Common Formulas

Distance

$d = rt$

d = distance traveled
t = time
r = rate

Temperature

$F = \dfrac{9}{5}C + 32$

F = degrees Fahrenheit
C = degrees Celsius

Simple Interest

$I = Prt$

I = interest
P = principal
r = annual interest rate
t = time in years

Compound Interest

$A = P\left(1 + \dfrac{r}{n}\right)^{nt}$

A = balance
P = principal
r = annual interest rate
n = compoundings per year
t = time in years

Coordinate Plane: Midpoint Formula

Midpoint of line segment joining (x_1, y_1) and (x_2, y_2)

$\left(\dfrac{x_1 + x_2}{2}, \dfrac{y_1 + y_2}{2}\right)$

Coordinate Plane: Distance Formula

d = distance between points (x_1, y_1) and (x_2, y_2)

$d = \sqrt{(x_2 - x_1)^2 + (y_2 - y_1)^2}$

Quadratic Formula

Solutions of $ax^2 + bx + c = 0$

$x = \dfrac{-b \pm \sqrt{b^2 - 4ac}}{2a}$

Rules of Exponents

(Assume $a \neq 0$ and $b \neq 0$.)

$a^0 = 1$

$a^m \cdot a^n = a^{m+n}$

$(ab)^m = a^m \cdot b^m$

$(a^m)^n = a^{mn}$

$\dfrac{a^m}{a^n} = a^{m-n}$

$\left(\dfrac{a}{b}\right)^m = \dfrac{a^m}{b^m}$

$a^{-n} = \dfrac{1}{a^n}$

$\left(\dfrac{a}{b}\right)^{-n} = \dfrac{b^n}{a^n}$

Basic Rules of Algebra

Commutative Property of Addition

$a + b = b + a$

Commutative Property of Multiplication

$ab = ba$

Associative Property of Addition

$(a + b) + c = a + (b + c)$

Associative Property of Multiplication

$(ab)c = a(bc)$

Left Distributive Property

$a(b + c) = ab + ac$

Right Distributive Property

$(a + b)c = ac + bc$

Additive Identity Property

$a + 0 = 0 + a = a$

Multiplicative Identity Property

$a \cdot 1 = 1 \cdot a = a$

Additive Inverse Property

$a + (-a) = 0$

Multiplicative Inverse Property

$a \cdot \dfrac{1}{a} = 1, \quad a \neq 0$

Properties of Equality

Addition Property of Equality

If $a = b$, then $a + c = b + c$.

Multiplication Property of Equality

If $a = b$, then $ac = bc$.

Cancellation Property of Addition

If $a + c = b + c$, then $a = b$.

Cancellation Property of Multiplication

If $ac = bc$, and $c \neq 0$, then $a = b$.

Zero Factor Property

If $ab = 0$, then $a = 0$ or $b = 0$.

Intermediate Algebra

FIFTH EDITION

Intermediate Algebra
FIFTH EDITION

Ron Larson
The Pennsylvania State University
The Behrend College

With the assistance of
Kimberly Nolting
Hillsborough Community College

BROOKS/COLE
CENGAGE Learning™

Australia • Brazil • Japan • Korea • Mexico • Singapore • Spain • United Kingdom • United States

BROOKS/COLE
CENGAGE Learning™

Intermediate Algebra, Fifth Edition
Ron Larson

Publisher: Charlie Van Wagner

Associate Development Editor: Laura Localio

Assistant Editor: Shaun Williams

Editorial Assistant: Rebecca Dashiell

Senior Media Editor: Maureen Ross

Executive Marketing Manager: Joe Rogove

Marketing Coordinator: Angela Kim

Marketing Communications Manager:
    Katherine Malatesta

Project Manager, Editorial Production: Carol Merrigan

Art & Design Manager: Jill Haber

Senior Manufacturing Coordinator: Diane Gibbons

Text Designer: Jerilyn Bockorick

Photo Researcher: Sue McDermott Barlow

Copy Editor: Craig Kirkpatrick

Cover Designer: Irene Morris

Cover Image: Frank Schwere/Stone+/Getty Images

Compositor: Larson Texts, Inc.

TI is a registered trademark of Texas Instruments, Inc.

Chapter Opener and Contents Photo Credits:
    p. 1: © Janine Wiedel Photolibrary/Alamy;
    p. 57: © Stock Connection Blue/Alamy; p. 125:
    John Kelly/Getty Images; p. 217: © Richard G.
    Bingham II/Alamy; p. 351: Kris Timken/Blend
    Images/JupiterImages; p. 368: Antonio Scorza/
    AFP/Getty Images; p. 369: © Digital Vision/
    Alamy; p. 448: Purestock/Getty Images; p. 449:
    © James Marshall/The Image Works; p. 511:
    JupiterImages/Banana Stock/Alamy; p. 576: Scott
    T. Baxter/Photodisc/Punchstock; p. 577: Ariel
    Skelley/Blend Images/Getty Images; p. 655:
    Purestock/Punchstock; p. 706: © SuperStock/
    Alamy; p. 707: UpperCut Images/Alamy.

For product information and technology assistance, contact us at
Cengage Learning Customer & Sales Support, 1-800-354-9706.
For permission to use material from this text or product, submit all requests online at **www.cengage.com/permissions.**
Further permissions questions can be e-mailed to
permissionrequest@cengage.com.

Library of Congress Control Number: 2008931148

Student Edition

ISBN-13: 978-0-547-10217-7

ISBN-10: 0-547-10217-8

Annotated Instructor's Edition

ISBN-13: 978-0-547-10220-7

ISBN-10: 0-547-10220-8

Brooks/Cole
10 Davis Drive
Belmont, CA 94002-3098
USA

Cengage Learning is a leading provider of customized learning solutions with office locations around the globe, including Singapore, the United Kingdom, Australia, Mexico, Brazil, and Japan. Locate your local office at **www.cengage.com/global.**

Cengage Learning products are represented in Canada by Nelson Education, Ltd.

For your course and learning solutions, visit **www.cengage.com.**

Purchase any of our products at your local college store or at our preferred online store **www.ichapters.com.**

Printed in the United States of America
2 3 4 5 6 7 12 11 10 09

Contents

7 Radicals and Complex Numbers 449

8 Quadratic Equations, Functions, and Inequalities 511

9 Exponential and Logarithmic Functions 577

*Appendices B, C, D, E, and F are available on the textbook website. Go to www.cengage.com/math/larson/algebra and link to Intermediate Algebra, Fifth Edition.

A Word from the Author

Welcome to *Intermediate Algebra*, Fifth Edition. In this revision I've focused on laying the groundwork for student success. Each chapter begins with study strategies to help the student do well in the course. Each chapter ends with an interactive summary of what they've learned to prepare them for the chapter test. Throughout the chapter, I've reinforced the skills needed to be successful and check to make sure the student understands the concepts being taught.

In order to address the diverse needs and abilities of students, I offer a straightforward approach to the presentation of difficult concepts. In the Fifth Edition, the emphasis is on helping students learn a variety of techniques—symbolic, numeric, and visual—for solving problems. I am committed to providing students with a successful and meaningful course of study.

Each chapter opens with a *Smart Study Strategy* that will help organize and improve the quality of studying. Mathematics requires students to remember every detail. These study strategies will help students organize, learn, and remember all the details. Each strategy has been student tested.

To improve the usefulness of the text as a study tool, I have a pair of features at the beginning of each section: *What You Should Learn* lists the main objectives that students will encounter throughout the section, and *Why You Should Learn It* provides a motivational explanation for learning the given objectives. To help keep students focused as they read the section, each objective presented in *What You Should Learn* is restated in the margin at the point where the concept is introduced.

In this edition, *Study Tip* features provide hints, cautionary notes, and words of advice for students as they learn the material. *Technology: Tip* features provide point-of-use instruction for using a graphing calculator, whereas *Technology: Discovery* features encourage students to explore mathematical concepts using their graphing or scientific calculators. All technology features are highlighted and can easily be omitted without loss of continuity in coverage of material.

The chapter summary feature *What Did You Learn?* highlights important mathematical vocabulary (*Key Terms*) and primary concepts (*Key Concepts*) from the chapter. For easy reference, the *Key Terms* are correlated to the chapter by page number and the *Key Concepts* by section number.

As students proceed through each chapter, they have many opportunities to assess their understanding and practice skills. A set of *Exercises*, located at the end of each section, correlates to the *Examples* found within the section. *Mid-Chapter Quizzes* and *Chapter Tests* offer students self-assessment tools halfway through and at the conclusion of each chapter. *Review Exercises*, organized by section, restate the *What You Should Learn* objectives so that students may refer back to the appropriate topic discussion when working through the exercises. In addition, the *Concept Check* exercises that precede each exercise set, and the *Cumulative Tests* that follow Chapters 4, 7, and 10, give students more opportunities to revisit and review previously learned concepts.

To show students the practical uses of algebra, I highlight the connections between the mathematical concepts and the real world in the multitude of applications found throughout the text. I believe that students can overcome their difficulties in mathematics if they are encouraged and supported throughout the learning process. Too often, students become frustrated and lose interest in the material when they cannot follow the text. With this in mind, every effort has been made to write a readable text that can be understood by every student. I hope that your students find this approach engaging and effective.

Ron Larson

Features

Study Skills in Action

Keeping a Positive Attitude

A student's experiences during the first three weeks in a math course often determine whether the student sticks with it or not. You can get yourself off to a good start by immediately acquiring a positive attitude and the study behaviors to support it.

Using Study Strategies

In each *Study Skills in Action* feature, you will learn a new study strategy that will help you progress through the course. Each strategy will help you:

- set up good study habits;
- organize information into smaller pieces;
- create review tools;
- memorize important definitions and rules;
- learn the math at hand.

Kimberly Nolting
VP, Academic Success Press
expert in developmental education

Daily Planner	Tuesday 9 September
	Study time for math:
	Review notes: 30 min.
	Rework notes: 1 hr.
	Homework: 2 hrs.

Daily Planner	Wednesday 10 September
	Important information:
	Tutoring Center hours:
	7:00 a.m. to 11:00 p.m.
	Instructor's office hours:
	M 4:30 to 6:00 p.m.
	Th 9:00 to 10:30 a.m.

Smart Study Strategy

Create a Positive Study Environment

1 ▶ After the first math class, set aside time for reviewing your notes and the textbook, reworking your notes, and completing homework.

2 ▶ Find a productive study environment on campus. Most colleges have a tutoring center where students can study and receive assistance as needed.

3 ▶ Set up a place for studying at home that is comfortable, but not too comfortable. It needs to be away from all potential distractions.

4 ▶ Make at least two other *collegial friends* in class. Collegial friends are students who study well together, help each other out when someone gets sick, and keep each other's attitudes positive.

5 ▶ Meet with your instructor at least once during the first two weeks. Ask the instructor what he or she advises for study strategies in the class. This will help you and let the instructor know that you really want to do well.

xxii

Chapter 1
Fundamentals of Algebra

1.1 The Real Number System
1.2 Operations with Real Numbers
1.3 Properties of Real Numbers
1.4 Algebraic Expressions
1.5 Constructing Algebraic Expressions

IT WORKED FOR ME!

"I get distracted very easily. If I study at home my video games call out to me. My instructor suggested studying on campus before going home or to work. I didn't like the idea at first, but tried it anyway. After a few times I realized that it was the best thing for me— I got things done and it took less time. I also did better on my next test."

Caleb
Music

1

Chapter Opener

Each chapter opener presents a study skill essential to success in mathematics. Following is a *Smart Study Strategy,* which gives concrete ways that students can help themselves with the study skill. In each chapter, there is a *Smart Study Strategy* note in the side column pointing out an appropriate time to use this strategy. Quotes from real students who have successfully used the strategy are given in *It Worked for Me!*

Section Opener

Every section begins with a list of learning objectives called *What You Should Learn*. Each objective is restated in the margin at the point where it is covered. *Why You Should Learn It* provides a motivational explanation for learning the given objectives.

3.6 Relations and Functions

What You Should Learn

1 ▶ Identify the domains and ranges of relations.
2 ▶ Determine if relations are functions by inspection.
3 ▶ Use function notation and evaluate functions.
4 ▶ Identify the domains and ranges of functions.

Why You Should Learn It
Functions can be used to model and solve real-life problems. For instance, in Exercise 87 on page 195, a function is used to find the cost of producing a video game.

1 ▶ Identify the domains and ranges of relations.

Relations

Many everyday occurrences involve two quantities that are paired or matched with each other by some rule of correspondence. The mathematical term for such a correspondence is **relation**.

Definition of Relation

A **relation** is any set of ordered pairs. The set of first components in the ordered pairs is the **domain** of the relation. The set of second components is the **range** of the relation.

EXAMPLE 1 Analyzing a Relation

Find the domain and range of the relation $\{(0, 1), (1, 3), (2, 5), (3, 5), (0, 3)\}$.

Solution
The domain is the set of all first components of the relation, and the range is the set of all second components.

Domain: $\{0, 1, 2, 3\}$

$\{(0, 1), (1, 3), (2, 5), (3, 5), (0, 3)\}$

Range: $\{1, 3, 5\}$

A graphical representation of this relation is shown in Figure 3.60.

Study Tip
When you write the domain or range of a relation, it is not necessary to list repeated components more than once.

Domain

Range

Figure 3.60

✓ **CHECKPOINT** Now try Exercise 1.

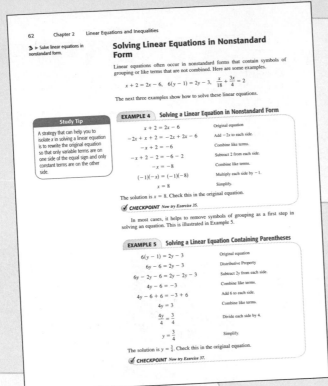

Examples

Each example has been carefully chosen to illustrate a particular mathematical concept or problem-solving technique. The examples cover a wide variety of problems and are titled for easy reference. Many examples include detailed, step-by-step solutions with side comments, which explain the key steps of the solution process.

Checkpoints

Each example is followed by a checkpoint exercise. After working through an example, students can try the checkpoint exercise in the exercise set to check their understanding of the concepts presented in the example. Checkpoint exercises are marked with a ✓ in the exercise set for easy reference.

Applications

A wide variety of real-life applications are integrated throughout the text in examples and exercises. These applications demonstrate the relevance of algebra in the real world. Many of the applications use current, real data. The icon 🌐 indicates an example involving a real-life application.

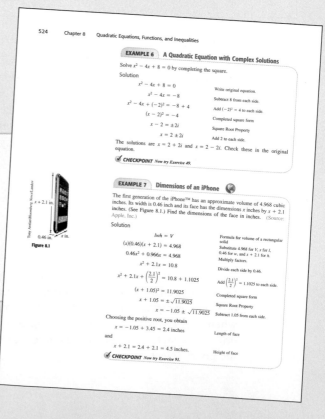

Problem Solving

This text provides many opportunities for students to sharpen their problem-solving skills. In both the examples and the exercises, students are asked to apply verbal, numerical, analytical, and graphical approaches to problem solving. In the spirit of the AMATYC and NCTM standards, students are taught a five-step strategy for solving applied problems, which begins with constructing a verbal model and ends with checking the answer.

Geometry

The Fifth Edition continues to provide coverage and integration of geometry in examples and exercises. The icon ▲ indicates an exercise involving geometry.

Graphics

Visualization is a critical problem-solving skill. To encourage the development of this skill, students are shown how to use graphs to reinforce algebraic and numeric solutions and to interpret data. The numerous figures in examples and exercises throughout the text were computer-generated for accuracy.

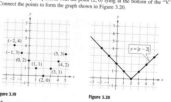

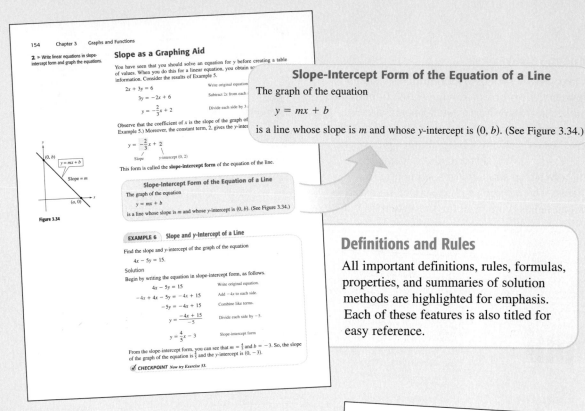

Slope-Intercept Form of the Equation of a Line

The graph of the equation

$$y = mx + b$$

is a line whose slope is m and whose y-intercept is $(0, b)$. (See Figure 3.34.)

Definitions and Rules

All important definitions, rules, formulas, properties, and summaries of solution methods are highlighted for emphasis. Each of these features is also titled for easy reference.

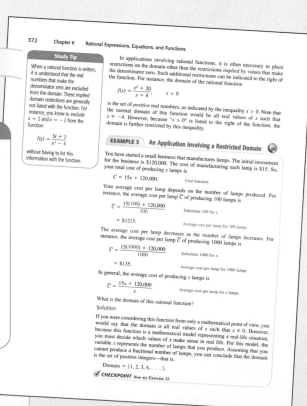

Study Tip

When a rational function is written, it is understood that the real numbers that make the denominator zero are excluded from the domain. These *implied* domain restrictions are generally not listed with the function. For instance, you know to exclude $x = 2$ and $x = -2$ from the function

$$f(x) = \frac{3x + 2}{x^2 - 4}$$

without having to list this information with the function.

Study Tips

Study Tips offer students specific point-of-use suggestions for studying algebra, as well as pointing out common errors and discussing alternative solution methods. They appear in the margins.

Technology Tips

Point-of-use instructions for using graphing calculators appear in the margins. These features encourage the use of graphing technology as a tool for visualization of mathematical concepts, for verification of other solution methods, and for facilitation of computations. The *Technology: Tips* can easily be omitted without loss of continuity in coverage. Answers to questions posed within these features are located in the back of the Annotated Instructor's Edition.

Technology: Tip

Most scientific and graphing calculators automatically switch to scientific notation to show large or small numbers that exceed the display range.

To *enter* numbers in scientific notation, your calculator should have a key labeled $\boxed{EE}$ or $\boxed{EXP}$. Consult the user's guide for your calculator for specific instructions.

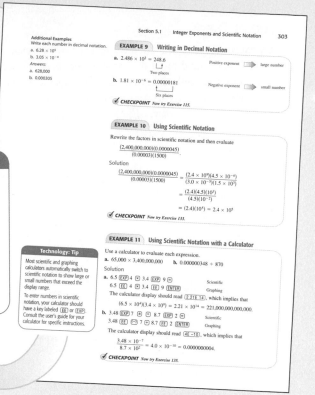

Technology: Discovery

Technology: Discovery features invite students to engage in active exploration of mathematical concepts and discovery of mathematical relationships through the use of scientific or graphing calculators. These activities encourage students to utilize their critical thinking skills and help them develop an intuitive understanding of theoretical concepts. *Technology: Discovery* features can easily be omitted without loss of continuity in coverage. Answers to questions posed within these features are located in the back of the Annotated Instructor's Edition.

Technology: Discovery

Rewrite each system of equations in slope-intercept form and graph the equations using a graphing calculator. What is the relationship between the slopes of the two lines and the number of points of intersection?

a. $\begin{cases} 2x + 4y = 8 \\ 4x - 3y = -6 \end{cases}$

b. $\begin{cases} -x + 5y = 15 \\ 2x - 10y = -7 \end{cases}$

c. $\begin{cases} x - y = 9 \\ 2x - 2y = 18 \end{cases}$

See Technology Answers.

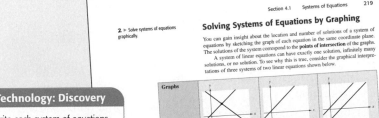

Concept Check

Each exercise set is preceded by four exercises that check students' understanding of the main concepts of the section. These exercises could be completed in class to make sure that students are ready to start the exercise set.

Section 8.1   Solving Quadratic Equations: Factoring and Special Forms   517

Concept Check

1. Explain the Zero-Factor Property and how it can be used to solve a quadratic equation. If a and b are factors such that $ab = 0$, then $a = 0$ or $b = 0$. To solve a quadratic equation using the Zero-Factor Property, write the quadratic expression equal to zero and factor. Then each factor can be set equal to zero to find the solution(s).

2. Determine whether the following statement is true or false. Justify your answer.
The only solution of the equation $x^2 = 25$ is $x = 5$.
False. Another solution is $x = -5$.
$x^2 = (-5)^2 = 25$

3. Does the equation $4x^2 + 9 = 0$ have two real solutions or two complex solutions? Explain your reasoning. Two complex solutions. When the squared expression is isolated on one side of the equation, the other side is negative. When the square root of each side is taken, the square root of the negative number is imaginary.

4. Is the equation $x^6 - 6x^3 + 9 = 0$ of quadratic form? Explain your reasoning. Yes. For an equation to be of quadratic form, the exponent of the algebraic expression in the first term must be twice the exponent of the same algebraic expression in the second term.

8.1 EXERCISES
Go to pages 570–571 to record your assignments.

Developing Skills

In Exercises 1–20, solve the equation by factoring. See Example 1.
1. $x^2 - 15x + 54 = 0$ 6, 9
2. $x^2 + 15x + 44 = 0$ −11, −4
3. $x^2 - x - 30 = 0$ −5, 6
4. $x^2 - 2x - 48 = 0$ −6, 8
5. $x^2 + 4x = 45$ −9, 5
6. $x^2 - 7x = 18$ −2, 9
7. $x^2 - 16x + 64 = 0$ 8
8. $x^2 + 60x + 900 = 0$ −30
9. $9x^2 - 10x - 16 = 0$ $-\frac{8}{9}$, 2
10. $8x^2 - 10x + 3 = 0$ $\frac{1}{2}$, $\frac{3}{4}$
11. $4x^2 - 12x = 0$ 0, 3
12. $25y^2 - 75y = 0$ 0, 3
13. $u(u - 9) - 12(u - 9) = 0$ 9, 12
14. $16x(x - 8) - 12(x - 8) = 0$ $\frac{3}{4}$, 8
15. $2x(x - 5) + 9(x - 5) = 0$ $-\frac{9}{2}$, 5
16. $3(4 - x) - 2x(4 - x) = 0$ $\frac{3}{2}$, 4
17. $(y - 4)(y - 3) = 6$ 1, 6
18. $(5 + u)(2 + u) = 4$ −6, −1
19. $2x(3x + 2) = 5 - 6x^2$ $-\frac{5}{6}$, $\frac{1}{2}$
20. $(2z + 1)(2z - 1) = -4z^2 - 5z + 2$ -1, $\frac{1}{8}$

In Exercises 21–42, solve the equation by using the Square Root Property. See Example 2.
21. $x^2 = 49$ ±7
22. $p^2 = 169$ ±13
23. $6x^2 = 54$ ±3
24. $5t^2 = 5$ ±1
25. $25x^2 = 16$ ±$\frac{4}{5}$
26. $9z^2 = 121$ ±$\frac{11}{3}$

27. $\frac{w^2}{4} = 49$ ±14
28. $\frac{x^2}{6} = 24$ ±12
29. $4x^2 - 25 = 0$ ±$\frac{5}{2}$
30. $16y^2 - 121 = 0$ ±$\frac{11}{4}$
31. $4u^2 - 225 = 0$ ±$\frac{15}{2}$
32. $16x^2 - 1 = 0$ ±$\frac{1}{4}$
33. $(x + 4)^2 = 64$ −12, 4
34. $(m - 12)^2 = 400$ −8, 32
35. $(x - 3)^2 = 0.25$ 2.5, 3.5
36. $(x + 2)^2 = 0.81$ −2.9, −1.1
37. $(x - 2)^2 = 7$ $2 ± \sqrt{7}$
38. $(y + 4)^2 = 27$ $-4 ± 3\sqrt{3}$
39. $(2x + 1)^2 = 50$ $-\frac{1}{2} ± \frac{5\sqrt{2}}{2}$
40. $(3x - 5)^2 = 48$ $\frac{5}{3} ± \frac{4\sqrt{3}}{3}$
41. $(9m - 2)^2 - 108 = 0$ $\frac{2}{9} ± \frac{2\sqrt{3}}{3}$
42. $(5x + 11)^2 - 300 = 0$ $-\frac{11}{5} ± 2\sqrt{3}$

In Exercises 43–64, solve the equation by using the Square Root Property. See Example 3.
43. $x^2 = -36$ ±6i
44. $x^2 = -16$ ±4i
45. $x^2 + 4 = 0$ ±2i
46. $p^2 + 9 = 0$ ±3i
47. $9x^2 + 17 = 0$ $±\frac{\sqrt{17}}{3}i$
48. $25x^2 + 4 = 0$ $±\frac{2}{5}i$
49. $(t - 3)^2 = -25$ 3 ± 5i
50. $(x + 5)^2 = -81$ −5 ± 9i
51. $(3z + 4)^2 + 144 = 0$ $-\frac{4}{3} ± 4i$
52. $(2y - 3)^2 + 25 = 0$ $\frac{3}{2} ± \frac{5}{2}i$

---

112. $x^{2/3} + 3x^{1/3} - 10 = 0$ −125, 8
113. $2x^{2/3} - 7x^{1/3} + 5 = 0$ 1, $\frac{125}{8}$
114. $5x^{2/3} - 13x^{1/3} + 6 = 0$ $\frac{27}{125}$, 8
115. $x^{2/5} - 3x^{1/5} + 2 = 0$ 1, 32
116. $x^{2/5} + 5x^{1/5} + 6 = 0$ −243, −32
117. $2x^{2/5} - 7x^{1/5} + 3 = 0$ $\frac{1}{32}$, 243
118. $2x^{2/5} + 3x^{1/5} + 1 = 0$ −1, $-\frac{1}{32}$
119. $x^{1/3} - x^{1/6} - 6 = 0$ 729
120. $x^{1/3} + 2x^{1/6} - 3 = 0$ 1

124. $\frac{2}{x^2} - \frac{x}{...}$
125. $4x^{-2} - x^{-1} - 5 = 0$ −1, $\frac{4}{5}$
126. $2x^{-2} - x^{-1} - 1 = 0$ −2, 1
127. $(x^2 - 3x)^2 - 2(x^2 - 3x) - 8 = 0$ ±1, 2, 4
128. $(x^2 - 6x)^2 - 2(x^2 - 6x) - 35 = 0$ ±1, 5, 7
129. $16\left(\frac{x-1}{x-8}\right)^2 + 8\left(\frac{x-1}{x-8}\right) + 1 = 0$ $\frac{17}{5}$
130. $9\left(\frac{x+2}{x+3}\right)^2 - 6\left(\frac{x+2}{x+3}\right) + 1 = 0$ $-\frac{5}{2}$

Solving Problems

131. *Unisphere* The Unisphere is the world's largest man-made globe. It was built as the symbol of the 1964–1965 New York World's Fair. A sphere with the same diameter as the Unisphere globe would have a surface area of 45,239 square feet. What is the diameter of the Unisphere? (Source: The World's Fair and Exposition Information and Reference Guide) 120 feet

© Rudy Sulgan/CORBIS

Designing the Unisphere was an engineering challenge that at one point involved simultaneously solving 670 equations.

132. *Geometry* The surface area S of a basketball is $900/\pi$ square inches. Find the radius r of the basketball. $\frac{15}{\pi} \approx 4.77$ inches

Free-Falling Object In Exercises 133–136, find the time required for an object to reach the ground when it is dropped from a height of s_0 feet. The height h (in feet) is given by $h = -16t^2 + s_0$, where t measures the time (in seconds) after the object is released.
133. $s_0 = 256$ 4 seconds
134. $s_0 = 48$ $\frac{\sqrt{3}}{...} \approx 1.73$ seconds
135. $s_0 = 128$ $2\sqrt{2} \approx 2.83$ seconds
136. $s_0 = 500$ $\frac{5\sqrt{5}}{2} \approx 5.59$ seconds

137. *Free-Falling Object* The height h (in feet) of an object thrown vertically upward from the top of a tower 144 feet tall is given by $h = 144 + 128t - 16t^2$, where t measures the time in seconds from the time when the object is released. How long does it take for the object to reach the ground? 9 seconds

138. *Profit* The monthly profit P (in dollars) a company makes depends on the amount x (in dollars) the company spends on advertising according to the model
$$P = 800 + 120x - \frac{1}{2}x^2.$$
Find the amount spent on advertising that will yield a monthly profit of $8000. $120

Exercises

The exercise sets are grouped into three categories: *Developing Skills, Solving Problems,* and *Explaining Concepts.* The exercise sets offer a diverse variety of computational, conceptual, and applied problems to accommodate many learning styles. Designed to build competence, skill, and understanding, each exercise set is graded in difficulty to allow students to gain confidence as they progress. Detailed solutions to all odd-numbered exercises are given in the *Student Solutions Guide,* and answers to all odd-numbered exercises are given in the back of the student text. Answers are located in place in the Annotated Instructor's Edition.

---

520   Chapter 8   Quadratic Equations, Functions, and Inequalities

Compound Interest The amount A in an account after 2 years when a principal of P dollars is invested at annual interest rate r compounded annually is given by $A = P(1 + r)^2$. In Exercises 139 and 140, find r.
139. $P = \$1500$, $A = \$1685.40$ 6%
140. $P = \$5000$, $A = \$5724.50$ 7%

National Health Expenditures In Exercises 141 and 142, the national expenditures for health care in the United States from 1997 through 2006 are given by
$$y = 4.95t^2 + 876, \quad 7 \le t \le 16.$$
In this model, y represents the expenditures (in billions of dollars) and t represents the year, with $t = 7$ corresponding to 1997 (see figure). (Source: U.S. Centers for Medicare & Medicaid Services)

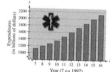

Year ($7 \leftrightarrow 1997$)

Figure for 141 and 142

141. Algebraically determine the year when expenditures were approximately $1500 billion. Graphically confirm the result. 2001
142. Algebraically determine the year when expenditures were approximately $1850 billion. Graphically confirm the result. 2004

Explaining Concepts

143. For a quadratic equation $ax^2 + bx + c = 0$, where a, b, and c are real numbers with $a \ne 0$, explain why b and c can equal zero, but a cannot. If $a = 0$, the equation would not be quadratic because it would be of degree 1, not 2.

144. Is it possible for a quadratic equation of the form $x^2 = m$ to have one real solution and one complex solution? Explain your reasoning. No. Complex solutions always occur in complex conjugate pairs.

145. Describe the steps you would use to solve a quadratic equation when using the Square Root Property. See Additional Answers.

146. Describe a procedure for solving an equation of quadratic form. Give an example. To solve an equation of quadratic form, determine an algebraic expression u such that substitution yields the quadratic equation $au^2 + bu + c = 0$. Solve this quadratic equation for u and then, through back-substitution, find the solution of the original equation.

Cumulative Review

In Exercises 147–150, solve the inequality and sketch the solution on the real number line. See Additional Answers.
147. $3x - 8 > 4$ $x > 4$
148. $4 - 5x \ge 12$ $x \le -\frac{8}{5}$
149. $2x - 6 \le 9$ $x \le \frac{15}{2}$
150. $x - 4 < 6$ or $x + 3 > 8$ $-\infty \le x \le \infty$

In Exercises 151 and 152, solve the system of linear equations.
151. $x + y - z = 4$
$2x + y + 2z = 10$
$x - 3y - 4z = -7$
$(3, 2, 1)$
152. $2x - y + z = -6$
$x + 5y - z = 7$
$-x - 2y - 3z = 8$
$(-1, 1, -3)$

In Exercises 153–158, combine the radical expressions, if possible, and simplify.
153. $5\sqrt{3} - 2\sqrt{3}$ $3\sqrt{3}$
154. $8\sqrt{27} + 4\sqrt{27}$ $36\sqrt{3}$
155. $16\sqrt[3]{y} - 9\sqrt[3]{x}$ $16\sqrt[3]{y} - 9\sqrt[3]{x}$
156. $12\sqrt{x - 1} + 6\sqrt{x - 1}$ $18\sqrt{x - 1}$
157. $\sqrt{16m^4n^3} + m\sqrt{m^2n}$ $(4n + 1)m^2\sqrt{n}$
158. $x^2y\sqrt[3]{32x^3} + x^3\sqrt[3]{2x^6y^4} - y\sqrt[3]{162x^{10}}$ 0

Cumulative Review

Each exercise set (except those in Chapter 1) is followed by exercises that cover concepts from previous sections. This serves as a review for students and also helps students connect old concepts with new concepts.

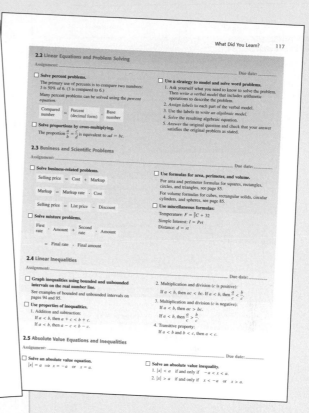

What Did You Learn? (Chapter Summary)

The *What Did You Learn?* at the end of each chapter has been reorganized and expanded in the Fifth Edition. The *Plan for Test Success* provides a place for students to plan their studying for a test and includes a checklist of things to review. Students are also able to check off the *Key Terms* and *Key Concepts* of the chapter as these are reviewed. A space to record assignments for each section of the chapter is also provided.

Review Exercises

The *Review Exercises* at the end of each chapter contain skill-building and application exercises that are first ordered by section, and then grouped according to the objectives stated within *What You Should Learn*. This organization allows students to easily identify the appropriate sections and concepts for study and review.

Mid-Chapter Quiz

Each chapter contains a *Mid-Chapter Quiz*. Answers to all questions in the *Mid-Chapter Quiz* are given in the back of the student text and are located in place in the Annotated Instructor's Edition.

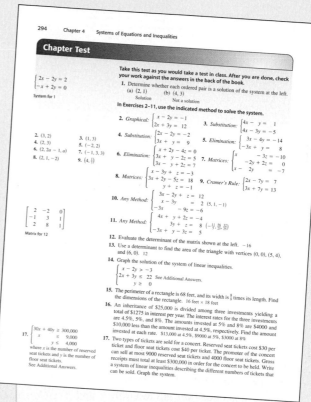

Chapter Test

Each chapter ends with a *Chapter Test*. Answers to all questions in the *Chapter Test* are given in the back of the student text and are located in place in the Annotated Instructor's Edition.

Cumulative Test

The *Cumulative Tests* that follow Chapters 4, 7, and 10 provide a comprehensive self-assessment tool that helps students check their mastery of previously covered material. Answers to all questions in the *Cumulative Tests* are given in the back of the student text and are located in place in the Annotated Instructor's Edition.

Intermediate Algebra, Fifth Edition, by Ron Larson is accompanied by a comprehensive supplements package, which includes resources for both students and instructors. All items are keyed to the text.

Printed Resources

For Students

Student Solutions Manual by Gerry Fitch, Louisiana State University
(0547140193)

- Detailed, step-by-step solutions to all odd-numbered exercises in the section exercise sets and in the review exercises
- Detailed, step-by-step solutions to all Mid-Chapter Quiz, Chapter Test, and Cumulative Test questions

For Instructors

Annotated Instructor's Edition
(0547102208)

- Includes answers in place for Exercise sets, Review Exercises, Mid-Chapter Quizzes, Chapter Tests, and Cumulative Tests
- Additional Answers section in the back of the text lists those answers that contain large graphics or lengthy exposition
- Answers to the Technology: Tip and Technology: Discovery questions are provided in the back of the book
- Annotations at point of use that offer strategies and suggestions for teaching the course and point out common student errors

Complete Solutions Manual by Gerry Fitch, Louisiana State University
(0547140266)

- Chapter and Final Exam test forms with answer key
- Individual test items and answers for Chapters 1–11
- Notes to the instructor including tips and strategies on student assessment, cooperative learning, classroom management, study skills, and problem solving

Technology Resources

For Students

Website *(www.cengage.com/math/larson/algebra)*

Instructional DVDs by Dana Mosely to accompany Larson, Developmental Math Series, 5e (05471402074)

Personal Tutor   An easy-to-use and effective live, online tutoring service. *Whiteboard Simulations* and Practice Area promote real-time visual interaction.

For Instructors

Power Lecture CD-ROM with Diploma® (0547140207)   This CD-ROM provides the instructor with dynamic media tools for teaching. Create, deliver, and customize tests (both print and online) in minutes with Diploma® computerized testing featuring algorithmic equations. Easily build solution sets for homework or exams using *Solution Builder*'s online solutions manual. Microsoft® PowerPoint® lecture slides, figures from the book, and Test Bank, in electronic format, are also included on this CD-ROM.

WebAssign   Instant feedback and ease of use are just two reasons why WebAssign is the most widely used homework system in higher education. WebAssign's homework delivery system allows you to assign, collect, grade, and record homework assignments via the web. And now, this proven system has been enhanced to include links to textbook sections, video examples, and problem-specific tutorials.

Website *(www.cengage.com/math/larson/algebra)*

Solution Builder   This online tool lets instructors build customized solution sets in three simple steps and then print and hand out in class or post to a password-protected class website.

Acknowledgments

I would like to thank the many people who have helped me revise the various editions of this text. Their encouragement, criticisms, and suggestions have been invaluable.

Reviewers

Tom Anthony, Central Piedmont Community College; Tina Cannon, Chattanooga State Technical Community College; LeAnne Conaway, Harrisburg Area Community College and Penn State University; Mary Deas, Johnson County Community College; Jeremiah Gilbert, San Bernadino Valley College; Jason Pallett, Metropolitan Community College-Longview; Laurence Small, L.A. Pierce College; Dr. Azar Raiszadeh, Chattanooga State Technical Community College; Patrick Ward, Illinois Central College.

My thanks to Kimberly Nolting, Hillsborough Community College, for her contributions to this project. My thanks also to Robert Hostetler, The Behrend College, The Pennsylvania State University, and Patrick M. Kelly, Mercyhurst College, for their significant contributions to previous editions of this text.

I would also like to thank the staff of Larson Texts, Inc., who assisted in preparing the manuscript, rendering the art package, and typesetting and proofreading the pages and the supplements.

On a personal level, I am grateful to my spouse, Deanna Gilbert Larson, for her love, patience, and support. Also, a special thanks goes to R. Scott O'Neil.

If you have suggestions for improving this text, please feel free to write to me. Over the past two decades I have received many useful comments from both instructors and students, and I value these comments very much.

Ron Larson

Study Skills in Action

Keeping a Positive Attitude

A student's experiences during the first three weeks in a math course often determine whether the student sticks with it or not. You can get yourself off to a good start by immediately acquiring a positive attitude and the study behaviors to support it.

Using Study Strategies

In each *Study Skills in Action* feature, you will learn a new study strategy that will help you progress through the course. Each strategy will help you:

- set up good study habits;
- organize information into smaller pieces;
- create review tools;
- memorize important definitions and rules;
- learn the math at hand.

Kimberly Nolting
VP, Academic Success Press
expert in developmental education

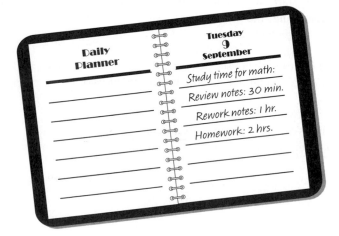

Smart Study Strategy

Create a Positive Study Environment

1 ► After the first math class, set aside time for reviewing your notes and the textbook, reworking your notes, and completing homework.

2 ► Find a productive study environment on campus. Most colleges have a tutoring center where students can study and receive assistance as needed.

3 ► Set up a place for studying at home that is comfortable, but not too comfortable. It needs to be away from all potential distractions.

4 ► Make at least two other *collegial friends* in class. Collegial friends are students who study well together, help each other out when someone gets sick, and keep each other's attitudes positive.

5 ► Meet with your instructor at least once during the first two weeks. Ask the instructor what he or she advises for study strategies in the class. This will help you and let the instructor know that you really want to do well.

Chapter 1
Fundamentals of Algebra

IT WORKED FOR ME!

"I get distracted very easily. If I study at home my video games call out to me. My instructor suggested studying on campus before going home or to work. I didn't like the idea at first, but tried it anyway. After a few times I realized that it was the best thing for me—I got things done and it took less time. I also did better on my next test."

Caleb
Music

1.1 The Real Number System

Duomo/CORBIS

What You Should Learn

1 ▶ Understand the set of real numbers and the subsets of real numbers.

2 ▶ Use the real number line to order real numbers.

3 ▶ Use the real number line to find the distance between two real numbers.

4 ▶ Determine the absolute value of a real number.

Why You Should Learn It

Inequality symbols can be used to represent many real-life situations, such as bicycling speeds (see Exercise 87 on page 10).

1 ▶ Understand the set of real numbers and the subsets of real numbers.

Sets and Real Numbers

This chapter introduces the basic definitions, operations, and rules that form the fundamental concepts of algebra. Section 1.1 begins with real numbers and their representation on the real number line. Sections 1.2 and 1.3 discuss operations and properties of real numbers, and Sections 1.4 and 1.5 discuss algebraic expressions.

The formal term that is used in mathematics to refer to a collection of objects is the word **set.** For instance, the set

$$\{1, 2, 3\}$$ A set with three members

contains the three numbers 1, 2, and 3. Note that the members of the set are enclosed in braces { }. Parentheses ( ) and brackets [ ] are used to represent other ideas.

The set of numbers that is used in arithmetic is called the set of **real numbers.** The term *real* distinguishes real numbers from *imaginary* or *complex* numbers—a type of number that you will study later in this text.

If all members of a set A are also members of a set B, then A is a **subset** of B. One of the most commonly used subsets of real numbers is the set of **natural numbers** or **positive integers.**

$$\{1, 2, 3, 4, . . .\}$$ The set of positive integers

Note that the three dots indicate that the pattern continues. For instance, the set also contains the numbers 5, 6, 7, and so on.

Positive integers can be used to describe many quantities in everyday life. For instance, you might be taking four classes this term, or you might be paying 240 dollars a month for rent. But even in everyday life, positive integers cannot describe some concepts accurately. For instance, you could have a zero balance in your checking account. To describe such a quantity, you need to expand the set of positive integers to include zero, forming the set of **whole numbers.** To describe a quantity such as $-5°$, you need to expand the set of whole numbers to include **negative integers.** This expanded set is called the set of **integers.** The set of integers is also a *subset* of the set of real numbers.

$$\underbrace{\{. . ., -3, -2, -1,}_{\text{Negative integers}} \overbrace{0, \underbrace{1, 2, 3, . . .\}}_{\text{Positive integers}}}^{\text{Whole numbers}}$$ The set of integers

Technology: Tip

You can use a calculator to round decimals. For instance, to round 0.2846 to three decimal places on a scientific calculator, enter

FIX 3 .2846 = .

On a graphing calculator, enter

round (.2846, 3) ENTER .

Consult the user's manual for your graphing calculator for specific keystrokes or instructions. Then, use your calculator to round 0.38174 to four decimal places.

See Technology Answers.

Even with the set of integers, there are still many quantities in everyday life that you cannot describe accurately. The costs of many items are not in whole dollar amounts, but in parts of dollars, such as $1.19 and $39.98. You might work $8\frac{1}{2}$ hours, or you might miss the first *half* of a movie. To describe such quantities, you can expand the set of integers to include **fractions.** The expanded set is called the set of **rational numbers.** Formally, a real number is **rational** if it can be written as the ratio p/q of two integers, where $q \neq 0$ (the symbol $\neq$ means **does not equal**). Here are some examples of rational numbers.

$$2 = \frac{2}{1}, \quad \frac{1}{3} = 0.333 \ldots, \quad \frac{1}{8} = 0.125, \quad \text{and} \quad \frac{125}{111} = 1.126126 \ldots$$

The decimal representation of a rational number is either **terminating** or **repeating.** For instance, the decimal representation of $\frac{1}{4} = 0.25$ is terminating, and the decimal representation of

$$\frac{4}{11} = 0.363636 \ldots = 0.\overline{36}$$

is repeating. (The overbar symbol over 36 indicates which digits repeat.) A real number that cannot be written as a ratio of two integers is **irrational.** For instance,

$$\sqrt{2} = 1.4142135 \ldots \quad \text{and} \quad \pi = 3.1415926 \ldots$$

are irrational.

The decimal representation of an irrational number neither terminates nor repeats. When you perform calculations using decimal representations of nonterminating, nonrepeating decimals, you usually use a decimal approximation that has been **rounded** to a certain number of decimal places. The rounding rule used in this text is to round up if the succeeding digit is 5 or more, or to round down if the succeeding digit is 4 or less. For example, to one decimal place, 7.35 would *round up* to 7.4. Similarly, to two decimal places, 2.364 would *round down* to 2.36. Rounded to four decimal places, the decimal approximations of the rational number $\frac{2}{3}$ and the irrational number π are

$$\frac{2}{3} \approx 0.6667 \quad \text{and} \quad \pi \approx 3.1416.$$

The symbol $\approx$ means **is approximately equal to.** Figure 1.1 shows several commonly used subsets of real numbers and their relationships to each other.

Figure 1.1   Subsets of Real Numbers

EXAMPLE 1 Classifying Real Numbers

Which of the numbers in the set $\left\{-7, -\sqrt{3}, -1, -\frac{1}{5}, 0, \frac{3}{4}, \sqrt{2}, \pi, 5\right\}$ are (a) natural numbers, (b) integers, (c) rational numbers, and (d) irrational numbers?

Solution

a. Natural numbers: $\{5\}$

b. Integers: $\{-7, -1, 0, 5\}$

c. Rational numbers: $\left\{-7, -1, -\frac{1}{5}, 0, \frac{3}{4}, 5\right\}$

d. Irrational numbers: $\left\{-\sqrt{3}, \sqrt{2}, \pi\right\}$

✓ **CHECKPOINT** *Now try Exercise 1.*

2 ▶ Use the real number line to order real numbers.

The Real Number Line

The picture that represents the real numbers is called the **real number line.** It consists of a horizontal line with a point (the **origin**) labeled 0. Numbers to the left of zero are **negative** and numbers to the right of zero are **positive,** as shown in Figure 1.2.

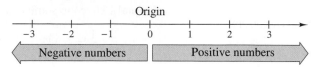

Figure 1.2   The Real Number Line

Zero is neither positive nor negative. So, to describe a real number that might be positive or zero, you can use the term **nonnegative real number.**

Each point on the real number line corresponds to exactly one real number, and each real number corresponds to exactly one point on the real number line, as shown in Figure 1.3. When you draw the point (on the real number line) that corresponds to a real number, you are **plotting** the real number.

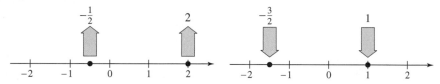

Each point on the real number line corresponds to a real number.

Each real number corresponds to a point on the real number line.

Figure 1.3

EXAMPLE 2   **Plotting Points on the Real Number Line**

Plot the real numbers on the real number line.

a. $-\dfrac{5}{3}$ **b.** 2.3 **c.** $\dfrac{9}{4}$ **d.** -0.3

Solution

All four points are shown in Figure 1.4.

a. The point representing the real number $-\dfrac{5}{3} = -1.666\ldots$ lies between -2 and -1, but closer to -2, on the real number line.

b. The point representing the real number 2.3 lies between 2 and 3, but closer to 2, on the real number line.

c. The point representing the real number $\dfrac{9}{4} = 2.25$ lies between 2 and 3, but closer to 2, on the real number line. Note that the point representing $\dfrac{9}{4}$ lies slightly to the left of the point representing 2.3.

d. The point representing the real number -0.3 lies between -1 and 0, but closer to 0, on the real number line.

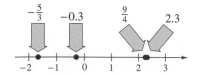

Figure 1.4

✓ **CHECKPOINT** *Now try Exercise 13.*

The real number line provides a way of comparing any two real numbers. For instance, if you choose any two (different) numbers on the real number line, one of the numbers must be to the left of the other. You can describe this by saying that the number to the left is **less than** the number to the right, or that the number to the right is **greater than** the number to the left, as shown in Figure 1.5.

$$a < b$$

Figure 1.5   *a* is to the left of *b*.

Order on the Real Number Line

If the real number *a* lies to the left of the real number *b* on the real number line, then *a* is **less than** *b*, which is written as

$a < b$.

This relationship can also be described by saying that *b* is **greater than** *a* and writing $b > a$. The expression $a \leq b$ means that *a* is **less than or equal to** *b*, and the expression $b \geq a$ means that *b* is **greater than or equal to** *a*. The symbols $<, >, \leq$, and $\geq$ are called **inequality symbols.**

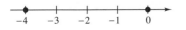

Figure 1.6

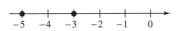

Figure 1.7

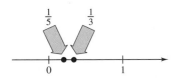

Figure 1.8

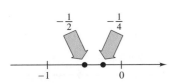

Figure 1.9

When asked to **order** two numbers, you are simply being asked to say which of the two numbers is greater.

EXAMPLE 3   Ordering Real Numbers

Place the correct inequality symbol ($<$ or $>$) between each pair of numbers.

a. -4 ___ 0 **b.** -3 ___ -5 **c.** $\frac{1}{5}$ ___ $\frac{1}{3}$ **d.** $-\frac{1}{4}$ ___ $-\frac{1}{2}$

Solution

a. Because -4 lies to the left of 0 on the real number line, as shown in Figure 1.6, you can say that -4 is *less than* 0, and write $-4 < 0$.

b. Because -3 lies to the right of -5 on the real number line, as shown in Figure 1.7, you can say that -3 is *greater than* -5, and write $-3 > -5$.

c. Because $\frac{1}{5}$ lies to the left of $\frac{1}{3}$ on the real number line, as shown in Figure 1.8, you can say that $\frac{1}{5}$ is *less than* $\frac{1}{3}$, and write $\frac{1}{5} < \frac{1}{3}$.

d. Because $-\frac{1}{4}$ lies to the right of $-\frac{1}{2}$ on the real number line, as shown in Figure 1.9, you can say that $-\frac{1}{4}$ is *greater than* $-\frac{1}{2}$, and write $-\frac{1}{4} > -\frac{1}{2}$.

✓ **CHECKPOINT** *Now try Exercise 19.*

One effective way to order two fractions such as $\frac{5}{12}$ and $\frac{9}{23}$ is to compare their decimal equivalents. Because $\frac{5}{12} = 0.41\overline{6}$ and $\frac{9}{23} \approx 0.391$, you can write

$$\frac{5}{12} > \frac{9}{23}.$$

3 ▶ Use the real number line to find the distance between two real numbers.

Distance on the Real Number Line

Once you know how to represent real numbers as points on the real number line, it is natural to talk about the **distance between two real numbers.** Specifically, if a and b are two real numbers such that $a \le b$, then the distance between a and b is defined as $b - a$.

Distance Between Two Real Numbers

If a and b are two real numbers such that $a \le b$, then the **distance between a and b** is given by

Distance between a and $b = b - a$.

Note from this definition that if $a = b$, the distance between a and b is zero. If $a \ne b$, then the distance between a and b is positive.

EXAMPLE 4   Finding the Distance Between Two Real Numbers

Find the distance between each pair of real numbers.

a. -2 and 3 **b.** 0 and 4 **c.** -4 and 0 **d.** 1 and $-\dfrac{1}{2}$

Solution

a. Because $-2 \le 3$, the distance between -2 and 3 is

$$3 - (-2) = 3 + 2 = 5. \qquad \text{See Figure 1.10.}$$

b. Because $0 \le 4$, the distance between 0 and 4 is

$$4 - 0 = 4. \qquad \text{See Figure 1.11.}$$

c. Because $-4 \le 0$, the distance between -4 and 0 is

$$0 - (-4) = 0 + 4 = 4. \qquad \text{See Figure 1.12.}$$

d. Because $-\dfrac{1}{2} \le 1$, let $a = -\dfrac{1}{2}$ and $b = 1$. So, the distance between 1 and $-\dfrac{1}{2}$ is

$$1 - \left(-\frac{1}{2}\right) = 1 + \frac{1}{2} = 1\frac{1}{2}. \qquad \text{See Figure 1.13.}$$

Study Tip

Recall that when you subtract a negative number, as in Example 4(a), you add the opposite of the second number to the first. Because the opposite of -2 is 2, you add 2 to 3.

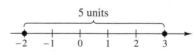

Figure 1.10

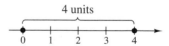

Figure 1.11

Figure 1.12

Figure 1.13

 **CHECKPOINT** *Now try Exercise 29.*

4 ▶ Determine the absolute value of a real number.

Absolute Value

Two real numbers are called **opposites** of each other if they lie the same distance from, but on opposite sides of, 0 on the real number line. For instance, -2 is the opposite of 2 (see Figure 1.14).

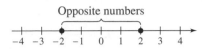

Figure 1.14

The opposite of a negative number is called a **double negative** (see Figure 1.15).

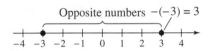

Figure 1.15

Opposite numbers are also referred to as **additive inverses** because their sum is zero. For instance, $3 + (-3) = 0$. In general, you have the following.

Opposites and Additive Inverses

Let a be a real number.

1. $-a$ is the opposite of a.

2. $-(-a) = a$ Double negative

3. $a + (-a) = 0$ Additive inverse

The distance between a real number a and 0 (the origin) is called the **absolute value** of a. Absolute value is denoted by double vertical bars $| \; |$. For example,

$$|5| = \text{“distance between 5 and 0”} = 5$$

and

$$|-8| = \text{“distance between } -8 \text{ and 0”} = 8.$$

Be sure you see from the following definition that the absolute value of a real number is never negative. For instance, if $a = -3$, then $|-3| = -(-3) = 3$. Moreover, the only real number whose absolute value is zero is 0. That is, $|0| = 0$.

Definition of Absolute Value

If a is a real number, then the **absolute value** of a is

$$|a| = \begin{cases} a, & \text{if } a \geq 0 \\ -a, & \text{if } a < 0 \end{cases}.$$

Commuting to and from work can be a good example for illustrating distance and absolute value. Assuming that the same route is used, the distance traveled to work is the same as the distance traveled back home, and the real number value is never negative.

EXAMPLE 5   **Finding Absolute Values**

a. $|-10| = 10$ The absolute value of -10 is 10.

b. $\left|\dfrac{3}{4}\right| = \dfrac{3}{4}$ The absolute value of $\frac{3}{4}$ is $\frac{3}{4}$.

c. $|-3.2| = 3.2$ The absolute value of -3.2 is 3.2.

d. $-|-6| = -(6) = -6$ The opposite of $|-6|$ is -6.

Note that part (d) does not contradict the fact that the absolute value of a number cannot be negative. The expression $-|-6|$ calls for the *opposite* of an absolute value, and so it must be negative.

 CHECKPOINT *Now try Exercise 41.*

For any two real numbers a and b, exactly one of the following orders must be true: $a < b$, $a = b$, or $a > b$. This property of real numbers is called the **Law of Trichotomy.** In words, this property tells you that if a and b are any two real numbers, then a is less than b, a is equal to b, or a is greater than b.

EXAMPLE 6   **Comparing Real Numbers**

Place the correct symbol ($<$, $>$, or $=$) between each pair of real numbers.

a. $|-2|$ ⬜ 1 **b.** $|-4|$ ⬜ $|4|$ **c.** $|12|$ ⬜ $|-15|$

d. $|-3|$ ⬜ -3 **e.** 2 ⬜ $-|-2|$ **f.** $-|-3|$ ⬜ -3

Solution

a. $|-2| > 1$, because $|-2| = 2$ and 2 is greater than 1.

b. $|-4| = |4|$, because $|-4| = 4$ and $|4| = 4$.

c. $|12| < |-15|$, because $|12| = 12$, $|-15| = 15$, and 12 is less than 15.

d. $|-3| > -3$, because $|-3| = 3$ and 3 is greater than -3.

e. $2 > -|-2|$, because $-|-2| = -2$ and 2 is greater than -2.

f. $-|-3| = -3$, because $-|-3| = -3$ and -3 is equal to -3.

 CHECKPOINT *Now try Exercise 55.*

When the distance between the two real numbers a and b was defined as $b - a$, the definition included the restriction $a \le b$. Using absolute value, you can generalize this definition. That is, if a and b are *any* two real numbers, then the distance between a and b is given by

Distance between a and $b = |b - a| = |a - b|$.

For instance, the distance between -2 and 1 is given by

$|-2 - 1| = |-3| = 3$.   Distance between -2 and 1

You could also find the distance between -2 and 1 as follows.

$|1 - (-2)| = |3| = 3$ Distance between -2 and 1

Smart Study Strategy

Go to page xxii for ways to *Create a Positive Study Environment.*

Concept Check

1. Two real numbers are plotted on the real number line. How can you tell which number is greater?
The number on the right is greater than the number on the left.

2. How are the numbers connected by each brace related?

The numbers connected by each brace are opposites.

3. Is the number 7 a rational number? Explain why or why not.
Yes. The number 7 can be written as $\frac{7}{1}$.

4. The distance between a number b and 0 is 6. Explain what you know about the number b.
The number b could be 6 or -6 because each number is 6 units from zero.

1.1 EXERCISES

Go to pages 50–51 to record your assignments.

Developing Skills

In Exercises 1–4, which of the real numbers in the set are (a) natural numbers, (b) integers, (c) rational numbers, and (d) irrational numbers? **See Example 1.**
See Additional Answers.

1. $\left\{-6, -\sqrt{6}, -\frac{4}{3}, 0, \frac{5}{8}, 1, \sqrt{2}, 2, \pi, 6\right\}$

2. $\left\{-\frac{10}{3}, -\pi, -\sqrt{3}, -1, 0, \frac{2}{5}, \sqrt{3}, \frac{5}{2}, 5, 101\right\}$

3. $\left\{-4.2, \sqrt{4}, -\frac{1}{9}, 0, \frac{3}{11}, \sqrt{11}, 5.\overline{5}, 5.543\right\}$

4. $\left\{-\sqrt{25}, -\sqrt{6}, -0.\overline{1}, -\frac{5}{3}, 0, 0.85, 3, 110\right\}$

In Exercises 5–8, use an overbar symbol to rewrite the decimal using the smallest number of digits possible.

5. $0.2222 \ldots$ $0.\overline{2}$

6. $1.5555 \ldots$ $1.\overline{5}$

7. $2.121212 \ldots$ $2.\overline{12}$

8. $0.436436436 \ldots$ $0.\overline{436}$

In Exercises 9–12, list all members of the set.

9. The integers between -5.8 and 3.2
$-5, -4, -3, -2, -1, 0, 1, 2, 3$

10. The even integers between -2.1 and 10.5
$-2, 0, 2, 4, 6, 8, 10$

11. The odd integers between π and 10 $5, 7, 9$

12. The prime numbers between 4 and 25
$5, 7, 11, 13, 17, 19, 23$

In Exercises 13 and 14, plot the real numbers on the real number line. **See Example 2.**   See Additional Answers.

13. (a) 3   (b) $\frac{5}{2}$ (c) $-\frac{7}{2}$ (d) -5.2

14. (a) 8   (b) $\frac{4}{3}$ (c) -6.75 (d) $-\frac{9}{2}$

In Exercises 15–18, approximate the two numbers and order them.

15.
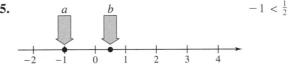
$-1 < \frac{1}{2}$

16.

$-\frac{3}{2} < \frac{7}{2}$

17.
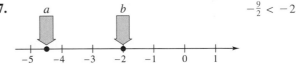
$-\frac{9}{2} < -2$

18.
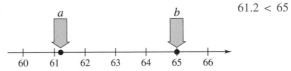
$61.2 < 65$

In Exercises 19–28, place the correct inequality symbol ($<$ or $>$) between the pair of numbers. **See Example 3.**

19. $\frac{4}{5}$ $<$ 1

20. 2 $>$ $\frac{5}{3}$

21. -5 $<$ 2

22. 9 $>$ -1

23. -5 $<$ -2

24. -8 $<$ -3

25. $\frac{5}{8}$ $>$ $\frac{1}{2}$

26. $\frac{3}{2}$ $<$ $\frac{5}{2}$

27. $-\frac{2}{3}$ $>$ $-\frac{10}{3}$

28. $-\frac{5}{3}$ $<$ $-\frac{3}{2}$

In Exercises 29–40, find the distance between the pair of real numbers. **See Example 4.**

✓ **29.** 4 and 10   6   **30.** 75 and 20   55

31. -12 and 7   19   **32.** -54 and 32   86

33. 18 and -32 50   **34.** 14 and -6 20

35. -8 and 0   8   **36.** 0 and 125   125

37. 0 and 35   35   **38.** -35 and 0   35

39. -6 and -9 3   **40.** -12 and -7 5

In Exercises 41–54, evaluate the expression. **See Example 5.**

✓ **41.** $|10|$ 10   **42.** $|62|$ 62

43. $|-225|$ 225   **44.** $|-14|$ 14

45. $-|-85|$ -85 **46.** $-|-36.5|$ -36.5

47. $-|16|$ -16 **48.** $-|-25|$ -25

49. $-\left|-\frac{3}{4}\right|$ $-\frac{3}{4}$ **50.** $-\left|\frac{3}{8}\right|$ $-\frac{3}{8}$

51. $-|3.5|$ -3.5 **52.** $|-1.4|$ 1.4

53. $|-\pi|$ π **54.** $-|\pi|$ $-\pi$

In Exercises 55–62, place the correct symbol ($<$, $>$, or $=$) between the pair of real numbers. **See Example 6.**

✓ **55.** $|-6|$ $>$ $|2|$ **56.** $|-2|$ $=$ $|2|$

57. $|47|$ $>$ $|-27|$ **58.** $|150|$ $<$ $|-310|$

59. $|-1.8|$ $=$ $|1.8|$ **60.** $|12.5|$ $>$ $-|-25|$

61. $\left|-\frac{3}{4}\right|$ $>$ $-\left|\frac{4}{5}\right|$ **62.** $-\left|-\frac{7}{3}\right|$ $<$ $-\left|\frac{1}{3}\right|$

In Exercises 63–72, find the opposite and the absolute value of the number.

63. 34 $-34, 34$ **64.** 225 $-225, 225$

65. -160 160, 160   **66.** -52 52, 52

67. $-\frac{3}{11}$ $\frac{3}{11}, \frac{3}{11}$ **68.** $\frac{7}{32}$ $-\frac{7}{32}, \frac{7}{32}$

69. $\frac{5}{4}$ $-\frac{5}{4}, \frac{5}{4}$ **70.** $\frac{4}{3}$ $-\frac{4}{3}, \frac{4}{3}$

71. 4.7 $-4.7, 4.7$ **72.** -0.4 0.4, 0.4

In Exercises 73–82, plot the number and its opposite on the real number line. Determine the distance of each from 0.   See Additional Answers.

73. -7 7   **74.** -4 4

75. 5   5   **76.** 6   6

77. $-\frac{3}{5}$ $\frac{3}{5}$ **78.** $\frac{7}{4}$ $\frac{7}{4}$

79. $\frac{5}{3}$ $\frac{5}{3}$ **80.** $-\frac{3}{4}$ $\frac{3}{4}$

81. -4.25 4.25   **82.** 3.5   3.5

In Exercises 83–90, write the statement using inequality notation.

83. x is negative.   **84.** y is more than 25.
$x < 0$ $y > 25$

85. u is at least 16.   **86.** x is nonnegative.
$u \geq 16$ $x \geq 0$

87. A bicycle racer's speed s is at least 16 miles per hour and at most 28 miles per hour.
$16 \leq s \leq 28$

88. The tire pressure p is at least 30 pounds per square inch and no more than 35 pounds per square inch.
$30 \leq p \leq 35$

89. The price p is less than \$225. $p < 225$

90. The average a will exceed 5000. $a > 5000$

In Exercises 91–94, find two possible values of a.

91. $|a| = 4$ $-4, 4$ **92.** $-|a| = -7$ $7, -7$

93. The distance between a and 3 is 5. $-2, 8$

94. The distance between a and -1 is 6. $-7, 5$

96. True. The distance between zero and the number b is the same as the distance between zero and the opposite of b.

_____ **Explaining Concepts** _____

✖ *True or False?*   In Exercises 95 and 96, decide whether the statement is true or false. Explain your reasoning.

95. Every real number is either rational or irrational.
True. If a number can be written as the ratio of two integers, it is rational. If not, the number is irrational.

96. The distance between a number b and its opposite is equal to the distance between 0 and twice the number b.

97. ✎ Describe the difference between the rational numbers 0.15 and $0.\overline{15}$. 0.15 is a terminating rational number and $0.\overline{15}$ is a repeating rational number.

98. ✎ Is there a difference between saying that a real number is positive and saying that a real number is nonnegative? Explain your answer.
Yes. The nonnegative real numbers include zero.

The symbol ✖ indicates an exercise that can be used as a group discussion problem.

1.2 Operations with Real Numbers

© JupiterImages/Pixland/Alamy

Why You Should Learn It

Real numbers can be used to represent many real-life quantities, such as the net profits for Columbia Sportswear Company (see Exercise 136 on page 21).

1 ▶ Add, subtract, multiply, and divide real numbers.

What You Should Learn

1 ▶ Add, subtract, multiply, and divide real numbers.

2 ▶ Write repeated multiplication in exponential form and evaluate exponential expressions.

3 ▶ Use order of operations to evaluate expressions.

4 ▶ Evaluate expressions using a calculator and order of operations.

Operations with Real Numbers

There are four basic operations of arithmetic: addition, subtraction, multiplication, and division.

The result of adding two real numbers is the **sum** of the two numbers, and the two real numbers are the **terms** of the sum. The rules for adding real numbers are as follows.

> ### Addition of Two Real Numbers
> 1. To **add** two real numbers with *like signs*, add their absolute values and attach the common sign to the result.
> 2. To **add** two real numbers with *unlike signs*, subtract the smaller absolute value from the greater absolute value and attach the sign of the number with the greater absolute value.

EXAMPLE 1   **Adding Integers**

a. $-84 + 14 = -(84 - 14)$ Use negative sign.

$= -70$ Subtract absolute values.

b. $-138 + (-62) = -(138 + 62)$ Use common sign.

$= -200$ Add absolute values.

 **CHECKPOINT** *Now try Exercise 7.*

EXAMPLE 2   **Adding Decimals**

a. $-26.41 + (-0.53) = -(26.41 + 0.53)$ Use common sign.

$= -26.94$ Add absolute values.

b. $3.2 + (-0.4) = +(3.2 - 0.4)$ Use positive sign.

$= 2.8$ Subtract absolute values.

 **CHECKPOINT** *Now try Exercise 9.*

The result of subtracting two real numbers is the **difference** of the two numbers. Subtraction of two real numbers is defined in terms of addition, as follows.

Subtraction of Two Real Numbers

To **subtract** the real number b from the real number a, add the opposite of b to a. That is, $a - b = a + (-b)$.

EXAMPLE 3   Subtracting Integers

Find each difference.

a. $9 - 21$ **b.** $-15 - 8$

Solution

a. $9 - 21 = 9 + (-21)$ Add opposite of 21.

$= -(21 - 9) = -12$ Use negative sign and subtract absolute values.

b. $-15 - 8 = -15 + (-8)$ Add opposite of 8.

$= -(15 + 8) = -23$ Use common sign and add absolute values.

✓ **CHECKPOINT** *Now try Exercise 11.*

EXAMPLE 4   Subtracting Decimals

Find each difference.

a. $-2.5 - (-2.7)$ **b.** $-7.02 - 13.8$

Solution

a. $-2.5 - (-2.7) = -2.5 + 2.7$ Add opposite of -2.7.

$= +(2.7 - 2.5) = 0.2$ Use positive sign and subtract absolute values.

b. $-7.02 - 13.8 = -7.02 + (-13.8)$ Add opposite of 13.8.

$= -(7.02 + 13.8) = -20.82$ Use common sign and add absolute values.

✓ **CHECKPOINT** *Now try Exercise 13.*

EXAMPLE 5   Evaluating an Expression

Evaluate $-13 - 7 + 11 - (-4)$.

Solution

$-13 - 7 + 11 - (-4) = -13 + (-7) + 11 + 4$ Add opposites.

$= -20 + 15$ Add two numbers at a time.

$= -5$ Add.

✓ **CHECKPOINT** *Now try Exercise 19.*

To add or subtract fractions, it is useful to recognize the equivalent forms of fractions, as illustrated below.

$$\frac{a}{b} = \frac{-a}{-b} = -\frac{-a}{b} = -\frac{a}{-b} \qquad \text{All are positive.}$$

$$-\frac{a}{b} = \frac{-a}{b} = \frac{a}{-b} = -\frac{-a}{-b} \qquad \text{All are negative.}$$

Study Tip

Here is an alternative method for adding and subtracting fractions with unlike denominators ($b \neq 0$ and $d \neq 0$).

$$\frac{a}{b} + \frac{c}{d} = \frac{ad + bc}{bd}$$

$$\frac{a}{b} - \frac{c}{d} = \frac{ad - bc}{bd}$$

For example,

$$\frac{1}{6} + \frac{3}{8} = \frac{1(8) + 6(3)}{6(8)}$$

$$= \frac{8 + 18}{48}$$

$$= \frac{26}{48}$$

$$= \frac{13}{24}.$$

Note that an additional step is needed to simplify the fraction after the numerators have been added.

Addition and Subtraction of Fractions

1. *Like Denominators:* The sum and difference of two fractions with like denominators ($c \neq 0$) are:

$$\frac{a}{c} + \frac{b}{c} = \frac{a + b}{c} \qquad \frac{a}{c} - \frac{b}{c} = \frac{a - b}{c}$$

2. *Unlike Denominators:* To add or subtract two fractions with unlike denominators, first rewrite the fractions so that they have the same denominator and then apply the first rule.

To find the **least common denominator (LCD)** for two or more fractions, find the **least common multiple (LCM)** of their denominators. For instance, the LCM of 6 and 8 is 24. To see this, consider all multiples of 6 (6, 12, 18, 24, 30, 36, 42, 48, . . .) and all multiples of 8 (8, 16, 24, 32, 40, 48, . . .). The numbers 24 and 48 are common multiples, and the number 24 is the smallest of the common multiples. To add $\frac{1}{6}$ and $\frac{3}{8}$, proceed as follows.

$$\frac{1}{6} + \frac{3}{8} = \frac{1(4)}{6(4)} + \frac{3(3)}{8(3)} = \frac{4}{24} + \frac{9}{24} = \frac{4 + 9}{24} = \frac{13}{24}$$

EXAMPLE 6   Adding and Subtracting Fractions

a. $\dfrac{5}{17} + \dfrac{9}{17} = \dfrac{5 + 9}{17}$ — Add numerators.

$= \dfrac{14}{17}$ — Simplify.

b. $\dfrac{3}{8} - \dfrac{5}{12} = \dfrac{3(3)}{8(3)} - \dfrac{5(2)}{12(2)}$ — Least common denominator is 24.

$= \dfrac{9}{24} - \dfrac{10}{24}$ — Simplify.

$= \dfrac{9 - 10}{24}$ — Subtract numerators.

$= -\dfrac{1}{24}$ — Simplify.

✓ **CHECKPOINT** *Now try Exercise 21.*

Study Tip

A quick way to convert the mixed number $1\frac{4}{5}$ into the fraction $\frac{9}{5}$ is to multiply the whole number by the denominator of the fraction and add the result to the numerator, as follows.

$$1\frac{4}{5} = \frac{1(5) + 4}{5} = \frac{9}{5}$$

EXAMPLE 7   **Adding Mixed Numbers**

Find the sum of $1\frac{4}{5}$ and $\frac{11}{7}$.

Solution

$$1\frac{4}{5} + \frac{11}{7} = \frac{9}{5} + \frac{11}{7} \qquad \text{Write } 1\frac{4}{5} \text{ as } \frac{9}{5}.$$

$$= \frac{9(7)}{5(7)} + \frac{11(5)}{7(5)} \qquad \text{Least common denominator is 35.}$$

$$= \frac{63}{35} + \frac{55}{35} \qquad \text{Simplify.}$$

$$= \frac{63 + 55}{35} = \frac{118}{35} \qquad \text{Add numerators and simplify.}$$

✓ **CHECKPOINT** *Now try Exercise 29.*

Multiplication of two real numbers can be described as *repeated addition*. For instance, 7×3 can be described as $3 + 3 + 3 + 3 + 3 + 3 + 3$. Multiplication is denoted in a variety of ways. For instance, 7×3, $7 \cdot 3$, $7(3)$, and $(7)(3)$ all denote the product "7 times 3." The result of multiplying two real numbers is their **product,** and each of the two numbers is a **factor** of the product.

Multiplication of Two Real Numbers

1. To multiply two real numbers with *like signs*, find the product of their absolute values. The product is *positive*.

2. To multiply two real numbers with *unlike signs*, find the product of their absolute values, and attach a minus sign. The product is *negative*.

3. The product of zero and any other real number is zero.

EXAMPLE 8   **Multiplying Integers**

Study Tip

To find the product of two or more numbers, first find the product of their absolute values. If there is an *even* number of negative factors, as in Example 8(c), the product is positive. If there is an *odd* number of negative factors, as in Example 8(a), the product is negative.

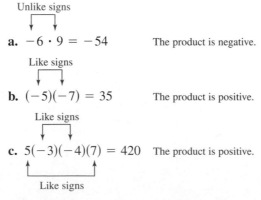

Unlike signs

a. $-6 \cdot 9 = -54$ The product is negative.

Like signs

b. $(-5)(-7) = 35$ The product is positive.

Like signs

c. $5(-3)(-4)(7) = 420$ The product is positive.

Like signs

✓ **CHECKPOINT** *Now try Exercise 45.*

Study Tip

When operating with fractions, you should check to see whether your answers can be simplified by dividing out factors that are common to the numerator and denominator. For instance, the fraction $\frac{4}{6}$ can be written in simplified form as

$$\frac{4}{6} = \frac{\overset{1}{\cancel{2}} \cdot 2}{\underset{1}{\cancel{2}} \cdot 3} = \frac{2}{3}.$$

Note that dividing out a common factor is the division of a number by itself, and what remains is a factor of 1.

Multiplication of Two Fractions

The product of the two fractions a/b and c/d is given by

$$\frac{a}{b} \cdot \frac{c}{d} = \frac{ac}{bd}, \quad b \neq 0, \quad d \neq 0.$$

EXAMPLE 9   Multiplying Fractions

Find the product.

$$\left(-\frac{3}{8}\right)\left(\frac{11}{6}\right)$$

Solution

$$\left(-\frac{3}{8}\right)\left(\frac{11}{6}\right) = -\frac{3(11)}{8(6)} \qquad \text{Multiply numerators and denominators.}$$

$$= -\frac{\cancel{3}(11)}{8(2)(\cancel{3})} \qquad \text{Factor and divide out common factor.}$$

$$= -\frac{11}{16} \qquad \text{Simplify.}$$

✓ **CHECKPOINT**   *Now try Exercise 57.*

The **reciprocal** of a nonzero real number a is defined as the number by which a must be multiplied to obtain 1. For instance, the reciprocal of 3 is $\frac{1}{3}$ because

$$3\left(\frac{1}{3}\right) = 1.$$

Similarly, the reciprocal of $-\frac{4}{5}$ is $-\frac{5}{4}$ because

$$-\frac{4}{5}\left(-\frac{5}{4}\right) = 1.$$

In general, the reciprocal of a/b is b/a. Note that the reciprocal of a positive number is positive, and the reciprocal of a negative number is negative.

Study Tip

Division by 0 is not defined because 0 has no reciprocal. If 0 had a reciprocal value b, then you would obtain the *false* result

$\dfrac{1}{0} = b$ The reciprocal of zero is b.

$1 = b \cdot 0$ Multiply each side by 0.

$1 = 0.$ False result, $1 \neq 0$

Division of Two Real Numbers

To divide the real number a by the nonzero real number b, multiply a by the reciprocal of b. That is,

$$a \div b = a \cdot \frac{1}{b}, \quad b \neq 0.$$

The result of dividing two real numbers is the **quotient** of the numbers. The number a is the **dividend** and the number b is the **divisor.** When the division is expressed as a/b or $\dfrac{a}{b}$, a is the **numerator** and b is the **denominator.**

EXAMPLE 10   Division of Real Numbers

a. $-30 \div 5 = -30 \cdot \dfrac{1}{5}$ Invert divisor and multiply.

$\qquad\qquad = -\dfrac{30}{5}$ Multiply.

$\qquad\qquad = -\dfrac{6 \cdot \cancel{5}}{\cancel{5}}$ Factor and divide out common factor.

$\qquad\qquad = -6$ Simplify.

b. $\dfrac{5}{16} \div 2\dfrac{3}{4} = \dfrac{5}{16} \div \dfrac{11}{4}$ Write $2\frac{3}{4}$ as $\frac{11}{4}$.

$\qquad\qquad = \dfrac{5}{16} \cdot \dfrac{4}{11}$ Invert divisor and multiply.

$\qquad\qquad = \dfrac{5(4)}{16(11)}$ Multiply.

$\qquad\qquad = \dfrac{5}{44}$ Simplify.

✓ **CHECKPOINT** *Now try Exercise 71.*

---

2 ▶ Write repeated multiplication in exponential form and evaluate exponential expressions.

Positive Integer Exponents

Repeated multiplication can be written in what is called **exponential form.**

Repeated Multiplication		Exponential Form
$\underbrace{7 \cdot 7 \cdot 7 \cdot 7}_{\text{4 factors of 7}}$	$=$	7^4
$\underbrace{\left(-\frac{3}{4}\right)\left(-\frac{3}{4}\right)\left(-\frac{3}{4}\right)}_{\text{3 factors of } -\frac{3}{4}}$	$=$	$\left(-\dfrac{3}{4}\right)^3$

Technology: Discovery

When a negative number is raised to a power, the use of parentheses is very important. To discover why, use a calculator to evaluate $(-4)^4$ and -4^4. Write a statement explaining the results. Then use a calculator to evaluate $(-4)^3$ and -4^3. If necessary, write a new statement explaining your discoveries.

Exponential Notation

Let n be a positive integer and let a be a real number. Then the product of n factors of a is given by

$$a^n = \underbrace{a \cdot a \cdot a \cdots a}_{n \text{ factors}}.$$

In the exponential form a^n, a is the **base** and n is the **exponent.** Writing the exponential form a^n is called **"raising a to the nth power."**

When a number, say 5, is raised to the *first* power, you would usually write 5 rather than 5^1. Raising a number to the *second* power is called **squaring** the number. Raising a number to the *third* power is called **cubing** the number.

EXAMPLE 11   **Evaluating Exponential Expressions**

a. $(-3)^4 = (-3)(-3)(-3)(-3) = 81$ Negative sign is part of the base.

b. $-3^4 = -(3)(3)(3)(3) = -81$ Negative sign is not part of the base.

c. $\left(\dfrac{2}{5}\right)^3 = \left(\dfrac{2}{5}\right)\left(\dfrac{2}{5}\right)\left(\dfrac{2}{5}\right) = \dfrac{8}{125}$

d. $(-5)^3 = (-5)(-5)(-5) = -125$ Negative raised to odd power.

e. $(-5)^4 = (-5)(-5)(-5)(-5) = 625$ Negative raised to even power.

 CHECKPOINT *Now try Exercise 91.*

In parts (d) and (e) of Example 11, note that when a negative number is raised to an *odd* power, the result is *negative,* and when a negative number is raised to an *even* power, the result is *positive.*

3 ▶ Use order of operations to evaluate expressions.

Order of Operations

One of your goals in studying this book is to learn to communicate about algebra by reading and writing information about numbers. One way to help avoid confusion when communicating algebraic ideas is to establish an **order of operations.** This is done by giving priorities to different operations. First priority is given to exponents, second priority is given to multiplication and division, and third priority is given to addition and subtraction. To distinguish between operations with the same priority, use the *Left-to-Right Rule.*

Study Tip

The order of operations for multiplication applies when multiplication is written with the symbol $\times$ or $\cdot$. When multiplication is implied by parentheses, it has a higher priority than the Left-to-Right Rule. For instance,

$$8 \div 4(2) = 8 \div 8 = 1$$

but

$$8 \div 4 \cdot 2 = 2 \cdot 2 = 4.$$

Order of Operations

To evaluate an expression involving more than one operation, use the following order.

1. First do operations that occur within symbols of grouping.
2. Then evaluate powers.
3. Then do multiplications and divisions from left to right.
4. Finally, do additions and subtractions from left to right.

EXAMPLE 12   **Order of Operations Without Symbols of Grouping**

a. $20 - 2 \cdot 3^2 = 20 - 2 \cdot 9$ Evaluate power.

$\quad\quad\quad\quad\quad = 20 - 18 = 2$ Multiply, then subtract.

b. $5 - 6 - 2 = (5 - 6) - 2$ Left-to-Right Rule

$\quad\quad\quad\quad = -1 - 2 = -3$ Subtract.

c. $8 \div 2 \cdot 2 = (8 \div 2) \cdot 2$ Left-to-Right Rule

$\quad\quad\quad\quad = 4 \cdot 2 = 8$ Divide, then multiply.

 CHECKPOINT *Now try Exercise 105.*

When you want to change the established order of operations, you must use parentheses or other symbols of grouping. Part (d) in the next example shows that a fraction bar acts as a symbol of grouping.

EXAMPLE 13   Order of Operations with Symbols of Grouping

a. $7 - 3(4 - 2) = 7 - 3(2)$ Subtract within symbols of grouping.

$ = 7 - 6 = 1$ Multiply, then subtract.

b. $4 - 3(2)^3 = 4 - 3(8)$ Evaluate power.

$ = 4 - 24 = -20$ Multiply, then subtract.

c. $1 - [4 - (5 - 3)] = 1 - (4 - 2)$ Subtract within symbols of grouping.

$ = 1 - 2 = -1$ Subtract within symbols of grouping, then subtract.

d. $\dfrac{2 \cdot 5^2 - 10}{3^2 - 4} = (2 \cdot 5^2 - 10) \div (3^2 - 4)$ Rewrite using parentheses.

$\phantom{\dfrac{2 \cdot 5^2 - 10}{3^2 - 4}} = (50 - 10) \div (9 - 4)$ Evaluate powers and multiply within symbols of grouping.

$\phantom{\dfrac{2 \cdot 5^2 - 10}{3^2 - 4}} = 40 \div 5 = 8$ Subtract within symbols of grouping, then divide.

✓ **CHECKPOINT** *Now try Exercise 109.*

4 ► Evaluate expressions using a calculator and order of operations.

Calculators and Order of Operations

When using your own calculator, be sure that you are familiar with the use of each of the keys. Two possible keystroke sequences are given in Example 14: one for a standard *scientific* calculator, and one for a *graphing* calculator.

Technology: Tip

Be sure you see the difference between the change sign key $\boxed{+/-}$ and the subtraction key $\boxed{-}$ on a scientific calculator. Also notice the difference between the negation key $\boxed{(-)}$ and the subtraction key $\boxed{-}$ on a graphing calculator.

Technology: Discovery

To discover if your calculator performs the established order of operations, evaluate $7 + 5 \cdot 3 - 2^4 \div 4$ exactly as it appears. If your calculator performs the established order of operations, it will display 18.

EXAMPLE 14   Evaluating Expressions on a Calculator

a. To evaluate the expression $7 - (5 \cdot 3)$, use the following keystrokes.

Keystrokes	Display	
7 ⊟ ⦅ 5 ⊠ 3 ⦆ ⊟	−8	Scientific
7 ⊟ ⦅ 5 ⊠ 3 ⦆ ENTER	−8	Graphing

b. To evaluate the expression $(-3)^2 + 4$, use the following keystrokes.

Keystrokes	Display	
3 +/− x² ⊞ 4 ⊟	13	Scientific
⦅ (−) 3 ⦆ x² ⊞ 4 ENTER	13	Graphing

c. To evaluate the expression $5/(4 + 3 \cdot 2)$, use the following keystrokes.

Keystrokes	Display	
5 ÷ ⦅ 4 ⊞ 3 ⊠ 2 ⦆ ⊟	0.5	Scientific
5 ÷ ⦅ 4 ⊞ 3 ⊠ 2 ⦆ ENTER	.5	Graphing

✓ **CHECKPOINT** *Now try Exercise 125.*

Concept Check

1. Is the reciprocal of every nonzero integer an integer?

2. Can the sum of two real numbers be less than either number? If so, give an example.

3. Explain how to subtract one real number from another.

4. If $a > 0$, state the values of n such that $(-a)^n = -a^n$.

1.2 EXERCISES

Go to pages 50–51 to record your assignments.

Developing Skills

In Exercises 1–38, evaluate the expression. *See Examples 1–7.*

1. $13 + 32$

2. $16 + 84$

3. $-8 + 12$

4. $-5 + 9$

5. $-6.4 + 3.7$

6. $-5.1 + 0.9$

✓ **7.** $13 + (-6)$

8. $12 + (-10)$

✓ **9.** $12.6 + (-38.5)$

10. $10.4 + (-43.5)$

✓ **11.** $-8 - 12$

12. $-3 - 17$

✓ **13.** $-21.5 - (-6.3)$

14. $-13.2 - 9.6$

15. $4 - (-11) + 9$

16. $-17 + 6 - (-24)$

17. $5.3 - 2.2 - 6.9$

18. $46.08 - 35.1 - 16.25$

✓ **19.** $15 - 6 + 31 + (-18)$

20. $6 + 26 - 17 + (-10)$

✓ **21.** $\frac{3}{8} + \frac{7}{8}$

22. $\frac{5}{6} + \frac{7}{6}$

23. $\frac{3}{4} - \frac{1}{4}$

24. $\frac{5}{9} - \frac{1}{9}$

25. $\frac{3}{5} + \left(-\frac{1}{2}\right)$

26. $\frac{6}{7} + \left(-\frac{3}{7}\right)$

27. $\frac{5}{8} + \frac{1}{4} - \frac{5}{6}$

28. $\frac{3}{10} - \frac{5}{2} + \frac{1}{5}$

✓ **29.** $3\frac{1}{2} + 4\frac{3}{8}$

30. $5\frac{3}{4} + 7\frac{3}{8}$

31. $10\frac{5}{8} - 6\frac{1}{4}$

32. $8\frac{1}{2} - 4\frac{2}{3}$

33. $85 - |-25|$

34. $-36 + |-8|$

35. $-(-11.325) + |34.625|$

36. $|-16.25| - 54.78$

37. $-\left|-6\frac{7}{8}\right| - 8\frac{1}{4}$

38. $-\left|-15\frac{2}{3}\right| - 12\frac{1}{3}$

In Exercises 39–44, write the expression as a multiplication problem.

39. $9 + 9 + 9 + 9$

40. $(-15) + (-15) + (-15) + (-15)$

41. $\frac{1}{4} + \frac{1}{4} + \frac{1}{4} + \frac{1}{4} + \frac{1}{4} + \frac{1}{4}$

42. $\frac{2}{3} + \frac{2}{3} + \frac{2}{3} + \frac{2}{3}$

43. $\left(-\frac{1}{5}\right) + \left(-\frac{1}{5}\right) + \left(-\frac{1}{5}\right) + \left(-\frac{1}{5}\right)$

44. $\left(-\frac{5}{22}\right) + \left(-\frac{5}{22}\right) + \left(-\frac{5}{22}\right)$

In Exercises 45–62, find the product. *See Examples 8 and 9.*

✓ **45.** $5(-6)$

46. $3(-9)$

47. $(-8)(-6)$

48. $(-4)(-7)$

49. $2(4)(-5)$

50. $3(-7)(10)$

51. $(-1)(12)(-3)$

52. $(-2)(-6)(4)$

53. $\left(-\frac{5}{8}\right)\left(-\frac{4}{5}\right)$

54. $\left(-\frac{4}{7}\right)\left(-\frac{4}{5}\right)$

55. $-\frac{3}{2}\left(\frac{8}{5}\right)$

56. $\left(\frac{10}{13}\right)\left(-\frac{3}{5}\right)$

✓ **57.** $\frac{1}{2}\left(\frac{1}{6}\right)$

58. $\frac{1}{3}\left(\frac{2}{3}\right)$

59. $-\frac{9}{8}\left(\frac{16}{27}\right)\left(\frac{1}{2}\right)$

60. $\frac{2}{3}\left(-\frac{18}{5}\right)\left(-\frac{5}{6}\right)$

61. $\frac{1}{3}\left(-\frac{3}{4}\right)(2)$

62. $\frac{2}{5}(-3)\left(\frac{10}{9}\right)$

In Exercises 63–68, find the reciprocal.

63. 6

64. 4

65. $\frac{2}{3}$

66. $\frac{9}{5}$

67. $-\frac{9}{7}$

68. $-\frac{2}{13}$

In Exercises 69–82, evaluate the expression. *See Example 10.*

69. $\dfrac{-18}{-3}$

70. $-\dfrac{30}{-15}$

✓ **71.** $-48 \div 16$

72. $-72 \div 12$

73. $63 \div (-7)$ **74.** $-27 \div (-9)$

75. $-\frac{4}{5} \div \frac{8}{25}$ **76.** $-\frac{11}{12} \div \frac{5}{24}$

77. $\left(-\frac{1}{3}\right) \div \left(-\frac{5}{6}\right)$ **78.** $\left(-\frac{3}{8}\right) \div \left(-\frac{4}{3}\right)$

79. $-4\frac{1}{4} \div -5\frac{5}{8}$ **80.** $-3\frac{5}{6} \div -2\frac{2}{3}$

81. $4\frac{1}{8} \div 4\frac{1}{2}$ **82.** $26\frac{2}{3} \div 10\frac{5}{6}$

In Exercises 83–88, write the expression using exponential notation.

83. $(-7) \cdot (-7) \cdot (-7)$

84. $(-4)(-4)(-4)(-4)(-4)(-4)$

85. $\left(\frac{1}{4}\right) \cdot \left(\frac{1}{4}\right) \cdot \left(\frac{1}{4}\right) \cdot \left(\frac{1}{4}\right)$

86. $\left(\frac{5}{8}\right) \cdot \left(\frac{5}{8}\right) \cdot \left(\frac{5}{8}\right) \cdot \left(\frac{5}{8}\right)$

87. $-(7 \cdot 7 \cdot 7)$

88. $-(5 \cdot 5 \cdot 5 \cdot 5 \cdot 5 \cdot 5)$

In Exercises 89–102, evaluate the exponential expression. *See Example 11.*

89. 2^5 **90.** 5^3

91. $(-2)^4$ **92.** $(-3)^3$

93. -4^3 **94.** -6^4

95. $\left(\frac{4}{5}\right)^3$ **96.** $\left(\frac{2}{3}\right)^4$

97. $-\left(-\frac{1}{2}\right)^5$ **98.** $\left(-\frac{3}{4}\right)^3$

99. $(0.3)^3$ **100.** $(0.2)^4$

101. $5(-0.4)^3$ **102.** $-3(0.8)^2$

In Exercises 103–124, evaluate the expression. *See Examples 12 and 13.*

103. $16 - 6 - 10$ **104.** $18 - 12 + 4$

105. $24 - 5 \cdot 2^2$ **106.** $18 + 3^2 - 12$

107. $28 \div 4 + 3 \cdot 5$ **108.** $6 \cdot 7 - 6^2 \div 4$

109. $14 - 2(8 - 4)$ **110.** $21 - 5(7 - 5)$

111. $17 - 5(16 \div 4^2)$ **112.** $72 - 8(6^2 \div 9)$

113. $5^2 - 2[9 - (18 - 8)]$ **114.** $8 \cdot 3^2 - 4(12 + 3)$

115. $5^3 + |-14 + 4|$ **116.** $|(-2)^5| - (25 + 7)$

117. $\dfrac{6 + 8(3)}{7 - 12}$ **118.** $\dfrac{9 + 6(2)}{3 + 4}$

119. $\dfrac{4^2 - 5}{11} - 7$ **120.** $\dfrac{5^3 - 50}{-15} + 27$

121. $\dfrac{6 \cdot 2^2 - 12}{3^2 + 3}$ **122.** $\dfrac{7^2 - 2(11)}{5^2 + 8(-2)}$

123. $\dfrac{3 + \frac{3}{4}}{\frac{1}{8}}$ **124.** $\dfrac{6 - \frac{2}{3}}{\frac{4}{9}}$

In Exercises 125–130, evaluate the expression using a calculator. Round your answer to two decimal places. *See Example 14.*

125. $5.6[13 - 2.5(-6.3)]$ **126.** $6.9[6.1(-4.2) + 16]$

127. $5^6 - 3(400)$

128. $300(1.09)^{10} + (-156.24)$

129. $\dfrac{500}{(1.055)^{20}}$ **130.** $5(100 - 3.6^4) \div 4.1$

Solving Problems

Circle Graphs In Exercises 131 and 132, find the unknown fractional part of the circle graph.

131.   **132.**

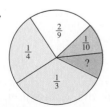   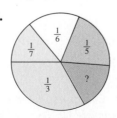

133. *Account Balance* During one month, you made the following transactions in your non-interest-bearing checking account. Find the balance at the end of the month.

NUMBER OR CODE	DATE	TRANSACTION DESCRIPTION	PAYMENT AMOUNT	✓ FEE	DEPOSIT AMOUNT	BALANCE $2618.68
	3/1	Pay			$1236 45	
2154	3/3	Magazine	$ 25 62			
2155	3/6	Insurance	$455 00			
	3/12	Withdrawal	$ 125 00			
2156	3/15	Mortgage	$ 715 95			

Figure for 133

134. *Profit* The midyear financial statement of a clothing company showed a profit of $1,345,298.55. At the close of the year, the financial statement showed a profit for the year of $867,132.87. Find the profit (or loss) of the company for the second 6 months of the year.

135. *Stock Values*   On Monday you purchased $500 worth of stock. The values of the stock during the remainder of the week are shown in the bar graph.

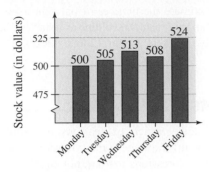

(a) Use the graph to complete the table.

Day	Daily gain or loss
Tuesday	
Wednesday	
Thursday	
Friday	

(b) Find the sum of the daily gains and losses. Interpret the result in the context of the problem. How could you determine this sum from the graph?

136. *Net Profit*   The net profits for Columbia Sportswear (in millions of dollars) for the years 2001 to 2006 are shown in the bar graph. Use the graph to create a table that shows the yearly gains or losses.   (Source: Columbia Sportswear Company)

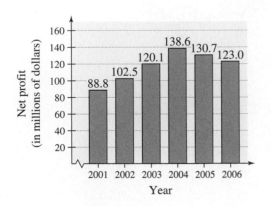

137. *Savings Plan*

(a) You save $50 per month for 18 years. How much money has been set aside during the 18 years?

(b) If the money in part (a) is deposited in a savings account earning 4% interest compounded monthly, the total amount in the account after 18 years will be

$$50\left[\left(1 + \frac{0.04}{12}\right)^{216} - 1\right]\left(1 + \frac{12}{0.04}\right).$$

Use a calculator to determine this amount.

(c) How much of the amount in part (b) is earnings from interest?

138. *Savings Plan*

(a) You save $60 per month for 30 years. How much money has been set aside during the 30 years?

(b) If the money in part (a) is deposited in a savings account earning 3% interest compounded monthly, the total amount in the account after 30 years will be

$$60\left[\left(1 + \frac{0.03}{12}\right)^{360} - 1\right]\left(1 + \frac{12}{0.03}\right).$$

Use a calculator to determine this amount.

(c) How much of the amount in part (b) is earnings from interest?

▲ *Geometry*   In Exercises 139–142, find the area of the figure. (The area *A* of a rectangle is given by $A = $ length · width, and the area *A* of a triangle is given by $A = \frac{1}{2} \cdot$ base · height.)

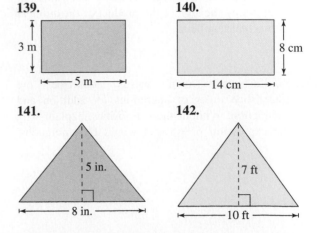

Volume In Exercises 143 and 144, use the following information. A bale of hay is a rectangular solid weighing approximately 50 pounds. It has a length of 42 inches, a width of 18 inches, and a height of 14 inches. (The volume V of a rectangular solid is given by $V = $ length $\cdot$ width $\cdot$ height.)

143. Find the volume of a bale of hay in cubic feet if 1728 cubic inches equals 1 cubic foot.

144. Approximate the number of bales in a ton of hay. Then approximate the volume of a stack of baled hay in cubic feet that weighs 12 tons. (2000 lb = 1 ton)

Explaining Concepts

True or False? In Exercises 145–149, determine whether the statement is true or false. Justify your answer.

145. The reciprocal of every nonzero rational number is a rational number.

146. The product of two fractions is the product of the numerators over the LCD.

147. If a negative real number is raised to the 12th power, the result will be positive.

148. If a negative real number is raised to the 11th power, the result will be positive.

149. $a \div b = b \div a$

150. Are the expressions $(2^2)^3$ and $2^{(2^3)}$ equal? Explain.

151. ✎ In your own words, describe the rules for determining the sign of the product or the quotient of two real numbers.

152. ✎ In your own words, describe the established order of operations for addition and subtraction. Without these priorities, explain why the expression $6 - 5 - 2$ would be ambiguous.

153. ✎ Decide which expressions are equal to 27 when you follow the standard order of operations. For the expressions that are not equal to 27, see if you can discover a way to insert symbols of grouping that make the expression equal to 27. Discuss the value of symbols of grouping in mathematical communication.

(a) $40 - 10 + 3$

(b) $5^2 + \frac{1}{2} \cdot 4$

(c) $8 \cdot 3 + 30 \div 2$

(d) $75 \div 2 + 1 + 2$

Error Analysis In Exercises 154–156, describe and correct the error.

154. $\dfrac{2}{3} + \dfrac{3}{2} = \dfrac{2+3}{3+2} = 1$

155. $\dfrac{5+12}{5} = \dfrac{\cancel{5}+12}{\cancel{5}} = 12$

156. $3 \cdot 4^2 = 12^2$

1.3 Properties of Real Numbers

What You Should Learn

1 ▶ Identify and use the properties of real numbers.

2 ▶ Develop additional properties of real numbers.

Basic Properties of Real Numbers

The following list summarizes the basic properties of addition and multiplication. Although the examples involve real numbers, these properties can also be applied to algebraic expressions.

Why You Should Learn It

Understanding the properties of real numbers will help you to understand and use the properties of algebra.

1 ▶ Identify and use the properties of real numbers.

Properties of Real Numbers

Let a, b, and c represent real numbers, variables, or algebraic expressions.

Property	*Example*
Commutative Property of Addition:	
$a + b = b + a$	$3 + 5 = 5 + 3$
Commutative Property of Multiplication:	
$ab = ba$	$2 \cdot 7 = 7 \cdot 2$
Associative Property of Addition:	
$(a + b) + c = a + (b + c)$	$(4 + 2) + 3 = 4 + (2 + 3)$
Associative Property of Multiplication:	
$(ab)c = a(bc)$	$(2 \cdot 5) \cdot 7 = 2 \cdot (5 \cdot 7)$
Distributive Property:	
$a(b + c) = ab + ac$	$4(7 + 3) = 4 \cdot 7 + 4 \cdot 3$
$(a + b)c = ac + bc$	$(2 + 5)3 = 2 \cdot 3 + 5 \cdot 3$
$a(b - c) = ab - ac$	$6(5 - 3) = 6 \cdot 5 - 6 \cdot 3$
$(a - b)c = ac - bc$	$(7 - 2)4 = 7 \cdot 4 - 2 \cdot 4$
Additive Identity Property:	
$a + 0 = 0 + a = a$	$9 + 0 = 0 + 9 = 9$
Multiplicative Identity Property:	
$a \cdot 1 = 1 \cdot a = a$	$-5 \cdot 1 = 1 \cdot (-5) = -5$
Additive Inverse Property:	
$a + (-a) = 0$	$3 + (-3) = 0$
Multiplicative Inverse Property:	
$a \cdot \dfrac{1}{a} = 1, \quad a \neq 0$	$8 \cdot \dfrac{1}{8} = 1$

Study Tip

The operations of subtraction and division are not listed at the right because they do not have many of the properties of real numbers. For instance, subtraction and division are not commutative or associative. To see this, consider the following.

$4 - 3 \neq 3 - 4$

$15 \div 5 \neq 5 \div 15$

$8 - (6 - 2) \neq (8 - 6) - 2$

$20 \div (4 \div 2) \neq (20 \div 4) \div 2$

EXAMPLE 1   **Identifying Properties of Real Numbers**

Identify the property of real numbers illustrated by each statement.

a. $4(a + 3) = 4 \cdot a + 4 \cdot 3$ **b.** $6 \cdot \dfrac{1}{6} = 1$

c. $-3 + (2 + b) = (-3 + 2) + b$

d. $(b + 8) + 0 = b + 8$

Solution

a. This statement illustrates the Distributive Property.

b. This statement illustrates the Multiplicative Inverse Property.

c. This statement illustrates the Associative Property of Addition.

d. This statement illustrates the Additive Identity Property, where $(b + 8)$ is an algebraic expression.

✔ **CHECKPOINT** *Now try Exercise 5.*

The properties of real numbers make up the third component of what is called a **mathematical system.** These three components are a *set of numbers* (Section 1.1), *operations* with the set of numbers (Section 1.2), and *properties* of the operations with the numbers (Section 1.3).

| Set of Numbers | ⟹ | Operations with the Numbers | ⟹ | Properties of the Operations |

Note that the properties of real numbers can be applied to variables and algebraic expressions as well as to real numbers.

EXAMPLE 2   **Using the Properties of Real Numbers**

Complete each statement using the specified property of real numbers.

a. Multiplicative Identity Property: $(4a)1 = $

b. Associative Property of Addition: $(b + 8) + 3 = $

c. Additive Inverse Property: $0 = 5c + $

d. Distributive Property: $7 \cdot b + 7 \cdot 5 = $

Solution

a. By the Multiplicative Identity Property, $(4a)1 = 4a$.

b. By the Associative Property of Addition, $(b + 8) + 3 = b + (8 + 3)$.

c. By the Additive Inverse Property, $0 = 5c + (-5c)$.

d. By the Distributive Property, $7 \cdot b + 7 \cdot 5 = 7(b + 5)$.

✔ **CHECKPOINT** *Now try Exercise 21.*

To help you understand each property of real numbers, try stating the properties in your own words. For instance, the Associative Property of Addition can be stated as follows: *When three real numbers are added, it makes no difference which two are added first.*

2 ▶ Develop additional properties of real numbers.

Additional Properties of Real Numbers

Once you have determined the basic properties (or *axioms*) of a mathematical system, you can go on to develop other properties. These additional properties are **theorems,** and the formal arguments that justify the theorems are **proofs.**

Additional Properties of Real Numbers

Let a, b, and c be real numbers, variables, or algebraic expressions.

Properties of Equality

Addition Property of Equality: If $a = b$, then $a + c = b + c$.

Multiplication Property of Equality: If $a = b$, then $ac = bc$.

Cancellation Property of Addition: If $a + c = b + c$, then $a = b$.

Cancellation Property of Multiplication: If $ac = bc$ and $c \neq 0$, then $a = b$.

Properties of Zero

Multiplication Property of Zero: $0 \cdot a = 0$

Division Property of Zero: $\dfrac{0}{a} = 0, \quad a \neq 0$

Division by Zero Is Undefined: $\dfrac{a}{0}$ is undefined.

Properties of Negation

Multiplication by -1: $(-1)(a) = -a,$
    $(-1)(-a) = a$

Placement of Negative Signs: $(-a)(b) = -(ab) = (a)(-b)$

Product of Two Opposites: $(-a)(-b) = ab$

Study Tip

When the properties of real numbers are used in practice, the process is usually less formal than it would appear from the list of properties on this page. For instance, the steps shown at the right are less formal than those shown in Examples 5 and 6 on page 27. The importance of the properties is that they can be used to justify the steps of a solution. They do not always need to be listed for every step of the solution.

In Section 2.1, you will see that the properties of equality are useful for solving equations, as shown below. Note that the Addition and Multiplication Properties of Equality can be used to subtract the same nonzero quantity from each side of an equation or to divide each side of an equation by the same nonzero quantity.

$5x + 4 = -2x + 18$	Original equation
$5x + 4 - 4 = -2x + 18 - 4$	Subtract 4 from each side.
$5x = -2x + 14$	Simplify.
$5x + 2x = -2x + 2x + 14$	Add $2x$ to each side.
$7x = 14$	Simplify.
$\dfrac{7x}{7} = \dfrac{14}{7}$	Divide each side by 7.
$x = 2$	Simplify.

Each of the additional properties in the list on the preceding page can be proved by using the basic properties of real numbers.

EXAMPLE 3   Proof of the Cancellation Property of Addition

Prove that if $a + c = b + c$, then $a = b$. (Use the Addition Property of Equality.)

Solution

Notice how each step is justified from the preceding step by means of a property of real numbers.

$$a + c = b + c \qquad \text{Write original equation.}$$
$$(a + c) + (-c) = (b + c) + (-c) \qquad \text{Addition Property of Equality}$$
$$a + [c + (-c)] = b + [c + (-c)] \qquad \text{Associative Property of Addition}$$
$$a + 0 = b + 0 \qquad \text{Additive Inverse Property}$$
$$a = b \qquad \text{Additive Identity Property}$$

✓ **CHECKPOINT** *Now try Exercise 59.*

EXAMPLE 4   Proof of a Property of Negation

Prove that $(-1)a = -a$. (You may use any of the properties of equality and properties of zero.)

Solution

At first glance, it is a little difficult to see what you are being asked to prove. However, a good way to start is to consider carefully the definitions of the three numbers in the equation.

$$a = \text{given real number}$$
$$-1 = \text{the additive inverse of } 1$$
$$-a = \text{the additive inverse of } a$$

Now, by showing that $(-1)a$ has the same properties as the additive inverse of a, you will be showing that $(-1)a$ must be the additive inverse of a.

$$(-1)a + a = (-1)a + (1)(a) \qquad \text{Multiplicative Identity Property}$$
$$= (-1 + 1)a \qquad \text{Distributive Property}$$
$$= (0)a \qquad \text{Additive Inverse Property}$$
$$= 0 \qquad \text{Multiplication Property of Zero}$$

Because $(-1)a + a = 0$, you can use the fact that $-a + a = 0$ to conclude that $(-1)a + a = -a + a$. From this, you can complete the proof as follows.

$$(-1)a + a = -a + a \qquad \text{Shown in first part of proof}$$
$$(-1)a = -a \qquad \text{Cancellation Property of Addition}$$

✓ **CHECKPOINT** *Now try Exercise 61.*

The list of additional properties of real numbers on page 25 forms a very important part of algebra. Knowing the names of the properties is useful, but knowing how to use each property is extremely important. The next two examples show how several of the properties can be used to solve equations. (You will study these techniques in detail in Section 2.1.)

EXAMPLE 5   Applying the Properties of Real Numbers

In the solution of the equation

$b + 2 = 6$

identify the property of real numbers that justifies each step.

Solution

$$b + 2 = 6 \qquad \text{Original equation}$$

Solution Step	Property
$(b + 2) + (-2) = 6 + (-2)$	Addition Property of Equality
$b + [2 + (-2)] = 6 - 2$	Associative Property of Addition
$b + 0 = 4$	Additive Inverse Property
$b = 4$	Additive Identity Property

✓ **CHECKPOINT** *Now try Exercise 63.*

EXAMPLE 6   Applying the Properties of Real Numbers

In the solution of the equation

$3x = 15$

identify the property of real numbers that justifies each step.

Solution

$$3x = 15 \qquad \text{Original equation}$$

Solution Step	Property
$\left(\frac{1}{3}\right)3x = \left(\frac{1}{3}\right)15$	Multiplication Property of Equality
$\left(\frac{1}{3} \cdot 3\right)x = 5$	Associative Property of Multiplication
$(1)(x) = 5$	Multiplicative Inverse Property
$x = 5$	Multiplicative Identity Property

✓ **CHECKPOINT** *Now try Exercise 65.*

—————— **Concept Check** ——————

1. Does every real number have an additive inverse? Explain.

2. Does every real number have a multiplicative inverse? Explain.

3. Are subtraction and division commutative? If not, show a counterexample.

4. Are subtraction and division associative? If not, show a counterexample.

1.3 EXERCISES

Go to pages 50–51 to record your assignments.

—————— **Developing Skills** ——————

In Exercises 1–20, identify the property of real numbers illustrated by the statement. *See Example 1.*

1. $18 - 18 = 0$

2. $5 + 0 = 5$

3. $\frac{1}{12} \cdot 12 = 1$

4. $52 \cdot 1 = 52$

 5. $13 + 12 = 12 + 13$

6. $(-4 \cdot 10) \cdot 8 = -4(10 \cdot 8)$

7. $3 + (12 - 9) = (3 + 12) - 9$

8. $(5 + 10)(8) = 8(5 + 10)$

9. $(8 - 5)(10) = 8 \cdot 10 - 5 \cdot 10$

10. $7(9 + 15) = 7 \cdot 9 + 7 \cdot 15$

11. $10(2x) = (10 \cdot 2)x$

12. $1 \cdot 9k = 9k$

13. $10x \cdot \dfrac{1}{10x} = 1$

14. $0 + 4x = 4x$

15. $2x - 2x = 0$

16. $4 + (3 - x) = (4 + 3) - x$

17. $3(2 + x) = 3 \cdot 2 + 3x$

18. $3(6 + b) = 3 \cdot 6 + 3 \cdot b$

19. $(x + 1) - (x + 1) = 0$

20. $6(x + 3) = 6 \cdot x + 6 \cdot 3$

In Exercises 21–28, complete the statement using the specified property of real numbers. *See Example 2.*

21. Commutative Property of Multiplication:
$15(-3) =$

22. Associative Property of Addition:
$6 + (5 + y) =$

23. Distributive Property:
$5(6 + z) =$

24. Distributive Property:
$(8 - y)(4) =$

25. Commutative Property of Addition:
$25 + (-x) =$

26. Additive Inverse Property:
$13x + (-13x) =$

27. Multiplicative Identity Property:
$(x + 8) \cdot 1 =$

28. Additive Identity Property:
$(8x) + 0 =$

In Exercises 29–40, give (a) the additive inverse and (b) the multiplicative inverse of the quantity.

29. 10

30. 18

31. -19

32. -37

33. $\frac{1}{2}$

34. $\frac{3}{4}$

35. $-\frac{5}{8}$

36. $-\frac{1}{5}$

37. $6z, \quad z \neq 0$

38. $2y, \quad y \neq 0$

39. $x - 2, \quad x \neq 2$

40. $y - 7, \quad y \neq 7$

In Exercises 41–44, rewrite the expression using the Associative Property of Addition or the Associative Property of Multiplication.

41. $32 + (4 + y)$

42. $15 + (3 - x)$

43. $9(6m)$

44. $11(4n)$

In Exercises 45–50, rewrite the expression using the Distributive Property.

45. $20(2 + 5)$

46. $-3(4 - 8)$

47. $(x + 6)(-2)$

48. $(z - 10)(12)$

49. $-6(2y - 5)$

50. $-4(10 - b)$

In Exercises 51–54, use the Distributive Property to simplify the expression.

51. $7x + 2x$

52. $8x - 6x$

53. $\dfrac{7x}{8} - \dfrac{5x}{8}$

54. $\dfrac{3x}{5} + \dfrac{x}{5}$

In Exercises 55–58, the right side of the statement does not equal the left side. Change the right side so that it *does* equal the left side.

55. $3(x + 5) \neq 3x + 5$

56. $4(x + 2) \neq 4x + 2$

57. $-2(x + 8) \neq -2x + 16$

58. $-9(x + 4) \neq -9x + 36$

In Exercises 59–62, use the basic properties of real numbers to prove the statement. *See Examples 3 and 4.*

✔ **59.** If $ac = bc$ and $c \neq 0$, then $a = b$.

60. If $a + c = b + c$, then $a = b$.

✔ **61.** $a = (a + b) + (-b)$

62. $a + (-a) = 0$

In Exercises 63–66, identify the property of real numbers that justifies each step. *See Examples 5 and 6.*

✔ **63.**
$$x + 5 = 3$$
$$(x + 5) + (-5) = 3 + (-5)$$
$$x + [5 + (-5)] = -2$$
$$x + 0 = -2$$
$$x = -2$$

64.
$$x - 8 = 20$$
$$(x - 8) + 8 = 20 + 8$$
$$x + (-8 + 8) = 28$$
$$x + 0 = 28$$
$$x = 28$$

✔ **65.**
$$2x - 5 = 6$$
$$(2x - 5) + 5 = 6 + 5$$
$$2x + (-5 + 5) = 11$$
$$2x + 0 = 11$$
$$2x = 11$$
$$\tfrac{1}{2}(2x) = \tfrac{1}{2}(11)$$
$$\left(\tfrac{1}{2} \cdot 2\right)x = \tfrac{11}{2}$$
$$1 \cdot x = \tfrac{11}{2}$$
$$x = \tfrac{11}{2}$$

66.
$$3x + 4 = 10$$
$$(3x + 4) + (-4) = 10 + (-4)$$
$$3x + [4 + (-4)] = 6$$
$$3x + 0 = 6$$
$$3x = 6$$
$$\tfrac{1}{3}(3x) = \tfrac{1}{3}(6)$$
$$\left(\tfrac{1}{3} \cdot 3\right)x = 2$$
$$1 \cdot x = 2$$
$$x = 2$$

Mental Math In Exercises 67–72, use the Distributive Property to perform the arithmetic mentally. For example, you work in an industry in which the wage is $14 per hour with "time and a half" for overtime. So, your hourly wage for overtime is

$$14(1.5) = 14\left(1 + \frac{1}{2}\right) = 14 + 7 = \$21.$$

67. $16(1.75) = 16\left(2 - \tfrac{1}{4}\right)$

68. $15\left(1\tfrac{2}{3}\right) = 15\left(2 - \tfrac{1}{3}\right)$

69. $7(62) = 7(60 + 2)$

70. $5(51) = 5(50 + 1)$

71. $9(6.98) = 9(7 - 0.02)$

72. $12(19.95) = 12(20 - 0.05)$

Solving Problems

73. ▲ *Geometry* The figure shows two adjoining rectangles. Demonstrate the Distributive Property by filling in the blanks to write the total area of the two rectangles in two ways.

$$\boxed{}\left(\boxed{} + \boxed{}\right) = \boxed{} + \boxed{}$$

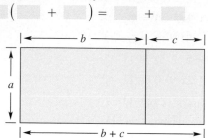

74. ▲ *Geometry* The figure shows two adjoining rectangles. Demonstrate the "subtraction version" of the Distributive Property by filling in the blanks to write the area of the left rectangle in two ways.

$$\boxed{}\left(\boxed{} - \boxed{}\right) = \boxed{} - \boxed{}$$

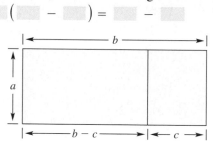

▲ *Geometry* In Exercises 75 and 76, write the expression for the perimeter of the triangle shown in the figure. Use the properties of real numbers to simplify the expression.

75.

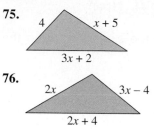

76.

77.

78.

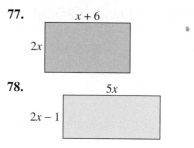

▲ *Geometry* In Exercises 77 and 78, write and simplify the expression for (a) the perimeter and (b) the area of the rectangle.

Explaining Concepts

79. What is the additive inverse of a real number? Give an example of the Additive Inverse Property.

80. What is the multiplicative inverse of a real number? Give an example of the Multiplicative Inverse Property.

81. ✎ In your own words, give a verbal description of the Commutative Property of Addition.

82. ✎ Explain how the Addition Property of Equality can be used to allow you to subtract the same number from each side of an equation.

83. You define a new mathematical operation using the symbol $\odot$. This operation is defined as $a \odot b = 2 \cdot a + b$. Give examples to show that this operation is neither commutative nor associative.

84. You define a new mathematical operation using the symbol $\ddagger$. This operation is defined as $a \ddagger b = a - (b + 1)$. Give examples to show that this operation is neither commutative nor associative.

Mid-Chapter Quiz

Take this quiz as you would take a quiz in class. After you are done, check your work against the answers in the back of the book.

In Exercises 1 and 2, plot the two real numbers on the real number line and place the correct inequality symbol ($<$ or $>$) between the two numbers.

1. -4.5 ░░░ -6 **2.** $\frac{3}{4}$ ░░░ $\frac{3}{2}$

In Exercises 3 and 4, find the distance between the two real numbers.

3. -15 and 7 **4.** -8.75 and -2.25

In Exercises 5 and 6, evaluate the expression.

5. $|-7.6|$ **6.** $-|9.8|$

In Exercises 7–16, evaluate the expression. Write fractions in simplest form.

7. $32 + (-18)$ **8.** $-12 - (-17)$

9. $\frac{3}{4} + \frac{7}{4}$ **10.** $\frac{2}{3} - \frac{1}{6}$

11. $(-3)(2)(-10)$ **12.** $\left(-\frac{4}{5}\right)\left(\frac{15}{32}\right)$

13. $\frac{7}{12} \div \frac{5}{6}$ **14.** $\left(-\frac{3}{2}\right)^3$

15. $3 - 2^2 + 25 \div 5$ **16.** $\dfrac{18 - 2(3 + 4)}{6^2 - (12 \cdot 2 + 10)}$

In Exercises 17 and 18, identify the property of real numbers illustrated by each statement.

17. (a) $8(u - 5) = 8 \cdot u - 8 \cdot 5$ (b) $10x - 10x = 0$

18. (a) $(7 + y) - z = 7 + (y - z)$ (b) $2x \cdot 1 = 2x$

19. During one month, you made the following transactions in your non-interest-bearing checking account. Find the balance at the end of the month.

NUMBER OR CODE	DATE	TRANSACTION DESCRIPTION	PAYMENT AMOUNT	✓	FEE	DEPOSIT AMOUNT	BALANCE $1406.98
2103	1/5	Car Payment	$375 03				
2104	1/7	Phone	$ 59 20				
	1/8	Withdrawal	$225 00				
	1/12	Deposit				$320 45	

20. You deposit $45 in a retirement account twice each month. How much will you deposit in the account in 8 years?

21. Determine the unknown fractional part of the circle graph at the left. Explain how you were able to make this determination.

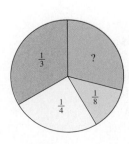

Figure for 21

1.4 Algebraic Expressions

William Thomas Cain/Getty Images

What You Should Learn

1 ▶ Identify the terms and coefficients of algebraic expressions.

2 ▶ Simplify algebraic expressions by combining like terms and removing symbols of grouping.

3 ▶ Evaluate algebraic expressions by substituting values for the variables.

Algebraic Expressions

One of the basic characteristics of algebra is the use of letters (or combinations of letters) to represent numbers. The letters used to represent the numbers are called **variables,** and combinations of letters and numbers are called **algebraic expressions.** Here are some examples.

$$3x, \quad x + 2, \quad \frac{x}{x^2 + 1}, \quad 2x - 3y, \quad 2x^3 - y^2$$

Why You Should Learn It

Many real-life quantities can be determined by evaluating algebraic expressions. For instance, in Example 9 on page 36, you will evaluate an algebraic expression to find yearly revenues of gambling industries.

1 ▶ Identify the terms and coefficients of algebraic expressions.

> ### Algebraic Expression
>
> A collection of letters (called **variables**) and real numbers (called **constants**) combined using the operations of addition, subtraction, multiplication, or division is called an **algebraic expression.**

Study Tip

It is important to understand the difference between a *term* and a *factor*. Terms are separated by addition, whereas factors are separated by multiplication. For instance, the expression $4x(x + 2)$ has three factors: 4, x, and $(x + 2)$.

The **terms** of an algebraic expression are those parts that are separated by *addition.* For example, the algebraic expression $x^2 - 3x + 6$ has three terms: x^2, $-3x$, and 6. Note that $-3x$ is a term, rather than $3x$, because

$$x^2 - 3x + 6 = x^2 + (-3x) + 6. \qquad \text{Think of subtraction as a form of addition.}$$

The terms x^2 and $-3x$ are the **variable terms** of the expression, and 6 is the **constant term.** The numerical factor of a term is called the **coefficient.** For instance, the coefficient of the variable term $-3x$ is -3, and the coefficient of the variable term x^2 is 1. Example 1 identifies the terms and coefficients of three different algebraic expressions.

EXAMPLE 1   Identifying Terms and Coefficients

Algebraic Expression	Terms	Coefficients
a. $5x - \dfrac{1}{3}$	$5x, \quad -\dfrac{1}{3}$	$5, \quad -\dfrac{1}{3}$
b. $4y + 6x - 9$	$4y, \quad 6x, \quad -9$	$4, \quad 6, \quad -9$
c. $\dfrac{2}{x} + 5x^4 - y$	$\dfrac{2}{x}, \quad 5x^4, \quad -y$	$2, \quad 5, \quad -1$

✓ **CHECKPOINT** *Now try Exercise 1.*

2 ▶ Simplify algebraic expressions by combining like terms and removing symbols of grouping.

Simplifying Algebraic Expressions

In an algebraic expression, two terms are said to be **like terms** if they are both constant terms or if they have the same variable factor. For example, $2x^2y$, $-x^2y$, and $\frac{1}{2}(x^2y)$ are like terms because they have the same variable factor x^2y. Note that $4x^2y$ and $-x^2y^2$ are not like terms because their variable factors x^2y and x^2y^2 are different.

One way to **simplify** an algebraic expression is to combine like terms.

EXAMPLE 2   Combining Like Terms

Simplify each expression by combining like terms.

a. $2x + 3x - 4$ **b.** $-3 + 5 + 2y - 7y$ **c.** $5x + 3y - 4x$

Solution

a. $2x + 3x - 4 = (2 + 3)x - 4$ ⟶ Distributive Property

$\qquad = 5x - 4$ ⟶ Simplest form

b. $-3 + 5 + 2y - 7y = (-3 + 5) + (2 - 7)y$ ⟶ Distributive Property

$\qquad = 2 - 5y$ ⟶ Simplest form

c. $5x + 3y - 4x = 3y + 5x - 4x$ ⟶ Commutative Property

$\qquad = 3y + (5x - 4x)$ ⟶ Associative Property

$\qquad = 3y + (5 - 4)x$ ⟶ Distributive Property

$\qquad = 3y + x$ ⟶ Simplest form

 CHECKPOINT *Now try Exercise 25.*

Study Tip

As you gain experience with the rules of algebra, you may want to combine some of the steps in your work. For instance, you might feel comfortable listing only the following steps to solve Example 2(c).

$5x + 3y - 4x = 3y + (5x - 4x)$

$\qquad = 3y + x$

EXAMPLE 3   Combining Like Terms

Simplify each expression by combining like terms.

a. $7x + 7y - 4x - y$ **b.** $2x^2 + 3x - 5x^2 - x$

c. $3xy^2 - 4x^2y^2 + 2xy^2 + x^2y^2$

Solution

a. $7x + 7y - 4x - y = (7x - 4x) + (7y - y)$ ⟶ Group like terms.

$\qquad = 3x + 6y$ ⟶ Combine like terms.

b. $2x^2 + 3x - 5x^2 - x = (2x^2 - 5x^2) + (3x - x)$ ⟶ Group like terms.

$\qquad = -3x^2 + 2x$ ⟶ Combine like terms.

c. $3xy^2 - 4x^2y^2 + 2xy^2 + x^2y^2$

$\qquad = (3xy^2 + 2xy^2) + (-4x^2y^2 + x^2y^2)$ ⟶ Group like terms.

$\qquad = 5xy^2 - 3x^2y^2$ ⟶ Combine like terms.

 CHECKPOINT *Now try Exercise 33.*

Another way to simplify an algebraic expression is to remove symbols of grouping. Remove the innermost symbols first and combine like terms. Repeat this process as needed to remove all the symbols of grouping.

A set of parentheses preceded by a *minus* sign can be removed by changing the sign of each term inside the parentheses. For instance,

$$3x - (2x - 7) = 3x - 2x + 7.$$

This is equivalent to using the Distributive Property with a multiplier of -1. That is,

$$3x - (2x - 7) = 3x + (-1)(2x - 7) = 3x - 2x + 7.$$

A set of parentheses preceded by a *plus* sign can be removed without changing the signs of the terms inside the parentheses. For instance,

$$3x + (2x - 7) = 3x + 2x - 7.$$

EXAMPLE 4   Removing Symbols of Grouping

Simplify each expression.

a. $3(x - 5) - (2x - 7)$ **b.** $-4(x^2 + 4) + x^2(x + 4)$

Solution

a. $3(x - 5) - (2x - 7) = 3x - 15 - 2x + 7$ Distributive Property

$\qquad\qquad\qquad\qquad = (3x - 2x) + (-15 + 7)$ Group like terms.

$\qquad\qquad\qquad\qquad = x - 8$ Combine like terms.

b. $-4(x^2 + 4) + x^2(x + 4) = -4x^2 - 16 + x^2 \cdot x + 4x^2$ Distributive Property

$\qquad\qquad\qquad\qquad\qquad = -4x^2 - 16 + x^3 + 4x^2$ Exponential form

$\qquad\qquad\qquad\qquad\qquad = x^3 + (4x^2 - 4x^2) - 16$ Group like terms.

$\qquad\qquad\qquad\qquad\qquad = x^3 + 0 - 16$ Combine like terms.

$\qquad\qquad\qquad\qquad\qquad = x^3 - 16$ Additive Identity Property

✓ **CHECKPOINT** *Now try Exercise 53.*

Study Tip

The exponential notation described in Section 1.2 can also be used when the base is a variable or an algebraic expression. For instance, in Example 4(b), $x^2 \cdot x$ can be written as

$x^2 \cdot x = x \cdot x \cdot x$

$\quad = x^3.$ 3 factors of x

EXAMPLE 5   Removing Symbols of Grouping

a. $5x - 2x[3 + 2(x - 7)] = 5x - 2x(3 + 2x - 14)$ Distributive Property

$\qquad\qquad\qquad\qquad = 5x - 2x(2x - 11)$ Combine like terms.

$\qquad\qquad\qquad\qquad = 5x - 4x^2 + 22x$ Distributive Property

$\qquad\qquad\qquad\qquad = -4x^2 + 27x$ Combine like terms.

b. $-3x(5x^4) + 2x^5 = -15x \cdot x^4 + 2x^5$ Multiply.

$\qquad\qquad\qquad = -15x^5 + 2x^5$ Exponential form

$\qquad\qquad\qquad = -13x^5$ Combine like terms.

✓ **CHECKPOINT** *Now try Exercise 65.*

3 ▶ Evaluate algebraic expressions by substituting values for the variables.

Evaluating Algebraic Expressions

To **evaluate** an algebraic expression, substitute numerical values for each of the variables in the expression. Note that you must substitute the value for *each* occurrence of the variable.

EXAMPLE 6   Evaluating Algebraic Expressions

Evaluate each algebraic expression when $x = -2$.

a. $5 + x^2$ **b.** $5 - x^2$

Solution

a. When $x = -2$, the expression $5 + x^2$ has a value of

$$5 + (-2)^2 = 5 + 4 = 9.$$

b. When $x = -2$, the expression $5 - x^2$ has a value of

$$5 - (-2)^2 = 5 - 4 = 1.$$

✓ **CHECKPOINT** *Now try Exercise 73.*

EXAMPLE 7   Evaluating Algebraic Expressions

Evaluate each algebraic expression when $x = 2$ and $y = -1$.

a. $|y - x|$ **b.** $x^2 - 2xy + y^2$

Solution

a. When $x = 2$ and $y = -1$, the expression $|y - x|$ has a value of

$$|(-1) - (2)| = |-3| = 3.$$

b. When $x = 2$ and $y = -1$, the expression $x^2 - 2xy + y^2$ has a value of

$$2^2 - 2(2)(-1) + (-1)^2 = 4 + 4 + 1 = 9.$$

✓ **CHECKPOINT** *Now try Exercise 81.*

EXAMPLE 8   Evaluating an Algebraic Expression

Evaluate $\dfrac{2xy}{x + 1}$ when $x = -4$ and $y = -3$.

Solution

When $x = -4$ and $y = -3$, the expression $(2xy)/(x + 1)$ has a value of

$$\frac{2(-4)(-3)}{-4 + 1} = \frac{24}{-3} = -8.$$

✓ **CHECKPOINT** *Now try Exercise 85.*

EXAMPLE 9 **Using a Mathematical Model**

The yearly revenues (in billions of dollars) for gambling industries in the United States for the years 1999 to 2005 can be modeled by

$$\text{Revenue} = 0.157t^2 - 1.19t + 11.0, \quad 9 \le t \le 15$$

where t represents the year, with $t = 9$ corresponding to 1999. Create a table that shows the revenue for each of these years. (Source: 2005 Service Annual Survey)

Solution

To create a table of values that shows the revenues (in billions of dollars) for the gambling industries for the years 1999 to 2005, evaluate the expression $0.157t^2 - 1.19t + 11.0$ for each integer value of t from $t = 9$ to $t = 15$.

Year	1999	2000	2001	2002	2003	2004	2005
t	9	10	11	12	13	14	15
Revenue	13.0	14.8	16.9	19.3	22.1	25.1	28.5

 CHECKPOINT *Now try Exercises 101 and 102.*

Technology: Tip

Most graphing calculators can be used to evaluate an algebraic expression for several values of x and display the results in a table. For instance, to evaluate $2x^2 - 3x + 2$ when x is 0, 1, 2, 3, 4, 5, and 6, you can use the following steps.

1. Enter the expression into the graphing calculator.

2. Set the minimum value of the table to 0.

3. Set the table step or table increment to 1.

4. Display the table.

The results are shown below.

X	Y₁
0	2
1	1
2	4
3	11
4	22
5	37
6	56

Consult the user's guide for your graphing calculator for specific instructions. Then complete a table for the expression $-4x^2 + 5x - 8$ when x is 0, 1, 2, 3, 4, 5, and 6.

Concept Check

1. Explain the difference between terms and factors in an algebraic expression.

2. Explain how you can use the Distributive Property to simplify the expression $5x + 3x$.

3. Explain how to combine like terms in an algebraic expression. Give an example.

4. Explain the difference between simplifying an algebraic expression and evaluating an algebraic expression.

1.4 EXERCISES

Go to pages 50–51 to record your assignments.

Developing Skills

In Exercises 1–14, identify the terms and coefficients of the algebraic expression. *See Example 1.*

1. $10x + 5$

2. $4 + 17y$

3. $12 - 6x^2$

4. $-16t^2 + 48$

5. $-3y^2 + 2y - 8$

6. $9t^2 + 2t + 10$

7. $1.2a - 4a^3$

8. $25z^3 - 4.8z^2$

9. $4x^2 - 3y^2 - 5x + 21$

10. $7a^2 + 4a - b^2 + 19$

11. $xy - 5x^2y + 2y^2$

12. $14u^2 + 25uv - 3v^2$

13. $\frac{1}{4}x^2 - \frac{3}{8}x + 5$

14. $\frac{2}{3}y + 8z + \frac{5}{6}$

In Exercises 15–20, identify the property of algebra illustrated by the statement.

15. $4 - 3x = -3x + 4$

16. $(10 + x) - y = 10 + (x - y)$

17. $-5(2x) = (-5 \cdot 2)x$

18. $(x - 2)(3) = 3(x - 2)$

19. $(5 - 2)x = 5x - 2x$

20. $7y + 2y = (7 + 2)y$

In Exercises 21–24, use the indicated property to rewrite the expression.

21. Distributive Property
$5(x + 6) =$

22. Distributive Property
$6x + 6 =$

23. Commutative Property of Multiplication
$5(x + 6) =$

24. Commutative Property of Addition
$6x + 6 =$

In Exercises 25–40, simplify the expression by combining like terms. *See Examples 2 and 3.*

25. $3x + 4x$

26. $18z + 14z$

27. $-2x^2 + 4x^2$

28. $20a^2 - 5a^2$

29. $7x - 11x$

30. $-23t + 11t$

31. $9y - 5y + 4y$

32. $8y + 7y - y$

33. $3x - 2y + 5x + 20y$

34. $-2a + 4b - 7a - b$

35. $7x^2 - 2x - x^2$

36. $9y + y^2 - 6y$

37. $-3z^4 + 6z - z + 8 + z^4 - 4z^2$

38. $-5y^3 + 3y - 6y^2 + 8y^3 + y - 4$

39. $x^2 + 2xy - 2x^2 + xy + y$

40. $3a - 5ab + 9a^2 + 4ab - a$

In Exercises 41–52, use the Distributive Property to simplify the expression.

41. $4(2x^2 + x - 3)$

42. $8(z^3 - 4z^2 + 2)$

43. $-3(6y^2 - y - 2)$

44. $-5(-x^2 + 2y + 1)$

45. $-(3x^2 - 2x + 4)$

46. $-(-5t^2 + 8t - 10)$

47. $x(5x + 2)$

48. $y(-y + 10)$

49. $3x(17 - 4x)$

50. $5y(2y - 1)$

51. $-5t(7 - 2t)$

52. $-6x(9x - 4)$

In Exercises 53–72, simplify the expression. *See Examples 4 and 5.*

✓ **53.** $10(x - 3) + 2x - 5$ **54.** $3(x + 1) + x - 6$

55. $x - (5x + 9)$ **56.** $y - (3y - 1)$

57. $5a - (4a - 3)$ **58.** $7x - (2x + 5)$

59. $-3(3y - 1) + 2(y - 5)$

60. $5(a + 6) - 4(2a - 1)$

61. $-3(y^2 - 2) + y^2(y + 3)$

62. $x(x^2 - 5) - 4(4 - x)$

63. $x(x^2 + 3) - 3(x + 4)$

64. $5(x + 1) - x(2x + 6)$

✓ **65.** $9a - [7 - 5(7a - 3)]$

66. $12b - [9 - 7(5b - 6)]$

67. $3[2x - 4(x - 8)]$

68. $4[5 - 3(x^2 + 10)]$

69. $8x + 3x[10 - 4(3 - x)]$

70. $5y - y[9 + 6(y - 2)]$

71. $2[3(b - 5) - (b^2 + b + 3)]$

72. $5[3(z + 2) - (z^2 + z - 2)]$

In Exercises 73–90, evaluate the expression for the specified values of the variable(s). If not possible, state the reason. *See Examples 6–8.*

Expression	Values
✓ **73.** $5 - 3x$	(a) $x = \frac{2}{3}$
	(b) $x = 5$
74. $\frac{3}{2}x - 2$	(a) $x = 6$
	(b) $x = -3$
75. $10 - 4x^2$	(a) $x = -1$
	(b) $x = \frac{1}{2}$
76. $3y^2 + 10$	(a) $y = -2$
	(b) $y = \frac{1}{2}$
77. $y^2 - y + 5$	(a) $y = 2$
	(b) $y = -2$
78. $2x^2 + 5x - 3$	(a) $x = 2$
	(b) $x = -3$
79. $\frac{1}{x^2} + 3$	(a) $x = 0$
	(b) $x = 3$
80. $5 - \frac{3}{x}$	(a) $x = 0$
	(b) $x = -6$
✓ **81.** $3x + 2y$	(a) $x = 1, \quad y = 5$
	(b) $x = -6, \quad y = -9$
82. $6x - 5y$	(a) $x = -2, \quad y = -3$
	(b) $x = 1, \quad y = 1$
83. $x^2 - xy + y^2$	(a) $x = 2, \quad y = -1$
	(b) $x = -3, \quad y = -2$
84. $y^2 + xy - x^2$	(a) $x = 5, \quad y = 2$
	(b) $x = -3, \quad y = 3$
✓ **85.** $\frac{x}{y^2 - x}$	(a) $x = 4, \quad y = 2$
	(b) $x = 3, \quad y = 3$
86. $\frac{x}{x - y}$	(a) $x = 0, \quad y = 10$
	(b) $x = 4, \quad y = 4$
87. $|y - x|$	(a) $x = 2, \quad y = 5$
	(b) $x = -2, \quad y = -2$
88. $|x^2 - y|$	(a) $x = 0, \quad y = -2$
	(b) $x = 3, \quad y = -15$
89. Distance traveled: rt	(a) $r = 40, \quad t = 5\frac{1}{4}$
	(b) $r = 35, \quad t = 4$
90. Simple interest: Prt	(a) $P = \$7000,$ $r = 0.065, \quad t = 10$
	(b) $P = \$4200,$ $r = 0.07, \quad t = 9$

Solving Problems

▲ *Geometry* In Exercises 91–94, find the volume of the rectangular solid by evaluating the expression *lwh* for the dimensions given in the figure.

91.

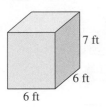

7 ft
6 ft
6 ft

92.

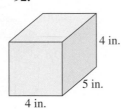

4 in.
5 in.
4 in.

93.

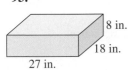

8 in.
18 in.
27 in.

94.

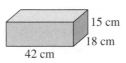

15 cm
18 cm
42 cm

In Exercises 95–98, evaluate the expression $0.01p + 0.05n + 0.10d + 0.25q$ to find the value of the given number of pennies *p*, nickels *n*, dimes *d*, and quarters *q*.

95. 11 pennies, 7 nickels, 3 quarters

96. 8 pennies, 13 nickels, 6 dimes

97. 43 pennies, 27 nickels, 17 dimes, 15 quarters

98. 111 pennies, 22 nickels, 2 dimes, 42 quarters

▲ *Geometry* In Exercises 99 and 100, write and simplify an expression for the area of the figure. Then evaluate the expression for the given value of the variable.

99. $b = 15$

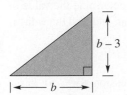

$b - 3$
b

100. $h = 12$

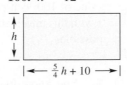

h
$\frac{5}{4}h + 10$

Using a Model In Exercises 101 and 102, use the following model, which approximates the annual sales (in millions of dollars) of sports equipment in the United States from 2001 to 2006 (see figure), where *t* represents the year, with $t = 1$ corresponding to 2001. (Source: National Sporting Goods Association)

$$\text{Sales} = 607.6t + 20{,}737, \quad 1 \le t \le 6$$

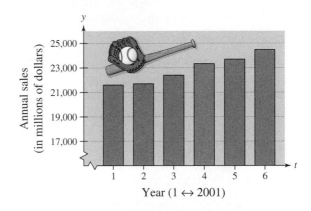

Year (1 ↔ 2001)

✔ **101.** Graphically approximate the sales of sports equipment in 2005. Then use the model to confirm your estimate algebraically.

✔ **102.** Use the model and a calculator to complete the table showing the sales from 2001 to 2006.

Year	2001	2002	2003
Sales			

Year	2004	2005	2006
Sales			

Using a Model In Exercises 103 and 104, use the following model, which approximates the total yearly disbursements (in billions of dollars) of Federal Family Education Loans (FFEL) in the United States from 1999 to 2005 (see figure), where *t* represents the year, with $t = 9$ corresponding to 1999. (Source: U.S. Department of Education)

$$\text{Disbursements} = 0.322t^2 - 3.75t + 27.6, \quad 9 \le t \le 15$$

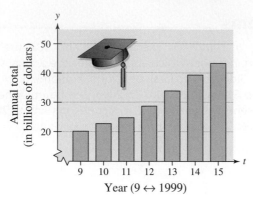

Figure for 103 and 104

103. Graphically approximate the total amount of FFEL disbursements in the year 2000. Then use the model to confirm your estimate algebraically.

104. Use the model and a calculator to complete the table showing the total yearly disbursements (in billions of dollars) from 1999 to 2005. Round each amount to the nearest tenth.

Year	1999	2000	2001	2002
Amount				

Year	2003	2004	2005
Amount			

105. ▲ *Geometry* The roof shown in the figure is made up of two trapezoids and two triangles. Find the total area of the roof. [For a trapezoid, area $= \frac{1}{2}h(b_1 + b_2)$, where b_1 and b_2 are the lengths of the bases and h is the height.]

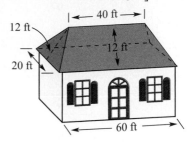

106. *Exploration*

(a) A convex polygon with n sides has

$$\frac{n(n-3)}{2}, n \geq 4$$

diagonals. Verify the formula for a square, a pentagon, and a hexagon.

(b) Explain why the formula in part (a) will always yield a natural number.

Explaining Concepts

107. ✎ Is it possible to evaluate the expression

$$\frac{x+2}{y-3}$$

when $x = 5$ and $y = 3$? Explain.

108. ✎ Is it possible to evaluate the expression $3x + 5y - 18z$ when $x = 10$ and $y = 8$? Explain.

109. ✎ State the procedure for simplifying an algebraic expression by removing a set of parentheses preceded by a minus sign, such as the parentheses in $a - (b + c)$. Then give an example.

110. ✎ How can a factor be part of a term in an algebraic expression? Explain and give an example.

111. ✎ How can an algebraic term be part of a factor in an algebraic expression? Give an example.

112. ✎ You know that the expression $180 - 10x$ has a value of 100. Is it possible to determine the value of x with this information? Explain and find the value if possible.

113. ✎ You know that the expression $8y - 5x$ has a value of 14. Is it possible to determine the values of x and y with this information? Explain and find the values if possible.

1.5 Constructing Algebraic Expressions

Clive Brunskill/Getty Images

Why You Should Learn It

Translating verbal phrases into algebraic expressions enables you to model real-life problems. For instance, in Exercise 73 on page 49, you will write an algebraic expression that models the area of a soccer field.

1 ▶ Translate verbal phrases into algebraic expressions, and vice versa.

What You Should Learn

1 ▶ Translate verbal phrases into algebraic expressions, and vice versa.

2 ▶ Construct algebraic expressions with hidden products.

Translating Phrases

In this section, you will study ways to *construct* algebraic expressions. When you translate a verbal sentence or phrase into an algebraic expression, watch for key words and phrases that indicate the four different operations of arithmetic.

Translating Key Words and Phrases

Key Words and Phrases	Verbal Description	Algebraic Expression
Addition: Sum, plus, greater than, increased by, more than, exceeds, total of	The sum of 5 and x Seven more than y	$5 + x$ $y + 7$
Subtraction: Difference, minus, less than, decreased by, subtracted from, reduced by, the remainder	b is subtracted from 4. Three less than z	$4 - b$ $z - 3$
Multiplication: Product, multiplied by, twice, times, percent of	Two times x	$2x$
Division: Quotient, divided by, ratio, per	The ratio of x and 8	$\dfrac{x}{8}$

EXAMPLE 1   Translating Verbal Phrases

Verbal Description	Algebraic Expression
a. Seven more than three times x	$3x + 7$
b. Four less than the product of 6 and n	$6n - 4$
c. The quotient of x and 3, decreased by 6	$\dfrac{x}{3} - 6$

✓ **CHECKPOINT** *Now try Exercise 1.*

EXAMPLE 2   **Translating Verbal Phrases**

Verbal Description	Algebraic Expression
a. Eight added to the product of 2 and n	$2n + 8$
b. Four times the sum of y and 9	$4(y + 9)$
c. The difference of a and 7, all divided by 9	$\dfrac{a - 7}{9}$

✓ **CHECKPOINT** *Now try Exercise 7.*

In Examples 1 and 2, the verbal description specified the name of the variable. In most real-life situations, however, the variables are not specified and it is your task to assign variables to the *appropriate* quantities.

EXAMPLE 3   **Translating Verbal Phrases**

Verbal Description	Label	Algebraic Expression
a. The sum of 7 and a number	The number $= x$	$7 + x$
b. Four decreased by the product of 2 and a number	The number $= n$	$4 - 2n$
c. Seven less than twice the sum of a number and 5	The number $= y$	$2(y + 5) - 7$

✓ **CHECKPOINT** *Now try Exercise 19.*

A good way to learn algebra is to do it forward and backward. For instance, the next example translates algebraic expressions into verbal form. Keep in mind that other key words could be used to describe the operations in each expression.

Study Tip

When you write a verbal model or construct an algebraic expression, watch out for statements that may be interpreted in more than one way. For instance, the statement "The sum of x and 1 divided by 5" is ambiguous because it could mean

$$\frac{x + 1}{5} \text{ or } x + \frac{1}{5}.$$

Notice in Example 4(b) that the verbal description for

$$\frac{3 + x}{4}$$

contains the phrase "all divided by 4."

EXAMPLE 4   **Translating Expressions into Verbal Phrases**

Without using a variable, write a verbal description for each expression.

a. $5x - 10$

b. $\dfrac{3 + x}{4}$

c. $2(3x + 4)$

d. $\dfrac{4}{x - 2}$

Solution

a. 10 less than the product of 5 and a number

b. The sum of 3 and a number, all divided by 4

c. Twice the sum of 3 times a number and 4

d. Four divided by a number reduced by 2

✓ **CHECKPOINT** *Now try Exercise 25.*

2 ▶ Construct algebraic expressions with hidden products.

Constructing Mathematical Models

Translating a verbal phrase into a mathematical model is critical in problem solving. The next four examples will demonstrate three steps for creating a mathematical model.

1. Construct a verbal model that represents the problem situation.
2. Assign labels to all quantities in the verbal model.
3. Construct a mathematical model (algebraic expression).

EXAMPLE 5   **Constructing a Mathematical Model**

A cash register contains x quarters. Write an algebraic expression for this amount of money in dollars.

Solution

Verbal Model:	Value of coin	·	Number of coins		

Labels:	Value of coin $= 0.25$	(dollars per quarter)
	Number of coins $= x$	(quarters)

Expression: $0.25x$ (dollars)

✓ **CHECKPOINT** *Now try Exercise 41.*

EXAMPLE 6   **Constructing a Mathematical Model**

A cash register contains n nickels and d dimes. Write an algebraic expression for this amount of money in cents.

Solution

Verbal Model:	Value of nickel	·	Number of nickels	+	Value of dime	·	Number of dimes

Labels:	Value of nickel $= 5$	(cents per nickel)
	Number of nickels $= n$	(nickels)
	Value of dime $= 10$	(cents per dime)
	Number of dimes $= d$	(dimes)

Expression: $5n + 10d$ (cents)

✓ **CHECKPOINT** *Now try Exercise 45.*

In Example 6, the final expression $5n + 10d$ is measured in cents. This makes "sense" as described below.

$$\frac{5 \text{ cents}}{\text{nickel}} \cdot n \text{ nickels} + \frac{10 \text{ cents}}{\text{dime}} \cdot d \text{ dimes}$$

Note that the nickels and dimes "divide out," leaving cents as the unit of measure for each term. This technique is called *unit analysis*, and it can be very helpful in determining the final unit of measure.

EXAMPLE 7   Constructing a Mathematical Model

A person riding a bicycle travels at a constant rate of 12 miles per hour. Write an algebraic expression showing how far the person can ride in *t* hours.

Solution

For this problem, use the formula Distance = (Rate)(Time).

Verbal Model: Rate · Time

Labels: Rate = 12   (miles per hour)
             Time = *t*   (hours)

Expression: 12*t*   (miles)

✓ **CHECKPOINT** *Now try Exercise 47.*

Using unit analysis, you can see that the expression in Example 7 has *miles* as its unit of measure.

$$12\frac{\text{miles}}{\cancel{\text{hour}}} \cdot t \ \cancel{\text{hours}}$$

When translating verbal phrases involving percents, be sure you write the percent *in decimal form.*

Percent	*Decimal Form*
4%	0.04
62%	0.62
140%	1.40
25%	0.25

Remember that when you find a percent of a number, you multiply. For instance, 25% of 78 is given by

$$0.25(78) = 19.5. \qquad \text{25\% of 78}$$

EXAMPLE 8   Constructing a Mathematical Model

A person adds *k* liters of fluid containing 55% antifreeze to a car radiator. Write an algebraic expression that indicates how much antifreeze was added.

Solution

Verbal Model: Percent antifreeze · Number of liters

Labels: Percent of antifreeze = 0.55   (in decimal form)
             Number of liters = *k*   (liters)

Expression: 0.55*k*   (liters)

Note that the algebraic expression uses the decimal form of 55%. That is, you compute with 0.55 rather than 55%.

✓ **CHECKPOINT** *Now try Exercise 51.*

When assigning labels to *two* unknown quantities, hidden operations are often involved. For example, two numbers add up to 18 and one of the numbers is assigned the variable x. What expression can you use to represent the second number? Let's try a specific case first, then apply it to a general case.

Specific Case: If the first number is 7, the second number is $18 - 7 = 11$.

General Case: If the first number is x, the second number is $18 - x$.

The strategy of using a *specific* case to help determine the general case is often helpful in applications. Observe the use of this strategy in the next example.

EXAMPLE 9   Using Specific Cases to Model General Cases

a. A person's weekly salary is d dollars. Write an expression for the person's annual salary.

b. A person's annual salary is y dollars. Write an expression for the person's monthly salary.

Solution

a. *Specific Case:* If the weekly salary is $300, the annual salary is 52(300) dollars.

General Case: If the weekly salary is d dollars, the annual salary is $52 \cdot d$ or $52d$ dollars.

b. *Specific Case:* If the annual salary is $24,000, the monthly salary is $24{,}000 \div 12$ dollars.

General Case: If the annual salary is y dollars, the monthly salary is $y \div 12$ or $y/12$ dollars.

✓ **CHECKPOINT** *Now try Exercise 57.*

Study Tip

You can check that your algebraic expressions are correct for even, odd, or consecutive integers by substituting an integer for n. For instance, by letting $n = 5$, you can see that $2n = 2(5) = 10$ is an even integer,
$2n - 1 = 2(5) - 1 = 9$
is an odd integer, and
$2n + 1 = 2(5) + 1 = 11$
is an odd integer.

In mathematics, it is useful to know how to represent certain types of integers algebraically. For instance, consider the set $\{2, 4, 6, 8, \ldots\}$ of *even* integers. Because every even integer has 2 as a factor,

$$2 = 2 \cdot 1, \quad 4 = 2 \cdot 2, \quad 6 = 2 \cdot 3, \quad 8 = 2 \cdot 4, \ldots$$

it follows that any integer n multiplied by 2 is sure to be the *even* number $2n$. Moreover, if $2n$ is even, then $2n - 1$ and $2n + 1$ are sure to be *odd* integers.

Two integers are called **consecutive integers** if they differ by 1. For any integer n, its next two larger consecutive integers are $n + 1$ and $(n + 1) + 1$ or $n + 2$. So, you can denote three consecutive integers by n, $n + 1$, and $n + 2$. These results are summarized below.

Labels for Integers

Let n represent an integer. Then even integers, odd integers, and consecutive integers can be represented as follows.

1. $2n$ denotes an *even* integer for $n = 1, 2, 3, \ldots$.

2. $2n - 1$ and $2n + 1$ denote *odd* integers for $n = 1, 2, 3, \ldots$.

3. $\{n, n + 1, n + 2, \ldots\}$ denotes a set of *consecutive* integers.

EXAMPLE 10   Constructing a Mathematical Model

Write an expression for the following phrase.

"The sum of two consecutive integers, the first of which is n"

Solution

The first integer is n. The next consecutive integer is $n + 1$. So the sum of two consecutive integers is

$$n + (n + 1) = 2n + 1.$$

 CHECKPOINT *Now try Exercise 59.*

Sometimes an expression may be written directly from a diagram using a common geometric formula, as shown in the next example.

EXAMPLE 11   Constructing a Mathematical Model

Write expressions for the perimeter and area of the rectangle shown in Figure 1.16.

$2w$ in.

$(w + 12)$ in.

Figure 1.16

Solution

For the perimeter of the rectangle, use the formula

Perimeter $= 2(\text{Length}) + 2(\text{Width})$.

Verbal Model: $2 \cdot$ Length $+ 2 \cdot$ Width

Labels: Length $= w + 12$ (inches)
             Width $= 2w$ (inches)

Expression: $2(w + 12) + 2(2w) = 2w + 24 + 4w$
$$= 6w + 24 \qquad \text{(inches)}$$

For the area of the rectangle, use the formula

Area $= (\text{Length})(\text{Width})$.

Verbal Model: Length $\cdot$ Width

Labels: Length $= w + 12$ (inches)
             Width $= 2w$ (inches)

Expression: $(w + 12)(2w) = 2w^2 + 24w$ (square inches)

 CHECKPOINT *Now try Exercise 69.*

Concept Check

1. The phrase *reduced by* implies what operation?

2. The word *ratio* indicates what operation?

3. A car travels at a constant rate of 45 miles per hour for t hours. The algebraic expression for the distance traveled is $45t$. What is the unit of measure of the algebraic expression?

4. Let n represent an integer. Is the expression $n(n + 1)$ even or odd? Explain.

1.5 EXERCISES

Go to pages 50–51 to record your assignments.

Developing Skills

In Exercises 1–24, translate the verbal phrase into an algebraic expression. *See Examples 1–3.*

✓ 1. The sum of 23 and a number n

2. Twelve more than a number n

3. The sum of 12 and twice a number n

4. The total of 25 and three times a number n

5. Six less than a number n

6. Fifteen decreased by three times a number n

✓ 7. Four times a number n minus 10

8. The product of a number y and 10 is decreased by 35.

9. Half of a number n

10. Seven-fifths of a number n

11. The quotient of a number x and 6

12. The ratio of y and 3

13. Eight times the ratio of N and 5

14. Fifteen times the ratio of x and 32

15. The number c is quadrupled and the product is increased by 10.

16. The number u is tripled and the product is increased by 250.

17. Thirty percent of the list price L

18. Twenty-five percent of the bill B

✓ 19. The sum of a number and 5, divided by 10

20. The sum of 7 and twice a number x, all divided by 8

21. The absolute value of the difference between a number and 8

22. The absolute value of the quotient of a number and 4

23. The product of 3 and the square of a number is decreased by 4.

24. The sum of 10 and one-fourth the square of a number

In Exercises 25–40, write a verbal description of the algebraic expression without using the variable. *See Example 4.*

✓ 25. $t - 2$

26. $5 - x$

27. $y + 50$

28. $2y + 3$

29. $2 - 3x$

30. $7y - 4$

31. $\dfrac{z}{2}$

32. $\dfrac{y}{8}$

33. $\dfrac{4}{5}x$

34. $\dfrac{2}{3}t$

35. $8(x - 5)$

36. $(y + 6)4$

37. $\dfrac{x + 10}{3}$

38. $\dfrac{3 - n}{9}$

39. $y^2 - 3$

40. $x^2 + 2$

In Exercises 41–64, write an algebraic expression that represents the specified quantity in the verbal statement, and simplify if possible. *See Examples 5–10.*

 41. The amount of money (in dollars) represented by n quarters

42. The amount of money (in dollars) represented by x nickels

43. The amount of money (in dollars) represented by m dimes

44. The amount of money (in dollars) represented by y pennies

✓ **45.** The amount of money (in cents) represented by m nickels and n dimes

46. The amount of money (in cents) represented by m dimes and n quarters

✓ **47.** The distance traveled in t hours at an average speed of 55 miles per hour

48. The distance traveled in 5 hours at an average speed of r miles per hour

49. The time required to travel 320 miles at an average speed of r miles per hour

50. The average rate of speed when traveling 320 miles in t hours

✓ **51.** The amount of antifreeze in a cooling system containing y gallons of coolant that is 45% antifreeze

52. The amount of water in q quarts of a food product that is 65% water

53. The amount of wage tax due for a taxable income of I dollars that is taxed at the rate of 1.25%

54. The amount of sales tax on a purchase valued at L dollars if the tax rate is 5.5%

55. The sale price of a coat that has a list price of L dollars if it is a "20% off" sale

56. The total bill for a meal that cost C dollars if you plan on leaving a 15% tip

✓ **57.** The total hourly wage for an employee when the base pay is \$8.25 per hour plus 60 cents for each of q units produced per hour

58. The total hourly wage for an employee when the base pay is \$11.65 per hour plus 80 cents for each of q units produced per hour

✓ **59.** The sum of a number n and five times the number

60. The sum of three consecutive integers, the first of which is n

61. The sum of three consecutive odd integers, the first of which is $2n + 1$

62. The sum of three consecutive even integers, the first of which is $2n$

63. The product of two consecutive even integers, divided by 4

64. The absolute value of the difference of two consecutive integers, divided by 2

Solving Problems

▲ *Geometry*   In Exercises 65–68, write an expression for the area of the figure. Simplify the expression.

65.

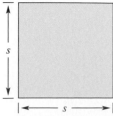

66.

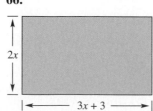

67.

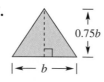

68.

▲ *Geometry* In Exercises 69–72, write expressions for the perimeter and area of the region. Simplify the expressions. *See Example 11.*

69.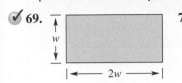

70.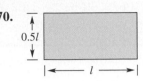

71.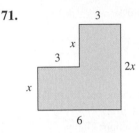

72.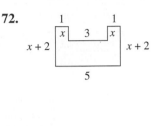

73. ▲ *Geometry* Write an expression for the area of the soccer field shown in the figure. What is the unit of measure for the area?

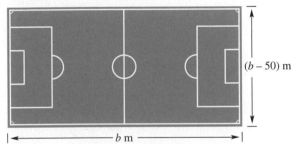

74. ▲ *Geometry* Write an expression for the area of the advertising banner shown in the figure. What is the unit of measure for the area?

75. *Finding a Pattern* Complete the table. The third row contains the differences between consecutive entries of the second row. Describe the pattern of the third row.

n	0	1	2	3	4	5
$5n - 3$						
Differences						

76. *Finding a Pattern* Complete the table. The third row contains the differences between consecutive entries of the second row. Describe the pattern of the third row.

n	0	1	2	3	4	5
$3n + 1$						
Differences						

77. *Finding a Pattern* Using the results of Exercises 75 and 76, guess the third-row difference that would result in a similar table if the algebraic expression were $an + b$.

78. *Think About It* Find a and b such that the expression $an + b$ would yield the following table.

n	0	1	2	3	4	5
$an + b$	3	7	11	15	19	23

Explaining Concepts

79. Which are equivalent to $4x$?

   (a) x multiplied by 4

   (b) x increased by 4

   (c) the product of x and 4

   (d) the ratio of 4 and x

80. ✎ If n is an integer, how are the integers $2n - 1$ and $2n + 1$ related? Explain.

81. ✎ When a statement is translated into an algebraic expression, explain why it may be helpful to use a specific case before writing the expression.

82. ✎ When each phrase is translated into an algebraic expression, is order important? Explain.

   (a) y multiplied by 5

   (b) 5 decreased by y

   (c) y divided by 5

   (d) the sum of 5 and y

What Did You Learn?

Use these two pages to help prepare for a test on this chapter. Check off the key terms and key concepts you know. You can also use this section to record your assignments.

Plan for Test Success

Date of test: [/ /]   **Study dates and times:** [/ /] at [:] A.M./P.M.

[/ /] at [:] A.M./P.M.

Things to review:

☐ Key Terms, *p. 50*
☐ Key Concepts, *pp. 50–51*
☐ Your class notes
☐ Your assignments

☐ Study Tips, *pp. 2, 6, 7, 13, 14, 15, 17, 23, 25, 32, 33, 34, 42, 43, 45*
☐ Technology Tips, *pp. 3, 18, 36*
☐ Mid-Chapter Quiz, *p. 31*

☐ Review Exercises, *pp. 52–54*
☐ Chapter Test, *p. 55*
☐ Video Explanations Online
☐ Tutorial Online

Key Terms

☐ set, *p. 2*
☐ subset, *p. 2*
☐ real numbers, *p. 2*
☐ natural numbers, *p. 2*
☐ positive integers, *p. 2*
☐ whole numbers, *p. 2*
☐ negative integers, *p. 2*
☐ integers, *p. 2*
☐ fractions, *p. 3*
☐ rational numbers, *p. 3*
☐ irrational numbers, *p. 3*
☐ real number line, *p. 4*
☐ origin, *p. 4*
☐ nonnegative real number, *p. 4*
☐ inequality symbols, *p. 5*

☐ opposites, *p. 7*
☐ additive inverses, *p. 7*
☐ absolute value, *p. 7*
☐ sum, *p. 11*
☐ difference, *p. 12*
☐ least common denominator, *p. 13*
☐ least common multiple, *p. 13*
☐ product, *p. 14*
☐ factor, *p. 14*
☐ reciprocal, *p. 15*
☐ quotient, *p. 15*
☐ dividend, *p. 15*
☐ divisor, *p. 15*
☐ numerator, *p. 15*
☐ denominator, *p. 15*

☐ exponential form, *p. 16*
☐ base, *p. 16*
☐ exponent, *p. 16*
☐ order of operations, *p. 16*
☐ variables, *p. 32*
☐ algebraic expressions, *p. 32*
☐ variable terms, *p. 32*
☐ constant term, *p. 32*
☐ coefficient, *p. 32*
☐ like terms, *p. 33*
☐ simplify, *p. 33*
☐ evaluate, *p. 35*
☐ consecutive integers, *p. 45*

Key Concepts

1.1 The Real Number System

Assignment: _____ Due date: _____

☐ **Use the real number line to order real numbers.**

If the real number a lies to the left of the real number b on the real number line, then a is less than b, which is written as $a < b$. This relationship can also be described by saying that b is greater than a and writing $b > a$.

☐ **Use the real number line to find the distance between two real numbers.**

If a and b are two real numbers such that $a \leq b$, then the distance between a and b is given by

$b - a$.

☐ **Use properties of opposites and additive inverses.**

Let a be a real number.
1. $-a$ is the opposite of a.
2. $-(-a) = a$ (The double negative of a is a.)
3. $a + (-a) = 0$ (a and $-a$ are additive inverses.)

☐ **Determine the absolute value of a real number.**

If a is a real number, then the absolute value of a is

$$|a| = \begin{cases} a, & \text{if } a \geq 0 \\ -a, & \text{if } a < 0 \end{cases}.$$

1.2 Operations with Real Numbers

Assignment: _____ Due date: _____

☐ **Perform operations on real numbers.**

1. *To add with like signs:* Add the absolute values and attach the common sign.
2. *To add with unlike signs:* Subtract the smaller absolute value from the greater absolute value and attach the sign of the number with the greater absolute value.
3. *To subtract b from a:* Add $-b$ to a.
4. *To multiply two real numbers:* With like signs, the product is positive. With unlike signs, the product is negative. The product of zero and any other real number is zero.
5. *To divide a by b:* Multiply a by the reciprocal of b.
6. *To add fractions:* Write the fractions so that they have the same denominator. Add the numerators over the common denominator.

7. *To multiply fractions:* Multiply the numerators and the denominators.
8. *To raise a to the nth power (n is an integer):*

$$a^n = \underbrace{a \cdot a \cdot a \cdots a}_{n \text{ factors}}$$

☐ **Use order of operations to evaluate expressions.**

1. First do operations within symbols of grouping.
2. Then evaluate powers.
3. Then do multiplications and divisions from left to right.
4. Finally, do additions and subtractions from left to right.

1.3 Properties of Real Numbers

Assignment: _____ Due date: _____

☐ **Use properties of real numbers.**

Let a, b, and c represent real numbers, variables, or algebraic expressions.

Commutative Properties:

$a + b = b + a$ $ab = ba$

Associative Properties:

$(a + b) + c = a + (b + c)$ $(ab)c = a(bc)$

Distributive Properties:

$a(b + c) = ab + ac$ $a(b - c) = ab - ac$

$(a + b)c = ac + bc$ $(a - b)c = ac - bc$

Identity Properties:

$a + 0 = 0 + a = a$ $a \cdot 1 = 1 \cdot a = a$

Inverse Properties:

$a + (-a) = 0$ $a \cdot \dfrac{1}{a} = 1, a \neq 0$

See page 25 for additional properties of real numbers.

1.4 Algebraic Expressions

Assignment: _____ Due date: _____

☐ **Identify the terms and coefficients of algebraic expressions.**

The *terms* of an algebraic expression are those parts that are separated by addition.

The *coefficient* of a term is its numerical factor.

☐ **Simplify algebraic expressions.**

To simplify an algebraic expression, remove the symbols of grouping and combine like terms.

☐ **Evaluate algebraic expressions.**

To evaluate an algebraic expression, substitute numerical values for each of the variables in the expression and simplify.

1.5 Constructing Algebraic Expressions

Assignment: _____ Due date: _____

☐ **Translate verbal phrases into algebraic expressions, and vice versa.**

Addition: sum, plus, greater than, increased by, more than, exceeds, total of

Subtraction: difference, minus, less than, decreased by, subtracted from, reduced by, the remainder

Multiplication: product, multiplied by, twice, times, percent of

Division: quotient, divided by, ratio, per

☐ **Write labels for integers.**

The following expressions are useful ways to denote integers.

1. $2n$ denotes an even integer for $n = 1, 2, 3, \ldots$.
2. $2n - 1$ and $2n + 1$ denote odd integers for $n = 1, 2, 3, \ldots$.
3. $\{n, n + 1, n + 2, \ldots\}$ denotes a set of consecutive integers.

Review Exercises

1.1 The Real Number System

1 ▶ Understand the set of real numbers and the subsets of real numbers.

In Exercises 1 and 2, which of the real numbers in the set are (a) natural numbers, (b) integers, (c) rational numbers, and (d) irrational numbers?

1. $\left\{\frac{3}{5}, -4, 0, \sqrt{2}, 52, -\frac{1}{8}, \sqrt{9}\right\}$

2. $\left\{98, -141, -\frac{7}{8}, 3.99, -\sqrt{12}, -\frac{54}{11}\right\}$

In Exercises 3 and 4, list all members of the set.

3. The natural numbers between -2.3 and 6.1

4. The even integers between -5.5 and 2.5

2 ▶ Use the real number line to order real numbers.

In Exercises 5 and 6, plot the real numbers on the real number line.

5. (a) 4   (b) -3 (c) $\frac{3}{4}$ (d) -2.4
6. (a) -9 (b) 7   (c) $-\frac{3}{2}$ (d) 5.25

In Exercises 7–10, place the correct inequality symbol (< or >) between the numbers.

7. -5 ____ 3
8. -2 ____ -8
9. $-\frac{8}{5}$ ____ $-\frac{2}{5}$
10. 8.4 ____ -3.2

3 ▶ Use the real number line to find the distance between two real numbers.

In Exercises 11–14, find the distance between the pair of real numbers.

11. 11 and -3

12. 4 and -13
13. -13.5 and -6.2
14. -8.4 and -0.3

4 ▶ Determine the absolute value of a real number.

In Exercises 15–18, evaluate the expression.

15. $|-5|$
16. $|6|$
17. $-|-7.2|$
18. $|-3.6|$

1.2 Operations with Real Numbers

1 ▶ Add, subtract, multiply, and divide real numbers.

In Exercises 19–40, evaluate the expression. If it is not possible, state the reason. Write all fractions in simplest form.

19. $15 + (-4)$
20. $-12 + 3$
21. $340 - 115 + 5$
22. $-154 + 86 - 240$
23. $-63.5 + 21.7$
24. $14.35 - 10.3$
25. $\frac{4}{21} + \frac{7}{21}$
26. $\frac{21}{16} - \frac{13}{16}$
27. $-\frac{5}{6} + 1$
28. $3 + \frac{4}{9}$
29. $8\frac{3}{4} - 6\frac{5}{8}$
30. $-2\frac{9}{10} + 5\frac{3}{20}$
31. $-7 \cdot 4$
32. $9 \cdot (-5)$
33. $120(-5)(7)$
34. $(-16)(-15)(-4)$
35. $\frac{3}{8} \cdot \left(-\frac{2}{15}\right)$
36. $\frac{5}{21} \cdot \frac{21}{5}$
37. $\frac{-56}{-4}$
38. $\frac{85}{0}$
39. $-\frac{7}{15} \div \left(-\frac{7}{30}\right)$
40. $-\frac{2}{3} \div \frac{4}{15}$

In Exercises 41 and 42, write the expression as a repeated addition problem.

41. $7(-3)$

42. $5\left(\frac{2}{3}\right)$

In Exercises 43 and 44, write the expression as a multiplication problem.

43. $8 + 8 + 8 + 8 + 8 + 8 + 8 + 8$
44. $(-5) + (-5) + (-5) + (-5) + (-5)$

2 ▶ Write repeated multiplication in exponential form and evaluate exponential expressions.

In Exercises 45 and 46, write the expression using exponential notation.

45. $6 \cdot 6 \cdot 6 \cdot 6 \cdot 6 \cdot 6 \cdot 6$
46. $\left(-\frac{1}{2}\right)\left(-\frac{1}{2}\right)\left(-\frac{1}{2}\right)$

In Exercises 47–52, evaluate the exponential expression.

47. $(-6)^4$
48. $-(-3)^4$
49. -4^2
50. 2^5
51. $-\left(-\frac{1}{2}\right)^3$
52. $-\left(\frac{2}{3}\right)^4$

3 ▶ Use order of operations to evaluate expressions.

In Exercises 53–56, evaluate the expression.

53. $120 - (5^2 \cdot 4)$
54. $45 - 45 \div 3^2$
55. $8 + 3[6^2 - 2(7 - 4)]$
56. $2^4 - [10 + 6(1 - 3)^2]$

4 ▶ Evaluate expressions using a calculator and order of operations.

In Exercises 57 and 58, evaluate the expression using a calculator. Round your answer to two decimal places.

57. $7(408.2^2 - 39.5 \div 0.3)$
58. $-59[4.6^3 + 5.8(-13.4)]$

59. *Total Charge* You purchased an entertainment system and made a down payment of \$395 plus nine monthly payments of \$45 each. What is the total amount you paid for the system?

60. *Savings Plan* You deposit \$80 per month in a savings account for 10 years. The account earns 2% interest compounded monthly. The total amount in the account after 10 years will be

$$80\left[\left(1 + \frac{0.02}{12}\right)^{120} - 1\right]\left(1 + \frac{12}{0.02}\right).$$

Use a calculator to determine this amount.

1.3 Properties of Real Numbers

1 ▶ Identify and use the properties of real numbers.

In Exercises 61–70, identify the property of real numbers illustrated by the statement.

61. $13 - 13 = 0$
62. $7\left(\frac{1}{7}\right) = 1$
63. $7(9 + 3) = 7 \cdot 9 + 7 \cdot 3$
64. $15(4) = 4(15)$
65. $5 + (4 - y) = (5 + 4) - y$

66. $6(4z) = (6 \cdot 4)z$
67. $(u - v)(2) = 2(u - v)$

68. $xy \cdot 1 = xy$
69. $8(x - y) = 8 \cdot x - 8 \cdot y$
70. $xz - yz = (x - y)z$

In Exercises 71–74, rewrite the expression by using the Distributive Property.

71. $-(-u + 3v)$
72. $-5(2x - 4y)$
73. $-a(8 - 3a)$
74. $x(3x + 4y)$

1.4 Algebraic Expressions

1 ▶ Identify the terms and coefficients of algebraic expressions.

In Exercises 75–78, identify the terms and coefficients of the algebraic expression.

75. $4y^3 - y^2 + \dfrac{17}{2}y$

76. $\dfrac{x}{3} + 2xy^2 + \dfrac{1}{5}$

77. $52 - 1.2x^3 + \dfrac{1}{x}$

78. $2ab^2 + a^2b^2 - \dfrac{1}{a}$

2 ▶ Simplify algebraic expressions by combining like terms and removing symbols of grouping.

In Exercises 79–88, simplify the expression.

79. $6x + 3x$ 80. $10y - 7y$

81. $3u - 2v + 7v - 3u$

82. $9m - 4n + m - 3n$

83. $5(x - 4) + 10$

84. $15 - 7(z + 2)$

85. $3x - (y - 2x)$

86. $30x - (10x + 80)$

87. $3[b + 5(b - a)]$

88. $-2t[8 - (6 - t)] + 5t$

3 ▶ Evaluate algebraic expressions by substituting values for the variables.

In Exercises 89–92, evaluate the algebraic expression for the specified values of the variable(s). If not possible, state the reason.

	Expression	*Values*

89. $x^2 - 2x - 3$ (a) $x = 3$
                       (b) $x = 0$

90. $\dfrac{x}{y + 2}$ (a) $x = 0, \quad y = 3$
                 (b) $x = 5, \quad y = -2$

91. $y^2 - 2y + 4x$ (a) $x = 4, \quad y = -1$
                          (b) $x = -2, \quad y = 2$

	Expression	*Values*

92. $|2x - x^2| - 2y$ (a) $x = -6, \quad y = 3$
                           (b) $x = 4, \quad y = -5$

1.5 Constructing Algebraic Expressions

1 ▶ Translate verbal phrases into algebraic expressions, and vice versa.

In Exercises 93–96, translate the verbal phrase into an algebraic expression.

93. Twelve decreased by twice the number n

94. One hundred increased by the product of 15 and a number x

95. The sum of the square of a number y and 49

96. Three times the absolute value of the difference of a number n and 3, all divided by 5

In Exercises 97–100, write a verbal description of the algebraic expression without using the variable.

97. $2y + 7$

98. $5u - 3$

99. $\dfrac{x - 5}{4}$

100. $4(a - 1)$

2 ▶ Construct algebraic expressions with hidden products.

In Exercises 101–104, write an algebraic expression that represents the quantity in the verbal statement, and simplify if possible.

101. The amount of income tax on a taxable income of I dollars when the tax rate is 18%

102. The distance traveled when you travel 8 hours at the average speed of r miles per hour

103. The area of a rectangle whose length is l units and whose width is 5 units less than the length

104. The sum of two consecutive integers, the first of which is n

Chapter Test

Take this test as you would take a test in class. After you are done, check your work against the answers in the back of the book.

1. Place the correct inequality symbol ($<$ or $>$) between each pair of numbers.

    (a) $\frac{5}{2}$ _____ $|-3|$ (b) $-\frac{2}{3}$ _____ $-\frac{3}{2}$

2. Find the distance between -4.4 and 6.9.

In Exercises 3–10, evaluate the expression.

3. $-14 + 9 - 15$

4. $\frac{2}{3} + \left(-\frac{7}{6}\right)$

5. $-2(225 - 150)$

6. $(-3)(4)(-5)$

7. $\left(-\frac{7}{16}\right)\left(-\frac{8}{21}\right)$

8. $\frac{5}{18} \div \frac{15}{8}$

9. $\left(-\frac{3}{5}\right)^3$

10. $\dfrac{4^2 - 6}{5} + 13$

11. Identify the property of real numbers illustrated by each statement.

    (a) $(-3 \cdot 5) \cdot 6 = -3(5 \cdot 6)$

    (b) $3y \cdot \dfrac{1}{3y} = 1$

12. Rewrite the expression $-6(2x - 1)$ using the Distributive Property.

In Exercises 13–16, simplify the expression.

13. $3x^2 - 2x - 5x^2 + 7x - 1$

14. $x(x + 2) - 2(x^2 + x - 13)$

15. $a(5a - 4) - 2(2a^2 - 2a)$

16. $4t - [3t - (10t + 7)]$

17. Explain the meaning of "evaluating an algebraic expression." Evaluate the expression $7 + (x - 3)^2$ for each value of x.

    (a) $x = -1$ (b) $x = 3$

18. An electrician wants to divide 102 inches of wire into 17 pieces with equal lengths. How long should each piece be?

19. A *cord* of wood is a pile 4 feet high, 4 feet wide, and 8 feet long. The volume of a rectangular solid is its length times its width times its height. Find the number of cubic feet in 5 cords of wood.

20. Translate the phrase into an algebraic expression.

    "The product of a number n and 5, decreased by 8"

21. Write an algebraic expression for the sum of two consecutive even integers, the first of which is $2n$.

22. Write expressions for the perimeter and area of the rectangle shown at the left. Simplify the expressions and evaluate them when $l = 45$.

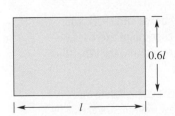

Figure for 22

Improving Your Memory

Have you ever driven on a highway for ten minutes when all of a sudden you kind of woke up and wondered where the last ten miles had gone? The car was on autopilot. The same thing happens to many college students as they sit through back-to-back classes. The longer students sit through classes on autopilot, the more likely they will "crash" when it comes to studying outside of class on their own.

While on autopilot, you do not process and retain new information effectively. Your memory can be improved by learning how to focus during class and while studying on your own.

Kimberly Nolting

VP, Academic Success Press
expert in developmental education

Memorize these models from Section 2.3:

$$\text{Selling price} = \text{Cost} + \text{Markup}$$
$$\$24 = \$18 + \$6$$

$$\text{Markup} = \text{Markup rate} \cdot \text{Cost}$$
$$\$8 = 25\% \cdot 32$$

$$\text{Selling price} = \text{List price} - \text{Discount}$$
$$\$350 = \$395 - \$45$$

$$\text{First rate} \cdot \text{Amount} + \text{Second rate} \cdot \text{Amount}$$
$$= \text{Final rate} \cdot \text{Final amount}$$

$$\text{Distance} = \text{Rate} \cdot \text{Time}$$
$$180 \text{ mi} = 60 \text{ mi/h} \cdot 3 \text{ h}$$

$$\text{Per hour work rate} = \frac{1}{\text{Total hours to complete a job}}$$

Smart Study Strategy

Keep Your Mind Focused

During class
- When you sit down at your desk, get all other issues out of your mind by reviewing your notes from the last class and focusing just on math.
- Repeat in your mind what you are writing in your notes.
- When the math is particularly difficult, ask your instructor for another example.

While completing homework
- Before doing homework, review the concept boxes and examples. Talk through the examples out loud.
- Complete homework as though you were also preparing for a quiz. Memorize the different types of problems, formulas, rules, and so on.

Between classes
- Review the concept boxes and check your memory using the checkpoint exercises, Concept Check exercises, and the What Did You Learn? section.

Preparing for a test
- Review all your notes that pertain to the upcoming test. Review examples of each type of problem that could appear on the test.

Chapter 2
Linear Equations and Inequalities

IT WORKED FOR ME!

"I have to work and I have two kids. The only time I sit still is in class and it is hard to stay awake. I went to talk to a counselor and she had lots of suggestions about organization. I found more time to study my math and made sure that I was really learning while I did my homework. Things made more sense in class and it was easier to understand and remember everything."

Tamika
Nursing

2.1 Linear Equations

John Coletti/Getty Images

What You Should Learn

1 ▶ Check solutions of equations.

2 ▶ Solve linear equations in standard form.

3 ▶ Solve linear equations in nonstandard form.

Why You Should Learn It

Linear equations are used in many real-life applications. For instance, in Example 9 on page 64, a linear equation is used to model the number of bridges in the United States.

1 ▶ Check solutions of equations.

Introduction

An **equation** is a statement that equates two algebraic expressions. Some examples are $x = 4$, $4x + 3 = 15$, $2x - 8 = 2(x - 4)$, and $x^2 - 16 = 0$.

Solving an equation involving a variable means finding all values of the variable for which the equation is true. Such values are **solutions** and are said to **satisfy** the equation. For instance, $x = 3$ is a solution of $4x + 3 = 15$ because $4(3) + 3 = 15$ is a true statement.

The **solution set** of an equation is the set of all solutions of the equation. Sometimes, an equation will have the set of all real numbers as its solution set. Such an equation is an **identity.** For instance, the equation

$$2x - 8 = 2(x - 4) \qquad \text{Identity}$$

is an identity because the equation is true for all real values of x. Try values such as 0, 1, -2, and 5 in this equation to see that each one is a solution.

An equation whose solution set is not the entire set of real numbers is called a **conditional equation.** For instance, the equation

$$x^2 - 16 = 0 \qquad \text{Conditional equation}$$

is a conditional equation because it has only two solutions, $x = 4$ and $x = -4$. Example 1 shows how to **check** whether a given value is a solution.

EXAMPLE 1   Checking a Solution of an Equation

Determine whether $x = -3$ is a solution of $-3x - 5 = 4x + 16$.

Solution

$$\begin{aligned}
-3x - 5 &= 4x + 16 & &\text{Write original equation.} \\
-3(-3) - 5 &\stackrel{?}{=} 4(-3) + 16 & &\text{Substitute } -3 \text{ for } x. \\
9 - 5 &\stackrel{?}{=} -12 + 16 & &\text{Simplify.} \\
4 &= 4 & &\text{Solution checks. } \checkmark
\end{aligned}$$

Because each side turns out to be the same number, you can conclude that $x = -3$ *is* a solution of the original equation. Try checking to see whether $x = -2$ is a solution.

 **CHECKPOINT** *Now try Exercise 1.*

Study Tip

When checking a solution, you should write a question mark over the equal sign to indicate that you are uncertain whether the "equation" is true for a given value of the variable.

It is helpful to think of an equation as having two sides that are in balance. Consequently, when you try to solve an equation, you must be careful to maintain that balance by performing the same operation(s) on each side.

Two equations that have the same set of solutions are **equivalent equations.** For instance, the equations $x = 3$ and $x - 3 = 0$ are equivalent equations because both have only one solution—the number 3. When any one of the four techniques in the following list is applied to an equation, the resulting equation is equivalent to the original equation.

Forming Equivalent Equations: Properties of Equality

An equation can be transformed into an *equivalent equation* using one or more of the following procedures.

	Original Equation	*Equivalent Equation*
1. *Simplify Either Side:* Remove symbols of grouping, combine like terms, or simplify fractions on one or both sides of the equation.	$4x - x = 8$	$3x = 8$
2. *Apply the Addition Property of Equality:* Add (or subtract) the same quantity to (from) *each* side of the equation.	$x - 3 = 5$	$x = 8$
3. *Apply the Multiplication Property of Equality:* Multiply (or divide) *each* side of the equation by the same *nonzero* quantity.	$3x = 12$	$x = 4$
4. *Interchange Sides:* Interchange the two sides of the equation.	$7 = x$	$x = 7$

When solving an equation, you can use any of the four techniques for forming equivalent equations to eliminate terms or factors in the equation. For example, to solve the equation

$$x + 4 = 2$$

you need to remove the term 4 from the left side. This is accomplished by subtracting 4 from each side.

$x + 4 = 2$	Write original equation.
$x + 4 - 4 = 2 - 4$	Subtract 4 from each side.
$x + 0 = -2$	Combine like terms.
$x = -2$	Simplify.

Although this solution method requires you to subtract 4 from each side, you can just as easily add -4 to each side. Both techniques are legitimate—which one you decide to use is a matter of personal preference.

2 ▸ Solve linear equations in standard form.

Solving Linear Equations in Standard Form

The most common type of equation in one variable is a *linear equation*.

> ### Definition of Linear Equation
> A **linear equation** in one variable x is an equation that can be written in the standard form
>
> $$ax + b = 0 \qquad \text{Standard form}$$
>
> where a and b are real numbers with $a \neq 0$.

A linear equation in one variable is also called a **first-degree equation** because its variable has an implied exponent of 1. Some examples of linear equations in the standard form $ax + b = 0$ are $3x + 2 = 0$ and $5x - 4 = 0$.

Remember that to *solve* an equation in x means to find the values of x that satisfy the equation. For a linear equation in the standard form

$$ax + b = 0$$

the goal is to **isolate** x by rewriting the standard equation in the form

$$x = \boxed{\text{a number}}.$$

Beginning with the original equation, you write a sequence of equivalent equations, each having the same solution as the original equation.

EXAMPLE 2 Solving a Linear Equation in Standard Form

Solve $4x - 12 = 0$. Then check the solution.

Solution

$4x - 12 = 0$	Write original equation.
$4x - 12 + 12 = 0 + 12$	Add 12 to each side.
$4x = 12$	Combine like terms.
$\dfrac{4x}{4} = \dfrac{12}{4}$	Divide each side by 4.
$x = 3$	Simplify.

The solution is $x = 3$. You can check this as follows.

Check

$4x - 12 = 0$	Write original equation.
$4(3) - 12 \stackrel{?}{=} 0$	Substitute 3 for x.
$12 - 12 \stackrel{?}{=} 0$	Simplify.
$0 = 0$	Solution checks. ✓

✓ **CHECKPOINT** *Now try Exercise 23.*

Study Tip

Be sure you see that solving an equation such as the one in Example 2 has two important parts. The first part is *finding* the solution(s). The second part is *checking* that each solution you find actually satisfies the original equation. You can improve your accuracy in algebra by developing the habit of checking each solution.

You know that $x = 3$ is a solution of the equation in Example 2, but at this point you might be asking, "How can I be sure that the equation does not have other solutions?" The answer is that a linear equation in one variable always has *exactly one* solution. You can show this with the following steps.

$$ax + b = 0$$ Original equation, with $a \neq 0$

$$ax + b - b = 0 - b$$ Subtract b from each side.

$$ax = -b$$ Combine like terms.

$$\frac{ax}{a} = \frac{-b}{a}$$ Divide each side by a.

$$x = -\frac{b}{a}$$ Simplify.

It is clear that the last equation has only one solution, $x = -b/a$. Because the last equation is equivalent to the original equation, you can conclude that every linear equation in one variable written in standard form has exactly one solution.

EXAMPLE 3   **Solving a Linear Equation in Standard Form**

Solve $2x + 2 = 0$. Then check the solution.

Solution

$$2x + 2 = 0$$ Write original equation.

$$2x + 2 - 2 = 0 - 2$$ Subtract 2 from each side.

$$2x = -2$$ Combine like terms.

$$\frac{2x}{2} = \frac{-2}{2}$$ Divide each side by 2.

$$x = -1$$ Simplify.

The solution is $x = -1$. You can check this as follows.

Check

$$2(-1) + 2 \overset{?}{=} 0$$ Substitute -1 for x in original equation.

$$-2 + 2 \overset{?}{=} 0$$ Simplify.

$$0 = 0$$ Solution checks. ✓

✓ **CHECKPOINT** *Now try Exercise 25.*

As you gain experience in solving linear equations, you will probably be able to perform some of the solution steps in your head. For instance, you might solve the equation given in Example 3 by performing two of the steps mentally and writing only three steps, as follows.

$$2x + 2 = 0$$ Write original equation.

$$2x = -2$$ Subtract 2 from each side.

$$x = -1$$ Divide each side by 2.

3 ▶ Solve linear equations in nonstandard form.

Solving Linear Equations in Nonstandard Form

Linear equations often occur in nonstandard forms that contain symbols of grouping or like terms that are not combined. Here are some examples.

$$x + 2 = 2x - 6, \quad 6(y - 1) = 2y - 3, \quad \frac{x}{18} + \frac{3x}{4} = 2$$

The next three examples show how to solve these linear equations.

Study Tip

A strategy that can help you to isolate x in solving a linear equation is to rewrite the original equation so that only variable terms are on one side of the equal sign and only constant terms are on the other side.

EXAMPLE 4   Solving a Linear Equation in Nonstandard Form

$x + 2 = 2x - 6$	Original equation
$-2x + x + 2 = -2x + 2x - 6$	Add $-2x$ to each side.
$-x + 2 = -6$	Combine like terms.
$-x + 2 - 2 = -6 - 2$	Subtract 2 from each side.
$-x = -8$	Combine like terms.
$(-1)(-x) = (-1)(-8)$	Multiply each side by -1.
$x = 8$	Simplify.

The solution is $x = 8$. Check this in the original equation.

✓ **CHECKPOINT** *Now try Exercise 35.*

In most cases, it helps to remove symbols of grouping as a first step in solving an equation. This is illustrated in Example 5.

EXAMPLE 5   Solving a Linear Equation Containing Parentheses

$6(y - 1) = 2y - 3$	Original equation
$6y - 6 = 2y - 3$	Distributive Property
$6y - 2y - 6 = 2y - 2y - 3$	Subtract $2y$ from each side.
$4y - 6 = -3$	Combine like terms.
$4y - 6 + 6 = -3 + 6$	Add 6 to each side.
$4y = 3$	Combine like terms.
$\dfrac{4y}{4} = \dfrac{3}{4}$	Divide each side by 4.
$y = \dfrac{3}{4}$	Simplify.

The solution is $y = \frac{3}{4}$. Check this in the original equation.

✓ **CHECKPOINT** *Now try Exercise 37.*

If a linear equation contains fractions, you should first *clear the equation of fractions* by multiplying each side of the equation by the least common denominator (LCD) of the fractions.

EXAMPLE 6   **Solving a Linear Equation Containing Fractions**

$$\frac{x}{18} + \frac{3x}{4} = 2$$ Original equation

$$36\left(\frac{x}{18} + \frac{3x}{4}\right) = 36(2)$$ Multiply each side by LCD of 36.

$$36 \cdot \frac{x}{18} + 36 \cdot \frac{3x}{4} = 36(2)$$ Distributive Property

$$2x + 27x = 72$$ Simplify.

$$29x = 72$$ Combine like terms.

$$\frac{29x}{29} = \frac{72}{29}$$ Divide each side by 29.

$$x = \frac{72}{29}$$ Simplify.

The solution is $x = \frac{72}{29}$. Check this in the original equation.

✓ **CHECKPOINT** *Now try Exercise 47.*

The next example shows how to solve a linear equation involving decimals. The procedure is basically the same as for a linear equation involving integers, but the arithmetic can be messier.

EXAMPLE 7   **Solving a Linear Equation Involving Decimals**

Solve $0.12x + 0.09(5000 - x) = 513$.

Solution

$$0.12x + 0.09(5000 - x) = 513$$ Write original equation.

$$0.12x + 450 - 0.09x = 513$$ Distributive Property

$$0.03x + 450 = 513$$ Combine like terms.

$$0.03x + 450 - 450 = 513 - 450$$ Subtract 450 from each side.

$$0.03x = 63$$ Combine like terms.

$$\frac{0.03x}{0.03} = \frac{63}{0.03}$$ Divide each side by 0.03.

$$x = 2100$$ Simplify.

The solution is $x = 2100$. Check this in the original equation.

✓ **CHECKPOINT** *Now try Exercise 53.*

Study Tip

Avoid the temptation to divide an equation by x. You may obtain an incorrect solution, as in the following example.

$$7x = -4x \quad \text{Original equation}$$

$$\frac{7x}{x} = -\frac{4x}{x} \quad \begin{array}{l}\text{Divide each side}\\\text{by } x.\end{array}$$

$$7 = -4 \quad \text{False statement}$$

The false statement indicates that there is no solution, but when the equation is solved correctly, the solution is $x = 0$.

$$7x = -4x$$

$$7x + 4x = -4x + 4x$$

$$11x = 0$$

$$\frac{11x}{11} = \frac{0}{11}$$

$$x = 0$$

Some equations in nonstandard form have *no solution* or *infinitely many solutions*. Two such cases are illustrated in Example 8.

EXAMPLE 8   Solving Equations: Special Cases

Solve each equation.

a. $2x - 4 = 2(x - 3)$ **b.** $3x + 2 + 2(x - 6) = 5(x - 2)$

Solution

a. $2x - 4 = 2(x - 3)$ Write original equation.

$\quad\ \ 2x - 4 = 2x - 6$ Distributive Property

$\quad\quad\quad -4 \neq -6$ Subtract $2x$ from each side.

Because -4 does not equal -6, you can conclude that the original equation has no solution.

b. $3x + 2 + 2(x - 6) = 5(x - 2)$ Write original equation.

$\quad\ \ 3x + 2 + 2x - 12 = 5x - 10$ Distributive Property

$\quad\quad\quad\quad\ \ 5x - 10 = 5x - 10$ Combine like terms.

$\quad\quad\quad\quad\quad\quad -10 = -10$ Subtract $5x$ from each side.

Because the last equation is true for any value of x, the equation is an identity, and you can conclude that the original equation has infinitely many solutions.

 **CHECKPOINT** *Now try Exercise 29.*

EXAMPLE 9   Bridges

The bar graph in Figure 2.1 shows the numbers y of bridges (in thousands) in the United States from 2000 to 2005. An equation that models the data is

$$y = 1.34t + 588.3, \qquad 0 \le t \le 5$$

where t represents the year, with $t = 0$ corresponding to 2000. Determine when the number of bridges reached 591,000. (Source: U.S. Federal Highway Administration)

Solution

Let $y = 591$ and solve the resulting equation for t.

$\quad\quad\quad y = 1.34t + 588.3$ Write original equation.

$\quad\ \ 591 = 1.34t + 588.3$ Substitute 591 for y.

$\quad\quad\ \ 2.7 = 1.34t$ Subtract 588.3 from each side.

$\quad\quad\quad 2 \approx t$ Divide each side by 1.34.

Because $t = 0$ corresponds to 2000, it follows that $t = 2$ corresponds to 2002. So, the number of bridges reached 591,000 in 2002. The bar graph in Figure 2.1 supports this answer.

 CHECKPOINT *Now try Exercise 69.*

Figure 2.1

_____ **Concept Check** _____

1. Explain how the process of evaluating an algebraic expression can be used to check a solution of an equation.

2. Explain the difference between a conditional equation and an identity.

3. The following equations are equivalent. Explain what must be true about a and b.
$$3x + 7 = 13$$
$$3x + 7 + a = 13 + b$$

4. Describe two procedures you can apply to the equation $30 = 3(x + 8)$ to obtain the equivalent equation $3x + 24 = 30$.

2.1 EXERCISES

Go to pages 116–117 to record your assignments.

_____ **Developing Skills** _____

In Exercises 1–6, determine whether each value of the variable is a solution of the equation. *See Example 1.*

Equation	Values

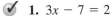 **1.** $3x - 7 = 2$ (a) $x = 0$
                                            (b) $x = 3$

2. $5x + 9 = 4$ (a) $x = -1$
                        (b) $x = 2$

3. $x + 8 = 3x$ (a) $x = 4$
                        (b) $x = -4$

4. $10x - 3 = 7x$.   (a) $x = 0$
                              (b) $x = -1$

5. $\frac{1}{4}x = 3$ (a) $x = -4$
                              (b) $x = 12$

6. $3(y + 2) = y - 5$ (a) $y = -\frac{3}{2}$
                              (b) $y = -5.5$

In Exercises 7–10, identify the equation as a conditional equation, an identity, or an equation with no solution.

7. $6(x + 3) = 6x + 3$

8. $3x + 11 - 6x = 11 - 3x$

9. $\frac{2}{3}x + 4 = \frac{1}{3}x + 12$

10. $\frac{4}{3}(2x - 6) = \frac{8}{3}x - 8$

In Exercises 11 and 12, justify each step of the solution.

11. $3x + 15 = 0$
           $3x + 15 - 15 = 0 - 15$
                  $3x = -15$
                  $\dfrac{3x}{3} = \dfrac{-15}{3}$
                  $x = -5$

12. $7x - 21 = 0$
           $7x - 21 + 21 = 0 + 21$
                  $7x = 21$
                  $\dfrac{7x}{7} = \dfrac{21}{7}$
                  $x = 3$

In Exercises 13–20, determine whether the two equations are equivalent. Explain your reasoning.

13. $3x = 10, \ 4x = x + 10$

14. $5x = 22, \ 4x = 22 - x$

15. $x + 5 = 12, \ 2x + 15 = 24$

16. $x - 3 = 8, \ 3x - 6 = 24$

17. $3(4 - 2t) = 5, \ 12 - 6t = 5$

18. $(3 + 2^2)z = 16, \ 7z = 16$

19. $2x - 7 = 3, \ x = 3$

20. $6 - 5x = -4, \ x = -4$

In Exercises 21–58, solve the equation. If there is exactly one solution, check your answer. If not, describe the solution. *See Examples 2–8.*

21. $x - 3 = 0$

22. $x + 8 = 0$

✓ **23.** $3x - 12 = 0$

24. $-14x - 28 = 0$

✓ **25.** $6x + 4 = 0$

26. $8z - 10 = 0$

27. $3t + 8 = -2$

28. $10 - 6x = -5$

✓ **29.** $4y - 3 = 4y$

30. $24 - 2x = -2x$

31. $-9y - 4 = -9y$

32. $6a + 2 = 6a$

33. $7 - 8x = 13x$

34. $2s - 16 = 34s$

✓ **35.** $3x - 1 = 2x + 14$

36. $9y + 4 = 12y - 2$

✓ **37.** $8(x - 8) = 24$

38. $6(x + 2) = 30$

39. $3(x - 4) = 7x + 6$

40. $-2(t + 3) = 9 - 5t$

41. $4(2x - 3) = 8x - 12$

42. $9x + 6 = 3(3x + 2)$

43. $12(x + 3) = 7(x + 3)$

44. $-5(x - 10) = 6(x - 10)$

45. $7(x + 6) = 3(2x + 14) + x$

46. $5(x + 8) = 4(2x + 10) - 3x$

✓ **47.** $t - \frac{2}{5} = \frac{3}{2}$

48. $z + \frac{1}{15} = -\frac{3}{10}$

49. $\frac{t}{5} - \frac{t}{2} = 1$

50. $\frac{t}{6} + \frac{t}{8} = 1$

51. $\frac{8x}{5} - \frac{x}{4} = -3$

52. $\frac{11x}{6} + \frac{1}{3} = 2x$

✓ **53.** $0.3x + 1.5 = 8.4$

54. $16.3 - 0.2x = 7.1$

55. $1.2(x - 3) = 10.8$

56. $6.5(1 - 2x) = 13$

57. $\frac{2}{3}(2x - 4) = \frac{1}{2}(x + 3) - 4$

58. $\frac{3}{4}(6 - x) = \frac{1}{3}(4x + 5) + 2$

Solving Problems

59. *Number Problem*   The sum of two consecutive integers is 251. Find the integers.

60. *Number Problem*   The sum of two consecutive even integers is 626. Find the integers.

61. *Car Repair*   The bill for the repair of your car was $257. The cost for parts was $162. The cost for labor was $38 per hour. How many hours did the repair work take?

62. *Appliance Repair*   The bill for the repair of your refrigerator was $187. The cost for parts was $74. The cost for the service call and the first half hour of service was $50. The additional cost for labor was $21 per half hour. How many hours did the repair work take?

63. *Work Rate*   Two people can complete a task in t hours, where t must satisfy the equation

$$\frac{t}{10} + \frac{t}{15} = 1.$$

Find the required time t.

64. *Work Rate*   Two people can complete a task in t hours, where t must satisfy the equation

$$\frac{t}{12} + \frac{t}{20} = 1.$$

Find the required time t.

65. *Height*   Consider the fountain shown in the figure. The initial velocity of the stream of the water is 48 feet per second. The velocity v of the water at any time t (in seconds) is given by $v = 48 - 32t$. Find the time for a drop of water to travel from the base to the maximum height of the fountain.   (*Hint:* The maximum height is reached when $v = 0$.)

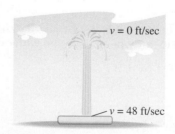

$v = 0$ ft/sec

$v = 48$ ft/sec

66. *Height* The velocity v of an object projected vertically upward with an initial velocity of 64 feet per second is given by $v = 64 - 32t$, where t is time in seconds. When does the object reach its maximum height?

67. ▲ *Geometry* The length of a rectangle is t times its width (see figure). So, the perimeter P is given by $P = 2w + 2(tw)$, where w is the width of the rectangle. The perimeter of the rectangle is 1000 meters.

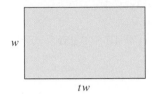

w

tw

(a) Complete the table of widths, lengths, and areas of the rectangle for the specified values of t.

t	1	1.5	2
Width			
Length			
Area			

t	3	4	5
Width			
Length			
Area			

(b) Use the table to write a short paragraph describing the relationship among the width, length, and area of a rectangle that has a *fixed* perimeter.

68. ▲ *Geometry* Repeat parts (a) and (b) of Exercise 67 for a rectangle with a fixed perimeter of 60 inches and a length that is t inches greater than the width (see figure). Use the values $t = 0, 1, 2, 3, 5$, and 10 to create a table given that $P = 2w + 2(t + w)$.

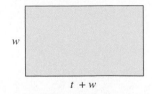

w

$t + w$

69. *Using a Model* The average annual expenditures per student y (in dollars) for primary and secondary public schools in the United States from 1998 to 2005 can be approximated by the model $y = 355.3t + 3725$, $8 \le t \le 15$, where t represents the year, with $t = 8$ corresponding to 1998 (see figure). According to this model, during which year did the expenditures reach $7633.30? Explain how to answer the question graphically, numerically, and algebraically. (Source: National Education Association)

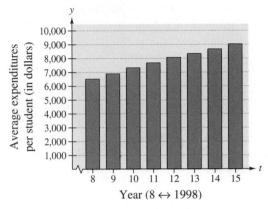

70. *Using a Model* The annual sales y (in billions of dollars) of food and beverages at full-service restaurants in the United States from 1998 to 2005 can be approximated by the model $y = 6.37t + 67.5$, $8 \le t \le 15$, where t represents the year, with $t = 8$ corresponding to 1998 (see figure). According to this model, in what year were the annual sales about $125 billion? Explain how to answer the question graphically, numerically, and algebraically. (Source: National Restaurant Association)

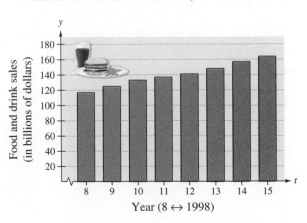

────────────────────── **Explaining Concepts** ──────────────────────

71. *True or False?* Multiplying each side of an equation by zero yields an equivalent equation. Justify your answer.

72. *True or False?* Subtracting zero from each side of an equation yields an equivalent equation. Justify your answer.

73. ✎ Can an identity be written as a linear equation in standard form? Explain.

74. ✎ Explain how you can determine whether a first-degree equation in one variable is linear from its solution. Describe two types of solutions of first-degree equations in one variable that are *not* linear equations.

✎ In Exercises 75–78, classify the equation as an *identity*, a *conditional equation*, or an *equation with no solution*. Discuss real-life situations that could be represented by the equation, or could be used to show that the equation has no solution.

75. $x + 0.20x = 50.16$

76. $3(x - 1) = 3x + 5$

77. $0.25(40 + x) = 10 + 0.25x$

78. $5w + 3 = 28$

────────────────────── **Cumulative Review** ──────────────────────

In Exercises 79–82, evaluate the expression.

79. $\frac{2}{5} + \frac{4}{5}$

80. $\frac{5}{6} - \frac{2}{3}$

81. $-5 - (-3)$

82. $-12 - (6 - 5)$

In Exercises 83–86, evaluate the expression for the specified values of the variable.

83. $8 + 7x; x = 2, x = 3$

84. $\dfrac{2x}{x + 1}; x = 1, x = 5$

85. $x^2 - 1; x = -4, x = 3$

86. $|3x - 7|; x = -1, x = 1$

In Exercises 87–90, translate the verbal phrase into an algebraic expression.

87. Eight less than a number n

88. The ratio of n and four

89. Twice the sum of n and three

90. Five less than half of a number n

2.2 Linear Equations and Problem Solving

Joe Raedle/Getty Images

Why You Should Learn It

Percents appear in many real-life situations. For instance, in Exercise 68 on page 76, a percent is used to show how price inflation affects the cost of a new car.

1 ▶ Use mathematical modeling to write algebraic equations representing real-life situations.

What You Should Learn

1 ▶ Use mathematical modeling to write algebraic equations representing real-life situations.

2 ▶ Solve percent problems using the percent equation.

3 ▶ Use ratios to compare unit prices for products.

4 ▶ Solve proportions.

Mathematical Modeling

In this section you will see how algebra can be used to solve problems that occur in real-life situations. This process is called **mathematical modeling,** and its basic steps are as follows.

Verbal description ⟹ Verbal model ⟹ Assign labels ⟹ Algebraic equation

EXAMPLE 1   **Mathematical Modeling**

Write an algebraic equation that represents the following problem. Then solve the equation and answer the question.

You have accepted a job at an annual salary of $40,830. This salary includes a year-end bonus of $750. You are paid twice a month. What will your gross pay be for each paycheck?

Solution

Because there are 12 months in a year and you will be paid twice a month, it follows that you will receive 24 paychecks during the year. Construct an algebraic equation for this problem as follows. Begin with a verbal model, then assign labels, and finally form an algebraic equation.

Verbal Model: $\boxed{\text{Income for year}} = 24 \times \boxed{\text{Amount of each paycheck}} + \boxed{\text{Bonus}}$

Labels: Income for year = 40,830 (dollars)
Amount of each paycheck = x (dollars)
Bonus = 750 (dollars)

Equation:

$40,830 = 24x + 750$	Original equation
$40,080 = 24x$	Subtract 750 from each side.
$\dfrac{40,080}{24} = \dfrac{24x}{24}$	Divide each side by 24.
$1670 = x$	Simplify.

Each paycheck will be $1670. Check this in the original statement of the problem.

✓ **CHECKPOINT** *Now try Exercise 3.*

2 ▶ Solve percent problems using the percent equation.

Percent Problems

Rates that describe increases, decreases, and discounts are often given as percents. **Percent** means *per hundred*, so 40% means 40 per hundred or, equivalently, $\frac{40}{100}$. The word *per* occurs in many other rates, such as price per ounce, miles per gallon, revolutions per minute, and cost per share. In applications involving percents, you need to convert the percent number to decimal or fractional form before performing any arithmetic operations. Some examples are listed below.

Percent	10%	$12\frac{1}{2}\%$	20%	25%	$33\frac{1}{3}\%$	50%	$66\frac{2}{3}\%$	75%
Decimal	0.1	0.125	0.2	0.25	$0.\overline{3}$	0.5	$0.\overline{6}$	0.75
Fraction	$\frac{1}{10}$	$\frac{1}{8}$	$\frac{1}{5}$	$\frac{1}{4}$	$\frac{1}{3}$	$\frac{1}{2}$	$\frac{2}{3}$	$\frac{3}{4}$

The primary use of percents is to compare two numbers. For example, you can compare 3 and 6 by saying that 3 is 50% of 6. In this statement, 6 is the **base number,** and 3 is the number being compared with the base number. The following model, which is called the **percent equation,** is helpful.

Verbal Model: Compared number = Percent (decimal form) · Base number

Labels: Compared number = a
Percent = p (decimal form)
Base number = b

Equation: $a = p \cdot b$ Percent equation

Remember to convert p to a decimal value before multiplying by b.

EXAMPLE 2   Solving a Percent Problem

The number 15.6 is 26% of what number?

Solution

Verbal Model: Compared number = Percent (decimal form) · Base number

Labels: Compared number = 15.6
Percent = 0.26   (decimal form)
Base number = b

Equation: $15.6 = 0.26b$ Original equation

$\dfrac{15.6}{0.26} = b$ Divide each side by 0.26.

$60 = b$ Simplify.

Check that 15.6 is 26% of 60 by multiplying 60 by 0.26 to get 15.6.

✓ **CHECKPOINT** *Now try Exercise 23.*

EXAMPLE 3   Solving a Percent Problem

The number 28 is what percent of 80?

Solution

*Verbal
Model:* $\boxed{\text{Compared number}} = \boxed{\text{Percent (decimal form)}} \cdot \boxed{\text{Base number}}$

Labels: Compared number = 28
Percent = p (decimal form)
Base number = 80

Equation: $28 = p(80)$ Original equation

$\dfrac{28}{80} = p$ Divide each side by 80.

$0.35 = p$ Simplify.

So, 28 is 35% of 80. Check this solution by multiplying 80 by 0.35 to obtain 28.

✓ **CHECKPOINT** *Now try Exercise 29.*

In most real-life applications, the base number b and the compared number a are much more disguised than in Examples 2 and 3. It sometimes helps to think of a as the "new" amount and b as the "original" amount.

EXAMPLE 4   A Percent Application

A real estate agency receives a commission of $13,812.50 for the sale of a $212,500 house. What percent commission is this?

Solution

A commission is a percent of the sale price paid to the agency for their services. To determine the percent commission, start with a verbal model.

*Verbal
Model:* $\boxed{\text{Commission}} = \boxed{\text{Percent (decimal form)}} \cdot \boxed{\text{Sale price}}$

Labels: Commission = 13,812.50   (dollars)
Percent = p (decimal form)
Sale price = 212,500   (dollars)

Equation: $13,812.50 = p(212,500)$ Original equation

$\dfrac{13,812.50}{212,500} = p$ Divide each side by 212,500.

$0.065 = p$ Simplify.

The real estate agency receives a commission of 6.5%. Check this solution by multiplying $212,500 by 0.065 to obtain $13,812.50.

✓ **CHECKPOINT** *Now try Exercise 65.*

3 ▶ Use ratios to compare unit prices for products.

Ratios and Unit Prices

You know that a percent compares a number with 100. A **ratio** is a more generic rate form that compares one number with another. If a and b represent two quantities, then a/b is called the ratio of a to b. Note the *order* implied by a ratio. The ratio of a to b means a/b, whereas the ratio of b to a means b/a.

> **Study Tip**
>
> Conversions for common units of measure can be found on our website *www.cengage.com/math/larson/algebra*.

EXAMPLE 5   Using a Ratio

Find the ratio of 4 feet to 8 inches using the same units.

Solution

Because the units of feet and inches are not the same, you must first convert 4 feet into its equivalent in inches or convert 8 inches into its equivalent in feet. You can convert 4 feet to 48 inches (by multiplying 4 by 12) to obtain

$$\frac{4 \text{ feet}}{8 \text{ inches}} = \frac{48 \text{ inches}}{8 \text{ inches}} = \frac{48}{8} = \frac{6}{1}.$$

Or, you can convert 8 inches to $\frac{8}{12}$ feet (by dividing 8 by 12) to obtain

$$\frac{4 \text{ feet}}{8 \text{ inches}} = \frac{4 \text{ feet}}{\frac{8}{12} \text{ feet}} = 4 \div \frac{8}{12} = 4 \cdot \frac{12}{8} = \frac{6}{1}.$$

 CHECKPOINT *Now try Exercise 35.*

> **Study Tip**
>
> If the denominator of a rate is 1 unit, the rate is called a *unit rate.*

The **unit price** of an item is the quotient of the total price divided by the total units. That is,

$$\text{Unit price} = \frac{\text{Total price}}{\text{Total units}}.$$

To state unit prices, use the word "per." For instance, the unit price for a brand of coffee might be 7.59 dollars *per* pound.

EXAMPLE 6   Comparing Unit Prices

Which is the better buy, a 12-ounce box of breakfast cereal for $2.79 or a 16-ounce box of the same cereal for $3.59?

Solution

The unit price for the 12-ounce box is

$$\text{Unit price} = \frac{\text{Total price}}{\text{Total units}} = \frac{\$2.79}{12 \text{ ounces}} = \$0.23 \text{ per ounce}.$$

The unit price for the 16-ounce box is approximately

$$\text{Unit price} = \frac{\text{Total price}}{\text{Total units}} = \frac{\$3.59}{16 \text{ ounces}} \approx \$0.22 \text{ per ounce}.$$

The 16-ounce box has a slightly lower unit price, and so it is the better buy.

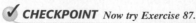 **CHECKPOINT** *Now try Exercise 87.*

4 ▶ Solve proportions.

Solving Proportions

A **proportion** is a statement that equates two ratios. For example, if the ratio of a to b is the same as the ratio of c to d, you can write the proportion as

$$\frac{a}{b} = \frac{c}{d}.$$

In typical problems, you know three of the values and need to find the fourth. The quantities a and d are called the **extremes** of the proportion, and the quantities b and c are called the **means** of the proportion. In a proportion, the product of the extremes is equal to the product of the means. This is done by **cross-multiplying.** That is, if

$$\frac{a}{b} = \frac{c}{d}$$

then $ad = bc$.

Proportions are often used in geometric applications involving similar triangles. Similar triangles have the same shape, but they may differ in size. The corresponding sides of similar triangles are proportional.

EXAMPLE 7   **Solving a Proportion in Geometry**

The triangles shown in Figure 2.2 are similar triangles. Use this fact to find the length of the unknown side x of the larger triangle.

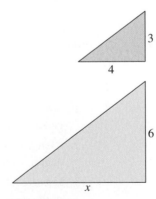

3

4

6

x

Figure 2.2

Study Tip

The proportion in Example 7 could also be written as

$$\frac{4}{3} = \frac{x}{6}.$$

After cross-multiplying, you would obtain the same equation

$$24 = 3x.$$

Solution

$$\frac{4}{x} = \frac{3}{6} \qquad \text{Set up proportion.}$$

$$4 \cdot 6 = x \cdot 3 \qquad \text{Cross-multiply.}$$

$$24 = 3x \qquad \text{Simplify.}$$

$$8 = x \qquad \text{Divide each side by 3.}$$

So, the length of the unknown side of the larger triangle is 8 units. Check this in the original statement of the problem.

 **CHECKPOINT** *Now try Exercise 91.*

Study Tip

You can write a proportion in several ways. Just be sure to put like quantities in similar positions on each side of the proportion.

EXAMPLE 8 **Gasoline Cost**

You are driving from New York City to Phoenix, a trip of 2450 miles. You begin the trip with a full tank of gas. After traveling 424 miles, you refill the tank for $40.00. How much should you plan to spend on gasoline for the entire trip?

Solution

Verbal Model: $\dfrac{\text{Cost for trip}}{\text{Cost for tank}} = \dfrac{\text{Miles for trip}}{\text{Miles for tank}}$

Labels:
Cost of gas for entire trip $= x$ (dollars)
Cost of gas for tank $= 40$ (dollars)
Miles for entire trip $= 2450$ (miles)
Miles for tank $= 424$ (miles)

Proportion:

$$\frac{x}{40} = \frac{2450}{424}$$ Original proportion

$$x \cdot 424 = 40 \cdot 2450$$ Cross-multiply.

$$424x = 98{,}000$$ Simplify.

$$x \approx 231.13$$ Divide each side by 424.

You should plan to spend approximately $231.13 for gasoline on the trip. Check this in the original statement of the problem.

✓ **CHECKPOINT** *Now try Exercise 97.*

The following list summarizes a strategy for modeling and solving real-life problems.

Strategy for Solving Word Problems

1. Ask yourself what you need to know to solve the problem. Then *write a verbal model* that includes arithmetic operations to describe the problem.

2. *Assign labels* to each part of the verbal model—numbers to the known quantities and letters (or expressions) to the variable quantities.

3. Use the labels to *write an algebraic model* based on the verbal model.

4. *Solve* the resulting algebraic equation.

5. *Answer* the original question and check that your answer satisfies the original problem as stated.

In previous mathematics courses, you studied several other problem-solving strategies, such as *drawing a diagram, making a table, looking for a pattern,* and *solving a simpler problem.* Each of these strategies can also help you to solve problems in algebra.

Concept Check

1. Explain the meaning of the word *percent*.

2. Define the term *ratio*. Give an example of a ratio.

3. If $\dfrac{a}{b} = \dfrac{c}{d}$, does this mean that $a = c$ and $b = d$? Explain.

4. If you know the means of a proportion, do you have enough information to find the extremes? Explain.

2.2 EXERCISES

Go to pages 116–117 to record your assignments.

Developing Skills

Mathematical Modeling In Exercises 1–4, construct a verbal model and write an algebraic equation that represents the problem. Solve the equation. **See Example 1.**

1. Find a number such that the sum of the number and 24 is 68.

2. Find a number such that the difference of the number and 18 is 27.

✔ 3. You have accepted a job offer at an annual salary of $37,120. This salary includes a year-end bonus of $2800. You are paid every 2 weeks. What will your gross pay be for each paycheck?

4. You have a job on an assembly line for which you are paid $10 per hour plus $0.75 per unit assembled. Find the number of units produced in an eight-hour day if your earnings for the day are $146.

In Exercises 5–12, complete the table showing the equivalent forms of various percents.

	Percent	Parts out of 100	Decimal	Fraction
5.	30%			
6.	75%			
7.			0.075	
8.			0.08	

	Percent	Parts out of 100	Decimal	Fraction
9.				$\frac{2}{3}$
10.				$\frac{1}{8}$
11.		100		
12.		42		

In Exercises 13–34, solve using a percent equation. **See Examples 2 and 3.**

13. What is 35% of 250?

14. What is 65% of 800?

15. What is 42.5% of 816?

16. What is 70.2% of 980?

17. What is $12\frac{1}{2}\%$ of 1024?

18. What is $33\frac{1}{3}\%$ of 816?

19. What is 0.4% of 150,000?

20. What is 0.1% of 8925?

21. What is 250% of 32?

22. What is 300% of 16?

✔ 23. 84 is 24% of what number?

24. 416 is 65% of what number?

25. 42 is 120% of what number?

26. 168 is 350% of what number?

27. 22 is 0.8% of what number?

28. 18 is 2.4% of what number?

✔ 29. 496 is what percent of 800?

30. 1650 is what percent of 5000?

31. 2.4 is what percent of 480?

32. 3.3 is what percent of 220?

33. 2100 is what percent of 1200?

34. 900 is what percent of 500?

In Exercises 35–42, write the verbal expression as a ratio. Use the same units in both the numerator and denominator, and simplify. *See Example 5.*

✓ **35.** 120 meters to 180 meters

36. 12 ounces to 20 ounces

37. 36 inches to 48 inches

38. 125 centimeters to 2 meters

39. 40 milliliters to 1 liter

40. 1 pint to 1 gallon

41. 5 pounds to 24 ounces

42. 45 minutes to 2 hours

In Exercises 43–52, solve the proportion. *See Example 7.*

43. $\dfrac{x}{6} = \dfrac{2}{3}$

44. $\dfrac{t}{4} = \dfrac{3}{2}$

45. $\dfrac{y}{36} = \dfrac{6}{7}$

46. $\dfrac{5}{16} = \dfrac{x}{4}$

47. $\dfrac{5}{4} = \dfrac{t}{6}$

48. $\dfrac{7}{8} = \dfrac{x}{2}$

49. $\dfrac{y}{6} = \dfrac{y-2}{4}$

50. $\dfrac{a}{5} = \dfrac{a+4}{8}$

51. $\dfrac{z-3}{3} = \dfrac{z+8}{12}$

52. $\dfrac{y+1}{10} = \dfrac{y-1}{6}$

Solving Problems

53. *College Enrollment*   In the fall of 2006, Penn State University admitted 20,181 applicants to attend the University Park campus. Of those admitted, approximately 40% accepted enrollment. How many students were enrolled? (Source: Penn State University)

54. *Pension Fund*   Your employer withholds $6\frac{1}{2}\%$ of your monthly gross income of $3800 for your retirement. Determine the amount withheld each month.

55. *Passing Grade*   There are 40 students in your class. On one test, 95% of the students received passing grades. How many students failed the test?

56. *Elections*   There are 255 members of an on-campus organization. In the election for officers, 60% of the members voted. How many members did not vote?

57. *Company Layoff*   Because of slumping sales, a small company laid off 25 of its 160 employees. What percent of the work force was laid off?

58. *Monthly Rent*   You spend $748 of your monthly income of $3400 for rent. What percent of your monthly income is your monthly rent payment?

59. *Gratuity*   You want to leave a 15% tip for a meal that costs $32.60. How much should you leave?

60. *Gratuity*   You want to leave a 20% tip for a meal that costs $49.24. How much should you leave?

61. *Gratuity*   A customer left $25 for a meal that cost $20.66. Determine the tip percent.

62. *Gratuity*   A customer left $60 for a meal that cost $47.24. Determine the tip percent.

63. *Gratuity*   A customer gave a taxi driver $9 for a ride that cost $8.20. Determine the tip percent.

64. *Gratuity*   A customer gave a taxi driver $21 for a ride that cost $18.80. Determine the tip percent.

✓ **65.** *Real Estate Commission*   A real estate agency receives a commission of $12,250 for the sale of a $175,000 house. What percent commission is this?

66. *Real Estate Commission*   A real estate agency receives a commission of $24,225 for the sale of a $285,000 house. What percent commission is this?

67. *Quality Control*   A quality control engineer reported that 1.5% of a sample of parts were defective. The engineer found three defective parts. How large was the sample?

68. *Price Inflation*   A new car costs $29,750, which is approximately 115% of what a comparable car cost 3 years ago. What did the car cost 3 years ago?

69. *Floor Space*   You are planning to build a tool shed, but you are undecided about the size. The two sizes you are considering are 12 feet by 15 feet and 16 feet by 20 feet. The floor space of the larger is what percent of the floor space of the smaller? The floor space of the smaller is what percent of the floor space of the larger?

70. ▲ *Geometry*   The floor of a rectangular room that measures 10 feet by 12 feet is partially covered by a circular rug with a radius of 4 feet (see figure). What percent of the floor is covered by the rug? (*Hint:* The area of a circle is $A = \pi r^2$.)

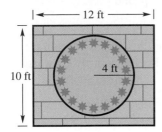

71. *Population*   The populations in 2006 of the four largest cities in Texas are shown in the circle graph. What percent of the total population is each city's population? (Source: U.S. Census Bureau)

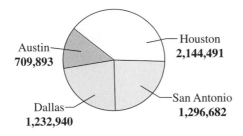

72. *Population*   The populations in 2006 of the four largest cities in California are shown in the circle graph. What percent of the total population is each city's population? (Source: U.S. Census Bureau)

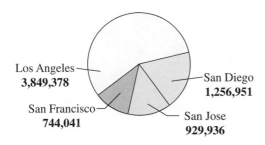

Graphical Estimation   In Exercises 73–76, use the bar graph to answer the questions. The graph shows the wholesale prices per 100 pounds of selected meats at the beginnings of 1995, 2000, and 2005. (Source: U.S. Department of Agriculture)

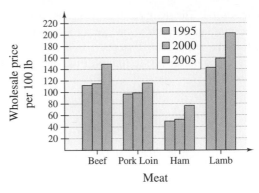

Figure for 73–76

73. Approximate the increase in the wholesale price of beef from 1995 to 2005. Use this estimate to approximate the percent increase.

74. Approximate the increase in the wholesale price of pork loin from 1995 to 2005. Use this estimate to approximate the percent increase.

75. In 2000, the wholesale price of ham was about what percent of the wholesale price of lamb?

76. In 2005, the wholesale price of ham was about what percent of the wholesale price of beef?

77. *Income Tax*   You have $12.50 of state tax withheld from your paycheck per week when your gross pay is $625. Find the ratio of tax to gross pay.

78. *Price-Earnings Ratio*   The **price-earnings ratio** is the ratio of the price of a stock to its earnings. Find the price-earnings ratio of a stock that sells for $56.25 per share and earns $6.25 per share.

79. *Compression Ratio*   The **compression ratio** of a cylinder is the ratio of its expanded volume to its compressed volume (see figure). The expanded volume of one cylinder of a small diesel engine is 425 cubic centimeters, and its compressed volume is 20 cubic centimeters. Find the compression ratio.

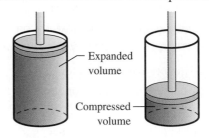

80. *Gear Ratio* The **gear ratio** of two gears is the ratio of the number of teeth in one gear to the number of teeth in the other gear. Two gears in a gearbox have 60 teeth and 40 teeth (see figure). Find the gear ratio of A to B.

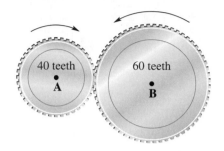

81. ▲ *Geometry* Find the ratio of the area of the smaller circle to the area of the larger circle in the figure. (*Hint:* The area of a circle is $A = \pi r^2$.)

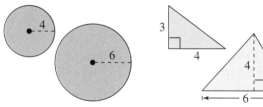

Figure for 81   Figure for 82

82. ▲ *Geometry* Find the ratio of the area of the smaller triangle to the area of the larger triangle in the figure.

Unit Prices In Exercises 83–86, find the unit price (in dollars per ounce) of the product.

83. A 20-ounce can of pineapple for $1.10

84. A 64-ounce bottle of juice for $1.89

85. A one-pound, four-ounce loaf of bread for $2.29

86. A one-pound, six-ounce box of cereal for $5.19

Consumer Awareness In Exercises 87–90, use unit prices to determine the better buy. *See Example 6.*

✓ **87.** (a) A $14\frac{1}{2}$-ounce bag of chips for $2.32
    (b) A $5\frac{1}{2}$-ounce bag of chips for $0.99

88. (a) A $10\frac{1}{2}$-ounce package of cookies for $1.79
    (b) A 16-ounce package of cookies for $2.39

89. (a) A four-ounce tube of toothpaste for $1.69
    (b) A six-ounce tube of toothpaste for $2.39

90. (a) A two-pound package of hamburger for $4.79
    (b) A three-pound package of hamburger for $6.99

▲ *Geometry* In Exercises 91–94, the triangles are similar. Solve for the length x by using the fact that corresponding sides of similar triangles are proportional. *See Example 7.*

✓ **91.**   **92.**

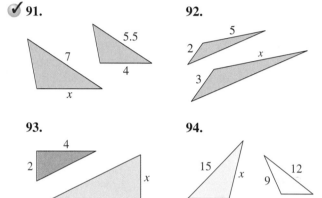

93.   **94.**

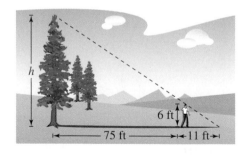

95. ▲ *Geometry* A man who is 6 feet tall walks directly toward the tip of the shadow of a tree. When the man is 75 feet from the tree, he starts forming his own shadow beyond the shadow of the tree (see figure). The length of the shadow of the tree beyond this point is 11 feet. Find the height h of the tree.

96. ▲ *Geometry* Find the length l of the shadow of a man who is 6 feet tall and is standing 15 feet from a streetlight that is 20 feet high (see figure on page 79).

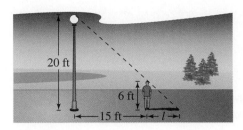

Figure for 96

✓ **97.** *Fuel Usage*   A tractor uses 5 gallons of diesel fuel to plow for 105 minutes. Assuming conditions remain the same, determine the number of gallons of fuel used in 6 hours.

98. *Spring Length*   A force of 32 pounds stretches a spring 6 inches. Determine the number of pounds of force required to stretch it 1.25 feet.

99. *Property Tax*   The tax on a property with an assessed value of $110,000 is $1650. Find the tax on a property with an assessed value of $160,000.

100. *Recipe*   Three cups of flour are required to make one batch of cookies. How many cups are required to make $3\frac{1}{2}$ batches?

101. *Quality Control*   A quality control engineer finds one defective unit in a sample of 75. At this rate, what is the expected number of defective units in a shipment of 200,000?

102. *Quality Control*   A quality control engineer finds 3 defective units in a sample of 120. At this rate, what is the expected number of defective units in a shipment of 5000?

103. *Quality Control*   A quality control inspector finds one color defect in a sample of 40 units. At this rate, what is the expected number of color defects in a shipment of 235?

104. *Quality Control*   A quality control inspector finds 4 scratch defects in a sample of 25 units. At this rate, what is the expected number of scratch defects in a shipment of 520?

105. *Public Opinion Poll*   In a public opinion poll, 870 people from a sample of 1500 indicate they will vote for the Republican candidate. Assuming this poll to be a correct indicator of the electorate, how many votes can the candidate expect to receive out of 80,000 votes cast?

106. *Public Opinion Poll*   In a public opinion poll, 1530 citizens from a sample of 1800 indicate they are in favor of a new tax legislation proposed by city council. Assuming this poll to be a correct indicator of the electorate, how many votes for the new tax legislation would you expect out of 110,000 votes cast?

Explaining Concepts

107. ✎   Explain how to change percents to decimals and decimals to percents. Give examples.

108. ✎   Is it true that $\frac{1}{2}\% = 50\%$? Explain.

109. ✎   In your own words, describe the meaning of *mathematical modeling*. Give an example.

110. ✎   During a year of financial difficulties, your company reduces your salary by 7%. What percent increase in this reduced salary is required to raise your salary to the amount it was prior to the reduction? Why isn't the percent increase the same as the percent of the reduction?

Cumulative Review

In Exercises 111–114, evaluate the expression.

111. $-\frac{4}{15} \cdot \frac{15}{16}$ **112.** $\frac{3}{8} \div \frac{5}{16}$

113. $(12 - 15)^3$ **114.** $\left(-\frac{5}{8}\right)^2$

In Exercises 115–118, identify the property of real numbers illustrated by the statement.

115. $5 + x = x + 5$

116. $3x \cdot \dfrac{1}{3x} = 1$

117. $6(x - 2) = 6x - 6 \cdot 2$

118. $3 + (4 + x) = (3 + 4) + x$

In Exercises 119–124, solve the equation.

119. $2x - 5 = x + 9$ **120.** $6x + 8 = 8 - 2x$

121. $2x + \dfrac{3}{2} = \dfrac{3}{2}$ **122.** $-\dfrac{x}{10} = 1000$

123. $-0.35x = 70$ **124.** $0.60x = 24$

2.3 Business and Scientific Problems

©Scott Speakes/CORBIS

Why You Should Learn It

Mathematical models can be used to solve a wide variety of real-life problems. For instance, you can find how much interest you earn by investing in bonds. See Exercise 83 on page 91.

1 ▶ Use mathematical models to solve business-related problems.

What You Should Learn

1 ▶ Use mathematical models to solve business-related problems.
2 ▶ Use mathematical models to solve mixture problems.
3 ▶ Use mathematical models to solve classic rate problems.
4 ▶ Use formulas to solve application problems.

Rates in Business Problems

Many business problems can be represented by mathematical models involving the sum of a fixed term and a variable term. The variable term is often a *hidden product* in which one of the factors is a percent or some other type of rate. Watch for these occurrences in the discussions and examples that follow.

The **markup** on a consumer item is the difference between the **cost** a retailer pays for an item and the **price** at which the retailer sells the item. A verbal model for this relationship is as follows.

$$\boxed{\text{Selling price}} = \boxed{\text{Cost}} + \boxed{\text{Markup}}$$ Markup is a hidden product.

The markup is the hidden product of the **markup rate** and the cost.

$$\boxed{\text{Markup}} = \boxed{\text{Markup rate}} \cdot \boxed{\text{Cost}}$$

EXAMPLE 1 **Finding the Markup Rate**

The Granger Collection

In 1874, Levi Strauss designed the first pair of blue jeans. Today, billions of pairs of jeans are sold each year throughout the world.

A clothing store sells a pair of jeans for $42. The cost of the jeans is $16.80. What is the markup rate?

Solution

Verbal Model: $\boxed{\text{Selling price}} = \boxed{\text{Cost}} + \boxed{\text{Markup}}$

Labels: Selling price = 42   (dollars)
             Cost = 16.80   (dollars)
             Markup rate = p (percent in decimal form)
             Markup = $p(16.80)$ (dollars)

Equation: $42 = 16.80 + p(16.80)$ Original equation

              $25.2 = p(16.80)$ Subtract 16.80 from each side.

              $\dfrac{25.2}{16.80} = p$ Divide each side by 16.80.

              $1.5 = p$ Simplify.

Because $p = 1.5$, it follows that the markup rate is 150%. Check this in the original statement of the problem.

 **CHECKPOINT** *Now try Exercise 1.*

The model for a **discount** is similar to that for a markup.

Selling price	=	List price	−	Discount

Discount is a hidden product.

The discount is the hidden product of the **discount rate** and the list price.

EXAMPLE 2   Finding the Discount and the Discount Rate

A DVD/VCR combination unit is marked down from its list price of $310 to a sale price of $217. What is the discount rate?

Solution

Verbal Model: Discount = Discount rate · List price

Labels:
Discount = $310 − 217 = 93$ (dollars)
List price = 310   (dollars)
Discount rate = p (percent in decimal form)

Equation:
$$93 = p(310)$$ Original equation

$$\frac{93}{310} = p$$ Divide each side by 310.

$$0.30 = p$$ Simplify.

The discount rate is 30%. Check this in the original statement of the problem.

✓ **CHECKPOINT** *Now try Exercise 9.*

EXAMPLE 3   Finding the Hours of Labor

An auto repair bill of $338 lists $170 for parts and the rest for labor. It took 6 hours to repair the auto. What is the hourly rate for labor?

Solution

Verbal Model: Total bill = Price of parts + Price of labor

Labels:
Total bill = 338   (dollars)
Price of parts = 170   (dollars)
Hours of labor = 6   (hours)
Hourly rate for labor = x (dollars per hour)
Price of labor = $6x$ (dollars)

Equation:
$$338 = 170 + 6x$$ Original equation
$$168 = 6x$$ Subtract 170 from each side.

$$\frac{168}{6} = x$$ Divide each side by 6.

$$28 = x$$ Simplify.

The hourly rate for labor is $28 per hour. Check this in the original problem.

✓ **CHECKPOINT** *Now try Exercise 29.*

2 ▶ Use mathematical models to solve mixture problems.

Rates in Mixture Problems

Many real-life problems involve combinations of two or more quantities that make up new or different quantities. Such problems are called **mixture problems.** They are usually composed of the sum of two or more "hidden products" that involve rate factors. Here is the generic form of the verbal model for mixture problems.

$$\boxed{\text{First rate}} \cdot \boxed{\text{Amount}} + \boxed{\text{Second rate}} \cdot \boxed{\text{Amount}} = \boxed{\text{Final rate}} \cdot \boxed{\text{Final amount}}$$

Study Tip

When you set up a verbal model, be sure to check that you are working with *the same type of units* in each part of the model. For instance, in Example 4 note that each of the three parts of the verbal model measures cost. (If two parts measured cost and the other part measured pounds, you would know that the model was incorrect.)

EXAMPLE 4   A Mixture Problem

A nursery wants to mix two types of lawn seed. Type A sells for $10 per pound and type B sells for $15 per pound. To obtain 20 pounds of a mixture at $12 per pound, how many pounds of each type of seed are needed?

Solution

The rates are the unit prices for each type of seed.

Verbal Model: $\boxed{\text{Total cost of \$10 seed}} + \boxed{\text{Total cost of \$15 seed}} = \boxed{\text{Total cost of \$12 seed}}$

Labels: Unit price of type A = 10   (dollars per pound)
    Pounds of $10 seed = x   (pounds)
    Unit price of type B = 15   (dollars per pound)
    Pounds of $15 seed = 20 − x   (pounds)
    Unit price of mixture = 12   (dollars per pound)
    Pounds of $12 seed = 20   (pounds)

Equation: $10x + 15(20 - x) = 12(20)$ Original equation

$\qquad\quad 10x + 300 - 15x = 240$ Distributive Property

$\qquad\qquad\quad 300 - 5x = 240$ Combine like terms.

$\qquad\qquad\qquad\quad -5x = -60$ Subtract 300 from each side.

$\qquad\qquad\qquad\qquad x = 12$ Divide each side by −5.

The mixture should contain 12 pounds of the $10 seed and $20 - x = 20 - 12 = 8$ pounds of the $15 seed.

✓ **CHECKPOINT** *Now try Exercise 37.*

Remember that when you have found a solution, you should always go back to the original statement of the problem and check to see that the solution makes sense—both algebraically and from a practical point of view. For instance, you can check the result of Example 4 as follows.

$$\overbrace{\left(\begin{array}{c}\$10\text{ per}\\\text{pound}\end{array}\right)\left(\begin{array}{c}12\\\text{pounds}\end{array}\right)}^{\$10\text{ seed}} + \overbrace{\left(\begin{array}{c}\$15\text{ per}\\\text{pound}\end{array}\right)\left(\begin{array}{c}8\\\text{pounds}\end{array}\right)}^{\$15\text{ seed}} \stackrel{?}{=} \overbrace{\left(\begin{array}{c}\$12\text{ per}\\\text{pound}\end{array}\right)\left(\begin{array}{c}20\\\text{pounds}\end{array}\right)}^{\$12\text{ seed}}$$

$$\$120 + \$120 = \$240 \quad \text{Solution checks.} \checkmark$$

3 ▶ Use mathematical models to solve classic rate problems.

Classic Rate Problems

Time-dependent problems such as distance traveled at a given speed and work done at a specified rate are classic types of **rate problems.** The distance-rate-time problem fits the verbal model

$$\boxed{\text{Distance}} = \boxed{\text{Rate}} \cdot \boxed{\text{Time}}.$$

For instance, if you travel at a constant (or average) rate of 55 miles per hour for 45 minutes, the total distance you travel is given by

$$\left(55\,\frac{\text{miles}}{\text{hour}}\right)\left(\frac{45}{60}\,\text{hour}\right) = 41.25 \text{ miles.}$$

As with all problems involving applications, be sure to check that the units in the verbal model make sense. In this problem the rate is given in *miles per hour*. For the solution to be given in *miles*, you must convert the time (from minutes) to *hours*. In the model, you can think of the hours as dividing out, as follows.

$$\left(55\,\frac{\text{miles}}{\cancel{\text{hour}}}\right)\left(\frac{45}{60}\,\cancel{\text{hour}}\right) = 41.25 \text{ miles}$$

EXAMPLE 5   Distance-Rate Problem

Students are traveling in two cars to a football game 150 miles away. The first car leaves on time and travels at an average speed of 48 miles per hour. The second car starts $\frac{1}{2}$ hour later and travels at an average speed of 58 miles per hour. At these speeds, how long will it take the second car to catch up to the first car?

Solution

Verbal Model: $\boxed{\text{Distance of first car}} = \boxed{\text{Distance of second car}}$

Labels: Time for first car $= t$ (hours)
Distance of first car $= 48t$ (miles)
Time for second car $= t - \frac{1}{2}$ (hours)
Distance of second car $= 58\left(t - \frac{1}{2}\right)$ (miles)

Equation:

$48t = 58\left(t - \frac{1}{2}\right)$	Original equation
$48t = 58t - 29$	Distributive Property
$48t - 58t = 58t - 58t - 29$	Subtract $58t$ from each side.
$-10t = -29$	Combine like terms.
$\dfrac{-10t}{-10} = \dfrac{-29}{-10}$	Divide each side by -10.
$t = 2.9$	Simplify.

After the first car travels for 2.9 hours, the second car catches up to it. So, it takes the second car $t - 0.5 = 2.9 - 0.5 = 2.4$ hours to catch up to the first car.

✓ **CHECKPOINT** *Now try Exercise 55.*

In work-rate problems, the **rate of work** is the *reciprocal* of the time needed to do the entire job. For instance, if it takes 5 hours to complete a job, then the per hour work rate is $\frac{1}{5}$ job per hour. In general,

$$\boxed{\text{Per hour work rate}} = \frac{1}{\boxed{\text{Total hours to complete a job}}}.$$

The next example illustrates the model for solving *mixture* problems using two rates of work.

EXAMPLE 6   Work-Rate Problem

Consider two machines in a paper manufacturing plant. Machine 1 can complete one job (2000 pounds of paper) in 4 hours. Machine 2 is newer and can complete one job in $2\frac{1}{2}$ hours. How long will it take the two machines working together to complete one job?

Solution

Verbal Model: $\boxed{\dfrac{\text{Work}}{\text{done}}} = \boxed{\dfrac{\text{Portion done}}{\text{by machine 1}}} + \boxed{\dfrac{\text{Portion done}}{\text{by machine 2}}}$

Labels: Work done by both machines $= 1$ (job)

        Time for each machine $= t$ (hours)

        Per hour work rate for machine 1 $= \frac{1}{4}$ (job per hour)

        Per hour work rate for machine 2 $= \frac{2}{5}$ (job per hour)

Equation:

$$1 = \left(\frac{1}{4}\right)(t) + \left(\frac{2}{5}\right)(t) \qquad \text{Rate} \cdot \text{time} + \text{rate} \cdot \text{time}$$

$$1 = \left(\frac{1}{4} + \frac{2}{5}\right)(t) \qquad \text{Distributive Property}$$

$$1 = \left(\frac{5}{20} + \frac{8}{20}\right)(t) \qquad \text{Least common denominator is 20.}$$

$$1 = \left(\frac{13}{20}\right)(t) \qquad \text{Simplify.}$$

$$1 \div \frac{13}{20} = t \qquad \text{Divide each side by } \frac{13}{20}.$$

$$\frac{20}{13} = t \qquad \text{Simplify.}$$

It will take $\frac{20}{13}$ hours (or about 1.54 hours) for both machines to complete the job. Check this solution in the original statement of the problem.

✓ **CHECKPOINT** *Now try Exercise 61.*

Note in Example 6 that the "2000 pounds of paper" is unnecessary information. The 2000 pounds is represented as one job. This type of unnecessary information in an applied problem is sometimes called a *red herring*. The 150 miles given in Example 5 is also a red herring.

4 ▶ Use formulas to solve application problems.

Formulas

Many common types of geometric, scientific, and investment problems use ready-made equations called **formulas.** Knowing formulas such as those in the following lists will help you translate and solve a wide variety of real-life problems involving perimeter, circumference, area, volume, temperature, interest, and distance.

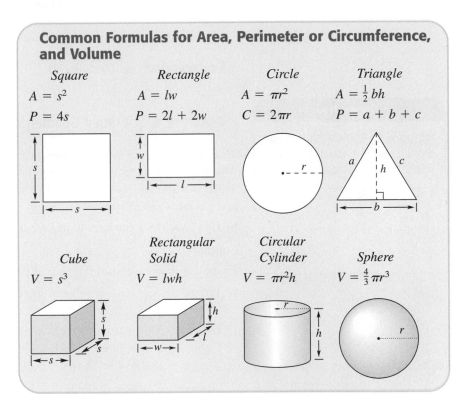

Common Formulas for Area, Perimeter or Circumference, and Volume

Square
$A = s^2$
$P = 4s$

Rectangle
$A = lw$
$P = 2l + 2w$

Circle
$A = \pi r^2$
$C = 2\pi r$

Triangle
$A = \frac{1}{2}bh$
$P = a + b + c$

Cube
$V = s^3$

Rectangular Solid
$V = lwh$

Circular Cylinder
$V = \pi r^2 h$

Sphere
$V = \frac{4}{3}\pi r^3$

Miscellaneous Common Formulas

Temperature: F = degrees Fahrenheit, C = degrees Celsius

$$F = \frac{9}{5}C + 32$$

Simple Interest: I = interest, P = principal, r = interest rate, t = time

$$I = Prt$$

Distance: d = distance traveled, r = rate, t = time

$$d = rt$$

Smart Study Strategy

Go to page 56 for ways to *Keep Your Mind Focused.*

When working with applied problems, you often need to rewrite one of the common formulas, as shown in the next example.

EXAMPLE 7   Rewriting a Formula

In the perimeter formula $P = 2l + 2w$, solve for w.

Solution

$$P = 2l + 2w$$ Original formula

$$P - 2l = 2w$$ Subtract $2l$ from each side.

$$\frac{P - 2l}{2} = w$$ Divide each side by 2.

✔ **CHECKPOINT** *Now try Exercise 65.*

Study Tip

When solving problems such as the one in Example 8, you may find it helpful to draw and label a diagram.

EXAMPLE 8   Using a Geometric Formula

The city plans to put sidewalks along the two streets that border your corner lot, which is 250 feet long on one side and has an area of 30,000 square feet. Each lot owner is to pay $1.50 per foot of sidewalk bordering his or her lot.

a. Find the width of your lot.

b. How much will you have to pay for the sidewalks put on your lot?

Solution

Figure 2.3 shows a labeled diagram of your lot.

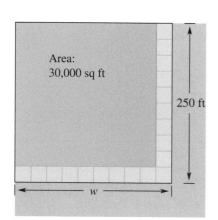

Area: 30,000 sq ft

250 ft

w

Figure 2.3

a. *Verbal Model:* | Area | = | Length | · | Width |

Labels: Area of lot = 30,000   (square feet)
Length of lot = 250   (feet)
Width of lot = w (feet)

Equation: $30,000 = 250 \cdot w$ Original equation

$$\frac{30,000}{250} = w$$ Divide each side by 250.

$$120 = w$$ Simplify.

Your lot is 120 feet wide.

b. *Verbal Model:* | Cost | = | Rate per foot | · | Length of sidewalk |

Labels: Cost of sidewalks = C (dollars)
Rate per foot = 1.50   (dollars per foot)
Total length of sidewalk = 120 + 250   (feet)

Equation: $C = 1.50(120 + 250)$ Original equation

$$C = 1.50 \cdot 370$$ Add within parentheses.

$$C = 555$$ Multiply.

You will have to pay $555 to have the sidewalks put on your lot.

✔ **CHECKPOINT** *Now try Exercise 77.*

EXAMPLE 9 **Converting Temperature**

The average daily low temperature in January in Chicago, Illinois is 14°F. In January, what is Chicago's average daily low temperature in degrees Celsius? (Source: U.S. National Oceanic and Atmospheric Administration)

Solution

Verbal Model: Fahrenheit temperature $= \dfrac{9}{5} \cdot$ Celsius temperature $+ 32$

Labels: Fahrenheit temperature $= 14$ (degrees Fahrenheit)
Celsius temperature $= C$ (degrees Celsius)

Equation:

$14 = \frac{9}{5}C + 32$ Original equation

$-18 = \frac{9}{5}C$ Subtract 32 from each side.

$-10 = C$ Multiply each side by $\frac{5}{9}$.

The average daily low temperature in January in Chicago is $-10°C$.

✓ **CHECKPOINT** *Now try Exercise 81.*

EXAMPLE 10 **Simple Interest**

A deposit of $8000 earned $300 in interest in 6 months.

a. What was the annual interest rate for this account?

b. At this rate, how long would it take to earn $800 in total interest?

Solution

a. *Verbal Model:* Interest $=$ Principal $\cdot$ Rate $\cdot$ Time

Labels: Interest $= 300$ (dollars)
Principal $= 8000$ (dollars)
Annual interest rate $= r$ (percent in decimal form)
Time $= \frac{1}{2}$ (year)

Equation: $300 = 8000(r)\left(\dfrac{1}{2}\right)$ Original equation

$\dfrac{300}{4000} = r \implies 0.075 = r$ Divide each side by 4000.

The annual interest rate is $r = 0.075$ or 7.5%.

b. Using the same verbal model as in part (a) with t representing time, you obtain the following equation.

$800 = 8000(0.075)(t)$ Original equation

$\dfrac{800}{600} = t \implies \dfrac{4}{3} = t$ Divide each side by 600.

So, it would take $\frac{4}{3}$ years, or $\frac{4}{3} \times 12 = 16$ months.

✓ **CHECKPOINT** *Now try Exercise 83.*

Technology: Tip

You can use a graphing calculator to solve simple interest problems by using the program found at our website *www.cengage.com/math/larson/algebra*. Use the program and the guess, check, and revise method to find P when $I = \$3330$, $r = 6\%$, and $t = 3$ years.

_____ **Concept Check** _____

1. Explain the difference between markup and markup rate.

2. The concentration of solution 1 is 10%. The concentration of solution 2 is 20%. You mix equal amounts of solution 1 and solution 2. Will the final solution have a concentration of 30%? Explain.

3. It takes you t hours to complete a task. What portion of the task can you complete in 1 hour?

4. If you double the height of a triangle, does the area double? Explain.

2.3 EXERCISES

Go to pages 116–117 to record your assignments.

_____ **Developing Skills** _____

In Exercises 1–8, find the missing quantities. (Assume that the markup rate is a percent based on the cost.) **See Example 1.**

	Cost	Selling Price	Markup	Markup Rate
✓ 1.	$45.97	$64.33		
2.	$62.40	$96.72		
3.		$250.80	$98.80	
4.		$623.72	$221.32	
5.		$26,922.50	$4672.50	
6.		$16,440.50	$3890.50	
7.	$225.00			85.2%
8.	$732.00			$33\frac{1}{3}\%$

In Exercises 9–16, find the missing quantities. (Assume that the discount rate is a percent based on the list price.) **See Example 2.**

	List Price	Sale Price	Discount	Discount Rate
✓ 9.	$49.95	$25.74		
10.	$119.00	$79.73		
11.	$300.00		$189.00	
12.	$345.00		$134.55	
13.		$27.00		40%
14.		$19.90		20%
15.		$831.96	$323.54	
16.		$257.32	$202.18	

_____ **Solving Problems** _____

17. *Markup*   The selling price of a jacket in a department store is $157.14. The cost of the jacket to the store is $130.95. What is the markup?

18. *Markup*   A shoe store sells a pair of shoes for $89.95. The cost of the shoes to the store is $46.50. What is the markup?

19. *Markup Rate*   A jewelry store sells a pair of earrings for $84. The cost of the earrings to the store is $46.67. What is the markup rate?

20. *Markup Rate*   A department store sells a sweater for $60. The cost of the sweater to the store is $35. What is the markup rate?

21. *Discount*   A shoe store sells a pair of athletic shoes for $75. The shoes go on sale for $45. What is the discount?

22. *Discount*   A bakery sells a dozen rolls for $2.25. You can buy a dozen day-old rolls for $0.75. What is the discount?

23. *Discount Rate*   An auto store sells a pair of car mats for $20. On sale, the car mats sell for $16. What is the discount rate?

24. *Discount Rate*   A department store sells a beach towel for $32. On sale, the beach towel sells for $24. What is the discount rate?

25. *Long-Distance Rate* The weekday rate for a telephone call is $0.75 for the first minute plus $0.55 for each additional minute. Determine the length of a call that costs $5.15. What would have been the cost of the call if it had been made during the weekend, when there is a 60% discount?

26. *Long-Distance Rate* The weekday rate for a telephone call is $0.65 for the first minute plus $0.40 for each additional minute. Determine the length of a call that costs $5.45. What would have been the cost of the call if it had been made during the weekend, when there is a 40% discount?

27. *Cost* An auto store gives the list price of a tire as $79.42. During a promotional sale, the store is selling four tires for the price of three. The store needs a markup on cost of 10% during the sale. What is the cost to the store of each tire?

28. *Price* The produce manager of a supermarket pays $22.60 for a 100-pound box of bananas. The manager estimates that 10% of the bananas will spoil before they are sold. At what price per pound should the bananas be sold to give the supermarket an average markup rate on cost of 30%?

✓ 29. *Labor* An automobile repair bill of $216.37 lists $136.37 for parts and the rest for labor. The labor rate is $32 per hour. How many hours did it take to repair the automobile?

30. *Labor* The bill for the repair of an automobile is $380. Included in this bill is a charge of $275 for parts, and the remainder of the bill is for labor. The charge for labor is $35 per hour. How many hours were spent in repairing the automobile?

31. *Labor* The bill for the repair of an automobile is $648. Included in this bill is a charge of $315 for parts, and the remainder of the bill is for labor. It took 9 hours to repair the automobile. What was the charge per hour for labor?

32. *Labor* An appliance repair store charges $60 for the first $\frac{1}{2}$ hour of a service call. The bill for a 3-hour service call is $185. What was the charge per hour after the first $\frac{1}{2}$ hour?

Mixture Problems In Exercises 33–36, determine the numbers of units of solutions 1 and 2 needed to obtain a final solution of the specified amount and concentration. *See Example 4.*

Concentration of Solution 1	Concentration of Solution 2	Concentration of Final Solution	Amount of Final Solution
33. 20%	60%	40%	100 gal
34. 50%	75%	60%	10 L
35. 15%	60%	45%	24 qt
36. 45%	85%	70%	600 ml

Phil Schermeister/Peter Arnold Inc.

✓ 37. *Seed Mixture* A nursery wants to mix two types of lawn seed. Type 1 sells for $12 per pound, and type 2 sells for $20 per pound. To obtain 100 pounds of a mixture at $14 per pound, how many pounds of each type of seed are needed?

38. *Nut Mixture* A grocer mixes two kinds of nuts costing $3.88 per pound and $4.88 per pound to make 100 pounds of a mixture costing $4.28 per pound. How many pounds of each kind of nut are in the mixture?

39. *Ticket Sales* Ticket sales for a play total $2200. There are three times as many adult tickets sold as children's tickets. The prices of the tickets for adults and children are $6 and $4, respectively. Find the number of children's tickets sold.

40. *Ticket Sales* Ticket sales for a spaghetti dinner total $1350. There are four times as many adult tickets sold as children's tickets. The prices of the tickets for adults and children are $6 and $3, respectively. Find the number of children's tickets sold.

41. *Antifreeze Mixture* The cooling system on a truck contains 5 gallons of coolant that is 40% antifreeze. How much must be withdrawn and replaced with 100% antifreeze to bring the coolant in the system to 50% antifreeze?

42. *Fuel Mixture* You mix gasoline and oil to obtain $2\frac{1}{2}$ gallons of mixture for an engine. The mixture is 40 parts gasoline and 1 part two-cycle oil. How much gasoline must be added to bring the mixture to 50 parts gasoline and 1 part oil?

Distance In Exercises 43–48, determine the unknown distance, rate, or time. *See Example 5.*

Distance, d	Rate, r	Time, t
43.	650 mi/hr	$3\frac{1}{2}$ hr
44.	60 ft/sec	$6\frac{1}{4}$ sec
45. 1000 km	110 km/hr	
46. 250 ft	32 ft/sec	
47. 385 mi		7 hr
48. 828 ft		9 sec

49. *Travel Distance* You ride your bike at an average speed of 10 miles per hour for $2\frac{1}{2}$ hours. How far do you ride?

50. *Travel Distance* You ride your bike at an average speed of 15 miles per hour for $3\frac{1}{3}$ hours. How far do you ride?

51. *Travel Time* You ride your bike at an average speed of 16 miles per hour. How long will it take you to ride 20 miles?

52. *Travel Time* You ride your bike at an average speed of 28 feet per second. How long will it take you to ride 252 feet?

53. *Travel Rate* It takes you $\frac{2}{3}$ of an hour to jog 4 miles. What is your jogging rate?

54. *Travel Rate* It takes you $\frac{7}{12}$ of an hour to jog 5250 meters. What is your jogging rate?

✓ 55. *Distance* Two planes leave Chicago's O'Hare International Airport at approximately the same time and fly in opposite directions. How far apart are the planes after $1\frac{1}{3}$ hours if their average speeds are 480 miles per hour and 600 miles per hour?

56. *Distance* Two trucks leave a depot at approximately the same time and travel the same route. How far apart are the trucks after $4\frac{1}{2}$ hours if their average speeds are 52 miles per hour and 56 miles per hour?

57. *Travel Time* Determine the time required for a space shuttle to travel a distance of 5000 miles in orbit when its average speed is 17,500 miles per hour.

58. *Speed of Light* The distance between the Sun and Earth is 93,000,000 miles, and the speed of light is 186,282.397 miles per second. Determine the time required for light to travel from the Sun to Earth.

59. *Travel Time* On the first part of a 317-mile trip, a sales representative averaged 58 miles per hour. The sales representative averaged only 52 miles per hour on the remainder of the trip because of an increased volume of traffic (see figure). The total time of the trip was 5 hours and 45 minutes. Find the amount of driving time at each speed.

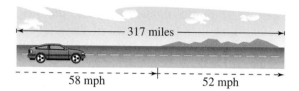

60. *Travel Time* Two cars start at the same location and travel in the same direction at average speeds of 30 miles per hour and 45 miles per hour. How much time must elapse before the two cars are 5 miles apart?

✓ 61. *Work-Rate Problem* You can complete a typing project in 5 hours, and your friend can complete it in 8 hours.

(a) What fractional part of the project can be accomplished by each person in 1 hour?

(b) How long will it take both of you to complete the project working together?

62. *Work-Rate Problem* You can mow a lawn in 3 hours, and your friend can mow it in 4 hours.

(a) What fractional part of the lawn can each of you mow in 1 hour?

(b) How long will it take both of you to mow the lawn working together?

63. *Work-Rate Problem* It takes 30 minutes for a pump to empty a water tank. A larger pump can empty the tank in half the time. How long would it take to empty the tank with both pumps operating?

64. *Work-Rate Problem* It takes 90 minutes to mow a lawn with a push mower. It takes half the time using a riding mower. How long would it take to mow the lawn using both mowers?

In Exercises 65–74, solve for the specified variable. *See Example 7.*

✓ **65.** Solve for R.

Ohm's Law: $E = IR$

66. Solve for r.

Simple Interest: $A = P + Prt$

67. Solve for L.

Discount: $S = L - rL$

68. Solve for C.

Markup: $S = C + rC$

69. Solve for a.

Free-Falling Body: $h = 48t + \dfrac{1}{2}at^2$

70. Solve for a.

Free-Falling Body: $h = 18t + \dfrac{1}{2}at^2$

71. Solve for a.

Free-Falling Body: $h = 36t + \dfrac{1}{2}at^2 + 50$

72. Solve for a.

Free-Falling Body: $h = -15t + \dfrac{1}{2}at^2 + 9.5$

73. Solve for h.

Surface Area of a Circular Cylinder:

$S = 2\pi r^2 + 2\pi rh$

74. Solve for b.

Area of a Trapezoid: $A = \dfrac{1}{2}(a + b)h$

75. ▲ *Geometry*   Find the volume of the circular cylinder shown in the figure.

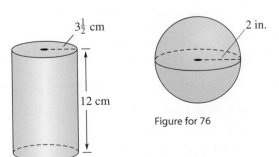

Figure for 75

76. ▲ *Geometry*   Find the volume of the sphere shown in the figure.

Figure for 76

✓ **77.** ▲ *Geometry*   A rectangular picture frame has a perimeter of 3 feet. The width of the frame is 0.62 times its height. Find the height of the frame.

78. ▲ *Geometry*   A rectangular stained glass window has a perimeter of 18 feet. The height of the window is 1.25 times its width. Find the width of the window.

79. ▲ *Geometry*   A "Slow Moving Vehicle" sign has the shape of an equilateral triangle. The sign has a perimeter of 129 centimeters. Find the length of each side.

80. ▲ *Geometry*   The length of a rectangle is three times its width. The perimeter of the rectangle is 64 inches. Find the dimensions of the rectangle.

✓ **81.** *Meteorology*   The average daily high temperature in August in Denver, Colorado is 86°F. In August, what is Denver's average daily high temperature in degrees Celsius? (Source: U.S. National Oceanic and Atmospheric Administration)

82. *Meteorology*   The average daily low temperature in December in Kansas City, Kansas is 23°F. In December, what is Kansas City's average daily low temperature in degrees Celsius? (Source: U.S. National Oceanic and Atmospheric Administration)

✓ **83.** *Simple Interest*   Find the interest on a $5000 bond that pays an annual percentage rate of $6\frac{1}{2}\%$ for 6 years.

84. *Simple Interest*   Find the annual interest rate on a certificate of deposit that accumulated $400 interest in 2 years on a principal of $2500.

85. *Simple Interest*   The interest on a savings account is 7%. Find the principal required to earn $500 in interest in 2 years.

86. *Simple Interest*   The interest on a bond is $5\frac{1}{2}\%$. Find the principal required to earn $900 in interest in 3 years.

87. *Average Wage*   The average hourly wage y (in dollars) for custodians at public schools in the United States from 2000 through 2005 can be approximated by the model $y = 0.264t + 11.49$, for $0 \le t \le 5$, where t represents the year, with $t = 0$ corresponding to 2000 (see figure). (Source: Educational Research Service)

(a) Use the graph to determine the year in which the average hourly wage was $11.96. Is the result the same when you use the model?

(b) What was the average annual hourly raise for custodians during this six-year period? Explain how you arrived at your answer.

88. *Average Wage*   The average hourly wage y (in dollars) for cafeteria workers at public schools in the United States from 2000 through 2005 can be approximated by the model $y = 0.261t + 9.12$, for $0 \le t \le 5$, where t represents the year, with $t = 0$ corresponding to 2000 (see figure). (Source: Educational Research Service)

(a) Use the graph to determine the year in which the average hourly wage was $10.18. Is the result the same when you use the model?

(b) What was the average annual hourly raise for cafeteria workers during this six-year period? Explain how you determined your answer.

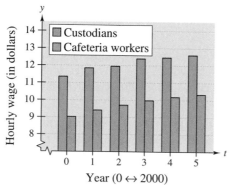

Figure for 87 and 88

Explaining Concepts

89. ✎   Explain how to find the sale price of an item when you are given the list price and the discount rate.

90. ✎   If the sides of a square are doubled, does the perimeter double? Explain.

91. ✎   If the sides of a square are doubled, does the area double? Explain.

92. ✎   If you forget the formula for the volume of a right circular cylinder, how can you derive it?

Cumulative Review

In Exercises 93–96, give (a) the additive inverse and (b) the multiplicative inverse of the quantity.

93. 21

94. -34

95. $-5x$

96. $8m$

In Exercises 97–100, simplify the expression.

97. $2x(x - 4) + 3$

98. $4x - 5(1 - x)$

99. $x^2(x - 4) - 2x^2$

100. $x^2(2x + 1) + 2x(3 - x)$

In Exercises 101–104, solve using a percent equation.

101. 52 is 40% of what number?

102. 72 is 48% of what number?

103. 117 is what percent of 900?

104. 287 is what percent of 350?

Mid-Chapter Quiz

Take this quiz as you would take a quiz in class. After you are done, check your work against the answers in the back of the book.

In Exercises 1–8, solve the equation and check the result. (If it is not possible, state the reason.)

1. $4x - 8 = 0$
2. $-3(z - 2) = 0$
3. $2(y + 3) = 18 - 4y$
4. $5t + 7 = 7(t + 1) - 2t$
5. $\frac{1}{4}x + 6 = \frac{3}{2}x - 1$
6. $\frac{2b}{5} + \frac{b}{2} = 3$
7. $\frac{4 - x}{5} + 5 = \frac{5}{2}$
8. $3x + \frac{11}{12} = \frac{5}{16}$

In Exercises 9 and 10, solve the equation and round your answer to two decimal places.

9. $0.25x + 6.2 = 4.45x + 3.9$
10. $0.42x + 6 = 5.25x - 0.80$

11. Write the decimal 0.45 as a fraction and as a percent.

12. 500 is 250% of what number?

13. Find the unit price (in dollars per ounce) of a 12-ounce box of cereal that sells for $4.85.

14. A quality control engineer for a manufacturer finds one defective unit in a sample of 150. At this rate, what is the expected number of defective units in a shipment of 750,000?

15. A store is offering a discount of 25% on a computer with a list price of $1080. A mail-order catalog has the same computer for $799 plus $14.95 for shipping. Which is the better buy?

16. Last week you earned $616. Your regular hourly wage is $12.25 for the first 40 hours, and your overtime hourly wage is $18. How many hours of overtime did you work?

17. Fifty gallons of a 30% acid solution is obtained by combining solutions that are 25% acid and 50% acid. How much of each solution is required?

18. On the first part of a 300-mile trip, a sales representative averaged 62 miles per hour. The sales representative averaged 46 miles per hour on the remainder of the trip because of an increased volume of traffic. The total time of the trip was 6 hours. Find the amount of driving time at each speed.

19. You can paint a room in 3 hours, and your friend can paint it in 5 hours. How long will it take both of you to paint the room together?

20. The accompanying figure shows three squares. The perimeters of squares I and II are 20 inches and 32 inches, respectively. Find the area of square III.

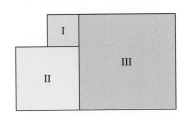

Figure for 20

2.4 Linear Inequalities

© Jose Fuste Raga/CORBIS

Why You Should Learn It

You can use linear inequalities to model and solve real-life problems. For instance, you will use inequalities to analyze the consumption of wind energy in Exercises 119 and 120 on page 106.

1 ▶ Sketch the graphs of inequalities.

What You Should Learn

1 ▶ Sketch the graphs of inequalities.

2 ▶ Identify the properties of inequalities that can be used to create equivalent inequalities.

3 ▶ Solve linear inequalities.

4 ▶ Solve compound inequalities.

5 ▶ Solve application problems involving inequalities.

Intervals on the Real Number Line

In this section you will study **algebraic inequalities,** which are inequalities that contain one or more variable terms. Some examples are

$$x \le 4, \quad x \ge -3, \quad x + 2 < 7, \quad \text{and} \quad 4x - 6 < 3x + 8.$$

As with an equation, you **solve** an inequality in the variable x by finding all values of x for which the inequality is true. Such values are called **solutions** and are said to *satisfy* the inequality. The set of all solutions of an inequality is the **solution set** of the inequality. The **graph** of an inequality is obtained by plotting its solution set on the real number line. Often, these graphs are intervals—either bounded or unbounded.

Bounded Intervals on the Real Number Line

Let a and b be real numbers such that $a < b$. The following intervals on the real number line are called **bounded intervals.** The numbers a and b are the **endpoints** of each interval. A bracket indicates that the endpoint is included in the interval, and a parenthesis indicates that the endpoint is excluded.

Notation	Interval Type	Inequality	Graph
$[a, b]$	Closed	$a \le x \le b$	
(a, b)	Open	$a < x < b$	
$[a, b)$		$a \le x < b$	
$(a, b]$		$a < x \le b$	

The **length** of the interval $[a, b]$ is the distance between its endpoints: $b - a$. The lengths of $[a, b]$, (a, b), $[a, b)$, and $(a, b]$ are the same. The reason that these four types of intervals are called "bounded" is that each has a finite length. An interval that *does not* have a finite length is **unbounded** (or **infinite**).

Unbounded Intervals on the Real Number Line

Let a and b be real numbers. The following intervals on the real number line are called **unbounded intervals**.

Notation	Interval Type	Inequality	Graph
$[a, \infty)$		$x \geq a$	
(a, ∞)	Open	$x > a$	
$(-\infty, b]$		$x \leq b$	
$(-\infty, b)$	Open	$x < b$	
$(-\infty, \infty)$	Entire real line		

The symbols ∞ (**positive infinity**) and $-\infty$ (**negative infinity**) do not represent real numbers. They are simply convenient symbols used to describe the unboundedness of an interval such as $(-5, \infty)$. This is read as the interval from -5 to infinity.

EXAMPLE 1   Graphing Inequalities

Sketch the graph of each inequality.

a. $-3 < x \leq 1$ **b.** $0 < x < 2$

c. $-3 < x$ **d.** $x \leq 2$

Solution

a. The graph of $-3 < x \leq 1$ is a bounded interval.

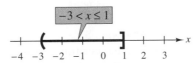

b. The graph of $0 < x < 2$ is a bounded interval.

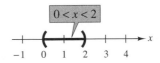

c. The graph of $-3 < x$ is an unbounded interval.

d. The graph of $x \leq 2$ is an unbounded interval.

✓ **CHECKPOINT** *Now try Exercise 11.*

2 ▶ Identify the properties of inequalities that can be used to create equivalent inequalities.

Properties of Inequalities

Solving a linear inequality is much like solving a linear equation. You isolate the variable by using the **properties of inequalities.** These properties are similar to the properties of equality, but there are two important exceptions. *When each side of an inequality is multiplied or divided by a negative number, the direction of the inequality symbol must be reversed.* Here is an example.

$$-2 < 5 \qquad \text{Original inequality}$$

$$(-3)(-2) > (-3)(5) \qquad \text{Multiply each side by } -3 \text{ and reverse the inequality.}$$

$$6 > -15 \qquad \text{Simplify.}$$

Two inequalities that have the same solution set are **equivalent inequalities.** The following list of operations can be used to create equivalent inequalities.

Properties of Inequalities

1. *Addition and Subtraction Properties*

Adding the same quantity to, or subtracting the same quantity from, each side of an inequality produces an equivalent inequality.

If $a < b$, then $a + c < b + c$.

If $a < b$, then $a - c < b - c$.

2. *Multiplication and Division Properties: Positive Quantities*

Multiplying or dividing each side of an inequality by a positive quantity produces an equivalent inequality.

If $a < b$ and c is positive, then $ac < bc$.

If $a < b$ and c is positive, then $\dfrac{a}{c} < \dfrac{b}{c}$.

3. *Multiplication and Division Properties: Negative Quantities*

Multiplying or dividing each side of an inequality by a negative quantity produces an equivalent inequality in which the inequality symbol is reversed.

If $a < b$ and c is negative, then $ac > bc$.   Reverse inequality.

If $a < b$ and c is negative, then $\dfrac{a}{c} > \dfrac{b}{c}$.   Reverse inequality.

4. *Transitive Property*

Consider three quantities for which the first quantity is less than the second, and the second is less than the third. It follows that the first quantity must be less than the third quantity.

If $a < b$ and $b < c$, then $a < c$.

These properties remain true if the symbols $<$ and $>$ are replaced by $\leq$ and $\geq$. Moreover, a, b, and c can represent real numbers, variables, or expressions. Note that you cannot multiply or divide each side of an inequality by zero.

3 ▶ Solve linear inequalities.

Solving a Linear Inequality

An inequality in one variable is a **linear inequality** if it can be written in one of the following forms.

$$ax + b \leq 0, \quad ax + b < 0, \quad ax + b \geq 0, \quad ax + b > 0$$

The solution set of a linear inequality can be written in set notation. For the solution $x > 1$, the set notation is $\{x \mid x > 1\}$ and is read "the set of all x such that x is greater than 1."

As you study the following examples, *remember that when you multiply or divide an inequality by a negative number, you must reverse the inequality symbol.*

Study Tip

Checking the solution set of an inequality is not as simple as checking the solution set of an equation. (There are usually too many x-values to substitute back into the original inequality.) You can, however, get an indication of the validity of a solution set by substituting a few convenient values of x. For instance, in Example 2, try checking that $x = 0$ satisfies the original inequality, whereas $x = 4$ does not.

EXAMPLE 2   **Solving a Linear Inequality**

$$x + 6 < 9 \qquad\qquad \text{Original inequality}$$

$$x + 6 - 6 < 9 - 6 \qquad\qquad \text{Subtract 6 from each side.}$$

$$x < 3 \qquad\qquad \text{Combine like terms.}$$

The solution set consists of all real numbers that are less than 3. The solution set in interval notation is $(-\infty, 3)$ and in set notation is $\{x \mid x < 3\}$. The graph is shown in Figure 2.4.

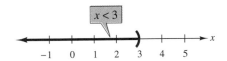

Figure 2.4

✔ **CHECKPOINT** *Now try Exercise 37.*

EXAMPLE 3   **Solving a Linear Inequality**

$$8 - 3x \leq 20 \qquad\qquad \text{Original inequality}$$

$$8 - 8 - 3x \leq 20 - 8 \qquad\qquad \text{Subtract 8 from each side.}$$

$$-3x \leq 12 \qquad\qquad \text{Combine like terms.}$$

$$\frac{-3x}{-3} \geq \frac{12}{-3} \qquad\qquad \text{Divide each side by } -3 \text{ and reverse the inequality symbol.}$$

$$x \geq -4 \qquad\qquad \text{Simplify.}$$

The solution set in interval notation is $[-4, \infty)$ and in set notation is $\{x \mid x \geq -4\}$. The graph is shown in Figure 2.5.

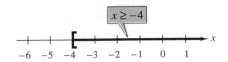

Figure 2.5

✔ **CHECKPOINT** *Now try Exercise 41.*

Most graphing calculators can graph a linear inequality. Consult your user's guide for specific instructions. The graph below shows the solution of the inequality in Example 4. Notice that the graph representing the solution interval lies above the x-axis.

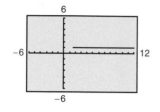

EXAMPLE 4   Solving a Linear Inequality

$7x - 3 > 3(x + 1)$	Original inequality
$7x - 3 > 3x + 3$	Distributive Property
$7x - 3x - 3 > 3x - 3x + 3$	Subtract $3x$ from each side.
$4x - 3 > 3$	Combine like terms.
$4x - 3 + 3 > 3 + 3$	Add 3 to each side.
$4x > 6$	Combine like terms.
$\dfrac{4x}{4} > \dfrac{6}{4}$	Divide each side by 4.
$x > \dfrac{3}{2}$	Simplify.

The solution set consists of all real numbers that are greater than $\frac{3}{2}$. The solution set in interval notation is $\left(\frac{3}{2}, \infty\right)$ and in set notation is $\left\{x \mid x > \frac{3}{2}\right\}$. The graph is shown in Figure 2.6.

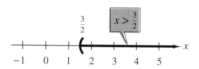

Figure 2.6

✔ **CHECKPOINT** *Now try Exercise 79.*

An inequality can be cleared of fractions in the same way an equation can be cleared of fractions—by multiplying each side by the least common denominator. This is shown in Example 5.

Figure 2.7

EXAMPLE 5   Solving a Linear Inequality

$\dfrac{2x}{3} + 12 < \dfrac{x}{6} + 18$	Original inequality
$6 \cdot \left(\dfrac{2x}{3} + 12\right) < 6 \cdot \left(\dfrac{x}{6} + 18\right)$	Multiply each side by LCD of 6.
$4x + 72 < x + 108$	Distributive Property
$4x - x < 108 - 72$	Subtract x and 72 from each side.
$3x < 36$	Combine like terms.
$x < 12$	Divide each side by 3.

The solution set consists of all real numbers that are less than 12. The solution set in interval notation is $(-\infty, 12)$ and in set notation is $\{x \mid x < 12\}$. The graph is shown in Figure 2.7.

✔ **CHECKPOINT** *Now try Exercise 55.*

4 ▶ Solve compound inequalities.

Solving a Compound Inequality

Two inequalities joined by the word *and* or the word *or* constitute a **compound inequality.** When two inequalities are joined by the word *and*, the solution set consists of all real numbers that satisfy *both* inequalities. The solution set for the compound inequality $-4 \le 5x - 2$ *and* $5x - 2 < 7$ can be written more simply as the **double inequality**

$$-4 \le 5x - 2 < 7.$$

A compound inequality formed by the word *and* is called **conjunctive** and is the only kind that has the potential to form a double inequality. A compound inequality joined by the word *or* is called **disjunctive** and cannot be re-formed into a double inequality.

EXAMPLE 6   **Solving a Double Inequality**

Solve the double inequality $-7 \le 5x - 2 < 8$.

Solution

$-7 \le 5x - 2 < 8$	Write original inequality.
$-7 + 2 \le 5x - 2 + 2 < 8 + 2$	Add 2 to all three parts.
$-5 \le 5x < 10$	Combine like terms.
$\dfrac{-5}{5} \le \dfrac{5x}{5} < \dfrac{10}{5}$	Divide each part by 5.
$-1 \le x < 2$	Simplify.

The solution set consists of all real numbers that are greater than or equal to -1 and less than 2. The solution set in interval notation is $[-1, 2)$ and in set notation is $\{x \mid -1 \le x < 2\}$. The graph is shown in Figure 2.8.

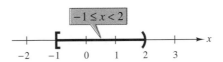

Figure 2.8

✓ **CHECKPOINT** *Now try Exercise 61.*

The double inequality in Example 6 could have been solved in two parts, as follows.

$-7 \le 5x - 2$	and	$5x - 2 < 8$
$-5 \le 5x$		$5x < 10$
$-1 \le x$		$x < 2$

The solution set consists of all real numbers that satisfy both inequalities. In other words, the solution set is the set of all values of x for which

$$-1 \le x < 2.$$

EXAMPLE 7   **Solving a Conjunctive Inequality**

Solve the compound inequality $-1 \leq 2x - 3$ and $2x - 3 < 5$.

Solution

Begin by writing the conjunctive inequality as a double inequality.

$$-1 \leq 2x - 3 < 5 \qquad \text{Write as double inequality.}$$

$$-1 + 3 \leq 2x - 3 + 3 < 5 + 3 \qquad \text{Add 3 to all three parts.}$$

$$2 \leq 2x < 8 \qquad \text{Combine like terms.}$$

$$\frac{2}{2} \leq \frac{2x}{2} < \frac{8}{2} \qquad \text{Divide each part by 2.}$$

$$1 \leq x < 4 \qquad \text{Solution set (See Figure 2.9.)}$$

The solution set in interval notation is $[1, 4)$ and in set notation is $\{x \mid 1 \leq x < 4\}$.

✓ **CHECKPOINT** *Now try Exercise 71.*

Figure 2.9

EXAMPLE 8   **Solving a Disjunctive Inequality**

Solve the compound inequality $-3x + 6 \leq 2$ or $-3x + 6 \geq 7$.

Solution

$-3x + 6 \leq 2$	or	$-3x + 6 \geq 7$	Write original inequality.

$$-3x + 6 - 6 \leq 2 - 6 \qquad -3x + 6 - 6 \geq 7 - 6 \qquad \text{Subtract 6 from all parts.}$$

$$-3x \leq -4 \qquad \qquad -3x \geq 1 \qquad \text{Combine like terms.}$$

$$\frac{-3x}{-3} \geq \frac{-4}{-3} \qquad \qquad \frac{-3x}{-3} \leq \frac{1}{-3} \qquad \begin{array}{l}\text{Divide all parts by} \\ -3 \text{ and reverse both} \\ \text{inequality symbols.}\end{array}$$

$$x \geq \frac{4}{3} \qquad \qquad x \leq -\frac{1}{3} \qquad \text{Solution set (See Figure 2.10.)}$$

The solution set in set notation is $\left\{x \mid x \leq -\frac{1}{3} \text{ or } x \geq \frac{4}{3}\right\}$.

✓ **CHECKPOINT** *Now try Exercise 75.*

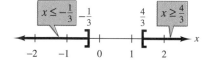

Figure 2.10

Compound inequalities can be written using *symbols*. For compound inequalities, the word *and* is represented by the symbol ∩, which is read as **intersection.** The word *or* is represented by the symbol ∪, which is read as **union.** Graphical representations are shown in Figure 2.11. If A and B are sets, then x is in $A \cap B$ if it is in both A and B. Similarly, x is in $A \cup B$ if it is in A, B, or both A and B.

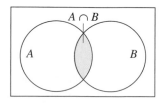

Intersection of two sets
Figure 2.11

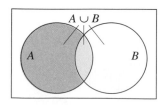

Union of two sets

EXAMPLE 9　**Writing a Solution Set Using Union**

A solution set is shown on the number line in Figure 2.12.

a. Write the solution set as a compound inequality.

b. Write the solution set using the union symbol.

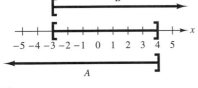

Figure 2.12

Solution

a. As a compound inequality, you can write the solution set as $x \leq -1$ or $x > 2$.

b. Using set notation, you can write the left interval as $A = \{x \mid x \leq -1\}$ and the right interval as $B = \{x \mid x > 2\}$. So, using the union symbol, the entire solution set can be written as $A \cup B$.

✔ **CHECKPOINT** *Now try Exercise 83.*

EXAMPLE 10　**Writing a Solution Set Using Intersection**

Write the compound inequality using the intersection symbol.

$$-3 \leq x \leq 4$$

Figure 2.13

Solution

Consider the two sets $A = \{x \mid x \leq 4\}$ and $B = \{x \mid x \geq -3\}$. These two sets overlap, as shown on the number line in Figure 2.13. The compound inequality $-3 \leq x \leq 4$ consists of all numbers that are in both $x \leq 4$ and $x \geq -3$, which means that it can be written as $A \cap B$.

✔ **CHECKPOINT** *Now try Exercise 89.*

5 ▶ Solve application problems involving inequalities.

Applications

Linear inequalities in real-life problems arise from statements that involve phrases such as "at least," "no more than," "minimum value," and so on. Study the meanings of the key phrases in the next example.

EXAMPLE 11　**Translating Verbal Statements**

Verbal Statement	*Inequality*	
a. x is at most 3.	$x \leq 3$	"at most" means "less than or equal to."
b. x is no more than 3.	$x \leq 3$	
c. x is at least 3.	$x \geq 3$	"at least" means "greater than or equal to."
d. x is no less than 3.	$x \geq 3$	
e. x is more than 3.	$x > 3$	
f. x is less than 3.	$x < 3$	
g. x is a minimum of 3.	$x \geq 3$	
h. x is at least 2, but less than 7.	$2 \leq x < 7$	
i. x is greater than 2, but no more than 7.	$2 < x \leq 7$	

✔ **CHECKPOINT** *Now try Exercise 97.*

To solve real-life problems involving inequalities, you can use the same "verbal-model approach" you use with equations.

Figure 2.14

EXAMPLE 12 **Finding the Maximum Width of a Package**

An overnight delivery service will not accept any package with a combined length and girth (perimeter of a cross section perpendicular to the length) exceeding 132 inches. Consider a rectangular box that is 68 inches long and has square cross sections. What is the maximum acceptable width of such a box?

Solution

First make a sketch (see Figure 2.14). The length of the box is 68 inches, and because a cross section is square, the width and height are each x inches long.

Verbal Model:	Length + Girth ≤ 132 inches	
Labels:	Width of a side $= x$	(inches)
	Length $= 68$	(inches)
	Girth $= 4x$	(inches)

Inequality: $68 + 4x \leq 132$

$$4x \leq 64$$

$$x \leq 16$$

The width of the box can be at most 16 inches.

 CHECKPOINT *Now try Exercise 113.*

EXAMPLE 13 **Comparing Costs**

Company A rents a subcompact car for $240 per week with no extra charge for mileage. Company B rents a similar car for $100 per week plus an additional 25 cents for each mile driven. How many miles must you drive in a week so that the rental fee of Company B is more than that of Company A?

Solution

Verbal Model:	Weekly cost for Company B	>	Weekly cost for Company A

Labels:	Number of miles driven in one week $= m$	(miles)
	Weekly cost for Company A $= 240$	(dollars)
	Weekly cost for Company B $= 100 + 0.25m$	(dollars)

Inequality: $100 + 0.25m > 240$

$$0.25m > 140$$

$$m > 560$$

So, the car from Company B is more expensive if you drive more than 560 miles in a week. The table shown at the left helps confirm this conclusion.

 CHECKPOINT *Now try Exercise 117.*

Miles driven	Company A	Company B
520	$240.00	$230.00
530	$240.00	$232.50
540	$240.00	$235.00
550	$240.00	$237.50
560	$240.00	$240.00
570	$240.00	$242.50

_____ Concept Check _____

1. Is dividing each side of an inequality by 5 the same as multiplying each side by $\frac{1}{5}$? Explain.

2. State whether each inequality is equivalent to $x > 3$. Explain your reasoning in each case.

(a) $x < 3$ (b) $3 < x$

(c) $-x < -3$ (d) $-3 < x$

3. Describe two types of situations involving application of properties of inequalities for which you must reverse the inequality symbol.

4. Explain the distinction between using the word *or* and using the word *and* to form a compound inequality.

2.4 EXERCISES

Go to pages 116–117 to record your assignments.

_____ Developing Skills _____

In Exercises 1–4, determine whether each value of x satisfies the inequality.

Inequality		*Values*	

1. $7x - 10 > 0$ (a) $x = 3$ (b) $x = -2$

   (c) $x = \frac{5}{2}$ (d) $x = \frac{1}{2}$

2. $3x + 2 < \dfrac{7x}{5}$ (a) $x = 0$ (b) $x = 4$

   (c) $x = -4$ (d) $x = -1$

3. $0 < \dfrac{x + 4}{5} < 2$ (a) $x = 10$ (b) $x = -4$

   (c) $x = 0$ (d) $x = 6$

4. $-3 < \dfrac{2 - x}{2} \le 3$ (a) $x = 0$ (b) $x = 7$

   (c) $x = 9$ (d) $x = -1$

In Exercises 5–10, match the inequality with its graph. [The graphs are labeled (a), (b), (c), (d), (e), and (f).]

(a)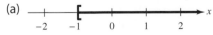

(b)

(c)

(d)

(e)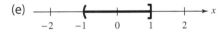

(f)

5. $x \ge -1$ **6.** $-1 < x \le 1$

7. $x \le -1$ or $x \ge 2$ **8.** $x < -1$ or $x \ge 1$

9. $-2 < x < 1$ **10.** $x < 2$

In Exercises 11–24, sketch the graph of the inequality. *See Example 1.*

✓ **11.** $x \le 4$ **12.** $x > -6$

13. $x > 3.5$ **14.** $x \le -2.5$

15. $-5 < x \le 3$ **16.** $-1 < x \le 5$

17. $4 > x \ge 1$ **18.** $9 \ge x \ge 3$

19. $\frac{3}{2} \ge x > 0$ **20.** $-\frac{15}{4} < x < -\frac{5}{2}$

21. $x < -5$ or $x \ge -1$ **22.** $x \le -4$ or $x > 0$

23. $x \le 3$ or $x > 7$ **24.** $x \le -1$ or $x \ge 1$

25. Write an inequality equivalent to $5 - \frac{1}{3}x > 8$ by multiplying each side by -3.

26. Write an inequality equivalent to $5 - \frac{1}{3}x > 8$ by adding $\frac{1}{3}x$ to each side.

In Exercises 27–34, determine whether the inequalities are equivalent.

27. $3x - 2 < 12, \quad 3x < 10$

28. $6x + 7 \geq 11, \quad 6x \geq 18$

29. $-5(x + 12) > 25, \quad x + 12 > -5$

30. $-4(5 - x) < 32, \quad 5 - x < -8$

31. $7x - 6 \leq 3x + 12, \quad 4x \leq 18$

32. $11 - 3x \geq 7x + 1, \quad 10 \geq 10x$

33. $3x > 5x, \quad 3 > 5$

34. $4x > -8x, \quad -4 < 8$

In Exercises 35–82, solve the inequality and sketch the solution on the real number line. *See Examples 2–8.*

35. $x - 4 \geq 0$

36. $x + 1 < 0$

37. $x + 7 \leq 9$

38. $z - 4 > 0$

39. $2x < 8$

40. $3x \geq 12$

41. $-9x \geq 36$

42. $-6x \leq 24$

43. $-\frac{3}{4}x < -6$

44. $-\frac{1}{5}x > -2$

45. $5 - x \leq -2$

46. $1 - y \geq -5$

47. $2x - 5.3 > 9.8$

48. $1.6x + 4 \leq 12.4$

49. $5 - 3x < 7$

50. $12 - 5x > 5$

51. $3x - 11 > -x + 7$

52. $21x - 11 \leq 6x + 19$

53. $-3x + 7 < 8x - 13$

54. $6x - 1 > 3x - 11$

55. $\frac{x}{4} > 2 - \frac{x}{2}$

56. $\frac{x}{6} - 1 \leq \frac{x}{4}$

57. $\frac{x - 4}{3} + 3 \leq \frac{x}{8}$

58. $\frac{x + 3}{6} + \frac{x}{8} \geq 1$

59. $\frac{3x}{5} - 4 < \frac{2x}{3} - 3$

60. $\frac{4x}{7} + 1 > \frac{x}{2} + \frac{5}{7}$

61. $0 < 2x - 5 < 9$

62. $-6 \leq 3x - 9 < 0$

63. $8 < 6 - 2x \leq 12$

64. $-10 \leq 4 - 7x < 10$

65. $-1 < -0.2x < 1$

66. $-2 < -0.5s \leq 0$

67. $-3 < \frac{2x - 3}{2} < 3$

68. $0 \leq \frac{x - 5}{2} < 4$

69. $1 > \frac{x - 4}{-3} > -2$

70. $-\frac{2}{3} < \frac{x - 4}{-6} \leq \frac{1}{3}$

71. $2x - 4 \leq 4$ and $2x + 8 > 6$

72. $7 + 4x < -5 + x$ and $2x + 10 \leq -2$

73. $8 - 3x > 5$ and $x - 5 \geq 10$

74. $9 - x \leq 3 + 2x$ and $3x - 7 \leq -22$

75. $7.2 - 1.1x > 1$ or $1.2x - 4 > 2.7$

76. $0.4x - 3 \leq 8.1$ or $4.2 - 1.6x \leq 3$

77. $7x + 11 < 3 + 4x$ or $\frac{5}{2}x - 1 \geq 9 - \frac{3}{2}x$

78. $3x + 10 \leq -x - 6$ or $\frac{1}{2}x + 5 < \frac{5}{2}x - 4$

79. $-3(y + 10) \geq 4(y + 10)$

80. $2(4 - z) \geq 8(1 + z)$

81. $-4 \leq 2 - 3(x + 2) < 11$

82. $16 < 4(y + 2) - 5(2 - y) \leq 24$

In Exercises 83–88, write the solution set as a compound inequality. Then write the solution using set notation and the union or intersection symbol. *See Example 9.*

83.

84.

85.

86.

87.

88.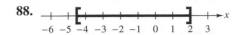

In Exercises 89–94, write the compound inequality using set notation and the union or intersection symbol. *See Example 10.*

89. $-7 \le x < 0$

90. $2 < x < 8$

91. $-\frac{9}{2} < x \le -\frac{3}{2}$

92. $-\frac{4}{5} \le x < \frac{1}{5}$

93. $x < 0$ or $x \ge \frac{2}{3}$

94. $-3 > x$ or $x > 8$

In Exercises 95–100, rewrite the statement using inequality notation. *See Example 11.*

95. x is nonnegative.   **96.** y is more than -2.

97. z is at least 8.   **98.** m is at least 4.

99. n is at least 10, but no more than 16.

100. x is at least 450, but no more than 500.

In Exercises 101–104, write a verbal description of the inequality.

101. $x \ge \frac{5}{2}$

102. $t < 4$

103. $3 \le y < 5$

104. $0 < z \le \pi$

Solving Problems

105. *Budget*   A student group has $4500 budgeted for a field trip. The cost of transportation for the trip is $1900. To stay within the budget, all other costs C must be no more than what amount?

106. *Budget*   You have budgeted $1800 per month for your total expenses. The cost of rent per month is $600 and the cost of food is $350. To stay within your budget, all other costs C must be no more than what amount?

107. *Meteorology*   Miami's average temperature is greater than the average temperature in Washington, DC, and the average temperature in Washington, DC is greater than the average temperature in New York City. How does the average temperature in Miami compare with the average temperature in New York City?

108. *Elevation*   The elevation (above sea level) of San Francisco is less than the elevation of Dallas, and the elevation of Dallas is less than the elevation of Denver. How does the elevation of San Francisco compare with the elevation of Denver?

109. *Operating Costs*   A utility company has a fleet of vans. The annual operating cost per van is $C = 0.35m + 2900$, where m is the number of miles traveled by a van in a year. What is the maximum number of miles that will yield an annual operating cost that is no more than $12,000?

110. *Operating Costs*   A fuel company has a fleet of trucks. The annual operating cost per truck is $C = 0.58m + 7800$, where m is the number of miles traveled by a truck in a year. What is the maximum number of miles that will yield an annual operating cost that is less than $25,000?

Cost, Revenue, and Profit   In Exercises 111 and 112, the revenue R from selling x units and the cost C of producing x units of a product are given. In order to obtain a profit, the revenue must be greater than the cost. For what values of x will this product produce a profit?

111. $R = 89.95x$ **112.** $R = 105.45x$

  $C = 61x + 875$ $C = 78x + 25{,}850$

113. ▲ *Geometry*   The width of a rectangle is 22 meters. The perimeter of the rectangle must be at least 90 meters and not more than 120 meters. Find the interval for the length x (see figure on page 106).

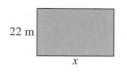
22 m

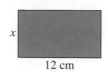
x

x

12 cm

Figure for 113   Figure for 114

114. ▲ *Geometry* The length of a rectangle is 12 centimeters. The perimeter of the rectangle must be at least 30 centimeters and not more than 42 centimeters. Find the interval for the width x.

115. *Number Problem* Four times a number n must be at least 12 and no more than 30. What interval represents the possible values of n?

116. *Number Problem* What interval represents the values of n for which $\frac{1}{3}n$ is no more than 9?

✔ **117.** *Hourly Wage* Your company requires you to select one of two payment plans. One plan pays a straight $12.50 per hour. The second plan pays $8.00 per hour plus $0.75 per unit produced per hour. Write an inequality for the number of units that must be produced per hour so that the second option yields the greater hourly wage. Solve the inequality.

118. *Monthly Wage* Your company requires you to select one of two payment plans. One plan pays a straight $3000 per month. The second plan pays $1000 per month plus a commission of 4% of your gross sales. Write an inequality for the gross sales per month for which the second option yields the greater monthly wage. Solve the inequality.

Energy In Exercises 119 and 120, use the equation $y = 21.8t - 160$, for $9 \le t \le 15$, which models the annual consumption of energy produced by wind (in trillions of British thermal units) in the United States from 1999 to 2005. In this model, t represents the year, with $t = 9$ corresponding to 1999. (Source: U.S. Energy Information Administration)

119. During which years was the consumption of energy produced by wind less than 100 trillion Btu?

120. During which years was the consumption of energy produced by wind greater than 130 trillion Btu?

Explaining Concepts

121. ✎ Describe any differences between properties of equalities and properties of inequalities.

122. If $-3 \le x \le 10$, then $-x$ must be in what interval? Explain.

123. Discuss whether the solution set of a linear inequality is a *bounded* interval or an *unbounded* interval.

124. Two linear inequalities are joined by the word *or* to form a compound inequality. Discuss whether the solution set is a bounded interval.

In Exercises 125–128, let a and b be real numbers such that $a < b$. Use a and b to write a compound algebraic inequality in x with the given type of solution. Explain your reasoning.

125. A bounded interval

126. Two unbounded intervals

127. The set of all real numbers

128. No solution

Cumulative Review

In Exercises 129–132, place the correct symbol ($<$, $>$, or $=$) between the pair of real numbers.

129. $|4|$ ___ $|-5|$

130. $|-4|$ ___ $|-6|$

131. $|-7|$ ___ $|7|$

132. $-|5|$ ___ $-(5)$

In Exercises 133–136, determine whether each value of the variable is a solution of the equation.

133. $3x = 27$; $x = 6, x = 9$

134. $x - 14 = 8$; $x = 6, x = 22$

135. $7x - 5 = 7 + x$; $x = 2, x = 6$

136. $2 + 5x = 8x - 13$; $x = 3, x = 5$

In Exercises 137–140, solve the equation.

137. $2x - 17 = 0$

138. $x - 17 = 4$

139. $32x = -8$

140. $14x + 5 = 2 - x$

2.5 Absolute Value Equations and Inequalities

Ronnie Kaufman/CORBIS

Why You Should Learn It

Absolute value equations and inequalities can be used to model and solve real-life problems. For instance, in Exercise 92 on page 115, you will use an absolute value inequality to describe the normal body temperature range.

1 ▶ Solve absolute value equations.

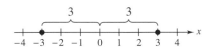

Figure 2.15

What You Should Learn

1 ▶ Solve absolute value equations.
2 ▶ Solve inequalities involving absolute value.

Solving Equations Involving Absolute Value

Consider the **absolute value equation**

$$|x| = 3.$$

The only solutions of this equation are $x = -3$ and $x = 3$, because these are the only two real numbers whose distance from zero is 3. (See Figure 2.15.) In other words, the absolute value equation $|x| = 3$ has exactly two solutions:

$$x = -3 \quad \text{and} \quad x = 3.$$

Solving an Absolute Value Equation

Let x be a variable or an algebraic expression and let a be a real number such that $a \geq 0$. The solutions of the equation $|x| = a$ are given by $x = -a$ and $x = a$. That is,

$$|x| = a \quad \Longrightarrow \quad x = -a \quad \text{or} \quad x = a.$$

EXAMPLE 1   Solving Absolute Value Equations

Solve each absolute value equation.

a. $|x| = 10$ **b.** $|x| = 0$ **c.** $|y| = -1$

Solution

a. This equation is equivalent to the two linear equations

$$x = -10 \quad \text{and} \quad x = 10. \quad \text{Equivalent linear equations}$$

So, the absolute value equation has two solutions: $x = -10$ and $x = 10$.

b. This equation is equivalent to the two linear equations

$$x = -(0) = 0 \quad \text{and} \quad x = 0. \quad \text{Equivalent linear equations}$$

Because both equations are the same, you can conclude that the absolute value equation has only one solution: $x = 0$.

c. This absolute value equation has *no solution* because it is not possible for the absolute value of a real number to be negative.

✓ **CHECKPOINT** *Now try Exercise 13.*

Study Tip

The strategy for solving an absolute value equation is to *rewrite* the equation in *equivalent forms* that can be solved by previously learned methods. This is a common strategy in mathematics. That is, when you encounter a new type of problem, you try to rewrite the problem so that it can be solved by techniques you already know.

EXAMPLE 2   **Solving an Absolute Value Equation**

Solve $|3x + 4| = 10$.

Solution

$\|3x + 4\| = 10$	Write original equation.
$3x + 4 = -10$ or $3x + 4 = 10$	Equivalent equations
$3x + 4 - 4 = -10 - 4$ $3x + 4 - 4 = 10 - 4$	Subtract 4 from each side.
$3x = -14$ $3x = 6$	Combine like terms.
$x = -\dfrac{14}{3}$ $x = 2$	Divide each side by 3.

Check

$$|3x + 4| = 10 \qquad\qquad |3x + 4| = 10$$
$$\left|3\left(-\tfrac{14}{3}\right) + 4\right| \overset{?}{=} 10 \qquad |3(2) + 4| \overset{?}{=} 10$$
$$|-14 + 4| \overset{?}{=} 10 \qquad\qquad |6 + 4| \overset{?}{=} 10$$
$$|-10| = 10 \checkmark \qquad\qquad |10| = 10 \checkmark$$

✔ **CHECKPOINT** *Now try Exercise 21.*

When solving absolute value equations, remember that it is possible that they have no solution. For instance, the equation $|3x + 4| = -10$ has no solution because the absolute value of a real number cannot be negative. Do not make the mistake of trying to solve such an equation by writing the "equivalent" linear equations as $3x + 4 = -10$ and $3x + 4 = 10$. These equations have solutions, but they are both extraneous.

The equation in the next example is not given in the **standard form**

$$|ax + b| = c, \quad c \geq 0.$$

Notice that the first step in solving such an equation is to write it in standard form.

EXAMPLE 3   **An Absolute Value Equation in Nonstandard Form**

Solve $|2x - 1| + 3 = 8$.

Solution

$\|2x - 1\| + 3 = 8$	Write original equation.
$\|2x - 1\| = 5$	Write in standard form.
$2x - 1 = -5$ or $2x - 1 = 5$	Equivalent equations
$2x = -4$ $2x = 6$	Add 1 to each side.
$x = -2$ $x = 3$	Divide each side by 2.

The solutions are $x = -2$ and $x = 3$. Check these in the original equation.

✔ **CHECKPOINT** *Now try Exercise 27.*

If two algebraic expressions are equal in absolute value, they must either be equal to each other or be the *opposites* of each other. So, you can solve equations of the form $|ax + b| = |cx + d|$ by forming the two linear equations

Expressions equal

Expressions opposite

$$ax + b = cx + d \quad \text{and} \quad ax + b = -(cx + d).$$

EXAMPLE 4   Solving an Equation Involving Two Absolute Values

Solve $|3x - 4| = |7x - 16|$.

Solution

$|3x - 4| = |7x - 16|$	Write original equation.

$3x - 4 = 7x - 16 \quad \text{or} \quad 3x - 4 = -(7x - 16)$ Equivalent equations

$-4x - 4 = -16 \qquad\qquad 3x - 4 = -7x + 16$

$-4x = -12 \qquad\qquad\quad 10x = 20$

$x = 3 \qquad\qquad\qquad\quad x = 2$ Solutions

The solutions are $x = 3$ and $x = 2$. Check these in the original equation.

 **CHECKPOINT** *Now try Exercise 35.*

Study Tip

When solving an equation of the form

$$|ax + b| = |cx + d|$$

it is possible that one of the resulting equations will not have a solution. Note this occurrence in Example 5.

EXAMPLE 5   Solving an Equation Involving Two Absolute Values

Solve $|x + 5| = |x + 11|$.

Solution

By equating the expression $(x + 5)$ to the opposite of $(x + 11)$, you obtain

$x + 5 = -(x + 11)$	Equivalent equation
$x + 5 = -x - 11$	Distributive Property
$2x + 5 = -11$	Add x to each side.
$2x = -16$	Subtract 5 from each side.
$x = -8.$	Divide each side by 2.

However, by setting the two expressions equal to each other, you obtain

$x + 5 = x + 11$	Equivalent equation
$x = x + 6$	Subtract 5 from each side.
$0 = 6$	Subtract x from each side.

which is a false statement. So, the original equation has only one solution: $x = -8$. Check this solution in the original equation.

 **CHECKPOINT** *Now try Exercise 37.*

2 ▶ Solve inequalities involving absolute value.

Solving Inequalities Involving Absolute Value

To see how to solve inequalities involving absolute value, consider the following comparisons.

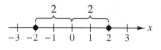

$|x| = 2$

$x = -2$ and $x = 2$

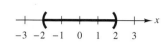

$|x| < 2$

$-2 < x < 2$

$|x| > 2$

$x < -2$ or $x > 2$

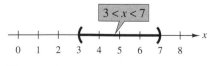

These comparisons suggest the following rules for solving inequalities involving absolute value.

Solving an Absolute Value Inequality

Let x be a variable or an algebraic expression and let a be a real number such that $a > 0$.

1. The solutions of $|x| < a$ are all values of x that lie between $-a$ and a. That is,

$|x| < a$ if and only if $-a < x < a$.

2. The solutions of $|x| > a$ are all values of x that are less than $-a$ or greater than a. That is,

$|x| > a$ if and only if $x < -a$ or $x > a$.

These rules are also valid if $<$ is replaced by $\leq$ and $>$ is replaced by $\geq$.

EXAMPLE 6   Solving an Absolute Value Inequality

Solve $|x - 5| < 2$.

Solution

$$|x - 5| < 2 \qquad\qquad \text{Write original inequality.}$$

$$-2 < x - 5 < 2 \qquad\qquad \text{Equivalent double inequality}$$

$$-2 + 5 < x - 5 + 5 < 2 + 5 \qquad\qquad \text{Add 5 to all three parts.}$$

$$3 < x < 7 \qquad\qquad \text{Combine like terms.}$$

The solution set consists of all real numbers that are greater than 3 and less than 7. The solution set in interval notation is $(3, 7)$ and in set notation is $\{x \mid 3 < x < 7\}$. The graph of this solution set is shown in Figure 2.16.

Figure 2.16

✓ CHECKPOINT *Now try Exercise 53.*

To verify the solution of an absolute value inequality, you must check values in the solution set and outside of the solution set. In Example 6 you can check that $x = 4$ is in the solution set and that $x = 2$ and $x = 8$ are not in the solution set.

<table>
<tr><td>

Study Tip

In Example 6, note that an absolute value inequality of the form $|x| < a$ (or $|x| \leq a$) can be solved with a double inequality. An inequality of the form $|x| > a$ (or $|x| \geq a$) cannot be solved with a double inequality. Instead, you must solve two separate inequalities, as demonstrated in Example 7.

</td></tr>
</table>

EXAMPLE 7   Solving an Absolute Value Inequality

Solve $|3x - 4| \geq 5$.

Solution

$|3x - 4| \geq 5$			Write original inequality.
$3x - 4 \leq -5$	or	$3x - 4 \geq 5$	Equivalent inequalities
$3x - 4 + 4 \leq -5 + 4$		$3x - 4 + 4 \geq 5 + 4$	Add 4 to all parts.
$3x \leq -1$		$3x \geq 9$	Combine like terms.
$\dfrac{3x}{3} \leq \dfrac{-1}{3}$		$\dfrac{3x}{3} \geq \dfrac{9}{3}$	Divide each side by 3.
$x \leq -\dfrac{1}{3}$		$x \geq 3$	Simplify.

The solution set consists of all real numbers that are less than or equal to $-\frac{1}{3}$ or greater than or equal to 3. The solution set in interval notation is $\left(-\infty, -\frac{1}{3}\right] \cup [3, \infty)$ and in set notation is $\left\{x \mid x \leq -\frac{1}{3} \text{ or } x \geq 3\right\}$. The graph is shown in Figure 2.17.

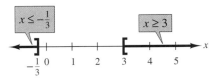

Figure 2.17

✓ **CHECKPOINT** *Now try Exercise 59.*

EXAMPLE 8   Solving an Absolute Value Inequality

$\left|2 - \dfrac{x}{3}\right| \leq 0.01$	Original inequality
$-0.01 \leq 2 - \dfrac{x}{3} \leq 0.01$	Equivalent double inequality
$-2.01 \leq -\dfrac{x}{3} \leq -1.99$	Subtract 2 from all three parts.
$-2.01(-3) \geq -\dfrac{x}{3}(-3) \geq -1.99(-3)$	Multiply all three parts by -3 and reverse both inequality symbols.
$6.03 \geq x \geq 5.97$	Simplify.
$5.97 \leq x \leq 6.03$	Solution set in standard form

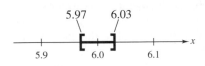

Figure 2.18

The solution set consists of all real numbers that are greater than or equal to 5.97 *and* less than or equal to 6.03. The solution set in interval notation is $[5.97, 6.03]$ and in set notation is $\{x \mid 5.97 \leq x \leq 6.03\}$. The graph is shown in Figure 2.18.

✓ **CHECKPOINT** *Now try Exercise 65.*

Technology: Tip

Most graphing calculators can graph absolute value inequalities. Consult your user's guide for specific instructions. The graph below shows the solution of the inequality in Example 6 on page 110, which is $|x - 5| < 2$. Notice that the graph representing the solution interval lies above the *x*-axis.

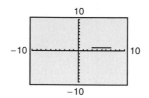

EXAMPLE 9 **Production**

The estimated daily production at an oil refinery is given by the absolute value inequality $|x - 200{,}000| \leq 25{,}000$, where *x* is measured in barrels of oil. Solve the inequality to determine the maximum and minimum production levels.

Solution

$$|x - 200{,}000| \leq 25{,}000 \qquad \text{Write original inequality.}$$

$$-25{,}000 \leq x - 200{,}000 \leq 25{,}000 \qquad \text{Equivalent double inequality}$$

$$175{,}000 \leq x \leq 225{,}000 \qquad \text{Add 200,000 to all three parts.}$$

So, the oil refinery produces a maximum of 225,000 barrels of oil and a minimum of 175,000 barrels of oil per day.

 CHECKPOINT *Now try Exercise 89.*

EXAMPLE 10 **Creating a Model**

To test the accuracy of a rattlesnake's "pit-organ sensory system," a biologist blindfolds a rattlesnake and presents the snake with a warm "target." Of 36 strikes, the snake is on target 17 times. Let *A* represent the number of degrees by which the snake is off target. Then $A = 0$ represents a strike that is aimed directly at the target. Positive values of *A* represent strikes to the right of the target, and negative values of *A* represent strikes to the left of the target. Use the diagram shown in Figure 2.19 to write an absolute value inequality that describes the interval in which the 36 strikes occurred.

Solution

From the diagram, you can see that in the 36 strikes, the snake is never off by more than 15 degrees in either direction. As a compound inequality, this can be represented by $-15 \leq A \leq 15$. As an absolute value inequality, this interval can be represented by $|A| \leq 15$.

 CHECKPOINT *Now try Exercise 91.*

Figure 2.19

Concept Check

1. In your own words, explain how to solve an absolute value equation. Illustrate your explanation with an example.

2. In the equation $|x| = b$, b is a positive real number. How many solutions does this equation have? Explain.

3. In the inequality $|x| < b$, b is a positive real number. Describe the solution set of this inequality.

4. Can you use a double inequality to solve the inequality $|5x - 4| \geq 3$? Explain your reasoning.

2.5 EXERCISES

Go to pages 116–117 to record your assignments.

Developing Skills

In Exercises 1–4, determine whether the value is a solution of the equation.

Equation	Value
1. $\|4x + 5\| = 10$	$x = -3$
2. $\|2x - 16\| = 10$	$x = 3$
3. $\|6 - 2w\| = 2$	$w = 4$
4. $\left\|\frac{1}{2}t + 4\right\| = 8$	$t = 6$

In Exercises 5–8, transform the absolute value equation into two linear equations.

5. $|x - 10| = 17$

6. $|7 - 2t| = 5$

7. $|4x + 1| = \frac{1}{2}$

8. $|22k + 6| = 9$

In Exercises 9–12, write the absolute value equation in standard form.

9. $|3x| + 7 = 8$

10. $|5x| - 6 = -3$

11. $3|2x| - 1 = 5$

12. $\frac{1}{4}|3x + 1| = 4$

In Exercises 13–40, solve the equation. (Some equations have no solution.) **See Examples 1–5.**

13. $|x| = 4$

14. $|x| = 3$

15. $|t| = -45$

16. $|s| = 16$

17. $|h| = 0$

18. $|x| = -82$

19. $|5x| = 15$

20. $\left|\frac{1}{3}x\right| = 2$

21. $|x + 1| = 5$

22. $|x + 5| = 7$

23. $\left|\frac{2s + 3}{5}\right| = 5$

24. $\left|\frac{7a + 6}{4}\right| = 2$

25. $|4 - 3x| = 0$

26. $|3x - 2| = -5$

27. $|5x - 3| + 8 = 22$

28. $|5 - 2x| + 10 = 6$

29. $\left|\frac{x - 2}{3}\right| + 6 = 6$

30. $\left|\frac{x - 2}{5}\right| + 4 = 4$

31. $-2|7 - 4x| = -16$

32. $4|5x + 1| = 24$

33. $3|2x - 5| + 4 = 7$

34. $2|4 - 3x| - 6 = -2$

35. $|x + 8| = |2x + 1|$

36. $|10 - 3x| = |x + 7|$

37. $|3x + 1| = |3x - 3|$

38. $|2x + 7| = |2x + 9|$

39. $|4x - 10| = 2|2x + 3|$

40. $3|2 - 3x| = |9x + 21|$

Think About It   In Exercises 41 and 42, write an absolute value equation that represents the verbal statement.

41. The distance between x and 4 is 9.

42. The distance between -3 and t is 5.

In Exercises 43–46, determine whether the x-value is a solution of the inequality.

Inequality	Value
43. $\|x\| < 3$	$x = 2$
44. $\|x\| \leq 5$	$x = -7$
45. $\|x - 7\| \geq 3$	$x = 9$
46. $\|x - 3\| > 5$	$x = 16$

In Exercises 47–50, transform the absolute value inequality into a double inequality or two separate inequalities.

47. $|y + 5| < 3$

48. $|6x + 7| \leq 5$

49. $|7 - 2h| \geq 9$

50. $|8 - x| > 25$

In Exercises 51–70, solve the inequality. *See Examples 6–8.*

51. $|y| < 4$ **52.** $|x| < 6$

✓ **53.** $|x| \geq 6$ **54.** $|y| \geq 4$

55. $|2x| < 14$ **56.** $|4z| \leq 9$

57. $\left|\dfrac{y}{3}\right| \leq \dfrac{1}{3}$ **58.** $\left|\dfrac{t}{5}\right| < \dfrac{3}{5}$

✓ **59.** $|x + 6| > 10$ **60.** $|y - 2| \leq 4$

61. $|2x - 1| \leq 7$ **62.** $|6t + 15| \geq 30$

63. $|3x + 10| < -1$ **64.** $|4x - 5| > -3$

✓ **65.** $\dfrac{|y - 16|}{4} < 30$ **66.** $\dfrac{|a + 6|}{2} \geq 16$

67. $|0.2x - 3| < 4$ **68.** $|1.5t - 8| \leq 16$

69. $\left|\dfrac{3x - 2}{4}\right| + 5 \geq 5$ **70.** $\left|\dfrac{2x - 4}{5}\right| - 9 \leq 3$

In Exercises 71–76, use a graphing calculator to solve the inequality.

71. $|3x + 2| < 4$ **72.** $|2x - 1| \leq 3$

73. $|2x + 3| > 9$ **74.** $|7r - 3| > 11$

75. $|x - 5| + 3 \leq 5$ **76.** $|a + 1| - 4 < 0$

In Exercises 77–80, match the inequality with its graph. [The graphs are labeled (a), (b), (c), and (d).]

(a)

(b)

(c)

(d)

77. $|x - 4| \leq 4$ **78.** $|x - 4| < 1$

79. $\frac{1}{2}|x - 4| > 4$ **80.** $|2(x - 4)| \geq 4$

In Exercises 81–84, write an absolute value inequality that represents the interval.

81.

82.

83.

84.

In Exercises 85–88, write an absolute value inequality that represents the verbal statement.

85. The set of all real numbers x whose distance from 0 is less than 3.

86. The set of all real numbers x whose distance from 0 is more than 2.

87. The set of all real numbers x for which the distance from 0 to 3 less than twice x is more than 5.

88. The set of all real numbers x for which the distance from 0 to 5 more than half of x is less than 13.

The symbol 🖩 indicates an exercise in which you are instructed to use a graphing calculator.

Solving Problems

89. *Speed Skating* In the 2006 Winter Olympics, each skater in the 500-meter short track speed skating final had a time that satisfied the inequality $|t - 42.238| \leq 0.412$, where t is the time in seconds. Sketch the graph of the solution of the inequality. What are the fastest and slowest times?

90. *Time Study* A time study was conducted to determine the length of time required to perform a task in a manufacturing process. The times required by approximately two-thirds of the workers in the study satisfied the inequality

$$\left| \frac{t - 15.6}{1.9} \right| \leq 1$$

where t is time in minutes. Sketch the graph of the solution of the inequality. What are the maximum and minimum times?

91. *Accuracy of Measurements* In woodshop class, you must cut several pieces of wood to within $\frac{3}{16}$ inch of the teacher's specifications. Let $(s - x)$ represent the difference between the specification s and the measured length x of a cut piece.

(a) Write an absolute value inequality that describes the values of x that are within specifications.

(b) The length of one piece of wood is specified to be $s = 5\frac{1}{8}$ inches. Describe the acceptable lengths for this piece.

92. *Body Temperature* Physicians generally consider an adult's body temperature x to be normal if it is within 1°F of the temperature 98.6°F.

(a) Write an absolute value inequality that describes the values of x that are considered normal.

(b) Describe the range of body temperatures that are considered normal.

Explaining Concepts

93. The graph of the inequality $|x - 3| < 2$ can be described as *all real numbers that are within two units of 3*. Give a similar description of $|x - 4| < 1$.

94. Write an absolute value inequality to represent all the real numbers that are more than $|a|$ units from b. Then write an example showing the solution of the inequality for sample values of a and b.

95. Complete $|2x - 6| \leq$ ___ so that the solution is $0 \leq x \leq 6$.

96. ✎ Describe and correct the error. Explain how you can recognize that the solution is wrong without solving the inequality.

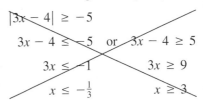

$$|3x - 4| \geq -5$$
$$3x - 4 \leq -5 \quad \text{or} \quad 3x - 4 \geq 5$$
$$3x \leq -1 \qquad\qquad 3x \geq 9$$
$$x \leq -\tfrac{1}{3} \qquad\qquad x \geq 3$$

Cumulative Review

In Exercises 97 and 98, translate the verbal phrase into an algebraic expression.

97. Four times the sum of a number n and 3

98. Eight less than two times a number n

In Exercises 99 and 100, find the missing quantities.

	Cost	Selling Price	Markup	Markup Rate
99.	$80.00			40%
100.	$74.00			62%

In Exercises 101–104, solve the inequality.

101. $x - 7 > 13$

102. $x + 7 \leq 13$

103. $4x + 11 \geq 27$

104. $-4 < x + 2 < 12$

What Did You Learn?

Use these two pages to help prepare for a test on this chapter. Check off the key terms and key concepts you know. You can also use this section to record your assignments.

Plan for Test Success

Date of test: ☐ / /

Study dates and times: ☐ / / at ☐ : A.M./P.M.

☐ / / at ☐ : A.M./P.M.

Things to review:

☐ Key Terms, *p. 116*
☐ Key Concepts, *pp. 116–117*
☐ Your class notes
☐ Your assignments

☐ Study Tips, *pp. 58, 60, 62, 63, 64,*
   69, 70, 72, 73, 74, 81, 82, 83, 84,
   86, 95, 97, 98, 107, 109, 111
☐ Technology Tips, *pp. 70, 87, 98, 112*
☐ Mid-Chapter Quiz, *p. 93*

☐ Review Exercises, *pp. 118–122*
☐ Chapter Test, *p. 123*
☐ Video Explanations Online
☐ Tutorial Online

Key Terms

☐ equation, *p. 58*
☐ solutions (equation), *p. 58*
☐ solution set, *p. 58*
☐ identity, *p. 58*
☐ conditional equation, *p. 58*
☐ equivalent equations, *p. 59*
☐ linear equation, *p. 60*
☐ first-degree equation, *p. 60*
☐ mathematical modeling, *p. 69*
☐ percent, *p. 70*
☐ percent equation, *p. 70*
☐ ratio, *p. 72*
☐ unit price, *p. 72*
☐ proportion, *p. 73*

☐ cross-multiplying, *p. 73*
☐ markup, *p. 80*
☐ markup rate, *p. 80*
☐ discount, *p. 81*
☐ discount rate, *p. 81*
☐ mixture problems, *p. 82*
☐ rate problems, *p. 83*
☐ rate of work, *p. 84*
☐ algebraic inequalities, *p. 94*
☐ solutions (inequality), *p. 94*
☐ solution set of an inequality, *p. 94*
☐ graph of an inequality, *p. 94*
☐ bounded intervals, *p. 94*
☐ endpoints of an interval, *p. 94*

☐ length of an interval, *p. 94*
☐ unbounded (infinite) intervals, *p. 95*
☐ positive infinity, *p. 95*
☐ negative infinity, *p. 95*
☐ equivalent inequalities, *p. 96*
☐ linear inequality, *p. 97*
☐ compound inequality, *p. 99*
☐ double inequality, *p. 99*
☐ intersection, *p. 100*
☐ union, *p. 100*
☐ absolute value equation, *p. 107*
☐ standard form of an absolute value
   equation, *p. 108*

Key Concepts

2.1 Linear Equations

Assignment: _____ Due date: _____

☐ **Form equivalent equations.**

An equation can be transformed into an equivalent equation using one or more of the following procedures:

1. Simplify either side.
2. Apply the Addition Property of Equality.
3. Apply the Multiplication Property of Equality.
4. Interchange sides.

☐ **Solve linear equations in standard form.**

To solve a linear equation in the standard form $ax + b = 0$, where a and b are real numbers and $a \neq 0$, isolate x by rewriting the standard equation in the form

$x = $ [a number] .

☐ **Solve linear equations in nonstandard form.**

To solve a linear equation in nonstandard form, rewrite the equation in the form

$x = $ [a number] .

2.2 Linear Equations and Problem Solving

Assignment: _____ Due date: _____

☐ **Solve percent problems.**

The primary use of percents is to compare two numbers: 3 is 50% of 6. (3 is compared to 6.)

Many percent problems can be solved using the *percent equation*:

| Compared number | = | Percent (decimal form) | · | Base number |

☐ **Solve proportions by cross-multiplying.**

The proportion $\frac{a}{b} = \frac{c}{d}$ is equivalent to $ad = bc$.

☐ **Use a strategy to model and solve word problems.**

1. Ask yourself what you need to know to solve the problem. Then *write a verbal model* that includes arithmetic operations to describe the problem.
2. *Assign labels* to each part of the verbal model.
3. Use the labels to *write an algebraic model*.
4. *Solve* the resulting algebraic equation.
5. *Answer* the original question and check that your answer satisfies the original problem as stated.

2.3 Business and Scientific Problems

Assignment: _____ Due date: _____

☐ **Solve business-related problems.**

| Selling price | = | Cost | + | Markup |

| Markup | = | Markup rate | · | Cost |

| Selling price | = | List price | − | Discount |

☐ **Solve mixture problems.**

| First rate | · | Amount | + | Second rate | · | Amount |

= Final rate · Final amount

☐ **Use formulas for area, perimeter, and volume.**

For area and perimeter formulas for squares, rectangles, circles, and triangles, see page 85.

For volume formulas for cubes, rectangular solids, circular cylinders, and spheres, see page 85.

☐ **Use miscellaneous formulas:**

Temperature: $F = \frac{9}{5}C + 32$

Simple Interest: $I = Prt$

Distance: $d = rt$

2.4 Linear Inequalities

Assignment: _____ Due date: _____

☐ **Graph inequalities using bounded and unbounded intervals on the real number line.**

See examples of bounded and unbounded intervals on pages 94 and 95.

☐ **Use properties of inequalities.**

1. Addition and subtraction:

   If $a < b$, then $a + c < b + c$.

   If $a < b$, then $a - c < b - c$.

2. Multiplication and division (c is positive):

   If $a < b$, then $ac < bc$. If $a < b$, then $\frac{a}{c} < \frac{b}{c}$.

3. Multiplication and division (c is negative):

   If $a < b$, then $ac > bc$.

   If $a < b$, then $\frac{a}{c} > \frac{b}{c}$.

4. Transitive property:

   If $a < b$ and $b < c$, then $a < c$.

2.5 Absolute Value Equations and Inequalities

Assignment: _____ Due date: _____

☐ **Solve an absolute value equation.**

$|x| = a \implies x = -a$ or $x = a$.

☐ **Solve an absolute value inequality.**

1. $|x| < a$ if and only if $-a < x < a$.
2. $|x| > a$ if and only if $x < -a$ or $x > a$.

Review Exercises

2.1 Linear Equations

1 ▶ Check solutions of equations.

In Exercises 1–4, determine whether each value of the variable is a solution of the equation.

Equation	Values

1. $45 - 7x = 3$ (a) $x = 3$ (b) $x = 6$

2. $3(3 + 4x) = 15x$ (a) $x = 3$ (b) $x = -2$

3. $\dfrac{x}{7} + \dfrac{x}{5} = 12$ (a) $x = 28$ (b) $x = 35$

4. $\dfrac{t + 2}{6} = \dfrac{7}{2}$ (a) $t = -12$ (b) $t = 19$

2 ▶ Solve linear equations in standard form.

In Exercises 5–8, solve the equation and check the result.

5. $3x + 21 = 0$ **6.** $-4x + 64 = 0$

7. $5x - 120 = 0$ **8.** $7x - 49 = 0$

3 ▶ Solve linear equations in nonstandard form.

In Exercises 9–30, solve the equation and check the result. (Some equations have no solution.)

9. $x + 4 = 9$

10. $x - 7 = 3$

11. $-3x = 36$

12. $11x = 44$

13. $-\frac{1}{8}x = 3$

14. $\frac{1}{10}x = 5$

15. $5x + 4 = 19$

16. $3 - 2x = 9$

17. $17 - 7x = 3$

18. $3 + 6x = 51$

19. $7x - 5 = 3x + 11$

20. $9 - 2x = 4x - 7$

21. $3(2y - 1) = 9 + 3y$

22. $-2(x + 4) = 2x - 7$

23. $4y - 6(y - 5) = 2$

24. $7x + 2(7 - x) = 8$

25. $4(3x - 5) = 6(2x + 3)$

26. $8(x - 2) = 4(2x + 3)$

27. $\frac{4}{5}x - \frac{1}{10} = \frac{3}{2}$

28. $\frac{1}{4}s + \frac{3}{8} = \frac{5}{2}$

29. $1.4t + 2.1 = 0.9t$

30. $2.5x - 6.2 = 3.7x - 5.8$

31. *Number Problem* Find two consecutive positive integers whose sum is 115.

32. *Number Problem* Find two consecutive even integers whose sum is 74.

2.2 Linear Equations and Problem Solving

1 ▶ Use mathematical modeling to write algebraic equations representing real-life situations.

Mathematical Modeling **In Exercises 33 and 34, construct a verbal model and write an algebraic equation that represents the problem. Solve the equation.**

33. An internship pays $320 per week plus an additional $75 for a training session. The total pay for the internship and training is $2635. How many weeks long is the internship?

34. You have a job as a salesperson for which you are paid $6 per hour plus $1.25 per sale made. Find the number of sales made in an eight-hour day if your earnings for the day are $88.

2 ▶ Solve percent problems using the percent equation.

In Exercises 35–38, complete the table showing the equivalent forms of various percents.

	Percent	Parts out of 100	Decimal	Fraction
35.	68%			
36.			0.35	
37.		60		
38.				$\frac{1}{6}$

In Exercises 39–44, solve using a percent equation.

39. What is 130% of 50?

40. What is 0.8% of 2450?

41. 645 is $21\frac{1}{2}$% of what number?

42. 498 is 83% of what number?

43. 250 is what percent of 200?

44. 162.5 is what percent of 6500?

45. *Real Estate Commission* A real estate agency receives a commission of $9000 for the sale of a $150,000 house. What percent commission is this?

46. *Pension Fund* Your employer withholds $216 of your gross income each month for your retirement. Determine the percent of your total monthly gross income of $3200 that is withheld for retirement.

47. *Quality Control* A quality control engineer reported that 1.6% of a sample of parts were defective. The engineer found six defective parts. How large was the sample?

48. *Sales Tax* The state sales tax on an item you purchase is $6\frac{1}{4}$%. If the item costs $44, how much sales tax will you pay?

3 ▶ Use ratios to compare unit prices for products.

Consumer Awareness **In Exercises 49 and 50, use unit prices to determine the better buy.**

49. (a) A 39-ounce can of coffee for $9.79

(b) An 8-ounce can of coffee for $1.79

50. (a) A 3.5-pound bag of sugar for $3.08

(b) A 2.2-pound bag of sugar for $1.87

51. *Income Tax* You have $9.90 of state tax withheld from your paycheck per week when your gross pay is $396. Find the ratio of tax to gross pay.

52. *Price-Earnings Ratio* The ratio of the price of a stock to its earnings is called the price-earnings ratio. Find the price-earnings ratio of a stock that sells for $46.75 per share and earns $5.50 per share.

4 ▶ Solve proportions.

In Exercises 53–56, solve the proportion.

53. $\frac{7}{8} = \frac{y}{4}$

54. $\frac{x}{16} = \frac{5}{12}$

55. $\frac{b}{6} = \frac{5+b}{15}$

56. $\frac{x+1}{3} = \frac{x-1}{2}$

▲ *Geometry* **In Exercises 57 and 58, the triangles are similar. Solve for the length x by using the fact that corresponding sides of similar triangles are proportional.**

57.

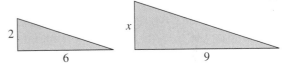

58.

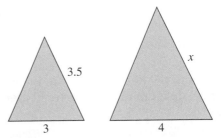

59. *Property Tax* The tax on a property with an assessed value of $105,000 is $1680. Find the tax on a property with an assessed value of $125,000.

60. *Masonry* You use $1\frac{1}{2}$ bags of mortar mix to lay 42 bricks. How many bags will you use to lay 336 bricks?

61. ▲ *Geometry*   You want to measure the height of a flagpole. To do this, you measure the flagpole's shadow and find that it is 30 feet long. You also measure the height of a five-foot lamp post and find its shadow to be 3 feet long (see figure). Find the height *h* of the flagpole.

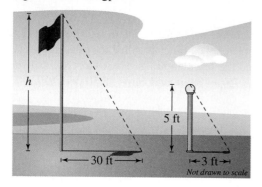

62. ▲ *Geometry*   You want to measure the height of a silo. To do this, you measure the silo's shadow and find that it is 20 feet long. You are 6 feet tall and your shadow is $1\frac{1}{2}$ feet long. Find the height of the silo.

2.3 Business and Scientific Problems

1 ▶ Use mathematical models to solve business-related problems.

In Exercises 63 and 64, find the missing quantities. (Assume that the markup rate is a percent based on the cost.)

	Cost	Selling Price	Markup	Markup Rate
63.	$99.95	$149.93		
64.	$23.50	$31.33		

In Exercises 65 and 66, find the missing quantities. (Assume that the discount rate is a percent based on the list price.)

	List Price	Sale Price	Discount	Discount Rate
65.	$71.95	$53.96		
66.	$559.95	$279.98		

67. *Amount Financed*   You buy a motorcycle for $2795 plus 6% sales tax. Find the amount of sales tax and the total bill. You make a down payment of $800. Find the amount financed.

68. *Labor*   An automobile repair bill of $366.44 lists $208.80 for parts, a 5% tax on parts, and the rest for labor. The labor rate is $32 per hour. How many hours did it take to repair the automobile?

2 ▶ Use mathematical models to solve mixture problems.

69. *Mixture Problem*   Determine the number of liters of a 30% saline solution and the number of liters of a 60% saline solution that are required to obtain 10 liters of a 50% saline solution.

70. *Mixture Problem*   Determine the number of gallons of a 20% bleach solution and the number of gallons of a 50% bleach solution that are required to obtain 8 gallons of a 40% bleach solution.

3 ▶ Use mathematical models to solve classic rate problems.

71. *Distance*   Determine the distance an Air Force jet can travel in $2\frac{1}{3}$ hours when its average speed is 1500 miles per hour.

72. *Travel Time*   Determine the time for a migrating bird to fly 185 kilometers at an average speed of 66 kilometers per hour.

73. *Speed*   A truck driver traveled at an average speed of 48 miles per hour on a 100-mile trip to pick up a load of freight. On the return trip with the truck fully loaded, the average speed was 40 miles per hour. Find the average speed for the round trip.

74. *Speed*   For 2 hours of a 400-mile trip, your average speed is 40 miles per hour. Determine the average speed that must be maintained for the remainder of the trip if you want the average speed for the entire trip to be 50 miles per hour.

75. *Work-Rate Problem*   Find the time for two people working together to complete a task if it takes them 4.5 hours and 6 hours working individually.

76. *Work-Rate Problem*   Find the time for two people working together to complete half a task if it takes them 8 hours and 10 hours to complete the entire task working individually.

4 ▶ Use formulas to solve application problems.

77. *Simple Interest* Find the interest on a $1000 corporate bond that matures in 4 years and has an 8.5% interest rate.

78. *Simple Interest* Find the annual interest rate on a certificate of deposit that pays $37.50 per year in interest on a principal of $500.

79. *Simple Interest* The interest on a savings account is 9.5%. Find the principal required to earn $20,000 in interest in 4 years.

80. *Simple Interest* A corporation borrows 3.25 million dollars for 2 years at an annual interest rate of 12% to modernize one of its manufacturing facilities. What is the total principal and interest that must be repaid?

81. *Simple Interest* An inheritance of $50,000 is divided into two investments earning 8.5% and 10% simple interest. (The 10% investment has a greater risk.) Your objective is to obtain a total annual interest income of $4700 from the investments. What is the smallest amount you can invest at 10% in order to meet your objective?

82. *Simple Interest* You invest $1000 in a certificate of deposit that has an annual interest rate of 7%. After 6 months, the interest is computed and added to the principal. During the second 6 months, the interest is computed using the original investment plus the interest earned during the first 6 months. What is the total interest earned during the first year of the investment?

83. ▲ *Geometry* The perimeter of the rectangle shown in the figure is 64 feet. Find its dimensions.

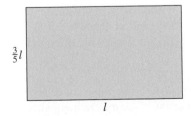

84. ▲ *Geometry* The area of the triangle shown in the figure is 60 square meters. Solve for x.

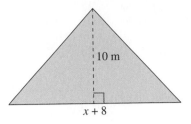

Figure for 84

85. *Meteorology* The average daily temperature in Boise, Idaho is 10.5°C. What is Boise's average daily temperature in degrees Fahrenheit? (Source: U.S. National Oceanic and Atmospheric Administration)

86. *Meteorology* The temperature reached 101°F on September 10, 2007 at Raleigh-Durham International Airport. What was the temperature in degrees Celsius? (Source: U.S. National Oceanic and Atmospheric Administration)

2.4 Linear Inequalities

1 ▶ Sketch the graphs of inequalities.

In Exercises 87–90, sketch the graph of the inequality.

87. $-3 \le x < 1$

88. $-2.5 \le x < 4$

89. $-7 < x$

90. $x \ge -2$

3 ▶ Solve linear inequalities.

In Exercises 91–102, solve the inequality and sketch the solution on the real number line.

91. $x - 5 \le -1$

92. $x + 8 > 5$

93. $-6x < -24$

94. $-16x \ge -48$

95. $5x + 3 > 18$

96. $3x - 11 \le 7$

97. $8x + 1 \ge 10x - 11$

98. $12 - 3x < 4x - 2$

99. $\frac{1}{3} - \frac{1}{2}y < 12$

100. $\dfrac{x}{4} - 2 < \dfrac{3x}{8} + 5$

101. $-4(3 - 2x) \le 3(2x - 6)$

102. $3(2 - y) \ge 2(1 + y)$

4 ▶ Solve compound inequalities.

In Exercises 103–108, solve the compound inequality and sketch the solution on the real number line.

103. $-6 \le 2x + 8 < 4$

104. $-13 \le 3 - 4x < 13$

105. $5 > \dfrac{x + 1}{-3} > 0$

106. $12 \ge \dfrac{x - 3}{2} > 1$

107. $5x - 4 < 6$ and $3x + 1 > -8$

108. $6 - 2x \le 1$ or $10 - 4x > -6$

5 ▶ Solve application problems involving inequalities.

109. *Earnings*   A country club waiter earns $6 per hour plus tips of at least 15% of the restaurant tab from each table served. What total amount of restaurant tabs assures the waiter of making at least $150 in a five-hour shift?

110. *Long-Distance Charges*   The cost of an international long-distance telephone call is $0.99 for the first minute and $0.49 for each additional minute. Your prepaid calling card has $22.50 left to pay for a call. How many minutes can you talk?

2.5 Absolute Value Equations and Inequalities

1 ▶ Solve absolute value equations.

In Exercises 111–118, solve the equation.

111. $|x| = 6$

112. $|x| = -4$

113. $|4 - 3x| = 8$

114. $|2x + 3| = 7$

115. $|5x + 4| - 10 = -6$

116. $|x - 2| - 2 = 4$

117. $|3x - 4| = |x + 2|$

118. $|5x + 6| = |2x - 1|$

2 ▶ Solve inequalities involving absolute value.

In Exercises 119–126, solve the inequality.

119. $|x - 4| > 3$

120. $|t + 3| > 2$

121. $|3x| < 12$

122. $\left|\dfrac{t}{3}\right| < 1$

123. $|2x - 7| < 15$

124. $|4x - 1| > 7$

125. $|b + 2| - 6 > 1$

126. $|2y - 1| + 4 < -1$

▦ In Exercises 127 and 128, use a graphing calculator to solve the inequality.

127. $|4(x - 3)| \ge 8$

128. $|5(1 - x)| \le 25$

In Exercises 129 and 130, write an absolute value inequality that represents the interval.

129.

130.

131. *Temperature*   The storage temperature of a computer must satisfy the inequality

$$|t - 78.3| \le 38.3$$

where t is the temperature in degrees Fahrenheit. Sketch the graph of the solution of the inequality. What are the maximum and minimum temperatures?

132. *Temperature*   The operating temperature of a computer must satisfy the inequality

$$|t - 77| \le 27$$

where t is the temperature in degrees Fahrenheit. Sketch the graph of the solution of the inequality. What are the maximum and minimum temperatures?

Chapter Test

$2.49   12 oz

$2.99   15 oz

Figure for 8

Take this test as you would take a test in class. After you are done, check your work against the answers in the back of the book.

In Exercises 1–4, solve the equation.

1. $6x - 5 = 19$

2. $5x - 6 = 7x - 12$

3. $15 - 7(1 - x) = 3(x + 8)$

4. $\dfrac{2x}{3} = \dfrac{x}{2} + 4$

5. What is 125% of 3200?

6. 32 is what percent of 8000?

7. A store is offering a 20% discount on all items in its inventory. Find the list price on a tractor that has a sale price of $8900.

8. Which of the packages at the left is a better buy? Explain your reasoning.

9. The tax on a property with an assessed value of $110,000 is $1650. What is the tax on a property with an assessed value of $145,000?

10. The bill (including parts and labor) for the repair of a home appliance was $165. The cost for parts was $85. The labor rate was $16 per half hour. How many hours were spent in repairing the appliance?

11. A pet store owner mixes two types of dog food costing $2.60 per pound and $3.80 per pound to make 40 pounds of a mixture costing $3.35 per pound. How many pounds of each kind of dog food are in the mixture?

12. Two cars start at the same location and travel in the same direction at average speeds of 40 miles per hour and 55 miles per hour. How much time must elapse before the two cars are 10 miles apart?

13. The interest on a savings account is 7.5%. Find the principal required to earn $300 in interest in 2 years.

14. Solve each equation.
    (a) $|3x - 6| = 9$ (b) $|3x - 5| = |6x - 1|$ (c) $|9 - 4x| + 4 = 1$

15. Solve each inequality and sketch the solution on the real number line.

    (a) $3x + 12 \geq -6$

    (b) $9 - 5x < 5 - 3x$

    (c) $0 \leq \dfrac{1 - x}{4} < 2$

    (d) $-7 < 4(2 - 3x) \leq 20$

16. Rewrite the statement "t is at least 8" using inequality notation.

17. Solve each inequality.

    (a) $|x - 3| \leq 2$ (b) $|5x - 3| > 12$ (c) $\left| \dfrac{x}{4} + 2 \right| < 0.2$

18. A utility company has a fleet of vans. The annual operating cost per van is

    $$C = 0.37m + 2700$$

    where m is the number of miles traveled by a van in a year. What is the maximum number of miles that will yield an annual operating cost that is less than or equal to $11,950?

Study Skills in Action

Reading Your Textbook Like a Manual

Many students avoid opening their textbooks for the same reason many people avoid opening their checkbooks—anxiety and frustration. The truth? Not opening your math textbook will cause more anxiety and frustration! Your textbook is a manual designed to help you master skills and understand and remember concepts. It contains many features and resources that can help you be successful in your course.

For more information about reading a textbook, refer to the *Math Study Skills Workbook* (Nolting, 2008).

Kimberly Nolting

VP, Academic Success Press
expert in developmental education

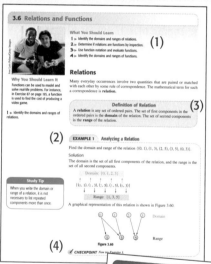

Smart Study Strategy

Use the Features of Your Textbook

To review what you learned in a previous class:

- Read the list of What You Should Learn (1) at the beginning of the section. If you cannot remember how to perform a skill, review the appropriate example (2) in the section.

- Read and understand the contents of all tinted concept boxes (3)—these contain important definitions and rules.

To prepare for homework:

- Complete the checkpoint exercises (4) in the section. If you have difficulty with a checkpoint exercise, reread the example or seek help from a peer or instructor.

To review for quizzes and tests:

- Make use of the What Did You Learn? (5) feature. Check off the key terms (6) and key concepts (7) you know, and review those you do not know.

- Complete the Review Exercises. Then take the Mid-Chapter Quiz, Chapter Test, or Cumulative Test, as appropriate.

Chapter 3
Graphs and Functions

IT WORKED FOR ME!

"I used to be really afraid of math, so reading the textbook was torture. I have learned that it just takes different strategies to read the textbook. It's my resource book when I do homework. I lug it to class because it helps me follow along. I'm not afraid of math anymore because I know how to study it—finally."

Jodie
Elementary education

3.1 The Rectangular Coordinate System

Don Johnston/Johnston Images/JupiterImages

Why You Should Learn It

A rectangular coordinate system can be used to represent relationships between two variables. For instance, Exercise 97 on page 138 shows the relationship between the speed of a car and fuel efficiency.

1 ▶ Plot points on a rectangular coordinate system.

What You Should Learn

1 ▶ Plot points on a rectangular coordinate system.
2 ▶ Determine whether ordered pairs are solutions of equations.
3 ▶ Use the Distance Formula to find the distance between two points.
4 ▶ Use the Midpoint Formula to find the midpoints of line segments.

The Rectangular Coordinate System

Just as you can represent real numbers by points on the real number line, you can represent ordered pairs of real numbers by points in a plane. This plane is called a **rectangular coordinate system** or the **Cartesian plane,** after the French mathematician René Descartes.

A rectangular coordinate system is formed by two real number lines intersecting at a right angle, as shown in Figure 3.1. The horizontal number line is usually called the **x-axis,** and the vertical number line is usually called the **y-axis.** (The plural of axis is *axes*.) The point of intersection of these two axes is the **origin.** The axes separate the plane into four regions known as **quadrants.**

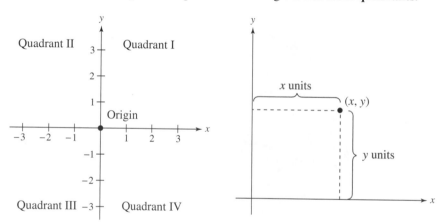

Figure 3.1   **Figure 3.2**

Each point in the plane corresponds to an **ordered pair** (x, y) of real numbers x and y, called the **coordinates** of the point. The first number (or **x-coordinate**) tells how far to the left or right the point is from the vertical axis, and the second number (or **y-coordinate**) tells how far up or down the point is from the horizontal axis, as shown in Figure 3.2.

A positive x-coordinate implies that the point lies to the *right* of the vertical axis; a negative x-coordinate implies that the point lies to the *left* of the vertical axis; and an x-coordinate of zero implies that the point lies *on* the vertical axis. Similarly, a positive y-coordinate implies that the point lies *above* the horizontal axis; a negative y-coordinate implies that the point lies *below* the horizontal axis; and a y-coordinate of zero implies that the point lies *on* the horizontal axis.

Locating a given point in a plane is called **plotting** the point. Example 1 shows how this is done.

EXAMPLE 1 Plotting Points on a Rectangular Coordinate System

Plot the points $(-2, 1)$, $(4, 0)$, $(3, -1)$, $(4, 3)$, $(0, 0)$, and $(-1, -3)$ on a rectangular coordinate system.

Solution

The point $(-2, 1)$ is two units to the *left* of the vertical axis and one unit *above* the horizontal axis.

Two units to the left   One unit above the
of the vertical axis   horizontal axis

$$(-2, 1)$$

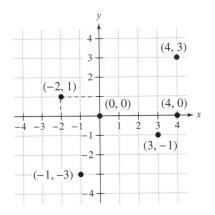

Figure 3.3

Similarly, the point $(4, 0)$ is four units to the *right* of the vertical axis and *on* the horizontal axis. (It is on the horizontal axis because its y-coordinate is 0.) The other four points can be plotted in a similar way, as shown in Figure 3.3.

✔ **CHECKPOINT** *Now try Exercise 1.*

In Example 1, you were given the coordinates of several points and asked to plot the points on a rectangular coordinate system. Example 2 looks at the reverse problem. That is, given the points on a rectangular coordinate system, you are asked to determine their coordinates.

EXAMPLE 2 Finding Coordinates of Points

Determine the coordinates of each of the points shown in Figure 3.4.

Solution

Point A lies two units to the *right* of the vertical axis and one unit *below* the horizontal axis. So, point A must be given by the ordered pair $(2, -1)$. The coordinates of the other five points can be determined in a similar way. The results are summarized as follows.

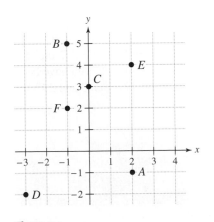

Figure 3.4

Point	Position	Coordinates
A	Two units *right*, one unit *down*	$(2, -1)$
B	One unit *left*, five units *up*	$(-1, 5)$
C	Zero units *right* (or *left*), three units *up*	$(0, 3)$
D	Three units *left*, two units *down*	$(-3, -2)$
E	Two units *right*, four units *up*	$(2, 4)$
F	One unit *left*, two units *up*	$(-1, 2)$

✔ **CHECKPOINT** *Now try Exercise 9.*

In Example 2, note that point $A(2, -1)$ and point $F(-1, 2)$ are different points. The order in which the numbers appear in an ordered pair is important.

As a consumer today, you are presented almost daily with vast amounts of data given in various forms. Data is given in *numerical form* using lists and tables and in *graphical form* using scatter plots, line graphs, circle graphs, and bar graphs. Graphical forms are more visual and make wide use of Descartes's rectangular coordinate system to show the relationship between two variables. Today, Descartes's ideas are commonly used in virtually every scientific and business-related field.

In the next example, data is represented graphically by points plotted on a rectangular coordinate system. This type of graph is called a **scatter plot.**

EXAMPLE 3   **Representing Data Graphically**

The populations (in millions) of California for the years 1990 through 2005 are shown in the table. Sketch a scatter plot of the data.   (Source: U.S. Census Bureau)

Year	1990	1991	1992	1993	1994	1995	1996	1997
Population	30.0	30.5	31.0	31.3	31.5	31.7	32.0	32.5

Year	1998	1999	2000	2001	2002	2003	2004	2005
Population	33.0	33.5	34.0	34.5	35.0	35.5	35.8	36.1

Solution

To sketch a scatter plot, begin by choosing which variable will be plotted on the horizontal axis and which will be plotted on the vertical axis. For this data, it seems natural to plot the years on the horizontal axis (which means that the population must be plotted on the vertical axis). Next, use the data in the table to form ordered pairs. For instance, the first three ordered pairs are (1990, 30.0), (1991, 30.5), and (1992, 31.0). All 16 points are shown in Figure 3.5. Note that the break in the *x*-axis indicates that the numbers between 0 and 1990 have been omitted. The break in the *y*-axis indicates that the numbers between 0 and 29 have been omitted.

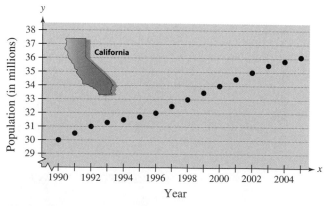

Figure 3.5

✓ *CHECKPOINT* *Now try Exercise 43.*

2 ▶ Determine whether ordered pairs are solutions of equations.

Ordered Pairs as Solutions

In Example 3, the relationship between the year and the population was given by a **table of values.** In mathematics, the relationship between the variables x and y is often given by an equation, from which you can construct a table of values.

Technology: Discovery

In the table of values in Example 4, successive x-values differ by 1. How do the successive y-values differ? Use the *table* feature of your graphing calculator to create a table of values for each of the following equations. If successive x-values differ by 1, how do the successive y-values differ?

a. $y = x + 2$

b. $y = 2x + 2$

c. $y = 4x + 2$

d. $y = -x + 2$

Describe the pattern.

| **EXAMPLE 4** | **Constructing a Table of Values** |

Construct a table of values for $y = 3x + 2$. Then plot the solution points on a rectangular coordinate system. Choose x-values of $-3, -2, -1, 0, 1, 2,$ and 3.

Solution

For each x-value, you must calculate the corresponding y-value. For example, if you choose $x = 1$, then the y-value is

$$y = 3(1) + 2 = 5.$$

The ordered pair $(x, y) = (1, 5)$ is a **solution point** (or **solution**) of the equation.

Choose x	**Calculate y from $y = 3x + 2$**	**Solution point**
$x = -3$	$y = 3(-3) + 2 = -7$	$(-3, -7)$
$x = -2$	$y = 3(-2) + 2 = -4$	$(-2, -4)$
$x = -1$	$y = 3(-1) + 2 = -1$	$(-1, -1)$
$x = 0$	$y = 3(0) + 2 = 2$	$(0, 2)$
$x = 1$	$y = 3(1) + 2 = 5$	$(1, 5)$
$x = 2$	$y = 3(2) + 2 = 8$	$(2, 8)$
$x = 3$	$y = 3(3) + 2 = 11$	$(3, 11)$

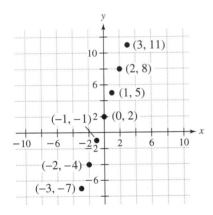

Figure 3.6

Once you have constructed a table of values, you can get a visual idea of the relationship between the variables x and y by plotting the solution points on a rectangular coordinate system, as shown in Figure 3.6.

✓ **CHECKPOINT** *Now try Exercise 49.*

In many places throughout this course, you will see that approaching a problem in different ways can help you understand the problem better. In Example 4, for instance, solutions of an equation are arrived at in three ways.

Three Approaches to Problem Solving

1. **Algebraic Approach** Use algebra to find several solutions.

2. **Numerical Approach** Construct a table that shows several solutions.

3. **Graphical Approach** Draw a graph that shows several solutions.

Smart Study Strategy

Go to page 124 for ways to *Use the Features of Your Textbook.*

Guidelines for Verifying Solutions

To verify that an ordered pair (x, y) is a solution of an equation with variables x and y, use the steps below.

1. Substitute the values of x and y into the equation.
2. Simplify each side of the equation.
3. If each side simplifies to the same number, the ordered pair is a solution. If the two sides yield different numbers, the ordered pair is not a solution.

EXAMPLE 5   **Verifying Solutions of an Equation**

Which of the ordered pairs are solutions of $x^2 - 2y = 6$?

a. $(2, 1)$ **b.** $(0, -3)$ **c.** $(-2, -5)$ **d.** $\left(1, -\frac{5}{2}\right)$

Solution

a. For the ordered pair $(2, 1)$, substitute $x = 2$ and $y = 1$ into the equation.

$$(2)^2 - 2(1) \overset{?}{=} 6 \qquad \text{Substitute 2 for } x \text{ and 1 for } y.$$

$$2 \neq 6 \qquad \text{Not a solution } ✗$$

Because the substitution does not satisfy the original equation, you can conclude that the ordered pair $(2, 1)$ *is not* a solution of the original equation.

b. For the ordered pair $(0, -3)$, substitute $x = 0$ and $y = -3$ into the equation.

$$(0)^2 - 2(-3) \overset{?}{=} 6 \qquad \text{Substitute 0 for } x \text{ and } -3 \text{ for } y.$$

$$6 = 6 \qquad \text{Solution } ✓$$

Because the substitution satisfies the original equation, you can conclude that the ordered pair $(0, -3)$ *is* a solution of the original equation.

c. For the ordered pair $(-2, -5)$, substitute $x = -2$ and $y = -5$ into the equation.

$$(-2)^2 - 2(-5) \overset{?}{=} 6 \qquad \text{Substitute } -2 \text{ for } x \text{ and } -5 \text{ for } y.$$

$$14 \neq 6 \qquad \text{Not a solution } ✗$$

Because the substitution does not satisfy the original equation, you can conclude that the ordered pair $(-2, -5)$ *is not* a solution of the original equation.

d. For the ordered pair $\left(1, -\frac{5}{2}\right)$, substitute $x = 1$ and $y = -\frac{5}{2}$ into the equation.

$$(1)^2 - 2\left(-\tfrac{5}{2}\right) \overset{?}{=} 6 \qquad \text{Substitute 1 for } x \text{ and } -\tfrac{5}{2} \text{ for } y.$$

$$6 = 6 \qquad \text{Solution } ✓$$

Because the substitution satisfies the original equation, you can conclude that the ordered pair $\left(1, -\frac{5}{2}\right)$ *is* a solution of the original equation.

✓ **CHECKPOINT** *Now try Exercise 55.*

3 ▸ Use the Distance Formula to find the distance between two points.

The Distance Formula

You know from Section 1.1 that the distance d between two points a and b on the real number line is simply

$$d = |b - a|.$$

The same "absolute value rule" is used to find the distance between two points that lie on the same *vertical or horizontal line* in the coordinate plane, as shown in Example 6.

EXAMPLE 6   **Finding Horizontal and Vertical Distances**

a. Find the distance between the points $(2, -2)$ and $(2, 4)$.

b. Find the distance between the points $(-3, -2)$ and $(2, -2)$.

Solution

a. Because the x-coordinates are equal, you can visualize a vertical line through the points $(2, -2)$ and $(2, 4)$, as shown in Figure 3.7. The distance between these two points is the absolute value of the difference of their y-coordinates.

Vertical distance $= |4 - (-2)|$	Subtract y-coordinates.
$= |6|$	Simplify.
$= 6$	Evaluate absolute value.

b. Because the y-coordinates are equal, you can visualize a horizontal line through the points $(-3, -2)$ and $(2, -2)$, as shown in Figure 3.7. The distance between these two points is the absolute value of the difference of their x-coordinates.

Horizontal distance $= |2 - (-3)|$	Subtract x-coordinates.
$= |5|$	Simplify.
$= 5$	Evaluate absolute value.

✓ **CHECKPOINT** *Now try Exercise 61.*

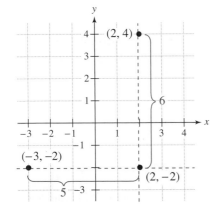

Figure 3.7

In Figure 3.7, note that the horizontal distance between the points $(-3, -2)$ and $(2, -2)$ is the absolute value of the difference of the x-coordinates, and the vertical distance between the points $(2, -2)$ and $(2, 4)$ is the absolute value of the difference of the y-coordinates.

The technique applied in Example 6 can be used to develop a general formula for finding the distance between two points in the plane. This general formula will work for any two points, even if they do not lie on the same vertical or horizontal line. To develop the formula, you use the **Pythagorean Theorem,** which states that for a right triangle, the hypotenuse c and sides a and b are related by the formula

$$a^2 + b^2 = c^2 \qquad \text{Pythagorean Theorem}$$

as shown in Figure 3.8. (The converse is also true. That is, if $a^2 + b^2 = c^2$, then the triangle is a right triangle.)

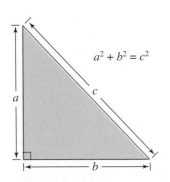

Figure 3.8

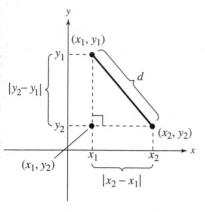

Figure 3.9   Distance Between Two Points

To develop a general formula for the distance between two points, let (x_1, y_1) and (x_2, y_2) represent two points that do not lie on the same horizontal or vertical line in the plane. With these two points, a right triangle can be formed, as shown in Figure 3.9. Note that the third vertex of the triangle is (x_1, y_2). Because (x_1, y_1) and (x_1, y_2) lie on the same vertical line, the length of the vertical side of the triangle is $|y_2 - y_1|$. Similarly, the length of the horizontal side is $|x_2 - x_1|$. By the Pythagorean Theorem, the square of the distance between (x_1, y_1) and (x_2, y_2) is

$$d^2 = |x_2 - x_1|^2 + |y_2 - y_1|^2.$$

Because the distance d must be positive, you can choose the positive square root and write

$$d = \sqrt{|x_2 - x_1|^2 + |y_2 - y_1|^2}.$$

Finally, replacing $|x_2 - x_1|^2$ and $|y_2 - y_1|^2$ by the equivalent expressions $(x_2 - x_1)^2$ and $(y_2 - y_1)^2$ yields the **Distance Formula.**

The Distance Formula

The distance d between two points (x_1, y_1) and (x_2, y_2) is

$$d = \sqrt{(x_2 - x_1)^2 + (y_2 - y_1)^2}.$$

Note that for the special case in which the two points lie on the same vertical or horizontal line, the Distance Formula still works. For instance, applying the Distance Formula to the points $(2, -2)$ and $(2, 4)$ produces

$$d = \sqrt{(2 - 2)^2 + [4 - (-2)]^2} = \sqrt{6^2} = 6$$

which is the same result obtained in Example 6(a).

EXAMPLE 7   **Finding the Distance Between Two Points**

Find the distance between the points $(-1, 2)$ and $(2, 4)$, as shown in Figure 3.10.

Solution

Let $(x_1, y_1) = (-1, 2)$ and let $(x_2, y_2) = (2, 4)$. Then apply the Distance Formula.

$$d = \sqrt{(x_2 - x_1)^2 + (y_2 - y_1)^2} \qquad \text{Distance Formula}$$

$$d = \sqrt{[2 - (-1)]^2 + (4 - 2)^2} \qquad \text{Substitute coordinates of points.}$$

$$= \sqrt{3^2 + 2^2} \qquad \text{Simplify.}$$

$$= \sqrt{13} \approx 3.61 \qquad \text{Simplify and use a calculator.}$$

✓ CHECKPOINT *Now try Exercise 69.*

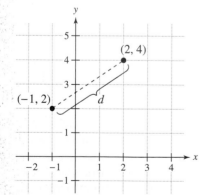

Figure 3.10

When the Distance Formula is used, it does not matter which point is (x_1, y_1) and which is (x_2, y_2), because the result will be the same. For instance, in Example 7, let $(x_1, y_1) = (2, 4)$ and let $(x_2, y_2) = (-1, 2)$. Then

$$d = \sqrt{[(-1) - 2]^2 + (2 - 4)^2} = \sqrt{(-3)^2 + (-2)^2} = \sqrt{13} \approx 3.61.$$

The Distance Formula has many applications in mathematics. In the next example, you can use the Distance Formula and the converse of the Pythagorean Theorem to verify that three points form the vertices of a right triangle.

EXAMPLE 8   Verifying a Right Triangle

Show that the points $(1, 2)$, $(3, 1)$, and $(4, 3)$ are vertices of a right triangle.

Solution

The three points are plotted in Figure 3.11. Using the Distance Formula, you can find the lengths of the three sides of the triangle.

$$d_1 = \sqrt{(3 - 1)^2 + (1 - 2)^2} = \sqrt{4 + 1} = \sqrt{5}$$
$$d_2 = \sqrt{(4 - 3)^2 + (3 - 1)^2} = \sqrt{1 + 4} = \sqrt{5}$$
$$d_3 = \sqrt{(4 - 1)^2 + (3 - 2)^2} = \sqrt{9 + 1} = \sqrt{10}$$

Because

$$d_1^2 + d_2^2 = \left(\sqrt{5}\right)^2 + \left(\sqrt{5}\right)^2 = 5 + 5 = 10$$

and

$$d_3^2 = \left(\sqrt{10}\right)^2 = 10$$

you can conclude from the converse of the Pythagorean Theorem that the triangle is a right triangle.

✓ **CHECKPOINT** *Now try Exercise 79.*

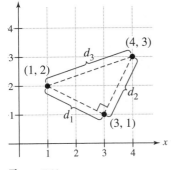

Figure 3.11

Three or more points are **collinear** if they all lie on the same line. You can use the Distance Formula to determine whether points are collinear, as shown in Example 9.

EXAMPLE 9   Collinear Points

Determine whether the set of points is collinear.

$$\{A(2, 6), B(5, 2), C(8, -2)\}$$

Solution

Let d_1 equal the distance between A and B, let d_2 equal the distance between B and C, and let d_3 equal the distance between A and C. You can find these distances as follows.

$$d_1 = \sqrt{(5 - 2)^2 + (2 - 6)^2} = \sqrt{9 + 16} = \sqrt{25} = 5$$
$$d_2 = \sqrt{(8 - 5)^2 + (-2 - 2)^2} = \sqrt{9 + 16} = \sqrt{25} = 5$$
$$d_3 = \sqrt{(8 - 2)^2 + (-2 - 6)^2} = \sqrt{36 + 64} = \sqrt{100} = 10$$

So, the set of points is collinear because $d_1 + d_2 = 5 + 5 = 10 = d_3$. These three points are plotted in Figure 3.12. From the figure, the points appear to lie on the same line.

✓ **CHECKPOINT** *Now try Exercise 83.*

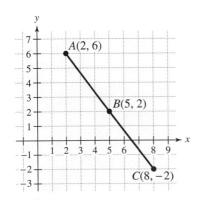

Figure 3.12

When you use coordinate geometry to solve real-life problems, you can place the coordinate system in any way that is convenient to find the solution of the problem, as shown in Example 10.

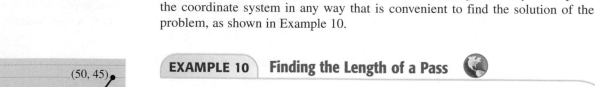

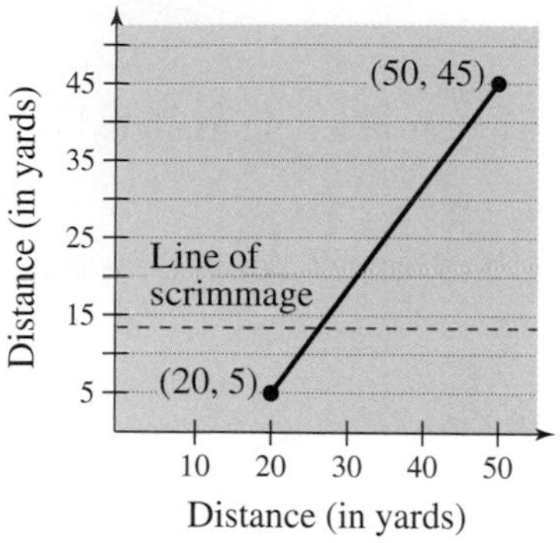

EXAMPLE 10 Finding the Length of a Pass

A football quarterback throws a pass from the five-yard line, 20 yards from the sideline. The pass is caught by a wide receiver on the 45-yard line, 50 yards from the same sideline, as shown in Figure 3.13. How long is the pass?

Solution

The length of the pass is the distance between points $(20, 5)$ and $(50, 45)$.

$$d = \sqrt{(50 - 20)^2 + (45 - 5)^2}$$
Substitute coordinates of points into Distance Formula.

$$= \sqrt{900 + 1600} = 50$$
Simplify.

So, the pass is 50 yards long.

✓ **CHECKPOINT** *Now try Exercise 95.*

Figure 3.13

4 ▶ Use the Midpoint Formula to find the midpoints of line segments.

The Midpoint Formula

The **midpoint** of a line segment that joins two points is the point that divides the segment into two equal parts. To find the midpoint of the line segment that joins two points in a coordinate plane, you can simply find the average values of the respective coordinates of the two endpoints using the **Midpoint Formula.**

The Midpoint Formula

The midpoint of the line segment joining the points (x_1, y_1) and (x_2, y_2) is given by the Midpoint Formula

$$\text{Midpoint} = \left(\frac{x_1 + x_2}{2}, \frac{y_1 + y_2}{2} \right).$$

EXAMPLE 11 Finding the Midpoint of a Line Segment

Find the midpoint of the line segment joining the points $(-5, -3)$ and $(9, 3)$, as shown in Figure 3.14.

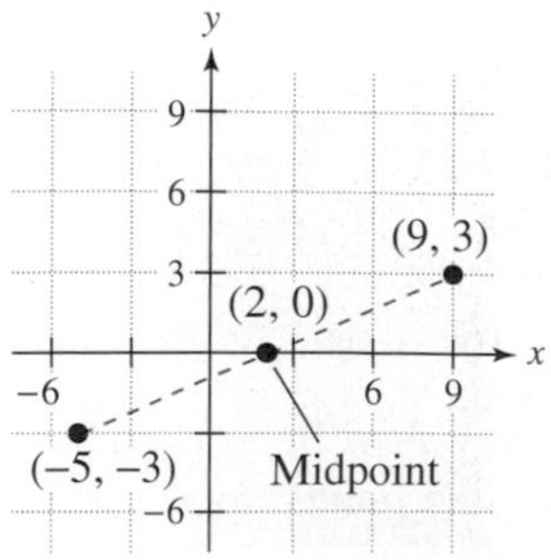

Solution

Let $(x_1, y_1) = (-5, -3)$ and let $(x_2, y_2) = (9, 3)$.

$$\text{Midpoint} = \left(\frac{x_1 + x_2}{2}, \frac{y_1 + y_2}{2} \right)$$
Midpoint Formula

$$= \left(\frac{-5 + 9}{2}, \frac{-3 + 3}{2} \right)$$
Substitute for x_1, y_1, x_2, and y_2.

$$= (2, 0)$$
Simplify.

✓ **CHECKPOINT** *Now try Exercise 89.*

Figure 3.14

_____ **Concept Check** _____

1. Discuss the significance of the word *ordered* when referring to an ordered pair (x, y).

2. When plotting the point (x, y), what does the x-coordinate measure? What does the y-coordinate measure?

3. What is the x-coordinate of any point on the y-axis? What is the y-coordinate of any point on the x-axis?

4. Point C is in Quadrant I and point D is in Quadrant III. Is it possible for the distance from C to D to be -5? Explain.

3.1 EXERCISES

Go to pages 208–209 to record your assignments.

_____ **Developing Skills** _____

In Exercises 1–8, plot the points on a rectangular coordinate system. *See Example 1.*

1. $(4, 3), (-5, 3), (3, -5)$ 2. $(-2, 5), (-2, -5), (3, 5)$

3. $(-8, -2), (6, -2), (5, 0)$

4. $(0, 4), (0, 0), (-7, 0)$ 5. $\left(\frac{5}{2}, -2\right), \left(-2, \frac{1}{4}\right), \left(\frac{3}{2}, -\frac{7}{2}\right)$

6. $\left(-\frac{2}{3}, 3\right), \left(\frac{1}{4}, -\frac{5}{4}\right), \left(-5, -\frac{7}{4}\right)$

7. $\left(\frac{3}{2}, 1\right), (4, -3), \left(-\frac{4}{3}, \frac{7}{3}\right)$ 8. $(-3, -5), \left(\frac{9}{4}, \frac{3}{4}\right), \left(\frac{5}{2}, -2\right)$

In Exercises 9–12, determine the coordinates of the points. *See Example 2.*

9.
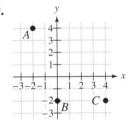

10.
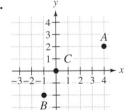

11.
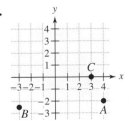

12.
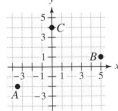

In Exercises 13–20, plot the points and connect them with line segments to form the figure. (*Note:* A *rhombus* is a parallelogram whose sides are all of the same length.)

13. *Square:* $(2, 4), (5, 1), (2, -2), (-1, 1)$

14. *Rectangle:* $(7, 0), (9, 1), (4, 6), (6, 7)$

15. *Triangle:* $(-1, 2), (2, 0), (3, 5)$

16. *Triangle:* $(-1, 3), (-2, -2), (3, 8)$

17. *Parallelogram:* $(4, 0), (6, -2), (0, -4), (-2, -2)$

18. *Parallelogram:* $(-1, 1), (0, 4), (4, -2), (5, 1)$

19. *Rhombus:* $(0, 0), (3, 2), (2, 3), (5, 5)$

20. *Rhombus:* $(-3, -3), (-2, -1), (-1, -2), (0, 0)$

In Exercises 21–28, find the coordinates of the point.

21. The point is located one unit to the right of the y-axis and four units above the x-axis.

22. The point is located five units to the left of the y-axis and two units above the x-axis.

23. The point is located 10 units to the right of the y-axis and four units below the x-axis.

24. The point is located three units to the left of the y-axis and two units below the x-axis.

25. The point is on the positive x-axis 10 units from the origin.

26. The point is on the negative y-axis five units from the origin.

27. The coordinates of the point are equal, and the point is located in the third quadrant eight units to the left of the y-axis.

28. The coordinates of the point are equal in magnitude and opposite in sign, and the point is located seven units to the right of the y-axis.

In Exercises 29–42, determine the quadrant in which the point is located without plotting it. (x and y are real numbers.)

29. $(-3, -5)$ **30.** $(4, -2)$

31. $\left(-\frac{8}{9}, \frac{3}{4}\right)$ **32.** $\left(\frac{5}{11}, \frac{3}{8}\right)$

33. $(-9.5, -12.13)$ **34.** $(-6.2, 8.05)$

35. $(x, y), \quad x > 0, y < 0$ **36.** $(x, y), \quad x > 0, y > 0$

37. $(x, 4)$ **38.** $(x, -6)$

39. $(-3, y)$ **40.** $(10, y)$

41. $(x, y), \quad xy > 0$ **42.** $(x, y), \quad xy < 0$

In Exercises 43–46, sketch a scatter plot of the points whose coordinates are shown in the table. *See Example 3.*

✔ 43. *Exam Scores*   The table shows the times x in hours invested in studying for five different algebra exams and the resulting scores y.

x	5	2	3	6.5	4
y	81	71	88	92	86

44. *Net Sales*   The net sales y (in billions of dollars) of Wal-Mart for the years 2003 through 2007 are shown in the table. The time in years is given by x. (Source: Wal-Mart 2007 Annual Report)

x	2003	2004	2005	2006	2007
y	226.5	252.8	281.5	308.9	345.0

45. *Meteorology*   The table shows the monthly normal temperatures y (in degrees Fahrenheit) for Duluth, Minnesota, for each month of the year, with $x = 1$ representing January. (Source: National Climatic Data Center)

x	1	2	3	4	5	6
y	7	12	24	39	51	60

x	7	8	9	10	11	12
y	66	64	54	44	28	13

46. *Fuel Efficiency*   The table shows various speeds x of a car in miles per hour and the corresponding approximate fuel efficiencies y in miles per gallon.

x	50	55	60	65	70
y	35	33.8	32.2	30	27.5

Shifting a Graph   In Exercises 47 and 48, the figure is shifted to a new location in the plane. Find the coordinates of the vertices of the figure in its new location.

47.

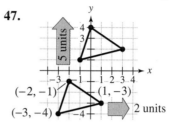

48.

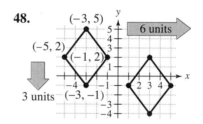

In Exercises 49–52, complete the table of values. Then plot the solution points on a rectangular coordinate system. *See Example 4.*

✔ 49.

x	−2	0	2	4	6
y = 5x + 3					

50.

x	−3	0	3	6	9
y = 6x − 7					

51.

x	−4	0	3	5	10
y = |2x − 7| + 2					

52.

x	−5	−1	0	3	8
y = |−3x + 1| − 5					

In Exercises 53 and 54, use the *table* feature of a graphing calculator to complete the table of values.

53.

x		-2	0	2	4	6
$y = x^2 + 2x + 5$						

54.

x		-2	0	2	4	6
$y = 4x^2 + x - 2$						

In Exercises 55–60, determine whether each ordered pair is a solution of the equation. *See Example 5.*

55. $4y - 2x + 1 = 0$

(a) $(0, 0)$

(b) $\left(\frac{1}{2}, 0\right)$

(c) $\left(-3, -\frac{7}{4}\right)$

(d) $\left(1, -\frac{3}{4}\right)$

56. $5x - 2y + 50 = 0$

(a) $(-10, 0)$

(b) $\left(\frac{4}{5}, -27\right)$

(c) $\left(-9, \frac{5}{2}\right)$

(d) $(20, -2)$

57. $y = \frac{7}{8}x + 3$

(a) $\left(\frac{8}{7}, 4\right)$

(b) $(8, 10)$

(c) $(0, 0)$

(d) $(-16, 14)$

58. $y = \frac{5}{8}x - 2$

(a) $(0, 0)$

(b) $(8, 3)$

(c) $(16, -7)$

(d) $\left(-\frac{8}{5}, 3\right)$

59. $x^2 + 3y = -5$

(a) $(3, -2)$

(b) $(-2, -3)$

(c) $(3, -5)$

(d) $(4, -7)$

60. $y^2 - 4x = 8$

(a) $(0, 6)$

(b) $(-4, 2)$

(c) $(-1, 3)$

(d) $(7, 6)$

In Exercises 61–68, plot the points and find the distance between them. State whether the points lie on a horizontal or a vertical line. *See Example 6.*

61. $(3, -2), (3, 5)$

62. $(-2, 8), (-2, 1)$

63. $(3, 2), (10, 2)$

64. $(-120, -2), (130, -2)$

65. $\left(-3, \frac{3}{2}\right), \left(-3, \frac{9}{4}\right)$

66. $\left(\frac{3}{4}, 1\right), \left(\frac{3}{4}, -10\right)$

67. $\left(-4, \frac{1}{3}\right), \left(\frac{5}{2}, \frac{1}{3}\right)$

68. $\left(\frac{1}{2}, \frac{7}{8}\right), \left(\frac{11}{2}, \frac{7}{8}\right)$

In Exercises 69–78, find the distance between the points. *See Example 7.*

69. $(1, 3), (5, 6)$

70. $(3, 10), (15, 5)$

71. $(3, 7), (4, 5)$

72. $(5, 2), (8, 3)$

73. $(-3, 0), (4, -3)$

74. $(0, -5), (2, -8)$

75. $(-2, -3), (4, 2)$

76. $(-5, 4), (10, -3)$

77. $\left(3, \frac{3}{4}\right), \left(7, -\frac{1}{4}\right)$

78. $\left(\frac{1}{2}, 1\right), \left(\frac{3}{2}, 2\right)$

▲ *Geometry*   In Exercises 79–82, show that the points are vertices of a right triangle. *See Example 8.*

79.

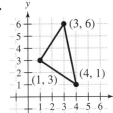

80.

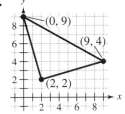

81.

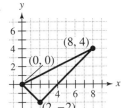

82.

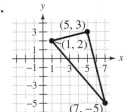

In Exercises 83–86, use the Distance Formula to determine whether the three points are collinear. *See Example 9.*

83. $(2, 3), (2, 6), (6, 3)$

84. $(1, 4), (4, -2), (2, 1)$

85. $(8, 3), (5, 2), (2, 1)$

86. $(2, 4), (1, 1), (0, -2)$

▲ *Geometry*   In Exercises 87 and 88, find the perimeter of the triangle with the given vertices.

87. $(-2, 0), (0, 5), (1, 0)$

88. $(-5, -2), (-1, 4), (3, -1)$

In Exercises 89–92, find the midpoint of the line segment joining the points, and then plot the points and the midpoint. *See Example 11.*

✓ **89.** $(-2, 0)$, $(4, 8)$

90. $(9, 3)$, $(-3, 7)$

91. $(1, 6)$, $(6, 3)$

92. $(2, 7)$, $(9, -1)$

Solving Problems

93. *Numerical Interpretation* For a handyman to install x windows in your home, the cost y is given by $y = 150x + 425$. Use x-values of 1, 2, 3, 4, and 5 to construct a table of values for y. Then use the table to help describe the relationship between the number of windows x and the cost of installation y.

94. *Numerical Interpretation* When an employee works x hours of overtime in a week, the employee's weekly pay y is given by $y = 18x + 480$. Use x-values of 0, 1, 2, 3, and 4 to construct a table of values for y. Then use the table to help describe the relationship between the number of overtime hours x and the weekly pay y.

✓ **95.** *Football Pass* A football quarterback throws a pass from the 10-yard line, 10 yards from the sideline. The pass is caught by a wide receiver on the 40-yard line, 35 yards from the same sideline, as shown in the figure. How long is the pass?

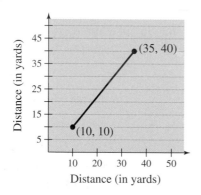

96. *Soccer Pass* A soccer player passes the ball from a point that is 18 yards from the endline and 12 yards from the sideline. The pass is received by a teammate who is 42 yards from the same endline and 50 yards from the same sideline, as shown in the figure. How long is the pass?

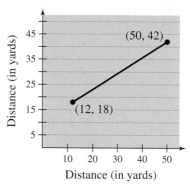

Figure for 96

97. *Fuel Efficiency* The scatter plot shows the speed of a car in kilometers per hour and the amount of fuel, in liters used per 100 kilometers traveled, that the car needs to maintain that speed. From the graph, how would you describe the relationship between the speed of the car and the fuel used? What is the approximate fuel efficiency if the car is traveling at 120 kilometers per hour?

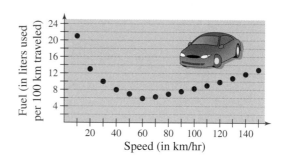

98. *Heating a Home*   A family wishes to study the relationship between the temperature outside and the amount of natural gas used to heat their house. For 16 months, the family observes the average temperature in degrees Fahrenheit for the month and the average amount of natural gas in hundreds of cubic feet used in that month. The scatter plot shows this data over the 16 months. Does the graph suggest that there is a strong relationship between the temperature outside and the amount of natural gas used? If a month were to have an average temperature of 45°F, about how much natural gas would you expect this family to use on average to heat their house for that month?

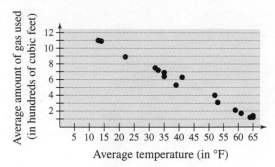

Figure for 98

99. *Net Sales*   Apple Inc. had net sales of $8279 million in 2004 and $19,315 million in 2006. Use the Midpoint Formula to estimate the net sales in 2005. (Source: Apple Inc.)

100. *Net Profit*   Staples Inc. had a net profit of $708.4 million in 2004 and $948.2 million in 2006. Use the Midpoint Formula to estimate the net profit in 2005.   (Source: Staples Inc.)

Explaining Concepts

101. State the Pythagorean Theorem and give examples of its use.

102. ✎   Explain why the ordered pair $(-3, 4)$ is not a solution point of the equation $y = 4x + 15$.

103. ✎   When points are plotted on the rectangular coordinate system, is it true that the scales on the x- and y-axes must be the same? Explain.

104. *Conjecture*   Plot the points $(2, 1)$, $(-3, 5)$, and $(7, -3)$ on a rectangular coordinate system. Then change the sign of the x-coordinate of each point and plot the three new points on the same rectangular coordinate system. What conjecture can you make about the location of a point when the sign of the x-coordinate is changed?

105. *Conjecture*   Plot the points $(2, 1)$, $(-3, 5)$, and $(7, -3)$ on a rectangular coordinate system. Then change the sign of the y-coordinate of each point and plot the three new points on the same rectangular coordinate system. What conjecture can you make about the location of a point when the sign of the y-coordinate is changed?

Cumulative Review

In Exercises 106–111, solve the proportion.

106. $\dfrac{3}{4} = \dfrac{x}{28}$ **107.** $\dfrac{5}{6} = \dfrac{y}{36}$

108. $\dfrac{a}{27} = \dfrac{4}{9}$ **109.** $\dfrac{m}{49} = \dfrac{5}{7}$

110. $\dfrac{z + 1}{10} = \dfrac{z}{9}$ **111.** $\dfrac{n}{16} = \dfrac{n - 3}{8}$

In Exercises 112–115, find the missing quantities.

List Price	Sale Price	Discount	Discount Rate
112. $80.00	$52.00		
113. $55.00			35%
114. $112.50		$31.50	
115.	$134.42	$124.08	

3.2 Graphs of Equations

© Thinkstock/CORBIS

Why You Should Learn It

Graphs of equations can help you to see relationships between real-life quantities. For instance, in Exercise 96 on page 147, you will sketch a graph that shows the changing value of a limousine over time.

1 ▶ Sketch graphs of equations using the point-plotting method.

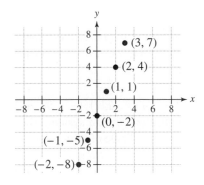

Figure 3.15

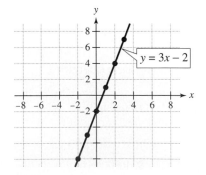

Figure 3.16

What You Should Learn

1 ▶ Sketch graphs of equations using the point-plotting method.

2 ▶ Find and use x- and y-intercepts as aids to sketching graphs.

3 ▶ Use a pattern to write an equation for an application problem, and sketch its graph.

The Graph of an Equation

In Section 3.1, you saw that the solutions of an equation in x and y can be represented by points on a rectangular coordinate system. The set of *all* solution points of an equation is called its **graph.** In this section, you will study a basic technique for sketching the graph of an equation—the **point-plotting method.**

EXAMPLE 1   Sketching the Graph of an Equation

To sketch the graph of $3x - y = 2$, first solve the equation for y.

$3x - y = 2$	Write original equation.
$-y = -3x + 2$	Subtract $3x$ from each side.
$y = 3x - 2$	Divide each side by -1.

Next, create a table of values. The choice of x-values to use in the table is somewhat arbitrary. However, the more x-values you choose, the easier it will be to recognize a pattern.

x	-2	-1	0	1	2	3
$y = 3x - 2$	-8	-5	-2	1	4	7
Solution point	$(-2, -8)$	$(-1, -5)$	$(0, -2)$	$(1, 1)$	$(2, 4)$	$(3, 7)$

Now, plot the solution points, as shown in Figure 3.15. It appears that all six points lie on a line, so complete the sketch by drawing a line through the points, as shown in Figure 3.16.

✓ **CHECKPOINT** *Now try Exercise 11.*

The equation in Example 1 is an example of a **linear equation** in two variables—the variables are raised to the first power and the graph of the equation is a line. By drawing a line through the plotted points, you are implying that every point on this line is a solution point of the given equation.

Technology: Discovery

Use a graphing calculator to graph each equation, and then answer the questions.

  i. $y = 3x + 2$

 ii. $y = 4 - x$

iii. $y = x^2 + 3x$

iv. $y = x^2 - 5$

  v. $y = |x - 4|$

 vi. $y = |x + 1|$

a. Which of the graphs are lines?

b. Which of the graphs are U-shaped?

c. Which of the graphs are V-shaped?

d. Describe the graph of the equation $y = x^2 + 7$ before you graph it. Use a graphing calculator to confirm your answer.

The Point-Plotting Method of Sketching a Graph

1. If possible, rewrite the equation by isolating one of the variables.
2. Make a table of values showing several solution points.
3. Plot these points on a rectangular coordinate system.
4. Connect the points with a smooth curve or line.

EXAMPLE 2   **Sketching the Graph of a Nonlinear Equation**

Sketch the graph of $-x^2 + 2x + y = 0$.

Solution

Begin by solving the equation for y.

$$-x^2 + 2x + y = 0 \qquad \text{Write original equation.}$$

$$2x + y = x^2 \qquad \text{Add } x^2 \text{ to each side.}$$

$$y = x^2 - 2x \qquad \text{Subtract } 2x \text{ from each side.}$$

Next, create a table of values.

x	-2	-1	0	1	2	3	4
$y = x^2 - 2x$	8	3	0	-1	0	3	8
Solution point	$(-2, 8)$	$(-1, 3)$	$(0, 0)$	$(1, -1)$	$(2, 0)$	$(3, 3)$	$(4, 8)$

Study Tip

Example 2 shows three common ways to represent the relationship between two variables. The equation $y = x^2 - 2x$ is the *analytical* or *algebraic* representation, the table of values is the *numerical* representation, and the graph in Figure 3.18 is the *graphical* representation. You will see and use analytical, numerical, and graphical representations throughout this course.

Now, plot the seven solution points, as shown in Figure 3.17. Finally, connect the points with a smooth curve, as shown in Figure 3.18.

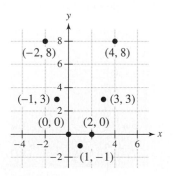

Figure 3.17

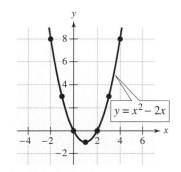

Figure 3.18

✓ **CHECKPOINT** *Now try Exercise 19.*

The graph of the equation given in Example 2 is called a *parabola*. You will study this type of graph in detail in a later chapter.

Example 3 examines the graph of an equation that involves an absolute value. Remember that the absolute value of a number is its distance from zero on the real number line. For instance, $|-5| = 5$, $|2| = 2$, and $|0| = 0$.

| EXAMPLE 3 | The Graph of an Absolute Value Equation |

Sketch the graph of $y = |x - 2|$.

Solution

This equation is already written in a form with y isolated on the left. So, begin by creating a table of values. Be sure that you understand how the absolute value is evaluated. For instance, when $x = -2$, the value of y is

$$y = |-2 - 2| \qquad \text{Substitute } -2 \text{ for } x.$$
$$= |-4| \qquad \text{Simplify.}$$
$$= 4 \qquad \text{Simplify.}$$

and when $x = 3$, the value of y is

$$y = |3 - 2| \qquad \text{Substitute } 3 \text{ for } x.$$
$$= |1| \qquad \text{Simplify.}$$
$$= 1. \qquad \text{Simplify.}$$

x	-2	-1	0	1	2	3	4	5
$y = \|x - 2\|$	4	3	2	1	0	1	2	3
Solution point	$(-2, 4)$	$(-1, 3)$	$(0, 2)$	$(1, 1)$	$(2, 0)$	$(3, 1)$	$(4, 2)$	$(5, 3)$

Next, plot the solution points, as shown in Figure 3.19. It appears that the points lie in a "V-shaped" pattern, with the point $(2, 0)$ lying at the bottom of the "V." Connect the points to form the graph shown in Figure 3.20.

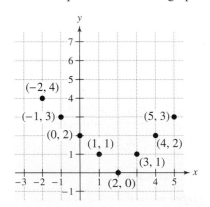

Figure 3.19

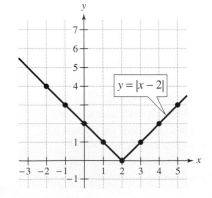

Figure 3.20

✓ **CHECKPOINT** *Now try Exercise 27.*

2 ▶ Find and use *x*- and *y*-intercepts as aids to sketching graphs.

Intercepts: Aids to Sketching Graphs

Solution points having zero as the *y*-coordinate or zero as the *x*-coordinate are especially useful. Such points are called **intercepts** because they are the points at which the graph intersects the *x*- and *y*-axes.

Study Tip

When you create a table of values for a graph, include any intercepts you have found. You should also include points to the left and right of the intercepts of the graph. This helps to give a more complete view of the graph.

Definition of Intercepts

The point $(a, 0)$ is called an ***x*-intercept** of the graph of an equation if it is a solution point of the equation. To find the *x*-intercept(s), let $y = 0$ and solve the equation for *x*.

The point $(0, b)$ is called a ***y*-intercept** of the graph of an equation if it is a solution point of the equation. To find the *y*-intercept(s), let $x = 0$ and solve the equation for *y*.

Some texts denote the *x*-intercept as the *x*-coordinate of the point $(a, 0)$ rather than the point itself. Unless it is necessary to make a distinction, this text will use the term "intercept" to mean either the point or the coordinate.

EXAMPLE 4   Finding the Intercepts of a Graph

Find the intercepts and sketch the graph of $y = 2x - 3$.

Solution

Find the *x*-intercept by letting $y = 0$ and solving for *x*.

$y = 2x - 3$	Write original equation.
$0 = 2x - 3$	Substitute 0 for *y*.
$3 = 2x$	Add 3 to each side.
$\frac{3}{2} = x$	Solve for *x*.

Find the *y*-intercept by letting $x = 0$ and solving for *y*.

$y = 2x - 3$	Write original equation.
$y = 2(0) - 3$	Substitute 0 for *x*.
$y = -3$	Solve for *y*.

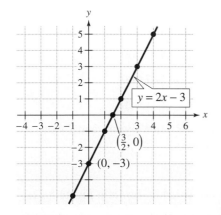

Figure 3.21

So, the graph has one *x*-intercept, which occurs at the point $\left(\frac{3}{2}, 0\right)$, and one *y*-intercept, which occurs at the point $(0, -3)$. To sketch the graph of the equation, create a table of values. Finally, use the solution points given in the table to sketch the graph of the equation, as shown in Figure 3.21.

x	-1	0	1	$\frac{3}{2}$	2	3	4
$y = 2x - 3$	-5	-3	-1	0	1	3	5
Solution point	$(-1, -5)$	$(0, -3)$	$(1, -1)$	$\left(\frac{3}{2}, 0\right)$	$(2, 1)$	$(3, 3)$	$(4, 5)$

✓ **CHECKPOINT** *Now try Exercise 57.*

3 ▶ Use a pattern to write an equation for an application problem, and sketch its graph.

Real-Life Application of Graphs

Newspapers and news magazines frequently use graphs to show real-life relationships between variables. Example 5 shows how such a graph can help you visualize the concept of **straight-line depreciation.**

EXAMPLE 5   **Straight-Line Depreciation: Finding the Pattern**

Your small business buys a new printing press for $65,000. For income tax purposes, you decide to depreciate the printing press over a 10-year period. At the end of 10 years, the value of the printing press is expected to be $5000.

  a. Find an equation that relates the depreciated value of the printing press to the number of years since it was purchased.

  b. Sketch the graph of the equation.

  c. What is the y-intercept of the graph and what does it represent?

Solution

  a. The total depreciation over the 10-year period is

$$\$65{,}000 - 5000 = \$60{,}000.$$

  Because the same amount is depreciated each year, it follows that the annual depreciation is

$$\frac{\$60{,}000}{10} = \$6000.$$

  So, after 1 year, the value of the printing press is

$$\text{Value after 1 year} = \$65{,}000 - (1)6000 = \$59{,}000.$$

  By similar reasoning, you can see that the values after 2, 3, and 4 years are

$$\text{Value after 2 years} = \$65{,}000 - (2)6000 = \$53{,}000$$

$$\text{Value after 3 years} = \$65{,}000 - (3)6000 = \$47{,}000$$

$$\text{Value after 4 years} = \$65{,}000 - (4)6000 = \$41{,}000.$$

  Let y represent the value of the printing press after t years and follow the pattern determined for the first 4 years to obtain

$$y = 65{,}000 - 6000t.$$

  b. A sketch of the graph of the depreciation equation is shown in Figure 3.22.

  c. To find the y-intercept of the graph, let $t = 0$ and solve the equation for y.

$y = 65{,}000 - 6000t$	Write original equation.
$y = 65{,}000 - 6000(0)$	Substitute 0 for t.
$y = 65{,}000$	Simplify.

  So, the y-intercept is $(0, 65{,}000)$, which corresponds to the initial value of the printing press.

 CHECKPOINT *Now try Exercise 95.*

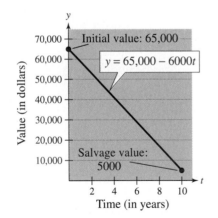

Straight-Line Depreciation
Figure 3.22

_____ **Concept Check** _____

1. A table of values has been used to plot the solution points shown below for a given equation. Describe what else needs to be done to complete the graph of the equation.

2. Is enough information given in the figure below to determine the x- and y-intercepts? Explain.

3. *True or False?* To find the x-intercept(s) of the graph of an equation algebraically, you can substitute 0 for x in the equation and then solve for y. Justify your reasoning.

4. The value of a piece of equipment is said to have *straight-line depreciation*. Describe the pattern of change in the value from one year to the next.

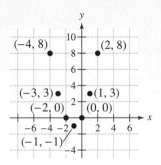

Figure for 1 and 2

3.2 EXERCISES

Go to pages 208–209 to record your assignments.

_____ **Developing Skills** _____

In Exercises 1–6, match the equation with its graph. [The graphs are labeled (a), (b), (c), (d), (e), and (f).]

(a)

(b)

(c)
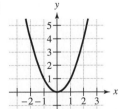

(d)
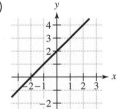

(e)
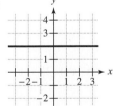

(f)
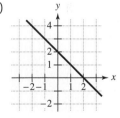

1. $y = 2$

2. $y = 2 + x$

3. $y = 2 - x$

4. $y = x^2$

5. $y = x^2 - 4$

6. $y = |x|$

In Exercises 7–30, sketch the graph of the equation. *See Examples 1–3.*

7. $y = 3x$

8. $y = -2x$

9. $y = 4 - x$

10. $y = x - 7$

11. $2x - y = 3$

12. $3x - y = -2$

13. $3x + 2y = 2$

14. $2y + 5x = 6$

15. $y = -x^2$

16. $y = x^2$

17. $y = x^2 - 3$

18. $y = 4 - x^2$

19. $-x^2 - 3x + y = 0$

20. $-x^2 + x + y = 0$

21. $x^2 - 2x - y = 1$

22. $x^2 + 3x - y = 4$

23. $y = |x|$

24. $y = -|x|$

25. $y = |x| + 3$

26. $y = |x| - 1$

27. $y = |x + 3|$

28. $y = |x - 1|$

29. $y = x^3$

30. $y = -x^3$

In Exercises 31–44, find the x- and y-intercepts (if any) of the graph of the equation. *See Example 4.*

31. $y = 6x - 3$

32. $y = 4 - 3x$

33. $y = 12 - \frac{2}{5}x$

34. $y = \frac{3}{4}x + 15$

35. $x + 2y = 10$

36. $3x - 2y = 12$

37. $4x - y + 3 = 0$

38. $2x + 3y - 8 = 0$

39. $y = |x| - 1$

40. $y = |x| + 4$

41. $y = -|x + 5|$

42. $y = |x - 4|$

43. $y = |x - 1| - 3$

44. $y = |x + 3| - 1$

In Exercises 45–50, graphically estimate the x- and y-intercepts (if any) of the graph.

45. $2x + 3y = 6$

46. $3x - y + 9 = 0$

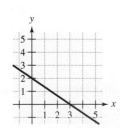

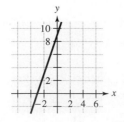

47. $y = x^2 + 3$

48. $y = -x^2 + 4x$

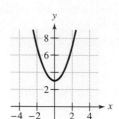

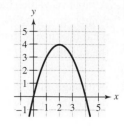

49. $y = -2$

50. $x = 3$

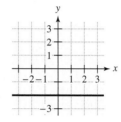

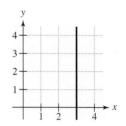

In Exercises 51–56, use a graphing calculator to graph the equation. Approximate the x- and y-intercepts (if any).

51. $y = 8 - 4x$

52. $y = 5x - 20$

53. $y = (x - 1)(x - 6)$

54. $y = (x + 2)(x - 3)$

55. $y = |4x + 6| - 2$

56. $y = |2x - 4| + 1$

In Exercises 57–92, sketch the graph of the equation and show the coordinates of three solution points (including x- and y-intercepts). *See Example 4.*

57. $y = 3 - x$

58. $y = x - 3$

59. $y = 2x - 3$

60. $y = -4x + 8$

61. $4x + y = 3$

62. $y - 2x = -4$

63. $2x - 3y = 6$

64. $3x - 2y = 8$

65. $3x + 4y = 12$

66. $4x + 5y = 10$

67. $x + 5y = 10$

68. $x + 3y = 15$

69. $5x - y = 10$

70. $7x - y = -21$

71. $y = x^2 - 9$

72. $y = x^2 - 16$

73. $y = 9 - x^2$

74. $y = 16 - x^2$

75. $y = 1 - x^2$

76. $y = 1 + x^2$

77. $y = x^2 - 4$

78. $y = x^2 - 25$

79. $y = x(x - 2)$ **80.** $y = x(x + 2)$

81. $y = -x(x + 4)$ **82.** $y = -x(x - 6)$

83. $y = |x| - 3$ **84.** $y = |x| - 5$

85. $y = |x| + 2$ **86.** $y = |x| + 4$

87. $y = |x + 2|$ **88.** $y = |x + 4|$

89. $y = |x - 3|$ **90.** $y = |x - 4|$

91. $y = -|x| + |x + 1|$

92. $y = |x| + |x - 2|$

93. *Straight-Line Depreciation* A manufacturing plant purchases a new molding machine for $230,000. The depreciated value y after t years is given by

$$y = 230,000 - 25,000t, \quad 0 \le t \le 8.$$

Sketch a graph of this model.

94. *Straight-Line Depreciation* A manufacturing plant purchases a new computer system for $20,000. The depreciated value y after t years is given by

$$y = 20,000 - 3000t, \quad 0 \le t \le 6.$$

Sketch a graph of this model.

✓ **95.** *Straight-Line Depreciation* Your company purchases a new delivery van for $40,000. For tax purposes, the van will be depreciated over a seven-year period. At the end of 7 years, the value of the van is expected to be $5000.

(a) Find an equation that relates the depreciated value of the van to the number of years since it was purchased.

(b) Sketch the graph of the equation.

(c) What is the y-intercept of the graph and what does it represent?

96. *Straight-Line Depreciation* Your company purchases a new limousine for $65,000. For tax purposes, the limousine will be depreciated over a 10-year period. At the end of 10 years, the value of the limousine is expected to be $10,000.

(a) Find an equation that relates the depreciated value of the limousine to the number of years since it was purchased.

(b) Sketch the graph of the equation.

(c) What is the y-intercept of the graph and what does it represent?

97. *Hooke's Law* The force F (in pounds) required to stretch a spring x inches from its natural length is given by

$$F = \frac{4}{3}x, \quad 0 \le x \le 12.$$

(a) Use the model to complete the table.

x	0	3	6	9	12
F					

(b) Sketch a graph of the model.

(c) Determine the required change in F if x is doubled. Explain your reasoning.

98. *Energy* The table shows the yearly world production P of natural gas (in quadrillion Btu) for the years 1998 through 2005. (Source: Energy Information Administration)

Year	1998	1999	2000	2001
P	86.0	87.9	91.3	93.7

Year	2002	2003	2004	2005
P	96.7	98.9	102.0	105.3

A model for this data is

$$P = 2.75t + 63.5$$

where t represents the year, with $t = 8$ corresponding to 1998.

(a) Sketch a graph of the model and plot the data on the same set of coordinate axes.

(b) How well does the model represent the data? Explain your reasoning.

(c) Use the model to predict the total world production of natural gas in the year 2006.

Explaining Concepts

99. *Misleading Graphs*   Graphs can help you visualize relationships between two variables, but they can also be misused to imply results that are not correct. The two graphs below represent the same data points. Why do the graphs appear different? Identify ways in which the graphs could be misleading.

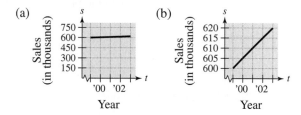

(a)   (b)

100. *Exploration*   Graph the equations $y = x^2 + 1$ and $y = -(x^2 + 1)$ on the same set of coordinate axes. Explain how the graph of an equation changes when the expression for y is multiplied by -1. Justify your answer by giving additional examples.

101. A company's profits decrease rapidly for a time, but then begin decreasing at a lower rate. Sketch an example of a graph representing such a situation, showing the profit y in terms of the time t.

102. Discuss the possible numbers of x- and y-intercepts of the graph of $y = ax + 5$, where a is a positive integer.

103. Discuss the possible numbers of x- and y-intercepts of a horizontal line in the rectangular coordinate system.

104. ✎   The graph shown represents the distance d in miles that a person drives during a 10-minute trip from home to work.

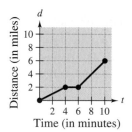

(a) How far is the person's home from the person's place of work? Explain.

(b) Describe the trip for time $4 < t < 6$. Explain.

(c) During what time interval is the person's speed greatest? Explain.

Cumulative Review

In Exercises 105–108, find the reciprocal.

105. 7

106. $\frac{1}{7}$

107. $\frac{4}{5}$

108. $\frac{8}{5}$

In Exercises 109–112, solve the equation and check the result.

109. $x - 8 = 0$

110. $x + 11 = 24$

111. $4x + 15 = 23$

112. $13 - 6x = -5$

In Exercises 113 and 114, plot the points and connect them with line segments to form a triangle.

113. $(0, 5), (3, 5), (3, 7)$

114. $(2, 4), (6, 2), (2, 2)$

3.3 Slope and Graphs of Linear Equations

Gary Conner/PhotoEdit Inc.

Why You Should Learn It

Slopes of lines can be used to describe rates of change. For instance, in Exercise 87 on page 161, you will use slope to describe the average rate of change in the tuition and fees paid by college students.

1 ▶ Determine the slope of a line through two points.

What You Should Learn

1 ▶ Determine the slope of a line through two points.
2 ▶ Write linear equations in slope-intercept form and graph the equations.
3 ▶ Use slopes to determine whether two lines are parallel, perpendicular, or neither.
4 ▶ Use slopes to describe rates of change in real-life problems.

The Slope of a Line

The **slope** of a nonvertical line is the number of units the line rises or falls vertically for each unit of horizontal change from left to right. For example, the line in Figure 3.23 rises two units for each unit of horizontal change from left to right, and so this line has a slope of $m = 2$.

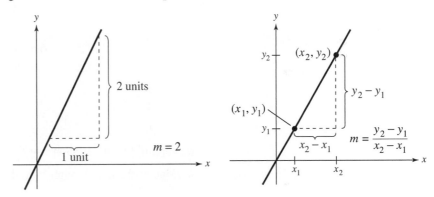

Figure 3.23   **Figure 3.24**

Definition of the Slope of a Line

The **slope** m of the nonvertical line passing through the points (x_1, y_1) and (x_2, y_2) is

$$m = \frac{y_2 - y_1}{x_2 - x_1} = \frac{\text{Change in } y}{\text{Change in } x} = \frac{\text{Rise}}{\text{Run}}$$

where $x_1 \neq x_2$. (See Figure 3.24.)

When the formula for slope is used, the *order of subtraction* is important. Given two points on a line, you can label either of them (x_1, y_1) and the other (x_2, y_2). However, once this has been done, you must form the numerator and denominator using the same order of subtraction.

$$m = \frac{y_2 - y_1}{x_2 - x_1} \qquad m = \frac{y_1 - y_2}{x_1 - x_2} \qquad m = \frac{y_2 - y_1}{x_1 - x_2}$$

Correct   Correct   Incorrect

For instance, the slope of the line passing through the points $(-1, 3)$ and $(4, -2)$ is

$$m = \frac{y_2 - y_1}{x_2 - x_1} = \frac{-2 - 3}{4 - (-1)} = \frac{-5}{5} = -1$$

or

$$m = \frac{y_1 - y_2}{x_1 - x_2} = \frac{3 - (-2)}{-1 - 4} = \frac{5}{-5} = -1.$$

EXAMPLE 1   Finding the Slope of a Line Through Two Points

Find the slope of the line passing through each pair of points.

a. $(1, 2)$ and $(4, 5)$ **b.** $(-1, 4)$ and $(2, 1)$

Solution

a. Let $(x_1, y_1) = (1, 2)$ and $(x_2, y_2) = (4, 5)$. The slope of the line through these points is

$$m = \frac{y_2 - y_1}{x_2 - x_1} \qquad \text{Difference in } y\text{-values}$$
$$\qquad \qquad \text{Difference in } x\text{-values}$$

$$= \frac{5 - 2}{4 - 1}$$

$$= 1.$$

The graph of the line is shown in Figure 3.25.

b. The slope of the line through $(-1, 4)$ and $(2, 1)$ is

$$m = \frac{1 - 4}{2 - (-1)}$$

$$= \frac{-3}{3}$$

$$= -1.$$

The graph of the line is shown in Figure 3.26.

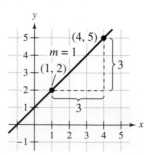

Positive Slope
Figure 3.25

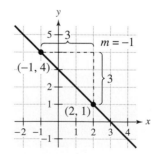

Negative Slope
Figure 3.26

✓ **CHECKPOINT** *Now try Exercise 9.*

EXAMPLE 2   **Finding the Slope of a Line Through Two Points**

Find the slope of the line passing through each pair of points.

a. $(1, 4)$ and $(3, 4)$ **b.** $(3, 1)$ and $(3, 3)$

Solution

a. Let $(x_1, y_1) = (1, 4)$ and $(x_2, y_2) = (3, 4)$. The slope of the line through these points is

$$m = \frac{y_2 - y_1}{x_2 - x_1}$$ ⇐ Difference in y-values
⇐ Difference in x-values

$$= \frac{4 - 4}{3 - 1}$$

$$= \frac{0}{2} = 0.$$

The graph of the line is shown in Figure 3.27.

b. The slope of the line through $(3, 1)$ and $(3, 3)$ is undefined. Applying the formula for slope, you have

$$\frac{3 - 1}{3 - 3} = \frac{2}{0}.$$ Division by 0 is undefined.

Because division by zero is not defined, the slope of a vertical line is not defined. The graph of the line is shown in Figure 3.28.

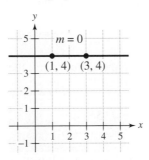

Zero Slope
Figure 3.27

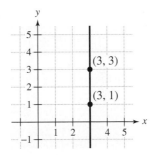

Slope is undefined.
Figure 3.28

✓ **CHECKPOINT** *Now try Exercise 11.*

From the slopes of the lines shown in Examples 1 and 2, you can make the following generalizations about the slope of a line.

Slope of a Line

1. A line with positive slope $(m > 0)$ *rises* from left to right.

2. A line with negative slope $(m < 0)$ *falls* from left to right.

3. A line with zero slope $(m = 0)$ is *horizontal*.

4. A line with undefined slope is *vertical*.

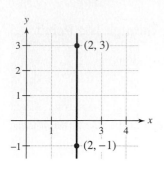

Vertical line
Undefined slope

Figure 3.29

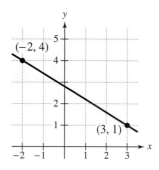

Line falls
Negative slope

Figure 3.30

| EXAMPLE 3 | Using Slope to Describe Lines |

Describe the lines through (a) $(2, -1)$ and $(2, 3)$ and (b) $(-2, 4)$ and $(3, 1)$.

Solution

a. Let $(x_1, y_1) = (2, -1)$ and $(x_2, y_2) = (2, 3)$.

$$m = \frac{3 - (-1)}{2 - 2} = \frac{4}{0} \qquad \text{Undefined slope (See Figure 3.29.)}$$

Because the slope is undefined, the line is vertical.

b. Let $(x_1, y_1) = (-2, 4)$ and $(x_2, y_2) = (3, 1)$.

$$m = \frac{1 - 4}{3 - (-2)} = -\frac{3}{5} < 0 \qquad \text{Negative slope (See Figure 3.30.)}$$

Because the slope is negative, the line falls from left to right.

✓ **CHECKPOINT** *Now try Exercise 19.*

| EXAMPLE 4 | Using Slope to Describe Lines |

Describe the lines through (a) $(1, 3)$ and $(4, 3)$ and (b) $(-1, 1)$ and $(2, 5)$.

Solution

a. Let $(x_1, y_1) = (1, 3)$ and $(x_2, y_2) = (4, 3)$.

$$m = \frac{3 - 3}{4 - 1} = \frac{0}{3} = 0 \qquad \text{Zero slope (See Figure 3.31.)}$$

Because the slope is zero, the line is horizontal.

b. Let $(x_1, y_1) = (-1, 1)$ and $(x_2, y_2) = (2, 5)$.

$$m = \frac{5 - 1}{2 - (-1)} = \frac{4}{3} > 0 \qquad \text{Positive slope (See Figure 3.32.)}$$

Because the slope is positive, the line rises from left to right.

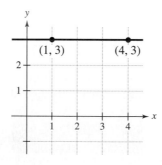

Horizontal line
Zero slope
Figure 3.31

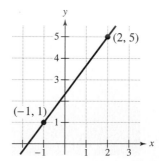

Line rises
Positive slope
Figure 3.32

✓ **CHECKPOINT** *Now try Exercise 23.*

Any two points on a nonvertical line can be used to calculate its slope. This is demonstrated in the next example.

EXAMPLE 5 · Finding the Slope of a Line

Sketch the graph of the line given by $2x + 3y = 6$. Then find the slope of the line. (Choose two different pairs of points on the line and show that the same slope is obtained from either pair.)

Solution

Begin by solving the equation for y.

$$2x + 3y = 6 \qquad \text{Write original equation.}$$

$$3y = -2x + 6 \qquad \text{Subtract } 2x \text{ from each side.}$$

$$y = \frac{-2x + 6}{3} \qquad \text{Divide each side by 3.}$$

$$y = -\frac{2}{3}x + 2 \qquad \text{Simplify.}$$

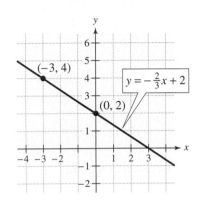

(a)

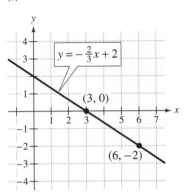

(b)

Figure 3.33

Then construct a table of values, as shown below.

x	-3	0	3	6
$y = -\frac{2}{3}x + 2$	4	2	0	-2
Solution point	$(-3, 4)$	$(0, 2)$	$(3, 0)$	$(6, -2)$

From the solution points shown in the table, sketch the graph of the line. (See Figure 3.33.) To calculate the slope of the line using two different sets of points, first use the points $(-3, 4)$ and $(0, 2)$, as shown in Figure 3.33(a), and obtain

$$m = \frac{2 - 4}{0 - (-3)} = -\frac{2}{3}.$$

Next, use the points $(3, 0)$ and $(6, -2)$, as shown in Figure 3.33(b), and obtain

$$m = \frac{-2 - 0}{6 - 3} = -\frac{2}{3}.$$

Try some other pairs of points on the line to see that you obtain a slope of $m = -\frac{2}{3}$ regardless of which two points you use.

✓ **CHECKPOINT** *Now try Exercise 25.*

Technology: Tip

Setting the viewing window on a graphing calculator affects the appearance of a line's slope. When you are using a graphing calculator, you cannot judge whether a slope is steep or shallow unless you use a *square setting.* See Appendix A for more information on setting a viewing window.

2 ▶ Write linear equations in slope-intercept form and graph the equations.

Slope as a Graphing Aid

You have seen that you should solve an equation for y before creating a table of values. When you do this for a linear equation, you obtain some very useful information. Consider the results of Example 5.

$$2x + 3y = 6 \qquad \text{Write original equation.}$$

$$3y = -2x + 6 \qquad \text{Subtract } 2x \text{ from each side.}$$

$$y = -\frac{2}{3}x + 2 \qquad \text{Divide each side by 3 and simplify.}$$

Observe that the coefficient of x is the slope of the graph of this equation. (See Example 5.) Moreover, the constant term, 2, gives the y-intercept of the graph.

$$y = -\frac{2}{3}x + 2$$

Slope y-intercept $(0, 2)$

This form is called the **slope-intercept form** of the equation of the line.

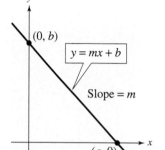

Figure 3.34

> ### Slope-Intercept Form of the Equation of a Line
>
> The graph of the equation
>
> $$y = mx + b$$
>
> is a line whose slope is m and whose y-intercept is $(0, b)$. (See Figure 3.34.)

EXAMPLE 6   Slope and y-Intercept of a Line

Find the slope and y-intercept of the graph of the equation

$$4x - 5y = 15.$$

Solution

Begin by writing the equation in slope-intercept form, as follows.

$$4x - 5y = 15 \qquad \text{Write original equation.}$$

$$-4x + 4x - 5y = -4x + 15 \qquad \text{Add } -4x \text{ to each side.}$$

$$-5y = -4x + 15 \qquad \text{Combine like terms.}$$

$$y = \frac{-4x + 15}{-5} \qquad \text{Divide each side by } -5.$$

$$y = \frac{4}{5}x - 3 \qquad \text{Slope-intercept form}$$

From the slope-intercept form, you can see that $m = \frac{4}{5}$ and $b = -3$. So, the slope of the graph of the equation is $\frac{4}{5}$ and the y-intercept is $(0, -3)$.

✓ **CHECKPOINT** *Now try Exercise 53.*

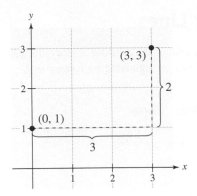

Figure 3.35

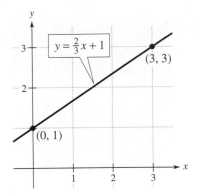

Figure 3.36

So far, you have been plotting several points in order to sketch the equation of a line. However, now that you can recognize equations of lines (linear equations), you don't have to plot as many points—two points are enough. (You might remember from geometry that *two points are all that are necessary to determine a line.*)

EXAMPLE 7   Using the Slope and y-Intercept to Sketch a Line

Use the slope and y-intercept to sketch the graph of

$$y = \frac{2}{3}x + 1.$$

Solution

The equation is already in slope-intercept form.

$$y = mx + b$$

$$y = \frac{2}{3}x + 1 \qquad\qquad \text{Slope-intercept form}$$

So, the slope of the line is

$$m = \frac{2}{3} = \frac{\text{Change in } y}{\text{Change in } x}$$

and the y-intercept is $(0, b) = (0, 1)$. You can sketch the graph of the line as follows. First, plot the y-intercept. Then, using a slope of $\frac{2}{3}$, locate a second point on the line by moving three units to the right and two units upward (or two units upward and three units to the right), as shown in Figure 3.35. Finally, draw a line through the two points to form the graph shown in Figure 3.36.

✓ **CHECKPOINT** *Now try Exercise 57.*

EXAMPLE 8   Using the Slope and y-Intercept to Sketch a Line

Use the slope and y-intercept to sketch the graph of $12x + 3y = 6$.

Solution

Begin by writing the equation in slope-intercept form.

$$12x + 3y = 6 \qquad\qquad \text{Write original equation.}$$

$$3y = -12x + 6 \qquad\qquad \text{Subtract } 12x \text{ from each side.}$$

$$y = \frac{-12x + 6}{3} \qquad\qquad \text{Divide each side by 3.}$$

$$y = -4x + 2 \qquad\qquad \text{Slope-intercept form}$$

So, the slope of the line is $m = -4$ and the y-intercept is $(0, b) = (0, 2)$. You can sketch the graph of the line as follows. First, plot the y-intercept. Then, using a slope of -4, locate a second point on the line by moving one unit to the right and four units downward (or four units downward and one unit to the right). Finally, draw a line through the two points to form the graph shown in Figure 3.37.

✓ **CHECKPOINT** *Now try Exercise 61.*

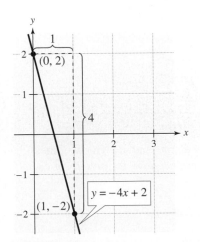

Figure 3.37

3 ▶ Use slopes to determine whether two lines are parallel, perpendicular, or neither.

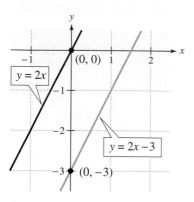

Figure 3.38

Parallel and Perpendicular Lines

You know from geometry that two lines in a plane are *parallel* if they do not intersect. What this means in terms of their slopes is suggested by Example 9.

EXAMPLE 9   Lines That Have the Same Slope

On the same set of coordinate axes, sketch the lines given by

$$y = 2x \quad \text{and} \quad y = 2x - 3.$$

Solution

For the line given by $y = 2x$, the slope is $m = 2$ and the y-intercept is $(0, 0)$. For the line given by $y = 2x - 3$, the slope is also $m = 2$ and the y-intercept is $(0, -3)$. The graphs of these two lines are shown in Figure 3.38.

✓ **CHECKPOINT** *Now try Exercise 75.*

In Example 9, notice that the two lines have the same slope *and* appear to be parallel. The following rule states that this is always the case.

> ### Parallel Lines
> Two distinct nonvertical lines are parallel if and only if they have the same slope.

The phrase "if and only if" is used in mathematics as a way to write two statements in one. The first statement says that *if two distinct nonvertical lines have the same slope, they must be parallel.* The second statement says that *if two distinct nonvertical lines are parallel, they must have the same slope.*

Another rule from geometry is that two lines in a plane are perpendicular if they intersect at right angles. In terms of their slopes, this means that two nonvertical lines are perpendicular if their slopes are negative reciprocals of each other. For instance, the negative reciprocal of 5 is $-\frac{1}{5}$, so the lines

$$y = 5x + 2 \quad \text{and} \quad y = -\frac{1}{5}x - 4$$

are perpendicular to each other, as shown in Figure 3.39.

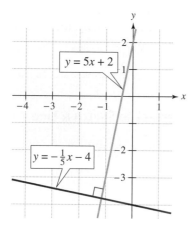

Figure 3.39

> ### Perpendicular Lines
> Consider two nonvertical lines whose slopes are m_1 and m_2. The two lines are perpendicular if and only if their slopes are *negative reciprocals* of each other. That is,
> $$m_1 = -\frac{1}{m_2}, \quad \text{or equivalently,} \quad m_1 \cdot m_2 = -1.$$

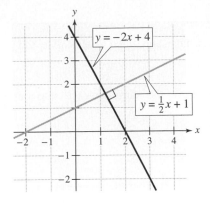

Figure 3.40

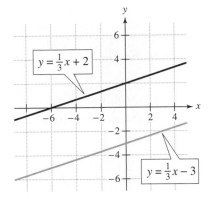

Figure 3.41

4 ▶ Use slopes to describe rates of change in real-life problems.

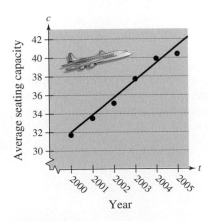

Figure 3.42

EXAMPLE 10   **Parallel or Perpendicular?**

Are the pairs of lines parallel, perpendicular, or neither?

a. $y = -2x + 4$, $y = \frac{1}{2}x + 1$ **b.** $y = \frac{1}{3}x + 2$, $y = \frac{1}{3}x - 3$

Solution

a. The first line has a slope of $m_1 = -2$, and the second line has a slope of $m_2 = \frac{1}{2}$. Because these slopes are negative reciprocals of each other, the two lines must be perpendicular, as shown in Figure 3.40.

b. Each of these two lines has a slope of $m = \frac{1}{3}$. So, the two lines must be parallel, as shown in Figure 3.41.

✓ **CHECKPOINT** *Now try Exercise 77.*

Slope as a Rate of Change

In real-life problems, slope can describe a **constant rate of change** or an **average rate of change.** In such cases, units of measure are used, such as miles per hour.

EXAMPLE 11   **Slope as a Rate of Change**

In 2000, the average seating capacity of a commuter aircraft was 31.7. By 2005, the average seating capacity had risen to 40.4. Find the average rate of change in average seating capacity of a commuter aircraft from 2000 to 2005.   (Source: Regional Airline Association and AvStat Associates)

Solution

Let c represent the average seating capacity and let t represent the year. The two given data points are represented by (t_1, c_1) and (t_2, c_2).

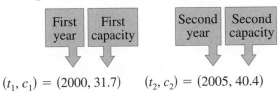

$$(t_1, c_1) = (2000, 31.7) \qquad (t_2, c_2) = (2005, 40.4)$$

Now use the formula for slope to find the average rate of change.

$$\text{Rate of change} = \frac{c_2 - c_1}{t_2 - t_1} \qquad \text{Slope formula}$$

$$= \frac{40.4 - 31.7}{2005 - 2000} \qquad \text{Substitute values.}$$

$$= \frac{8.7}{5} = 1.74 \qquad \text{Simplify.}$$

From 2000 through 2005, the average rate of change in the average seating capacity of a commuter aircraft was 1.74. The exact change in seating capacity varied from one year to the next, as shown in the scatter plot in Figure 3.42.

✓ **CHECKPOINT** *Now try Exercise 87.*

―――――――――――――――― **Concept Check** ――――――――――――――――

1. In your own words, give the interpretations of a negative slope, a zero slope, and a positive slope.

2. In the form $y = mx + b$, what does m represent? What does b represent?

3. Two distinct lines have undefined slopes. Are the lines parallel?

4. Do parallel lines have the same average rate of change? Explain.

3.3 EXERCISES

Go to pages 208–209 to record your assignments.

―――――――――――――――― **Developing Skills** ――――――――――――――――

In Exercises 1–6, estimate the slope of the line from its graph.

1.
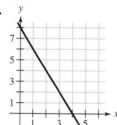

2.
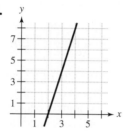

3.
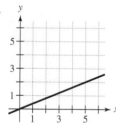

4.
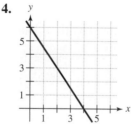

5.
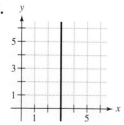

6.
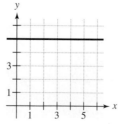

In Exercises 7 and 8, identify the line that has each slope m.

7. (a) $m = \frac{3}{4}$
   (b) $m = 0$
   (c) $m = -3$

8. (a) $m = -\frac{5}{2}$
   (b) m is undefined.
   (c) $m = 2$

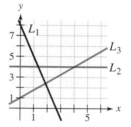
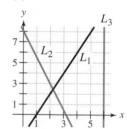

In Exercises 9–12, find the slope of the line passing through each pair of points (if possible). *See Examples 1 and 2.*

9. $(0, 0), (4, 8)$

10. $(0, 0), (-3, -12)$

11. $(2, 4), (7, 4)$

12. $(-1, 4), (-1, -3)$

In Exercises 13–24, plot the points and find the slope (if possible) of the line passing through them. State whether the line rises, falls, is horizontal, or is vertical. *See Examples 3 and 4.*

13. $(-4, 3), (-2, 5)$

14. $(7, 1), (4, -5)$

15. $(-5, -3), (-5, 4)$

16. $(9, 2), (-9, 2)$

17. $(2, -5), (7, -5)$

18. $(-3, 4), (-3, 8)$

☑ 19. $\left(\frac{3}{4}, 2\right), \left(5, -\frac{5}{2}\right)$ **20.** $\left(\frac{1}{2}, -1\right), \left(3, \frac{2}{3}\right)$

21. $\left(\frac{3}{4}, \frac{1}{4}\right), \left(-\frac{3}{2}, \frac{1}{8}\right)$ **22.** $\left(-\frac{3}{2}, -\frac{1}{2}\right), \left(\frac{4}{3}, -\frac{3}{2}\right)$

☑ 23. $(4.2, -1), (-4.2, 6)$ **24.** $(3.4, 0), (3.4, 1)$

In Exercises 25–30, sketch the graph of the line. Then find the slope of the line. *See Example 5.*

☑ 25. $y = 2x - 1$
26. $y = 3x + 2$
27. $y = -\frac{1}{2}x + 4$
28. $y = \frac{3}{4}x - 5$
29. $3x - 9y = 18$
30. $6x + 2y = 12$

In Exercises 31 and 32, solve for x so that the line through the points has the given slope.

31. $(4, 5), (x, 7)$
   $m = -\frac{2}{3}$
32. $(x, 3), (6, 6)$
   $m = \frac{3}{4}$

In Exercises 33 and 34, solve for y so that the line through the points has the given slope.

33. $(-3, y), (9, 3)$
   $m = \frac{3}{2}$
34. $(-3, 20), (2, y)$
   $m = -6$

In Exercises 35–42, a point on a line and the slope of the line are given. Find two additional points on the line. (There are many correct answers.)

35. $(5, 2)$ **36.** $(-4, 3)$
   $m = 0$ m is undefined.

37. $(3, -4)$ **38.** $(-1, -5)$
   $m = 3$ $m = 2$

39. $(-2, -3)$ **40.** $(-1, 6)$
   $m = -1$ $m = -3$

41. $(-5, 0)$ **42.** $(-1, 1)$
   $m = \frac{4}{3}$ $m = -\frac{3}{4}$

In Exercises 43–50, write the equation of the line in slope-intercept form.

43. $6x - 3y = 9$ **44.** $12x - 4y = 4$

45. $4y - x = -4$ **46.** $3x - 2y = -10$

47. $2x + 5y - 3 = 0$ **48.** $8x - 6y + 1 = 0$

49. $x = 2y - 4$ **50.** $x = -\frac{3}{2}y + \frac{2}{3}$

In Exercises 51–56, find the slope and y-intercept of the line. *See Example 6.*

51. $y = 3x - 2$
52. $y = 4 - 2x$
☑ 53. $4x - 6y = 24$
54. $4x + 8y = -1$
55. $5x + 3y - 2 = 0$
56. $6y - 5x + 18 = 0$

In Exercises 57–66, write the equation of the line in slope-intercept form, and then use the slope and y-intercept to sketch the line. *See Examples 7 and 8.*

☑ 57. $x + y = 0$ **58.** $x - y = 0$

59. $3x - y - 2 = 0$ **60.** $2x - y - 3 = 0$

☑ 61. $3x + 2y - 2 = 0$ **62.** $x - 2y - 2 = 0$

63. $x - 4y + 2 = 0$ **64.** $8x + 6y - 3 = 0$

65. $0.2x - 0.8y - 4 = 0$ **66.** $0.5x + 0.6y - 3 = 0$

In Exercises 67–70, sketch the graph of a line through the point (3, 2) having the given slope.

67. $m = -\frac{1}{3}$

68. $m = \frac{3}{2}$

69. m is undefined.

70. $m = 0$

In Exercises 71–74, plot the x- and y-intercepts and sketch the line.

71. $3x - 5y - 15 = 0$

72. $3x + 5y + 15 = 0$

73. $-4x - 2y + 16 = 0$

74. $-5x + 2y - 20 = 0$

In Exercises 75–78, determine whether the two lines are parallel, perpendicular, or neither. *See Examples 9 and 10.*

✓ **75.** L_1: $y = \frac{1}{2}x - 2$
    L_2: $y = \frac{1}{2}x + 3$

76. L_1: $y = 3x - 2$
    L_2: $y = 3x + 1$

✓ **77.** L_1: $y = \frac{3}{4}x - 3$
    L_2: $y = -\frac{4}{3}x + 1$

78. L_1: $y = -\frac{2}{3}x - 5$
    L_2: $y = \frac{3}{2}x + 1$

In Exercises 79–82, determine whether the lines L_1 and L_2 passing through the pair of points are parallel, perpendicular, or neither.

79. L_1: $(0, 4), (2, 8)$
    L_2: $(0, -1), (3, 5)$

80. L_1: $(3, 4), (-2, 3)$
    L_2: $(0, -3), (2, -1)$

81. L_1: $(0, 2), (6, -2)$
    L_2: $(2, 0), (8, 4)$

82. L_1: $(3, 2), (-1, -2)$
    L_2: $(2, 0), (3, -1)$

Solving Problems

83. *Road Grade*   When driving down a mountain road, you notice warning signs indicating an "8% grade." This means that the slope of the road is $-\frac{8}{100}$. Over a stretch of the road, your elevation drops by 2000 feet (see figure). What is the horizontal change in your position?

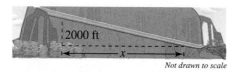

2000 ft

x

Not drawn to scale

84. *Ramp*   A loading dock ramp rises 4 feet above the ground. The ramp has a slope of $\frac{1}{10}$. What is the horizontal length of the ramp?

85. *Roof Pitch*   The slope, or pitch, of a roof (see figure) is such that it rises (or falls) 3 feet for every 4 feet of horizontal distance. Determine the maximum height of the attic of the house for a 30-foot wide house.

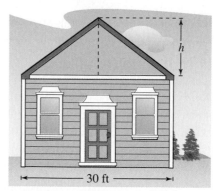

h

30 ft

Figure for 85 and 86

86. *Roof Pitch*   The slope, or pitch, of a roof (see figure) is such that it rises (or falls) 4 feet for every 5 feet of horizontal distance. Determine the maximum height of the attic of the house for a 30-foot wide house.

87. *Education*   The average annual amount of tuition and fees y paid by an in-state student attending a 4-year university in the United States from 2001 to 2006 can be approximated by the model

$$y = 690.9t + 6505, \quad 1 \le t \le 6$$

where t represents the year, with $t = 1$ corresponding to 2001. (Source: U.S. National Center for Education Statistics)

(a) Use the model to complete the table.

t	1	2	3
y			

t	4	5	6
y			

(b) Graph the model on a rectangular coordinate system.

(c) Find the average rate of change of tuition and fees from 2001 to 2006.

(d) Use the model to predict the amount of tuition and fees that would be paid in the year 2015.

88. *Simple Interest*   An inheritance of $8000 is invested in an account that pays $7\frac{1}{2}\%$ simple interest. The amount of money in the account after t years is given by the model

$$y = 8000 + 600t, \quad t \ge 0.$$

(a) Use the model to complete the table.

t	0	1	2
y			

t	3	4	5
y			

(b) Graph the equation on a rectangular coordinate system.

(c) Find the average rate of change of the amount in the account over 5 years.

Explaining Concepts

89. ✎   Can any pair of points on a line be used to calculate the slope of the line? Explain.

90. ✎   The slopes of two lines are -3 and $\frac{3}{2}$. Which is steeper? Explain.

91. ✎   What is the relationship between the x-intercept of the graph of the line $y = mx + b$ and the solution to the equation $mx + b = 0$? Explain.

92. ✎   Is it possible for two lines with positive slopes to be perpendicular to each other? Explain.

93. ✎   Lines L_1 and L_2 are perpendicular. Line L_1 has a y-intercept of $(0, 3)$. Is it true that the y-intercept of line L_2 must equal $\left(0, -\frac{1}{3}\right)$? Explain.

Cumulative Review

In Exercises 94–97, rewrite the statement using inequality notation.

94. x is negative

95. m is at least -3

96. z is at least 85, but no more than 100

97. n is less than 20, but no less than 16

In Exercises 98–105, solve the equation.

98. $|x| = 8$

99. $|g| = -4$

100. $|4h| = 24$

101. $\left|\frac{1}{5}m\right| = 2$

102. $|x + 4| = 5$

103. $|2t - 3| = 11$

104. $|6b + 8| = -2b$

105. $|n - 3| = |2n + 9|$

3.4 Equations of Lines

Why You Should Learn It

Linear equations can be used to model and solve real-life problems. For instance, in Exercise 90 on page 171, a linear equation is used to model the relationship between the price of soft drinks and the demand for that product.

1 ▶ Write equations of lines using point-slope form.

What You Should Learn

1 ▶ Write equations of lines using point-slope form.
2 ▶ Write equations of horizontal, vertical, parallel, and perpendicular lines.
3 ▶ Use linear models to solve application problems.

The Point-Slope Form of the Equation of a Line

In Sections 3.1 through 3.3, you have been studying analytic (or coordinate) geometry. Analytic geometry uses a coordinate plane to give visual representations of algebraic concepts, such as equations or functions.

There are two basic types of problems in analytic geometry.

1. Given an equation, sketch its graph: Algebra ⟹ Geometry
2. Given a graph, write its equation: Geometry ⟹ Algebra

In Section 3.3, you worked primarily with the first type of problem. In this section, you will study the second type. Specifically, you will learn how to write the equation of a line when you are given its slope and a point on the line. Before you learn a general formula for doing this, consider the following example.

EXAMPLE 1   Writing an Equation of a Line

Write an equation of the line with slope $\frac{4}{3}$ that passes through the point $(-2, 1)$.

Solution

Begin by sketching the line, as shown in Figure 3.43. You know that the slope of a line is the same through any two points on the line. So, to find an equation of the line, let (x, y) represent any point on the line. Using the representative point (x, y) and the point $(-2, 1)$, it follows that the slope of the line is

$$m = \frac{y - 1}{x - (-2)}.$$

◀ Difference in y-values
◀ Difference in x-values

Because the slope of the line is $m = \frac{4}{3}$, this equation can be rewritten as follows.

$$\frac{4}{3} = \frac{y - 1}{x + 2} \qquad \text{Slope formula}$$

$$4(x + 2) = 3(y - 1) \qquad \text{Cross-multiply.}$$

$$4x + 8 = 3y - 3 \qquad \text{Distributive Property}$$

$$4x - 3y = -11 \qquad \text{Subtract 8 and } 3y \text{ from each side.}$$

An equation of the line is $4x - 3y = -11$.

 CHECKPOINT *Now try Exercise 5.*

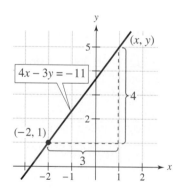

Figure 3.43

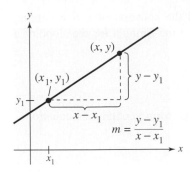

Figure 3.44

The procedure in Example 1 can be used to derive a *formula* for the equation of a line, given its slope and a point on the line. In Figure 3.44, let (x_1, y_1) be a given point on the line whose slope is m. If (x, y) is any *other* point on the line, it follows that

$$\frac{y - y_1}{x - x_1} = m.$$

This equation in variables x and y can be rewritten in the form

$$y - y_1 = m(x - x_1)$$

which is called the **point-slope form** of the equation of a line.

Point-Slope Form of the Equation of a Line

The **point-slope form** of the equation of the line that passes through the point (x_1, y_1) and has a slope of m is

$$y - y_1 = m(x - x_1).$$

EXAMPLE 2   **The Point-Slope Form of the Equation of a Line**

Write an equation of the line that passes through the point $(2, -3)$ and has slope $m = -2$.

Solution

Use the point-slope form with $(x_1, y_1) = (2, -3)$ and $m = -2$.

$$y - y_1 = m(x - x_1) \qquad \text{Point-slope form}$$
$$y - (-3) = -2(x - 2) \qquad \text{Substitute } y_1 = -3, \ x_1 = 2, \text{ and } m = -2.$$
$$y + 3 = -2x + 4 \qquad \text{Simplify.}$$
$$y = -2x + 1 \qquad \text{Subtract 3 from each side.}$$

So, an equation of the line is $y = -2x + 1$. Note that this is the slope-intercept form of the equation. The graph of the line is shown in Figure 3.45.

✓ **CHECKPOINT** *Now try Exercise 19.*

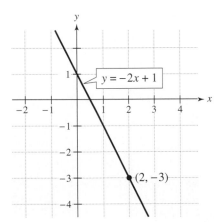

Figure 3.45

In Example 2, note that it was stated that $y = -2x + 1$ is "an" equation of the line rather than "the" equation of the line. The reason for this is that every equation can be written in many equivalent forms. For instance,

$$y = -2x + 1, \quad 2x + y = 1, \quad \text{and} \quad 2x + y - 1 = 0$$

are all equations of the line in Example 2. The first of these equations $(y = -2x + 1)$ is the slope-intercept form

$$y = mx + b \qquad \text{Slope-intercept form}$$

and it provides the most information about the line. The last of these equations $(2x + y - 1 = 0)$ is the **general form** of the equation of a line.

$$ax + by + c = 0 \qquad \text{General form}$$

The point-slope form can be used to find the equation of a line passing through two points (x_1, y_1) and (x_2, y_2). First, use the formula for the slope of a line passing through two points.

$$m = \frac{y_2 - y_1}{x_2 - x_1}$$

Then, substitute this value for m in the point-slope form to obtain the equation

$$y - y_1 = \frac{y_2 - y_1}{x_2 - x_1}(x - x_1). \qquad \text{Two-point form}$$

This is sometimes called the **two-point form** of the equation of a line.

EXAMPLE 3   An Equation of a Line Passing Through Two Points

Write the general form of the equation of the line that passes through the points $(4, 2)$ and $(-2, 3)$.

Solution

Let $(x_1, y_1) = (4, 2)$ and $(x_2, y_2) = (-2, 3)$. Then apply the formula for the slope of a line passing through two points, as follows.

$$m = \frac{y_2 - y_1}{x_2 - x_1}$$

$$= \frac{3 - 2}{-2 - 4}$$

$$= -\frac{1}{6}$$

Now, using the point-slope form, you can find the equation of the line.

$y - y_1 = m(x - x_1)$	Point-slope form
$y - 2 = -\dfrac{1}{6}(x - 4)$	Substitute $y_1 = 2$, $x_1 = 4$, and $m = -\frac{1}{6}$.
$6(y - 2) = -(x - 4)$	Multiply each side by 6.
$6y - 12 = -x + 4$	Distributive Property
$x + 6y - 12 = 4$	Add x to each side.
$x + 6y - 16 = 0$	Subtract 4 from each side.

The general form of the equation of the line is $x + 6y - 16 = 0$. The graph of this line is shown in Figure 3.46.

✓ **CHECKPOINT** *Now try Exercise 39.*

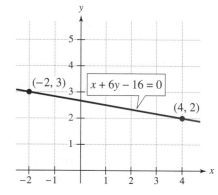

Figure 3.46

In Example 3, it does not matter which of the two points is labeled (x_1, y_1) and which is labeled (x_2, y_2). Try switching these labels to

$$(x_1, y_1) = (-2, 3) \quad \text{and} \quad (x_2, y_2) = (4, 2)$$

and reworking the problem to see that you obtain the same equation.

2 ▶ Write equations of horizontal, vertical, parallel, and perpendicular lines.

Other Equations of Lines

Recall from Section 3.3 that a horizontal line has a slope of zero. From the slope-intercept form of the equation of a line, you can see that a horizontal line has an equation of the form

$$y = (0)x + b \quad \text{or} \quad y = b. \qquad \text{Horizontal line}$$

This is consistent with the fact that each point on a horizontal line through $(0, b)$ has a y-coordinate of b, as shown in Figure 3.47. Similarly, each point on a vertical line through $(a, 0)$ has an x-coordinate of a, as shown in Figure 3.48. Because a vertical line has an undefined slope, it has an equation of the form

$$x = a. \qquad \text{Vertical line}$$

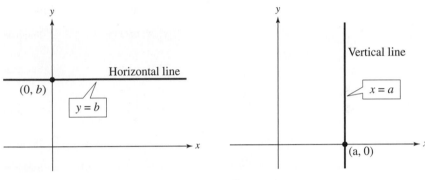

Figure 3.47   **Figure 3.48**

EXAMPLE 4   **Writing Equations of Horizontal and Vertical Lines**

Write an equation for each line.

a. Vertical line through $(-2, 4)$

b. Horizontal line through $(0, 6)$

c. Line passing through $(-2, 3)$ and $(3, 3)$

d. Line passing through $(-1, 2)$ and $(-1, 3)$

Solution

a. Because the line is vertical and passes through the point $(-2, 4)$, you know that every point on the line has an x-coordinate of -2. So, the equation is $x = -2$.

b. Because the line is horizontal and passes through the point $(0, 6)$, you know that every point on the line has a y-coordinate of 6. So, the equation of the line is $y = 6$.

c. Because both points have the same y-coordinate, the line through $(-2, 3)$ and $(3, 3)$ is horizontal. So, its equation is $y = 3$.

d. Because both points have the same x-coordinate, the line through $(-1, 2)$ and $(-1, 3)$ is vertical. So, its equation is $x = -1$.

✓ **CHECKPOINT** *Now try Exercise 59.*

In Section 3.3, you learned that parallel lines have the same slope and perpendicular lines have slopes that are negative reciprocals of each other. You can use these facts to write an equation of a line parallel or perpendicular to a given line.

EXAMPLE 5   Parallel and Perpendicular Lines

Write equations of the lines that pass through the point $(3, -2)$ and are (a) parallel and (b) perpendicular to the line $x - 4y = 6$, as shown in Figure 3.49.

Solution

By writing the given line in slope-intercept form, $y = \frac{1}{4}x - \frac{3}{2}$, you can see that it has a slope of $\frac{1}{4}$. So, a line parallel to it must also have a slope of $\frac{1}{4}$ and a line perpendicular to it must have a slope of -4.

a. $y - y_1 = m(x - x_1)$ Point-slope form

$y - (-2) = \frac{1}{4}(x - 3)$ Substitute $y_1 = -2$, $x_1 = 3$, and $m = \frac{1}{4}$.

$y = \frac{1}{4}x - \frac{11}{4}$ Equation of parallel line

b. $y - y_1 = m(x - x_1)$ Point-slope form

$y - (-2) = -4(x - 3)$ Substitute $y_1 = -2$, $x_1 = 3$, and $m = -4$.

$y = -4x + 10$ Equation of perpendicular line

✓ **CHECKPOINT** *Now try Exercise 65.*

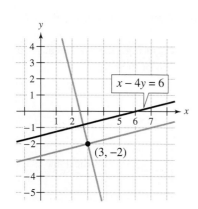

$x - 4y = 6$

$(3, -2)$

Figure 3.49

The equation of a vertical line cannot be written in slope-intercept form because the slope of a vertical line is undefined. However, *every* line has an equation that can be written in the general form $ax + by + c = 0$, where a and b are not both zero.

Study Tip

The slope-intercept form of the equation of a line is better suited for *sketching a line.* On the other hand, the point-slope form of the equation of a line is better suited for *creating the equation of a line,* given its slope and a point on the line.

Summary of Equations of Lines

1. Slope of a line through (x_1, y_1) and (x_2, y_2): $m = \dfrac{y_2 - y_1}{x_2 - x_1}$

2. General form of equation of line: $ax + by + c = 0$

3. Equation of vertical line: $x = a$

4. Equation of horizontal line: $y = b$

5. Slope-intercept form of equation of line: $y = mx + b$

6. Point-slope form of equation of line: $y - y_1 = m(x - x_1)$

7. Parallel lines (equal slopes): $m_1 = m_2$

8. Perpendicular lines (negative reciprocal slopes): $m_1 = -\dfrac{1}{m_2}$

3 ▶ Use linear models to solve application problems.

Application

EXAMPLE 6   **Total Sales**

Texas Instruments, Inc. had total sales of $12.6 billion in 2004 and $14.3 billion in 2006.   (Source: Texas Instruments, Inc.)

a. Using only this information, write a linear equation that models the sales (in billions of dollars) in terms of the year.

b. Interpret the meaning of the slope in the context of the problem.

c. Predict the sales for 2007.

Solution

a. Let $t = 4$ represent 2004. Then the two given values are represented by the data points $(4, 12.6)$ and $(6, 14.3)$. The slope of the line through these points is

$$m = \frac{y_2 - y_1}{t_2 - t_1}$$

$$= \frac{14.3 - 12.6}{6 - 4}$$

$$= 0.85.$$

Using the point-slope form, you can find the equation that relates the sales y and the year t to be

$y - y_1 = m(t - t_1)$	Point-slope form
$y - 12.6 = 0.85(t - 4)$	Substitute for y_1, m, and t_1.
$y - 12.6 = 0.85t - 3.4$	Distributive Property
$y = 0.85t + 9.2.$	Write in slope-intercept form.

b. The slope of the equation in part (a) indicates that the total sales for Texas Instruments, Inc. increased by $0.85 billion each year.

c. Using the equation from part (a), you can predict the sales for 2007 $(t = 7)$ to be

$y = 0.85t + 9.2$	Equation from part (a)
$y = 0.85(7) + 9.2$	Substitute 7 for t.
$y = 15.15.$	Simplify.

So, the predicted sales for 2007 are $15.2 billion. The graph of this equation is shown in Figure 3.50.

✓ **CHECKPOINT** *Now try Exercise 79.*

The estimation method illustrated in Example 6 is called **linear extrapolation.** Note in Figure 3.51(a) that for linear extrapolation, the estimated point lies to the right of the given points. When the estimated point lies *between* two given points, as in Figure 3.51(b), the procedure is called **linear interpolation.**

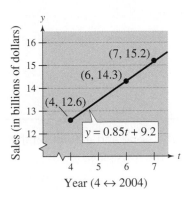

Figure 3.50

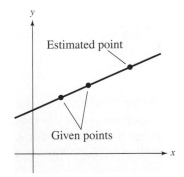

(a) Linear Extrapolation

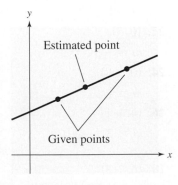

(b) Linear Interpolation

Figure 3.51

Concept Check

1. The equation $y - 1 = 4(x - 1)$ is written in point-slope form. What is the slope? By observation, what point does the equation of the line pass through?

2. Is the general form of an equation of a line equivalent to the point-slope form? Explain.

3. A horizontal line passes through the point $(-4, 5)$. Do you have enough information to write an equation of the line? If so, what is the equation of the line?

4. A line passes through the point $(3, 2)$. Do you have enough information to write an equation of the line? Explain.

3.4 EXERCISES

Go to pages 208–209 to record your assignments.

Developing Skills

In Exercises 1–4, match the equation with its graph. [The graphs are labeled (a), (b), (c), and (d).]

(a)
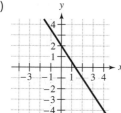

(b)
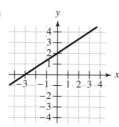

(c)
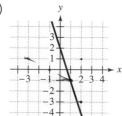

(d)
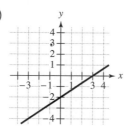

1. $y = \frac{2}{3}x + 2$

2. $y = \frac{2}{3}x - 2$

3. $y = -\frac{3}{2}x + 2$

4. $y = -3x + 2$

In Exercises 5–10, write an equation of the line that passes through the point and has the specified slope. **See Example 1.**

 5. $(2, -3)$
   $m = 3$

6. $(-1, 6)$
   $m = -4$

7. $(-3, 1)$
   $m = -\frac{1}{2}$

8. $(6, 9)$
   $m = \frac{2}{3}$

9. $\left(\frac{3}{4}, -1\right)$
   $m = \frac{4}{5}$

10. $\left(-2, \frac{3}{2}\right)$
    $m = -\frac{1}{6}$

In Exercises 11–18, state the slope and a point on the graph for each equation in point-slope form.

11. $y - 1 = 2(x - 3)$

12. $y - 4 = 3(x - 2)$

13. $y + 1 = -5(x - 8)$

14. $y - 6 = 4(x + 1)$

15. $y + 3 = \frac{1}{2}(x + 6)$

16. $y + 7 = -\frac{2}{5}(x - 5)$

17. $y = -8x$

18. $y = 9x$

In Exercises 19–34, use the point-slope form of the equation of a line to write an equation of the line that passes through the point and has the specified slope. When possible, write the equation in slope-intercept form. **See Example 2.**

 19. $(0, 0)$
   $m = -\frac{1}{2}$

20. $(0, 0)$
    $m = \frac{1}{5}$

21. $(0, -4)$
    $m = 3$

22. $(0, 5)$
    $m = -3$

23. $(0, 6)$
    $m = -\frac{3}{4}$

24. $(0, -8)$
    $m = \frac{2}{3}$

25. $(-2, 8)$
    $m = -2$

26. $(4, -1)$
    $m = 3$

27. $(-4, -7)$
    $m = \frac{5}{4}$

28. $(6, -8)$
    $m = -\frac{2}{3}$

29. $\left(-2, \frac{7}{2}\right)$
$m = -4$

30. $\left(1, -\frac{3}{2}\right)$
$m = 1$

31. $\left(\frac{3}{4}, \frac{5}{2}\right)$
$m = \frac{4}{3}$

32. $\left(-\frac{5}{2}, \frac{1}{2}\right)$
$m = -\frac{2}{5}$

33. $(2, -1)$
$m = 0$

34. $(-8, 5)$
$m = 0$

In Exercises 35–38, write an equation of the line that passes through the point and has the specified y-intercept.

35. $(5, 2)$
$b = 3$

36. $(3, -2)$
$b = -2$

37. $(-1, 6)$
$b = -\frac{2}{3}$

38. $(-2, -4)$
$b = \frac{1}{4}$

In Exercises 39–54, write the general form of the equation of the line that passes through the two points. *See Example 3.*

✓ **39.** $(0, 0), (3, -4)$

40. $(0, 0), (5, 2)$

41. $(0, 4), (4, 0)$

42. $(0, -2), (2, 0)$

43. $(1, 4), (5, 6)$

44. $(6, 3), (3, 2)$

45. $(-5, 2), (5, -2)$

46. $(-2, -3), (-4, -6)$

47. $\left(\frac{3}{2}, 3\right), \left(\frac{9}{2}, 4\right)$

48. $\left(4, \frac{7}{3}\right), \left(-1, \frac{1}{3}\right)$

49. $\left(10, \frac{1}{2}\right), \left(\frac{3}{2}, \frac{7}{4}\right)$

50. $\left(-4, \frac{3}{5}\right), \left(\frac{3}{4}, -\frac{2}{5}\right)$

51. $(5, 9), (8, -1.4)$

52. $(2, -8), (6, 2.3)$

53. $(2, 0.6), (8, -4.2)$

54. $(-5, 0.6), (3, -3.4)$

In Exercises 55–58, write the slope-intercept form of the equation of the line that passes through the two points.

55. $(-1, 1), (1, 7)$

56. $(0, 10), (5, 0)$

57. $(-2, 3), (4, 3)$

58. $(-6, -3), (4, 3)$

In Exercises 59–64, write an equation of the line. *See Example 4.*

✓ **59.** Vertical line through $(-1, 5)$

60. Vertical line through $(2, -3)$

61. Horizontal line through $(0, -5)$

62. Horizontal line through $(-4, 6)$

63. Line through $(-7, 2)$ and $(-7, -1)$

64. Line through $(6, 4)$ and $(-9, 4)$

In Exercises 65–74, write equations of the lines that pass through the point and are (a) parallel and (b) perpendicular to the given line. *See Example 5.*

✓ **65.** $(2, 1)$
$6x - 2y = 3$

66. $(-3, 4)$
$x + 6y = 12$

67. $(-5, 4)$
$5x + 4y = 24$

68. $(6, -4)$
$3x + 10y = 24$

69. $(5, -3)$
$4x - y - 3 = 0$

70. $(-5, -10)$
$2x + 5y - 12 = 0$

71. $\left(\frac{2}{3}, \frac{4}{3}\right)$
$x - 5 = 0$

72. $\left(\frac{5}{8}, \frac{9}{4}\right)$
$-5x + 4y = 0$

73. $(-1, 2)$
$y + 5 = 0$

74. $(3, -4)$
$x - 10 = 0$

In Exercises 75–78, write the **intercept form** of the equation of the line with intercepts $(a, 0)$ and $(0, b)$. The equation is given by
$$\frac{x}{a} + \frac{y}{b} = 1, \; a \neq 0, \; b \neq 0.$$

75. x-intercept: $(3, 0)$
y-intercept: $(0, 2)$

76. x-intercept: $(-6, 0)$
y-intercept: $(0, 2)$

77. x-intercept: $\left(-\frac{5}{6}, 0\right)$
    y-intercept: $\left(0, -\frac{7}{3}\right)$

78. x-intercept: $\left(-\frac{8}{3}, 0\right)$
    y-intercept: $(0, -4)$

Solving Problems

79. *Cost*   The cost C (in dollars) of producing x units of a product is shown in the table. Find a linear model to represent the data. Estimate the cost of producing 400 units.

x	0	50	100	150	200
C	5000	6000	7000	8000	9000

80. *Temperature Conversion*   The relationship between the Fahrenheit F and Celsius C temperature scales is shown in the table. Find a linear model to represent the data. Estimate the Fahrenheit temperature when the Celsius temperature is 18 degrees.

C	0	5	10	15	20
F	32	41	50	59	68

81. *Sales*   The total sales for a new camera equipment store were $200,000 for the second year and $500,000 for the fifth year. Find a linear model to represent the data. Estimate the total sales for the sixth year.

Scott Olson/Getty Images

82. *Sales*   The total sales for a new sportswear store were $150,000 for the third year and $250,000 for the fifth year. Find a linear model to represent the data. Estimate the total sales for the sixth year.

83. *Sales Commission*   The salary for a sales representative is $1500 per month plus a commission of total monthly sales. The table shows the relationship between the salary S and total monthly sales M. Write an equation of the line giving the salary S in terms of the monthly sales M. What is the commission rate?

M	0	1000	2000	3000	4000
S	1500	1530	1560	1590	1620

84. *Reimbursed Expenses*   A sales representative is reimbursed $140 per day for lodging and meals plus an amount per mile driven. The table below shows the relationship between the daily cost C to the company and the number of miles driven x. Write an equation giving the daily cost C to the company in terms of x, the number of miles driven. How much is the sales representative reimbursed per mile?

x	50	100	150	200	250
C	161	182	203	224	245

85. *Discount*   A store is offering a 30% discount on all items in its inventory.

(a) Write an equation of the line giving the sale price S for an item in terms of its list price L.

(b) 🖩 Use a graphing calculator to graph the line in part (a) and estimate graphically the sale price of an item that has a list price of $500. Confirm your estimate algebraically.

(c) 🖩 Use the graph in part (b) to estimate graphically the list price when the sale price of an item is $210. Confirm your estimate algebraically.

86. *Reimbursed Expenses* A sales representative is reimbursed $150 per day for lodging and meals plus $0.40 per mile driven.

(a) Write an equation of the line giving the daily cost C to the company in terms of x, the number of miles driven.

(b) ▦ Use a graphing calculator to graph the line in part (a) and estimate graphically the daily cost to the company when the representative drives 230 miles. Confirm your estimate algebraically.

(c) ▦ Use the graph in part (b) to estimate graphically the number of miles driven when the daily cost to the company is $200. Confirm your estimate algebraically.

87. *Straight-Line Depreciation* A small business purchases a photocopier for $7400. After 4 years, its depreciated value will be $1500.

(a) Assuming straight-line depreciation, write an equation of the line giving the value V of the copier in terms of time t in years.

(b) Use the equation in part (a) to find the value of the copier after 2 years.

88. *Straight-Line Depreciation* A business purchases a van for $27,500. After 5 years, its depreciated value will be $12,000.

(a) Assuming straight-line depreciation, write an equation of the line giving the value V of the van in terms of time t in years.

(b) Use the equation in part (a) to find the value of the van after 2 years.

89. *Education* A small college had an enrollment of 1500 students in 1995. During the next 10 years, the enrollment increased by approximately 60 students per year.

(a) Write an equation of the line giving the enrollment N in terms of the year t. (Let $t = 5$ correspond to the year 1995.)

(b) *Linear Extrapolation* Use the equation in part (a) to predict the enrollment in the year 2015.

(c) *Linear Interpolation* Use the equation in part (a) to estimate the enrollment in 2000.

90. *Demand* When soft drinks sold for $0.80 per cup at football games, approximately 6000 cups were sold. When the price was raised to $1.00 per cup, the demand dropped to 4000. Assume that the relationship between the price p and demand d is linear.

(a) Write an equation of the line giving the demand d in terms of the price p.

(b) *Linear Extrapolation* Use the equation in part (a) to predict the number of cups of soft drinks sold if the price is raised to $1.10.

(c) *Linear Interpolation* Use the equation in part (a) to estimate the number of cups of soft drinks sold if the price is $0.90.

91. *Data Analysis* The table shows the expected number of additional years of life E for a person of age A. (Source: U.S. National Center for Health Statistics)

A	Birth	10	20	40	60	80
E	77.5	68.2	58.4	39.5	22.2	9.0

(a) Sketch a scatter plot of the data.

(b) Use a straightedge to sketch the "best-fitting" line through the points.

(c) Find an equation of the line you sketched in part (b).

(d) Use the equation in part (c) to estimate the expected number of additional years of life of a person who is 30 years old.

92. *Data Analysis* The table shows the total revenue R (in billions of dollars) of Verizon Wireless for the years 2002 through 2006, where t represents the year, with $t = 2$ corresponding to 2002. (Source: Verizon Wireless)

t	2	3	4	5	6
R	19.5	22.5	27.7	32.3	38.0

(a) Sketch a scatter plot of the data.

(b) Use a straightedge to sketch the "best-fitting" line through the points.

(c) Find an equation of the line you sketched in part (b).

(d) Use the equation in part (c) to estimate the revenue for 2007.

93. *Depth Markers* A swimming pool is 40 feet long, 20 feet wide, 4 feet deep at the shallow end, and 9 feet deep at the deep end. Position the side of the pool on a rectangular coordinate system as shown in the figure and find an equation of the line representing the edge of the inclined bottom of the pool. Use this equation to determine the distances from the deep end at which markers must be placed to indicate each one-foot change in the depth of the pool.

94. *Carpentry* A carpenter uses a wedge-shaped block of wood to support heavy objects. The block of wood is 12 inches long, 2 inches wide, 6 inches high at the tall end, and 2 inches high at the short end. Position the side of the block on a rectangular coordinate system as shown in the figure and find an equation of the line representing the edge of the slanted top of the block. Use this equation to determine the distances from the tall end at which marks must be made to indicate each one-inch change in the height of the block.

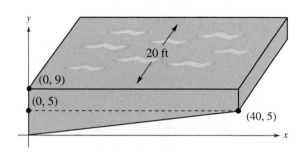

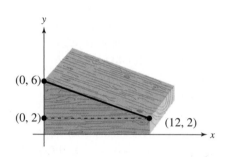

Explaining Concepts

95. Write, from memory, the point-slope form, the slope-intercept form, and the general form of an equation of a line.

96. In the equation $y = 3x + 5$, what does the 3 represent? What does the 5 represent?

97. ✎ In the equation of a vertical line, the variable y is missing. Explain why.

98. ✎ Can any pair of points on a line be used to determine the equation of the line? Explain.

Cumulative Review

In Exercises 99–102, sketch the graph of the equation.

99. $y = 4x$

100. $y = 2x - 3$

101. $y = x^2 - 1$

102. $y = -|x + 1|$

In Exercises 103–110, solve for a so that the line through the points has the given slope.

103. $(1, 2), (a, 4)$
   $m = 2$

104. $(0, 1), (2, a)$
   $m = 3$

105. $(-4, a), (-2, 3)$
   $m = \frac{1}{2}$

106. $(a, 3), (6, 3)$
   $m = 0$

107. $(5, 0), (0, a)$
   $m = -\frac{3}{5}$

108. $(0, a), (4, -6)$
   $m = -\frac{2}{3}$

109. $(-7, -2), (-1, a)$
   $m = -1$

110. $(9, -a), (-3, 4)$
   $m = -\frac{4}{5}$

Mid-Chapter Quiz

Take this quiz as you would take a quiz in class. After you are done, check your work against the answers in the back of the book.

1. Determine the quadrant(s) in which the point $(x, 4)$ must be located if x is a real number. Explain your reasoning.

2. Determine whether each ordered pair is a solution point of the equation $4x - 3y = 10$.
   (a) $(2, 1)$ (b) $(1, -2)$ (c) $(2.5, 0)$ (d) $\left(2, -\frac{2}{3}\right)$

In Exercises 3 and 4, plot the points on a rectangular coordinate system, find the distance between them, and determine the coordinates of the midpoint of the line segment joining the two points.

3. $(-1, 5), (3, 2)$ 　　　　　　　4. $(-4, 3), (6, -7)$

In Exercises 5–7, sketch the graph of the equation and show the coordinates of three solution points (including x- and y-intercepts). (There are many correct answers.)

5. $3x + y - 6 = 0$ 　6. $y = 6x - x^2$ 　7. $y = |x - 2| - 3$

In Exercises 8–10, determine the slope of the line (if possible) through the two points. State whether the line rises, falls, is horizontal, or is vertical.

8. $(-3, 8), (7, 8)$ 　9. $(3, 0), (6, 5)$ 　10. $(-2, 7), (4, -1)$

In Exercises 11 and 12, write the equation of the line in slope-intercept form. Find the slope and y-intercept and use them to sketch the graph of the equation.

11. $3x + 6y = 6$ 　　　　　　　12. $6x - 4y = 12$

In Exercises 13 and 14, determine whether the lines are parallel, perpendicular, or neither.

13. $y = 3x + 2, y = -\frac{1}{3}x - 4$

14. L_1: $(4, 3), (-2, -9)$; L_2: $(0, -5), (5, 5)$

15. Write the general form of the equation of a line that passes through the point $(6, -1)$ and has a slope of $\frac{1}{2}$.

16. Your company purchases a new printing press for $124,000. For tax purposes, the printing press will be depreciated over a 10-year period. At the end of 10 years, the salvage value of the printing press is expected to be $4000. Find an equation that relates the depreciated value of the printing press to the number of years since it was purchased. Then sketch the graph of the equation.

3.5 Graphs of Linear Inequalities

Elsa/Getty Images

Why You Should Learn It

Linear inequalities in two variables can be used to model and solve real-life problems. For instance, in Exercise 69 on page 182, a linear inequality is used to find the numbers of wins and ties a hockey team needs to reach the playoffs.

1 ▶ Verify solutions of linear inequalities in two variables.

What You Should Learn

1 ▶ Verify solutions of linear inequalities in two variables.

2 ▶ Sketch graphs of linear inequalities in two variables.

Linear Inequalities in Two Variables

A **linear inequality in two variables,** x and y, is an inequality that can be written in one of the forms below (where a and b are not both zero).

$$ax + by < c, \quad ax + by > c, \quad ax + by \le c, \quad ax + by \ge c$$

Here are some examples.

$$4x - 3y < 7, \quad x - y > -3, \quad x \le 2, \quad y \ge -4$$

An ordered pair (x_1, y_1) is a **solution** of a linear inequality in x and y if the inequality is true when x_1 and y_1 are substituted for x and y, respectively. For instance, the ordered pair $(3, 2)$ is a solution of the inequality $x - y > 0$ because $3 - 2 > 0$ is a true statement.

EXAMPLE 1   Verifying Solutions of Linear Inequalities

Determine whether each point is a solution of $2x - 3y \ge -2$.

a. $(0, 0)$ **b.** $(0, 1)$

Solution

To determine whether a point (x_1, y_1) is a solution of the inequality, substitute the coordinates of the point into the inequality.

a. $2x - 3y \ge -2$ Write original inequality.

$2(0) - 3(0) \overset{?}{\ge} -2$ Substitute 0 for x and 0 for y.

$0 \ge -2$ Inequality is satisfied. ✓

Because the inequality is satisfied, the point $(0, 0)$ *is* a solution.

b. $2x - 3y \ge -2$ Write original inequality.

$2(0) - 3(1) \overset{?}{\ge} -2$ Substitute 0 for x and 1 for y.

$-3 \not\ge -2$ Inequality is not satisfied. ✗

Because the inequality is not satisfied, the point $(0, 1)$ *is not* a solution.

✓ **CHECKPOINT** *Now try Exercise 1.*

2 ▶ Sketch graphs of linear inequalities in two variables.

The Graph of a Linear Inequality in Two Variables

The **graph** of a linear inequality is the collection of all solution points of the inequality. To sketch the graph of a linear inequality such as

$$4x - 3y < 12 \qquad \text{Original linear inequality}$$

begin by sketching the graph of the *corresponding linear equation*

$$4x - 3y = 12. \qquad \text{Corresponding linear equation}$$

Use *dashed* lines for the inequalities $<$ and $>$ and *solid* lines for the inequalities $\leq$ and $\geq$. The graph of the equation separates the plane into two regions, called **half-planes.** In each half-plane, one of the following *must* be true.

1. All points in the half-plane are solutions of the inequality.

2. No point in the half-plane is a solution of the inequality.

So, you can determine whether the points in an entire half-plane satisfy the inequality by simply testing *one* point in the region. This graphing procedure is summarized as follows.

Study Tip

The graph of the corresponding equation is a dashed line for an inequality involving $<$ or $>$ because the points on the line are *not* solutions of the inequality. It is a solid line for an inequality involving $\leq$ or $\geq$ because the points on the line *are* solutions of the inequality. In either case, to test a half-plane you should choose a test point that is *not* on the line of the corresponding equation.

Sketching the Graph of a Linear Inequality in Two Variables

1. Replace the inequality symbol by an equal sign, and sketch the graph of the resulting equation. (Use a dashed line for $<$ or $>$, and a solid line for $\leq$ or $\geq$.)

2. Test one point in one of the half-planes formed by the graph in Step 1.

  a. If the point satisfies the inequality, shade the entire half-plane to denote that every point in the region satisfies the inequality.

  b. If the point does not satisfy the inequality, then shade the other half-plane.

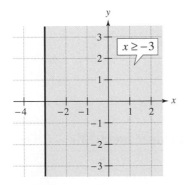

Figure 3.52

EXAMPLE 2   Sketching the Graphs of Linear Inequalities

Sketch the graph of each linear inequality.

a. $x \geq -3$ **b.** $y < 4$

Solution

a. The graph of the corresponding equation $x = -3$ is a vertical line. The line is solid because the inequality symbol is $\geq$. By testing the point $(0, 0)$, you can see that the points satisfying the inequality are those in the half-plane to the right of the line $x = -3$. The graph is shown in Figure 3.52.

b. The graph of the corresponding equation $y = 4$ is a horizontal line. The line is dashed because the inequality symbol is $<$. By testing the point $(0, 0)$, you can see that the points satisfying the inequality are those in the half-plane below the line $y = 4$. The graph is shown in Figure 3.53.

 **CHECKPOINT** *Now try Exercise 17.*

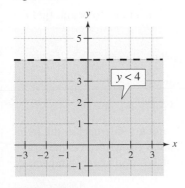

Figure 3.53

EXAMPLE 3   Sketching the Graph of a Linear Inequality

Sketch the graph of the linear inequality $x + y > 3$.

Solution

The graph of the corresponding equation $x + y = 3$ is a line. To begin, find the x-intercept by letting $y = 0$ and solving for x.

$$x + 0 = 3 \quad \Longrightarrow \quad x = 3 \qquad \text{Substitute 0 for } y \text{ and solve for } x.$$

Find the y-intercept by letting $x = 0$ and solving for y.

$$0 + y = 3 \quad \Longrightarrow \quad y = 3 \qquad \text{Substitute 0 for } x \text{ and solve for } y.$$

So, the graph has an x-intercept at the point $(3, 0)$ and a y-intercept at the point $(0, 3)$. Plot these points and connect them with a dashed line. Because the origin $(0, 0)$ does not satisfy the inequality, the graph consists of the half-plane lying above the line, as shown in Figure 3.54.

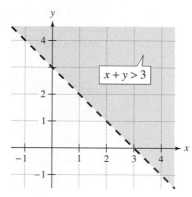

Figure 3.54

✓ **CHECKPOINT** *Now try Exercise 27.*

EXAMPLE 4   Sketching the Graph of a Linear Inequality

Sketch the graph of the linear inequality $2x + y \le 2$.

Solution

The graph of the corresponding equation $2x + y = 2$ is a line. To begin, find the x-intercept by letting $y = 0$ and solving for x.

$$2x + 0 = 2 \quad \Longrightarrow \quad x = 1 \qquad \text{Substitute 0 for } y \text{ and solve for } x.$$

Find the y-intercept by letting $x = 0$ and solving for y.

$$2(0) + y = 2 \quad \Longrightarrow \quad y = 2 \qquad \text{Substitute 0 for } x \text{ and solve for } y.$$

So, the graph has an x-intercept at the point $(1, 0)$ and a y-intercept at the point $(0, 2)$. Plot these points and connect them with a solid line. Because the origin $(0, 0)$ satisfies the inequality, the graph consists of the half-plane lying on or below the line, as shown in Figure 3.55.

✓ **CHECKPOINT** *Now try Exercise 29.*

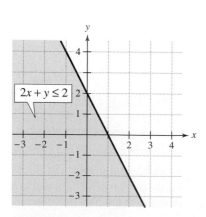

Figure 3.55

For a linear inequality in two variables, you can sometimes simplify the graphing procedure by writing the inequality in *slope-intercept form.* For instance, by writing $x + y > 1$ in the form

$$y > -x + 1 \qquad \text{Slope-intercept form}$$

you can see that the solution points lie *above* the line $y = -x + 1$, as shown in Figure 3.56. So, when $y > mx + b$ (or $y \geq mx + b$), shade above the line $y = mx + b$. Similarly, by writing the inequality $4x - 3y > 12$ in the form

$$y < \frac{4}{3}x - 4 \qquad \text{Slope-intercept form}$$

you can see that the solutions lie *below* the line $y = \frac{4}{3}x - 4$, as shown in Figure 3.57. So, when $y < mx + b$ (or $y \leq mx + b$), shade below the line $y = mx + b$.

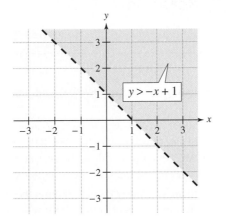

Figure 3.56

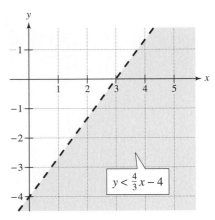

Figure 3.57

EXAMPLE 5   Sketching the Graph of a Linear Inequality

Use the slope-intercept form of a linear equation as an aid in sketching the graph of the inequality $2x - 3y \leq 15$.

Solution

To begin, rewrite the inequality in slope-intercept form.

$$2x - 3y \leq 15 \qquad \text{Write original inequality.}$$

$$-3y \leq -2x + 15 \qquad \text{Subtract } 2x \text{ from each side.}$$

$$y \geq \frac{2}{3}x - 5 \qquad \text{Divide each side by } -3 \text{ and reverse the inequality symbol.}$$

From this form, you can conclude that the solution is the half-plane lying *on or above* the line

$$y = \frac{2}{3}x - 5.$$

The graph is shown in Figure 3.58. To verify the solution, test any point in the shaded region.

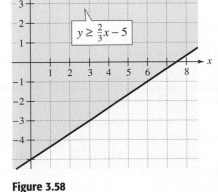

Figure 3.58

✓ **CHECKPOINT** *Now try Exercise 31.*

Most graphing calculators can graph inequalities in two variables. Consult the user's guide for your graphing calculator for specific instructions. The graph of $y \leq \frac{1}{2}x - 3$ is shown below.

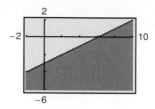

Try using a graphing calculator to graph each inequality.

a. $y \geq -\frac{2}{3}x + 1$ **b.** $3x - 2y \leq 5$

EXAMPLE 6   **An Application: Working to Meet a Budget**

Your budget requires you to earn *at least* $160 per week. You work two part-time jobs. One is tutoring, which pays $10 per hour, and the other is at a fast-food restaurant, which pays $8 per hour. Let x represent the number of hours tutoring and let y represent the number of hours worked at the fast-food restaurant.

a. Write an inequality that represents the numbers of hours you can work at each job in order to meet your budget requirements.

b. Graph the inequality and identify at least two ordered pairs (x, y) that represent numbers of hours you can work at each job in order to meet your budget requirements.

Solution

a. *Verbal Model:* $10 \cdot \boxed{\text{Number of hours tutoring}} + 8 \cdot \boxed{\text{Number of hours at fast-food restaurant}} \geq 160$

Labels: Number of hours tutoring $= x$ (hours)
Number of hours at fast-food restaurant $= y$ (hours)

Inequality: $10x + 8y \geq 160$

b. Rewrite the inequality in slope-intercept form.

$10x + 8y \geq 160$ Write original inequality.

$8y \geq -10x + 160$ Subtract $10x$ from each side.

$y \geq -1.25x + 20$ Divide each side by 8.

Graph the corresponding equation $y = -1.25x + 20$ and shade the half-plane lying above the line, as shown in Figure 3.59. From the graph, you can see that two solutions that yield the desired weekly earnings of at least $160 are $(8, 10)$ and $(12, 15)$. (There are many other solutions.)

✓ **CHECKPOINT** *Now try Exercise 67.*

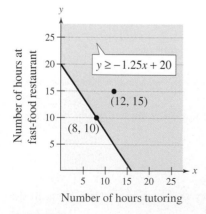

Figure 3.59

_____ **Concept Check** _____

1. *True or False?*   To determine whether the point $(2, 4)$ is a solution of the inequality $2x + 3y > 12$, you must graph the inequality. Justify your answer.

2. *True or False?*   Any point in the Cartesian plane can be used as a test point to determine the half-plane that represents the solution of a linear inequality. Justify your answer.

3. How does the solution of $x - y > 1$ differ from the solution of $x - y \geq 1$?

4. Explain how you can use *slope-intercept form* to graph a linear inequality without using a test point.

3.5 EXERCISES

Go to pages 208–209 to record your assignments.

_____ **Developing Skills** _____

In Exercises 1–8, determine whether each point is a solution of the inequality. *See Example 1.*

	Inequality		*Points*	
✓ **1.**	$x - 2y < 4$	(a) $(0, 0)$	(b) $(2, -1)$	
		(c) $(3, 4)$	(d) $(5, 1)$	
2.	$x + y < 3$	(a) $(0, 6)$	(b) $(4, 0)$	
		(c) $(0, -2)$	(d) $(1, 1)$	
3.	$3x + y \geq 10$	(a) $(1, 3)$	(b) $(-3, 1)$	
		(c) $(3, 1)$	(d) $(2, 15)$	
4.	$-3x + 5y \geq 6$	(a) $(2, 8)$	(b) $(-10, -3)$	
		(c) $(0, 0)$	(d) $(3, 3)$	
5.	$y > 0.2x - 1$	(a) $(0, 2)$	(b) $(6, 0)$	
		(c) $(4, -1)$	(d) $(-2, 7)$	
6.	$y < -3.5x + 7$	(a) $(1, 5)$	(b) $(5, -1)$	
		(c) $(-1, 4)$	(d) $\left(0, \frac{4}{3}\right)$	
7.	$y \leq 3 - |x|$	(a) $(-1, 4)$	(b) $(2, -2)$	
		(c) $(6, 0)$	(d) $(5, -2)$	
8.	$y \geq |x - 3|$	(a) $(0, 0)$	(b) $(1, 2)$	
		(c) $(4, 10)$	(d) $(5, -1)$	

In Exercises 9–14, match the inequality with its graph. [The graphs are labeled (a), (b), (c), (d), (e), and (f).]

(a)
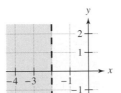

(b)
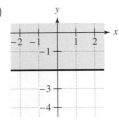

(c)
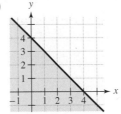

(d)
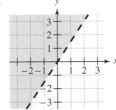

(e)
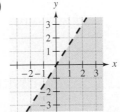

(f)
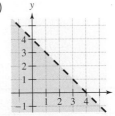

9. $y \geq -2$

10. $x < -2$

11. $3x - 2y < 0$

12. $3x - 2y > 0$

13. $x + y < 4$

14. $x + y \leq 4$

In Exercises 15 and 16, complete the graph of the inequality by shading the correct half-plane.

15. $3x - y > 2$

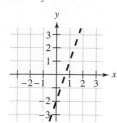

16. $3x + 4y \leq 4$

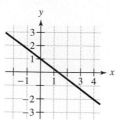

In Exercises 17–44, sketch the graph of the solution of the linear inequality. *See Examples 2–5.*

✓ 17. $x \geq 6$

18. $x < -3$

19. $y < 5$

20. $y > 2$

21. $y > \frac{1}{2}x$

22. $y \leq 2x$

23. $y \geq 3 - x$

24. $y > x + 6$

25. $y \leq x + 2$

26. $y \leq 1 - x$

✓ 27. $x + y \geq 4$

28. $x + y \leq 5$

✓ 29. $x - 2y \geq 6$

30. $3x + y \leq 9$

✓ 31. $3x + 2y \geq 2$

32. $3x + 5y \leq 15$

33. $5x + 4y < 20$

34. $4x - 7y > 21$

35. $x - 3y - 9 < 0$

36. $x + 4y + 12 > 0$

37. $3x - 2 \leq 5x + y$

38. $2x - 2y \geq 8 + 2y$

39. $0.2x + 0.3y < 2$

40. $0.25x - 0.75y > 6$

41. $y - 1 > -\frac{1}{2}(x - 2)$

42. $y - 2 < -\frac{2}{3}(x - 3)$

43. $\frac{x}{3} + \frac{y}{4} \leq 1$

44. $\frac{x}{2} + \frac{y}{6} \geq 1$

In Exercises 45–52, use a graphing calculator to graph the solution of the inequality.

45. $y \geq \frac{3}{4}x - 1$

46. $y \leq 9 - \frac{3}{2}x$

47. $y \leq -\frac{2}{3}x + 6$

48. $y \geq \frac{1}{4}x + 3$

49. $x - 2y - 4 \geq 0$

50. $2x + 4y - 3 \leq 0$

51. $2x + 3y - 12 \leq 0$

52. $x - 3y + 9 \geq 0$

In Exercises 53–58, write an inequality for the shaded region shown in the figure.

53.

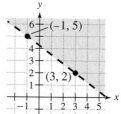

54.

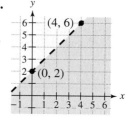

55.

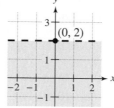

56.

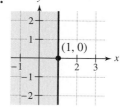

57.

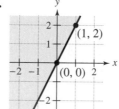

58.

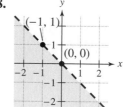

Solving Problems

59. ▲ *Geometry* The perimeter of a rectangle of length x and width y cannot exceed 500 inches.

(a) Write a linear inequality for this constraint.

(b) ▦ Use a graphing calculator to graph the solution of the inequality.

60. ▲ *Geometry* The perimeter of a rectangle of length x and width y must be at least 100 centimeters.

(a) Write a linear inequality for this constraint.

(b) ▦ Use a graphing calculator to graph the solution of the inequality.

61. *Storage Space*   A warehouse for storing chairs and tables has 1000 square feet of floor space. Each chair requires 10 square feet of floor space and each table requires 15 square feet.

   (a) Write a linear inequality for this space constraint, where x is the number of chairs and y is the number of tables stored.

   (b) Sketch a graph of the solution of the inequality.

62. *Storage Space*   A warehouse for storing desks and filing cabinets has 2000 square feet of floor space. Each desk requires 15 square feet of floor space and each filing cabinet requires 6 square feet.

   (a) Write a linear inequality for this space constraint, where x is the number of desks and y is the number of filing cabinets stored.

   (b) Sketch a graph of the solution of the inequality.

63. *Consumerism*   You and some friends go out for pizza. Together you have \$32. You want to order two large pizzas with cheese at \$10 each. Each additional topping costs \$0.60 and each small soft drink costs \$1.00.

   (a) Write an inequality that represents the numbers of toppings x and drinks y that your group can afford. (Assume there is no sales tax.)

   (b) Sketch a graph of the solution of the inequality.

   (c) What are the coordinates for an order of six soft drinks and two large pizzas with cheese, each with three additional toppings? Is this a solution of the inequality?

64. *Consumerism*   You and some friends go out for pizza. Together you have \$48. You want to order three large pizzas with cheese at \$9 each. Each additional topping costs \$1 and each soft drink costs \$1.50.

   (a) Write an inequality that represents the numbers of toppings x and drinks y that your group can afford. (Assume there is no sales tax.)

   (b) Sketch a graph of the solution of the inequality.

   (c) What are the coordinates for an order of eight soft drinks and three large pizzas with cheese, each with two additional toppings? Is this a solution of the inequality?

65. *Nutrition*   A dietitian is asked to design a special diet supplement using two foods. Each ounce of food X contains 12 units of protein and each ounce of food Y contains 16 units of protein. The minimum daily requirement in the diet is 250 units of protein.

   (a) Write an inequality that represents the different numbers of units of food X and food Y required.

   (b) Sketch a graph of the solution of the inequality. From the graph, find several ordered pairs with positive integer coordinates that are solutions of the inequality.

66. *Nutrition*   A dietitian is asked to design a special diet supplement using two foods. Each ounce of food X contains 30 units of calcium and each ounce of food Y contains 20 units of calcium. The minimum daily requirement in the diet is 300 units of calcium.

   (a) Write an inequality that represents the different numbers of units of food X and food Y required.

   (b) Sketch a graph of the solution of the inequality. From the graph, find several ordered pairs with positive integer coordinates that are solutions of the inequality.

67. *Weekly Pay*   You have two part-time jobs. One is at a grocery store, which pays \$11 per hour, and the other is mowing lawns, which pays \$9 per hour. Between the two jobs, you want to earn at least \$240 a week.

   (a) Write an inequality that shows the different numbers of hours you can work at each job.

   (b) Sketch the graph of the solution of the inequality. From the graph, find several ordered pairs with positive integer coordinates that are solutions of the inequality.

68. *Weekly Pay*   You have two part-time jobs. One is at a candy store, which pays \$8 per hour, and the other is providing childcare, which pays \$10 per hour. Between the two jobs, you want to earn at least \$160 a week.

   (a) Write an inequality that shows the different numbers of hours you can work at each job.

(b) Sketch the graph of the solution of the inequality. From the graph, find several ordered pairs with positive integer coordinates that are solutions of the inequality.

69. *Sports* Your hockey team needs at least 70 points for the season in order to advance to the playoffs. Your team finishes with w wins, each worth 2 points, and t ties, each worth 1 point.

(a) Write a linear inequality that shows the different numbers of wins and ties your team must record in order to advance to the playoffs.

(b) Sketch the graph of the solution of the inequality. From the graph, find several ordered pairs with positive integer coordinates that are solutions of the inequality.

70. *Exercise* The maximum heart rate r (in beats per minute) of a person in normal health is related to the person's age A (in years). The relationship between r and A is given by $r \le 220 - A$. (Source: American Heart Association)

(a) Sketch a graph of the solution of the inequality.

(b) Physicians recommend that during a workout a person strive to increase his or her heart rate to 75% of the maximum rate for the person's age. Sketch the graph of $r = 0.75(220 - A)$ on the same set of coordinate axes used in part (a).

─────────────── **Explaining Concepts** ───────────────

71. ✎ What is meant by saying that (x_1, y_1) is a solution of a linear inequality in x and y?

72. ✎ Explain the difference between graphing the solution of the inequality $x \le 3$ on the real number line and graphing it on a rectangular coordinate system.

73. The graph of a particular inequality consists of the half-plane above a dashed line. Explain how to modify the inequality to represent all of the points in the plane *other* than the half-plane represented by the graph.

74. Discuss how you could write a double inequality whose solution set is the graph of a line in the plane. Give an example.

75. The origin *cannot* be used as the test point to determine which half-plane to shade in graphing a linear inequality. What is the y-intercept of the graph of the corresponding non-vertical linear equation? Explain.

76. Every point in the solution set of the linear inequality $y \le ax + b$ is a solution of the linear inequality $y < cx + d$. What can you say about the values of a, b, c, and d?

─────────────── **Cumulative Review** ───────────────

In Exercises 77–80, transform the absolute value equation into two linear equations.

77. $|x| = 6$

78. $|x + 2| = 3$

79. $|2x + 3| = 9$

80. $|8 - 3x| = 10$

In Exercises 81–84, plot the points on a rectangular coordinate system.

81. $(0, 0), (-3, 4), (-4, -2)$

82. $(5, -1), (-7, 4), (3, 3)$

83. $(1, 1), \left(\frac{3}{4}, \frac{3}{4}\right), \left(\frac{5}{4}, \frac{5}{4}\right)$

84. $\left(\frac{1}{2}, \frac{3}{2}\right), \left(2, \frac{1}{2}\right), \left(\frac{1}{2}, \frac{7}{2}\right)$

In Exercises 85 and 86, write equations of the lines that pass through the point and are (a) parallel and (b) perpendicular to the given line.

85. $(3, 0); y = -2x + 1$ **86.** $(5, 7); 2x - 3y = -6$

3.6 Relations and Functions

Jose Luis Pelaez/Getty Images

What You Should Learn

1 ▶ Identify the domains and ranges of relations.

2 ▶ Determine if relations are functions by inspection.

3 ▶ Use function notation and evaluate functions.

4 ▶ Identify the domains and ranges of functions.

Why You Should Learn It

Functions can be used to model and solve real-life problems. For instance, in Exercise 87 on page 195, a function is used to find the cost of producing a video game.

1 ▶ Identify the domains and ranges of relations.

Relations

Many everyday occurrences involve two quantities that are paired or matched with each other by some rule of correspondence. The mathematical term for such a correspondence is **relation.**

Definition of Relation

A **relation** is any set of ordered pairs. The set of first components in the ordered pairs is the **domain** of the relation. The set of second components is the **range** of the relation.

EXAMPLE 1   **Analyzing a Relation**

Find the domain and range of the relation $\{(0, 1), (1, 3), (2, 5), (3, 5), (0, 3)\}$.

Solution

The domain is the set of all first components of the relation, and the range is the set of all second components.

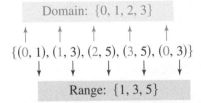

Domain: $\{0, 1, 2, 3\}$

$\{(0, 1), (1, 3), (2, 5), (3, 5), (0, 3)\}$

Range: $\{1, 3, 5\}$

A graphical representation of this relation is shown in Figure 3.60.

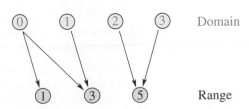

Figure 3.60

✓ **CHECKPOINT** *Now try Exercise 1.*

Study Tip

When you write the domain or range of a relation, it is not necessary to list repeated components more than once.

2 ▶ Determine if relations are functions by inspection.

Functions

In modeling real-life situations, you will work with a special type of relation called a function. A **function** is a relation in which no two ordered pairs have the same first component and different second components. For instance, (2, 3) and (2, 4) could not be ordered pairs of a function.

> ## Definition of a Function
>
> A **function** f from a set A to a set B is a rule of correspondence that assigns to each element x in the set A exactly one element y in the set B. The set A is called the **domain** (or set of inputs) of the function f, and the set B contains the **range** (or set of outputs) of the function.

The rule of correspondence for a function establishes a set of "input-output" ordered pairs of the form (x, y), where x is an input and y is the corresponding output. In some cases, the rule may generate only a finite set of ordered pairs, whereas in other cases the rule may generate an infinite set of ordered pairs.

EXAMPLE 2   Input-Output Ordered Pairs for Functions

Write a set of ordered pairs that represents the rule of correspondence.

a. Winners of the Super Bowl in 2004, 2005, 2006, and 2007

b. The squares of all real numbers

Solution

a. For the function that pairs the year from 2004 to 2007 with the winner of the Super Bowl, each ordered pair is of the form (year, winner).

$$\{(2004, \text{Patriots}), (2005, \text{Patriots}), (2006, \text{Steelers}), (2007, \text{Colts})\}$$

b. For the function that pairs each real number with its square, each ordered pair is of the form (x, x^2).

$$\{\text{All points } (x, x^2), \text{ where } x \text{ is a real number}\}$$

 CHECKPOINT *Now try Exercise 5.*

Study Tip

In Example 2, the set in part (a) has only a finite number of ordered pairs, whereas the set in part (b) has an infinite number of ordered pairs.

Functions are commonly represented in four ways.

1. *Verbally* by a sentence that describes how the input variable is related to the output variable.

2. *Numerically* by a table or a list of ordered pairs that matches input values with output values.

3. *Graphically* by points on a graph in a coordinate plane in which the input values are represented by the horizontal axis and the output values are represented by the vertical axis.

4. *Algebraically* by an equation in two variables.

A function has certain characteristics that distinguish it from a relation. To determine whether or not a relation is a function, use the following list of characteristics of a function.

Characteristics of a Function

1. Each element in the domain A must be matched with an element in the range, which is contained in the set B.

2. Some elements in set B may not be matched with any element in the domain A.

3. Two or more elements of the domain may be matched with the same element in the range.

4. No element of the domain is matched with two different elements in the range.

EXAMPLE 3   Testing for Functions

Decide whether or not the description represents a function.

a. The input value x is any of the 50 states in the United States and the output value y is the number of governors of that state.

b. $\{(1, 1), (2, 4), (3, 9), (4, 16), (5, 25), (6, 36)\}$

c.

Input, x	Output, y
-1	7
3	2
4	0
3	4

d. Let $A = \{a, b, c\}$ and let $B = \{1, 2, 3, 4, 5\}$.

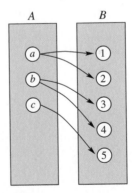

Solution

a. This set of ordered pairs *does* represent a function. Regardless of the input value x, the output value is always 1.

b. This set of ordered pairs *does* represent a function. No input value is matched with *two* output values.

c. This table *does not* represent a function. The input value 3 is matched with *two* different output values, 2 and 4.

d. This diagram *does not* represent a function. The element a in set A is matched with *two* elements in set B. This is also true of element b.

✓ **CHECKPOINT** *Now try Exercise 23.*

Representing functions by sets of ordered pairs is a common practice in the study of *discrete mathematics*, which mainly involves finite sets of data or finite subsets of the set of real numbers. In algebra, however, it is more common to represent functions by equations or formulas involving two variables. For instance,

$$y = x^2 \qquad \text{Squaring function}$$

represents the variable y as a function of the variable x. The variable x is the **independent variable** and the variable y is the **dependent variable.** In this context, the domain of the function is the set of all *allowable* real values for the independent variable x, and the range of the function is the *resulting* set of all values taken on by the dependent variable y.

EXAMPLE 4   **Testing for Functions Represented by Equations**

Which of the equations represent y as a function of x?

a. $y = x^2 + 1$ **b.** $x - y^2 = 2$ **c.** $-2x + 3y = 4$

Solution

a. For the equation

$$y = x^2 + 1$$

just one value of y corresponds to each value of x. For instance, when $x = 1$, the value of y is

$$y = 1^2 + 1 = 2.$$

So, y *is* a function of x.

b. By writing the equation $x - y^2 = 2$ in the form

$$y^2 = x - 2$$

you can see that two values of y correspond to some values of x. For instance, when $x = 3$,

$$y^2 = 3 - 2$$
$$y^2 = 1$$
$$y = 1 \quad \text{or} \quad y = -1.$$

So, the solution points $(3, 1)$ and $(3, -1)$ show that y *is not* a function of x.

c. By writing the equation $-2x + 3y = 4$ in the form

$$y = \frac{2}{3}x + \frac{4}{3}$$

you can see that just one value of y corresponds to each value of x. For instance, when $x = 2$, the value of y is $\frac{4}{3} + \frac{4}{3} = \frac{8}{3}$. So, y *is* a function of x.

✓ *CHECKPOINT Now try Exercise 29.*

An equation that defines y as a function of x may or may not also define x as a function of y. For instance, the equation in part (a) of Example 4 does not define x as a function of y, but the equation in part (c) does.

3 ▶ Use function notation and evaluate functions.

Function Notation

When an equation is used to represent a function, it is convenient to name the function so that it can be easily referenced. For example, the function $y = x^2 + 1$ in Example 4(a) can be given the name "f" and written in **function notation** as

$$f(x) = x^2 + 1.$$

Function Notation

In the notation $f(x)$:

f is the **name** of the function.

x is the **domain** (or input) value.

$f(x)$ is the **range** (or output) value y for a given x.

The symbol $f(x)$ is read as *the value of f at x* or simply *f of x.*

The process of finding the value of $f(x)$ for a given value of x is called **evaluating a function.** This is accomplished by substituting a given x-value (input) into the equation to obtain the value of $f(x)$ (output). Here is an example.

Function	*x-Value*	*Function Value*
$f(x) = 3 - 4x$	$x = -1$	$f(-1) = 3 - 4(-1)$
		$= 3 + 4$
		$= 7$

Although f is often used as a convenient function name and x as the independent variable, you can use other letters. For instance, the equations

$$f(x) = 2x^2 + 5, \quad f(t) = 2t^2 + 5, \quad \text{and} \quad g(s) = 2s^2 + 5$$

all define the same function. In fact, the letters used are simply "placeholders" and this same function is well described by the form

$$f(\quad) = 2(\quad)^2 + 5$$

where the parentheses are used in place of a letter. To evaluate $f(-2)$, simply place -2 in each set of parentheses, as follows.

$$f(-2) = 2(-2)^2 + 5$$
$$= 2(4) + 5$$
$$= 8 + 5$$
$$= 13$$

When evaluating a function, you are not restricted to substituting only numerical values into the parentheses. For instance, the value of $f(3x)$ is

$$f(3x) = 2(3x)^2 + 5$$
$$= 2(9x^2) + 5$$
$$= 18x^2 + 5.$$

Study Tip

Note that

$$g(x + 1) \neq g(x) + g(1).$$

In general, $g(a + b)$ is not equal to $g(a) + g(b)$.

EXAMPLE 5   Evaluating a Function

Let $g(x) = 3x - 4$. Find each value of the function.

a. $g(1)$ **b.** $g(-2)$ **c.** $g(y)$ **d.** $g(x + 1)$ **e.** $g(x) + g(1)$

Solution

a. Replacing x with 1 produces $g(1) = 3(1) - 4 = 3 - 4 = -1$.

b. Replacing x with -2 produces $g(-2) = 3(-2) - 4 = -6 - 4 = -10$.

c. Replacing x with y produces $g(y) = 3(y) - 4 = 3y - 4$.

d. Replacing x with $(x + 1)$ produces

$$g(x + 1) = 3(x + 1) - 4 = 3x + 3 - 4 = 3x - 1.$$

e. Using the result of part (a) for $g(1)$, you have

$$g(x) + g(1) = (3x - 4) + (-1) = 3x - 4 - 1 = 3x - 5.$$

✓ **CHECKPOINT** *Now try Exercise 43.*

Sometimes a function is defined by more than one equation, each of which is given a portion of the domain. Such a function is called a **piecewise-defined function.** To evaluate a piecewise-defined function f for a given value of x, first determine the portion of the domain in which the x-value lies and then use the corresponding equation to evaluate f. This is illustrated in Example 6.

EXAMPLE 6   A Piecewise-Defined Function

Let $f(x) = \begin{cases} x^2 + 1, & \text{if } x < 0 \\ x - 2, & \text{if } x \geq 0 \end{cases}$. Find each value of the function.

a. $f(-1)$ **b.** $f(0)$ **c.** $f(-2)$ **d.** $f(-3) + f(4)$

Solution

a. Because $x = -1 < 0$, use $f(x) = x^2 + 1$ to obtain

$$f(-1) = (-1)^2 + 1 = 1 + 1 = 2.$$

b. Because $x = 0 \geq 0$, use $f(x) = x - 2$ to obtain

$$f(0) = 0 - 2 = -2.$$

c. Because $x = -2 < 0$, use $f(x) = x^2 + 1$ to obtain

$$f(-2) = (-2)^2 + 1 = 4 + 1 = 5.$$

d. Because $x = -3 < 0$, use $f(x) = x^2 + 1$ to obtain

$$f(-3) = (-3)^2 + 1 = 9 + 1 = 10.$$

Because $x = 4 \geq 0$, use $f(x) = x - 2$ to obtain

$$f(4) = 4 - 2 = 2.$$

So, $f(-3) + f(4) = 10 + 2 = 12.$

✓ **CHECKPOINT** *Now try Exercise 59.*

4 ▶ Identify the domains and ranges of functions.

Finding the Domain and Range of a Function

The domain of a function may be explicitly described along with the function, or it may be *implied* by the expression used to define the function. The **implied domain** is the set of all real numbers (inputs) that yield real number values for the function. For instance, the function given by

$$f(x) = \frac{1}{x - 3} \qquad \text{Domain: all } x \neq 3$$

has an implied domain that consists of all real values of x other than $x = 3$. The value $x = 3$ is excluded from the domain because division by zero is undefined. Another common type of implied domain is one used to avoid even roots of negative numbers. For instance, the function given by

$$f(x) = \sqrt{x} \qquad \text{Domain: all } x \geq 0$$

is defined only for $x \geq 0$. So, its implied domain is the set of all real numbers x such that $x \geq 0$. More is said about the domains of square root functions in Chapter 7.

EXAMPLE 7   Finding the Domain of a Function

Find the domain of each function.

a. $f(x) = \sqrt{2x - 6}$ **b.** $g(x) = \dfrac{4x}{(x - 1)(x + 5)}$

Solution

a. The domain of f consists of all real numbers x such that $2x - 6 \geq 0$. Solving this inequality yields

$$2x - 6 \geq 0 \qquad \text{Original inequality}$$
$$2x \geq 6 \qquad \text{Add 6 to each side.}$$
$$x \geq 3. \qquad \text{Divide each side by 2.}$$

So, the domain consists of all real numbers x such that $x \geq 3$.

b. The domain of g consists of all real numbers x such that the denominator is not equal to zero. The denominator is equal to zero when either factor of the denominator is zero.

First Factor
$$x - 1 = 0 \qquad \text{Set the first factor equal to zero.}$$
$$x = 1 \qquad \text{Add 1 to each side.}$$

Second Factor
$$x + 5 = 0 \qquad \text{Set the second factor equal to zero.}$$
$$x = -5 \qquad \text{Subtract 5 from each side.}$$

So, the domain consists of all real numbers x such that $x \neq 1$ and $x \neq -5$.

✓ **CHECKPOINT** *Now try Exercise 69.*

EXAMPLE 8 | Finding the Domain and Range of a Function

Find the domain and range of each function.

a. f: $\{(-3, 0), (-1, 2), (0, 4), (2, 4), (4, -1)\}$
b. Area of a circle: $A = \pi r^2$

Solution

a. The domain of f consists of all first coordinates in the set of ordered pairs. The range consists of all second coordinates in the set of ordered pairs. So, the domain and range are as follows.

$$\text{Domain} = \{-3, -1, 0, 2, 4\} \qquad \text{Range} = \{0, 2, 4, -1\}$$

b. For the area of a circle, you must choose positive values for the radius r. So, the domain is the set of all real numbers r such that $r > 0$. The range is therefore the set of all real numbers A such that $A > 0$.

✓ **CHECKPOINT** *Now try Exercise 79.*

Note in Example 8(b) that the domain of a function can be implied by a physical context. For instance, from the equation $A = \pi r^2$, you would have no strictly mathematical reason to restrict r to positive values. However, because you know that this function represents the area of a circle, you can conclude that the radius must be positive.

EXAMPLE 9 | Geometry: The Dimensions of a Container

You work in the marketing department of a soft drink company, where you are experimenting with a new can for iced tea that is slightly narrower and taller than a standard can. For your experimental can, the ratio of the height to the radius is 4, as shown in Figure 3.61.

a. Write the volume of the can as a function of the radius r. Find the domain of the function.

b. Write the volume of the can as a function of the height h. Find the domain of the function.

Solution

The volume of a circular cylinder is given by the formula $V = \pi r^2 h$.

a. Because the ratio of the height of the can to the radius is 4, $h = 4r$. Substitute this value of h into the formula to obtain

$$V(r) = \pi r^2 h = \pi r^2 (4r) = 4\pi r^3. \qquad \text{Write } V \text{ as a function of } r.$$

The domain is the set of all real numbers r such that $r > 0$.

b. You know that $h = 4r$ from part (a), so you can determine that $r = h/4$. Substitute this value of r into the formula to obtain

$$V(h) = \pi r^2 h = \pi \left(\frac{h}{4}\right)^2 h = \frac{\pi h^3}{16}. \qquad \text{Write } V \text{ as a function of } h.$$

The domain is the set of all real numbers h such that $h > 0$.

$\dfrac{h}{r} = 4$

Figure 3.61

✓ **CHECKPOINT** *Now try Exercise 83.*

Concept Check

1. Explain how a set of ordered pairs can be a relation but not a function.

2. Can a set of ordered pairs be a function but not a relation? Explain.

3. Use the following terms to make a list of domain-related terms and a list of range-related terms.

   y, second component, x, input, independent variable, dependent variable, output, first component, $f(x)$

   Domain:

   Range:

4. What is the meaning of the notation $f(2)$? Explain how to carry out the process suggested by this notation.

3.6 EXERCISES

Go to pages 208–209 to record your assignments.

Developing Skills

In Exercises 1–4, find the domain and the range of the relation. Then draw a graphical representation of the relation. *See Example 1.*

✓ 1. $\{(-2, 0), (0, 1), (1, 4), (0, -1)\}$

2. $\{(3, 10), (4, 5), (6, -2), (8, 3)\}$

3. $\{(0, 0), (4, -3), (2, 8), (5, 5), (6, 5)\}$

4. $\{(-3, 6), (-3, 2), (-3, 5)\}$

In Exercises 5–10, write a set of ordered pairs that represents the rule of correspondence. *See Example 2.*

✓ 5. The cubes of all positive integers less than 8

6. The cubes of all integers greater than -2 and less than 5

7. The winners of the World Series from 2004 to 2007

8. The men inaugurated as president of the United States in 1989, 1993, 1997, 2001, and 2005.

9. The fuel used by a vehicle on a trip is a function of the driving time in hours. Fuel is used at a rate of 3 gallons per hour on trips of 3 hours, 1 hour, 2 hours, 8 hours, and 7 hours.

10. The time it takes a court stenographer to transcribe a testimony is a function of the number of words. Working at a rate of 120 words per minute, the stenographer transcribes testimonies of 360 words, 600 words, 1200 words, and 2040 words.

In Exercises 11–22, determine whether the relation is a function. *See Example 3.*

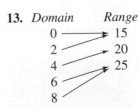

11.

Domain	Range
-2	5
-1	6
0	7
1	8
2	

12.

Domain	Range
-2	3
-1	4
0	5
1	
2	

13.

Domain	Range
0	15
2	20
4	25
6	
8	

14.

Domain	Range
100	25
200	30
300	40
400	45
500	

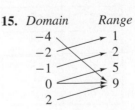

15.

Domain	Range
-4	1
-2	2
-1	5
0	9
2	

16. *Domain* *Range*

Year
3-year-olds in
Head Start
(in percent)

2001   34
2002   35
2003   36
2004
2005

(Source: U.S. Administration for Children and Families)

17. *Domain* *Range*

Desperate Housewives
ABC → Grey's Anatomy
Lost

American Idol
FOX → House
24

18. *Domain* *Range*

Desperate Housewives
Grey's Anatomy → ABC
Lost

American Idol
House → FOX
24

19.

Input value, x	Output value, y
0	0
1	1
2	4
3	9
4	16

20.

Input value, x	Output value, y
0	1
1	8
2	12
1	15
0	20

21.

Input value, x	Output value, y
4	2
7	4
9	6
7	8
4	10

22.

Input value, x	Output value, y
0	5
2	5
4	5
6	5
8	5

In Exercises 23 and 24, determine which sets of ordered pairs represent functions from *A* to *B*. *See Example 3.*

23. $A = \{0, 1, 2, 3\}$ and $B = \{-2, -1, 0, 1, 2\}$

  (a) $\{(0, 1), (1, -2), (2, 0), (3, 2)\}$

  (b) $\{(0, -1), (2, 2), (1, -2), (3, 0), (1, 1)\}$

  (c) $\{(0, 0), (1, 0), (2, 0), (3, 0)\}$

  (d) $\{(0, 2), (3, 0), (1, 1)\}$

24. $A = \{1, 2, 3\}$ and $B = \{9, 10, 11, 12\}$

  (a) $\{(1, 10), (3, 11), (3, 12), (2, 12)\}$

  (b) $\{(1, 10), (2, 11), (3, 12)\}$

  (c) $\{(1, 10), (1, 9), (3, 11), (2, 12)\}$

  (d) $\{(3, 9), (2, 9), (1, 12)\}$

In Exercises 25–28, determine whether the description represents a function. *See Example 3.*

25. The input value x is any of 16 homes on a street and the output value y is the number of vehicles in the driveway of the home.

26. The input value x is any of 15 football games played in a day when one game ended in a tie. The output value y is the winner of game x.

27. The input value x is any of the integers from 1 to 20. The output value y is the set of all factors of the integer.

28. The input value x is any of the positive even integers. The output value y is one-half the value of x.

In Exercises 29–32, show that both ordered pairs are solutions of the equation, and explain why this implies that y is not a function of x. *See Example 4.*

29. $x^2 + y^2 = 25$; $(0, 5), (0, -5)$

30. $x^2 + 4y^2 = 16$; $(0, 2), (0, -2)$

31. $|y| = x + 2$; $(1, 3), (1, -3)$

32. $|y - 2| = x$; $(2, 4), (2, 0)$

In Exercises 33–38, explain why the equation represents y as a function of x. *See Example 4.*

33. $y = 10x + 12$

34. $y = 3 - 8x$

35. $3x + 7y - 2 = 0$

36. $x - 9y + 3 = 0$

37. $y = x(x - 10)$

38. $y = (x + 2)^2 + 3$

In Exercises 39–42, determine whether the equation represents y as a function of x. *See Example 4.*

39. $y^2 = x$

40. $y = x^2$

41. $y = |x|$

42. $|y| = x$

In Exercises 43–48, fill in each blank and simplify. *See Example 5.*

✓ **43.** $f(x) = 3x + 5$

   (a) $f(2) = 3(\rule{2cm}{0.3mm}) + 5$

   (b) $f(-2) = 3(\rule{2cm}{0.3mm}) + 5$

   (c) $f(k) = 3(\rule{2cm}{0.3mm}) + 5$

   (d) $f(k + 1) = 3(\rule{2cm}{0.3mm}) + 5$

44. $f(x) = 6 - 2x$

   (a) $f(3) = 6 - 2(\rule{2cm}{0.3mm})$

   (b) $f(-4) = 6 - 2(\rule{2cm}{0.3mm})$

   (c) $f(n) = 6 - 2(\rule{2cm}{0.3mm})$

   (d) $f(n - 2) = 6 - 2(\rule{2cm}{0.3mm})$

45. $f(x) = 3 - x^2$

   (a) $f(0) = 3 - (\rule{2cm}{0.3mm})^2$

   (b) $f(-3) = 3 - (\rule{2cm}{0.3mm})^2$

   (c) $f(m) = 3 - (\rule{2cm}{0.3mm})^2$

   (d) $f(2t) = 3 - (\rule{2cm}{0.3mm})^2$

46. $f(x) = \sqrt{x + 8}$

   (a) $f(1) = \sqrt{(\rule{2cm}{0.3mm}) + 8}$

   (b) $f(-4) = \sqrt{(\rule{2cm}{0.3mm}) + 8}$

   (c) $f(h) = \sqrt{(\rule{2cm}{0.3mm}) + 8}$

   (d) $f(h - 8) = \sqrt{(\rule{2cm}{0.3mm}) + 8}$

47. $f(x) = \dfrac{x}{x + 2}$

   (a) $f(3) = \dfrac{(\rule{2cm}{0.3mm})}{(\rule{2cm}{0.3mm}) + 2}$

   (b) $f(-4) = \dfrac{(\rule{2cm}{0.3mm})}{(\rule{2cm}{0.3mm}) + 2}$

   (c) $f(s) = \dfrac{(\rule{2cm}{0.3mm})}{(\rule{2cm}{0.3mm}) + 2}$

   (d) $f(s - 2) = \dfrac{(\rule{2cm}{0.3mm})}{(\rule{2cm}{0.3mm}) + 2}$

48. $f(x) = \dfrac{2x}{x - 7}$

   (a) $f(2) = \dfrac{2(\rule{2cm}{0.3mm})}{(\rule{2cm}{0.3mm}) - 7}$

   (b) $f(-3) = \dfrac{2(\rule{2cm}{0.3mm})}{(\rule{2cm}{0.3mm}) - 7}$

   (c) $f(t) = \dfrac{2(\rule{2cm}{0.3mm})}{(\rule{2cm}{0.3mm}) - 7}$

   (d) $f(t + 5) = \dfrac{2(\rule{2cm}{0.3mm})}{(\rule{2cm}{0.3mm}) - 7}$

In Exercises 49–64, evaluate the function as indicated, and simplify. *See Examples 5 and 6.*

49. $f(x) = 12x - 7$

   (a) $f(3)$ (b) $f\left(\frac{3}{2}\right)$

   (c) $f(a) + f(1)$ (d) $f(a + 1)$

50. $f(x) = 3 - 7x$

   (a) $f(-1)$ (b) $f\left(\frac{1}{2}\right)$

   (c) $f(t) + f(-2)$ (d) $f(2t - 3)$

51. $g(x) = 2 - 4x + x^2$

   (a) $g(4)$ (b) $g(0)$

   (c) $g(2y)$ (d) $g(4) + g(6)$

52. $h(x) = x^2 - 2x$

   (a) $h(2)$ (b) $h(0)$

   (c) $h(1) - h(-4)$ (d) $h(4t)$

53. $f(x) = \sqrt{x + 5}$

   (a) $f(-1)$ (b) $f(4)$

   (c) $f(z - 5)$ (d) $f(5z)$

54. $h(x) = \sqrt{2x - 3}$

   (a) $h(4)$ (b) $h(2)$

   (c) $h(4n)$ (d) $h(n + 2)$

55. $g(x) = 8 - |x - 4|$

   (a) $g(0)$ (b) $g(8)$

   (c) $g(16) - g(-1)$ (d) $g(x - 2)$

56. $g(x) = 2|x + 1| - 2$

   (a) $g(2)$ (b) $g(-1)$

   (c) $g(-4)$ (d) $g(3) + g(-5)$

57. $f(x) = \dfrac{3x}{x - 5}$

   (a) $f(0)$ (b) $f\left(\frac{5}{3}\right)$

   (c) $f(2) - f(-1)$ (d) $f(x + 4)$

58. $f(x) = \dfrac{x + 2}{x - 3}$

   (a) $f(-3)$ (b) $f\left(-\frac{3}{2}\right)$

   (c) $f(4) + f(8)$ (d) $f(x - 5)$

59. $f(x) = \begin{cases} x + 8, & \text{if } x < 0 \\ 10 - 2x, & \text{if } x \geq 0 \end{cases}$

   (a) $f(4)$ (b) $f(-10)$

   (c) $f(0)$ (d) $f(6) - f(-2)$

60. $f(x) = \begin{cases} -x, & \text{if } x \leq 0 \\ 6 - 3x, & \text{if } x > 0 \end{cases}$

   (a) $f(0)$ (b) $f\left(-\frac{3}{2}\right)$

   (c) $f(4)$ (d) $f(-2) + f(25)$

61. $h(x) = \begin{cases} 4 - x^2, & \text{if } x \leq 2 \\ x - 2, & \text{if } x > 2 \end{cases}$

   (a) $h(2)$ (b) $h\left(-\frac{3}{2}\right)$

   (c) $h(5)$ (d) $h(-3) + h(7)$

62. $f(x) = \begin{cases} x^2, & \text{if } x < 1 \\ x^2 - 3x + 2, & \text{if } x \geq 1 \end{cases}$

   (a) $f(1)$ (b) $f(-1)$

   (c) $f(2)$ (d) $f(-3) + f(3)$

63. $f(x) = 2x + 5$

   (a) $\dfrac{f(x + 2) - f(2)}{x}$ (b) $\dfrac{f(x - 3) - f(3)}{x}$

64. $f(x) = 3x + 4$

   (a) $\dfrac{f(x + 1) - f(1)}{x}$ (b) $\dfrac{f(x - 5) - f(5)}{x}$

In Exercises 65–74, find the domain of the function. *See Example 7.*

65. $f(x) = x^2 + x - 2$

66. $h(x) = 3x^2 - x$

67. $f(t) = \dfrac{t + 3}{t(t + 2)}$

68. $g(s) = \dfrac{s - 2}{(s - 6)(s - 10)}$

69. $g(x) = \sqrt{x + 4}$

70. $f(x) = \sqrt{2 - x}$

71. $f(x) = \sqrt{2x - 1}$

72. $G(x) = \sqrt{8 - 3x}$

73. $f(t) = |t - 4|$

74. $f(x) = |x + 3|$

In Exercises 75–82, find the domain and range of the function. *See Example 8.*

75. f: $\{(0, 0), (2, 1), (4, 8), (6, 27)\}$

76. f: $\{(-3, 4), (-1, 3), (2, 0), (5, 3)\}$

77. f: $\left\{\left(-3, -\frac{17}{2}\right), \left(-1, -\frac{5}{2}\right), (4, 2), (10, 2)\right\}$

78. f: $\left\{\left(\frac{1}{2}, 4\right), \left(\frac{3}{4}, 5\right), (1, 6), \left(\frac{5}{4}, 7\right)\right\}$

79. Circumference of a circle: $C = 2\pi r$

80. Area of a square with side s: $A = s^2$

81. Area of a circle with radius r: $A = \pi r^2$

82. Volume of a sphere with radius r: $V = \frac{4}{3}\pi r^3$

Solving Problems

83. ▲ *Geometry* Write the perimeter P of a square as a function of the length x of one of its sides.

84. ▲ *Geometry* Write the surface area S of a cube as a function of the length x of one of its edges.

85. ▲ *Geometry* Write the volume V of a cube as a function of the length x of one of its edges.

86. ▲ *Geometry* Write the length L of the diagonal of a square as a function of the length x of one of its sides.

87. *Cost* The cost of producing a video game is $1.95 per unit with fixed costs of $8000. Write the total cost C as a function of x, the number of units produced.

88. *Cost* The cost of producing a software program is $3.25 per unit with fixed costs of $495. Write the total cost C as a function of x, the number of units produced.

89. *Distance* An airplane flies at a speed of 120 miles per hour. Write the distance d traveled by the airplane as a function of time t in hours.

90. *Distance* A car travels for 4 hours on a highway at a steady speed. Write the distance d traveled by the car as a function of its speed s in miles per hour.

91. *Distance* A train travels at a speed of 65 miles per hour. Write the distance d traveled by the train as a function of time t in hours. Then find d when the value of t is 4.

92. *Distance* A migrating bird flies at a steady speed for 8 hours. Write the distance d traveled by the bird as a function of its speed s in miles per hour. Then find d when the value of s is 35.

93. ▲ *Geometry* Strips of width x are cut from the four sides of a square that is 32 inches on a side (see figure). Write the area A of the remaining square as a function of x.

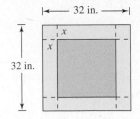

94. ▲ *Geometry* Strips of width x are cut from two adjacent sides of a square that is 32 inches on a side (see figure). Write the area A of the remaining square as a function of x.

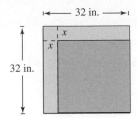

95. ▲ *Geometry* An open box is to be made from a square piece of material 24 inches on a side by cutting equal squares from the corners and turning up the sides (see figure). Write the volume V of the box as a function of x.

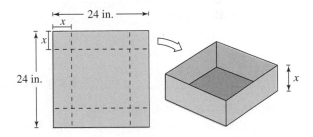

96. ▲ *Geometry* You use the method of Exercise 95 to make a box from a rectangular piece of material that is 24 inches long and 20 inches wide. Write the volume of the box as a function of x.

97. *Profit* The marketing department of a business has determined that the profit from selling x units of a product is approximated by the model

$$P(x) = 50\sqrt{x} - 0.5x - 500.$$

Find (a) $P(1600)$ and (b) $P(2500)$.

98. *Safe Load* A solid rectangular beam has a height of 6 inches and a width of 4 inches. The safe load S of the beam with the load at the center is a function of its length L and is approximated by the model

$$S(L) = \frac{128,160}{L}$$

where S is measured in pounds and L is measured in feet. Find (a) $S(12)$ and (b) $S(16)$.

Wages   In Exercises 99 and 100, use the following information. A wage earner is paid $12.00 per hour for regular time and time-and-a-half for overtime. The weekly wage function is

$$W(h) = \begin{cases} 12h, & 0 \le h \le 40 \\ 18(h - 40) + 480, & h > 40, \end{cases}$$

where h represents the number of hours worked in a week.

99. (a) Evaluate $W(30)$, $W(40)$, $W(45)$, and $W(50)$.

   (b) Could you use values of h for which $h < 0$ in this model? Why or why not?

100. (a) Evaluate $W(20)$, $W(25)$, $W(35)$, and $W(55)$.

   (b) Describe the domain implied by the situation.

Data Analysis   In Exercises 101 and 102, use the graph, which shows the numbers of students (in millions) enrolled in public and private post-secondary degree-granting institutions in the United States.   (Source: U.S. National Center for Education Statistics)

101. Is the enrollment in public post-secondary schools a function of the year? Explain. Let $f(x)$ represent the number of students in these schools in year x. Approximate $f(2004)$.

102. Is the enrollment in private post-secondary schools a function of the year? Explain. Let $g(x)$ represent the number of students in these schools in year x. Approximate $g(2003)$.

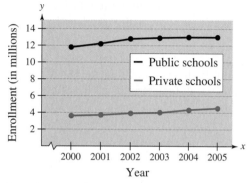

Figure for 101 and 102

Explaining Concepts

In Exercises 103 and 104, determine whether the statement uses the word *function* in a way that is mathematically correct. Explain.

103. The sales tax on clothes is 5.5% of the selling price, so the sales tax on clothes is a function of the price.

104. A shoe store sells shoes at prices from $19.99 to $189.99 per pair, so the daily revenue of the store is a function of the number of pairs of shoes sold.

105. ✎   Set A is a relation but not a function. Is it possible that a subset of set A is a function? Explain.

106. ✎   Set B is a function. Is it possible that a subset of set B is a relation but not a function? Explain.

107. Make a list of real-life situations that are relations. In each situation, identify the domain and the range and discuss whether the relation is also a function.

108. Make a list of real-life situations that are functions. In each situation, identify the domain and range and discuss whether the function represents a finite or an infinite number of ordered pairs.

Cumulative Review

In Exercises 109–112, identify the property of real numbers illustrated by the statement.

109. $(6x) \cdot 1 = 6x$

110. $0(3x + 17) + 12 = 12$

111. $2a \cdot \dfrac{1}{2a} = 1$

112. $8 - 2(x + 7) = 8 - 2x - 2(7)$

In Exercises 113–116, sketch the graph of the line.

113. $y = -x$

114. $y = \frac{1}{2}x + 3$

115. $x - 4y = 16$

116. $5x - 3y = 0$

In Exercises 117–120, sketch the graph of the solution of the inequality.

117. $x \le -2$

118. $y > 3$

119. $y > \frac{3}{4}x - 1$

120. $2x + 3y \le 6$

3.7 Graphs of Functions

Charles Gullung/Getty Images

Why You Should Learn It

Graphs of functions can help you visualize relationships between variables in real-life situations. For instance, in Exercise 77 on page 206, you will use a graphing calculator to represent the sales of petroleum and coal products in the United States.

1 ▶ Sketch graphs of functions on rectangular coordinate systems.

What You Should Learn

1 ▶ Sketch graphs of functions on rectangular coordinate systems.

2 ▶ Identify the graphs of basic functions.

3 ▶ Use the Vertical Line Test to determine if graphs represent functions.

4 ▶ Use vertical and horizontal shifts and reflections to sketch graphs of functions.

The Graph of a Function

Consider a function f whose domain and range are the set of real numbers. The **graph** of f is the set of ordered pairs $(x, f(x))$, where x is in the domain of f.

$x = x$-coordinate of the ordered pair

$f(x) = y$-coordinate of the ordered pair

Figure 3.62 shows a typical graph of such a function.

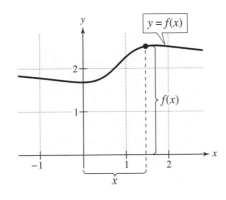

Figure 3.62

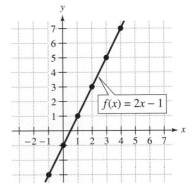

Figure 3.63

Study Tip

In Example 1, the (implied) domain of the function is the set of all real numbers. When writing the equation of a function, you may choose to restrict its domain by writing a condition to the right of the equation. For instance, the domain of the function

$$f(x) = 4x + 5, \quad x \geq 0$$

is the set of all nonnegative real numbers (all $x \geq 0$).

EXAMPLE 1   **Sketching the Graph of a Function**

Sketch the graph of $f(x) = 2x - 1$.

Solution

One way to sketch the graph is to begin by making a table of values.

x	-1	0	1	2	3	4
$f(x)$	-3	-1	1	3	5	7

Next, plot the six points shown in the table. Finally, connect the points with a line, as shown in Figure 3.63.

 **CHECKPOINT** *Now try Exercise 1.*

$f(x) = x + 3, x < 0$

$f(x) = -x + 4, x \geq 0$

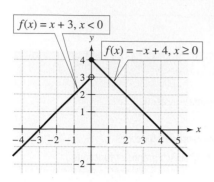

Figure 3.64

EXAMPLE 2   **Sketching the Graph of a Piecewise-Defined Function**

Sketch the graph of $f(x) = \begin{cases} x + 3, & x < 0 \\ -x + 4, & x \geq 0 \end{cases}$. Then determine its domain and range.

Solution

Begin by graphing $f(x) = x + 3$ for $x < 0$, as shown in Figure 3.64. You will recognize that this is the graph of the line $y = x + 3$ with the restriction that the x-values are negative. Because $x = 0$ is not in the domain, the right endpoint of the line is an open dot. Next, graph $f(x) = -x + 4$ for $x \geq 0$ on the same set of coordinate axes, as shown in Figure 3.64. This is the graph of the line $y = -x + 4$ with the restriction that the x-values are nonnegative. Because $x = 0$ is in the domain, the left endpoint of the line is a solid dot. From the graph, you can see that the domain is $-\infty \leq x \leq \infty$ and the range is $y \leq 4$.

✓ CHECKPOINT *Now try Exercise 25.*

2 ▶ Identify the graphs of basic functions.

Graphs of Basic Functions

To become good at sketching the graphs of functions, it helps to be familiar with the graphs of some basic functions. The functions shown in Figure 3.65, and variations of them, occur frequently in applications.

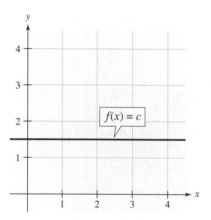

(a) Constant function

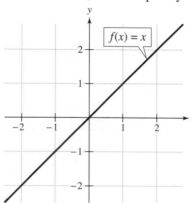

(b) Identity function

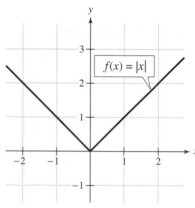

(c) Absolute value function

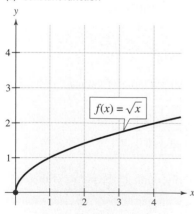

(d) Square root function

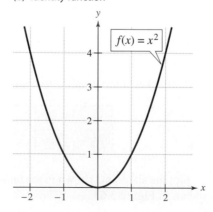

(e) Squaring function

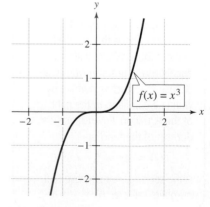

(f) Cubing function

Figure 3.65

3 ▶ Use the Vertical Line Test to determine if graphs represent functions.

The Vertical Line Test

By the definition of a function, at most one y-value corresponds to a given x-value. This implies that any vertical line can intersect the graph of a function at most once.

Vertical Line Test for Functions

A set of points on a rectangular coordinate system is the graph of y as a function of x if and only if no vertical line intersects the graph at more than one point.

EXAMPLE 3   **Using the Vertical Line Test**

Determine whether each equation represents y as a function of x.

a. $y = x^2 - 3x + \frac{1}{4}$

b. $x = y^2 - 1$

c. $x = y^3$

Solution

a. From the graph of the equation in Figure 3.66, you can see that every vertical line intersects the graph at most once. So, by the Vertical Line Test, the equation *does* represent y as a function of x.

b. From the graph of the equation in Figure 3.67, you can see that a vertical line intersects the graph twice. So, by the Vertical Line Test, the equation *does not* represent y as a function of x.

c. From the graph of the equation in Figure 3.68, you can see that every vertical line intersects the graph at most once. So, by the Vertical Line Test, the equation *does* represent y as a function of x.

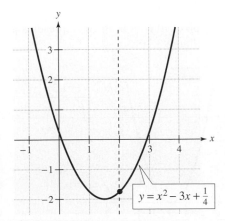

Graph of a function of x.
Vertical line intersects once.
Figure 3.66

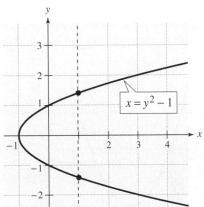

Not a graph of a function of x.
Vertical line intersects twice.
Figure 3.67

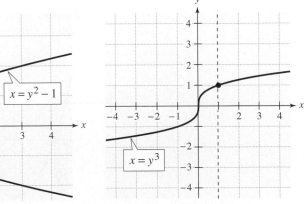

Graph of a function of x.
Vertical line intersects once.
Figure 3.68

✓ **CHECKPOINT** *Now try Exercise 33.*

4 ▶ Use vertical and horizontal shifts and reflections to sketch graphs of functions.

Transformations of Graphs of Functions

Many functions have graphs that are simple transformations of the basic graphs shown in Figure 3.65 on page 198. The following list summarizes the various types of **vertical** and **horizontal shifts** of the graphs of functions.

Technology: Discovery

Use a graphing calculator to display the graphs of $y = x^2 + c$, where c is equal to $-2, 0, 2,$ and 4. What conclusions can you make?

Use a graphing calculator to display the graphs of $y = (x + c)^2$, where c is equal to $-3, -1, 0, 1,$ and 3. What conclusions can you make?

Vertical and Horizontal Shifts

Let c be a positive real number. **Vertical** and **horizontal shifts** of the graph of the function $y = f(x)$ are represented as follows.

1. Vertical shift c units *upward*: $h(x) = f(x) + c$
2. Vertical shift c units *downward*: $h(x) = f(x) - c$
3. Horizontal shift c units to the *right*: $h(x) = f(x - c)$
4. Horizontal shift c units to the *left*: $h(x) = f(x + c)$

Note that for a vertical transformation, the addition of a positive number c yields a shift upward (in the positive direction) and the subtraction of a positive number c yields a shift downward (in the negative direction). For a horizontal transformation, replacing x with $x + c$ yields a shift to the left (in the negative direction) and replacing x with $x - c$ yields a shift to the right (in the positive direction).

EXAMPLE 4   Shifts of the Graphs of Functions

Use the graph of $f(x) = x^2$ to sketch the graph of each function.

a. $g(x) = x^2 - 2$ **b.** $h(x) = (x + 3)^2$

Solution

a. Relative to the graph of $f(x) = x^2$, the graph of $g(x) = x^2 - 2$ represents a shift of two units *downward*, as shown in Figure 3.69.

b. Relative to the graph of $f(x) = x^2$, the graph of $h(x) = (x + 3)^2$ represents a shift of three units to the *left*, as shown in Figure 3.70.

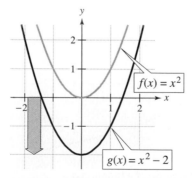

Vertical Shift: Two Units Downward
Figure 3.69

Horizontal Shift: Three Units Left
Figure 3.70

 **CHECKPOINT** *Now try Exercise 47(a).*

Some graphs can be obtained from *combinations* of vertical and horizontal shifts, as shown in part (b) of the next example.

EXAMPLE 5   **Shifts of the Graphs of Functions**

Use the graph of $f(x) = x^3$ to sketch the graph of each function.

a. $g(x) = x^3 + 2$

b. $h(x) = (x - 1)^3 + 2$

Solution

a. Relative to the graph of $f(x) = x^3$, the graph of $g(x) = x^3 + 2$ represents a shift of two units *upward*, as shown in Figure 3.71.

b. Relative to the graph of $f(x) = x^3$, the graph of $h(x) = (x - 1)^3 + 2$ represents a shift of one unit to the *right*, followed by a shift of two units *upward*, as shown in Figure 3.72.

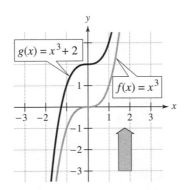

Vertical Shift: Two Units Upward

Figure 3.71

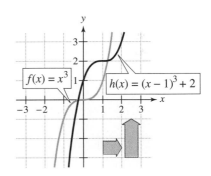

Horizontal Shift: One Unit Right
Vertical Shift: Two Units Upward

Figure 3.72

✓ **CHECKPOINT** *Now try Exercise 47(e).*

The second basic type of transformation is a **reflection.** For instance, if you imagine that the x-axis represents a mirror, then the graph of

$$h(x) = -x^2$$

is the mirror image (or reflection) of the graph of

$$f(x) = x^2$$

as shown in Figure 3.73.

Figure 3.73   Reflection

Reflections in the Coordinate Axes

Reflections of the graph of $y = f(x)$ are represented as follows.

1. Reflection in the x-axis: $h(x) = -f(x)$

2. Reflection in the y-axis: $h(x) = f(-x)$

EXAMPLE 6 **Reflections of the Graphs of Functions**

Use the graph of $f(x) = \sqrt{x}$ to sketch the graph of each function.

a. $g(x) = -\sqrt{x}$ **b.** $h(x) = \sqrt{-x}$

Solution

a. Relative to the graph of

$$f(x) = \sqrt{x}$$

the graph of

$$g(x) = -\sqrt{x} = -f(x)$$

represents a *reflection in the x-axis*, as shown in Figure 3.74.

b. Relative to the graph of

$$f(x) = \sqrt{x}$$

the graph of

$$h(x) = \sqrt{-x} = f(-x)$$

represents a *reflection in the y-axis*, as shown in Figure 3.75.

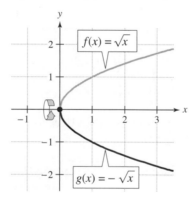

Reflection in *x*-Axis

Figure 3.74

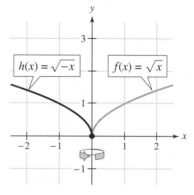

Reflection in *y*-Axis

Figure 3.75

✓ **CHECKPOINT** *Now try Exercise 47(f).*

EXAMPLE 7 **Graphical Reasoning**

Identify the basic function and any transformation shown in Figure 3.76. Write the equation for the graphed function.

Solution

In Figure 3.76, the basic function is $f(x) = |x|$. The transformation is a reflection in the *x*-axis and a vertical shift one unit downward. The equation of the graphed function is $f(x) = -|x| - 1$.

✓ **CHECKPOINT** *Now try Exercise 69.*

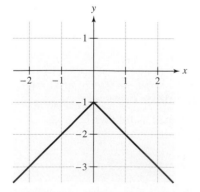

Figure 3.76

_____ Concept Check _____

1. In a rectangular coordinate system, does the graph of $x = 3$ represent the graph of a function? Explain.

2. Describe the relationship between the graphs of $f(x)$ and $g(x) = f(x - 2)$.

3. Describe the relationship between the graphs of $f(x)$ and $g(x) = f(-x)$.

4. Identify the basic function in $g(x) = (x - 1)^2 + 1$. What transformations are used to obtain g?

3.7 EXERCISES

Go to pages 208–209 to record your assignments.

_____ Developing Skills _____

In Exercises 1–28, sketch the graph of the function. Then determine its domain and range. *See Examples 1 and 2.*

✓ 1. $f(x) = 2x - 7$

2. $f(x) = 1 - 3x$

3. $g(x) = \frac{1}{2}x^2$

4. $h(x) = \frac{1}{4}x^2 - 1$

5. $f(x) = -(x - 1)^2$

6. $g(x) = (x + 2)^2 + 3$

7. $h(x) = x^2 - 6x + 8$

8. $f(x) = -x^2 - 2x + 1$

9. $C(x) = \sqrt{x} + 2$

10. $Q(x) = 4 - \sqrt{x}$

11. $f(t) = \sqrt{t - 2}$

12. $h(x) = \sqrt{4 - x}$

13. $G(x) = 8$

14. $H(x) = -4$

15. $g(s) = s^3 + 1$

16. $f(x) = x^3 - 4$

17. $f(x) = |x + 3|$

18. $g(x) = |x - 1|$

19. $K(s) = |s - 4| + 1$

20. $Q(t) = 1 - |t + 1|$

21. $f(x) = 6 - 3x, \quad 0 \le x \le 2$

22. $f(x) = \frac{1}{3}x - 2, \quad 6 \le x \le 12$

23. $h(x) = x^2 - 2x, \quad -2 \le x \le 6$

24. $h(x) = 6x - x^2, \quad 0 \le x \le 6$

✓ 25. $h(x) = \begin{cases} 2x + 3, & x < 0 \\ 3 - x, & x \ge 0 \end{cases}$

26. $f(x) = \begin{cases} x + 6, & x < 0 \\ 6 - 2x, & x \ge 0 \end{cases}$

27. $f(x) = \begin{cases} 3 - x, & x < -3 \\ x^2 + x, & x \ge -3 \end{cases}$

28. $h(x) = \begin{cases} 4 - x^2, & x \le 2 \\ x - 2, & x > 2 \end{cases}$

In Exercises 29–32, use a graphing calculator to graph the function and find its domain and range.

29. $g(x) = -x^4 - 3$

30. $f(x) = 3x^3 - 4$

31. $f(x) = \sqrt{x^2 + 1}$

32. $h(x) = \sqrt{4 - x^2}$

In Exercises 33–38, use the Vertical Line Test to determine whether y is a function of x. **See Example 3.**

✓ **33.** $y = \frac{1}{3}x^3$

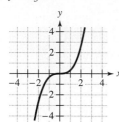

34. $y = x^2 - 2x$

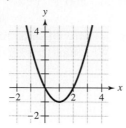

35. $y = -(x - 3)^2$

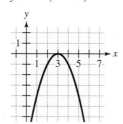

36. $-2x + y^2 = 6$

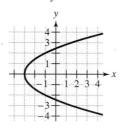

37. $x^2 + y^2 = 16$

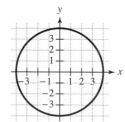

38. $|y| = x$

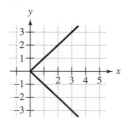

In Exercises 39–42, sketch a graph of the equation. Use the Vertical Line Test to determine whether y is a function of x.

39. $-2x + 3y = 12$

40. $y = x^2 + 2$

41. $y^2 = x + 1$

42. $x = y^4$

In Exercises 43–46, match the function with its graph. [The graphs are labeled (a), (b), (c), and (d).]

(a)

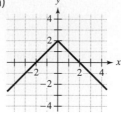

(b)

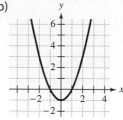

(c)

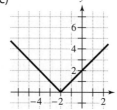

(d)

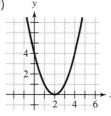

43. $f(x) = x^2 - 1$

44. $f(x) = (x - 2)^2$

45. $f(x) = 2 - |x|$

46. $f(x) = |x + 2|$

In Exercises 47 and 48, identify the transformation of f, and sketch a graph of the function h. **See Examples 4–6.**

47. $f(x) = x^2$

✓ (a) $h(x) = x^2 + 2$ (b) $h(x) = x^2 - 4$

(c) $h(x) = (x + 2)^2$ (d) $h(x) = (x - 4)^2$

✓ (e) $h(x) = (x - 3)^2 + 1$ ✓ (f) $h(x) = -x^2 + 4$

48. $f(x) = x^3$

(a) $h(x) = x^3 + 3$ (b) $h(x) = x^3 - 5$

(c) $h(x) = (x - 3)^3$ (d) $h(x) = (x + 2)^3$

(e) $h(x) = 2 - (x - 1)^3$ (f) $h(x) = -x^3$

In Exercises 49–54, identify the transformation of the graph of $f(x) = |x|$, and sketch the graph of h.

49. $h(x) = |x - 3|$

50. $h(x) = |x + 3|$

51. $h(x) = |x| - 4$

52. $h(x) = |-x|$

53. $h(x) = -|x|$

54. $h(x) = 4 - |x|$

In Exercises 55–60, use the graph of $f(x) = x^2$ to write a function that represents the graph.

55.

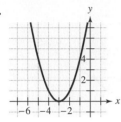

56.

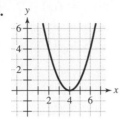

65.

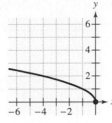

66.

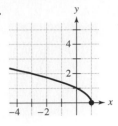

57.

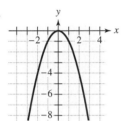

58.

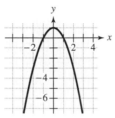

In Exercises 67 and 68, use a graphing calculator to graph f. Decide how to alter the function to produce each of the transformation descriptions. Graph each transformation in the same viewing window with f; confirm that each transformation moved f as described.

a. The graph of f shifted to the left three units

b. The graph of f shifted downward five units

c. The graph of f shifted upward one unit

d. The graph of f shifted to the right two units

59.

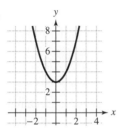

60.

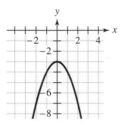

67. $f(x) = (x - 1)^2 + 1$ **68.** $f(x) = \sqrt{x + 3}$

In Exercises 69–74, identify the basic function and any transformation shown in the graph. Write the equation for the graphed function. *See Example 7.*

In Exercises 61–66, use the graph of $f(x) = \sqrt{x}$ to write a function that represents the graph.

61.

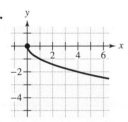

62.

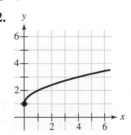

69.

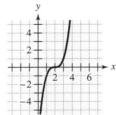

70.

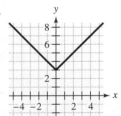

63.

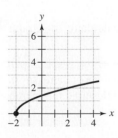

64.

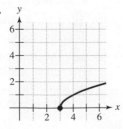

71.

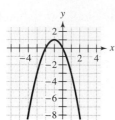

72.

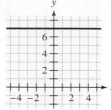

73.

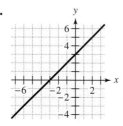

74.

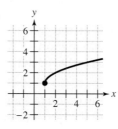

75. Use the graph of f to sketch each graph.

(a) $y = f(x) + 2$

(b) $y = -f(x)$

(c) $y = f(x - 2)$

(d) $y = f(x + 2)$

(e) $y = f(x) - 1$

(f) $y = f(-x)$

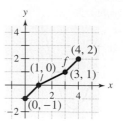

76. Use the graph of f to sketch each graph.

(a) $y = f(x) - 1$

(b) $y = f(x + 1)$

(c) $y = f(x - 1)$

(d) $y = -f(x - 2)$

(e) $y = f(-x)$

(f) $y = f(x) + 2$

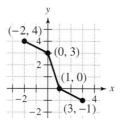

Solving Problems

77. 🖩 *Petroleum and Coal Sales*   For the years 2001 through 2005, the sales S (in billions of dollars) of petroleum and coal products in the United States can be modeled by

$$S(t) = 38.943t^2 - 95.04t + 523.2, \quad 1 \le t \le 5$$

where $t = 1$ represents 2001.   (Source: U.S. Census Bureau)

(a) Use a graphing calculator to graph the model over the specified domain.

(b) Use the graph to approximate the year in which sales were $600 billion.

78. 🖩 *Electric Power Sales*   For the years 2001 through 2005, the sales S (in billions of kilowatt-hours) of electric power in the United States can be modeled by

$$S(t) = 4.86t^2 + 32.3t + 3356, \quad 1 \le t \le 5$$

where $t = 1$ represents 2001.   (Source: Edison Electric Institute)

(a) Use a graphing calculator to graph the model over the specified domain.

(b) Use the graph to approximate the sales in 2004.

79. 🔺 *Geometry*   The perimeter of a rectangle is 200 meters.

(a) Show algebraically that the area of the rectangle is given by $A = l(100 - l)$, where l is its length.

(b) 🖩 Use a graphing calculator to graph the area function.

(c) 🖩 Use the graph to determine the value of l that yields the largest value of A. Interpret the result.

80. 🔺 *Geometry*   The length and width of a rectangular flower garden are 40 feet and 30 feet, respectively. A walkway of uniform width x surrounds the garden.

(a) Write the outside perimeter y of the walkway as a function of x.

(b) 🖩 Use a graphing calculator to graph the function for the perimeter.

(c) 🖩 Determine the slope of the graph in part (b). For each additional one-foot increase in the width of the walkway, determine the increase in its outside perimeter.

81. *Population*   For the years 1990 through 2005, the population P (in millions) of the United States can be modeled by

$$P(t) = -0.025t^2 + 3.53t + 248.9, \quad 0 \le t \le 15$$

where $t = 0$ represents 1990.   (Source: U.S. Census Bureau)

(a) 🖩 Use a graphing calculator to graph the model over the specified domain.

(b) Is P a function of t? Explain.

(c) Find $P(5)$ and $P(10)$.

(d) In the transformation of the population model

$$P_1(t) = -0.025(t - 10)^2 + 3.53(t - 10)$$
$$+ 248.9$$

what calendar year corresponds to $t = 10$? Explain.

(e) 🖩 Use a graphing calculator to graph P_1 over the appropriate domain.

82. *Graphical Reasoning*   An electronically controlled thermostat in a home is programmed to lower the temperature automatically during the night. The temperature T, in degrees Fahrenheit, is given in terms of t, the time on a 24-hour clock (see figure).

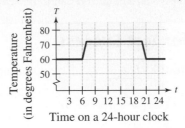

(a) Explain why T is a function of t.

(b) Find $T(4)$ and $T(15)$.

(c) The thermostat is reprogrammed to produce a temperature H, where $H(t) = T(t - 1)$. Explain how this changes the temperature in the house.

(d) The thermostat is reprogrammed to produce a temperature H, where $H(t) = T(t) - 1$. Explain how this changes the temperature in the house.

Explaining Concepts

83. ✎ In your own words, explain how to use the Vertical Line Test.

84. ✎ Describe the four types of shifts of the graph of a function.

85. ✎ Explain the change in the range of the function $f(x) = 2x$ if the domain is changed from $0 \le x \le 2$ to $0 \le x \le 4$.

86. ✎ Is it possible for the graph of a function to be a scatter plot rather than one or more continuous lines or a curve? If so, when would this be appropriate?

Cumulative Review

In Exercises 87–90, write a verbal description of the algebraic expression.

87. $4x + 1$

88. $x(x - 2)$

89. $\dfrac{2n}{3}$

90. $x^2 + 6$

In Exercises 91–94, solve for y in terms of x.

91. $2x + y = 4$

92. $3x - 6y = 12$

93. $-4x + 3y + 3 = 0$

94. $3x + 4y - 5 = 0$

In Exercises 95–98, determine whether the point is a solution of the inequality.

95. $y < 2x + 1, \quad (0, 1)$

96. $y \le 4 - |2x|, \quad (-2, -1)$

97. $2x - 3y > 2y, \quad (6, 2)$

98. $4x + 1 > x - y, \quad (-1, 3)$

What Did You Learn?

Use these two pages to help prepare for a test on this chapter. Check off the key terms and key concepts you know. You can also use this section to record your assignments.

Plan for Test Success

Date of test: ☐ / /

Study dates and times: ☐ / / at ☐ : A.M./P.M.

☐ / / at ☐ : A.M./P.M.

Things to review:

☐ Key Terms, *p. 208*
☐ Key Concepts, *pp. 208–209*
☐ Your class notes
☐ Your assignments

☐ Study Tips, *pp. 126, 141, 143, 166, 175, 176, 183, 184, 188, 197, 199*
☐ Technology Tips, *pp. 153, 164, 178, 202*
☐ Mid-Chapter Quiz, *p. 173*

☐ Review Exercises, *pp. 210–214*
☐ Chapter Test, *p. 215*
☐ Video Explanations Online
☐ Tutorial Online

Key Terms

☐ rectangular coordinate system, *p. 126*
☐ *x*- and *y*-axes, *p. 126*
☐ origin, *p. 126*
☐ quadrants, *p. 126*
☐ ordered pair, *p. 126*
☐ *x*- and *y*-coordinates, *p. 126*
☐ scatter plot, *p. 128*
☐ solution (of an equation), *p. 129*
☐ Pythagorean Theorem, *p. 131*
☐ Distance Formula, *p. 132*
☐ collinear, *p. 133*
☐ Midpoint Formula, *p. 134*

☐ graph of an equation, *p. 140*
☐ linear equation, *p. 140*
☐ *x*-intercept, *y*-intercept, *p. 143*
☐ slope, *p. 149*
☐ slope-intercept form, *p. 154*
☐ constant rate of change, *p. 157*
☐ average rate of change, *p. 157*
☐ point-slope form, *p. 163*
☐ general form, *p. 163*
☐ two-point form, *p. 164*
☐ linear extrapolation, *p. 167*
☐ linear interpolation, *p. 167*
☐ linear inequality, *p. 174*

☐ graph of a linear inequality, *p. 175*
☐ half-plane, *p. 175*
☐ relation, *p. 183*
☐ domain, *pp. 183, 184, 187*
☐ range, *pp. 183, 184, 187*
☐ function, *p. 184*
☐ independent variable, *p. 186*
☐ dependent variable, *p. 186*
☐ function notation, *p. 187*
☐ piecewise-defined function, *p. 188*
☐ implied domain, *p. 189*
☐ graph of a function, *p. 197*
☐ Vertical Line Test, *p. 199*

Key Concepts

3.1 The Rectangular Coordinate System

Assignment: _____ Due date: _____

☐ **Check ordered pairs as solutions of equations.**
1. Substitute the ordered pair coordinates into the equation.
2. Simplify each side of the equation.
3. Check whether a true equation results.

☐ **Use the Distance Formula:**
The distance between two points (x_1, y_1) and (x_2, y_2) is
$d = \sqrt{(x_2 - x_1)^2 + (y_2 - y_1)^2}$.

3.2 Graphs of Equations

Assignment: _____ Due date: _____

☐ **Use the point-plotting method to sketch a graph**
1. Rewrite the equation by isolating one variable.
2. Make a table of values.
3. Plot the resulting points.
4. Connect the points with a smooth curve or line.

☐ **Use the intercepts to aid in sketching a graph.**
To find *x*-intercepts: Let $y = 0$ and solve for *x*.
To find *y*-intercepts: Let $x = 0$ and solve for *y*.

3.3 Slope and Graphs of Linear Equations

Assignment: _____ Due date: _____

☐ **Determine the slope of a line through two points.**

Slope through two points (x_1, y_1) and (x_2, y_2):

$$m = \frac{y_2 - y_1}{x_2 - x_1} = \frac{\text{Change in } y}{\text{Change in } x} = \frac{\text{Rise}}{\text{Run}}$$

Slope Characteristics:

$m > 0$: ↗ $m < 0$: ↘

$m = 0$: ↔ m undefined: ↕

☐ **Use the slope-intercept form.**

In the slope-intercept form of the equation of a line
$$y = mx + b$$
the slope of the graph is m and the y-intercept is $(0, b)$.

☐ **Identify parallel and perpendicular lines.**

Two distinct nonvertical lines are *parallel* if and only if they have the same slope.

Two nonvertical lines are *perpendicular* if and only if their slopes are negative reciprocals of each other.

3.4 Equations of Lines

Assignment: _____ Due date: _____

☐ **Use point-slope form to write equations of lines.**

The point-slope form of the equation of a line with slope m, passing through point (x_1, y_1), is
$$y - y_1 = m(x - x_1).$$

☐ **Write equations of horizontal and vertical lines.**

The equation of a *horizontal line* has the form $y = b$.

The equation of a *vertical line* has the form $x = a$.

3.5 Graphs of Linear Inequalities

Assignment: _____ Due date: _____

☐ **Graph a linear inequality in two variables.**

1. Form the corresponding linear equation. Graph the equation using a dashed line for $<$ or $>$, or a solid line for $\leq$ or $\geq$.

2. Test a point in one of the half-planes.

   a. If the point satisfies the inequality, shade its half-plane.

   b. If not, shade the other half-plane.

3.6 Relations and Functions

Assignment: _____ Due date: _____

☐ **Determine if relations are functions.**

A *relation* is any set of ordered pairs. A *function* is a relation in which no two ordered pairs have the same first component and different second components.

☐ **Use function notation.**

$f(x)$ is read as *the value of f at x* or simply f *of x*.

Independent variable, *input*, and x are terms associated with a function's domain. *Dependent variable*, *output*, y, and $f(x)$ are terms associated with a function's range.

☐ **Identify the domain and range of a function.**

A function's *implied domain* is the set of all real numbers that yield real number values for the function. The *range* consists of the function values that result from all of the values of the domain.

3.7 Graphs of Functions

Assignment: _____ Due date: _____

☐ **Use the Vertical Line Test.**

A vertical line can intersect the graph of a function at no more than one point.

☐ **Graph functions using shifts of basic functions.**

Vertical shift of c units: $h(x) = f(x) \pm c$

Horizontal shift of c units: $h(x) = f(x \pm c)$

☐ **Graph functions using reflections of basic functions.**

Reflection in x-axis: $h(x) = -f(x)$

Reflection in y-axis: $h(x) = f(-x)$

Review Exercises

3.1 The Rectangular Coordinate System

1 ▶ Plot points on a rectangular coordinate system.

In Exercises 1 and 2, plot the points on a rectangular coordinate system.

1. $(0, 2), \left(-4, \frac{1}{2}\right), (2, -3)$
2. $\left(1, -\frac{3}{2}\right), \left(-2, 2\frac{3}{4}\right), (5, 10)$

In Exercises 3–6, determine the quadrant in which the point is located without plotting it. (*x* and *y* are real numbers.)

3. $(6, 4)$

4. $(-4.8, -2)$

5. $(4, y)$

6. $(x, y), \quad xy > 0$

2 ▶ Determine whether ordered pairs are solutions of equations.

In Exercises 7 and 8, determine whether each ordered pair is a solution of the equation.

7. $y = 4 - \frac{1}{2}x$
   (a) $(2, 3)$ (b) $(-1, 5)$
   (c) $(-6, 1)$ (d) $(8, 0)$

8. $3x - 2y + 18 = 0$
   (a) $(3, 10)$ (b) $(0, 9)$
   (c) $(-4, 3)$ (d) $(-8, 0)$

3 ▶ Use the Distance Formula to find the distance between two points.

In Exercises 9–12, plot the points and find the distance between them.

9. $(4, 3), (4, 8)$ 10. $(2, -5), (6, -5)$
11. $(-5, -1), (1, 2)$ 12. $(-3, 3), (6, -1)$

In Exercises 13 and 14, use the Distance Formula to determine whether the three points are collinear.

13. $(-3, -2), (1, 0), (5, 2)$
14. $(-5, 7), (-1, 2), (3, -4)$

4 ▶ Use the Midpoint Formula to find the midpoints of line segments.

In Exercises 15–18, find the midpoint of the line segment joining the points, and then plot the points and the midpoint.

15. $(1, 4), (7, 2)$ 16. $(-1, 3), (5, 5)$
17. $(5, -2), (-3, 5)$ 18. $(1, 6), (6, 1)$

3.2 Graphs of Equations

1 ▶ Sketch graphs of equations using the point-plotting method.

In Exercises 19–24, sketch the graph of the equation.

19. $3y - 2x - 3 = 0$ 20. $3x + 4y + 12 = 0$
21. $y = x^2 - 1$ 22. $y = (x + 3)^2$
23. $y = |x| - 2$ 24. $y = |x - 3|$

2 ▶ Find and use *x*- and *y*-intercepts as aids to sketching graphs.

In Exercises 25–30, find the *x*- and *y*-intercepts of the graph of the equation. Then sketch the graph of the equation and show the coordinates of three solution points (including *x*- and *y*-intercepts).

25. $8x - 2y = -4$ 26. $5x + 4y = 10$

27. $y = 5 - |x|$ 28. $y = |x| + 4$

29. $y = |2x + 1| - 5$ 30. $y = |3 - 6x| - 15$

▦ In Exercises 31–36, use a graphing calculator to graph the equation. Approximate the *x*- and *y*-intercepts (if any).

31. $y = (x + 1)^2 - 9$ 32. $y = \frac{1}{4}(x - 2)^3$

33. $y = -|x - 4| - 7$ 34. $y = 3 - |x - 3|$

35. $y = \sqrt{3 - x}$ 36. $y = x - 2\sqrt{x}$

3 ▸ Use a pattern to write an equation for an application problem, and sketch its graph.

37. *Straight-Line Depreciation* Your family purchases a new SUV for $35,000. For financing purposes, the SUV will be depreciated over a five-year period. At the end of 5 years, the value of the SUV is expected to be $15,000.

(a) Find an equation that relates the depreciated value of the SUV to the number of years since it was purchased.

(b) Sketch the graph of the equation.

(c) What is the *y*-intercept of the graph and what does it represent?

38. *Straight-Line Depreciation* A company purchases a new computer system for $20,000. For tax purposes, the computer system will be depreciated over a six-year period. At the end of the 6 years, the value of the system is expected to be $2000.

(a) Find an equation that relates the depreciated value of the computer system to the number of years since it was purchased.

(b) Sketch the graph of the equation.

(c) What is the *y*-intercept of the graph and what does it represent?

3.3 Slope and Graphs of Linear Equations

1 ▸ Determine the slope of a line through two points.

In Exercises 39–44, find the slope of the line through the points.

39. $(4, 2), (-3, -1)$ **40.** $(-2, 5), (3, -8)$

41. $\left(-6, \frac{3}{4}\right), \left(4, \frac{3}{4}\right)$ **42.** $(7, 2), (7, 8)$

43. $(0, 6), (8, 0)$ **44.** $(0, 0), \left(\frac{7}{2}, 6\right)$

In Exercises 45–48, a point on a line and the slope of the line are given. Find two additional points on the line. (There are many correct answers.)

45. $(2, -4)$ **46.** $(3, 1)$
  $m = -3$ $m = 6$

47. $\left(-4, \frac{1}{2}\right)$ **48.** $\left(-3, -\frac{3}{2}\right)$
  $m = \frac{1}{4}$ $m = -\frac{1}{3}$

49. *Ramp* A loading dock ramp rises 4 feet above the ground. The ramp has a slope of $\frac{1}{12}$. What is the length of the ramp?

50. *Road Grade* When driving down a mountain road, you notice a warning sign indicating a "9% grade." This means that the slope of the road is $-\frac{9}{100}$. Over a stretch of road, your elevation drops by 1500 feet. What is the horizontal change in your position?

2 ▸ Write linear equations in slope-intercept form and graph the equations.

In Exercises 51–54, write the equation of the line in slope-intercept form and use the slope and *y*-intercept to sketch the line.

51. $5x - 2y - 4 = 0$ **52.** $x - 3y - 6 = 0$

53. $6x + 2y + 5 = 0$ **54.** $y - 6 = 0$

3 ▸ Use slopes to determine whether two lines are parallel, perpendicular, or neither.

In Exercises 55–58, determine whether the lines are parallel, perpendicular, or neither.

55. L_1: $y = \frac{3}{2}x + 1$ **56.** L_1: $y = 2x - 5$
  L_2: $y = \frac{2}{3}x - 1$ L_2: $y = 2x + 3$

57. L_1: $y = \frac{3}{2}x - 2$ **58.** L_1: $y = -0.3x - 2$
  L_2: $y = -\frac{2}{3}x + 1$ L_2: $y = 0.3x + 1$

4 ▸ Use slopes to describe rates of change in real-life problems.

59. *Consumer Awareness* In May of 1998, the average cost of regular unleaded gasoline was $1.09 per gallon. By May of 2007, the average cost had risen to $3.13 per gallon. Find the average rate of change in the cost of a gallon of regular unleaded gasoline from 1998 to 2007. (Source: U.S. Bureau of Labor Statistics)

60. *Consumer Awareness* In January of 2000, the average hourly earnings of manufacturing employees was \$13.33 per hour. By January of 2007, the average earnings had risen to \$16.17 per hour. Find the average rate of change in the earnings per hour of manufacturing employees from 2000 to 2007. (Source: U.S. Bureau of Labor Statistics)

3.4 Equations of Lines

1 ▸ Write equations of lines using point-slope form.

In Exercises 61–64, write the general form of the equation of the line that passes through the point and has the specified slope.

61. $(1, -4)$
$m = -4$

62. $(-5, -5)$
$m = 3$

63. $(-6, 5)$
$m = \frac{1}{4}$

64. $(3, 7)$
$m = -\frac{3}{5}$

In Exercises 65–68, write the slope-intercept form of the equation of the line that passes through the two points.

65. $(-6, 0), (0, -3)$

66. $(0, 10), (6, 8)$

67. $(-2, -3), (4, 6)$

68. $(-10, 2), (4, -7)$

2 ▸ Write equations of horizontal, vertical, parallel, and perpendicular lines.

In Exercises 69–72, write an equation of the line.

69. Horizontal line through $(5, -9)$

70. Vertical line through $(-4, -6)$

71. Line through $(-5, -2)$ and $(-5, 3)$

72. Line through $(1, 8)$ and $(15, 8)$

In Exercises 73 and 74, write equations of the lines that pass through the point and are (a) parallel and (b) perpendicular to the given line.

73. $\left(\frac{3}{5}, -\frac{4}{5}\right)$
$3x + y = 2$

74. $(-1, 5)$
$2x + 4y = 1$

3 ▸ Use linear models to solve application problems.

75. *Annual Salary* Your annual salary in 2002 was \$32,000. During the next 5 years, your annual salary increased by approximately \$1050 per year.

(a) Write an equation of the line giving the annual salary S in terms of the year t. (Let $t = 2$ correspond to the year 2002.)

(b) *Linear Extrapolation* Use the equation in part (a) to predict your annual salary in the year 2010.

(c) *Linear Interpolation* Use the equation in part (a) to estimate your annual salary in 2005.

76. *Rent* The rent for a two-bedroom apartment was \$525 per month in 2003. During the next 5 years, the rent increased by approximately \$45 per year.

(a) Write an equation of the line giving the rent R in terms of the year t. (Let $t = 3$ correspond to the year 2003.)

(b) *Linear Extrapolation* Use the equation in part (a) to predict the rent in the year 2011.

(c) *Linear Interpolation* Use the equation in part (a) to estimate the rent in 2007.

3.5 Graphs of Linear Inequalities

1 ▸ Verify solutions of linear inequalities in two variables.

In Exercises 77 and 78, determine whether each point is a solution of the inequality.

77. $5x - 8y \geq 12$
(a) $(-1, 2)$ (b) $(3, -1)$
(c) $(4, 0)$ (d) $(0, 3)$

78. $2x + 4y - 14 < 0$
(a) $(0, 0)$ (b) $(3, 2)$
(c) $(1, 3)$ (d) $(-4, 1)$

2 ▸ Sketch graphs of linear inequalities in two variables.

In Exercises 79–86, sketch the graph of the solution of the linear inequality.

79. $y > -2$
80. $x \leq 5$
81. $x - 2 \geq 0$
82. $y - 4 < 0$
83. $2x + y < 1$
84. $3x - 4y > 2$
85. $-(x - 1) \leq 4y - 2$
86. $(y - 3) \geq 2(x - 5)$

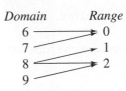 **In Exercises 87–90, use a graphing calculator to graph the solution of the inequality.**

87. $y \le 12 - \frac{3}{2}x$

88. $y \le \frac{1}{3}x + 1$

89. $x + y \ge 0$

90. $4x - 3y \ge 2$

91. ▲ *Geometry*   The perimeter of a rectangle of length x and width y cannot exceed 800 feet.

    (a) Write a linear inequality for this constraint.

    (b) Use a graphing calculator to graph the solution of the inequality.

92. *Weekly Pay*   You have two part-time jobs. One is at a grocery store, which pays \$8.50 per hour, and the other is mowing lawns, which pays \$10 per hour. Between the two jobs, you want to earn at least \$250 a week.

    (a) Write an inequality that shows the different numbers of hours you can work at each job.

    (b) Use a graphing calculator to graph the solution of the inequality. From the graph, find several ordered pairs with positive integer coordinates that are solutions of the inequality.

3.6 Relations and Functions

1 ▶ Identify the domains and ranges of relations.

In Exercises 93 and 94, find the domain and the range of the relation. Then draw a graphical representation of the relation.

93. $\{(-3, 4), (-1, 0), (0, 1), (1, 4), (-3, 5)\}$

94. $\{(-2, 4), (-1, 1), (0, 0), (1, 1), (2, 4)\}$

2 ▶ Determine if relations are functions by inspection.

In Exercises 95–98, determine whether the relation is a function.

95.

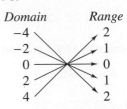

Domain	Range
6	0
7	1
8	2
9	

96.

Domain	Range
-4	2
-2	1
0	0
2	1
4	2

97.

Input value, x	Output value, y
1	10
2	10
3	10
4	10
5	10

98.

Input value, x	Output value, y
3	0
6	3
9	8
6	12
0	2

3 ▶ Use function notation and evaluate functions.

In Exercises 99–104, evaluate the function as indicated, and simplify.

99. $f(t) = \sqrt{5 - t}$

    (a) $f(-4)$ (b) $f(5)$

    (c) $f(3)$ (d) $f(5z)$

100. $g(x) = |x + 4|$

    (a) $g(0)$ (b) $g(-8)$

    (c) $g(2) - g(-5)$ (d) $g(x - 2)$

101. $f(x) = \begin{cases} -3x, & x \le 0 \\ 1 - x^2, & x > 0 \end{cases}$

    (a) $f(3)$ (b) $f\left(-\frac{2}{3}\right)$

    (c) $f(0)$ (d) $f(4) - f(3)$

102. $h(x) = \begin{cases} x^3, & x \le 1 \\ (x - 1)^2 + 1, & x > 1 \end{cases}$

    (a) $h(1)$ (b) $h\left(-\frac{1}{2}\right)$

    (c) $h(-2)$ (d) $h(4) - h(3)$

103. $f(x) = 3 - 2x$

    (a) $\dfrac{f(x + 2) - f(2)}{x}$ (b) $\dfrac{f(x - 3) - f(3)}{x}$

104. $f(x) = 7x + 10$

(a) $\dfrac{f(x+1) - f(1)}{x}$ (b) $\dfrac{f(x-5) - f(5)}{x}$

4 ▶ Identify the domains and ranges of functions.

In Exercises 105–108, find the domain of the function.

105. $h(x) = 4x^2 - 7$

106. $g(s) = \dfrac{s+1}{(s-1)(s+5)}$

107. $f(x) = \sqrt{3x + 10}$

108. $f(x) = |x - 6| + 10$

109. ▲ *Geometry*   A wire 150 inches long is cut into four pieces to form a rectangle whose shortest side has a length of x. Write the area A of the rectangle as a function of x. What is the domain of the function?

110. ▲ *Geometry*   A wire 240 inches long is cut into four pieces to form a rectangle whose shortest side has a length of x. Write the area A of the rectangle as a function of x. What is the domain of the function?

3.7 Graphs of Functions

1 ▶ Sketch graphs of functions on rectangular coordinate systems.

In Exercises 111–116, sketch the graph of the function. Then determine its domain and range.

111. $y = 4 - (x - 3)^2$

112. $h(x) = 9 - (x - 2)^2$

113. $g(x) = 6 - 3x, \quad -2 \le x < 4$

114. $h(x) = x(4 - x), \quad 0 \le x \le 4$

115. $f(x) = \begin{cases} 2 - (x - 1)^2, & x < 1 \\ 2 + (x - 1)^2, & x \ge 1 \end{cases}$

116. $f(x) = \begin{cases} 2x, & x \le 0 \\ x^2 + 1, & x > 0 \end{cases}$

2 ▶ Identify the graphs of basic functions.

In Exercises 117–120, match the function with its graph. [The graphs are labeled (a), (b), (c), and (d).]

(a)
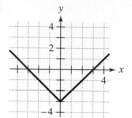

(b)
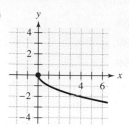

(c)
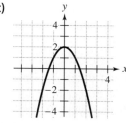

(d)
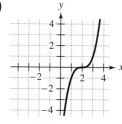

117. $f(x) = -x^2 + 2$ **118.** $f(x) = |x| - 3$

119. $f(x) = -\sqrt{x}$ **120.** $f(x) = (x - 2)^3$

3 ▶ Use the Vertical Line Test to determine if graphs represent functions.

In Exercises 121 and 122, use the Vertical Line Test to determine if the graph represents y as a function of x.

121. $9y^2 = 4x^3$
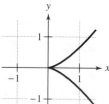

122. $y = x^3 - 3x^2$
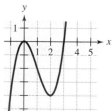

4 ▶ Use vertical and horizontal shifts and reflections to sketch graphs of functions.

In Exercises 123–126, identify the transformation of the graph of $f(x) = \sqrt{x}$, and sketch the graph of h.

123. $h(x) = -\sqrt{x}$

124. $h(x) = \sqrt{x} + 3$

125. $h(x) = \sqrt{x - 2} - 1$

126. $h(x) = 1 - \sqrt{x + 4}$

Chapter Test

Take this test as you would take a test in class. After you are done, check your work against the answers in the back of the book.

1. Determine the quadrant in which the point (x, y) lies if $x > 0$ and $y < 0$.

2. Plot the points $(7, -2)$ and $(3, 1)$. Then find the distance between them and the coordinates of the midpoint of the line segment joining the two points.

3. Find the x- and y-intercepts of the graph of the equation $y = -3(x + 1)$.

4. Sketch the graph of the equation $y = |x - 2|$.

5. Find the slope (if possible) of the line passing through each pair of points.
   (a) $(-4, 7), (2, 3)$ (b) $(3, -2), (3, 6)$

6. Sketch the graph of the line passing through the point $(-3, 5)$ with slope $m = \frac{4}{3}$.

7. Find the x- and y-intercepts of the graph of $2x + 5y - 10 = 0$. Use the results to sketch the graph.

8. Find an equation of the line through the points $(2, -6)$ and $(8, 3)$.

9. Find an equation of the vertical line through the point $(-2, 4)$.

10. Write equations of the lines that pass through the point $(-2, 3)$ and are (a) parallel and (b) perpendicular to the line $3x - 5y = 4$.

11. Sketch the graph of the inequality $x + 4y \leq 8$.

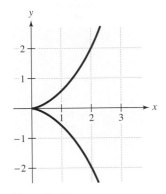

Figure for 12

12. The graph of $y^2(4 - x) = x^3$ is shown at the left. Does the graph represent y as a function of x? Explain your reasoning.

13. Determine whether the relation represents a function. Explain.
    (a) $\{(2, 4), (-6, 3), (3, 3), (1, -2)\}$ (b) $\{(0, 0), (1, 5), (-2, 1), (0, -4)\}$

14. Evaluate $g(x) = x/(x - 3)$ as indicated, and simplify.
    (a) $g(2)$ (b) $g\left(\frac{7}{2}\right)$ (c) $g(x + 2)$

15. Find the domain of each function.
    (a) $h(t) = \sqrt{9 - t}$ (b) $f(x) = \dfrac{x + 1}{x - 4}$

16. Sketch the graph of the function $g(x) = \sqrt{2 - x}$.

17. Describe the transformation of the graph of $f(x) = x^2$ that would produce the graph of $g(x) = -(x - 2)^2 + 1$.

18. After 4 years, the value of a \$26,000 car will have depreciated to \$10,000. Write the value V of the car as a linear function of t, the number of years since the car was purchased. When will the car be worth \$16,000?

19. Use the graph of $f(x) = |x|$ to write a function that represents each graph.

(a)
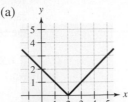

(b)
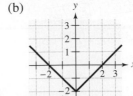

(c)
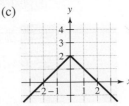

Viewing Math as a Foreign Language

Learning math requires more than just completing homework problems. For instance, learning the material in a chapter may require using approaches similar to those used for learning a foreign language (Nolting, 2008) in that you must:

- understand and memorize vocabulary words;
- understand and memorize mathematical rules (as you would memorize grammatical rules); and
- apply rules to mathematical expressions or equations (like creating sentences using correct grammar rules).

You should understand the vocabulary words and rules in a chapter as well as memorize and say them out loud. Strive to speak the mathematical language with fluency, just as a student learning a foreign language must.

Kimberly Nolting

VP, Academic Success Press
expert in developmental education

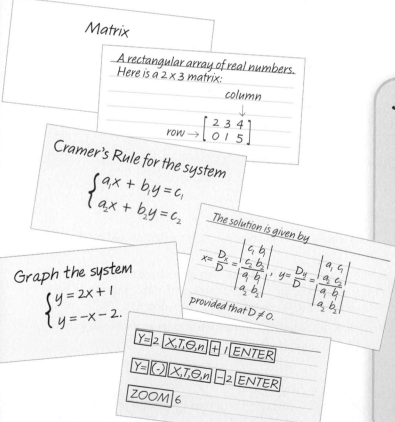

Matrix

A rectangular array of real numbers.
Here is a 2 × 3 matrix:

$$\text{row} \rightarrow \begin{bmatrix} 2 & 3 & 4 \\ 0 & 1 & 5 \end{bmatrix}$$

column

Cramer's Rule for the system

$$\begin{cases} a_1 x + b_1 y = c_1 \\ a_2 x + b_2 y = c_2 \end{cases}$$

The solution is given by

$$x = \frac{D_x}{D} = \frac{\begin{vmatrix} c_1 & b_1 \\ c_2 & b_2 \end{vmatrix}}{\begin{vmatrix} a_1 & b_1 \\ a_2 & b_2 \end{vmatrix}}, \quad y = \frac{D_y}{D} = \frac{\begin{vmatrix} a_1 & c_1 \\ a_2 & c_2 \end{vmatrix}}{\begin{vmatrix} a_1 & b_1 \\ a_2 & b_2 \end{vmatrix}}$$

provided that $D \neq 0$.

Graph the system

$$\begin{cases} y = 2x + 1 \\ y = -x - 2. \end{cases}$$

Y= 2 X,T,θ,n + 1 ENTER
Y= (-) X,T,θ,n − 2 ENTER
ZOOM 6

Smart Study Strategy

Make Note Cards

Invest in three different colors of 4 × 6 note cards. Use one color for each of the following: vocabulary words; rules; and graphing keystrokes.

1 ▶ Write vocabulary words on note cards, one word per card. Write the definition and an example on the other side. If possible, put definitions in your own words.

2 ▶ Write rules on note cards, one per card. Include an example and an explanation on the other side.

3 ▶ Write each kind of calculation on a separate note card. Include the keystrokes required to perform the calculation on the other side.

Use the note cards as references while doing your homework. Quiz yourself once a day.

Chapter 4
Systems of Equations and Inequalities

IT WORKED FOR ME!

"When I trained to be a tutor we learned how note cards can be used to help review and memorize math concepts. I started using them in my own math class and it really helps me, especially when I am caught with short amounts of time on campus. I don't have to pull out all of my books to study. I just keep my note cards in my purse."

Di Iris
Computer science

4.1 Systems of Equations

What You Should Learn

1 ▶ Determine if ordered pairs are solutions of systems of equations.

2 ▶ Solve systems of equations graphically.

3 ▶ Solve systems of equations algebraically using the method of substitution.

4 ▶ Use systems of equations to model and solve real-life problems.

David Lassman/The Image Works

Why You Should Learn It

Systems of equations can be used to model and solve real-life problems. For instance, in Exercise 116 on page 230, a system of equations is used to model the numbers of morning and evening newspapers in the United States.

1 ▶ Determine if ordered pairs are solutions of systems of equations.

Systems of Equations

Many problems in business and science involve **systems of equations.** These systems consist of two or more equations involving two or more variables.

$$\begin{cases} ax + by = c & \text{Equation 1} \\ dx + ey = f & \text{Equation 2} \end{cases}$$

A **solution** of such a system is an ordered pair (x, y) of real numbers that satisfies *each* equation in the system. When you find the set of all solutions of the system of equations, you are **solving the system of equations.**

EXAMPLE 1 Checking Solutions of a System of Equations

Check whether each ordered pair is a solution of the system of equations.

$$\begin{cases} x + y = 6 & \text{Equation 1} \\ 2x - 5y = -2 & \text{Equation 2} \end{cases}$$

a. $(3, 3)$ **b.** $(4, 2)$

Solution

a. To determine whether the ordered pair $(3, 3)$ is a solution of the system of equations, substitute 3 for x and 3 for y in *each* of the equations.

$3 + 3 = 6$ ✓   Substitute 3 for x and 3 for y in Equation 1.

$2(3) - 5(3) \neq -2$ ✗   Substitute 3 for x and 3 for y in Equation 2.

Because the check fails in Equation 2, you can conclude that the ordered pair $(3, 3)$ *is not* a solution of the original system of equations.

b. By substituting 4 for x and 2 for y in each of the original equations, you can determine that the ordered pair $(4, 2)$ is a solution of *both* equations.

$4 + 2 = 6$ ✓   Substitute 4 for x and 2 for y in Equation 1.

$2(4) - 5(2) = -2$ ✓   Substitute 4 for x and 2 for y in Equation 2.

So, $(4, 2)$ *is* a solution of the original system of equations.

✓ *CHECKPOINT* Now try Exercise 1.

2 ▶ Solve systems of equations graphically.

Solving Systems of Equations by Graphing

You can gain insight about the location and number of solutions of a system of equations by sketching the graph of each equation in the same coordinate plane. The solutions of the system correspond to the **points of intersection** of the graphs.

A system of linear equations can have exactly one solution, infinitely many solutions, or no solution. To see why this is true, consider the graphical interpretations of three systems of two linear equations shown below.

Graphs			
Graphical interpretation	The two lines intersect.	The two lines coincide (are identical).	The two lines are parallel.
Intersection	Single point of intersection	Infinitely many points of intersection	No point of intersection
Slopes of lines	Slopes are not equal.	Slopes are equal.	Slopes are equal.
Number of solutions	Exactly one solution	Infinitely many solutions	No solution
Type of system	**Consistent system**	**Dependent (consistent) system**	**Inconsistent system**

Technology: Discovery

Rewrite each system of equations in slope-intercept form and graph the equations using a graphing calculator. What is the relationship between the slopes of the two lines and the number of points of intersection?

a. $\begin{cases} 2x + 4y = 8 \\ 4x - 3y = -6 \end{cases}$

b. $\begin{cases} -x + 5y = 15 \\ 2x - 10y = -7 \end{cases}$

c. $\begin{cases} x - y = 9 \\ 2x - 2y = 18 \end{cases}$

A system of equations is *consistent* if it has at least one solution and *dependent* if its graphs coincide. So, a linear system of equations is either consistent with one solution, dependent (consistent with infinitely many solutions), or *inconsistent* (there is no solution).

You can see from the graphs above that a comparison of the slopes of two lines gives useful information about the number of solutions of the corresponding system of equations. For instance:

Consistent systems with one solution have lines with different slopes.
Dependent (consistent) systems have lines with equal slopes and equal y-intercepts.
Inconsistent systems have lines with equal slopes, but different y-intercepts.

So, when solving a system of equations graphically, it is helpful to know the slopes of the lines. Writing these linear equations in the slope-intercept form

$$y = mx + b \qquad \text{Slope-intercept form}$$

allows you to identify the slopes quickly.

EXAMPLE 2   The Graphical Method of Solving a System

Use the graphical method to solve the system of equations.

$$\begin{cases} 2x + 3y = 7 & \text{Equation 1} \\ 2x - 5y = -1 & \text{Equation 2} \end{cases}$$

Solution

Because both equations in the system are linear, you know they have graphs that are straight lines. To sketch these lines, first write each equation in slope-intercept form, as follows.

$$y = -\frac{2}{3}x + \frac{7}{3} \qquad \text{Equation 1}$$

$$y = \frac{2}{5}x + \frac{1}{5} \qquad \text{Equation 2}$$

Because their slopes are not equal, you can conclude that the graphs intersect at a single point. The lines corresponding to these two equations are shown in Figure 4.1. From this figure, it appears that the two lines intersect at the point $(2, 1)$. You can check these coordinates as follows.

Check

Substitute in 1st Equation

$$2x + 3y = 7$$
$$2(2) + 3(1) \stackrel{?}{=} 7$$
$$4 + 3 \stackrel{?}{=} 7$$
$$7 = 7 \checkmark$$

Substitute in Final Equation

$$2x - 5y = -1$$
$$2(2) - 5(1) \stackrel{?}{=} -1$$
$$4 - 5 \stackrel{?}{=} -1$$
$$-1 = -1 \checkmark$$

Because both equations in the system are satisfied, the point $(2, 1)$ is the solution of the system.

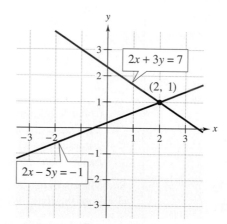

Figure 4.1

✓ **CHECKPOINT** *Now try Exercise 35.*

3 ▶ Solve systems of equations algebraically using the method of substitution.

Solving Systems of Equations by Substitution

Solving a system of equations graphically is limited by the ability to sketch an accurate graph. An accurate solution is difficult to obtain if one or both coordinates of a solution point are fractional or irrational. One analytic way to determine an exact solution of a system of two equations in two variables is to convert the system to *one* equation in *one* variable by an appropriate substitution.

EXAMPLE 3   The Method of Substitution: One-Solution Case

Solve the system of equations.

$$\begin{cases} -x + y = 3 & \text{Equation 1} \\ 3x + y = -1 & \text{Equation 2} \end{cases}$$

Solution

Begin by solving for y in Equation 1.

$$y = x + 3 \qquad \text{Revised Equation 1}$$

Next, substitute this expression for y in Equation 2.

$3x + y = -1$	Equation 2
$3x + (x + 3) = -1$	Substitute $x + 3$ for y.
$4x + 3 = -1$	Combine like terms.
$4x = -4$	Subtract 3 from each side.
$x = -1$	Divide each side by 4.

At this point, you know that the x-coordinate of the solution is -1. To find the y-coordinate, *back-substitute* the x-value into the revised Equation 1.

$y = x + 3$	Revised Equation 1
$y = -1 + 3$	Substitute -1 for x.
$y = 2$	Simplify.

The solution is $(-1, 2)$. Check this in the original system of equations.

✔ **CHECKPOINT** *Now try Exercise 53.*

Study Tip

The term **back-substitute** implies that you work backwards. After finding a value for one of the variables, substitute that value back into one of the equations in the original (or revised) system to find the value of the other variable.

When using substitution, it does not matter which variable you solve for first. You will obtain the same solution regardless. When making your choice, you should choose the variable that is easier to work with. For instance, in the system

$$\begin{cases} 3x - 2y = 1 & \text{Equation 1} \\ x + 4y = 3 & \text{Equation 2} \end{cases}$$

it is easier to begin by solving for x in the second equation. But in the system

$$\begin{cases} 2x + y = 5 & \text{Equation 1} \\ 3x - 2y = 11 & \text{Equation 2} \end{cases}$$

it is easier to begin by solving for y in the first equation.

The steps for using the method of substitution to solve a system of two equations involving two variables are summarized as follows.

The Method of Substitution

1. Solve one of the equations for one variable in terms of the other.
2. Substitute the expression obtained in Step 1 in the other equation to obtain an equation in one variable.
3. Solve the equation obtained in Step 2.
4. Back-substitute the solution from Step 3 in the expression obtained in Step 1 to find the value of the other variable.
5. Check the solution to see that it satisfies *both* of the original equations.

EXAMPLE 4   The Method of Substitution: No-Solution Case

To solve the system of equations

$$\begin{cases} 2x - 2y = 0 & \text{Equation 1} \\ x - y = 1 & \text{Equation 2} \end{cases}$$

begin by solving for y in Equation 2.

$$y = x - 1 \qquad \text{Revised Equation 2}$$

Next, substitute this expression for y in Equation 1.

$$2x - 2(x - 1) = 0 \qquad \text{Substitute } x - 1 \text{ for } y \text{ in Equation 1.}$$
$$2x - 2x + 2 = 0 \qquad \text{Distributive Property}$$
$$2 = 0 \qquad \text{False statement}$$

Because the substitution process produces a false statement ($2 = 0$), you can conclude that the original system of equations is inconsistent and has no solution. You can check your solution graphically, as shown in Figure 4.2.

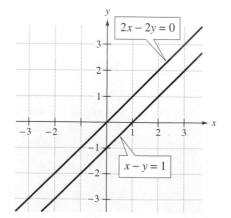

Figure 4.2

✓ **CHECKPOINT** *Now try Exercise 61.*

Study Tip

Figure 4.2 shows that the two lines in Example 4 are parallel, so the system has no solution. You can also see that the system has no solution by writing each equation in slope-intercept form, as follows.

$$y = x$$
$$y = x - 1$$

Because the slopes are equal and the y-intercepts are different, the system has no solution.

4 ▶ Use systems of equations to model and solve real-life problems.

Applications

To model a real-life situation with a system of equations, you can use the same basic problem-solving strategy that has been used throughout the text.

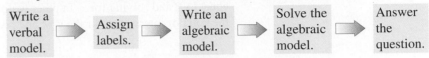

Write a verbal model. ⇒ Assign labels. ⇒ Write an algebraic model. ⇒ Solve the algebraic model. ⇒ Answer the question.

After answering the question, remember to check the answer in the original statement of the problem.

EXAMPLE 5   A Mixture Problem

A roofing contractor buys 30 bundles of shingles and four rolls of roofing paper for $528. In a second purchase (at the same prices), the contractor pays $140 for eight bundles of shingles and one roll of roofing paper. Find the price per bundle of shingles and the price per roll of roofing paper.

Solution

Verbal Model:

$$30\left(\begin{array}{c}\text{Price of}\\\text{a bundle}\end{array}\right) + 4\left(\begin{array}{c}\text{Price of}\\\text{a roll}\end{array}\right) = 528$$

$$8\left(\begin{array}{c}\text{Price of}\\\text{a bundle}\end{array}\right) + 1\left(\begin{array}{c}\text{Price of}\\\text{a roll}\end{array}\right) = 140$$

Labels: Price of a bundle of shingles $= x$ (dollars)
Price of a roll of roofing paper $= y$ (dollars)

System:
$$\begin{cases} 30x + 4y = 528 & \text{Equation 1} \\ 8x + y = 140 & \text{Equation 2} \end{cases}$$

Solving the second equation for y produces $y = 140 - 8x$, and substituting this expression for y in the first equation produces the following.

$30x + 4(140 - 8x) = 528$	Substitute $140 - 8x$ for y.
$30x + 560 - 32x = 528$	Distributive Property
$-2x = -32$	Combine like terms.
$x = 16$	Divide each side by -2.

Back-substituting 16 for x in revised Equation 2 produces

$$y = 140 - 8(16) = 12.$$

So, you can conclude that the contractor pays $16 per bundle of shingles and $12 per roll of roofing paper. Check this in the original statement of the problem.

Check

1st Equation	*Final Equation*	
$30(16) + 4(12) \stackrel{?}{=} 528$	$8(16) + 12 \stackrel{?}{=} 140$	Substitute 16 for x and 12 for y.
$480 + 48 = 528$	$128 + 12 = 140$	Solution checks. ✓

✓ **CHECKPOINT** *Now try Exercise 95.*

The total cost C of producing x units of a product usually has two components—the initial cost and the cost per unit. When enough units have been sold so that the total revenue R equals the total cost, the sales are said to have reached the **break-even point.** You can find this break-even point by setting C equal to R and solving for x. In other words, the break-even point corresponds to the point of intersection of the cost and revenue graphs.

EXAMPLE 6   Break-Even Analysis

A small business invests $14,000 to produce a new energy bar. Each bar costs $0.80 to produce and sells for $1.50. How many bars must be sold before the business breaks even?

Solution

Verbal Model:

| Total cost | $=$ | Cost per bar | $\cdot$ | Number of bars | $+$ | Initial cost |

| Total revenue | $=$ | Price per bar | $\cdot$ | Number of bars |

Labels: Total cost $= C$ (dollars)
Cost per bar $= 0.80$ (dollars per bar)
Number of bars $= x$ (bars)
Initial cost $= 14,000$ (dollars)
Total revenue $= R$ (dollars)
Price per bar $= 1.50$ (dollars per bar)

System:
$$\begin{cases} C = 0.80x + 14,000 & \text{Equation 1} \\ R = 1.50x & \text{Equation 2} \end{cases}$$

Because the break-even point occurs when $R = C$, you have

$1.50x = 0.80x + 14,000$ $R = C$

$0.7x = 14,000$ Subtract $0.80x$ from each side.

$x = 20,000.$ Divide each side by 0.7.

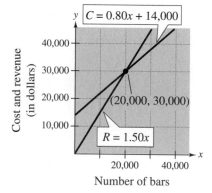

Figure 4.3

So, it follows that the business must sell 20,000 bars before it breaks even. Profit P (or loss) for the business can be determined by the equation $P = R - C$. Note in Figure 4.3 that sales less than the break-even point correspond to a loss for the business, whereas sales greater than the break-even point correspond to a profit for the business. The following table helps confirm this conclusion.

Units, x	0	5000	10,000	15,000	20,000	25,000
Revenue, R	$0	$7500	$15,000	$22,500	$30,000	$37,500
Cost, C	$14,000	$18,000	$22,000	$26,000	$30,000	$34,000
Profit, P	$-$14,000	$-$10,500	$-$7000	$-$3500	$0	$3500

✓ **CHECKPOINT** *Now try Exercise 97.*

EXAMPLE 7   Investment

A total of $12,000 is invested in two funds paying 6% and 8% simple interest. The annual interest is $880. How much of the $12,000 is invested in each fund?

Solution

Verbal Model:

$$\boxed{\text{Amount in 6\% fund}} + \boxed{\text{Amount in 8\% fund}} = 12{,}000$$

$$\boxed{6\%} \cdot \boxed{\text{Amount in 6\% fund}} + \boxed{8\%} \cdot \boxed{\text{Amount in 8\% fund}} = 880$$

Labels: Amount in 6% fund = x (dollars)
            Amount in 8% fund = y (dollars)

System: $\begin{cases} x + y = 12{,}000 & \text{Equation 1} \\ 0.06x + 0.08y = 880 & \text{Equation 2} \end{cases}$

Begin by solving for x in the first equation.

$$x + y = 12{,}000 \qquad\qquad \text{Equation 1}$$

$$x = 12{,}000 - y \qquad\qquad \text{Revised Equation 1}$$

Substituting this expression for x in the second equation produces the following.

$$0.06x + 0.08y = 880 \qquad\qquad \text{Equation 2}$$

$$0.06(12{,}000 - y) + 0.08y = 880 \qquad\qquad \text{Substitute } 12{,}000 - y \text{ for } x.$$

$$720 - 0.06y + 0.08y = 880 \qquad\qquad \text{Distributive Property}$$

$$720 + 0.02y = 880 \qquad\qquad \text{Simplify.}$$

$$0.02y = 160 \qquad\qquad \text{Combine like terms.}$$

$$y = 8000 \qquad\qquad \text{Divide each side by 0.02.}$$

Back-substitute this value for y in revised Equation 1.

$$x = 12{,}000 - y \qquad\qquad \text{Revised Equation 1}$$

$$x = 12{,}000 - 8000 \qquad\qquad \text{Substitute 8000 for } y.$$

$$x = 4000 \qquad\qquad \text{Simplify.}$$

So, $4000 is invested in the fund paying 6% and $8000 is invested in the fund paying 8%. Check this in the original statement of the problem, as follows.

Check

Substitute in 1st Equation

$$x + y = 12{,}000$$

$$4000 + 8000 \stackrel{?}{=} 12{,}000$$

$$12{,}000 = 12{,}000 \ \checkmark$$

Substitute in Final Equation

$$0.06x + 0.08y = 880$$

$$0.06(4000) + 0.08(8000) \stackrel{?}{=} 880$$

$$880 = 880 \ \checkmark$$

 **CHECKPOINT** *Now try Exercise 101.*

Concept Check

1. You are checking a solution of a system of linear equations. How can you tell if this solution is *not* valid?

2. Describe the geometric properties of the graph of each possible type of solution set of a system of linear equations in two variables.

3. Describe the *simplest* way to carry out the first step in solving the following system of equations by substitution.

$$\begin{cases} y + 18 = 12x \\ 7x + 5y - 4 = 0 \end{cases}$$

4. What does it mean to *back-substitute* when solving a system of equations?

4.1 EXERCISES

Go to pages 288–289 to record your assignments.

Developing Skills

In Exercises 1–8, determine whether each ordered pair is a solution of the system of equations. *See Example 1.*

 1. $\begin{cases} x + 2y = 9 \\ -2x + 3y = 10 \end{cases}$
   (a) $(1, 4)$
   (b) $(3, -1)$

2. $\begin{cases} 5x - 4y = 34 \\ x - 2y = 8 \end{cases}$
   (a) $(0, 3)$
   (b) $(6, -1)$

3. $\begin{cases} -2x + 7y = 46 \\ 3x + y = 0 \end{cases}$
   (a) $(-3, 2)$
   (b) $(-2, 6)$

4. $\begin{cases} -5x - 2y = 23 \\ x + 4y = -19 \end{cases}$
   (a) $(-3, -4)$
   (b) $(3, 7)$

5. $\begin{cases} 4x - 5y = 12 \\ 3x + 2y = -2.5 \end{cases}$
   (a) $(8, 4)$
   (b) $\left(\frac{1}{2}, -2\right)$

6. $\begin{cases} 2x - y = 1.5 \\ 4x - 2y = 3 \end{cases}$
   (a) $\left(0, -\frac{3}{2}\right)$
   (b) $\left(2, \frac{5}{2}\right)$

7. $\begin{cases} -3x + 2y = -19 \\ 5x - y = 27 \end{cases}$
   (a) $(5, -2)$
   (b) $(3, 4)$

8. $\begin{cases} 7x + 2y = -1 \\ 3x - 6y = -21 \end{cases}$
   (a) $(-2, -4)$
   (b) $(-1, 3)$

In Exercises 9–14, state the number of solutions of the system of linear equations without solving the system.

9. $\begin{cases} y = 4x \\ y = 4x + 1 \end{cases}$

10. $\begin{cases} y = 3x + 2 \\ y = -3x + 2 \end{cases}$

11. $\begin{cases} y = 2x - 5 \\ y = 5x - 2 \end{cases}$

12. $\begin{cases} y = 3x - 8 \\ y = 3x + 8 \end{cases}$

13. $\begin{cases} y = \frac{1}{2}x \\ 2y = x \end{cases}$

14. $\begin{cases} y = \frac{2}{3}x + 1 \\ 3y = 2x + 3 \end{cases}$

In Exercises 15–22, determine whether the system is consistent or inconsistent.

15. $\begin{cases} x + 2y = 6 \\ x + 2y = 3 \end{cases}$

16. $\begin{cases} x - 2y = 3 \\ 2x - 4y = 7 \end{cases}$

17. $\begin{cases} 2x - 3y = -12 \\ -8x + 12y = -12 \end{cases}$

18. $\begin{cases} -5x + 8y = 8 \\ 7x - 4y = 14 \end{cases}$

19. $\begin{cases} -x + 4y = 7 \\ 3x - 12y = -21 \end{cases}$

20. $\begin{cases} 3x + 8y = 28 \\ -4x + 9y = 1 \end{cases}$

21. $\begin{cases} 5x - 3y = 1 \\ 6x - 4y = -3 \end{cases}$

22. $\begin{cases} 9x + 6y = 10 \\ -6x - 4y = 3 \end{cases}$

In Exercises 23–26, use a graphing calculator to graph the equations in the system. Use the graphs to determine whether the system is consistent or inconsistent. If the system is consistent, determine the number of solutions.

23. $\begin{cases} \frac{1}{3}x - \frac{1}{2}y = 1 \\ -2x + 3y = 6 \end{cases}$

24. $\begin{cases} x + y = 5 \\ x - y = 5 \end{cases}$

25. $\begin{cases} -2x + 3y = 6 \\ x - y = -1 \end{cases}$ **26.** $\begin{cases} 2x - 4y = 9 \\ x - 2y = 4.5 \end{cases}$

33. $\begin{cases} 0.5x + 0.5y = 1.5 \\ -x + y = 1 \end{cases}$ **34.** $\begin{cases} 0.25x - 0.5y = -1 \\ -0.5x + y = 2 \end{cases}$

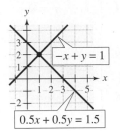

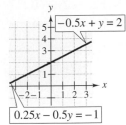

In Exercises 27–34, use the graphs of the equations to determine whether the system has any solutions. Find any solutions that exist.

27. $\begin{cases} x + y = 4 \\ x + y = -1 \end{cases}$ **28.** $\begin{cases} -x + y = 5 \\ x + 2y = 4 \end{cases}$

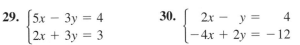

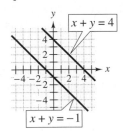

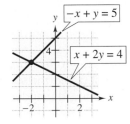

In Exercises 35–48, use the graphical method to solve the system of equations. *See Example 2.*

✓ **35.** $\begin{cases} y = -x + 3 \\ y = x + 1 \end{cases}$ **36.** $\begin{cases} y = 2x - 4 \\ y = -\frac{1}{2}x + 1 \end{cases}$

29. $\begin{cases} 5x - 3y = 4 \\ 2x + 3y = 3 \end{cases}$ **30.** $\begin{cases} 2x - y = 4 \\ -4x + 2y = -12 \end{cases}$

37. $\begin{cases} x - y = 2 \\ x + y = 2 \end{cases}$ **38.** $\begin{cases} x - y = 0 \\ x + y = 4 \end{cases}$

39. $\begin{cases} 3x - 4y = 5 \\ x = 3 \end{cases}$ **40.** $\begin{cases} 5x + 2y = 24 \\ y = 2 \end{cases}$

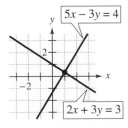

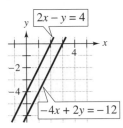

41. $\begin{cases} 4x + 5y = 20 \\ \frac{4}{5}x + y = 4 \end{cases}$ **42.** $\begin{cases} -x + 3y = 7 \\ 2x - 6y = 6 \end{cases}$

43. $\begin{cases} 2x - 5y = 20 \\ 4x - 5y = 40 \end{cases}$ **44.** $\begin{cases} 5x + 3y = 24 \\ x - 2y = 10 \end{cases}$

45. $\begin{cases} x + y = 3 \\ 3x + 3y = 6 \end{cases}$ **46.** $\begin{cases} 4x - 3y = -3 \\ 8x - 6y = -6 \end{cases}$

31. $\begin{cases} -2x + 3y = 6 \\ 8x - 12y = -24 \end{cases}$ **32.** $\begin{cases} 2x - 3y = 6 \\ 4x + 3y = 12 \end{cases}$

47. $\begin{cases} 4x + 5y = 7 \\ 2x - 3y = 9 \end{cases}$ **48.** $\begin{cases} 7x + 4y = 6 \\ 5x - 3y = -25 \end{cases}$

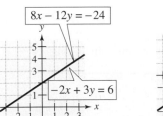

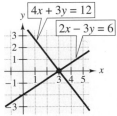

In Exercises 49–52, use a graphing calculator to graph the equations and approximate any solutions of the system of equations.

49. $\begin{cases} 2x - y = 7.7 \\ -x - 3y = 1.4 \end{cases}$ **50.** $\begin{cases} 3x + 2y = 7.8 \\ -x + 3y = 15 \end{cases}$

51. $\begin{cases} 3.4x - 5.6y = 10.2 \\ 5.8x + 1.4y = -33.6 \end{cases}$

52. $\begin{cases} -2.3x + 7.9y = 88.3 \\ -5.3x - 2.7y = -16.5 \end{cases}$

In Exercises 53–78, solve the system of equations by the method of substitution. *See Examples 3 and 4.*

✓ 53. $\begin{cases} x - 2y = 0 \\ 3x + 2y = 8 \end{cases}$
54. $\begin{cases} x - y = 0 \\ 5x - 2y = 6 \end{cases}$

55. $\begin{cases} x = 4 \\ x - 2y = -2 \end{cases}$
56. $\begin{cases} y = 2 \\ x - 6y = -6 \end{cases}$

57. $\begin{cases} x + y = 3 \\ 2x - y = 0 \end{cases}$
58. $\begin{cases} -x + y = 5 \\ x - 4y = 0 \end{cases}$

59. $\begin{cases} x + y = 2 \\ x - 4y = 12 \end{cases}$
60. $\begin{cases} x - 2y = -1 \\ x - 5y = 2 \end{cases}$

✓ 61. $\begin{cases} 3x + y = 8 \\ 3x + y = 6 \end{cases}$
62. $\begin{cases} x - 3y = 12 \\ -2x + 6y = -18 \end{cases}$

63. $\begin{cases} x + 6y = 19 \\ x - 7y = -7 \end{cases}$
64. $\begin{cases} x - 5y = -6 \\ 4x - 3y = 10 \end{cases}$

65. $\begin{cases} 8x + 5y = 100 \\ 9x - 10y = 50 \end{cases}$
66. $\begin{cases} x + 4y = 300 \\ x - 2y = 0 \end{cases}$

67. $\begin{cases} -13x + 16y = 10 \\ 5x + 16y = -26 \end{cases}$
68. $\begin{cases} 2x + 5y = 29 \\ 5x + 2y = 13 \end{cases}$

69. $\begin{cases} 4x - 14y = -15 \\ 18x - 12y = 9 \end{cases}$
70. $\begin{cases} 5x - 24y = -12 \\ 17x - 24y = 36 \end{cases}$

71. $\begin{cases} \frac{1}{5}x + \frac{1}{2}y = 8 \\ x + y = 20 \end{cases}$
72. $\begin{cases} \frac{1}{2}x + \frac{3}{4}y = 10 \\ \frac{3}{2}x - y = 4 \end{cases}$

73. $\begin{cases} \frac{1}{8}x + \frac{1}{2}y = 1 \\ \frac{3}{5}x + y = \frac{3}{5} \end{cases}$
74. $\begin{cases} \frac{1}{8}x - \frac{1}{4}y = \frac{3}{4} \\ -\frac{1}{4}x + \frac{3}{4}y = -1 \end{cases}$

75. $\begin{cases} 8x + 6y = 6 \\ 12x + 9y = 6 \end{cases}$
76. $\begin{cases} 12x - 14y = 15 \\ 18x - 21y = 10 \end{cases}$

77. $\begin{cases} \frac{1}{9}x - \frac{2}{3}y = 2 \\ \frac{2}{3}x - 4y = 6 \end{cases}$
78. $\begin{cases} \frac{3}{7}x + \frac{4}{5}y = 3 \\ x - \frac{28}{15}y = 6 \end{cases}$

In Exercises 79–82, use a graphing calculator to graph the equations in the system. The graphs appear to be parallel, yet you can find from the slope-intercept forms of the lines that the slopes are not equal and the graphs intersect. Find the point of intersection of the two lines.

79. $\begin{cases} x - 100y = -200 \\ 3x - 275y = 198 \end{cases}$
80. $\begin{cases} 35x - 33y = 0 \\ 12x - 11y = 92 \end{cases}$

81. $\begin{cases} 3x - 25y = 50 \\ 9x - 100y = 50 \end{cases}$
82. $\begin{cases} x + 40y = 80 \\ 2x + 150y = 195 \end{cases}$

Think About It In Exercises 83–86, determine the value of a and the value of b for which the system of equations is dependent.

83. $\begin{cases} y = \frac{2}{3}x + 1 \\ 3y = ax + b \end{cases}$
84. $\begin{cases} 4y = 5x \\ 2y = ax + b \end{cases}$

85. $\begin{cases} 6x - 8y - 48 = 0 \\ \frac{3}{4}x - ay + b = 0 \end{cases}$
86. $\begin{cases} \frac{1}{2}y = -\frac{3}{4}x + \frac{9}{4} \\ ax = -\frac{5}{6}y + b \end{cases}$

Think About It In Exercises 87–94, write a system of equations having the given solution. (There are many correct answers.)

87. $(4, 5)$
88. No solution

89. $(-1, -2)$
90. $\left(\frac{1}{2}, 3\right)$

91. Infinitely many solutions, including $\left(4, -\frac{1}{2}\right)$ and $(-1, 1)$

92. $(-2, 6)$

93. No solution

94. Infinitely many solutions, including $(-3, -1)$ and $(0, 3)$

Solving Problems

✓ **95.** *Hay Mixture*   A farmer wants to mix two types of hay. The first type sells for $125 per ton and the second type sells for $75 per ton. The farmer wants a total of 100 tons of hay at a cost of $90 per ton. How many tons of each type of hay should be used in the mixture?

96. *Seed Mixture*   Ten pounds of mixed birdseed sells for $6.97 per pound. The mixture is obtained from two kinds of birdseed, with one variety priced at $5.65 per pound and the other at $8.95 per pound. How many pounds of each variety of birdseed are used in the mixture?

✓ **97.** *Break-Even Analysis*   A small business invests $8000 in equipment to produce a new candy bar. Each bar costs $1.20 to produce and is sold for $2.00. How many candy bars must be sold before the business breaks even?

98. *Break-Even Analysis*   A business invests $50,000 in equipment to produce a new hand-held video game. Each game costs $19.25 to produce and is sold for $35.95. How many hand-held video games must be sold before the business breaks even?

99. *Break-Even Analysis*   You are setting up a small business and have invested $10,000 to produce hand cream that will sell for $3.25 a bottle. Each bottle can be produced for $1.65. How many bottles of hand cream must you sell to break even?

100. *Break-Even Analysis*   You are setting up a small guitar business and have made an initial investment of $30,000. The unit cost of the guitar you are producing is $26.50, and the selling price is $76.50. How many guitars must you sell to break even?

© Jim Craigmyle/CORBIS

✓ **101.** *Investment*   A total of $12,000 is invested in two bonds that pay 8.5% and 10% simple interest. The annual interest is $1140. How much is invested in each bond?

102. *Investment*   A total of $25,000 is invested in two funds paying 8% and 8.5% simple interest. The annual interest is $2060. How much is invested in each fund?

Number Problems   In Exercises 103–110, find two positive integers that satisfy the given requirements.

103. The sum of the two numbers is 80 and their difference is 18.

104. The sum of the two numbers is 93 and their difference is 31.

105. The sum of the larger number and three times the smaller number is 51 and their difference is 3.

106. The sum of the larger number and twice the smaller number is 61 and their difference is 7.

107. The sum of the two numbers is 52 and the larger number is 8 less than twice the smaller number.

108. The sum of the two numbers is 160 and the larger number is three times the smaller number.

109. The difference of twice the smaller number and the larger number is 13 and the sum of the smaller number and twice the larger number is 114.

110. The difference of the numbers is 86 and the larger number is three times the smaller number.

▲ *Geometry*   In Exercises 111–114, find the dimensions of the rectangle meeting the specified conditions.

	Perimeter	Condition
111.	50 feet	The length is 5 feet greater than the width.
112.	320 inches	The width is 20 inches less than the length.
113.	68 yards	The width is $\frac{7}{10}$ of the length.
114.	90 meters	The length is $1\frac{1}{2}$ times the width.

115. 🖩 *Graphical Analysis*   From 1991 through 2005, the population of South Carolina grew at a faster rate than that of Kentucky. The two populations can be modeled by

$$S = 49.2t + 3516 \qquad \text{South Carolina}$$

$$K = 31.3t + 3719 \qquad \text{Kentucky}$$

where S and K are the populations in thousands and t represents the year, with $t = 1$ corresponding to 1991. Use a graphing calculator to approximate the year in which the population of South Carolina first exceeded the population of Kentucky. (Source: U.S. Census Bureau)

116. 🖩 *Graphical Analysis*   From 1995 to 2005, the number of daily morning newspapers in the United States increased, while the number of daily evening newspapers decreased. Models that represent the circulations of the two types of daily papers are

$$M = 15.6t + 593 \qquad \text{Morning}$$

$$E = -24.0t + 985 \qquad \text{Evening}$$

where M and E are the numbers of newspapers and t is the year, with $t = 5$ corresponding to 1995. Use a graphing calculator to approximate the year in which the number of morning papers first exceeded the number of evening papers. (Source: Editor & Publisher Co.)

Explaining Concepts

117. ✎   When solving a system of equations by substitution, how do you recognize when the system has no solution?

118. When solving a system of equations by substitution, what type of result do you suppose might indicate that the system has an infinite number of solutions? (Consider the special cases for solutions of linear equations.)

119. ✎   Describe any advantages of the method of substitution over the graphical method of solving a system of equations.

120. ✎   Is it possible for a consistent system of linear equations to have exactly two solutions? Explain.

121. ✎   In a consistent system of three linear equations in two variables, exactly two of the equations are dependent. How many solutions does the system have? Explain.

122. ✎   Is it possible for a system of three linear equations in two variables to be inconsistent if two of the equations are dependent? Explain.

123. ✎   In the graph of a system of three linear equations in two variables, all three lines have different slopes and exactly two of the lines have the same y-intercept. How many solutions does the system have? Explain.

124. You want to create several systems of equations with relatively simple solutions that students can use for practice. Discuss how to create a system of equations that has a given solution. Illustrate your method by creating a system of linear equations that has one of the following solutions: $(1, 4)$, $(-2, 5)$, $(-3, 1)$, or $(4, 2)$.

Cumulative Review

In Exercises 125–128, solve the equation and check the result.

125. $x - 4 = 1$

126. $6x - 2 = -7$

127. $3x - 12 = x + 2$

128. $4x + 21 = 4(x + 5)$

In Exercises 129–132, sketch the graph of the function.

129. $f(x) = x^2 - 6$

130. $g(x) = 5$

131. $h(t) = 3t - 1$

132. $s(t) = |t - 2|$

In Exercises 133 and 134, show that both ordered pairs are solutions of the equation, and explain why this implies that y is not a function of x.

133. $-x + y^2 = 0$; $(4, 2)$, $(4, -2)$

134. $|y| = x + 4$; $(3, 7)$, $(3, -7)$

4.2 Linear Systems in Two Variables

© JupiterImages/Comstock Images/Alamy

What You Should Learn

1 ▶ Solve systems of linear equations algebraically using the method of elimination.

2 ▶ Use systems of linear equations to model and solve real-life problems.

Why You Should Learn It

Systems of linear equations can be used to find the best-fitting line that models a data set. For instance, in Exercise 85 on page 240, a system of linear equations is used to find the best-fitting line to model the hourly wages for retail employees in the United States.

1 ▶ Solve systems of linear equations algebraically using the method of elimination.

The Method of Elimination

In Section 4.1, you studied two ways of solving a system of equations—substitution and graphing. Now you will study a third way—the **method of elimination.**

The key step in the method of elimination is to obtain two equations in which, for one of the variables, the coefficients are opposites. Then, by *adding* the two equations, you eliminate this variable and get a single equation in one variable. Notice how this is accomplished in Example 1. When the two equations are added, the y-terms are eliminated.

EXAMPLE 1   **The Method of Elimination**

Solve the system of linear equations.

$$\begin{cases} 3x + 2y = 4 & \text{Equation 1} \\ 5x - 2y = 8 & \text{Equation 2} \end{cases}$$

Solution

Begin by noting that the coefficients of y are opposites. By adding the two equations, you can eliminate y.

$$
\begin{array}{ll}
3x + 2y = 4 & \text{Equation 1} \\
\underline{5x - 2y = 8} & \text{Equation 2} \\
8x = 12 & \text{Add equations.}
\end{array}
$$

So, $x = \frac{3}{2}$. By back-substituting this value in the first equation, you can solve for y, as follows.

$$
\begin{array}{ll}
3x + 2y = 4 & \text{Equation 1} \\
3\left(\dfrac{3}{2}\right) + 2y = 4 & \text{Substitute } \frac{3}{2} \text{ for } x. \\
2y = -\dfrac{1}{2} & \text{Subtract } \frac{9}{2} \text{ from each side.} \\
y = -\dfrac{1}{4} & \text{Divide each side by 2.}
\end{array}
$$

The solution is $\left(\frac{3}{2}, -\frac{1}{4}\right)$. Check this solution in the original system.

✓ **CHECKPOINT** *Now try Exercise 1.*

The steps for solving a system of linear equations by the method of elimination are summarized as follows.

> ## The Method of Elimination
>
> 1. Obtain coefficients for x (or y) that are opposites by multiplying all terms of one or both equations by suitable constants.
> 2. Add the equations to eliminate one variable, and solve the resulting equation.
> 3. Back-substitute the value obtained in Step 2 in either of the original equations and solve for the other variable.
> 4. Check your solution in *both* of the original equations.

Example 2 shows how Step 1 above is used in the method of elimination.

EXAMPLE 2   The Method of Elimination

Solve the system of linear equations.

$$\begin{cases} 4x - 5y = 13 & \text{Equation 1} \\ 3x - y = 7 & \text{Equation 2} \end{cases}$$

Solution

To obtain coefficients of y that are opposites, multiply Equation 2 by -5.

$$\begin{cases} 4x - 5y = 13 \\ 3x - y = 7 \end{cases} \implies \begin{array}{ll} 4x - 5y = 13 & \text{Equation 1} \\ \underline{-15x + 5y = -35} & \text{Multiply Equation 2 by } -5. \\ -11x = -22 & \text{Add equations.} \end{array}$$

So, $x = 2$. Back-substitute this value in Equation 2 and solve for y.

$$3x - y = 7 \qquad\qquad \text{Equation 2}$$

$$3(2) - y = 7 \qquad\qquad \text{Substitute 2 for } x.$$

$$y = -1 \qquad\qquad \text{Solve for } y.$$

The solution is $(2, -1)$. Check this in the original system of equations, as follows.

Substitute in 1st Equation

$$4x - 5y = 13 \qquad\qquad \text{Equation 1}$$

$$4(2) - 5(-1) \overset{?}{=} 13 \qquad\qquad \text{Substitute 2 for } x \text{ and } -1 \text{ for } y.$$

$$8 + 5 = 13 \qquad\qquad \text{Solution checks. } \checkmark$$

Substitute in Final Equation

$$3x - y = 7 \qquad\qquad \text{Equation 2}$$

$$3(2) - (-1) \overset{?}{=} 7 \qquad\qquad \text{Substitute 2 for } x \text{ and } -1 \text{ for } y.$$

$$6 + 1 = 7 \qquad\qquad \text{Solution checks. } \checkmark$$

✓ **CHECKPOINT** *Now try Exercise 19.*

Example 3 shows how the method of elimination can be used to determine that a system of linear equations has no solution, while Example 4 shows how the method of elimination works with a system that has infinitely many solutions.

EXAMPLE 3   The Method of Elimination: No-Solution Case

To solve the system of linear equations

$$\begin{cases} 3x + 9y = 8 & \text{Equation 1} \\ 2x + 6y = 7 & \text{Equation 2} \end{cases}$$

obtain coefficients of x that are opposites by multiplying Equation 1 by 2 and Equation 2 by -3.

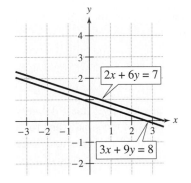

$$\begin{cases} 3x + 9y = 8 \\ 2x + 6y = 7 \end{cases} \implies \begin{aligned} 6x + 18y &= 16 &&\text{Multiply Equation 1 by 2.} \\ -6x - 18y &= -21 &&\text{Multiply Equation 2 by } -3. \\ \hline 0 &= -5 &&\text{False statement} \end{aligned}$$

Because $0 = -5$ is a false statement, you can conclude that the system is inconsistent and has no solution. You can confirm this by graphing the two lines, as shown in Figure 4.4. Because the lines are parallel, the system is inconsistent.

Figure 4.4

✔ **CHECKPOINT** *Now try Exercise 27.*

EXAMPLE 4   The Method of Elimination: Infinitely Many Solutions Case

To solve the system of linear equations

$$\begin{cases} -2x + 6y = 3 & \text{Equation 1} \\ 4x - 12y = -6 & \text{Equation 2} \end{cases}$$

obtain coefficients of x that are opposites by multiplying Equation 1 by 2.

$$\begin{cases} -2x + 6y = 3 \\ 4x - 12y = -6 \end{cases} \implies \begin{aligned} -4x + 12y &= 6 &&\text{Multiply Equation 1 by 2.} \\ 4x - 12y &= -6 &&\text{Equation 2} \\ \hline 0 &= 0 &&\text{Add equations.} \end{aligned}$$

Because $0 = 0$ is a true statement, the system has infinitely many solutions. You can confirm this by graphing the two lines, as shown in Figure 4.5. So, the solution set consists of all points (x, y) lying on the line $-2x + 6y = 3$.

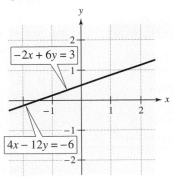

Figure 4.5

✔ **CHECKPOINT** *Now try Exercise 35.*

2 ▶ Use systems of linear equations to model and solve real-life problems.

Applications

To determine whether a real-life problem can be solved using a system of linear equations, consider these questions:

1. Does the problem involve more than one unknown quantity?

2. Are there two (or more) equations or conditions to be satisfied?

If one or both of these conditions are present, the appropriate mathematical model for the problem may be a system of linear equations.

EXAMPLE 5   A Mixture Problem

A company with two stores buys six large delivery vans and five small delivery vans. The first store receives four of the large vans and two of the small vans for a total cost of $200,000. The second store receives two of the large vans and three of the small vans for a total cost of $160,000. What is the cost of each type of van?

Solution

The two unknowns in this problem are the costs of the two types of vans.

Verbal Model: $4 \left(\dfrac{\text{Cost of}}{\text{large van}} \right) + 2 \left(\dfrac{\text{Cost of}}{\text{small van}} \right) = \$200,000$

$2 \left(\dfrac{\text{Cost of}}{\text{large van}} \right) + 3 \left(\dfrac{\text{Cost of}}{\text{small van}} \right) = \$160,000$

Labels: Cost of large van = x (dollars)
Cost of small van = y (dollars)

System: $\begin{cases} 4x + 2y = 200{,}000 & \text{Equation 1} \\ 2x + 3y = 160{,}000 & \text{Equation 2} \end{cases}$

To solve this system of linear equations, use the method of elimination. To obtain coefficients of x that are opposites, multiply Equation 2 by -2.

$\begin{cases} 4x + 2y = 200{,}000 \\ 2x + 3y = 160{,}000 \end{cases}$ ⟹

$\begin{aligned} 4x + 2y &= 200{,}000 && \text{Equation 1} \\ -4x - 6y &= -320{,}000 && \text{Multiply Equation 2 by } -2. \\ \hline -4y &= -120{,}000 && \text{Add equations.} \\ y &= 30{,}000 && \text{Divide each side by } -4. \end{aligned}$

So, the cost of each small van is $y = \$30,000$. Back-substitute this value in Equation 1 to find the cost of each large van.

$\begin{aligned} 4x + 2y &= 200{,}000 && \text{Equation 1} \\ 4x + 2(30{,}000) &= 200{,}000 && \text{Substitute 30,000 for } y. \\ 4x &= 140{,}000 && \text{Simplify.} \\ x &= 35{,}000 && \text{Divide each side by 4.} \end{aligned}$

The cost of each large van is $x = \$35,000$. Check this solution in the original statement of the problem.

 CHECKPOINT *Now try Exercise 63.*

EXAMPLE 6   **An Application Involving Two Speeds**

You take a motorboat trip on a river (18 miles upstream and 18 miles downstream). You run the motor at the same speed going up and down the river, but because of the river's current, the trip upstream takes $1\frac{1}{2}$ hours and the trip downstream takes only 1 hour. Determine the speed of the current.

Solution

Verbal Model:

Boat speed − Current speed = Upstream speed

Boat speed + Current speed = Downstream speed

Labels: Boat speed (in still water) $= x$ (miles per hour)
Current speed $= y$ (miles per hour)
Upstream speed $= 18/1.5 = 12$ (miles per hour)
Downstream speed $= 18/1 = 18$ (miles per hour)

System: $\begin{cases} x - y = 12 & \text{Equation 1} \\ x + y = 18 & \text{Equation 2} \end{cases}$

To solve this system of linear equations, use the method of elimination.

$$\begin{array}{ll} x - y = 12 & \text{Equation 1} \\ \underline{x + y = 18} & \text{Equation 2} \\ 2x \quad\;\; = 30 & \text{Add equations.} \end{array}$$

So, the speed of the boat in still water is 15 miles per hour. Back-substitution yields $y = 3$. So, the speed of the current is 3 miles per hour.

✔ **CHECKPOINT** *Now try Exercise 67.*

EXAMPLE 7   **Data Analysis: Best-Fitting Line**

The slope and y-intercept of the line $y = mx + b$ that best fits the three noncollinear points $(1, 0), (2, 1),$ and $(3, 4)$ are given by the solution of the system of linear equations below.

$$\begin{cases} 3b + \;\;6m = \;\;5 & \text{Equation 1} \\ 6b + 14m = 14 & \text{Equation 2} \end{cases}$$

Solve this system. Then find the equation of the best-fitting line.

Solution

To solve this system of linear equations, use the method of elimination.

$$\begin{cases} 3b + \;\;6m = \;\;5 \\ 6b + 14m = 14 \end{cases} \implies \begin{array}{ll} -6b - 12m = -10 & \text{Multiply Equation 1 by } -2. \\ \underline{\;\;6b + 14m = \quad 14} & \text{Equation 2} \\ \qquad\quad 2m = \quad 4 & \text{Add equations.} \end{array}$$

So, $m = 2$. Back-substitution yields $b = -\frac{7}{3}$. So, the equation of the best-fitting line is $y = 2x - \frac{7}{3}$. Figure 4.6 shows the points and the best-fitting line.

✔ **CHECKPOINT** *Now try Exercise 81.*

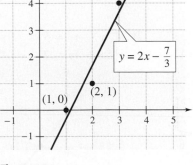

Figure 4.6

Concept Check

1. In a system of linear equations, how do you obtain coefficients for x (or y) that are opposites?

2. If a system of linear equations has coefficients of y that are opposites, how do you eliminate the y variable?

3. Explain what is meant by an *inconsistent* system of linear equations.

4. Both $(1, 3)$ and $(5, 5)$ are solutions of a system of two linear equations in two variables. How many solutions does the system have? Explain.

4.2 EXERCISES

Go to pages 288–289 to record your assignments.

Developing Skills

In Exercises 1–12, solve the system of linear equations by the method of elimination. Identify and label each line with its equation, and label the point of intersection (if any). *See Examples 1–4.*

 1. $\begin{cases} 2x + y = 4 \\ x - y = 2 \end{cases}$

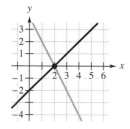

2. $\begin{cases} x + 3y = 2 \\ -x + 2y = 3 \end{cases}$

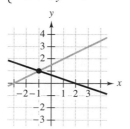

3. $\begin{cases} -x + 2y = 1 \\ x - y = 2 \end{cases}$

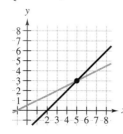

4. $\begin{cases} x + y = 0 \\ -3x - 3y = 0 \end{cases}$

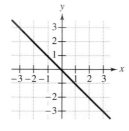

5. $\begin{cases} 3x + y = 3 \\ 2x - y = 7 \end{cases}$

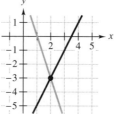

6. $\begin{cases} -x + 2y = 2 \\ 3x + y = 15 \end{cases}$

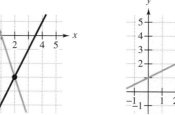

7. $\begin{cases} -2x + 4y = 8 \\ 0.4x - 0.8y = 2.4 \end{cases}$

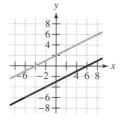

8. $\begin{cases} 3x + 4y = 2 \\ 0.6x + 0.8y = 1.6 \end{cases}$

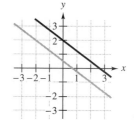

9. $\begin{cases} x - 3y = 5 \\ -2x + 6y = -10 \end{cases}$

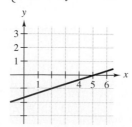

10. $\begin{cases} x - 4y = 5 \\ 5x + 4y = 7 \end{cases}$

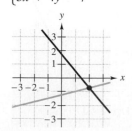

11. $\begin{cases} 2x - 8y = -11 \\ 5x + 3y = \quad 7 \end{cases}$

12. $\begin{cases} 3x + 4y = \quad 0 \\ 9x - 5y = 17 \end{cases}$

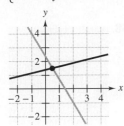

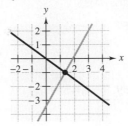

32. $\begin{cases} 0.02x - 0.05y = -0.19 \\ 0.03x + 0.04y = \quad 0.52 \end{cases}$

33. $\begin{cases} 0.7u - \quad v = -0.4 \\ 0.3u - 0.8v = \quad 0.2 \end{cases}$

34. $\begin{cases} 0.15x - 0.35y = -0.5 \\ -0.12x + 0.25y = \quad 0.1 \end{cases}$

✓ **35.** $\begin{cases} 5x + \quad 7y = 25 \\ x + 1.4y = \quad 5 \end{cases}$

36. $\begin{cases} 6b - 1.25m = -2 \\ -24b + \quad 5m = -1 \end{cases}$

In Exercises 13–40, solve the system of linear equations by the method of elimination. *See Examples 1–4.*

13. $\begin{cases} 6x - 6y = 25 \\ \quad 3y = 11 \end{cases}$

14. $\begin{cases} x + 3y = 4 \\ 2x \quad = 2 \end{cases}$

15. $\begin{cases} x + y = 0 \\ x - y = 4 \end{cases}$

16. $\begin{cases} x - y = 12 \\ x + y = \quad 2 \end{cases}$

17. $\begin{cases} 3x - 2y = 8 \\ 2x + 2y = 7 \end{cases}$

18. $\begin{cases} -x + 2y = 12 \\ \quad x + 6y = 20 \end{cases}$

✓ **19.** $\begin{cases} 5x + 2y = \quad 7 \\ 3x - \quad y = 13 \end{cases}$

20. $\begin{cases} 4x + 3y = \quad 8 \\ \quad x - 2y = 13 \end{cases}$

21. $\begin{cases} \quad x - 3y = 2 \\ 3x - 7y = 4 \end{cases}$

22. $\begin{cases} 2s - \quad t = \quad 9 \\ 3s + 4t = -14 \end{cases}$

23. $\begin{cases} 4x + 3y = -10 \\ 3x - \quad y = -14 \end{cases}$

24. $\begin{cases} 7r - \quad s = -25 \\ 2r + 5s = \quad 14 \end{cases}$

25. $\begin{cases} 2u + 3v = \quad 8 \\ 3u + 4v = 13 \end{cases}$

26. $\begin{cases} 4x - 3y = 25 \\ -3x + 8y = 10 \end{cases}$

✓ **27.** $\begin{cases} 12x - \quad 5y = 2 \\ -24x + 10y = 6 \end{cases}$

28. $\begin{cases} -2x + 3y = \quad 9 \\ 6x - 9y = -27 \end{cases}$

29. $\begin{cases} \frac{2}{3}r - \quad s = \quad 0 \\ 10r + 4s = 19 \end{cases}$

30. $\begin{cases} x - \quad y = -\frac{1}{2} \\ 4x - 48y = -35 \end{cases}$

31. $\begin{cases} 0.05x - 0.03y = 0.21 \\ \quad x + \quad y = 9 \end{cases}$

37. $\begin{cases} \frac{1}{2}x - \frac{1}{3}y = 1 \\ \frac{1}{4}x - \frac{1}{9}y = \frac{2}{3} \end{cases}$

38. $\begin{cases} \frac{1}{3}x = 4 - \frac{1}{4}y \\ \frac{3}{5}x - \frac{4}{5}y = 0 \end{cases}$

39. $\begin{cases} \frac{1}{5}x + \frac{1}{5}y = 4 \\ \frac{2}{3}x - \quad y = \frac{8}{3} \end{cases}$

40. $\begin{cases} \frac{2}{3}x - 4 = \frac{1}{2}y \\ x - 3y = \frac{1}{3} \end{cases}$

In Exercises 41–48, solve the system of linear equations by any convenient method.

41. $\begin{cases} x + 7y = -6 \\ x - 5y = \quad 18 \end{cases}$

42. $\begin{cases} 4x + y = -2 \\ -6x + y = \quad 18 \end{cases}$

43. $\begin{cases} y = \quad 5x - \quad 3 \\ y = -2x + 11 \end{cases}$

44. $\begin{cases} 3y = 2x + 21 \\ \quad x = 50 - 4y \end{cases}$

45. $\begin{cases} 2x - y = \quad 20 \\ -x + y = -5 \end{cases}$

46. $\begin{cases} 3x - 2y = -20 \\ 5x + 6y = \quad 32 \end{cases}$

47. $\begin{cases} \frac{3}{2}x + 2y = 12 \\ \frac{1}{4}x + \quad y = \quad 4 \end{cases}$

48. $\begin{cases} x + 2y = 4 \\ \frac{1}{2}x + \frac{1}{3}y = 1 \end{cases}$

In Exercises 49–54, decide whether the system is consistent or inconsistent.

49. $\begin{cases} 4x - \quad 5y = \quad 3 \\ -8x + 10y = -6 \end{cases}$

50. $\begin{cases} 4x - \quad 5y = \quad 3 \\ -8x + 10y = 14 \end{cases}$

51. $\begin{cases} -2x + 5y = 3 \\ 5x + 2y = 8 \end{cases}$

52. $\begin{cases} x + 10y = 12 \\ -2x + \quad 5y = \quad 2 \end{cases}$

53. $\begin{cases} -10x + 15y = 25 \\ 2x - 3y = -24 \end{cases}$ **54.** $\begin{cases} 4x - 5y = 28 \\ -2x + 2.5y = -14 \end{cases}$

In Exercises 55 and 56, determine the value of k such that the system of linear equations is inconsistent.

55. $\begin{cases} 5x - 10y = 40 \\ -2x + ky = 30 \end{cases}$ **56.** $\begin{cases} 12x - 18y = 5 \\ -18x + ky = 10 \end{cases}$

In Exercises 57–60, find a system of linear equations that has the given solution. (There are many correct answers.)

57. $(3, -6)$ **58.** $(-2, 7)$

59. $\left(-\frac{1}{2}, 4\right)$ **60.** $\left(\frac{2}{3}, -\frac{3}{4}\right)$

Solving Problems

61. *Break-Even Analysis*   To open a small business, you need an initial investment of $85,000. Your costs each week will be about $7400. Your projected weekly revenue is $8300. How many weeks will it take to break even?

62. *Break-Even Analysis*   To open a small business, you need an initial investment of $250,000. Your costs each week will be about $8650. Your projected weekly revenue is $9950. How many weeks will it take to break even?

✓ 63. *Comparing Costs*   A band charges $500 to play for 4 hours plus $50 for each additional hour. A DJ costs $300 to play for 4 hours plus $75 for each additional hour. After how many hours will the cost of the DJ exceed the cost of the band?

64. *Comparing Costs*   An SUV costs $26,445 and costs an average of $0.18 per mile to maintain. A hybrid model of the SUV costs $31,910 and costs an average of $0.13 per mile to maintain. After how many miles will the cost of the gas-only SUV exceed the cost of the hybrid?

65. *Investment*   A total of $20,000 is invested in two bonds that pay 8% and 9.5% simple interest. The annual interest is $1780. How much is invested in each bond?

66. *Investment*   A total of $15,000 is invested in two funds paying 4% and 5% simple interest. The annual interest is $700. How much is invested in each fund?

✓ 67. *Average Speed*   A van travels for 2 hours at an average speed of 40 miles per hour. How much longer must the van travel at an average speed of 55 miles per hour so that the average speed for the total trip will be 50 miles per hour?

68. *Average Speed*   A truck travels for 4 hours at an average speed of 42 miles per hour. How much longer must the truck travel at an average speed of 55 miles per hour so that the average speed for the total trip will be 50 miles per hour?

69. *Air Speed*   An airplane flying into a headwind travels 1800 miles in 3 hours and 36 minutes. On the return flight, the same distance is traveled in 3 hours (see figure). Find the speed of the plane in still air and the speed of the wind, assuming that both remain constant throughout the round trip.

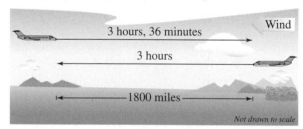

70. *Air Speed*   An airplane flying into a headwind travels 3000 miles in 6 hours and 15 minutes. On the return flight, the same distance is traveled in 5 hours (see figure). Find the speed of the plane in still air and the speed of the wind, assuming that both remain constant throughout the round trip.

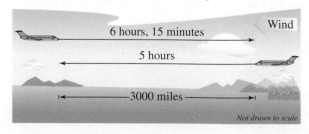

71. *Ticket Sales*   Five hundred tickets were sold for a fundraising dinner. The receipts totaled $3400.00. Adult tickets were $7.50 each and children's tickets were $4.00 each. How many tickets of each type were sold?

72. *Ticket Sales*   A fundraising dinner was held on two consecutive nights. On the first night, 100 adult tickets and 175 children's tickets were sold, for a total of $937.50. On the second night, 200 adult tickets and 316 children's tickets were sold, for a total of $1790.00. Find the price of each type of ticket.

73. *Gasoline Mixture*   Twelve gallons of regular unleaded gasoline plus 8 gallons of premium unleaded gasoline cost $76.48. Premium unleaded gasoline costs $0.11 more per gallon than regular unleaded. Find the price per gallon for each grade of gasoline.

74. *Gasoline Mixture*   The total cost of 8 gallons of regular unleaded gasoline and 12 gallons of premium unleaded gasoline is $71.84. Premium unleaded gasoline costs $0.17 more per gallon than regular unleaded. Find the price per gallon for each grade of gasoline.

75. *Alcohol Mixture*   How many liters of a 40% alcohol solution must be mixed with a 65% solution to obtain 20 liters of a 50% solution?

76. *Alcohol Mixture*   How many liters of a 20% alcohol solution must be mixed with a 60% solution to obtain 40 liters of a 35% solution?

77. *Alcohol Mixture*   How many fluid ounces of a 50% alcohol solution must be mixed with a 90% solution to obtain 32 fluid ounces of a 75% solution?

78. *Acid Mixture*   Five gallons of a 45% acid solution is obtained by mixing a 90% solution with a 30% solution. How many gallons of each solution must be used to obtain the desired mixture?

79. *Acid Mixture*   Thirty liters of a 46% acid solution is obtained by mixing a 40% solution with a 70% solution. How many liters of each solution must be used to obtain the desired mixture?

80. *Acid Mixture*   Fifty gallons of a 60% acid solution is obtained by mixing an 80% solution with a 50% solution. How many gallons of each solution must be used to obtain the desired mixture?

✓ 81. *Nut Mixture*   Ten pounds of mixed nuts sells for $6.87 per pound. The mixture is obtained from two kinds of nuts, peanuts priced at $5.70 per pound and cashews at $8.70 per pound. How many pounds of each variety of nut are used in the mixture?

82. *Nut Mixture*   Thirty pounds of mixed nuts sells for $6.30 per pound. The mixture is obtained from two kinds of nuts, walnuts priced at $6.50 per pound and peanuts at $5.90 per pound. How many pounds of each variety of nut are used in the mixture?

83. *Best-Fitting Line*   The slope and y-intercept of the line $y = mx + b$ that best fits the three noncollinear points $(0, 0)$, $(1, 1)$, and $(2, 3)$ are given by the solution of the following system of linear equations.
$$\begin{cases} 5m + 3b = 7 \\ 3m + 3b = 4 \end{cases}$$

(a) Solve the system and find the equation of the best-fitting line.

(b) Plot the three points and sketch the graph of the best-fitting line.

84. *Best-Fitting Line*   The slope and y-intercept of the line $y = mx + b$ that best fits the three noncollinear points $(0, 4)$, $(1, 2)$, and $(2, 1)$ are given by the solution of the following system of linear equations.
$$\begin{cases} 3b + 3m = 7 \\ 3b + 5m = 4 \end{cases}$$

(a) Solve the system and find the equation of the best-fitting line.

(b) Plot the three points and sketch the graph of the best-fitting line.

85. *Hourly Wages* The average hourly wages (in December) for those employed in the retail trade industry in the United States for selected years from 2002 through 2006 are shown in the table. (Source: U.S. Bureau of Labor Statistics)

Year, t	2002	2004	2006
Wage, y	$11.84	$12.24	$12.67

(a) Plot the data shown in the table. Let t represent the year, with $t = 2$ corresponding to 2002.

(b) The line $y = mx + b$ that best fits the data is given by the solution of the following system.

$$\begin{cases} 3b + 15m = 37.38 \\ 15b + 125m = 197.3 \end{cases}$$

Solve the system and find the equation of the best-fitting line. Sketch the graph of the line on the same set of coordinate axes used in part (a).

(c) Interpret the meaning of the slope of the line in the context of this problem.

86. *Hourly Wages* The average hourly wages (in December) for those employed in the manufacturing industry in the United States for selected years from 2002 through 2006 are shown in the table. (Source: U.S. Bureau of Labor Statistics)

Year, t	2002	2004	2006
Wage, y	$15.53	$16.35	$16.95

(a) Plot the data shown in the table. Let t represent the year, with $t = 2$ corresponding to 2002.

(b) The line $y = mx + b$ that best fits the data is given by the solution of the following system.

$$\begin{cases} 3b + 15m = 49.905 \\ 15b + 125m = 267.275 \end{cases}$$

Solve the system and find the equation of the best-fitting line. Sketch the graph of the line on the same set of coordinate axes used in part (a).

(c) Interpret the meaning of the slope of the line in the context of this problem.

Explaining Concepts

87. When solving a system by elimination, how do you recognize that it has infinitely many solutions?

88. In your own words, explain how to solve a system of linear equations by elimination.

89. How can you recognize that a system of linear equations has no solution? Give an example.

90. Under what conditions might substitution be better than elimination for solving a system of linear equations?

Cumulative Review

In Exercises 91–94, write the general form of the equation of the line that passes through the two points.

91. $(0, 0), (4, 2)$

92. $(1, 2), (6, 3)$

93. $(-1, 2), (5, 2)$

94. $(-3, 3), (8, -6)$

In Exercises 95–98, determine if the set of ordered pairs is a function.

95. $\{(0, 0), (2, 1), (4, 2), (6, 3)\}$

96. $\{(0, 2), (1, 4), (4, 1), (0, 4)\}$

97. $\{(-4, 5), (-1, 0), (3, -2), (3, -4)\}$

98. $\{(-3, 1), (-1, 3), (1, 3), (3, 1)\}$

In Exercises 99 and 100, determine whether each ordered pair is a solution of the system of equations.

99. $\begin{cases} 3x - 4y = 10 \\ 2x + 6y = -2 \end{cases}$

(a) $(2, -1)$

(b) $(-1, 0)$

100. $\begin{cases} -4x - y = 2 \\ 2x + 7y = 38 \end{cases}$

(a) $(2, -10)$

(b) $(-2, 6)$

4.3 Linear Systems in Three Variables

Frank Whitney/Getty Images

Why You Should Learn It

Systems of linear equations in three variables can be used to model and solve real-life problems. For instance, in Exercise 47 on page 251, a system of linear equations can be used to determine a chemical mixture for a pesticide.

1 ▶ Solve systems of linear equations in row-echelon form using back-substitution.

What You Should Learn

1 ▶ Solve systems of linear equations in row-echelon form using back-substitution.

2 ▶ Solve systems of linear equations using the method of Gaussian elimination.

3 ▶ Solve application problems using the method of Gaussian elimination.

Row-Echelon Form

The method of elimination can be applied to a system of linear equations in more than two variables. In fact, this method easily adapts to computer use for solving systems of linear equations with dozens of variables.

When the method of elimination is used to solve a system of linear equations, the goal is to rewrite the system in a form to which back-substitution can be applied. For instance, consider the following two systems of linear equations.

$$\begin{cases} x - 2y + 2z = 9 \\ -x + 3y \quad\;\; = -4 \\ 2x - 5y + z = 10 \end{cases} \qquad \begin{cases} x - 2y + 2z = 9 \\ \quad\;\; y + 2z = 5 \\ \quad\quad\quad\; z = 3 \end{cases}$$

Which of these two systems do you think is easier to solve? After comparing the two systems, it should be clear that it is easier to solve the system on the right because the value of z is already shown and back-substitution will readily yield the values of x and y. The system on the right is said to be in **row-echelon form,** which means that it has a "stair-step" pattern with leading coefficients of 1.

EXAMPLE 1   **Using Back-Substitution**

In the following system of linear equations, you know the value of z from Equation 3.

$$\begin{cases} x - 2y + 2z = 9 & \text{Equation 1} \\ \quad\;\; y + 2z = 5 & \text{Equation 2} \\ \quad\quad\quad\; z = 3 & \text{Equation 3} \end{cases}$$

To solve for y, substitute $z = 3$ in Equation 2 to obtain

$$y + 2(3) = 5 \quad \Longrightarrow \quad y = -1. \qquad \text{Substitute 3 for } z.$$

Finally, substitute $y = -1$ and $z = 3$ in Equation 1 to obtain

$$x - 2(-1) + 2(3) = 9 \quad \Longrightarrow \quad x = 1. \qquad \text{Substitute } -1 \text{ for } y \text{ and 3 for } z.$$

The solution is $x = 1$, $y = -1$, and $z = 3$, which can also be written as the **ordered triple** $(1, -1, 3)$. Check this in the original system of equations.

✓ **CHECKPOINT** *Now try Exercise 3.*

Study Tip

When checking a solution, remember that the solution must satisfy each equation in the original system.

2 ▶ Solve systems of linear equations using the method of Gaussian elimination.

The Method of Gaussian Elimination

Two systems of equations are **equivalent systems** if they have the same solution set. To solve a system that is not in row-echelon form, first convert it to an *equivalent* system that is in row-echelon form. To see how this is done, take another look at the method of elimination, as applied to a system of two linear equations.

EXAMPLE 2   The Method of Elimination

Solve the system of linear equations.

$$\begin{cases} 3x - 2y = -1 & \text{Equation 1} \\ x - y = 0 & \text{Equation 2} \end{cases}$$

Solution

$$\begin{cases} x - y = 0 \\ 3x - 2y = -1 \end{cases}$$ Interchange the two equations in the system.

$$\begin{aligned} -3x + 3y &= 0 \\ \underline{3x - 2y} &= -1 \\ y &= -1 \end{aligned}$$ Multiply new Equation 1 by -3 and add it to new Equation 2.

$$\begin{cases} x - y = 0 \\ y = -1 \end{cases}$$ New system in row-echelon form

Using back-substitution, you can determine that the solution is $(-1, -1)$. Check the solution in each equation in the original system, as follows.

Equation 1	*Equation 2*
$3x - 2y \overset{?}{=} -1$	$x - y \overset{?}{=} 0$
$3(-1) - 2(-1) = -1$ ✓	$(-1) - (-1) = 0$ ✓

✓ **CHECKPOINT** *Now try Exercise 7.*

Rewriting a system of linear equations in row-echelon form usually involves a chain of equivalent systems, each of which is obtained by using one of the three basic row operations. This process is called **Gaussian elimination.**

Operations That Produce Equivalent Systems

Each of the following **row operations** on a system of linear equations produces an *equivalent* system of linear equations.

1. Interchange two equations.
2. Multiply one of the equations by a nonzero constant.
3. Add a multiple of one of the equations to another equation to replace the latter equation.

| **EXAMPLE 3** | **Using Gaussian Elimination to Solve a System** |

Solve the system of linear equations.

$$\begin{cases} x - 2y + 2z = 9 & \text{Equation 1} \\ -x + 3y = -4 & \text{Equation 2} \\ 2x - 5y + z = 10 & \text{Equation 3} \end{cases}$$

Solution

Because the leading coefficient of the first equation is 1, you can begin by keeping the x in the upper left position and eliminating the other x terms from the first column, as follows.

$$\begin{cases} x - 2y + 2z = 9 \\ y + 2z = 5 \\ 2x - 5y + z = 10 \end{cases}$$

Adding the first equation to the second equation produces a new second equation.

$$\begin{cases} x - 2y + 2z = 9 \\ y + 2z = 5 \\ {-}y - 3z = -8 \end{cases}$$

Adding -2 times the first equation to the third equation produces a new third equation.

Now that the x terms are eliminated from all but the first row, work on the second column. (You need to eliminate y from the third equation.)

$$\begin{cases} x - 2y + 2z = 9 \\ y + 2z = 5 \\ {-}z = -3 \end{cases}$$

Adding the second equation to the third equation produces a new third equation.

Finally, you need a coefficient of 1 for z in the third equation.

$$\begin{cases} x - 2y + 2z = 9 \\ y + 2z = 5 \\ z = 3 \end{cases}$$

Multiplying the third equation by -1 produces a new third equation.

This is the same system that was solved in Example 1, and, as in that example, you can conclude by back-substitution that the solution is

$$x = 1, \quad y = -1, \quad \text{and} \quad z = 3.$$

The solution can be written as the ordered triple

$$(1, -1, 3).$$

You can check the solution by substituting 1 for x, -1 for y, and 3 for z in each equation of the original system, as follows.

Check

Equation 1: $\quad x - 2y + 2z \stackrel{?}{=} 9$

$$1 - 2(-1) + 2(3) = 9 \checkmark$$

Equation 2: $\quad -x + 3y \stackrel{?}{=} -4$

$$-(1) + 3(-1) = -4 \checkmark$$

Equation 3: $\quad 2x - 5y + z \stackrel{?}{=} 10$

$$2(1) - 5(-1) + 3 = 10 \checkmark$$

✓ **CHECKPOINT** *Now try Exercise 11.*

EXAMPLE 4   Using Gaussian Elimination to Solve a System

Solve the system of linear equations.

$$\begin{cases} 4x + y - 3z = 11 & \text{Equation 1} \\ 2x - 3y + 2z = 9 & \text{Equation 2} \\ x + y + z = -3 & \text{Equation 3} \end{cases}$$

Solution

$$\begin{cases} x + y + z = -3 \\ 2x - 3y + 2z = 9 \\ 4x + y - 3z = 11 \end{cases}$$

Interchange the first and third equations.

$$\begin{cases} x + y + z = -3 \\ -5y = 15 \\ 4x + y - 3z = 11 \end{cases}$$

Adding -2 times the first equation to the second equation produces a new second equation.

$$\begin{cases} x + y + z = -3 \\ -5y = 15 \\ -3y - 7z = 23 \end{cases}$$

Adding -4 times the first equation to the third equation produces a new third equation.

$$\begin{cases} x + y + z = -3 \\ y = -3 \\ -3y - 7z = 23 \end{cases}$$

Multiplying the second equation by $-\frac{1}{5}$ produces a new second equation.

$$\begin{cases} x + y + z = -3 \\ y = -3 \\ -7z = 14 \end{cases}$$

Adding 3 times the second equation to the third equation produces a new third equation.

$$\begin{cases} x + y + z = -3 \\ y = -3 \\ z = -2 \end{cases}$$

Multiplying the third equation by $-\frac{1}{7}$ produces a new third equation.

Now you can see that $z = -2$ and $y = -3$. Moreover, by back-substituting these values in Equation 1, you can determine that $x = 2$. So, the solution is

$$x = 2, \quad y = -3, \quad \text{and} \quad z = -2$$

which can be written as the ordered triple $(2, -3, -2)$. You can check this solution as follows.

Check

Equation 1: $4x + y - 3z \stackrel{?}{=} 11$
$$4(2) + (-3) - 3(-2) \stackrel{?}{=} 11$$
$$11 = 11 \checkmark$$

Equation 2: $2x - 3y + 2z \stackrel{?}{=} 9$
$$2(2) - 3(-3) + 2(-2) \stackrel{?}{=} 9$$
$$9 = 9 \checkmark$$

Equation 3: $x + y + z \stackrel{?}{=} -3$
$$(2) + (-3) + (-2) \stackrel{?}{=} -3$$
$$-3 = -3 \checkmark$$

✓ **CHECKPOINT** *Now try Exercise 15.*

The next example involves an inconsistent system—one that has no solution. The key to recognizing an inconsistent system is that at some stage in the elimination process, you obtain a false statement such as $0 = 6$. Watch for such statements as you do the exercises for this section.

EXAMPLE 5   An Inconsistent System

Solve the system of linear equations.

$$\begin{cases} x - 3y + z = 1 & \text{Equation 1} \\ 2x - y - 2z = 2 & \text{Equation 2} \\ x + 2y - 3z = -1 & \text{Equation 3} \end{cases}$$

Solution

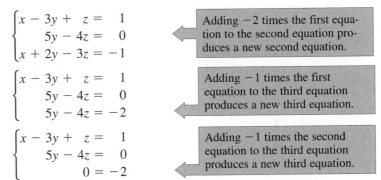

$$\begin{cases} x - 3y + z = 1 \\ 5y - 4z = 0 \\ x + 2y - 3z = -1 \end{cases}$$
Adding -2 times the first equation to the second equation produces a new second equation.

$$\begin{cases} x - 3y + z = 1 \\ 5y - 4z = 0 \\ 5y - 4z = -2 \end{cases}$$
Adding -1 times the first equation to the third equation produces a new third equation.

$$\begin{cases} x - 3y + z = 1 \\ 5y - 4z = 0 \\ 0 = -2 \end{cases}$$
Adding -1 times the second equation to the third equation produces a new third equation.

Because the third "equation" is a false statement, you can conclude that this system is inconsistent and therefore has no solution. Moreover, because this system is equivalent to the original system, you can conclude that the original system also has no solution.

 CHECKPOINT *Now try Exercise 17.*

As with a system of linear equations in two variables, the number of solutions of a system of linear equations in more than two variables must fall into one of three categories.

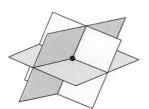

Solution: one point
Figure 4.7

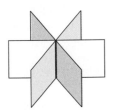

Solution: one line

Solution: one plane
Figure 4.8

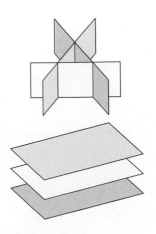

Solution: none
Figure 4.9

The Number of Solutions of a Linear System

For a system of linear equations, exactly one of the following is true.

1. There is exactly one solution.
2. There are infinitely many solutions.
3. There is no solution.

The graph of a system of three linear equations in three variables consists of *three planes*. When these planes intersect in a single point, the system has exactly one solution. (See Figure 4.7.) When the three planes intersect in a line or a plane, the system has infinitely many solutions. (See Figure 4.8.) When the three planes have no point in common, the system has no solution. (See Figure 4.9.)

> **EXAMPLE 6**   **A System with Infinitely Many Solutions**

Solve the system of linear equations.

$$\begin{cases} x + y - 3z = -1 & \text{Equation 1} \\ \quad\quad y - z = 0 & \text{Equation 2} \\ -x + 2y \quad\quad = 1 & \text{Equation 3} \end{cases}$$

Solution

Begin by rewriting the system in row-echelon form.

$$\begin{cases} x + y - 3z = -1 \\ \quad\quad y - z = 0 \\ \quad\quad 3y - 3z = 0 \end{cases}$$

> Adding the first equation to the third equation produces a new third equation.

$$\begin{cases} x + y - 3z = -1 \\ \quad\quad y - z = 0 \\ \quad\quad\quad 0 = 0 \end{cases}$$

> Adding -3 times the second equation to the third equation produces a new third equation.

This means that Equation 3 depends on Equations 1 and 2 in the sense that it gives no additional information about the variables. So, the original system is equivalent to the system

$$\begin{cases} x + y - 3z = -1 \\ \quad\quad y - z = 0. \end{cases}$$

In the last equation, solve for y in terms of z to obtain $y = z$. Back-substituting for y in the previous equation produces $x = 2z - 1$. Finally, letting $z = a$, where a is any real number, you can see that there are an infinite number of solutions to the original system, all of the form

$$x = 2a - 1, \ y = a, \text{ and } z = a.$$

So, every ordered triple of the form

$$(2a - 1, a, a), \quad a \text{ is a real number}$$

is a solution of the system.

✓ **CHECKPOINT** *Now try Exercise 27.*

In Example 6, there are other ways to write the same infinite set of solutions. For instance, letting $x = b$, the solutions could have been written as

$$\left(b, \frac{1}{2}(b + 1), \frac{1}{2}(b + 1) \right), \quad b \text{ is a real number.}$$

To convince yourself that this description produces the same set of solutions, consider the comparison shown below.

Substitution	*Solution*	
$a = 0$	$(2(0)-1, 0, 0) = (-1, 0, 0)$	
$b = -1$	$\left(-1, \frac{1}{2}(-1 + 1), \frac{1}{2}(-1 + 1)\right) = (-1, 0, 0)$	Same solution
$a = 1$	$(2(1)-1, 1, 1) = (1, 1, 1)$	
$b = 1$	$\left(1, \frac{1}{2}(1 + 1), \frac{1}{2}(1 + 1)\right) = (1, 1, 1)$	Same solution

Study Tip

When comparing descriptions of an infinite solution set, keep in mind that there is more than one way to describe the set.

3 ▶ Solve application problems using the method of Gaussian elimination.

Applications

EXAMPLE 7 **Vertical Motion**

The height at time t of an object that is moving in a (vertical) line with constant acceleration a is given by the **position equation**

$$s = \frac{1}{2}at^2 + v_0 t + s_0.$$

The height s is measured in feet, the acceleration a is measured in feet per second squared, the time t is measured in seconds, v_0 is the initial velocity (at time $t = 0$), and s_0 is the initial height. Find the values of a, v_0, and s_0, if $s = 164$ feet at 1 second, $s = 180$ feet at 2 seconds, and $s = 164$ feet at 3 seconds.

Solution

By substituting the three values of t and s into the position equation, you obtain three linear equations in a, v_0, and s_0.

When $t = 1$, $s = 164$: $\frac{1}{2}a(1)^2 + v_0(1) + s_0 = 164$

When $t = 2$, $s = 180$: $\frac{1}{2}a(2)^2 + v_0(2) + s_0 = 180$

When $t = 3$, $s = 164$: $\frac{1}{2}a(3)^2 + v_0(3) + s_0 = 164$

By multiplying the first and third equations by 2, this system can be rewritten as

$$\begin{cases} a + 2v_0 + 2s_0 = 328 & \text{Equation 1} \\ 2a + 2v_0 + s_0 = 180 & \text{Equation 2} \\ 9a + 6v_0 + 2s_0 = 328 & \text{Equation 3} \end{cases}$$

and you can apply Gaussian elimination to obtain

$$\begin{cases} a + 2v_0 + 2s_0 = 328 & \text{Equation 1} \\ \quad\; -2v_0 - 3s_0 = -476 & \text{Equation 2} \\ \quad\qquad\quad 2s_0 = 232. & \text{Equation 3} \end{cases}$$

From the third equation, $s_0 = 116$, so back-substitution in Equation 2 yields

$$-2v_0 - 3(116) = -476$$

$$-2v_0 = -128$$

$$v_0 = 64.$$

Finally, back-substituting $s_0 = 116$ and $v_0 = 64$ in Equation 1 yields

$$a + 2(64) + 2(116) = 328$$

$$a = -32.$$

So, the position equation for this object is $s = -16t^2 + 64t + 116$.

✓ **CHECKPOINT** *Now try Exercise 37.*

© Peter Brogden/Alamy

The "Big Shot" zero-gravity ride looks like part of a needle atop the Stratosphere Tower in Las Vegas. The "Big Shot" lets riders experience zero gravity by allowing them to free-fall after first catapulting them upward to a height of nearly 1100 feet above the ground.

EXAMPLE 8   A Geometry Application

The sum of the measures of two angles of a triangle is twice the measure of the third angle. The measure of the first angle is 18° more than the measure of the third angle. Find the measures of the three angles.

Solution

Let x, y, and z represent the measures of the first, second, and third angles, respectively. The sum of the measures of the three angles of a triangle is 180°. From the given information, you can write a system of equations as follows.

$$\begin{cases} x + y + z = 180 & \text{Equation 1} \\ x + y = 2z & \text{Equation 2} \\ x = z + 18 & \text{Equation 3} \end{cases}$$

By rewriting this system with the variable terms on the left side, you obtain

$$\begin{cases} x + y + z = 180 & \text{Equation 1} \\ x + y - 2z = 0 & \text{Equation 2} \\ x \quad - z = 18. & \text{Equation 3} \end{cases}$$

Using Gaussian elimination to solve this system yields $x = 78$, $y = 42$, and $z = 60$. So, the measures of the three angles are 78°, 42°, and 60°, respectively.

Check

Equation 1: $78 + 42 + 60 = 180$ ✓

Equation 2: $78 + 42 - 2(60) = 0$ ✓

Equation 3: $78 - 60 = 18$ ✓

✓ **CHECKPOINT** *Now try Exercise 41.*

EXAMPLE 9   Grades of Paper

A paper manufacturer sells a 50-pound package that consists of three grades of computer paper. Grade A costs \$6.00 per pound, grade B costs \$4.50 per pound, and grade C costs \$3.50 per pound. Half of the 50-pound package consists of the two cheaper grades. The cost of the 50-pound package is \$252.50. How many pounds of each grade of paper are there in the 50-pound package?

Solution

Let A, B, and C represent the numbers of pounds of grade A, grade B, and grade C paper, respectively. From the given information, you can write a system of equations as follows.

$$\begin{cases} A + B + C = 50 & \text{Equation 1} \\ 6A + 4.50B + 3.50C = 252.50 & \text{Equation 2} \\ B + C = 25 & \text{Equation 3} \end{cases}$$

Using Gaussian elimination to solve this system yields $A = 25$, $B = 15$, and $C = 10$. So, there are 25 pounds of grade A paper, 15 pounds of grade B paper, and 10 pounds of grade C paper in the 50-pound package. Check this solution.

✓ **CHECKPOINT** *Now try Exercise 49.*

Concept Check

1. How can the process of Gaussian elimination help you to solve a system of equations? In general, after applying Gaussian elimination in a system of equations, what are the next steps you take to find the solution of the system?

2. Describe the three row operations that you can use to produce an equivalent system of equations while applying Gaussian elimination.

3. Give an example of a system of three linear equations in three variables that is in row-echelon form.

4. Show how to use back-substitution to solve the system of equations you wrote in Concept Check 3.

4.3 EXERCISES

Go to pages 288–289 to record your assignments.

Developing Skills

In Exercises 1 and 2, determine whether each ordered triple is a solution of the system of linear equations.

1. $\begin{cases} x + 3y + 2z = 1 \\ 5x - y + 3z = 16 \\ -3x + 7y + z = -14 \end{cases}$

   (a) $(0, 3, -2)$ (b) $(12, 5, -13)$
   (c) $(1, -2, 3)$ (d) $(-2, 5, -3)$

2. $\begin{cases} 3x - y + 4z = -10 \\ -x + y + 2z = 6 \\ 2x - y + z = -8 \end{cases}$

   (a) $(-2, 4, 0)$ (b) $(0, -3, 10)$
   (c) $(1, -1, 5)$ (d) $(7, 19, -3)$

In Exercises 3–6, use back-substitution to solve the system of linear equations. *See Example 1.*

3. $\begin{cases} x - 2y + 4z = 4 \\ 3y - z = 2 \\ z = -5 \end{cases}$

4. $\begin{cases} 5x + 4y - z = 0 \\ 10y - 3z = 11 \\ z = 3 \end{cases}$

5. $\begin{cases} x - 2y + 4z = 4 \\ y = 3 \\ y + z = 2 \end{cases}$

6. $\begin{cases} x = 10 \\ 3x + 2y = 2 \\ x + y + 2z = 0 \end{cases}$

In Exercises 7 and 8, perform the row operation and write the equivalent system of linear equations. *See Example 2.*

7. Add Equation 1 to Equation 2.

   $\begin{cases} x - 2y = 8 & \text{Equation 1} \\ -x + 3y = 6 & \text{Equation 2} \end{cases}$

   What did this operation accomplish?

8. Add -2 times Equation 1 to Equation 3.

   $\begin{cases} x - 2y + 3z = 5 & \text{Equation 1} \\ -x + y + 5z = 4 & \text{Equation 2} \\ 2x - 3z = 0 & \text{Equation 3} \end{cases}$

   What did this operation accomplish?

In Exercises 9 and 10, determine whether the two systems of linear equations are equivalent. Give reasons for your answer.

9. $\begin{cases} x + 3y - z = 6 \\ 2x - y + 2z = 1 \\ 3x + 2y - z = 2 \end{cases}$ $\begin{cases} x + 3y - z = 6 \\ -7y + 4z = -11 \\ -7y + 2z = -16 \end{cases}$

10. $\begin{cases} x - 2y + 3z = 9 \\ -x + 3y = -4 \\ 2x - 5y + 5z = 17 \end{cases}$ $\begin{cases} x - 2y + 3z = 9 \\ y + 3z = 5 \\ -y - z = -1 \end{cases}$

In Exercises 11–34, solve the system of linear equations.
See Examples 3–6.

11.
$$\begin{cases} x \ \ + z = 4 \\ \quad y \quad = 2 \\ 4x \ \ + z = 7 \end{cases}$$

12.
$$\begin{cases} x + y \quad = 6 \\ 3x - y \quad = 2 \\ \quad z = 3 \end{cases}$$

13.
$$\begin{cases} x + y + z = 6 \\ 2x - y + z = 3 \\ 3x \quad - z = 0 \end{cases}$$

14.
$$\begin{cases} x + y + z = 2 \\ -x + 3y + 2z = 8 \\ 4x + y \quad = 4 \end{cases}$$

15.
$$\begin{cases} x + y + z = -3 \\ 4x + y - 3z = 11 \\ 2x - 3y + 2z = 9 \end{cases}$$

16.
$$\begin{cases} x - y + 2z = -4 \\ 3x + y - 4z = -6 \\ 2x + 3y - 4z = 4 \end{cases}$$

17.
$$\begin{cases} x + 2y + 6z = 5 \\ -x + y - 2z = 3 \\ x - 4y - 2z = 1 \end{cases}$$

18.
$$\begin{cases} x + 6y + 2z = 9 \\ 3x - 2y + 3z = -1 \\ 5x - 5y + 2z = 7 \end{cases}$$

19.
$$\begin{cases} 2x \quad + 2z = 2 \\ 5x + 3y \quad = 4 \\ \quad 3y - 4z = 4 \end{cases}$$

20.
$$\begin{cases} x + y + 8z = 3 \\ 2x + y + 11z = 4 \\ x \quad + 3z = 0 \end{cases}$$

21.
$$\begin{cases} \quad 6y + 4z = -12 \\ 3x + 3y \quad = 9 \\ 2x \quad - 3z = 10 \end{cases}$$

22.
$$\begin{cases} 2x - 4y + z = 0 \\ 3x \quad + 2z = -1 \\ -6x + 3y + 2z = -10 \end{cases}$$

23.
$$\begin{cases} 2x + y + 3z = 1 \\ 2x + 6y + 8z = 3 \\ 6x + 8y + 18z = 5 \end{cases}$$

24.
$$\begin{cases} 3x - y - 2z = 5 \\ 2x + y + 3z = 6 \\ 6x - y - 4z = 9 \end{cases}$$

25.
$$\begin{cases} \quad y + z = 5 \\ 2x \quad + 4z = 4 \\ 2x - 3y \quad = -14 \end{cases}$$

26.
$$\begin{cases} 5x + 2y \quad = -8 \\ \quad z = 5 \\ 3x - y + z = 9 \end{cases}$$

27.
$$\begin{cases} 2x \quad + z = 1 \\ \quad 5y - 3z = 2 \\ 6x + 20y - 9z = 11 \end{cases}$$

28.
$$\begin{cases} 2x + y - z = 4 \\ \quad y + 3z = 2 \\ 3x + 2y \quad = 4 \end{cases}$$

29.
$$\begin{cases} 2x \quad + 3z = 4 \\ 5x + y + z = 2 \\ 11x + 3y - 3z = 0 \end{cases}$$

30.
$$\begin{cases} 3x + y + z = 2 \\ 4x \quad + 2z = 1 \\ 5x - y + 3z = 0 \end{cases}$$

31.
$$\begin{cases} 0.2x + 1.3y + 0.6z = 0.1 \\ 0.1x \quad + 0.3z = 0.7 \\ 2x + 10y + 8z = 8 \end{cases}$$

32.
$$\begin{cases} 0.3x - 0.1y + 0.2z = 0.35 \\ 2x + y - 2z = -1 \\ 2x + 4y + 3z = 10.5 \end{cases}$$

33.
$$\begin{cases} x + 4y - 2z = 2 \\ -3x + y + z = -2 \\ 5x + 7y - 5z = 6 \end{cases}$$

34.
$$\begin{cases} x - 2y - z = 3 \\ 2x + y - 3z = 1 \\ x + 8y - 3z = -7 \end{cases}$$

In Exercises 35 and 36, find a system of linear equations in three variables with integer coefficients that has the given point as a solution. (There are many correct answers.)

35. $(4, -3, 2)$

36. $(5, 7, -10)$

Solving Problems

Vertical Motion In Exercises 37–40, find the position equation $s = \frac{1}{2}at^2 + v_0t + s_0$ for an object that has the indicated heights at the specified times. *See Example 7.*

37. $s = 128$ feet at $t = 1$ second
$s = 80$ feet at $t = 2$ seconds
$s = 0$ feet at $t = 3$ seconds

38. $s = 48$ feet at $t = 1$ second
$s = 64$ feet at $t = 2$ seconds
$s = 48$ feet at $t = 3$ seconds

39. $s = 32$ feet at $t = 1$ second
$s = 32$ feet at $t = 2$ seconds
$s = 0$ feet at $t = 3$ seconds

40. $s = 10$ feet at $t = 0$ seconds
$s = 54$ feet at $t = 1$ second
$s = 46$ feet at $t = 3$ seconds

41. ▲ *Geometry* The sum of the measures of two angles of a triangle is twice the measure of the third angle. The measure of the second angle is 28° less than the measure of the third angle. Find the measures of the three angles.

42. ▲ *Geometry* The measure of one angle of a triangle is two-thirds the measure of a second angle, and the measure of the second angle is 12° greater than the measure of the third angle. Find the measures of the three angles.

43. *Investment* An inheritance of $80,000 is divided among three investments yielding a total of $8850 in simple interest in 1 year. The interest rates for the three investments are 6%, 10%, and 15%. The amount invested at 10% is $750 more than the amount invested at 15%. Find the amount invested at each rate.

44. *Investment* An inheritance of $16,000 is divided among three investments yielding a total of $940 in simple interest in 1 year. The interest rates for the three investments are 5%, 6%, and 7%. The amount invested at 6% is $3000 less than the amount invested at 5%. Find the amount invested at each rate.

45. *Investment* You receive a total of $1150 in simple interest in 1 year from three investments. The interest rates for the three investments are 6%, 8%, and 9%. The 8% investment is twice the 6% investment, and the 9% investment is $1000 less than the 6% investment. What is the amount of each investment?

46. *Investment* You receive a total of $620 in simple interest in 1 year from three investments. The interest rates for the three investments are 5%, 7%, and 8%. The 5% investment is twice the 7% investment, and the 7% investment is $1500 less than the 8% investment. What is the amount of each investment?

47. *Chemical Mixture* A mixture of 12 gallons of chemical A, 16 gallons of chemical B, and 26 gallons of chemical C is required to kill a destructive crop insect. Commercial spray X contains one, two, and two parts of these chemicals. Spray Y contains only chemical C. Spray Z contains only chemicals A and B in equal amounts. How much of each type of commercial spray is needed to obtain the desired mixture?

48. *Fertilizer Mixture* A mixture of 5 pounds of fertilizer A, 13 pounds of fertilizer B, and 4 pounds of fertilizer C provides the optimal nutrients for a plant. Commercial brand X contains equal parts of fertilizer B and fertilizer C. Brand Y contains one part of fertilizer A and two parts of fertilizer B. Brand Z contains two parts of fertilizer A, five parts of fertilizer B, and two parts of fertilizer C. How much of each fertilizer brand is needed to obtain the desired mixture?

49. *Hot Dogs* A vendor sells three sizes of hot dogs at prices of $1.50, $2.50, and $3.25. On a day when the vendor had a total revenue of $289.25 from sales of 143 hot dogs, four times as many $1.50 hot dogs were sold as $3.25 hot dogs. How many hot dogs were sold at each price?

50. *Coffee* A coffee manufacturer sells a 10-pound package that consists of three flavors of coffee. Vanilla flavored coffee costs $6 per pound, Hazelnut flavored coffee costs $6.50 per pound, and French Roast flavored coffee costs $7 per pound. The package contains the same amount of Hazelnut coffee as French Roast coffee. The cost of the 10-pound package is $66. How many pounds of each type of coffee are in the package?

51. *Mixture Problem* A chemist needs 12 gallons of a 20% acid solution. It is mixed from three solutions whose concentrations are 10%, 15%, and 25%. How many gallons of each solution will satisfy each condition?

(a) Use 4 gallons of the 25% solution.

(b) Use as little as possible of the 25% solution.

(c) Use as much as possible of the 25% solution.

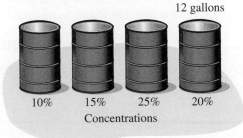

12 gallons

10%   15%   25%   20%
Concentrations

52. *Mixture Problem* A chemist needs 10 liters of a 25% acid solution. It is mixed from three solutions whose concentrations are 10%, 20%, and 50%. How many liters of each solution will satisfy each condition?

(a) Use 2 liters of the 50% solution.

(b) Use as little as possible of the 50% solution.

(c) Use as much as possible of the 50% solution.

53. *School Orchestra* The table shows the percents of each section of the North High School orchestra that were chosen to participate in the city orchestra, the county orchestra, and the state orchestra. Thirty members of the city orchestra, 17 members of the county orchestra, and 10 members of the state orchestra are from North High. How many members are in each section of North High's orchestra?

Orchestra	String	Wind	Percussion
City orchestra	40%	30%	50%
County orchestra	20%	25%	25%
State orchestra	10%	15%	25%

54. *Sports* The table shows the percents of each unit of the North High School football team that were chosen for academic honors, as city all-stars, and as county all-stars. Of all the players on the football team, 5 were awarded with academic honors, 13 were named city all-stars, and 4 were named county all-stars. How many members of each unit are there on the football team?

	Defense	Offense	Special teams
Academic honors	0%	10%	20%
City all-stars	10%	20%	50%
County all-stars	10%	0%	20%

Explaining Concepts

55. ✎ You apply Gaussian elimination to a system of three equations in the variables x, y, and z. From the row-echelon form, the solution $(1, -3, 4)$ is apparent *without* applying back-substitution or any other calculations. Explain why.

56. ✎ A system of three linear equations in three variables has an infinite number of solutions. Is it possible that the graphs of two of the three equations are parallel planes? Explain.

57. ✎ Two ways that a system of three linear equations in three variables can have no solution are shown in Figure 4.9 on page 245. Describe the graph for a third type of situation that results in no solution.

58. ✎ Describe the graphs and numbers of solutions possible for a system of three linear equations in three variables in which at least two of the equations are dependent.

59. ✎ Describe the graphs and numbers of solutions possible for a system of three linear equations in three variables if each pair of equations is consistent and *not* dependent.

60. Write a system of four linear equations in four unknowns, and use Gaussian elimination with back-substitution to solve it.

Cumulative Review

In Exercises 61–64, identify the terms and coefficients of the algebraic expression.

61. $3x + 2$

62. $4x^2 + 5x - 4$

63. $14t^5 - t + 25$

64. $5s^2 + 3st + 2t^2$

In Exercises 65–68, solve the system of linear equations by the method of elimination.

65. $\begin{cases} 2x + 3y = 17 \\ 4y = 12 \end{cases}$

66. $\begin{cases} x - 2y = 11 \\ 3x + 3y = 6 \end{cases}$

67. $\begin{cases} 3x - 4y = -30 \\ 5x + 4y = 14 \end{cases}$

68. $\begin{cases} 3x + 5y = 1 \\ 4x + 15y = 5 \end{cases}$

Mid-Chapter Quiz

Take this quiz as you would take a quiz in class. After you are done, check your work against the answers in the back of the book.

1. Determine whether each ordered pair is a solution of the system of linear equations: (a) $(1, -2)$ (b) $(10, 4)$

$$\begin{cases} 5x - 12y = 2 \\ 2x + 1.5y = 26 \end{cases}$$

In Exercises 2–4, graph the equations in the system. Use the graphs to determine the number of solutions of the system.

2. $\begin{cases} -6x + 9y = 9 \\ 2x - 3y = 6 \end{cases}$ 3. $\begin{cases} x - 2y = -4 \\ 3x - 2y = 4 \end{cases}$ 4. $\begin{cases} 0.5x - 1.5y = 7 \\ -2x + 6y = -28 \end{cases}$

In Exercises 5–7, use the graphical method to solve the system of equations.

5. $\begin{cases} x - y = 0 \\ 2x = 8 \end{cases}$ 6. $\begin{cases} 2x + 7y = 16 \\ 3x + 2y = 24 \end{cases}$ 7. $\begin{cases} 4x - y = 9 \\ x - 3y = 16 \end{cases}$

In Exercises 8–10, solve the system of equations by the method of substitution.

8. $\begin{cases} 2x - 3y = 4 \\ y = 2 \end{cases}$ 9. $\begin{cases} 5x - y = 32 \\ 6x - 9y = 18 \end{cases}$ 10. $\begin{cases} 6x - 2y = 2 \\ 9x - 3y = 1 \end{cases}$

In Exercises 11–14, use elimination or Gaussian elimination to solve the linear system.

11. $\begin{cases} x + 10y = 18 \\ 5x + 2y = 42 \end{cases}$ 12. $\begin{cases} x - 3y = 6 \\ 3x + y = 8 \end{cases}$

13. $\begin{cases} a + b + c = 1 \\ 4a + 2b + c = 2 \\ 9a + 3b + c = 4 \end{cases}$ 14. $\begin{cases} x + 4z = 17 \\ -3x + 2y - z = -20 \\ x - 5y + 3z = 19 \end{cases}$

In Exercises 15 and 16, write a system of linear equations having the given solution. (There are many correct answers.)

15. $(10, -12)$ 16. $(2, -5, 10)$

17. Twenty gallons of a 30% brine solution is obtained by mixing a 20% solution with a 50% solution. How many gallons of each solution are required?

18. In a triangle, the measure of one angle is 14° less than twice the measure of a second angle. The measure of the third angle is 30° greater than the measure of the second angle. Find the measures of the three angles.

4.4 Matrices and Linear Systems

© South West Images Scotland/Alamy

What You Should Learn

1 ▶ Determine the orders of matrices.

2 ▶ Form coefficient and augmented matrices and form linear systems from augmented matrices.

3 ▶ Perform elementary row operations to solve systems of linear equations in matrix form.

4 ▶ Use matrices and Gaussian elimination with back-substitution to solve systems of linear equations.

Why You Should Learn It

Systems of linear equations that model real-life situations can be solved using matrices. For instance, in Exercise 85 on page 265, the numbers of computer parts a company produces can be found using a matrix.

1 ▶ Determine the orders of matrices.

Matrices

In this section, you will study a streamlined technique for solving systems of linear equations. This technique involves the use of a rectangular array of real numbers called a **matrix.** (The plural of matrix is *matrices.*) Here is an example of a matrix.

$$
\begin{array}{c}

\begin{array}{cccc}
\text{Column} & \text{Column} & \text{Column} & \text{Column} \\
1 & 2 & 3 & 4
\end{array} \\
\begin{array}{c}
\text{Row 1} \\
\text{Row 2} \\
\text{Row 3}
\end{array}
\begin{bmatrix}
3 & -2 & 4 & 1 \\
0 & 1 & -1 & 2 \\
2 & 0 & -3 & 0
\end{bmatrix}
\end{array}
$$

This matrix has three rows and four columns, which means that its **order** is 3×4, which is read as "3 by 4." Each number in the matrix is an **entry** of the matrix.

EXAMPLE 1   Orders of Matrices

Determine the order of each matrix.

a. $\begin{bmatrix} 1 & -2 & 4 \\ 0 & 1 & -2 \end{bmatrix}$ **b.** $\begin{bmatrix} 0 & 0 \\ 0 & 0 \end{bmatrix}$ **c.** $\begin{bmatrix} 1 & -3 \\ -2 & 0 \\ 4 & -2 \end{bmatrix}$

Solution

a. This matrix has two rows and three columns, so the order is 2×3.

b. This matrix has two rows and two columns, so the order is 2×2.

c. This matrix has three rows and two columns, so the order is 3×2.

 **CHECKPOINT** *Now try Exercise 1.*

Smart Study Strategy

Go to page 216 for ways to *Make Note Cards.*

Study Tip

The order of a matrix is always given as *row by column.* A matrix with the same number of rows as columns is called a **square matrix.** For instance, the 2×2 matrix in Example 1(b) is square.

2 ▶ Form coefficient and augmented matrices and form linear systems from augmented matrices.

Augmented and Coefficient Matrices

A matrix derived from a system of linear equations (each written in standard form with the constant term on the right) is the **augmented matrix** of the system. Moreover, the matrix derived from the coefficients of the system (but not including the constant terms) is the **coefficient matrix** of the system. Here is an example.

System

$$\begin{cases} x - 4y + 3z = 5 \\ -x + 3y - z = -3 \\ 2x - 4z = 6 \end{cases}$$

Coefficient Matrix

$$\begin{bmatrix} 1 & -4 & 3 \\ -1 & 3 & -1 \\ 2 & 0 & -4 \end{bmatrix}$$

Augmented Matrix

$$\begin{bmatrix} 1 & -4 & 3 & \vdots & 5 \\ -1 & 3 & -1 & \vdots & -3 \\ 2 & 0 & -4 & \vdots & 6 \end{bmatrix}$$

> **Study Tip**
>
> Note the use of 0 for the missing y-variable in the third equation, and also note the fourth column of constant terms in the augmented matrix.

When forming either the coefficient matrix or the augmented matrix of a system, you should begin by vertically aligning the variables in the equations.

Given System

$$\begin{cases} x + 3y = 9 \\ -y + 4z = -2 \\ x - 5z = 0 \end{cases}$$

Align Variables

$$\begin{cases} x + 3y = 9 \\ -y + 4z = -2 \\ x - 5z = 0 \end{cases}$$

Form Augmented Matrix

$$\begin{bmatrix} 1 & 3 & 0 & \vdots & 9 \\ 0 & -1 & 4 & \vdots & -2 \\ 1 & 0 & -5 & \vdots & 0 \end{bmatrix}$$

EXAMPLE 2   Forming Coefficient and Augmented Matrices

Form the coefficient matrix and the augmented matrix for each system.

a. $\begin{cases} -x + 5y = 2 \\ 7x - 2y = -6 \end{cases}$ **b.** $\begin{cases} 3x + 2y - z = 1 \\ x + 2z = -3 \\ -2x - y = 4 \end{cases}$

Solution

| | *System* | *Coefficient Matrix* | *Augmented Matrix* |

a. $\begin{cases} -x + 5y = 2 \\ 7x - 2y = -6 \end{cases}$ $\begin{bmatrix} -1 & 5 \\ 7 & -2 \end{bmatrix}$ $\begin{bmatrix} -1 & 5 & \vdots & 2 \\ 7 & -2 & \vdots & -6 \end{bmatrix}$

b. $\begin{cases} 3x + 2y - z = 1 \\ x + 2z = -3 \\ -2x - y = 4 \end{cases}$ $\begin{bmatrix} 3 & 2 & -1 \\ 1 & 0 & 2 \\ -2 & -1 & 0 \end{bmatrix}$ $\begin{bmatrix} 3 & 2 & -1 & \vdots & 1 \\ 1 & 0 & 2 & \vdots & -3 \\ -2 & -1 & 0 & \vdots & 4 \end{bmatrix}$

✓ **CHECKPOINT** *Now try Exercise 11.*

EXAMPLE 3   Forming Linear Systems from Their Matrices

Write the system of linear equations represented by each matrix.

a. $\begin{bmatrix} 3 & -5 & \vdots & 4 \\ -1 & 2 & \vdots & 0 \end{bmatrix}$ **b.** $\begin{bmatrix} 1 & 3 & \vdots & 2 \\ 0 & 1 & \vdots & -3 \end{bmatrix}$ **c.** $\begin{bmatrix} 2 & 0 & -8 & \vdots & 1 \\ -1 & 1 & 1 & \vdots & 2 \\ 5 & -1 & 7 & \vdots & 3 \end{bmatrix}$

Solution

a. $\begin{cases} 3x - 5y = 4 \\ -x + 2y = 0 \end{cases}$ **b.** $\begin{cases} x + 3y = 2 \\ y = -3 \end{cases}$ **c.** $\begin{cases} 2x - 8z = 1 \\ -x + y + z = 2 \\ 5x - y + 7z = 3 \end{cases}$

✓ **CHECKPOINT** *Now try Exercise 17.*

3 ▶ Perform elementary row operations to solve systems of linear equations in matrix form.

Elementary Row Operations

In Section 4.3, you studied three operations that can be used on a system of linear equations to produce an equivalent system: (1) interchange two equations, (2) multiply an equation by a nonzero constant, and (3) add a multiple of an equation to another equation. In matrix terminology, these three operations correspond to **elementary row operations.**

Elementary Row Operations

Any of the following **elementary row operations** performed on an augmented matrix will produce a matrix that is row-equivalent to the original matrix. Two matrices are **row-equivalent** if one can be obtained from the other by a sequence of elementary row operations.

1. Interchange two rows.
2. Multiply a row by a nonzero constant.
3. Add a multiple of a row to another row.

EXAMPLE 4   Performing Elementary Row Operations

a. Interchange the first and second rows.

Original Matrix

$$\begin{bmatrix} 0 & 1 & 3 & 4 \\ -1 & 2 & 0 & 3 \\ 2 & -3 & 4 & 1 \end{bmatrix}$$

New Row-Equivalent Matrix

$$\begin{matrix} R_2 \\ R_1 \end{matrix} \begin{bmatrix} -1 & 2 & 0 & 3 \\ 0 & 1 & 3 & 4 \\ 2 & -3 & 4 & 1 \end{bmatrix}$$

b. Multiply the first row by $\frac{1}{2}$.

Original Matrix

$$\begin{bmatrix} 2 & -4 & 6 & -2 \\ 1 & 3 & -3 & 0 \\ 5 & -2 & 1 & 2 \end{bmatrix}$$

New Row-Equivalent Matrix

$$\tfrac{1}{2}R_1 \rightarrow \begin{bmatrix} 1 & -2 & 3 & -1 \\ 1 & 3 & -3 & 0 \\ 5 & -2 & 1 & 2 \end{bmatrix}$$

c. Add -2 times the first row to the third row.

Original Matrix

$$\begin{bmatrix} 1 & 2 & -4 & 3 \\ 0 & 3 & -2 & -1 \\ 2 & 1 & 5 & -2 \end{bmatrix}$$

New Row-Equivalent Matrix

$$\begin{bmatrix} 1 & 2 & -4 & 3 \\ 0 & 3 & -2 & -1 \\ -2R_1 + R_3 \rightarrow & 0 & -3 & 13 & -8 \end{bmatrix}$$

d. Add 6 times the first row to the second row.

Original Matrix

$$\begin{bmatrix} 1 & 2 & 2 & -4 \\ -6 & -11 & 3 & 18 \\ 0 & 0 & 4 & 7 \end{bmatrix}$$

New Row-Equivalent Matrix

$$6R_1 + R_2 \rightarrow \begin{bmatrix} 1 & 2 & 2 & -4 \\ 0 & 1 & 15 & -6 \\ 0 & 0 & 4 & 7 \end{bmatrix}$$

✓ **CHECKPOINT** *Now try Exercise 31.*

In Section 4.3, Gaussian elimination was used with back-substitution to solve systems of linear equations. Example 5 demonstrates the matrix version of Gaussian elimination. The two methods are essentially the same. The basic difference is that with matrices you do not need to keep writing the variables.

EXAMPLE 5   Solving a System of Linear Equations

Linear System

$$\begin{cases} x - 2y + 2z = 9 \\ -x + 3y \quad\quad = -4 \\ 2x - 5y + z = 10 \end{cases}$$

Associated Augmented Matrix

$$\left[\begin{array}{ccc:c} 1 & -2 & 2 & 9 \\ -1 & 3 & 0 & -4 \\ 2 & -5 & 1 & 10 \end{array}\right]$$

Add the first equation to the second equation.

$$\begin{cases} x - 2y + 2z = 9 \\ y + 2z = 5 \\ 2x - 5y + z = 10 \end{cases}$$

Add the first row to the second row.

$$R_1 + R_2 \rightarrow \left[\begin{array}{ccc:c} 1 & -2 & 2 & 9 \\ 0 & 1 & 2 & 5 \\ 2 & -5 & 1 & 10 \end{array}\right]$$

Add -2 times the first equation to the third equation.

$$\begin{cases} x - 2y + 2z = 9 \\ y + 2z = 5 \\ -y - 3z = -8 \end{cases}$$

Add -2 times the first row to the third row.

$$-2R_1 + R_3 \rightarrow \left[\begin{array}{ccc:c} 1 & -2 & 2 & 9 \\ 0 & 1 & 2 & 5 \\ 0 & -1 & -3 & -8 \end{array}\right]$$

Add the second equation to the third equation.

$$\begin{cases} x - 2y + 2z = 9 \\ y + 2z = 5 \\ -z = -3 \end{cases}$$

Add the second row to the third row.

$$R_2 + R_3 \rightarrow \left[\begin{array}{ccc:c} 1 & -2 & 2 & 9 \\ 0 & 1 & 2 & 5 \\ 0 & 0 & -1 & -3 \end{array}\right]$$

Multiply the third equation by -1.

$$\begin{cases} x - 2y + 2z = 9 \\ y + 2z = 5 \\ z = 3 \end{cases}$$

Multiply the third row by -1.

$$-R_3 \rightarrow \left[\begin{array}{ccc:c} 1 & -2 & 2 & 9 \\ 0 & 1 & 2 & 5 \\ 0 & 0 & 1 & 3 \end{array}\right]$$

At this point, you can use back-substitution to find that the solution is $x = 1$, $y = -1$, and $z = 3$. The solution can be written as the ordered triple $(1, -1, 3)$.

✓ **CHECKPOINT** *Now try Exercise 53.*

Definition of Row-Echelon Form of a Matrix

A matrix in **row-echelon form** has the following properties.

1. All rows consisting entirely of zeros occur at the bottom of the matrix.
2. For each row that does not consist entirely of zeros, the first nonzero entry is 1 (called a **leading 1**).
3. For two successive (nonzero) rows, the leading 1 in the higher row is farther to the left than the leading 1 in the lower row.

4 ▶ Use matrices and Gaussian elimination with back-substitution to solve systems of linear equations.

Solving a System of Linear Equations

> ### Gaussian Elimination with Back-Substitution
>
> To use matrices and Gaussian elimination to solve a system of linear equations, use the following steps.
>
> 1. Write the augmented matrix of the system of linear equations.
> 2. Use elementary row operations to rewrite the augmented matrix in row-echelon form.
> 3. Write the system of linear equations corresponding to the matrix in row-echelon form, and use back-substitution to find the solution.

When you perform Gaussian elimination with back-substitution, you should operate from *left to right by columns*, using elementary row operations to obtain zeros in all entries directly below the leading 1's.

EXAMPLE 6   Gaussian Elimination with Back-Substitution

Solve the system of linear equations.

$$\begin{cases} 2x - 3y = -2 \\ x + 2y = 13 \end{cases}$$

Solution

$$\begin{bmatrix} 2 & -3 & \vdots & -2 \\ 1 & 2 & \vdots & 13 \end{bmatrix}$$ Augmented matrix for system of linear equations

$$\begin{matrix} R_2 \\ R_1 \end{matrix} \begin{bmatrix} 1 & 2 & \vdots & 13 \\ 2 & -3 & \vdots & -2 \end{bmatrix}$$ First column has leading 1 in upper left corner.

$$-2R_1 + R_2 \rightarrow \begin{bmatrix} 1 & 2 & \vdots & 13 \\ 0 & -7 & \vdots & -28 \end{bmatrix}$$ First column has a zero under its leading 1.

$$-\tfrac{1}{7}R_2 \rightarrow \begin{bmatrix} 1 & 2 & \vdots & 13 \\ 0 & 1 & \vdots & 4 \end{bmatrix}$$ Second column has leading 1 in second row.

The system of linear equations that corresponds to the (row-echelon) matrix is

$$\begin{cases} x + 2y = 13 \\ y = 4. \end{cases}$$

Using back-substitution, you can find that the solution of the system is $x = 5$ and $y = 4$, which can be written as the ordered pair $(5, 4)$. Check this solution in the original system, as follows.

Check

Equation 1: $2(5) - 3(4) = -2$ ✓

Equation 2: $5 + 2(4) = 13$ ✓

✓ **CHECKPOINT** *Now try Exercise 55.*

EXAMPLE 7   **Gaussian Elimination with Back-Substitution**

Solve the system of linear equations.

$$\begin{cases} 3x + 3y & = 9 \\ 2x & - 3z = 10 \\ & 6y + 4z = -12 \end{cases}$$

Solution

$$\begin{bmatrix} 3 & 3 & 0 & \vdots & 9 \\ 2 & 0 & -3 & \vdots & 10 \\ 0 & 6 & 4 & \vdots & -12 \end{bmatrix}$$ Augmented matrix for system of linear equations

$$\frac{1}{3}R_1 \rightarrow \begin{bmatrix} 1 & 1 & 0 & \vdots & 3 \\ 2 & 0 & -3 & \vdots & 10 \\ 0 & 6 & 4 & \vdots & -12 \end{bmatrix}$$ First column has leading 1 in upper left corner.

$$-2R_1 + R_2 \rightarrow \begin{bmatrix} 1 & 1 & 0 & \vdots & 3 \\ 0 & -2 & -3 & \vdots & 4 \\ 0 & 6 & 4 & \vdots & -12 \end{bmatrix}$$ First column has zeros under its leading 1.

$$-\frac{1}{2}R_2 \rightarrow \begin{bmatrix} 1 & 1 & 0 & \vdots & 3 \\ 0 & 1 & \frac{3}{2} & \vdots & -2 \\ 0 & 6 & 4 & \vdots & -12 \end{bmatrix}$$ Second column has leading 1 in second row.

$$-6R_2 + R_3 \rightarrow \begin{bmatrix} 1 & 1 & 0 & \vdots & 3 \\ 0 & 1 & \frac{3}{2} & \vdots & -2 \\ 0 & 0 & -5 & \vdots & 0 \end{bmatrix}$$ Second column has zero under its leading 1.

$$-\frac{1}{5}R_3 \rightarrow \begin{bmatrix} 1 & 1 & 0 & \vdots & 3 \\ 0 & 1 & \frac{3}{2} & \vdots & -2 \\ 0 & 0 & 1 & \vdots & 0 \end{bmatrix}$$ Third column has leading 1 in third row.

The system of linear equations that corresponds to this (row-echelon) matrix is

$$\begin{cases} x + y & = 3 \\ y + \frac{3}{2}z = -2 \\ z = 0. \end{cases}$$

Using back-substitution, you can find that the solution is

$$x = 5, \quad y = -2, \quad \text{and} \quad z = 0$$

which can be written as the ordered triple $(5, -2, 0)$. Check this in the original system, as follows.

Check

Equation 1: $3(5) + 3(-2)$ $= \quad 9$ ✓

Equation 2: $2(5)$ $- 3(0) = \quad 10$ ✓

Equation 3: $6(-2) + 4(0) = -12$ ✓

 **CHECKPOINT** *Now try Exercise 65.*

EXAMPLE 8   **A System with No Solution**

Solve the system of linear equations.

$$\begin{cases} 6x - 10y = -4 \\ 9x - 15y = 5 \end{cases}$$

Solution

$$\begin{bmatrix} 6 & -10 & \vdots & -4 \\ 9 & -15 & \vdots & 5 \end{bmatrix}$$ Augmented matrix for system of linear equations

$$\tfrac{1}{6}R_1 \rightarrow \begin{bmatrix} 1 & -\tfrac{5}{3} & \vdots & -\tfrac{2}{3} \\ 9 & -15 & \vdots & 5 \end{bmatrix}$$ First column has leading 1 in upper left corner.

$$-9R_1 + R_2 \rightarrow \begin{bmatrix} 1 & -\tfrac{5}{3} & \vdots & -\tfrac{2}{3} \\ 0 & 0 & \vdots & 11 \end{bmatrix}$$ First column has a zero under its leading 1.

The "equation" that corresponds to the second row of this matrix is $0 = 11$. Because this is a false statement, the system of equations has no solution.

 CHECKPOINT *Now try Exercise 59.*

EXAMPLE 9   **A System with Infinitely Many Solutions**

Solve the system of linear equations.

$$\begin{cases} 12x - 6y = -3 \\ -8x + 4y = 2 \end{cases}$$

Solution

$$\begin{bmatrix} 12 & -6 & \vdots & -3 \\ -8 & 4 & \vdots & 2 \end{bmatrix}$$ Augmented matrix for system of linear equations

$$\tfrac{1}{12}R_1 \rightarrow \begin{bmatrix} 1 & -\tfrac{1}{2} & \vdots & -\tfrac{1}{4} \\ -8 & 4 & \vdots & 2 \end{bmatrix}$$ First column has leading 1 in upper left corner.

$$8R_1 + R_2 \rightarrow \begin{bmatrix} 1 & -\tfrac{1}{2} & \vdots & -\tfrac{1}{4} \\ 0 & 0 & \vdots & 0 \end{bmatrix}$$ First column has a zero under its leading 1.

Because the second row of the matrix is all zeros, the system of equations has an infinite number of solutions, represented by all points (x, y) on the line

$$x - \frac{1}{2}y = -\frac{1}{4}.$$

Because this line can be written as

$$x = \frac{1}{2}y - \frac{1}{4}$$

you can write the solution set as

$$\left(\frac{1}{2}a - \frac{1}{4}, a \right),$$ where a is any real number.

 **CHECKPOINT** *Now try Exercise 63.*

EXAMPLE 10   **Investment Portfolio**

You have a portfolio totaling $219,000 and want to invest in municipal bonds, blue-chip stocks, and growth or speculative stocks. The municipal bonds pay 6% annually. Over a five-year period, you expect blue-chip stocks to return 10% annually and growth stocks to return 15% annually. You want a combined annual return of 8%, and you also want to have only one-fourth of the portfolio invested in stocks. How much should be allocated to each type of investment?

Solution

Let M, B, and G represent the amounts invested in municipal bonds, blue-chip stocks, and growth stocks, respectively. This situation is represented by the following system.

$$\begin{cases} M + B + G = 219{,}000 & \text{Equation 1: Total investment is \$219,000.} \\ 0.06M + 0.10B + 0.15G = 17{,}520 & \text{Equation 2: Combined annual return is 8\%.} \\ B + G = 54{,}750 & \text{Equation 3: } \tfrac{1}{4} \text{ of investment is allocated} \\ & \text{to stocks.} \end{cases}$$

Form the augmented matrix for this system of equations, and then use elementary row operations to obtain the row-echelon form of the matrix.

$$\begin{bmatrix} 1 & 1 & 1 & \vdots & 219{,}000 \\ 0.06 & 0.10 & 0.15 & \vdots & 17{,}520 \\ 0 & 1 & 1 & \vdots & 54{,}750 \end{bmatrix}$$ Augmented matrix for system of linear equations

$-0.06R_1 + R_2 \longrightarrow \begin{bmatrix} 1 & 1 & 1 & \vdots & 219{,}000 \\ 0 & 0.04 & 0.09 & \vdots & 4{,}380 \\ 0 & 1 & 1 & \vdots & 54{,}750 \end{bmatrix}$ First column has zeros under its leading 1.

$25R_2 \longrightarrow \begin{bmatrix} 1 & 1 & 1 & \vdots & 219{,}000 \\ 0 & 1 & 2.25 & \vdots & 109{,}500 \\ 0 & 1 & 1 & \vdots & 54{,}750 \end{bmatrix}$ Second column has leading 1 in second row.

$-R_2 + R_3 \longrightarrow \begin{bmatrix} 1 & 1 & 1 & \vdots & 219{,}000 \\ 0 & 1 & 2.25 & \vdots & 109{,}500 \\ 0 & 0 & -1.25 & \vdots & -54{,}750 \end{bmatrix}$ Second column has zero under its leading 1.

$-0.8R_3 \longrightarrow \begin{bmatrix} 1 & 1 & 1 & \vdots & 219{,}000 \\ 0 & 1 & 2.25 & \vdots & 109{,}500 \\ 0 & 0 & 1 & \vdots & 43{,}800 \end{bmatrix}$ Third column has leading 1 in third row and matrix is in row-echelon form.

From the row-echelon form, you can see that $G = 43{,}800$. By back-substituting G into the revised second equation, you can determine the value of B.

$$B + 2.25(43{,}800) = 109{,}500 \quad \Longrightarrow \quad B = 10{,}950$$

By back-substituting B and G into Equation 1, you can solve for M.

$$M + 10{,}950 + 43{,}800 = 219{,}000 \quad \Longrightarrow \quad M = 164{,}250$$

So, you should invest $164,250 in municipal bonds, $10,950 in blue-chip stocks, and $43,800 in growth or speculative stocks. Check this solution by substituting these values into the original system of equations.

 **CHECKPOINT**   *Now try Exercise 81.*

——————————————— **Concept Check** ———————————————

1. A matrix contains exactly four entries. What are the possible orders of the matrix? State the numbers of rows and columns in each possible order.

3. What is the primary difference between performing row operations on a system of equations and performing elementary row operations?

2. For a given system of equations, which has more entries, the coefficient matrix or the augmented matrix? Explain.

4. After using matrices to perform Gaussian elimination, what steps are generally needed to find the solution of the original system of equations?

4.4 EXERCISES

Go to pages 288–289 to record your assignments.

——————————————— **Developing Skills** ———————————————

In Exercises 1–10, determine the order of the matrix. *See Example 1.*

✓ 1. $\begin{bmatrix} 3 & -2 \\ -4 & 0 \\ 2 & -7 \\ -1 & -3 \end{bmatrix}$

2. $\begin{bmatrix} 3 & 4 \\ 2 & -1 \\ 8 & 10 \\ -6 & -6 \\ 12 & 50 \end{bmatrix}$

3. $\begin{bmatrix} -2 & 5 \\ 0 & -1 \end{bmatrix}$

4. $\begin{bmatrix} 5 & -8 & 32 \\ 7 & 15 & 28 \end{bmatrix}$

5. $\begin{bmatrix} 4 \\ -2 \\ 0 \\ 1 \end{bmatrix}$

6. $\begin{bmatrix} 4 & 0 & -5 \\ -1 & 8 & 9 \\ 0 & -3 & 4 \end{bmatrix}$

7. $\begin{bmatrix} 5 \end{bmatrix}$

8. $\begin{bmatrix} 1 & -1 & 2 & 3 \end{bmatrix}$

9. $\begin{bmatrix} 13 & 12 & -9 & 0 \end{bmatrix}$

10. $\begin{bmatrix} 6 \\ -13 \\ 22 \end{bmatrix}$

In Exercises 11–16, form (a) the coefficient matrix and (b) the augmented matrix for the system of linear equations. *See Example 2.*

✓ 11. $\begin{cases} 4x - 5y = -2 \\ -x + 8y = 10 \end{cases}$

12. $\begin{cases} 8x + 3y = 25 \\ 3x - 9y = 12 \end{cases}$

13. $\begin{cases} x + y = 0 \\ 5x - 2y - 2z = 12 \\ 2x + 4y + z = 5 \end{cases}$

14. $\begin{cases} 9x - 3y + z = 13 \\ 12x - 8z = 5 \\ 3x + 4y - z = 6 \end{cases}$

15. $\begin{cases} 5x + y - 3z = 7 \\ 2y + 4z = 12 \end{cases}$

16. $\begin{cases} 10x + 6y - 8z = -4 \\ -4x - 7y = 9 \end{cases}$

In Exercises 17–24, write the system of linear equations represented by the augmented matrix. (Use variables x, y, z, and w.) *See Example 3.*

✓ 17. $\begin{bmatrix} 4 & 3 & \vdots & 8 \\ 1 & -2 & \vdots & 3 \end{bmatrix}$

18. $\begin{bmatrix} 9 & -4 & \vdots & 0 \\ 6 & 1 & \vdots & -4 \end{bmatrix}$

19. $\begin{bmatrix} 1 & 0 & 2 & \vdots & -10 \\ 0 & 3 & -1 & \vdots & 5 \\ 4 & 2 & 0 & \vdots & 3 \end{bmatrix}$

20. $\begin{bmatrix} 4 & -1 & 3 & \vdots & 5 \\ 2 & 0 & -2 & \vdots & -1 \\ -1 & 6 & 0 & \vdots & 3 \end{bmatrix}$

21. $\begin{bmatrix} 5 & 8 & 2 & 0 & \vdots & -1 \\ -2 & 15 & 5 & 1 & \vdots & 9 \\ 1 & 6 & -7 & 0 & \vdots & -3 \end{bmatrix}$

22. $\begin{bmatrix} 0 & 1 & -5 & 8 & \vdots & 10 \\ 2 & 4 & -1 & 0 & \vdots & 15 \\ 1 & 1 & 7 & 9 & \vdots & -8 \end{bmatrix}$

23. $\begin{bmatrix} 13 & 1 & 4 & -2 & \vdots & -4 \\ 5 & 4 & 0 & -1 & \vdots & 0 \\ 1 & 2 & 6 & 8 & \vdots & 5 \\ -10 & 12 & 3 & 1 & \vdots & -2 \end{bmatrix}$

24. $\begin{bmatrix} 7 & 3 & -2 & 4 & \vdots & 2 \\ -1 & 0 & 4 & -1 & \vdots & 6 \\ 8 & 3 & 0 & 0 & \vdots & -4 \\ 0 & 2 & -4 & 3 & \vdots & 12 \end{bmatrix}$

In Exercises 25–30, describe the elementary row operation used to transform the first matrix into the second matrix. *See Examples 4 and 5.*

25. $\begin{bmatrix} 0 & 3 & -2 \\ 2 & 5 & -7 \end{bmatrix} \Rightarrow \begin{bmatrix} 2 & 5 & -7 \\ 0 & 3 & -2 \end{bmatrix}$

26. $\begin{bmatrix} 1 & 6 & 7 \\ 0 & 1 & 2 \\ 0 & -3 & -4 \end{bmatrix} \Rightarrow \begin{bmatrix} 1 & 6 & 7 \\ 0 & 1 & 2 \\ 0 & 0 & 2 \end{bmatrix}$

27. $\begin{bmatrix} -3 & 6 & 9 \\ 5 & 6 & 7 \end{bmatrix} \Rightarrow \begin{bmatrix} 1 & -2 & -3 \\ 5 & 6 & 7 \end{bmatrix}$

28. $\begin{bmatrix} \frac{1}{3} & 1 & 4 \\ -7 & 2 & 5 \end{bmatrix} \Rightarrow \begin{bmatrix} 1 & 3 & 12 \\ -7 & 2 & 5 \end{bmatrix}$

29. $\begin{bmatrix} 1 & 3 & 2 \\ -3 & 4 & 2 \\ -5 & 6 & -7 \end{bmatrix} \Rightarrow \begin{bmatrix} 1 & 3 & 2 \\ -3 & 4 & 2 \\ 0 & 21 & 3 \end{bmatrix}$

30. $\begin{bmatrix} 0 & -2 & 6 \\ 2 & 5 & 3 \end{bmatrix} \Rightarrow \begin{bmatrix} 2 & 5 & 3 \\ 0 & -2 & 6 \end{bmatrix}$

In Exercises 31–36, fill in the entries of the row-equivalent matrix formed by performing the indicated elementary row operation. *See Example 4.*

31. $\begin{bmatrix} 1 & 1 & -4 & 2 \\ 0 & 0 & 8 & 3 \\ 0 & 4 & 5 & 5 \end{bmatrix} \begin{matrix} \\ R_3 \\ R_2 \end{matrix} \begin{bmatrix} \ & \ & \ & \ \\ \ & \ & \ & \ \\ \ & \ & \ & \ \end{bmatrix}$

32. $\begin{bmatrix} 0 & 0 & -5 & 2 \\ 0 & -7 & -3 & 3 \\ 1 & 4 & 5 & 4 \end{bmatrix} \begin{matrix} R_3 \\ \\ R_1 \end{matrix} \begin{bmatrix} \ & \ & \ & \ \\ \ & \ & \ & \ \\ \ & \ & \ & \ \end{bmatrix}$

33. $\begin{bmatrix} 9 & -18 & 27 \\ 3 & 4 & 5 \end{bmatrix} \begin{matrix} \frac{1}{9}R_1 \rightarrow \end{matrix} \begin{bmatrix} \ & \ & \ \\ \ & \ & \ \end{bmatrix}$

34. $\begin{bmatrix} 1 & 21 & 7 \\ 0 & -7 & 14 \end{bmatrix} \begin{matrix} -\frac{1}{7}R_2 \rightarrow \end{matrix} \begin{bmatrix} \ & \ & \ \\ \ & \ & \ \end{bmatrix}$

35. $\begin{bmatrix} 1 & 4 & 3 \\ 2 & 8 & 6 \end{bmatrix} \begin{matrix} -2R_1 + R_2 \rightarrow \end{matrix} \begin{bmatrix} \ & \ & \ \\ \ & \ & \ \end{bmatrix}$

36. $\begin{bmatrix} 1 & 4 & 5 \\ 4 & -7 & 3 \end{bmatrix} \begin{matrix} -4R_1 + R_2 \rightarrow \end{matrix} \begin{bmatrix} \ & \ & \ \\ \ & \ & \ \end{bmatrix}$

In Exercises 37–42, convert the matrix to row-echelon form. (There are many correct answers.)

37. $\begin{bmatrix} 1 & 2 & 3 \\ 2 & -1 & -4 \end{bmatrix}$

38. $\begin{bmatrix} 1 & 3 & 6 \\ -4 & -9 & 3 \end{bmatrix}$

39. $\begin{bmatrix} 4 & 6 & 1 \\ -2 & 2 & 5 \end{bmatrix}$

40. $\begin{bmatrix} 3 & 2 & 6 \\ 2 & 3 & -3 \end{bmatrix}$

41. $\begin{bmatrix} 1 & 1 & 0 & 5 \\ -2 & -1 & 2 & -10 \\ 3 & 6 & 7 & 14 \end{bmatrix}$

42. $\begin{bmatrix} 1 & 2 & -1 & 3 \\ 3 & 7 & -5 & 14 \\ -2 & -1 & -3 & 8 \end{bmatrix}$

In Exercises 43–46, use the matrix capabilities of a graphing calculator to write the matrix in row-echelon form. (There are many correct answers.)

43. $\begin{bmatrix} 1 & -1 & -1 & 1 \\ 4 & -4 & 1 & 8 \\ -6 & 8 & 18 & 0 \end{bmatrix}$

44. $\begin{bmatrix} 1 & -3 & 0 & -7 \\ -3 & 10 & 1 & 23 \\ 4 & -10 & 2 & -24 \end{bmatrix}$

45. $\begin{bmatrix} 1 & 1 & -1 & 3 \\ 2 & 1 & 2 & 5 \\ 3 & 2 & 1 & 8 \end{bmatrix}$

46. $\begin{bmatrix} 1 & -3 & -2 & -8 \\ 1 & 3 & -2 & 17 \\ 1 & 2 & -2 & -5 \end{bmatrix}$

In Exercises 47–52, write the system of linear equations represented by the augmented matrix. Then use back-substitution to find the solution. (Use variables x, y, and z.)

47. $\begin{bmatrix} 1 & -2 & \vdots & 4 \\ 0 & 1 & \vdots & -3 \end{bmatrix}$ **48.** $\begin{bmatrix} 1 & 5 & \vdots & 0 \\ 0 & 1 & \vdots & -1 \end{bmatrix}$

49. $\begin{bmatrix} 1 & 5 & \vdots & 3 \\ 0 & 1 & \vdots & -2 \end{bmatrix}$ **50.** $\begin{bmatrix} 1 & 5 & -3 & \vdots & 0 \\ 0 & 1 & 0 & \vdots & 6 \\ 0 & 0 & 1 & \vdots & -5 \end{bmatrix}$

51. $\begin{bmatrix} 1 & -1 & 2 & \vdots & 4 \\ 0 & 1 & -1 & \vdots & 2 \\ 0 & 0 & 1 & \vdots & -2 \end{bmatrix}$

52. $\begin{bmatrix} 1 & 2 & -2 & \vdots & -1 \\ 0 & 1 & 1 & \vdots & 9 \\ 0 & 0 & 1 & \vdots & -3 \end{bmatrix}$

In Exercises 53–78, use matrices to solve the system of linear equations. *See Examples 5–9.*

✓ 53. $\begin{cases} x + 2y = 7 \\ 3x - 7y = 8 \end{cases}$ **54.** $\begin{cases} 2x + 6y = 16 \\ 2x + 3y = 7 \end{cases}$

✓ 55. $\begin{cases} 6x - 4y = 2 \\ 5x + 2y = 7 \end{cases}$ **56.** $\begin{cases} x - 3y = 5 \\ -2x + 6y = -10 \end{cases}$

57. $\begin{cases} 12x + 10y = -14 \\ 4x - 3y = -11 \end{cases}$ **58.** $\begin{cases} -x - 5y = -10 \\ 2x - 3y = 7 \end{cases}$

✓ 59. $\begin{cases} -x + 2y = 1.5 \\ 2x - 4y = 3 \end{cases}$ **60.** $\begin{cases} 2x - y = -0.1 \\ 3x + 2y = 1.6 \end{cases}$

61. $\begin{cases} x - 2y - z = 6 \\ y + 4z = 5 \\ 4x + 2y + 3z = 8 \end{cases}$ **62.** $\begin{cases} x - 3z = -2 \\ 3x + y - 2z = 5 \\ 2x + 2y + z = 4 \end{cases}$

✓ 63. $\begin{cases} x + y - 5z = 3 \\ x - 2z = 1 \\ 2x - y - z = 0 \end{cases}$ **64.** $\begin{cases} 2y + z = 3 \\ -4y - 2z = 0 \\ x + y + z = 2 \end{cases}$

✓ 65. $\begin{cases} 2x + 4y = 10 \\ 2x + 2y + 3z = 3 \\ -3x + y + 2z = -3 \end{cases}$

66. $\begin{cases} 2x - y + 3z = 24 \\ 2y - z = 14 \\ 7x - 5y = 6 \end{cases}$ **67.** $\begin{cases} x - 3y + 2z = 8 \\ 2y - z = -4 \\ x + z = 3 \end{cases}$

68. $\begin{cases} 2x + 3z = 3 \\ 4x - 3y + 7z = 5 \\ 8x - 9y + 15z = 9 \end{cases}$

69. $\begin{cases} -2x - 2y - 15z = 0 \\ x + 2y + 2z = 18 \\ 3x + 3y + 22z = 2 \end{cases}$

70. $\begin{cases} 2x + 4y + 5z = 5 \\ x + 3y + 3z = 2 \\ 2x + 4y + 4z = 2 \end{cases}$ **71.** $\begin{cases} 2x + 4z = 1 \\ x + y + 3z = 0 \\ x + 3y + 5z = 0 \end{cases}$

72. $\begin{cases} 3x + y - 2z = 2 \\ 6x + 2y - 4z = 1 \\ -3x - y + 2z = 1 \end{cases}$ **73.** $\begin{cases} x + 3y = 2 \\ 2x + 6y = 4 \\ 2x + 5y + 4z = 3 \end{cases}$

76. $\begin{cases} 2x + 2y + z = 8 \\ 2x + 3y + z = 7 \\ 6x + 8y + 3z = 22 \end{cases}$

74. $\begin{cases} 4x + 3y = 10 \\ 2x - y = 10 \\ -2x + z = -9 \end{cases}$

77. $\begin{cases} 2x + y - 2z = 4 \\ 3x - 2y + 4z = 6 \\ -4x + y + 6z = 12 \end{cases}$

78. $\begin{cases} 3x + 3y + z = 4 \\ 2x + 6y + z = 5 \\ -x - 3y + 2z = -5 \end{cases}$

75. $\begin{cases} 4x - y + z = 4 \\ -6x + 3y - 2z = -5 \\ 2x + 5y - z = 7 \end{cases}$

Solving Problems

79. *Investment*   A corporation borrows $1,500,000 to expand its line of clothing. Some of the money is borrowed at 8%, some at 9%, and the remainder at 12%. The annual interest payment to the lenders is $133,000. The amount borrowed at 8% is four times the amount borrowed at 12%. How much is borrowed at each rate?

80. *Investment*   An inheritance of $25,000 is divided among three investments yielding a total of $1890 in simple interest per year. The interest rates for the three investments are 5%, 7%, and 10%. The 5% and 7% investments are $2000 and $3000 less than the 10% investment, respectively. Find the amount placed in each investment.

✓**81.** *Ticket Sales*   A theater owner wants to sell 1500 total tickets at his three theaters for a total revenue of $10,050. Tickets cost $1.50 at Theater A, $7.50 at Theater B, and $8.50 at Theater C. Theaters B and C each have twice as many seats as Theater A. How many tickets must be sold at each theater to reach the owner's goal?

82. *Nut Mixture*   A grocer wishes to mix three kinds of nuts to obtain 50 pounds of a mixture priced at $4.10 per pound. Peanuts cost $3.00 per pound, pecans cost $4.00 per pound, and cashews cost $6.00 per pound. Three-quarters of the mixture is composed of peanuts and pecans. How many pounds of each variety should the grocer use?

83. *Number Problem*   The sum of three positive numbers is 33. The second number is 3 greater than the first, and the third is four times the first. Find the three numbers.

84. *Number Problem*   The sum of three positive numbers is 24. The second number is 4 greater than the first, and the third is three times the first. Find the three numbers.

85. *Production*   A company produces computer chips, resistors, and transistors. Each computer chip requires 2 units of copper, 2 units of zinc, and 1 unit of glass. Each resistor requires 1 unit of copper, 3 units of zinc, and 2 units of glass. Each transistor requires 3 units of copper, 2 units of zinc, and 2 units of glass. There are 70 units of copper, 80 units of zinc, and 55 units of glass available for use. Find the numbers of computer chips, resistors, and transistors the company can produce.

86. *Production*   A gourmet baked goods company specializes in chocolate muffins, chocolate cookies, and chocolate brownies. Each muffin requires 2 units of chocolate, 3 units of flour, and 2 units of sugar. Each cookie requires 1 unit of chocolate, 1 unit of flour, and 1 unit of sugar. Each brownie requires 2 units of chocolate, 1 unit of flour, and 1.5 units of sugar. There are 550 units of chocolate, 525 units of flour, and 500 units of sugar available for use. Find the numbers of chocolate muffins, chocolate cookies, and chocolate brownies the company can produce.

Investment Portfolio   In Exercises 87 and 88, consider an investor with a portfolio totaling $500,000 that is to be allocated among the following types of investments: certificates of deposit, municipal bonds, blue-chip stocks, and growth or speculative stocks. Use the given conditions to find expressions for the amounts that can be invested in each type of stock. Then find the other amounts when the amount invested in growth stocks is $100,000.

87. The certificates of deposit pay 10% annually, and the municipal bonds pay 8% annually. Over a five-year period, the investor expects the blue-chip stocks to return 12% annually and the growth stocks to return 13% annually. The investor wants a combined annual return of 10% and also wants to have only one-fourth of the portfolio invested in stocks.

88. The certificates of deposit pay 9% annually, and the municipal bonds pay 5% annually. Over a five-year period, the investor expects the blue-chip stocks to return 12% annually and the growth stocks to return 14% annually. The investor wants a combined annual return of 10% and also wants to have only one-fourth of the portfolio invested in stocks.

Explaining Concepts

89. The entries in a matrix consist of the whole numbers from 1 to 15. The matrix has more than one row and there are more columns than rows. What is the order of the matrix? Explain.

90. Give an example of a matrix in *row-echelon form*. There are many correct answers.

91. ✎ Describe the row-echelon form of an augmented matrix that corresponds to a system of linear equations that is inconsistent.

92. ✎ Describe the row-echelon form of an augmented matrix that corresponds to a system of linear equations that has an infinite number of solutions.

93. An augmented matrix in row-echelon form represents a system of three variables in three equations that has exactly one solution. The matrix has six nonzero entries, and three of them are in the last column. Discuss the possible entries in the first three columns of this matrix.

94. ✎ An augmented matrix in row-echelon form represents a system of three variables in three equations with exactly one solution. What is the smallest number of nonzero entries that this matrix can have? Explain.

Cumulative Review

In Exercises 95–98, evaluate the expression.

95. $6(-7)$

96. $45 \div (-5)$

97. $5(4) - 3(-2)$

98. $\dfrac{(-45) - (-20)}{-5}$

In Exercises 99 and 100, solve the system of linear equations.

99. $\begin{cases} x \qquad\quad = 4 \\ \quad 3y + 2z = -4 \\ x + y + z = 3 \end{cases}$

100. $\begin{cases} \quad x - 2y - 3z = 4 \\ \quad 2x + 2y + z = -4 \\ -2x \qquad + z = 0 \end{cases}$

4.5 Determinants and Linear Systems

Keven R. Morris/CORBIS

What You Should Learn

1 ▶ Find determinants of 2×2 matrices and 3×3 matrices.

2 ▶ Use determinants and Cramer's Rule to solve systems of linear equations.

3 ▶ Use determinants to find areas of triangles, to test for collinear points, and to find equations of lines.

Why You Should Learn It

You can use determinants and matrices to model and solve real-life problems. For instance, in Exercise 71 on page 277, you can use a matrix to estimate the area of a region of land.

1 ▶ Find determinants of 2×2 matrices and 3×3 matrices.

The Determinant of a Matrix

Associated with each square matrix is a real number called its **determinant.** The use of determinants arose from special number patterns that occur during the solution of systems of linear equations. For instance, the system

$$\begin{cases} a_1x + b_1y = c_1 \\ a_2x + b_2y = c_2 \end{cases}$$

has a solution given by

$$x = \frac{b_2c_1 - b_1c_2}{a_1b_2 - a_2b_1} \qquad \text{and} \qquad y = \frac{a_1c_2 - a_2c_1}{a_1b_2 - a_2b_1}$$

provided that $a_1b_2 - a_2b_1 \neq 0$. Note that the denominator of each fraction is the same. This denominator is called the **determinant** of the coefficient matrix of the system.

Coefficient Matrix

$$A = \begin{bmatrix} a_1 & b_1 \\ a_2 & b_2 \end{bmatrix}$$

Determinant

$$\det(A) = a_1b_2 - a_2b_1$$

The determinant of the matrix A can also be denoted by vertical bars on both sides of the matrix, as indicated in the following definition.

Study Tip

Note that $\det(A)$ and $|A|$ are used interchangeably to represent the determinant of A. Although vertical bars are also used to denote the absolute value of a real number, the context will show which use is intended.

Definition of the Determinant of a 2×2 Matrix

$$\det(A) = |A| = \begin{vmatrix} a_1 & b_1 \\ a_2 & b_2 \end{vmatrix} = a_1b_2 - a_2b_1$$

A convenient method for remembering the formula for the determinant of a 2×2 matrix is shown in the diagram below.

$$\det(A) = \begin{vmatrix} a_1 & b_1 \\ a_2 & b_2 \end{vmatrix} = a_1b_2 - a_2b_1$$

Note that the determinant is given by the difference of the products of the two diagonals of the matrix.

EXAMPLE 1   **The Determinant of a 2 × 2 Matrix**

Find the determinant of each matrix.

a. $A = \begin{bmatrix} 2 & -3 \\ 1 & 4 \end{bmatrix}$ **b.** $B = \begin{bmatrix} -1 & 2 \\ 2 & -4 \end{bmatrix}$ **c.** $C = \begin{bmatrix} 1 & 3 \\ 2 & 5 \end{bmatrix}$

Solution

a. $\det(A) = \begin{vmatrix} 2 & -3 \\ 1 & 4 \end{vmatrix} = 2(4) - 1(-3) = 8 + 3 = 11$

b. $\det(B) = \begin{vmatrix} -1 & 2 \\ 2 & -4 \end{vmatrix} = (-1)(-4) - 2(2) = 4 - 4 = 0$

c. $\det(C) = \begin{vmatrix} 1 & 3 \\ 2 & 5 \end{vmatrix} = 1(5) - 2(3) = 5 - 6 = -1$

✓ **CHECKPOINT** *Now try Exercise 1.*

Notice in Example 1 that the determinant of a matrix can be positive, zero, or negative.

One way to evaluate the determinant of a 3 × 3 matrix, called **expanding by minors,** allows you to write the determinant of a 3 × 3 matrix in terms of three 2 × 2 determinants. The **minor** of an entry in a 3 × 3 matrix is the determinant of the 2 × 2 matrix that remains after deletion of the row and column in which the entry occurs. Here are three examples.

Determinant	Entry	Minor of Entry	Value of Minor
$\begin{vmatrix} 1 & -1 & 3 \\ 0 & 2 & 5 \\ -2 & 4 & -7 \end{vmatrix}$	1	$\begin{vmatrix} 2 & 5 \\ 4 & -7 \end{vmatrix}$	$2(-7) - 4(5) = -34$
$\begin{vmatrix} 1 & -1 & 3 \\ 0 & 2 & 5 \\ -2 & 4 & -7 \end{vmatrix}$	-1	$\begin{vmatrix} 0 & 5 \\ -2 & -7 \end{vmatrix}$	$0(-7) - (-2)(5) = 10$
$\begin{vmatrix} 1 & -1 & 3 \\ 0 & 2 & 5 \\ -2 & 4 & -7 \end{vmatrix}$	3	$\begin{vmatrix} 0 & 2 \\ -2 & 4 \end{vmatrix}$	$0(4) - (-2)(2) = 4$

Expanding by Minors

$$\det(A) = \begin{vmatrix} a_1 & b_1 & c_1 \\ a_2 & b_2 & c_2 \\ a_3 & b_3 & c_3 \end{vmatrix}$$

$$= a_1(\text{minor of } a_1) - b_1(\text{minor of } b_1) + c_1(\text{minor of } c_1)$$

$$= a_1 \begin{vmatrix} b_2 & c_2 \\ b_3 & c_3 \end{vmatrix} - b_1 \begin{vmatrix} a_2 & c_2 \\ a_3 & c_3 \end{vmatrix} + c_1 \begin{vmatrix} a_2 & b_2 \\ a_3 & b_3 \end{vmatrix}$$

This pattern is called **expanding by minors** along the first row. A similar pattern can be used to expand by minors along any row or column.

$$\begin{bmatrix} + & - & + \\ - & + & - \\ + & - & + \end{bmatrix}$$

Figure 4.10 Sign Pattern for a
3×3 Matrix

The *signs* of the terms used in expanding by minors follow the alternating pattern shown in Figure 4.10. For instance, the signs used to expand by minors along the second row are $-, +, -$, as shown below.

$$\det(A) = \begin{vmatrix} a_1 & b_1 & c_1 \\ a_2 & b_2 & c_2 \\ a_3 & b_3 & c_3 \end{vmatrix}$$

$$= -a_2(\text{minor of } a_2) + b_2(\text{minor of } b_2) - c_2(\text{minor of } c_2)$$

EXAMPLE 2   Finding the Determinant of a 3 × 3 Matrix

Find the determinant of $A = \begin{bmatrix} -1 & 1 & 2 \\ 0 & 2 & 3 \\ 3 & 4 & 2 \end{bmatrix}$.

Solution

By expanding by minors along the *first column*, you obtain

$$\det(A) = \begin{vmatrix} -1 & 1 & 2 \\ 0 & 2 & 3 \\ 3 & 4 & 2 \end{vmatrix}$$

$$= (-1)\begin{vmatrix} 2 & 3 \\ 4 & 2 \end{vmatrix} - (0)\begin{vmatrix} 1 & 2 \\ 4 & 2 \end{vmatrix} + (3)\begin{vmatrix} 1 & 2 \\ 2 & 3 \end{vmatrix}$$

$$= (-1)(4 - 12) - (0)(2 - 8) + (3)(3 - 4)$$

$$= 8 - 0 - 3 = 5$$

✓ **CHECKPOINT** *Now try Exercise 13.*

EXAMPLE 3   Finding the Determinant of a 3 × 3 Matrix

Find the determinant of $A = \begin{bmatrix} 1 & 2 & 1 \\ 3 & 0 & 2 \\ 4 & 0 & -1 \end{bmatrix}$.

Solution

By expanding by minors along the *second column*, you obtain

$$\det(A) = \begin{vmatrix} 1 & 2 & 1 \\ 3 & 0 & 2 \\ 4 & 0 & -1 \end{vmatrix}$$

$$= -(2)\begin{vmatrix} 3 & 2 \\ 4 & -1 \end{vmatrix} + (0)\begin{vmatrix} 1 & 1 \\ 4 & -1 \end{vmatrix} - (0)\begin{vmatrix} 1 & 1 \\ 3 & 2 \end{vmatrix}$$

$$= -(2)(-3 - 8) + 0 - 0 = 22$$

✓ **CHECKPOINT** *Now try Exercise 17.*

A zero entry in a matrix will always yield a zero term when expanding by minors. So, when you are evaluating the determinant of a matrix, choose to expand along the row or column that has the most zero entries.

2 ▶ Use determinants and Cramer's Rule to solve systems of linear equations.

Cramer's Rule

So far in this chapter, you have studied four methods for solving a system of linear equations: graphing, substitution, elimination with equations, and elimination with matrices. You will now learn one more method, called **Cramer's Rule,** which is named after Gabriel Cramer (1704–1752). This rule uses determinants to write the solution of a system of linear equations.

In Cramer's Rule, the value of a variable is expressed as the quotient of two determinants of the coefficient matrix of the system. The numerator is the determinant of the matrix formed by using the column of constants as replacements for the coefficients of the variable. In the definition below, note the notation for the different determinants.

Study Tip

Cramer's Rule is not as general as the elimination method because Cramer's Rule requires that the coefficient matrix of the system be square *and* that the system have exactly one solution.

Cramer's Rule

1. For the system of linear equations

$$\begin{cases} a_1x + b_1y = c_1 \\ a_2x + b_2y = c_2 \end{cases}$$

the solution is given by

$$x = \frac{D_x}{D} = \frac{\begin{vmatrix} c_1 & b_1 \\ c_2 & b_2 \end{vmatrix}}{\begin{vmatrix} a_1 & b_1 \\ a_2 & b_2 \end{vmatrix}}, \qquad y = \frac{D_y}{D} = \frac{\begin{vmatrix} a_1 & c_1 \\ a_2 & c_2 \end{vmatrix}}{\begin{vmatrix} a_1 & b_1 \\ a_2 & b_2 \end{vmatrix}}$$

provided that $D \neq 0$.

2. For the system of linear equations

$$\begin{cases} a_1x + b_1y + c_1z = d_1 \\ a_2x + b_2y + c_2z = d_2 \\ a_3x + b_3y + c_3z = d_3 \end{cases}$$

the solution is given by

$$x = \frac{D_x}{D} = \frac{\begin{vmatrix} d_1 & b_1 & c_1 \\ d_2 & b_2 & c_2 \\ d_3 & b_3 & c_3 \end{vmatrix}}{\begin{vmatrix} a_1 & b_1 & c_1 \\ a_2 & b_2 & c_2 \\ a_3 & b_3 & c_3 \end{vmatrix}}, \qquad y = \frac{D_y}{D} = \frac{\begin{vmatrix} a_1 & d_1 & c_1 \\ a_2 & d_2 & c_2 \\ a_3 & d_3 & c_3 \end{vmatrix}}{\begin{vmatrix} a_1 & b_1 & c_1 \\ a_2 & b_2 & c_2 \\ a_3 & b_3 & c_3 \end{vmatrix}},$$

$$z = \frac{D_z}{D} = \frac{\begin{vmatrix} a_1 & b_1 & d_1 \\ a_2 & b_2 & d_2 \\ a_3 & b_3 & d_3 \end{vmatrix}}{\begin{vmatrix} a_1 & b_1 & c_1 \\ a_2 & b_2 & c_2 \\ a_3 & b_3 & c_3 \end{vmatrix}}, D \neq 0$$

EXAMPLE 4   Using Cramer's Rule for a 2 × 2 System

Use Cramer's Rule to solve the system of linear equations.

$$\begin{cases} 4x - 2y = 10 \\ 3x - 5y = 11 \end{cases}$$

Solution

The determinant of the coefficient matrix is

$$D = \begin{vmatrix} 4 & -2 \\ 3 & -5 \end{vmatrix} = -20 - (-6) = -14.$$

$$x = \frac{D_x}{D} = \frac{\begin{vmatrix} 10 & -2 \\ 11 & -5 \end{vmatrix}}{-14} = \frac{(-50) - (-22)}{-14} = \frac{-28}{-14} = 2$$

$$y = \frac{D_y}{D} = \frac{\begin{vmatrix} 4 & 10 \\ 3 & 11 \end{vmatrix}}{-14} = \frac{44 - 30}{-14} = \frac{14}{-14} = -1$$

The solution is $(2, -1)$. Check this in the original system of equations.

✓ **CHECKPOINT** *Now try Exercise 37.*

EXAMPLE 5   Using Cramer's Rule for a 3 × 3 System

Use Cramer's Rule to solve the system of linear equations.

$$\begin{cases} -x + 2y - 3z = 1 \\ 2x \qquad + z = 0 \\ 3x - 4y + 4z = 2 \end{cases}$$

Solution

The determinant of the coefficient matrix is $D = 10$.

$$x = \frac{D_x}{D} = \frac{\begin{vmatrix} 1 & 2 & -3 \\ 0 & 0 & 1 \\ 2 & -4 & 4 \end{vmatrix}}{10} = \frac{8}{10} = \frac{4}{5}$$

$$y = \frac{D_y}{D} = \frac{\begin{vmatrix} -1 & 1 & -3 \\ 2 & 0 & 1 \\ 3 & 2 & 4 \end{vmatrix}}{10} = \frac{-15}{10} = -\frac{3}{2}$$

$$z = \frac{D_z}{D} = \frac{\begin{vmatrix} -1 & 2 & 1 \\ 2 & 0 & 0 \\ 3 & -4 & 2 \end{vmatrix}}{10} = \frac{-16}{10} = -\frac{8}{5}$$

The solution is $\left(\frac{4}{5}, -\frac{3}{2}, -\frac{8}{5}\right)$. Check this in the original system of equations.

✓ **CHECKPOINT** *Now try Exercise 47.*

Study Tip

When using Cramer's Rule, remember that the method *does not* apply if the determinant of the coefficient matrix is zero.

3 ▶ Use determinants to find areas of triangles, to test for collinear points, and to find equations of lines.

Applications of Determinants

In addition to Cramer's Rule, determinants have many other practical applications. For instance, you can use a determinant to find the area of a triangle whose vertices are given by three points on a rectangular coordinate system.

Area of a Triangle

The area of a triangle with vertices (x_1, y_1), (x_2, y_2), and (x_3, y_3) is

$$\text{Area} = \pm\frac{1}{2}\begin{vmatrix} x_1 & y_1 & 1 \\ x_2 & y_2 & 1 \\ x_3 & y_3 & 1 \end{vmatrix}$$

where the symbol $(\pm)$ indicates that the appropriate sign should be chosen to yield a positive area.

EXAMPLE 6   Finding the Area of a Triangle

Find the area of the triangle whose vertices are $(2, 0)$, $(1, 3)$, and $(3, 2)$, as shown in Figure 4.11.

Solution

Choose $(x_1, y_1) = (2, 0)$, $(x_2, y_2) = (1, 3)$, and $(x_3, y_3) = (3, 2)$. To find the area of the triangle, evaluate the determinant by expanding by minors along the first row.

$$\begin{vmatrix} x_1 & y_1 & 1 \\ x_2 & y_2 & 1 \\ x_3 & y_3 & 1 \end{vmatrix} = \begin{vmatrix} 2 & 0 & 1 \\ 1 & 3 & 1 \\ 3 & 2 & 1 \end{vmatrix}$$

$$= 2\begin{vmatrix} 3 & 1 \\ 2 & 1 \end{vmatrix} - 0\begin{vmatrix} 1 & 1 \\ 3 & 1 \end{vmatrix} + 1\begin{vmatrix} 1 & 3 \\ 3 & 2 \end{vmatrix}$$

$$= 2(1) - 0 + 1(-7)$$

$$= -5$$

Using this value, you can conclude that the area of the triangle is

$$\text{Area} = -\frac{1}{2}\begin{vmatrix} 2 & 0 & 1 \\ 1 & 3 & 1 \\ 3 & 2 & 1 \end{vmatrix}$$

$$= -\frac{1}{2}(-5) = \frac{5}{2}.$$

 CHECKPOINT *Now try Exercise 59.*

To see the benefit of the "determinant formula," try finding the area of the triangle in Example 6 by using the standard formula:

$$\text{Area} = \frac{1}{2}(\text{Base})(\text{Height}).$$

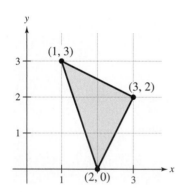

Figure 4.11

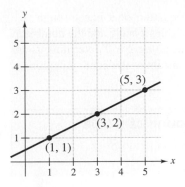

Figure 4.12

Suppose the three points in Example 6 had been on the same line. What would have happened in applying the area formula to three such points? The answer is that the determinant would have been zero. Consider, for instance, the three collinear points $(1, 1)$, $(3, 2)$, and $(5, 3)$, as shown in Figure 4.12. The area of the "triangle" that has these three points as vertices is

$$\frac{1}{2}\begin{vmatrix} 1 & 1 & 1 \\ 3 & 2 & 1 \\ 5 & 3 & 1 \end{vmatrix} = \frac{1}{2}\left(1\begin{vmatrix} 2 & 1 \\ 3 & 1 \end{vmatrix} - 1\begin{vmatrix} 3 & 1 \\ 5 & 1 \end{vmatrix} + 1\begin{vmatrix} 3 & 2 \\ 5 & 3 \end{vmatrix} \right)$$

$$= \frac{1}{2}[-1 - (-2) + (-1)]$$

$$= 0.$$

This result is generalized as follows.

Test for Collinear Points

Three points (x_1, y_1), (x_2, y_2), and (x_3, y_3) are collinear (lie on the same line) if and only if

$$\begin{vmatrix} x_1 & y_1 & 1 \\ x_2 & y_2 & 1 \\ x_3 & y_3 & 1 \end{vmatrix} = 0.$$

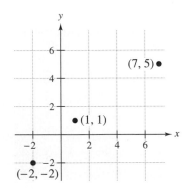

Figure 4.13

EXAMPLE 7   Testing for Collinear Points

Determine whether the points $(-2, -2)$, $(1, 1)$, and $(7, 5)$ are collinear. (See Figure 4.13.)

Solution

Letting $(x_1, y_1) = (-2, -2)$, $(x_2, y_2) = (1, 1)$, and $(x_3, y_3) = (7, 5)$, you have

$$\begin{vmatrix} x_1 & y_1 & 1 \\ x_2 & y_2 & 1 \\ x_3 & y_3 & 1 \end{vmatrix} = \begin{vmatrix} -2 & -2 & 1 \\ 1 & 1 & 1 \\ 7 & 5 & 1 \end{vmatrix}$$

$$= -2\begin{vmatrix} 1 & 1 \\ 5 & 1 \end{vmatrix} - (-2)\begin{vmatrix} 1 & 1 \\ 7 & 1 \end{vmatrix} + 1\begin{vmatrix} 1 & 1 \\ 7 & 5 \end{vmatrix}$$

$$= -2(-4) - (-2)(-6) + 1(-2)$$

$$= -6.$$

Because the value of this determinant *is not* zero, you can conclude that the three points *do not* lie on the same line and so are not collinear.

 **CHECKPOINT** *Now try Exercise 73.*

As a good review, look at how the slope can be used to verify the result in Example 7. Label the points $A(-2, -2)$, $B(1, 1)$, and $C(7, 5)$. Because the slopes from A to B and from A to C are different, the points are not collinear.

You can also use determinants to find the equation of a line through two points. In this case, the first row consists of the variables x and y and the number 1. By expanding by minors along the first row, the resulting 2×2 determinants are the coefficients of the variables x and y and the constant of the linear equation, as shown in Example 8.

Two-Point Form of the Equation of a Line

An equation of the line passing through the distinct points (x_1, y_1) and (x_2, y_2) is given by

$$\begin{vmatrix} x & y & 1 \\ x_1 & y_1 & 1 \\ x_2 & y_2 & 1 \end{vmatrix} = 0.$$

EXAMPLE 8   Finding an Equation of a Line

Find an equation of the line passing through $(-2, 1)$ and $(3, -2)$.

Solution

Applying the determinant formula for the equation of a line produces

$$\begin{vmatrix} x & y & 1 \\ -2 & 1 & 1 \\ 3 & -2 & 1 \end{vmatrix} = 0.$$

To evaluate this determinant, you can expand by minors along the first row to obtain the following.

$$x\begin{vmatrix} 1 & 1 \\ -2 & 1 \end{vmatrix} - y\begin{vmatrix} -2 & 1 \\ 3 & 1 \end{vmatrix} + 1\begin{vmatrix} -2 & 1 \\ 3 & -2 \end{vmatrix} = 0$$

$$3x + 5y + 1 = 0$$

So, an equation of the line is $3x + 5y + 1 = 0$.

✓ **CHECKPOINT** *Now try Exercise 79.*

Note that this method of finding the equation of a line works for all lines, including horizontal and vertical lines, as shown below.

Vertical Line Through
(2, 0) and (2, 2):

$$\begin{vmatrix} x & y & 1 \\ 2 & 0 & 1 \\ 2 & 2 & 1 \end{vmatrix} = 0$$

$$-2x - 0y + 4 = 0$$

$$-2x = -4$$

$$x = 2$$

Horizontal Line Through
(−3, 4) and (2, 4):

$$\begin{vmatrix} x & y & 1 \\ -3 & 4 & 1 \\ 2 & 4 & 1 \end{vmatrix} = 0$$

$$0x + 5y - 20 = 0$$

$$5y = 20$$

$$y = 4$$

_____ **Concept Check** _____

1. The determinant of a matrix can be represented by vertical bars, similar to the vertical bars used for absolute value. Does this mean that every determinant is nonnegative?

2. Is it possible to find the determinant of a 2×3 matrix? Explain.

3. If one column of a 3×3 matrix is all zeros, what is the determinant of the matrix? Explain.

4. Can Cramer's Rule be used to solve any system of linear equations? Explain.

4.5 EXERCISES

Go to pages 288–289 to record your assignments.

_____ **Developing Skills** _____

In Exercises 1–12, find the determinant of the matrix. *See Example 1.*

1. $\begin{bmatrix} 2 & 1 \\ 3 & 4 \end{bmatrix}$

2. $\begin{bmatrix} -3 & 1 \\ 5 & 2 \end{bmatrix}$

3. $\begin{bmatrix} 5 & 2 \\ -6 & 3 \end{bmatrix}$

4. $\begin{bmatrix} 2 & -2 \\ 4 & 3 \end{bmatrix}$

5. $\begin{bmatrix} -4 & 0 \\ 9 & 0 \end{bmatrix}$

6. $\begin{bmatrix} 4 & -3 \\ 0 & 0 \end{bmatrix}$

7. $\begin{bmatrix} 3 & -3 \\ -6 & 6 \end{bmatrix}$

8. $\begin{bmatrix} -2 & 3 \\ 6 & -9 \end{bmatrix}$

9. $\begin{bmatrix} -7 & 6 \\ \frac{1}{2} & 3 \end{bmatrix}$

10. $\begin{bmatrix} \frac{2}{3} & \frac{5}{6} \\ 14 & -2 \end{bmatrix}$

11. $\begin{bmatrix} 0.4 & 0.7 \\ 0.7 & 0.4 \end{bmatrix}$

12. $\begin{bmatrix} -1.2 & 4.5 \\ 0.4 & -0.9 \end{bmatrix}$

In Exercises 13–30, evaluate the determinant of the matrix. Expand by minors along the row or column that appears to make the computation easiest. *See Examples 2 and 3.*

13. $\begin{bmatrix} 2 & 3 & -1 \\ 6 & 0 & 0 \\ 4 & 1 & 1 \end{bmatrix}$

14. $\begin{bmatrix} 10 & 2 & -4 \\ 8 & 0 & -2 \\ 4 & 0 & 2 \end{bmatrix}$

15. $\begin{bmatrix} 1 & 1 & 2 \\ 3 & 1 & 0 \\ -2 & 0 & 3 \end{bmatrix}$

16. $\begin{bmatrix} 2 & 1 & 3 \\ 1 & 4 & 4 \\ 1 & 0 & 2 \end{bmatrix}$

17. $\begin{bmatrix} 2 & 4 & 6 \\ 0 & 3 & 1 \\ 0 & 0 & -5 \end{bmatrix}$

18. $\begin{bmatrix} 2 & 3 & 1 \\ 0 & 5 & -2 \\ 0 & 0 & -2 \end{bmatrix}$

19. $\begin{bmatrix} -2 & 2 & 3 \\ 1 & -1 & 0 \\ 0 & 1 & 4 \end{bmatrix}$

20. $\begin{bmatrix} -2 & 3 & 0 \\ 3 & 1 & -4 \\ 0 & 4 & 2 \end{bmatrix}$

21. $\begin{bmatrix} 1 & 4 & -2 \\ 3 & 6 & -6 \\ -2 & 1 & 4 \end{bmatrix}$

22. $\begin{bmatrix} 2 & -1 & 0 \\ 4 & 2 & 1 \\ 4 & 2 & 1 \end{bmatrix}$

23. $\begin{bmatrix} 2 & -2 & 7 \\ 1 & -3 & -2 \\ -2 & 6 & 4 \end{bmatrix}$

24. $\begin{bmatrix} 6 & 8 & -7 \\ 0 & 0 & 0 \\ 4 & -6 & 22 \end{bmatrix}$

25. $\begin{bmatrix} 2 & -5 & 0 \\ 4 & 7 & 0 \\ -7 & 25 & 3 \end{bmatrix}$

26. $\begin{bmatrix} 8 & 7 & 6 \\ -4 & 0 & 0 \\ 5 & 1 & 4 \end{bmatrix}$

27. $\begin{bmatrix} 0.1 & 0.2 & 0.3 \\ -0.3 & 0.2 & 0.2 \\ 5 & 4 & 4 \end{bmatrix}$

28. $\begin{bmatrix} -0.4 & 0.4 & 0.3 \\ 0.2 & 0.2 & 0.2 \\ 0.3 & 0.2 & 0.2 \end{bmatrix}$

29. $\begin{bmatrix} x & y & 1 \\ 3 & 1 & 1 \\ -2 & 0 & 1 \end{bmatrix}$

30. $\begin{bmatrix} x & y & 1 \\ -2 & -2 & 1 \\ 1 & 5 & 1 \end{bmatrix}$

In Exercises 31–36, use a graphing calculator to evaluate the determinant of the matrix.

31. $\begin{bmatrix} 5 & -3 & 2 \\ 7 & 5 & -7 \\ 0 & 6 & -1 \end{bmatrix}$

32. $\begin{bmatrix} 3 & -1 & 2 \\ 1 & -1 & 2 \\ -2 & 3 & 10 \end{bmatrix}$

33. $\begin{bmatrix} -\frac{1}{2} & -1 & 6 \\ 8 & -\frac{1}{4} & -4 \\ 1 & 2 & 1 \end{bmatrix}$

34. $\begin{bmatrix} \frac{1}{2} & \frac{3}{2} & \frac{1}{2} \\ 4 & 8 & 10 \\ -2 & -6 & 12 \end{bmatrix}$

35. $\begin{bmatrix} 0.6 & 0.4 & -0.6 \\ 0.1 & 0.5 & -0.3 \\ 8 & -2 & 12 \end{bmatrix}$

36. $\begin{bmatrix} 0.4 & 0.3 & 0.3 \\ -0.2 & 0.6 & 0.6 \\ 3 & 1 & 1 \end{bmatrix}$

In Exercises 37–52, use Cramer's Rule to solve the system of linear equations. (If not possible, state the reason.) *See Examples 4 and 5.*

37. $\begin{cases} x + 2y = 5 \\ -x + y = 1 \end{cases}$

38. $\begin{cases} 2x - y = -10 \\ 3x + 2y = -1 \end{cases}$

39. $\begin{cases} 3x + 4y = -2 \\ 5x + 3y = 4 \end{cases}$

40. $\begin{cases} 3x + 2y = -3 \\ 4x + 5y = -11 \end{cases}$

41. $\begin{cases} 20x + 8y = 11 \\ 12x - 24y = 21 \end{cases}$

42. $\begin{cases} 13x - 6y = 17 \\ 26x - 12y = 8 \end{cases}$

43. $\begin{cases} -0.4x + 0.8y = 1.6 \\ 2x - 4y = 5 \end{cases}$

44. $\begin{cases} -0.4x + 0.8y = 1.6 \\ 0.2x + 0.3y = 2.2 \end{cases}$

45. $\begin{cases} 3u + 6v = 5 \\ 6u + 14v = 11 \end{cases}$

46. $\begin{cases} 3x_1 + 2x_2 = 1 \\ 2x_1 + 10x_2 = 6 \end{cases}$

47. $\begin{cases} 4x - y + z = -5 \\ 2x + 2y + 3z = 10 \\ 5x - 2y + 6z = 1 \end{cases}$

48. $\begin{cases} 4x - 2y + 3z = -2 \\ 2x + 2y + 5z = 16 \\ 8x - 5y - 2z = 4 \end{cases}$

49. $\begin{cases} 4a + 3b + 4c = 1 \\ 4a - 6b + 8c = 8 \\ -a + 9b - 2c = -7 \end{cases}$

50. $\begin{cases} 2x + 3y + 5z = 4 \\ 3x + 5y + 9z = 7 \\ 5x + 9y + 17z = 13 \end{cases}$

51. $\begin{cases} 5x - 3y + 2z = 2 \\ 2x + 2y - 3z = 3 \\ x - 7y + 8z = -4 \end{cases}$

52. $\begin{cases} 5x + 4y - 6z = -10 \\ -4x + 2y + 3z = -1 \\ 8x + 4y + 12z = 2 \end{cases}$

In Exercises 53–56, solve the system of linear equations using a graphing calculator and Cramer's Rule. *See Examples 4 and 5.*

53. $\begin{cases} -3x + 10y = 22 \\ 9x - 3y = 0 \end{cases}$

54. $\begin{cases} 3x + 7y = 3 \\ 7x + 25y = 11 \end{cases}$

55. $\begin{cases} 3x - 2y + 3z = 8 \\ x + 3y + 6z = -3 \\ x + 2y + 9z = -5 \end{cases}$

56. $\begin{cases} 6x + 4y - 8z = -22 \\ -2x + 2y + 3z = 13 \\ -2x + 2y - z = 5 \end{cases}$

In Exercises 57 and 58, solve the equation.

57. $\begin{vmatrix} -3x & x \\ 4 & 3 \end{vmatrix} = 26$

58. $\begin{vmatrix} -8 & x \\ 6 & -x \end{vmatrix} = 6$

Solving Problems

Area of a Triangle   In Exercises 59–66, use a determinant to find the area of the triangle with the given vertices. *See Example 6.*

59. $(0, 3), (4, 0), (8, 5)$

60. $(2, 0), (0, 5), (6, 3)$

61. $(-3, 4), (1, -2), (6, 1)$

62. $(-2, -3), (2, -3), (0, 4)$

63. $(-2, 1), (3, -1), (1, 6)$

64. $(-1, 4), (-4, 0), (1, 3)$

65. $\left(0, \frac{1}{2}\right), \left(\frac{5}{2}, 0\right), (4, 3)$

66. $\left(\frac{1}{4}, 0\right), \left(0, \frac{3}{4}\right), (8, -2)$

Area of a Region   In Exercises 67–70, find the area of the shaded region of the figure.

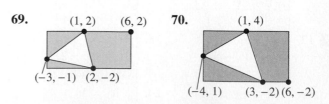

67. $(3, 5)$ $(5, 4)$ $(-1, 2)$ $(4, 0)$

68. $(-1, 2)$ $(5, 2)$ $(0, 0)$ $(4, -2)$

69. $(1, 2)$ $(6, 2)$ $(-3, -1)$ $(2, -2)$

70. $(1, 4)$ $(-4, 1)$ $(3, -2)$ $(6, -2)$

71. *Area of a Region*   A large region of forest has been infested with gypsy moths. The region is roughly triangular, as shown in the figure. Find the area of this region. (*Note:* The measurements in the figure are in miles.)

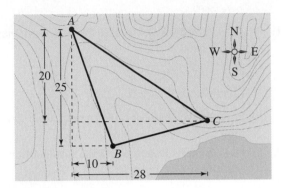

72. *Area of a Region*   You have purchased a triangular tract of land, as shown in the figure. What is the area of this tract of land? (*Note:* The measurements in the figure are in feet.)

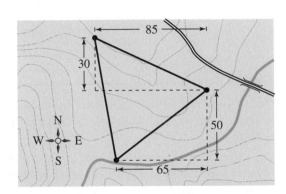

Collinear Points   In Exercises 73–78, determine whether the points are collinear. *See Example 7.*

✓ **73.** $(-1, 11), (0, 8), (2, 2)$

74. $(-1, -1), (1, 9), (2, 13)$

75. $(2, -4), (5, 2), (10, 10)$

76. $(1, 8), (3, 2), (6, -7)$

77. $\left(-2, \frac{1}{3}\right), (2, 1), \left(3, \frac{1}{5}\right)$

78. $\left(0, \frac{1}{2}\right), \left(1, \frac{7}{6}\right), \left(9, \frac{13}{2}\right)$

Equation of a Line   In Exercises 79–86, use a determinant to find the equation of the line through the points. *See Example 8.*

✓ **79.** $(-2, -1), (4, 2)$

80. $(-1, 3), (2, -6)$

81. $(10, 7), (-2, -7)$

82. $(-8, 3), (4, 6)$

83. $\left(-2, \frac{3}{2}\right), (3, -3)$

84. $\left(-\frac{1}{2}, 3\right), \left(\frac{5}{2}, 1\right)$

85. $(2, 3.6), (8, 10)$

86. $(3, 1.6), (5, -2.2)$

87. *Electrical Networks*   Laws that deal with electrical currents are known as *Kirchhoff's Laws*. When Kirchhoff's Laws are applied to the electrical network shown in the figure, the currents I_1, I_2, and I_3 are the solution of the system

$$\begin{cases} I_1 - I_2 + I_3 = 0 \\ 3I_1 + 2I_2 \quad\quad = 7 \\ \quad\quad 2I_2 + 4I_3 = 8. \end{cases}$$

Find the currents.

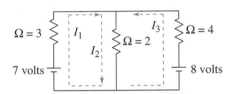

© Jose Luis Pelaez, Inc./CORBIS

88. *Force*   When three forces are applied to a beam, Newton's Laws suggest that the forces F_1, F_2, and F_3 are the solution of the system

$$\begin{cases} 3F_1 + F_2 - F_3 = 2 \\ F_1 - 2F_2 + F_3 = 0 \\ 4F_1 - F_2 + F_3 = 0. \end{cases}$$

Find the forces.

89. *Electrical Networks* When Kirchhoff's Laws are applied to the electrical network shown in the figure, the currents I_1, I_2, and I_3 are the solution of the system

$$\begin{cases} I_1 + I_2 - I_3 = 0 \\ I_1 + 2I_3 = 12 \\ I_1 - 2I_2 = -4. \end{cases}$$

Find the currents.

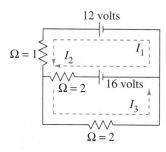

90. *Electrical Networks* When Kirchhoff's Laws are applied to the electrical network shown in the figure, the currents I_1, I_2, and I_3 are the solution of the system

$$\begin{cases} I_1 - I_2 + I_3 = 0 \\ I_2 + 4I_3 = 8 \\ 4I_1 + I_2 = 16. \end{cases}$$

Find the currents.

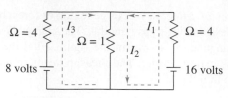

Figure for 90

91. (a) Use Cramer's Rule to solve the system of linear equations.

$$\begin{cases} kx + 3ky = 2 \\ (2 + k)x + ky = 5 \end{cases}$$

(b) For what values of k can Cramer's Rule not be used?

92. (a) Use Cramer's Rule to solve the system of linear equations.

$$\begin{cases} kx + (1 - k)y = 1 \\ (1 - k)x + ky = 3 \end{cases}$$

(b) For what value(s) of k will the system be inconsistent?

Explaining Concepts

93. ✎ Explain the difference between a square matrix and its determinant.

94. ✎ What is meant by the minor of an entry of a square matrix?

95. ✎ If two rows of a 3×3 matrix have identical entries, what is the value of the determinant? Explain.

96. ✎ What conditions must be met in order to use Cramer's Rule to solve a system of linear equations?

Cumulative Review

In Exercises 97–100, sketch the graph of the solution of the linear inequality.

97. $4x - 2y < 0$ **98.** $2x + 8y \geq 0$

99. $-x + 3y > 12$ **100.** $-3x - y \leq 2$

101. Given a function $f(x)$, describe how the graph of $h(x) = f(x) + c$ compares with the graph of $f(x)$ for a positive real number c.

102. Given a function $f(x)$, describe how the graph of $h(x) = f(x + c)$ compares with the graph of $f(x)$ for a positive real number c.

In Exercises 103–106, use the graph of f to sketch the graph.

103. $f(x) - 2$

104. $f(x - 2)$

105. $f(-x)$

106. $f(x - 1) + 3$

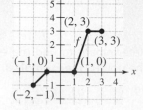

4.6 Systems of Linear Inequalities

Bonnie Kamin/PhotoEdit

Why You Should Learn It

Systems of linear inequalities can be used to model and solve real-life problems. For instance, in Exercise 61 on page 286, a system of linear inequalities can be used to analyze the compositions of dietary supplements.

1 ▶ Solve systems of linear inequalities in two variables.

What You Should Learn

1 ▶ Solve systems of linear inequalities in two variables.

2 ▶ Use systems of linear inequalities to model and solve real-life problems.

Systems of Linear Inequalities in Two Variables

You have already graphed linear inequalities in two variables. However, many practical problems in business, science, and engineering involve **systems of linear inequalities.** This type of system arises in problems that have *constraint* statements that contain phrases such as "more than," "less than," "at least," "no more than," "a minimum of," and "a maximum of." A **solution** of a system of linear inequalities in x and y is a point (x, y) that satisfies each inequality in the system.

To sketch the graph of a system of inequalities in two variables, first sketch (on the same coordinate system) the graph of each individual inequality. The **solution set** is the region that is *common* to every graph in the system.

EXAMPLE 1   Graphing a System of Linear Inequalities

Sketch the graph of the system of linear inequalities.

$$\begin{cases} 2x - y \le 5 \\ x + 2y \ge 2 \end{cases}$$

Solution

Begin by rewriting each inequality in slope-intercept form. Then sketch the line for the corresponding equation of each inequality. See Figures 4.14–4.16.

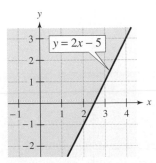

Graph of $2x - y \le 5$ is all points on and above $y = 2x - 5$.

Figure 4.14

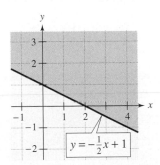

Graph of $x + 2y \ge 2$ is all points on and above $y = -\frac{1}{2}x + 1$.

Figure 4.15

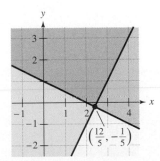

Graph of system is the purple wedge-shaped region.

Figure 4.16

✔ **CHECKPOINT** *Now try Exercise 13.*

In Figure 4.16, note that the two borderlines of the region

$$y = 2x - 5 \quad \text{and} \quad y = -\frac{1}{2}x + 1$$

intersect at the point $\left(\frac{12}{5}, -\frac{1}{5}\right)$. Such a point is called a **vertex** of the region. The region shown in the figure has only one vertex. Some regions, however, have several vertices. When you are sketching the graph of a system of linear inequalities, it is helpful to find and label any vertices of the region.

Graphing a System of Linear Inequalities

1. Sketch the line that corresponds to each inequality. (Use dashed lines for inequalities with $<$ or $>$ and solid lines for inequalities with $\leq$ or $\geq$.)

2. Lightly shade the half-plane that is the graph of each linear inequality. (Colored pencils may help distinguish different half-planes.)

3. The graph of the system is the intersection of the half-planes. (If you use colored pencils, it is the region that is selected with *every* color.)

EXAMPLE 2   Graphing a System of Linear Inequalities

Sketch the graph of the system of linear inequalities: $\begin{cases} y < 4 \\ y > 1 \end{cases}$.

Solution

The graph of the first inequality is the half-plane below the horizontal line

$y = 4.$ Upper boundary

The graph of the second inequality is the half-plane above the horizontal line

$y = 1.$ Lower boundary

The graph of the system is the horizontal band that lies *between* the two horizontal lines (where $y < 4$ *and* $y > 1$), as shown in Figure 4.17.

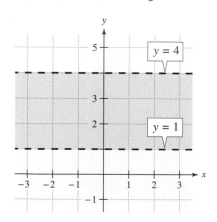

Figure 4.17

✓ **CHECKPOINT** *Now try Exercise 11.*

| EXAMPLE 3 | Graphing a System of Linear Inequalities |

Sketch the graph of the system of linear inequalities, and label the vertices.

$$\begin{cases} x - y < 2 \\ x > -2 \\ y \le 3 \end{cases}$$

Solution

Begin by sketching the half-planes represented by the three linear inequalities. The graph of

$$x - y < 2$$

is the half-plane lying above the line $y = x - 2$, the graph of

$$x > -2$$

is the half-plane lying to the right of the line $x = -2$, and the graph of

$$y \le 3$$

is the half-plane lying on and below the line $y = 3$. As shown in Figure 4.18, the region that is common to all three of these half-planes is a triangle. The vertices of the triangle are found as follows.

Vertex A: $(-2, -4)$	*Vertex B:* $(5, 3)$	*Vertex C:* $(-2, 3)$
Solution of the system	Solution of the system	Solution of the system
$\begin{cases} x - y = 2 \\ x = -2 \end{cases}$	$\begin{cases} x - y = 2 \\ y = 3 \end{cases}$	$\begin{cases} x = -2 \\ y = 3 \end{cases}$

✔ **CHECKPOINT** *Now try Exercise 31.*

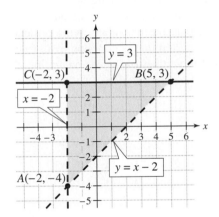

Figure 4.18

For the triangular region shown in Figure 4.18, each point of intersection of a pair of boundary lines corresponds to a vertex. With more complicated regions, two border lines can sometimes intersect at a point that is not a vertex of the region, as shown in Figure 4.19. To keep track of which points of intersection are actually vertices of the region, you should sketch the region and refer to your sketch as you find each point of intersection.

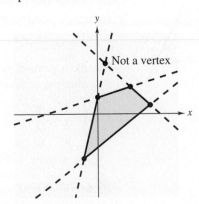

Figure 4.19

EXAMPLE 4   Graphing a System of Linear Inequalities

Sketch the graph of the system of linear inequalities, and label the vertices.

$$\begin{cases} x + y \le 5 \\ 3x + 2y \le 12 \\ x \ge 0 \\ y \ge 0 \end{cases}$$

Solution

Begin by sketching the half-planes represented by the four linear inequalities. The graph of $x + y \le 5$ is the half-plane lying on and below the line $y = -x + 5$. The graph of $3x + 2y \le 12$ is the half-plane lying on and below the line $y = -\frac{3}{2}x + 6$. The graph of $x \ge 0$ is the half-plane lying on and to the right of the y-axis, and the graph of $y \ge 0$ is the half-plane lying on and above the x-axis. As shown in Figure 4.20, the region that is common to all four of these half-planes is a four-sided polygon. The vertices of the region are found as follows.

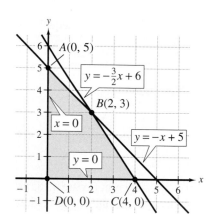

Figure 4.20

Vertex A: $(0, 5)$	Vertex B: $(2, 3)$	Vertex C: $(4, 0)$	Vertex D: $(0, 0)$
Solution of the system	Solution of the system	Solution of the system	Solution of the system
$\begin{cases} x + y = 5 \\ x = 0 \end{cases}$	$\begin{cases} x + y = 5 \\ 3x + 2y = 12 \end{cases}$	$\begin{cases} 3x + 2y = 12 \\ y = 0 \end{cases}$	$\begin{cases} x = 0 \\ y = 0 \end{cases}$

✓ **CHECKPOINT** *Now try Exercise 43.*

EXAMPLE 5   Finding the Boundaries of a Region

Find a system of inequalities that defines the region shown in Figure 4.21.

Solution

Three of the boundaries of the region are horizontal or vertical—they are easy to find. To find the diagonal boundary line, you can use the techniques of Section 3.4 to find the equation of the line passing through the points $(4, 4)$ and $(6, 0)$. Use the formula for slope to find $m = -2$, and then use the point-slope form with point $(6, 0)$ and $m = -2$ to obtain

$$y - 0 = -2(x - 6).$$

So, the equation is $y = -2x + 12$. The system of linear inequalities that describes the region is as follows.

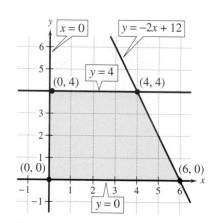

Figure 4.21

$$\begin{cases} y \le 4 & \text{Region lies on and below line } y = 4. \\ y \ge 0 & \text{Region lies on and above } x\text{-axis.} \\ x \ge 0 & \text{Region lies on and to the right of } y\text{-axis.} \\ y \le -2x + 12 & \text{Region lies on and below line } y = -2x + 12. \end{cases}$$

✓ **CHECKPOINT** *Now try Exercise 51.*

Technology: Tip

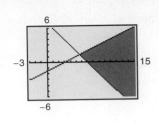

A graphing calculator can be used to graph a system of linear inequalities. The graph of

$$\begin{cases} 4y < 2x - 6 \\ x + y \geq 7 \end{cases}$$

is shown at the left. The grey shaded region, in which all points satisfy both inequalities, is the solution of the system. Try using a graphing calculator to graph

$$\begin{cases} 3x + y < 1 \\ -2x - 2y < 8. \end{cases}$$

2 ▶ Use systems of linear inequalities to model and solve real-life problems.

Application

EXAMPLE 6   Nutrition

The minimum daily requirements for the liquid portion of a diet are 300 calories, 36 units of vitamin A, and 90 units of vitamin C. A cup of dietary drink X provides 60 calories, 12 units of vitamin A, and 10 units of vitamin C. A cup of dietary drink Y provides 60 calories, 6 units of vitamin A, and 30 units of vitamin C. Write a system of linear inequalities that describes how many cups of each drink should be consumed each day to meet the minimum daily requirements for calories and vitamins.

Solution

Begin by letting x and y represent the following.

$x =$ number of cups of dietary drink X

$y =$ number of cups of dietary drink Y

To meet the minimum daily requirements, the following inequalities must be satisfied.

$$\begin{cases} 60x + 60y \geq 300 & \text{Calories} \\ 12x + 6y \geq 36 & \text{Vitamin A} \\ 10x + 30y \geq 90 & \text{Vitamin C} \\ x \geq 0 \\ y \geq 0 \end{cases}$$

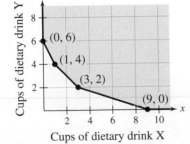

Figure 4.22

The last two inequalities are included because x and y cannot be negative. The graph of this system of inequalities is shown in Figure 4.22.

✔ **CHECKPOINT** *Now try Exercise 57.*

Concept Check

1. What is a system of linear inequalities in two variables?

2. Explain when you should use dashed lines and when you should use solid lines in sketching a system of linear inequalities.

3. Does the point of intersection of each pair of boundary lines correspond to a vertex? Explain.

4. Is it possible for a system of linear inequalities to have no solution? Explain.

4.6 EXERCISES

Go to pages 288–289 to record your assignments.

Developing Skills

In Exercises 1–6, match the system of linear inequalities with its graph. [The graphs are labeled (a), (b), (c), (d), (e), and (f).]

(a)

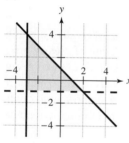

(b)

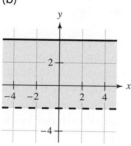

(c)

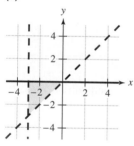

(d)

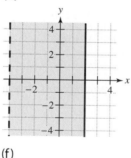

(e)

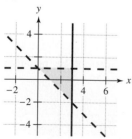

(f)

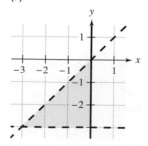

1. $\begin{cases} x > -4 \\ x \le 2 \end{cases}$

2. $\begin{cases} y \le 4 \\ y > -2 \end{cases}$

3. $\begin{cases} y < x \\ y > -3 \\ x \le 0 \end{cases}$

4. $\begin{cases} y > x \\ x > -3 \\ y \le 0 \end{cases}$

5. $\begin{cases} y > -1 \\ x \ge -3 \\ y \le -x + 1 \end{cases}$

6. $\begin{cases} x \le 3 \\ y < 1 \\ y > -x + 1 \end{cases}$

In Exercises 7–10, determine if each ordered pair is a solution of the system of linear inequalities.

7. $\begin{cases} 2x - y > 4 \\ x + 3y \le 6 \end{cases}$

   (a) $(2, 0)$

   (b) $(4, -2)$

8. $\begin{cases} x + 4y > 4 \\ 3x + 2y \ge -6 \end{cases}$

   (a) $(0, 2)$

   (b) $(-1, 1)$

9. $\begin{cases} -x + y < -2 \\ 4x + y < -3 \end{cases}$

   (a) $(-3, 4)$

   (b) $(-1, -3)$

10. $\begin{cases} 5x - 3y \le 12 \\ -3x + 5y \ge -15 \end{cases}$

    (a) $(3, 1)$

    (b) $(-5, -6)$

In Exercises 11–44, sketch a graph of the solution of the system of linear inequalities. *See Examples 1–4.*

 11. $\begin{cases} x < 3 \\ x > -2 \end{cases}$

12. $\begin{cases} y > -1 \\ y \le 2 \end{cases}$

13. $\begin{cases} x + y \le 3 \\ x - 1 \le 1 \end{cases}$

14. $\begin{cases} x + y \ge 2 \\ x - y \le 2 \end{cases}$

15. $\begin{cases} 2x - 4y \le 6 \\ x + y \ge 2 \end{cases}$

16. $\begin{cases} 4x + 10y \le 5 \\ x - y \le 4 \end{cases}$

17. $\begin{cases} x + 2y \le 6 \\ x - 2y \le 0 \end{cases}$

18. $\begin{cases} 2x + y \le 0 \\ x - y \le 8 \end{cases}$

19. $\begin{cases} x - 2y > 4 \\ 2x + y > 6 \end{cases}$

20. $\begin{cases} 3x + y < 6 \\ x + 2y > 2 \end{cases}$

21. $\begin{cases} x + y > -1 \\ x + y < 3 \end{cases}$

22. $\begin{cases} x - y > 2 \\ x - y < -4 \end{cases}$

23. $\begin{cases} y \geq \frac{4}{3}x + 1 \\ y \leq 5x - 2 \end{cases}$

24. $\begin{cases} y \geq \frac{1}{2}x + \frac{1}{2} \\ y \leq 4x - \frac{1}{2} \end{cases}$

25. $\begin{cases} y > x - 2 \\ y > -\frac{1}{3}x + 5 \end{cases}$

26. $\begin{cases} y > x - 4 \\ y > \frac{2}{3}x + \frac{1}{3} \end{cases}$

27. $\begin{cases} y \geq 3x - 3 \\ y \leq -x + 1 \end{cases}$

28. $\begin{cases} y \geq 2x - 3 \\ y \leq 3x + 1 \end{cases}$

29. $\begin{cases} x + 2y \leq -4 \\ y \geq x + 5 \end{cases}$

30. $\begin{cases} x + y \leq -3 \\ y \geq 3x - 4 \end{cases}$

✓ 31. $\begin{cases} x + y \leq 4 \\ x \geq 0 \\ y \geq 0 \end{cases}$

32. $\begin{cases} 2x + y \leq 6 \\ x \geq 0 \\ y \geq 0 \end{cases}$

33. $\begin{cases} 4x - 2y > 8 \\ x \geq 0 \\ y \leq 0 \end{cases}$

34. $\begin{cases} 2x - 6y > 6 \\ x \leq 0 \\ y \leq 0 \end{cases}$

35. $\begin{cases} y > -5 \\ x \leq 2 \\ y \leq x + 2 \end{cases}$

36. $\begin{cases} y \geq -1 \\ x < 3 \\ y \geq x - 1 \end{cases}$

37. $\begin{cases} x + y \leq 1 \\ -x + y \leq 1 \\ y \geq 0 \end{cases}$

38. $\begin{cases} 3x + 2y < 6 \\ x - 3y \geq 1 \\ y \geq 0 \end{cases}$

39. $\begin{cases} x + y \leq 5 \\ x - 2y \geq 2 \\ y \geq 3 \end{cases}$

40. $\begin{cases} 2x + y \geq 2 \\ x - 3y \leq 2 \\ y \leq 1 \end{cases}$

41. $\begin{cases} -3x + 2y < 6 \\ x - 4y > -2 \\ 2x + y < 3 \end{cases}$

42. $\begin{cases} x + 2y > 14 \\ -2x + 3y > 15 \\ x + 3y < 3 \end{cases}$

✓ 43. $\begin{cases} x \geq 1 \\ x - 2y \leq 3 \\ 3x + 2y \geq 9 \\ x + y \leq 6 \end{cases}$

44. $\begin{cases} x + y \leq 4 \\ x + y \geq -1 \\ x - y \geq -2 \\ x - y \leq 2 \end{cases}$

In Exercises 45–50, use a graphing calculator to graph the solution of the system of linear inequalities.

45. $\begin{cases} 2x - 3y \leq 6 \\ y \leq 4 \end{cases}$

46. $\begin{cases} 6x + 3y \geq 12 \\ y \leq 4 \end{cases}$

47. $\begin{cases} 2x - 2y \leq 5 \\ y \leq 6 \end{cases}$

48. $\begin{cases} 2x + 3y \geq 12 \\ y \geq 2 \end{cases}$

49. $\begin{cases} 2x + y \leq 2 \\ y \geq -4 \end{cases}$

50. $\begin{cases} 4x - 3y \geq -3 \\ y \geq -1 \end{cases}$

In Exercises 51–56, write a system of linear inequalities that describes the shaded region. *See Example 5.*

✓ 51.
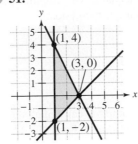

52.
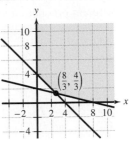

53.
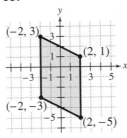

54.
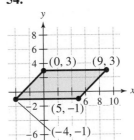

55.
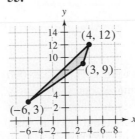

56.
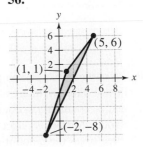

Solving Problems

✓ **57.** *Production* A furniture company can sell all the tables and chairs it produces. Each table requires 1 hour in the assembly center and $1\frac{1}{3}$ hours in the finishing center. Each chair requires $1\frac{1}{2}$ hours in the assembly center and $\frac{3}{4}$ hour in the finishing center. The company's assembly center is available 12 hours per day, and its finishing center is available 16 hours per day. Write a system of linear inequalities describing the different production levels. Graph the system.

58. *Production* An electronics company can sell all the HD TVs and DVD players it produces. Each HD TV requires 3 hours on the assembly line and $1\frac{1}{4}$ hours on the testing line. Each DVD player requires $2\frac{1}{2}$ hours on the assembly line and 1 hour on the testing line. The company's assembly line is available 20 hours per day, and its testing line is available 16 hours per day. Write a system of linear inequalities describing the different production levels. Graph the system.

59. *Investment* A person plans to invest up to $40,000 in two different interest-bearing accounts, account X and account Y. Account X is to contain at least $10,000. Moreover, account Y should have at least twice the amount in account X. Write a system of linear inequalities describing the various amounts that can be deposited in each account. Graph the system.

60. *Investment* A person plans to invest up to $25,000 in two different interest-bearing accounts, account X and account Y. Account Y is to contain at least $4000. Moreover, account X should have at least three times the amount in account Y. Write a system of linear inequalities describing the various amounts that can be deposited in each account. Graph the system.

61. *Nutrition* A dietitian is asked to design a special dietary supplement using two different foods. Each ounce of food X contains 20 units of calcium, 15 units of iron, and 10 units of vitamin B. Each ounce of food Y contains 10 units of calcium, 10 units of iron, and 20 units of vitamin B. The minimum daily requirements in the diet are 280 units of calcium, 160 units of iron, and 180 units of vitamin B. Write a system of linear inequalities describing the different amounts of food X and food Y that can be used in the diet. Use a graphing calculator to graph the system.

62. *Nutrition* A veterinarian is asked to design a special canine dietary supplement using two different dog foods. Each ounce of food X contains 12 units of calcium, 8 units of iron, and 6 units of protein. Each ounce of food Y contains 10 units of calcium, 10 units of iron, and 8 units of protein. The minimum daily requirements of the diet are 200 units of calcium, 100 units of iron, and 120 units of protein. Write a system of linear inequalities describing the different amounts of dog food X and dog food Y that can be used. Use a graphing calculator to graph the system.

63. *Ticket Sales* Two types of tickets are to be sold for a concert. General admission tickets cost $30 per ticket and stadium seat tickets cost $45 per ticket. The promoter of the concert must sell at least 15,000 tickets, including at least 8000 general admission tickets and at least 4000 stadium seat tickets. Moreover, the gross receipts must total at least $525,000 in order for the concert to be held. Write a system of linear inequalities describing the different numbers of tickets that can be sold. Use a graphing calculator to graph the system.

64. 🖩 *Ticket Sales* For a concert event, there are $30 reserved seat tickets and $20 general admission tickets. There are 2000 reserved seats available, and fire regulations limit the number of paid ticket holders to 3000. The promoter must take in at least $75,000 in ticket sales. Write a system of linear inequalities describing the different numbers of tickets that can be sold. Use a graphing calculator to graph the system.

65. ▲ *Geometry* The figure shows a cross section of a roped-off swimming area at a beach. Write a system of linear inequalities describing the cross section. (Each unit in the coordinate system represents 1 foot.)

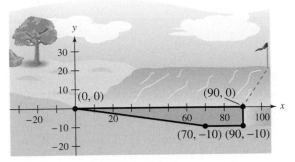

66. ▲ *Geometry* The figure shows the chorus platform on a stage. Write a system of linear inequalities describing the part of the audience that can see the full chorus. (Each unit in the coordinate system represents 1 meter.)

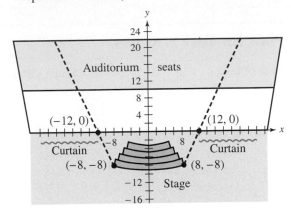

Explaining Concepts

67. ✎ Explain the meaning of the term *half-plane*. Give an example of an inequality whose graph is a half-plane.

68. ✎ Explain how you can check any single point (x_1, y_1) to determine whether the point is a solution of a system of linear inequalities.

69. ✎ Explain how to determine the vertices of the solution region for a system of linear inequalities.

70. ✎ Describe the difference between the solution set of a system of linear equations and the solution set of a system of linear inequalities.

Cumulative Review

In Exercises 71–76, find the x- and y-intercepts (if any) of the graph of the equation.

71. $y = 4x + 2$

72. $y = 8 - 3x$

73. $-x + 3y = -3$

74. $3x - 6y = 12$

75. $y = |x + 2|$

76. $y = |x - 1| - 2$

In Exercises 77–80, evaluate the function as indicated, and simplify.

77. $f(x) = 3x - 7$
(a) $f(-1)$
(b) $f\left(\frac{2}{3}\right)$

78. $f(x) = x^2 + x$
(a) $f(3)$
(b) $f(-2)$

79. $f(x) = 3x - x^2$
(a) $f(0)$ (b) $f(2m)$

80. $f(x) = \dfrac{x + 3}{x - 1}$
(a) $f(8)$ (b) $f(k - 2)$

What Did You Learn?

Use these two pages to help prepare for a test on this chapter. Check off the key terms and key concepts you know. You can also use this section to record your assignments.

Plan for Test Success

Date of test: ☐ / / ☐   **Study dates and times:** ☐ / / ☐ at ☐ : ☐ A.M./P.M.

                                                                ☐ / / ☐ at ☐ : ☐ A.M./P.M.

Things to review:

☐ Key Terms, *p. 288*
☐ Key Concepts, *pp. 288–289*
☐ Your class notes
☐ Your assignments

☐ Study Tips, *pp. 221, 222, 234, 241, 246, 254, 255, 256, 257, 267, 270, 271*
☐ Technology Tips, *pp. 234, 257, 268, 283*

☐ Mid-Chapter Quiz, *p. 253*
☐ Review Exercises, *pp. 290–293*
☐ Chapter Test, *p. 294*
☐ Video Explanations Online
☐ Tutorial Online

Key Terms

☐ systems of equations, *p. 218*
☐ solution of a system of equations, *p. 218*
☐ points of intersection, *p. 219*
☐ consistent system, *p. 219*
☐ dependent system, *p. 219*
☐ inconsistent system, *p. 219*
☐ break-even point, *p. 224*
☐ method of elimination, *p. 231*
☐ row-echelon form, *pp. 241, 257*
☐ ordered triple, *p. 241*

☐ equivalent systems, *p. 242*
☐ Gaussian elimination, *p. 242*
☐ row operations, *p. 242*
☐ position equation, *p. 247*
☐ matrix, *p. 254*
☐ order (of a matrix), *p. 254*
☐ entry (of a matrix), *p. 254*
☐ square matrix, *p. 254*
☐ augmented matrix, *p. 255*
☐ coefficient matrix, *p. 255*

☐ elementary row operations, *p. 256*
☐ row-equivalent matrices, *p. 256*
☐ determinant, *p. 267*
☐ expanding by minors, *p. 268*
☐ minor (of an entry), *p. 268*
☐ Cramer's Rule, *p. 270*
☐ system of linear inequalities, *p. 279*
☐ solution of a system of linear inequalities, *p. 279*
☐ vertex, *p. 280*

Key Concepts

4.1 Systems of Equations

Assignment: _____ Due date: _____

☐ **Solve systems of equations graphically.**

A system of equations can have one solution, infinitely many solutions, or no solution.

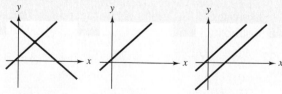

Consistent system (one solution)

Dependent (consistent) system (infinitely many solutions)

Inconsistent system (no solution)

☐ **Solve systems of equations algebraically using the method of substitution.**

1. Solve one equation for one variable in terms of the other variable.
2. Substitute the expression obtained in Step 1 in the other equation to obtain an equation in one variable.
3. Solve the equation obtained in Step 2.
4. Back-substitute the Step 3 solution into the expression found in Step 1 to find the value of the other variable.
5. Check the solution to see that it satisfies both of the original equations.

4.2 Linear Systems in Two Variables

Assignment: _____ Due date: _____

☐ **Solve systems of linear equations algebraically using the method of elimination.**

1. Obtain coefficients for x (or y) that are opposites by multiplying one or both equations by suitable constants.
2. Add the equations to eliminate one variable. Solve the resulting equation.
3. Back-substitute the value from Step 2 to find the value of the other variable.
4. Check the solution in both original equations.

☐ **Use systems of linear equations to model and solve real-life problems.**

1. Does the problem involve more than one unknown quantity?
2. Are there two (or more) equations or conditions to be satisfied?

If one or both of these conditions occur, the appropriate mathematical model for the problem may be a system of linear equations.

4.3 Linear Systems in Three Variables

Assignment: _____ Due date: _____

☐ **Solve systems of linear equations in row-echelon form using back-substitution.**

A system of equations in *row-echelon form* has a stair-step pattern with leading coefficients of 1. You can use back-substitution to solve a system in row-echelon form.

☐ **Use Gaussian elimination to write a system of linear equations in row-echelon form.**

Two systems of equations are *equivalent systems* if they have the same solution set.

Each of the following *row operations* produces an equivalent system of linear equations.

1. Interchange two equations.
2. Multiply one of the equations by a nonzero constant.
3. Add a multiple of one of the equations to another equation to replace the latter equation.

Gaussian elimination is the process of forming a chain of equivalent systems by performing one row operation at a time to obtain an equivalent system in row-echelon form.

4.4 Matrices and Linear Systems

Assignment: _____ Due date: _____

☐ **Perform elementary row operations on a matrix.**

1. Interchange two rows.
2. Multiply a row by a nonzero constant.
3. Add a multiple of a row to another row.

☐ **Use Gaussian elimination with back-substitution.**

1. Write the augmented matrix of the system of equations.
2. Use elementary row operations to rewrite the augmented matrix in row-echelon form.
3. Write the system of equations corresponding to the matrix in row-echelon form. Then use back-substitution to find the solution.

4.5 Determinants and Linear Systems

Assignment: _____ Due date: _____

☐ **Find the determinant of a 2×2 matrix.**

$$\begin{vmatrix} a_1 & b_1 \\ a_2 & b_2 \end{vmatrix} = a_1 b_2 - a_2 b_1$$

☐ **Use expanding by minors to find the determinant of a 3×3 matrix.**

$$\begin{vmatrix} a_1 & b_1 & c_1 \\ a_2 & b_2 & c_2 \\ a_3 & b_3 & c_3 \end{vmatrix} = a_1 \begin{vmatrix} b_2 & c_2 \\ b_3 & c_3 \end{vmatrix} - b_1 \begin{vmatrix} a_2 & c_2 \\ a_3 & c_3 \end{vmatrix} + c_1 \begin{vmatrix} a_2 & b_2 \\ a_3 & b_3 \end{vmatrix}$$

☐ **Use Cramer's Rule to solve a system.**

See page 270.

4.6 Systems of Linear Inequalities

Assignment: _____ Due date: _____

☐ **Graph a system of linear inequalities.**

1. Sketch a dashed or solid line corresponding to each inequality.
2. Shade the half-plane for each inequality.
3. The intersection of all half-planes represents the system.

Review Exercises

4.1 Systems of Equations

1 ▶ Determine if ordered pairs are solutions of systems of equations.

In Exercises 1–4, determine whether each ordered pair is a solution of the system of equations.

1. $\begin{cases} 3x + 7y = 2 \\ 5x + 6y = 9 \end{cases}$
   (a) $(3, 4)$

   (b) $(3, -1)$

2. $\begin{cases} -2x + 5y = 21 \\ 9x - y = 13 \end{cases}$
   (a) $(2, 5)$

   (b) $(-2, 4)$

3. $\begin{cases} 26x + 13y = 26 \\ 20x + 10y = 30 \end{cases}$
   (a) $(4, -5)$

   (b) $(7, 12)$

4. $\begin{cases} 8x + 6y = 10 \\ 12x + 9y = 15 \end{cases}$
   (a) $(3, -4)$

   (b) $(2, -1)$

2 ▶ Solve systems of equations graphically.

In Exercises 5–12, use the graphical method to solve the system of equations.

5. $\begin{cases} x + y = 2 \\ x - y = 2 \end{cases}$

6. $\begin{cases} 2x - 3y = -3 \\ y = x \end{cases}$

7. $\begin{cases} x - y = 3 \\ -x + y = 1 \end{cases}$

8. $\begin{cases} x + y = -1 \\ 3x + 2y = 0 \end{cases}$

9. $\begin{cases} 2x - y = 0 \\ -x + y = 4 \end{cases}$

10. $\begin{cases} x = y + 3 \\ x = y + 1 \end{cases}$

11. $\begin{cases} 2x + y = 4 \\ -4x - 2y = -8 \end{cases}$

12. $\begin{cases} 3x - 2y = 6 \\ -6x + 4y = 12 \end{cases}$

In Exercises 13 and 14, determine the number of solutions of the system of linear equations without solving the system.

13. $y = 3x - 3$
    $y = 3x + 2$

14. $y = 2x + 1$
    $y = 3x + 1$

In Exercises 15 and 16, use a graphing calculator to graph the equations and approximate any solutions of the system of equations.

15. $\begin{cases} 5x - 3y = 3 \\ 2x + 2y = 14 \end{cases}$

16. $\begin{cases} 0.2x - 1.2y = 10 \\ 2x - 3y = 4.5 \end{cases}$

3 ▶ Solve systems of equations algebraically using the method of substitution.

In Exercises 17–26, solve the system of equations by the method of substitution.

17. $\begin{cases} 2x - 3y = -1 \\ x + 4y = 16 \end{cases}$

18. $\begin{cases} 3x - 7y = 10 \\ -2x + y = -14 \end{cases}$

19. $\begin{cases} -5x + 2y = 4 \\ 10x - 4y = 7 \end{cases}$

20. $\begin{cases} 5x + 2y = 3 \\ 2x + 3y = 10 \end{cases}$

21. $\begin{cases} 3x - 7y = 5 \\ 5x - 9y = -5 \end{cases}$

22. $\begin{cases} 24x - 4y = 20 \\ 6x - y = 5 \end{cases}$

23. $\begin{cases} -x + y = 6 \\ 15x + y = -10 \end{cases}$

24. $\begin{cases} -3x + y = 6 \\ 2x + 3y = -3 \end{cases}$

25. $\begin{cases} -3x - 3y = 3 \\ x + y = -1 \end{cases}$

26. $\begin{cases} x + y = 9 \\ x + y = 0 \end{cases}$

4 ▶ Use systems of equations to model and solve real-life problems.

27. *Break-Even Analysis* A small business invests $25,000 in equipment to produce a one-time-use camera. Each camera costs $4.45 to produce and sells for $8.95. How many one-time-use cameras must be sold before the business breaks even?

28. *Seed Mixture* Fifteen pounds of mixed birdseed sells for $8.85 per pound. The mixture is obtained from two kinds of birdseed, with one variety priced at $7.05 per pound and the other at $9.30 per pound. How many pounds of each variety of birdseed are used in the mixture?

4.2 Linear Systems in Two Variables

1 ▶ Solve systems of linear equations algebraically using the method of elimination.

In Exercises 29–36, solve the system of linear equations by the method of elimination.

29. $\begin{cases} x + y = 0 \\ 2x + y = 0 \end{cases}$

30. $\begin{cases} 4x + y = 1 \\ x - y = 4 \end{cases}$

31. $\begin{cases} 2x - y = 2 \\ 6x + 8y = 39 \end{cases}$

32. $\begin{cases} 3x + 2y = 11 \\ x - 3y = -11 \end{cases}$

33. $\begin{cases} 4x + y = -3 \\ -4x + 3y = 23 \end{cases}$

34. $\begin{cases} -3x + 5y = -23 \\ 2x - 5y = 22 \end{cases}$

35. $\begin{cases} 0.2x + 0.3y = 0.14 \\ 0.4x + 0.5y = 0.20 \end{cases}$

36. $\begin{cases} 0.1x + 0.5y = -0.17 \\ -0.3x - 0.2y = -0.01 \end{cases}$

2 ▶ Use systems of linear equations to model and solve real-life problems.

37. *Acid Mixture*   Forty gallons of a 60% acid solution is obtained by mixing a 75% solution with a 50% solution. How many gallons of each solution must be used to obtain the desired mixture?

38. *Alcohol Mixture*   Fifty gallons of a 90% alcohol solution is obtained by mixing a 100% solution with a 75% solution. How many gallons of each solution must be used to obtain the desired mixture?

39. *Average Speed*   A bus travels for 3 hours at an average speed of 50 miles per hour. How much longer must the bus travel at an average speed of 60 miles per hour so that the average speed for the total trip will be 52 miles per hour?

40. *Homeland Security*   The numbers of aircraft checked by the U.S. Customs and Border Protection Agency in the years from 2002 to 2005 are shown in the table. (Source: U.S. Customs and Border Protection Agency)

Year	2002	2003	2004	2005
Aircraft, y	769	790	824	866

(a) Plot the data shown in the table. Let x represent the year, with $x = 2$ corresponding to 2002.

(b) The line $y = mx + b$ that best fits the data is given by the solution of the system below. Solve the system and find the equation of the best-fitting line. Sketch the graph of the line on the same set of coordinate axes used in part (a).

$$\begin{cases} 4b + 4m = 2924 \\ 16b + 24m = 11{,}956 \end{cases}$$

(c) Interpret the meaning of the slope of the line in the context of this problem.

4.3 Linear Systems in Three Variables

1 ▶ Solve systems of linear equations in row-echelon form using back-substitution.

In Exercises 41–44, use back-substitution to solve the system of linear equations.

41. $\begin{cases} x = 3 \\ x + 2y = 7 \\ -3x - y + 4z = 9 \end{cases}$

42. $\begin{cases} 2x + 3y = 9 \\ 4x - 6z = 12 \\ y = 5 \end{cases}$

43. $\begin{cases} x + 2y = 6 \\ 3y = 9 \\ x + 2z = 12 \end{cases}$

44. $\begin{cases} 3x - 2y + 5z = -10 \\ 3y = 18 \\ 6x - 4y = -6 \end{cases}$

2 ▶ Solve systems of linear equations using the method of Gaussian elimination.

In Exercises 45–48, solve the system of linear equations.

45. $\begin{cases} -x + y + 2z = 1 \\ 2x + 3y + z = -2 \\ 5x + 4y + 2z = 4 \end{cases}$

46. $\begin{cases} 2x + 3y + z = 10 \\ 2x - 3y - 3z = 22 \\ 4x - 2y + 3z = -2 \end{cases}$

47. $\begin{cases} x - y - z = 1 \\ -2x + y + 3z = -5 \\ 3x + 4y - z = 6 \end{cases}$

48. $\begin{cases} -3x + y + 2z = -13 \\ -x - y + z = 0 \\ 2x + 2y - 3z = -1 \end{cases}$

3 ▶ Solve application problems using the method of Gaussian elimination.

49. *Investment* An inheritance of $20,000 is divided among three investments yielding a total of $1780 in interest per year. The interest rates for the three investments are 7%, 9%, and 11%. The amounts invested at 9% and 11% are $3000 and $1000 less than the amount invested at 7%, respectively. Find the amount invested at each rate.

50. *Vertical Motion* Find the position equation

$$s = \frac{1}{2}at^2 + v_0t + s_0$$

for an object that has the indicated heights at the specified times.

$s = 192$ feet at $t = 1$ second

$s = 152$ feet at $t = 2$ seconds

$s = 80$ feet at $t = 3$ seconds

4.4 Matrices and Linear Systems

1 ▶ Determine the orders of matrices.

In Exercises 51–54, determine the order of the matrix.

51. $[3 \quad 6 \quad -7 \quad -8]$ **52.** $\begin{bmatrix} 1 & 5 \\ 3 & -4 \end{bmatrix}$

53. $\begin{bmatrix} 5 & 7 & 9 \\ 11 & -12 & 0 \end{bmatrix}$ **54.** $\begin{bmatrix} 15 \\ 13 \\ -9 \end{bmatrix}$

2 ▶ Form coefficient and augmented matrices and form linear systems from augmented matrices.

In Exercises 55 and 56, form (a) the coefficient matrix and (b) the augmented matrix for the system of linear equations.

55. $\begin{cases} 7x - 5y = 11 \\ x - y = -5 \end{cases}$

56. $\begin{cases} x + 2y + z = 4 \\ 3x \quad\quad - z = 2 \\ -x + 5y - 2z = -6 \end{cases}$

In Exercises 57 and 58, write the system of linear equations represented by the matrix. (Use variables x, y, and z.)

57. $\begin{bmatrix} 4 & -1 & 0 & \vdots & 2 \\ 6 & 3 & 2 & \vdots & 1 \\ 0 & 1 & 4 & \vdots & 0 \end{bmatrix}$

58. $\begin{bmatrix} 7 & 8 & \vdots & -26 \\ 4 & -9 & \vdots & -12 \end{bmatrix}$

3 ▶ Perform elementary row operations to solve systems of linear equations in matrix form.

In Exercises 59–62, use matrices and elementary row operations to solve the system.

59. $\begin{cases} 5x + 4y = 2 \\ -x + y = -22 \end{cases}$ **60.** $\begin{cases} 2x - 5y = 2 \\ 3x - 7y = 1 \end{cases}$

61. $\begin{cases} 0.2x - 0.1y = 0.07 \\ 0.4x - 0.5y = -0.01 \end{cases}$ **62.** $\begin{cases} 2x + y = 0.3 \\ 3x - y = -1.3 \end{cases}$

4 ▶ Use matrices and Gaussian elimination with back-substitution to solve systems of linear equations.

In Exercises 63–68, use matrices to solve the system of linear equations.

63. $\begin{cases} x + 4y + 4z = 7 \\ -3x + 2y + 3z = 0 \\ 4x \quad\quad - 2z = -2 \end{cases}$

64. $\begin{cases} -x + 3y - z = -4 \\ 2x \quad\quad + 6z = 14 \\ -3x - y + z = 10 \end{cases}$

65. $\begin{cases} 2x_1 + 3x_2 + 3x_3 = 3 \\ 6x_1 + 6x_2 + 12x_3 = 13 \\ 12x_1 + 9x_2 - x_3 = 2 \end{cases}$

66. $\begin{cases} -x_1 + 2x_2 + 3x_3 = 4 \\ 2x_1 - 4x_2 - x_3 = -13 \\ 3x_1 + 2x_2 - 4x_3 = -1 \end{cases}$

67. $\begin{cases} x \quad\quad - 4z = 17 \\ -2x + 4y + 3z = -14 \\ 5x - y + 2z = -3 \end{cases}$

68. $\begin{cases} 2x + 3y - 5z = 3 \\ -x + 2y \quad\quad = 3 \\ 3x + 5y + 2z = 15 \end{cases}$

4.5 Determinants and Linear Systems

1 ▶ Find determinants of 2×2 matrices and 3×3 matrices.

In Exercises 69–74, find the determinant of the matrix using any appropriate method.

69. $\begin{bmatrix} 9 & 8 \\ 10 & 10 \end{bmatrix}$
70. $\begin{bmatrix} -3.4 & 1.2 \\ -5 & 2.5 \end{bmatrix}$

71. $\begin{bmatrix} 8 & 6 & 3 \\ 6 & 3 & 0 \\ 3 & 0 & 2 \end{bmatrix}$
72. $\begin{bmatrix} 7 & -1 & 10 \\ -3 & 0 & -2 \\ 12 & 1 & 1 \end{bmatrix}$

73. $\begin{bmatrix} 8 & 3 & 2 \\ 1 & -2 & 4 \\ 6 & 0 & 5 \end{bmatrix}$
74. $\begin{bmatrix} 4 & 0 & 10 \\ 0 & 10 & 0 \\ 10 & 0 & 34 \end{bmatrix}$

2 ▶ Use determinants and Cramer's Rule to solve systems of linear equations.

In Exercises 75–78, use Cramer's Rule to solve the system of linear equations. (If not possible, state the reason.)

75. $\begin{cases} 7x + 12y = 63 \\ 2x + 3y = 15 \end{cases}$
76. $\begin{cases} 12x + 42y = -17 \\ 30x - 18y = 19 \end{cases}$

77. $\begin{cases} -x + y + 2z = 1 \\ 2x + 3y + z = -2 \\ 5x + 4y + 2z = 4 \end{cases}$
78. $\begin{cases} 2x_1 + x_2 + 2x_3 = 4 \\ 2x_1 + 2x_2 = 5 \\ 2x_1 - x_2 + 6x_3 = 2 \end{cases}$

3 ▶ Use determinants to find areas of triangles, to test for collinear points, and to find equations of lines.

Area of a Triangle **In Exercises 79–82, use a determinant to find the area of the triangle with the given vertices.**

79. $(1, 0), (5, 0), (5, 8)$
80. $(-6, 0), (6, 0), (0, 5)$
81. $(1, 2), (4, -5), (3, 2)$
82. $\left(\frac{3}{2}, 1\right), \left(4, -\frac{1}{2}\right), (4, 2)$

Collinear Points **In Exercises 83 and 84, determine whether the points are collinear.**

83. $(1, 2), (5, 0), (10, -2)$
84. $(-4, 3), (1, 1), (6, -1)$

Equation of a Line **In Exercises 85 and 86, use a determinant to find the equation of the line through the points.**

85. $(-4, 0), (4, 4)$
86. $\left(-\frac{5}{2}, 3\right), \left(\frac{7}{2}, 1\right)$

4.6 Systems of Linear Inequalities

1 ▶ Solve systems of linear inequalities in two variables.

In Exercises 87–90, sketch a graph of the solution of the system of linear inequalities.

87. $\begin{cases} x + y < 5 \\ x > 2 \\ y \geq 0 \end{cases}$
88. $\begin{cases} \frac{1}{2}x + y > 4 \\ x < 6 \\ y < 3 \end{cases}$

89. $\begin{cases} x + 2y \leq 160 \\ 3x + y \leq 180 \\ x \geq 0 \\ y \geq 0 \end{cases}$
90. $\begin{cases} 2x + 3y \leq 24 \\ 2x + y \leq 16 \\ x \geq 0 \\ y \geq 0 \end{cases}$

2 ▶ Use systems of linear inequalities to model and solve real-life problems.

91. *Soup Distribution* A charitable organization can purchase up to 500 cartons of soup to be divided between a soup kitchen and a homeless shelter in the Chicago area. These two organizations need at least 150 cartons and 220 cartons, respectively. Write a system of linear inequalities describing the various numbers of cartons that can go to each organization. Graph the system.

92. *Inventory Costs* A warehouse operator has up to 24,000 square feet of floor space in which to store two products. Each unit of product x requires 20 square feet of floor space and costs \$12 per day to store. Each unit of product y requires 30 square feet of floor space and costs \$8 per day to store. The total storage cost per day cannot exceed \$12,400. Write a system of linear inequalities describing the various ways the two products can be stored. Graph the system.

Chapter Test

$$\begin{cases} 2x - 2y = 2 \\ -x + 2y = 0 \end{cases}$$

System for 1

Take this test as you would take a test in class. After you are done, check your work against the answers in the back of the book.

1. Determine whether each ordered pair is a solution of the system at the left.
   (a) $(2, 1)$ (b) $(4, 3)$

In Exercises 2–11, use the indicated method to solve the system.

2. *Graphical:* $\begin{cases} x - 2y = -1 \\ 2x + 3y = 12 \end{cases}$ 3. *Substitution:* $\begin{cases} 4x - y = 1 \\ 4x - 3y = -5 \end{cases}$

4. *Substitution:* $\begin{cases} 2x - 2y = -2 \\ 3x + y = 9 \end{cases}$ 5. *Elimination:* $\begin{cases} 3x - 4y = -14 \\ -3x + y = 8 \end{cases}$

6. *Elimination:* $\begin{cases} x + 2y - 4z = 0 \\ 3x + y - 2z = 5 \\ 3x - y + 2z = 7 \end{cases}$ 7. *Matrices:* $\begin{cases} x - 3z = -10 \\ -2y + 2z = 0 \\ x - 2y = -7 \end{cases}$

8. *Matrices:* $\begin{cases} x - 3y + z = -3 \\ 3x + 2y - 5z = 18 \\ y + z = -1 \end{cases}$ 9. *Cramer's Rule:* $\begin{cases} 2x - 7y = 7 \\ 3x + 7y = 13 \end{cases}$

10. *Any Method:* $\begin{cases} 3x - 2y + z = 12 \\ x - 3y = 2 \\ -3x - 9z = -6 \end{cases}$

11. *Any Method:* $\begin{cases} 4x + y + 2z = -4 \\ 3y + z = 8 \\ -3x + y - 3z = 5 \end{cases}$

$$\begin{bmatrix} 2 & -2 & 0 \\ -1 & 3 & 1 \\ 2 & 8 & 1 \end{bmatrix}$$

Matrix for 12

12. Evaluate the determinant of the matrix shown at the left.

13. Use a determinant to find the area of the triangle with vertices $(0, 0)$, $(5, 4)$, and $(6, 0)$.

14. Graph the solution of the system of linear inequalities.
    $$\begin{cases} x - 2y > -3 \\ 2x + 3y \le 22 \\ y \ge 0 \end{cases}$$

15. The perimeter of a rectangle is 68 feet, and its width is $\frac{8}{9}$ times its length. Find the dimensions of the rectangle.

16. An inheritance of $25,000 is divided among three investments yielding a total of $1275 in interest per year. The interest rates for the three investments are 4.5%, 5%, and 8%. The amounts invested at 5% and 8% are $4000 and $10,000 less than the amount invested at 4.5%, respectively. Find the amount invested at each rate.

17. Two types of tickets are sold for a concert. Reserved seat tickets cost $30 per ticket and floor seat tickets cost $40 per ticket. The promoter of the concert can sell at most 9000 reserved seat tickets and 4000 floor seat tickets. Gross receipts must total at least $300,000 in order for the concert to be held. Write a system of linear inequalities describing the different numbers of tickets that can be sold. Graph the system.

Cumulative Test: Chapters 1–4

Take this test as you would take a test in class. After you are done, check your work against the answers in the back of the book.

1. Place the correct symbol ($<$, $>$, or $=$) between the two real numbers.

   (a) -2 ___ -4 (b) $\frac{2}{3}$ ___ $\frac{1}{2}$ (c) -4.5 ___ $-|-4.5|$

2. Write an algebraic expression for the statement, "The number n is tripled and the product is decreased by 8."

In Exercises 3 and 4, perform the operations and simplify.

3. $t(3t - 1) - 2(t + 4)$

4. $4x(x + x^2) - 6(x^2 + 4)$

In Exercises 5–8, solve the equation or inequality.

5. $12 - 5(3 - x) = x + 3$

6. $1 - \dfrac{x + 2}{4} = \dfrac{7}{8}$

7. $|x - 2| \geq 3$

8. $-12 \leq 4x - 6 < 10$

9. Your annual automobile insurance premium is $1150. Because of a driving violation, your premium is increased by 20%. What is your new premium?

10. The triangles at the left are similar. Solve for x by using the fact that corresponding sides of similar triangles are proportional.

11. The revenue R from selling x units of a product is $R = 12.90x$. The cost C of producing x units is $C = 8.50x + 450$. To obtain a profit, the revenue must be greater than the cost. For what values of x will this product produce a profit? Explain your reasoning.

12. Does the equation $x - y^2 = 0$ represent y as a function of x?

13. Find the domain of the function $f(x) = \sqrt{x - 2}$.

14. Given $f(x) = x^2 - 2x$, find (a) $f(3)$ and (b) $f(-3c)$.

15. Find the slope of the line passing through $(-4, 0)$ and $(4, 6)$. Then find the distance between the points and the midpoint of the line segment joining the points.

16. Determine the equations of lines through the point $(-2, 1)$ (a) parallel to $2x - y = 1$ and (b) perpendicular to $3x + 2y = 5$.

In Exercises 17 and 18, graph the equation.

17. $4x + 3y - 12 = 0$

18. $y = 2 - (x - 3)^2$

In Exercises 19–21, use the indicated method to solve the system.

19. *Substitution:*

$$\begin{cases} x + y = 6 \\ 2x - y = 3 \end{cases}$$

20. *Elimination:*

$$\begin{cases} 2x + y = 6 \\ 3x - 2y = 16 \end{cases}$$

21. *Matrices:*

$$\begin{cases} 2x + y - 2z = 1 \\ x \quad\quad - z = 1 \\ 3x + 3y + z = 12 \end{cases}$$

9

4.5

13

x

Figure for 10

Using Different Approaches

In math, one small detail such as a misplaced parenthesis or exponent can lead to a wrong answer, even if you set up the problem correctly. One way you can minimize detail mistakes is to be aware of learning modalities. A learning modality is a preferred way of taking in information that is then transferred into the brain for processing. The three modalities are *visual*, *auditory*, and *kinesthetic*. The following are brief descriptions of these modalities.

- **Visual** You take in information more productively if you can see the information.
- **Auditory** You take in information more productively when you listen to an explanation and talk about it.
- **Kinesthetic** You take in information more productively if you can experience it or use physical activity in studying.

You may find that one approach, or even a combination of approaches, works best for you.

Kimberly Nolting

VP, Academic Success Press
expert in developmental education

$$4x^2(3x - 2x^3 + 1)$$
$$= 4x^2(3x) - 4x^2(2x^3) + 4x^2(1)$$
$$= 12x^3 \quad - 8x^5 \quad + 4x^2$$
$$= -8x^5 + 12x^3 + 4x^2$$

$$4x^2(3x - 2x^3 + 1)$$
↑   ↑

Second product

Smart Study Strategy

Use All Three Modalities

Design a way to solve a problem using all three learning modalities.

Visual Highlight or use a different color of pencil to indicate when you have simplified something in a problem. This is done in the example at the left.

Auditory Say each calculation step out loud.

"I need to multiply four x to the second power and two x to the third power. Since two x to the third power is negative, I must include the negative sign when I multiply by four x to the second power."

Kinesthetic When completing and checking your calculations, use your fingers to point to each step. Double-check signs, parentheses, exponents, and so on. For the given example, you can use your fingers to point to the products formed by the monomial and each term of the trinomial.

Chapter 5
Polynomials and Factoring

$$\frac{-b \pm \sqrt{b^2 - 4ac}}{2a}$$

IT WORKED FOR ME!

"I understand math but never did as well as I wanted to until this semester. My instructor pointed out that I set problems up correctly and know all the steps, but I mess up in the details. I took a learning modality inventory and I am a visual learner, but I discovered that if I talk out loud and move my finger over each little step, I catch my mistakes."

Katie
Major undecided

5.1 Integer Exponents and Scientific Notation

© Blend Images/Alamy

Why You Should Learn It

Scientific notation can be used to represent very large real-life quantities. For instance, in Exercise 152 on page 307, you will use scientific notation to determine the total amount of rice consumed in the world in 1 year.

1 ▶ Use the rules of exponents to simplify expressions.

What You Should Learn

1 ▶ Use the rules of exponents to simplify expressions.
2 ▶ Rewrite exponential expressions involving negative and zero exponents.
3 ▶ Write very large and very small numbers in scientific notation.

Rules of Exponents

Recall from Section 1.2 that *repeated multiplication* can be written in what is called **exponential form.** Let n be a positive integer and let a be a real number. Then the product of n factors of a is given by

$$a^n = \underbrace{a \cdot a \cdot a \cdots a}_{n \text{ factors}}.$$ a is the base and n is the exponent.

When multiplying two exponential expressions that have the *same base*, you add exponents. To see why this is true, consider the product $a^3 \cdot a^2$. Because the first expression represents three factors of a and the second represents two factors of a, the product of the two expressions represents five factors of a, as follows.

$$a^3 \cdot a^2 = \underbrace{(a \cdot a \cdot a)}_{3 \text{ factors}} \cdot \underbrace{(a \cdot a)}_{2 \text{ factors}} = \underbrace{(a \cdot a \cdot a \cdot a \cdot a)}_{5 \text{ factors}} = a^{3+2} = a^5$$

Rules of Exponents

Let m and n be positive integers, and let a and b represent real numbers, variables, or algebraic expressions.

	Rule	*Example*
1. Product:	$a^m \cdot a^n = a^{m+n}$	$x^5(x^4) = x^{5+4} = x^9$
2. Product-to-Power:	$(ab)^m = a^m \cdot b^m$	$(2x)^3 = 2^3(x^3) = 8x^3$
3. Power-to-Power:	$(a^m)^n = a^{mn}$	$(x^2)^3 = x^{2 \cdot 3} = x^6$
4. Quotient:	$\dfrac{a^m}{a^n} = a^{m-n}, m > n, a \neq 0$	$\dfrac{x^5}{x^3} = x^{5-3} = x^2, x \neq 0$
5. Quotient-to-Power:	$\left(\dfrac{a}{b}\right)^m = \dfrac{a^m}{b^m}, b \neq 0$	$\left(\dfrac{x}{4}\right)^2 = \dfrac{x^2}{4^2} = \dfrac{x^2}{16}$

The product rule and the product-to-power rule can be extended to three or more factors. For example,

$$a^m \cdot a^n \cdot a^k = a^{m+n+k} \quad \text{and} \quad (abc)^m = a^m b^m c^m.$$

EXAMPLE 1   Using Rules of Exponents

Simplify: **a.** $(x^2y^4)(3x)$ **b.** $-2(y^2)^3$ **c.** $(-2y^2)^3$ **d.** $(3x^2)(-5x)^3$

Solution

a. $(x^2y^4)(3x) = 3(x^2 \cdot x)(y^4) = 3(x^{2+1})(y^4) = 3x^3y^4$

b. $-2(y^2)^3 = (-2)(y^{2 \cdot 3}) = -2y^6$

c. $(-2y^2)^3 = (-2)^3(y^2)^3 = -8(y^{2 \cdot 3}) = -8y^6$

d. $(3x^2)(-5x)^3 = 3(-5)^3(x^2 \cdot x^3) = 3(-125)(x^{2+3}) = -375x^5$

✓ **CHECKPOINT** *Now try Exercise 1.*

EXAMPLE 2   Using Rules of Exponents

Simplify: **a.** $\dfrac{14a^5b^3}{7a^2b^2}$ **b.** $\left(\dfrac{x^2}{2y}\right)^3$ **c.** $\dfrac{x^n y^{3n}}{x^2 y^4}$ **d.** $\dfrac{(2a^2b^3)^2}{a^3b^2}$

Solution

a. $\dfrac{14a^5b^3}{7a^2b^2} = 2(a^{5-2})(b^{3-2}) = 2a^3b$

b. $\left(\dfrac{x^2}{2y}\right)^3 = \dfrac{(x^2)^3}{(2y)^3} = \dfrac{x^{2 \cdot 3}}{2^3 y^3} = \dfrac{x^6}{8y^3}$

c. $\dfrac{x^n y^{3n}}{x^2 y^4} = x^{n-2} y^{3n-4}$

d. $\dfrac{(2a^2b^3)^2}{a^3b^2} = \dfrac{2^2(a^{2 \cdot 2})(b^{3 \cdot 2})}{a^3b^2} = \dfrac{4a^4b^6}{a^3b^2} = 4(a^{4-3})(b^{6-2}) = 4ab^4$

✓ **CHECKPOINT** *Now try Exercise 11.*

2 ▶ Rewrite exponential expressions involving negative and zero exponents.

Integer Exponents

The definition of an exponent can be extended to include zero and negative integers. If a is a real number such that $a \neq 0$, then a^0 is defined as 1. Moreover, if m is an integer, then a^{-m} is defined as the reciprocal of a^m.

Definitions of Zero Exponents and Negative Exponents

Let a and b be real numbers such that $a \neq 0$ and $b \neq 0$, and let m be an integer.

1. $a^0 = 1$ **2.** $a^{-m} = \dfrac{1}{a^m}$ **3.** $\left(\dfrac{a}{b}\right)^{-m} = \left(\dfrac{b}{a}\right)^m$

These definitions are consistent with the rules of exponents given on page 298. For instance, consider the following.

$$x^0 \cdot x^m = x^{0+m} = x^m = 1 \cdot x^m$$

$(x^0$ is the same as 1$)$

EXAMPLE 3   **Zero Exponents and Negative Exponents**

Evaluate each expression.

a. 3^0 **b.** 3^{-2} **c.** $\left(\frac{3}{4}\right)^{-1}$

Solution

a. $3^0 = 1$ Definition of zero exponents

b. $3^{-2} = \dfrac{1}{3^2} = \dfrac{1}{9}$ Definition of negative exponents

c. $\left(\dfrac{3}{4}\right)^{-1} = \left(\dfrac{4}{3}\right)^1 = \dfrac{4}{3}$ Definition of negative exponents

✓ **CHECKPOINT** *Now try Exercise 21.*

Study Tip

Notice that by definition, $a^0 = 1$ for all real *nonzero* values of a. Zero cannot have a zero exponent, because the expression 0^0 is undefined.

Study Tip

Notice that the first five rules of exponents were first listed on page 298 for *positive* values of m and n, and the quotient rule included the restriction $m > n$.

Because the rules shown here allow the use of zero exponents and negative exponents, the restriction $m > n$ is no longer necessary for the quotient rule.

Summary of Rules of Exponents

Let m and n be integers, and let a and b represent real numbers, variables, or algebraic expressions. (All denominators and bases are nonzero.)

Product and Quotient Rules	*Example*
1. $a^m \cdot a^n = a^{m+n}$	$x^4(x^3) = x^{4+3} = x^7$
2. $\dfrac{a^m}{a^n} = a^{m-n}$	$\dfrac{x^3}{x} = x^{3-1} = x^2$

Power Rules

3. $(ab)^m = a^m \cdot b^m$ $(3x)^2 = 3^2(x^2) = 9x^2$

4. $(a^m)^n = a^{mn}$ $(x^3)^3 = x^{3\cdot3} = x^9$

5. $\left(\dfrac{a}{b}\right)^m = \dfrac{a^m}{b^m}$ $\left(\dfrac{x}{3}\right)^2 = \dfrac{x^2}{3^2} = \dfrac{x^2}{9}$

Zero and Negative Exponent Rules

6. $a^0 = 1$ $(x^2 + 1)^0 = 1$

7. $a^{-m} = \dfrac{1}{a^m}$ $x^{-2} = \dfrac{1}{x^2}$

8. $\left(\dfrac{a}{b}\right)^{-m} = \left(\dfrac{b}{a}\right)^m$ $\left(\dfrac{x}{3}\right)^{-2} = \left(\dfrac{3}{x}\right)^2 = \dfrac{3^2}{x^2} = \dfrac{9}{x^2}$

EXAMPLE 4   **Using Rules of Exponents**

a. $2x^{-1} = 2(x^{-1}) = 2\left(\dfrac{1}{x}\right) = \dfrac{2}{x}$ Use negative exponent rule and simplify.

b. $(2x)^{-1} = \dfrac{1}{(2x)^1} = \dfrac{1}{2x}$ Use negative exponent rule and simplify.

✓ **CHECKPOINT** *Now try Exercise 55.*

Study Tip

As you become accustomed to working with negative exponents, you will probably not write as many steps as shown in Example 5. For instance, to rewrite a fraction involving exponents, you might use the following simplified rule. *To move a factor from the numerator to the denominator or vice versa, change the sign of its exponent.* You can apply this rule to the expression in Example 5(a) by "moving" the factor x^{-2} to the numerator and changing the exponent to 2. That is,

$$\frac{3}{x^{-2}} = 3x^2.$$

Remember, you can move only *factors* in this manner, not terms.

EXAMPLE 5   Using Rules of Exponents

Rewrite each expression using only positive exponents. (Assume that $x \neq 0$.)

a. $\dfrac{3}{x^{-2}} = \dfrac{3}{\left(\dfrac{1}{x^2}\right)}$ Negative exponent rule

  $= 3\left(\dfrac{x^2}{1}\right) = 3x^2$ Invert divisor and multiply.

b. $\dfrac{1}{(3x)^{-2}} = \dfrac{1}{\left[\dfrac{1}{(3x)^2}\right]}$ Use negative exponent rule.

  $= \dfrac{1}{\left(\dfrac{1}{9x^2}\right)}$ Use product-to-power rule and simplify.

  $= (1)\left(\dfrac{9x^2}{1}\right) = 9x^2$ Invert divisor and multiply.

✔ **CHECKPOINT** *Now try Exercise 63.*

EXAMPLE 6   Using Rules of Exponents

Rewrite each expression using only positive exponents. (Assume that $x \neq 0$ and $y \neq 0$.)

a. $(-5x^{-3})^2 = (-5)^2 (x^{-3})^2$ Product-to-power rule

  $= 25x^{-6}$ Power-to-product rule

  $= \dfrac{25}{x^6}$ Negative exponent rule

b. $-\left(\dfrac{7x}{y^2}\right)^{-2} = -\left(\dfrac{y^2}{7x}\right)^2$ Negative exponent rule

  $= -\dfrac{(y^2)^2}{(7x)^2}$ Quotient-to-power rule

  $= -\dfrac{y^4}{49x^2}$ Power-to-power and product-to-power rules

c. $\dfrac{12x^2 y^{-4}}{6x^{-1} y^2} = 2(x^{2-(-1)})(y^{-4-2})$ Quotient rule

  $= 2x^3 y^{-6}$ Simplify.

  $= \dfrac{2x^3}{y^6}$ Negative exponent rule

✔ **CHECKPOINT** *Now try Exercise 73.*

> **EXAMPLE 7**   **Using Rules of Exponents**
>
> Rewrite each expression using only positive exponents. (Assume that $x \neq 0$ and $y \neq 0$.)
>
> **a.** $\left(\dfrac{8x^{-1}y^4}{4x^3y^2}\right)^{-3} = \left(\dfrac{2y^2}{x^4}\right)^{-3}$ Simplify.
>
> $= \left(\dfrac{x^4}{2y^2}\right)^{3}$ Negative exponent rule
>
> $= \dfrac{x^{12}}{2^3y^6} = \dfrac{x^{12}}{8y^6}$ Quotient-to-power rule
>
> **b.** $\dfrac{3xy^0}{x^2(5y)^0} = \dfrac{3x(1)}{x^2(1)} = \dfrac{3}{x}$ Zero exponent rule
>
> ✓ **CHECKPOINT** *Now try Exercise 75.*

3 ▶ Write very large and very small numbers in scientific notation.

Scientific Notation

Exponents provide an efficient way of writing and computing with very large and very small numbers. For instance, a drop of water contains more than 33 billion billion molecules—that is, 33 followed by 18 zeros. It is convenient to write such numbers in **scientific notation.** This notation has the form $c \times 10^n$, where $1 \leq c < 10$ and n is an integer. So, the number of molecules in a drop of water can be written in scientific notation as follows.

$$33{,}000{,}000{,}000{,}000{,}000{,}000 = 3.3 \times 10^{19}$$

19 places

The *positive* exponent 19 indicates that the number being written in scientific notation is *large* (10 or more) and that the decimal point has been moved 19 places. A *negative* exponent in scientific notation indicates that the number is *small* (less than 1).

> **EXAMPLE 8**   **Writing in Scientific Notation**
>
> Write each number in scientific notation.
>
> **a.** 0.0000684   **b.** 937,200,000
>
> **Solution**
>
> **a.** $0.0000684 = 6.84 \times 10^{-5}$ Small number ⟹ negative exponent
>
>   Five places
>
> **b.** $937{,}200{,}000.0 = 9.372 \times 10^{8}$ Large number ⟹ positive exponent
>
>   Eight places
>
> ✓ **CHECKPOINT** *Now try Exercise 101.*

 **EXAMPLE 9   Writing in Decimal Notation**

a. $2.486 \times 10^2 = 248.6$ Positive exponent ⟹ large number

                         Two places

b. $1.81 \times 10^{-6} = 0.00000181$ Negative exponent ⟹ small number

                         Six places

 CHECKPOINT *Now try Exercise 115.*

EXAMPLE 10   Using Scientific Notation

Rewrite the factors in scientific notation and then evaluate

$$\frac{(2,400,000,000)(0.0000045)}{(0.00003)(1500)}.$$

Solution

$$\frac{(2,400,000,000)(0.0000045)}{(0.00003)(1500)} = \frac{(2.4 \times 10^9)(4.5 \times 10^{-6})}{(3.0 \times 10^{-5})(1.5 \times 10^3)}$$

$$= \frac{(2.4)(4.5)(10^3)}{(4.5)(10^{-2})}$$

$$= (2.4)(10^5) = 2.4 \times 10^5$$

 CHECKPOINT *Now try Exercise 133.*

EXAMPLE 11   Using Scientific Notation with a Calculator

Use a calculator to evaluate each expression.

a. $65,000 \times 3,400,000,000$ **b.** $0.000000348 \div 870$

Solution

a. 6.5 $\boxed{\text{EXP}}$ 4 $\boxed{\times}$ 3.4 $\boxed{\text{EXP}}$ 9 $\boxed{=}$ Scientific

   6.5 $\boxed{\text{EE}}$ 4 $\boxed{\times}$ 3.4 $\boxed{\text{EE}}$ 9 $\boxed{\text{ENTER}}$ Graphing

The calculator display should read $\boxed{2.21\text{E }14}$, which implies that

$$(6.5 \times 10^4)(3.4 \times 10^9) = 2.21 \times 10^{14} = 221,000,000,000,000.$$

b. 3.48 $\boxed{\text{EXP}}$ 7 $\boxed{\pm}$ $\boxed{\div}$ 8.7 $\boxed{\text{EXP}}$ 2 $\boxed{=}$ Scientific

   3.48 $\boxed{\text{EE}}$ $\boxed{(-)}$ 7 $\boxed{\div}$ 8.7 $\boxed{\text{EE}}$ 2 $\boxed{\text{ENTER}}$ Graphing

The calculator display should read $\boxed{4\text{E }-10}$, which implies that

$$\frac{3.48 \times 10^{-7}}{8.7 \times 10^2} = 4.0 \times 10^{-10} = 0.0000000004.$$

 CHECKPOINT *Now try Exercise 135.*

Technology: Tip

Most scientific and graphing calculators automatically switch to scientific notation to show large or small numbers that exceed the display range.

To *enter* numbers in scientific notation, your calculator should have a key labeled $\boxed{\text{EE}}$ or $\boxed{\text{EXP}}$. Consult the user's guide for your calculator for specific instructions.

Concept Check

1. In your own words, describe how to simplify the expression.
   (a) $x^a \cdot x^b$ (b) $(xy)^m$
   (c) $(x^a)^b$ (d) $\dfrac{c^m}{c^n}$

2. Let x represent a real number such that $x > 1$. What can you say about the value of each of the expressions x^2, x^1, x^0, x^{-1}, and x^{-2}?

3. To write a decimal in scientific notation, you write the number as $c \times 10^n$, where $1 \le c < 10$. Explain in your own words how to write the number c for a given decimal.

4. Explain why you would not use scientific notation to list the ingredient amounts in a cooking recipe.

5.1 EXERCISES

Go to pages 360–361 to record your assignments.

Developing Skills

In Exercises 1–20, use the rules of exponents to simplify the expression (if possible). *See Examples 1 and 2.*

1. (a) $-3x^3 \cdot x^5$ (b) $(-3x)^2 \cdot x^5$
2. (a) $5^2 y^4 \cdot y^2$ (b) $(5y)^2 \cdot y^4$
3. (a) $(-5z^2)^3$ (b) $(-5z^4)^2$
4. (a) $(-5z^3)^2$ (b) $(-5z)^4$
5. (a) $(u^3 v)(2v^2)$ (b) $(-4u^4)(u^5 v)$
6. (a) $(6xy^7)(-x)$ (b) $(x^5 y^3)(2y^3)$
7. (a) $5u^2 \cdot (-3u^6)$ (b) $(2u)^4(4u)$

8. (a) $(3y)^3(2y^2)$ (b) $3y^3 \cdot 2y^2$
9. (a) $-(m^5 n)^3(-m^2 n^2)^2$ (b) $(-m^5 n)(m^2 n^2)$

10. (a) $-(m^3 n^2)(mn^3)$ (b) $-(m^3 n^2)^2(-mn^3)$

11. (a) $\dfrac{27m^5 n^6}{9mn^3}$ (b) $\dfrac{-18m^3 n^6}{-6mn^3}$

12. (a) $\dfrac{28x^2 y^3}{2xy^2}$ (b) $\dfrac{24xy^2}{8y}$

13. (a) $\left(\dfrac{3x}{4y}\right)^2$ (b) $\left(\dfrac{5u}{3v}\right)^3$

14. (a) $\left(\dfrac{2a}{3y}\right)^5$ (b) $-\left(\dfrac{2a}{3y}\right)^2$

15. (a) $-\dfrac{(-2x^2 y)^3}{9x^2 y^2}$ (b) $-\dfrac{(-2xy^3)^2}{6y^2}$

16. (a) $\dfrac{(-4xy)^3}{8xy^2}$ (b) $\dfrac{(-xy)^4}{-3(xy)^2}$

17. (a) $\left[\dfrac{(-5u^3 v)^2}{10u^2 v}\right]^2$ (b) $\left[\dfrac{-5(u^3 v)^2}{10u^2 v}\right]^2$

18. (a) $\left[\dfrac{(3x^2)(2x)^2}{(-2x)(6x)}\right]^2$ (b) $\left[\dfrac{(3x^2)(2x)^4}{(-2x)^2(6x)}\right]^2$

19. (a) $\dfrac{x^{2n+4}\, y^{4n}}{x^5\, y^{2n+1}}$ (b) $\dfrac{x^{6n}\, y^{n-7}}{x^{4n+2}\, y^5}$

20. (a) $\dfrac{x^{3n}\, y^{2n-1}}{x^n\, y^{n+3}}$ (b) $\dfrac{x^{4n-6}\, y^{n+10}}{x^{2n-5}\, y^{n-2}}$

In Exercises 21–50, evaluate the expression. *See Example 3.*

21. 5^{-2} 22. 2^{-4}
23. -10^{-3} 24. -20^{-2}
25. $(-3)^0$ 26. 25^0
27. $\dfrac{1}{4^{-3}}$ 28. $\dfrac{1}{-8^{-2}}$
29. $\dfrac{1}{(-2)^{-5}}$ 30. $-\dfrac{1}{6^{-2}}$
31. $\left(\dfrac{2}{3}\right)^{-1}$ 32. $\left(\dfrac{4}{5}\right)^{-3}$
33. $\left(\dfrac{3}{16}\right)^0$ 34. $\left(-\dfrac{5}{8}\right)^{-2}$
35. $27 \cdot 3^{-3}$ 36. $16 \cdot 4^{-4}$
37. $\dfrac{3^4}{3^{-2}}$ 38. $\dfrac{5^{-1}}{5^2}$
39. $\dfrac{10^3}{10^{-2}}$ 40. $\dfrac{10^{-5}}{10^{-6}}$
41. $(4^2 \cdot 4^{-1})^{-2}$ 42. $(5^3 \cdot 5^{-4})^{-3}$

43. $(2^{-3})^2$

44. $(-4^{-1})^{-2}$

45. $2^{-3} + 2^{-4}$

46. $4 - 3^{-2}$

47. $\left(\frac{3}{4} + \frac{5}{8}\right)^{-2}$

48. $\left(\frac{1}{2} - \frac{2}{3}\right)^{-1}$

49. $(5^0 - 4^{-2})^{-1}$

50. $(32 + 4^{-3})^0$

In Exercises 51–90, rewrite the expression using only positive exponents, and simplify. (Assume that any variables in the expression are nonzero.) *See Examples 4–7.*

51. $y^4 \cdot y^{-2}$

52. $x^{-2} \cdot x^{-5}$

53. $z^5 \cdot z^{-3}$

54. $t^{-1} \cdot t^{-6}$

✓ 55. $7x^{-4}$

56. $3y^{-3}$

57. $(4x)^{-3}$

58. $(5u)^{-2}$

59. $\frac{1}{x^{-6}}$

60. $\frac{4}{y^{-1}}$

61. $\frac{8a^{-6}}{6a^{-7}}$

62. $\frac{6u^{-2}}{15u^{-1}}$

✓ 63. $\frac{(4t)^0}{t^{-2}}$

64. $\frac{(5u)^{-4}}{(5u)^0}$

65. $(2x^2)^{-2}$

66. $(4a^{-2})^{-3}$

67. $(-3x^{-3}y^2)(4x^2y^{-5})$

68. $(5s^5t^{-5})(-6s^{-2}t^4)$

69. $(3x^2y^{-2})^{-2}$

70. $(-4y^{-3}z)^{-3}$

71. $\left(\frac{x}{10}\right)^{-1}$

72. $\left(\frac{4}{z}\right)^{-2}$

✓ 73. $\frac{6x^3y^{-3}}{12x^{-2}y}$

74. $\frac{2y^{-1}z^{-3}}{4yz^{-3}}$

✓ 75. $\left(\frac{3u^2v^{-1}}{3^3u^{-1}v^3}\right)^{-2}$

76. $\left(\frac{5^2x^3y^{-3}}{125xy}\right)^{-1}$

77. $\left(\frac{a^{-2}}{b^{-2}}\right)\left(\frac{b}{a}\right)^3$

78. $\left(\frac{a^{-3}}{b^{-3}}\right)\left(\frac{b}{a}\right)^3$

79. $(2x^3y^{-1})^{-3}(4xy^{-6})$

80. $(ab)^{-2}(a^2b^2)^{-1}$

81. $u^4(6u^{-3}v^0)(7v)^0$

82. $x^5(3x^0y^4)(7y)^0$

83. $[(x^{-4}y^{-6})^{-1}]^2$

84. $[(2x^{-3}y^{-2})^2]^{-2}$

85. $\frac{(2a^{-2}b^4)^3b}{(10a^3b)^2}$

86. $\frac{(5x^2y^{-5})^{-1}}{2x^{-5}y^4}$

87. $(u + v^{-2})^{-1}$

88. $x^{-2}(x^2 + y^2)$

89. $\frac{a + b}{b^{-1}a + 1}$

90. $\frac{u^{-1} - v^{-1}}{u^{-1} + v^{-1}}$

In Exercises 91–100, evaluate the expression when $x = -3$ and $y = 4$.

91. $x^2 \cdot x^{-3} \cdot x \cdot y$

92. $x^4 \cdot x^{-1} \cdot x^{-1} \cdot y$

93. $\frac{x^2}{y^{-2}}$

94. $\frac{y^2}{x^{-2}}$

95. $(x + y)^{-4}$

96. $(-x - y)^{-2}$

97. $\left(\frac{5x}{3y}\right)^{-1}$

98. $\left(\frac{3y}{12x}\right)^{-2}$

99. $(xy)^{-2}$

100. $(x^2y)^{-1}$

In Exercises 101–114, write the number in scientific notation. *See Example 8.*

✓ 101. 3,600,000

102. 98,100,000

103. 47,620,000

104. 841,000,000,000

105. 0.00031

106. 0.00625

107. 0.0000000381

108. 0.0000000000692

109. *Land Area of Earth*: 57,300,000 square miles

110. *Water Area of Earth*: 139,500,000 square miles

111. *Light Year*: 9,460,800,000,000 kilometers

112. *Thickness of a Soap Bubble*: 0.0000001 meter

113. *Relative Density of Hydrogen*: 0.0899 gram per milliliter

114. *One Micron (Millionth of a Meter)*: 0.00003937 inch

In Exercises 115–124, write the number in decimal notation. *See Example 9.*

✓ **115.** 7.2×10^8

116. 7.413×10^{11}

117. 1.359×10^{-7}

118. 8.6×10^{-9}

119. *2006 Merrill Lynch Revenues:* $\$3.4659 \times 10^{10}$ (Source: 2006 Merrill Lynch Annual Report)

120. *Number of Air Sacs in the Lungs:* 3.5×10^8

121. *Interior Temperature of the Sun:* 1.5×10^7 degrees Celsius

122. *Width of an Air Molecule:* 9.0×10^{-9} meter

123. *Charge of an Electron:* 4.8×10^{-10} electrostatic unit

124. *Width of a Human Hair:* 9.0×10^{-4} meter

In Exercises 125–134, evaluate the expression without a calculator. *See Example 10.*

125. $(2 \times 10^9)(3.4 \times 10^{-4})$

126. $(6.5 \times 10^6)(2 \times 10^4)$

127. $(5 \times 10^4)^2$

128. $(4 \times 10^6)^3$

129. $\dfrac{3.6 \times 10^{12}}{6 \times 10^5}$

130. $\dfrac{2.5 \times 10^{-3}}{5 \times 10^2}$

131. $(4,500,000)(2,000,000,000)$

132. $(62,000,000)(0.0002)$

✓ **133.** $\dfrac{64,000,000}{0.00004}$

134. $\dfrac{72,000,000,000}{0.00012}$

In Exercises 135–142, evaluate with a calculator. Write the answer in scientific notation, $c \times 10^n$, with c rounded to two decimal places. *See Example 11.*

✓ **135.** $\dfrac{(0.0000565)(2,850,000,000,000)}{0.00465}$

136. $\dfrac{(3,450,000,000)(0.000125)}{(52,000,000)(0.000003)}$

137. $\dfrac{1.357 \times 10^{12}}{(4.2 \times 10^2)(6.87 \times 10^{-3})}$

138. $\dfrac{(3.82 \times 10^5)^2}{(8.5 \times 10^4)(5.2 \times 10^{-3})}$

139. $(2.58 \times 10^6)^4$

140. $(8.67 \times 10^4)^7$

141. $\dfrac{(5,000,000)^3(0.000037)^2}{(0.005)^4}$

142. $\dfrac{(6,200,000)(0.005)^3}{(0.00035)^5}$

Solving Problems

143. *Distance*   The distance from Earth to the Sun is approximately 93 million miles. Write this distance in scientific notation.

144. *Stars*   A study by Australian astronomers estimated the number of stars within range of modern telescopes to be 70,000,000,000,000,000,000,000. Write this number in scientific notation. (Source: The Australian National University)

145. *Electrons*   A cube of copper with an edge of 1 centimeter has approximately 8.483×10^{22} free electrons. Write this real number in decimal notation.

146. *Lumber Consumption*   The total volume of the lumber consumed in the United States in 2005 was about 1.0862×10^{10} cubic feet. Write this volume in decimal notation. (Source: U.S. Forest Service)

147. *Light Year*   One light year (the distance light can travel in 1 year) is approximately 9.46×10^{15} meters. Approximate the time to the nearest minute for light to travel from the Sun to Earth if that distance is approximately 1.50×10^{11} meters.

148. *Masses of Earth and the Sun* The masses of Earth and the Sun are approximately 5.98×10^{24} kilograms and 1.99×10^{30} kilograms, respectively. The mass of the Sun is approximately how many times that of Earth?

149. *Distance* The star Alpha Andromeda is approximately 95 light years from Earth. Determine this distance in meters. (See Exercise 147 for the definition of a light year.)

150. *Metal Expansion* When the temperature of an iron steam pipe 200 feet long is increased by 75°C, the length of the pipe will increase by an amount $75(200)(1.1 \times 10^{-5})$. Find this amount and write the answer in decimal notation.

151. *Federal Debt* In 2005, the resident population of the United States was about 296 million people, and it would have cost each resident about $26,600 to pay off the federal debt. Use these two numbers to approximate the federal debt in 2005. (Source: U.S. Census Bureau and U.S. Office of Management and Budget)

152. *Rice Consumption* In 2005, the population of the world was about 6.451 billion people, and the average person consumed about 141.8 pounds of milled rice. Use these two numbers to approximate the total amount (in pounds) of milled rice consumed in the world in 2005. (Source: U.S. Census Bureau and U.S. Department of Agriculture)

Explaining Concepts

153. *Think About It* Discuss whether you feel that using scientific notation to multiply or divide very large or very small numbers makes the process *easier* or *more difficult*. Support your position with an example.

154. You multiply an expression by a^5. The product is a^{12}. What was the original expression? Explain how you found your answer.

True or False? In Exercises 155 and 156, determine whether the statement is true or false. Justify your reasoning.

155. The value of $\dfrac{1}{3^{-3}}$ is less than 1.

156. The expression 0.142×10^{10} is in scientific notation.

In Exercises 157–160, use the rules of exponents to explain why the statement is *false*.

157. $a^m \cdot b^n = ab^{m+n}$ ✗

158. $(ab)^m = a^m + b^m$ ✗

159. $(a^m)^n = a^{m+n}$ ✗

160. $\dfrac{a^m}{a^n} = a^m - a^n$ ✗

Cumulative Review

In Exercises 161–164, simplify the expression by combining like terms.

161. $3x + 4x - x$

162. $y - 3x + 4y - 2$

163. $a^2 + 2ab - b^2 + ab + 4b^2$

164. $x^2 + 5x^2y - 3x^2y + 4x^2$

In Exercises 165 and 166, sketch a graph of the solution of the system of linear inequalities.

165. $\begin{cases} x > 2 \\ x - y \le 0 \\ y < 0 \end{cases}$

166. $\begin{cases} x - 2y \le 6 \\ x + y \le 0 \\ y > 0 \end{cases}$

5.2 Adding and Subtracting Polynomials

Herbert Hartmann/Getty Images

What You Should Learn

1 ▶ Identify leading coefficients and degrees of polynomials.

2 ▶ Add and subtract polynomials using a horizontal format and a vertical format.

3 ▶ Use polynomials to model and solve real-life problems.

Why You Should Learn It

Polynomials can be used to model many aspects of the physical world. For instance, in Exercise 103 on page 315, a polynomial is used to model the height of an object thrown from the top of the Eiffel Tower.

1 ▶ Identify leading coefficients and degrees of polynomials.

Basic Definitions

A **polynomial in x** is an algebraic expression whose terms are all of the form ax^k, where a is any real number and k is a nonnegative integer. The following *are not* polynomials for the reasons stated.

- The expression $2x^{-1} + 5$ is not a polynomial because the exponent in $2x^{-1}$ is negative.
- The expression $x^3 + 3x^{1/2}$ is not a polynomial because the exponent in $3x^{1/2}$ is not an integer.

Definition of a Polynomial in x

Let $a_n, a_{n-1}, \ldots, a_2, a_1, a_0$ be real numbers and let n be a *nonnegative integer*. A **polynomial in x** is an expression of the form

$$a_n x^n + a_{n-1} x^{n-1} + \cdots + a_2 x^2 + a_1 x + a_0$$

where $a_n \neq 0$. The polynomial is of **degree n,** and the number a_n is called the **leading coefficient.** The number a_0 is called the **constant term.**

In the term ax^k, a is the **coefficient** and k is the degree of the term. Note that the degree of the term ax is 1 and the degree of a constant term is 0. Because a polynomial is an algebraic sum, the coefficients take on the signs between the terms. For instance, the polynomial

$$x^3 - 4x^2 + 3 = (1)x^3 + (-4)x^2 + (0)x + 3$$

has coefficients 1, $-4, 0$, and 3. A polynomial that is written in order of descending powers of the variable is said to be in **standard form.** A polynomial with only one term is a **monomial.** Polynomials with two *unlike* terms are called **binomials,** and those with three *unlike* terms are called **trinomials.** Some examples of polynomials are shown below.

Polynomial	Standard Form	Degree	Leading Coefficient
$5x^2 - 2x^7 + 4 - 2x$	$-2x^7 + 5x^2 - 2x + 4$	7	-2
$16 + x^3$	$x^3 + 16$	3	1
10	10	0	10

2 ▶ Add and subtract polynomials using a horizontal format and a vertical format.

Adding and Subtracting Polynomials

To add two polynomials, simply combine like terms. This can be done in either a horizontal or a vertical format, as shown in Examples 1 and 2.

Technology: Tip

You can use a graphing calculator to check the results of polynomial operations. For instance, in Example 1(a), you can graph the equations

$$y = (2x^3 + x^2 - 5) +$$
$$(x^2 + x + 6)$$

and

$$y = 2x^3 + 2x^2 + x + 1$$

in the same viewing window. The fact that the graphs coincide, as shown below, confirms that the two polynomials are equivalent.

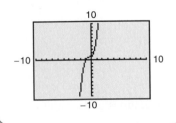

| EXAMPLE 1 | Adding Polynomials Horizontally |

a. $(2x^3 + x^2 - 5) + (x^2 + x + 6)$ Original polynomials

$\quad = (2x^3) + (x^2 + x^2) + (x) + (-5 + 6)$ Group like terms.

$\quad = 2x^3 + 2x^2 + x + 1$ Combine like terms.

b. $(3x^2 + 2x + 4) + (3x^2 - 6x + 3) + (-x^2 + 2x - 4)$

$\quad = (3x^2 + 3x^2 - x^2) + (2x - 6x + 2x) + (4 + 3 - 4)$

$\quad = 5x^2 - 2x + 3$

✔ **CHECKPOINT** *Now try Exercise 29.*

| EXAMPLE 2 | Using a Vertical Format to Add Polynomials |

Use a vertical format to find the sum.

$$(5x^3 + 2x^2 - x + 7) + (3x^2 - 4x + 7) + (-x^3 + 4x^2 - 8)$$

Solution

To use a vertical format, align the terms of the polynomials by their degrees.

$$\begin{array}{r} 5x^3 + 2x^2 - \ x + 7 \\ 3x^2 - 4x + 7 \\ -\ x^3 + 4x^2 \qquad - 8 \\ \hline 4x^3 + 9x^2 - 5x + 6 \end{array}$$

✔ **CHECKPOINT** *Now try Exercise 43.*

To subtract one polynomial from another, *add the opposite*. You can do this by changing the sign of each term of the polynomial that is being subtracted and then adding the resulting like terms.

| EXAMPLE 3 | Subtracting Polynomials Horizontally |

Use a horizontal format to subtract $x^3 + 2x^2 - x - 4$ from $3x^3 - 5x^2 + 3$.

Solution

$\quad (3x^3 - 5x^2 + 3) - (x^3 + 2x^2 - x - 4)$ Write original polynomials.

$\quad = (3x^3 - 5x^2 + 3) + (-x^3 - 2x^2 + x + 4)$ Add the opposite.

$\quad = (3x^3 - x^3) + (-5x^2 - 2x^2) + (x) + (3 + 4)$ Group like terms.

$\quad = 2x^3 - 7x^2 + x + 7$ Combine like terms.

✔ **CHECKPOINT** *Now try Exercise 53.*

Be especially careful to use the correct signs when subtracting one polynomial from another. One of the most common mistakes in algebra is forgetting to change signs correctly when subtracting one expression from another. Here is an example.

Wrong sign
↓
$$(x^2 - 2x + 3) - (x^2 + 2x - 2) \neq x^2 - 2x + 3 - x^2 + 2x - 2$$ Common error
↑
Wrong sign

The error illustrated above is forgetting to change two of the signs in the polynomial that is being subtracted. Be sure to add the opposite of *every* term of the subtracted polynomial.

EXAMPLE 4   Using a Vertical Format to Subtract Polynomials

Use a vertical format to find the difference.

$$(4x^4 - 2x^3 + 5x^2 - x + 8) - (3x^4 - 2x^3 + 3x - 4)$$

Solution

$$
\begin{array}{l}
(4x^4 - 2x^3 + 5x^2 - \ x + 8) \\
-(3x^4 - 2x^3 \qquad\quad + 3x - 4)
\end{array}
\quad\Longrightarrow\quad
\begin{array}{l}
\ 4x^4 - 2x^3 + 5x^2 - \ x + \ 8 \\
-3x^4 + 2x^3 \qquad\quad - 3x + \ 4 \\
\hline
\ \ x^4 \qquad\quad + 5x^2 - 4x + 12
\end{array}
$$

✔ **CHECKPOINT** *Now try Exercise 63.*

EXAMPLE 5   Combining Polynomials

Use a horizontal format to perform the indicated operations and simplify.

a. $(2x^2 - 7x + 2) - (4x^2 + 5x - 1) + (-x^2 + 4x + 4)$

b. $(-x^2 + 4x - 3) - [(4x^2 - 3x + 8) - (-x^2 + x + 7)]$

Solution

a. $(2x^2 - 7x + 2) - (4x^2 + 5x - 1) + (-x^2 + 4x + 4)$

$$= 2x^2 - 7x + 2 - 4x^2 - 5x + 1 - x^2 + 4x + 4$$

$$= (2x^2 - 4x^2 - x^2) + (-7x - 5x + 4x) + (2 + 1 + 4)$$

$$= -3x^2 - 8x + 7$$

b. $(-x^2 + 4x - 3) - [(4x^2 - 3x + 8) - (-x^2 + x + 7)]$

$$= (-x^2 + 4x - 3) - (4x^2 - 3x + 8 + x^2 - x - 7)$$

$$= (-x^2 + 4x - 3) - [(4x^2 + x^2) + (-3x - x) + (8 - 7)]$$

$$= (-x^2 + 4x - 3) - (5x^2 - 4x + 1)$$

$$= -x^2 + 4x - 3 - 5x^2 + 4x - 1$$

$$= (-x^2 - 5x^2) + (4x + 4x) + (-3 - 1)$$

$$= -6x^2 + 8x - 4$$

✔ **CHECKPOINT** *Now try Exercise 71.*

3 ▶ Use polynomials to model and solve real-life problems.

Applications

Function notation can be used to represent polynomials in a single variable. Such notation is useful for evaluating polynomials and is used in applications involving polynomials, as shown in Example 6.

There are many applications that involve polynomials. One commonly used second-degree polynomial is called a **position function.** This polynomial is a function of time and has the form

$$h(t) = -16t^2 + v_0 t + s_0 \qquad \text{Position function}$$

where the height h is measured in feet and the time t is measured in seconds.

This position function gives the height (above ground) of a free-falling object. The coefficient of t, v_0, is called the **initial velocity** of the object, and the constant term, s_0, is called the **initial height** of the object. If the initial velocity is positive, the object was projected upward (at $t = 0$); if the initial velocity is negative, the object was projected downward; and if the initial velocity is zero, the object was dropped.

EXAMPLE 6   Free-Falling Object

An object is thrown downward from the 86th floor observatory at the Empire State Building, which is 1050 feet high. The initial velocity is -15 feet per second. Use the position function

$$h(t) = -16t^2 - 15t + 1050$$

to find the height of the object when $t = 1$, $t = 4$, and $t = 7$. (See Figure 5.1.)

Solution

When $t = 1$, the height of the object is

$$h(1) = -16(1)^2 - 15(1) + 1050 \qquad \text{Substitute 1 for } t.$$

$$= -16 - 15 + 1050 \qquad \text{Simplify.}$$

$$= 1019 \text{ feet.}$$

When $t = 4$, the height of the object is

$$h(4) = -16(4)^2 - 15(4) + 1050 \qquad \text{Substitute 4 for } t.$$

$$= -256 - 60 + 1050 \qquad \text{Simplify.}$$

$$= 734 \text{ feet.}$$

When $t = 7$, the height of the object is

$$h(7) = -16(7)^2 - 15(7) + 1050 \qquad \text{Substitute 7 for } t.$$

$$= -784 - 105 + 1050 \qquad \text{Simplify.}$$

$$= 161 \text{ feet.}$$

✓ **CHECKPOINT** *Now try Exercise 97.*

Use your calculator to determine the height of the object in Example 6 when $t = 7.6457$. What can you conclude?

$t = 0$
$t = 1$
$t = 4$
1050 ft
$t = 7$

Figure 5.1

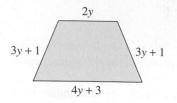

$2y$

$3y + 1$ $3y + 1$

$4y + 3$

Figure 5.2

In 2005, minority-owned businesses borrowed over 6 billion dollars.

Keith Brofsky/Photodisc/Getty Images

EXAMPLE 7 **Geometry: Perimeter**

Write an expression for the perimeter of the trapezoid shown in Figure 5.2. Then find the perimeter when $y = 5$.

Solution

To write an expression for the perimeter P of the trapezoid, find the sum of the lengths of the sides.

$$P = (2y) + (3y + 1) + (3y + 1) + (4y + 3)$$
$$= (2y + 3y + 3y + 4y) + (1 + 1 + 3)$$
$$= 12y + 5$$

To find the perimeter when $y = 5$, substitute 5 for y in the expression for the perimeter.

$$P = 12y + 5 \ = 12(5) + 5 \ = 60 + 5 = 65 \text{ units}$$

 CHECKPOINT *Now try Exercise 107.*

EXAMPLE 8 **Using Polynomial Models**

The numbers of small business loans granted by the U.S. Small Business Administration to African American owned businesses A and Hispanic American owned businesses H from 2001 to 2005 can be modeled by

$$A = 187.36t^2 + 83.0t + 1553, \quad 1 \le t \le 5 \qquad \text{African American loans}$$

$$H = 47.29t^2 + 1091.3t + 2304, \quad 1 \le t \le 5 \qquad \text{Hispanic American loans}$$

where t represents the year, with $t = 1$ corresponding to 2001. Find a polynomial that models the total number T of loans granted to both types of minority-owned businesses. Estimate the total number T of loans granted in 2005.   (Source: U.S. Small Business Administration)

Solution

The sum of the two polynomial models is as follows.

$$A + H = (187.36t^2 + 83.0t + 1553) + (47.29t^2 + 1091.3t + 2304)$$
$$= 234.65t^2 + 1174.3t + 3857$$

So, the polynomial that models the total number of loans granted is

$$T = A + H = 234.65t^2 + 1174.3t + 3857.$$

Using this model, and substituting $t = 5$, you can estimate the total number of small business loans granted to African American and Hispanic American owned businesses in 2005 to be

$$T = 234.65(5)^2 + 1174.3(5) + 3857 \ \approx 15,595.$$

 CHECKPOINT *Now try Exercise 106.*

Concept Check

1. Is a polynomial an algebraic expression? Explain.

2. Is the sum of two binomials always a binomial? Explain.

3. In your own words, define "like terms." What is the only factor of like terms that can differ?

4. Describe how to combine like terms. What operations are used?

5.2 EXERCISES

Go to pages 360–361 to record your assignments.

Developing Skills

In Exercises 1–12, write the polynomial in standard form, and find its degree and leading coefficient.

1. $4y + 16$

2. $50 - x$

3. $2x + x^2 - 6$

4. $12 + 4y - y^2$

5. $3x - 10x^2 + 5 - 42x^3$

6. $9x^3 - 2x^2 + 5x - 7$

7. $4 - 14t^4 + t^5 - 20t$

8. $6t + 4t^5 - t^2 + 3$

9. -4

10. 28

11. $v_0t - 16t^2$ (v_0 is constant.)

12. $48 - \frac{1}{2}at^2$ (a is constant.)

In Exercises 13–18, determine whether the polynomial is a monomial, a binomial, or a trinomial.

13. $12 - 5y^2$

14. $-6y + 3 + y^3$

15. $x^3 + 2x^2 - 4$

16. t^3

17. 5

18. $4m^3 + 4$

In Exercises 19–22, give an example of a polynomial in x that satisfies the conditions. (There are many correct answers.)

19. A monomial of degree 2

20. A trinomial of degree 4 and leading coefficient -2

21. A binomial of degree 3 and leading coefficient 8

22. A monomial of degree 0

In Exercises 23–26, state why the expression is not a polynomial.

23. $y^{-3} - 2$

24. $x^3 - 4x^{1/3}$

25. $6 - \sqrt{n}$

26. $\dfrac{2}{x - 4}$

In Exercises 27–42, use a horizontal format to find the sum. *See Example 1.*

27. $5 + (2 + 3x)$

28. $(6 - 2x) + 4x$

✓ 29. $(2x^2 - 3) + (5x^2 + 6)$

30. $(3x^2 + 2) + (4x^2 - 8)$

31. $(5y + 6) + (4y^2 - 6y - 3)$

32. $(3x^3 - 2x + 8) + (3x - 5)$

33. $(2 - 8y) + (-2y^4 + 3y + 2)$

34. $(z^3 + 6z - 2) + (3z^2 - 6z)$

35. $(x^3 + 9) + (2x^2 + 5) + (x^3 - 14)$

36. $(y^5 - 4y) + (3y - y^5) + (y^5 - 5)$

37. $(x^2 - 3x + 8) + (2x^2 - 4x) + 3x^2$

38. $(3a^2 + 5a) + (7 - a^2 - 5a) + (2a^2 + 8)$

39. $\left(\frac{2}{3}x^3 - 4x + 1\right) + \left(-\frac{3}{5} + 7x - \frac{1}{2}x^3\right)$

40. $\left(2 - \frac{1}{4}y^2 + y^4\right) + \left(\frac{1}{3}y^4 - \frac{3}{2}y^2 - 3\right)$

41. $(6.32t - 4.51t^2) + (7.2t^2 + 1.03t - 4.2)$

42. $(7.4x^3 + 0.26x - 15.88) + (3.5x^2 + 2.37 - 6.28x^3)$

In Exercises 43–50, use a vertical format to find the sum. *See Example 2.*

43.
$$\begin{array}{r} 5x^2 - 3x + 4 \\ -3x^2 \quad\; - 4 \\ \hline \end{array}$$

44.
$$\begin{array}{r} 3x^4 - 2x^2 - 9 \\ -5x^4 + \; x^2 \\ \hline \end{array}$$

45. $(4x^3 - 2x^2 + 8x) + (4x^2 + x - 6)$

46. $(4x^3 + 8x^2 - 5x + 3) + (x^3 - 3x^2 - 7)$

47. $(5p^2 - 4p + 2) + (-3p^2 + 2p - 7)$

48. $(16 - 32t) + (64 + 48t - 16t^2)$

49. $(2.5b - 3.6b^2) + (7.1 - 3.1b - 2.4b^2) + 6.6b^2$

50. $(2.9n^3 - 6.1n) + 1.6n + (12.2 + 3.1n - 5.3n^3)$

In Exercises 51–62, use a horizontal format to find the difference. *See Example 3.*

51. $(4 - y^3) - (4 + y^3)$

52. $(5y^4 - 2) - (3y^4 + 2)$

53. $(3x^2 - 2x + 1) - (2x^2 + x - 1)$

54. $(5q^2 - 3q + 5) - (4q^2 - 3q - 10)$

55. $(6t^3 - 12) - (-t^3 + t - 2)$

56. $(-2v^3 + v^2 - 4) - (-5v^3 - 10)$

57. $\left(\frac{1}{4}y^2 - 5y\right) - \left(12 + 4y - \frac{3}{2}y^2\right)$

58. $\left(12 - \frac{2}{3}x + \frac{1}{2}x^2\right) - \left(x^3 + 3x^2 - \frac{1}{6}x\right)$

59. $(10.4t^4 - 0.23t^5 + 1.3t^2) -$
$(2.6 - 7.35t + 6.7t^2 - 9.6t^5)$

60. $(u^3 - 9.75u^2 + 0.12u - 3) -$
$(0.7u^3 - 6.9u^2 - 4.83)$

61. Subtract $3x^3 - (x^2 + 5x)$ from $x^3 - 3x$.

62. Subtract $y^4 - (y^2 - 8y)$ from $y^2 + 3y^4$.

In Exercises 63–68, use a vertical format to find the difference. *See Example 4.*

63.
$$\begin{array}{r} x^2 - \; x + 3 \\ - \qquad (x - 2) \\ \hline \end{array}$$

64.
$$\begin{array}{r} 3t^4 - 5t^2 \\ -(-t^4 + 2t^2 - 14) \\ \hline \end{array}$$

65. $(2x^2 - 4x + 5) - (4x^2 + 5x - 6)$

66. $(4x^2 + 5x - 6) - (2x^2 - 4x + 5)$

67. $(6x^4 - 3x^7 + 4) - (8x^7 + 10x^5 - 2x^4 - 12)$

68. $(13x^3 - 9x^2 + 4x - 5) - (5x^3 + 7x + 3)$

In Exercises 69–84, perform the indicated operations and simplify. *See Example 5.*

69. $-(2x^3 - 3) + (4x^3 - 2x)$

70. $(2x^2 + 1) - (x^2 - 2x + 1)$

71. $(4x^5 - 10x^3 + 6x) - (8x^5 - 3x^3 + 11) +$
$(4x^5 + 5x^3 - x^2)$

72. $(15 - 2y + y^2) + (3y^2 - 6y + 1) -$
$(4y^2 - 8y + 16)$

73. $(5n^2 + 6) + [(2n - 3n^2) - (2n^2 + 2n + 6)]$

74. $(p^3 + 4) - [(p^2 + 4) + (3p - 9)]$

75. $(8x^3 - 4x^2 + 3x) -$
$[(x^3 - 4x^2 + 5) + (x - 5)]$

76. $(5x^4 - 3x^2 + 9) -$
$[(2x^4 + x^3 - 7x^2) - (x^2 + 6)]$

77. $3(4x^2 - 1) + (3x^3 - 7x^2 + 5)$

78. $(x^3 - 2x^2 - x) - 5(2x^3 + x^2 - 4x)$

79. $2(t^2 + 12) - 5(t^2 + 5) + 6(t^2 + 5)$

80. $-10(v + 2) + 8(v - 1) - 3(v - 9)$

81. $15v - 3(3v - v^2) + 9(8v + 3)$

82. $9(7x^2 - 3x + 3) - 4(15x + 2) - (3x^2 - 7x)$

83. $5s - [6s - (30s + 8)]$

84. $3x^2 - 2[3x + (9 - x^2)]$

In Exercises 85–90, perform the indicated operations and simplify. (Assume that all exponents represent positive integers.)

85. $(2x^{2r} - 6x^r - 3) + (3x^{2r} - 2x^r + 6)$

86. $(6x^{2r} - 5x^r + 4) + (2x^{2r} + 2x^r + 3)$

87. $(3x^{2m} + 2x^m - 8) - (x^{2m} - 4x^m + 3)$

88. $(x^{2m} - 6x^m + 4) - (2x^{2m} - 4x^m - 3)$

89. $(7x^{4n} - 3x^{2n} - 1) - (4x^{4n} + x^{3n} - 6x^{2n})$

90. $(-4x^{3n} + 5x^{2n} + x^n) - (x^{2n} + 9x^n - 14)$

 Graphical Reasoning   In Exercises 91 and 92, use a graphing calculator to graph the expressions for y_1 and y_2 in the same viewing window. What conclusion can you make?

91. $y_1 = (x^3 - 3x^2 - 2) - (x^2 + 1)$

$y_2 = x^3 - 4x^2 - 3$

92. $y_1 = \left(\frac{1}{2}x^3 + 2x\right) + (x^3 - x^2 - x + 1)$

$y_2 = \frac{3}{2}x^3 - x^2 + x + 1$

In Exercises 93–96, $f(x) = 4x^3 - 3x^2 + 7$ and $g(x) = 9 - x - x^2 - 5x^3$. Find $h(x)$.

93. $h(x) = f(x) + g(x)$

94. $h(x) = f(x) - g(x)$

95. $h(x) = -f(x) - g(x)$

96. $h(x) = -f(x) + g(x)$

Solving Problems

Free-Falling Object   In Exercises 97 and 98, find the height in feet of a free-falling object at the specified times using the position function. Then describe the vertical path of the object. *See Example 6.*

✓ 97. $h(t) = -16t^2 + 64$

    (a) $t = 0$ (b) $t = \frac{1}{2}$

    (c) $t = 1$ (d) $t = 2$

98. $h(t) = -16t^2 + 80t + 50$

    (a) $t = 0$ (b) $t = \frac{5}{2}$

    (c) $t = 4$ (d) $t = 5$

Free-Falling Object   In Exercises 99–102, use the position function to determine whether the free-falling object was dropped, was thrown upward, or was thrown downward. Also determine the height in feet of the object at time $t = 0$.

99. $h(t) = -16t^2 + 100$

100. $h(t) = -16t^2 + 50t$

101. $h(t) = -16t^2 + 40t + 12$

102. $h(t) = -16t^2 - 28t + 150$

103. *Free-Falling Object*   An object is thrown upward from the top of the Eiffel Tower, which is 984 feet tall. The initial velocity is 40 feet per second. Use the position function

$$h(t) = -16t^2 + 40t + 984$$

to find the heights of the object when $t = 1$, $t = 5$, and $t = 9$.

104. *Free-Falling Object*   An object is dropped from a hot-air balloon that is 200 feet above the ground (see figure). Use the position function

$$h(t) = -16t^2 + 200$$

to find the heights of the object when $t = 1$, $t = 2$, and $t = 3$.

200 ft

105. *Cost, Revenue, and Profit* A manufacturer can produce and sell x radios per week. The total cost C (in dollars) of producing the radios is given by

$$C = 8x + 15,000$$

and the total revenue R is given by

$$R = 14x.$$

Find the profit P obtained by selling 5000 radios per week. (*Note: P = R - C*)

✓ **106.** *Holiday Spending* One thousand U.S. adults were surveyed about holiday spending. The amount they planned on spending (in dollars) in 2004 through 2007 can be modeled by

$$P = 3.5t^2 - 85.5t + 1288, \quad 4 \le t \le 7$$

and the amount they actually spent can be modeled by

$$A = t^2 - 62.4t + 1207, \quad 4 \le t \le 7$$

where t represents the year, with $t = 4$ corresponding to 2004.

(a) Determine the polynomial that represents the amount saved per year.

(b) Use the polynomial from part (a) to find the amount saved in 2005.

▲ *Geometry* In Exercises 107 and 108, write and simplify an expression for the perimeter of the figure. **See Example 7.**

✓ **107.**

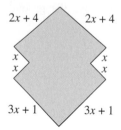

2x + 4   2x + 4

x
x   x
           x

3x + 1   3x + 1

108.

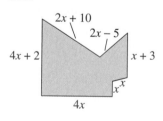

2x + 10

2x − 5

4x + 2   x + 3

x   x

4x

▲ *Geometry* In Exercises 109 and 110, find an expression for the area of the shaded region of the figure.

109.

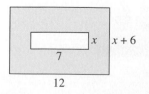

x   x + 6

7

12

110.

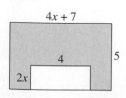

4x + 7

4   5

2x

▲ *Geometry* In Exercises 111 and 112, find an expression that represents the area of the entire region.

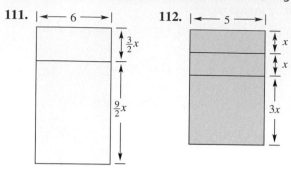

111. |← 6 →|

$\frac{3}{2}x$

$\frac{9}{2}x$

112. |← 5 →|

x

x

$3x$

113. *Stopping Distance* The total stopping distance of an automobile is the distance traveled during the driver's reaction time plus the distance traveled after the brakes are applied. In an experiment, these distances were measured (in feet) when an automobile was traveling at a speed of x miles per hour on dry, level pavement. The distance traveled during the reaction time was $R = 1.1x$, and the braking distance was $B = 0.0475x^2 - 0.001x$.

(a) Determine the polynomial that represents the total stopping distance T.

(b) 🖩 Use a graphing calculator to graph R, B, and T in the same viewing window.

(c) 🖩 Use the graph to estimate the total stopping distances when $x = 25$ and $x = 50$ miles per hour.

114. *Beverage Availability* The per capita availability (average available per person) of all beverage milks M and bottled water W in the U.S. from 2001 to 2005 can be approximated by the two polynomial models

$$M = -0.27t + 22.4, \quad 1 \le t \le 5 \text{ and}$$

$$W = 0.050t^2 + 1.45t + 16.8, \quad 1 \le t \le 5$$

where t represents the year, with $t = 1$ corresponding to 2001. Both M and W are measured in gallons. (Source: U.S. Department of Agriculture)

(a) Find a polynomial that represents the per capita availability of both beverage milks and bottled water during the time period.

(b) ⊞ During the given period, the per capita availability of beverage milks was decreasing and the per capita availability of bottled water was increasing (see figure). Use a graphing calculator to graph the model from part (a). Was the total per capita availability of beverage milks and bottled water increasing or decreasing over this time period?

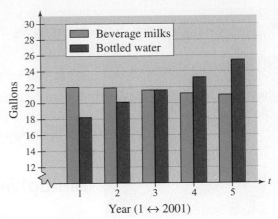

Figure for 114

Explaining Concepts

115. ✎ Explain the difference between the degree of a term of a polynomial in x and the degree of a polynomial.

116. What algebraic operation separates the terms of a polynomial? What operation separates the factors of a term?

117. ✎ Determine which of the two statements is always true. Is the other statement always false? Explain.

(a) A polynomial is a trinomial.

(b) A trinomial is a polynomial.

118. Can two third-degree polynomials be added to produce a second-degree polynomial? If so, give an example.

119. ✎ Is every trinomial a second-degree polynomial? If not, give an example of a trinomial that is not a second-degree polynomial.

120. ✎ Describe the method for subtracting polynomials.

Cumulative Review

In Exercises 121–130, consider the following matrices.

$$A = \begin{bmatrix} 3 & 2 \\ 1 & 4 \end{bmatrix}, B = \begin{bmatrix} 6 & -3 & \vdots & 0 \\ 2 & 1 & \vdots & 4 \end{bmatrix},$$

$$C = \begin{bmatrix} 2 & 0 & -2 \\ 4 & -1 & 1 \\ 0 & 4 & -6 \end{bmatrix}$$

121. What is the order of B?

122. What is the order of C?

123. Is A a square matrix?

124. Write B in row-echelon form.

125. Write C in row-echelon form.

126. Write the system of linear equations represented by B.

127. Solve the system of linear equations obtained in Exercise 126.

128. Find $|A|$.

129. Find $|C|$.

130. Find $|B|$.

5.3 Multiplying Polynomials

Paul Poplis/Getty Images

What You Should Learn

1 ▶ Use the Distributive Property and the FOIL Method to multiply polynomials.

2 ▶ Use special product formulas to multiply two binomials.

3 ▶ Use multiplication of polynomials in application problems.

Why You Should Learn It

Multiplication of polynomials can be used to model many aspects of the business world. For instance, in Exercise 108 on page 326, you will use a polynomial to model the total revenue from selling apple pies.

1 ▶ Use the Distributive Property and the FOIL Method to multiply polynomials.

Multiplying Polynomials

The simplest type of polynomial multiplication involves a monomial multiplier. The product is obtained by direct application of the Distributive Property. For instance, to multiply the monomial $3x$ by the polynomial $(2x^2 - 5x + 3)$, multiply *each* term of the polynomial by $3x$.

$$(3x)(2x^2 - 5x + 3) = (3x)(2x^2) - (3x)(5x) + (3x)(3)$$
$$= 6x^3 - 15x^2 + 9x$$

EXAMPLE 1 **Finding Products with Monomial Multipliers**

Multiply the polynomial by the monomial.

a. $(2x - 7)(3x)$ **b.** $4x^2(3x - 2x^3 + 1)$

Solution

a. $(2x - 7)(3x) = 2x(3x) - 7(3x)$ Distributive Property

$$= 6x^2 - 21x$$ Rules of exponents

b. $4x^2(3x - 2x^3 + 1)$

$$= 4x^2(3x) - 4x^2(2x^3) + 4x^2(1)$$ Distributive Property

$$= 12x^3 - 8x^5 + 4x^2$$ Rules of exponents

$$= -8x^5 + 12x^3 + 4x^2$$ Standard form

✓ **CHECKPOINT** *Now try Exercise 3.*

EXAMPLE 2 **Finding Products with Negative Monomial Multipliers**

a. $(-x)(5x^2 - x) = (-x)(5x^2) - (-x)(x)$ Distributive Property

$$= -5x^3 + x^2$$ Rules of exponents

b. $(-2x^2)(x^2 - 8x + 4)$

$$= (-2x^2)(x^2) - (-2x^2)(8x) + (-2x^2)(4)$$ Distributive Property

$$= -2x^4 + 16x^3 - 8x^2$$ Rules of exponents

✓ **CHECKPOINT** *Now try Exercise 7.*

Smart Study Strategy

Go to page 296 for ways to *Use All Three Modalities.*

To multiply two *binomials*, you can use both (left and right) forms of the Distributive Property. For example, if you treat the binomial $(2x + 7)$ as a single quantity, you can multiply $(3x - 2)$ by $(2x + 7)$ as follows.

$$(3x - 2)(2x + 7) = 3x(2x + 7) - 2(2x + 7)$$

$$= (3x)(2x) + (3x)(7) - (2)(2x) - (2)(7)$$

$$= 6x^2 + 21x - 4x - 14$$

Product of First terms	Product of Outer terms	Product of Inner terms	Product of Last terms

$$= 6x^2 + 17x - 14$$

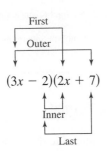

First
Outer

$(3x - 2)(2x + 7)$

Inner
Last

The FOIL Method

With practice, you should be able to multiply two binomials without writing out all of the steps shown above. In fact, the four products in the boxes above suggest that you can write the product of two binomials in just one step. This is called the **FOIL Method.** Note that the words *First*, *Outer*, *Inner*, and *Last* refer to the positions of the terms in the original product, as shown at the left.

EXAMPLE 3   **Multiplying Binomials (FOIL Method)**

Use the FOIL Method to find the product.

$$(x - 3)(x + 3)$$

Solution

$$
\overset{\text{F}}{}\quad\overset{\text{O}}{}\quad\overset{\text{I}}{}\quad\overset{\text{L}}{}
$$
$$(x - 3)(x + 3) = x^2 + 3x - 3x - 9$$

$$= x^2 - 9 \qquad \text{Combine like terms.}$$

✓ **CHECKPOINT** *Now try Exercise 15.*

EXAMPLE 4   **Multiplying Binomials (FOIL Method)**

Use the FOIL Method to find the product.

$$(3x + 4)(2x + 1)$$

Solution

$$
\overset{\text{F}}{}\quad\overset{\text{O}}{}\quad\overset{\text{I}}{}\quad\overset{\text{L}}{}
$$
$$(3x + 4)(2x + 1) = 6x^2 + 3x + 8x + 4$$

$$= 6x^2 + 11x + 4 \qquad \text{Combine like terms.}$$

✓ **CHECKPOINT** *Now try Exercise 23.*

To multiply two polynomials that have three or more terms, you can use the same basic principle as for multiplying monomials and binomials. That is, *each term of one polynomial must be multiplied by each term of the other polynomial.* This can be done using either a horizontal or a vertical format.

EXAMPLE 5   Multiplying Polynomials (Horizontal Format)

$(4x^2 - 3x - 1)(2x - 5)$

$= (4x^2 - 3x - 1)(2x) - (4x^2 - 3x - 1)(5)$ Distributive Property

$= 8x^3 - 6x^2 - 2x - (20x^2 - 15x - 5)$ Distributive Property

$= 8x^3 - 6x^2 - 2x - 20x^2 + 15x + 5$ Subtract (change signs).

$= 8x^3 - 26x^2 + 13x + 5$ Combine like terms.

✓ **CHECKPOINT** *Now try Exercise 33.*

EXAMPLE 6   Multiplying Polynomials (Vertical Format)

Write the polynomials in standard form and use a vertical format to find the product.

$(4x^2 + x - 2)(5 + 3x - x^2)$

Solution

$$
\begin{array}{r}
4x^2 + x - 2 \\
\times \qquad -x^2 + 3x + 5 \\
\hline
20x^2 + 5x - 10 \\
12x^3 + 3x^2 - 6x \\
-4x^4 - x^3 + 2x^2 \\
\hline
-4x^4 + 11x^3 + 25x^2 - x - 10
\end{array}
$$

Standard form

Standard form

⟸ $5(4x^2 + x - 2)$

⟸ $3x(4x^2 + x - 2)$

⟸ $-x^2(4x^2 + x - 2)$

Combine like terms.

✓ **CHECKPOINT** *Now try Exercise 45.*

EXAMPLE 7   An Area Model for Multiplying Polynomials

Show that $(2x + 1)(x + 2) = 2x^2 + 5x + 2$.

Solution

An appropriate area model for demonstrating the multiplication of two binomials would be $A = lw$, the area formula for a rectangle. Think of a rectangle whose sides are $x + 2$ and $2x + 1$. The area of this rectangle is

$(2x + 1)(x + 2)$.   Area = (length)(width)

Another way to find the area is to add the areas of the rectangular parts, as shown in Figure 5.3. There are two squares whose sides are x, five rectangles whose sides are x and 1, and two squares whose sides are 1. The total area of these nine rectangles is

$2x^2 + 5x + 2$.   Area = sum of rectangular areas

Because each method must produce the same area, you can conclude that

$(2x + 1)(x + 2) = 2x^2 + 5x + 2$.

✓ **CHECKPOINT** *Now try Exercise 111.*

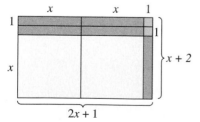

Figure 5.3

2 ▶ Use special product formulas to multiply two binomials.

Special Products

Special Products

Let u and v be real numbers, variables, or algebraic expressions. Then the following formulas are true.

| *Special Product* | *Example* |

Sum and Difference of Two Terms

$(u + v)(u - v) = u^2 - v^2$ $(3x - 4)(3x + 4) = 9x^2 - 16$

Square of a Binomial

$(u + v)^2 = u^2 + 2uv + v^2$ $(2x + 5)^2 = 4x^2 + 2(2x)(5) + 25$

$\qquad\qquad\qquad\qquad\qquad\qquad\qquad = 4x^2 + 20x + 25$

$(u - v)^2 = u^2 - 2uv + v^2$ $(x - 6)^2 = x^2 - 2(x)(6) + 36$

$\qquad\qquad\qquad\qquad\qquad\qquad\qquad = x^2 - 12x + 36$

Study Tip

When squaring a binomial, note that the resulting middle term, $\pm 2uv$, is always twice the product of the two terms.

EXAMPLE 8   **Product of the Sum and Difference of Two Terms**

a. $(3x - 2)(3x + 2) = (3x)^2 - 2^2$ $(u + v)(u - v) = u^2 - v^2$

$\qquad\qquad\qquad\quad = 9x^2 - 4$ Simplify.

b. $(6 + 5x)(6 - 5x) = 6^2 - (5x)^2 = 36 - 25x^2$

 **CHECKPOINT** *Now try Exercise 53.*

EXAMPLE 9   **Squaring a Binomial**

$\quad (2x - 7)^2 = (2x)^2 - 2(2x)(7) + 7^2$ Square of a binomial

$\qquad\qquad\quad = 4x^2 - 28x + 49$ Simplify.

 **CHECKPOINT** *Now try Exercise 69.*

EXAMPLE 10   **Cubing a Binomial**

$\quad (x + 4)^3 = (x + 4)^2(x + 4)$ Rules of exponents

$\qquad\qquad = (x^2 + 8x + 16)(x + 4)$ Square of a binomial

$\qquad\qquad = x^2(x + 4) + 8x(x + 4) + 16(x + 4)$ Distributive Property

$\qquad\qquad = x^3 + 4x^2 + 8x^2 + 32x + 16x + 64$ Distributive Property

$\qquad\qquad = x^3 + 12x^2 + 48x + 64$ Combine like terms.

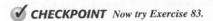 **CHECKPOINT** *Now try Exercise 83.*

3 ▶ Use multiplication of polynomials in application problems.

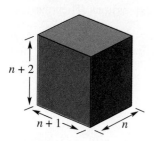

$n + 2$

$n + 1$ n

Figure 5.4

Applications

| EXAMPLE 11 | Geometry: Area and Volume |

The closed rectangular box shown in Figure 5.4 has sides whose lengths (in inches) are consecutive integers.

a. Write a polynomial function $V(n)$ that represents the volume of the box.

b. What is the volume if the length of the shortest side is 4 inches?

c. Write a polynomial function $A(n)$ for the area of the base of the box.

d. Write a polynomial function for the area of the base if the length and width increase by 3. That is, find $A(n + 3)$.

Solution

a. The volume can be represented by the following function.

$$V(n) = n(n + 1)(n + 2) = n(n^2 + 3n + 2) = n^3 + 3n^2 + 2n$$

b. If $n = 4$, the volume of the box is

$$V(4) = (4)^3 + 3(4)^2 + 2(4) = 64 + 48 + 8 = 120 \text{ cubic inches.}$$

c. $A(n) = (\text{Length})(\text{Width}) = n(n + 1) = n^2 + n$

d. $A(n + 3) = (n + 3)^2 + (n + 3) = n^2 + 6n + 9 + n + 3 = n^2 + 7n + 12$

 **CHECKPOINT** *Now try Exercise 99.*

| EXAMPLE 12 | Revenue | |

A software manufacturer has determined that the demand for its new video game is given by the equation $p = 50 - 0.001x$, where p is the price of the game (in dollars) and x is the number of units sold. The total revenue R from selling x units of a product is given by the equation $R = xp$. Find the revenue equation for the video game. Then find the revenue when 3000 units are sold.

Solution

$R = xp$		Revenue equation
$= x(50 - 0.001x)$		Substitute for p.
$= 50x - 0.001x^2$		Distributive Property

So, the revenue equation for the video game is $R = 50x - 0.001x^2$. To find the revenue when 3000 units are sold, substitute 3000 for x in the revenue equation.

$R = 50x - 0.001x^2$		Revenue equation
$= 50(3000) - 0.001(3000)^2$		Substitute 3000 for x.
$= 141,000$		Simplify.

So, the revenue when 3000 units are sold is $141,000.

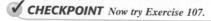 **CHECKPOINT** *Now try Exercise 107.*

_____ Concept Check _____

1. Is the product of two binomials a polynomial?

2. Explain the meaning of each letter of FOIL as it relates to multiplying two binomials.

3. Can the FOIL Method be used to multiply two trinomials? Explain.

4. How many terms does the square of a binomial have? Explain.

5.3 EXERCISES

Go to pages 360–361 to record your assignments.

_____ Developing Skills _____

In Exercises 1–14, perform the indicated multiplication(s). *See Examples 1 and 2.*

1. $(-2a^2)(-8a)$

2. $(-6n)(3n^2)$

✓ 3. $2y(5 - y)$

4. $4z(6 - 3z)$

5. $4x(2x^2 - 3x + 5)$

6. $3y(-3y^2 + 7y - 3)$

✓ 7. $-2m^2(7 - 4m + 2m^2)$

8. $-3a^2(8 - 2a - a^2)$

9. $-x^3(x^4 - 2x^3 + 5x - 6)$

10. $-y^4(7y^3 - 4y^2 + y - 4)$

11. $-3x(-5x)(5x + 2)$

12. $-7t(2t)(6 - 3t)$

13. $u^2v(3u^4 - 5u^2v + 6uv^3)$

14. $ab^3(2a - 9a^2b + 3b)$

In Exercises 15–32, multiply using the FOIL Method. *See Examples 3 and 4.*

✓ 15. $(x + 2)(x + 4)$

16. $(x - 5)(x - 3)$

17. $(x - 4)(x + 4)$

18. $(x + 7)(x - 1)$

19. $(x - 3)(x - 3)$

20. $(x - 6)(x + 6)$

21. $(2x - 3)(x + 5)$

22. $(3x + 1)(x - 4)$

✓ 23. $(5x - 2)(2x - 6)$

24. $(4x + 7)(3x + 7)$

25. $(2x^2 - 1)(x + 2)$

26. $(4x - 5)(2 - 3x^2)$

27. $\left(4y - \frac{1}{3}\right)(12y + 9)$

28. $\left(3z - \frac{3}{4}\right)(4z - 8)$

29. $(2x + y)(3x + 2y)$

30. $(2x - y)(3x - 2y)$

31. $(2t - 1)(t + 1) + (2t - 5)(t - 1)$

32. $(s - 3t)(s + t) - (s - 3t)(s - t)$

In Exercises 33–44, use a horizontal format to find the product. *See Example 5.*

✓ 33. $(x - 1)(x^2 - 4x + 6)$

34. $(z + 2)(z^2 - 4z + 4)$

35. $(3a + 2)(a^2 + 3a + 1)$

36. $(2t + 3)(t^2 - 5t + 1)$

37. $(2u^2 + 3u - 4)(4u + 5)$

38. $(2x^2 - 5x + 1)(3x - 4)$

39. $(x^3 - 3x + 2)(x - 2)$

40. $(x^2 - 5x - 2)(x^2 - 2)$

41. $(5x^2 + 2)(x^2 + 4x - 1)$

42. $(2x^2 - 3)(2x^2 - 2x + 3)$

43. $(t^2 + t - 2)(t^2 - t + 2)$

44. $(y^2 + 3y + 5)(2y^2 - 3y - 1)$

In Exercises 45–52, use a vertical format to find the product. *See Example 6.*

45.
$$7x^2 - 14x + 9$$
$$\times \quad\quad 2x + 1$$

46.
$$4x^4 - 6x^2 + 9$$
$$\times \quad\quad 2x^2 + 3$$

47. $(u - 2)(2u^2 + 5u + 3)$

48. $(p + 2)(-p^2 - 4p + 7)$

49. $(-x^2 + 2x - 1)(2x + 1)$

50. $(2s^2 - 5s + 6)(3s - 4)$

51. $(t^2 + t - 2)(t^2 - t + 2)$

52. $(y^2 + 3y + 5)(2y^2 - 3y - 1)$

In Exercises 53–82, use a special product formula to find the product. *See Examples 8 and 9.*

53. $(x + 2)(x - 2)$

54. $(x - 5)(x + 5)$

55. $(x - 8)(x + 8)$

56. $(x + 10)(x - 10)$

57. $(2 + 7y)(2 - 7y)$

58. $(4 + 3z)(4 - 3z)$

59. $(3 - 2x^2)(3 + 2x^2)$

60. $(6 - 5r^2)(6 + 5r^2)$

61. $(2a + 5b)(2a - 5b)$

62. $(5u + 12v)(5u - 12v)$

63. $(6x - 9y)(6x + 9y)$

64. $(8x - 5y)(8x + 5y)$

65. $\left(2x - \frac{1}{4}\right)\left(2x + \frac{1}{4}\right)$

66. $\left(\frac{2}{3}x + 7\right)\left(\frac{2}{3}x - 7\right)$

67. $(0.2t + 0.5)(0.2t - 0.5)$

68. $(4a - 0.1b)(4a + 0.1b)$

69. $(x + 5)^2$

70. $(x + 2)^2$

71. $(x - 10)^2$

72. $(u - 7)^2$

73. $(2x + 5)^2$

74. $(3x + 8)^2$

75. $(6x - 1)^2$

76. $(5 - 3z)^2$

77. $(2x - 7y)^2$

78. $(3m + 4n)^2$

79. $[(x + 2) + y]^2$

80. $[(x - 4) - y]^2$

81. $[u - (v - 3)][u + (v - 3)]$

82. $[z + (y + 1)][z - (y + 1)]$

In Exercises 83–86, simplify the expression. *See Example 10.*

83. $(k + 5)^3$

84. $(y - 2)^3$

85. $(u + v)^3$

86. $(u - v)^3$

In Exercises 87–92, simplify the expression. (Assume that all variables represent positive integers.)

87. $3x^r(5x^{2r} + 4x^{3r-1})$

88. $5x^r(4x^{r+2} - 3x^r)$

89. $(6x^m - 5)(2x^{2m} - 3)$

90. $(x^{3m} - x^{2m})(x^{2m} + 2x^{4m})$

91. $\left(x^{m-n}\right)^{m+n}$

92. $\left(y^{m+n}\right)^{m+n}$

In Exercises 93–96, use a graphing calculator to graph the expressions for y_1 and y_2 in the same viewing window. What can you conclude? Verify the conclusion algebraically.

93. $y_1 = (x + 1)(x^2 - x + 2)$
$y_2 = x^3 + x + 2$

94. $y_1 = (x - 3)^2$
$y_2 = x^2 - 6x + 9$

95. $y_1 = \left(x + \frac{1}{2}\right)\left(x - \frac{1}{2}\right)$
$y_2 = x^2 - \frac{1}{4}$

96. $y_1 = \left(\frac{1}{4}x^2 - 1\right)\left(\frac{1}{2}x^2 + 1\right)$
$y_2 = \frac{1}{8}x^4 - \frac{1}{4}x^2 - 1$

97. For the function $f(x) = x^2 - 2x$, find and simplify each of the following.
(a) $f(w + 2)$
(b) $f(a - 4) + f(3)$

98. For the function $f(x) = 2x^2 - 5x + 4$, find and simplify each of the following.
(a) $f(y + 2)$
(b) $f(1 + h) - f(1)$

Solving Problems

✓ **99.** ▲ *Geometry*   A closed rectangular box has sides of lengths n, $n + 2$, and $n + 4$ inches (see figure).

(a) Write a polynomial function $V(n)$ that represents the volume of the box.

(b) What is the volume if the length of the shortest side is 3 inches?

(c) Write a polynomial function $A(n)$ that represents the area of the base of the box.

(d) Write a polynomial function for the area of the base if the length and width increase by 5. Show that the polynomial function is $A(n + 5)$.

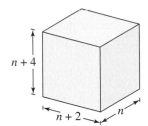
Figure for 99

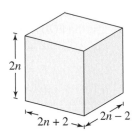
Figure for 100

100. ▲ *Geometry*   A closed rectangular box has sides of lengths $2n - 2$, $2n + 2$, and $2n$ inches (see figure).

(a) Write a polynomial function $V(n)$ that represents the volume of the box.

(b) What is the volume if the length of the shortest side is 4 inches?

(c) Write a polynomial function $A(n)$ that represents the area of the base of the box.

(d) Write a polynomial function for the area of the base if the length and width increase by 3. Show that the area of the base is not $A(n + 4)$.

▲ *Geometry*   In Exercises 101–104, write an expression for the area of the shaded region of the figure. Then simplify the expression.

101.

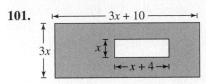

102.

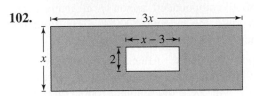

103.

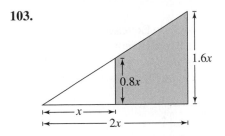

104.

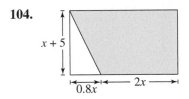

105. ▲ *Geometry*   The length of a rectangle is $\frac{5}{2}$ times its width $2w$. Find expressions for (a) the perimeter and (b) the area of the rectangle.

106. ▲ *Geometry*   The base of a triangle is $3x$ and its height is $x + 8$. Find an expression for the area A of the triangle.

✓ **107.** *Revenue*   A shop owner has determined that the demand for his daily newspapers is given by the equation $p = 175 - 0.02x$, where p is the price of the newspaper (in cents) and x is the number of papers sold. The total revenue R from selling x units of a product is given by the equation $R = xp$. Find the revenue equation for the shop owner's daily newspaper sales. Then find the revenue when 3000 newspapers are sold.

108. *Revenue*   A supermarket has determined that the demand for its apple pies is given by the equation $p = 20 - 0.015x$, where p is the price of the apple pie (in dollars) and x is the number of pies sold. The total revenue R from selling x units of a product is given by the equation $R = xp$. Find the revenue equation for the supermarket's apple pie sales. Then find the revenue when 50 apple pies are sold.

109. *Compound Interest*   After 2 years, an investment of $5000 compounded annually at interest rate r will yield the amount $5000(1 + r)^2$. Find this product.

110. *Compound Interest*   After 2 years, an investment of $1000 compounded annually at an interest rate of 5.5% will yield the amount $1000(1 + 0.055)^2$. Find this product.

▲ *Geometric Modeling*   In Exercises 111 and 112, use the area model to write two different expressions for the total area. Then equate the two expressions and name the algebraic property that is illustrated. ***See Example 7.***

☑ 111.

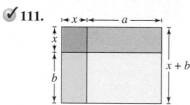

112.

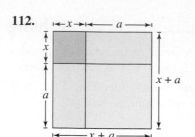

113. *Finding a Pattern*   Find each product.
(a) $(x - 1)(x + 1)$
(b) $(x - 1)(x^2 + x + 1)$
(c) $(x - 1)(x^3 + x^2 + x + 1)$

From the pattern formed by these products, can you predict the result of $(x - 1)(x^4 + x^3 + x^2 + x + 1)$?

114. Use the FOIL Method to verify each of the following.
(a) $(x + y)^2 = x^2 + 2xy + y^2$
(b) $(x - y)^2 = x^2 - 2xy + y^2$
(c) $(x - y)(x + y) = x^2 - y^2$

Explaining Concepts

115. ✎ Explain why an understanding of the Distributive Property is essential in multiplying polynomials.

116. Give an example of how to use the Distributive Property to multiply two binomials.

117. What is the degree of the product of two polynomials of degrees m and n?

118. *True or False?*   Determine whether the statement is true or false. Justify your answer.
(a) The product of two monomials is a monomial.

(b) The product of two binomials is a binomial.

Cumulative Review

In Exercises 119–124, graph the equation. Use the Vertical Line Test to determine whether y is a function of x.

119. $y = 5 - \frac{1}{2}x$ **120.** $y = \frac{3}{2}x - 2$

121. $y - 4x + 1 = 0$ **122.** $5x + 3y - 9 = 0$

123. $|y| + 2x = 0$ **124.** $|y| = 3 - x$

In Exercises 125–130, evaluate the expression.

125. 2^{-5}

126. $\dfrac{1}{5^{-2}}$

127. $\dfrac{4^2}{4^{-1}}$

128. $\dfrac{3^{-5}}{3^{-6}}$

129. $(6^3 + 3^{-6})^0$

130. $2^{-4} - 16^{-1}$

Mid-Chapter Quiz

Take this quiz as you would take a quiz in class. After you are done, check your work against the answers in the back of the book.

1. Determine the degree and leading coefficient of the polynomial
   $3 - 2x + 4x^3 - 2x^4$.

2. Explain why $x^{-3} + 2x^2 - 6$ is not a polynomial.

In Exercises 3–22, perform the indicated operations and simplify (use only positive exponents).

3. $(5y^2)(-y^4)(2y^3)$

4. $(-6x)(-3x^2)^2$

5. $(-5n^2)(-2n^3)$

6. $(3m^3)^2(-2m^4)$

7. $\dfrac{6x^{-7}}{(-2x^2)^{-3}}$

8. $\left(\dfrac{4y^2}{5x}\right)^{-2}$

9. $\left(\dfrac{3a^{-2}b^5}{9a^{-4}b^0}\right)^{-2}$

10. $\left(\dfrac{5x^0y^{-7}}{2x^{-2}y^4}\right)^{-3}$

11. Add $2t^3 + 3t^2 - 2$ to $t^3 + 9$.

12. $(3 - 7y) + (7y^2 + 2y - 3)$

13. $(7x^3 - 3x^2 + 1) - (x^2 - 2x^3)$

14. $(5 - u) - 2[3 - (u^2 + 1)]$

15. $7y(4 - 3y)$

16. $(k + 8)(k + 5)$

17. $(4x - y)(6x - 5y)$

18. $2z(z + 5) - 7(z + 5)$

19. $(6r + 5)(6r - 5)$

20. $(2x - 3)^2$

21. $(x + 1)(x^2 - x + 1)$

22. $(x^2 - 3x + 2)(x^2 + 5x - 10)$

23. Find the area of the shaded region of the figure.

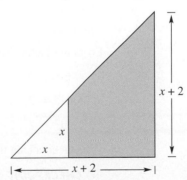

24. An object is thrown upward from the top of a 100-foot building with an initial velocity of 32 feet per second. Use the position function $h(t) = -16t^2 + 32t + 100$ to find the height of the object when $t = \frac{3}{2}$ and $t = 3$.

25. A manufacturer can produce and sell x T-shirts per week. The total cost C (in dollars) of producing the T-shirts is given by $C = 5x + 2250$, and the total revenue R is given by $R = 24x$. Find the profit P obtained by selling 1500 T-shirts per week. (*Hint: $P = R - C$.*)

5.4 Factoring by Grouping and Special Forms

What You Should Learn

1 ▶ Factor greatest common monomial factors from polynomials.

2 ▶ Factor polynomials by grouping terms.

3 ▶ Factor the difference of two squares and factor the sum or difference of two cubes.

4 ▶ Factor polynomials completely by repeated factoring.

Joel Sartore/National Geographic/Getty Images

Why You Should Learn It

In some cases, factoring a polynomial enables you to determine unknown quantities. For instance, in Exercise 138 on page 336, you will factor the expression for the area of a rectangular pig pen to find an expression for its length.

1 ▶ Factor greatest common monomial factors from polynomials.

Common Monomial Factors

In Section 5.3, you studied ways of multiplying polynomials. In this section and the next two sections, you will study the reverse process—**factoring polynomials.**

Use Distributive Property to multiply.

$$3x(4 - 5x) = 12x - 15x^2$$

Use Distributive Property to factor.

$$12x - 15x^2 = 3x(4 - 5x)$$

Notice that factoring changes a *sum of terms* into a *product of factors.*

To be efficient in factoring expressions, you need to understand the concept of the *greatest common factor* of two (or more) integers or terms. Recall from arithmetic that every integer can be factored into a product of prime numbers. The **greatest common factor** (or **GCF**) of two or more integers is the greatest integer that is a factor of each number.

EXAMPLE 1   **Finding the Greatest Common Factor**

Find the greatest common factor of $6x^5$, $30x^4$, and $12x^3$.

Solution

From the factorizations

$$6x^5 = 2 \cdot 3 \cdot x \cdot x \cdot x \cdot x \cdot x = (6x^3)(x^2)$$

$$30x^4 = 2 \cdot 3 \cdot 5 \cdot x \cdot x \cdot x \cdot x = (6x^3)(5x)$$

$$12x^3 = 2 \cdot 2 \cdot 3 \cdot x \cdot x \cdot x = (6x^3)(2)$$

you can conclude that the greatest common factor is $6x^3$.

✓ **CHECKPOINT** *Now try Exercise 15.*

Consider the three terms given in Example 1 as terms of the polynomial $6x^5 + 30x^4 + 12x^3$. The greatest common factor of these terms, $6x^3$, is the **greatest common monomial factor** of the polynomial. When you use the Distributive Property to remove this factor from each term of the polynomial, you are **factoring out** the greatest common monomial factor.

$$6x^5 + 30x^4 + 12x^3 = 6x^3(x^2) + 6x^3(5x) + 6x^3(2)$$ Factor each term.

$$= 6x^3(x^2 + 5x + 2)$$ Factor out common monomial factor.

> ### Greatest Common Monomial Factor
>
> If a polynomial in x with integer coefficients has a greatest common monomial factor of the form ax^n, the following statements must be true.
>
> 1. The coefficient a must be the greatest integer that *divides* each of the coefficients in the polynomial.
> 2. The variable factor x^n is the highest-powered variable factor that *is common* to all terms of the polynomial.

EXAMPLE 2   Factoring Out a Greatest Common Monomial Factor

Factor out the greatest common monomial factor from $24x^3 - 32x^2$.

Solution

For the terms $24x^3$ and $32x^2$, 8 is the greatest integer factor of 24 and 32, and x^2 is the highest-powered variable factor common to x^3 and x^2. So, the greatest common monomial factor of $24x^3$ and $32x^2$ is $8x^2$. You can factor the polynomial as follows.

$$24x^3 - 32x^2 = (8x^2)(3x) - (8x^2)(4)$$
$$= 8x^2(3x - 4)$$

 CHECKPOINT *Now try Exercise 21.*

The greatest common monomial factor of a polynomial is usually considered to have a positive coefficient. However, sometimes it is convenient to factor a negative number out of a polynomial. You can see how this is done in the next example.

EXAMPLE 3   A Negative Common Monomial Factor

Factor the polynomial $-3x^2 + 12x - 18$ in two ways.

a. Factor out a 3.   **b.** Factor out a -3.

Solution

a. By factoring out the common monomial factor of 3, you obtain

$$-3x^2 + 12x - 18 = 3(-x^2) + 3(4x) + 3(-6)$$
$$= 3(-x^2 + 4x - 6).$$

b. By factoring out the common monomial factor of -3, you obtain

$$-3x^2 + 12x - 18 = -3(x^2) + (-3)(-4x) + (-3)(6)$$
$$= -3(x^2 - 4x + 6).$$

Check these results by multiplying.

 **CHECKPOINT** *Now try Exercise 41.*

2 ▶ Factor polynomials by grouping terms.

Factoring by Grouping

Some expressions have common factors that are not simple monomials. For instance, the expression $x^2(2x - 3) + 4(2x - 3)$ has the common *binomial* factor $(2x - 3)$. Factoring out this common binomial produces

$$x^2(2x - 3) + 4(2x - 3) = (2x - 3)(x^2 + 4).$$

This type of factoring is part of a more general procedure called **factoring by grouping.**

EXAMPLE 4   Common Binomial Factors

Factor the expression $5x^2(6x - 5) - 2(6x - 5)$.

Solution

Each of the terms of this expression has a binomial factor of $(6x - 5)$. Factoring this binomial out of each term produces the following.

$$5x^2(6x - 5) - 2(6x - 5) = (6x - 5)(5x^2 - 2)$$

✓ **CHECKPOINT** *Now try Exercise 55.*

In Example 4, the original expression was already grouped so that it was easy to determine the common binomial factor. When you use the process of factoring by grouping, you do the grouping as well as the factoring.

EXAMPLE 5   Factoring By Grouping

Factor each polynomial by grouping.

a. $x^3 - 5x^2 + x - 5$ **b.** $4x^3 + 3x - 8x^2 - 6$

Solution

a. $x^3 - 5x^2 + x - 5 = (x^3 - 5x^2) + (x - 5)$ Group terms.

$\qquad\qquad\qquad\quad = x^2(x - 5) + 1(x - 5)$ Factor grouped terms.

$\qquad\qquad\qquad\quad = (x - 5)(x^2 + 1)$ Common binomial factor

b. $4x^3 + 3x - 8x^2 - 6 = 4x^3 - 8x^2 + 3x - 6$ Write in standard form.

$\qquad\qquad\qquad\quad = (4x^3 - 8x^2) + (3x - 6)$ Group terms.

$\qquad\qquad\qquad\quad = 4x^2(x - 2) + 3(x - 2)$ Factor grouped terms.

$\qquad\qquad\qquad\quad = (x - 2)(4x^2 + 3)$ Common binomial factor

✓ **CHECKPOINT** *Now try Exercise 65.*

Note that in Example 5(a) the polynomial is factored by grouping the first and second terms and the third and fourth terms. You could just as easily have grouped the first and third terms and the second and fourth terms, as follows.

$$x^3 - 5x^2 + x - 5 = (x^3 + x) - (5x^2 + 5)$$

$$= x(x^2 + 1) - 5(x^2 + 1) = (x^2 + 1)(x - 5)$$

3 ▸ Factor the difference of two squares and factor the sum or difference of two cubes.

Factoring Special Products

Some polynomials have special forms that you should learn to recognize so that you can factor them easily. One of the easiest special polynomial forms to recognize and to factor is the form $u^2 - v^2$, called a **difference of two squares.** This form arises from the special product $(u + v)(u - v)$ in Section 5.3.

Difference of Two Squares

Let u and v be real numbers, variables, or algebraic expressions. Then the expression $u^2 - v^2$ can be factored as follows.

$$u^2 - v^2 = (u + v)(u - v)$$

↑ Difference   ↑ Opposite signs ↑

To recognize perfect squares, look for coefficients that are squares of integers and for variables raised to *even* powers.

EXAMPLE 6   **Factoring the Difference of Two Squares**

Factor each polynomial.

a. $x^2 - 64$ **b.** $49x^2 - 81y^2$

Solution

a. $x^2 - 64 = x^2 - 8^2$ Write as difference of two squares.

$\qquad\qquad = (x + 8)(x - 8)$ Factored form

b. $49x^2 - 81y^2 = (7x)^2 - (9y)^2$ Write as difference of two squares.

$\qquad\qquad\qquad = (7x + 9y)(7x - 9y)$ Factored form

✓ **CHECKPOINT** *Now try Exercise 77.*

Remember that the rule $u^2 - v^2 = (u + v)(u - v)$ also applies when u and v are algebraic expressions.

EXAMPLE 7   **Factoring the Difference of Two Squares**

Factor the expression $(x + 2)^2 - 9$.

Solution

$\qquad (x + 2)^2 - 9 = (x + 2)^2 - 3^2$ Write as difference of two squares.

$\qquad\qquad\qquad = [(x + 2) + 3][(x + 2) - 3]$ Factored form

$\qquad\qquad\qquad = (x + 5)(x \quad 1)$ Simplify.

To check this result, write the original polynomial in standard form. Then multiply the factored form to see that you obtain the same standard form.

✓ **CHECKPOINT** *Now try Exercise 93.*

Sum or Difference of Two Cubes

Let u and v be real numbers, variables, or algebraic expressions. Then the expressions $u^3 + v^3$ and $u^3 - v^3$ can be factored as follows.

Like signs

1. $u^3 + v^3 = (u + v)(u^2 - uv + v^2)$

Unlike signs

Like signs

2. $u^3 - v^3 = (u - v)(u^2 + uv + v^2)$

Unlike signs

EXAMPLE 8   **Factoring Sums and Differences of Cubes**

Factor each polynomial.

a. $x^3 - 125$ **b.** $8y^3 + 1$ **c.** $y^3 - 27x^3$

Solution

a. This polynomial is the difference of two cubes because x^3 is the cube of x and 125 is the cube of 5.

$$x^3 - 125 = x^3 - 5^3 \qquad \text{Write as difference of two cubes.}$$
$$= (x - 5)(x^2 + 5x + 5^2) \qquad \text{Factored form}$$
$$= (x - 5)(x^2 + 5x + 25) \qquad \text{Simplify.}$$

b. This polynomial is the sum of two cubes because $8y^3$ is the cube of $2y$ and 1 is the cube of 1.

$$8y^3 + 1 = (2y)^3 + 1^3 \qquad \text{Write as sum of two cubes.}$$
$$= (2y + 1)[(2y)^2 - (2y)(1) + 1^2] \qquad \text{Factored form}$$
$$= (2y + 1)(4y^2 - 2y + 1) \qquad \text{Simplify.}$$

c. $y^3 - 27x^3 = y^3 - (3x)^3 \qquad \text{Write as difference of two cubes.}$
$$= (y - 3x)[y^2 + 3xy + (3x)^2] \qquad \text{Factored form}$$
$$= (y - 3x)(y^2 + 3xy + 9x^2) \qquad \text{Simplify.}$$

✓ **CHECKPOINT** *Now try Exercise 99.*

You can check the result of Example 8(a) by multiplying, as follows.

$$(x - 5)(x^2 + 5x + 25) = x(x^2 + 5x + 25) - 5(x^2 + 5x + 25)$$
$$= x(x^2) + x(5x) + x(25) - 5(x^2) - 5(5x) - 5(25)$$
$$= x^3 + 5x^2 + 25x - 5x^2 - 25x - 125$$
$$= x^3 - 125$$

4 ▶ Factor polynomials completely by repeated factoring.

Factoring Completely

Sometimes the difference of two squares can be hidden by the presence of a common monomial factor. Remember that with *all* factoring techniques, you should first factor out any common monomial factors.

EXAMPLE 9   Factoring Completely

Factor the polynomial $125x^2 - 80$ completely.

Solution

Because both terms have a common factor of 5, begin by factoring 5 out of the expression.

$$125x^2 - 80 = 5(25x^2 - 16) \qquad \text{Factor out common monomial factor.}$$
$$= 5[(5x)^2 - 4^2] \qquad \text{Write as difference of two squares.}$$
$$= 5(5x + 4)(5x - 4) \qquad \text{Factored form}$$

✔ **CHECKPOINT** *Now try Exercise 111.*

The expression $5(5x + 4)(5x - 4)$ is said to be **completely factored** because none of its factors can be further factored using integer coefficients.

EXAMPLE 10   Factoring Completely

Factor each polynomial completely: **a.** $x^4 - y^4$ **b.** $81m^4 - 1$

Solution

a. $x^4 - y^4 = (x^2)^2 - (y^2)^2$ Write as difference of two squares.
$$= (x^2 + y^2)(x^2 - y^2) \qquad \text{Factor as difference of two squares.}$$
$$= (x^2 + y^2)(x + y)(x - y) \qquad \text{Factor second difference of two squares.}$$

b. $81m^4 - 1 = (9m^2)^2 - 1^2$ Write as difference of two squares.
$$= (9m^2 + 1)(9m^2 - 1) \qquad \text{Factor as difference of two squares.}$$
$$= (9m^2 + 1)(3m + 1)(3m - 1) \qquad \text{Factor second difference of two squares.}$$

✔ **CHECKPOINT** *Now try Exercise 115.*

Study Tip

The sum of two squares, such as $9m^2 + 1$ in Example 10(b), cannot be factored further using integer coefficients. Such polynomials are called **prime** with respect to the integers. Some other prime polynomials are $x^2 + 4$ and $4x^2 + 9$.

EXAMPLE 11   Geometry: Area of a Rectangle

The area of a rectangle of width $4x$ is given by the polynomial $12x^2 + 32x$, as shown in Figure 5.5. Factor this expression to determine the length of the rectangle.

Solution

The polynomial for the area of the rectangle factors as follows.

$$12x^2 + 32x = 4x(3x + 8) \qquad \text{Factor out common monomial factor.}$$

So, the length of the rectangle is $3x + 8$.

✔ **CHECKPOINT** *Now try Exercise 137.*

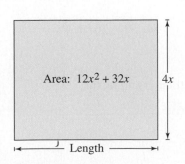

Area: $12x^2 + 32x$ $4x$

Length

Figure 5.5

_____ **Concept Check** _____

1. What is the greatest common monomial factor of the monomial terms ax^b and ax^c, where a, b, and c are integers and $b > c$?

2. You can sometimes factor out a common binomial from a polynomial expression of four or more terms. What procedure can you use to do this?

3. Which of the following polynomials *cannot* be factored by one of the formulas for special forms? Can this polynomial be factored at all? Explain why or why not.
   (a) $x^2 + y^2$ (b) $x^2 - y^2$
   (c) $x^3 + y^3$ (d) $x^3 - y^3$

4. Decide whether the following expression is factored completely. Explain your reasoning.
   $(4x^2 + 9)(3x^2 - 1)(x^2 - 1)$

5.4 EXERCISES

Go to pages 360–361 to record your assignments.

_____ **Developing Skills** _____

In Exercises 1–8, write the number as a product of prime factors.

1. 6
2. 10
3. 8
4. 12
5. 30
6. 40
7. 27
8. 54

In Exercises 9–20, find the greatest common factor of the expressions. *See Example 1.*

9. 16, 24
10. 18, 30
11. 6, 12, 16
12. 10, 15, 20
13. x^3, x^4
14. y^2, y^5
15. $3x^2, 12x$
16. $27x^4, 18x^3$
17. $16ab^2, 40a^2b^3$
18. $9x^3y, 24xy^2$
19. $9(x - 2)^2, 6(x - 2)^3$
20. $10(y + 3)^3, 25(y + 3)^4$

In Exercises 21–40, factor out the greatest common monomial factor. (Some of the polynomials have no common monomial factor.) *See Example 2.*

21. $4x + 4$
22. $7y - 7$
23. $6y - 20$
24. $9x + 30$
25. $24t^2 - 36$
26. $54x^2 - 36$
27. $x^2 + 9x$
28. $y^2 - 5y$
29. $8t^2 + 8t$
30. $12x^2 + 6x$
31. $11u^2 + 9$
32. $16 - 3y^3$
33. $3x^2y^2 - 15y$
34. $4uv + 6u^2v^2$
35. $28x^2 + 16x - 8$
36. $9 - 27y - 15y^2$
37. $45x^2 - 15x + 30$
38. $4x^2 + 16x + 24$
39. $14x^4y^3 + 21x^3y^2 + 9x^2$
40. $17x^5y^3 - xy^2 + 34y^2$

In Exercises 41–50, factor a negative real number out of the polynomial and then write the polynomial factor in standard form. *See Example 3.*

41. $7 - 14x$
42. $15 - 5x$
43. $6 - x$
44. $-5 - x$
45. $7 - y^2$
46. $4 - x^3$
47. $4 + x - x^2$
48. $1 + y - 3y^2$
49. $2y - 2 - 6y^2$
50. $9x - 9x^2 - 24$

In Exercises 51–54, fill in the missing factor.

51. $40y - 12 = 4(\quad\quad)$
52. $72z + 9 = 9(\quad\quad)$
53. $30x^2 + 25x = 5x(\quad\quad)$
54. $12x^2 - 30x = 6x(\quad\quad)$

In Exercises 55–64, factor the expression by factoring out the common binomial factor. *See Example 4.*

✓ **55.** $2y(y - 4) + 5(y - 4)$

56. $7t(s + 9) - 6(s + 9)$

57. $5x(3x + 2) - 3(3x + 2)$

58. $6(4t - 3) - 5t(4t - 3)$

59. $2(7a + 6) - 3a^2(7a + 6)$

60. $4(5y - 12) + 3y^2(5y - 12)$

61. $8t^3(4t - 1)^2 + 3(4t - 1)^2$

62. $2y^2(y^2 + 6)^3 + 7(y^2 + 6)^3$

63. $(x - 5)(4x + 9) - (3x + 4)(4x + 9)$

64. $(3x + 7)(2x - 1) + (x - 6)(2x - 1)$

In Exercises 65–76, factor the polynomial by grouping. *See Example 5.*

✓ **65.** $x^2 + 25x + x + 25$ **66.** $x^2 - 9x + x - 9$

67. $y^2 - 6y + 2y - 12$ **68.** $y^2 + 3y + 4y + 12$

69. $x^3 + 2x^2 + x + 2$ **70.** $t^3 - 11t^2 + t - 11$

71. $3a^3 - 12a^2 - 2a + 8$

72. $3s^3 + 6s^2 + 5s + 10$

73. $z^4 - 2z + 3z^3 - 6$

74. $4u^4 - 6u - 2u^3 + 3$

75. $5x^3 - 10x^2y + 7xy^2 - 14y^3$

76. $10u^4 - 8u^2v^3 - 12v^4 + 15u^2v$

In Exercises 77–98, factor the difference of two squares. *See Examples 6 and 7.*

✓ **77.** $x^2 - 9$ **78.** $y^2 - 4$

79. $1 - a^2$ **80.** $16 - b^2$

81. $16y^2 - 9$ **82.** $9z^2 - 36$

83. $81 - 4x^2$ **84.** $49 - 64x^2$

85. $4z^2 - y^2$ **86.** $9u^2 - v^2$

87. $36x^2 - 25y^2$ **88.** $100a^2 - 49b^2$

89. $u^2 - \frac{1}{16}$ **90.** $v^2 - \frac{9}{25}$

91. $\frac{4}{9}x^2 - \frac{16}{25}y^2$ **92.** $\frac{1}{4}x^2 - \frac{36}{49}y^2$

✓ **93.** $(x - 1)^2 - 16$ **94.** $(x - 3)^2 - 4$

95. $81 - (z + 5)^2$ **96.** $36 - (y - 6)^2$

97. $(2x + 5)^2 - (x - 4)^2$

98. $(3y - 1)^2 - (x + 6)^2$

In Exercises 99–110, factor the sum or difference of cubes. *See Example 8.*

✓ **99.** $x^3 - 8$ **100.** $t^3 - 1$

101. $y^3 + 64$ **102.** $z^3 + 125$

103. $8t^3 - 27$

104. $27s^3 + 64$

105. $27u^3 + 1$

106. $64v^3 - 125$

107. $64a^3 + b^3$

108. $m^3 - 8n^3$

109. $x^3 + 27y^3$

110. $u^3 + 125v^3$

In Exercises 111–120, factor the polynomial completely. *See Examples 9 and 10.*

✓ **111.** $8 - 50x^2$ **112.** $8y^2 - 18$

113. $8x^3 + 64$ **114.** $a^3 - 16a$

✓ **115.** $y^4 - 81$

116. $u^4 - 16$

117. $3x^4 - 300x^2$ **118.** $6x^5 + 30x^3$

119. $6x^6 - 48y^6$

120. $2u^6 + 54v^6$

In Exercises 121–126, factor the expression. (Assume that all exponents represent positive integers.)

121. $4x^{2n} - 25$

122. $81 - 16y^{4n}$

123. $2x^{3r} + 8x^r + 4x^{2r}$

124. $3x^{n+1} + 6x^n - 15x^{n+2}$

125. $4y^{m+n} + 7y^{2m+n} - y^{m+2n}$

126. $x^{2r+s} - 5x^{r+3s} + 10x^{2r+2s}$

Graphical Reasoning In Exercises 127–130, use a graphing calculator to graph y_1 and y_2 in the same viewing window. What can you conclude?

127. $y_1 = 3x - 6$
$\quad\ y_2 = 3(x - 2)$

128. $y_1 = x^3 - 2x^2$
$\quad\ y_2 = x^2(x - 2)$

129. $y_1 = x^2 - 4$
$\quad\ y_2 = (x + 2)(x - 2)$

130. $y_1 = x(x + 1) - 4(x + 1)$
$\quad\ y_2 = (x + 1)(x - 4)$

Think About It In Exercises 131 and 132, show all the different groupings that can be used to factor the polynomial completely. Carry out the various factorizations to show that they yield the same result.

131. $3x^3 + 4x^2 - 3x - 4$

132. $6x^3 - 8x^2 + 9x - 12$

Solving Problems

Revenue The revenue from selling x units of a product at a price of p dollars per unit is given by $R = xp$. In Exercises 133 and 134, factor the expression for revenue and determine an expression that gives the price in terms of x.

133. $R = 800x - 0.25x^2$ **134.** $R = 1000x - 0.4x^2$

135. *Simple Interest* The total amount of money accrued from a principal of P dollars invested at a simple interest rate r for t years is given by $P + Prt$. Factor this expression.

136. *Chemical Reaction* The rate of change of a chemical reaction is given by $kQx - kx^2$, where Q is the amount of the original substance, x is the amount of substance formed, and k is a constant of proportionality. Factor this expression.

137. ▲ *Geometry* The area of a rectangle of length l is given by the polynomial $45l - l^2$. Factor this expression to determine the width of the rectangle.

138. *Farming* A farmer has enough fencing to construct a rectangular pig pen that encloses an area given by $32w - w^2$, where w is the width (in feet) of the pen. Use factoring to find the length of the pen in terms of w.

139. ▲ *Geometry* The surface area of a rectangular solid of height h and square base with edge of length x is given by $2x^2 + 4xh$. Factor this expression.

140. ▲ *Geometry* The surface area of a right circular cylinder is given by $S = 2\pi r^2 + 2\pi rh$ (see figure). Factor this expression.

141. *Product Design* A washer on the drive train of a car has an inside radius of r centimeters and an outside radius of R centimeters (see figure). Find the area of one of the flat surfaces of the washer and write the area in factored form.

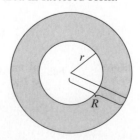

142. ▲ *Geometry* The cube shown in the figure is formed by solids I, II, III, and IV.

(a) Explain how you could determine each expression for volume.

	Volume
Entire cube	a^3
Solid I	$a^2(a - b)$
Solid II	$ab(a - b)$
Solid III	$b^2(a - b)$
Solid IV	b^3

(b) Add the volumes of solids I, II, and III. Factor the result to show that their total volume can be expressed as $(a - b)(a^2 + ab + b^2)$.

(c) Explain why the total volume of solids I, II, and III can also be expressed as $a^3 - b^3$. Then explain how the figure can be used as a geometric model for the *difference of two cubes* factoring pattern.

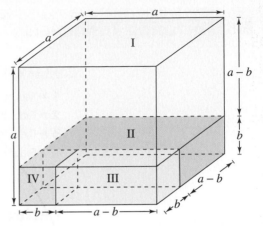

Figure for 142

Explaining Concepts

143. ✎ Explain what is meant by saying that a polynomial is in factored form.

144. ✎ Explain how the word *factor* can be used as a noun or as a verb.

145. ✎ How can you check your result after factoring a polynomial?

146. ✎ Describe a method for finding the greatest common factor of two (or more) monomials.

147. Give an example of using the Distributive Property to factor a polynomial.

148. Give an example of a polynomial that is prime with respect to the integers.

149. *Think About It* A binomial expression consists of a difference of two monomial terms that are squares of squares. Describe the polynomial factors of the complete factorization of the expression.

150. *Think About It* A binomial expression consists of a difference of two monomial terms that are squares of cubes. Describe the polynomial factors of the complete factorization of the expression.

Cumulative Review

In Exercises 151–154, find the determinant of the matrix.

151. $\begin{bmatrix} 3 & 4 \\ 2 & 1 \end{bmatrix}$

152. $\begin{bmatrix} -1 & 2 \\ -3 & 5 \end{bmatrix}$

153. $\begin{bmatrix} -1 & 3 & 0 \\ -2 & 0 & 6 \\ 0 & 4 & 2 \end{bmatrix}$

154. $\begin{bmatrix} 1 & 2 & 5 \\ -2 & 4 & 3 \\ -2 & 2 & 2 \end{bmatrix}$

In Exercises 155–158, use a special product formula to find the product.

155. $(x + 7)(x - 7)$

156. $(2x + 3)^2$

157. $(2x - 3)^2$

158. $[(x - 2) - y][(x - 2) + y]$

5.5 Factoring Trinomials

What You Should Learn

1 ▶ Recognize and factor perfect square trinomials.

2 ▶ Factor trinomials of the forms $x^2 + bx + c$ and $ax^2 + bx + c$.

3 ▶ Factor trinomials of the form $ax^2 + bx + c$ by grouping.

4 ▶ Factor polynomials using the guidelines for factoring.

Perfect Square Trinomials

Why You Should Learn It

The technique for factoring trinomials will help you in solving quadratic equations in Section 5.6.

1 ▶ Recognize and factor perfect square trinomials.

A **perfect square trinomial** is the square of a binomial. For instance,

$$x^2 + 6x + 9 = (x + 3)(x + 3) = (x + 3)^2$$

is the square of the binomial $(x + 3)$. Perfect square trinomials come in one of two forms: the middle term is either positive or negative.

> ### Perfect Square Trinomials
>
> Let u and v represent real numbers, variables, or algebraic expressions.
>
> **1.** $u^2 + 2uv + v^2 = (u + v)^2$ **2.** $u^2 - 2uv + v^2 = (u - v)^2$
>
>   Same sign   Same sign

To recognize a perfect square trinomial, remember that the first and last terms must be perfect squares and positive, and the middle term must be *twice* the product of u and v. (The middle term can be positive or negative.)

EXAMPLE 1   Factoring Perfect Square Trinomials

a. $x^2 - 4x + 4 = x^2 - 2(x)(2) + 2^2 = (x - 2)^2$

b. $16y^2 + 24y + 9 = (4y)^2 + 2(4y)(3) + 3^2 = (4y + 3)^2$

c. $9x^2 - 30xy + 25y^2 = (3x)^2 - 2(3x)(5y) + (5y)^2 = (3x - 5y)^2$

✓ **CHECKPOINT** *Now try Exercise 1.*

EXAMPLE 2   Factoring Out a Common Monomial Factor First

a. $3x^2 - 30x + 75 = 3(x^2 - 10x + 25)$ Factor out common monomial factor.

$ = 3(x - 5)^2$ Factor as perfect square trinomial.

b. $16y^3 + 80y^2 + 100y = 4y(4y^2 + 20y + 25)$ Factor out common monomial factor.

$ = 4y(2y + 5)^2$ Factor as perfect square trinomial.

✓ **CHECKPOINT** *Now try Exercise 13.*

2 ▶ Factor trinomials of the forms $x^2 + bx + c$ and $ax^2 + bx + c$.

Factoring Trinomials

To factor a trinomial of the form $x^2 + bx + c$, consider the following.

$$(x + m)(x + n) = x^2 + nx + mx + mn$$
$$= x^2 + \underbrace{(m + n)}_{\substack{\text{Sum of} \\ \text{terms}}}x + \underbrace{mn}_{\substack{\text{Product} \\ \text{of terms}}}$$

$$= x^2 + \boxed{b}\,x + \boxed{c}$$

From this, you can see that to factor a trinomial $x^2 + bx + c$ into a product of two binomials, you must find *two factors of c whose sum is b*. There are many different techniques for factoring trinomials. The most common technique is to use *guess, check, and revise* with mental math.

EXAMPLE 3   Factoring a Trinomial of the Form $x^2 + bx + c$

Factor the trinomial $x^2 + 3x - 4$.

Solution

You need to find two numbers whose product is -4 and whose sum is 3. Using mental math, you can determine that the numbers are 4 and -1.

The product of 4 and -1 is -4.

$$x^2 + 3x - 4 = (x + 4)(x - 1)$$

The sum of 4 and -1 is 3.

✓ **CHECKPOINT** *Now try Exercise 37.*

EXAMPLE 4   Factoring Trinomials of the Form $x^2 + bx + c$

Factor each trinomial.

a. $x^2 - 2x - 8$

b. $x^2 - 5x + 6$

Solution

a. You need to find two numbers whose product is -8 and whose sum is -2.

The product of -4 and 2 is -8.

$$x^2 - 2x - 8 = (x - 4)(x + 2)$$

The sum of -4 and 2 is -2.

b. You need to find two numbers whose product is 6 and whose sum is -5.

The product of -3 and -2 is 6.

$$x^2 - 5x + 6 = (x - 3)(x - 2)$$

The sum of -3 and -2 is -5.

✓ **CHECKPOINT** *Now try Exercise 41.*

Study Tip

Use a list to help you find the two numbers with the required product and sum. For Example 4(a):

Factors of -8	Sum
$1, -8$	-7
$-1, 8$	7
$2, -4$	-2
$-2, 4$	2

Because -2 is the required sum, the correct factorization is

$$x^2 - 2x - 8 = (x - 4)(x + 2).$$

When factoring a trinomial of the form $x^2 + bx + c$, if you have trouble finding two factors of c whose sum is b, it may be helpful to list all of the distinct pairs of factors, and then choose the appropriate pair from the list. For instance, consider the trinomial

$$x^2 - 2x - 24.$$

For this trinomial, $c = -24$ and $b = -2$. So, you need two factors of -24 whose sum is -2. Here is the complete list.

Factors of -24	Sum of Factors
$1, -24$	$1 - 24 = -23$
$-1, 24$	$-1 + 24 = 23$
$2, -12$	$2 - 12 = -10$
$-2, 12$	$-2 + 12 = 10$
$3, -8$	$3 - 8 = -5$
$-3, 8$	$-3 + 8 = 5$
$4, -6$	$4 - 6 = -2$ ⬅ Correct choice
$-4, 6$	$-4 + 6 = 2$

With experience, you will be able to narrow this list down *mentally* to only two or three possibilities whose sums can then be tested to determine the correct factorization. Here are some suggestions for narrowing down the list.

Guidelines for Factoring $x^2 + bx + c$

1. If c is *positive*, its factors have like signs that match the sign of b.
2. If c is *negative*, its factors have unlike signs.
3. If $|b|$ is small relative to $|c|$, first try those factors of c that are closest to each other in absolute value.
4. If $|b|$ is near $|c|$, first try those factors of c that are farthest from each other in absolute value.

Study Tip

With *any* factoring problem, remember that you can check your result by multiplying. For instance, in Example 5, you can check the result by multiplying $(x - 18)$ by $(x + 1)$ to see that you obtain $x^2 - 17x - 18$.

Remember that not all trinomials are factorable using integers. For instance, $x^2 - 2x - 4$ is not factorable using integers because there is no pair of factors of -4 whose sum is -2. Such non-factorable trinomials are called **prime polynomials**.

EXAMPLE 5   Factoring a Trinomial of the Form $x^2 + bx + c$

Factor $x^2 - 17x - 18$.

Solution

You need to find two numbers whose product is -18 and whose sum is -17. Because $|b| = |-17| = 17$ and $|c| = |-18| = 18$ are close in value, choose factors of -18 that are farthest from each other.

The product of -18 and 1 is -18.

$$x^2 - 17x - 18 = (x - 18)(x + 1)$$

The sum of -18 and 1 is -17.

✓ **CHECKPOINT** *Now try Exercise 45.*

To factor a trinomial whose leading coefficient is not 1, use the following pattern.

Factors of a

$$ax^2 + bx + c = (\quad x + \quad)(\quad x + \quad)$$

Factors of c

The goal is to find a combination of factors of a and c such that the outer and inner products add up to the middle term bx.

EXAMPLE 6   Factoring a Trinomial of the Form $ax^2 + bx + c$

Factor the trinomial $4x^2 + 5x - 6$.

Solution

First, observe that $4x^2 + 5x - 6$ has no common monomial factor. For this trinomial, $a = 4$, which factors as $(1)(4)$ or $(2)(2)$, and $c = -6$, which factors as $(-1)(6)$, $(1)(-6)$, $(-2)(3)$, or $(2)(-3)$. A test of the many possibilities is shown below.

Factors	$O + I$	
$(x + 1)(4x - 6)$	$-6x + 4x = -2x$	$-2x$ does not equal $5x$.
$(x - 1)(4x + 6)$	$6x - 4x = 2x$	$2x$ does not equal $5x$.
$(x + 6)(4x - 1)$	$-x + 24x = 23x$	$23x$ does not equal $5x$.
$(x - 6)(4x + 1)$	$x - 24x = -23x$	$-23x$ does not equal $5x$.
$(x - 2)(4x + 3)$	$3x - 8x = -5x$	$-5x$ does not equal $5x$.
$(x + 2)(4x - 3)$	$-3x + 8x = 5x$	$5x$ equals $5x$. ✔
$(2x + 1)(2x - 6)$	$-12x + 2x = -10x$	$-10x$ does not equal $5x$.
$(2x - 1)(2x + 6)$	$12x - 2x = 10x$	$10x$ does not equal $5x$.
$(2x + 2)(2x - 3)$	$-6x + 4x = -2x$	$-2x$ does not equal $5x$.
$(2x - 2)(2x + 3)$	$6x - 4x = 2x$	$2x$ does not equal $5x$.
$(x + 3)(4x - 2)$	$-2x + 12x = 10x$	$10x$ does not equal $5x$.
$(x - 3)(4x + 2)$	$2x - 12x = -10x$	$-10x$ does not equal $5x$.

So, you can conclude that the correct factorization is

$$4x^2 + 5x - 6 = (x + 2)(4x - 3).$$

Check this result by multiplying $(x + 2)$ by $(4x - 3)$.

✔ **CHECKPOINT** *Now try Exercise 67.*

The guidelines on the following page can help shorten the list of possible factorizations of a trinomial.

Study Tip

If the original trinomial has no common monomial factor, its binomial factors cannot have common monomial factors. So, in Example 6, you do not have to test factors, such as $(4x - 6)$, that have a common monomial factor of 2. Which of the other factors in Example 6 did not need to be tested?

> ### Guidelines for Factoring $ax^2 + bx + c$
>
> 1. If the trinomial has a common monomial factor, you should factor out the common factor before trying to find binomial factors.
> 2. Because the resulting trinomial has no common monomial factors, you do not have to test any binomial factors that have a common monomial factor.
> 3. Do not switch the signs of the factors of c unless the middle term $(O + I)$ is correct except in sign.

EXAMPLE 7   **Factoring a Trinomial of the Form $ax^2 + bx + c$**

Factor the trinomial $2x^2 - x - 21$.

Solution

First observe that $2x^2 - x - 21$ has no common monomial factor. For this trinomial, $a = 2$, which factors as $(1)(2)$, and $c = -21$, which factors as $(1)(-21)$, $(-1)(21)$, $(3)(-7)$, or $(-3)(7)$. Because b is small, avoid the large factors of -21, and test the smaller ones.

Factors	$O + I$	
$(2x + 3)(x - 7)$	$-14x + 3x = -11x$	$-11x$ does not equal $-x$.
$(2x + 7)(x - 3)$	$-6x + 7x = x$	x does not equal $-x$.

Because $(2x + 7)(x - 3)$ results in a middle term that is correct except in sign, you need only switch the signs of the factors of c to obtain the correct factorization.

$$2x^2 - x - 21 = (2x - 7)(x + 3) \qquad \text{Correct factorization}$$

Check this result by multiplying $(2x - 7)$ by $(x + 3)$.

✓ **CHECKPOINT** *Now try Exercise 75.*

Study Tip

Notice in Example 8 that a factorization such as $(2x + 2)(3x + 5)$ was not considered because $(2x + 2)$ has a common monomial factor of 2.

EXAMPLE 8   **Factoring a Trinomial of the Form $ax^2 + bx + c$**

Factor the trinomial $6x^2 + 19x + 10$.

Solution

First observe that $6x^2 + 19x + 10$ has no common monomial factor. For this trinomial, $a = 6$, which factors as $(1)(6)$ or $(2)(3)$, and $c = 10$, which factors as $(1)(10)$ or $(2)(5)$. You can test the potential factors as follows.

Factors	$O + I$	
$(x + 10)(6x + 1)$	$x + 60x = 61x$	$61x$ does not equal $19x$.
$(x + 2)(6x + 5)$	$5x + 12x = 17x$	$17x$ does not equal $19x$.
$(2x + 1)(3x + 10)$	$20x + 3x = 23x$	$23x$ does not equal $19x$.
$(2x + 5)(3x + 2)$	$4x + 15x = 19x$	$19x$ equals $19x$. ✓

So, the correct factorization is $6x^2 + 19x + 10 = (2x + 5)(3x + 2)$.

✓ **CHECKPOINT** *Now try Exercise 77.*

EXAMPLE 9   Factoring Completely

Factor the trinomial $8x^2y - 60xy + 28y$ completely.

Solution

Begin by factoring out the common monomial factor $4y$.

$$8x^2y - 60xy + 28y = 4y(2x^2 - 15x + 7)$$

Now, for the new trinomial $2x^2 - 15x + 7$, $a = 2$ and $c = 7$. The possible factorizations of this trinomial are as follows.

Factors	$O + I$	
$(2x - 7)(x - 1)$	$-2x - 7x = -9x$	$-9x$ does not equal $-15x$.
$(2x - 1)(x - 7)$	$-14x - x = -15x$	$-15x$ equals $-15x$. ✓

So, the complete factorization of the original trinomial is

$$8x^2y - 60xy + 28y = 4y(2x^2 - 15x + 7) = 4y(2x - 1)(x - 7).$$

Check this result by multiplying.

✔ **CHECKPOINT** *Now try Exercise 87.*

When factoring a trinomial with a negative leading coefficient, first factor -1 out of the trinomial, as demonstrated in Example 10.

EXAMPLE 10   A Trinomial with a Negative Leading Coefficient

Factor the trinomial $-3x^2 + 16x + 35$.

Solution

Begin by factoring (-1) out of the trinomial.

$$-3x^2 + 16x + 35 = (-1)(3x^2 - 16x - 35)$$

For the new trinomial $3x^2 - 16x - 35$, you have $a = 3$ and $c = -35$. Some possible factorizations of this trinomial are as follows.

Factors	$O + I$	
$(3x - 1)(x + 35)$	$105x - x = 104x$	$104x$ does not equal $-16x$.
$(3x - 35)(x + 1)$	$3x - 35x = -32x$	$-32x$ does not equal $-16x$.
$(3x - 7)(x + 5)$	$15x - 7x = 8x$	$8x$ does not equal $-16x$.
$(3x - 5)(x + 7)$	$21x - 5x = 16x$	$16x$ does not equal $-16x$.

Because $(3x - 5)(x + 7)$ results in a middle term that is correct except in sign, you need only switch the signs of the factors of c to obtain the correct factorization.

$(3x + 5)(x - 7)$	$-21x + 5x = -16x$	$-16x$ equals $-16x$. ✓

So, the correct factorization is

$$-3x^2 + 16x + 35 = (-1)(3x + 5)(x - 7) = (3x + 5)(-x + 7).$$

✔ **CHECKPOINT** *Now try Exercise 79.*

3 ▶ Factor trinomials of the form $ax^2 + bx + c$ by grouping.

Factoring Trinomials by Grouping (Optional)

So far in this section, you have been using *guess, check, and revise* to factor trinomials. An alternative technique is to use *factoring by grouping* to factor a trinomial. For instance, suppose you rewrite the trinomial $2x^2 + 7x - 15$ as

$$2x^2 + 7x - 15 = 2x^2 + 10x - 3x - 15.$$

Then, by grouping the first two terms and the third and fourth terms, you can factor the polynomial as follows.

$$2x^2 + 7x - 15 = 2x^2 + (10x - 3x) - 15 \qquad \text{Rewrite middle term.}$$

$$= (2x^2 + 10x) - (3x + 15) \qquad \text{Group terms.}$$

$$= 2x(x + 5) - 3(x + 5) \qquad \text{Factor out common monomial factor in each group.}$$

$$= (x + 5)(2x - 3) \qquad \text{Distributive Property}$$

Guidelines for Factoring $ax^2 + bx + c$ by Grouping

1. If necessary, write the trinomial in standard form.
2. Choose factors of the product ac that add up to b.
3. Use these factors to rewrite the middle term as a sum or difference.
4. Group and remove a common monomial factor from the first two terms and the last two terms.
5. If possible, factor out the common binomial factor.

EXAMPLE 11 **Factoring a Trinomial by Grouping**

Use factoring by grouping to factor the trinomial $3x^2 + 5x - 2$.

Solution

For the trinomial $3x^2 + 5x - 2$, $a = 3$ and $c = -2$, which implies that the product ac is -6. Now, because -6 factors as $(6)(-1)$, and $6 - 1 = 5 = b$, you can rewrite the middle term as $5x = 6x - x$. This produces the following result.

$$3x^2 + 5x - 2 = 3x^2 + (6x - x) - 2 \qquad \text{Rewrite middle term.}$$

$$= (3x^2 + 6x) - (x + 2) \qquad \text{Group terms.}$$

$$= 3x(x + 2) - (x + 2) \qquad \text{Factor out common monomial factor in first group.}$$

$$= (x + 2)(3x - 1) \qquad \text{Distributive Property}$$

So, the trinomial factors as $3x^2 + 5x - 2 = (x + 2)(3x - 1)$. Check this result as follows.

$$(x + 2)(3x - 1) = 3x^2 - x + 6x - 2 \qquad \text{FOIL Method}$$

$$= 3x^2 + 5x - 2 \qquad \text{Combine like terms.}$$

✓ **CHECKPOINT** *Now try Exercise 93.*

Study Tip

Factoring by grouping can be more efficient than the *guess, check, and revise* method, especially when the coefficients a and c have many factors.

4 ▶ Factor polynomials using the guidelines for factoring.

Summary of Factoring

Although the basic factoring techniques have been discussed one at a time, from this point on you must decide which technique to apply for any given problem. The guidelines below should assist you in this selection process.

Guidelines for Factoring Polynomials

1. Factor out any common factors.
2. Factor according to one of the special polynomial forms: difference of two squares, sum or difference of two cubes, or perfect square trinomials.
3. Factor trinomials, which have the form $ax^2 + bx + c$, using the methods for $a = 1$ and $a \neq 1$.
4. For polynomials with four terms, factor by grouping.
5. Check to see whether the factors themselves can be factored.
6. Check the results by multiplying the factors.

EXAMPLE 12   Factoring Polynomials

Factor each polynomial completely.

a. $3x^2 - 108$

b. $4x^3 - 32x^2 + 64x$

c. $6x^3 + 27x^2 - 15x$

d. $x^3 - 3x^2 - 4x + 12$

Solution

a. $3x^2 - 108 = 3(x^2 - 36)$ Factor out common factor.

$\qquad\qquad\quad = 3(x + 6)(x - 6)$ Difference of two squares

b. $4x^3 - 32x^2 + 64x = 4x(x^2 - 8x + 16)$ Factor out common factor.

$\qquad\qquad\qquad\quad = 4x(x - 4)^2$ Factor as perfect square trinomial.

c. $6x^3 + 27x^2 - 15x = 3x(2x^2 + 9x - 5)$ Factor out common factor.

$\qquad\qquad\qquad\quad = 3x(2x - 1)(x + 5)$ Factor.

d. $x^3 - 3x^2 - 4x + 12 = (x^3 - 3x^2) + (-4x + 12)$ Group terms.

$\qquad\qquad\qquad\quad = x^2(x - 3) - 4(x - 3)$ Factor out common factors.

$\qquad\qquad\qquad\quad = (x - 3)(x^2 - 4)$ Distributive Property

$\qquad\qquad\qquad\quad = (x - 3)(x + 2)(x - 2)$ Difference of two squares

✓ **CHECKPOINT** *Now try Exercise 99.*

Concept Check

1. When factoring $x^2 + bx + c$, explain how the last terms in each factor are related to b and c.

2. Is it possible to factor every trinomial into the product of two binomials? Explain.

3. When factoring $x^2 - 7x + 12$, why is it unnecessary to test $(x - 3)(x + 4)$ or $(x + 3)(x - 4)$?

4. Explain why $(2x + 2)$ cannot be one of the factors in the complete factorization of $2x^2 - 8x - 10$.

5.5 EXERCISES

Go to pages 360–361 to record your assignments.

Developing Skills

In Exercises 1–20, factor the perfect square trinomial. *See Examples 1 and 2.*

✓ 1. $x^2 + 4x + 4$

2. $z^2 + 6z + 9$

3. $a^2 - 10a + 25$

4. $y^2 - 14y + 49$

5. $25y^2 - 10y + 1$

6. $4z^2 + 28z + 49$

7. $9b^2 + 12b + 4$

8. $16a^2 - 24a + 9$

9. $u^2 + 8uv + 16v^2$

10. $x^2 - 14xy + 49y^2$

11. $36x^2 - 60xy + 25y^2$

12. $4y^2 + 20yz + 25z^2$

✓ 13. $5x^2 + 30x + 45$

14. $4x^2 - 32x + 64$

15. $3m^3 - 18m^2 + 27m$

16. $4m^3 + 16m^2 + 16m$

17. $20v^4 - 60v^3 + 45v^2$

18. $8y^3 + 24y^2 + 18y$

19. $\frac{1}{4}x^2 - \frac{2}{3}x + \frac{4}{9}$

20. $\frac{1}{9}x^2 + \frac{8}{15}x + \frac{16}{25}$

In Exercises 21–24, find two real numbers b such that the expression is a perfect square trinomial.

21. $x^2 + bx + 81$

22. $x^2 + bx + 49$

23. $4x^2 + bx + 9$

24. $16x^2 + bxy + 25y^2$

In Exercises 25–28, find a real number c such that the expression is a perfect square trinomial.

25. $x^2 + 8x + c$

26. $x^2 + 12x + c$

27. $y^2 - 6y + c$

28. $z^2 - 20z + c$

In Exercises 29–36, fill in the missing factor.

29. $a^2 + 6a + 8 = (a + 4)()$

30. $a^2 + 2a - 8 = (a + 4)()$

31. $y^2 - y - 20 = (y + 4)()$

32. $y^2 - 4y - 32 = (y + 4)()$

33. $x^2 + 10x + 24 = (x + 4)()$

34. $x^2 + 7x + 12 = (x + 4)()$

35. $z^2 - 6z + 8 = (z - 4)()$

36. $z^2 + 2z - 24 = (z - 4)()$

In Exercises 37–50, factor the trinomial. *See Examples 3–5.*

✓ 37. $x^2 + 6x + 5$

38. $x^2 + 7x + 10$

39. $x^2 - 5x + 6$

40. $x^2 - 10x + 24$

✓ 41. $y^2 + 7y - 30$

42. $m^2 - 3m - 10$

43. $t^2 - 6t - 16$

44. $x^2 + 4x - 12$

45. $x^2 - 20x + 96$

46. $y^2 - 35y + 300$

47. $x^2 - 2xy - 35y^2$

48. $u^2 - uv - 12v^2$

49. $x^2 + 30xy + 216y^2$

50. $a^2 - 21ab + 110b^2$

In Exercises 51–56, find all integers b such that the trinomial can be factored.

51. $x^2 + bx + 8$

52. $x^2 + bx + 10$

53. $x^2 + bx - 21$

54. $x^2 + bx - 7$

55. $x^2 + bx + 35$

56. $x^2 + bx - 38$

In Exercises 57–60, find two integers c such that the trinomial can be factored. (There are many correct answers.)

57. $x^2 + 6x + c$

58. $x^2 + 9x + c$

59. $x^2 - 3x + c$

60. $x^2 - 12x + c$

In Exercises 61–66, fill in the missing factor.

61. $5x^2 + 18x + 9 = (x + 3)()$

62. $5x^2 + 19x + 12 = (x + 3)()$

63. $5a^2 + 12a - 9 = (a + 3)()$

64. $5a^2 + 13a - 6 = (a + 3)()$

65. $2y^2 - 3y - 27 = (y + 3)()$

66. $3y^2 - y - 30 = (y + 3)()$

In Exercises 67–92, factor the trinomial, if possible. (*Note:* Some of the trinomials may be prime.) **See Examples 6–10.**

67. $6x^2 - 5x - 25$

68. $3x^2 - 16x - 12$

69. $10y^2 - 7y - 12$

70. $6x^2 - x - 15$

71. $12x^2 - 7x + 1$

72. $3y^2 - 10y + 8$

73. $5z^2 + 2z - 3$

74. $15x^2 + 4x - 3$

75. $2t^2 - 7t - 4$

76. $3z^2 - z - 4$

77. $6b^2 + 19b - 7$

78. $10x^2 - 24x - 18$

79. $-2x^2 - x + 6$

80. $-6x^2 + 5x - 6$

81. $-15d^2 + 19d - 6$

82. $-8k^2 - 2k + 3$

83. $2 + 5x - 12x^2$

84. $2 + x - 6x^2$

85. $4w^2 - 3w + 8$

86. $12x^2 + 32x - 12$

87. $60y^3 + 35y^2 - 50y$

88. $12x^2 + 42x^3 - 54x^4$

89. $10a^2 + 23ab + 6b^2$

90. $6u^2 - 5uv - 4v^2$

91. $24x^2 - 14xy - 3y^2$

92. $10x^2 + 9xy - 9y^2$

In Exercises 93–98, factor the trinomial by grouping. **See Example 11.**

93. $3x^2 + 10x + 8$

94. $2x^2 + 9x + 9$

95. $5x^2 - 12x - 9$

96. $7x^2 - 13x - 2$

97. $15x^2 - 11x + 2$

98. $12x^2 - 28x + 15$

In Exercises 99–114, factor the expression completely. **See Example 12.**

99. $3x^3 - 3x$

100. $20y^2 - 45$

101. $10t^3 + 2t^2 - 36t$

102. $16z^3 - 56z^2 + 49z$

103. $54x^3 - 2$

104. $3t^3 - 24$

105. $27a^3b^4 - 9a^2b^3 - 18ab^2$

106. $8m^3n + 20m^2n^2 - 48mn^3$

107. $x^3 + 2x^2 - 16x - 32$

108. $x^3 - 7x^2 - 4x + 28$

109. $49 - (r - 2)^2$

110. $(x + 7y)^2 - 4a^2$

111. $(x^2 - 10x + 25) - y^2$

112. $(a^2 - 2ab + b^2) - 16$

113. $x^8 - 1$

114. $x^4 - 16y^4$

In Exercises 115–120, factor the trinomial. (Assume that n represents a positive integer.)

115. $x^{2n} - 5x^n - 24$ **116.** $y^{2n} + y^n - 2$

117. $x^{2n} + 3x^n - 10$ **118.** $x^{2n} + 4x^n - 12$

119. $6y^{2n} + 13y^n + 6$ **120.** $3x^{2n} - 16x^n - 12$

Graphical Reasoning   In Exercises 121–124, use a graphing calculator to graph the two equations in the same viewing window. What can you conclude?

121. $y_1 = x^2 + 6x + 9$

$y_2 = (x + 3)^2$

122. $y_1 = x^2 - 8x + 16$

$y_2 = (x - 4)^2$

123. $y_1 = 4x^2 - 13x - 12$

$y_2 = (4x + 3)(x - 4)$

124. $y_1 = 3x^2 - 8x - 16$

$y_2 = (3x + 4)(x - 4)$

Solving Problems

Geometric Modeling   In Exercises 125–128, match the geometric factoring model with the correct factoring formula. [The models are labeled (a), (b), (c), and (d).]

(d)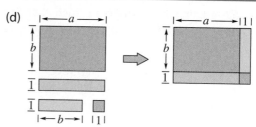

(a)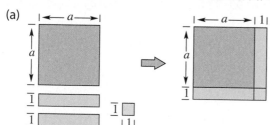

125. $a^2 - b^2 = (a + b)(a - b)$

126. $a^2 + 2a + 1 = (a + 1)^2$

127. $a^2 + 2ab + b^2 = (a + b)^2$

128. $ab + a + b + 1 = (a + 1)(b + 1)$

(b)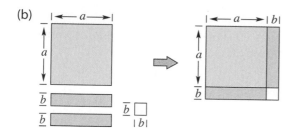

Geometry   In Exercises 129 and 130, write, in factored form, an expression for the area of the shaded region of the figure.

129.

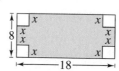

(c)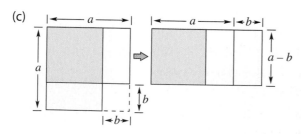

130.

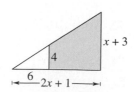

131. *Number Problem*   Let n be an integer.

    (a) Factor $8n^3 + 24n^2 + 16n$ so as to verify that it represents the product of three consecutive even integers. (*Hint:* Show that each factor has a common factor of 2.)

    (b) If $n = 10$, what are the three integers?

132. *Number Problem*   Let n be an integer.

    (a) Factor $8n^3 + 12n^2 - 2n - 3$ so as to verify that it represents the product of three consecutive odd integers.

    (b) If $n = 15$, what are the three integers?

Explaining Concepts

133. ✎   In your own words, explain how you would factor $x^2 - 5x + 6$.

134. ✎   Explain how you can check the factors of a trinomial. Give an example.

135. Give an example of a prime trinomial.

136. *Error Analysis*   Describe and correct the error.

$$9x^2 - 9x - 54 = (3x + 6)(3x - 9)$$
$$= 3(x + 2)(x - 3)$$

137. Is $x(x + 2) - 2(x + 2)$ completely factored? If not, show the complete factorization.

138. Is $(2x - 4)(x + 1)$ completely factored? If not, show the complete factorization.

139. ✎   Create five factoring problems that you think represent a fair test of a person's factoring skills. Discuss how it is possible to *create* polynomials that are factorable.

Cumulative Review

In Exercises 140–143, solve using a percent equation.

140. What is 68% of 250?

141. What is 125% of 340?

142. 34 is 5% of what number?

143. 725 is what percent of 2000?

Mixture Problems   In Exercises 144–147, determine the number of units of solutions 1 and 2 needed to obtain the desired amount and concentration of the final solution.

	Concentration of Solution 1	Concentration of Solution 2	Concentration of Final Solution	Amount of Final Solution
144.	30%	80%	50%	100 gal
145.	20%	60%	30%	10 L
146.	25%	65%	45%	40 qt
147.	60%	90%	85%	120 gal

5.6 Solving Polynomial Equations by Factoring

Carol Havens/CORBIS

What You Should Learn

1 ▶ Use the Zero-Factor Property to solve equations.
2 ▶ Solve quadratic equations by factoring.
3 ▶ Solve higher-degree polynomial equations by factoring.
4 ▶ Solve application problems by factoring.

Why You Should Learn It

Quadratic equations can be used to model and solve real-life problems. For instance, Exercise 105 on page 358 shows how a quadratic equation can be used to model the time it takes an object thrown from the Royal Gorge Bridge to reach the ground.

1 ▶ Use the Zero-Factor Property to solve equations.

The Zero-Factor Property

In the first five sections of this chapter, you have developed skills for *rewriting* (simplifying and factoring) polynomials. In this section you will use these skills, together with the **Zero-Factor Property,** to solve polynomial equations.

> ### Zero-Factor Property
> Let a and b be real numbers, variables, or algebraic expressions. If a and b are factors such that
>
> $$ab = 0$$
>
> then $a = 0$ or $b = 0$. This property also applies to three or more factors.

Study Tip

The Zero-Factor Property is basically a formal way of saying that the only way the product of two or more factors can be zero is if one (or more) of the factors is zero.

The Zero-Factor Property is the primary property for solving equations in algebra. For instance, to solve the equation

$$(x - 1)(x + 2) = 0 \qquad \text{Original equation}$$

you can use the Zero-Factor Property to conclude that either $(x - 1)$ or $(x + 2)$ must be zero. Setting the first factor equal to zero implies that $x = 1$ is a solution.

$$x - 1 = 0 \quad \Longrightarrow \quad x = 1 \qquad \text{First solution}$$

Similarly, setting the second factor equal to zero implies that $x = -2$ is a solution.

$$x + 2 = 0 \quad \Longrightarrow \quad x = -2 \qquad \text{Second solution}$$

So, the equation $(x - 1)(x + 2) = 0$ has exactly two solutions: $x = 1$ and $x = -2$. Check these solutions by substituting them into the original equation.

$$(x - 1)(x + 2) = 0 \qquad \text{Write original equation.}$$
$$(1 - 1)(1 + 2) \overset{?}{=} 0 \qquad \text{Substitute 1 for } x.$$
$$(0)(3) = 0 \qquad \text{First solution checks.} \ \checkmark$$
$$(-2 - 1)(-2 + 2) \overset{?}{=} 0 \qquad \text{Substitute } -2 \text{ for } x.$$
$$(-3)(0) = 0 \qquad \text{Second solution checks.} \ \checkmark$$

2 ▶ Solve quadratic equations by factoring.

Solving Quadratic Equations by Factoring

> ### Definition of Quadratic Equation
> A **quadratic equation** is an equation that can be written in the general form
>
> $$ax^2 + bx + c = 0 \qquad \text{Quadratic equation}$$
>
> where a, b, and c are real numbers with $a \neq 0$.

Here are some examples of quadratic equations.

$$x^2 - 2x - 3 = 0, \quad 2x^2 + x - 1 = 0, \quad x^2 - 5x = 0$$

In the next four examples, note how you can combine your factoring skills with the Zero-Factor Property to solve quadratic equations.

EXAMPLE 1   Solving a Quadratic Equation by Factoring

Solve $x^2 - x - 6 = 0$.

Solution

First, make sure that the right side of the equation is zero. Next, factor the left side of the equation. Finally, apply the Zero-Factor Property to find the solutions.

$x^2 - x - 6 = 0$	Write original equation.
$(x + 2)(x - 3) = 0$	Factor left side of equation.
$x + 2 = 0 \implies x = -2$	Set 1st factor equal to 0 and solve for x.
$x - 3 = 0 \implies x = 3$	Set 2nd factor equal to 0 and solve for x.

The equation has two solutions: $x = -2$ and $x = 3$.

Check

$(-2)^2 - (-2) - 6 \overset{?}{=} 0$	Substitute -2 for x in original equation.
$4 + 2 - 6 \overset{?}{=} 0$	Simplify.
$0 = 0$	Solution checks. ✔
$(3)^2 - 3 - 6 \overset{?}{=} 0$	Substitute 3 for x in original equation.
$9 - 3 - 6 \overset{?}{=} 0$	Simplify.
$0 = 0$	Solution checks. ✔

✔ **CHECKPOINT** *Now try Exercise 25.*

Study Tip

In Section 2.1, you learned that the basic idea in solving a linear equation is to *isolate the variable.* Notice in Example 1 that the basic idea in solving a quadratic equation is to factor the left side so that the equation can be converted into two linear equations.

Factoring and the Zero-Factor Property allow you to solve a quadratic equation by converting it into two *linear* equations, which you already know how to solve. This is a common strategy of algebra—to break down a given problem into simpler parts, each of which can be solved by previously learned methods.

In order for the Zero-Factor Property to be used, a polynomial equation *must* be written in **general form**. That is, the polynomial must be on one side of the equation and zero must be the only term on the other side of the equation. To write $x^2 - 3x = 10$ in general form, subtract 10 from each side of the equation.

$x^2 - 3x = 10$	Write original equation.
$x^2 - 3x - 10 = 10 - 10$	Subtract 10 from each side.
$x^2 - 3x - 10 = 0$	General form

To solve this equation, factor the left side as $(x - 5)(x + 2)$, then form the linear equations $x - 5 = 0$ and $x + 2 = 0$ to obtain $x = 5$ and $x = -2$, respectively.

Guidelines for Solving Quadratic Equations

1. Write the quadratic equation in general form.
2. Factor the left side of the equation.
3. Set each factor with a variable equal to zero.
4. Solve each linear equation.
5. Check each solution in the original equation.

EXAMPLE 2   Solving a Quadratic Equation by Factoring

Solve $2x^2 + 5x = 12$.

Solution

$2x^2 + 5x = 12$	Write original equation.
$2x^2 + 5x - 12 = 0$	Write in general form.
$(2x - 3)(x + 4) = 0$	Factor left side of equation.
$2x - 3 = 0 \implies x = \frac{3}{2}$	Set 1st factor equal to 0 and solve for x.
$x + 4 = 0 \implies x = -4$	Set 2nd factor equal to 0 and solve for x.

The solutions are $x = \frac{3}{2}$ and $x = -4$.

Check

$2\left(\frac{3}{2}\right)^2 + 5\left(\frac{3}{2}\right) \overset{?}{=} 12$	Substitute $\frac{3}{2}$ for x in original equation.
$\frac{9}{2} + \frac{15}{2} \overset{?}{=} 12$	Simplify.
$12 = 12$	Solution checks. ✓
$2(-4)^2 + 5(-4) \overset{?}{=} 12$	Substitute -4 for x in original equation.
$32 - 20 \overset{?}{=} 12$	Simplify.
$12 = 12$	Solution checks. ✓

✓ **CHECKPOINT** *Now try Exercise 29.*

Study Tip

Be sure you see that one side of an equation must be zero to apply the Zero-Factor Property. For instance, in Example 2, you cannot simply factor the left side to obtain $x(2x + 5) = 12$ and assume that $x = 12$ and $2x + 5 = 12$ yield correct solutions. In fact, neither of the resulting solutions satisfies the original equation.

In Examples 1 and 2, the original equations each involved a second-degree (quadratic) polynomial, and each had *two different* solutions. You will sometimes encounter second-degree polynomial equations that have only one (repeated) solution. This occurs when the left side of the general form of the equation is a perfect square trinomial, as shown in Example 3.

EXAMPLE 3   A Quadratic Equation with a Repeated Solution

Solve $x^2 - 2x + 16 = 6x$.

Solution

$x^2 - 2x + 16 = 6x$	Write original equation.
$x^2 - 8x + 16 = 0$	Write in general form.
$(x - 4)^2 = 0$	Factor.
$x - 4 = 0$ or $x - 4 = 0$	Set factors equal to 0.
$x = 4$	Solve for x.

Note that even though the left side of this equation has two factors, the factors are the same. So, the only solution of the equation is $x = 4$. This solution is called a **repeated solution.**

Check

$x^2 - 2x + 16 = 6x$	Write original equation.
$(4)^2 - 2(4) + 16 \overset{?}{=} 6(4)$	Substitute 4 for x.
$16 - 8 + 16 \overset{?}{=} 24$	Simplify.
$24 = 24$	Solution checks. ✓

✓ **CHECKPOINT** *Now try Exercise 37.*

EXAMPLE 4   Solving a Quadratic Equation by Factoring

Solve $(x + 3)(x + 6) = 4$.

Solution

Begin by multiplying the factors on the left side.

$(x + 3)(x + 6) = 4$	Write original equation.
$x^2 + 9x + 18 = 4$	Multiply factors.
$x^2 + 9x + 14 = 0$	Write in general form.
$(x + 2)(x + 7) = 0$	Factor.
$x + 2 = 0 \implies x = -2$	Set 1st factor equal to 0 and solve for x.
$x + 7 = 0 \implies x = -7$	Set 2nd factor equal to 0 and solve for x.

The equation has two solutions: $x = -2$ and $x = -7$. Check these in the original equation.

✓ **CHECKPOINT** *Now try Exercise 47.*

3 ▶ Solve higher-degree polynomial equations by factoring.

Solving Higher-Degree Equations by Factoring

EXAMPLE 5   Solving a Polynomial Equation with Three Factors

Solve $3x^3 = 15x^2 + 18x$.

Solution

$3x^3 = 15x^2 + 18x$	Write original equation.
$3x^3 - 15x^2 - 18x = 0$	Write in general form.
$3x(x^2 - 5x - 6) = 0$	Factor out common factor.
$3x(x - 6)(x + 1) = 0$	Factor.
$3x = 0 \implies x = 0$	Set 1st factor equal to 0.
$x - 6 = 0 \implies x = 6$	Set 2nd factor equal to 0.
$x + 1 = 0 \implies x = -1$	Set 3rd factor equal to 0.

The solutions are $x = 0$, $x = 6$, and $x = -1$. Check these three solutions.

✓ **CHECKPOINT** *Now try Exercise 65.*

Notice that the equation in Example 5 is a third-degree equation and has three solutions. This is not a coincidence. In general, a polynomial equation can have *at most* as many solutions as its degree. For instance, a second-degree equation can have zero, one, or two solutions. Notice that the equation in Example 6 is a fourth-degree equation and has four solutions.

EXAMPLE 6   Solving a Polynomial Equation with Four Factors

Solve $x^4 + x^3 - 4x^2 - 4x = 0$.

Solution

$x^4 + x^3 - 4x^2 - 4x = 0$	Write original equation.
$x(x^3 + x^2 - 4x - 4) = 0$	Factor out common factor.
$x[(x^3 + x^2) + (-4x - 4)] = 0$	Group terms.
$x[x^2(x + 1) - 4(x + 1)] = 0$	Factor grouped terms.
$x[(x + 1)(x^2 - 4)] = 0$	Distributive Property
$x(x + 1)(x + 2)(x - 2) = 0$	Difference of two squares
$x = 0 \implies x = 0$	
$x + 1 = 0 \implies x = -1$	
$x + 2 = 0 \implies x = -2$	
$x - 2 = 0 \implies x = 2$	

The solutions are $x = 0$, $x = -1$, $x = -2$, and $x = 2$. Check these four solutions.

✓ **CHECKPOINT** *Now try Exercise 75.*

4 ▶ Solve application problems by factoring.

Applications

EXAMPLE 7 Geometry: Dimensions of a Room

A rectangular room has an area of 192 square feet. The length of the room is 4 feet more than its width, as shown in Figure 5.6. Find the dimensions of the room.

Solution

Verbal Model:

| Length · Width = Area |

Labels: Length = $x + 4$ (feet)
Width = x (feet)
Area = 192   (square feet)

Equation:
$$(x + 4)x = 192$$
$$x^2 + 4x - 192 = 0$$
$$(x + 16)(x - 12) = 0$$
$$x = -16 \quad \text{or} \quad x = 12$$

Because the negative solution does not make sense, choose the positive solution $x = 12$. When the width of the room is 12 feet, the length of the room is

Length = $x + 4 = 12 + 4 = 16$ feet.

So, the dimensions of the room are 12 feet by 16 feet.

✔ **CHECKPOINT** *Now try Exercise 97.*

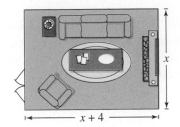

Figure 5.6

EXAMPLE 8 Free-Falling Object

A rock is dropped into a well from a height of 64 feet above the water. (See Figure 5.7.) The rock's height (in feet) relative to the water surface is given by the position function $h(t) = -16t^2 + 64$, where t is the time (in seconds) since the rock was dropped. How long does it take the rock to hit the water?

Solution

The water surface corresponds to a height of 0 feet. So, substitute a height of 0 for $h(t)$ in the equation, and solve for t.

$$0 = -16t^2 + 64 \qquad \text{Substitute 0 for } h(t).$$
$$16t^2 - 64 = 0 \qquad \text{Write in general form.}$$
$$16(t^2 - 4) = 0 \qquad \text{Factor out common factor.}$$
$$16(t + 2)(t - 2) = 0 \qquad \text{Difference of two squares}$$
$$t = -2 \quad \text{or} \quad t = 2 \qquad \text{Solutions using Zero-Factor Property}$$

Because a time of -2 seconds does not make sense, choose the positive solution $t = 2$, and conclude that the rock hits the water 2 seconds after it is dropped.

✔ **CHECKPOINT** *Now try Exercise 101.*

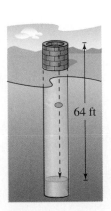

Figure 5.7

Concept Check

1. Fill in the blanks to complete the statement: In order to apply the Zero-Factor Property to an equation, one side of the equation must consist of a _____ of two or more _____, and the other side must consist of the number _____.

2. *True or False?* If $(2x - 5)(x + 4) = 1$, then $2x - 5 = 1$ or $x + 4 = 1$. Justify your answer.

3. Is it possible for a quadratic equation to have just one solution? Explain.

4. You want to solve an equation of the form $ax^2 + bx + c = d$, where a, b, c, and d are nonzero integers. What step(s) must you perform before you can apply the Zero-Factor Property?

5.6 EXERCISES

Go to pages 360–361 to record your assignments.

Developing Skills

In Exercises 1–12, use the Zero-Factor Property to solve the equation.

1. $x(x - 4) = 0$

2. $z(z + 6) = 0$

3. $(y - 3)(y + 10) = 0$

4. $(s - 7)(s + 4) = 0$

5. $25(a + 4)(a - 2) = 0$

6. $17(t - 3)(t + 8) = 0$

7. $(2t + 5)(3t + 1) = 0$

8. $(5x - 3)(2x - 8) = 0$

9. $4x(2x - 3)(2x + 25) = 0$

10. $\frac{1}{5}x(x - 2)(3x + 4) = 0$

11. $(x - 3)(2x + 1)(x + 4) = 0$

12. $(y - 39)(2y + 7)(y + 12) = 0$

In Exercises 13–78, solve the equation by factoring. *See Examples 1–6.*

13. $5y - y^2 = 0$

14. $3x^2 + 9x = 0$

15. $9x^2 + 15x = 0$

16. $4x^2 - 6x = 0$

17. $2x^2 = 32x$

18. $8x^2 = 5x$

19. $5y^2 = 15y$

20. $5x^2 = 7x$

21. $x^2 - 25 = 0$

22. $x^2 - 121 = 0$

23. $3y^2 - 48 = 0$

24. $5z^2 - 45 = 0$

25. $x^2 - 3x - 10 = 0$

26. $x^2 - x - 12 = 0$

27. $x^2 - 10x + 24 = 0$

28. $x^2 - 13x + 42 = 0$

29. $4x^2 + 15x = 25$

30. $14x^2 + 9x = -1$

31. $7 + 13x - 2x^2 = 0$

32. $11 + 32y - 3y^2 = 0$

33. $3y^2 - 2 = -y$

34. $-2x - 15 = -x^2$

35. $-13x + 36 = -x^2$

36. $x^2 - 15 = -2x$

37. $m^2 - 8m + 18 = 2$

38. $a^2 + 4a + 10 = 6$

39. $x^2 + 16x + 57 = -7$

40. $x^2 - 12x + 21 = -15$

41. $4z^2 - 12z + 15 = 6$

42. $16t^2 + 48t + 40 = 4$

43. $x(x + 2) - 10(x + 2) = 0$

44. $x(x - 15) + 3(x - 15) = 0$

45. $u(u - 3) + 3(u - 3) = 0$

46. $x(x + 10) - 2(x + 10) = 0$

47. $x(x - 5) = 36$

48. $s(s + 4) = 96$

49. $y(y + 6) = 72$

50. $x(x - 4) = 12$

51. $3t(2t - 3) = 15$

52. $3u(3u + 1) = 20$

53. $(a + 2)(a + 5) = 10$

54. $(x - 8)(x - 7) = 20$

55. $(x - 4)(x + 5) = 10$

56. $(u - 6)(u + 4) = -21$

57. $(t - 2)^2 = 16$

58. $(s + 4)^2 = 49$

59. $9 = (x + 2)^2$

60. $1 = (y + 3)^2$

61. $(x - 3)^2 - 25 = 0$ **62.** $1 - (x + 1)^2 = 0$

63. $81 - (x + 4)^2 = 0$ **64.** $(s + 5)^2 - 49 = 0$

65. $x^3 - 19x^2 + 84x = 0$ **66.** $x^3 + 18x^2 + 45x = 0$

67. $6t^3 = t^2 + t$ **68.** $3u^3 = 5u^2 + 2u$

69. $z^2(z + 2) - 4(z + 2) = 0$
70. $16(3 - u) - u^2(3 - u) = 0$
71. $a^3 + 2a^2 - 9a - 18 = 0$
72. $x^3 - 2x^2 - 4x + 8 = 0$
73. $c^3 - 3c^2 - 9c + 27 = 0$
74. $v^3 + 4v^2 - 4v - 16 = 0$
75. $x^4 - 3x^3 - x^2 + 3x = 0$
76. $x^4 + 2x^3 - 9x^2 - 18x = 0$
77. $8x^4 + 12x^3 - 32x^2 - 48x = 0$
78. $9x^4 - 15x^3 - 9x^2 + 15x = 0$

Graphical Reasoning   In Exercises 79–82, determine the x-intercepts of the graph and explain how the x-intercepts correspond to the solutions of the polynomial equation when $y = 0$.

79. $y = x^2 - 9$ **80.** $y = x^2 - 4x + 4$

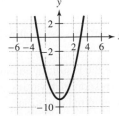   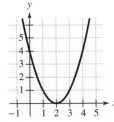

81. $y = x^3 - 6x^2 + 9x$ **82.** $y = x^3 - 3x^2 - x + 3$

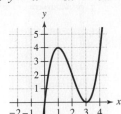   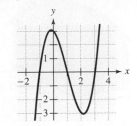

In Exercises 83–90, use a graphing calculator to graph the equation and find any x-intercepts of the graph. Verify algebraically that any x-intercepts are solutions of the polynomial equation when $y = 0$.

83. $y = x^2 + 5x$ **84.** $y = x^2 - 11x + 28$

85. $y = x^2 - 8x + 12$ **86.** $y = (x - 2)^2 - 9$

87. $y = 2x^2 + 5x - 12$ **88.** $y = x^3 - 9x$

89. $y = 2x^3 - 5x^2 - 12x$ **90.** $y = 2 + x - 2x^2 - x^3$

91. Let a and b be real numbers such that $a \neq 0$. Find the solutions of $ax^2 + bx = 0$.

92. Let a be a nonzero real number. Find the solutions of $ax^2 - ax = 0$.

Solving Problems

Think About It   In Exercises 93 and 94, find a quadratic equation with the given solutions.

93. $x = -2, \quad x = 6$
94. $x = -2, \quad x = 4$

95. *Number Problem*   The sum of a positive number and its square is 240. Find the number.

96. *Number Problem*   Find two consecutive positive integers whose product is 132.

97. *Geometry*   The rectangular floor of a storage shed has an area of 540 square feet. The length of the floor is 7 feet more than its width (see figure on next page). Find the dimensions of the floor.

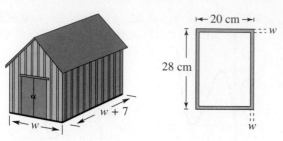

Figure for 97   Figure for 98

98. ▲ *Geometry* The outside dimensions of a picture frame are 28 centimeters and 20 centimeters (see figure). The area of the exposed part of the picture is 468 square centimeters. Find the width w of the frame.

99. ▲ *Geometry* A triangle has an area of 27 square inches. The height of the triangle is $1\frac{1}{2}$ times its base. Find the base and height of the triangle.

100. ▲ *Geometry* The height of a triangle is 2 inches less than its base. The area of the triangle is 60 square inches. Find the base and height of the triangle.

101. *Free-Falling Object* A hammer is dropped from a construction project 400 feet above the ground. The height h (in feet) of the hammer is modeled by the position equation $h = -16t^2 + 400$, where t is the time in seconds. How long does it take for the hammer to reach the ground?

102. *Free-Falling Object* A penny is dropped from the roof of a building 256 feet above the ground. The height h (in feet) of the penny after t seconds is modeled by the equation $h = -16t^2 + 256$. How long does it take for the penny to reach the ground?

103. *Free-Falling Object* An object falls from the roof of a building 80 feet above the ground toward a balcony 16 feet above the ground. The object's height h (in feet, relative to the ground) after t seconds is modeled by the equation $h = -16t^2 + 80$. How long does it take for the object to reach the balcony?

104. *Free-Falling Object* You throw a baseball upward with an initial velocity of 30 feet per second. The baseball's height h (in feet) relative to your glove after t seconds is modeled by the equation $h = -16t^2 + 30t$. How long does it take for the ball to reach your glove?

105. *Free-Falling Object* An object is thrown upward from the Royal Gorge Bridge in Colorado, 1053 feet above the Arkansas River, with an initial velocity of 48 feet per second. The height h (in feet) of the object is modeled by the position equation $h = -16t^2 + 48t + 1053$, where t is the time measured in seconds. How long does it take for the object to reach the river?

106. *Free-Falling Object* Your friend stands 96 feet above you on a cliff. You throw an object upward with an initial velocity of 80 feet per second. The height h (in feet) of the object after t seconds is modeled by the equation $h = -16t^2 + 80t$. How long does it take for the object to reach your friend on the way up? On the way down?

107. *Break-Even Analysis* The revenue R from the sale of x home theater systems is given by $R = 140x - x^2$. The cost of producing x systems is given by $C = 2000 + 50x$. How many home theater systems must be produced and sold in order to break even?

108. *Break-Even Analysis* The revenue R from the sale of x digital cameras is given by $R = 120x - x^2$. The cost of producing x digital cameras is given by $C = 1200 + 40x$. How many cameras must be produced and sold in order to break even?

109. *Investigation* Solve the equation $2(x + 3)^2 + (x + 3) - 15 = 0$ in the following two ways.

    (a) Let $u = x + 3$, and solve the resulting equation for u. Then find the corresponding values of x that are solutions of the original equation.

    (b) Expand and collect like terms in the original equation, and solve the resulting equation for x.

    (c) Which method is easier? Explain.

110. *Investigation* Solve each equation using both methods described in Exercise 109.

    (a) $3(x + 6)^2 - 10(x + 6) - 8 = 0$
    (b) $8(x + 2)^2 - 18(x + 2) + 9 = 0$

111. ▲ *Geometry* An open box is to be made from a rectangular piece of material that is 5 meters long and 4 meters wide. The box is made by cutting squares of dimension x from the corners and turning up the sides, as shown in the figure. The volume V of a rectangular solid is the product of its length, width, and height.

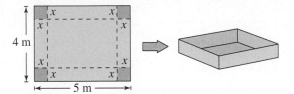

(a) Show algebraically that the volume of the box is given by $V = (5 - 2x)(4 - 2x)x$.

(b) Determine the values of x for which $V = 0$. Determine an appropriate domain for the function V in the context of this problem.

(c) Complete the table.

x	0.25	0.50	0.75	1.00	1.25	1.50	1.75
V							

(d) Use the table to determine x when $V = 3$. Verify the result algebraically.

(e) 🖩 Use a graphing calculator to graph the volume function. Use the graph to approximate the value of x that yields the box of greatest volume.

112. ▲ *Geometry* An open box with a square base stands 5 inches tall. The total surface area of the outside of the box is 525 square inches. What are the dimensions of the base?

Explaining Concepts

113. What is the maximum number of solutions of an nth-degree polynomial equation? Give an example of a third-degree equation that has only one real number solution.

114. What is the maximum number of first-degree factors that an nth-degree polynomial equation can have? Explain.

115. *Think About It* A quadratic equation has a repeated solution. Describe the x-intercept(s) of the graph of the equation formed by replacing 0 with y in the general form of the equation.

116. ✎ A third-degree polynomial equation has two solutions. What must be special about one of the solutions? Explain.

117. ✎ There are some polynomial equations that have real number solutions but cannot be solved by factoring. Explain how this can be.

118. 🖩 The polynomial equation $x^3 - x - 3 = 0$ *cannot* be solved algebraically using any of the techniques described in this book. It does, however, have one solution that is a real number.

(a) *Graphical Solution:* Use a graphing calculator to graph the equation and estimate the solution.

(b) *Numerical Solution:* Use the *table* feature of a graphing calculator to create a table and estimate the solution.

Cumulative Review

In Exercises 119–122, find the unit price (in dollars per ounce) of the product.

119. A 12-ounce soda for $0.75

120. A 12-ounce package of brown-and-serve rolls for $1.89

121. A 30-ounce can of pumpkin pie filling for $2.13

122. Turkey meat priced at $0.94 per pound

In Exercises 123–126, find the domain of the function.

123. $f(x) = \dfrac{x + 3}{x + 1}$

124. $f(x) = \dfrac{12}{x - 2}$

125. $g(x) = \sqrt{3 - x}$

126. $h(x) = \sqrt{x^2 - 4}$

What Did You Learn?

Use these two pages to help prepare for a test on this chapter. Check off the key terms and key concepts you know. You can also use this section to record your assignments.

Plan for Test Success

Date of test: ☐ / /

Study dates and times: ☐ / / at ☐ : A.M./P.M.

☐ / / at ☐ : A.M./P.M.

Things to review:

☐ Key Terms, *p. 360*
☐ Key Concepts, *pp. 360–361*
☐ Your class notes
☐ Your assignments

☐ Study Tips, *pp. 299, 300, 301, 320, 321, 322, 329, 330, 333, 339, 340, 341, 342, 344, 350, 351, 352*
☐ Technology Tips, *pp. 303, 309*
☐ Mid-Chapter Quiz, *p. 327*

☐ Review Exercises, *pp. 362–366*
☐ Chapter Test, *p. 367*
☐ Video Explanations Online
☐ Tutorial Online

Key Terms

☐ exponential form, *p. 298*
☐ scientific notation, *p. 302*
☐ polynomial in *x*, *p. 308*
☐ degree *n*, *p. 308*
☐ leading coefficient, *p. 308*
☐ constant term, *p. 308*
☐ standard form, *p. 308*

☐ monomial, *p. 308*
☐ binomial, *p. 308*
☐ trinomial, *p. 308*
☐ FOIL Method, *p. 319*
☐ factoring polynomials, *p. 328*
☐ greatest common monomial factor, *p. 328*

☐ factoring by grouping, *p. 330*
☐ completely factored, *p. 333*
☐ perfect square trinomial, *p. 338*
☐ Zero-Factor Property, *p. 350*
☐ quadratic equation, *p. 351*
☐ general form, *p. 352*
☐ repeated solution, *p. 353*

Key Concepts

5.1 Integer Exponents and Scientific Notation

Assignment: _____ Due date: _____

☐ **Use the rules of exponents to simplify expressions.**

$$a^m \cdot a^n = a^{m+n} \qquad (ab)^m = a^m \cdot b^m$$

$$(a^m)^n = a^{mn} \qquad \left(\frac{a}{b}\right)^m = \frac{a^m}{b^m}$$

$$\frac{a^m}{a^n} = a^{m-n} \qquad a^0 = 1$$

$$a^{-m} = \frac{1}{a^m} \qquad \left(\frac{a}{b}\right)^{-m} = \left(\frac{b}{a}\right)^m$$

☐ **Write very large and very small numbers in scientific notation.**

Examples: $1{,}230{,}000 = 1.23 \times 10^6$

$0.000123 = 1.23 \times 10^{-4}$

5.2 Adding and Subtracting Polynomials

Assignment: _____ Due date: _____

☐ **Identify leading coefficients and degrees of polynomials.**

A polynomial in x is an expression of the form

$$a_n x^n + a_{n-1} x^{n-1} + \cdots + a_2 x^2 + a_1 x + a_0$$

where $a_n \neq 0$. The polynomial is of degree n, and the number a_n is called the leading coefficient. The number a_0 is called the constant term.

☐ **Add and subtract polynomials.**

To *add* two polynomials, simply combine like terms.

To *subtract* one polynomial from another, add the opposite.

5.3 Multiplying Polynomials

Assignment: _____ Due date: _____

☐ **Use the Distributive Property and the FOIL Method to multiply polynomials.**

$$(x + a)(x + b) = \underbrace{x \cdot x}_{\text{First}} + \underbrace{b \cdot x}_{\text{Outer}} + \underbrace{a \cdot x}_{\text{Inner}} + \underbrace{a \cdot b}_{\text{Last}}$$

☐ **Use special product formulas to multiply two binomials.**

$$(u + v)(u - v) = u^2 - v^2$$

$$(u \pm v)^2 = u^2 \pm 2uv + v^2$$

5.4 Factoring by Grouping and Special Forms

Assignment: _____ Due date: _____

☐ **Factor greatest common monomial factors from polynomials.**

☐ **Factor polynomials by grouping terms.**

Example:

$$x^3 - 5x^2 + x - 5 = (x^3 - 5x^2) + (x - 5)$$

$$= x^2(x - 5) + 1(x - 5)$$

$$= (x - 5)(x^2 + 1)$$

☐ **Factor the difference of two squares.**

$$u^2 - v^2 = (u + v)(u - v)$$

☐ **Factor the sum or difference of two cubes.**

$$u^3 \pm v^3 = (u \pm v)(u^2 \pm uv + v^2)$$

5.5 Factoring Trinomials

Assignment: _____ Due date: _____

☐ **Recognize and factor perfect square trinomials.**

Let u and v represent real numbers, variables, or algebraic expressions.

$$u^2 \pm 2uv + v^2 = (u \pm v)^2$$

☐ **Factor trinomials of the form $x^2 + bx + c$.**

See page 340.

☐ **Factor trinomials of the form $ax^2 + bx + c$.**

See pages 342 and 344.

☐ **Factor polynomials using the guidelines for factoring.**

1. Factor out any common factor.
2. Factor according to one of the special polynomial forms.
3. Factor trinomials, which have the form $ax^2 + bx + c$, using methods for $a = 1$ and $a \neq 1$.
4. For polynomials with four terms, factor by grouping.
5. Check to see whether the factors themselves can be factored.
6. Check the results by multiplying the factors.

5.6 Solving Polynomial Equations by Factoring

Assignment: _____ Due date: _____

☐ **Use the Zero-Factor Property to solve equations.**

Let a and b be real numbers, variables, or algebraic expressions. If a and b are factors such that $ab = 0$, then $a = 0$ or $b = 0$. This property also applies to three or more factors.

☐ **Solve quadratic equations by factoring.**

1. Write the quadratic equation in general form.
2. Factor the left side of the equation.
3. Set each factor with a variable equal to zero.
4. Solve each linear equation.
5. Check each solution in the original equation.

Review Exercises

5.1 Integer Exponents and Scientific Notation

1 ▶ Use the rules of exponents to simplify expressions.

In Exercises 1–14, use the rules of exponents to simplify the expression (if possible).

1. $x^4 \cdot x^5$

2. $-3y^2 \cdot y^4$

3. $(u^2)^3$

4. $(v^4)^2$

5. $(-2z)^3$

6. $(-3y)^2(2)$

7. $-(u^2v)^2(-4u^3v)$

8. $(12x^2y)(3x^2y^4)^2$

9. $\dfrac{12z^5}{6z^2}$

10. $\dfrac{15m^3}{25m}$

11. $\dfrac{25g^4d^2}{80g^2d^2}$

12. $-\dfrac{-48u^8v^6}{(-2u^2v)^3}$

13. $\left(\dfrac{72x^4}{6x^2}\right)^2$

14. $\left(-\dfrac{y^2}{2}\right)^3$

2 ▶ Rewrite exponential expressions involving negative and zero exponents.

In Exercises 15–18, evaluate the expression.

15. $(2^3 \cdot 3^2)^{-1}$

16. $(2^{-2} \cdot 5^2)^{-2}$

17. $\left(\dfrac{3}{4}\right)^{-3}$

18. $\left(\dfrac{1}{3^{-2}}\right)^2$

In Exercises 19–30, rewrite the expression using only positive exponents, and simplify. (Assume that any variables in the expression are nonzero.)

19. $(6y^4)(2y^{-3})$

20. $4(-3x)^{-3}$

21. $\dfrac{4x^{-2}}{2x}$

22. $\dfrac{15t^5}{24t^{-3}}$

23. $(x^3y^{-4})^0$

24. $(5x^{-2}y^4)^{-2}$

25. $\dfrac{7a^6b^{-2}}{14a^{-1}b^4}$

26. $\dfrac{2u^0v^{-2}}{10u^{-1}v^{-3}}$

27. $\left(\dfrac{3x^{-1}y^2}{12x^5y^{-3}}\right)^{-1}$

28. $\left(\dfrac{4x^{-3}z^{-1}}{8x^4z}\right)^{-2}$

29. $u^3(5u^0v^{-1})(9u)^2$

30. $a^4(16a^{-2}b^4)(2b)^{-3}$

3 ▶ Write very large and very small numbers in scientific notation.

In Exercises 31–34, write the number in scientific notation.

31. 0.0000319

32. 0.0000008924

33. 17,350,000

34. 849,600,000

In Exercises 35–38, write the number in decimal form.

35. 1.95×10^6

36. 7.025×10^4

37. 2.05×10^{-5}

38. 6.118×10^{-8}

In Exercises 39–42, evaluate the expression without a calculator.

39. $(6 \times 10^3)^2$

40. $(3 \times 10^{-3})(8 \times 10^7)$

41. $\dfrac{3.5 \times 10^7}{7 \times 10^4}$

42. $\dfrac{1}{(6 \times 10^{-3})^2}$

5.2 Adding and Subtracting Polynomials

1 ▶ Identify leading coefficients and degrees of polynomials.

In Exercises 43–46, write the polynomial in standard form, and find its degree and leading coefficient.

43. $6x^3 - 4x + 5x^2 - x^4$

44. $2x^6 - 5x^3 + x^5 - 7$

45. $x^4 + 3x^5 - 4 - 6x$

46. $9x - 2x^3 + x^5 - 8x^7$

In Exercises 47 and 48, give an example of a polynomial in x that satisfies the conditions. (There are many correct answers.)

47. A trinomial of degree 5 and leading coefficient -6

48. A binomial of degree 3 and leading coefficient 4

2 ▶ Add and subtract polynomials using a horizontal format and a vertical format.

In Exercises 49–60, use a horizontal format to find the sum or difference.

49. $(10x + 8) + (x^2 + 3x)$

50. $(6 - 4u) + (2u^2 - 3u)$

51. $(5x^3 - 6x + 11) + (5 + 6x - x^2 - 8x^3)$

52. $(7 - 12x^2 + 8x^3) + (x^4 - 6x^3 + 7x^2 - 5)$

53. $(3y - 4) - (2y^2 + 1)$

54. $(x^2 - 5) - (3 - 6x)$

55. $(-x^3 - 3x) - 4(2x^3 - 3x + 1)$

56. $(7z^2 + 6z) - 3(5z^2 + 2z)$

57. $3y^2 - [2y + 3(y^2 + 5)]$

58. $(16a^3 + 5a) - 5[a + (2a^3 - 1)]$

59. $(3x^5 + 4x^2 - 8x + 12) - (2x^5 + x) + (3x^2 - 4x^3 - 9)$

60. $(7x^4 - 10x^2 + 4x) + (x^3 - 3x) - (3x^4 - 5x^2 + 1)$

In Exercises 61–64, use a vertical format to find the sum or difference.

61. $3x^2 + 5x$
    $\underline{-4x^2 -\ x + 6}$

62. $6x + 1$
    $\underline{x^2 - 4x}$

63. $3t - 5$
    $\underline{-(t^2 - t - 5)}$

64. $10y^2\qquad + 3$
    $\underline{-(y^2 + 4y - 9)}$

3 ▶ Use polynomials to model and solve real-life problems.

▲ *Geometry* **In Exercises 65 and 66, write and simplify an expression for the perimeter of the figure.**

65.   **66.**

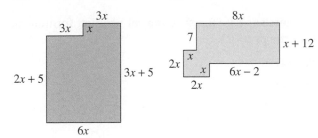

67. *Cost, Revenue, and Profit* A manufacturer can produce and sell x backpacks per week. The total cost C (in dollars) of producing the backpacks is given by $C = 16x + 3000$, and the total revenue R is given by $R = 35x$. Find the profit P obtained by selling 1200 backpacks per week.

68. *Cost, Revenue, and Profit* A manufacturer can produce and sell x notepads per week. The total cost C (in dollars) of producing the notepads is given by $C = 0.8x + 1000$, and the total revenue R is given by $R = 1.6x$. Find the profit P obtained by selling 5000 notepads per week.

5.3 Multiplying Polynomials

1 ▶ Use the Distributive Property and the FOIL Method to multiply polynomials.

In Exercises 69–82, perform the multiplication and simplify.

69. $(-2x)^3(x + 4)$

70. $(-4y)^2(y - 2)$

71. $3x(2x^2 - 5x + 3)$

72. $-2y(5y^2 - y - 4)$

73. $(x - 2)(x + 7)$

74. $(u + 5)(u - 8)$

75. $(5x + 3)(3x - 4)$

76. $(4x - 1)(2x - 5)$

77. $(4x^2 + 3)(6x^2 + 1)$

78. $(3y^2 + 2)(4y^2 - 5)$

79. $(2x^2 - 3x + 2)(2x + 3)$

80. $(5s^3 + 4s - 3)(4s - 5)$

81. $2u(u - 7) - (u + 1)(u - 7)$

82. $(3v + 2)(-5v) + 5v(3v + 2)$

2 ▶ Use special product formulas to multiply two binomials.

In Exercises 83–88, use a special product formula to find the product.

83. $(4x - 7)^2$

84. $(2x + 3y)^2$

85. $(6v + 9)(6v - 9)$

86. $(5x - 2y)(5x + 2y)$

87. $[(u - 3) + v][(u - 3) - v]$

88. $[(m - 5) + n]^2$

3 ▶ Use multiplication of polynomials in application problems.

▲ *Geometry* **In Exercises 89 and 90, write an expression for the area of the shaded region of the figure. Then simplify the expression.**

89.

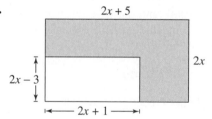

90.

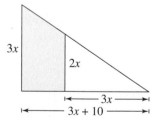

91. *Compound Interest* After 2 years, an investment of $1000 compounded annually at an interest rate of 6% will yield the amount $1000(1 + 0.06)^2$. Find this product.

92. *Compound Interest* After 2 years, an investment of $1500 compounded annually at an interest rate of r will yield the amount $1500(1 + r)^2$. Find this product.

5.4 Factoring by Grouping and Special Forms

1 ▶ Factor greatest common monomial factors from polynomials.

In Exercises 93–98, factor out the greatest common monomial factor.

93. $24x^2 - 18$

94. $14z^3 + 21$

95. $-3b^2 + b$

96. $-a^3 - 4a$

97. $6x^2 + 15x^3 - 3x$

98. $8y - 12y^2 + 24y^3$

2 ▶ Factor polynomials by grouping terms.

In Exercises 99–104, factor the polynomial by grouping.

99. $28(x + 5) - 70(x + 5)$

100. $(u - 9v)(u - v) + v(u - 9v)$

101. $v^3 - 2v^2 - v + 2$

102. $y^3 + 4y^2 - y - 4$

103. $t^3 + 3t^2 + 3t + 9$

104. $x^3 + 7x^2 + 3x + 21$

3 ▶ Factor the difference of two squares and factor the sum or difference of two cubes.

In Exercises 105–108, factor the difference of two squares.

105. $x^2 - 36$

106. $16y^2 - 49$

107. $(u + 6)^2 - 81$

108. $(y - 3)^2 - 16$

In Exercises 109–112, factor the sum or difference of two cubes.

109. $u^3 - 1$

110. $t^3 - 125$

111. $8x^3 + 27$

112. $64y^3 + 1$

4 ▶ Factor polynomials completely by repeated factoring.

In Exercises 113–116, factor the polynomial completely.

113. $x^3 - x$

114. $y^4 - 4y^2$

115. $24 + 3u^3$

116. $54 - 2x^3$

5.5 Factoring Trinomials

1 ▶ Recognize and factor perfect square trinomials.

In Exercises 117–120, factor the perfect square trinomial.

117. $x^2 - 18x + 81$

118. $y^2 + 16y + 64$

119. $4s^2 + 40st + 100t^2$

120. $9u^2 - 30uv + 25v^2$

2 ▶ Factor trinomials of the forms $x^2 + bx + c$ and $ax^2 + bx + c$.

In Exercises 121–126, factor the trinomial.

121. $x^2 + 2x - 35$

122. $x^2 - 12x + 32$

123. $2x^2 - 7x + 6$

124. $5x^2 + 11x - 12$

125. $18x^2 + 27x + 10$

126. $12x^2 - 13x - 14$

3 ▶ Factor trinomials of the form $ax^2 + bx + c$ by grouping.

In Exercises 127–132, factor the trinomial by grouping.

127. $4x^2 - 3x - 1$

128. $12x^2 - 7x + 1$

129. $5x^2 - 12x + 7$

130. $3u^2 + 7u - 6$

131. $7s^2 + 10s - 8$

132. $3x^2 - 13x - 10$

4 ▶ Factor polynomials using the guidelines for factoring.

In Exercises 133–140, factor the expression completely.

133. $4a - 64a^3$

134. $3b + 27b^3$

135. $z^3 + z^2 + 3z + 3$

136. $x^3 + 3x^2 - 4x - 12$

137. $\frac{1}{4}x^2 + xy + y^2$

138. $x^2 - \frac{2}{3}x + \frac{1}{9}$

139. $x^2 - 10x + 25 - y^2$

140. $u^6 - 8v^6$

5.6 Solving Polynomial Equations by Factoring

1 ▶ Use the Zero-Factor Property to solve equations.

In Exercises 141–146, use the Zero-Factor Property to solve the equation.

141. $4x(x - 2) = 0$

142. $-3x(2x + 6) = 0$

143. $(2x + 1)(x - 3) = 0$

144. $(x - 7)(3x - 8) = 0$

145. $(x + 10)(4x - 1)(5x + 9) = 0$

146. $3x(x + 8)(2x - 7) = 0$

2 ▶ Solve quadratic equations by factoring.

In Exercises 147–154, solve the quadratic equation by factoring.

147. $3s^2 - 2s - 8 = 0$

148. $5v^2 - 12v - 9 = 0$

149. $m(2m - 1) + 3(2m - 1) = 0$

150. $4w(2w + 8) - 7(2w + 8) = 0$

151. $z(5 - z) + 36 = 0$

152. $(x + 3)^2 - 25 = 0$

153. $v^2 - 100 = 0$

154. $x^2 - 121 = 0$

3 ▶ Solve higher-degree polynomial equations by factoring.

In Exercises 155–162, solve the polynomial equation by factoring.

155. $2y^4 + 2y^3 - 24y^2 = 0$

156. $9x^4 - 15x^3 - 6x^2 = 0$

157. $x^3 - 11x^2 + 18x = 0$

158. $x^3 + 20x^2 + 36x = 0$

159. $b^3 - 6b^2 - b + 6 = 0$

160. $q^3 + 3q^2 - 4q - 12 = 0$

161. $x^4 - 5x^3 - 9x^2 + 45x = 0$

162. $2x^4 + 6x^3 - 50x^2 - 150x = 0$

4 ▶ Solve application problems by factoring.

163. *Number Problem*   Find two consecutive positive odd integers whose product is 99.

164. *Number Problem*   Find two consecutive positive even integers whose product is 168.

165. ▲ *Geometry*   A rectangle has an area of 900 square inches. The length of the rectangle is $2\frac{1}{4}$ times its width. Find the dimensions of the rectangle.

166. ▲ *Geometry*   A rectangle has an area of 432 square inches. The width of the rectangle is $\frac{3}{4}$ times its length. Find the dimensions of the rectangle.

167. ▲ *Geometry*   A closed box with a square base stands 12 inches tall. The total surface area of the outside of the box is 512 square inches. What are the dimensions of the base?   (*Hint:* The surface area is given by $S = 2x^2 + 4xh$.)

168. ▲ *Geometry*   An open box with a square base stands 10 inches tall. The total surface area of the outside of the box is 225 square inches. What are the dimensions of the base? (*Hint:* The surface area is given by $S = x^2 + 4xh$.)

169. *Free-Falling Object*   An object is dropped from a weather balloon 3600 feet above the ground. The height h (in feet) of the object is modeled by the position equation $h = -16t^2 + 3600$, where t is the time (in seconds). How long will it take the object to reach the ground?

170. *Free-Falling Object*   An object is thrown upward from the Trump Tower in New York City, which is 664 feet tall, with an initial velocity of 45 feet per second. The height h (in feet) of the object is modeled by the position equation $h = -16t^2 + 45t + 664$, where t is the time (in seconds). How long will it take the object to reach the ground?

Chapter Test

Take this test as you would take a test in class. After you are done, check your work against the answers in the back of the book.

1. Determine the degree and leading coefficient of $3 - 4.5x + 8.2x^3$.

2. Explain why the following expression is not a polynomial.

$$\frac{4}{x^2 + 2}$$

In Exercises 3 and 4, rewrite each expression using only positive exponents, and simplify. (Assume that any variables in the expression are nonzero.)

3. (a) $\dfrac{2^{-1}x^5y^{-3}}{4x^{-2}y^2}$ (b) $\left(\dfrac{-2x^2y}{z^{-3}}\right)^{-2}$

4. (a) $\left(-\dfrac{2u^2}{v^{-1}}\right)^3\left(\dfrac{3v^2}{u^{-3}}\right)$ (b) $\dfrac{(-3x^2y^{-1})^4}{6x^2y^0}$

In Exercises 5–9, perform the indicated operations and simplify.

5. (a) $(5a^2 - 3a + 4) + (a^2 - 4)$ (b) $(16 - y^2) - (16 + 2y + y^2)$

6. (a) $-2(2x^4 - 5) + 4x(x^3 + 2x - 1)$ (b) $4t - [3t - (10t + 7)]$

7. (a) $-3x(x - 4)$ (b) $(2x - 3y)(x + 5y)$

8. (a) $(x - 1)[2x + (x - 3)]$ (b) $(2s - 3)(3s^2 - 4s + 7)$

9. (a) $(2w - 7)^2$ (b) $[4 - (a + b)][4 + (a + b)]$

In Exercises 10–15, factor the expression completely.

10. $18y^2 - 12y$ 11. $v^2 - \dfrac{16}{9}$

12. $x^3 - 3x^2 - 4x + 12$ 13. $9u^2 - 6u + 1$

14. $6x^2 - 26x - 20$ 15. $x^3 + 27$

In Exercises 16–19, solve the equation.

16. $(x - 3)(x + 2) = 14$ 17. $(y + 2)^2 - 9 = 0$

18. $12 + 5y - 3y^2 = 0$ 19. $2x^3 + 10x^2 + 8x = 0$

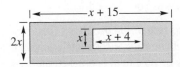

Figure for 20

20. Write an expression for the area of the shaded region in the figure. Then simplify the expression.

21. The area of a rectangle is 54 square centimeters. The length of the rectangle is $1\frac{1}{2}$ times its width. Find the dimensions of the rectangle.

22. The area of a triangle is 20 square feet. The height of the triangle is 2 feet more than twice its base. Find the base and height of the triangle.

23. The revenue R from the sale of x computer desks is given by $R = x^2 - 35x$. The cost C of producing x computer desks is given by $C = 150 + 12x$. How many computer desks must be produced and sold in order to break even?

Study Skills in Action

Using a Test-Taking Strategy

What do runners do before a race? They design a strategy for running their best. They make sure they get enough rest, eat sensibly, and get to the track early to warm up. In the same way, it is important for students to get a good night's sleep, eat a healthy meal, and get to class early to allow time to focus before a test.

The biggest difference between a runner's race and a math test is that a math student does not have to reach the finish line first! In fact, many students would increase their scores if they used all the test time instead of worrying about being the last student left in the class. This is why it is important to have a strategy for taking the test.

Kimberly Nolting

VP, Academic Success Press
expert in developmental education

These runners are focusing on their techniques, not on whether other runners are ahead of or behind them.

Smart Study Strategy

Use Ten Steps for Test-Taking

1 ▶ **Do a memory data dump.** As soon as you get the test, turn it over and write down anything that you still have trouble remembering sometimes (formulas, calculations, rules).

2 ▶ **Preview the test.** Look over the test and mark the questions you know how to do easily. These are the problems you should do first.

3 ▶ **Do a second memory data dump.** As you previewed the test, you may have remembered other information. Write this information on the back of the test.

4 ▶ **Develop a test progress schedule.** Based on how many points each question is worth, decide on a progress schedule. You should always have more than half the test done before half the time has elapsed.

5 ▶ **Answer the easiest problems first.** Solve the problems you marked while previewing the test.

6 ▶ **Skip difficult problems.** Skip the problems that you suspect will give you trouble.

7 ▶ **Review the skipped problems.** After solving all the problems that you know how to do easily, go back and reread the problems you skipped.

8 ▶ **Try your best at the remaining problems that confuse you.** Even if you cannot completely solve a problem, you may be able to get partial credit for a few correct steps.

9 ▶ **Review the test.** Look for any careless errors you may have made.

10▶ **Use all the allowed test time.** The test is not a race against the other students.

Chapter 6
Rational Expressions, Equations, and Functions

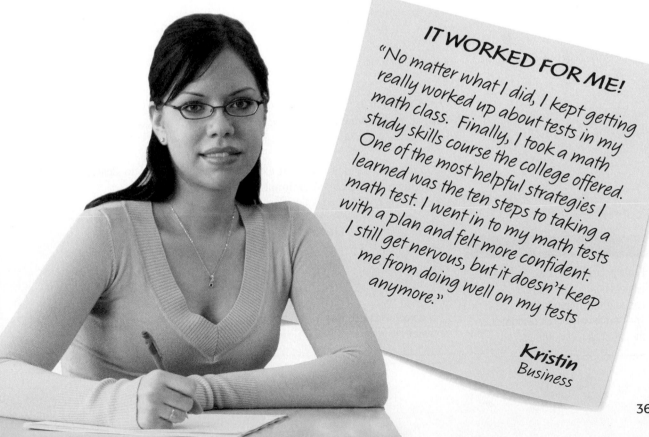

IT WORKED FOR ME!

"No matter what I did, I kept getting really worked up about tests in my math class. Finally, I took a math study skills course the college offered. One of the most helpful strategies I learned was the ten steps to taking a math test. I went in to my math tests with a plan and felt more confident. I still get nervous, but it doesn't keep me from doing well on my tests anymore."

Kristin
Business

6.1 Rational Expressions and Functions

Paul Barton/CORBIS

Why You Should Learn It

Rational expressions can be used to solve real-life problems. For instance, in Exercise 93 on page 381, you will find a rational expression that models the average cable television revenue per subscriber.

1 ▶ Find the domain of a rational function.

What You Should Learn

1 ▶ Find the domain of a rational function.
2 ▶ Simplify rational expressions.

The Domain of a Rational Function

A fraction whose numerator and denominator are polynomials is called a **rational expression.** Some examples are

$$\frac{3}{x+4}, \quad \frac{2x}{x^2-4x+4}, \quad \text{and} \quad \frac{x^2-5x}{x^2+2x-3}.$$

In Section 3.6, you learned that because division by zero is undefined, the denominator of a rational expression cannot be zero. So, in your work with rational expressions, you must assume that all real number values of the variable that make the denominator zero are excluded. For the three fractions above, $x = -4$ is excluded from the first fraction, $x = 2$ from the second, and both $x = 1$ and $x = -3$ from the third. The set of *usable* values of the variable is called the **domain** of the rational expression.

Definition of a Rational Expression

Let u and v be polynomials. The algebraic expression

$$\frac{u}{v}$$

is a **rational expression.** The **domain** of this rational expression is the set of all real numbers for which $v \neq 0$.

Like polynomials, rational expressions can be used to describe functions. Such functions are called **rational functions.**

Study Tip

Every polynomial is also a rational expression because you can consider the denominator to be 1. The domain of every polynomial is the set of all real numbers.

Definition of a Rational Function

Let $u(x)$ and $v(x)$ be polynomial functions. The function

$$f(x) = \frac{u(x)}{v(x)}$$

is a **rational function.** The **domain** of f is the set of all real numbers for which $v(x) \neq 0$.

EXAMPLE 1   Finding the Domains of Rational Functions

Find the domain of each rational function.

a. $f(x) = \dfrac{4}{x - 2}$ **b.** $g(x) = \dfrac{2x + 5}{8}$

Solution

a. The denominator is zero when $x - 2 = 0$ or $x = 2$. So, the domain is all real values of x such that $x \neq 2$. In interval notation, you can write the domain as

   Domain $= (-\infty, 2) \cup (2, \infty)$.

b. The denominator, 8, is never zero, so the domain is the set of *all* real numbers. In interval notation, you can write the domain as

   Domain $= (-\infty, \infty)$.

 **CHECKPOINT** *Now try Exercise 3.*

Technology: Discovery

Use a graphing calculator to graph the equation that corresponds to part (a) of Example 1. Then use the *trace* or *table* feature of the calculator to determine the behavior of the graph near $x = 2$. Graph the equation that corresponds to part (b) of Example 1. How does this graph differ from the graph in part (a)?

EXAMPLE 2   Finding the Domains of Rational Functions

Find the domain of each rational function.

a. $f(x) = \dfrac{5x}{x^2 - 16}$ **b.** $h(x) = \dfrac{3x - 1}{x^2 - 2x - 3}$

Study Tip

Remember that when interval notation is used, the symbol $\cup$ means *union* and the symbol $\cap$ means *intersection*.

Solution

a. The denominator is zero when $x^2 - 16 = 0$. Solving this equation by factoring, you find that the denominator is zero when $x = -4$ or $x = 4$. So, the domain is all real values of x such that $x \neq -4$ and $x \neq 4$. In interval notation, you can write the domain as

   Domain $= (-\infty, -4) \cup (-4, 4) \cup (4, \infty)$.

b. The denominator is zero when $x^2 - 2x - 3 = 0$. Solving this equation by factoring, you find that the denominator is zero when $x = 3$ or when $x = -1$. So, the domain is all real values of x such that $x \neq 3$ and $x \neq -1$. In interval notation, you can write the domain as

   Domain $= (-\infty, -1) \cup (-1, 3) \cup (3, \infty)$.

 **CHECKPOINT** *Now try Exercise 15.*

In applications involving rational functions, it is often necessary to place restrictions on the domain other than the restrictions *implied* by values that make the denominator zero. Such additional restrictions can be indicated to the right of the function. For instance, the domain of the rational function

$$f(x) = \frac{x^2 + 20}{x + 4}, \qquad x > 0$$

is the set of *positive* real numbers, as indicated by the inequality $x > 0$. Note that the normal domain of this function would be all real values of x such that $x \neq -4$. However, because "$x > 0$" is listed to the right of the function, the domain is further restricted by this inequality.

EXAMPLE 3   An Application Involving a Restricted Domain

You have started a small business that manufactures lamps. The initial investment for the business is $120,000. The cost of manufacturing each lamp is $15. So, your total cost of producing x lamps is

$$C = 15x + 120{,}000. \qquad \text{Cost function}$$

Your average cost per lamp depends on the number of lamps produced. For instance, the average cost per lamp $\overline{C}$ of producing 100 lamps is

$$\overline{C} = \frac{15(100) + 120{,}000}{100} \qquad \text{Substitute 100 for } x.$$

$$= \$1215. \qquad \text{Average cost per lamp for 100 lamps}$$

The average cost per lamp decreases as the number of lamps increases. For instance, the average cost per lamp $\overline{C}$ of producing 1000 lamps is

$$\overline{C} = \frac{15(1000) + 120{,}000}{1000} \qquad \text{Substitute 1000 for } x.$$

$$= \$135. \qquad \text{Average cost per lamp for 1000 lamps}$$

In general, the average cost of producing x lamps is

$$\overline{C} = \frac{15x + 120{,}000}{x}. \qquad \text{Average cost per lamp for } x \text{ lamps}$$

What is the domain of this rational function?

Solution

If you were considering this function from only a mathematical point of view, you would say that the domain is all real values of x such that $x \neq 0$. However, because this function is a mathematical model representing a real-life situation, you must decide which values of x make sense in real life. For this model, the variable x represents the number of lamps that you produce. Assuming that you cannot produce a fractional number of lamps, you can conclude that the domain is the set of positive integers—that is,

$$\text{Domain} = \{1, 2, 3, 4, \ldots\}.$$

 **CHECKPOINT** *Now try Exercise 31.*

2 ▶ Simplify rational expressions.

Simplifying Rational Expressions

As with numerical fractions, a rational expression is said to be in **simplified** (or **reduced**) **form** if its numerator and denominator have no common factors (other than ±1). To simplify rational expressions, you can apply the rule below.

> ### Simplifying Rational Expressions
>
> Let u, v, and w represent real numbers, variables, or algebraic expressions such that $v \neq 0$ and $w \neq 0$. Then the following is valid.
>
> $$\frac{uw}{vw} = \frac{u\cancel{w}}{v\cancel{w}} = \frac{u}{v}$$

Be sure you divide out only *factors*, not *terms*. For instance, consider the expressions below.

$$\frac{\cancel{2} \cdot 2}{\cancel{2}(x + 5)}$$ You *can* divide out the common factor 2.

$$\frac{3 + x}{3 + 2x}$$ You *cannot* divide out the common term 3.

Simplifying a rational expression requires two steps: (1) completely factor the numerator and denominator and (2) divide out any *factors* that are common to both the numerator and denominator. So, your success in simplifying rational expressions actually lies in your ability to *factor completely* the polynomials in both the numerator and denominator.

EXAMPLE 4   Simplifying a Rational Expression

Simplify the rational expression $\dfrac{2x^3 - 6x}{6x^2}$.

Solution

First note that the domain of the rational expression is all real values of x such that $x \neq 0$. Then, completely factor both the numerator and denominator.

$$\frac{2x^3 - 6x}{6x^2} = \frac{2x(x^2 - 3)}{2x(3x)}$$ Factor numerator and denominator.

$$= \frac{\cancel{2x}(x^2 - 3)}{\cancel{2x}(3x)}$$ Divide out common factor $2x$.

$$= \frac{x^2 - 3}{3x}$$ Simplified form

In simplified form, the domain of the rational expression is the same as that of the original expression—all real values of x such that $x \neq 0$.

 **CHECKPOINT** *Now try Exercise 43.*

EXAMPLE 5   Simplifying a Rational Expression

Simplify the rational expression $\frac{x^2 + 2x - 15}{3x - 9}$.

Solution

The domain of the rational expression is all real values of x such that $x \neq 3$.

$$\frac{x^2 + 2x - 15}{3x - 9} = \frac{(x + 5)(x - 3)}{3(x - 3)} \quad \text{Factor numerator and denominator.}$$

$$= \frac{(x + 5)(x - 3)}{3(x - 3)} \quad \text{Divide out common factor } (x - 3).$$

$$= \frac{x + 5}{3}, \ x \neq 3 \quad \text{Simplified form}$$

✓ **CHECKPOINT** *Now try Exercise 51.*

Dividing out common factors from the numerator and denominator of a rational expression can change the implied domain. For instance, in Example 5 the domain restriction $x \neq 3$ must be listed because it is no longer implied in the simplified expression. With this restriction, the new expression is equal to the original expression.

EXAMPLE 6   Simplifying a Rational Expression

Simplify the rational expression $\frac{x^3 - 16x}{x^2 - 2x - 8}$.

Solution

The domain of the rational expression is all real values of x such that $x \neq -2$ and $x \neq 4$.

$$\frac{x^3 - 16x}{x^2 - 2x - 8} = \frac{x(x^2 - 16)}{(x + 2)(x - 4)} \quad \text{Partially factor.}$$

$$= \frac{x(x + 4)(x - 4)}{(x + 2)(x - 4)} \quad \text{Factor completely.}$$

$$= \frac{x(x + 4)(x - 4)}{(x + 2)(x - 4)} \quad \text{Divide out common factor } (x - 4).$$

$$= \frac{x(x + 4)}{x + 2}, \ x \neq 4 \quad \text{Simplified form}$$

✓ **CHECKPOINT** *Now try Exercise 61.*

When you simplify a rational expression, keep in mind that you must list any domain restrictions that are no longer implied in the simplified expression. For instance, in Example 6 the restriction $x \neq 4$ is listed so that the domains agree for the original and simplified expressions. The example does not list $x \neq -2$ because this restriction is apparent by looking at either expression.

EXAMPLE 7 Simplification Involving a Change in Sign

Simplify the rational expression $\dfrac{2x^2 - 9x + 4}{12 + x - x^2}$.

Solution

The domain of the rational expression is all real values of x such that $x \neq -3$ and $x \neq 4$.

$$\frac{2x^2 - 9x + 4}{12 + x - x^2} = \frac{(2x - 1)(x - 4)}{(4 - x)(3 + x)} \qquad \text{Factor numerator and denominator.}$$

$$= \frac{(2x - 1)(x - 4)}{-(x - 4)(3 + x)} \qquad (4 - x) = -(x - 4)$$

$$= \frac{(2x - 1)\cancel{(x - 4)}}{-\cancel{(x - 4)}(3 + x)} \qquad \text{Divide out common factor } (x - 4).$$

$$= -\frac{2x - 1}{3 + x}, \quad x \neq 4 \qquad \text{Simplified form}$$

The simplified form is equivalent to the original expression for all values of x such that $x \neq 4$. Note that by implied restriction, $x = -3$ is excluded from the domains of both the original and simplified expressions.

✔ **CHECKPOINT** *Now try Exercise 65.*

In Example 7, be sure you see that when dividing the numerator and denominator by the common factor of $(x - 4)$, you keep the minus sign. In the simplified form of the fraction, this text uses the convention of moving the minus sign out in front of the fraction. However, this is a personal preference. All of the following forms are equivalent.

$$-\frac{2x - 1}{3 + x} = \frac{-(2x - 1)}{3 + x} = \frac{-2x + 1}{3 + x} = \frac{2x - 1}{-3 - x} = \frac{2x - 1}{-(3 + x)}$$

EXAMPLE 8 A Rational Expression Involving Two Variables

Simplify the rational expression $\dfrac{3xy + y^2}{2y}$.

Solution

The domain of the rational expression is all real values of y such that $y \neq 0$.

$$\frac{3xy + y^2}{2y} = \frac{y(3x + y)}{2y} \qquad \text{Factor numerator and denominator.}$$

$$= \frac{\cancel{y}(3x + y)}{2\cancel{y}} \qquad \text{Divide out common factor } y.$$

$$= \frac{3x + y}{2}, \quad y \neq 0 \qquad \text{Simplified form}$$

✔ **CHECKPOINT** *Now try Exercise 71.*

EXAMPLE 9 **A Rational Expression Involving Two Variables**

$$\frac{2x^2 + 2xy - 4y^2}{5x^3 - 5xy^2} = \frac{2(x - y)(x + 2y)}{5x(x - y)(x + y)}$$ Factor numerator and denominator.

$$= \frac{2\cancel{(x - y)}(x + 2y)}{5x\cancel{(x - y)}(x + y)}$$ Divide out common factor $(x - y)$.

$$= \frac{2(x + 2y)}{5x(x + y)}, \quad x \ne y$$ Simplified form

The domain of the original rational expression is all real values of x and y such that $x \ne 0$ and $x \ne \pm y$.

 CHECKPOINT *Now try Exercise 73.*

Study Tip

As you study the examples and work the exercises in this section and the next four sections, keep in mind that you are *rewriting expressions in simpler forms*. You are not solving equations. Equal signs are used in the steps of the simplification process only to indicate that the new form of the expression is *equivalent* to the original form.

EXAMPLE 10 **A Rational Expression Involving Two Variables**

$$\frac{4x^2y - y^3}{2x^2y - xy^2} = \frac{(2x - y)(2x + y)y}{(2x - y)xy}$$ Factor numerator and denominator.

$$= \frac{\cancel{(2x - y)}(2x + y)\cancel{y}}{\cancel{(2x - y)}x\cancel{y}}$$ Divide out common factors $(2x - y)$ and y.

$$= \frac{2x + y}{x}, \quad y \ne 0, \ y \ne 2x$$ Simplified form

The domain of the original rational expression is all real values of x and y such that $x \ne 0$, $y \ne 0$, and $y \ne 2x$.

 CHECKPOINT *Now try Exercise 75.*

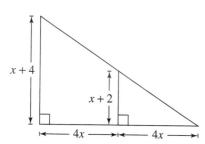

Figure 6.1

EXAMPLE 11 **Geometry: Area**

Find the ratio of the area of the shaded portion of the triangle to the total area of the triangle. (See Figure 6.1.)

Solution

The area of the shaded portion of the triangle is given by

$$\text{Area} = \tfrac{1}{2}(4x)(x + 2) = \tfrac{1}{2}(4x^2 + 8x) = 2x^2 + 4x.$$

The total area of the triangle is given by

$$\text{Area} = \tfrac{1}{2}(4x + 4x)(x + 4) = \tfrac{1}{2}(8x)(x + 4) = \tfrac{1}{2}(8x^2 + 32x) = 4x^2 + 16x.$$

So, the ratio of the area of the shaded portion of the triangle to the total area of the triangle is

$$\frac{2x^2 + 4x}{4x^2 + 16x} = \frac{2x(x + 2)}{4x(x + 4)} = \frac{x + 2}{2(x + 4)}, \quad x > 0.$$

CHECKPOINT *Now try Exercise 85.*

_____ Concept Check _____

1. Describe the process for finding the implied domain restrictions of a rational function.

2. Describe a situation in which you would need to indicate a domain restriction to the right of a rational function.

3. What expression(s) must you factor completely in order to simplify a rational function of the form $f(x) = \dfrac{u(x)}{v(x)}$, and why?

4. After factoring completely, what is one additional step that is sometimes needed to find common factors in the numerator and denominator of a rational expression?

6.1 EXERCISES

Go to pages 440–441 to record your assignments.

_____ Developing Skills _____

In Exercises 1–22, find the domain of the rational function. **See Examples 1 and 2.**

1. $f(x) = \dfrac{x^2 + 9}{4}$

2. $f(y) = \dfrac{y^2 - 3}{7}$

✓ **3.** $f(x) = \dfrac{4}{x - 3}$

4. $g(x) = \dfrac{-2}{x - 7}$

5. $f(x) = \dfrac{12x}{9 - x}$

6. $h(y) = \dfrac{2y}{1 - y}$

7. $g(x) = \dfrac{2x}{x + 10}$

8. $f(x) = \dfrac{4x}{x + 1}$

9. $h(x) = \dfrac{x}{x^2 + 4}$

10. $h(x) = \dfrac{4x}{x^2 + 16}$

11. $f(y) = \dfrac{y - 4}{y(y + 3)}$

12. $f(z) = \dfrac{z + 2}{z(z - 4)}$

13. $f(x) = \dfrac{x^2}{x(x - 1)}$

14. $g(x) = \dfrac{x^3}{x(x + 2)}$

✓ **15.** $f(t) = \dfrac{5t}{t^2 - 16}$

16. $f(x) = \dfrac{x}{x^2 - 4}$

17. $g(y) = \dfrac{y + 5}{y^2 - 3y}$

18. $g(t) = \dfrac{t - 6}{t^2 + 5t}$

19. $g(x) = \dfrac{x + 1}{x^2 - 5x + 6}$

20. $h(t) = \dfrac{3t^2}{t^2 - 2t - 3}$

21. $f(u) = \dfrac{u^2}{3u^2 - 2u - 5}$

22. $g(y) = \dfrac{y + 5}{4y^2 - 5y - 6}$

In Exercises 23–28, evaluate the rational function as indicated, and simplify. If not possible, state the reason.

23. $f(x) = \dfrac{4x}{x + 3}$

    (a) $f(1)$ (b) $f(-2)$

    (c) $f(-3)$ (d) $f(0)$

24. $f(x) = \dfrac{x - 5}{4x}$

    (a) $f(10)$ (b) $f(0)$

    (c) $f(-3)$ (d) $f(5)$

25. $g(x) = \dfrac{x^2 - 4x}{x^2 - 9}$

    (a) $g(0)$ (b) $g(4)$

    (c) $g(3)$ (d) $g(-3)$

26. $g(t) = \dfrac{t^2 + 4t}{t^2 - 4}$

    (a) $g(2)$ (b) $g(1)$

    (c) $g(-2)$ (d) $g(-4)$

27. $h(s) = \dfrac{s^2}{s^2 - s - 2}$

    (a) $h(10)$ (b) $h(0)$

    (c) $h(-1)$ (d) $h(2)$

28. $f(x) = \dfrac{x^3 + 1}{x^2 - 6x + 9}$

    (a) $f(-1)$ (b) $f(3)$

    (c) $f(-2)$ (d) $f(2)$

In Exercises 29–34, describe the domain. *See Example 3.*

29. ▲ *Geometry* A rectangle of length x inches has an area of 500 square inches. The perimeter P of the rectangle is given by

$$P = 2\left(x + \frac{500}{x}\right).$$

30. *Cost* The cost C in millions of dollars for the government to seize $p\%$ of an illegal drug as it enters the country is given by

$$C = \frac{528p}{100 - p}.$$

✔ **31.** *Inventory Cost* The inventory cost I when x units of a product are ordered from a supplier is given by

$$I = \frac{0.25x + 2000}{x}.$$

32. *Average Cost* The average cost $\overline{C}$ for a manufacturer to produce x units of a product is given by

$$\overline{C} = \frac{1.35x + 4570}{x}.$$

33. *Pollution Removal* The cost C in dollars of removing $p\%$ of the air pollutants in the stack emission of a utility company is given by the rational function

$$C = \frac{60{,}000p}{100 - p}.$$

34. *Consumer Awareness* The average cost of a movie rental $\overline{M}$ when you consider the cost of purchasing a DVD player and renting x DVDs at \$3.49 per movie is

$$\overline{M} = \frac{75 + 3.49x}{x}.$$

In Exercises 35–42, fill in the missing factor.

35. $\dfrac{5(\quad\quad\quad\quad)}{6(x + 3)} = \dfrac{5}{6}, \quad x \neq -3$

36. $\dfrac{7(\quad\quad\quad\quad)}{15(x - 10)} = \dfrac{7}{15}, \quad x \neq 10$

37. $\dfrac{3x(x + 16)^2}{2(\quad\quad\quad\quad)} = \dfrac{x}{2}, \quad x \neq -16$

38. $\dfrac{25x^2(x - 10)}{12(\quad\quad\quad\quad)} = \dfrac{5x}{12}, \quad x \neq 10, \quad x \neq 0$

39. $\dfrac{(x + 5)(\quad\quad\quad\quad)}{3x^2(x - 2)} = \dfrac{x + 5}{3x}, \quad x \neq 2$

40. $\dfrac{(3y - 7)(\quad\quad\quad\quad)}{y^2 - 4} = \dfrac{3y - 7}{y + 2}, \quad y \neq 2$

41. $\dfrac{8x(\quad\quad\quad\quad)}{x^2 - 2x - 15} = \dfrac{8x}{x - 5}, \quad x \neq -3$

42. $\dfrac{(3 - z)(\quad\quad\quad\quad)}{z^3 + 2z^2} = \dfrac{3 - z}{z^2}, \quad z \neq -2$

In Exercises 43–80, simplify the rational expression. *See Examples 4–10.*

✔ **43.** $\dfrac{5x}{25}$ **44.** $\dfrac{32y}{24}$

45. $\dfrac{12x^2}{12x}$ **46.** $\dfrac{15z^3}{15z^3}$

47. $\dfrac{18x^2y}{15xy^4}$ **48.** $\dfrac{24xz^4}{16x^3z}$

49. $\dfrac{3x^2 - 9x}{12x^2}$ **50.** $\dfrac{8x^3 + 4x^2}{20x}$

51. $\dfrac{x^2(x-8)}{x(x-8)}$

52. $\dfrac{a^2b(b-3)}{b^3(b-3)^2}$

71. $\dfrac{3xy^2}{xy^2+x}$

72. $\dfrac{x+3x^2y}{3xy+1}$

73. $\dfrac{y^2-64x^2}{5(3y+24x)}$

74. $\dfrac{x^2-25z^2}{2(3x+15z)}$

53. $\dfrac{2x-3}{4x-6}$

54. $\dfrac{5-x}{3x-15}$

55. $\dfrac{y^2-49}{2y-14}$

56. $\dfrac{x^2-36}{6-x}$

75. $\dfrac{5xy+3x^2y^2}{xy^3}$

76. $\dfrac{4u^2v-12uv^2}{18uv}$

57. $\dfrac{a+3}{a^2+6a+9}$

58. $\dfrac{u^2-12u+36}{u-6}$

77. $\dfrac{u^2-4v^2}{u^2+uv-2v^2}$

78. $\dfrac{x^2+4xy}{x^2-16y^2}$

59. $\dfrac{x^2-7x}{x^2-14x+49}$

60. $\dfrac{z^2+22z+121}{3z+33}$

79. $\dfrac{3m^2-12n^2}{m^2+4mn+4n^2}$

80. $\dfrac{x^2+xy-2y^2}{x^2+3xy+2y^2}$

In Exercises 81 and 82, complete the table. What can you conclude?

61. $\dfrac{y^3-4y}{y^2+4y-12}$

62. $\dfrac{x^3-4x}{x^2-5x+6}$

81.

x	-2	-1	0	1	2	3	4
$\dfrac{x^2-x-2}{x-2}$							
$x+1$							

63. $\dfrac{y^4-16y^2}{y^2+y-12}$

64. $\dfrac{x^4-25x^2}{x^2+2x-15}$

65. $\dfrac{3x^2-7x-20}{12+x-x^2}$

66. $\dfrac{2x^2+3x-5}{7-6x-x^2}$

82.

x	-2	-1	0	1	2	3	4
$\dfrac{x^2+5x}{x}$							
$x+5$							

67. $\dfrac{2x^2+19x+24}{2x^2-3x-9}$

68. $\dfrac{2y^2+13y+20}{2y^2+17y+30}$

69. $\dfrac{15x^2+7x-4}{25x^2-16}$

70. $\dfrac{56z^2-3z-20}{49z^2-16}$

Solving Problems

 Geometry   In Exercises 83–86, find the ratio of the area of the shaded portion to the total area of the figure. *See Example 11.*

83.
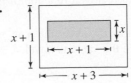

84.
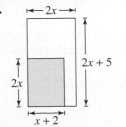

✓ **85.** 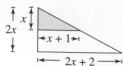 **86.**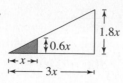

87. *Average Cost*   A machine shop has a setup cost of $2500 for the production of a new product. The cost of labor and material for producing each unit is $9.25.

(a) Write the total cost C as a function of x, the number of units produced.

(b) Write the average cost per unit $\overline{C} = C/x$ as a function of x, the number of units produced.

(c) Determine the domain of the function in part (b).

(d) Find the value of $\overline{C}(100)$.

88. *Average Cost*   A greeting card company has an initial investment of $60,000. The cost of producing one dozen cards is $6.50.

(a) Write the total cost C as a function of x, the number of dozens of cards produced.

(b) Write the average cost per dozen $\overline{C} = C/x$ as a function of x, the number of dozens of cards produced.

(c) Determine the domain of the function in part (b).

(d) Find the value of $\overline{C}(11,000)$.

89. *Distance Traveled*   A van starts on a trip and travels at an average speed of 45 miles per hour. Three hours later, a car starts on the same trip and travels at an average speed of 60 miles per hour.

(a) Find the distance each vehicle has traveled when the car has been on the road for t hours.

(b) Use the result of part (a) to write the distance between the van and the car as a function of t.

(c) Write the ratio of the distance the car has traveled to the distance the van has traveled as a function of t.

90. *Distance Traveled*   A car starts on a trip and travels at an average speed of 55 miles per hour. Two hours later, a second car starts on the same trip and travels at an average speed of 65 miles per hour.

(a) Find the distance each vehicle has traveled when the second car has been on the road for t hours.

(b) Use the result of part (a) to write the distance between the first car and the second car as a function of t.

(c) Write the ratio of the distance the second car has traveled to the distance the first car has traveled as a function of t.

91. ▲ *Geometry*   One swimming pool is circular and another is rectangular. The rectangular pool's width is three times its depth. Its length is 6 feet more than its width. The circular pool has a diameter that is twice the width of the rectangular pool, and it is 2 feet deeper. Find the ratio of the circular pool's volume to the rectangular pool's volume.

92. ▲ *Geometry*   A circular pool has a radius five times its depth. A rectangular pool has the same depth as the circular pool. Its width is 4 feet more than three times its depth and its length is 2 feet less than six times its depth. Find the ratio of the rectangular pool's volume to the circular pool's volume.

Cable TV Revenue   In Exercises 93 and 94, use the following polynomial models, which give the total basic cable television revenue R (in millions of dollars) and the number of basic cable subscribers S (in millions) for the years 2001 through 2005 (see figures).

$R = 1189.2t + 25{,}266, \quad 1 \le t \le 5$
$S = -0.35t + 67.1, \quad 1 \le t \le 5$

In these models, t represents the year, with $t = 1$ corresponding to 2001.   (Source: Kagan Research, LLC)

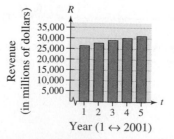

Figures for 93 and 94

93. Find a rational model that represents the average basic cable television revenue per subscriber during the years 2001 to 2005.

94. Use the model found in Exercise 93 to complete the table, which shows the average basic cable television revenue per subscriber.

Year	2001	2002	2003	2004	2005
Average revenue					

Explaining Concepts

95. ✎ How do you determine whether a rational expression is in simplified form?

96. ✎ Can you divide out common terms from the numerator and denominator of a rational expression? Explain.

97. Give an example of a rational function whose domain is the set of all real numbers and whose denominator is a second-degree polynomial function.

98. *Error Analysis* Describe the error.
$$\frac{2x^2}{x^2+4} = \frac{2\cancel{x^2}}{\cancel{x^2}+4} = \frac{2}{1+4} = \frac{2}{5}$$

99. A student writes the following incorrect solution for simplifying a rational expression. Discuss the student's errors and misconceptions, and construct a correct solution.

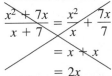
$$\frac{x^2+7x}{x+7} = \frac{x^2}{x} + \frac{7x}{7}$$
$$= x + x$$
$$= 2x$$

100. ✎ Is the following statement true? Explain.
$$\frac{6x-5}{5-6x} = -1$$

101. ✎ Explain how you can use a given polynomial function $f(x)$ to write a rational function $g(x)$ that is equivalent to $f(x)$, $x \neq 2$.

102. ✎ Is it possible for a rational function $f(x)$ (without added domain restrictions) to be undefined on an interval $[a, b]$, where a and b are real numbers such that $a < b$? Explain.

Cumulative Review

In Exercises 103–106, find the product.

103. $\frac{1}{4}\left(\frac{3}{4}\right)$

104. $\frac{2}{3}\left(-\frac{5}{6}\right)$

105. $\frac{1}{3}\left(\frac{3}{5}\right)(5)$

106. $\left(-\frac{3}{7}\right)\left(\frac{2}{5}\right)\left(-\frac{1}{6}\right)$

In Exercises 107–110, perform the indicated multiplication.

107. $(-2a^3)(-2a)$

108. $6x^2(-3x)$

109. $(-3b)(b^2 - 3b + 5)$

110. $ab^2(3a - 4ab + 6a^2b^2)$

6.2 Multiplying and Dividing Rational Expressions

Steven E. Frischling/Bloomberg News/Landov

What You Should Learn

1 ▶ Multiply rational expressions and simplify.

2 ▶ Divide rational expressions and simplify.

Why You Should Learn It

Multiplication and division of rational expressions can be used to solve real-life applications. For instance, Example 9 on page 386 shows how to divide rational expressions to find a model for the annual amount the average American spent on meals away from home from 2000 to 2006.

1 ▶ Multiply rational expressions and simplify.

Multiplying Rational Expressions

The rule for multiplying rational expressions is the same as the rule for multiplying numerical fractions. That is, you *multiply numerators, multiply denominators, and write the new fraction in simplified form.*

$$\frac{3}{4} \cdot \frac{7}{6} = \frac{21}{24} = \frac{\cancel{3} \cdot 7}{\cancel{3} \cdot 8} = \frac{7}{8}$$

Multiplying Rational Expressions

Let u, v, w, and z represent real numbers, variables, or algebraic expressions such that $v \neq 0$ and $z \neq 0$. Then the product of u/v and w/z is

$$\frac{u}{v} \cdot \frac{w}{z} = \frac{uw}{vz}.$$

In order to recognize common factors when simplifying the product, use factoring in the numerator and denominator, as demonstrated in Example 1.

EXAMPLE 1   Multiplying Rational Expressions

Multiply the rational expressions $\dfrac{4x^3y}{3xy^4} \cdot \dfrac{-6x^2y^2}{10x^4}$.

Solution

$$\frac{4x^3y}{3xy^4} \cdot \frac{-6x^2y^2}{10x^4} = \frac{(4x^3y) \cdot (-6x^2y^2)}{(3xy^4) \cdot (10x^4)}$$ Multiply numerators and denominators.

$$= \frac{-24x^5y^3}{30x^5y^4}$$ Simplify.

$$= \frac{-4\cancel{(6)}\cancel{(x^5)}\cancel{(y^3)}}{5\cancel{(6)}\cancel{(x^5)}\cancel{(y^3)}(y)}$$ Factor and divide out common factors.

$$= -\frac{4}{5y}, \ x \neq 0$$ Simplified form

✓ **CHECKPOINT** *Now try Exercise 11.*

EXAMPLE 2   **Multiplying Rational Expressions**

Multiply the rational expressions.

$$\frac{x}{5x^2 - 20x} \cdot \frac{x - 4}{2x^2 + x - 3}$$

Solution

$$\frac{x}{5x^2 - 20x} \cdot \frac{x - 4}{2x^2 + x - 3}$$

$$= \frac{x \cdot (x - 4)}{(5x^2 - 20x) \cdot (2x^2 + x - 3)} \qquad \text{Multiply numerators and denominators.}$$

$$= \frac{x(x - 4)}{5x(x - 4)(x - 1)(2x + 3)} \qquad \text{Factor.}$$

$$= \frac{\cancel{x}\cancel{(x - 4)}}{5\cancel{x}\cancel{(x - 4)}(x - 1)(2x + 3)} \qquad \text{Divide out common factors.}$$

$$= \frac{1}{5(x - 1)(2x + 3)}, \ x \neq 0, \ x \neq 4 \qquad \text{Simplified form}$$

✓ **CHECKPOINT** *Now try Exercise 23.*

Technology: Tip

You can use a graphing calculator to check your results when multiplying rational expressions. For instance, in Example 3, try graphing the equations

$$y_1 = \frac{4x^2 - 4x}{x^2 + 2x - 3} \cdot \frac{x^2 + x - 6}{4x}$$

and

$$y_2 = x - 2$$

in the same viewing window and use the *table* feature to create a table of values for the two equations. If the two graphs coincide, and the values of y_1 and y_2 are the same in the table except where a common factor has been divided out, as shown below, you can conclude that the solution checks.

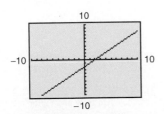

EXAMPLE 3   **Multiplying Rational Expressions**

Multiply the rational expressions.

$$\frac{4x^2 - 4x}{x^2 + 2x - 3} \cdot \frac{x^2 + x - 6}{4x}$$

Solution

$$\frac{4x^2 - 4x}{x^2 + 2x - 3} \cdot \frac{x^2 + x - 6}{4x}$$

$$= \frac{4x(x - 1)(x + 3)(x - 2)}{(x - 1)(x + 3)(4x)} \qquad \text{Multiply and factor.}$$

$$= \frac{\cancel{4x}\cancel{(x - 1)}\cancel{(x + 3)}(x - 2)}{\cancel{(x - 1)}\cancel{(x + 3)}\cancel{(4x)}} \qquad \text{Divide out common factors.}$$

$$= x - 2, \ x \neq 0, \ x \neq 1, \ x \neq -3 \qquad \text{Simplified form}$$

✓ **CHECKPOINT** *Now try Exercise 25.*

The rule for multiplying rational expressions can be extended to cover products involving expressions that are not in fractional form. To do this, rewrite each expression that is not in fractional form as a fraction whose denominator is 1. Here is a simple example.

$$\frac{x + 3}{x - 2} \cdot (5x) = \frac{x + 3}{x - 2} \cdot \frac{5x}{1} = \frac{(x + 3)(5x)}{x - 2} = \frac{5x(x + 3)}{x - 2}$$

In the next example, note how to divide out a factor that differs only in sign. The Distributive Property is used in the step in which $(y - x)$ is rewritten as $(-1)(x - y)$.

EXAMPLE 4   **Multiplying Rational Expressions**

Multiply the rational expressions.

$$\frac{x - y}{y^2 - x^2} \cdot \frac{x^2 - xy - 2y^2}{3x - 6y}$$

Solution

$$\frac{x - y}{y^2 - x^2} \cdot \frac{x^2 - xy - 2y^2}{3x - 6y}$$

$$= \frac{(x - y)(x - 2y)(x + y)}{(y + x)(y - x)(3)(x - 2y)} \qquad \text{Multiply and factor.}$$

$$= \frac{(x - y)(x - 2y)(x + y)}{(y + x)(-1)(x - y)(3)(x - 2y)} \qquad (y - x) = -1(x - y)$$

$$= \frac{\cancel{(x - y)}\cancel{(x - 2y)}\cancel{(x + y)}}{\cancel{(x + y)}(-1)\cancel{(x - y)}(3)\cancel{(x - 2y)}} \qquad \text{Divide out common factors.}$$

$$= -\frac{1}{3}, \ x \neq y, \ x \neq -y, \ x \neq 2y \qquad \text{Simplified form}$$

✓ **CHECKPOINT** *Now try Exercise 27.*

The rule for multiplying rational expressions can be extended to cover products of three or more expressions, as shown in Example 5.

EXAMPLE 5   **Multiplying Three Rational Expressions**

Multiply the rational expressions.

$$\frac{x^2 - 3x + 2}{x + 2} \cdot \frac{3x}{x - 2} \cdot \frac{2x + 4}{x^2 - 5x}$$

Solution

$$\frac{x^2 - 3x + 2}{x + 2} \cdot \frac{3x}{x - 2} \cdot \frac{2x + 4}{x^2 - 5x}$$

$$= \frac{(x - 1)(x - 2)(3)(x)(2)(x + 2)}{(x + 2)(x - 2)(x)(x - 5)} \qquad \text{Multiply and factor.}$$

$$= \frac{(x - 1)\cancel{(x - 2)}(3)\cancel{(x)}(2)\cancel{(x + 2)}}{\cancel{(x + 2)}\cancel{(x - 2)}\cancel{(x)}(x - 5)} \qquad \text{Divide out common factors.}$$

$$= \frac{6(x - 1)}{x - 5}, \ x \neq 0, \ x \neq 2, \ x \neq -2 \qquad \text{Simplified form}$$

✓ **CHECKPOINT** *Now try Exercise 31.*

2 ▸ Divide rational expressions and simplify.

Dividing Rational Expressions

To divide two rational expressions, multiply the first expression by the *reciprocal* of the second. That is, *invert the divisor and multiply.*

Dividing Rational Expressions

Let u, v, w, and z represent real numbers, variables, or algebraic expressions such that $v \neq 0$, $w \neq 0$, and $z \neq 0$. Then the quotient of u/v and w/z is

$$\frac{u}{v} \div \frac{w}{z} = \frac{u}{v} \cdot \frac{z}{w} = \frac{uz}{vw}.$$

Study Tip

Don't forget to add domain restrictions as needed in division problems. In Example 6, an implied domain restriction in the original expression is $x \neq 1$. Because this restriction is not implied by the final expression, it must be added as a written restriction.

EXAMPLE 6 **Dividing Rational Expressions**

Divide the rational expressions.

$$\frac{x}{x+3} \div \frac{4}{x-1}$$

Solution

$$\frac{x}{x+3} \div \frac{4}{x-1} = \frac{x}{x+3} \cdot \frac{x-1}{4} \qquad \text{Invert divisor and multiply.}$$

$$= \frac{x(x-1)}{(x+3)(4)} \qquad \text{Multiply numerators and denominators.}$$

$$= \frac{x(x-1)}{4(x+3)}, \ x \neq 1 \qquad \text{Simplify.}$$

✓ **CHECKPOINT** *Now try Exercise 37.*

EXAMPLE 7 **Dividing Rational Expressions**

$$\frac{2x}{3x-12} \div \frac{x^2-2x}{x^2-6x+8} \qquad \text{Original expressions}$$

$$= \frac{2x}{3x-12} \cdot \frac{x^2-6x+8}{x^2-2x} \qquad \text{Invert divisor and multiply.}$$

$$= \frac{(2)(x)(x-2)(x-4)}{(3)(x-4)(x)(x-2)} \qquad \text{Factor.}$$

$$= \frac{(2)(\cancel{x})(\cancel{x-2})(\cancel{x-4})}{(3)(\cancel{x-4})(\cancel{x})(\cancel{x-2})} \qquad \text{Divide out common factors.}$$

$$= \frac{2}{3}, \ x \neq 0, \ x \neq 2, \ x \neq 4 \qquad \text{Simplified form}$$

Remember that the original expression is equivalent to $\frac{2}{3}$ except for $x=0$, $x=2$, and $x=4$.

✓ **CHECKPOINT** *Now try Exercise 47.*

EXAMPLE 8   **Dividing Rational Expressions**

Divide the rational expressions.

$$\frac{x^2 - y^2}{2x + 2y} \div \frac{2x^2 - 3xy + y^2}{6x + 2y}$$

Solution

$$\frac{x^2 - y^2}{2x + 2y} \div \frac{2x^2 - 3xy + y^2}{6x + 2y}$$

$$= \frac{x^2 - y^2}{2x + 2y} \cdot \frac{6x + 2y}{2x^2 - 3xy + y^2} \qquad \text{Invert divisor and multiply.}$$

$$= \frac{(x + y)(x - y)(2)(3x + y)}{(2)(x + y)(2x - y)(x - y)} \qquad \text{Factor.}$$

$$= \frac{\cancel{(x + y)}\cancel{(x - y)}\cancel{(2)}(3x + y)}{\cancel{(2)}\cancel{(x + y)}(2x - y)\cancel{(x - y)}} \qquad \text{Divide out common factors.}$$

$$= \frac{3x + y}{2x - y}, \; x \neq y, x \neq -y, y \neq -3x \qquad \text{Simplified form}$$

✓ **CHECKPOINT** *Now try Exercise 49.*

EXAMPLE 9   **Amount Spent on Meals and Beverages**

The annual amount A (in millions of dollars) Americans spent on meals and beverages purchased away from home, and the population P (in millions) of the United States, for the years 2000 through 2006 can be modeled by

$$A = \frac{-8242.58t + 348{,}299.6}{-0.06t + 1}, \quad 0 \le t \le 6$$

and

$$P = 2.71t + 282.7, \quad 0 \le t \le 6$$

where t represents the year, with $t = 0$ corresponding to 2000. Find a model T for the amount Americans spent *per person* on meals and beverages.   (Source: U.S. Bureau of Economic Analysis and U.S. Census Bureau)

Solution

To find a model T for the amount Americans spent per person on meals and beverages, divide the total amount by the population.

$$T = \frac{-8242.58t + 348{,}299.6}{-0.06t + 1} \div (2.71t + 282.7) \qquad \begin{array}{l}\text{Divide amount spent by}\\ \text{population.}\end{array}$$

$$= \frac{-8242.58t + 348{,}299.6}{-0.06t + 1} \cdot \frac{1}{2.71t + 282.7} \qquad \text{Invert divisor and multiply.}$$

$$= \frac{-8242.58t + 348{,}299.6}{(-0.06t + 1)(2.71t + 282.7)}, \quad 0 \le t \le 6 \qquad \text{Model}$$

✓ **CHECKPOINT** *Now try Exercise 69.*

Concept Check

1. Why is factoring used in multiplying rational expressions?

2. In your own words, explain how to divide rational expressions.

3. Explain how to divide a rational expression by a polynomial.

4. In dividing rational expressions, explain how you can lose implied domain restrictions when you invert the divisor.

6.2 EXERCISES

Go to pages 440–441 to record your assignments.

Developing Skills

In Exercises 1–8, fill in the missing factor.

1. $\dfrac{7x^2}{3y()} = \dfrac{7}{3y}, \quad x \neq 0$

2. $\dfrac{14x(x-3)^2}{(x-3)()} = \dfrac{2x}{x-3}$

3. $\dfrac{3x(x+2)^2}{(x-4)()} = \dfrac{3x}{x-4}, \quad x \neq -2$

4. $\dfrac{(x+1)^3}{x()} = \dfrac{x+1}{x}, \quad x \neq -1$

5. $\dfrac{3u()}{7v(u+1)} = \dfrac{3u}{7v}, \quad u \neq -1$

6. $\dfrac{(3t+5)()}{5t^2(3t-5)} = \dfrac{3t+5}{t}, \quad t \neq \dfrac{5}{3}$

7. $\dfrac{13x()}{4-x^2} = \dfrac{13x}{x-2}, \quad x \neq -2$

8. $\dfrac{x^2()}{x^2-10x} = \dfrac{x^2}{10-x}, \quad x \neq 0$

In Exercises 9–36, multiply and simplify. *See Examples 1–5.*

9. $4x \cdot \dfrac{7}{12x}$

10. $\dfrac{8}{7y} \cdot (42y)$

11. $\dfrac{8s^3}{9s} \cdot \dfrac{6s^2}{32s}$

12. $\dfrac{3x^4}{7x} \cdot \dfrac{8x^2}{9}$

13. $16u^4 \cdot \dfrac{12}{8u^2}$

14. $18x^4 \cdot \dfrac{4}{15x}$

15. $\dfrac{8}{3+4x} \cdot (9+12x)$

16. $(6-4x) \cdot \dfrac{10}{3-2x}$

17. $\dfrac{8u^2v}{3u+v} \cdot \dfrac{u+v}{12u}$

18. $\dfrac{1-3xy}{4x^2y} \cdot \dfrac{46x^4y^2}{15-45xy}$

19. $\dfrac{12-r}{3} \cdot \dfrac{3}{r-12}$

20. $\dfrac{8-z}{8+z} \cdot \dfrac{z+8}{z-8}$

21. $\dfrac{(2x-3)(x+8)}{x^3} \cdot \dfrac{x}{3-2x}$

22. $\dfrac{x+14}{x^3(10-x)} \cdot \dfrac{x(x-10)}{5}$

✓ 23. $\dfrac{4r-12}{r-2} \cdot \dfrac{r^2-4}{r-3}$

24. $\dfrac{5y-20}{5y+15} \cdot \dfrac{2y+6}{y-4}$

✓ 25. $\dfrac{2t^2-t-15}{t+2} \cdot \dfrac{t^2-t-6}{t^2-6t+9}$

26. $\dfrac{y^2-16}{y^2+8y+16} \cdot \dfrac{3y^2-5y-2}{y^2-6y+8}$

✓ 27. $(4y^2-x^2) \cdot \dfrac{xy}{(x-2y)^2}$

28. $(u-2v)^2 \cdot \dfrac{u+2v}{2v-u}$

29. $\dfrac{x^2+2xy-3y^2}{(x+y)^2} \cdot \dfrac{x^2-y^2}{x+3y}$

30. $\dfrac{(x-2y)^2}{x+2y} \cdot \dfrac{x^2+7xy+10y^2}{x^2-4y^2}$

✓ 31. $\dfrac{x+5}{x-5} \cdot \dfrac{2x^2-9x-5}{3x^2+x-2} \cdot \dfrac{x^2-1}{x^2+7x+10}$

32. $\dfrac{t^2+4t+3}{2t^2-t-10} \cdot \dfrac{t}{t^2+3t+2} \cdot \dfrac{2t^2+4t^3}{t^2+3t}$

33. $\dfrac{9-x^2}{2x+3} \cdot \dfrac{4x^2+8x-5}{4x^2-8x+3} \cdot \dfrac{6x^4-2x^3}{8x^2+4x}$

34. $\dfrac{16x^2-1}{4x^2+9x+5} \cdot \dfrac{5x^2-9x-18}{x^2-12x+36} \cdot \dfrac{12+4x-x^2}{4x^2-13x+3}$

35. $\dfrac{x^3+3x^2-4x-12}{x^3-3x^2-4x+12} \cdot \dfrac{x^2-9}{x}$

36. $\dfrac{xu-yu+xv-yv}{xu+yu-xv-yv} \cdot \dfrac{xu+yu+xv+yv}{xu-yu-xv+yv}$

In Exercises 37–52, divide and simplify. *See Examples 6–8.*

✓ 37. $\dfrac{x}{x+2} \div \dfrac{3}{x+1}$ **38.** $\dfrac{x+3}{4} \div \dfrac{x-2}{x}$

39. $x^2 \div \dfrac{3x}{4}$ **40.** $\dfrac{u}{10} \div u^2$

41. $\dfrac{2x}{5} \div \dfrac{x^2}{15}$ **42.** $\dfrac{3y^2}{20} \div \dfrac{y}{15}$

43. $\dfrac{7xy^2}{10u^2v} \div \dfrac{21x^3}{45uv}$ **44.** $\dfrac{25x^2y}{60x^3y^2} \div \dfrac{5x^4y^3}{16x^2y}$

45. $\dfrac{3(a+b)}{4} \div \dfrac{(a+b)^2}{2}$

46. $\dfrac{x^2+9}{5(x+2)} \div \dfrac{x+3}{5(x^2-4)}$

✓ 47. $\dfrac{4x}{3x-3} \div \dfrac{x^2+2x}{x^2+x-2}$

48. $\dfrac{5x+5}{2x} \div \dfrac{x^2-3x}{x^2-2x-3}$

✓ 49. $\dfrac{(x^3y)^2}{(x+2y)^2} \div \dfrac{x^2y}{(x+2y)^3}$

50. $\dfrac{x^2-y^2}{2x^2-8x} \div \dfrac{(x-y)^2}{2xy}$

51. $\dfrac{x^2+2x-15}{x^2+11x+30} \div \dfrac{x^2-8x+15}{x^2+2x-24}$

52. $\dfrac{y^2+5y-14}{y^2+10y+21} \div \dfrac{y^2+5y+6}{y^2+7y+12}$

In Exercises 53–60, perform the operations and simplify. (In Exercises 59 and 60, n is a positive integer.)

53. $\left[\dfrac{x^2}{9} \cdot \dfrac{3(x+4)}{x^2+2x} \right] \div \dfrac{x}{x+2}$

54. $\left(\dfrac{x^2+6x+9}{x^2} \cdot \dfrac{2x+1}{x^2-9} \right) \div \dfrac{4x^2+4x+1}{x^2-3x}$

55. $\left[\dfrac{xy+y}{4x} \div (3x+3) \right] \div \dfrac{y}{3x}$

56. $\dfrac{3u^2-u-4}{u^2} \div \dfrac{3u^2+12u+4}{u^4-3u^3}$

57. $\dfrac{2x^2+5x-25}{3x^2+5x+2} \cdot \dfrac{3x^2+2x}{x+5} \div \left(\dfrac{x}{x+1} \right)^2$

58. $\dfrac{t^2-100}{4t^2} \cdot \dfrac{t^3-5t^2-50t}{t^4+10t^3} \div \dfrac{(t-10)^2}{5t}$

59. $x^3 \cdot \dfrac{x^{2n}-9}{x^{2n}+4x^n+3} \div \dfrac{x^{2n}-2x^n-3}{x}$

60. $\dfrac{x^{n+1} - 8x}{x^{2n} + 2x^n + 1} \cdot \dfrac{x^{2n} - 4x^n - 5}{x} \div x^n$

61. $y_1 = \dfrac{x^2 - 10x + 25}{x^2 - 25} \cdot \dfrac{x + 5}{2}$

$y_2 = \dfrac{x - 5}{2}, \quad x \neq \pm 5$

62. $y_1 = \dfrac{3x + 15}{x^4} \div \dfrac{x + 5}{x^2}$

$y_2 = \dfrac{3}{x^2}, \quad x \neq -5$

In Exercises 61 and 62, use a graphing calculator to graph the two equations in the same viewing window. Use the graphs and a table of values to verify that the expressions are equivalent. Verify the results algebraically.

Solving Problems

▲ *Geometry* In Exercises 63 and 64, write and simplify an expression for the area of the shaded region.

63.

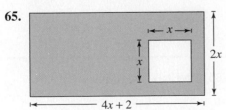

64.

Probability In Exercises 65–68, consider an experiment in which a marble is tossed into a rectangular box with dimensions $2x$ centimeters by $4x + 2$ centimeters. The probability that the marble will come to rest in the unshaded portion of the box is equal to the ratio of the unshaded area to the total area of the figure. Find the probability in simplified form.

65.

66.

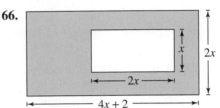

67.

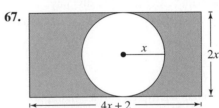

68.

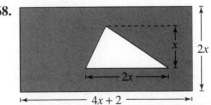

✔ **69.** *Employment* The number of jobs J (in millions) in Florida, and the population P (in millions) of Florida, for the years 2001 through 2006 can be modeled by

$$J = \dfrac{-0.696t + 8.94}{-0.092t + 1}, \quad 1 \le t \le 6 \quad \text{and}$$

$$P = 0.352t + 15.97, \quad 1 \le t \le 6$$

where t represents the year, with $t = 1$ corresponding to 2001. Find a model Y for the number of jobs per person during these years. (Source: U.S. Bureau of Economic Analysis)

70. *Per Capita Income*   The total annual amount I (in millions of dollars) of personal income earned in Alabama, and its population P (in millions), for the years 2001 through 2006 can be modeled by

$$I = \frac{-4.665t + 106.48}{-0.075t + 1}, \quad 1 \le t \le 6 \quad \text{and}$$

$$P = \frac{-0.467t + 4.46}{-0.107t + 1}, \quad 1 \le t \le 6$$

where t represents the year, with $t = 1$ corresponding to 2001. Find a model Y for the annual per capita income for these years.   (Source: U.S. Bureau of Economic Analysis)

Explaining Concepts

71. ✎   Describe how the operation of division is used in the process of simplifying a product of rational expressions.

72. ✎   In a quotient of two rational expressions, the denominator of the divisor is x. Describe a set of circumstances in which you will *not* need to list $x \ne 0$ as a domain restriction after dividing.

73. ✎   Explain what is missing in the following statement.

$$\frac{x - a}{x - b} \div \frac{x - a}{x - b} = 1$$

74. ✎   When two rational expressions are multiplied, the resulting expression is a polynomial. Explain how the total number of factors in the numerators of the expressions you multiplied compares to the total number of factors in the denominators.

75. *Error Analysis*   Describe and correct the errors.

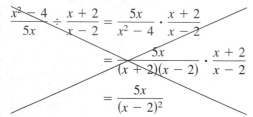

$$\frac{x^2 - 4}{5x} \div \frac{x + 2}{x - 2} = \frac{5x}{x^2 - 4} \cdot \frac{x + 2}{x - 2}$$

$$= \frac{5x}{(x + 2)(x - 2)} \cdot \frac{x + 2}{x - 2}$$

$$= \frac{5x}{(x - 2)^2}$$

76. ✎   Complete the table for the given values of x. Round your answers to five decimal places.

x	60	100	1000
$\dfrac{x - 10}{x + 10}$			
$\dfrac{x + 50}{x - 50}$			
$\dfrac{x - 10}{x + 10} \cdot \dfrac{x + 50}{x - 50}$			

x	10,000	100,000	1,000,000
$\dfrac{x - 10}{x + 10}$			
$\dfrac{x + 50}{x - 50}$			
$\dfrac{x - 10}{x + 10} \cdot \dfrac{x + 50}{x - 50}$			

What kind of pattern do you see? Try to explain what is going on. Can you see why?

Cumulative Review

In Exercises 77–80, evaluate the expression.

77. $\frac{1}{8} + \frac{3}{8} + \frac{5}{8}$

78. $\frac{3}{7} - \frac{2}{7}$

79. $\frac{3}{5} + \frac{4}{15}$

80. $\frac{7}{6} - \frac{9}{7}$

In Exercises 81–84, solve the equation by factoring.

81. $x^2 + 3x = 0$

82. $x^2 + 3x - 10 = 0$

83. $4x^2 - 25 = 0$

84. $x(x - 4) + 2(x - 12) = 0$

6.3 Adding and Subtracting Rational Expressions

© James Marshall/The Image Works

Why You Should Learn It

Addition and subtraction of rational expressions can be used to solve real-life applications. For instance, in Exercise 89 on page 399, you will find a rational expression that models the total number of people enrolled as undergraduate students.

1 ▶ Add or subtract rational expressions with like denominators, and simplify.

What You Should Learn

1 ▶ Add or subtract rational expressions with like denominators, and simplify.

2 ▶ Add or subtract rational expressions with unlike denominators, and simplify.

Adding or Subtracting with Like Denominators

As with numerical fractions, the procedure used to add or subtract two rational expressions depends on whether the expressions have *like* or *unlike* denominators. To add or subtract two rational expressions with *like* denominators, simply combine their numerators and place the result over the common denominator.

Adding or Subtracting with Like Denominators

If u, v, and w are real numbers, variables, or algebraic expressions, and $w \neq 0$, the following rules are valid.

1. $\dfrac{u}{w} + \dfrac{v}{w} = \dfrac{u + v}{w}$ Add fractions with like denominators.

2. $\dfrac{u}{w} - \dfrac{v}{w} = \dfrac{u - v}{w}$ Subtract fractions with like denominators.

EXAMPLE 1   **Adding and Subtracting with Like Denominators**

a. $\dfrac{x}{4} + \dfrac{5 - x}{4} = \dfrac{x + (5 - x)}{4} = \dfrac{5}{4}$ Add numerators.

b. $\dfrac{7}{2x - 3} - \dfrac{3x}{2x - 3} = \dfrac{7 - 3x}{2x - 3}$ Subtract numerators.

✔ **CHECKPOINT** *Now try Exercise 1.*

Study Tip

After adding or subtracting two (or more) rational expressions, check the resulting fraction to see if it can be simplified, as illustrated in Example 2.

EXAMPLE 2   **Subtracting Rational Expressions and Simplifying**

$$\frac{x}{x^2 - 2x - 3} - \frac{3}{x^2 - 2x - 3} = \frac{x - 3}{x^2 - 2x - 3}$$ Subtract numerators.

$$= \frac{(1)(x - 3)}{(x - 3)(x + 1)}$$ Factor.

$$= \frac{1}{x + 1}, \quad x \neq 3$$ Simplified form

✔ **CHECKPOINT** *Now try Exercise 17.*

The rules for adding and subtracting rational expressions with like denominators can be extended to sums and differences involving three or more rational expressions, as illustrated in Example 3.

EXAMPLE 3   Combining Three Rational Expressions

$$\frac{x^2 - 26}{x - 5} - \frac{2x + 4}{x - 5} + \frac{10 + x}{x - 5}$$ Original expressions

$$= \frac{(x^2 - 26) - (2x + 4) + (10 + x)}{x - 5}$$ Write numerator over common denominator.

$$= \frac{x^2 - x - 20}{x - 5}$$ Use the Distributive Property and combine like terms.

$$= \frac{(x - 5)(x + 4)}{x - 5}$$ Factor and divide out common factor.

$$= x + 4, \quad x \neq 5$$ Simplified form

✓ **CHECKPOINT** *Now try Exercise 21.*

---

2 ▶ Add or subtract rational expressions with unlike denominators, and simplify.

Adding or Subtracting with Unlike Denominators

The **least common multiple (LCM)** of two (or more) polynomials can be helpful when adding or subtracting rational expressions with *unlike* denominators. The least common multiple of two (or more) polynomials is the simplest polynomial that is a multiple of each of the original polynomials. This means that the LCM must contain all the *different* factors in each polynomial, with each factor raised to the greatest power of its occurrence in any one of the polynomials.

EXAMPLE 4   Finding Least Common Multiples

a. The least common multiple of

$$6x = 2 \cdot 3 \cdot x, \quad 2x^2 = 2 \cdot x^2, \quad \text{and} \quad 9x^3 = 3^2 \cdot x^3$$

is $2 \cdot 3^2 \cdot x^3 = 18x^3$.

b. The least common multiple of

$$x^2 - x = x(x - 1) \quad \text{and} \quad 2x - 2 = 2(x - 1)$$

is $2x(x - 1)$.

c. The least common multiple of

$$3x^2 + 6x = 3x(x + 2) \quad \text{and} \quad x^2 + 4x + 4 = (x + 2)^2$$

is $3x(x + 2)^2$.

✓ **CHECKPOINT** *Now try Exercise 23.*

To add or subtract rational expressions with *unlike* denominators, you must first rewrite the rational expressions so that they have a common denominator. You can always find a common denominator of two (or more) rational expressions by multiplying their denominators. However, if you use the **least common denominator (LCD),** which is the least common multiple of the denominators, you may have less simplifying to do. After the rational expressions have been written with a common denominator, you can simply add or subtract using the rules given at the beginning of this section.

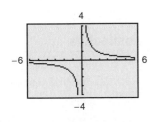
EXAMPLE 5   Adding with Unlike Denominators

Add the rational expressions: $\dfrac{7}{6x} + \dfrac{5}{8x}$.

Solution

By factoring the denominators, $6x = 2 \cdot 3 \cdot x$ and $8x = 2^3 \cdot x$, you can conclude that the least common denominator is $2^3 \cdot 3 \cdot x = 24x$.

$$\frac{7}{6x} + \frac{5}{8x} = \frac{7(4)}{6x(4)} + \frac{5(3)}{8x(3)}$$ Rewrite expressions using LCD of $24x$.

$$= \frac{28}{24x} + \frac{15}{24x}$$ Like denominators

$$= \frac{28 + 15}{24x} = \frac{43}{24x}$$ Add fractions and simplify.

✓ **CHECKPOINT** *Now try Exercise 51.*

EXAMPLE 6   Subtracting with Unlike Denominators

Subtract the rational expressions: $\dfrac{3}{x - 3} - \dfrac{5}{x + 2}$.

Solution

The only factors of the denominators are $x - 3$ and $x + 2$. So, the least common denominator is $(x - 3)(x + 2)$.

$$\frac{3}{x - 3} - \frac{5}{x + 2}$$ Write original expressions.

$$= \frac{3(x + 2)}{(x - 3)(x + 2)} - \frac{5(x - 3)}{(x - 3)(x + 2)}$$ Rewrite expressions using LCD of $(x - 3)(x + 2)$.

$$= \frac{3x + 6}{(x - 3)(x + 2)} - \frac{5x - 15}{(x - 3)(x + 2)}$$ Distributive Property

$$= \frac{3x + 6 - 5x + 15}{(x - 3)(x + 2)}$$ Subtract fractions and use the Distributive Property.

$$= \frac{-2x + 21}{(x - 3)(x + 2)}$$ Simplified form

✓ **CHECKPOINT** *Now try Exercise 63.*

EXAMPLE 7   Adding with Unlike Denominators

$$\frac{6x}{x^2 - 4} + \frac{3}{2 - x}$$ Original expressions

$$= \frac{6x}{(x + 2)(x - 2)} + \frac{3}{(-1)(x - 2)}$$ Factor denominators.

$$= \frac{6x}{(x + 2)(x - 2)} - \frac{3(x + 2)}{(x + 2)(x - 2)}$$ Rewrite expressions using LCD of $(x + 2)(x - 2)$.

$$= \frac{6x}{(x + 2)(x - 2)} - \frac{3x + 6}{(x + 2)(x - 2)}$$ Distributive Property

$$= \frac{6x - (3x + 6)}{(x + 2)(x - 2)}$$ Subtract.

$$= \frac{6x - 3x - 6}{(x + 2)(x - 2)}$$ Distributive Property

$$= \frac{3x - 6}{(x + 2)(x - 2)}$$ Simplify.

$$= \frac{3(x - 2)}{(x + 2)(x - 2)}$$ Factor and divide out common factor.

$$= \frac{3}{x + 2}, \quad x \neq 2$$ Simplified form

✓ **CHECKPOINT** *Now try Exercise 55.*

EXAMPLE 8   Subtracting with Unlike Denominators

$$\frac{x}{x^2 - 5x + 6} - \frac{1}{x^2 - x - 2}$$ Original expressions

$$= \frac{x}{(x - 3)(x - 2)} - \frac{1}{(x - 2)(x + 1)}$$ Factor denominators.

$$= \frac{x(x + 1)}{(x - 3)(x - 2)(x + 1)} - \frac{1(x - 3)}{(x - 3)(x - 2)(x + 1)}$$ Rewrite expressions using LCD of $(x - 3)(x - 2)(x + 1)$.

$$= \frac{x^2 + x}{(x - 3)(x - 2)(x + 1)} - \frac{x - 3}{(x - 3)(x - 2)(x + 1)}$$ Distributive Property

$$= \frac{(x^2 + x) - (x - 3)}{(x - 3)(x - 2)(x + 1)}$$ Subtract fractions.

$$= \frac{x^2 + x - x + 3}{(x - 3)(x - 2)(x + 1)}$$ Distributive Property

$$= \frac{x^2 + 3}{(x - 3)(x - 2)(x + 1)}$$ Simplified form

✓ **CHECKPOINT** *Now try Exercise 71.*

EXAMPLE 9   **Combining Three Rational Expressions**

$$\frac{4x}{x^2 - 16} + \frac{x}{x + 4} - \frac{2}{x} = \frac{4x}{(x + 4)(x - 4)} + \frac{x}{x + 4} - \frac{2}{x}$$

$$= \frac{4x(x)}{x(x + 4)(x - 4)} + \frac{x(x)(x - 4)}{x(x + 4)(x - 4)} - \frac{2(x + 4)(x - 4)}{x(x + 4)(x - 4)}$$

$$= \frac{4x^2 + x^2(x - 4) - 2(x^2 - 16)}{x(x + 4)(x - 4)}$$

$$= \frac{4x^2 + x^3 - 4x^2 - 2x^2 + 32}{x(x + 4)(x - 4)}$$

$$= \frac{x^3 - 2x^2 + 32}{x(x + 4)(x - 4)}$$

✓ **CHECKPOINT** *Now try Exercise 77.*

To add or subtract two rational expressions, you can use the LCD method or the basic definition

$$\frac{a}{b} \pm \frac{c}{d} = \frac{ad \pm bc}{bd}, \quad b \neq 0, d \neq 0.$$ Basic definition

This definition provides an efficient way of adding or subtracting two rational expressions that have no common factors in their denominators.

EXAMPLE 10   **Motor Vehicle and Parts Sales**

For the years 2000 through 2005, the total annual retail sales T (in billions of dollars) and the E-commerce annual retail sales E (in billions of dollars) of motor vehicles and vehicle parts in the United States can be modeled by

$$T = \frac{-57.6t + 800}{-0.085t + 1} \quad \text{and} \quad E = \frac{t^2 + 4.3}{0.03t^2 + 1}, \quad 0 \leq t \leq 5$$

where t represents the year, with $t = 0$ corresponding to 2000. Find a rational model N for the annual retail sales *not* from E-commerce during this time period. (Source: U.S. Department of Labor)

Solution

To find a model for N, find the difference of T and E.

$$N = \frac{-57.6t + 800}{-0.085t + 1} - \frac{t^2 + 4.3}{0.03t^2 + 1}$$ Subtract E from T.

$$= \frac{(-57.6t + 800)(0.03t^2 + 1) - (-0.085t + 1)(t^2 + 4.3)}{(-0.085t + 1)(0.03t^2 + 1)}$$ Basic definition

$$= \frac{-1.643t^3 + 23t^2 - 57.2345t + 795.7}{(-0.085t + 1)(0.03t^2 + 1)}$$ Use FOIL Method and combine like terms.

✓ **CHECKPOINT** *Now try Exercise 89.*

© Syracuse Newspapers/The Image Works

In 2005, E-commerce accounted for only about 2 percent of the sales of vehicles and vehicle parts in the United States.

_____ **Concept Check** _____

1. *True or False?* Two rational expressions with *like* denominators have a common denominator.

2. When adding or subtracting rational expressions, how do you rewrite each rational expression as an equivalent expression whose denominator is the LCD?

3. In your own words, describe how to add or subtract rational expressions with *like* denominators.

4. In your own words, describe how to add or subtract rational expressions with *unlike* denominators.

6.3 EXERCISES

Go to pages 440–441 to record your assignments.

_____ **Developing Skills** _____

In Exercises 1–22, combine and simplify. *See Examples 1–3.*

✓ 1. $\dfrac{5x}{6} + \dfrac{4x}{6}$

2. $\dfrac{7y}{12} + \dfrac{9y}{12}$

3. $\dfrac{2}{3a} - \dfrac{11}{3a}$

4. $\dfrac{6}{19x} - \dfrac{7}{19x}$

5. $\dfrac{x}{9} - \dfrac{x+2}{9}$

6. $\dfrac{4-y}{4} + \dfrac{3y}{4}$

7. $\dfrac{z^2}{3} + \dfrac{z^2-2}{3}$

8. $\dfrac{10x^2+1}{3} - \dfrac{10x^2}{3}$

9. $\dfrac{2x+5}{3x} + \dfrac{1-x}{3x}$

10. $\dfrac{16+z}{5z} - \dfrac{11-z}{5z}$

11. $\dfrac{3y-22}{y-6} - \dfrac{2y-16}{y-6}$

12. $\dfrac{5x-1}{x+4} + \dfrac{5-4x}{x+4}$

13. $\dfrac{2x-1}{x(x-3)} + \dfrac{1-x}{x(x-3)}$

14. $\dfrac{3-2n}{n(n+2)} - \dfrac{1-3n}{n(n+2)}$

15. $\dfrac{w}{w^2-4} + \dfrac{2}{w^2-4}$

16. $\dfrac{d}{d^2-36} - \dfrac{6}{d^2-36}$

✓ 17. $\dfrac{c}{c^2+3c-4} - \dfrac{1}{c^2+3c-4}$

18. $\dfrac{2v}{2v^2-5v-12} + \dfrac{3}{2v^2-5v-12}$

19. $\dfrac{3y}{3} - \dfrac{3y-3}{3} - \dfrac{7}{3}$

20. $\dfrac{-16u}{9} - \dfrac{27-16u}{9} + \dfrac{2}{9}$

✓ 21. $\dfrac{x^2-4x}{x-3} + \dfrac{10-4x}{x-3} - \dfrac{x-8}{x-3}$

22. $\dfrac{6-7z}{z+4} + \dfrac{z^2-14}{z+4} + \dfrac{4z-20}{z+4}$

In Exercises 23–34, find the least common multiple of the expressions. *See Example 4.*

✓ 23. $5x^2, 20x^3$

24. $14t^2, 42t^5$

25. $9y^3, 12y$

26. $18m^2, 45m$

27. $15x^2, 3(x+5)$

28. $6x^2, 15x(x-1)$

29. $63z^2(z+1), 14(z+1)^4$

30. $18y^3, 27y(y-3)^2$

31. $8t(t+2), 14(t^2-4)$

32. $6(x^2-4), 2x(x+2)$

33. $2y^2+y-1, 4y^2-2y$

34. $t^3+3t^2+9t, 2t^2(t^2-9)$

In Exercises 35–40, fill in the missing factor.

35. $\dfrac{7x^2}{4a()} = \dfrac{7}{4a}, \quad x \neq 0$

36. $\dfrac{8y^2}{(b+2)()} = \dfrac{2y}{b+2}, \quad y \neq 0$

37. $\dfrac{5r()}{3v(u+1)} = \dfrac{5r}{3v}, \quad u \neq -1$

38. $\dfrac{(3t+5)()}{10t^2(3t-5)} = \dfrac{3t+5}{2t}, \quad t \neq \dfrac{5}{3}$

39. $\dfrac{7y()}{4-x^2} = \dfrac{7y}{x-2}, \quad x \neq -2$

40. $\dfrac{4x^2()}{x^2-10x} = \dfrac{4x^2}{10-x}, \quad x \neq 0$

In Exercises 41–48, find the least common denominator of the two fractions and rewrite each fraction using the least common denominator.

41. $\dfrac{n+8}{3n-12}, \dfrac{10}{6n^2}$

42. $\dfrac{y-4}{2y+14}, \dfrac{3y}{10y^3}$

43. $\dfrac{2}{x^2(x-3)}, \dfrac{5}{x(x+3)}$

44. $\dfrac{5t}{2t(t-3)^2}, \dfrac{4}{t(t-3)}$

45. $\dfrac{v}{2v^2+2v}, \dfrac{4}{3v^2}$

46. $\dfrac{4x}{(x+5)^2}, \dfrac{x-2}{x^2-25}$

47. $\dfrac{x-8}{x^2-25}, \dfrac{9x}{x^2-10x+25}$

48. $\dfrac{3y}{y^2-y-12}, \dfrac{y-4}{y^2+3y}$

In Exercises 49–82, combine and simplify. *See Examples 5–9.*

49. $\dfrac{5}{4x} - \dfrac{3}{5}$

50. $\dfrac{10}{b} + \dfrac{1}{10b}$

✓ **51.** $\dfrac{7}{a} + \dfrac{14}{a^2}$

52. $\dfrac{1}{6u^2} - \dfrac{2}{9u}$

53. $25 + \dfrac{10}{x+4}$

54. $\dfrac{30}{x-6} - 4$

✓ **55.** $\dfrac{20}{x-4} + \dfrac{20}{4-x}$

56. $\dfrac{15}{2-t} - \dfrac{7}{t-2}$

57. $\dfrac{3x}{x-8} - \dfrac{6}{8-x}$

58. $\dfrac{1}{y-6} + \dfrac{y}{6-y}$

59. $\dfrac{3x}{3x-2} + \dfrac{2}{2-3x}$

60. $\dfrac{y}{5y-3} - \dfrac{3}{3-5y}$

61. $\dfrac{9}{5v} + \dfrac{3}{v-1}$

62. $\dfrac{3}{y-1} + \dfrac{5}{4y}$

✓ **63.** $\dfrac{x}{x+3} - \dfrac{5}{x-2}$

64. $\dfrac{1}{x+4} - \dfrac{1}{x+2}$

65. $\dfrac{12}{x^2-9} - \dfrac{2}{x-3}$

66. $\dfrac{12}{x^2-4} - \dfrac{3}{x+2}$

67. $\dfrac{3}{x-5} + \dfrac{2}{x+5}$

68. $\dfrac{7}{2x-3} + \dfrac{3}{2x+3}$

69. $\dfrac{4}{x^2} - \dfrac{4}{x^2+1}$

70. $\dfrac{3}{y^2-3} + \dfrac{2}{3y^2}$

✓ **71.** $\dfrac{x}{x^2-x-30} - \dfrac{1}{x+5}$

72. $\dfrac{x}{x^2-9} + \dfrac{3}{x^2-5x+6}$

73. $\dfrac{4}{x-4} + \dfrac{16}{(x-4)^2}$

74. $\dfrac{3}{x-2} - \dfrac{1}{(x-2)^2}$

75. $\dfrac{y}{x^2 + xy} - \dfrac{x}{xy + y^2}$ **76.** $\dfrac{5}{x + y} + \dfrac{5}{x^2 - y^2}$

✓ **77.** $\dfrac{4}{x} - \dfrac{2}{x^2} + \dfrac{4}{x + 3}$ **78.** $\dfrac{5}{2} - \dfrac{1}{2x} - \dfrac{3}{x + 1}$

79. $\dfrac{3u}{u^2 - 2uv + v^2} + \dfrac{2}{u - v} - \dfrac{u}{u - v}$

80. $\dfrac{1}{x - y} - \dfrac{3}{x + y} + \dfrac{3x - y}{x^2 - y^2}$

81. $\dfrac{x + 2}{x - 1} - \dfrac{2}{x + 6} - \dfrac{14}{x^2 + 5x - 6}$

82. $\dfrac{-2x - 10}{x^2 + 8x + 15} + \dfrac{2}{x + 3} + \dfrac{x}{x + 5}$

🖩 In Exercises 83 and 84, use a graphing calculator to graph the two equations in the same viewing window. Use the graphs to verify that the expressions are equivalent. Verify the results algebraically.

83. $y_1 = \dfrac{2}{x} + \dfrac{4}{x - 2},\quad y_2 = \dfrac{6x - 4}{x(x - 2)}$

84. $y_1 = \dfrac{x}{3} - \dfrac{2}{x + 3},\quad y_2 = \dfrac{x^2 + 3x - 6}{3x + 9}$

Solving Problems

85. *Work Rate*   After working together for t hours on a common task, two workers have completed fractional parts of the job equal to $t/4$ and $t/6$. What fractional part of the task has been completed?

86. *Work Rate*   After working together for t hours on a common task, two workers have completed fractional parts of the job equal to $t/3$ and $t/5$. What fractional part of the task has been completed?

87. *Rewriting a Fraction*   The fraction $4/(x^3 - x)$ can be rewritten as a sum of three fractions, as follows.

$$\frac{4}{x^3 - x} = \frac{A}{x} + \frac{B}{x + 1} + \frac{C}{x - 1}$$

The numbers A, B, and C are the solutions of the system

$$\begin{cases} A + B + C = 0 \\ \quad\;\; -B + C = 0 \\ -A \qquad\qquad = 4. \end{cases}$$

Solve the system and verify that the sum of the three resulting fractions is the original fraction.

88. *Rewriting a Fraction*   The fraction

$$\frac{x + 1}{x^3 - x^2}$$

can be rewritten as a sum of three fractions, as follows.

$$\frac{x + 1}{x^3 - x^2} = \frac{A}{x} + \frac{B}{x^2} + \frac{C}{x - 1}$$

The numbers A, B, and C are the solutions of the system

$$\begin{cases} A \qquad\quad + C = 0 \\ -A + B \qquad\;\; = 1 \\ \quad\;\; -B \qquad\quad = 1. \end{cases}$$

Solve the system and verify that the sum of the three resulting fractions is the original fraction.

Undergraduate Students   In Exercises 89 and 90, use the following models, which give the numbers (in thousands) of males M and females F enrolled as undergraduate students from 2000 through 2005.

$$M = \frac{1434.4t + 5797.28}{0.205t + 1},\; 0 \le t \le 5$$

and $F = \dfrac{1809.8t + 7362.51}{0.183t + 1},\; 0 \le t \le 5$

In these models, t represents the year, with $t = 0$ corresponding to 2000.   (Source: U.S. National Center for Education Statistics)

89. Find a rational model T for the total number of undergraduate students (in thousands) from 2000 through 2005.

Year	2000	2001	2002
Undergraduates (in thousands)			

90. Use the model you found in Exercise 89 to complete the table showing the total number of undergraduate students (rounded to the nearest thousand) each year from 2000 through 2005.

Year	2003	2004	2005
Undergraduates (in thousands)			

Explaining Concepts

91. *Error Analysis*   Describe the error.

$$\frac{x-1}{x+4} - \frac{4x-11}{x+4} = \frac{x-1-4x-11}{x+4}$$
$$= \frac{-3x-12}{x+4}$$
$$= \frac{-3(x+4)}{x+4}$$
$$= -3$$

92. *Error Analysis*   Describe the error.

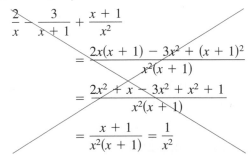

$$\frac{2}{x} - \frac{3}{x+1} + \frac{x+1}{x^2}$$
$$= \frac{2x(x+1) - 3x^2 + (x+1)^2}{x^2(x+1)}$$
$$= \frac{2x^2 + x - 3x^2 + x^2 + 1}{x^2(x+1)}$$
$$= \frac{x+1}{x^2(x+1)} = \frac{1}{x^2}$$

93. Is it possible for the least common denominator of two fractions to be the same as one of the fraction's denominators? If so, give an example.

94. Evaluate each expression at the given value of the variable in two different ways: (1) combine and simplify the rational expressions first and then evaluate the simplified expression at the given value of the variable, and (2) substitute the given value of the variable first and then simplify the resulting expression. Do you get the same result with each method? Discuss which method you prefer and why. List the advantages and/or disadvantages of each method.

(a) $\dfrac{1}{m-4} - \dfrac{1}{m+4} + \dfrac{3m}{m^2-16}$, $m = 2$

(b) $\dfrac{x-2}{x^2-9} + \dfrac{3x+2}{x^2-5x+6}$, $x = 4$

(c) $\dfrac{3y^2+16y-8}{y^2+2y-8} - \dfrac{y-1}{y-2} + \dfrac{y}{y+4}$, $y = 3$

Cumulative Review

In Exercises 95–98, find the sum or difference.

95. $5v + (4 - 3v)$

96. $(2v + 7) + (9v + 8)$

97. $(x^2 - 4x + 3) - (6 - 2x)$

98. $(5y + 2) - (2y^2 + 8y - 5)$

In Exercises 99–102, factor the trinomial, if possible.

99. $x^2 - 7x + 12$

100. $c^2 + 6c + 10$

101. $2a^2 - 9a - 18$

102. $6w^2 + 14w - 12$

6.4 Complex Fractions

David Forbert/SuperStock, Inc.

Why You Should Learn It

Complex fractions can be used to model real-life situations. For instance, in Exercise 64 on page 406, a complex fraction is used to model the annual percent rate for a home-improvement loan.

1 ▶ Simplify complex fractions using rules for dividing rational expressions.

What You Should Learn

1 ▶ Simplify complex fractions using rules for dividing rational expressions.

2 ▶ Simplify complex fractions having a sum or difference in the numerator and/or denominator.

Complex Fractions

Problems involving the division of two rational expressions are sometimes written as **complex fractions.** A complex fraction is a fraction that has a fraction in its numerator or denominator, or both. The rules for dividing rational expressions still apply. For instance, consider the following complex fraction.

$$\dfrac{\left(\dfrac{x+2}{3}\right)}{\left(\dfrac{x-2}{x}\right)} \longrightarrow$$

Numerator fraction

Main fraction line

Denominator fraction

To perform the division implied by this complex fraction, invert the denominator fraction (the divisor) and multiply, as follows.

$$\dfrac{\left(\dfrac{x+2}{3}\right)}{\left(\dfrac{x-2}{x}\right)} = \dfrac{x+2}{3} \cdot \dfrac{x}{x-2}$$

$$= \dfrac{x(x+2)}{3(x-2)}, \quad x \neq 0$$

Note that for complex fractions, you make the main fraction line slightly longer than the fraction lines in the numerator and denominator.

EXAMPLE 1   **Simplifying a Complex Fraction**

$$\dfrac{\left(\dfrac{5}{14}\right)}{\left(\dfrac{25}{8}\right)} = \dfrac{5}{14} \cdot \dfrac{8}{25} \qquad \text{Invert divisor and multiply.}$$

$$= \dfrac{\cancel{5} \cdot \cancel{2} \cdot 2 \cdot 2}{\cancel{2} \cdot 7 \cdot \cancel{5} \cdot 5} \qquad \text{Multiply, factor, and divide out common factors.}$$

$$= \dfrac{4}{35} \qquad \text{Simplified form}$$

✓ **CHECKPOINT** *Now try Exercise 1.*

EXAMPLE 2   **Simplifying a Complex Fraction**

Simplify the complex fraction.

$$\dfrac{\left(\dfrac{4y^3}{(5x)^2}\right)}{\left(\dfrac{(2y)^2}{10x^3}\right)}$$

Solution

$$\dfrac{\left(\dfrac{4y^3}{(5x)^2}\right)}{\left(\dfrac{(2y)^2}{10x^3}\right)} = \dfrac{4y^3}{25x^2} \cdot \dfrac{10x^3}{4y^2} \qquad \text{Invert divisor and multiply.}$$

$$= \dfrac{4y^2 \cdot y \cdot 2 \cdot 5x^2 \cdot x}{5 \cdot 5x^2 \cdot 4y^2} \qquad \text{Multiply and factor.}$$

$$= \dfrac{\cancel{4y^2} \cdot y \cdot 2 \cdot \cancel{5x^2} \cdot x}{5 \cdot \cancel{5x^2} \cdot \cancel{4y^2}} \qquad \text{Divide out common factors.}$$

$$= \dfrac{2xy}{5}, \quad x \neq 0, \; y \neq 0 \qquad \text{Simplified form}$$

✓ **CHECKPOINT** *Now try Exercise 5.*

> **Study Tip**
>
> Domain restrictions result from the values that make any denominator zero in a complex fraction. In Example 2, note that the original expression has three denominators: $(5x)^2$, $10x^3$, and $(2y)^2/10x^3$. The domain restrictions that result from these denominators are $x \neq 0$ and $y \neq 0$.

EXAMPLE 3   **Simplifying a Complex Fraction**

Simplify the complex fraction.

$$\dfrac{\left(\dfrac{x+1}{x+2}\right)}{\left(\dfrac{x+1}{x+5}\right)}$$

Solution

$$\dfrac{\left(\dfrac{x+1}{x+2}\right)}{\left(\dfrac{x+1}{x+5}\right)} = \dfrac{x+1}{x+2} \cdot \dfrac{x+5}{x+1} \qquad \text{Invert divisor and multiply.}$$

$$= \dfrac{(x+1)(x+5)}{(x+2)(x+1)} \qquad \text{Multiply numerators and denominators.}$$

$$= \dfrac{\cancel{(x+1)}(x+5)}{(x+2)\cancel{(x+1)}} \qquad \text{Divide out common factors.}$$

$$= \dfrac{x+5}{x+2}, \quad x \neq -1, \; x \neq -5 \qquad \text{Simplified form}$$

✓ **CHECKPOINT** *Now try Exercise 7.*

EXAMPLE 4 Simplifying a Complex Fraction

$$\frac{\left(\frac{x^2 + 4x + 3}{x - 2}\right)}{2x + 6} = \frac{\left(\frac{x^2 + 4x + 3}{x - 2}\right)}{\left(\frac{2x + 6}{1}\right)}$$ Rewrite denominator.

$$= \frac{x^2 + 4x + 3}{x - 2} \cdot \frac{1}{2x + 6}$$ Invert divisor and multiply.

$$= \frac{(x + 1)(x + 3)}{(x - 2)(2)(x + 3)}$$ Multiply and factor.

$$= \frac{(x + 1)\cancel{(x + 3)}}{(x - 2)(2)\cancel{(x + 3)}}$$ Divide out common factor.

$$= \frac{x + 1}{2(x - 2)}, \quad x \neq -3$$ Simplified form

✓ **CHECKPOINT** *Now try Exercise 11.*

2 ▶ Simplify complex fractions having a sum or difference in the numerator and/or denominator.

Complex Fractions with Sums or Differences

Complex fractions can have numerators and/or denominators that are sums or differences of fractions. One way to simplify such a complex fraction is to combine the terms so that the numerator and denominator each consist of a single fraction. Then divide by inverting the denominator and multiplying.

Study Tip

Another way of simplifying the complex fraction in Example 5 is to multiply the numerator and denominator by $3x$, the least common denominator of all the fractions in the numerator and denominator. This produces the same result, as shown below.

$$\frac{\left(\frac{x}{3} + \frac{2}{3}\right)}{\left(1 - \frac{2}{x}\right)} = \frac{\left(\frac{x}{3} + \frac{2}{3}\right)}{\left(1 - \frac{2}{x}\right)} \cdot \frac{3x}{3x}$$

$$= \frac{\frac{x}{3}(3x) + \frac{2}{3}(3x)}{(1)(3x) - \frac{2}{x}(3x)}$$

$$= \frac{x^2 + 2x}{3x - 6}$$

$$= \frac{x(x + 2)}{3(x - 2)}, \quad x \neq 0$$

EXAMPLE 5 Simplifying a Complex Fraction

$$\frac{\left(\frac{x}{3} + \frac{2}{3}\right)}{\left(1 - \frac{2}{x}\right)} = \frac{\left(\frac{x}{3} + \frac{2}{3}\right)}{\left(\frac{x}{x} - \frac{2}{x}\right)}$$

$$= \frac{\left(\frac{x + 2}{3}\right)}{\left(\frac{x - 2}{x}\right)}$$ Add fractions. Rewrite with least common denominators.

$$= \frac{x + 2}{3} \cdot \frac{x}{x - 2}$$ Invert divisor and multiply.

$$= \frac{x(x + 2)}{3(x - 2)}, \quad x \neq 0$$ Simplified form

✓ **CHECKPOINT** *Now try Exercise 25.*

Study Tip

In Example 6, you might wonder about the domain restrictions that result from the main denominator

$$\left(\frac{3}{x+2} + \frac{2}{x}\right)$$

of the original expression. By setting this expression equal to zero and solving for x, you can see that it leads to the domain restriction $x \neq -\frac{4}{5}$. Notice that this restriction is implied by the denominator of the simplified expression.

EXAMPLE 6   **Simplifying a Complex Fraction**

$$\frac{\left(\dfrac{2}{x+2}\right)}{\left(\dfrac{3}{x+2} + \dfrac{2}{x}\right)} = \frac{\left(\dfrac{2}{x+2}\right)(x)(x+2)}{\left(\dfrac{3}{x+2}\right)(x)(x+2) + \left(\dfrac{2}{x}\right)(x)(x+2)}$$ $x(x+2)$ is the least common denominator.

$$= \frac{2x}{3x + 2(x+2)}$$ Multiply and simplify.

$$= \frac{2x}{3x + 2x + 4}$$ Distributive Property

$$= \frac{2x}{5x + 4}, \quad x \neq -2, \quad x \neq 0$$ Simplify.

Notice that the numerator and denominator of the complex fraction were multiplied by $(x)(x+2)$, which is the least common denominator of the fractions in the original complex fraction.

✔ **CHECKPOINT**   *Now try Exercise 41.*

When simplifying a rational expression containing negative exponents, first rewrite the expression with positive exponents and then proceed with simplifying the expression. This is demonstrated in Example 7.

EXAMPLE 7   **Simplifying a Complex Fraction**

$$\frac{5 + x^{-2}}{8x^{-1} + x} = \frac{\left(5 + \dfrac{1}{x^2}\right)}{\left(\dfrac{8}{x} + x\right)}$$ Rewrite with positive exponents.

$$= \frac{\left(\dfrac{5x^2}{x^2} + \dfrac{1}{x^2}\right)}{\left(\dfrac{8}{x} + \dfrac{x^2}{x}\right)}$$ Rewrite with least common denominators.

$$= \frac{\left(\dfrac{5x^2 + 1}{x^2}\right)}{\left(\dfrac{x^2 + 8}{x}\right)}$$ Add fractions.

$$= \frac{5x^2 + 1}{x^2} \cdot \frac{x}{x^2 + 8}$$ Invert divisor and multiply.

$$= \frac{\cancel{x}(5x^2 + 1)}{\cancel{x}(x)(x^2 + 8)}$$ Divide out common factor.

$$= \frac{5x^2 + 1}{x(x^2 + 8)}$$ Simplified form

✔ **CHECKPOINT**   *Now try Exercise 45.*

_____ Concept Check _____

1. What kind of division problem can be represented by a complex fraction?

2. Describe a method for simplifying complex fractions that uses the process for the "Division of Two Real Numbers" on page 15.

3. Describe the method for simplifying complex fractions that involves the use of a least common denominator.

4. Explain how you can find the implied domain restrictions for a complex fraction.

6.4 EXERCISES

Go to pages 440–441 to record your assignments.

_____ Developing Skills _____

In Exercises 1–22, simplify the complex fraction. See Examples 1–4.

1. $\dfrac{\left(\dfrac{3}{16}\right)}{\left(\dfrac{9}{12}\right)}$

2. $\dfrac{\left(\dfrac{20}{21}\right)}{\left(\dfrac{8}{7}\right)}$

3. $\dfrac{\left(\dfrac{8x^2y}{3z^2}\right)}{\left(\dfrac{4xy}{9z^5}\right)}$

4. $\dfrac{\left(\dfrac{36x^4}{5y^4z^5}\right)}{\left(\dfrac{9xy^2}{20z^5}\right)}$

5. $\dfrac{\left(\dfrac{6x^3}{(5y)^2}\right)}{\left(\dfrac{(3x)^2}{15y^4}\right)}$

6. $\dfrac{\left(\dfrac{(3r)^3}{10t^4}\right)}{\left(\dfrac{9r}{(2t)^2}\right)}$

7. $\dfrac{\left(\dfrac{y}{3-y}\right)}{\left(\dfrac{y^2}{y-3}\right)}$

8. $\dfrac{\left(\dfrac{x}{x-4}\right)}{\left(\dfrac{x}{4-x}\right)}$

9. $\dfrac{\left(\dfrac{25x^2}{x-5}\right)}{\left(\dfrac{10x}{5+4x-x^2}\right)}$

10. $\dfrac{\left(\dfrac{5x}{x+7}\right)}{\left(\dfrac{10}{x^2+8x+7}\right)}$

11. $\dfrac{\left(\dfrac{x^2+3x-10}{x+4}\right)}{3x-6}$

12. $\dfrac{\left(\dfrac{x^2-2x-8}{x-1}\right)}{5x-20}$

13. $\dfrac{2x-14}{\left(\dfrac{x^2-9x+14}{x+3}\right)}$

14. $\dfrac{4x+16}{\left(\dfrac{x^2+9x+20}{x-1}\right)}$

15. $\dfrac{\left(\dfrac{6x^2-17x+5}{3x^2+3x}\right)}{\left(\dfrac{3x-1}{3x+1}\right)}$

16. $\dfrac{\left(\dfrac{6x^2-13x-5}{5x^2+5x}\right)}{\left(\dfrac{2x-5}{5x+1}\right)}$

17. $\dfrac{\left(\dfrac{16x^2+8x+1}{3x^2+8x-3}\right)}{\left(\dfrac{4x^2-3x-1}{x^2+6x+9}\right)}$

18. $\dfrac{\left(\dfrac{9x^2 - 24x + 16}{x^2 + 10x + 25}\right)}{\left(\dfrac{6x^2 - 5x - 4}{2x^2 + 3x - 35}\right)}$

19. $\dfrac{x^2 + x - 6}{x^2 - 4} \div \dfrac{x + 3}{x^2 + 4x + 4}$

20. $\dfrac{t^3 + t^2 - 9t - 9}{t^2 - 5t + 6} \div \dfrac{t^2 + 6t + 9}{t - 2}$

21. $\dfrac{\left(\dfrac{x^2 - 3x - 10}{x^2 - 4x + 4}\right)}{\left(\dfrac{21 + 4x - x^2}{x^2 - 5x - 14}\right)}$

22. $\dfrac{\left(\dfrac{x^2 + 5x + 6}{4x^2 - 20x + 25}\right)}{\left(\dfrac{x^2 - 5x - 24}{4x^2 - 25}\right)}$

In Exercises 23–44, simplify the complex fraction. *See Examples 5 and 6.*

23. $\dfrac{\left(1 + \dfrac{4}{y}\right)}{y}$

24. $\dfrac{x}{\left(\dfrac{3}{x} + 2\right)}$

✓ 25. $\dfrac{\left(\dfrac{4}{x} + 3\right)}{\left(\dfrac{4}{x} - 3\right)}$

26. $\dfrac{\left(\dfrac{1}{t} - 1\right)}{\left(\dfrac{1}{t} + 1\right)}$

27. $\dfrac{\left(\dfrac{x}{2}\right)}{\left(2 + \dfrac{3}{x}\right)}$

28. $\dfrac{\left(1 - \dfrac{2}{x}\right)}{\left(\dfrac{x}{2}\right)}$

29. $\dfrac{\left(3 + \dfrac{9}{x - 3}\right)}{\left(4 + \dfrac{12}{x - 3}\right)}$

30. $\dfrac{\left(4 + \dfrac{16}{x - 4}\right)}{\left(5 + \dfrac{20}{x - 4}\right)}$

31. $\dfrac{\left(\dfrac{3}{x^2} + \dfrac{1}{x}\right)}{\left(2 - \dfrac{4}{5x}\right)}$

32. $\dfrac{\left(16 - \dfrac{1}{x^2}\right)}{\left(\dfrac{1}{4x^2} - 4\right)}$

33. $\dfrac{\left(\dfrac{y}{x} - \dfrac{x}{y}\right)}{\left(\dfrac{x + y}{xy}\right)}$

34. $\dfrac{\left(\dfrac{x}{y} - \dfrac{y}{x}\right)}{\left(\dfrac{x - y}{xy}\right)}$

35. $\dfrac{\left(x - \dfrac{2y^2}{x - y}\right)}{x - 2y}$

36. $\dfrac{\left(x - \dfrac{6y^2}{x - y}\right)}{x - 3y}$

37. $\dfrac{\left(1 - \dfrac{1}{y}\right)}{\left(1 - 4y\over y - 3\right)}$

38. $\dfrac{\left(\dfrac{x + 1}{x + 2} - \dfrac{1}{x}\right)}{\left(\dfrac{2}{x + 2}\right)}$

39. $\dfrac{\left(\dfrac{10}{x + 1}\right)}{\left(\dfrac{1}{2x + 2} + \dfrac{3}{x + 1}\right)}$

40. $\dfrac{\left(\dfrac{2}{x + 5}\right)}{\left(\dfrac{2}{x + 5} + \dfrac{1}{4x + 20}\right)}$

✓ 41. $\dfrac{\left(\dfrac{1}{x} - \dfrac{1}{x + 1}\right)}{\left(\dfrac{1}{x + 1}\right)}$

42. $\dfrac{\left(\dfrac{5}{y} - \dfrac{6}{2y + 1}\right)}{\left(\dfrac{5}{2y + 1}\right)}$

43. $\dfrac{\left(\dfrac{x}{x - 3} - \dfrac{2}{3}\right)}{\left(\dfrac{10}{3x} + \dfrac{x^2}{x - 3}\right)}$

44. $\dfrac{\left(\dfrac{1}{2x} - \dfrac{6}{x + 5}\right)}{\left(\dfrac{x}{x - 5} + \dfrac{1}{x}\right)}$

In Exercises 45–52, simplify the expression. *See Example 7.*

45. $\dfrac{2y - y^{-1}}{10 - y^{-2}}$

46. $\dfrac{9x - x^{-1}}{3 + x^{-1}}$

51. $\dfrac{x^{-2} - y^{-2}}{(x + y)^2}$

52. $\dfrac{x - y}{x^{-2} - y^{-2}}$

47. $\dfrac{7x^2 + 2x^{-1}}{5x^{-3} + x}$

48. $\dfrac{3x^{-2} - x}{4x^{-1} + 6x}$

In Exercises 53 and 54, use the function to find and simplify the expression for

$$\frac{f(2 + h) - f(2)}{h}.$$

49. $\dfrac{x^{-1} + y^{-1}}{x^{-1} - y^{-1}}$

53. $f(x) = \dfrac{1}{x}$

54. $f(x) = \dfrac{x}{x - 1}$

50. $\dfrac{x^{-1} - y^{-1}}{x^{-2} - y^{-2}}$

Solving Problems

55. *Average of Two Numbers* Determine the average of two real numbers $x/5$ and $x/6$.

56. *Average of Two Numbers* Determine the average of two real numbers $2x/3$ and $3x/5$.

57. *Average of Two Numbers* Determine the average of two real numbers $2x/3$ and $x/4$.

58. *Average of Two Numbers* Determine the average of two real numbers $4/a^2$ and $2/a$.

59. *Average of Two Numbers* Determine the average of two real numbers $(b + 5)/4$ and $2/b$.

60. *Average of Two Numbers* Determine the average of two real numbers $5/2s$ and $(s + 1)/5$.

61. *Number Problem* Find three real numbers that divide the real number line between $x/9$ and $x/6$ into four equal parts (see figure).

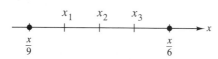

62. *Number Problem* Find two real numbers that divide the real number line between $x/3$ and $5x/4$ into three equal parts (see figure).

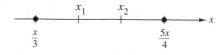

63. *Electrical Resistance* When two resistors of resistances R_1 and R_2 are connected in parallel, the total resistance is modeled by

$$\frac{1}{\left(\dfrac{1}{R_1} + \dfrac{1}{R_2}\right)}.$$

Simplify this complex fraction.

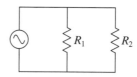

64. *Monthly Payment* The approximate annual percent interest rate r of a monthly installment loan is

$$r = \frac{\left[\dfrac{24(MN - P)}{N}\right]}{\left(P + \dfrac{MN}{12}\right)}$$

where N is the total number of payments, M is the monthly payment, and P is the amount financed.

(a) Simplify the expression.

(b) Approximate the annual percent interest rate for a four-year home-improvement loan of $15,000 with monthly payments of $350.

In Exercises 65 and 66, use the following models, which give the number N (in thousands) of cellular telephone subscribers and the annual revenue R (in millions of dollars) from cell phone subscriptions in the United States from 2000 through 2005.

$$N = \frac{6433.62t + 111,039.2}{-0.06t + 1}, \ 0 \leq t \leq 5$$

and $R = \dfrac{8123.73t + 60,227.5}{-0.04t + 1}, \ 0 \leq t \leq 5$

In these models, t represents the year, with $t = 0$ corresponding to 2000. (Source: Cellular Telecommunications and Internet Association)

© Strauss/Curtis/CORBIS

65. (a) Use a graphing calculator to graph the two models in the same viewing window.

(b) Find a model for the average monthly bill per subscriber. (*Note:* Modify the revenue model from years to months.)

66. (a) Use the model in Exercise 65 (b) to complete the table.

Year, t	0	1	2	3	4	5
Monthly bill						

(b) Use the model in Exercise 65(b) to predict the average monthly bill per subscriber in 2006, 2007, and 2008. Notice that, according to the model, N and R increase in these years, but the average bill decreases. Explain how this is possible.

Explaining Concepts

67. ✎ Is the simplified form of a complex fraction a complex fraction? Explain.

68. ✎ Describe the effect of multiplying two rational expressions by their least common denominator.

Error Analysis In Exercises 69 and 70, describe and correct the error.

69.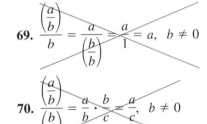

$$\frac{\left(\dfrac{a}{b}\right)}{b} = \frac{a}{\left(\dfrac{b}{b}\right)} = \frac{a}{1} = a, \ b \neq 0$$

70.

$$\frac{\left(\dfrac{a}{b}\right)}{\left(\dfrac{b}{c}\right)} = \frac{a}{b} \cdot \frac{b}{c} = \frac{a}{c}, \ b \neq 0$$

Cumulative Review

In Exercises 71 and 72, use the rules of exponents to simplify the expression.

71. $(2y)^3(3y)^2$

72. $\dfrac{27x^4y^2}{9x^3y}$

In Exercises 73 and 74, factor the trinomial.

73. $3x^2 + 5x - 2$

74. $x^2 + xy - 2y^2$

In Exercises 75–78, divide and simplify.

75. $\dfrac{x^2}{2} \div 4x$

76. $\dfrac{4x^3}{3} \div \dfrac{2x^2}{9}$

77. $\dfrac{(x + 1)^2}{x + 2} \div \dfrac{x + 1}{(x + 2)^3}$

78. $\dfrac{x^2 - 4x + 4}{x - 3} \div \dfrac{x^2 - 3x + 2}{x^2 - 6x + 9}$

Mid-Chapter Quiz

Take this quiz as you would take a quiz in class. After you are done, check your work against the answers in the back of the book.

1. Determine the domain of $f(x) = \dfrac{x}{x^2 + x}$.

2. Evaluate $h(x) = (x^2 - 9)/(x^2 - x - 2)$ for the indicated values of x, and simplify. If it is not possible, state the reason.

    (a) $h(-3)$ (b) $h(0)$ (c) $h(-1)$ (d) $h(5)$

In Exercises 3–8, simplify the rational expression.

3. $\dfrac{9y^2}{6y}$

4. $\dfrac{6u^4v^3}{15uv^3}$

5. $\dfrac{4x^2 - 1}{x - 2x^2}$

6. $\dfrac{(z + 3)^2}{2z^2 + 5z - 3}$

7. $\dfrac{5a^2b + 3ab^3}{a^2b^2}$

8. $\dfrac{2mn^2 - n^3}{2m^2 + mn - n^2}$

In Exercises 9–20, perform the indicated operations and simplify.

9. $\dfrac{11t^2}{6} \cdot \dfrac{9}{33t}$

10. $(x^2 + 2x) \cdot \dfrac{5}{x^2 - 4}$

11. $\dfrac{4}{3(x - 1)} \cdot \dfrac{12x}{6(x^2 + 2x - 3)}$

12. $\dfrac{32z^4}{5x^5y^5} \div \dfrac{80z^5}{25x^8y^6}$

13. $\dfrac{a - b}{9a + 9b} \div \dfrac{a^2 - b^2}{a^2 + 2a + 1}$

14. $\dfrac{5u}{3(u + v)} \cdot \dfrac{2(u^2 - v^2)}{3v} \div \dfrac{25u^2}{18(u - v)}$

15. $\dfrac{5x - 6}{x - 2} + \dfrac{2x - 5}{x - 2}$

16. $\dfrac{x}{x^2 - 9} - \dfrac{4(x - 3)}{x + 3}$

17. $\dfrac{x^2 + 2}{x^2 - x - 2} + \dfrac{1}{x + 1} - \dfrac{x}{x - 2}$

18. $\dfrac{\left(\dfrac{9t^2}{3 - t}\right)}{\left(\dfrac{6t}{t - 3}\right)}$

19. $\dfrac{\left(\dfrac{10}{x^2 + 2x}\right)}{\left(\dfrac{15}{x^2 + 3x + 2}\right)}$

20. $\dfrac{3x^{-1} - y^{-1}}{(x - y)^{-1}}$

21. You open a floral shop with a setup cost of $25,000. The cost of creating one dozen floral arrangements is $144.

    (a) Write the total cost C as a function of x, the number of floral arrangements (in dozens) created.

    (b) Write the average cost per dozen $\overline{C} = C/x$ as a function of x, the number of floral arrangements (in dozens) created.

    (c) Find the value of $\overline{C}(500)$.

22. Determine the average of three real numbers x, $x/2$, and $2x/3$.

6.5 Dividing Polynomials and Synthetic Division

What You Should Learn

1 ▶ Divide polynomials by monomials and write in simplest form.

2 ▶ Use long division to divide polynomials by polynomials.

3 ▶ Use synthetic division to divide polynomials by polynomials of the form $x - k$.

4 ▶ Use synthetic division to factor polynomials.

Why You Should Learn It

Division of polynomials is useful in higher-level mathematics when factoring and finding zeros of polynomials.

1 ▶ Divide polynomials by monomials and write in simplest form.

Dividing a Polynomial by a Monomial

To divide a polynomial by a monomial, *reverse* the procedure used to add or subtract two rational expressions. Here is an example.

$$2 + \frac{1}{x} = \frac{2x}{x} + \frac{1}{x} = \frac{2x + 1}{x} \qquad \text{Add fractions.}$$

$$\frac{2x + 1}{x} = \frac{2x}{x} + \frac{1}{x} = 2 + \frac{1}{x} \qquad \text{Divide by monomial.}$$

Dividing a Polynomial by a Monomial

Let u, v, and w represent real numbers, variables, or algebraic expressions such that $w \neq 0$.

$$\textbf{1.} \ \frac{u + v}{w} = \frac{u}{w} + \frac{v}{w} \qquad \textbf{2.} \ \frac{u - v}{w} = \frac{u}{w} - \frac{v}{w}$$

When dividing a polynomial by a monomial, remember to write the resulting expressions in simplest form, as illustrated in Example 1.

EXAMPLE 1   Dividing a Polynomial by a Monomial

Perform the division and simplify.

$$\frac{12x^2 - 20x + 8}{4x}$$

Solution

$$\frac{12x^2 - 20x + 8}{4x} = \frac{12x^2}{4x} - \frac{20x}{4x} + \frac{8}{4x} \qquad \begin{array}{l}\text{Divide each term in the}\\\text{numerator by } 4x.\end{array}$$

$$= \frac{3(4x)(x)}{4x} - \frac{5(4x)}{4x} + \frac{2(4)}{4x} \qquad \text{Divide out common factors.}$$

$$= 3x - 5 + \frac{2}{x} \qquad \text{Simplified form}$$

✓ **CHECKPOINT** *Now try Exercise 5.*

Long Division

In Section 6.1, you learned how to divide one polynomial by another by factoring and dividing out common factors. For instance, you can divide $x^2 - 2x - 3$ by $x - 3$ as follows.

$$(x^2 - 2x - 3) \div (x - 3) = \frac{x^2 - 2x - 3}{x - 3}$$ Write as a fraction.

$$= \frac{(x + 1)(x - 3)}{x - 3}$$ Factor numerator.

$$= \frac{(x + 1)\cancel{(x - 3)}}{\cancel{x - 3}}$$ Divide out common factor.

$$= x + 1, \quad x \neq 3$$ Simplified form

This procedure works well for polynomials that factor easily. For those that do not, you can use a more general procedure that follows a "long division algorithm" similar to the algorithm used for dividing positive integers, which is reviewed in Example 2.

EXAMPLE 2   Long Division Algorithm for Positive Integers

Use the long division algorithm to divide 6584 by 28.

Solution

Think $\frac{65}{28} \approx 2$.
Think $\frac{98}{28} \approx 3$.
Think $\frac{144}{28} \approx 5$.

$$\begin{array}{r} 235 \\ 28\overline{)6584} \\ \underline{56} \\ 98 \\ \underline{84} \\ 144 \\ \underline{140} \\ 4 \end{array}$$

Multiply 2 by 28.
Subtract and bring down 8.
Multiply 3 by 28.
Subtract and bring down 4.
Multiply 5 by 28.
Remainder

So, you have

$$6584 \div 28 = 235 + \frac{4}{28}$$

$$= 235 + \frac{1}{7}.$$

✔ **CHECKPOINT** *Now try Exercise 15.*

In Example 2, 6584 is the **dividend,** 28 is the **divisor,** 235 is the **quotient,** and 4 is the **remainder.**

In the next several examples, you will see how the long division algorithm can be extended to cover the division of one polynomial by another.

When you use long division to divide polynomials, follow the steps below.

Long Division of Polynomials

1. Write the dividend and divisor in descending powers of the variable.

2. Insert placeholders with zero coefficients for missing powers of the variable. (See Example 5.)

3. Perform the long division of the polynomials as you would with integers.

4. Continue the process until the degree of the remainder is less than that of the divisor.

EXAMPLE 3   Long Division Algorithm for Polynomials

Think $x^2/x = x$.

Think $3x/x = 3$.

$$
\begin{array}{r}
x + 3 \\
x - 1 \,\overline{)\, x^2 + 2x + 4} \\
\underline{x^2 - x} \\
3x + 4 \\
\underline{3x - 3} \\
7
\end{array}
$$

Multiply x by $(x - 1)$.
Subtract and bring down 4.
Multiply 3 by $(x - 1)$.
Subtract.

The remainder is a fractional part of the divisor, so you can write

$$
\underbrace{\frac{x^2 + 2x + 4}{x - 1}}_{\text{Dividend} \atop \text{Divisor}} = \overbrace{x + 3}^{\text{Quotient}} + \overbrace{\underbrace{\frac{7}{x - 1}}_{\text{Divisor}}}^{\text{Remainder}}.
$$

✓ **CHECKPOINT** *Now try Exercise 19.*

Study Tip

Note that in Example 3, the division process requires $3x - 3$ to be subtracted from $3x + 4$. The difference

$$
\begin{array}{r}
3x + 4 \\
-(3x - 3)
\end{array}
$$

is implied and written simply as

$$
\begin{array}{r}
3x + 4 \\
\underline{3x - 3} \\
7.
\end{array}
$$

You can check a long division problem by multiplying by the divisor. For instance, you can check the result of Example 3 as follows.

$$
\frac{x^2 + 2x + 4}{x - 1} \overset{?}{=} x + 3 + \frac{7}{x - 1}
$$

$$
(x - 1)\left(\frac{x^2 + 2x + 4}{x - 1}\right) \overset{?}{=} (x - 1)\left(x + 3 + \frac{7}{x - 1}\right)
$$

$$
x^2 + 2x + 4 \overset{?}{=} (x + 3)(x - 1) + 7
$$

$$
x^2 + 2x + 4 \overset{?}{=} (x^2 + 2x - 3) + 7
$$

$$
x^2 + 2x + 4 = x^2 + 2x + 4 \quad ✓
$$

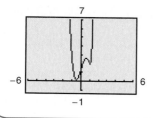

EXAMPLE 4   Writing in Standard Form Before Dividing

Divide $-13x^3 + 10x^4 + 8x - 7x^2 + 4$ by $3 - 2x$.

Solution

First write the divisor and dividend in standard polynomial form.

$$
\begin{array}{r}
-5x^3 - x^2 + 2x - 1 \\
-2x + 3 \overline{)\, 10x^4 - 13x^3 - 7x^2 + 8x + 4\,} \\
\underline{10x^4 - 15x^3} \\
2x^3 - 7x^2 \\
\underline{2x^3 - 3x^2} \\
-4x^2 + 8x \\
\underline{-4x^2 + 6x} \\
2x + 4 \\
\underline{2x - 3} \\
7
\end{array}
$$

Multiply $-5x^3$ by $(-2x + 3)$.
Subtract and bring down $-7x^2$.
Multiply $-x^2$ by $(-2x + 3)$.
Subtract and bring down $8x$.
Multiply $2x$ by $(-2x + 3)$.
Subtract and bring down 4.
Multiply -1 by $(-2x + 3)$.
Subtract.

This shows that

$$
\underbrace{\overbrace{10x^4 - 13x^3 - 7x^2 + 8x + 4}^{\text{Dividend}}}_{\underbrace{-2x + 3}_{\text{Divisor}}} = \overbrace{-5x^3 - x^2 + 2x - 1}^{\text{Quotient}} + \frac{\overset{\text{Remainder}}{7}}{\underbrace{-2x + 3}_{\text{Divisor}}}.
$$

✓ **CHECKPOINT** *Now try Exercise 25.*

When the dividend is missing one or more powers of x, the long division algorithm requires that you account for the missing powers, as shown in Example 5.

EXAMPLE 5   Accounting for Missing Powers of x

Divide $x^3 - 2$ by $x - 1$.

Solution

To account for the missing x^2- and x-terms, insert $0x^2$ and $0x$.

$$
\begin{array}{r}
x^2 + x + 1 \\
x - 1 \overline{)\, x^3 + 0x^2 + 0x - 2\,} \\
\underline{x^3 - x^2} \\
x^2 + 0x \\
\underline{x^2 - x} \\
x - 2 \\
\underline{x - 1} \\
-1
\end{array}
$$

Insert $0x^2$ and $0x$.
Multiply x^2 by $(x - 1)$.
Subtract and bring down $0x$.
Multiply x by $(x - 1)$.
Subtract and bring down -2.
Multiply 1 by $(x - 1)$.
Subtract.

So, you have

$$
\frac{x^3 - 2}{x - 1} = x^2 + x + 1 - \frac{1}{x - 1}.
$$

✓ **CHECKPOINT** *Now try Exercise 41.*

In each of the long division examples presented so far, the divisor has been a first-degree polynomial. The long division algorithm works just as well with polynomial divisors of degree two or more, as shown in Example 6.

EXAMPLE 6   A Second-Degree Divisor

Divide $x^4 + 6x^3 + 6x^2 - 10x - 3$ by $x^2 + 2x - 3$.

Solution

$$
\begin{array}{r}
x^2 + 4x + 1 \\
x^2 + 2x - 3 \overline{)\ x^4 + 6x^3 + 6x^2 - 10x - 3} \\
\underline{x^4 + 2x^3 - 3x^2} \\
4x^3 + 9x^2 - 10x \\
\underline{4x^3 + 8x^2 - 12x} \\
x^2 + 2x - 3 \\
\underline{x^2 + 2x - 3} \\
0
\end{array}
$$

Multiply x^2 by $(x^2 + 2x - 3)$.
Subtract and bring down $-10x$.
Multiply $4x$ by $(x^2 + 2x - 3)$.
Subtract and bring down -3.
Multiply 1 by $(x^2 + 2x - 3)$.
Subtract.

So, $x^2 + 2x - 3$ divides evenly into $x^4 + 6x^3 + 6x^2 - 10x - 3$. That is,

$$\frac{x^4 + 6x^3 + 6x^2 - 10x - 3}{x^2 + 2x - 3} = x^2 + 4x + 1, \ x \neq -3, x \neq 1.$$

✓ **CHECKPOINT** *Now try Exercise 49.*

Study Tip

If the remainder of a division problem is zero, the divisor is said to **divide evenly** into the dividend.

3 ▶ Use synthetic division to divide polynomials by polynomials of the form $x - k$.

Synthetic Division

There is a nice shortcut for division by polynomials of the form $x - k$. It is called **synthetic division** and is outlined for a third-degree polynomial as follows.

Synthetic Division of a Third-Degree Polynomial

Use synthetic division to divide $ax^3 + bx^2 + cx + d$ by $x - k$, as follows.

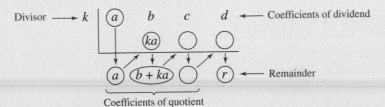

Vertical Pattern: Add terms.

Diagonal Pattern: Multiply by k.

Keep in mind that this algorithm for synthetic division works *only* for divisors of the form $x - k$. Remember that $x + k = x - (-k)$. Moreover, the degree of the quotient is always one less than the degree of the dividend.

EXAMPLE 7 **Using Synthetic Division**

Use synthetic division to divide $x^3 + 3x^2 - 4x - 10$ by $x - 2$.

Solution

The coefficients of the dividend form the top row of the synthetic division array. Because you are dividing by $x - 2$, write 2 at the top left of the array. To begin the algorithm, bring down the first coefficient. Then multiply this coefficient by 2, write the result in the second row, and add the two numbers in the second column. By continuing this pattern, you obtain the following.

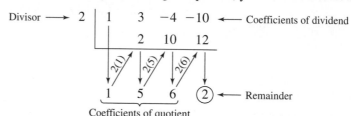

The bottom row shows the coefficients of the quotient. So, the quotient is

$$1x^2 + 5x + 6$$

and the remainder is 2. So, the result of the division problem is

$$\frac{x^3 + 3x^2 - 4x - 10}{x - 2} = x^2 + 5x + 6 + \frac{2}{x - 2}.$$

✓ **CHECKPOINT** *Now try Exercise 61.*

4 ▶ Use synthetic division to factor polynomials.

Factoring and Division

Synthetic division (or long division) can be used to factor polynomials. If the remainder in a synthetic division problem is zero, you know that the divisor divides *evenly* into the dividend.

EXAMPLE 8 **Factoring a Polynomial**

The polynomial $x^3 - 7x + 6$ can be factored completely using synthetic division. Because $x - 1$ is a factor of the polynomial, you can divide as follows.

Because the remainder is zero, the divisor divides evenly into the dividend:

$$\frac{x^3 - 7x + 6}{x - 1} = x^2 + x - 6.$$

From this result, you can factor the original polynomial as follows.

$$x^3 - 7x + 6 = (x - 1)(x^2 + x - 6) = (x - 1)(x + 3)(x - 2)$$

✓ **CHECKPOINT** *Now try Exercise 73.*

Concept Check

1. Consider the equation $1253 \div 12 = 104 + \dfrac{5}{12}$. Identify the dividend, divisor, quotient, and remainder.

2. Explain what it means for a divisor to divide *evenly* into a dividend.

3. Explain how you can check polynomial division.

4. For synthetic division, what form must the divisor have?

6.5 EXERCISES

Go to pages 440–441 to record your assignments.

Developing Skills

In Exercises 1–14, perform the division. *See Example 1.*

1. $(7x^3 - 2x^2) \div x$

2. $(3w^2 - 6w) \div w$

3. $(4x^2 - 2x) \div (-x)$

4. $(5y^3 + 6y^2 - 3y) \div (-y)$

5. $(m^4 + 2m^2 - 7) \div m$

6. $(x^3 + x - 2) \div x$

7. $\dfrac{50z^3 + 30z}{-5z}$

8. $\dfrac{18c^4 - 24c^2}{-6c}$

9. $\dfrac{4v^4 + 10v^3 - 8v^2}{4v^2}$

10. $\dfrac{6x^4 + 8x^3 - 18x^2}{3x^2}$

11. $\dfrac{4x^5 - 6x^4 + 12x^3 - 8x^2}{4x^2}$

12. $\dfrac{15x^{12} - 5x^9 + 30x^6}{5x^6}$

13. $(5x^2y - 8xy + 7xy^2) \div 2xy$

14. $(-14s^4t^2 + 7s^2t^2 - 18t) \div 2s^2t$

In Exercises 15–18, use the long division algorithm to perform the division. *See Example 2.*

15. Divide 1013 by 9.

16. Divide 3713 by 22.

17. $3235 \div 15$

18. $6344 \div 28$

In Exercises 19–56, perform the division. *See Examples 3–6.*

19. $\dfrac{x^2 - 8x + 15}{x - 3}$

20. $\dfrac{t^2 - 18t + 72}{t - 6}$

21. $(x^2 + 15x + 50) \div (x + 5)$

22. $(y^2 - 6y - 16) \div (y + 2)$

23. Divide $x^2 - 5x + 8$ by $x - 2$.

24. Divide $x^2 + 10x - 9$ by $x - 3$.

25. Divide $21 - 4x - x^2$ by $3 - x$.

26. Divide $5 + 4x - x^2$ by $1 + x$.

27. $\dfrac{5x^2 + 2x + 3}{x + 2}$

28. $\dfrac{2x^2 + 13x + 15}{x + 5}$

29. $\dfrac{12x^2 + 17x - 5}{3x + 2}$

30. $\dfrac{8x^2 + 2x + 3}{4x - 1}$

31. $(12 - 17t + 6t^2) \div (2t - 3)$

32. $(15 - 14u - 8u^2) \div (5 + 2u)$

33. Divide $2y^2 + 7y + 3$ by $2y + 1$.

34. Divide $10t^2 - 7t - 12$ by $2t - 3$.

35. $\dfrac{x^3 - 2x^2 + 4x - 8}{x - 2}$

36. $\dfrac{x^3 + 4x^2 + 7x + 28}{x + 4}$

37. $\dfrac{9x^3 - 3x^2 - 3x + 4}{3x + 2}$

38. $\dfrac{4y^3 + 12y^2 + 7y - 3}{2y + 3}$

39. $(2x + 9) \div (x + 2)$

40. $(12x - 5) \div (2x + 3)$

✓ **41.** $\dfrac{x^2 + 16}{x + 4}$

42. $\dfrac{y^2 + 8}{y + 2}$

43. $\dfrac{6z^2 + 7z}{5z - 1}$

44. $\dfrac{8y^2 - 2y}{3y + 5}$

45. $\dfrac{16x^2 - 1}{4x + 1}$

46. $\dfrac{25y^2 - 9}{5y - 3}$

47. $\dfrac{x^3 + 125}{x + 5}$

48. $\dfrac{x^3 - 27}{x - 3}$

✓ **49.** $(x^3 + 4x^2 + 7x + 7) \div (x^2 + 2x + 3)$

50. $(2x^3 + 2x^2 - 2x - 15) \div (2x^2 + 4x + 5)$

51. $(4x^4 - 3x^2 + x - 5) \div (x^2 - 3x + 2)$

52. $(8x^5 + 6x^4 - x^3 + 1) \div (2x^3 - x^2 - 3)$

53. Divide $x^4 - 1$ by $x - 1$.
54. Divide $x^6 - 1$ by $x - 1$.

55. $x^5 \div (x^2 + 1)$

56. $x^6 \div (x^3 - 1)$

In Exercises 57–60, simplify the expression.

57. $\dfrac{8u^2v}{2u} + \dfrac{3(uv)^2}{uv}$

58. $\dfrac{15x^3y}{10x^2} + \dfrac{3xy^2}{2y}$

59. $\dfrac{x^2 + 3x + 2}{x + 2} + (2x + 3)$

60. $\dfrac{x^2 + 2x - 3}{x - 1} - (3x - 4)$

In Exercises 61–72, use synthetic division to divide. *See Example 7.*

✓ **61.** $(x^2 + x - 6) \div (x - 2)$

62. $(x^2 + 5x - 6) \div (x + 6)$

63. $\dfrac{x^3 + 3x^2 - 1}{x + 4}$

64. $\dfrac{x^3 - 4x + 7}{x - 1}$

65. $\dfrac{x^4 - 4x^3 + x + 10}{x - 2}$

66. $\dfrac{2x^5 - 3x^3 + x}{x - 3}$

67. $\dfrac{5x^3 - 6x^2 + 8}{x - 4}$

68. $\dfrac{5x^3 + 6x + 8}{x + 2}$

69. $\dfrac{10x^4 - 50x^3 - 800}{x - 6}$

70. $\dfrac{x^5 - 13x^4 - 120x + 80}{x + 3}$

71. $\dfrac{0.1x^2 + 0.8x + 1}{x - 0.2}$

72. $\dfrac{x^3 - 0.8x + 2.4}{x + 0.1}$

In Exercises 73–80, completely factor the polynomial given one of its factors. *See Example 8.*

	Polynomial	Factor
✓ **73.**	$x^3 - x^2 - 14x + 24$	$x - 3$
74.	$x^3 + x^2 - 32x - 60$	$x + 5$
75.	$4x^3 - 3x - 1$	$x - 1$
76.	$9x^3 + 51x^2 + 88x + 48$	$x + 3$
77.	$x^4 + 7x^3 + 3x^2 - 63x - 108$	$x + 4$
78.	$x^4 - 6x^3 - 8x^2 + 96x - 128$	$x - 4$
79.	$15x^2 - 2x - 8$	$x - \frac{4}{5}$
80.	$18x^2 - 9x - 20$	$x + \frac{5}{6}$

In Exercises 81 and 82, find the constant c such that the denominator divides evenly into the numerator.

81. $\dfrac{x^3 + 2x^2 - 4x + c}{x - 2}$ **82.** $\dfrac{x^4 - 3x^2 + c}{x + 6}$

In Exercises 83 and 84, use a graphing calculator to graph the two equations in the same viewing window. Use the graphs to verify that the expressions are equivalent. Verify the results algebraically.

83. $y_1 = \dfrac{x + 4}{2x}$

$y_2 = \dfrac{1}{2} + \dfrac{2}{x}$

84. $y_1 = \dfrac{x^2 + 2}{x + 1}$

$y_2 = x - 1 + \dfrac{3}{x + 1}$

In Exercises 85 and 86, perform the division assuming that n is a positive integer.

85. $\dfrac{x^{3n} + 3x^{2n} + 6x^n + 8}{x^n + 2}$ **86.** $\dfrac{x^{3n} - x^{2n} + 5x^n - 5}{x^n - 1}$

Think About It   In Exercises 87 and 88, the divisor, quotient, and remainder are given. Find the dividend.

	Divisor	Quotient	Remainder
87.	$x - 6$	$x^2 + x + 1$	-4
88.	$x + 3$	$x^2 - 2x - 5$	8

Finding a Pattern   In Exercises 89 and 90, complete the table for the function. The first row is completed for Exercise 89. What conclusion can you draw as you compare the values of $f(k)$ with the remainders? (Use synthetic division to find the remainders.)

89. $f(x) = x^3 - x^2 - 2x$

90. $f(x) = 2x^3 - x^2 - 2x + 1$

k	$f(k)$	Divisor $(x - k)$	Remainder
-2	-8	$x + 2$	-8
-1			
0			
$\frac{1}{2}$			
1			
2			

Solving Problems

91. △ *Geometry*   The volume of a cube is $x^3 + 3x^2 + 3x + 1$. The height of the cube is $x + 1$. Use division to find the area of the base.

92. △ *Geometry*   A rectangular house has a volume of $x^3 + 55x^2 + 650x + 2000$ cubic feet (the space in the attic is not included). The height of the house is $x + 5$ feet (see figure). Find the number of square feet of floor space *on the first floor* of the house.

$x + 5$

Figure for 92

▲ *Geometry*   In Exercises 93 and 94, you are given the expression for the volume of the solid shown. Find the expression for the missing dimension.

93. $V = x^3 + 18x^2 + 80x + 96$

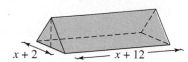

94. $V = 2h^3 + 3h^2 + h$

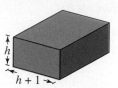

─────────── **Explaining Concepts** ───────────

95. *Error Analysis*   Describe and correct the error.

$$\frac{6x + 5y}{x} = \frac{6x + 5y}{x} = 6 + 5y$$

96. *Error Analysis*   Describe and correct the error.

$$\frac{x^2}{x+1} = \frac{x^2}{x} + \frac{x^2}{1} = x + x^2$$

97. Create a polynomial division problem and identify the dividend, divisor, quotient, and remainder.

98. *True or False?*   If the divisor divides evenly into the dividend, then the divisor and quotient are factors of the dividend. Justify your answer.

99. ⌨ ✎   Use a graphing calculator to graph each polynomial in the same viewing window using the standard setting. Use the *zero* or *root* feature to find the x-intercepts. What can you conclude about the polynomials? Verify your conclusion algebraically.

(a) $y = (x - 4)(x - 2)(x + 1)$

(b) $y = (x^2 - 6x + 8)(x + 1)$

(c) $y = x^3 - 5x^2 + 2x + 8$

100. ⌨ ✎   Use a graphing calculator to graph the function

$$f(x) = \frac{x^3 - 5x^2 + 2x + 8}{x - 2}.$$

Use the *zero* or *root* feature to find the x-intercepts. Why does this function have only two x-intercepts? To what other function does the graph of $f(x)$ appear to be equivalent? What is the difference between the two graphs?

─────────── **Cumulative Review** ───────────

In Exercises 101–106, solve the inequality.

101. $7 - 3x > 4 - x$ **102.** $2(x + 6) - 20 < 2$

103. $|x - 3| < 2$ **104.** $|x - 5| > 3$

105. $\left|\frac{1}{4}x - 1\right| \geq 3$ **106.** $\left|2 - \frac{1}{3}x\right| \leq 10$

In Exercises 107 and 108, determine the quadrants in which the point must be located.

107. $(-3, y)$, y is a real number.

108. $(x, 7)$, x is a real number.

109. Describe the location of the set of points whose x-coordinates are 0.

110. Find the coordinates of the point five units to the right of the y-axis and seven units below the x-axis.

6.6 Solving Rational Equations

Thinkstock/Getty Images

Why You Should Learn It

Rational equations can be used to model and solve real-life applications. For instance, in Exercise 86 on page 426, you will use a rational equation to determine the speeds of two runners.

1 ▶ Solve rational equations containing constant denominators.

What You Should Learn

1 ▶ Solve rational equations containing constant denominators.

2 ▶ Solve rational equations containing variable denominators.

Equations Containing Constant Denominators

In Section 2.1, you studied a strategy for solving equations that contain fractions with *constant* denominators. That procedure is reviewed here because it is the basis for solving more general equations involving fractions. Recall from Section 2.1 that you can "clear an equation of fractions" by multiplying each side of the equation by the least common denominator (LCD) of the fractions in the equation. Note how this is done in the next three examples.

> **Study Tip**
>
> A *rational equation* is an equation containing one or more rational expressions.

EXAMPLE 1   An Equation Containing Constant Denominators

Solve $\dfrac{3}{5} = \dfrac{x}{2} + 1$.

Solution

The least common denominator of the fractions is 10, so begin by multiplying each side of the equation by 10.

$$\frac{3}{5} = \frac{x}{2} + 1 \qquad \text{Write original equation.}$$

$$10\left(\frac{3}{5}\right) = 10\left(\frac{x}{2} + 1\right) \qquad \text{Multiply each side by LCD of 10.}$$

$$6 = 5x + 10 \qquad \text{Distribute and simplify.}$$

$$-4 = 5x \quad \Longrightarrow \quad -\frac{4}{5} = x \qquad \text{Subtract 10 from each side, then divide each side by 5.}$$

The solution is $x = -\frac{4}{5}$. You can check this in the original equation as follows.

Check

$$\frac{3}{5} \stackrel{?}{=} \frac{-4/5}{2} + 1 \qquad \text{Substitute } -\tfrac{4}{5} \text{ for } x \text{ in the original equation.}$$

$$\frac{3}{5} \stackrel{?}{=} -\frac{4}{5} \cdot \frac{1}{2} + 1 \qquad \text{Invert divisor and multiply.}$$

$$\frac{3}{5} = -\frac{2}{5} + 1 \qquad \text{Solution checks. } \checkmark$$

 **CHECKPOINT** *Now try Exercise 5.*

EXAMPLE 2   An Equation Containing Constant Denominators

Solve $\dfrac{x-3}{6} = 7 - \dfrac{x}{12}$.

Solution

The least common denominator of the fractions is 12, so begin by multiplying each side of the equation by 12.

$$\dfrac{x-3}{6} = 7 - \dfrac{x}{12}$$ Write original equation.

$$12\left(\dfrac{x-3}{6}\right) = 12\left(7 - \dfrac{x}{12}\right)$$ Multiply each side by LCD of 12.

$$2x - 6 = 84 - x$$ Distribute and simplify.

$$3x - 6 = 84$$ Add x to each side.

$$3x = 90 \quad \Longrightarrow \quad x = 30$$ Add 6 to each side, then divide each side by 3.

The solution is $x = 30$. Check this in the original equation.

✓ **CHECKPOINT** *Now try Exercise 11.*

EXAMPLE 3   An Equation That Has Two Solutions

Solve $\dfrac{x^2}{3} + \dfrac{x}{2} = \dfrac{5}{6}$.

Solution

The least common denominator of the fractions is 6, so begin by multiplying each side of the equation by 6.

$$\dfrac{x^2}{3} + \dfrac{x}{2} = \dfrac{5}{6}$$ Write original equation.

$$6\left(\dfrac{x^2}{3} + \dfrac{x}{2}\right) = 6\left(\dfrac{5}{6}\right)$$ Multiply each side by LCD of 6.

$$\dfrac{6x^2}{3} + \dfrac{6x}{2} = \dfrac{30}{6}$$ Distributive Property

$$2x^2 + 3x = 5$$ Simplify.

$$2x^2 + 3x - 5 = 0$$ Subtract 5 from each side.

$$(2x + 5)(x - 1) = 0$$ Factor.

$$2x + 5 = 0 \quad \Longrightarrow \quad x = -\tfrac{5}{2}$$ Set 1st factor equal to 0.

$$x - 1 = 0 \quad \Longrightarrow \quad x = 1$$ Set 2nd factor equal to 0.

The solutions are $x = -\tfrac{5}{2}$ and $x = 1$. Check these in the original equation.

✓ **CHECKPOINT** *Now try Exercise 15.*

2 ▶ Solve rational equations containing variable denominators.

Equations Containing Variable Denominators

In a rational expression, remember that the variable(s) cannot take on values that make the denominator zero. This is especially critical in solving rational equations that contain variable denominators.

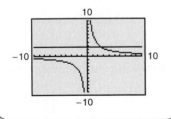

EXAMPLE 4   **An Equation Containing Variable Denominators**

Solve the equation.

$$\frac{7}{x} - \frac{1}{3x} = \frac{8}{3}$$

Solution

The least common denominator of the fractions is $3x$, so begin by multiplying each side of the equation by $3x$.

$$\frac{7}{x} - \frac{1}{3x} = \frac{8}{3} \qquad \text{Write original equation.}$$

$$3x\left(\frac{7}{x} - \frac{1}{3x}\right) = 3x\left(\frac{8}{3}\right) \qquad \text{Multiply each side by LCD of } 3x.$$

$$\frac{21x}{x} - \frac{3x}{3x} = \frac{24x}{3} \qquad \text{Distributive Property}$$

$$21 - 1 = 8x \qquad \text{Simplify.}$$

$$\frac{20}{8} = x \qquad \text{Combine like terms and divide each side by 8.}$$

$$\frac{5}{2} = x \qquad \text{Simplify.}$$

The solution is $x = \frac{5}{2}$. You can check this in the original equation as follows.

Check

$$\frac{7}{x} - \frac{1}{3x} = \frac{8}{3} \qquad \text{Write original equation.}$$

$$\frac{7}{5/2} - \frac{1}{3(5/2)} \overset{?}{=} \frac{8}{3} \qquad \text{Substitute } \frac{5}{2} \text{ for } x.$$

$$7\left(\frac{2}{5}\right) - \left(\frac{1}{3}\right)\left(\frac{2}{5}\right) \overset{?}{=} \frac{8}{3} \qquad \text{Invert divisors and multiply.}$$

$$\frac{14}{5} - \frac{2}{15} \overset{?}{=} \frac{8}{3} \qquad \text{Simplify.}$$

$$\frac{40}{15} \overset{?}{=} \frac{8}{3} \qquad \text{Combine like terms.}$$

$$\frac{8}{3} = \frac{8}{3} \qquad \text{Solution checks. ✓}$$

 **CHECKPOINT** *Now try Exercise 25.*

Throughout the text, the importance of checking solutions is emphasized. Up to this point, the main reason for checking has been to make sure that you did not make arithmetic errors in the solution process. In the next example, you will see that there is another reason for checking solutions in the *original* equation. That is, even with no mistakes in the solution process, it can happen that a "trial solution" does not satisfy the original equation. This type of solution is called an **extraneous solution.** An extraneous solution of an equation *must not* be listed as an actual solution.

EXAMPLE 5   An Equation with No Solution

Solve $\dfrac{5x}{x-2} = 7 + \dfrac{10}{x-2}$.

Solution

The least common denominator of the fractions is $x - 2$, so begin by multiplying each side of the equation by $x - 2$.

$$\frac{5x}{x-2} = 7 + \frac{10}{x-2} \qquad \text{Write original equation.}$$

$$(x-2)\left(\frac{5x}{x-2}\right) = (x-2)\left(7 + \frac{10}{x-2}\right) \qquad \text{Multiply each side by } x-2.$$

$$5x = 7(x-2) + 10 \qquad \text{Distribute and simplify.}$$

$$5x = 7x - 14 + 10 \qquad \text{Distributive Property}$$

$$5x = 7x - 4 \qquad \text{Combine like terms.}$$

$$-2x = -4 \qquad \text{Subtract } 7x \text{ from each side.}$$

$$x = 2 \qquad \text{Divide each side by } -2.$$

At this point, the solution appears to be $x = 2$. However, by performing a check, you can see that this "trial solution" is extraneous.

Check

$$\frac{5x}{x-2} = 7 + \frac{10}{x-2} \qquad \text{Write original equation.}$$

$$\frac{5(2)}{2-2} \stackrel{?}{=} 7 + \frac{10}{2-2} \qquad \text{Substitute 2 for } x.$$

$$\frac{10}{0} \stackrel{?}{=} 7 + \frac{10}{0} \qquad \text{Solution does not check. } \textbf{✗}$$

Because the check results in *division by zero*, you can conclude that 2 is extraneous. So, the original equation has no solution.

✓ **CHECKPOINT** *Now try Exercise 45.*

Notice that $x = 2$ is excluded from the domains of the two fractions in the original equation in Example 5. You may find it helpful when solving these types of equations to list the domain restrictions *before* beginning the solution process.

EXAMPLE 6   **Cross-Multiplying**

Solve $\dfrac{2x}{x + 4} = \dfrac{3}{x - 1}$.

Solution

The domain is all real values of x such that $x \neq -4$ and $x \neq 1$. You can use cross-multiplication to solve this equation.

$$\frac{2x}{x + 4} = \frac{3}{x - 1}$$ Write original equation.

$$2x(x - 1) = 3(x + 4)$$ Cross-multiply.

$$2x^2 - 2x = 3x + 12$$ Distributive Property

$$2x^2 - 5x - 12 = 0$$ Subtract $3x$ and 12 from each side.

$$(2x + 3)(x - 4) = 0$$ Factor.

$$2x + 3 = 0 \implies x = -\tfrac{3}{2}$$ Set 1st factor equal to 0.

$$x - 4 = 0 \implies x = 4$$ Set 2nd factor equal to 0.

The solutions are $x = -\tfrac{3}{2}$ and $x = 4$. Check these in the original equation.

✓ **CHECKPOINT** *Now try Exercise 47.*

EXAMPLE 7   **An Equation That Has Two Solutions**

Solve $\dfrac{3x}{x + 1} = \dfrac{12}{x^2 - 1} + 2$.

Solution

The domain is all real values of x such that $x \neq 1$ and $x \neq -1$. The least common denominator is $(x + 1)(x - 1) = x^2 - 1$.

$$(x^2 - 1)\left(\frac{3x}{x + 1}\right) = (x^2 - 1)\left(\frac{12}{x^2 - 1} + 2\right)$$ Multiply each side of original equation by LCD of $x^2 - 1$.

$$(x - 1)(3x) = 12 + 2(x^2 - 1)$$ Simplify.

$$3x^2 - 3x = 12 + 2x^2 - 2$$ Distributive Property

$$x^2 - 3x - 10 = 0$$ Subtract $2x^2$ and 10 from each side.

$$(x + 2)(x - 5) = 0$$ Factor.

$$x + 2 = 0 \implies x = -2$$ Set 1st factor equal to 0.

$$x - 5 = 0 \implies x = 5$$ Set 2nd factor equal to 0.

The solutions are $x = -2$ and $x = 5$. Check these in the original equation.

✓ **CHECKPOINT** *Now try Exercise 61.*

_____ **Concept Check** _____

1. What is a rational equation?

2. Describe how to solve a rational equation.

3. Explain the domain restrictions that may exist for a rational equation.

4. When can you use cross-multiplication to solve a rational equation? Explain.

6.6 EXERCISES

Go to pages 440–441 to record your assignments.

_____ **Developing Skills** _____

In Exercises 1–4, determine whether each value of x is a solution of the equation.

Equation   *Values*

1. $\dfrac{x}{3} - \dfrac{x}{5} = \dfrac{4}{3}$ (a) $x = 0$ (b) $x = -2$

                                   (c) $x = \frac{1}{8}$ (d) $x = 10$

2. $\dfrac{x}{4} + \dfrac{3}{4x} = 1$ (a) $x = -1$ (b) $x = 1$

                                   (c) $x = 3$ (d) $x = \frac{1}{2}$

3. $x = 4 + \dfrac{21}{x}$ (a) $x = 0$ (b) $x = -3$

                                   (c) $x = 7$ (d) $x = -1$

4. $5 - \dfrac{1}{x - 3} = 2$ (a) $x = \frac{10}{3}$ (b) $x = -\frac{1}{3}$

                                   (c) $x = 0$ (d) $x = 1$

In Exercises 5–22, solve the equation. *See Examples 1–3.*

5. $\dfrac{x}{6} - 1 = \dfrac{2}{3}$

6. $\dfrac{y}{8} + 7 = -\dfrac{1}{2}$

7. $\dfrac{1}{4} = \dfrac{z + 1}{8}$

8. $\dfrac{a}{2} = \dfrac{a + 2}{3}$

9. $\dfrac{x}{4} + \dfrac{x}{2} = \dfrac{2x}{3}$

10. $\dfrac{x}{4} - \dfrac{x}{6} = \dfrac{1}{4}$

11. $\dfrac{z + 2}{3} = 4 - \dfrac{z}{12}$

12. $\dfrac{2y - 9}{6} = 3y - \dfrac{3}{4}$

13. $\dfrac{x - 5}{5} + 3 = -\dfrac{x}{4}$

14. $\dfrac{4x - 2}{7} - \dfrac{5}{14} = 2x$

15. $\dfrac{t}{2} = 12 - \dfrac{3t^2}{2}$

16. $\dfrac{x^2}{2} - \dfrac{3x}{5} = -\dfrac{1}{10}$

17. $\dfrac{5y - 1}{12} + \dfrac{y}{3} = -\dfrac{1}{4}$

18. $\dfrac{z - 4}{9} - \dfrac{3z + 1}{18} = \dfrac{3}{2}$

19. $\dfrac{h + 2}{5} - \dfrac{h - 1}{9} = \dfrac{2}{3}$

20. $\dfrac{u - 2}{6} + \dfrac{2u + 5}{15} = 3$

21. $\dfrac{x + 5}{4} - \dfrac{3x - 8}{3} = \dfrac{4 - x}{12}$

22. $\dfrac{2x - 7}{10} - \dfrac{3x + 1}{5} = \dfrac{6 - x}{5}$

In Exercises 23–66, solve the equation. (Check for extraneous solutions.) *See Examples 4–7.*

23. $\dfrac{9}{25 - y} = -\dfrac{1}{4}$

24. $-\dfrac{6}{u + 3} = \dfrac{2}{3}$

25. $5 - \dfrac{12}{a} = \dfrac{5}{3}$

26. $\dfrac{5}{b} - 18 = 21$

27. $\dfrac{4}{x} - \dfrac{7}{5x} = -\dfrac{1}{2}$

28. $\dfrac{5}{3} = \dfrac{6}{7x} + \dfrac{2}{x}$

29. $\dfrac{12}{y + 5} + \dfrac{1}{2} = 2$

30. $\dfrac{7}{8} - \dfrac{16}{t - 2} = \dfrac{3}{4}$

31. $\dfrac{5}{x} = \dfrac{25}{3(x + 2)}$

32. $\dfrac{10}{x + 4} = \dfrac{15}{4(x + 1)}$

33. $\dfrac{8}{3x + 5} = \dfrac{1}{x + 2}$

34. $\dfrac{500}{3x + 5} = \dfrac{50}{x - 3}$

35. $\dfrac{3}{x+2} - \dfrac{1}{x} = \dfrac{1}{5x}$

36. $\dfrac{12}{x+5} + \dfrac{5}{x} = \dfrac{20}{x}$

37. $\dfrac{1}{2} = \dfrac{18}{x^2}$

38. $\dfrac{1}{4} = \dfrac{16}{z^2}$

39. $\dfrac{t}{4} = \dfrac{4}{t}$

40. $\dfrac{20}{u} = \dfrac{u}{5}$

41. $x + 1 = \dfrac{72}{x}$

42. $\dfrac{48}{x} = x - 2$

43. $y + \dfrac{18}{y} = 9$

44. $x - \dfrac{24}{x} = 5$

✓ 45. $\dfrac{4}{x(x-1)} + \dfrac{3}{x} = \dfrac{4}{x-1}$

46. $\dfrac{x-2}{2} - \dfrac{15}{2x} = 0$

✓ 47. $\dfrac{2x}{5} = \dfrac{x^2 - 5x}{5x}$

48. $\dfrac{3x}{4} = \dfrac{x^2 + 3x}{8x}$

49. $\dfrac{y+1}{y+10} = \dfrac{y-2}{y+4}$

50. $\dfrac{x-3}{x+1} = \dfrac{x-6}{x+5}$

51. $\dfrac{15}{x} + \dfrac{9x-7}{x+2} = 9$

52. $\dfrac{3z-2}{z+1} = 4 - \dfrac{z+2}{z-1}$

53. $\dfrac{2}{6q+5} - \dfrac{3}{4(6q+5)} = \dfrac{1}{28}$

54. $\dfrac{10}{x(x-2)} + \dfrac{4}{x} = \dfrac{5}{x-2}$

55. $\dfrac{4}{2x+3} + \dfrac{17}{5x-3} = 3$

56. $\dfrac{5}{3x+1} + \dfrac{3}{2x+2} = 2$

57. $\dfrac{2}{x-10} - \dfrac{3}{x-2} = \dfrac{6}{x^2 - 12x + 20}$

58. $\dfrac{5}{x+2} + \dfrac{2}{x^2 - 6x - 16} = -\dfrac{4}{x-8}$

59. $\dfrac{x+3}{x^2 - 9} + \dfrac{4}{3-x} - 2 = 0$

60. $1 - \dfrac{6}{4-x} = \dfrac{x+2}{x^2 - 16}$

✓ 61. $\dfrac{x}{x-2} + \dfrac{3x}{x-4} = -\dfrac{2(x-6)}{x^2 - 6x + 8}$

62. $\dfrac{2(x+1)}{x^2 - 4x + 3} + \dfrac{6x}{x-3} = \dfrac{3x}{x-1}$

63. $\dfrac{5}{x^2 + 4x + 3} + \dfrac{2}{x^2 + x - 6} = \dfrac{3}{x^2 - x - 2}$

64. $\dfrac{2}{x^2 + 2x - 8} - \dfrac{1}{x^2 + 9x + 20} = \dfrac{4}{x^2 + 3x - 10}$

65. $\dfrac{x}{3} = \dfrac{1 + \dfrac{4}{x}}{1 + \dfrac{2}{x}}$

66. $\dfrac{2x}{3} = \dfrac{1 + \dfrac{2}{x}}{1 + \dfrac{1}{x}}$

In Exercises 67–70, (a) use the graph to determine any *x*-intercepts of the graph and (b) set $y = 0$ and solve the resulting rational equation to confirm the result of part (a).

67. $y = \dfrac{x+2}{x-2}$

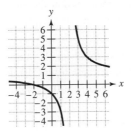

68. $y = \dfrac{2x}{x+4}$

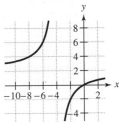

69. $y = x - \dfrac{1}{x}$

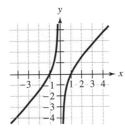

70. $y = x - \dfrac{2}{x} - 1$

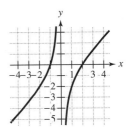

In Exercises 71–76, (a) use a graphing calculator to graph the equation and determine any *x*-intercepts of the graph and (b) set $y = 0$ and solve the resulting rational equation to confirm the result of part (a).

71. $y = \dfrac{x-4}{x+5}$

72. $y = \dfrac{1}{x} - \dfrac{3}{x+4}$

73. $y = x + 3 + \dfrac{7}{x}$

74. $y = 20\left(\dfrac{2}{x} - \dfrac{3}{x-1}\right)$

75. $y = (x+1) - \dfrac{6}{x}$

76. $y = \dfrac{x^2 + 9}{x}$

Think About It   In Exercises 77–80, if the exercise is an equation, solve it; if it is an expression, simplify it.

77. $\dfrac{16}{x^2 - 16} + \dfrac{x}{2x - 8} = \dfrac{1}{2}$

78. $\dfrac{5}{x + 3} + \dfrac{5}{3} + 3$

79. $\dfrac{16}{x^2 - 16} + \dfrac{x}{2x - 8} + \dfrac{1}{2}$

80. $\dfrac{5}{x + 3} + \dfrac{5}{3} = 3$

Solving Problems

81. *Number Problem*   Find a number such that the sum of the number and its reciprocal is $\frac{37}{6}$.

82. *Number Problem*   Find a number such that the sum of two times the number and three times its reciprocal is $\frac{203}{10}$.

83. *Painting*   A painter can paint a fence in 4 hours, while his partner can paint the fence in 6 hours. How long would it take to paint the fence if both worked together?

84. *Roofing*   A roofer requires 15 hours to shingle a roof, while an apprentice requires 21 hours. How long would it take to shingle the roof if both worked together?

85. *Wind Speed*   A plane has a speed of 300 miles per hour in still air. The plane travels a distance of 680 miles with a tail wind in the same time it takes to travel 520 miles into a head wind. Find the speed of the wind.

86. *Speed*   One person runs 2 miles per hour faster than a second person. The first person runs 5 miles in the same time the second person runs 4 miles. Find the speed of each person.

87. *Saves*   A hockey goalie has faced 799 shots and saved 707 of them. How many additional consecutive saves does the goalie need to obtain a save percent (in decimal form) of .900?

Nick Didlick/Getty Images

88. *Batting Average*   A softball player has been up to bat 47 times and has hit the ball safely 8 times. How many additional consecutive times must the player hit the ball safely to obtain a batting average of .250?

Explaining Concepts

89. ✎   Define the term *extraneous solution*. How do you identify an extraneous solution?

90. ✎   Explain how you can use a graphing calculator to estimate the solution of a rational equation.

91. ✎   Explain why the equation $\dfrac{n}{x} + n = \dfrac{n}{x}$ has no solution if n is any real nonzero number.

92. Does multiplying a rational equation by its LCD produce an equivalent equation? Explain.

Cumulative Review

In Exercises 93–96, factor the expression.

93. $x^2 - 81$

94. $x^2 - 121$

95. $4x^2 - \frac{1}{4}$

96. $49 - (x - 2)^2$

In Exercises 97 and 98, find the domain of the rational function.

97. $f(x) = \dfrac{2x^2}{5}$

98. $f(x) = \dfrac{4}{x - 6}$

6.7 Applications and Variation

NASA

Why You Should Learn It

You can use mathematical models in a wide variety of applications involving variation. For instance, in Exercise 64 on page 438, you will use direct variation to model the weight of a person on the moon.

1 ▶ Solve application problems involving rational equations.

What You Should Learn

1 ▶ Solve application problems involving rational equations.
2 ▶ Solve application problems involving direct variation.
3 ▶ Solve application problems involving inverse variation.
4 ▶ Solve application problems involving joint variation.

Rational Equation Applications

The first three examples in this section are types of application problems that you have seen earlier in the text. The difference now is that the variable appears in the denominator of a rational expression.

> **EXAMPLE 1**   Average Speeds
>
> You and your friend travel to separate colleges in the same amount of time. You drive 380 miles and your friend drives 400 miles. Your friend's average speed is 3 miles per hour faster than your average speed. What is your average speed and what is your friend's average speed?
>
> **Solution**
>
> Begin by setting your time equal to your friend's time. Then use an alternative version of the formula for distance that gives the time in terms of the distance and the rate.
>
> *Verbal Model:*
>
> | Your time | = | Your friend's time |
>
> $$\frac{\text{Your distance}}{\text{Your rate}} = \frac{\text{Friend's distance}}{\text{Friend's rate}}$$
>
> *Labels:*   Your distance = 380   (miles)
> Your rate = r (miles per hour)
> Friend's distance = 400   (miles)
> Friend's rate = $r + 3$ (miles per hour)
>
> *Equation:* $\dfrac{380}{r} = \dfrac{400}{r + 3}$ Original equation.
>
> $380(r + 3) = 400(r), \quad r \neq 0, r \neq -3$ Cross-multiply.
>
> $380r + 1140 = 400r$ Distributive Property
>
> $1140 = 20r \quad \Longrightarrow \quad 57 = r$ Simplify.
>
> Your average speed is 57 miles per hour and your friend's average speed is $57 + 3 = 60$ miles per hour. Check this in the original statement of the problem.
>
> ✓ **CHECKPOINT** *Now try Exercise 43.*

Study Tip

When determining the domain of a real-life problem, you must also consider the context of the problem. For instance, in Example 2, the time it takes to fill the tub with water could not be a negative number. The problem implies that the domain must be all real numbers greater than zero.

EXAMPLE 2 **A Work-Rate Problem**

With the cold water valve open, it takes 8 minutes to fill a washing machine tub. With both the hot and cold water valves open, it takes 5 minutes to fill the tub. How long will it take to fill the tub with only the hot water valve open?

Solution

Verbal Model:

$$\boxed{\text{Rate for cold water}} + \boxed{\text{Rate for hot water}} = \boxed{\text{Rate for warm water}}$$

Labels: Rate for cold water $= \dfrac{1}{8}$ (tub per minute)

Rate for hot water $= \dfrac{1}{t}$ (tub per minute)

Rate for warm water $= \dfrac{1}{5}$ (tub per minute)

Equation: $\dfrac{1}{8} + \dfrac{1}{t} = \dfrac{1}{5}$ Original equation

$5t + 40 = 8t$ Multiply each side by LCD of $40t$ and simplify.

$40 = 3t$ $\dfrac{40}{3} = t$ Simplify.

So, it takes $13\frac{1}{3}$ minutes to fill the tub with hot water. Check this solution.

✓ **CHECKPOINT** *Now try Exercise 47.*

EXAMPLE 3   **Cost-Benefit Model**

A utility company burns coal to generate electricity. The cost C (in dollars) of removing $p\%$ of the pollutants from smokestack emissions is modeled by

$$C = \frac{80{,}000p}{100 - p}, \quad 0 \le p < 100.$$

What percent of air pollutants in the stack emissions can be removed for $420,000?

Solution

To determine the percent of air pollutants in the stack emissions that can be removed for $420,000, substitute 420,000 for C in the model.

$$420{,}000 = \frac{80{,}000p}{100 - p}$$ Substitute 420,000 for C.

$$420{,}000(100 - p) = 80{,}000p$$ Cross-multiply.

$$42{,}000{,}000 - 420{,}000p = 80{,}000p$$ Distributive Property

$$42{,}000{,}000 = 500{,}000p$$ Add $420{,}000p$ to each side.

$$84 = p$$ Divide each side by 500,000.

So, 84% of air pollutants in the stack emissions can be removed for $420,000.

✓ **CHECKPOINT** *Now try Exercise 49.*

2 ▶ Solve application problems involving direct variation.

Direct Variation

In the mathematical model for **direct variation,** y is a *linear* function of x. Specifically,

$$y = kx.$$

To use this mathematical model in applications involving direct variation, you need to use given values of x and y to find the value of the constant k.

Direct Variation

The following statements are equivalent.

1. y **varies directly** as x.
2. y is **directly proportional** to x.
3. $y = kx$ for some constant k.

The number k is called the **constant of proportionality.**

EXAMPLE 4   **Direct Variation**

The total revenue R (in dollars) obtained from selling x ice show tickets is directly proportional to the number of tickets sold x. When 10,000 tickets are sold, the total revenue is $142,500.

a. Find a mathematical model that relates the total revenue R to the number of tickets sold x.

b. Find the total revenue obtained from selling 12,000 tickets.

Solution

a. Because the total revenue is directly proportional to the number of tickets sold, the linear model is $R = kx$. To find the value of the constant k, use the fact that $R = 142,500$ when $x = 10,000$. Substituting these values into the model produces

$$142,500 = k(10,000) \qquad \text{Substitute for } R \text{ and } x.$$

which implies that

$$k = \frac{142,500}{10,000} = 14.25.$$

So, the equation relating the total revenue to the total number of tickets sold is

$$R = 14.25x. \qquad \text{Direct variation model}$$

The graph of this equation is shown in Figure 6.2.

b. When $x = 12,000$, the total revenue is

$$R = 14.25(12,000) = \$171,000.$$

✓ **CHECKPOINT** *Now try Exercise 53.*

Figure with $R = 14.25x$; Revenue (in dollars) on vertical axis with marks 50,000; 100,000; 150,000; 200,000. Tickets sold on horizontal axis with marks 5000; 10,000; 15,000.

Figure 6.2

EXAMPLE 5   Direct Variation

Hooke's Law for springs states that the distance a spring is stretched (or compressed) is directly proportional to the force on the spring. A force of 20 pounds stretches a spring 5 inches.

a. Find a mathematical model that relates the distance the spring is stretched to the force applied to the spring.

b. How far will a force of 30 pounds stretch the spring?

Solution

a. For this problem, let d represent the distance (in inches) that the spring is stretched and let F represent the force (in pounds) that is applied to the spring. Because the distance d is directly proportional to the force F, the model is

$$d = kF.$$

To find the value of the constant k, use the fact that $d = 5$ when $F = 20$. Substituting these values into the model produces

$$5 = k(20)$$ Substitute 5 for d and 20 for F.

$$\frac{5}{20} = k$$ Divide each side by 20.

$$\frac{1}{4} = k.$$ Simplify.

So, the equation relating distance and force is

$$d = \frac{1}{4}F.$$ Direct variation model

b. When $F = 30$, the distance is

$$d = \frac{1}{4}(30) = 7.5 \text{ inches.}$$ See Figure 6.3.

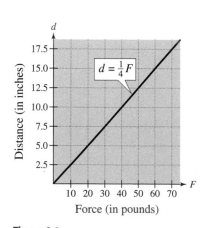

Figure 6.3

✔ CHECKPOINT   Now try Exercise 55.

In Example 5, you can get a clearer understanding of Hooke's Law by using the model $d = \frac{1}{4}F$ to create a table or a graph (see Figure 6.4). From the table or from the graph, you can see what it means for the distance to be "proportional to the force."

Force, F	10 lb	20 lb	30 lb	40 lb	50 lb	60 lb
Distance, d	2.5 in.	5.0 in.	7.5 in.	10.0 in.	12.5 in.	15.0 in.

In Examples 4 and 5, the direct variations are such that an *increase* in one variable corresponds to an *increase* in the other variable. There are, however, other applications of direct variation in which an increase in one variable corresponds to a *decrease* in the other variable. For instance, in the model $y = -2x$, an increase in x will yield a decrease in y.

Figure 6.4

Another type of direct variation relates one variable to a power of another.

> ## Direct Variation as *n*th Power
>
> The following statements are equivalent.
> 1. *y* **varies directly as the *n*th power** of *x*.
> 2. *y* is **directly proportional to the *n*th power** of *x*.
> 3. $y = kx^n$ for some constant *k*.

EXAMPLE 6   **Direct Variation as a Power**

The distance a ball rolls down an inclined plane is directly proportional to the square of the time it rolls. During the first second, a ball rolls down a plane a distance of 6 feet.

a. Find a mathematical model that relates the distance traveled to the time.

b. How far will the ball roll during the first 2 seconds?

Solution

a. Letting *d* be the distance (in feet) that the ball rolls and letting *t* be the time (in seconds), you obtain the model

$$d = kt^2.$$

Because $d = 6$ when $t = 1$, you obtain

$d = kt^2$	Write original equation.
$6 = k(1)^2 \implies 6 = k.$	Substitute 6 for *d* and 1 for *t*.

So, the equation relating distance to time is

$d = 6t^2.$	Direct variation as 2nd power model

The graph of this equation is shown in Figure 6.5.

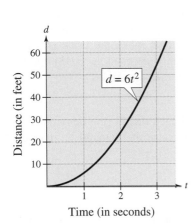

Figure 6.5

b. When $t = 2$, the distance traveled is

$d = 6(2)^2 = 6(4) = 24$ feet.	See Figure 6.6.

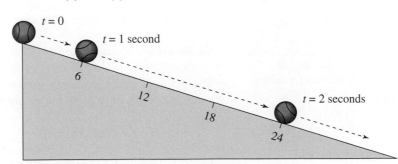

Figure 6.6

✔ **CHECKPOINT** *Now try Exercise 61.*

3 ▶ Solve application problems involving inverse variation.

Inverse Variation

A second type of variation is called **inverse variation.** With this type of variation, one of the variables is said to be inversely proportional to the other variable.

> ### Inverse Variation
>
> **1.** The following three statements are equivalent.
>
>   **a.** y **varies inversely** as x.
>
>   **b.** y is **inversely proportional** to x.
>
>   **c.** $y = \dfrac{k}{x}$ for some constant k.
>
> **2.** If $y = \dfrac{k}{x^n}$, then y is inversely proportional to the nth power of x.

EXAMPLE 7   Inverse Variation

The marketing department of a large company has found that the demand for one of its hand tools varies inversely as the price of the product. (When the price is low, more people are willing to buy the product than when the price is high.) When the price of the tool is $7.50, the monthly demand is 50,000 tools. Approximate the monthly demand if the price is reduced to $6.00.

Solution

Let x represent the number of tools that are sold each month (the demand), and let p represent the price per tool (in dollars). Because the demand is inversely proportional to the price, the model is

$$x = \frac{k}{p}.$$

By substituting $x = 50,000$ when $p = 7.50$, you obtain

$$50,000 = \frac{k}{7.50} \qquad \text{Substitute 50,000 for } x \text{ and 7.50 for } p.$$

$$375,000 = k. \qquad \text{Multiply each side by 7.50.}$$

So, the inverse variation model is $x = \dfrac{375,000}{p}$.

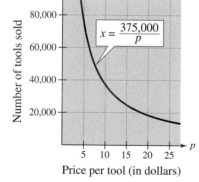

Figure 6.7

The graph of this equation is shown in Figure 6.7. To find the demand that corresponds to a price of $6.00, substitute 6 for p in the equation and obtain

$$x = \frac{375,000}{6} = 62,500 \text{ tools.}$$

So, if the price is lowered from $7.50 per tool to $6.00 per tool, you can expect the monthly demand to increase from 50,000 tools to 62,500 tools.

✓ **CHECKPOINT** *Now try Exercise 65.*

Some applications of variation involve problems with *both* direct and inverse variation in the same model. These types of models are said to have **combined variation.**

EXAMPLE 8   **Direct and Inverse Variation**

A computer hardware manufacturer determines that the demand for its USB flash drive is directly proportional to the amount spent on advertising and inversely proportional to the price of the flash drive. When $40,000 is spent on advertising and the price per unit is $20, the monthly demand is 10,000 flash drives.

a. If the amount of advertising were increased to $50,000, how much could the price be increased to maintain a monthly demand of 10,000 flash drives?

b. If you were in charge of the advertising department, would you recommend this increased expense in advertising?

Solution

a. Let x represent the number of flash drives that are sold each month (the demand), let a represent the amount spent on advertising (in dollars), and let p represent the price per unit (in dollars). Because the demand is directly proportional to the advertising expense and inversely proportional to the price, the model is

$$x = \frac{ka}{p}.$$

By substituting 10,000 for x when $a = 40,000$ and $p = 20$, you obtain

$$10,000 = \frac{k(40,000)}{20} \qquad \text{Substitute 10,000 for } x \text{, 40,000 for } a \text{, and 20 for } p.$$

$$200,000 = 40,000k \qquad \text{Multiply each side by 20.}$$

$$5 = k. \qquad \text{Divide each side by 40,000.}$$

So, the model is

$$x = \frac{5a}{p}. \qquad \text{Direct and inverse variation model}$$

To find the price that corresponds to a demand of 10,000 and an advertising expense of $50,000, substitute 10,000 for x and 50,000 for a into the model and solve for p.

$$10,000 = \frac{5(50,000)}{p} \qquad p = \frac{5(50,000)}{10,000} = \$25$$

So, the price increase would be $25 - $20 = $5.

b. The total revenue for selling 10,000 units at $20 each is $200,000, and the revenue for selling 10,000 units at $25 each is $250,000. So, increasing the advertising expense from $40,000 to $50,000 would increase the revenue by $50,000. This implies that you should recommend the increased expense in advertising.

 CHECKPOINT *Now try Exercise 69.*

4 ▶ Solve application problems involving joint variation.

Joint Variation

The model used in Example 8 involved both direct and inverse variation, and the word "and" was used to couple the two types of variation together. To describe two different *direct* variations in the same statement, the word "jointly" is used. For instance, the model $z = kxy$ can be described by saying that z is *jointly* proportional to x and y. So, in **joint variation,** one variable varies directly with the product of two variables.

Joint Variation

1. The following three statements are equivalent.

    a. z **varies jointly** as x and y.

    b. z is **jointly proportional** to x and y.

    c. $z = kxy$ for some constant k.

2. If $z = kx^n y^m$, then z is jointly proportional to the nth power of x and the mth power of y.

EXAMPLE 9 **Joint Variation**

The *simple interest* earned by a savings account is jointly proportional to the time and the principal. After one quarter (3 months), the interest for a principal of $6000 is $120. How much interest would a principal of $7500 earn in 5 months?

Solution

To begin, let I represent the interest earned (in dollars), let P represent the principal (in dollars), and let t represent the time (in years). Because the interest is jointly proportional to the time and the principal, the model is

$$I = ktP.$$

Because $I = 120$ when $P = 6000$ and $t = \frac{1}{4}$, you have

$$120 = k\left(\frac{1}{4}\right)(6000) \qquad \text{Substitute 120 for } I, \tfrac{1}{4} \text{ for } t, \text{ and 6000 for } P.$$

$$120 = 1500\,k \qquad\qquad \text{Simplify.}$$

$$0.08 = k. \qquad\qquad\quad \text{Divide each side by 1500.}$$

So, the model that relates interest to time and principal is

$$I = 0.08tP. \qquad\qquad \text{Joint variation model}$$

To find the interest earned on a principal of $7500 over a five-month period of time, substitute $P = 7500$ and $t = \frac{5}{12}$ into the model to obtain an interest of

$$I = 0.08\left(\frac{5}{12}\right)(7500) = \$250.$$

✓ **CHECKPOINT** *Now try Exercise 71.*

_____ **Concept Check** _____

1. In a problem, y varies directly as x and the constant of proportionality is positive. If one of the variables increases, how does the other change? Explain.

2. In a problem, y varies inversely as x and the constant of proportionality is positive. If one of the variables increases, how does the other change? Explain.

3. Are the following statements equivalent? Explain.
   (a) y varies directly as x.
   (b) y is directly proportional to the square of x.

4. Describe the difference between *combined variation* and *joint variation*.

6.7 EXERCISES   Go to pages 440–441 to record your assignments.

_____ **Developing Skills** _____

In Exercises 1–14, write a model for the statement.

1. I varies directly as V.
2. C varies directly as r.
3. V is directly proportional to t.
4. A is directly proportional to w.
5. u is directly proportional to the square of v.
6. s varies directly as the cube of t.
7. p varies inversely as d.
8. S varies inversely as the square of v.
9. A is inversely proportional to the fourth power of t.
10. P is inversely proportional to the square root of $1 + r$.
11. A varies jointly as l and w.
12. V varies jointly as h and the square of r.
13. *Boyle's Law* If the temperature of a gas is not allowed to change, its absolute pressure P is inversely proportional to its volume V.
14. *Newton's Law of Universal Gravitation* The gravitational attraction F between two particles of masses m_1 and m_2 is directly proportional to the product of the masses and inversely proportional to the square of the distance r between the particles.

In Exercises 15–20, write a verbal sentence using variation terminology to describe the formula.

15. *Area of a Triangle:* $A = \frac{1}{2}bh$

16. *Area of a Rectangle:* $A = lw$

17. *Volume of a Right Circular Cylinder:* $V = \pi r^2 h$

18. *Volume of a Sphere:* $V = \frac{4}{3}\pi r^3$

19. *Average Speed:* $r = \dfrac{d}{t}$

20. *Height of a Cylinder:* $h = \dfrac{V}{\pi r^2}$

In Exercises 21–32, find the constant of proportionality and write an equation that relates the variables.

21. s varies directly as t, and $s = 20$ when $t = 4$.

22. h is directly proportional to r, and $h = 28$ when $r = 12$.

23. F is directly proportional to the square of x, and $F = 500$ when $x = 40$.

24. M varies directly as the cube of n, and $M = 0.012$ when $n = 0.2$.

25. n varies inversely as m, and $n = 32$ when $m = 1.5$.

26. q is inversely proportional to p, and $q = \frac{3}{2}$ when $p = 50$.

27. g varies inversely as the square root of z, and $g = \frac{4}{5}$ when $z = 25$.

28. u varies inversely as the square of v, and $u = 40$ when $v = \frac{1}{2}$.

29. F varies jointly as x and y, and $F = 500$ when $x = 15$ and $y = 8$.

30. V varies jointly as h and the square of b, and $V = 288$ when $h = 6$ and $b = 12$.

31. d varies directly as the square of x and inversely with r, and $d = 3000$ when $x = 10$ and $r = 4$.

32. z is directly proportional to x and inversely proportional to the square root of y, and $z = 720$ when $x = 48$ and $y = 81$.

In Exercises 33–36, complete the table and plot the resulting points.

x	2	4	6	8	10
$y = kx^2$					

33. $k = 1$
34. $k = 2$
35. $k = \frac{1}{2}$
36. $k = \frac{1}{4}$

In Exercises 37–40, complete the table and plot the resulting points.

x	2	4	6	8	10
$y = \dfrac{k}{x^2}$					

37. $k = 2$
38. $k = 5$
39. $k = 10$
40. $k = 20$

In Exercises 41 and 42, determine whether the variation model is of the form $y = kx$ or $y = k/x$, and find k.

41.

x	10	20	30	40	50
y	$\frac{2}{5}$	$\frac{1}{5}$	$\frac{2}{15}$	$\frac{1}{10}$	$\frac{2}{25}$

42.

x	10	20	30	40	50
y	-3	-6	-9	-12	-15

Solving Problems

✓ 43. *Average Speeds*   You and a friend jog for the same amount of time. You jog 10 miles and your friend jogs 12 miles. Your friend's average speed is 1.5 miles per hour faster than yours. What are the average speeds of you and your friend?

44. *Current Speed*   A boat travels at a speed of 20 miles per hour in still water. It travels 48 miles upstream and then returns to the starting point in a total of 5 hours. Find the speed of the current.

45. *Partnership Costs*   A group plans to start a new business that will require $240,000 for start-up capital. The individuals in the group share the cost equally. If two additional people join the group, the cost per person will decrease by $4000. How many people are presently in the group?

46. *Partnership Costs*   A group of people share equally the cost of a $180,000 endowment. If they could find four more people to join the group, each person's share of the cost would decrease by $3750. How many people are presently in the group?

✓ 47. *Work Rate*   It takes a lawn care company 60 minutes to complete a job using only a riding mower, or 45 minutes using the riding mower and a push mower. How long does the job take using only the push mower?

© James Kirkus Photography

48. *Flow Rate*   It takes 3 hours to fill a pool using two pipes. It takes 5 hours to fill the pool using only the larger pipe. How long does it take to fill the pool using only the smaller pipe?

49. *Pollution Removal* The cost C in dollars of removing $p\%$ of the air pollutants in the stack emissions of a utility company is modeled by the equation below. Determine the percent of air pollutants in the stack emissions that can be removed for $680,000.

$$C = \frac{120,000p}{100 - p}$$

50. *Population Growth* A biologist starts a culture with 100 bacteria. The population P of the culture is approximated by the model below, where t is the time in hours. Find the time required for the population to increase to 800 bacteria.

$$P = \frac{500(1 + 3t)}{5 + t}$$

51. *Nail Sizes* The unit for determining the size of a nail is the *penny*. For example, 8d represents an 8-penny nail. The number N of finishing nails per pound can be modeled by

$$N = -139.1 + \frac{2921}{x}$$

where x is the size of the nail.

(a) What is the domain of the function?

(b) ▦ Use a graphing calculator to graph the function.

(c) Use the graph to determine the size of the finishing nail if there are 153 nails per pound.

(d) Verify the result of part (c) algebraically.

52. *Learning Curve* A psychologist observes that the number of lines N of a poem that a four-year-old child can memorize depends on the number x of short sessions spent on the task, according to the model

$$N = \frac{20x}{x + 1}.$$

(a) What is the domain of the function?

(b) ▦ Use a graphing calculator to graph the function.

(c) Use the graph to determine the number of sessions needed for a child to memorize 15 lines of the poem.

(d) Verify the result of part (c) algebraically.

53. *Revenue* The total revenue R is directly proportional to the number of units sold x. When 500 units are sold, the revenue is $4825. Find the revenue when 620 units are sold. Then interpret the constant of proportionality.

54. *Revenue* The total revenue R is directly proportional to the number of units sold x. When 25 units are sold, the revenue is $300. Find the revenue when 42 units are sold. Then interpret the constant of proportionality.

55. *Hooke's Law* A force of 50 pounds stretches a spring 5 inches.

(a) How far will a force of 20 pounds stretch the spring?

(b) What force is required to stretch the spring 1.5 inches?

56. *Hooke's Law* A force of 50 pounds stretches a spring 3 inches.

(a) How far will a force of 20 pounds stretch the spring?

(b) What force is required to stretch the spring 1.5 inches?

57. *Hooke's Law* A baby weighing $10\frac{1}{2}$ pounds compresses the spring of a baby scale 7 millimeters. Determine the weight of a baby that compresses the spring 12 millimeters.

58. *Hooke's Law* An apple weighing 14 ounces compresses the spring of a produce scale 3 millimeters. Determine the weight of a grapefruit that compresses the spring 5 millimeters.

59. *Free-Falling Object* The velocity v of a free-falling object is directly proportional to the time t (in seconds) that the object has fallen. The velocity of a falling object is -64 feet per second after the object has fallen for 2 seconds. Find the velocity of the object after it has fallen for a total of 4 seconds.

60. *Free-Falling Object* Neglecting air resistance, the distance d that an object falls varies directly as the square of the time t it has fallen. An object falls 64 feet in 2 seconds. Determine the distance it will fall in 6 seconds.

61. *Stopping Distance*   The stopping distance d of an automobile is directly proportional to the square of its speed s. On one road, a car requires 75 feet to stop from a speed of 30 miles per hour. How many feet does the car require to stop from a speed of 48 miles per hour on the same road?

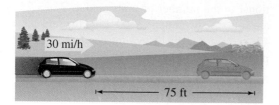

62. *Frictional Force*   The frictional force F (between the tires of a car and the road) that is required to keep a car on a curved section of a highway is directly proportional to the square of the speed s of the car. By what factor does the force F change when the speed of the car is doubled on the same curve?

63. *Power Generation*   The power P generated by a wind turbine varies directly as the cube of the wind speed w. The turbine generates 400 watts of power in a 20-mile-per-hour wind. Find the power it generates in a 30-mile-per-hour wind.

64. *Weight of an Astronaut*   A person's weight on the moon varies directly as his or her weight on Earth. An astronaut weighs 360 pounds on Earth, including heavy equipment. On the moon the astronaut weighs only 60 pounds with the equipment. If the first woman in space, Valentina Tereshkova, had landed on the moon and weighed 54 pounds with equipment, how much would she have weighed on Earth with her equipment?

65. *Demand*   A company has found that the daily demand x for its boxes of chocolates is inversely proportional to the price p. When the price is \$5, the demand is 800 boxes. Approximate the demand when the price is increased to \$6.

66. *Pressure*   When a person walks, the pressure P on each sole varies inversely as the area A of the sole. A person is trudging through deep snow, wearing boots that have a sole area of 29 square inches each. The sole pressure is 4 pounds per square inch. If the person was wearing snowshoes, each with an area 11 times that of their boot soles, what would be the pressure on each snowshoe? The constant of variation in this problem is the weight of the person. How much does the person weigh?

67. *Environment*   The graph shows the percent p of oil that remained in Chedabucto Bay, Nova Scotia, after an oil spill. The cleaning of the spill was left primarily to natural actions. After about a year, the percent that remained varied inversely as time. Find a model that relates p and t, where t is the number of years since the spill. Then use it to find the percent of oil that remained $6\frac{1}{2}$ years after the spill, and compare the result with the graph.

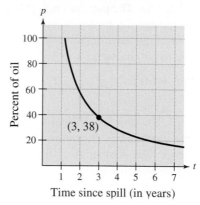

68. *Meteorology*   The graph shows the water temperature in relation to depth in the north central Pacific Ocean. At depths greater than 900 meters, the water temperature varies inversely with the water depth. Find a model that relates the temperature T to the depth d. Then use it to find the water temperature at a depth of 4385 meters, and compare the result with the graph.

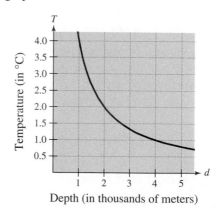

69. *Revenue*   The weekly demand for a company's frozen pizzas varies directly as the amount spent on advertising and inversely as the price per pizza. At \$5 per pizza, when \$500 is spent each week on ads, the demand is 2000 pizzas. If advertising is increased to \$600, what price will yield a demand of 2000 pizzas? Is this increase worthwhile in terms of revenue?

70. *Revenue* The monthly demand for a company's sports caps varies directly as the amount spent on advertising and inversely as the square of the price per cap. At $15 per cap, when $2500 is spent each week on ads, the demand is 300 caps. If advertising is increased to $3000, what price will yield a demand of 300 caps? Is this increase worthwhile in terms of revenue?

✓ **71.** *Simple Interest* The simple interest earned by an account varies jointly as the time and the principal. A principal of $600 earns $10 interest in 4 months. How much would $900 earn in 6 months?

72. *Simple Interest* The simple interest earned by an account varies jointly as the time and the principal. In 2 years, a principal of $5000 earns $650 interest. How much would $1000 earn in 1 year?

73. *Engineering* The load P that can be safely supported by a horizontal beam varies jointly as the product of the width W of the beam and the square of the depth D, and inversely as the length L (see figure).

(a) Write a model for the statement.

(b) How does P change when the width and length of the beam are both doubled?

(c) How does P change when the width and depth of the beam are doubled?

(d) How does P change when all three of the dimensions are doubled?

(e) How does P change when the depth of the beam is cut in half?

(f) A beam with width 3 inches, depth 8 inches, and length 120 inches can safely support 2000 pounds. Determine the safe load of a beam made from the same material if its depth is increased to 10 inches.

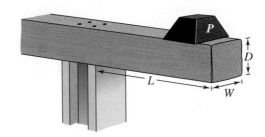

Explaining Concepts

True or False? In Exercises 74 and 75, determine whether the statement is true or false. Explain your reasoning.

74. In a situation involving combined variation, y can vary directly as x and inversely as x at the same time.

75. In a joint variation problem where z varies jointly as x and y, if x increases, then z and y must both increase.

76. ✎ If y varies directly as the square of x and x is doubled, how does y change? Use the rules of exponents to explain your answer.

77. ✎ If y varies inversely as the square of x and x is doubled, how does y change? Use the rules of exponents to explain your answer.

78. ✎ Describe a real-life problem for each type of variation (direct, inverse, and joint).

Cumulative Review

In Exercises 79–82, write the expression using exponential notation.

79. $(6)(6)(6)(6)$

80. $(-4)(-4)(-4)$

81. $\left(\frac{1}{5}\right)\left(\frac{1}{5}\right)\left(\frac{1}{5}\right)\left(\frac{1}{5}\right)\left(\frac{1}{5}\right)$

82. $-\left(-\frac{3}{4}\right)\left(-\frac{3}{4}\right)\left(-\frac{3}{4}\right)$

In Exercises 83–86, use synthetic division to divide.

83. $(x^2 - 5x - 14) \div (x + 2)$

84. $(3x^2 - 5x + 2) \div (x + 1)$

85. $\dfrac{4x^5 - 14x^4 + 6x^3}{x - 3}$

86. $\dfrac{x^5 - 3x^2 - 5x + 1}{x - 2}$

What Did You Learn?

Use these two pages to help prepare for a test on this chapter. Check off the key terms and key concepts you know. You can also use this section to record your assignments.

Plan for Test Success

Date of test: / /   **Study dates and times:** / / at : A.M./P.M.

 / / at : A.M./P.M.

Things to review:

☐ Key Terms, *p. 440*
☐ Key Concepts, *pp. 440–441*
☐ Your class notes
☐ Your assignments

☐ Study Tips, *pp. 370, 371, 372, 375, 376, 385, 391, 394, 401, 402, 403, 411, 413, 414, 419, 423, 428*
☐ Technology Tips, *pp. 374, 383, 393, 412, 421*

☐ Mid-Chapter Quiz, *p. 408*
☐ Review Exercises, *pp. 442–446*
☐ Chapter Test, *p. 447*
☐ Video Explanations Online
☐ Tutorial Online

Key Terms

☐ rational expression, *p. 370*
☐ rational function, *p. 370*
☐ domain (of a rational function), *p. 370*
☐ simplified form, *p. 373*
☐ least common multiple, *p. 392*
☐ least common denominator, *p. 393*

☐ complex fraction, *p. 400*
☐ dividend, *p. 410*
☐ divisor, *p. 410*
☐ quotient, *p. 410*
☐ remainder, *p. 410*
☐ synthetic division, *p. 413*
☐ extraneous solution, *p. 422*

☐ direct variation, *p. 429*
☐ constant of proportionality, *p. 429*
☐ inverse variation, *p. 432*
☐ combined variation, *p. 433*
☐ joint variation, *p. 434*

Key Concepts

6.1 Rational Expressions and Functions

Assignment: _____ Due date: _____

☐ **Find the domain of a rational function.**

The set of *usable* values of the variable (values that do not make the denominator zero) is called the **domain** of a rational expression.

☐ **Simplify a rational expression.**

Divide out common factors: $\dfrac{uw}{vw} = \dfrac{u\cancel{w}}{v\cancel{w}} = \dfrac{u}{v}, w \neq 0$

Domain restrictions of the original expression that are not implied by the simplified form must be listed.

6.2 Multiplying and Dividing Rational Expressions

Assignment: _____ Due date: _____

☐ **Multiply rational expressions.**

1. Multiply the numerators and the denominators.
$$\frac{u}{v} \cdot \frac{w}{z} = \frac{uw}{vz}$$
2. Factor the numerator and the denominator.
3. Simplify by dividing out the common factors.

☐ **Divide rational expressions.**

Invert the divisor and multiply using the steps for multiplying rational expressions.
$$\frac{u}{v} \div \frac{w}{z} = \frac{u}{v} \cdot \frac{z}{w} = \frac{uz}{vw}$$

6.3 Adding and Subtracting Rational Expressions

Assignment: _____ Due date:_____

☐ **Add or subtract rational expressions with like denominators.**

1. Combine the numerators: $\dfrac{u}{w} \pm \dfrac{v}{w} = \dfrac{u \pm v}{w}$

2. Simplify the resulting rational expression.

☐ **Add or subtract rational expressions with unlike denominators.**

Multiply the rational expressions by their LCD, then use the process for adding or subtracting rational expressions with like denominators.

6.4 Complex Fractions

Assignment: _____ Due date:_____

☐ **Simplify a complex fraction.**

When the numerator and denominator of the complex fraction each consist of a single fraction, use the rules for dividing rational expressions.

When a sum or difference is present in the numerator or denominator of the complex fraction, first combine the terms so that the numerator and the denominator each consist of a single fraction.

6.5 Dividing Polynomials and Synthetic Division

Assignment: _____ Due date:_____

☐ **Divide a polynomial by a monomial.**

Divide each term of the polynomial by the monomial, then simplify each fraction.

☐ **Use long division to divide a polynomial by a polynomial.**

See page 411.

☐ **Use synthetic division to divide a polynomial by a polynomial of the form $x - k$.**

See pages 413 and 414.

☐ **Use synthetic division to factor a polynomial.**

If the remainder in a synthetic division problem is zero, then the binomial divisor is a factor and the quotient is another factor.

6.6 Solving Rational Equations

Assignment: _____ Due date:_____

☐ **Solve a rational equation containing constant denominators.**

1. Multiply each side of the equation by the LCD of all the fractions in the equation.

2. Solve the resulting equation.

☐ **Solve a rational equation containing variable denominators.**

1. Determine the domain restrictions of the equation.

2. Multiply each side by the LCD of all the fractions in the equation.

3. Solve the resulting equation.

☐ **Solve a rational equation by cross-multiplying.**

This method can be used when each side of the equation consists of a single fraction.

6.7 Applications and Variation

Assignment: _____ Due date:_____

☐ **Solve application problems involving rational equations.**

☐ **Solve application problems involving direct variation.**

Direct variation: $y = kx$

Direct variation as nth power: $y = kx^n$

☐ **Solve application problems involving inverse variation.**

Inverse variation: $y = k/x$

Inverse variation as nth power: $y = k/x^n$

☐ **Solve application problems involving joint variation.**

Joint variation: $z = kxy$

Joint variation as nth and mth powers: $z = kx^n y^m$

Review Exercises

6.1 Rational Expressions and Functions

1 ▶ Find the domain of a rational function.

In Exercises 1–6, find the domain of the rational function.

1. $f(y) = \dfrac{3y}{y - 8}$

2. $g(t) = \dfrac{t + 4}{t + 12}$

3. $f(x) = \dfrac{2x}{x^2 + 1}$

4. $g(t) = \dfrac{t + 2}{t^2 + 4}$

5. $g(u) = \dfrac{u}{u^2 - 7u + 6}$

6. $f(x) = \dfrac{x - 12}{x(x^2 - 16)}$

7. ▲ *Geometry*   A rectangle with a width of w inches has an area of 36 square inches. The perimeter P of the rectangle is given by

$$P = 2\left(w + \dfrac{36}{w}\right).$$

Describe the domain of the function.

8. *Average Cost*   The average cost $\overline{C}$ for a manufacturer to produce x units of a product is given by

$$\overline{C} = \dfrac{15{,}000 + 0.75x}{x}.$$

Describe the domain of the function.

2 ▶ Simplify rational expressions.

In Exercises 9–16, simplify the rational expression.

9. $\dfrac{6x^4y^2}{15xy^2}$

10. $\dfrac{2(y^3z)^2}{28(yz^2)^2}$

11. $\dfrac{5b - 15}{30b - 120}$

12. $\dfrac{4a}{10a^2 + 26a}$

13. $\dfrac{9x - 9y}{y - x}$

14. $\dfrac{x + 3}{x^2 - x - 12}$

15. $\dfrac{x^2 - 5x}{2x^2 - 50}$

16. $\dfrac{x^2 + 3x + 9}{x^3 - 27}$

In Exercises 17 and 18, find the ratio of the area of the shaded region to the area of the whole figure.

17.

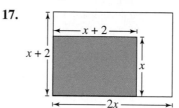

18.

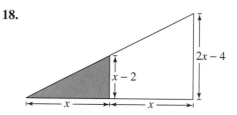

6.2 Multiplying and Dividing Rational Expressions

1 ▶ Multiply rational expressions and simplify.

In Exercises 19–26, multiply and simplify.

19. $\dfrac{4}{x} \cdot \dfrac{x^2}{12}$

20. $\dfrac{3}{y^3} \cdot 5y^3$

21. $\dfrac{7}{8} \cdot \dfrac{2x}{y} \cdot \dfrac{y^2}{14x^2}$

22. $\dfrac{15(x^2y)^3}{3y^3} \cdot \dfrac{12y}{x}$

23. $\dfrac{60z}{z + 6} \cdot \dfrac{z^2 - 36}{5}$

24. $\dfrac{x^2 - 16}{6} \cdot \dfrac{3}{x^2 - 8x + 16}$

25. $\dfrac{u}{u - 3} \cdot \dfrac{3u - u^2}{4u^2}$

26. $x^2 \cdot \dfrac{x + 1}{x^2 - x} \cdot \dfrac{(5x - 5)^2}{x^2 + 6x + 5}$

2 ▶ Divide rational expressions and simplify.

In Exercises 27–34, divide and simplify.

27. $24x^4 \div \dfrac{6x}{5}$

28. $\dfrac{8u^2}{3} \div \dfrac{u}{9}$

29. $25y^2 \div \dfrac{xy}{5}$

30. $\dfrac{6}{z^2} \div 4z^2$

31. $\dfrac{x^2 + 3x + 2}{3x^2 + x - 2} \div (x + 2)$

32. $\dfrac{x^2 - 14x + 48}{x^2 - 6x} \div (3x - 24)$

33. $\dfrac{x^2 - 7x}{x + 1} \div \dfrac{x^2 - 14x + 49}{x^2 - 1}$

34. $\dfrac{x^2 - x}{x + 1} \div \dfrac{5x - 5}{x^2 + 6x + 5}$

6.3 Adding and Subtracting Rational Expressions

1 ▶ Add or subtract rational expressions with like denominators, and simplify.

In Exercises 35–44, combine and simplify.

35. $\dfrac{4x}{5} + \dfrac{11x}{5}$

36. $\dfrac{7y}{12} - \dfrac{4y}{12}$

37. $\dfrac{15}{3x} - \dfrac{3}{3x}$

38. $\dfrac{4}{5x} + \dfrac{1}{5x}$

39. $\dfrac{8 - x}{4x} + \dfrac{5}{4x}$

40. $\dfrac{3}{5x} - \dfrac{x - 1}{5x}$

41. $\dfrac{2(3y + 4)}{2y + 1} + \dfrac{3 - y}{2y + 1}$

42. $\dfrac{4x - 2}{3x + 1} - \dfrac{x + 1}{3x + 1}$

43. $\dfrac{4x}{x + 2} + \dfrac{3x - 7}{x + 2} - \dfrac{9}{x + 2}$

44. $\dfrac{3}{2y - 3} - \dfrac{y - 10}{2y - 3} + \dfrac{5y}{2y - 3}$

2 ▶ Add or subtract rational expressions with unlike denominators, and simplify.

In Exercises 45–54, combine and simplify.

45. $\dfrac{3}{5x^2} + \dfrac{4}{10x}$

46. $\dfrac{3}{z} - \dfrac{5}{2z^2}$

47. $\dfrac{1}{x + 5} + \dfrac{3}{x - 12}$

48. $\dfrac{2}{x - 10} + \dfrac{3}{4 - x}$

49. $5x + \dfrac{2}{x - 3} - \dfrac{3}{x + 2}$

50. $4 - \dfrac{4x}{x + 6} + \dfrac{7}{x - 5}$

51. $\dfrac{6}{x - 5} - \dfrac{4x + 7}{x^2 - x - 20}$

52. $\dfrac{5}{x + 2} + \dfrac{25 - x}{x^2 - 3x - 10}$

53. $\dfrac{5}{x + 3} - \dfrac{4x}{(x + 3)^2} - \dfrac{1}{x - 3}$

54. $\dfrac{8}{y} - \dfrac{3}{y + 5} + \dfrac{4}{y - 2}$

In Exercises 55 and 56, use a graphing calculator to graph the two equations in the same viewing window. Use the graphs to verify that the expressions are equivalent. Verify the results algebraically.

55. $y_1 = \dfrac{1}{x} - \dfrac{3}{x + 3}$

$y_2 = \dfrac{3 - 2x}{x(x + 3)}$

56. $y_1 = \dfrac{5x}{x - 5} + \dfrac{7}{x + 1}$

$y_2 = \dfrac{5x^2 + 12x - 35}{x^2 - 4x - 5}$

6.4 Complex Fractions

1 ▶ Simplify complex fractions using rules for dividing rational expressions.

In Exercises 57–62, simplify the complex fraction.

57. $\dfrac{\left(\dfrac{6}{x}\right)}{\left(\dfrac{2}{x^3}\right)}$

58. $\dfrac{xy}{\left(\dfrac{5x^2}{2y}\right)}$

59. $\dfrac{\left(\dfrac{x}{x-2}\right)}{\left(\dfrac{2x}{2-x}\right)}$

60. $\dfrac{\left(\dfrac{y^2}{5-y}\right)}{\left(\dfrac{y}{y-5}\right)}$

61. $\dfrac{\left(\dfrac{6x^2}{x^2+2x-35}\right)}{\left(\dfrac{x^3}{x^2-25}\right)}$

62. $\dfrac{\left[\dfrac{24-18x}{(2-x)^2}\right]}{\left(\dfrac{60-45x}{x^2-4x+4}\right)}$

2 ▶ Simplify complex fractions having a sum or difference in the numerator and/or denominator.

In Exercises 63–68, simplify the complex fraction.

63. $\dfrac{3t}{\left(5-\dfrac{2}{t}\right)}$

64. $\dfrac{\left(\dfrac{1}{x}-\dfrac{1}{2}\right)}{2x}$

65. $\dfrac{\left(x-3+\dfrac{2}{x}\right)}{\left(1-\dfrac{2}{x}\right)}$

66. $\dfrac{3x-1}{\left(\dfrac{2}{x^2}+\dfrac{5}{x}\right)}$

67. $\dfrac{\left(\dfrac{1}{a^2-16}-\dfrac{1}{a}\right)}{\left(\dfrac{1}{a^2+4a}+4\right)}$

68. $\dfrac{\left(\dfrac{1}{x^2}-\dfrac{1}{y^2}\right)}{\left(\dfrac{1}{x}+\dfrac{1}{y}\right)}$

6.5 Dividing Polynomials and Synthetic Division

1 ▶ Divide polynomials by monomials and write in simplest form.

In Exercises 69–72, perform the division.

69. $(4x^3-x) \div (2x)$

70. $(10x+15) \div (5x)$

71. $\dfrac{3x^3y^2-x^2y^2+x^2y}{x^2y}$

72. $\dfrac{6a^3b^3+2a^2b-4ab^2}{2ab}$

2 ▶ Use long division to divide polynomials by polynomials.

In Exercises 73–78, perform the division.

73. $\dfrac{6x^3+2x^2-4x+2}{3x-1}$

74. $\dfrac{4x^4-x^3-7x^2+18x}{x-2}$

75. $\dfrac{x^4-3x^2+2}{x^2-1}$

76. $\dfrac{x^4-4x^3+3x}{x^2-1}$

77. $\dfrac{x^5-3x^4+x^2+6}{x^3-2x^2+x-1}$

78. $\dfrac{x^6+4x^5-3x^2+5x}{x^3+x^2-4x+3}$

3 ▶ Use synthetic division to divide polynomials by polynomials of the form $x-k$.

In Exercises 79–84, use synthetic division to divide.

79. $\dfrac{x^2+3x+5}{x+1}$

80. $\dfrac{2x^2+x-10}{x-2}$

81. $\dfrac{x^3+7x^2+3x-14}{x+2}$

82. $\dfrac{x^4 - 2x^3 - 15x^2 - 2x + 10}{x - 5}$

83. $(x^4 - 3x^2 - 25) \div (x - 3)$

84. $(2x^3 + 5x - 2) \div \left(x + \tfrac{1}{2}\right)$

4 ▶ Use synthetic division to factor polynomials.

In Exercises 85 and 86, completely factor the polynomial given one of its factors.

Polynomial	Factor
85. $x^3 + 2x^2 - 5x - 6$	$x - 2$
86. $2x^3 + x^2 - 2x - 1$	$x + 1$

6.6 Solving Rational Equations

1 ▶ Solve rational equations containing constant denominators.

In Exercises 87–92, solve the equation.

87. $\dfrac{x}{15} + \dfrac{3}{5} = 1$

88. $\dfrac{x}{6} + \dfrac{5}{3} = 3$

89. $\dfrac{3x}{8} = -15 + \dfrac{x}{4}$

90. $\dfrac{t + 1}{6} = \dfrac{1}{2} - 2t$

91. $\dfrac{x^2}{6} - \dfrac{x}{12} = \dfrac{1}{2}$

92. $\dfrac{x^2}{4} = -\dfrac{x}{12} + \dfrac{1}{6}$

2 ▶ Solve rational equations containing variable denominators.

In Exercises 93–108, solve the equation.

93. $8 - \dfrac{12}{t} = \dfrac{1}{3}$

94. $5 + \dfrac{2}{x} = \dfrac{1}{4}$

95. $\dfrac{2}{y} - \dfrac{1}{3y} = \dfrac{1}{3}$

96. $\dfrac{7}{4x} - \dfrac{6}{8x} = 1$

97. $r = 2 + \dfrac{24}{r}$

98. $\dfrac{2}{x} - \dfrac{x}{6} = \dfrac{2}{3}$

99. $8\left(\dfrac{6}{x} - \dfrac{1}{x + 5}\right) = 15$

100. $16\left(\dfrac{5}{x} - \dfrac{3}{x + 2}\right) = 4$

101. $\dfrac{3}{y + 1} - \dfrac{8}{y} = 1$

102. $\dfrac{4x}{x - 5} + \dfrac{2}{x} = -\dfrac{4}{x - 5}$

103. $\dfrac{2x}{x - 3} - \dfrac{3}{x} = 0$

104. $\dfrac{6x}{x - 3} = 9 + \dfrac{18}{x - 3}$

105. $\dfrac{12}{x^2 + x - 12} - \dfrac{1}{x - 3} = -1$

106. $\dfrac{3}{x - 1} + \dfrac{6}{x^2 - 3x + 2} = 2$

107. $\dfrac{5}{x^2 - 4} - \dfrac{6}{x - 2} = -5$

108. $\dfrac{3}{x^2 - 9} + \dfrac{4}{x + 3} = 1$

6.7 Applications and Variation

1 ▶ Solve application problems involving rational equations.

109. *Average Speeds* You and a friend ride bikes for the same amount of time. You ride 24 miles and your friend rides 15 miles. Your friend's average speed is 6 miles per hour slower than yours. What are the average speeds of you and your friend?

110. *Average Speed* You drive 220 miles to see a friend. The return trip takes 20 minutes less than the original trip, and your average speed is 5 miles per hour faster. What is your average speed on the return trip?

111. *Partnership Costs* A group of people starting a business agree to share equally in the cost of a $60,000 piece of machinery. If they could find two more people to join the group, each person's share of the cost would decrease by $5000. How many people are presently in the group?

112. *Work Rate*   One painter works $1\frac{1}{2}$ times as fast as another painter. It takes them 4 hours working together to paint a room. Find the time it takes each painter to paint the room working alone.

113. *Population Growth*   The Parks and Wildlife Commission introduces 80,000 fish into a large lake. The population P (in thousands) of the fish is approximated by the model

$$P = \frac{20(4 + 3t)}{1 + 0.05t}$$

where t is the time in years. Find the time required for the population to increase to 400,000 fish.

114. *Average Cost*   The average cost $\overline{C}$ of producing x units of a product is given by

$$\overline{C} = 1.5 + \frac{4200}{x}.$$

Determine the number of units that must be produced to obtain an average cost of $2.90 per unit.

2 ▶ Solve application problems involving direct variation.

115. *Hooke's Law*   A force of 100 pounds stretches a spring 4 inches. Find the force required to stretch the spring 6 inches.

116. *Stopping Distance*   The stopping distance d of an automobile is directly proportional to the square of its speed s. How will the stopping distance be changed by doubling the speed of the car?

3 ▶ Solve application problems involving inverse variation.

117. *Travel Time*   The travel time between two cities is inversely proportional to the average speed. A train travels between the cities in 3 hours at an average speed of 65 miles per hour. How long would it take to travel between the cities at an average speed of 80 miles per hour?

118. *Demand*   A company has found that the daily demand x for its cordless telephones is inversely proportional to the price p. When the price is $25, the demand is 1000 telephones. Approximate the demand when the price is increased to $28.

119. *Revenue*   The monthly demand for Brand X athletic shoes varies directly as the amount spent on advertising and inversely as the square of the price per pair. When $20,000 is spent on monthly advertising and the price per pair of shoes is $55, the demand is 900 pairs. If advertising is increased to $25,000, what price will yield a demand of 900 pairs? Is this increase worthwhile in terms of revenue?

120. *Revenue*   The seasonal demand for Ace brand sunglasses varies directly as the amount spent on advertising and inversely as the square of the price per pair. When $125,000 is spent on advertising and the price per pair is $35, the demand is 5000 pairs. If advertising is increased to $135,000, what price will yield a demand of 5000 pairs? Is this increase worthwhile in terms of revenue?

4 ▶ Solve application problems involving joint variation.

121. *Simple Interest*   The simple interest earned on a savings account is jointly proportional to the time and the principal. After three quarters (9 months), the interest for a principal of $12,000 is $675. How much interest would a principal of $8200 earn in 18 months?

122. *Cost*   The cost of constructing a wooden box with a square base varies jointly as the height of the box and the square of the width of the box. A box of height 16 inches and of width 6 inches costs $28.80. How much would a box of height 14 inches and of width 8 inches cost?

Chapter Test

Take this test as you would take a test in class. After you are done, check your work against the answers in the back of the book.

1. Find the domain of $f(x) = \dfrac{x + 1}{x^2 - 6x + 5}$.

In Exercises 2 and 3, simplify the rational expression.

2. $\dfrac{4 - 2x}{x - 2}$

3. $\dfrac{2a^2 - 5a - 12}{5a - 20}$

4. Find the least common multiple of x^2, $3x^3$, and $(x + 4)^2$.

Smart Study Strategy

Go to page 368 for ways to *Use Ten Steps for Test-Taking.*

In Exercises 5–18, perform the operation and simplify.

5. $\dfrac{4z^3}{5} \cdot \dfrac{25}{12z^2}$

6. $\dfrac{y^2 + 8y + 16}{2(y - 2)} \cdot \dfrac{8y - 16}{(y + 4)^3}$

7. $\dfrac{(2xy^2)^3}{15} \div \dfrac{12x^3}{21}$

8. $(4x^2 - 9) \div \dfrac{2x + 3}{2x^2 - x - 3}$

9. $\dfrac{3}{x - 3} + \dfrac{x - 2}{x - 3}$

10. $2x + \dfrac{1 - 4x^2}{x + 1}$

11. $\dfrac{5x}{x + 2} - \dfrac{2}{x^2 - x - 6}$

12. $\dfrac{3}{x} - \dfrac{5}{x^2} + \dfrac{2x}{x^2 + 2x + 1}$

13. $\dfrac{\left(\dfrac{3x}{x + 2}\right)}{\left(\dfrac{12}{x^3 + 2x^2}\right)}$

14. $\dfrac{\left(9x - \dfrac{1}{x}\right)}{\left(\dfrac{1}{x} - 3\right)}$

15. $\dfrac{3x^{-2} + y^{-1}}{(x + y)^{-1}}$

16. $\dfrac{6x^2 - 4x + 8}{2x}$

17. $\dfrac{t^4 + t^2 - 6t}{t^2 - 2}$

18. $\dfrac{2x^4 - 15x^2 - 7}{x - 3}$

In Exercises 19–21, solve the equation.

19. $\dfrac{3}{h + 2} = \dfrac{1}{6}$

20. $\dfrac{2}{x + 5} - \dfrac{3}{x + 3} = \dfrac{1}{x}$

21. $\dfrac{1}{x + 1} + \dfrac{1}{x - 1} = \dfrac{2}{x^2 - 1}$

22. Find a mathematical model that relates u and v if v varies directly as the square root of u, and $v = \frac{3}{2}$ when $u = 36$.

23. If the temperature of a gas is not allowed to change, the absolute pressure P of the gas is inversely proportional to its volume V, according to Boyle's Law. A large balloon is filled with 180 cubic meters of helium at atmospheric pressure (1 atm) at sea level. What is the volume of the helium if the balloon rises to an altitude at which the atmospheric pressure is 0.75 atm? (Assume that the temperature does not change.)

Study Skills in Action

Studying in a Group

Many students endure unnecessary frustration because they study by themselves. Studying in a group or with a partner has many benefits. First, the combined memory and comprehension of the members minimizes the likelihood of any member getting "stuck" on a particular problem. Second, discussing math often helps clarify unclear areas. Third, regular study groups keep many students from procrastinating. Finally, study groups often build a camaraderie that helps students stick with the course when it gets tough.

Kimberly Nolting

VP, Academic Success Press
expert in developmental education

These students are keeping each other motivated.

Smart Study Strategy

Form a Weekly Study Group

1 ▶ Set up the group.

- Select students who are just as dedicated to doing well in the math class as you are.

- Find a regular meeting place on campus that has minimal distractions. Try to find a place that has a white board.

- Compare schedules, and select at least one time a week to meet, allowing at least 1.5 hours for study time.

2 ▶ Organize the study time. If you are unsure about how to structure your time during the first few study sessions, try using the guidelines at the right.

3 ▶ Set up rules for the group. Consider using the following rules:

- Members must attend regularly, be on time, and participate.

- The sessions will focus on the key math concepts, not on the needs of one student.

- Students who skip classes will not be allowed to participate in the study group.

- Students who keep the group from being productive will be asked to leave the group.

4 ▶ Inform the instructor. Let the instructor know about your study group. Ask for advice about maintaining a productive group.

> - Review and compare notes - 20 minutes
>
> - Identify and review the key rules, definitions, etc. - 20 minutes
>
> - Demonstrate at least one homework problem for each key concept - 40 minutes
>
> - Make small talk (saving this until the end improves your chances of getting through all the math) - 10 minutes

Chapter 7
Radicals and Complex Numbers

IT WORKED FOR ME!

"I learned about study groups by accident last semester. I started studying in the learning center. When I saw someone else from my class come into the center, I mentioned something to the tutor and she suggested asking her over to study with us. We did, and it actually turned into a group of three or four students every session. I learned a lot more and enjoyed it more too."

Jeremy
Graphic design

7.1 Radicals and Rational Exponents

Jeffrey Blackman/Index Stock

What You Should Learn

1 ▶ Determine the *n*th roots of numbers and evaluate radical expressions.

2 ▶ Use the rules of exponents to evaluate or simplify expressions with rational exponents.

3 ▶ Use a calculator to evaluate radical expressions.

4 ▶ Evaluate radical functions and find the domains of radical functions.

Why You Should Learn It

Algebraic equations often involve rational exponents. For instance, in Exercise 167 on page 459, you will use an equation involving a rational exponent to find the depreciation rate for a truck.

1 ▶ Determine the *n*th roots of numbers and evaluate radical expressions.

Roots and Radicals

A **square root** of a number is defined as one of its two equal factors. For example, 5 is a square root of 25 because 5 is one of the two equal factors of 25. In a similar way, a **cube root** of a number is one of its three equal factors.

Number	Equal Factors	Root	Type
$9 = 3^2$	$3 \cdot 3$	3	Square root
$25 = (-5)^2$	$(-5)(-5)$	-5	Square root
$-27 = (-3)^3$	$(-3)(-3)(-3)$	-3	Cube root
$64 = 4^3$	$4 \cdot 4 \cdot 4$	4	Cube root
$16 = 2^4$	$2 \cdot 2 \cdot 2 \cdot 2$	2	Fourth root

Definition of *n*th Root of a Number

Let a and b be real numbers and let n be an integer such that $n \geq 2$. If

$$a = b^n$$

then b is an **nth root of a.** If $n = 2$, the root is a **square root.** If $n = 3$, the root is a **cube root.**

Some numbers have more than one *n*th root. For example, both 5 and -5 are square roots of 25. To avoid ambiguity about which root you are referring to, the **principal *n*th root** of a number is defined in terms of a **radical symbol** $\sqrt[n]{}$. So the *principal square root* of 25, written as $\sqrt{25}$, is the positive root, 5.

Principal *n*th Root of a Number

Let a be a real number that has at least one (real number) nth root. The **principal nth root of a** is the nth root that has the same sign as a, and it is denoted by the **radical symbol**

$$\sqrt[n]{a}. \qquad \text{Principal } n\text{th root}$$

The positive integer n is the **index** of the radical, and the number a is the **radicand.** If $n = 2$, omit the index and write $\sqrt{a}$ rather than $\sqrt[2]{a}$.

Study Tip

In the definition at the right, "the nth root that has the same sign as a" means that the principal nth root of a is positive if a is positive and negative if a is negative. For example, $\sqrt{4} = 2$ and $\sqrt[3]{-8} = -2$. Furthermore, to denote the negative square root of a number, you must use a negative sign in front of the radical. For example, $-\sqrt{4} = -2$.

EXAMPLE 1   **Finding Roots of Numbers**

Find each root.

a. $\sqrt{36}$ **b.** $-\sqrt{36}$ **c.** $\sqrt{-4}$ **d.** $\sqrt[3]{8}$ **e.** $\sqrt[3]{-8}$

Solution

a. $\sqrt{36} = 6$ because $6 \cdot 6 = 6^2 = 36$.

b. $-\sqrt{36} = -6$ because $6 \cdot 6 = 6^2 = 36$. So, $(-1)(\sqrt{36}) = (-1)(6) = -6$.

c. $\sqrt{-4}$ is not real because there is no real number that when multiplied by itself yields -4.

d. $\sqrt[3]{8} = 2$ because $2 \cdot 2 \cdot 2 = 2^3 = 8$.

e. $\sqrt[3]{-8} = -2$ because $(-2)(-2)(-2) = (-2)^3 = -8$.

 CHECKPOINT *Now try Exercise 1.*

Properties of *n*th Roots

Property	*Example*
1. If a is a positive real number and n is even, then a has exactly two (real) nth roots, which are denoted by $\sqrt[n]{a}$ and $-\sqrt[n]{a}$.	The two real square roots of 81 are $\sqrt{81} = 9$ and $-\sqrt{81} = -9$.
2. If a is any real number and n is odd, then a has only one (real) nth root, which is denoted by $\sqrt[n]{a}$.	$\sqrt[3]{27} = 3$ $\sqrt[3]{-64} = -4$
3. If a is a negative real number and n is even, then a has no (real) nth root.	$\sqrt{-64}$ is not a real number.

Numbers such as 1, 4, 9, 16, 49, and 81 are called **perfect squares** because they have rational square roots. Similarly, numbers such as 1, 8, 27, 64, and 125 are called **perfect cubes** because they have rational cube roots.

EXAMPLE 2   **Classifying Perfect *n*th Powers**

State whether each number is a perfect square, a perfect cube, both, or neither.

a. 81   **b.** -125 **c.** 64   **d.** 32

Solution

a. 81 is a perfect square because $9^2 = 81$. It is not a perfect cube.

b. -125 is a perfect cube because $(-5)^3 = -125$. It is not a perfect square.

c. 64 is a perfect square because $8^2 = 64$, and it is also a perfect cube because $4^3 = 64$.

d. 32 is not a perfect square or a perfect cube. (It is, however, a perfect fifth power, because $2^5 = 32$.)

 **CHECKPOINT** *Now try Exercise 9.*

Raising a number to the nth power and taking the principal nth root of a number can be thought of as *inverse* operations. Here are some examples.

$$\left(\sqrt{4}\right)^2 = (2)^2 = 4 \quad \text{and} \quad \sqrt{4} = \sqrt{2^2} = 2$$

$$\left(\sqrt[3]{27}\right)^3 = (3)^3 = 27 \quad \text{and} \quad \sqrt[3]{27} = \sqrt[3]{3^3} = 3$$

$$\left(\sqrt[4]{16}\right)^4 = (2)^4 = 16 \quad \text{and} \quad \sqrt[4]{16} = \sqrt[4]{2^4} = 2$$

$$\left(\sqrt[5]{-243}\right)^5 = (-3)^5 = -243 \quad \text{and} \quad \sqrt[5]{-243} = \sqrt[5]{(-3)^5} = -3$$

Inverse Properties of nth Powers and nth Roots

Let a be a real number, and let n be an integer such that $n \geq 2$.

Property	Example
1. If a has a principal nth root, then $\left(\sqrt[n]{a}\right)^n = a.$	$\left(\sqrt{5}\right)^2 = 5$
2. If n is odd, then $\sqrt[n]{a^n} = a.$	$\sqrt[3]{5^3} = 5$
If n is even, then $\sqrt[n]{a^n} = |a|.$	$\sqrt{(-5)^2} = |-5| = 5$

EXAMPLE 3   Evaluating Radical Expressions

Evaluate each radical expression.

a. $\sqrt[3]{4^3}$ **b.** $\sqrt[3]{(-2)^3}$ **c.** $\left(\sqrt{7}\right)^2$

d. $\sqrt{(-3)^2}$ **e.** $\sqrt{-3^2}$

Solution

a. Because the index of the radical is odd, you can write

$$\sqrt[3]{4^3} = 4.$$

b. Because the index of the radical is odd, you can write

$$\sqrt[3]{(-2)^3} = -2.$$

c. Because the radicand is positive, $\sqrt{7}$ is real and you can write

$$\left(\sqrt{7}\right)^2 = 7.$$

d. Because the index of the radical is even, you must include absolute value signs, and write

$$\sqrt{(-3)^2} = |-3| = 3.$$

e. Because $\sqrt{-3^2} = \sqrt{-9}$ is an even root of a negative number, its value is not a real number.

✓ **CHECKPOINT** *Now try Exercise 39.*

Study Tip

In parts (d) and (e) of Example 3, notice that the two expressions inside the radical are different. In part (d), the negative sign is part of the base of the exponential expression. In part (e), the negative sign is not part of the base.

2 ▶ Use the rules of exponents to evaluate or simplify expressions with rational exponents.

Rational Exponents

So far in the text you have worked with algebraic expressions involving only integer exponents. Next you will see that algebraic expressions may also contain **rational exponents.**

Definition of Rational Exponents

Let a be a real number, and let n be an integer such that $n \geq 2$. If the principal nth root of a exists, then $a^{1/n}$ is defined as

$$a^{1/n} = \sqrt[n]{a}.$$

If m is a positive integer that has no common factor with n, then

$$a^{m/n} = (a^{1/n})^m = \left(\sqrt[n]{a}\right)^m \quad \text{and} \quad a^{m/n} = (a^m)^{1/n} = \sqrt[n]{a^m}.$$

It does not matter in which order the two operations are performed, provided the nth root exists. Here is an example.

$$8^{2/3} = \left(\sqrt[3]{8}\right)^2 = 2^2 = 4 \qquad \text{Cube root, then second power}$$

$$8^{2/3} = \sqrt[3]{8^2} = \sqrt[3]{64} = 4 \qquad \text{Second power, then cube root}$$

The rules of exponents that were listed in Section 5.1 also apply to rational exponents (provided the roots indicated by the denominators exist). These rules are listed below, with different examples.

Summary of Rules of Exponents

Let r and s be rational numbers, and let a and b be real numbers, variables, or algebraic expressions. (All denominators and bases are nonzero.)

Product and Quotient Rules	*Example*
1. $a^r \cdot a^s = a^{r+s}$	$4^{1/2}(4^{1/3}) = 4^{5/6}$
2. $\dfrac{a^r}{a^s} = a^{r-s}$	$\dfrac{x^2}{x^{1/2}} = x^{2-(1/2)} = x^{3/2}$
Power Rules	
3. $(ab)^r = a^r \cdot b^r$	$(2x)^{1/2} = 2^{1/2}(x^{1/2})$
4. $(a^r)^s = a^{rs}$	$(x^3)^{1/2} = x^{3/2}$
5. $\left(\dfrac{a}{b}\right)^r = \dfrac{a^r}{b^r}$	$\left(\dfrac{x}{3}\right)^{2/3} = \dfrac{x^{2/3}}{3^{2/3}}$
Zero and Negative Exponent Rules	
6. $a^0 = 1$	$(3x)^0 = 1$
7. $a^{-r} = \dfrac{1}{a^r}$	$4^{-3/2} = \dfrac{1}{4^{3/2}} = \dfrac{1}{(2)^3} = \dfrac{1}{8}$
8. $\left(\dfrac{a}{b}\right)^{-r} = \left(\dfrac{b}{a}\right)^r$	$\left(\dfrac{x}{4}\right)^{-1/2} = \left(\dfrac{4}{x}\right)^{1/2} = \dfrac{2}{x^{1/2}}$

Study Tip

The numerator of a rational exponent denotes the *power* to which the base is raised, and the denominator denotes the *root* to be taken.

Power

Root

$a^{m/n} = \left(\sqrt[n]{a}\right)^m$

Technology: Discovery

Use a calculator to evaluate each pair of expressions below.

a. $5.6^{3.5} \cdot 5.6^{0.4}$ and $5.6^{3.9}$

b. $\dfrac{3.4^{4.6}}{3.4^{3.1}}$ and $3.4^{1.5}$

c. $6.2^{-0.75}$ and $\dfrac{1}{6.2^{0.75}}$

What rule is illustrated by each pair of expressions? Use your calculator to illustrate some of the other rules of exponents.

EXAMPLE 4 **Evaluating Expressions with Rational Exponents**

Evaluate each expression.

a. $8^{4/3}$　　　**b.** $(4^2)^{3/2}$　　**c.** $25^{-3/2}$

d. $\left(\dfrac{64}{125}\right)^{2/3}$　　**e.** $-16^{1/2}$　　**f.** $(-16)^{1/2}$

Solution

a. $8^{4/3} = (8^{1/3})^4 = \left(\sqrt[3]{8}\right)^4 = 2^4 = 16$ 　　　　Root is 3. Power is 4.

b. $(4^2)^{3/2} = 4^{2 \cdot (3/2)} = 4^{6/2} = 4^3 = 64$ 　　　Root is 2. Power is 3.

c. $25^{-3/2} = \dfrac{1}{25^{3/2}} = \dfrac{1}{(\sqrt{25})^3} = \dfrac{1}{5^3} = \dfrac{1}{125}$ 　　Root is 2. Power is 3.

d. $\left(\dfrac{64}{125}\right)^{2/3} = \dfrac{64^{2/3}}{125^{2/3}} = \dfrac{\left(\sqrt[3]{64}\right)^2}{\left(\sqrt[3]{125}\right)^2} = \dfrac{4^2}{5^2} = \dfrac{16}{25}$ 　Root is 3. Power is 2.

e. $-16^{1/2} = -\sqrt{16} = -(4) = -4$ 　　　　Root is 2. Power is 1.

f. $(-16)^{1/2} = \sqrt{-16}$ is not a real number. 　　Root is 2. Power is 1.

 CHECKPOINT *Now try Exercise 75.*

Study Tip

In parts (e) and (f) of Example 4, be sure that you see the distinction between the expressions $-16^{1/2}$ and $(-16)^{1/2}$.

EXAMPLE 5 **Using Rules of Exponents**

Rewrite each expression using rational exponents.

a. $x\sqrt[4]{x^3}$ 　　**b.** $\dfrac{\sqrt[3]{x^2}}{\sqrt{x^3}}$ 　　**c.** $\sqrt[3]{x^2 y}$

Solution

a. $x\sqrt[4]{x^3} = x(x^{3/4}) = x^{1+(3/4)} = x^{7/4}$

b. $\dfrac{\sqrt[3]{x^2}}{\sqrt{x^3}} = \dfrac{x^{2/3}}{x^{3/2}} = x^{(2/3)-(3/2)} = x^{-5/6} = \dfrac{1}{x^{5/6}}$

c. $\sqrt[3]{x^2 y} = (x^2 y)^{1/3} = (x^2)^{1/3} y^{1/3} = x^{2/3} y^{1/3}$

 CHECKPOINT *Now try Exercise 91.*

EXAMPLE 6 **Using Rules of Exponents**

Use rules of exponents to simplify each expression.

a. $\sqrt{\sqrt[3]{x}}$ 　　**b.** $\dfrac{(2x-1)^{4/3}}{\sqrt[3]{2x-1}}$

Solution

a. $\sqrt{\sqrt[3]{x}} = \sqrt{x^{1/3}} = (x^{1/3})^{1/2} = x^{(1/3)(1/2)} = x^{1/6}$

b. $\dfrac{(2x-1)^{4/3}}{\sqrt[3]{2x-1}} = \dfrac{(2x-1)^{4/3}}{(2x-1)^{1/3}} = (2x-1)^{(4/3)-(1/3)} = (2x-1)^{3/3} = 2x-1$

CHECKPOINT *Now try Exercise 107.*

3 ▸ Use a calculator to evaluate radical expressions.

Radicals and Calculators

There are two methods of evaluating radicals on most calculators. For square roots, you can use the *square root key* $\boxed{\sqrt{}}$ or $\boxed{\sqrt{x}}$. For other roots, you can first convert the radical to exponential form and then use the *exponential key* $\boxed{y^x}$ or $\boxed{\wedge}$.

EXAMPLE 7   Evaluating Roots with a Calculator

Evaluate each expression. Round the result to three decimal places.

a. $\sqrt{5}$ **b.** $\sqrt[5]{25}$ **c.** $\sqrt[3]{-4}$ **d.** $(-1.4)^{3/2}$

Solution

a. 5 $\boxed{\sqrt{x}}$ Scientific

$\boxed{\sqrt{}}$ 5 $\boxed{\text{ENTER}}$ Graphing

The display is 2.236067977. Rounded to three decimal places, $\sqrt{5} \approx 2.236$.

b. First rewrite the expression as $\sqrt[5]{25} = 25^{1/5}$. Then use one of the following keystroke sequences.

25 $\boxed{y^x}$ $\boxed{(}$ 1 $\boxed{\div}$ 5 $\boxed{)}$ $\boxed{=}$ Scientific

25 $\boxed{\wedge}$ $\boxed{(}$ 1 $\boxed{\div}$ 5 $\boxed{)}$ $\boxed{\text{ENTER}}$ Graphing

The display is 1.903653939. Rounded to three decimal places, $\sqrt[5]{25} \approx 1.904$.

c. If your calculator does not have a cube root key, use the fact that

$$\sqrt[3]{-4} = \sqrt[3]{(-1)(4)} = \sqrt[3]{-1}\sqrt[3]{4} = -\sqrt[3]{4} = -4^{1/3}$$

and attach the negative sign as the last keystroke.

4 $\boxed{y^x}$ $\boxed{(}$ 1 $\boxed{\div}$ 3 $\boxed{)}$ $\boxed{=}$ $\boxed{+/-}$ Scientific

$\boxed{\sqrt[3]{}}$ $\boxed{(-)}$ 4 $\boxed{)}$ $\boxed{\text{ENTER}}$ Graphing

The display is -1.587401052. Rounded to three decimal places, $\sqrt[3]{-4} \approx -1.587$.

d. 1.4 $\boxed{+/-}$ $\boxed{y^x}$ $\boxed{(}$ 3 $\boxed{\div}$ 2 $\boxed{)}$ $\boxed{=}$ Scientific

$\boxed{(}$ $\boxed{(-)}$ 1.4 $\boxed{)}$ $\boxed{\wedge}$ $\boxed{(}$ 3 $\boxed{\div}$ 2 $\boxed{)}$ $\boxed{\text{ENTER}}$ Graphing

The display should indicate an error because an even root of a negative number is not real.

✓ **CHECKPOINT** *Now try Exercise 129.*

---

4 ▸ Evaluate radical functions and find the domains of radical functions.

Radical Functions

A **radical function** is a function that contains a radical such as

$$f(x) = \sqrt{x} \quad \text{or} \quad g(x) = \sqrt[3]{x}.$$

When evaluating a radical function, note that the radical symbol is a symbol of grouping.

Technology: Discovery

Consider the function $f(x) = x^{2/3}$.

a. What is the domain of the function?

b. Use your graphing calculator to graph each equation, in order.

$$y_1 = x^{(2 \div 3)}$$

$$y_2 = (x^2)^{1/3} \quad \text{Power, then root}$$

$$y_3 = (x^{1/3})^2 \quad \text{Root, then power}$$

c. Are the graphs all the same? Are their domains all the same?

d. On your graphing calculator, which of the forms properly represent the function $f(x) = x^{m/n}$?

$$y_1 = x^{(m \div n)}$$

$$y_2 = (x^m)^{1/n}$$

$$y_3 = (x^{1/n})^m$$

e. Explain how the domains of $f(x) = x^{2/3}$ and $g(x) = x^{-2/3}$ differ.

EXAMPLE 8   Evaluating Radical Functions

Evaluate each radical function when $x = 4$.

a. $f(x) = \sqrt[3]{x - 31}$ **b.** $g(x) = \sqrt{16 - 3x}$

Solution

a. $f(4) = \sqrt[3]{4 - 31} = \sqrt[3]{-27} = -3$

b. $g(4) = \sqrt{16 - 3(4)} = \sqrt{16 - 12} = \sqrt{4} = 2$

 CHECKPOINT *Now try Exercise 143.*

The **domain** of the radical function $f(x) = \sqrt[n]{x}$ is the set of all real numbers such that x has a principal nth root.

Domain of a Radical Function

Let n be an integer that is greater than or equal to 2.

1. If n is odd, the domain of $f(x) = \sqrt[n]{x}$ is the set of all real numbers.
2. If n is even, the domain of $f(x) = \sqrt[n]{x}$ is the set of all nonnegative real numbers.

EXAMPLE 9   Finding the Domains of Radical Functions

Describe the domains of (a) $f(x) = \sqrt[3]{x}$ and (b) $f(x) = \sqrt{x^3}$.

Solution

a. The domain of $f(x) = \sqrt[3]{x}$ is the set of all real numbers because for any real number x, the expression $\sqrt[3]{x}$ is a real number.

b. The domain of $f(x) = \sqrt{x^3}$ is the set of all nonnegative real numbers. For instance, 1 is in the domain but -1 is not because $\sqrt{(-1)^3} = \sqrt{-1}$ is not a real number.

 CHECKPOINT *Now try Exercise 149.*

EXAMPLE 10   Finding the Domain of a Radical Function

Find the domain of $f(x) = \sqrt{2x - 1}$.

Solution

The domain of f consists of all x such that $2x - 1 \geq 0$. Using the methods described in Section 2.4, you can solve this inequality as follows.

$2x - 1 \geq 0$	Write original inequality.
$2x \geq 1$	Add 1 to each side.
$x \geq \frac{1}{2}$	Divide each side by 2.

So, the domain is the set of all real numbers x such that $x \geq \frac{1}{2}$.

 CHECKPOINT *Now try Exercise 155.*

Study Tip

In general, when the index n of a radical function is even, the domain of the function includes all real values for which the expression under the radical is greater than or equal to zero.

Concept Check

1. Describe all values of n and x for which x has each number of real nth roots.

   (a) 2   (b) 1   (c) 0

2. Describe two ways of performing two operations to evaluate the expression $32^{2/5}$.

3. You rewrite a radical expression using a rational exponent. What part of the rational exponent corresponds to the index of the radical?

4. If n is even, what must be true about the radicand for the nth root to be a real number?

7.1 EXERCISES

Go to pages 502–503 to record your assignments.

Developing Skills

In Exercises 1–8, find the root if it exists. *See Example 1.*

 1. $\sqrt{64}$

2. $-\sqrt{100}$

3. $-\sqrt{49}$

4. $\sqrt{-25}$

5. $\sqrt[3]{-27}$

6. $\sqrt[3]{-64}$

7. $\sqrt{-1}$

8. $-\sqrt{-1}$

In Exercises 9–14, state whether the number is a perfect square, a perfect cube, or neither. *See Example 2.*

9. 49

10. -27

11. 1728

12. 964

13. 96

14. 225

In Exercises 15–22, find all of the square roots of the perfect square.

15. 25

16. 121

17. $\frac{9}{16}$

18. $\frac{25}{36}$

19. $\frac{1}{49}$

20. $\frac{81}{100}$

21. 0.16

22. 0.25

In Exercises 23–30, find all of the cube roots of the perfect cube.

23. 8

24. -8

25. $-\frac{1}{27}$

26. $\frac{27}{64}$

27. $\frac{1}{1000}$

28. $-\frac{8}{125}$

29. 0.001

30. -0.008

In Exercises 31–38, determine whether the square root is a rational or an irrational number.

31. $\sqrt{3}$

32. $\sqrt{6}$

33. $\sqrt{16}$

34. $\sqrt{25}$

35. $\sqrt{\frac{4}{25}}$

36. $\sqrt{\frac{2}{7}}$

37. $\sqrt{\frac{3}{16}}$

38. $\sqrt{\frac{1}{49}}$

In Exercises 39–68, evaluate the radical expression without using a calculator. If not possible, state the reason. *See Example 3.*

39. $\sqrt{8^2}$

40. $-\sqrt{10^2}$

41. $\sqrt{(-10)^2}$

42. $\sqrt{(-12)^2}$

43. $\sqrt{-9^2}$

44. $\sqrt{-12^2}$

45. $-\sqrt{\left(\frac{2}{3}\right)^2}$

46. $\sqrt{\left(\frac{3}{4}\right)^2}$

47. $\sqrt{-\left(\frac{3}{10}\right)^2}$

48. $\sqrt{\left(-\frac{3}{5}\right)^2}$

49. $\left(\sqrt{5}\right)^2$

50. $-\left(\sqrt{10}\right)^2$

51. $-\left(\sqrt{23}\right)^2$

52. $\left(-\sqrt{18}\right)^2$

53. $\sqrt[3]{5^3}$

54. $\sqrt[3]{(-7)^3}$

55. $\sqrt[3]{10^3}$

56. $\sqrt[3]{4^3}$

57. $-\sqrt[3]{(-6)^3}$

58. $-\sqrt[3]{9^3}$

59. $\sqrt[3]{\left(-\frac{1}{4}\right)^3}$

60. $-\sqrt[3]{\left(\frac{1}{5}\right)^3}$

61. $\left(\sqrt[3]{11}\right)^3$

62. $\left(\sqrt[3]{-6}\right)^3$

63. $\left(-\sqrt[3]{24}\right)^3$

64. $\left(\sqrt[3]{21}\right)^3$

65. $\sqrt[4]{3^4}$

66. $\sqrt[5]{(-2)^5}$

67. $-\sqrt[4]{2^4}$

68. $-\sqrt[4]{-5^4}$

In Exercises 69–72, fill in the missing description.

Radical Form	Rational Exponent Form
69. $\sqrt{36} = 6$	
70. $\sqrt[3]{27^2} = 9$	
71.	$256^{3/4} = 64$
72.	$125^{1/3} = 5$

In Exercises 73–88, evaluate without using a calculator. **See Example 4.**

73. $25^{1/2}$ **74.** $49^{1/2}$

✓ **75.** $-36^{1/2}$ **76.** $-121^{1/2}$

77. $32^{-2/5}$ **78.** $81^{-3/4}$

79. $(-27)^{-2/3}$ **80.** $(-243)^{-3/5}$

81. $\left(\frac{8}{27}\right)^{2/3}$ **82.** $\left(\frac{256}{625}\right)^{1/4}$

83. $\left(\frac{121}{9}\right)^{-1/2}$ **84.** $\left(\frac{27}{1000}\right)^{-4/3}$

85. $(-3^3)^{2/3}$ **86.** $(8^2)^{3/2}$

87. $-(4^4)^{3/4}$ **88.** $(-2^3)^{5/3}$

In Exercises 89–106, rewrite the expression using rational exponents. **See Example 5.**

89. $\sqrt{t}$ **90.** $\sqrt[3]{x}$

✓ **91.** $x\sqrt[3]{x^6}$ **92.** $t\sqrt[5]{t^2}$

93. $u^2\sqrt[3]{u}$ **94.** $y\sqrt[4]{y^2}$

95. $\dfrac{\sqrt{x}}{\sqrt{x^3}}$ **96.** $\dfrac{\sqrt[3]{x^2}}{\sqrt[3]{x^4}}$

97. $\dfrac{\sqrt[4]{t}}{\sqrt{t^5}}$ **98.** $\dfrac{\sqrt[3]{x^4}}{\sqrt{x^3}}$

99. $\sqrt[3]{x^2} \cdot \sqrt[3]{x^7}$ **100.** $\sqrt[5]{z^3} \cdot \sqrt[5]{z^2}$

101. $\sqrt[4]{y^3} \cdot \sqrt[3]{y}$ **102.** $\sqrt[6]{x^5} \cdot \sqrt[3]{x^4}$

103. $\sqrt[4]{x^3 y}$ **104.** $\sqrt[3]{u^4 v^2}$

105. $z^2\sqrt{y^5 z^4}$ **106.** $x^2\sqrt[3]{xy^4}$

In Exercises 107–128, simplify the expression. **See Example 6.**

✓ **107.** $3^{1/4} \cdot 3^{3/4}$ **108.** $2^{2/5} \cdot 2^{3/5}$

109. $(2^{1/2})^{2/3}$ **110.** $(4^{1/3})^{9/4}$

111. $\dfrac{2^{1/5}}{2^{6/5}}$ **112.** $\dfrac{5^{-3/4}}{5}$

113. $(c^{3/2})^{1/3}$ **114.** $(k^{-1/3})^{3/2}$

115. $\dfrac{18y^{4/3}z^{-1/3}}{24y^{-2/3}z}$ **116.** $\dfrac{a^{3/4} \cdot a^{1/2}}{a^{5/2}}$

117. $(3x^{-1/3}y^{3/4})^2$ **118.** $(-2u^{3/5}v^{-1/5})^3$

119. $\left(\dfrac{x^{1/4}}{x^{1/6}}\right)^3$ **120.** $\left(\dfrac{3m^{1/6}n^{1/3}}{4n^{-2/3}}\right)^2$

121. $\sqrt{\sqrt[4]{y}}$ **122.** $\sqrt[3]{\sqrt{2x}}$

123. $\sqrt[4]{\sqrt{x^3}}$ **124.** $\sqrt[5]{\sqrt[3]{y^4}}$

125. $\dfrac{(x+y)^{3/4}}{\sqrt[4]{x+y}}$ **126.** $\dfrac{(a-b)^{1/3}}{\sqrt[3]{a-b}}$

127. $\dfrac{(3u-2v)^{2/3}}{\sqrt{(3u-2v)^3}}$ **128.** $\dfrac{\sqrt[4]{2x+y}}{(2x+y)^{3/2}}$

In Exercises 129–142, use a calculator to evaluate the expression. Round your answer to four decimal places. If not possible, state the reason. **See Example 7.**

✓ **129.** $\sqrt{35}$ **130.** $\sqrt{-23}$

131. $315^{2/5}$ **132.** $962^{2/3}$

133. $82^{-3/4}$ **134.** $382.5^{-3/2}$

135. $\sqrt[4]{212}$ **136.** $\sqrt[3]{-411}$

137. $\sqrt[3]{545^2}$ **138.** $\sqrt[5]{-35^3}$

139. $\dfrac{8 - \sqrt{35}}{2}$ **140.** $\dfrac{-5 + \sqrt{3215}}{10}$

141. $\dfrac{3 + \sqrt{17}}{9}$ **142.** $\dfrac{7 - \sqrt{241}}{12}$

In Exercises 143–148, evaluate the function for each indicated x-value, if possible, and simplify. **See Example 8.**

✓ **143.** $f(x) = \sqrt{2x + 9}$
  (a) $f(0)$ (b) $f(8)$ (c) $f(-6)$ (d) $f(36)$

144. $g(x) = \sqrt{5x - 6}$
  (a) $g(0)$ (b) $g(2)$ (c) $g(30)$ (d) $g\left(\frac{7}{5}\right)$

145. $g(x) = \sqrt[3]{x + 1}$
  (a) $g(7)$ (b) $g(26)$ (c) $g(-9)$ (d) $g(-65)$

146. $f(x) = \sqrt[3]{2x - 1}$
  (a) $f(0)$ (b) $f(-62)$ (c) $f(-13)$ (d) $f(63)$

147. $f(x) = \sqrt[4]{x-3}$

  (a) $f(19)$ (b) $f(1)$ (c) $f(84)$ (d) $f(4)$

148. $g(x) = \sqrt[4]{x+1}$

  (a) $g(0)$ (b) $g(15)$ (c) $g(-82)$ (d) $g(80)$

In Exercises 149–158, describe the domain of the function. *See Examples 9 and 10.*

✓ **149.** $f(x) = 3\sqrt{x}$ **150.** $h(x) = \sqrt[4]{x}$

151. $g(x) = \sqrt{4-9x}$ **152.** $g(x) = \sqrt{10-2x}$

153. $f(x) = \sqrt[3]{x^4}$ **154.** $f(x) = \sqrt{-x}$

✓ **155.** $h(x) = \sqrt{2x+9}$ **156.** $f(x) = \sqrt{3x-5}$

157. $g(x) = \dfrac{2}{\sqrt[4]{x}}$ **158.** $g(x) = \dfrac{10}{\sqrt[3]{x}}$

In Exercises 159–162, describe the domain of the function. Then check your answer by using a graphing calculator to graph the function.

159. $y = \dfrac{5}{\sqrt[4]{x^3}}$

160. $y = 4\sqrt[3]{x}$

161. $g(x) = 2x^{3/5}$

162. $h(x) = 5x^{2/3}$

In Exercises 163–166, perform the multiplication. Use a graphing calculator to confirm your result.

163. $x^{1/2}(2x-3)$

164. $x^{4/3}(3x^2 - 4x + 5)$

165. $y^{-1/3}(y^{1/3} + 5y^{4/3})$

166. $(x^{1/2} - 3)(x^{1/2} + 3)$

Solving Problems

Mathematical Modeling   In Exercises 167 and 168, use the formula for the *declining balances method*

$$r = 1 - \left(\frac{S}{C}\right)^{1/n}$$

to find the depreciation rate r. In the formula, n is the useful life of the item (in years), S is the salvage value (in dollars), and C is the original cost (in dollars).

167. A \$75,000 truck depreciates over an eight-year period, as shown in the graph. Find r. (Round your answer to three decimal places.)

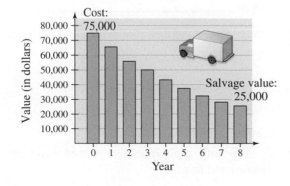

168. A \$125,000 stretch limousine depreciates over a 10-year period, as shown in the graph. Find r. (Round your answer to three decimal places.)

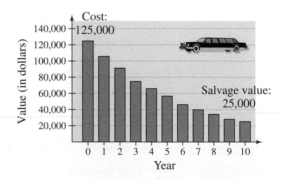

169. ▲ *Geometry*   Find the dimensions of a piece of carpet for a classroom with 529 square feet of floor space, assuming the floor is square.

170. ▲ *Geometry*   Find the dimensions of a square mirror with an area of 1024 square inches.

171. ▲ *Geometry*   The length D of a diagonal of a rectangular solid of length l, width w, and height h is represented by $D = \sqrt{l^2 + w^2 + h^2}$. Approximate to two decimal places the length of the diagonal D of the bed of the pickup truck shown in the figure.

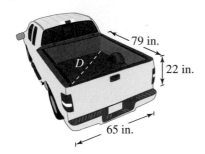

79 in.

D

22 in.

65 in.

172. *Velocity*   A stream of water moving at a rate of v feet per second can carry a particle of a certain type if its diameter is at most $0.03\sqrt{v}$ inches.

(a) Find the largest particle size that can be carried by a stream flowing at the rate of $\frac{3}{4}$ foot per second. Round your answer to three decimal places.

(b) Find the largest particle size that can be carried by a stream flowing at the rate of $\frac{3}{16}$ foot per second. Round your answer to three decimal places.

Explaining Concepts

173. ✎   In your own words, explain what the nth root of a number is.

174. ✎   Explain how you can determine the domain of a function that has the form of a fraction with radical expressions in both the numerator and the denominator.

175. ✎   Is it true that $\sqrt{2} = 1.414$? Explain.

176. Given that x represents a real number, state the conditions on n for each of the following.

(a) $\sqrt[n]{x^n} = x$

(b) $\sqrt[n]{x^n} = |x|$

177. *Investigation*   Find all possible "last digits" of perfect squares. (For instance, the last digit of 81 is 1 and the last digit of 64 is 4.) Is it possible that 4,322,788,986 is a perfect square?

178. ✎   Use what you know about the domains of radical functions to write a set of rules for the domain of a rational exponent function of the form $f(x) = x^{1/n}$. Can the same rules be used for a function of the form $f(x) = x^{m/n}$? Explain.

Cumulative Review

In Exercises 179–182, solve the equation.

179. $\dfrac{a}{5} = \dfrac{a-3}{2}$

180. $\dfrac{x}{3} - \dfrac{3x}{4} = \dfrac{5x}{12}$

181. $\dfrac{2}{u+4} = \dfrac{5}{8}$

182. $\dfrac{6}{b} + 22 = 24$

In Exercises 183–186, write a model for the statement.

183. s is directly proportional to the square of t.

184. r varies inversely as the fourth power of x.

185. a varies jointly as b and c.

186. x is directly proportional to y and inversely proportional to z.

7.2 Simplifying Radical Expressions

Fundamental Photographs

Why You Should Learn It

Algebraic equations often involve radicals. For instance, in Exercise 76 on page 467, you will use a radical equation to find the period of a pendulum.

1 ▶ Use the Product and Quotient Rules for Radicals to simplify radical expressions.

What You Should Learn

1 ▶ Use the Product and Quotient Rules for Radicals to simplify radical expressions.
2 ▶ Use rationalization techniques to simplify radical expressions.
3 ▶ Use the Pythagorean Theorem in application problems.

Simplifying Radicals

In this section, you will study ways to simplify radicals. For instance, the expression $\sqrt{12}$ can be simplified as

$$\sqrt{12} = \sqrt{4 \cdot 3} = \sqrt{4}\sqrt{3} = 2\sqrt{3}.$$

This rewritten form is based on the following rules for multiplying and dividing radicals.

Product and Quotient Rules for Radicals

Let u and v be real numbers, variables, or algebraic expressions. If the nth roots of u and v are real, the following rules are true.

1. $\sqrt[n]{uv} = \sqrt[n]{u}\,\sqrt[n]{v}$ Product Rule for Radicals

2. $\sqrt[n]{\dfrac{u}{v}} = \dfrac{\sqrt[n]{u}}{\sqrt[n]{v}}, \quad v \neq 0$ Quotient Rule for Radicals

---

You can use the Product Rule for Radicals to *simplify* square root expressions by finding the largest perfect square factor and removing it from the radical, as follows.

$$\sqrt{48} = \sqrt{16 \cdot 3} = \sqrt{16}\sqrt{3} = 4\sqrt{3}$$

This process is called **removing perfect square factors from the radical.**

EXAMPLE 1   **Removing Constant Factors from Radicals**

Simplify each radical by removing as many factors as possible.

a. $\sqrt{75}$ **b.** $\sqrt{72}$ **c.** $\sqrt{162}$

Solution

a. $\sqrt{75} = \sqrt{25 \cdot 3} = \sqrt{25}\sqrt{3} = 5\sqrt{3}$ 25 is a perfect square factor of 75.

b. $\sqrt{72} = \sqrt{36 \cdot 2} = \sqrt{36}\sqrt{2} = 6\sqrt{2}$ 36 is a perfect square factor of 72.

c. $\sqrt{162} = \sqrt{81 \cdot 2} = \sqrt{81}\sqrt{2} = 9\sqrt{2}$ 81 is a perfect square factor of 162.

✓ **CHECKPOINT** *Now try Exercise 1.*

When removing *variable* factors from a square root radical, remember that it is not valid to write $\sqrt{x^2} = x$ *unless* you happen to know that x is nonnegative. Without knowing anything about x, the only way you can simplify $\sqrt{x^2}$ is to include absolute value signs when you remove x from the radical.

$$\sqrt{x^2} = |x| \qquad \text{Restricted by absolute value signs}$$

When simplifying the expression $\sqrt{x^3}$, it is not necessary to include absolute value signs because the domain does not include negative numbers.

$$\sqrt{x^3} = \sqrt{x^2(x)} = x\sqrt{x} \qquad \text{Restricted by domain of radical}$$

EXAMPLE 2   Removing Variable Factors from Radicals

Simplify each radical expression.

a. $\sqrt{25x^2}$ **b.** $\sqrt{12x^3}$ **c.** $\sqrt{144x^4}$ **d.** $\sqrt{72x^3y^2}$

Solution

a. $\sqrt{25x^2} = \sqrt{5^2x^2} = \sqrt{5^2}\sqrt{x^2}$ Product Rule for Radicals

$\qquad = 5|x|$ $\sqrt{x^2} = |x|$

b. $\sqrt{12x^3} = \sqrt{2^2x^2(3x)} = \sqrt{2^2}\sqrt{x^2}\sqrt{3x}$ Product Rule for Radicals

$\qquad = 2x\sqrt{3x}$ $\sqrt{2^2}\sqrt{x^2} = 2x, \quad x \geq 0$

c. $\sqrt{144x^4} = \sqrt{12^2(x^2)^2} = \sqrt{12^2}\sqrt{(x^2)^2}$ Product Rule for Radicals

$\qquad = 12x^2$ $\sqrt{12^2}\sqrt{(x^2)^2} = 12|x^2| = 12x^2$

d. $\sqrt{72x^3y^2} = \sqrt{6^2x^2y^2} \cdot \sqrt{2x}$ Product Rule for Radicals

$\qquad = \sqrt{6^2}\sqrt{x^2}\sqrt{y^2} \cdot \sqrt{2x}$ Product Rule for Radicals

$\qquad = 6x|y|\sqrt{2x}$ $\sqrt{6^2}\sqrt{x^2}\sqrt{y^2} = 6x|y|, x \geq 0$

✓ CHECKPOINT *Now try Exercise 19.*

You can use the inverse properties of nth powers and nth roots described in Section 7.1 to remove perfect nth powers from nth root radicals.

EXAMPLE 3   Removing Factors from Radicals

Simplify each radical expression.

a. $\sqrt[3]{40}$ **b.** $\sqrt[4]{x^5}$

Solution

a. $\sqrt[3]{40} = \sqrt[3]{8(5)} = \sqrt[3]{2^3} \cdot \sqrt[3]{5}$ Product Rule for Radicals

$\qquad = 2\sqrt[3]{5}$ $\sqrt[3]{2^3} = 2$

b. $\sqrt[4]{x^5} = \sqrt[4]{x^4(x)} = \sqrt[4]{x^4}\sqrt[4]{x}$ Product Rule for Radicals

$\qquad = x\sqrt[4]{x}$ $\sqrt[4]{x^4} = x, \quad x \geq 0$

✓ CHECKPOINT *Now try Exercise 29.*

EXAMPLE 4   Removing Factors from Radicals

Simplify each radical expression.

a. $\sqrt[5]{486x^7}$ **b.** $\sqrt[3]{128x^3y^5}$

Solution

a. $\sqrt[5]{486x^7} = \sqrt[5]{243x^5(2x^2)} = \sqrt[5]{3^5x^5} \cdot \sqrt[5]{2x^2}$ Product Rule for Radicals

$\qquad = 3x\sqrt[5]{2x^2}$ $\sqrt[5]{3^5}\sqrt[5]{x^5} = 3x$

b. $\sqrt[3]{128x^3y^5} = \sqrt[3]{64x^3y^3(2y^2)} = \sqrt[3]{4^3x^3y^3} \cdot \sqrt[3]{2y^2}$ Product Rule for Radicals

$\qquad = 4xy\sqrt[3]{2y^2}$ $\sqrt[3]{4^3}\sqrt[3]{x^3}\sqrt[3]{y^3} = 4xy$

 **CHECKPOINT** *Now try Exercise 35.*

EXAMPLE 5   Removing Factors from Radicals

Simplify each radical expression.

a. $\sqrt{\dfrac{81}{25}}$ **b.** $\dfrac{\sqrt{56x^2}}{\sqrt{8}}$

Solution

a. $\sqrt{\dfrac{81}{25}} = \dfrac{\sqrt{81}}{\sqrt{25}} = \dfrac{9}{5}$ Quotient Rule for Radicals

b. $\dfrac{\sqrt{56x^2}}{\sqrt{8}} = \sqrt{\dfrac{56x^2}{8}}$ Quotient Rule for Radicals

$\qquad = \sqrt{7x^2}$ Simplify.

$\qquad = \sqrt{7} \cdot \sqrt{x^2}$ Product Rule for Radicals

$\qquad = \sqrt{7}|x|$ $\sqrt{x^2} = |x|$

 **CHECKPOINT** *Now try Exercise 43.*

EXAMPLE 6   Removing Factors from Radicals

Simplify $-\sqrt[3]{\dfrac{y^5}{27x^3}}$.

Solution

$-\sqrt[3]{\dfrac{y^5}{27x^3}} = -\dfrac{\sqrt[3]{y^3y^2}}{\sqrt[3]{27x^3}}$ Quotient Rule for Radicals

$\qquad = -\dfrac{\sqrt[3]{y^3} \cdot \sqrt[3]{y^2}}{\sqrt[3]{27} \cdot \sqrt[3]{x^3}}$ Product Rule for Radicals

$\qquad = -\dfrac{y\sqrt[3]{y^2}}{3x}$ Simplify.

 **CHECKPOINT** *Now try Exercise 49.*

2 ▶ Use rationalization techniques to simplify radical expressions.

Rationalization Techniques

Removing factors from radicals is only one of two techniques used to simplify radicals. Three conditions must be met in order for a radical expression to be in simplest form. These three conditions are summarized as follows.

> ## Simplifying Radical Expressions
>
> A radical expression is said to be in *simplest form* if all three of the statements below are true.
>
> **1.** All possible nth powered factors have been removed from each radical.
>
> **2.** No radical contains a fraction.
>
> **3.** No denominator of a fraction contains a radical.

To meet the last two conditions, you can use a second technique for simplifying radical expressions called **rationalizing the denominator.** This involves multiplying both the numerator and the denominator by a *rationalizing factor* that creates a perfect nth power in the denominator.

Study Tip

When rationalizing a denominator, remember that for square roots you want a perfect square in the denominator, for cube roots you want a perfect cube, and so on. For instance, to find the rationalizing factor needed to create a perfect square in the denominator of Example 7(c), you can write the prime factorization of 18.

$$18 = 2 \cdot 3 \cdot 3$$
$$= 2 \cdot 3^2$$

From its prime factorization, you can see that 3^2 is a square root factor of 18. You need one more factor of 2 to create a perfect square in the denominator:

$$2 \cdot (2 \cdot 3^2) = 2 \cdot 2 \cdot 3^2$$
$$= 2^2 \cdot 3^2$$
$$= 4 \cdot 9 = 36.$$

EXAMPLE 7   Rationalizing the Denominator

Rationalize the denominator in each expression.

a. $\sqrt{\dfrac{3}{5}}$ **b.** $\dfrac{4}{\sqrt[3]{9}}$ **c.** $\dfrac{8}{3\sqrt{18}}$

Solution

a. $\sqrt{\dfrac{3}{5}} = \dfrac{\sqrt{3}}{\sqrt{5}} = \dfrac{\sqrt{3}}{\sqrt{5}} \cdot \dfrac{\sqrt{5}}{\sqrt{5}} = \dfrac{\sqrt{15}}{\sqrt{5^2}} = \dfrac{\sqrt{15}}{5}$ Multiply by $\sqrt{5}/\sqrt{5}$ to create a perfect square in the denominator.

b. $\dfrac{4}{\sqrt[3]{9}} = \dfrac{4}{\sqrt[3]{9}} \cdot \dfrac{\sqrt[3]{3}}{\sqrt[3]{3}} = \dfrac{4\sqrt[3]{3}}{\sqrt[3]{27}} = \dfrac{4\sqrt[3]{3}}{3}$ Multiply by $\sqrt[3]{3}/\sqrt[3]{3}$ to create a perfect cube in the denominator.

c. $\dfrac{8}{3\sqrt{18}} = \dfrac{8}{3\sqrt{18}} \cdot \dfrac{\sqrt{2}}{\sqrt{2}} = \dfrac{8\sqrt{2}}{3\sqrt{36}} = \dfrac{8\sqrt{2}}{3\sqrt{6^2}}$ Multiply by $\sqrt{2}/\sqrt{2}$ to create a perfect square in the denominator.

$= \dfrac{8\sqrt{2}}{3(6)} = \dfrac{4\sqrt{2}}{9}$

✓ **CHECKPOINT** *Now try Exercise 55.*

EXAMPLE 8   Rationalizing the Denominator

a. $\sqrt{\dfrac{8x}{12y^5}} = \sqrt{\dfrac{(4)(2)x}{(4)(3)y^5}} = \sqrt{\dfrac{2x}{3y^5}} = \dfrac{\sqrt{2x}}{\sqrt{3y^5}} \cdot \dfrac{\sqrt{3y}}{\sqrt{3y}} = \dfrac{\sqrt{6xy}}{\sqrt{3^2y^6}} = \dfrac{\sqrt{6xy}}{3|y^3|}$

b. $\sqrt[3]{\dfrac{54x^6y^3}{5z^2}} = \dfrac{\sqrt[3]{(3^3)(2)(x^6)(y^3)}}{\sqrt[3]{5z^2}} \cdot \dfrac{\sqrt[3]{25z}}{\sqrt[3]{25z}} = \dfrac{3x^2y\sqrt[3]{50z}}{\sqrt[3]{5^3z^3}} = \dfrac{3x^2y\sqrt[3]{50z}}{5z}$

✓ **CHECKPOINT** *Now try Exercise 69.*

3 ▶ Use the Pythagorean Theorem in application problems.

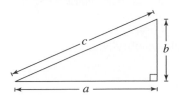

Figure 7.1

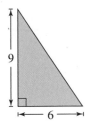

Figure 7.2

Applications of Radicals

Radicals commonly occur in applications involving right triangles. Recall that a right triangle is one that contains a right (or 90°) angle, as shown in Figure 7.1. The relationship among the three sides of a right triangle is described by the **Pythagorean Theorem,** which states that if a and b are the lengths of the legs and c is the length of the hypotenuse, then

$$c = \sqrt{a^2 + b^2} \quad \text{and} \quad a = \sqrt{c^2 - b^2}. \quad \text{Pythagorean Theorem: } a^2 + b^2 = c^2$$

EXAMPLE 9   The Pythagorean Theorem

Find the length of the hypotenuse of the right triangle shown in Figure 7.2.

Solution

Because you know that $a = 6$ and $b = 9$, you can use the Pythagorean Theorem to find c, as follows.

$c = \sqrt{a^2 + b^2}$	Pythagorean Theorem
$= \sqrt{6^2 + 9^2}$	Substitute 6 for a and 9 for b.
$= \sqrt{117}$	Simplify.
$= \sqrt{9}\sqrt{13}$	Product Rule for Radicals
$= 3\sqrt{13}$	Simplify.

✓ **CHECKPOINT** *Now try Exercise 73.*

EXAMPLE 10   An Application of the Pythagorean Theorem

A softball diamond has the shape of a square with 60-foot sides, as shown in Figure 7.3. The catcher is 5 feet behind home plate. How far does the catcher have to throw to reach second base?

Solution

In Figure 7.3, let x be the hypotenuse of a right triangle with 60-foot sides. So, by the Pythagorean Theorem, you have the following.

$x = \sqrt{60^2 + 60^2}$	Pythagorean Theorem
$= \sqrt{7200}$	Simplify.
$= \sqrt{3600}\sqrt{2}$	Product Rule for Radicals
$= 60\sqrt{2}$	Simplify.
≈ 84.9 feet	Use a calculator.

So, the distance from home plate to second base is approximately 84.9 feet. Because the catcher is 5 feet behind home plate, the catcher must make a throw of

$$x + 5 \approx 84.9 + 5 = 89.9 \text{ feet.}$$

✓ **CHECKPOINT** *Now try Exercise 77.*

Figure 7.3

_____ **Concept Check** _____

1. When is $\sqrt{x^2} \neq x$? Explain.

2. Explain why $\sqrt{8}$ is not in simplest form.

3. Describe the three conditions that characterize a simplified radical expression.

4. Describe how you would simplify $\dfrac{1}{\sqrt{3}}$.

7.2 EXERCISES

Go to pages 502–503 to record your assignments.

_____ **Developing Skills** _____

In Exercises 1–18, simplify the radical. (Do not use a calculator.) *See Example 1.*

1. $\sqrt{18}$

2. $\sqrt{27}$

3. $\sqrt{45}$

4. $\sqrt{125}$

5. $\sqrt{96}$

6. $\sqrt{84}$

7. $\sqrt{153}$

8. $\sqrt{147}$

9. $\sqrt{1183}$

10. $\sqrt{1176}$

11. $\sqrt{0.04}$

12. $\sqrt{0.25}$

13. $\sqrt{0.0072}$

14. $\sqrt{0.0027}$

15. $\sqrt{\frac{60}{3}}$

16. $\sqrt{\frac{208}{4}}$

17. $\sqrt{\frac{13}{25}}$

18. $\sqrt{\frac{15}{36}}$

In Exercises 19–54, simplify the radical expression. *See Examples 2–6.*

19. $\sqrt{9x^5}$

20. $\sqrt{64x^3}$

21. $\sqrt{48y^4}$

22. $\sqrt{32x}$

23. $\sqrt{117y^6}$

24. $\sqrt{160x^8}$

25. $\sqrt{120x^2y^3}$

26. $\sqrt{125u^4v^6}$

27. $\sqrt{192a^5b^7}$

28. $\sqrt{363x^{10}y^9}$

29. $\sqrt[3]{48}$

30. $\sqrt[3]{54}$

31. $\sqrt[3]{112}$

32. $\sqrt[4]{112}$

33. $\sqrt[3]{40x^5}$

34. $\sqrt[3]{81a^7}$

35. $\sqrt[4]{324y^6}$

36. $\sqrt[5]{160x^8}$

37. $\sqrt[3]{x^4y^3}$

38. $\sqrt[3]{a^5b^6}$

39. $\sqrt[4]{4x^4y^6}$

40. $\sqrt[4]{128u^4v^7}$

41. $\sqrt[5]{32x^5y^6}$

42. $\sqrt[3]{16x^4y^5}$

43. $\sqrt[3]{\frac{35}{64}}$

44. $\sqrt[4]{\frac{5}{16}}$

45. $\dfrac{\sqrt{39y^2}}{\sqrt{3}}$

46. $\dfrac{\sqrt{56w^3}}{\sqrt{2}}$

47. $\sqrt{\dfrac{32a^4}{b^2}}$

48. $\sqrt{\dfrac{18x^2}{z^6}}$

49. $\sqrt[5]{\dfrac{32x^2}{y^5}}$

50. $\sqrt[3]{\dfrac{16z^3}{y^6}}$

51. $\sqrt[3]{\dfrac{54a^4}{b^9}}$

52. $\sqrt[4]{\dfrac{3u^2}{16v^8}}$

53. $-\sqrt[3]{\dfrac{3w^4}{8z^3}}$

54. $-\sqrt[4]{\dfrac{42y^7}{81x^4}}$

In Exercises 55–72, rationalize the denominator and simplify further, if possible. *See Examples 7 and 8.*

55. $\sqrt{\frac{1}{3}}$

56. $\sqrt{\frac{1}{5}}$

57. $\dfrac{1}{\sqrt{7}}$

58. $\dfrac{12}{\sqrt{3}}$

59. $\sqrt[4]{\frac{5}{4}}$

60. $\sqrt[3]{\frac{9}{25}}$

61. $\dfrac{6}{\sqrt[3]{32}}$

62. $\dfrac{10}{\sqrt[5]{16}}$

63. $\dfrac{1}{\sqrt{y}}$

64. $\dfrac{2}{\sqrt{3c}}$

65. $\sqrt{\dfrac{4}{x}}$

66. $\sqrt{\dfrac{4}{x^3}}$

67. $\dfrac{1}{x\sqrt{2}}$

68. $\dfrac{1}{3x\sqrt{x}}$

69. $\dfrac{6}{\sqrt{3b^3}}$

70. $\dfrac{1}{\sqrt{xy}}$

71. $\sqrt[3]{\dfrac{2x}{3y}}$

72. $\sqrt[3]{\dfrac{20x^2}{9y^2}}$

Solving Problems

 Geometry   In Exercises 73 and 74, find the length of the hypotenuse of the right triangle. *See Example 9.*

73.

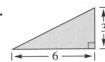

74.

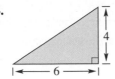

75. *Frequency*   The frequency *f* in cycles per second of a vibrating string is given by

$$f = \frac{1}{100}\sqrt{\frac{400 \times 10^6}{5}}.$$

Use a calculator to approximate this number. (Round the result to two decimal places.)

76. *Period of a Pendulum*   The time *t* (in seconds) for a pendulum of length *L* (in feet) to go through one complete cycle (its period) is given by

$$t = 2\pi\sqrt{\frac{L}{32}}.$$

Find the period of a pendulum whose length is 4 feet. (Round your answer to two decimal places.)

77. ▲ *Geometry*   A ladder is to reach a window that is 26 feet high. The ladder is placed 10 feet from the base of the wall (see figure). How long must the ladder be?

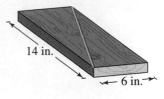

Figure for 77   Figure for 78

78. ▲ *Geometry*   A string is attached to opposite corners of a piece of wood that is 6 inches wide and 14 inches long (see figure). How long must the string be?

Explaining Concepts

79. Give an example of multiplying two radicals.

80. Enter any positive real number into your calculator and find its square root. Then repeatedly take the square root of the result.

$$\sqrt{x}, \ \sqrt{\sqrt{x}}, \ \sqrt{\sqrt{\sqrt{x}}}, \dots$$

What real number does the display appear to be approaching?

81. Square the real number $5/\sqrt{3}$ and note that the radical is eliminated from the denominator. Is this equivalent to rationalizing the denominator? Why or why not?

82. Let *u* be a positive real number. Explain why $\sqrt[3]{u} \cdot \sqrt[4]{u} \neq \sqrt[12]{u}$.

83. Explain how to find a perfect *n*th root factor in the radicand of an *n*th root radical.

Cumulative Review

In Exercises 84 and 85, sketch the graphs of the equations and approximate any solutions of the system of linear equations.

84. $\begin{cases} 3x + 2y = -4 \\ y = 3x + 7 \end{cases}$

85. $\begin{cases} 2x + 3y = 12 \\ 4x - y = 10 \end{cases}$

In Exercises 86 and 87, solve the system by the method of substitution.

86. $\begin{cases} x - 3y = -2 \\ 7y - 4x = 6 \end{cases}$

87. $\begin{cases} y = x + 2 \\ y - x = 8 \end{cases}$

In Exercises 88 and 89, solve the system by the method of elimination.

88. $\begin{cases} 1.5x - 3 = -2y \\ 3x + 4y = 6 \end{cases}$

89. $\begin{cases} x + 4y + 3z = 2 \\ 2x + y + z = 10 \\ -x + y + 2z = 8 \end{cases}$

7.3 Adding and Subtracting Radical Expressions

© James Marshall/The Image Works

Why You Should Learn It

Radical expressions can be used to model and solve real-life problems. For instance, Example 6 on page 470 shows how to find a radical expression that models the out-of-pocket expense of attending college.

1 ▶ Use the Distributive Property to add and subtract like radicals.

What You Should Learn

1 ▶ Use the Distributive Property to add and subtract like radicals.

2 ▶ Use radical expressions in application problems.

Adding and Subtracting Radical Expressions

Two or more radical expressions are called **like radicals** if they have the same index and the same radicand. For instance, the expressions $\sqrt{2}$ and $3\sqrt{2}$ are like radicals, whereas the expressions $\sqrt{3}$ and $\sqrt[3]{3}$ are not. Two radical expressions that are like radicals can be added or subtracted by adding or subtracting their coefficients.

EXAMPLE 1   **Combining Radical Expressions**

Simplify each expression by combining like radicals.

a. $\sqrt{7} + 5\sqrt{7} - 2\sqrt{7}$

b. $6\sqrt{x} - \sqrt[3]{4} - 5\sqrt{x} + 2\sqrt[3]{4}$

c. $3\sqrt[3]{x} + 2\sqrt[3]{x} + \sqrt{x} - 8\sqrt{x}$

Solution

a. $\sqrt{7} + 5\sqrt{7} - 2\sqrt{7} = (1 + 5 - 2)\sqrt{7}$ Distributive Property

$\qquad\qquad\qquad\qquad\quad = 4\sqrt{7}$ Simplify.

b. $6\sqrt{x} - \sqrt[3]{4} - 5\sqrt{x} + 2\sqrt[3]{4}$

$\quad = \left(6\sqrt{x} - 5\sqrt{x}\right) + \left(-\sqrt[3]{4} + 2\sqrt[3]{4}\right)$ Group like radicals.

$\quad = (6 - 5)\sqrt{x} + (-1 + 2)\sqrt[3]{4}$ Distributive Property

$\quad = \sqrt{x} + \sqrt[3]{4}$ Simplify.

c. $3\sqrt[3]{x} + 2\sqrt[3]{x} + \sqrt{x} - 8\sqrt{x}$

$\quad = (3 + 2)\sqrt[3]{x} + (1 - 8)\sqrt{x}$ Distributive Property

$\quad = 5\sqrt[3]{x} - 7\sqrt{x}$ Simplify.

✓ **CHECKPOINT** *Now try Exercise 13.*

Before concluding that two radicals cannot be combined, you should first rewrite them in simplest form. This is illustrated in Examples 2 and 3.

Study Tip

It is important to realize that the expression $\sqrt{a} + \sqrt{b}$ is not equal to $\sqrt{a + b}$. For instance, you may be tempted to add $\sqrt{6} + \sqrt{3}$ and get $\sqrt{9} = 3$. But remember, you cannot add unlike radicals. So, $\sqrt{6} + \sqrt{3}$ cannot be simplified further.

EXAMPLE 2 **Simplifying Before Combining Radical Expressions**

Simplify each expression by combining like radicals.

a. $\sqrt{45x} + 3\sqrt{20x}$

b. $5\sqrt{x^3} - x\sqrt{4x}$

Solution

a. $\sqrt{45x} + 3\sqrt{20x} = 3\sqrt{5x} + 6\sqrt{5x}$ Simplify radicals.

$= 9\sqrt{5x}$ Combine like radicals.

b. $5\sqrt{x^3} - x\sqrt{4x} = 5x\sqrt{x} - 2x\sqrt{x}$ Simplify radicals.

$= 3x\sqrt{x}$ Combine like radicals.

✓ **CHECKPOINT** *Now try Exercise 25.*

EXAMPLE 3 **Simplifying Before Combining Radical Expressions**

Simplify each expression by combining like radicals.

a. $\sqrt[3]{54y^5} + 4\sqrt[3]{2y^2}$

b. $\sqrt[3]{6x^4} + \sqrt[3]{48x} - \sqrt[3]{162x^4}$

Solution

a. $\sqrt[3]{54y^5} + 4\sqrt[3]{2y^2} = 3y\sqrt[3]{2y^2} + 4\sqrt[3]{2y^2}$ Simplify radicals.

$= (3y + 4)\sqrt[3]{2y^2}$ Distributive Property

b. $\sqrt[3]{6x^4} + \sqrt[3]{48x} - \sqrt[3]{162x^4}$ Write original expression.

$= x\sqrt[3]{6x} + 2\sqrt[3]{6x} - 3x\sqrt[3]{6x}$ Simplify radicals.

$= (x + 2 - 3x)\sqrt[3]{6x}$ Distributive Property

$= (2 - 2x)\sqrt[3]{6x}$ Combine like terms.

✓ **CHECKPOINT** *Now try Exercise 35.*

It may be necessary to rationalize denominators before combining radicals.

EXAMPLE 4 **Rationalizing Denominators Before Simplifying**

$\sqrt{7} - \dfrac{5}{\sqrt{7}} = \sqrt{7} - \left(\dfrac{5}{\sqrt{7}} \cdot \dfrac{\sqrt{7}}{\sqrt{7}}\right)$ Multiply by $\sqrt{7}/\sqrt{7}$ to remove the radical from the denominator.

$= \sqrt{7} - \dfrac{5\sqrt{7}}{7}$ Simplify.

$= \left(1 - \dfrac{5}{7}\right)\sqrt{7}$ Distributive Property

$= \dfrac{2}{7}\sqrt{7}$ Simplify.

✓ **CHECKPOINT** *Now try Exercise 47.*

2 ▶ Use radical expressions in application problems.

Applications

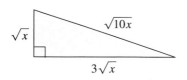

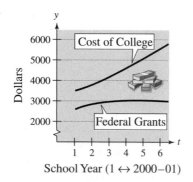

Figure 7.4

Geometry: Perimeter of a Triangle

Write and simplify an expression for the perimeter of the triangle shown in Figure 7.4.

Solution

$P = a + b + c$	Formula for perimeter of a triangle
$= \sqrt{x} + 3\sqrt{x} + \sqrt{10x}$	Substitute.
$= (1 + 3)\sqrt{x} + \sqrt{10x}$	Distributive Property
$= 4\sqrt{x} + \sqrt{10x}$	Simplify.

✓ **CHECKPOINT** *Now try Exercise 61.*

EXAMPLE 6   **Out-of-Pocket Expense**

The cost C of attending a four-year public college and the aid A per recipient (for eligible students) from federal grants from 2000 to 2006 can be modeled by the equations

$$C = 3783 + 697.7t - 989.2\sqrt{t}, \quad 1 \le t \le 6 \quad \text{College cost}$$

$$A = 1473 - 329.1t + 1431.1\sqrt{t}, \quad 1 \le t \le 6 \quad \text{Grant amount}$$

where t represents the school year, with $t = 1$ corresponding to 2000–01. (See Figure 7.5.) Find a radical expression that models the out-of-pocket expense E incurred by a college student from 2000 to 2006. Estimate the out-of-pocket expense E of a college student during the 2004–05 school year. (Source: *Annual Survey of Colleges,* The College Board)

Solution

The difference of the cost and the aid gives the out-of-pocket expense.

$$C - A = (3783 + 697.7t - 989.2\sqrt{t}) - (1473 - 329.1t + 1431.1\sqrt{t})$$

$$= 3783 + 697.7t - 989.2\sqrt{t} - 1473 + 329.1t - 1431.1\sqrt{t}$$

$$= (3783 - 1473) + (697.7t + 329.1t) - (989.2\sqrt{t} + 1431.1\sqrt{t})$$

$$= 2310 + 1026.8t - 2420.3\sqrt{t}$$

So, the radical expression that models the out-of-pocket expense incurred by a college student is

$$E = C - A$$

$$= 2310 + 1026.8t - 2420.3\sqrt{t}.$$

Using this model, substitute $t = 5$ to estimate the out-of-pocket expense of a college student during the 2004–05 school year.

$$E = 2310 + 1026.8(5) - 2420.3\sqrt{5} \approx 2032$$

✓ **CHECKPOINT** *Now try Exercise 67.*

From the graph, what can you deduce about the out-of-pocket expense of a college student?

Figure 7.5

_____ **Concept Check** _____

1. Explain what it means for two radical expressions to be like radicals.

2. Explain how to add or subtract like radicals.

3. Is $\sqrt{2} + \sqrt{18}$ in simplest form? Explain.

4. Is $\sqrt{2} - \dfrac{1}{\sqrt{2}}$ in simplest form? Explain.

7.3 EXERCISES

Go to pages 502–503 to record your assignments.

_____ **Developing Skills** _____

In Exercises 1–46, combine the radical expressions, if possible. *See Examples 1–3.*

1. $3\sqrt{2} - \sqrt{2}$

2. $6\sqrt{5} - 2\sqrt{5}$

3. $2\sqrt{6} + 5\sqrt{6}$

4. $3\sqrt{7} + 2\sqrt{7}$

5. $8\sqrt{5} + 9\sqrt[3]{5}$

6. $3\sqrt[3]{3} + 6\sqrt[3]{3}$

7. $9\sqrt[3]{5} - 6\sqrt[3]{5}$

8. $8\sqrt[4]{5} - 2\sqrt[3]{5}$

9. $4\sqrt[3]{y} + 9\sqrt[3]{y}$

10. $13\sqrt{x} + \sqrt{x}$

11. $15\sqrt[4]{s} - \sqrt[4]{s}$

12. $9\sqrt[4]{t} - 3\sqrt[4]{t}$

✓ 13. $8\sqrt{2} + 6\sqrt{2} - 5\sqrt{2}$

14. $2\sqrt{6} + 8\sqrt{6} - 3\sqrt{6}$

15. $\sqrt[4]{5} - 6\sqrt[4]{13} + 3\sqrt[4]{5} - \sqrt[4]{13}$

16. $9\sqrt[3]{17} + 7\sqrt[3]{2} - 4\sqrt[3]{17} + \sqrt[3]{2}$

17. $9\sqrt[3]{7} - \sqrt{3} + 4\sqrt[3]{7} + 2\sqrt{3}$

18. $5\sqrt{7} - 8\sqrt[4]{11} + \sqrt{7} + 9\sqrt[4]{11}$

19. $8\sqrt{27} - 3\sqrt{3}$

20. $9\sqrt{50} - 4\sqrt{2}$

21. $3\sqrt{45} + 7\sqrt{20}$

22. $5\sqrt{12} + 16\sqrt{27}$

23. $2\sqrt[3]{54} + 12\sqrt[3]{16}$

24. $4\sqrt[4]{48} - \sqrt[4]{243}$

✓ 25. $5\sqrt{9x} - 3\sqrt{x}$

26. $4\sqrt{y} + 2\sqrt{16y}$

27. $3\sqrt{x+1} + 10\sqrt{x+1}$

28. $7\sqrt{2a-3} - 4\sqrt{2a-3}$

29. $\sqrt{25y} + \sqrt{64y}$

30. $\sqrt[3]{16t^4} - \sqrt[3]{54t^4}$

31. $10\sqrt[3]{z} - \sqrt[3]{z^4}$

32. $5\sqrt[3]{24u^2} + 2\sqrt[3]{81u^5}$

33. $\sqrt{5a} + 2\sqrt{45a^3}$

34. $4\sqrt{3x^3} - \sqrt{12x}$

✓ 35. $\sqrt[3]{6x^4} + \sqrt[3]{48x}$

36. $\sqrt[3]{54x} - \sqrt[3]{2x^4}$

37. $\sqrt{9x-9} + \sqrt{x-1}$

38. $\sqrt{4y+12} + \sqrt{y+3}$

39. $\sqrt{x^3 - x^2} + \sqrt{4x-4}$

40. $\sqrt{9x-9} - \sqrt{x^3 - x^2}$

41. $2\sqrt[3]{a^4b^2} + 3a\sqrt[3]{ab^2}$

42. $3y\sqrt[4]{2x^5y^3} - x\sqrt[4]{162xy^7}$

43. $\sqrt{4r^7s^5} + 3r^2\sqrt{r^3s^5} - 2rs\sqrt{r^5s^3}$

44. $x\sqrt[3]{27x^5y^2} - x^2\sqrt[3]{x^2y^2} + z\sqrt[3]{x^8y^2}$

45. $\sqrt[3]{128x^9y^{10}} - 2x^2y\sqrt[3]{16x^3y^7}$

46. $5\sqrt[3]{320x^5y^8} + 2x\sqrt[3]{135x^2y^8}$

In Exercises 47–56, perform the addition or subtraction and simplify your answer. *See Example 4.*

✓ 47. $\sqrt{5} - \dfrac{3}{\sqrt{5}}$

48. $\sqrt{10} + \dfrac{5}{\sqrt{10}}$

49. $\sqrt{32} + \sqrt{\dfrac{1}{2}}$

50. $\sqrt{\dfrac{1}{5}} - \sqrt{45}$

51. $\sqrt{12y} - \dfrac{y}{\sqrt{3y}}$

52. $\dfrac{x}{\sqrt{3x}} + \sqrt{27x}$

53. $\dfrac{2}{\sqrt{3x}} + \sqrt{3x}$

54. $2\sqrt{7x} - \dfrac{4}{\sqrt{7x}}$

55. $\sqrt{7y^3} - \sqrt{\dfrac{9}{7y^3}}$

56. $\sqrt{\dfrac{4}{3x^3}} + \sqrt{3x^3}$

In Exercises 57–60, place the correct symbol ($<$, $>$, or $=$) between the numbers.

57. $\sqrt{7} + \sqrt{18}$ ▢ $\sqrt{7 + 18}$

58. $\sqrt{10} - \sqrt{6}$ ▢ $\sqrt{10 - 6}$

59. 5 ▢ $\sqrt{9^2 - 4^2}$

60. 5 ▢ $\sqrt{3^2 + 4^2}$

Solving Problems

▲ *Geometry*   In Exercises 61–64, write a simplified expression for the perimeter of the figure. ***See Example 5.***

✓ 61.

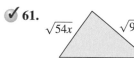

62.

63.

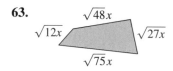

64.

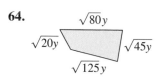

65. ▲ *Geometry*   The foundation of a house is 40 feet long and 30 feet wide. The height of the attic is 5 feet (see figure).

(a) Use the Pythagorean Theorem to find the length of the hypotenuse of each of the two right triangles formed by the roof line. (Assume there is no overhang.)

(b) Use the result of part (a) to determine the total area of the roof.

66. ▲ *Geometry*   The four corners are cut from a four-foot-by-eight-foot sheet of plywood, as shown in the figure. Find the perimeter of the remaining piece of plywood.

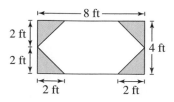

67. *Immigration* Legal permanent residents are immigrants who have received a "green card," which grants them the right to live in the United States. The number of immigrants from Colombia C (in thousands) and South America S (in thousands) who received green cards from 2001 to 2006 can be modeled by the equations

$$C = -118 + 286.7\sqrt{t} - 195.1t + 42.8\sqrt{t^3},$$
$$1 \leq t \leq 6$$

$$S = -182 + 565.5\sqrt{t} - 409.2t + 94.6\sqrt{t^3},$$
$$1 \leq t \leq 6$$

where t represents the year, with $t = 1$ corresponding to 2001. Find a radical expression that models the total number T of legal permanent residents from South America, excluding Colombia, from 2001 to 2006. Estimate T in 2003. (Source: U.S. Department of Homeland Security)

68. *Immigration* The number of immigrants from Africa A (in thousands) who received green cards from 2001 to 2006 can be modeled by the equation

$$A = -812 + 2138.9\sqrt{t} - 1881t + 694.2\sqrt{t^3}$$
$$- 89.8t^2, \quad 1 \leq t \leq 6$$

where t represents the year, with $t = 1$ corresponding to 2001. Use the South American model from Exercise 67 to find a radical expression that models the total number T of legal permanent residents from Africa and South America from 2001 to 2006. Estimate T in 2005. (Source: U.S. Department of Homeland Security)

Explaining Concepts

69. Will the sum of two radicals always be a radical? Give an example to support your answer.

70. Will the difference of two radicals always be a radical? Give an example to support your answer.

71. ✎ Is $\sqrt{2x} + \sqrt{2x}$ equal to $\sqrt{8x}$? Explain.

72. ✎ Explain how adding two monomials compares to adding two radicals.

73. You are an algebra instructor, and one of your students hands in the following work. Find and correct the errors, and discuss how you can help your student avoid such errors in the future.

(a) $7\sqrt{3} + 4\sqrt{2} = 11\sqrt{5}$

(b) $3\sqrt[3]{k} - 6\sqrt{k} = -3\sqrt{k}$

Cumulative Review

In Exercises 74–79, combine the rational expressions and simplify.

74. $\dfrac{2x + 1}{3x} + \dfrac{3 - 4x}{3x}$

75. $\dfrac{7z - 2}{2z} - \dfrac{4z + 1}{2z}$

76. $\dfrac{4m + 6}{m + 2} - \dfrac{3m + 4}{m + 2}$

77. $\dfrac{2x + 3}{x - 3} + \dfrac{6 - 5x}{x - 3}$

78. $\dfrac{4}{x - 4} + \dfrac{2x}{x + 1}$

79. $\dfrac{2v}{v - 5} - \dfrac{3}{5 - v}$

In Exercises 80–83, simplify the complex fraction.

80. $\dfrac{\left(\dfrac{2}{3}\right)}{\left(\dfrac{4}{15}\right)}$

81. $\dfrac{\left(\dfrac{27a^3}{4b^2c}\right)}{\left(\dfrac{9ac^2}{10b^2}\right)}$

82. $\dfrac{\left(\dfrac{x^2 + 2x - 8}{x - 8}\right)}{2x + 8}$

83. $\dfrac{3w - 9}{\left(\dfrac{w^2 - 10w + 21}{w + 1}\right)}$

Mid-Chapter Quiz

Take this quiz as you would take a quiz in class. After you are done, check your work against the answers in the back of the book.

In Exercises 1–4, evaluate the expression.

1. $\sqrt{225}$

2. $\sqrt[4]{\frac{81}{16}}$

3. $49^{1/2}$

4. $(-27)^{2/3}$

In Exercises 5 and 6, evaluate the function as indicated, if possible, and simplify.

5. $f(x) = \sqrt{3x - 5}$

    (a) $f(0)$ (b) $f(2)$ (c) $f(10)$

6. $g(x) = \sqrt{9 - x}$

    (a) $g(-7)$ (b) $g(5)$ (c) $g(9)$

In Exercises 7 and 8, describe the domain of the function.

7. $g(x) = \dfrac{12}{\sqrt[3]{x}}$

8. $h(x) = \sqrt{3x + 10}$

In Exercises 9–14, simplify the expression.

9. $\sqrt{27x^2}$

10. $\sqrt[4]{32x^8}$

11. $\sqrt{\dfrac{4u^3}{9}}$

12. $\sqrt[3]{\dfrac{16}{u^6}}$

13. $\sqrt{125x^3y^2z^4}$

14. $2a\sqrt[3]{16a^3b^5}$

In Exercises 15 and 16, rationalize the denominator and simplify further, if possible.

15. $\dfrac{24}{\sqrt{12}}$

16. $\dfrac{21x^2}{\sqrt{7x}}$

In Exercises 17–22, combine the radical expressions, if possible.

17. $2\sqrt{3} - 4\sqrt{7} + \sqrt{3}$

18. $\sqrt{200y} - 3\sqrt{8y}$

19. $5\sqrt{12} + 2\sqrt{3} - \sqrt{75}$

20. $\sqrt{25x + 50} - \sqrt{x + 2}$

21. $6x\sqrt[3]{5x^2} + 2\sqrt[3]{40x^4}$

22. $3\sqrt{x^3y^4z^5} + 2xy^2\sqrt{xz^5} - xz^2\sqrt{xy^4z}$

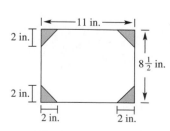

Figure for 23

23. The four corners are cut from an $8\frac{1}{2}$-inch-by-11-inch sheet of paper, as shown in the figure at the left. Find the perimeter of the remaining piece of paper.

7.4 Multiplying and Dividing Radical Expressions

Paul A. Souders/CORBIS

Why You Should Learn It

Multiplication of radicals is often used in real-life applications. For instance, in Exercise 111 on page 481, you will multiply two radical expressions to find the area of the cross section of a wooden beam.

1 ▶ Use the Distributive Property or the FOIL Method to multiply radical expressions.

What You Should Learn

1 ▶ Use the Distributive Property or the FOIL Method to multiply radical expressions.
2 ▶ Determine the products of conjugates.
3 ▶ Simplify quotients involving radicals by rationalizing the denominators.

Multiplying Radical Expressions

To multiply radical expressions, you can use the Product Rule for Radicals from Section 7.2, given by $\sqrt[n]{uv} = \sqrt[n]{u}\sqrt[n]{v}$, where u and v are real numbers whose nth roots are also real numbers. When the expressions you are multiplying involve sums or differences, you can also use the Distributive Property or the FOIL Method.

EXAMPLE 1 **Multiplying Radical Expressions**

Find each product and simplify.

a. $\sqrt{6} \cdot \sqrt{3}$ **b.** $\sqrt[3]{5} \cdot \sqrt[3]{16}$

Solution

a. $\sqrt{6} \cdot \sqrt{3} = \sqrt{6 \cdot 3} = \sqrt{18} = \sqrt{9 \cdot 2} = 3\sqrt{2}$
b. $\sqrt[3]{5} \cdot \sqrt[3]{16} = \sqrt[3]{5 \cdot 16} = \sqrt[3]{80} = \sqrt[3]{8 \cdot 10} = 2\sqrt[3]{10}$

✓ **CHECKPOINT** *Now try Exercise 1.*

EXAMPLE 2 **Multiplying Radical Expressions**

Find each product and simplify.

a. $\sqrt{3}\left(2 + \sqrt{5}\right)$ **b.** $\sqrt{2}\left(4 - \sqrt{8}\right)$ **c.** $\sqrt{6}\left(\sqrt{12} - \sqrt{3}\right)$

Solution

a. $\sqrt{3}\left(2 + \sqrt{5}\right) = 2\sqrt{3} + \sqrt{3}\sqrt{5}$ Distributive Property
$\qquad\qquad\qquad = 2\sqrt{3} + \sqrt{15}$ Product Rule for Radicals

b. $\sqrt{2}\left(4 - \sqrt{8}\right) = 4\sqrt{2} - \sqrt{2}\sqrt{8}$ Distributive Property
$\qquad\qquad\qquad = 4\sqrt{2} - \sqrt{16} = 4\sqrt{2} - 4$ Product Rule for Radicals

c. $\sqrt{6}\left(\sqrt{12} - \sqrt{3}\right) = \sqrt{6}\sqrt{12} - \sqrt{6}\sqrt{3}$ Distributive Property
$\qquad\qquad\qquad = \sqrt{72} - \sqrt{18}$ Product Rule for Radicals
$\qquad\qquad\qquad = 6\sqrt{2} - 3\sqrt{2} = 3\sqrt{2}$ Find perfect square factors.

✓ **CHECKPOINT** *Now try Exercise 9.*

Smart Study Strategy

Go to page 448 for ways to *Form a Weekly Study Group.*

In Example 2, the Distributive Property was used to multiply radical expressions. In Example 3, note how the FOIL Method is used.

EXAMPLE 3 Using the FOIL Method

$$
\begin{matrix}
& \text{F} & \text{O} & \text{I} & \text{L}
\end{matrix}
$$

a. $\left(2\sqrt{7} - 4\right)\left(\sqrt{7} + 1\right) = 2\left(\sqrt{7}\right)^2 + 2\sqrt{7} - 4\sqrt{7} - 4$ FOIL Method

$$= 2(7) + (2 - 4)\sqrt{7} - 4 \qquad \text{Combine like radicals.}$$

$$= 10 - 2\sqrt{7} \qquad \text{Simplify.}$$

b. $\left(3 - \sqrt{x}\right)\left(1 + \sqrt{x}\right) = 3 + 3\sqrt{x} - \sqrt{x} - \left(\sqrt{x}\right)^2$ FOIL Method

$$= 3 + 2\sqrt{x} - x, \quad x \geq 0 \qquad \text{Combine like radicals and simplify.}$$

✔ **CHECKPOINT** *Now try Exercise 23.*

2 ▶ Determine the products of conjugates.

Conjugates

The expressions $3 + \sqrt{6}$ and $3 - \sqrt{6}$ are called **conjugates** of each other. Notice that they differ only in the sign between the terms. The product of two conjugates is the difference of two squares, which is given by the special product formula $(a + b)(a - b) = a^2 - b^2$. Here are some other examples.

Expression	Conjugate	Product
$1 - \sqrt{3}$	$1 + \sqrt{3}$	$(1)^2 - \left(\sqrt{3}\right)^2 = 1 - 3 = -2$
$\sqrt{5} + \sqrt{2}$	$\sqrt{5} - \sqrt{2}$	$\left(\sqrt{5}\right)^2 - \left(\sqrt{2}\right)^2 = 5 - 2 = 3$
$\sqrt{10} - 3$	$\sqrt{10} + 3$	$\left(\sqrt{10}\right)^2 - (3)^2 = 10 - 9 = 1$
$\sqrt{x} + 2$	$\sqrt{x} - 2$	$\left(\sqrt{x}\right)^2 - (2)^2 = x - 4, x \geq 0$

EXAMPLE 4 Multiplying Conjugates

Find the conjugate of the expression and multiply the expression by its conjugate.

a. $2 - \sqrt{5}$ **b.** $\sqrt{3} + \sqrt{x}$

Solution

a. The conjugate of $2 - \sqrt{5}$ is $2 + \sqrt{5}$.

$$\left(2 - \sqrt{5}\right)\left(2 + \sqrt{5}\right) = 2^2 - \left(\sqrt{5}\right)^2 \qquad \text{Special product formula}$$

$$= 4 - 5 = -1 \qquad \text{Simplify.}$$

b. The conjugate of $\sqrt{3} + \sqrt{x}$ is $\sqrt{3} - \sqrt{x}$.

$$\left(\sqrt{3} + \sqrt{x}\right)\left(\sqrt{3} - \sqrt{x}\right) = \left(\sqrt{3}\right)^2 - \left(\sqrt{x}\right)^2 \qquad \text{Special product formula}$$

$$= 3 - x, \quad x \geq 0 \qquad \text{Simplify.}$$

✔ **CHECKPOINT** *Now try Exercise 57.*

3 ▶ Simplify quotients involving radicals by rationalizing the denominators.

Dividing Radical Expressions

To simplify a *quotient* involving radicals, you rationalize the denominator. For single-term denominators, you can use the rationalization process described in Section 7.2. To rationalize a denominator involving two terms, multiply both the numerator and denominator by the *conjugate of the denominator.*

EXAMPLE 5   Simplifying Quotients Involving Radicals

Simplify (a) $\dfrac{\sqrt{3}}{1 - \sqrt{5}}$ and (b) $\dfrac{4}{2 - \sqrt{3}}$.

Solution

a. $\dfrac{\sqrt{3}}{1 - \sqrt{5}} = \dfrac{\sqrt{3}}{1 - \sqrt{5}} \cdot \dfrac{1 + \sqrt{5}}{1 + \sqrt{5}}$ Multiply numerator and denominator by conjugate of denominator.

$= \dfrac{\sqrt{3}(1 + \sqrt{5})}{1^2 - (\sqrt{5})^2}$ Special product formula

$= \dfrac{\sqrt{3} + \sqrt{15}}{1 - 5}$ Simplify.

$= -\dfrac{\sqrt{3} + \sqrt{15}}{4}$ Simplify.

b. $\dfrac{4}{2 - \sqrt{3}} = \dfrac{4}{2 - \sqrt{3}} \cdot \dfrac{2 + \sqrt{3}}{2 + \sqrt{3}}$ Multiply numerator and denominator by conjugate of denominator.

$= \dfrac{4(2 + \sqrt{3})}{2^2 - (\sqrt{3})^2}$ Special product formula

$= \dfrac{8 + 4\sqrt{3}}{4 - 3}$ Simplify.

$= 8 + 4\sqrt{3}$ Simplify.

✓ **CHECKPOINT** *Now try Exercise 75.*

EXAMPLE 6   Simplifying a Quotient Involving Radicals

$\dfrac{5\sqrt{2}}{\sqrt{7} + \sqrt{2}} = \dfrac{5\sqrt{2}}{\sqrt{7} + \sqrt{2}} \cdot \dfrac{\sqrt{7} - \sqrt{2}}{\sqrt{7} - \sqrt{2}}$ Multiply numerator and denominator by conjugate of denominator.

$= \dfrac{5(\sqrt{14} - \sqrt{4})}{(\sqrt{7})^2 - (\sqrt{2})^2}$ Special product formula

$= \dfrac{5(\sqrt{14} - 2)}{7 - 2}$ Simplify.

$= \dfrac{\cancel{5}(\sqrt{14} - 2)}{\cancel{5}}$ Divide out common factor.

$= \sqrt{14} - 2$ Simplest form

✓ **CHECKPOINT** *Now try Exercise 81.*

EXAMPLE 7   **Dividing Radical Expressions**

Perform each division and simplify.

a. $6 \div \left(\sqrt{x} - 2 \right)$

b. $\left(2 - \sqrt{3} \right) \div \left(\sqrt{6} + \sqrt{2} \right)$

Solution

a. $\dfrac{6}{\sqrt{x} - 2} = \dfrac{6}{\sqrt{x} - 2} \cdot \dfrac{\sqrt{x} + 2}{\sqrt{x} + 2}$

Multiply numerator and denominator by conjugate of denominator.

$= \dfrac{6\left(\sqrt{x} + 2 \right)}{\left(\sqrt{x} \right)^2 - 2^2}$

Special product formula

$= \dfrac{6\sqrt{x} + 12}{x - 4}$

Simplify.

b. $\dfrac{2 - \sqrt{3}}{\sqrt{6} + \sqrt{2}} = \dfrac{2 - \sqrt{3}}{\sqrt{6} + \sqrt{2}} \cdot \dfrac{\sqrt{6} - \sqrt{2}}{\sqrt{6} - \sqrt{2}}$

Multiply numerator and denominator by conjugate of denominator.

$= \dfrac{2\sqrt{6} - 2\sqrt{2} - \sqrt{18} + \sqrt{6}}{\left(\sqrt{6} \right)^2 - \left(\sqrt{2} \right)^2}$

FOIL Method and special product formula

$= \dfrac{3\sqrt{6} - 2\sqrt{2} - 3\sqrt{2}}{6 - 2}$

Simplify.

$= \dfrac{3\sqrt{6} - 5\sqrt{2}}{4}$

Simplify.

✓ **CHECKPOINT** *Now try Exercise 85.*

EXAMPLE 8   **Dividing Radical Expressions**

Perform the division and simplify.

$$1 \div \left(\sqrt{x} - \sqrt{x + 1} \right)$$

Solution

$\dfrac{1}{\sqrt{x} - \sqrt{x + 1}} = \dfrac{1}{\sqrt{x} - \sqrt{x + 1}} \cdot \dfrac{\sqrt{x} + \sqrt{x + 1}}{\sqrt{x} + \sqrt{x + 1}}$

Multiply numerator and denominator by conjugate of denominator.

$= \dfrac{\sqrt{x} + \sqrt{x + 1}}{\left(\sqrt{x} \right)^2 - \left(\sqrt{x + 1} \right)^2}$

Special product formula

$= \dfrac{\sqrt{x} + \sqrt{x + 1}}{x - (x + 1)}$

Simplify.

$= \dfrac{\sqrt{x} + \sqrt{x + 1}}{-1}$

Combine like terms.

$= -\sqrt{x} - \sqrt{x + 1}$

Simplify.

✓ **CHECKPOINT** *Now try Exercise 97.*

Concept Check

1. Give an example of a product of expressions involving radicals in which the Distributive Property can be used to perform the multiplication.

2. Give an example of a product of expressions involving radicals in which the FOIL Method can be used to perform the multiplication.

3. Write a rule that can be used to find the conjugate of the expression $a + b$, where at least one of the expressions a and b is a radical expression.

4. Is the number $\dfrac{3}{1 + \sqrt{5}}$ in simplest form? If not, explain the steps for writing it in simplest form.

7.4 EXERCISES

Go to pages 502–503 to record your assignments.

Developing Skills

In Exercises 1–50, multiply and simplify. *See Examples 1–3.*

1. $\sqrt{2} \cdot \sqrt{8}$ 2. $\sqrt{6} \cdot \sqrt{18}$

3. $\sqrt{3} \cdot \sqrt{15}$ 4. $\sqrt{5} \cdot \sqrt{10}$

5. $\sqrt[3]{12} \cdot \sqrt[3]{6}$ 6. $\sqrt[3]{9} \cdot \sqrt[3]{3}$

7. $\sqrt[4]{8} \cdot \sqrt[4]{2}$ 8. $\sqrt[4]{54} \cdot \sqrt[4]{3}$

9. $\sqrt{7}(3 - \sqrt{7})$ 10. $\sqrt{3}(4 + \sqrt{3})$

11. $\sqrt{2}(\sqrt{20} + 8)$ 12. $\sqrt{7}(\sqrt{14} + 3)$

13. $\sqrt{6}(\sqrt{12} - \sqrt{3})$ 14. $\sqrt{10}(\sqrt{5} + \sqrt{6})$

15. $4\sqrt{3}(\sqrt{3} - \sqrt{5})$ 16. $3\sqrt{5}(\sqrt{5} - \sqrt{2})$

17. $\sqrt{y}(\sqrt{y} + 4)$ 18. $\sqrt{x}(5 - \sqrt{x})$

19. $\sqrt{a}(4 - \sqrt{a})$ 20. $\sqrt{z}(\sqrt{z} + 5)$

21. $\sqrt[3]{4}(\sqrt[3]{2} - 7)$ 22. $\sqrt[3]{9}(\sqrt[3]{3} + 2)$

23. $(\sqrt{5} + 3)(\sqrt{3} - 5)$

24. $(\sqrt{7} + 6)(\sqrt{2} + 6)$

25. $(\sqrt{20} + 2)^2$ 26. $(4 - \sqrt{20})^2$

27. $(\sqrt[3]{6} - 3)(\sqrt[3]{4} + 3)$

28. $(\sqrt[3]{9} + 5)(\sqrt[3]{12} - 5)$

29. $(\sqrt{3} + 2)(\sqrt{3} - 2)$ 30. $(3 - \sqrt{5})(3 + \sqrt{5})$

31. $(6 - \sqrt{7})(6 + \sqrt{7})$ 32. $(\sqrt{8} - 5)(\sqrt{8} + 5)$

33. $(\sqrt{5} - \sqrt{3})(\sqrt{5} - \sqrt{3})$

34. $(\sqrt{2} + \sqrt{7})(\sqrt{2} + \sqrt{7})$

35. $(10 + \sqrt{2x})^2$ 36. $(5 - \sqrt{3v})^2$

37. $(9\sqrt{x} + 2)(5\sqrt{x} - 3)$

38. $(16\sqrt{u} - 3)(\sqrt{u} - 1)$

39. $(2\sqrt{2x} - \sqrt{5})(2\sqrt{2x} + \sqrt{5})$

40. $(\sqrt{7} - 3\sqrt{3t})(\sqrt{7} + 3\sqrt{3t})$

41. $(\sqrt[3]{2x} + 5)^2$

42. $(\sqrt[3]{3x} - 4)^2$

43. $(\sqrt[3]{y} + 2)(\sqrt[3]{y^2} - 5)$

44. $(\sqrt[3]{2y} + 10)(\sqrt[3]{4y^2} - 10)$

45. $(\sqrt[3]{t} + 1)(\sqrt[3]{t^2} + 4\sqrt[3]{t} - 3)$

46. $(\sqrt[3]{x} - 2)(\sqrt[3]{x^2} - 2\sqrt[3]{x} + 1)$

47. $\sqrt{x^3y^4}(2\sqrt{xy^2} - \sqrt{x^3y})$

48. $3\sqrt{xy^3}(\sqrt{x^3y} + 2\sqrt{xy^2})$

49. $2\sqrt[3]{x^4y^5}(\sqrt[3]{8x^{12}y^4} + \sqrt[3]{16xy^9})$

50. $\sqrt[4]{8x^3y^5}(\sqrt[4]{4x^5y^7} - \sqrt[4]{3x^7y^6})$

In Exercises 51–56, complete the statement.

51. $5x\sqrt{3} + 15\sqrt{3} = 5\sqrt{3}(\quad\quad\quad)$

52. $x\sqrt{7} - x^2\sqrt{7} = x\sqrt{7}(\quad\quad\quad)$

53. $4\sqrt{12} - 2x\sqrt{27} = 2\sqrt{3}(\quad\quad\quad)$

54. $5\sqrt{50} + 10y\sqrt{8} = 5\sqrt{2}(\quad\quad\quad)$

55. $6u^2 + \sqrt{18u^3} = 3u(\quad\quad\quad)$

56. $12s^3 - \sqrt{32s^4} = 4s^2(\quad\quad\quad)$

In Exercises 57–70, find the conjugate of the expression. Then multiply the expression by its conjugate and simplify. *See Example 4.*

✓ **57.** $2 + \sqrt{5}$

58. $\sqrt{2} - 9$

59. $\sqrt{11} - \sqrt{3}$

60. $\sqrt{10} + \sqrt{7}$

61. $\sqrt{15} + 3$

62. $\sqrt{14} - 3$

63. $\sqrt{x} - 3$

64. $\sqrt{t} + 7$

65. $\sqrt{2u} - \sqrt{3}$

66. $\sqrt{5a} + \sqrt{2}$

67. $2\sqrt{2} + \sqrt{4}$

68. $4\sqrt{3} + \sqrt{2}$

69. $\sqrt{x} + \sqrt{y}$

70. $3\sqrt{u} + \sqrt{3v}$

In Exercises 71–74, evaluate the function as indicated and simplify.

71. $f(x) = x^2 - 6x + 1$
   (a) $f(2 - \sqrt{3})$ (b) $f(3 - 2\sqrt{2})$

72. $g(x) = x^2 + 8x + 11$
   (a) $g(-4 + \sqrt{5})$ (b) $g(-4\sqrt{2})$

73. $f(x) = x^2 - 2x - 2$
   (a) $f(1 + \sqrt{3})$ (b) $f(3 - \sqrt{3})$

74. $g(x) = x^2 - 4x + 1$
   (a) $g(1 + \sqrt{5})$ (b) $g(2 - \sqrt{3})$

In Exercises 75–98, simplify the expression. *See Examples 5–8.*

✓ **75.** $\dfrac{6}{\sqrt{11} - 2}$

76. $\dfrac{8}{\sqrt{7} + 3}$

77. $\dfrac{7}{\sqrt{3} + 5}$

78. $\dfrac{5}{9 - \sqrt{6}}$

79. $\dfrac{3}{2\sqrt{10} - 5}$

80. $\dfrac{4}{3\sqrt{5} - 1}$

✓ **81.** $\dfrac{2}{\sqrt{6} + \sqrt{2}}$

82. $\dfrac{10}{\sqrt{9} + \sqrt{5}}$

83. $\dfrac{10}{2\sqrt{3} - \sqrt{7}}$

84. $\dfrac{12}{2\sqrt{5} + \sqrt{8}}$

✓ **85.** $(\sqrt{7} + 2) \div (\sqrt{7} - 2)$

86. $(5 - \sqrt{3}) \div (3 + \sqrt{3})$

87. $(\sqrt{x} - 5) \div (2\sqrt{x} - 1)$

88. $(2\sqrt{t} + 1) \div (2\sqrt{t} - 1)$

89. $\dfrac{3x}{\sqrt{15} - \sqrt{3}}$

90. $\dfrac{5y}{\sqrt{12} + \sqrt{10}}$

91. $\dfrac{\sqrt{5t}}{\sqrt{5} - \sqrt{t}}$

92. $\dfrac{\sqrt{2x}}{\sqrt{x} - \sqrt{2}}$

93. $\dfrac{8a}{\sqrt{3a} + \sqrt{a}}$

94. $\dfrac{7z}{\sqrt{5z} - \sqrt{z}}$

95. $\dfrac{3(x - 4)}{x^2 - \sqrt{x}}$

96. $\dfrac{6(y + 1)}{y^2 + \sqrt{y}}$

✓ **97.** $\dfrac{\sqrt{u + v}}{\sqrt{u - v} - \sqrt{u}}$

98. $\dfrac{z}{\sqrt{u + z} - \sqrt{u}}$

In Exercises 99–102, use a graphing calculator to graph the functions in the same viewing window. Use the graphs to verify that the expressions are equivalent. Verify your results algebraically.

99. $y_1 = \dfrac{10}{\sqrt{x}+1}$

$y_2 = \dfrac{10(\sqrt{x}-1)}{x-1}, \quad x \neq 1$

100. $y_1 = \dfrac{4x}{\sqrt{x}+4}$

$y_2 = \dfrac{4x(\sqrt{x}-4)}{x-16}, \quad x \neq 16$

101. $y_1 = \dfrac{2\sqrt{3x}}{2-\sqrt{3x}}$

$y_2 = \dfrac{2(2\sqrt{3x}+3x)}{4-3x}$

102. $y_1 = \dfrac{\sqrt{2x}+6}{\sqrt{2x}-2}$

$y_2 = \dfrac{x+6+4\sqrt{2x}}{x-2}$

Rationalizing Numerators   In the study of calculus, students sometimes rewrite an expression by rationalizing the numerator. In Exercises 103–110, rationalize the numerator. (*Note:* The results will not be in simplest radical form.)

103. $\dfrac{\sqrt{2}}{7}$

104. $\dfrac{\sqrt{3}}{3}$

105. $\dfrac{\sqrt{10}}{\sqrt{3x}}$

106. $\dfrac{\sqrt{5}}{\sqrt{7x}}$

107. $\dfrac{\sqrt{7}+\sqrt{3}}{5}$

108. $\dfrac{\sqrt{2}-\sqrt{5}}{4}$

109. $\dfrac{\sqrt{y}-5}{\sqrt{3}}$

110. $\dfrac{\sqrt{x}+6}{\sqrt{2}}$

Solving Problems

111. ▲ *Geometry* The width w and height h of the strongest rectangular beam that can be cut from a log of diameter 24 inches (see figure) are given by

$$w = 8\sqrt{3} \quad \text{and} \quad h = \sqrt{24^2 - (8\sqrt{3})^2}.$$

Find the area of a rectangular cross section of the beam, and write the area in simplest form.

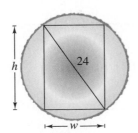

112. *Basketball* The area of the circular cross-section of a basketball is 70 square inches. The area enclosed by a basketball hoop is about 254 square inches. Find the ratio of the diameter of the basketball to the diameter of the hoop.

113. *Force*   The force required to slide a steel block weighing 500 pounds across a milling machine is

$$\dfrac{500k}{\dfrac{1}{\sqrt{k^2+1}}+\dfrac{k^2}{\sqrt{k^2+1}}}$$

where k is the friction constant (see figure). Simplify this expression.

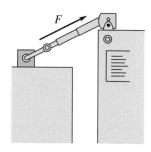

114. The ratio of the width of the Temple of Hephaestus to its height (see figure) is approximately

$$\dfrac{w}{h}\approx\dfrac{2}{\sqrt{5}-1}.$$

This number is called the **golden section.** Early Greeks believed that the most aesthetically pleasing rectangles were those whose sides had this ratio.

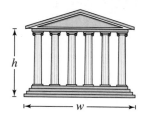

(a) Rationalize the denominator for this expression. Approximate your answer, rounded to two decimal places.

(b) Use the Pythagorean Theorem, a straightedge, and a compass to construct a rectangle whose sides have the golden section as their ratio.

Explaining Concepts

115. ✎   Let a and b be integers, but not perfect squares. Describe the circumstances (if any) for which each expression represents a rational number. Explain.

(a) $a\sqrt{b}$

(b) $\sqrt{a}\sqrt{b}$

116. ✎   Given that a and b are positive integers, what type of number is the product of the expression $\sqrt{a}+\sqrt{b}$ and its conjugate? Explain.

117. ✎   Find the conjugate of $\sqrt{a}+\sqrt{b}$. Multiply the conjugates. Next, find the conjugate of $\sqrt{b}+\sqrt{a}$. Multiply the conjugates. Explain how changing the order of the terms affects the conjugate and the product of the conjugates.

118. ✎   Rationalize the denominators of $\dfrac{1}{\sqrt{a}+\sqrt{b}}$ and $\dfrac{1}{\sqrt{b}+\sqrt{a}}$. Explain how changing the order of the terms in the denominator affects the rationalized form of the quotient.

Cumulative Review

In Exercises 119–122, solve the equation. If there is exactly one solution, check your answer. If not, describe the solution.

119. $3x-18=0$

120. $7t-4=4t+8$

121. $3x-4=3x$

122. $3(2x+5)=6x+15$

In Exercises 123–126, solve the equation by factoring.

123. $x^2-144=0$

124. $4x^2-25=0$

125. $x^2+2x-15=0$ **126.** $6x^2-x-12=0$

In Exercises 127–130, simplify the radical expression.

127. $\sqrt{32x^2y^5}$

128. $\sqrt[3]{32x^2y^5}$

129. $\sqrt[4]{32x^2y^5}$

130. $\sqrt[5]{32x^2y^5}$

7.5 Radical Equations and Applications

Jeff Greenberg/The Image Works

Why You Should Learn It

Radical equations can be used to model and solve real-life applications. For instance, in Exercise 106 on page 492, a radical equation is used to model the total monthly cost of daily flights between Chicago and Denver.

1 ▶ Solve a radical equation by raising each side to the nth power.

What You Should Learn

1 ▶ Solve a radical equation by raising each side to the nth power.
2 ▶ Solve application problems involving radical equations.

Solving Radical Equations

Solving equations involving radicals is somewhat like solving equations that contain fractions—first try to eliminate the radicals and obtain a polynomial equation. Then, solve the polynomial equation using the standard procedures. The following property plays a key role.

Raising Each Side of an Equation to the nth Power

Let u and v be real numbers, variables, or algebraic expressions, and let n be a positive integer. If $u = v$, then it follows that

$$u^n = v^n.$$

This is called **raising each side of an equation to the nth power.**

To use this property to solve a radical equation, first try to isolate one of the radicals on one side of the equation. When using this property to solve radical equations, it is critical that you check your solutions in the original equation.

Technology: Tip

To use a graphing calculator to check the solution in Example 1, graph

$$y = \sqrt{x} - 8$$

as shown below. Notice that the graph crosses the x-axis at $x = 64$, which confirms the solution that was obtained algebraically.

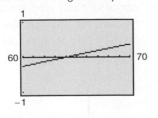

EXAMPLE 1 **Solving an Equation Having One Radical**

Solve $\sqrt{x} - 8 = 0$.

Solution

$\sqrt{x} - 8 = 0$	Write original equation.
$\sqrt{x} = 8$	Isolate radical.
$\left(\sqrt{x}\right)^2 = 8^2$	Square each side.
$x = 64$	Simplify.

Check

$\sqrt{64} - 8 \overset{?}{=} 0$	Substitute 64 for x in original equation.
$8 - 8 = 0$	Solution checks. ✓

So, the equation has one solution: $x = 64$.

✓ **CHECKPOINT** *Now try Exercise 11.*

Checking solutions of a radical equation is especially important because raising each side of an equation to the *n*th power to remove the radical(s) often introduces *extraneous* solutions.

EXAMPLE 2   Solving an Equation Having One Radical

$$\sqrt{3x} + 6 = 0$$ Original equation

$$\sqrt{3x} = -6$$ Isolate radical.

$$\left(\sqrt{3x}\right)^2 = (-6)^2$$ Square each side.

$$3x = 36$$ Simplify.

$$x = 12$$ Divide each side by 3.

Check

$$\sqrt{3(12)} + 6 \overset{?}{=} 0$$ Substitute 12 for *x* in original equation.

$$6 + 6 \neq 0$$ Solution does not check. ✗

The solution $x = 12$ is an extraneous solution. So, the original equation has no solution. You can also check this graphically, as shown in Figure 7.6. Notice that the graph does not cross the *x*-axis and so has no *x*-intercept.

✓ **CHECKPOINT** *Now try Exercise 21.*

Figure 7.6

EXAMPLE 3   Solving an Equation Having One Radical

$$\sqrt[3]{2x + 1} - 2 = 3$$ Original equation

$$\sqrt[3]{2x + 1} = 5$$ Isolate radical.

$$\left(\sqrt[3]{2x + 1}\right)^3 = 5^3$$ Cube each side.

$$2x + 1 = 125$$ Simplify.

$$2x = 124$$ Subtract 1 from each side.

$$x = 62$$ Divide each side by 2.

Check

$$\sqrt[3]{2(62) + 1} - 2 \overset{?}{=} 3$$ Substitute 62 for *x* in original equation.

$$\sqrt[3]{125} - 2 \overset{?}{=} 3$$ Simplify.

$$5 - 2 = 3$$ Solution checks. ✓

So, the equation has one solution: $x = 62$. You can also check the solution graphically by determining the point of intersection of the graphs of $y = \sqrt[3]{2x + 1} - 2$ (left side of equation) and $y = 3$ (right side of equation), as shown in Figure 7.7.

✓ **CHECKPOINT** *Now try Exercise 27.*

Figure 7.7

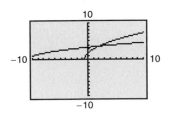
EXAMPLE 4   Solving an Equation Having Two Radicals

Solve $\sqrt{5x + 3} = \sqrt{x + 11}$.

Solution

$\sqrt{5x + 3} = \sqrt{x + 11}$	Write original equation.
$\left(\sqrt{5x + 3}\right)^2 = \left(\sqrt{x + 11}\right)^2$	Square each side.
$5x + 3 = x + 11$	Simplify.
$4x + 3 = 11$	Subtract x from each side.
$4x = 8$	Subtract 3 from each side.
$x = 2$	Divide each side by 4.

Check

$\sqrt{5x + 3} = \sqrt{x + 11}$	Write original equation.
$\sqrt{5(2) + 3} \overset{?}{=} \sqrt{2 + 11}$	Substitute 2 for x.
$\sqrt{13} = \sqrt{13}$	Solution checks. ✓

So, the equation has one solution: $x = 2$.

✓ **CHECKPOINT** *Now try Exercise 31.*

EXAMPLE 5   Solving an Equation Having Two Radicals

Solve $\sqrt[4]{3x} + \sqrt[4]{2x - 5} = 0$.

Solution

$\sqrt[4]{3x} + \sqrt[4]{2x - 5} = 0$	Write original equation.
$\sqrt[4]{3x} = -\sqrt[4]{2x - 5}$	Isolate radicals.
$\left(\sqrt[4]{3x}\right)^4 = \left(-\sqrt[4]{2x - 5}\right)^4$	Raise each side to fourth power.
$3x = 2x - 5$	Simplify.
$x = -5$	Subtract $2x$ from each side.

Check

$\sqrt[4]{3x} + \sqrt[4]{2x - 5} = 0$	Write original equation.
$\sqrt[4]{3(-5)} + \sqrt[4]{2(-5) - 5} \overset{?}{=} 0$	Substitute -5 for x.
$\sqrt[4]{-15} + \sqrt[4]{-15} \neq 0$	Solution does not check. ✗

The solution does not check because it yields fourth roots of negative radicands. So, this equation has no solution. Try checking this graphically. If you graph both sides of the equation, you will discover that the graphs do not intersect.

✓ **CHECKPOINT** *Now try Exercise 33.*

In the next example you will see that squaring each side of the equation results in a quadratic equation. Remember that you must check the solutions in the *original* radical equation.

EXAMPLE 6   An Equation That Converts to a Quadratic Equation

Solve $\sqrt{x} + 2 = x$.

Solution

$$\sqrt{x} + 2 = x \qquad \text{Write original equation.}$$
$$\sqrt{x} = x - 2 \qquad \text{Isolate radical.}$$
$$\left(\sqrt{x}\right)^2 = (x - 2)^2 \qquad \text{Square each side.}$$
$$x = x^2 - 4x + 4 \qquad \text{Simplify.}$$
$$-x^2 + 5x - 4 = 0 \qquad \text{Write in general form.}$$
$$(-1)(x - 4)(x - 1) = 0 \qquad \text{Factor.}$$
$$x - 4 = 0 \implies x = 4 \qquad \text{Set 1st factor equal to 0.}$$
$$x - 1 = 0 \implies x = 1 \qquad \text{Set 2nd factor equal to 0.}$$

Check

First Solution

$$\sqrt{4} + 2 \stackrel{?}{=} 4$$
$$2 + 2 = 4$$

Second Solution

$$\sqrt{1} + 2 \stackrel{?}{=} 1$$
$$1 + 2 \neq 1$$

From the check you can see that $x = 1$ is an extraneous solution. So, the only solution is $x = 4$.

✓ **CHECKPOINT** *Now try Exercise 39.*

When an equation contains two radicals, it may not be possible to isolate both. In such cases, you may have to raise each side of the equation to a power at *two* different stages in the solution.

EXAMPLE 7   Repeatedly Squaring Each Side of an Equation

$$\sqrt{3t + 1} = 2 - \sqrt{3t} \qquad \text{Original equation}$$
$$\left(\sqrt{3t + 1}\right)^2 = \left(2 - \sqrt{3t}\right)^2 \qquad \text{Square each side (1st time).}$$
$$3t + 1 = 4 - 4\sqrt{3t} + 3t \qquad \text{Simplify.}$$
$$-3 = -4\sqrt{3t} \qquad \text{Isolate radical.}$$
$$(-3)^2 = \left(-4\sqrt{3t}\right)^2 \qquad \text{Square each side (2nd time).}$$
$$9 = 16(3t) \qquad \text{Simplify.}$$
$$\frac{3}{16} = t \qquad \text{Divide each side by 48 and simplify.}$$

The solution is $t = \frac{3}{16}$. Check this in the original equation.

✓ **CHECKPOINT** *Now try Exercise 47.*

2 ▶ Solve application problems involving radical equations.

Applications

EXAMPLE 8   Electricity

The amount of power consumed by an electrical appliance is given by $I = \sqrt{P/R}$, where I is the current measured in amps, R is the resistance measured in ohms, and P is the power measured in watts. Find the power used by an electric heater for which $I = 10$ amps and $R = 16$ ohms.

Solution

$$10 = \sqrt{\frac{P}{16}}$$ Substitute 10 for I and 16 for R in original equation.

$$10^2 = \left(\sqrt{\frac{P}{16}}\right)^2$$ Square each side.

$$100 = \frac{P}{16} \implies 1600 = P$$ Simplify and multiply each side by 16.

So, the solution is $P = 1600$ watts. Check this in the original equation.

✓ **CHECKPOINT** *Now try Exercise 97.*

EXAMPLE 9   An Application of the Pythagorean Theorem

The distance between a house on shore and a playground on shore is 40 meters. The distance between the playground and a house on an island is 50 meters. (See Figure 7.8.) What is the distance between the two houses?

Solution

From Figure 7.8, you can see that the distances form a right triangle. So, you can use the Pythagorean Theorem to find the distance between the two houses.

$$c = \sqrt{a^2 + b^2}$$ Pythagorean Theorem

$$50 = \sqrt{40^2 + b^2}$$ Substitute 40 for a and 50 for c.

$$50 = \sqrt{1600 + b^2}$$ Simplify.

$$50^2 = (\sqrt{1600 + b^2})^2$$ Square each side.

$$2500 = 1600 + b^2$$ Simplify.

$$0 = b^2 - 900$$ Write in general form.

$$0 = (b + 30)(b - 30)$$ Factor.

$$b + 30 = 0 \implies b = -30$$ Set 1st factor equal to 0.

$$b - 30 = 0 \implies b = 30$$ Set 2nd factor equal to 0.

Choose the positive solution to obtain a distance of 30 meters. Check this solution in the original equation.

✓ **CHECKPOINT** *Now try Exercise 85.*

Figure 7.8

EXAMPLE 10 **Velocity of a Falling Object**

The velocity of a free-falling object can be determined from the equation $v = \sqrt{2gh}$, where v is the velocity measured in feet per second, $g = 32$ feet per second per second, and h is the distance (in feet) the object has fallen. Find the height from which a rock has been dropped when it strikes the ground with a velocity of 50 feet per second.

Solution

$v = \sqrt{2gh}$	Write original equation.
$50 = \sqrt{2(32)h}$	Substitute 50 for v and 32 for g.
$50^2 = \left(\sqrt{64h}\right)^2$	Square each side.
$2500 = 64h$	Simplify.
$39 \approx h$	Divide each side by 64.

Check

$v = \sqrt{2gh}$	Write original equation.
$50 \overset{?}{\approx} \sqrt{2(32)(39)}$	Substitute 50 for v, 32 for g, and 39 for h.
$50 \overset{?}{\approx} \sqrt{2496}$	Simplify.
$50 \approx 49.96$	Solution checks. ✓

So, the height from which the rock has been dropped is approximately 39 feet.

✓ **CHECKPOINT** *Now try Exercise 99.*

Study Tip

When checking a solution with a value that has been rounded, the check will not result in an equality. If the solution is valid, the expressions on each side of the equal sign will be *approximately* equal to each other, as shown in Example 10.

EXAMPLE 11 **Market Research**

The marketing department at a publisher determines that the demand for a book depends on the price of the book in accordance with the formula $p = 40 - \sqrt{0.0001x + 1}$, $x \geq 0$, where p is the price per book in dollars and x is the number of books sold at the given price. (See Figure 7.9.) The publisher sets the price at $12.95. How many copies can the publisher expect to sell?

Solution

$p = 40 - \sqrt{0.0001x + 1}$	Write original equation.
$12.95 = 40 - \sqrt{0.0001x + 1}$	Substitute 12.95 for p.
$\sqrt{0.0001x + 1} = 27.05$	Isolate radical.
$0.0001x + 1 = 731.7025$	Square each side.
$0.0001x = 730.7025$	Subtract 1 from each side.
$x = 7,307,025$	Divide each side by 0.0001.

So, the publisher can expect to sell about 7.3 million copies.

✓ **CHECKPOINT** *Now try Exercise 105.*

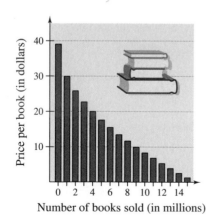

Figure 7.9

Concept Check

1. Is $1 - x\sqrt{5} = x^2$ a radical equation? Explain your reasoning.

2. One reason for checking a solution in the original equation is to discover errors made when solving the equation. Describe another reason.

3. In your own words, describe the steps used to solve $\sqrt{x} + 2 = x$.

4. The graphs of $f(x) = x - 1$ and $g(x) = \sqrt{x + 5}$ are shown at the right. Explain how you can use the graphs to solve $x - 1 = \sqrt{x + 5}$.

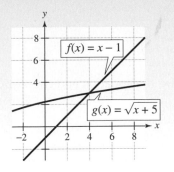

Figure for 4

7.5 EXERCISES

Go to pages 502–503 to record your assignments.

Developing Skills

In Exercises 1–4, determine whether each value of x is a solution of the equation.

Equation	Values of x
1. $\sqrt{x} - 10 = 0$	(a) $x = -4$ (b) $x = -100$
	(c) $x = \sqrt{10}$ (d) $x = 100$
2. $\sqrt{3x} - 6 = 0$	(a) $x = \frac{2}{3}$ (b) $x = 2$
	(c) $x = 12$ (d) $x = -\frac{1}{3}\sqrt{6}$
3. $\sqrt[3]{x} - 4 = 4$	(a) $x = -60$ (b) $x = 68$
	(c) $x = 20$ (d) $x = 0$
4. $\sqrt[4]{2x} + 2 = 6$	(a) $x = 128$ (b) $x = 2$
	(c) $x = -2$ (d) $x = 0$

In Exercises 5–54, solve the equation and check your solution(s). (Some of the equations have no solution.) **See Examples 1–7.**

5. $\sqrt{x} = 12$

6. $\sqrt{x} = 5$

7. $\sqrt{y} = 7$

8. $\sqrt{t} = 4$

9. $\sqrt[3]{z} = 3$

10. $\sqrt[4]{x} = 3$

✓ 11. $\sqrt{y} - 7 = 0$

12. $\sqrt{t} - 13 = 0$

13. $\sqrt{u} + 13 = 0$

14. $\sqrt{y} + 15 = 0$

15. $\sqrt{x} - 8 = 0$

16. $\sqrt{x} - 10 = 0$

17. $\sqrt{10x} = 30$

18. $\sqrt{8x} = 6$

19. $\sqrt{-3x} = 9$

20. $\sqrt{-4y} = 4$

✓ 21. $\sqrt{5t} - 2 = 0$

22. $10 - \sqrt{6x} = 0$

23. $\sqrt{3y + 1} = 4$

24. $\sqrt{3 - 2x} = 2$

25. $\sqrt{9 - 2x} = -9$

26. $\sqrt{2t - 7} = -5$

✓ 27. $\sqrt[3]{y - 3} + 4 = 6$

28. $\sqrt[4]{6a - 11} + 8 = -5$

29. $6\sqrt[4]{x + 3} = 15$

30. $4\sqrt[3]{x + 4} = 7$

✓ 31. $\sqrt{x + 3} = \sqrt{2x - 1}$

32. $\sqrt{3t + 1} = \sqrt{t + 15}$

✓ 33. $\sqrt{3y - 5} - 3\sqrt{y} = 0$

34. $\sqrt{2u + 10} - 2\sqrt{u} = 0$

35. $\sqrt[3]{3x - 4} = \sqrt[3]{x + 10}$

36. $2\sqrt[3]{10 - 3x} = \sqrt[3]{2 - x}$

37. $\sqrt[3]{2x + 15} - \sqrt[3]{x} = 0$

38. $\sqrt[4]{2x} + \sqrt[4]{x + 3} = 0$

✓ 39. $\sqrt{x^2 - 2} = x + 4$

40. $\sqrt{x^2 - 4} = x - 2$

41. $\sqrt{2x} = x - 4$

42. $\sqrt{x} = 6 - x$

43. $\sqrt{8x + 1} = x + 2$

44. $\sqrt{3x + 7} = x + 3$

45. $\sqrt{3x + 4} = \sqrt{4x + 3}$

46. $\sqrt{2x - 7} = \sqrt{3x - 12}$

✓ 47. $\sqrt{z + 2} = 1 + \sqrt{z}$

48. $\sqrt{2x + 5} = 7 - \sqrt{2x}$

49. $\sqrt{2t + 3} = 3 - \sqrt{2t}$

50. $\sqrt{x} + \sqrt{x + 2} = 2$

51. $\sqrt{x+5} - \sqrt{x} = 1$

52. $\sqrt{x+1} = 2 - \sqrt{x}$

53. $\sqrt{x-6} + 3 = \sqrt{x+9}$

54. $\sqrt{x+3} - \sqrt{x-1} = 1$

In Exercises 55–62, solve the equation and check your solution(s).

55. $t^{3/2} = 8$

56. $v^{2/3} = 25$

57. $3y^{1/3} = 18$

58. $2x^{3/4} = 54$

59. $(x+4)^{2/3} = 4$

60. $(u-2)^{4/3} = 81$

61. $(2x+5)^{1/3} + 3 = 0$

62. $(x-6)^{3/2} - 27 = 0$

In Exercises 63–72, use a graphing calculator to graph each side of the equation in the same viewing window. Use the graphs to approximate the solution(s). Verify your answer algebraically.

63. $\sqrt{x} = 2(2-x)$

64. $\sqrt{2x+3} = 4x - 3$

65. $\sqrt{x^2+1} = 5 - 2x$

66. $\sqrt{8-3x} = x$

67. $\sqrt{x+3} = 5 - \sqrt{x}$

68. $\sqrt[3]{5x-8} = 4 - \sqrt[3]{x}$

69. $3\sqrt[4]{x} = 9 - x$

70. $\sqrt[3]{x+4} = \sqrt{6-x}$

71. $\sqrt{15-4x} = 2x$

72. $\dfrac{4}{\sqrt{x}} = 3\sqrt{x} - 4$

In Exercises 73–76, use the given function to find the indicated value of x.

73. For $f(x) = \sqrt{x} - \sqrt{x-9}$,
find x such that $f(x) = 1$.

74. For $g(x) = \sqrt{x} + \sqrt{x-5}$,
find x such that $g(x) = 5$.

75. For $h(x) = \sqrt{x-2} - \sqrt{4x+1}$,
find x such that $h(x) = -3$.

76. For $f(x) = \sqrt{2x+7} - \sqrt{x+15}$,
find x such that $f(x) = -1$.

In Exercises 77–80, find the x-intercept(s) of the graph of the function without graphing the function.

77. $f(x) = \sqrt{x+5} - 3 + \sqrt{x}$

78. $f(x) = \sqrt{6x+7} - 2 - \sqrt{2x+3}$

79. $f(x) = \sqrt{3x-2} - 1 - \sqrt{2x-3}$

80. $f(x) = \sqrt{5x+6} - 1 - \sqrt{3x+3}$

Solving Problems

▲ *Geometry* In Exercises 81–84, find the length x of the unknown side of the right triangle. (Round your answer to two decimal places.)

81.

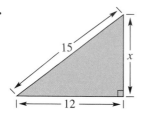

82.

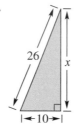

83.

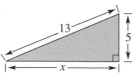

84.

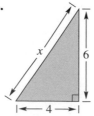

✓ **85.** ▲ *Plasma TV* The screen of a plasma television has a diagonal of 50 inches and a width of 43.75 inches. Draw a diagram of the plasma television and find the length of the screen.

86. ▲ *Basketball* A basketball court is 50 feet wide and 94 feet long. Draw a diagram of the basketball court and find the length of a diagonal of the court.

87. ▲ *Ladders* An extension ladder is placed against the side of a house such that the base of the ladder is 2 meters from the base of the house and the ladder reaches 6 meters up the side of the house. How far is the ladder extended?

88. ▲ *Guy Wires* A guy wire on a 100-foot radio tower is attached to the top of the tower and to an anchor 50 feet from the base of the tower. Find the length of the guy wire.

89. ▲ *Ladders* A ladder is 17 feet long, and the bottom of the ladder is 8 feet from the side of a house. How far does the ladder reach up the side of the house?

90. ▲ *Construction* A 10-foot plank is used to brace a basement wall during construction of a home. The plank is nailed to the wall 6 feet above the floor. Find the slope of the plank.

91. ▲ *Geometry* Determine the length and width of a rectangle with a perimeter of 92 inches and a diagonal of 34 inches.

92. ▲ *Geometry* Determine the length and width of a rectangle with a perimeter of 68 inches and a diagonal of 26 inches.

93. ▲ *Geometry* The lateral surface area of a cone (see figure) is given by $S = \pi r \sqrt{r^2 + h^2}$. Solve the equation for h. Then find the height of a cone with a lateral surface area of $364\pi\sqrt{2}$ square centimeters and a radius of 14 centimeters.

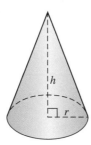

94. ▲ *Geometry* The slant height l of a truncated pyramid (see figure) is given by

$$l = \sqrt{h^2 + \tfrac{1}{4}(b_2 - b_1)^2}.$$

Solve the equation for h. Then find the height of a truncated pyramid when $l = 2\sqrt{26}$ inches, $b_2 = 8$ inches, and $b_1 = 4$ inches.

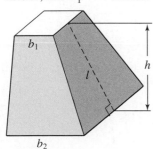

95. ▦ ▲ *Geometry* Write a function that gives the radius r of a circle in terms of the circle's area A. Use a graphing calculator to graph this function.

96. ▦ ▲ *Geometry* Write a function that gives the radius r of a sphere in terms of the sphere's volume V. Use a graphing calculator to graph this function.

Height In Exercises 97 and 98, use the formula $t = \sqrt{d/16}$, which gives the time t in seconds for a free-falling object to fall d feet.

✓ **97.** A construction worker drops a nail from a building and observes it strike a water puddle after approximately 2 seconds. Estimate the height from which the nail was dropped.

98. A farmer drops a stone down a well and hears it strike the water after approximately 4.5 seconds. Estimate the depth of the well.

Free-Falling Object In Exercises 99–102, use the equation for the velocity of a free-falling object, $v = \sqrt{2gh}$, as described in Example 10.

✓ **99.** A cliff diver dives from a height of 80 feet. Estimate the velocity of the diver when the diver strikes the water.

100. A coin is dropped from a hot air balloon that is 250 feet above the ground. Estimate the velocity of the coin when the coin strikes the ground.

101. An egg strikes the ground with a velocity of 50 feet per second. Estimate to two decimal places the height from which the egg was dropped.

102. A stone strikes the water with a velocity of 130 feet per second. Estimate to two decimal places the height from which the stone was dropped.

Period of a Pendulum In Exercises 103 and 104, the time t (in seconds) for a pendulum of length L (in feet) to go through one complete cycle (its period) is given by $t = 2\pi\sqrt{L/32}$.

103. How long is the pendulum of a grandfather clock with a period of 1.5 seconds?

104. How long is the pendulum of a mantel clock with a period of 0.75 second?

✓ **105.** *Demand* The demand equation for a sweater is given by

$$p = 50 - \sqrt{0.8(x - 1)}$$

where x is the number of units demanded per day and p is the price per sweater. Find the demand when the price is set at $30.02.

106. *Airline Passengers*   An airline offers daily flights between Chicago and Denver. The total monthly cost C (in millions of dollars) of these flights is

$$C = \sqrt{0.2x + 1}, \quad x \geq 0$$

where x is measured in thousands of passengers (see figure). The total cost of the flights for June is 2.5 million dollars. Approximately how many passengers flew in June?

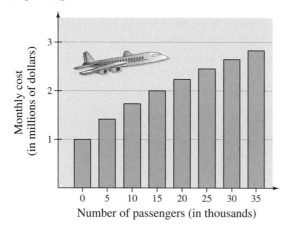

Number of passengers (in thousands)

107. *Killer Whales*   The weight w (in pounds) of a killer whale can be modeled by

$$w = 280 + 325\sqrt{t}, \quad 0 \leq t \leq 144$$

where t represents the age (in months) of the killer whale.

(a) ⌨ Use a graphing calculator to graph the model.

(b) At what age did the killer whale weigh about 3400 pounds?

108. *Consumer Spending*   The average movie ticket price p (in dollars) consumers paid in theaters in the United States for the years 1997 through 2006 can be modeled by

$$p = 0.518 + 1.52\sqrt{t}, \quad 7 \leq t \leq 16$$

where t represents the year, with $t = 7$ corresponding to 1997.   (Source: Veronis, Suhler & Associates Inc.)

(a) ⌨ Use a graphing calculator to graph the model.

(b) In what year did the average movie ticket price in theaters reach $5.80?

Explaining Concepts

109. *Error Analysis*   Describe the error.

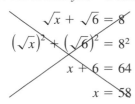

$$\sqrt{x} + \sqrt{6} = 8$$
$$(\sqrt{x})^2 + (\sqrt{6})^2 = 8^2$$
$$x + 6 = 64$$
$$x = 58$$

110. ✎   Does raising each side of an equation to the nth power always yield an equivalent equation? Explain.

111. *Exploration*   The solution of the equation $x + \sqrt{x - a} = b$ is $x = 20$. Discuss how to find a and b. (There are many correct values for a and b.)

112. ✎   Explain how you can tell that $\sqrt{x - 9} = -4$ has no solution without solving the equation.

Cumulative Review

In Exercises 113–116, determine whether the two lines are parallel, perpendicular, or neither.

113. $L_1: y = 4x + 2$
$L_2: y = 4x - 1$

114. $L_1: y = 3x - 8$
$L_2: y = -3x - 8$

115. $L_1: y = -x + 5$
$L_2: y = x - 3$

116. $L_1: y = 2x$
$L_2: y = \frac{1}{2}x + 4$

In Exercises 117 and 118, use matrices to solve the system of linear equations.

117. $\begin{cases} 4x - y = 10 \\ -7x - 2y = -25 \end{cases}$ **118.** $\begin{cases} 3x - 2y = 5 \\ 6x - 5y = 14 \end{cases}$

In Exercises 119–122, simplify the expression.

119. $a^{3/5} \cdot a^{1/5}$ **120.** $\dfrac{m^2}{m^{2/3}}$

121. $\left(\dfrac{x^{1/2}}{x^{1/8}}\right)^4$ **122.** $\dfrac{(a + b)^{3/4}}{\sqrt[4]{a + b}}$

7.6 Complex Numbers

What You Should Learn

1 ▶ Write square roots of negative numbers in i-form and perform operations on numbers in i-form.

2 ▶ Determine the equality of two complex numbers.

3 ▶ Add, subtract, and multiply complex numbers.

4 ▶ Use complex conjugates to write the quotient of two complex numbers in standard form.

Why You Should Learn It

Understanding complex numbers can help you in Section 8.3 to identify quadratic equations that have no real solutions.

1 ▶ Write square roots of negative numbers in i-form and perform operations on numbers in i-form.

The Imaginary Unit i

In Section 7.1, you learned that a negative number has no *real* square root. For instance, $\sqrt{-1}$ is not real because there is no real number x such that $x^2 = -1$. So, as long as you are dealing only with real numbers, the equation $x^2 = -1$ has no solution. To overcome this deficiency, mathematicians have expanded the set of numbers by including the **imaginary unit i,** defined as

$$i = \sqrt{-1}. \qquad \text{Imaginary unit}$$

This number has the property that $i^2 = -1$. So, the imaginary unit i is a solution of the equation $x^2 = -1$.

The Square Root of a Negative Number

Let c be a positive real number. Then the square root of $-c$ is given by

$$\sqrt{-c} = \sqrt{c(-1)} = \sqrt{c}\sqrt{-1} = \sqrt{c}\,i.$$

When writing $\sqrt{-c}$ in the **i-form,** $\sqrt{c}\,i$, note that i is outside the radical.

EXAMPLE 1 Writing Numbers in i-Form

Write each number in i-form.

a. $\sqrt{-36}$ **b.** $\sqrt{-\dfrac{16}{25}}$ **c.** $\sqrt{-54}$ **d.** $\dfrac{\sqrt{-48}}{\sqrt{-3}}$

Solution

a. $\sqrt{-36} = \sqrt{36(-1)} = \sqrt{36}\sqrt{-1} = 6i$

b. $\sqrt{-\dfrac{16}{25}} = \sqrt{\dfrac{16}{25}(-1)} = \sqrt{\dfrac{16}{25}}\sqrt{-1} = \dfrac{4}{5}i$

c. $\sqrt{-54} = \sqrt{54(-1)} = \sqrt{54}\sqrt{-1} = 3\sqrt{6}\,i$

d. $\dfrac{\sqrt{-48}}{\sqrt{-3}} = \dfrac{\sqrt{48}\sqrt{-1}}{\sqrt{3}\sqrt{-1}} = \dfrac{\sqrt{48}\,i}{\sqrt{3}\,i} = \sqrt{\dfrac{48}{3}} = \sqrt{16} = 4$

✓ **CHECKPOINT** *Now try Exercise 1.*

Technology: Discovery

Use a calculator to evaluate each radical. Does one result in an error message? Explain why.

a. $\sqrt{121}$

b. $\sqrt{-121}$

c. $-\sqrt{121}$

To perform operations with square roots of negative numbers, you must *first* write the numbers in *i*-form. You can then add, subtract, and multiply as follows.

$$ai + bi = (a + b)i \qquad \text{Addition}$$

$$ai - bi = (a - b)i \qquad \text{Subtraction}$$

$$(ai)(bi) = ab(i^2) = ab(-1) = -ab \qquad \text{Multiplication}$$

EXAMPLE 2   Operations with Square Roots of Negative Numbers

Perform each operation.

a. $\sqrt{-9} + \sqrt{-49}$ **b.** $\sqrt{-32} - 2\sqrt{-2}$

Solution

a. $\sqrt{-9} + \sqrt{-49} = \sqrt{9}\sqrt{-1} + \sqrt{49}\sqrt{-1}$ Product Rule for Radicals

$\qquad\qquad\qquad\quad = 3i + 7i$ Write in *i*-form.

$\qquad\qquad\qquad\quad = 10i$ Simplify.

b. $\sqrt{-32} - 2\sqrt{-2} = \sqrt{32}\sqrt{-1} - 2\sqrt{2}\sqrt{-1}$ Product Rule for Radicals

$\qquad\qquad\qquad\quad = 4\sqrt{2}i - 2\sqrt{2}i$ Write in *i*-form.

$\qquad\qquad\qquad\quad = 2\sqrt{2}i$ Simplify.

✓ **CHECKPOINT** *Now try Exercise 19.*

EXAMPLE 3   Multiplying Square Roots of Negative Numbers

Find each product.

a. $\sqrt{-15}\sqrt{-15}$ **b.** $\sqrt{-5}\left(\sqrt{-45} - \sqrt{-4}\right)$

Solution

a. $\sqrt{-15}\sqrt{-15} = \left(\sqrt{15}i\right)\left(\sqrt{15}i\right)$ Write in *i*-form.

$\qquad\qquad\qquad = \left(\sqrt{15}\right)^2 i^2$ Multiply.

$\qquad\qquad\qquad = 15(-1)$ $i^2 = -1$

$\qquad\qquad\qquad = -15$ Simplify.

b. $\sqrt{-5}\left(\sqrt{-45} - \sqrt{-4}\right) = \sqrt{5}i\left(3\sqrt{5}i - 2i\right)$ Write in *i*-form.

$\qquad\qquad\qquad\qquad = \left(\sqrt{5}i\right)\left(3\sqrt{5}i\right) - \left(\sqrt{5}i\right)(2i)$ Distributive Property

$\qquad\qquad\qquad\qquad = 3(5)(-1) - 2\sqrt{5}(-1)$ Multiply.

$\qquad\qquad\qquad\qquad = -15 + 2\sqrt{5}$ Simplify.

✓ **CHECKPOINT** *Now try Exercise 27.*

When multiplying square roots of negative numbers, always write them in *i*-form *before multiplying*. If you do not do this, you can obtain incorrect answers. For instance, in Example 3(a) be sure you see that

$$\sqrt{-15}\sqrt{-15} \neq \sqrt{(-15)(-15)} = \sqrt{225} = 15.$$

2 ▶ Determine the equality of two complex numbers.

Complex Numbers

A number of the form $a + bi$, where a and b are real numbers, is called a **complex number.** The real number a is called the **real part** of the complex number $a + bi$, and the number bi is called the **imaginary part.**

> ### Definition of Complex Number
> If a and b are real numbers, the number $a + bi$ is a **complex number,** and it is said to be written in **standard form.** If $b = 0$, the number $a + bi = a$ is a real number. If $b \neq 0$, the number $a + bi$ is called an **imaginary number.** A number of the form bi, where $b \neq 0$, is called a **pure imaginary number.**

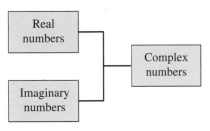

Figure 7.10

A number cannot be both real and imaginary. For instance, the numbers -2, 0, 1, $\frac{1}{2}$, and $\sqrt{2}$ are real numbers, and the numbers $-3i$, $2 + 4i$, and $-1 + i$ are imaginary numbers. The diagram shown in Figure 7.10 shows that the real numbers and the imaginary numbers make up the complex numbers.

Two complex numbers $a + bi$ and $c + di$, in standard form, are equal if and only if $a = c$ and $b = d$.

EXAMPLE 4   **Equality of Two Complex Numbers**

To determine whether the complex numbers $\sqrt{9} + \sqrt{-48}$ and $3 - 4\sqrt{3}i$ are equal, begin by writing the first number in standard form.

$$\sqrt{9} + \sqrt{-48} = \sqrt{3^2} + \sqrt{4^2(3)(-1)} = 3 + 4\sqrt{3}i$$

The two numbers are not equal because their imaginary parts differ in sign.

✔ **CHECKPOINT** *Now try Exercise 43.*

EXAMPLE 5   **Equality of Two Complex Numbers**

To find values of x and y that satisfy the equation $3x - \sqrt{-25} = -6 + 3yi$, begin by writing the left side of the equation in standard form.

$$3x - 5i = -6 + 3yi \qquad \text{Each side is in standard form.}$$

For these two numbers to be equal, their real parts must be equal to each other and their imaginary parts must be equal to each other.

Real Parts	*Imaginary Parts*
$3x = -6$	$3yi = -5i$
$x = -2$	$3y = -5$
	$y = -\frac{5}{3}$

So, $x = -2$ and $y = -\frac{5}{3}$.

✔ **CHECKPOINT** *Now try Exercise 51.*

3 ▶ Add, subtract, and multiply complex numbers.

Operations with Complex Numbers

To add or subtract two complex numbers, you add (or subtract) the real and imaginary parts separately. This is similar to combining like terms of a polynomial.

$$(a + bi) + (c + di) = (a + c) + (b + d)i \qquad \text{Addition of complex numbers}$$

$$(a + bi) - (c + di) = (a - c) + (b - d)i \qquad \text{Subtraction of complex numbers}$$

Study Tip

Note in part (b) of Example 6 that the sum of two complex numbers can be a real number.

EXAMPLE 6   Adding and Subtracting Complex Numbers

a. $(3 - i) + (-2 + 4i) = (3 - 2) + (-1 + 4)i = 1 + 3i$

b. $3i + (5 - 3i) = 5 + (3 - 3)i = 5$

c. $4 - (-1 + 5i) + (7 + 2i) = [4 - (-1) + 7] + (-5 + 2)i = 12 - 3i$

d. $(6 + 3i) + (2 - \sqrt{-8}) - \sqrt{-4} = (6 + 3i) + (2 - 2\sqrt{2}i) - 2i$

$$= (6 + 2) + (3 - 2\sqrt{2} - 2)i$$

$$= 8 + (1 - 2\sqrt{2})i$$

✓ **CHECKPOINT** *Now try Exercise 55.*

The Commutative, Associative, and Distributive Properties of real numbers are also valid for complex numbers, as is the FOIL Method.

EXAMPLE 7   Multiplying Complex Numbers

Perform each operation and write the result in standard form.

a. $(7i)(-3i)$ **b.** $(1 - i)(\sqrt{-9})$

c. $(2 - i)(4 + 3i)$ **d.** $(3 + 2i)(3 - 2i)$

Solution

a. $(7i)(-3i) = -21i^2$ Multiply.

$$= -21(-1) = 21 \qquad\qquad i^2 = -1$$

b. $(1 - i)(\sqrt{-9}) = (1 - i)(3i)$ Write in *i*-form.

$$= 3i - 3(i^2) \qquad\qquad \text{Distributive Property}$$

$$= 3i - 3(-1) = 3 + 3i \qquad i^2 = -1$$

c. $(2 - i)(4 + 3i) = 8 + 6i - 4i - 3i^2$ FOIL Method

$$= 8 + 6i - 4i - 3(-1) \qquad i^2 = -1$$

$$= 11 + 2i \qquad\qquad \text{Combine like terms.}$$

d. $(3 + 2i)(3 - 2i) = 3^2 - (2i)^2$ Special product formula

$$= 9 - 4i^2 \qquad\qquad \text{Simplify.}$$

$$= 9 - 4(-1) = 13 \qquad i^2 = -1$$

✓ **CHECKPOINT** *Now try Exercise 71.*

4 ▶ Use complex conjugates to write the quotient of two complex numbers in standard form.

Complex Conjugates

In Example 7(d), note that the product of two complex numbers can be a real number. This occurs with pairs of complex numbers of the form $a + bi$ and $a - bi$, called **complex conjugates.** In general, the product of complex conjugates has the following form.

$$(a + bi)(a - bi) = a^2 - (bi)^2 = a^2 - b^2i^2 = a^2 - b^2(-1) = a^2 + b^2$$

Here are some examples.

Complex Number	Complex Conjugate	Product
$4 - 5i$	$4 + 5i$	$4^2 + 5^2 = 41$
$3 + 2i$	$3 - 2i$	$3^2 + 2^2 = 13$
$-2 = -2 + 0i$	$-2 = -2 - 0i$	$(-2)^2 + 0^2 = 4$
$i = 0 + i$	$-i = 0 - i$	$0^2 + 1^2 = 1$

To write the quotient of $a + bi$ and $c + di$ (where $d \neq 0$) in standard form, multiply the numerator and denominator by the *complex conjugate of the denominator*, as shown in Example 8.

EXAMPLE 8   **Writing Quotients of Complex Numbers in Standard Form**

a. $\dfrac{2 - i}{4i} = \dfrac{2 - i}{4i} \cdot \dfrac{(-4i)}{(-4i)}$ Multiply numerator and denominator by complex conjugate of denominator.

$= \dfrac{-8i + 4i^2}{-16i^2}$ Multiply fractions.

$= \dfrac{-8i + 4(-1)}{-16(-1)}$ $i^2 = -1$

$= \dfrac{-8i - 4}{16}$ Simplify.

$= -\dfrac{1}{4} - \dfrac{1}{2}i$ Write in standard form.

b. $\dfrac{5}{3 - 2i} = \dfrac{5}{3 - 2i} \cdot \dfrac{3 + 2i}{3 + 2i}$ Multiply numerator and denominator by complex conjugate of denominator.

$= \dfrac{5(3 + 2i)}{(3 - 2i)(3 + 2i)}$ Multiply fractions.

$= \dfrac{5(3 + 2i)}{3^2 + 2^2}$ Product of complex conjugates

$= \dfrac{15 + 10i}{13}$ Simplify.

$= \dfrac{15}{13} + \dfrac{10}{13}i$ Write in standard form.

✓ **CHECKPOINT** *Now try Exercise 123.*

| EXAMPLE 9 | Writing a Quotient of Complex Numbers in Standard Form |

$$\frac{8 - i}{8 + i} = \frac{8 - i}{8 + i} \cdot \frac{8 - i}{8 - i}$$ Multiply numerator and denominator by complex conjugate of denominator.

$$= \frac{64 - 16i + i^2}{8^2 + 1^2}$$ Multiply fractions.

$$= \frac{64 - 16i + (-1)}{8^2 + 1^2}$$ $i^2 = -1$

$$= \frac{63 - 16i}{65}$$ Simplify.

$$= \frac{63}{65} - \frac{16}{65}i$$ Write in standard form.

✓ **CHECKPOINT** *Now try Exercise 135.*

| EXAMPLE 10 | Writing a Quotient of Complex Numbers in Standard Form |

$$\frac{2 + 3i}{4 - 2i} = \frac{2 + 3i}{4 - 2i} \cdot \frac{4 + 2i}{4 + 2i}$$ Multiply numerator and denominator by complex conjugate of denominator.

$$= \frac{8 + 16i + 6i^2}{4^2 + 2^2}$$ Multiply fractions.

$$= \frac{8 + 16i + 6(-1)}{4^2 + 2^2}$$ $i^2 = -1$

$$= \frac{2 + 16i}{20} = \frac{1}{10} + \frac{4}{5}i$$ Write in standard form.

✓ **CHECKPOINT** *Now try Exercise 137.*

| EXAMPLE 11 | Verifying a Complex Solution of an Equation |

Show that $x = 2 + i$ is a solution of the equation $x^2 - 4x + 5 = 0$.

Solution

$$x^2 - 4x + 5 = 0$$ Write original equation.

$$(2 + i)^2 - 4(2 + i) + 5 \stackrel{?}{=} 0$$ Substitute $2 + i$ for x.

$$4 + 4i + i^2 - 8 - 4i + 5 \stackrel{?}{=} 0$$ Expand.

$$i^2 + 1 \stackrel{?}{=} 0$$ Combine like terms.

$$(-1) + 1 \stackrel{?}{=} 0$$ $i^2 = -1$

$$0 = 0$$ Solution checks. ✓

So, $x = 2 + i$ is a solution of the original equation.

✓ **CHECKPOINT** *Now try Exercise 145.*

Concept Check

1. Write (in words) the steps you can use to write the square root of a negative number in i-form.

2. Describe the values of a and b for which the complex number $a + bi$ is (a) a real number, (b) an imaginary number, and (c) a pure imaginary number. Then (d) explain what you must do to show that two complex numbers are equal.

3. Explain how adding two complex numbers is similar to adding two binomials. Then explain how multiplying two complex numbers is similar to multiplying two binomials.

4. Explain how you can use a complex conjugate to write the quotient of two complex numbers in standard form.

7.6 EXERCISES

Go to pages 502–503 to record your assignments.

Developing Skills

In Exercises 1–18, write the number in i-form. **See Example 1.**

✔ 1. $\sqrt{-4}$

2. $\sqrt{-9}$

3. $-\sqrt{-144}$

4. $\sqrt{-49}$

5. $\sqrt{-\frac{4}{25}}$

6. $\sqrt{-\frac{9}{64}}$

7. $-\sqrt{-\frac{36}{121}}$

8. $-\sqrt{-\frac{9}{25}}$

9. $\sqrt{-8}$

10. $\sqrt{-75}$

11. $\sqrt{-7}$

12. $\sqrt{-15}$

13. $\dfrac{\sqrt{-12}}{\sqrt{-3}}$

14. $\dfrac{\sqrt{-45}}{\sqrt{-5}}$

15. $\sqrt{-\frac{18}{25}}$

16. $\sqrt{-\frac{20}{49}}$

17. $\sqrt{-0.09}$

18. $\sqrt{-0.0004}$

In Exercises 19–42, perform the operation(s) and write the result in standard form. **See Examples 2 and 3.**

✔ 19. $\sqrt{-16} + \sqrt{-36}$

20. $\sqrt{-25} - \sqrt{-9}$

21. $\sqrt{-9} - \sqrt{-1}$

22. $\sqrt{-81} + \sqrt{-64}$

23. $\sqrt{-50} - \sqrt{-8}$

24. $\sqrt{-500} + \sqrt{-45}$

25. $\sqrt{-48} + \sqrt{-12} - \sqrt{-27}$

26. $\sqrt{-32} - \sqrt{-18} + \sqrt{-50}$

✔ 27. $\sqrt{-12}\sqrt{-2}$

28. $\sqrt{-25}\sqrt{-6}$

29. $\sqrt{-18}\sqrt{-3}$

30. $\sqrt{-7}\sqrt{-7}$

31. $\sqrt{-0.16}\sqrt{-1.21}$

32. $\sqrt{-0.49}\sqrt{-1.44}$

33. $\sqrt{-3}\left(\sqrt{-3} + \sqrt{-4}\right)$

34. $\sqrt{-12}\left(\sqrt{-3} - \sqrt{-12}\right)$

35. $\sqrt{-5}\left(\sqrt{-16} - \sqrt{-10}\right)$

36. $\sqrt{-3}\left(\sqrt{-24} + \sqrt{-27}\right)$

37. $\sqrt{-2}\left(3 - \sqrt{-8}\right)$

38. $\sqrt{-9}\left(1 + \sqrt{-16}\right)$

39. $\left(\sqrt{-16}\right)^2$

40. $\left(\sqrt{-2}\right)^2$

41. $\left(\sqrt{-4}\right)^3$

42. $\left(\sqrt{-5}\right)^3$

In Exercises 43–46, determine whether the complex numbers are equal. **See Example 4.**

✔ 43. $\sqrt{1} + \sqrt{-25}$ and $1 + 5i$

44. $\sqrt{16} + \sqrt{-9}$ and $4 - 3i$

45. $\sqrt{27} - \sqrt{-8}$ and $3\sqrt{3} + 2\sqrt{2}\,i$

46. $\sqrt{18} - \sqrt{-12}$ and $3\sqrt{2} - 2\sqrt{3}\,i$

In Exercises 47–54, determine the values of a and b that satisfy the equation. **See Examples 4 and 5.**

47. $3 - 4i = a + bi$

48. $-8 + 6i = a + bi$

49. $5 - 4i = (a + 3) + (b - 1)i$

50. $-10 + 12i = 2a + (5b - 3)i$

✔ 51. $-4 - \sqrt{-8} = a + bi$

52. $\sqrt{-36} - 3 = a + bi$

53. $\sqrt{a} + \sqrt{-49} = 8 + bi$

54. $\sqrt{100} + \sqrt{b} = a + 2\sqrt{3}\,i$

In Exercises 55–70, perform the operation(s) and write the result in standard form. **See Example 6.**

✔ 55. $(4 - 3i) + (6 + 7i)$

56. $(-10 + 2i) + (4 - 7i)$

57. $(-4 - 7i) + (-10 - 33i)$

58. $(15 + 10i) - (2 + 10i)$

59. $13i - (14 - 7i)$ **60.** $17i + (9 - 14i)$

61. $(30 - i) - (18 + 6i) + 3i^2$

62. $(4 + 6i) + (15 + 24i) - 10i^2$

63. $6 - (3 - 4i) + 2i$

64. $22 + (-5 + 8i) + 10i$

65. $\left(\frac{4}{3} + \frac{1}{3}i\right) + \left(\frac{5}{6} + \frac{7}{6}i\right)$ **66.** $\left(\frac{4}{5} + \frac{2}{5}i\right) + \left(\frac{3}{10} - \frac{3}{10}i\right)$

67. $(0.05 + 2.50i) - (6.2 + 11.8i)$

68. $(1.8 + 4.3i) - (0.8 - 0.7i)$

69. $15i - (3 - 25i) + \sqrt{-81}$

70. $(-1 + i) - \sqrt{2} - \sqrt{-2}$

In Exercises 71–98, perform the operation and write the result in standard form. *See Example 7.*

✔ 71. $(3i)(12i)$ **72.** $(-5i)(4i)$

73. $(3i)(-8i)$ **74.** $(-2i)(-10i)$

75. $(-5i)(-i)(\sqrt{-49})$ **76.** $(10i)(\sqrt{-36})(-5i)$

77. $(-3i)^3$ **78.** $(8i)^2$

79. $(-3i)^2$ **80.** $(2i)^4$

81. $-5(13 + 2i)$ **82.** $10(8 - 6i)$

83. $4i(-3 - 5i)$ **84.** $-3i(10 - 15i)$

85. $(9 - 2i)(\sqrt{-4})$ **86.** $(11 + 3i)(\sqrt{-25})$

87. $(4 + 3i)(-7 + 4i)$ **88.** $(3 + 5i)(2 + 15i)$

89. $(-7 + 7i)(4 - 2i)$ **90.** $(3 + 5i)(2 - 15i)$

91. $\left(-2 + \sqrt{-5}\right)\left(-2 - \sqrt{-5}\right)$

92. $\left(-3 - \sqrt{-12}\right)\left(4 - \sqrt{-12}\right)$

93. $(3 - 4i)^2$ **94.** $(7 + i)^2$

95. $(2 + 5i)^2$ **96.** $(8 - 3i)^2$

97. $(3 + i)^3$ **98.** $(2 - 2i)^3$

In Exercises 99–108, simplify the expression.

99. i^7 **100.** i^{11}

101. i^{24} **102.** i^{35}

103. i^{42} **104.** i^{64}

105. i^9 **106.** i^{71}

107. $(-i)^6$ **108.** $(-i)^4$

In Exercises 109–122, multiply the number by its complex conjugate and simplify.

109. $2 + i$ **110.** $3 + 2i$

111. $-2 - 8i$ **112.** $10 - 3i$

113. $5 - \sqrt{6}i$ **114.** $-4 + \sqrt{2}i$

115. $10i$ **116.** 20

117. -12 **118.** $-12i$

119. $1 + \sqrt{-3}$ **120.** $-3 - \sqrt{-5}$

121. $1.5 + \sqrt{-0.25}$ **122.** $3.2 - \sqrt{-0.04}$

In Exercises 123–138, write the quotient in standard form. *See Examples 8–10.*

✔ 123. $\dfrac{20}{2i}$ **124.** $\dfrac{-5}{-3i}$

125. $\dfrac{2 + i}{-5i}$ **126.** $\dfrac{1 + i}{3i}$

127. $\dfrac{4}{1 - i}$ **128.** $\dfrac{20}{3 + i}$

129. $\dfrac{7i + 14}{7i}$ **130.** $\dfrac{6i + 3}{3i}$

131. $\dfrac{-12}{2 + 7i}$ **132.** $\dfrac{15}{2(1 - i)}$

133. $\dfrac{3i}{5 + 2i}$ **134.** $\dfrac{4i}{5 - 3i}$

✔ 135. $\dfrac{5 - i}{5 + i}$ **136.** $\dfrac{9 + i}{9 - i}$

✔ 137. $\dfrac{4 + 5i}{3 - 7i}$ **138.** $\dfrac{5 + 3i}{7 - 4i}$

In Exercises 139–144, perform the operation by first writing each quotient in standard form.

139. $\dfrac{5}{3 + i} + \dfrac{1}{3 - i}$ **140.** $\dfrac{1}{1 - 2i} + \dfrac{4}{1 + 2i}$

141. $\dfrac{3i}{1 + i} + \dfrac{2}{2 + 3i}$ **142.** $\dfrac{i}{4 - 3i} - \dfrac{5}{2 + i}$

143. $\dfrac{1+i}{i} - \dfrac{3}{5-2i}$

144. $\dfrac{3-2i}{i} - \dfrac{1}{7+i}$

In Exercises 145–148, determine whether each number is a solution of the equation. **See Example 11.**

✓ **145.** $x^2 + 2x + 5 = 0$
  (a) $x = -1 + 2i$ (b) $x = -1 - 2i$

146. $x^2 - 4x + 13 = 0$
  (a) $x = 2 - 3i$ (b) $x = 2 + 3i$

147. $x^3 + 4x^2 + 9x + 36 = 0$
  (a) $x = -4$ (b) $x = -3i$

148. $x^3 - 8x^2 + 25x - 26 = 0$
  (a) $x = 2$ (b) $x = 3 - 2i$

149. *Cube Roots* The principal cube root of 125, $\sqrt[3]{125}$, is 5. Evaluate the expression x^3 for each value of x.
  (a) $x = \dfrac{-5 + 5\sqrt{3}\,i}{2}$

  (b) $x = \dfrac{-5 - 5\sqrt{3}\,i}{2}$

150. *Cube Roots* The principal cube root of 27, $\sqrt[3]{27}$, is 3. Evaluate the expression x^3 for each value of x.
  (a) $x = \dfrac{-3 + 3\sqrt{3}\,i}{2}$

  (b) $x = \dfrac{-3 - 3\sqrt{3}\,i}{2}$

151. *Pattern Recognition* Compare the results of Exercises 149 and 150. Use the results to list possible cube roots of (a) 1, (b) 8, and (c) 64. Verify your results algebraically.

152. *Algebraic Properties* Consider the complex number $1 + 5i$.
  (a) Find the additive inverse of the number.

  (b) Find the multiplicative inverse of the number.

In Exercises 153–156, perform the operations.

153. $(a + bi) + (a - bi)$
154. $(a + bi)(a - bi)$
155. $(a + bi) - (a - bi)$
156. $(a + bi)^2 + (a - bi)^2$

Explaining Concepts

157. Look back at Exercises 153–156. Based on your results, write a general rule for each exercise about operations on complex conjugates of the form $a + bi$ and $a - bi$.

158. *True or False?* Some numbers are both real and imaginary. Justify your answer.

159. *Error Analysis* Describe and correct the error.
$$\sqrt{-3}\sqrt{-3} = \sqrt{(-3)(-3)} = \sqrt{9} = 3$$

160. ✎ Explain why the Product Rule for Radicals cannot be used to produce the second expression in Exercise 159.

161. ✎ The denominator of a quotient is a pure imaginary number of the form bi. How can you use the complex conjugate of bi to write the quotient in standard form? Can you use the number i instead of the conjugate of bi? Explain.

162. The polynomial $x^2 + 1$ is prime *with respect to the integers*. It is not, however, prime *with respect to the complex numbers*. Show how $x^2 + 1$ can be factored using complex numbers.

Cumulative Review

In Exercises 163–166, use the Zero-Factor Property to solve the equation.

163. $(x - 5)(x + 7) = 0$ **164.** $z(z - 2) = 0$

165. $3y(y - 3)(y + 4) = 0$
166. $(3x - 2)(4x + 1)(x + 9) = 0$

In Exercises 167–170, solve the equation and check your solution.

167. $\sqrt{x} = 9$ **168.** $\sqrt[3]{t} = 8$
169. $\sqrt{x} - 5 = 0$ **170.** $\sqrt{2x + 3} - 7 = 0$

What Did You Learn?

Use these two pages to help prepare for a test on this chapter. Check off the key terms and key concepts you know. You can also use this section to record your assignments.

Plan for Test Success

Date of test: ☐ / / ☐   **Study dates and times:** ☐ / / ☐ at ☐ : ☐ A.M./P.M.

☐ / / ☐ at ☐ : ☐ A.M./P.M.

Things to review:

☐ Key Terms, *p. 502*
☐ Key Concepts, *pp. 502–503*
☐ Your class notes
☐ Your assignments

☐ Study Tips, *pp. 450, 451, 452, 453, 454, 456, 461, 463, 464, 468, 487, 488, 494, 496*
☐ Technology Tips, *pp. 455, 483, 485*
☐ Mid-Chapter Quiz, *p. 474*

☐ Review Exercises, *pp. 504–506*
☐ Chapter Test, *p. 507*
☐ Video Explanations Online
☐ Tutorial Online

Key Terms

☐ square root, *p. 450*
☐ cube root, *p. 450*
☐ nth root of *a*, *p. 450*
☐ principal nth root of *a*, *p. 450*
☐ radical symbol, *p. 450*
☐ index, *p. 450*
☐ radicand, *p. 450*
☐ perfect square, *p. 451*
☐ perfect cube, *p. 451*

☐ rational exponent, *p. 453*
☐ radical function, *p. 455*
☐ domain (of a radical function), *p. 456*
☐ rationalizing the denominator, *p. 464*
☐ Pythagorean Theorem, *p. 465*
☐ like radicals, *p. 468*
☐ conjugates, *p. 476*
☐ imaginary unit *i*, *p. 493*
☐ *i*-form, *p. 493*

☐ complex number, *p. 495*
☐ real part, *p. 495*
☐ standard form (of a complex number), *p. 495*
☐ imaginary part, *p. 495*
☐ imaginary number, *p. 495*
☐ complex conjugates, *p. 497*

Key Concepts

7.1 Radicals and Rational Exponents

Assignment: _____ Due date: _____

☐ **Use the properties of nth roots.**
1. If *a* is a positive real number and *n* is even, then *a* has exactly two (real) nth roots, which are denoted by $\sqrt[n]{a}$ and $-\sqrt[n]{a}$.
2. If *a* is any real number and *n* is odd, then *a* has only one (real) nth root, denoted by $\sqrt[n]{a}$.
3. If *a* is a negative real number and *n* is even, then *a* has no (real) nth root.

☐ **Use the inverse properties of nth powers and nth roots.**
Let *a* be a real number, and let *n* be an integer such that $n \geq 2$.
1. If *a* has a principal nth root, then $\left(\sqrt[n]{a}\right)^n = a$.
2. If *n* is odd, then $\sqrt[n]{a^n} = a$.
   If *n* is even, then $\sqrt[n]{a^n} = |a|$.

☐ **Understand rational exponents.**
1. $a^{1/n} = \sqrt[n]{a}$
2. $a^{m/n} = (a^{1/n})^m = \left(\sqrt[n]{a}\right)^m$
3. $a^{m/n} = (a^m)^{1/n} = \sqrt[n]{a^m}$

☐ **Use the rules of exponents.**
See page 453 for the rules of exponents as they apply to rational exponents.

☐ **Find the domain of a radical function.**
Let *n* be an integer that is greater than or equal to 2.
1. If *n* is odd, the domain of $f(x) = \sqrt[n]{x}$ is the set of all real numbers.
2. If *n* is even, the domain of $f(x) = \sqrt[n]{x}$ is the set of all nonnegative real numbers.

7.2 Simplifying Radical Expressions

Assignment: _____ Due date: _____

☐ **Use the Product and Quotient Rules for Radicals.**

Let u and v be real numbers, variables, or algebraic expressions. If the nth roots of u and v are real, the following rules are true.

1. $\sqrt[n]{uv} = \sqrt[n]{u}\sqrt[n]{v}$

2. $\sqrt[n]{\dfrac{u}{v}} = \dfrac{\sqrt[n]{u}}{\sqrt[n]{v}}, v \neq 0$

☐ **Simplify radical expressions.**

A radical expression is said to be in *simplest form* if all three of the statements below are true.

1. All possible nth powered factors have been removed from each radical.
2. No radical contains a fraction.
3. No denominator of a fraction contains a radical.

7.3 Adding and Subtracting Radical Expressions

Assignment: _____ Due date: _____

☐ **Add and subtract radical expressions.**

1. *Like radicals* have the same index and the same radicand. Combine like radicals by combining their coefficients.
2. Before concluding that two radicals cannot be combined, simplify each radical expression to see if they become like radicals.

Example:

$\sqrt{4x} + \sqrt{9x}$ Sum of *unlike* radicals

$= 2\sqrt{x} + 3\sqrt{x}$ Simplify radicals.

$= 5\sqrt{x}$ Combine like radicals.

7.4 Multiplying and Dividing Radical Expressions

Assignment: _____ Due date: _____

☐ **Multiply radical expressions.**

Multiply radical expressions by using the Distributive Property or the FOIL Method.

☐ **Divide radical expressions.**

To simplify a quotient involving radicals, rationalize the denominator by multiplying both the numerator and denominator by the conjugate of the denominator.

7.5 Radical Equations and Applications

Assignment: _____ Due date: _____

☐ **Raise each side of an equation to the nth power.**

Let u and v be real numbers, variables, or algebraic expressions, and let n be a positive integer. If $u = v$, then it follows that $u^n = v^n$.

Example:

$\sqrt{x} = 9$ Let $u = \sqrt{x}, v = 9$.

$\left(\sqrt{x}\right)^2 = 9^2$ $u^2 = v^2$

$x = 81$ Simplify.

7.6 Complex Numbers

Assignment: _____ Due date: _____

☐ **Find the square root of a negative number.**

Let c be a positive real number. Then the square root of $-c$ is given by $\sqrt{-c} = \sqrt{c(-1)} = \sqrt{c}\sqrt{-1} = \sqrt{c}\,i$.

When writing $\sqrt{-c}$ in the i-form, $\sqrt{c}\,i$, note that i is outside the radical.

☐ **Perform operations with complex numbers.**

If a and b are real numbers, $a + bi$ is a complex number. To add complex numbers, add the real and imaginary parts separately. Use the FOIL Method or the Distributive Property to multiply complex numbers. To write the quotient of two complex numbers in standard form, multiply the numerator and denominator by the complex conjugate of the denominator, and simplify.

Review Exercises

7.1 Radicals and Rational Exponents

1 ▶ Determine the *n*th roots of numbers and evaluate radical expressions.

In Exercises 1–10, evaluate the radical expression without using a calculator. If not possible, state the reason.

1. $-\sqrt{81}$

2. $\sqrt{-16}$

3. $-\sqrt[3]{64}$

4. $\sqrt[3]{-8}$

5. $-\sqrt{\left(\frac{3}{4}\right)^2}$

6. $\sqrt{\left(-\frac{9}{13}\right)^2}$

7. $\sqrt[3]{-\left(\frac{1}{5}\right)^3}$

8. $-\sqrt[3]{\left(-\frac{27}{64}\right)^3}$

9. $\sqrt{-2^2}$

10. $-\sqrt{-3^2}$

2 ▶ Use the rules of exponents to evaluate or simplify expressions with rational exponents.

In Exercises 11–14, fill in the missing description.

Radical Form	Rational Exponent Form
11. $\sqrt[3]{27} = 3$	
12. $\sqrt[3]{0.125} = 0.5$	
13.	$216^{1/3} = 6$
14.	$16^{1/4} = 2$

In Exercises 15–20, evaluate without using a calculator.

15. $27^{4/3}$

16. $16^{3/4}$

17. $(-25)^{3/2}$

18. $-(4^3)^{2/3}$

19. $8^{-4/3}$

20. $243^{-2/5}$

In Exercises 21–32, rewrite the expression using rational exponents.

21. $x^{3/4} \cdot x^{-1/6}$

22. $a^{2/3} \cdot a^{3/5}$

23. $z\sqrt[3]{z^2}$

24. $x^2\sqrt[4]{x^3}$

25. $\dfrac{\sqrt[4]{x^3}}{\sqrt{x^4}}$

26. $\dfrac{\sqrt{x^3}}{\sqrt[3]{x^2}}$

27. $\sqrt[3]{a^3b^2}$

28. $\sqrt[4]{m^3n^8}$

29. $\sqrt[4]{\sqrt{x}}$

30. $\sqrt{\sqrt[3]{x^4}}$

31. $\dfrac{(3x+2)^{2/3}}{\sqrt[3]{3x+2}}$

32. $\dfrac{\sqrt[5]{3x+6}}{(3x+6)^{4/5}}$

3 ▶ Use a calculator to evaluate radical expressions.

In Exercises 33–36, use a calculator to evaluate the expression. Round the answer to four decimal places.

33. $75^{-3/4}$

34. $158^{7/3}$

35. $\sqrt{13^2 - 4(2)(7)}$

36. $\dfrac{-3.7 + \sqrt{15.8}}{2(2.3)}$

4 ▶ Evaluate radical functions and find the domains of radical functions.

In Exercises 37–40, evaluate the function as indicated, if possible, and simplify.

37. $f(x) = \sqrt{x-2}$
  (a) $f(-7)$ (b) $f(51)$

38. $f(x) = \sqrt{6x-5}$
  (a) $f(5)$ (b) $f(-1)$

39. $g(x) = \sqrt[3]{2x-1}$
  (a) $g(0)$ (b) $g(14)$

40. $g(x) = \sqrt[4]{x+5}$
  (a) $g(-4)$ (b) $g(76)$

In Exercises 41 and 42, describe the domain of the function.

41. $f(x) = \sqrt{9-2x}$

42. $g(x) = \sqrt[3]{x+2}$

7.2 Simplifying Radical Expressions

1 ▶ Use the Product and Quotient Rules for Radicals to simplify radical expressions.

In Exercises 43–48, simplify the radical expression.

43. $\sqrt{36u^5v^2}$

44. $\sqrt{24x^3y^4}$

45. $\sqrt{0.25x^4y}$

46. $\sqrt{0.16s^6t^3}$

47. $\sqrt[3]{48a^3b^4}$

48. $\sqrt[4]{48u^4v^6}$

2 ▶ Use rationalization techniques to simplify radical expressions.

In Exercises 49–52, rationalize the denominator and simplify further, if possible.

49. $\sqrt{\dfrac{5}{6}}$

50. $\dfrac{4y}{\sqrt{10z}}$

51. $\dfrac{2}{\sqrt[3]{2x}}$

52. $\sqrt[3]{\dfrac{16t}{s^2}}$

3 ▶ Use the Pythagorean Theorem in application problems.

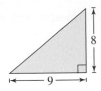 *Geometry*   In Exercises 53 and 54, find the length of the hypotenuse of the right triangle.

53.

54.

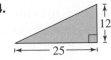

7.3 Adding and Subtracting Radical Expressions

1 ▶ Use the Distributive Property to add and subtract like radicals.

In Exercises 55–62, combine the radical expressions, if possible.

55. $2\sqrt{24} + 7\sqrt{6} - \sqrt{54}$

56. $9\sqrt{50} - 5\sqrt{8} + \sqrt{48}$

57. $5\sqrt{x} - \sqrt[3]{x} + 9\sqrt{x} - 8\sqrt[3]{x}$

58. $\sqrt{3x} - \sqrt[4]{6x^2} + 2\sqrt[4]{6x^2} - 4\sqrt{3x}$

59. $10\sqrt[4]{y + 3} - 3\sqrt[4]{y + 3}$

60. $5\sqrt[3]{x - 3} + 4\sqrt[3]{x - 3}$

61. $2x\sqrt[3]{24x^2y} - \sqrt[3]{3x^5y}$

62. $4xy^2\sqrt[4]{243x} + 2y^2\sqrt[4]{48x^5}$

2 ▶ Use radical expressions in application problems.

▲ *Dining Hall*   In Exercises 63 and 64, a campus dining hall is undergoing renovations. The four corners of the hall are to be walled off and used as storage units (see figure).

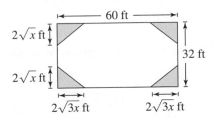

63. Find the perimeter of one of the storage units.

64. Find the perimeter of the newly designed dining hall.

7.4 Multiplying and Dividing Radical Expressions

1 ▶ Use the Distributive Property or the FOIL Method to multiply radical expressions.

In Exercises 65–70, multiply and simplify.

65. $\sqrt{15} \cdot \sqrt{20}$ 66. $\sqrt{36} \cdot \sqrt{60}$

67. $\sqrt{10}\left(\sqrt{2} + \sqrt{5}\right)$

68. $\sqrt{12}\left(\sqrt{6} - \sqrt{8}\right)$

69. $\left(\sqrt{3} - \sqrt{x}\right)\left(\sqrt{3} + \sqrt{x}\right)$

70. $\left(4 - 3\sqrt{2}\right)^2$

2 ▶ Determine the products of conjugates.

In Exercises 71–74, find the conjugate of the expression. Then multiply the expression by its conjugate and simplify.

71. $3 - \sqrt{7}$

72. $\sqrt{6} + 9$

73. $\sqrt{x} + 20$

74. $9 - \sqrt{2y}$

3 ▶ Simplify quotients involving radicals by rationalizing the denominators.

In Exercises 75–78, rationalize the denominator of the expression and simplify.

75. $\dfrac{\sqrt{2} - 1}{\sqrt{3} - 4}$

76. $\dfrac{2 + \sqrt{20}}{3 + \sqrt{5}}$

77. $\left(\sqrt{x} + 10\right) \div \left(\sqrt{x} - 10\right)$

78. $\left(3\sqrt{s} + 4\right) \div \left(\sqrt{s} + 2\right)$

7.5 Radical Equations and Applications

1 ▶ Solve a radical equation by raising each side to the *n*th power.

In Exercises 79–88, solve the equation and check your solution(s).

79. $\sqrt{2x} - 8 = 0$ 80. $\sqrt{4x} + 6 = 9$

81. $\sqrt[4]{3x - 1} + 6 = 3$

82. $\sqrt[3]{5x - 7} - 3 = -1$

83. $\sqrt[3]{5x + 2} - \sqrt[3]{7x - 8} = 0$

84. $\sqrt[4]{9x - 2} - \sqrt[4]{8x} = 0$

85. $\sqrt{2(x + 5)} = x + 5$

86. $y - 2 = \sqrt{y + 4}$

87. $\sqrt{1 + 6x} = 2 - \sqrt{6x}$

88. $\sqrt{2 + 9b} + 1 = 3\sqrt{b}$

2 ▶ Solve application problems involving radical equations.

89. ▲ *Geometry* Determine the length and width of a rectangle with a perimeter of 46 inches and a diagonal of 17 inches.

90. ▲ *Geometry* Determine the length and width of a rectangle with a perimeter of 82 inches and a diagonal of 29 inches.

91. *Period of a Pendulum* The time t (in seconds) for a pendulum of length L (in feet) to go through one complete cycle (its period) is given by

$$t = 2\pi\sqrt{\frac{L}{32}}.$$

How long is the pendulum of a grandfather clock with a period of 1.9 seconds?

92. *Height* The time t (in seconds) for a free-falling object to fall d feet is given by $t = \sqrt{d}/16$. A child drops a pebble from a bridge and observes it strike the water after approximately 4 seconds. Estimate the height from which the pebble was dropped.

Free-Falling Object **In Exercises 93 and 94, the velocity of a free-falling object can be determined from the equation $v = \sqrt{2gh}$, where v is the velocity (in feet per second), $g = 32$ feet per second per second, and h is the distance (in feet) the object has fallen.**

93. Find the height from which a brick has been dropped when it strikes the ground with a velocity of 64 feet per second.

94. Find the height from which a wrench has been dropped when it strikes the ground with a velocity of 112 feet per second.

7.6 Complex Numbers

1 ▶ Write square roots of negative numbers in *i*-form and perform operations on numbers in *i*-form.

In Exercises 95–100, write the number in *i*-form.

95. $\sqrt{-48}$

96. $\sqrt{-0.16}$

97. $10 - 3\sqrt{-27}$

98. $3 + 2\sqrt{-500}$

99. $\frac{3}{4} - 5\sqrt{-\frac{3}{25}}$

100. $-0.5 + 3\sqrt{-1.21}$

In Exercises 101–104, perform the operation(s) and write the result in standard form.

101. $\sqrt{-81} + \sqrt{-36}$

102. $\sqrt{-121} - \sqrt{-84}$

103. $\sqrt{-10}\left(\sqrt{-4} - \sqrt{-7}\right)$

104. $\sqrt{-5}\left(\sqrt{-10} + \sqrt{-15}\right)$

2 ▶ Determine the equality of two complex numbers.

In Exercises 105–108, determine the values of a and b that satisfy the equation.

105. $12 - 5i = (a + 2) + (b - 1)i$

106. $-48 + 9i = (a - 5) + (b + 10)i$

107. $\sqrt{-49} + 4 = a + bi$

108. $-3 - \sqrt{-4} = a + bi$

3 ▶ Add, subtract, and multiply complex numbers.

In Exercises 109–114, perform the operation and write the result in standard form.

109. $(-4 + 5i) - (-12 + 8i)$

110. $(-6 + 3i) + (-1 + i)$

111. $(4 - 3i)(4 + 3i)$

112. $(12 - 5i)(2 + 7i)$

113. $(6 - 5i)^2$

114. $(2 - 9i)^2$

4 ▶ Use complex conjugates to write the quotient of two complex numbers in standard form.

In Exercises 115–120, write the quotient in standard form.

115. $\dfrac{7}{3i}$

116. $\dfrac{4}{5i}$

117. $\dfrac{-3i}{4 - 6i}$

118. $\dfrac{5i}{2 + 9i}$

119. $\dfrac{3 - 5i}{6 + i}$

120. $\dfrac{2 + i}{1 - 9i}$

Chapter Test

Take this test as you would take a test in class. After you are done, check your work against the answers in the back of the book.

In Exercises 1 and 2, evaluate each expression without using a calculator.

1. (a) $16^{3/2}$

(b) $\sqrt{5}\sqrt{20}$

2. (a) $125^{-2/3}$

(b) $\sqrt{3}\sqrt{12}$

3. For $f(x) = \sqrt{9 - 5x}$, find $f(-8)$ and $f(0)$.

4. Find the domain of $g(x) = \sqrt{7x - 3}$.

In Exercises 5–7, simplify each expression.

5. (a) $\left(\dfrac{x^{1/2}}{x^{1/3}}\right)^2$

(b) $5^{1/4} \cdot 5^{7/4}$

6. (a) $\sqrt{\dfrac{32}{9}}$

(b) $\sqrt[3]{24}$

7. (a) $\sqrt{24x^3}$

(b) $\sqrt[4]{16x^5y^8}$

In Exercises 8 and 9, rationalize the denominator of the expression and simplify.

8. $\dfrac{2}{\sqrt[3]{9y}}$

9. $\dfrac{10}{\sqrt{6} - \sqrt{2}}$

10. Subtract: $6\sqrt{18x} - 3\sqrt{32x}$

11. Multiply and simplify: $\sqrt{5}\left(\sqrt{15x} + 3\right)$

12. Expand: $\left(4 - \sqrt{2x}\right)^2$

13. Factor: $7\sqrt{27} + 14y\sqrt{12} = 7\sqrt{3}\left(\quad\right)$

In Exercises 14–16, solve the equation.

14. $\sqrt{6z} + 5 = 17$

15. $\sqrt{x^2 - 1} = x - 2$

16. $\sqrt{x} - x + 6 = 0$

In Exercises 17–20, perform the operation(s) and simplify.

17. $(2 + 3i) - \sqrt{-25}$

18. $(3 - 5i)^2$

19. $\sqrt{-16}\left(1 + \sqrt{-4}\right)$

20. $(3 - 2i)(1 + 5i)$

21. Write $\dfrac{5 - 2i}{3 + i}$ in standard form.

22. The velocity v (in feet per second) of an object is given by $v = \sqrt{2gh}$, where $g = 32$ feet per second per second and h is the distance (in feet) the object has fallen. Find the height from which a rock has been dropped when it strikes the ground with a velocity of 96 feet per second.

Cumulative Test: Chapters 5–7

Take this test as you would take a test in class. After you are done, check your work against the answers in the back of the book.

In Exercises 1–3, simplify the expression.

1. $(-2x^5y^{-2}z^0)^{-1}$

2. $\dfrac{12s^5t^{-2}}{20s^{-2}t^{-1}}$

3. $\left(\dfrac{2x^{-4}y^3}{3x^5y^{-3}z^0}\right)^{-2}$

4. Evaluate $(5 \times 10^3)^2$ without using a calculator.

In Exercises 5–8, perform the operation(s) and simplify.

5. $(x^5 + 2x^3 + x^2 - 10x) - (2x^3 - x^2 + x - 4)$

6. $-3(3x^3 - 4x^2 + x) + 3x(2x^2 + x - 1)$

7. $(x + 8)(3x - 2)$

8. $(3x + 2)(3x^2 - x + 1)$

In Exercises 9–12, factor the polynomial completely.

9. $2x^2 - 11x + 15$

10. $9x^2 - 144$

11. $y^3 - 3y^2 - 9y + 27$

12. $8t^3 - 40t^2 + 50t$

In Exercises 13 and 14, solve the polynomial equation.

13. $3x^2 + x - 24 = 0$

14. $6x^3 - 486x = 0$

In Exercises 15–20, perform the operation(s) and simplify.

15. $\dfrac{x^2 + 8x + 16}{18x^2} \cdot \dfrac{2x^4 + 4x^3}{x^2 - 16}$

16. $\dfrac{x^2 + 4x}{2x^2 - 7x + 3} \div \dfrac{x^2 - 16}{x - 3}$

17. $\dfrac{5x}{x + 2} - \dfrac{2}{x^2 - x - 6}$

18. $\dfrac{2}{x} - \dfrac{x}{x^3 + 3x^2} + \dfrac{1}{x + 3}$

19. $\dfrac{\left(\dfrac{3x}{x + 2}\right)}{\left(\dfrac{12 \cdot}{x^3 + 2x^2}\right)}$

20. $\dfrac{\left(\dfrac{x}{y} - \dfrac{y}{x}\right)}{\left(\dfrac{x - y}{xy}\right)}$

21. Use synthetic division to divide $2x^3 + 7x^2 - 5$ by $x + 4$.

22. Use long division to divide $4x^4 - 6x^3 + x - 4$ by $2x - 1$.

In Exercises 23 and 24, solve the rational equation.

23. $\dfrac{1}{x} + \dfrac{4}{10 - x} = 1$

24. $\dfrac{x - 3}{x} + 1 = \dfrac{x - 4}{x - 6}$

In Exercises 25–28, simplify the expression.

25. $\sqrt{24x^2y^3}$

26. $\sqrt[3]{80a^{15}b^8}$

27. $(12a^{-4}b^6)^{1/2}$

28. $\left(\dfrac{t^{1/2}}{t^{1/4}}\right)^2$

29. Add: $10\sqrt{20x} + 3\sqrt{125x}$

30. Expand: $\left(\sqrt{2x} - 3\right)^2$

31. Rationalize the denominator and simplify the result: $\dfrac{3}{\sqrt{10} - \sqrt{x}}$

In Exercises 32–35, solve the radical equation.

32. $\sqrt{x - 5} - 6 = 0$

33. $\sqrt{3 - x} + 10 = 11$

34. $\sqrt{x + 5} - \sqrt{x - 7} = 2$

35. $\sqrt{x - 4} = \sqrt{x + 7} - 1$

In Exercises 36–38, perform the operation and simplify.

36. $\sqrt{-2}\left(\sqrt{-8} + 3\right)$

37. $(-4 + 11i) - (3 - 5i)$

38. $(5 + 2i)^2$

39. Write $\dfrac{2 + 3i}{6 - 2i}$ in standard form.

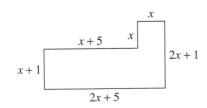

Figure for 40

40. Write and simplify an expression for the perimeter of the figure shown at the left.

41. You can mow a lawn in 2 hours and your friend can mow it in 3.5 hours. What fractional part of the lawn can each of you mow in 1 hour? How long will it take both of you to mow the lawn together?

42. The game commission introduces 50 deer into newly acquired state game lands. The population P of the herd is approximated by the model

$$P = \dfrac{10(5 + 3t)}{1 + 0.004t}$$

where t is the time in years. Find the time required for the population to increase to 250 deer.

43. A guy wire on a 180-foot cell phone tower is attached to the top of the tower and to an anchor 90 feet from the base of the tower. Find the length of the guy wire.

44. The time t (in seconds) for a free-falling object to fall d feet is given by

$$t = \sqrt{\dfrac{d}{16}}.$$

A construction worker drops a nail from a building and observes it strike a water puddle after 5 seconds. Estimate the height from which the nail was dropped.

Managing Test Anxiety

Test anxiety is different from the typical nervousness that usually occurs during tests. It interferes with the thinking process. After leaving the classroom, have you suddenly been able to recall what you could not remember during the test? It is likely that this was a result of test anxiety. Test anxiety is a learned reaction or response—no one is born with it. The good news is that most students can learn to manage test anxiety.

It is important to get as much information as you can into your long-term memory and to practice retrieving the information before you take a test. The more you practice retrieving information, the easier it will be during the test.

Kimberly Nolting

VP, Academic Success Press
expert in developmental education

Smart Study Strategy

Make Mental Cheat Sheets

No, we are not asking you to cheat! Just prepare as if you were going to and then memorize the information you've gathered.

1 ▶ Write down important information on note cards. This can include:

- formulas

- examples of problems you find difficult

- concepts that always trip you up

2 ▶ Memorize the information on the note cards. Flash through the cards, placing the ones containing information you know in one stack and the ones containing information you do not know in another stack. Keep working on the information you do not know.

3 ▶ As soon as you receive your test, turn it over and write down all the information you remember, starting with things you have the greatest difficulty remembering. Having this information available should boost your confidence and free up mental energy for focusing on the test.

Do not wait until the night before the test to make note cards. Make them after you study each section. Then review them two or three times a week.

Completing the square

To complete the square for the expression $x^2 + bx$, add $(b/2)^2$.

$$x^2 + bx + \left(\frac{b}{2}\right)^2 = \left(x + \frac{b}{2}\right)^2$$

Quadratic Formula:

$$x = \frac{-b \pm \sqrt{b^2 - 4ac}}{2a}$$

Use $x = -b/(2a)$ to find the vertex and axis of a parabola.

Chapter 8
Quadratic Equations, Functions, and Inequalities

IT WORKED FOR ME!

"I used to waste time reviewing for a test by just reading through my notes over and over again. Then I learned how to make note cards and keep working through them, saving the important ones for my mental cheat sheets. Now a couple of my friends and I go through our notes and make mental cheat sheets together. We are all doing well in our math class."

Stephanie
Associate of Arts
transfer degree

511

8.1 Solving Quadratic Equations: Factoring and Special Forms

Chris Whitehead/Getty Images

What You Should Learn

1 ▶ Solve quadratic equations by factoring.

2 ▶ Solve quadratic equations by the Square Root Property.

3 ▶ Solve quadratic equations with complex solutions by the Square Root Property.

4 ▶ Use substitution to solve equations of quadratic form.

Why You Should Learn It

Quadratic equations can be used to model and solve real-life problems. For instance, in Exercises 141 and 142 on page 520, you will use a quadratic equation to determine national health care expenditures in the United States.

1 ▶ Solve quadratic equations by factoring.

Solving Quadratic Equations by Factoring

In this chapter, you will study methods for solving quadratic equations and equations of quadratic form. To begin, let's review the method of factoring that you studied in Section 5.6.

Remember that the first step in solving a quadratic equation by factoring is to write the equation in general form. Next, factor the left side. Finally, set each factor equal to zero and solve for x. Be sure to check each solution in the original equation.

EXAMPLE 1   Solving Quadratic Equations by Factoring

a.

$x^2 + 5x = 24$	Original equation
$x^2 + 5x - 24 = 0$	Write in general form.
$(x + 8)(x - 3) = 0$	Factor.
$x + 8 = 0 \implies x = -8$	Set 1st factor equal to 0.
$x - 3 = 0 \implies x = 3$	Set 2nd factor equal to 0.

b.

$3x^2 = 4 - 11x$	Original equation
$3x^2 + 11x - 4 = 0$	Write in general form.
$(3x - 1)(x + 4) = 0$	Factor.
$3x - 1 = 0 \implies x = \dfrac{1}{3}$	Set 1st factor equal to 0.
$x + 4 = 0 \implies x = -4$	Set 2nd factor equal to 0.

c.

$9x^2 + 12 = 3 + 12x + 5x^2$	Original equation
$4x^2 - 12x + 9 = 0$	Write in general form.
$(2x - 3)(2x - 3) = 0$	Factor.
$2x - 3 = 0 \implies x = \dfrac{3}{2}$	Set factor equal to 0.

Check each solution in its original equation.

 **CHECKPOINT** *Now try Exercise 1.*

Study Tip

In Example 1(c), the quadratic equation produces two identical solutions. This is called a **double** or **repeated solution.**

2 ▸ Solve quadratic equations by the Square Root Property.

The Square Root Property

Consider the following equation, where $d > 0$ and u is an algebraic expression.

$$u^2 = d \qquad\qquad \text{Original equation}$$

$$u^2 - d = 0 \qquad\qquad \text{Write in general form.}$$

$$\left(u + \sqrt{d}\,\right)\left(u - \sqrt{d}\,\right) = 0 \qquad\qquad \text{Factor.}$$

$$u + \sqrt{d} = 0 \quad\Longrightarrow\quad u = -\sqrt{d} \qquad \text{Set 1st factor equal to 0.}$$

$$u - \sqrt{d} = 0 \quad\Longrightarrow\quad u = \sqrt{d} \qquad \text{Set 2nd factor equal to 0.}$$

Because the solutions differ only in sign, they can be written together using a "plus or minus sign": $u = \pm\sqrt{d}$. This form of the solution is read as "u is equal to plus or minus the square root of d." Now you can use the **Square Root Property** to solve an equation of the form $u^2 = d$ *without* going through the steps of factoring.

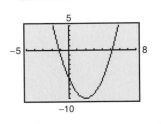

Technology: Tip

To check graphically the solutions of an equation written in general form, graph the left side of the equation and locate its *x*-intercepts. For instance, in Example 2(b), write the equation as

$$(x - 2)^2 - 10 = 0$$

and then use a graphing calculator to graph

$$y = (x - 2)^2 - 10$$

as shown below. You can use the *zoom* and *trace* features or the *zero* or *root* feature to approximate the *x*-intercepts of the graph to be $x \approx 5.16$ and $x \approx -1.16$.

> ### Square Root Property
>
> The equation $u^2 = d$, where $d > 0$, has exactly two solutions:
>
> $$u = \sqrt{d} \quad \text{and} \quad u = -\sqrt{d}.$$
>
> These solutions can also be written as $u = \pm\sqrt{d}$. This solution process is also called **extracting square roots.**

EXAMPLE 2   Square Root Property

a. $3x^2 = 15$ Original equation

$\quad x^2 = 5$ Divide each side by 3.

$\quad x = \pm\sqrt{5}$ Square Root Property

The solutions are $x = \sqrt{5}$ and $x = -\sqrt{5}$. Check these in the original equation.

b. $(x - 2)^2 = 10$ Original equation

$\quad x - 2 = \pm\sqrt{10}$ Square Root Property

$\quad x = 2 \pm\sqrt{10}$ Add 2 to each side.

The solutions are $x = 2 + \sqrt{10} \approx 5.16$ and $x = 2 - \sqrt{10} \approx -1.16$.

c. $(3x - 6)^2 - 8 = 0$ Original equation

$\quad (3x - 6)^2 = 8$ Add 8 to each side.

$\quad 3x - 6 = \pm 2\sqrt{2}$ Square Root Property and rewrite $\sqrt{8}$ as $2\sqrt{2}$.

$\quad 3x = 6 \pm 2\sqrt{2}$ Add 6 to each side.

$\quad x = 2 \pm \dfrac{2\sqrt{2}}{3}$ Divide each side by 3.

The solutions are $x = 2 + 2\sqrt{2}/3 \approx 2.94$ and $x = 2 - 2\sqrt{2}/3 \approx 1.06$.

✓ **CHECKPOINT** *Now try Exercise 23.*

3 ▶ Solve quadratic equations with complex solutions by the Square Root Property.

Quadratic Equations with Complex Solutions

Prior to Section 7.6, the only solutions you could find were real numbers. But now that you have studied complex numbers, it makes sense to look for other types of solutions. For instance, although the quadratic equation $x^2 + 1 = 0$ has no solutions that are real numbers, it does have two solutions that are complex numbers: i and $-i$. To check this, substitute i and $-i$ for x.

$$(i)^2 + 1 = -1 + 1 = 0 \qquad \text{Solution checks.} \checkmark$$

$$(-i)^2 + 1 = -1 + 1 = 0 \qquad \text{Solution checks.} \checkmark$$

One way to find complex solutions of a quadratic equation is to extend the Square Root Property to cover the case in which d is a negative number.

Technology: Discovery

Solve each quadratic equation below algebraically. Then use a graphing calculator to check the solutions. Which equations have real solutions and which have complex solutions? Which graphs have x-intercepts and which have no x-intercepts? Compare the type(s) of solution(s) of each quadratic equation with the x-intercept(s) of the graph of the equation.

a. $y = 2x^2 + 3x - 5$

b. $y = 2x^2 + 4x + 2$

c. $y = x^2 + 4$

d. $y = (x + 7)^2 + 2$

> ### Square Root Property (Complex Square Root)
>
> The equation $u^2 = d$, where $d < 0$, has exactly two solutions:
> $$u = \sqrt{|d|}\,i \quad \text{and} \quad u = -\sqrt{|d|}\,i.$$
> These solutions can also be written as $u = \pm\sqrt{|d|}\,i$.

EXAMPLE 3   Square Root Property

a. $x^2 + 8 = 0$ Original equation

$\qquad x^2 = -8$ Subtract 8 from each side.

$\qquad x = \pm\sqrt{8}\,i = \pm 2\sqrt{2}\,i$ Square Root Property

The solutions are $x = 2\sqrt{2}\,i$ and $x = -2\sqrt{2}\,i$. Check these in the original equation.

b. $(x - 4)^2 = -3$ Original equation

$\qquad x - 4 = \pm\sqrt{3}\,i$ Square Root Property

$\qquad x = 4 \pm \sqrt{3}\,i$ Add 4 to each side.

The solutions are $x = 4 + \sqrt{3}\,i$ and $x = 4 - \sqrt{3}\,i$. Check these in the original equation.

c. $2(3x - 5)^2 + 32 = 0$ Original equation

$\qquad 2(3x - 5)^2 = -32$ Subtract 32 from each side.

$\qquad (3x - 5)^2 = -16$ Divide each side by 2.

$\qquad 3x - 5 = \pm 4i$ Square Root Property

$\qquad 3x = 5 \pm 4i$ Add 5 to each side.

$\qquad x = \dfrac{5}{3} \pm \dfrac{4}{3}i$ Divide each side by 3.

The solutions are $x = 5/3 + 4/3i$ and $x = 5/3 - 4/3i$. Check these in the original equation.

✓ **CHECKPOINT** *Now try Exercise 45.*

4 ▶ Use substitution to solve equations of quadratic form.

Equations of Quadratic Form

Both the factoring method and the Square Root Property can be applied to nonquadratic equations that are of **quadratic form.** An equation is said to be of quadratic form if it has the form

$$au^2 + bu + c = 0$$

where u is an algebraic expression. Here are some examples.

Equation	Written in Quadratic Form
$x^4 + 5x^2 + 4 = 0$	$(x^2)^2 + 5(x^2) + 4 = 0$
$x - 5\sqrt{x} + 6 = 0$	$(\sqrt{x})^2 - 5(\sqrt{x}) + 6 = 0$
$2x^{2/3} + 5x^{1/3} - 3 = 0$	$2(x^{1/3})^2 + 5(x^{1/3}) - 3 = 0$
$18 + 2x^2 + (x^2 + 9)^2 = 8$	$(x^2 + 9)^2 + 2(x^2 + 9) - 8 = 0$

To solve an equation of quadratic form, it helps to make a substitution and rewrite the equation in terms of u, as demonstrated in Examples 4 and 5.

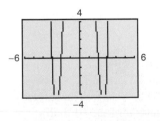
EXAMPLE 4   **Solving an Equation of Quadratic Form**

Solve $x^4 - 13x^2 + 36 = 0$.

Solution

Begin by writing the original equation in quadratic form, as follows.

$$x^4 - 13x^2 + 36 = 0 \qquad \text{Write original equation.}$$

$$(x^2)^2 - 13(x^2) + 36 = 0 \qquad \text{Write in quadratic form.}$$

Next, let $u = x^2$ and substitute u into the equation written in quadratic form. Then, factor and solve the equation.

$$u^2 - 13u + 36 = 0 \qquad \text{Substitute } u \text{ for } x^2.$$

$$(u - 4)(u - 9) = 0 \qquad \text{Factor.}$$

$$u - 4 = 0 \quad \Longrightarrow \quad u = 4 \qquad \text{Set 1st factor equal to 0.}$$

$$u - 9 = 0 \quad \Longrightarrow \quad u = 9 \qquad \text{Set 2nd factor equal to 0.}$$

At this point you have found the "u-solutions." To find the "x-solutions," replace u with x^2 and solve for x.

$$u = 4 \quad \Longrightarrow \quad x^2 = 4 \quad \Longrightarrow \quad x = \pm 2$$

$$u = 9 \quad \Longrightarrow \quad x^2 = 9 \quad \Longrightarrow \quad x = \pm 3$$

The solutions are $x = 2$, $x = -2$, $x = 3$, and $x = -3$. Check these in the original equation.

✓ **CHECKPOINT** *Now try Exercise 101.*

Be sure you see in Example 4 that the u-solutions of 4 and 9 represent only a temporary step. They are not solutions of the original equation and cannot be substituted into the original equation.

EXAMPLE 5   Solving Equations of Quadratic Form

a. $x - 5\sqrt{x} + 6 = 0$ Original equation

This equation is of quadratic form with $u = \sqrt{x}$.

$(\sqrt{x})^2 - 5(\sqrt{x}) + 6 = 0$ Write in quadratic form.

$u^2 - 5u + 6 = 0$ Substitute u for $\sqrt{x}$.

$(u - 2)(u - 3) = 0$ Factor.

$u - 2 = 0 \implies u = 2$ Set 1st factor equal to 0.

$u - 3 = 0 \implies u = 3$ Set 2nd factor equal to 0.

Now, using the u-solutions of 2 and 3, you obtain the x-solutions as follows.

$u = 2 \implies \sqrt{x} = 2 \implies x = 4$

$u = 3 \implies \sqrt{x} = 3 \implies x = 9$

b. $x^{2/3} - x^{1/3} - 6 = 0$ Original equation

This equation is of quadratic form with $u = x^{1/3}$.

$(x^{1/3})^2 - (x^{1/3}) - 6 = 0$ Write in quadratic form.

$u^2 - u - 6 = 0$ Substitute u for $x^{1/3}$.

$(u + 2)(u - 3) = 0$ Factor.

$u + 2 = 0 \implies u = -2$ Set 1st factor equal to 0.

$u - 3 = 0 \implies u = 3$ Set 2nd factor equal to 0.

Now, using the u-solutions of -2 and 3, you obtain the x-solutions as follows.

$u = -2 \implies x^{1/3} = -2 \implies x = -8$

$u = 3 \implies x^{1/3} = 3 \implies x = 27$

 **CHECKPOINT** *Now try Exercise 107.*

EXAMPLE 6   Surface Area of a Softball

The surface area of a sphere of radius r is given by $S = 4\pi r^2$. The surface area of a softball is $144/\pi$ square inches. Find the diameter d of the softball.

Solution

$\dfrac{144}{\pi} = 4\pi r^2$ Substitute $144/\pi$ for S.

$\dfrac{36}{\pi^2} = r^2 \implies \pm\sqrt{\dfrac{36}{\pi^2}} = r$ Divide each side by 4π and use Square Root Property.

Choose the positive root to obtain $r = 6/\pi$. The diameter of the softball is

$d = 2r = 2\left(\dfrac{6}{\pi}\right) = \dfrac{12}{\pi} \approx 3.82$ inches.

 **CHECKPOINT** *Now try Exercise 131.*

Concept Check

1. Explain the Zero-Factor Property and how it can be used to solve a quadratic equation.

2. Determine whether the following statement is true or false. Justify your answer.

   The only solution of the equation $x^2 = 25$ is $x = 5$.

3. Does the equation $4x^2 + 9 = 0$ have two real solutions or two complex solutions? Explain your reasoning.

4. Is the equation $x^6 - 6x^3 + 9 = 0$ of quadratic form? Explain your reasoning.

8.1 EXERCISES

Go to pages 570–571 to record your assignments.

Developing Skills

In Exercises 1–20, solve the equation by factoring. *See Example 1.*

1. $x^2 - 15x + 54 = 0$ 2. $x^2 + 15x + 44 = 0$

3. $x^2 - x - 30 = 0$ 4. $x^2 - 2x - 48 = 0$

5. $x^2 + 4x = 45$ 6. $x^2 - 7x = 18$

7. $x^2 - 16x + 64 = 0$

8. $x^2 + 60x + 900 = 0$

9. $9x^2 - 10x - 16 = 0$

10. $8x^2 - 10x + 3 = 0$

11. $4x^2 - 12x = 0$ 12. $25y^2 - 75y = 0$

13. $u(u - 9) - 12(u - 9) = 0$

14. $16x(x - 8) - 12(x - 8) = 0$

15. $2x(x - 5) + 9(x - 5) = 0$

16. $3(4 - x) - 2x(4 - x) = 0$

17. $(y - 4)(y - 3) = 6$

18. $(5 + u)(2 + u) = 4$

19. $2x(3x + 2) = 5 - 6x^2$

20. $(2z + 1)(2z - 1) = -4z^2 - 5z + 2$

In Exercises 21–42, solve the equation by using the Square Root Property. *See Example 2.*

21. $x^2 = 49$ 22. $p^2 = 169$

23. $6x^2 = 54$ 24. $5t^2 = 5$

25. $25x^2 = 16$ 26. $9z^2 = 121$

27. $\dfrac{w^2}{4} = 49$ 28. $\dfrac{x^2}{6} = 24$

29. $4x^2 - 25 = 0$ 30. $16y^2 - 121 = 0$

31. $4u^2 - 225 = 0$ 32. $16x^2 - 1 = 0$

33. $(x + 4)^2 = 64$

34. $(m - 12)^2 = 400$

35. $(x - 3)^2 = 0.25$

36. $(x + 2)^2 = 0.81$

37. $(x - 2)^2 = 7$ 38. $(y + 4)^2 = 27$

39. $(2x + 1)^2 = 50$ 40. $(3x - 5)^2 = 48$

41. $(9m - 2)^2 - 108 = 0$ 42. $(5x + 11)^2 - 300 = 0$

In Exercises 43–64, solve the equation by using the Square Root Property. *See Example 3.*

43. $z^2 = -36$ 44. $x^2 = -16$

45. $x^2 + 4 = 0$ 46. $p^2 + 9 = 0$

47. $9u^2 + 17 = 0$ 48. $25x^2 + 4 = 0$

49. $(t - 3)^2 = -25$

50. $(x + 5)^2 = -81$

51. $(3z + 4)^2 + 144 = 0$

52. $(2y - 3)^2 + 25 = 0$

53. $(4m + 1)^2 = -80$ **54.** $(6y - 5)^2 = -8$

55. $36(t + 3)^2 = -100$ **56.** $4(x - 4)^2 = -169$

57. $(x - 1)^2 = -27$ **58.** $(2x + 3)^2 = -54$

59. $(x + 1)^2 + 0.04 = 0$ **60.** $(y - 5)^2 + 6.25 = 0$

61. $\left(c - \frac{2}{3}\right)^2 + \frac{1}{9} = 0$ **62.** $\left(u + \frac{5}{8}\right)^2 + \frac{49}{16} = 0$

63. $\left(x + \frac{7}{3}\right)^2 = -\frac{38}{9}$ **64.** $\left(y - \frac{5}{8}\right)^2 = -\frac{5}{4}$

In Exercises 65–80, find all real and complex solutions of the quadratic equation.

65. $2x^2 - 5x = 0$ **66.** $4t^2 + 20t = 0$

67. $2x^2 + 5x - 12 = 0$ **68.** $3x^2 + 8x - 16 = 0$

69. $x^2 - 900 = 0$ **70.** $z^2 - 256 = 0$

71. $x^2 + 900 = 0$ **72.** $z^2 + 256 = 0$

73. $\frac{2}{3}x^2 = 6$ **74.** $\frac{1}{3}x^2 = 4$

75. $(p - 2)^2 - 108 = 0$ **76.** $(y + 12)^2 - 400 = 0$

77. $(p - 2)^2 + 108 = 0$ **78.** $(y + 12)^2 + 400 = 0$

79. $(x + 2)^2 + 18 = 0$ **80.** $(x + 2)^2 - 18 = 0$

In Exercises 81–90, use a graphing calculator to graph the function. Use the graph to approximate any x-intercepts. Set $y = 0$ and solve the resulting equation. Compare the result with the x-intercepts of the graph.

81. $y = x^2 - 9$
82. $y = 5x - x^2$
83. $y = x^2 - 2x - 15$

84. $y = x^2 + 3x - 40$
85. $y = 4 - (x - 3)^2$
86. $y = 4(x + 1)^2 - 9$
87. $y = 2x^2 - x - 6$
88. $y = 4x^2 - x - 14$
89. $y = 3x^2 - 13x - 10$
90. $y = 5x^2 + 9x - 18$

In Exercises 91–96, use a graphing calculator to graph the function and observe that the graph has no x-intercepts. Set $y = 0$ and solve the resulting equation. Of what type are the solutions of the equation?

91. $y = x^2 + 7$
92. $y = x^2 + 5$
93. $y = (x - 4)^2 + 2$
94. $y = (x + 2)^2 + 3$
95. $y = (x + 3)^2 + 5$
96. $y = (x - 2)^2 + 3$

In Exercises 97–100, solve for y in terms of x. Let f and g be functions representing, respectively, the positive square root and the negative square root. Use a graphing calculator to graph f and g in the same viewing window.

97. $x^2 + y^2 = 4$ **98.** $x^2 - y^2 = 4$

99. $x^2 + 4y^2 = 4$ **100.** $x - y^2 = 0$

In Exercises 101–130, solve the equation of quadratic form. (Find all real and complex solutions.) *See Examples 4 and 5.*

101. $x^4 - 5x^2 + 4 = 0$
102. $x^4 - 10x^2 + 25 = 0$
103. $x^4 - 5x^2 + 6 = 0$
104. $x^4 - 10x^2 + 21 = 0$
105. $(x^2 - 4)^2 + 2(x^2 - 4) - 3 = 0$
106. $(x^2 - 1)^2 + (x^2 - 1) - 6 = 0$
107. $x - 3\sqrt{x} - 4 = 0$

108. $x - \sqrt{x} - 6 = 0$

109. $x - 7\sqrt{x} + 10 = 0$

110. $x - 11\sqrt{x} + 24 = 0$

111. $x^{2/3} - x^{1/3} - 6 = 0$

112. $x^{2/3} + 3x^{1/3} - 10 = 0$

113. $2x^{2/3} - 7x^{1/3} + 5 = 0$

114. $5x^{2/3} - 13x^{1/3} + 6 = 0$

115. $x^{2/5} - 3x^{1/5} + 2 = 0$

116. $x^{2/5} + 5x^{1/5} + 6 = 0$

117. $2x^{2/5} - 7x^{1/5} + 3 = 0$

118. $2x^{2/5} + 3x^{1/5} + 1 = 0$

119. $x^{1/3} - x^{1/6} - 6 = 0$

120. $x^{1/3} + 2x^{1/6} - 3 = 0$

121. $x^{1/2} - 3x^{1/4} + 2 = 0$

122. $x^{1/2} - 5x^{1/4} + 6 = 0$

123. $\dfrac{1}{x^2} - \dfrac{3}{x} + 2 = 0$

124. $\dfrac{1}{x^2} - \dfrac{1}{x} - 6 = 0$

125. $4x^{-2} - x^{-1} - 5 = 0$

126. $2x^{-2} - x^{-1} - 1 = 0$

127. $(x^2 - 3x)^2 - 2(x^2 - 3x) - 8 = 0$

128. $(x^2 - 6x)^2 - 2(x^2 - 6x) - 35 = 0$

129. $16\left(\dfrac{x-1}{x-8}\right)^2 + 8\left(\dfrac{x-1}{x-8}\right) + 1 = 0$

130. $9\left(\dfrac{x+2}{x+3}\right)^2 - 6\left(\dfrac{x+2}{x+3}\right) + 1 = 0$

Solving Problems

131. *Unisphere* The Unisphere is the world's largest man-made globe. It was built as the symbol of the 1964–1965 New York World's Fair. A sphere with the same diameter as the Unisphere globe would have a surface area of 45,239 square feet. What is the diameter of the Unisphere? (Source: The World's Fair and Exposition Information and Reference Guide)

© Rudy Sulgan/CORBIS

Designing the Unisphere was an engineering challenge that at one point involved simultaneously solving 670 equations.

132. ▲ *Geometry* The surface area S of a basketball is $900/\pi$ square inches. Find the radius r of the basketball.

Free-Falling Object In Exercises 133–136, find the time required for an object to reach the ground when it is dropped from a height of s_0 feet. The height h (in feet) is given by $h = -16t^2 + s_0$, where t measures the time (in seconds) after the object is released.

133. $s_0 = 256$ **134.** $s_0 = 48$

135. $s_0 = 128$ **136.** $s_0 = 500$

137. *Free-Falling Object* The height h (in feet) of an object thrown vertically upward from the top of a tower 144 feet tall is given by $h = 144 + 128t - 16t^2$, where t measures the time in seconds from the time when the object is released. How long does it take for the object to reach the ground?

138. *Profit* The monthly profit P (in dollars) a company makes depends on the amount x (in dollars) the company spends on advertising according to the model

$$P = 800 + 120x - \frac{1}{2}x^2.$$

Find the amount spent on advertising that will yield a monthly profit of $8000.

Compound Interest   The amount A in an account after 2 years when a principal of P dollars is invested at annual interest rate r compounded annually is given by $A = P(1 + r)^2$. In Exercises 139 and 140, find r.

139. $P = \$1500$, $A = \$1685.40$

140. $P = \$5000$, $A = \$5724.50$

National Health Expenditures   In Exercises 141 and 142, the national expenditures for health care in the United States from 1997 through 2006 are given by

$$y = 4.95t^2 + 876, \quad 7 \leq t \leq 16.$$

In this model, y represents the expenditures (in billions of dollars) and t represents the year, with $t = 7$ corresponding to 1997 (see figure).   (Source: U.S. Centers for Medicare & Medicaid Services)

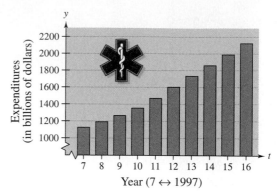

Figure for 141 and 142

141. Algebraically determine the year when expenditures were approximately $1500 billion. Graphically confirm the result.

142. Algebraically determine the year when expenditures were approximately $1850 billion. Graphically confirm the result.

Explaining Concepts

143. ✎   For a quadratic equation $ax^2 + bx + c = 0$, where a, b, and c are real numbers with $a \neq 0$, explain why b and c can equal 0, but a cannot.

144. Is it possible for a quadratic equation of the form $x^2 = m$ to have one real solution and one complex solution? Explain your reasoning.

145. ✎   Describe the steps you would use to solve a quadratic equation when using the Square Root Property.

146. ✎   Describe a procedure for solving an equation of quadratic form. Give an example.

Cumulative Review

In Exercises 147–150, solve the inequality and sketch the solution on the real number line.

147. $3x - 8 > 4$

148. $4 - 5x \geq 12$

149. $2x - 6 \leq 9 - x$

150. $x - 4 < 6$ or $x + 3 > 8$

In Exercises 151 and 152, solve the system of linear equations.

151. $x + y - z = 4$
$2x + y + 2z = 10$
$x - 3y - 4z = -7$

152. $2x - y + z = -6$
$x + 5y - z = 7$
$-x - 2y - 3z = 8$

In Exercises 153–158, combine the radical expressions, if possible, and simplify.

153. $5\sqrt{3} - 2\sqrt{3}$

154. $8\sqrt{27} + 4\sqrt{27}$

155. $16\sqrt[3]{y} - 9\sqrt[3]{x}$

156. $12\sqrt{x-1} + 6\sqrt{x-1}$

157. $\sqrt{16m^4n^3} + m\sqrt{m^2n}$

158. $x^2y\sqrt[4]{32x^2} + x\sqrt[4]{2x^6y^4} - y\sqrt[4]{162x^{10}}$

8.2 Completing the Square

© Sean Cayton/The Image Works

Why You Should Learn It

You can use techniques such as completing the square to solve quadratic equations that model real-life situations. For instance, in Exercise 90 on page 527, you will find the dimensions of an outdoor enclosure of a kennel by completing the square.

1 ▶ Rewrite quadratic expressions in completed square form.

What You Should Learn

1 ▶ Rewrite quadratic expressions in completed square form.

2 ▶ Solve quadratic equations by completing the square.

Constructing Perfect Square Trinomials

Consider the quadratic equation

$$(x - 2)^2 = 10. \qquad \text{Completed square form}$$

You know from Example 2(b) in the preceding section that this equation has two solutions: $x = 2 + \sqrt{10}$ and $x = 2 - \sqrt{10}$. Suppose you were given the equation in its general form

$$x^2 - 4x - 6 = 0. \qquad \text{General form}$$

How could you solve this form of the quadratic equation? You could try factoring, but after attempting to do so you would find that the left side of the equation is not factorable using integer coefficients.

In this section, you will study a technique for rewriting an equation in a completed square form. This technique is called **completing the square.** Note that prior to completing the square, the coefficient of the second-degree term must be 1.

Completing the Square

To **complete the square** for the expression $x^2 + bx$, add $(b/2)^2$, which is the square of half the coefficient of x. Consequently,

$$x^2 + bx + \left(\frac{b}{2}\right)^2 = \left(x + \frac{b}{2}\right)^2.$$

$$\underbrace{\qquad\qquad}_{(\text{half})^2}$$

EXAMPLE 1   **Constructing a Perfect Square Trinomial**

What term should be added to $x^2 - 8x$ so that it becomes a perfect square trinomial? To find this term, notice that the coefficient of the x-term is -8. Take half of this coefficient and square the result to get $(-4)^2 = 16$. Add this term to the expression to make it a perfect square trinomial.

$$x^2 - 8x + (-4)^2 = x^2 - 8x + 16 \qquad \text{Add } (-4)^2 = 16 \text{ to the expression.}$$

You can then rewrite the expression as the square of a binomial, $(x - 4)^2$.

✓ **CHECKPOINT** *Now try Exercise 3.*

2 ▶ Solve quadratic equations by completing the square.

Solving Equations by Completing the Square

Completing the square can be used to solve quadratic equations. When using this procedure, remember to *preserve the equality* by adding the same constant to each side of the equation.

EXAMPLE 2 Completing the Square: Leading Coefficient Is 1

Solve $x^2 + 12x = 0$ by completing the square.

Solution

$$x^2 + 12x = 0 \qquad \text{Write original equation.}$$

$$x^2 + 12x + 6^2 = 36 \qquad \text{Add } 6^2 = 36 \text{ to each side.}$$

$$\left(\tfrac{12}{2}\right)^2$$

$$(x + 6)^2 = 36 \qquad \text{Completed square form}$$

$$x + 6 = \pm\sqrt{36} \qquad \text{Square Root Property}$$

$$x = -6 \pm 6 \qquad \text{Subtract 6 from each side.}$$

$$x = -6 + 6 \text{ or } x = -6 - 6 \qquad \text{Separate solutions.}$$

$$x = 0 \qquad\qquad x = -12 \qquad \text{Simplify.}$$

The solutions are $x = 0$ and $x = -12$. Check these in the original equation.

✓ **CHECKPOINT** *Now try Exercise 17.*

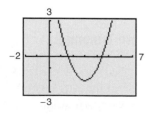
EXAMPLE 3 Completing the Square: Leading Coefficient Is 1

Solve $x^2 - 6x + 7 = 0$ by completing the square.

Solution

$$x^2 - 6x + 7 = 0 \qquad \text{Write original equation.}$$

$$x^2 - 6x = -7 \qquad \text{Subtract 7 from each side.}$$

$$x^2 - 6x + (-3)^2 = -7 + 9 \qquad \text{Add } (-3)^2 = 9 \text{ to each side.}$$

$$\left(-\tfrac{6}{2}\right)^2$$

$$(x - 3)^2 = 2 \qquad \text{Completed square form}$$

$$x - 3 = \pm\sqrt{2} \qquad \text{Square Root Property}$$

$$x = 3 \pm \sqrt{2} \qquad \text{Add 3 to each side.}$$

$$x = 3 + \sqrt{2} \text{ or } x = 3 - \sqrt{2} \qquad \text{Separate solutions.}$$

The solutions are $x = 3 + \sqrt{2} \approx 4.41$ and $x = 3 - \sqrt{2} \approx 1.59$. Check these in the original equation.

✓ **CHECKPOINT** *Now try Exercise 33.*

If the leading coefficient of a quadratic equation is not 1, you must divide each side of the equation by this coefficient *before* completing the square.

EXAMPLE 4   **Completing the Square: Leading Coefficient Is Not 1**

$$2x^2 - x - 2 = 0 \qquad\qquad \text{Original equation}$$

$$2x^2 - x = 2 \qquad\qquad \text{Add 2 to each side.}$$

$$x^2 - \frac{1}{2}x = 1 \qquad\qquad \text{Divide each side by 2.}$$

$$x^2 - \frac{1}{2}x + \left(-\frac{1}{4}\right)^2 = 1 + \frac{1}{16} \qquad\qquad \text{Add } \left(-\frac{1}{4}\right)^2 = \frac{1}{16} \text{ to each side.}$$

$$\left(x - \frac{1}{4}\right)^2 = \frac{17}{16} \qquad\qquad \text{Completed square form}$$

$$x - \frac{1}{4} = \pm\frac{\sqrt{17}}{4} \qquad\qquad \text{Square Root Property}$$

$$x = \frac{1}{4} \pm \frac{\sqrt{17}}{4} \qquad\qquad \text{Add } \frac{1}{4} \text{ to each side.}$$

The solutions are $x = \frac{1}{4} + \frac{\sqrt{17}}{4} \approx 1.28$ and $x = \frac{1}{4} - \frac{\sqrt{17}}{4} \approx -0.78$.

 **CHECKPOINT** *Now try Exercise 61.*

EXAMPLE 5   **Completing the Square: Leading Coefficient Is Not 1**

$$3x^2 - 6x + 1 = 0 \qquad\qquad \text{Original equation}$$

$$3x^2 - 6x = -1 \qquad\qquad \text{Subtract 1 from each side.}$$

$$x^2 - 2x = -\frac{1}{3} \qquad\qquad \text{Divide each side by 3.}$$

$$x^2 - 2x + (-1)^2 = -\frac{1}{3} + 1 \qquad\qquad \text{Add } (-1)^2 = 1 \text{ to each side.}$$

$$(x - 1)^2 = \frac{2}{3} \qquad\qquad \text{Completed square form}$$

$$x - 1 = \pm\sqrt{\frac{2}{3}} \qquad\qquad \text{Square Root Property}$$

$$x - 1 = \pm\frac{\sqrt{6}}{3} \qquad\qquad \text{Rationalize the denominator.}$$

$$x = 1 \pm \frac{\sqrt{6}}{3} \qquad\qquad \text{Add 1 to each side.}$$

The solutions are $x = 1 + \sqrt{6}/3 \approx 1.82$ and $x = 1 - \sqrt{6}/3 \approx 0.18$.

 **CHECKPOINT** *Now try Exercise 59.*

EXAMPLE 6   A Quadratic Equation with Complex Solutions

Solve $x^2 - 4x + 8 = 0$ by completing the square.

Solution

$x^2 - 4x + 8 = 0$	Write original equation.
$x^2 - 4x = -8$	Subtract 8 from each side.
$x^2 - 4x + (-2)^2 = -8 + 4$	Add $(-2)^2 = 4$ to each side.
$(x - 2)^2 = -4$	Completed square form
$x - 2 = \pm 2i$	Square Root Property
$x = 2 \pm 2i$	Add 2 to each side.

The solutions are $x = 2 + 2i$ and $x = 2 - 2i$. Check these in the original equation.

✓ **CHECKPOINT** *Now try Exercise 49.*

EXAMPLE 7   Dimensions of an iPhone

The first generation of the iPhone™ has an approximate volume of 4.968 cubic inches. Its width is 0.46 inch and its face has the dimensions x inches by $x + 2.1$ inches. (See Figure 8.1.) Find the dimensions of the face in inches.   (Source: Apple, Inc.)

Solution

$lwh = V$	Formula for volume of a rectangular solid
$(x)(0.46)(x + 2.1) = 4.968$	Substitute 4.968 for V, x for l, 0.46 for w, and $x + 2.1$ for h.
$0.46x^2 + 0.966x = 4.968$	Multiply factors.
$x^2 + 2.1x = 10.8$	Divide each side by 0.46.
$x^2 + 2.1x + \left(\dfrac{2.1}{2}\right)^2 = 10.8 + 1.1025$	Add $\left(\dfrac{2.1}{2}\right)^2 = 1.1025$ to each side.
$(x + 1.05)^2 = 11.9025$	Completed square form
$x + 1.05 = \pm\sqrt{11.9025}$	Square Root Property
$x = -1.05 \pm \sqrt{11.9025}$	Subtract 1.05 from each side.

Choosing the positive root, you obtain

$$x = -1.05 + 3.45 = 2.4 \text{ inches} \qquad \text{Length of face}$$

and

$$x + 2.1 = 2.4 + 2.1 = 4.5 \text{ inches.} \qquad \text{Height of face}$$

✓ **CHECKPOINT** *Now try Exercise 91.*

$x + 2.1$ in.

0.46 in. x in.

Figure 8.1

_____ **Concept Check** _____

1. What is a perfect square trinomial?

2. What term must be added to $x^2 + 5x$ to complete the square? Explain how you found the term.

3. When using the method of completing the square to solve $2x^2 - 7x = 6$, what is the first step? Is the resulting equation equivalent to the original equation? Explain.

4. Is it possible for a quadratic equation to have no real number solution? If so, give an example.

8.2 EXERCISES

Go to pages 570–571 to record your assignments.

_____ **Developing Skills** _____

In Exercises 1–16, add a term to the expression so that it becomes a perfect square trinomial. *See Example 1.*

1. $x^2 + 8x +$

2. $x^2 + 12x +$

✓ **3.** $y^2 - 20y +$

4. $y^2 - 2y +$

5. $x^2 + 14x +$

6. $x^2 - 24x +$

7. $t^2 + 5t +$

8. $u^2 + 7u +$

9. $x^2 - 9x +$

10. $y^2 - 11y +$

11. $a^2 - \frac{1}{3}a +$

12. $y^2 + \frac{4}{3}y +$

13. $y^2 + \frac{8}{5}y +$

14. $x^2 - \frac{9}{5}x +$

15. $r^2 - 0.4r +$

16. $s^2 + 4.6s +$

In Exercises 17–32, solve the equation first by completing the square and then by factoring. *See Examples 2–5.*

✓ **17.** $x^2 - 20x = 0$

18. $x^2 + 32x = 0$

19. $x^2 + 6x = 0$

20. $t^2 - 10t = 0$

21. $y^2 - 5y = 0$

22. $t^2 - 9t = 0$

23. $t^2 - 8t + 7 = 0$

24. $y^2 - 4y + 4 = 0$

25. $x^2 + 7x + 12 = 0$

26. $z^2 + 3z - 10 = 0$

27. $x^2 - 3x - 18 = 0$

28. $a^2 + 12a + 32 = 0$

29. $2u^2 - 12u + 18 = 0$

30. $3x^2 - 3x - 6 = 0$

31. $4x^2 + 4x - 15 = 0$

32. $6a^2 - 23a + 15 = 0$

In Exercises 33–72, solve the equation by completing the square. Give the solutions in exact form and in decimal form rounded to two decimal places. (The solutions may be complex numbers.) *See Examples 2–6.*

✓ **33.** $x^2 - 4x - 3 = 0$

34. $x^2 - 6x + 7 = 0$

35. $x^2 + 4x - 3 = 0$

36. $x^2 + 6x + 7 = 0$

37. $x^2 + 6x = 7$

38. $x^2 - 4x = -3$

39. $x^2 - 12x = -10$

40. $x^2 - 4x = -9$

41. $x^2 + 8x + 7 = 0$

42. $x^2 + 10x + 9 = 0$

43. $x^2 - 10x + 21 = 0$

44. $x^2 - 10x + 24 = 0$

45. $y^2 + 5y + 3 = 0$

46. $y^2 + 8y + 9 = 0$

47. $x^2 + 10 = 6x$

48. $x^2 + 23 = 10x$

✓ **49.** $z^2 + 4z + 13 = 0$

50. $z^2 - 6z + 18 = 0$

51. $-x^2 + x - 1 = 0$

52. $1 - x - x^2 = 0$

53. $a^2 + 7a + 11 = 0$

65. $5x^2 - 3x + 10 = 0$

54. $y^2 + 5y + 9 = 0$

66. $7u^2 - 8u - 3 = 0$ **67.** $x\left(x - \dfrac{2}{3}\right) = 14$

55. $x^2 - \frac{2}{3}x - 3 = 0$ **56.** $x^2 + \frac{4}{5}x - 1 = 0$

68. $2x\left(x + \dfrac{4}{3}\right) = 5$

57. $v^2 + \frac{3}{4}v - 2 = 0$

69. $0.1x^2 + 0.5x = -0.2$

58. $u^2 - \frac{2}{3}u + 5 = 0$

70. $0.2x^2 + 0.1x = -0.5$

✔ **59.** $2x^2 + 8x + 3 = 0$ **60.** $3x^2 - 24x - 5 = 0$

71. $0.75x^2 + 1.25x + 1.5 = 0$

✔ **61.** $3x^2 + 9x + 5 = 0$ **62.** $5x^2 - 15x + 7 = 0$

72. $0.625x^2 - 0.875x + 0.25 = 0$

In Exercises 73–78, find the real solutions.

73. $\dfrac{x}{2} - \dfrac{1}{x} = 1$ **74.** $\dfrac{x}{2} + \dfrac{5}{x} = 4$

63. $4y^2 + 4y - 9 = 0$

75. $\dfrac{x^2}{8} = \dfrac{x + 3}{2}$ **76.** $\dfrac{x^2 + 2}{24} = \dfrac{x - 1}{3}$

77. $\sqrt{2x + 1} = x - 3$ **78.** $\sqrt{3x - 2} = x - 2$

64. $4z^2 - 3z + 2 = 0$

In Exercises 79–86, use a graphing calculator to graph the function. Use the graph to approximate any x-intercepts of the graph. Set $y = 0$ and solve the resulting equation. Compare the result with the x-intercepts of the graph.

79. $y = x^2 + 4x - 1$

80. $y = x^2 + 6x - 4$

81. $y = x^2 - 2x - 5$

82. $y = 2x^2 - 6x - 5$

83. $y = \frac{1}{3}x^2 + 2x - 6$

84. $y = \frac{1}{2}x^2 - 3x + 1$

85. $y = x - 2\sqrt{x} + 1$

86. $y = \sqrt{x} - x + 2$

Solving Problems

87. ▲ *Geometric Modeling*

(a) Find the area of the two adjoining rectangles and large square in the figure.

(b) Find the area of the small square in the lower right-hand corner of the figure and add it to the area found in part (a).

(c) Find the dimensions and the area of the entire figure after adjoining the small square in the lower right-hand corner of the figure. Note that you have shown geometrically the technique of completing the square.

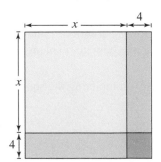

Figure for 87   Figure for 88

88. ▲ *Geometric Modeling* Repeat Exercise 87 for the model shown in the figure.

89. ▲ *Geometry* You have 200 meters of fencing to enclose two adjacent rectangular corrals (see figure). The total area of the enclosed region is 1400 square meters. What are the dimensions of each corral? (The corrals are the same size.)

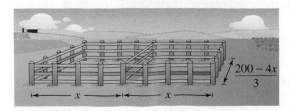

90. ▲ *Geometry* A kennel is adding a rectangular outdoor enclosure along one side of the kennel wall (see figure). The other three sides of the enclosure will be formed by a fence. The kennel has 111 feet of fencing and plans to use 1215 square feet of land for the enclosure. What are the dimensions of the enclosure?

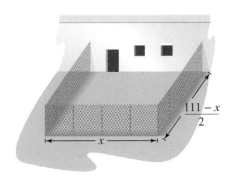

☑ 91. ▲ *Geometry* An open box with a rectangular base of x inches by $x + 4$ inches has a height of 6 inches (see figure). The volume of the box is 840 cubic inches. Find the dimensions of the box.

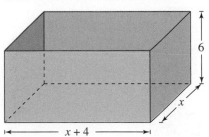

92. ▲ *Geometry* An open box with a rectangular base of $2x$ inches by $6x - 2$ inches has a height of 9 inches (see figure). The volume of the box is 1584 cubic inches. Find the dimensions of the box.

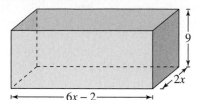

93. *Revenue* The revenue R (in dollars) from selling x pairs of running shoes is given by

$$R = x\left(80 - \frac{1}{2}x\right).$$

Find the number of pairs of running shoes that must be sold to produce a revenue of \$2750.

94. *Revenue* The revenue R (in dollars) from selling x golf clubs is given by

$$R = x\left(150 - \frac{1}{10}x\right).$$

Find the number of golf clubs that must be sold to produce a revenue of \$15,033.60.

Explaining Concepts

95. ✎ Explain the use of the Square Root Property when solving a quadratic equation by the method of completing the square.

96. *True or False?* If you solve a quadratic equation by completing the square and obtain solutions that are rational numbers, then you could have solved the equation by factoring. Justify your answer.

97. ✎ Consider the quadratic equation $(x - 1)^2 = d$.

(a) What value(s) of d will produce a quadratic equation that has exactly one (repeated) solution?

(b) Describe the value(s) of d that will produce two different solutions, both of which are *rational* numbers.

(c) Describe the value(s) of d that will produce two different solutions, both of which are *irrational* numbers.

(d) Describe the value(s) of d that will produce two different solutions, both of which are *complex* numbers.

98. ✎ You teach an algebra class and one of your students hands in the following solution. Find and correct the error(s). Discuss how to explain the error(s) to your student.

Solve $x^2 + 6x - 13 = 0$ by completing the square.

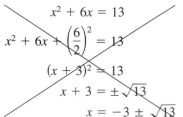

Cumulative Review

In Exercises 99–106, perform the operation and simplify the expression.

99. $3\sqrt{5}\sqrt{500}$

100. $2\sqrt{2x^2}\sqrt{27x}$

101. $(3 + \sqrt{2})(3 - \sqrt{2})$

102. $(\sqrt[3]{6} - 2)(\sqrt[3]{4} + 1)$

103. $(3 + \sqrt{2})^2$

104. $(2\sqrt{x} - 5)^2$

105. $\dfrac{8}{\sqrt{10}}$

106. $\dfrac{5}{\sqrt{12} - 2}$

In Exercises 107 and 108, rewrite the expression using the specified rule, where a and b are nonnegative real numbers.

107. Product Rule: $\sqrt{ab} = $ _____.

108. Quotient Rule: $\sqrt{\dfrac{a}{b}} = $ _____.

8.3 The Quadratic Formula

© Peter Turnley/CORBIS

What You Should Learn

1 ▶ Derive the Quadratic Formula by completing the square for a general quadratic equation.

2 ▶ Use the Quadratic Formula to solve quadratic equations.

3 ▶ Determine the types of solutions of quadratic equations using the discriminant.

4 ▶ Write quadratic equations from solutions of the equations.

Why You Should Learn It

Knowing the Quadratic Formula can be helpful in solving quadratic equations that model real-life situations. For instance, in Exercise 99 on page 536, you will use a quadratic equation that models the depth of a river after a heavy rain begins.

1 ▶ Derive the Quadratic Formula by completing the square for a general quadratic equation.

The Quadratic Formula

A fourth technique for solving a quadratic equation involves the **Quadratic Formula.** This formula is derived by completing the square for a general quadratic equation.

$$ax^2 + bx + c = 0 \qquad \text{General form, } a \neq 0$$

$$ax^2 + bx = -c \qquad \text{Subtract } c \text{ from each side.}$$

$$x^2 + \frac{b}{a}x = -\frac{c}{a} \qquad \text{Divide each side by } a.$$

$$x^2 + \frac{b}{a}x + \left(\frac{b}{2a}\right)^2 = -\frac{c}{a} + \left(\frac{b}{2a}\right)^2 \qquad \text{Add } \left(\frac{b}{2a}\right)^2 \text{ to each side.}$$

$$\left(x + \frac{b}{2a}\right)^2 = \frac{b^2 - 4ac}{4a^2} \qquad \text{Simplify.}$$

$$x + \frac{b}{2a} = \pm\sqrt{\frac{b^2 - 4ac}{4a^2}} \qquad \text{Square Root Property}$$

$$x = -\frac{b}{2a} \pm \frac{\sqrt{b^2 - 4ac}}{2|a|} \qquad \text{Subtract } \frac{b}{2a} \text{ from each side.}$$

$$x = \frac{-b \pm \sqrt{b^2 - 4ac}}{2a} \qquad \text{Simplify.}$$

Notice in the derivation of the Quadratic Formula that, because $\pm 2|a|$ represents the same numbers as $\pm 2a$, you can omit the absolute value bars.

Study Tip

The Quadratic Formula is one of the most important formulas in algebra, and you should memorize it. It helps to try to memorize a verbal statement of the rule. For instance, you might try to remember the following verbal statement of the Quadratic Formula: "The opposite of *b*, plus or minus the square root of *b* squared minus 4*ac*, all divided by 2*a*."

> ## The Quadratic Formula
>
> The solutions of $ax^2 + bx + c = 0$, $a \neq 0$, are given by the **Quadratic Formula**
>
> $$x = \frac{-b \pm \sqrt{b^2 - 4ac}}{2a}.$$

2 ▶ Use the Quadratic Formula to solve quadratic equations.

Solving Equations by the Quadratic Formula

When using the Quadratic Formula, remember that *before* the formula can be applied, you must first write the quadratic equation in general form in order to determine the values of a, b, and c.

EXAMPLE 1   **The Quadratic Formula: Two Distinct Solutions**

$x^2 + 6x = 16$	Original equation
$x^2 + 6x - 16 = 0$	Write in general form.
$x = \dfrac{-b \pm \sqrt{b^2 - 4ac}}{2a}$	Quadratic Formula
$x = \dfrac{-6 \pm \sqrt{6^2 - 4(1)(-16)}}{2(1)}$	Substitute 1 for a, 6 for b, and -16 for c.
$x = \dfrac{-6 \pm \sqrt{100}}{2}$	Simplify.
$x = \dfrac{-6 \pm 10}{2}$	Simplify.
$x = 2$ or $x = -8$	Solutions

The solutions are $x = 2$ and $x = -8$. Check these in the original equation.

 CHECKPOINT *Now try Exercise 5.*

Study Tip

In Example 1, the solutions are rational numbers, which means that the equation could have been solved by factoring. Try solving the equation by factoring.

Study Tip

If the leading coefficient of a quadratic equation is negative, you should begin by multiplying each side of the equation by -1, as shown in Example 2. This will produce a positive leading coefficient, which is easier to work with.

EXAMPLE 2   **The Quadratic Formula: Two Distinct Solutions**

$-x^2 - 4x + 8 = 0$	Leading coefficient is negative.
$x^2 + 4x - 8 = 0$	Multiply each side by -1.
$x = \dfrac{-b \pm \sqrt{b^2 - 4ac}}{2a}$	Quadratic Formula
$x = \dfrac{-4 \pm \sqrt{4^2 - 4(1)(-8)}}{2(1)}$	Substitute 1 for a, 4 for b, and -8 for c.
$x = \dfrac{-4 \pm \sqrt{48}}{2}$	Simplify.
$x = \dfrac{-4 \pm 4\sqrt{3}}{2}$	Simplify.
$x = -2 \pm 2\sqrt{3}$	Solutions

The solutions are $x = -2 + 2\sqrt{3}$ and $x = -2 - 2\sqrt{3}$. Check these in the original equation.

 **CHECKPOINT** *Now try Exercise 15.*

EXAMPLE 3   **The Quadratic Formula: One Repeated Solution**

$18x^2 - 24x + 8 = 0$	Original equation
$9x^2 - 12x + 4 = 0$	Divide each side by 2.
$x = \dfrac{-b \pm \sqrt{b^2 - 4ac}}{2a}$	Quadratic Formula
$x = \dfrac{-(-12) \pm \sqrt{(-12)^2 - 4(9)(4)}}{2(9)}$	Substitute 9 for a, -12 for b, and 4 for c.
$x = \dfrac{12 \pm \sqrt{144 - 144}}{18}$	Simplify.
$x = \dfrac{12 \pm \sqrt{0}}{18}$	Simplify.
$x = \dfrac{2}{3}$	Solution

The only solution is $x = \frac{2}{3}$. Check this in the original equation.

✓ **CHECKPOINT** *Now try Exercise 9.*

EXAMPLE 4   **The Quadratic Formula: Complex Solutions**

$2x^2 - 4x + 5 = 0$	Original equation
$x = \dfrac{-b \pm \sqrt{b^2 - 4ac}}{2a}$	Quadratic Formula
$x = \dfrac{-(-4) \pm \sqrt{(-4)^2 - 4(2)(5)}}{2(2)}$	Substitute 2 for a, -4 for b, and 5 for c.
$x = \dfrac{4 \pm \sqrt{-24}}{4}$	Simplify.
$x = \dfrac{4 \pm 2\sqrt{6}\,i}{4}$	Write in i-form.
$x = \dfrac{2\left(2 \pm \sqrt{6}\,i\right)}{2 \cdot 2}$	Factor numerator and denominator.
$x = \dfrac{\cancel{2}\left(2 \pm \sqrt{6}\,i\right)}{\cancel{2} \cdot 2}$	Divide out common factor.
$x = 1 \pm \dfrac{\sqrt{6}}{2}\,i$	Solutions

The solutions are $x = 1 + \dfrac{\sqrt{6}}{2}\,i$ and $x = 1 - \dfrac{\sqrt{6}}{2}\,i$. Check these in the original equation.

✓ **CHECKPOINT** *Now try Exercise 21.*

3 ▶ Determine the types of solutions of quadratic equations using the discriminant.

The Discriminant

The radicand in the Quadratic Formula, $b^2 - 4ac$, is called the **discriminant** because it allows you to "discriminate" among different types of solutions.

Study Tip

By reexamining Examples 1 through 4, you can see that the equations with rational or repeated solutions could have been solved by *factoring*. In general, quadratic equations (with integer coefficients) for which the discriminant is either zero or a perfect square are factorable using integer coefficients. Consequently, a quick test of the discriminant will help you decide which solution method to use to solve a quadratic equation.

Using the Discriminant

Let a, b, and c be rational numbers such that $a \neq 0$. The discriminant of the quadratic equation $ax^2 + bx + c = 0$ is given by $b^2 - 4ac$, and can be used to classify the solutions of the equation as follows.

Discriminant	Solution Type
1. Perfect square	Two distinct rational solutions (Example 1)
2. Positive nonperfect square	Two distinct irrational solutions (Example 2)
3. Zero	One repeated rational solution (Example 3)
4. Negative number	Two distinct complex solutions (Example 4)

Technology: Discovery

Use a graphing calculator to graph each equation.

a. $y = x^2 - x + 2$

b. $y = 2x^2 - 3x - 2$

c. $y = x^2 - 2x + 1$

d. $y = x^2 - 2x - 10$

Describe the solution type of each equation and check your results with those shown in Example 5. Why do you think the discriminant is used to determine solution types?

EXAMPLE 5   Using the Discriminant

Determine the type of solution(s) for each quadratic equation.

a. $x^2 - x + 2 = 0$

b. $2x^2 - 3x - 2 = 0$

c. $x^2 - 2x + 1 = 0$

d. $x^2 - 2x - 1 = 9$

Solution

Equation	Discriminant	Solution Type
a. $x^2 - x + 2 = 0$	$b^2 - 4ac = (-1)^2 - 4(1)(2)$ $= 1 - 8 = -7$	Two distinct complex solutions
b. $2x^2 - 3x - 2 = 0$	$b^2 - 4ac = (-3)^2 - 4(2)(-2)$ $= 9 + 16 = 25$	Two distinct rational solutions
c. $x^2 - 2x + 1 = 0$	$b^2 - 4ac = (-2)^2 - 4(1)(1)$ $= 4 - 4 = 0$	One repeated rational solution
d. $x^2 - 2x - 1 = 9$	$b^2 - 4ac = (-2)^2 - 4(1)(-10)$ $= 4 + 40 = 44$	Two distinct irrational solutions

✓ **CHECKPOINT** *Now try Exercise 41.*

Summary of Methods for Solving Quadratic Equations

Method	*Example*

1. Factoring

$$3x^2 + x = 0$$

$$x(3x + 1) = 0 \implies x = 0 \quad \text{and} \quad x = -\frac{1}{3}$$

2. Square Root Property

$$(x + 2)^2 = 7$$

$$x + 2 = \pm\sqrt{7} \implies x = -2 + \sqrt{7} \quad \text{and} \quad x = -2 - \sqrt{7}$$

3. Completing the square

$$x^2 + 6x = 2$$

$$x^2 + 6x + 3^2 = 2 + 9$$

$$(x + 3)^2 = 11 \implies x = -3 + \sqrt{11} \quad \text{and} \quad x = -3 - \sqrt{11}$$

4. Quadratic Formula

$$3x^2 - 2x + 2 = 0 \implies x = \frac{-(-2) \pm \sqrt{(-2)^2 - 4(3)(2)}}{2(3)} = \frac{1}{3} \pm \frac{\sqrt{5}}{3}i$$

4 ▶ Write quadratic equations from solutions of the equations.

Writing Quadratic Equations from Solutions

Using the Zero-Factor Property, you know that the equation $(x + 5)(x - 2) = 0$ has two solutions, $x = -5$ and $x = 2$. You can use the Zero-Factor Property in reverse to find a quadratic equation given its solutions. This process is demonstrated in Example 6.

Reverse of Zero-Factor Property

Let a and b be real numbers, variables, or algebraic expressions. If $a = 0$ or $b = 0$, then a and b are factors such that $ab = 0$.

Technology: Tip

A program for several models of graphing calculators that uses the Quadratic Formula to solve quadratic equations can be found at our website, *www.cengage.com/ math/larson/algebra*. This program will display real solutions to quadratic equations.

EXAMPLE 6   Writing a Quadratic Equation from Its Solutions

Write a quadratic equation that has the solutions $x = 4$ and $x = -7$. Using the solutions $x = 4$ and $x = -7$, you can write the following.

$x = 4$	and	$x = -7$ Solutions
$x - 4 = 0$		$x + 7 = 0$ Obtain zero on one side of each equation.
		$(x - 4)(x + 7) = 0$ Reverse of Zero-Factor Property
		$x^2 + 3x - 28 = 0$ Foil Method

So, a quadratic equation that has the solutions $x = 4$ and $x = -7$ is

$$x^2 + 3x - 28 = 0.$$

This is not the only quadratic equation with the solutions $x = 4$ and $x = -7$. You can obtain other quadratic equations with these solutions by multiplying $x^2 + 3x - 28 = 0$ by any nonzero real number.

✓ **CHECKPOINT** *Now try Exercise 65.*

_____ Concept Check _____

1. What method is used to derive the Quadratic Formula from a general quadratic equation?

2. To solve the quadratic equation $3x^2 = 3 - x$ using the Quadratic Formula, what are the values of a, b, and c?

3. The discriminant of a quadratic equation is -25. What type of solution(s) does the equation have?

4. Describe the steps you would use to write a quadratic equation that has the solutions $x = 3$ and $x = -2$.

8.3 EXERCISES

Go to pages 570–571 to record your assignments.

_____ Developing Skills _____

In Exercises 1–4, write the quadratic equation in general form.

1. $2x^2 = 7 - 2x$

2. $7x^2 + 15x = 5$

3. $x(10 - x) = 5$

4. $x(2x + 9) = 12$

In Exercises 5–14, solve the equation first by using the Quadratic Formula and then by factoring. *See Examples 1–4.*

✓ 5. $x^2 - 11x + 28 = 0$

6. $x^2 - 12x + 27 = 0$

7. $x^2 + 6x + 8 = 0$

8. $x^2 + 9x + 14 = 0$

✓ 9. $16x^2 + 8x + 1 = 0$

10. $9x^2 + 12x + 4 = 0$

11. $4x^2 + 12x + 9 = 0$

12. $10x^2 - 11x + 3 = 0$

13. $x^2 - 5x - 300 = 0$

14. $x^2 + 20x - 300 = 0$

In Exercises 15–40, solve the equation by using the Quadratic Formula. (Find all real *and* complex solutions.) *See Examples 1–4.*

✓ 15. $x^2 - 2x - 4 = 0$

16. $x^2 - 2x - 6 = 0$

17. $t^2 + 4t + 1 = 0$

18. $y^2 + 6y - 8 = 0$

19. $x^2 - 10x + 23 = 0$

20. $u^2 - 12u + 29 = 0$

✓ 21. $2x^2 + 3x + 3 = 0$

22. $2x^2 - 2x + 3 = 0$

23. $3v^2 - 2v - 1 = 0$

24. $4x^2 + 6x + 1 = 0$

25. $2x^2 + 4x - 3 = 0$

26. $x^2 - 8x + 19 = 0$

27. $-4x^2 - 6x + 3 = 0$

28. $-5x^2 - 15x + 10 = 0$

29. $8x^2 - 6x + 2 = 0$

30. $6x^2 + 3x - 9 = 0$

31. $-4x^2 + 10x + 12 = 0$

32. $-15x^2 - 10x + 25 = 0$

33. $9x^2 = 1 + 9x$

34. $7x^2 = 3 - 5x$

35. $2x - 3x^2 = 3 - 7x^2$

36. $x - x^2 = 1 - 6x^2$

37. $x^2 - 0.4x - 0.16 = 0$

38. $x^2 + 0.6x - 0.41 = 0$

39. $2.5x^2 + x - 0.9 = 0$

40. $0.09x^2 - 0.12x - 0.26 = 0$

In Exercises 41–48, use the discriminant to determine the type of solution(s) of the quadratic equation. *See Example 5.*

✓ **41.** $x^2 + x + 1 = 0$

42. $x^2 + x - 1 = 0$

43. $3x^2 - 2x - 5 = 0$

44. $5x^2 + 7x + 3 = 0$

45. $9x^2 - 24x + 16 = 0$

46. $2x^2 + 10x + 6 = 0$

47. $3x^2 - x + 2 = 0$

48. $4x^2 - 16x + 16 = 0$

In Exercises 49–64, solve the quadratic equation by using the most convenient method. (Find all real *and* complex solutions.)

49. $z^2 - 169 = 0$

50. $t^2 = 144$

51. $5y^2 + 15y = 0$

52. $12u^2 + 30u = 0$

53. $25(x - 3)^2 - 36 = 0$

54. $9(x + 4)^2 + 16 = 0$

55. $2y(y - 18) + 3(y - 18) = 0$

56. $4y(y + 7) - 5(y + 7) = 0$

57. $x^2 + 8x + 25 = 0$

58. $y^2 + 21y + 108 = 0$

59. $3x^2 - 13x + 169 = 0$ **60.** $2x^2 - 15x + 225 = 0$

61. $25x^2 + 80x + 61 = 0$ **62.** $14x^2 + 11x - 40 = 0$

63. $7x(x + 2) + 5 = 3x(x + 1)$

64. $5x(x - 1) - 7 = 4x(x - 2)$

In Exercises 65–74, write a quadratic equation having the given solutions. *See Example 6.*

✓ **65.** $5, -2$

66. $-2, 3$

67. $1, 7$

68. $2, 8$

69. $1 + \sqrt{2}, 1 - \sqrt{2}$

70. $-3 + \sqrt{5}, -3 - \sqrt{5}$

71. $5i, -5i$

72. $2i, -2i$

73. 12

74. -4

In Exercises 75–80, use a graphing calculator to graph the function. Use the graph to approximate any *x*-intercepts of the graph. Set $y = 0$ and solve the resulting equation. Compare the result with the *x*-intercepts of the graph.

75. $y = 3x^2 - 6x + 1$ **76.** $y = x^2 + x + 1$

77. $y = x^2 - 4x + 3$ **78.** $y = 5x^2 - 18x + 6$

79. $y = -0.03x^2 + 2x - 0.4$

80. $y = 3.7x^2 - 10.2x + 3.2$

In Exercises 81–84, use a graphing calculator to determine the number of real solutions of the quadratic equation. Verify your answer algebraically.

81. $2x^2 - 5x + 5 = 0$ **82.** $3x^2 - 7x - 6 = 0$

83. $\frac{1}{5}x^2 + \frac{6}{5}x - 8 = 0$ **84.** $\frac{1}{3}x^2 - 5x + 25 = 0$

In Exercises 85–88, determine all real values of *x* for which the function has the indicated value.

85. $f(x) = 2x^2 - 7x + 1, f(x) = -3$

86. $f(x) = 3x^2 - 7x + 4, f(x) = 0$

87. $g(x) = 2x^2 - 3x + 16, g(x) = 14$

88. $h(x) = 6x^2 + x + 10, h(x) = -2$

In Exercises 89–92, solve the equation.

89. $\frac{x^2}{4} - \frac{2x}{3} = 1$ **90.** $\frac{x^2 - 9x}{6} = \frac{x - 1}{2}$

91. $\sqrt{x + 3} = x - 1$ **92.** $\sqrt{2x - 3} = x - 2$

Think About It In Exercises 93 and 94, describe the values of c such that the equation has (a) two real number solutions, (b) one real number solution, and (c) two complex number solutions.

93. $x^2 - 6x + c = 0$

94. $x^2 + 2x + c = 0$

—————————————— **Solving Problems** ——————————————

95. ▲ *Geometry* A rectangle has a width of x inches, a length of $x + 6.3$ inches, and an area of 58.14 square inches. Find its dimensions.

96. ▲ *Geometry* A rectangle has a length of $x + 1.5$ inches, a width of x inches, and an area of 18.36 square inches. Find its dimensions.

97. *Free-Falling Object* A stone is thrown vertically upward at a velocity of 40 feet per second from a bridge that is 50 feet above the level of the water (see figure). The height h (in feet) of the stone at time t (in seconds) after it is thrown is

$$h = -16t^2 + 40t + 50.$$

(a) Find the time when the stone is again 50 feet above the water.

(b) Find the time when the stone strikes the water.

(c) Does the stone reach a height of 80 feet? Use the determinant to justify your answer.

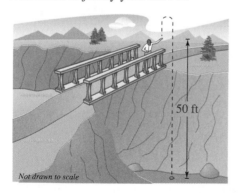

50 ft

Not drawn to scale

98. *Free-Falling Object* A stone is thrown vertically upward at a velocity of 20 feet per second from a bridge that is 40 feet above the level of the water. The height h (in feet) of the stone at time t (in seconds) after it is thrown is

$$h = -16t^2 + 20t + 40.$$

(a) Find the time when the stone is again 40 feet above the water.

(b) Find the time when the stone strikes the water.

(c) Does the stone reach a height of 50 feet? Use the determinant to justify your answer.

99. *Depth of a River* The depth d (in feet) of a river is given by

$$d = -0.25t^2 + 1.7t + 3.5, \ 0 \le t \le 7$$

where t is the time (in hours) after a heavy rain begins. When is the river 6 feet deep?

100. *Baseball* The path of a baseball after it has been hit is given by

$$h = -0.003x^2 + 1.19x + 5.2$$

where h is the height (in feet) of the baseball and x is the horizontal distance (in feet) of the ball from home plate. The ball hits the top of the outfield fence that is 10 feet high. How far is the outfield fence from home plate?

101. ▦ *Fuel Economy*   The fuel economy y (in miles per gallon) of a car is given by

$$y = -0.013x^2 + 1.25x + 5.6, \; 5 \le x \le 75$$

where x is the speed (in miles per hour) of the car.

(a) Use a graphing calculator to graph the model.

(b) Use the graph in part (a) to find the speeds at which you can travel and have a fuel economy of 32 miles per gallon. Verify your results algebraically.

102. ▦ *Cellular Phone Subscribers*   The number s (in millions) of cellular phone subscribers in the United States for the years 1997 through 2006 can be modeled by

$$s = 0.510t^2 + 7.74t - 23.5, \; 7 \le t \le 16$$

where $t = 7$ corresponds to 1997.

(Source: Cellular Telecommunications & Internet Association)

(a) Use a graphing calculator to graph the model.

(b) Use the graph in part (a) to determine the year in which there were 145 million cellular phone subscribers. Verify your answer algebraically.

103. *Exploration*   Determine the two solutions, x_1 and x_2, of each quadratic equation. Use the values of x_1 and x_2 to fill in the boxes.

Equation	x_1, x_2	$x_1 + x_2$	$x_1 x_2$
(a) $x^2 - x - 6 = 0$			
(b) $2x^2 + 5x - 3 = 0$			
(c) $4x^2 - 9 = 0$			
(d) $x^2 - 10x + 34 = 0$			

104. *Think About It*   Consider a general quadratic equation $ax^2 + bx + c = 0$ whose solutions are x_1 and x_2. Use the results of Exercise 103 to determine a relationship among the coefficients a, b, and c, and the sum $(x_1 + x_2)$ and product $(x_1 x_2)$ of the solutions.

Explaining Concepts

In Exercises 105–108, which method of solving the quadratic equation would be most convenient? Explain your reasoning.

105. $(x - 3)^2 = 25$

106. $x^2 + 8x - 12 = 0$

107. $2x^2 - 9x + 12 = 0$

108. $8x^2 - 40x = 0$

109. ✎ Explain how the discriminant of $ax^2 + bx + c = 0$ is related to the number of x-intercepts of the graph of $y = ax^2 + bx + c$.

110. *Error Analysis*   Describe and correct the student's error in writing a quadratic equation that has solutions $x = 2$ and $x = 4$.

$$\overbrace{(x + 2)(x + 4)} = 0$$
$$x^2 + 6x + 8 = 0$$

Cumulative Review

In Exercises 111–114, use the Distance Formula to determine whether the three points are collinear.

111. $(-1, 11), (2, 2), (1, 5)$

112. $(-2, 4), (3, -3), (5, -1)$

113. $(-6, -2), (-3, -4), (3, -4)$

114. $(-4, 7), (0, 4), (8, -2)$

In Exercises 115–118, sketch the graph of the function.

115. $f(x) = (x - 1)^2$

116. $f(x) = \frac{1}{2}x^2$

117. $f(x) = (x - 2)^2 + 4$

118. $f(x) = (x + 3)^2 - 1$

Mid-Chapter Quiz

Take this quiz as you would take a quiz in class. After you are done, check your work against the answers in the back of the book.

In Exercises 1–8, solve the quadratic equation by the specified method.

1. Factoring:

$2x^2 - 72 = 0$

2. Factoring:

$2x^2 + 3x - 20 = 0$

3. Square Root Property:

$3x^2 = 36$

4. Square Root Property:

$(u - 3)^2 - 16 = 0$

5. Completing the square:

$m^2 + 7m + 2 = 0$

6. Completing the square:

$2y^2 + 6y - 5 = 0$

7. Quadratic Formula:

$x^2 + 4x - 6 = 0$

8. Quadratic Formula:

$6v^2 - 3v - 4 = 0$

In Exercises 9–16, solve the equation by using the most convenient method. (Find all real *and* complex solutions.)

9. $x^2 + 5x + 7 = 0$

10. $36 - (t - 4)^2 = 0$

11. $x(x - 10) + 3(x - 10) = 0$

12. $x(x - 3) = 10$

13. $4b^2 - 12b + 9 = 0$

14. $3m^2 + 10m + 5 = 0$

15. $x - 4\sqrt{x} - 21 = 0$

16. $x^4 + 7x^2 + 12 = 0$

In Exercises 17 and 18, solve the equation of quadratic form. (Find all real *and* complex solutions.)

17. $x - 4\sqrt{x} + 3 = 0$

18. $x^4 - 14x^2 + 24 = 0$

In Exercises 19 and 20, use a graphing calculator to graph the function. Use the graph to approximate any *x*-intercepts of the graph. Set *y* = 0 and solve the resulting equation. Compare the results with the *x*-intercepts of the graph.

19. $y = \frac{1}{2}x^2 - 3x - 1$

20. $y = x^2 + 0.45x - 4$

21. The revenue R from selling x handheld video games is given by

$R = x(180 - 1.5x).$

Find the number of handheld video games that must be sold to produce a revenue of $5400.

22. A rectangle has a length of x meters, a width of $100 - x$ meters, and an area of 2275 square meters. Find its dimensions.

8.4 Graphs of Quadratic Functions

Joaquin Palting/Photodisc/Getty Images

Why You Should Learn It

Real-life situations can be modeled by graphs of quadratic functions. For instance, in Exercise 101 on page 547, a quadratic equation is used to model the maximum height of a diver.

1 ▶ Determine the vertices of parabolas by completing the square.

What You Should Learn

1 ▶ Determine the vertices of parabolas by completing the square.

2 ▶ Sketch parabolas.

3 ▶ Write the equation of a parabola given the vertex and a point on the graph.

4 ▶ Use parabolas to solve application problems.

Graphs of Quadratic Functions

In this section, you will study graphs of quadratic functions of the form

$$f(x) = ax^2 + bx + c. \qquad \text{Quadratic function}$$

Figure 8.2 shows the graph of a simple quadratic function, $f(x) = x^2$.

> ### Graphs of Quadratic Functions
>
> The graph of $f(x) = ax^2 + bx + c$, $a \neq 0$, is a **parabola.** The completed square form
>
> $$f(x) = a(x - h)^2 + k \qquad \text{Standard form}$$
>
> is the **standard form** of the function. The **vertex** of the parabola occurs at the point (h, k), and the vertical line passing through the vertex is the **axis** of the parabola.

Every parabola is *symmetric* about its axis, which means that if it were folded along its axis, the two parts would match.

If a is positive, the graph of $f(x) = ax^2 + bx + c$ opens upward, and if a is negative, the graph opens downward, as shown in Figure 8.3. Observe in Figure 8.3 that the y-coordinate of the vertex identifies the minimum function value if $a > 0$ and the maximum function value if $a < 0$.

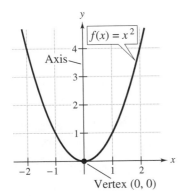

Figure 8.2

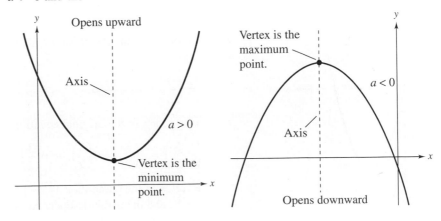

Figure 8.3

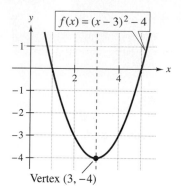

Figure 8.4

Finding the Vertex by Completing the Square

Find the vertex of the parabola given by $f(x) = x^2 - 6x + 5$.

Solution

Begin by writing the function in standard form.

$$f(x) = x^2 - 6x + 5 \qquad \text{Original function}$$

$$f(x) = x^2 - 6x + (-3)^2 - (-3)^2 + 5 \qquad \text{Complete the square.}$$

$$f(x) = (x^2 - 6x + 9) - 9 + 5 \qquad \text{Regroup terms.}$$

$$f(x) = (x - 3)^2 - 4 \qquad \text{Standard form}$$

From the standard form, you can see that the vertex of the parabola occurs at the point $(3, -4)$, as shown in Figure 8.4. The minimum value of the function is $f(3) = -4$.

✓ **CHECKPOINT** *Now try Exercise 9.*

Study Tip

When a number is added to a function and then that same number is subtracted from the function, the value of the function remains unchanged. Notice in Example 1 that $(-3)^2$ is added to the function to complete the square and then $(-3)^2$ is subtracted from the function so that the value of the function remains the same.

In Example 1, the vertex of the graph was found by *completing the square*. Another approach to finding the vertex is to complete the square once for a general function and then use the resulting formula to find the vertex.

$$f(x) = ax^2 + bx + c \qquad \text{Quadratic function}$$

$$= a\left(x^2 + \frac{b}{a}x\right) + c \qquad \text{Factor } a \text{ out of first two terms.}$$

$$= a\left[x^2 + \frac{b}{a}x + \left(\frac{b}{2a}\right)^2\right] + c - \left(\frac{b}{4a}\right)^2 \qquad \text{Complete the square.}$$

$$= a\left(x + \frac{b}{2a}\right)^2 + c - \frac{b^2}{4a} \qquad \text{Standard form}$$

From this form you can see that the vertex occurs when $x = -b/(2a)$.

Finding the Vertex Using a Formula

Find the vertex of the parabola given by $f(x) = 3x^2 - 9x$.

Solution

From the original function, it follows that $a = 3$ and $b = -9$. So, the x-coordinate of the vertex is

$$x = \frac{-b}{2a} = \frac{-(-9)}{2(3)} = \frac{3}{2}.$$

Substitute $\frac{3}{2}$ for x in the original equation to find the y-coordinate.

$$f\left(-\frac{b}{2a}\right) = f\left(\frac{3}{2}\right) = 3\left(\frac{3}{2}\right)^2 - 9\left(\frac{3}{2}\right) = -\frac{27}{4}$$

So, the vertex of the parabola is $\left(\frac{3}{2}, -\frac{27}{4}\right)$, the minimum value of the function is $f\left(\frac{3}{2}\right) = -\frac{27}{4}$, and the parabola opens upward, as shown in Figure 8.5.

✓ **CHECKPOINT** *Now try Exercise 19.*

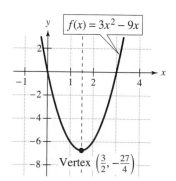

Figure 8.5

2 ▶ Sketch parabolas.

Sketching a Parabola

To obtain an accurate sketch of a parabola, the following guidelines are useful.

> ### Sketching a Parabola
>
> **1.** Determine the vertex and axis of the parabola by completing the square or by using the formula $x = -b/(2a)$.
>
> **2.** Plot the vertex, axis, x- and y-intercepts, and a few additional points on the parabola. (Using the symmetry about the axis can reduce the number of points you need to plot.)
>
> **3.** Use the fact that the parabola opens *upward* if $a > 0$ and opens *downward* if $a < 0$ to complete the sketch.

> ### Study Tip
>
> The x- and y-intercepts are useful points to plot. Another convenient fact is that the x-coordinate of the vertex lies halfway between the x-intercepts. Keep this in mind as you study the examples and do the exercises in this section.

EXAMPLE 3   **Sketching a Parabola**

To sketch the parabola given by $y = x^2 + 6x + 8$, begin by writing the equation in standard form.

$y = x^2 + 6x + 8$	Write original equation.
$y = (x^2 + 6x + 3^2 - 3^2) + 8$	Complete the square.

$$\left(\tfrac{6}{2}\right)^2$$

$y = (x^2 + 6x + 9) - 9 + 8$	Regroup terms.
$y = (x + 3)^2 - 1$	Standard form

The vertex occurs at the point $(-3, -1)$ and the axis is the line $x = -3$. After plotting this information, calculate a few additional points on the parabola, as shown in the table. Note that the y-intercept is $(0, 8)$ and the x-intercepts are solutions to the equation

$$x^2 + 6x + 8 = (x + 4)(x + 2) = 0.$$

x	-5	-4	-3	-2	-1
$y = (x + 3)^2 - 1$	3	0	-1	0	3
Solution point	$(-5, 3)$	$(-4, 0)$	$(-3, -1)$	$(-2, 0)$	$(-1, 3)$

The graph of the parabola is shown in Figure 8.6. Note that the parabola opens upward because the leading coefficient (in general form) is positive.

The graph of the parabola in Example 3 can also be obtained by shifting the graph of $y = x^2$ to the left three units and downward one unit, as discussed in Section 3.7.

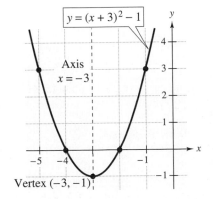

Figure 8.6

✔ **CHECKPOINT** *Now try Exercise 47.*

3 ▶ Write the equation of a parabola given the vertex and a point on the graph.

Writing the Equation of a Parabola

To write the equation of a parabola with a vertical axis, use the fact that its standard equation has the form $y = a(x - h)^2 + k$, where (h, k) is the vertex.

EXAMPLE 4   Writing the Equation of a Parabola

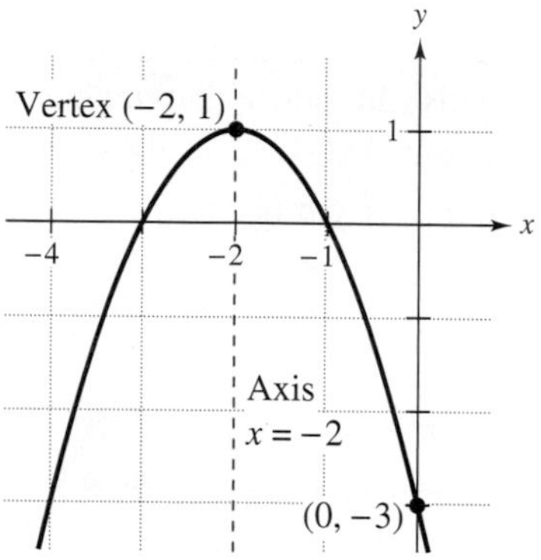

Write the equation of the parabola with vertex $(-2, 1)$ and y-intercept $(0, -3)$, as shown in Figure 8.7.

Solution

Because the vertex occurs at $(h, k) = (-2, 1)$, the equation has the form

$$y = a(x - h)^2 + k \qquad \text{Standard form}$$
$$y = a[x - (-2)]^2 + 1 \qquad \text{Substitute } -2 \text{ for } h \text{ and } 1 \text{ for } k.$$
$$y = a(x + 2)^2 + 1. \qquad \text{Simplify.}$$

To find the value of a, use the fact that the y-intercept is $(0, -3)$.

$$y = a(x + 2)^2 + 1 \qquad \text{Write standard form.}$$
$$-3 = a(0 + 2)^2 + 1 \qquad \text{Substitute 0 for } x \text{ and } -3 \text{ for } y.$$
$$-1 = a \qquad \text{Simplify.}$$

So, the standard form of the equation of the parabola is $y = -(x + 2)^2 + 1$.

✓ **CHECKPOINT** *Now try Exercise 83.*

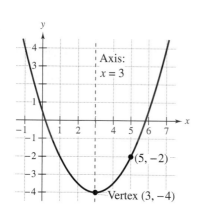

Vertex $(-2, 1)$

Axis
$x = -2$

$(0, -3)$

Figure 8.7

EXAMPLE 5   Writing the Equation of a Parabola

Write the equation of the parabola with vertex $(3, -4)$ and that passes through the point $(5, -2)$, as shown in Figure 8.8.

Solution

Because the vertex occurs at $(h, k) = (3, -4)$, the equation has the form

$$y = a(x - h)^2 + k \qquad \text{Standard form}$$
$$y = a(x - 3)^2 + (-4) \qquad \text{Substitute 3 for } h \text{ and } -4 \text{ for } k.$$
$$y = a(x - 3)^2 - 4. \qquad \text{Simplify.}$$

To find the value of a, use the fact that the parabola passes through the point $(5, -2)$.

$$y = a(x - 3)^2 - 4 \qquad \text{Write standard form.}$$
$$-2 = a(5 - 3)^2 - 4 \qquad \text{Substitute 5 for } x \text{ and } -2 \text{ for } y.$$
$$\frac{1}{2} = a \qquad \text{Simplify.}$$

So, the standard form of the equation of the parabola is $y = \frac{1}{2}(x - 3)^2 - 4$.

✓ **CHECKPOINT** *Now try Exercise 91.*

Axis:
$x = 3$

$(5, -2)$

Vertex $(3, -4)$

Figure 8.8

4 ▶ Use parabolas to solve application problems.

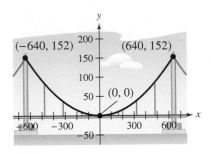

Figure 8.9

Application

| EXAMPLE 6 | **Golden Gate Bridge** |

Each cable of the Golden Gate Bridge is suspended (in the shape of a parabola) between two towers that are 1280 meters apart. The top of each tower is 152 meters above the roadway. The cables touch the roadway at the midpoint between the towers. (See Figure 8.9.)

a. Write an equation that models the cables of the bridge.

b. Find the height of the suspension cables over the roadway at a distance of 320 meters from the center of the bridge.

Solution

a. From Figure 8.9, you can see that the vertex of the parabola occurs at $(0, 0)$. So, the equation has the form

$$y = a(x - h)^2 + k \qquad \text{Standard form}$$

$$y = a(x - 0)^2 + 0 \qquad \text{Substitute 0 for } h \text{ and 0 for } k.$$

$$y = ax^2. \qquad \text{Simplify.}$$

To find the value of a, use the fact that the parabola passes through the point $(640, 152)$.

$$y = ax^2 \qquad \text{Write standard form.}$$

$$152 = a(640)^2 \qquad \text{Substitute 640 for } x \text{ and 152 for } y.$$

$$\frac{19}{51{,}200} = a \qquad \text{Simplify.}$$

So, an equation that models the cables of the bridge is

$$y = \frac{19}{51{,}200}x^2.$$

b. To find the height of the suspension cables over the roadway at a distance of 320 meters from the center of the bridge, evaluate the equation from part (a) for $x = 320$.

$$y = \frac{19}{51{,}200}x^2 \qquad \text{Write original equation.}$$

$$y = \frac{19}{51{,}200}(320)^2 \qquad \text{Substitute 320 for } x.$$

$$y = 38 \qquad \text{Simplify.}$$

So, the height of the suspension cables over the roadway is 38 meters.

✓ **CHECKPOINT** *Now try Exercise 107.*

Concept Check
---

1. In your own words, describe the graph of the quadratic function $f(x) = ax^2 + bx + c$.

2. Explain how to find the vertex of the graph of a quadratic function.

3. Explain how to find any x- or y-intercepts of the graph of a quadratic function.

4. Explain how to determine whether the graph of a quadratic function opens upward or downward.

8.4 EXERCISES

Go to pages 570–571 to record your assignments.

Developing Skills

In Exercises 1–6, match the equation with its graph. [The graphs are labeled (a), (b), (c), (d), (e), and (f).]

(a)
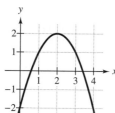

(b)
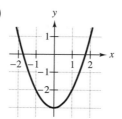

(c)
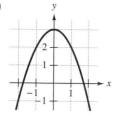

(d)
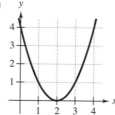

(e)
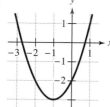

(f)
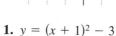

1. $y = (x + 1)^2 - 3$
2. $y = -(x + 1)^2$
3. $y = x^2 - 3$
4. $y = -x^2 + 3$
5. $y = (x - 2)^2$
6. $y = 2 - (x - 2)^2$

In Exercises 7–18, write the equation of the parabola in standard form and find the vertex of its graph. *See Example 1.*

7. $y = x^2 - 2x$
8. $y = x^2 + 2x$
✓ 9. $y = x^2 - 4x + 7$
10. $y = x^2 + 6x - 5$
11. $y = x^2 + 6x + 5$
12. $y = x^2 - 4x + 5$
13. $y = -x^2 + 6x - 10$
14. $y = -x^2 + 4x - 8$
15. $y = -x^2 - 8x + 5$
16. $y = -x^2 - 10x + 10$
17. $y = 2x^2 + 6x + 2$
18. $y = 3x^2 - 3x - 9$

In Exercises 19–24, find the vertex of the graph of the function by using the formula $x = -b/(2a)$. *See Example 2.*

✓ 19. $f(x) = x^2 - 8x + 15$
20. $f(x) = x^2 + 4x + 1$
21. $g(x) = -x^2 - 2x + 1$
22. $h(x) = -x^2 + 14x - 14$
23. $y = 4x^2 + 4x + 4$
24. $y = 9x^2 - 12x$

In Exercises 25–34, state whether the graph opens upward or downward, and find the vertex.

25. $y = 2(x - 0)^2 + 2$

26. $y = -3(x + 5)^2 - 3$

27. $y = 4 - (x - 10)^2$

28. $y = 2(x - 12)^2 + 3$

29. $y = x^2 - 6$

30. $y = -(x + 1)^2$

31. $y = -(x - 3)^2$

32. $y = x^2 - 6x$

33. $y = -x^2 + 6x$

34. $y = -x^2 - 5$

In Exercises 35–46, find the x- and y-intercepts of the graph.

35. $y = 25 - x^2$

36. $y = x^2 - 49$

37. $y = x^2 - 9x$

38. $y = x^2 + 4x$

39. $y = -x^2 - 6x + 7$

40. $y = -x^2 + 4x - 5$

41. $y = 4x^2 - 12x + 9$

42. $y = 10 - x - 2x^2$

43. $y = x^2 - 3x + 3$

44. $y = x^2 - 3x - 10$

45. $y = -2x^2 - 6x + 5$

46. $y = -4x^2 + 20x + 3$

In Exercises 47–70, sketch the parabola. Identify the vertex and any x-intercepts. Use a graphing calculator to verify your results. *See Example 3.*

✓ **47.** $g(x) = x^2 - 4$

48. $h(x) = x^2 - 9$

49. $f(x) = -x^2 + 4$

50. $f(x) = -x^2 + 9$

51. $f(x) = x^2 - 3x$

52. $g(x) = x^2 - 4x$

53. $y = -x^2 + 3x$

54. $y = -x^2 + 4x$

55. $y = (x - 4)^2$

56. $y = -(x + 4)^2$

57. $y = x^2 - 9x - 18$

58. $y = x^2 + 4x + 2$

59. $y = -(x^2 + 6x + 5)$

60. $y = -x^2 + 2x + 8$

61. $q(x) = -x^2 + 6x - 7$

62. $g(x) = x^2 + 4x + 7$

63. $y = -2x^2 - 12x - 21$

64. $y = -3x^2 + 6x - 5$

65. $y = \frac{1}{2}(x^2 - 2x - 3)$

66. $y = -\frac{1}{2}(x^2 - 6x + 7)$

67. $y = \frac{1}{5}(3x^2 - 24x + 38)$

68. $y = \frac{1}{5}(2x^2 - 4x + 7)$

69. $f(x) = 5 - \frac{1}{3}x^2$

70. $f(x) = \frac{1}{3}x^2 - 2$

In Exercises 71–78, identify the transformation of the graph of $f(x) = x^2$, and sketch a graph of h.

71. $h(x) = x^2 - 1$

72. $h(x) = x^2 + 3$

73. $h(x) = (x + 2)^2$

74. $h(x) = (x - 4)^2$

75. $h(x) = -(x + 5)^2$

76. $h(x) = -x^2 - 6$

77. $h(x) = -(x - 2)^2 - 3$

78. $h(x) = -(x + 1)^2 + 5$

In Exercises 79–82, use a graphing calculator to approximate the vertex of the graph. Verify the result algebraically.

79. $y = \frac{1}{6}(2x^2 - 8x + 11)$

80. $y = -\frac{1}{4}(4x^2 - 20x + 13)$

81. $y = -0.7x^2 - 2.7x + 2.3$

82. $y = 0.6x^2 + 4.8x + 10.4$

In Exercises 83–86, write an equation of the parabola. *See Example 4.*

83.

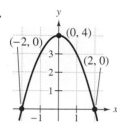

84.

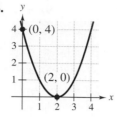

85.

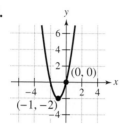

86.

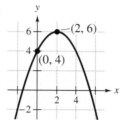

In Exercises 87–94, write an equation of the parabola $y = a(x - h)^2 + k$ that satisfies the given conditions. *See Example 5.*

87. Vertex: $(2, 1)$; $a = 1$

88. Vertex: $(-3, -3)$; $a = 1$

89. Vertex: $(2, -4)$; Point on the graph: $(0, 0)$

90. Vertex: $(-2, -4)$; Point on the graph: $(0, 0)$

91. Vertex: $(-2, -1)$; Point on the graph: $(1, 8)$

92. Vertex: $(4, 2)$; Point on the graph: $(2, -4)$

93. Vertex: $(-1, 1)$; Point on the graph: $(-4, 7)$

94. Vertex: $(5, 2)$; Point on the graph: $(10, 3)$

Solving Problems

95. *Path of a Ball* The height y (in feet) of a ball thrown by a child is given by

$$y = -\frac{1}{12}x^2 + 2x + 4$$

where x is the horizontal distance (in feet) from where the ball is thrown.

(a) How high is the ball when it leaves the child's hand?

(b) How high is the ball when it reaches its maximum height?

(c) How far from the child does the ball strike the ground?

96. *Path of a Ball* Repeat Exercise 95 if the path of the ball is modeled by

$$y = -\frac{1}{20}x^2 + 2x + 5.$$

97. *Path of an Object* A child launches a toy rocket from a table. The height y (in feet) of the rocket is given by

$$y = -\frac{1}{5}x^2 + 6x + 3$$

where x is the horizontal distance (in feet) from where the rocket is launched.

(a) Determine the height from which the rocket is launched.

(b) How high is the rocket at its maximum height?

(c) How far from where it is launched does the rocket land?

98. *Path of an Object* You use a fishing rod to cast a lure into the water. The height y (in feet) of the lure is given by

$$y = -\frac{1}{90}x^2 + \frac{1}{5}x + 9$$

where x is the horizontal distance (in feet) from the point where the lure is released.

(a) Determine the height from which the lure is released.

(b) How high is the lure at its maximum height?

(c) How far from its release point does the lure land?

99. *Path of a Golf Ball*   The height y (in yards) of a golf ball hit by a professional golfer is given by

$$y = -\frac{1}{480}x^2 + \frac{1}{2}x$$

where x is the horizontal distance (in yards) from where the ball is hit.

(a) How high is the ball when it is hit?

(b) How high is the ball at its maximum height?

(c) How far from where the ball is hit does it strike the ground?

100. *Path of a Softball*   The height y (in feet) of a softball that you hit is given by

$$y = -\frac{1}{70}x^2 + 2x + 2$$

where x is the horizontal distance (in feet) from where you hit the ball.

(a) How high is the ball when you hit it?

(b) How high is the ball at its maximum height?

(c) How far from where you hit the ball does it strike the ground?

101. *Path of a Diver*   The path of a diver is given by

$$y = -\frac{4}{9}x^2 + \frac{24}{9}x + 10$$

where y is the height in feet and x is the horizontal distance from the end of the diving board in feet. What is the maximum height of the diver?

102. *Path of a Diver*   Repeat Exercise 101 if the path of the diver is modeled by

$$y = -\frac{4}{3}x^2 + \frac{10}{3}x + 10.$$

103. 🖩 *Cost*   The cost C of producing x units of a product is given by

$$C = 800 - 10x + \frac{1}{4}x^2, \quad 0 < x < 40.$$

Use a graphing calculator to graph this function and approximate the value of x at which C is minimum.

104. 🖩 ▲ *Geometry*   The area A of a rectangle is given by the function

$$A = \frac{2}{\pi}(80x - 2x^2), \quad 0 < x < 40$$

where x is the length of the base of the rectangle in feet. Use a graphing calculator to graph the function and to approximate the value of x when A is maximum.

105. 🖩 *Graphical Estimation*   A bridge is to be constructed over a gorge with the main supporting arch being a parabola. The equation of the parabola is

$$y = 6[80 - (x^2/2400)]$$

where x and y are measured in feet. Use a graphing calculator to graph the equation and approximate the maximum height of the arch (relative to its base). Verify the maximum height algebraically.

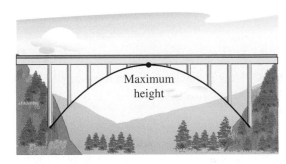

106. 🖩 *Graphical Estimation*   The profit P (in thousands of dollars) for a landscaping company is given by

$$P = 230 + 20s - \frac{1}{2}s^2$$

where s is the amount (in hundreds of dollars) spent on advertising. Use a graphing calculator to graph the profit function and approximate the amount of advertising that yields a maximum profit. Verify the maximum profit algebraically.

107. *Roller Coaster Design*   A structural engineer must design a parabolic arc for the bottom of a roller coaster track. The vertex of the parabola is placed at the origin, and the parabola must pass through the points $(-30, 15)$ and $(30, 15)$ (see figure). Find an equation of the parabolic arc.

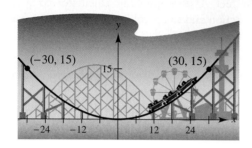

108. *Highway Design*   A highway department engineer must design a parabolic arc to create a turn in a freeway around a park. The vertex of the parabola is placed at the origin, and the parabola must connect with roads represented by the equations

$$y = -0.4x - 100, \quad x < -500$$

and

$$y = 0.4x - 100, \quad x > 500$$

(see figure). Find an equation of the parabolic arc.

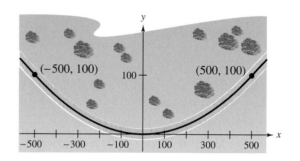

Explaining Concepts

109. ✎ How is the discriminant related to the graph of a quadratic function?

110. ✎ Is it possible for the graph of a quadratic function to have two y-intercepts? Explain.

111. ✎ Explain how to determine the maximum (or minimum) value of a quadratic function.

112. ✎ The domain of a quadratic function is the set of real numbers. Explain how to find the range.

Cumulative Review

In Exercises 113–120, find the slope-intercept form of the equation of the line through the two points.

113. $(0, 0), (4, -2)$ **114.** $(0, 0), (100, 75)$

115. $(-1, -2), (3, 6)$ **116.** $(1, 5), (6, 0)$

117. $\left(\frac{3}{2}, 8\right), \left(\frac{11}{2}, \frac{5}{2}\right)$ **118.** $(0, 2), (7.3, 15.4)$

119. $(0, 8), (5, 8)$ **120.** $(-3, 2), (-3, 5)$

In Exercises 121–124, write the number in i-form.

121. $\sqrt{-64}$ **122.** $\sqrt{-32}$

123. $\sqrt{-0.0081}$ **124.** $\sqrt{-\frac{20}{16}}$

8.5 Applications of Quadratic Equations

Lon C. Diehl/PhotoEdit, Inc.

What You Should Learn

1 ▶ Use quadratic equations to solve application problems.

Why You Should Learn It

Quadratic equations are used in a wide variety of real-life problems. For instance, in Exercise 42 on page 558, a quadratic equation is used to model the height of a baseball after you hit the ball.

1 ▶ Use quadratic equations to solve application problems.

Applications of Quadratic Equations

EXAMPLE 1   **An Investment Problem**

A car dealer buys a fleet of cars from a car rental agency for a total of $120,000. The dealer regains this $120,000 investment by selling all but four of the cars at an average profit of $2500 each. How many cars has the dealer sold, and what is the average price per car?

Solution

Although this problem is stated in terms of average price and average profit per car, you can use a model that assumes that each car has sold for the same price.

Verbal Model:

$$\boxed{\text{Selling price per car}} = \boxed{\text{Cost per car}} + \boxed{\text{Profit per car}}$$

Labels:

Number of cars sold $= x$	(cars)
Number of cars bought $= x + 4$	(cars)
Selling price per car $= 120{,}000/x$	(dollars per car)
Cost per car $= 120{,}000/(x + 4)$	(dollars per car)
Profit per car $= 2500$	(dollars per car)

Equation:

$$\frac{120{,}000}{x} = \frac{120{,}000}{x + 4} + 2500$$

$$120{,}000(x + 4) = 120{,}000x + 2500x(x + 4), \ x \neq 0, \ x \neq -4$$

$$120{,}000x + 480{,}000 = 120{,}000x + 2500x^2 + 10{,}000x$$

$$0 = 2500x^2 + 10{,}000x - 480{,}000$$

$$0 = x^2 + 4x - 192$$

$$0 = (x - 12)(x + 16)$$

$$x - 12 = 0 \implies x = 12$$

$$x + 16 = 0 \implies x = -16$$

Choosing the positive value, it follows that the dealer sold 12 cars at an average price of $120{,}000/12 = \$10{,}000$ per car. Check this in the original statement.

 **CHECKPOINT** *Now try Exercise 1.*

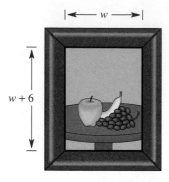

|←—— w ——→|

$w + 6$

Figure 8.10

EXAMPLE 2   Geometry: Dimensions of a Picture

A picture is 6 inches taller than it is wide and has an area of 216 square inches, as shown in Figure 8.10. What are the dimensions of the picture?

Solution

Verbal Model:

| Area of picture | = | Width | · | Height |

Labels: Picture width $= w$ (inches)
                 Picture height $= w + 6$ (inches)
                 Area $= 216$ (square inches)

Equation: $216 = w(w + 6)$

                   $0 = w^2 + 6w - 216$

                   $0 = (w + 18)(w - 12)$

               $w + 18 = 0 \implies w = -18$

               $w - 12 = 0 \implies w = 12$

Choosing the positive value of w, you can conclude that the picture is $w = 12$ inches wide and $w + 6 = 12 + 6 = 18$ inches tall. Check these solutions.

✔ **CHECKPOINT** *Now try Exercise 15.*

EXAMPLE 3   An Interest Problem

The formula $A = P(1 + r)^2$ represents the amount of money A in an account in which P dollars is deposited for 2 years at an annual interest rate of r (in decimal form). Find the interest rate if a deposit of $6000 increases to $6933.75 over a two-year period.

Solution

$A = P(1 + r)^2$	Write given formula.
$6933.75 = 6000(1 + r)^2$	Substitute 6933.75 for A and 6000 for P.
$1.155625 = (1 + r)^2$	Divide each side by 6000.
$\pm 1.075 = 1 + r$	Square Root Property
$0.075 = r$	Choose positive solution.

The annual interest rate is $r = 0.075 = 7.5\%$.

Check

$A = P(1 + r)^2$	Write given formula.
$6933.75 \overset{?}{=} 6000(1 + 0.075)^2$	Substitute 6933.75 for A, 6000 for P, and 0.075 for r.
$6933.75 \overset{?}{=} 6000(1.155625)$	Simplify.
$6933.75 = 6933.75$	Solution checks. ✔

✔ **CHECKPOINT** *Now try Exercise 23.*

EXAMPLE 4   **Reduced Rates**

A ski club charters a bus for a ski trip at a cost of $720. When four nonmembers accept invitations from the club to go on the trip, the bus fare per skier decreases by $6. How many club members are going on the trip?

Solution

Verbal Model: | Fare per skier | · | Number of skiers | = | $720 |

Labels: Number of ski club members $= x$ (people)

Number of skiers $= x + 4$ (people)

Original fare per skier $= \dfrac{720}{x}$ (dollars per person)

New fare per skier $= \dfrac{720}{x} - 6$ (dollars per person)

Equation: $\left(\dfrac{720}{x} - 6\right)(x + 4) = 720$ Original equation

$\left(\dfrac{720 - 6x}{x}\right)(x + 4) = 720$ Rewrite 2nd factor.

$(720 - 6x)(x + 4) = 720x, \ x \neq 0$ Multiply each side by x.

$720x + 2880 - 6x^2 - 24x = 720x$ Multiply factors.

$-6x^2 - 24x + 2880 = 0$ Subtract 720x from each side.

$x^2 + 4x - 480 = 0$ Divide each side by -6.

$(x + 24)(x - 20) = 0$ Factor left side of equation.

$x + 24 = 0 \quad \Longrightarrow \quad x = -24$ Set 1st factor equal to 0.

$x - 20 = 0 \quad \Longrightarrow \quad x = 20$ Set 2nd factor equal to 0.

Choosing the positive value of x, you can conclude that 20 ski club members are going on the trip. Check this solution in the original statement of the problem, as follows.

Check

Original fare per skier for 20 ski club members:

$\dfrac{720}{x} = \dfrac{720}{20} = \$36$ Substitute 20 for x.

New fare per skier with 4 nonmembers:

$\dfrac{720}{x + 4} = \dfrac{720}{24} = \$30$

Decrease in fare per skier with 4 nonmembers:

$36 - 30 = \$6$ Solution checks. ✓

 CHECKPOINT *Now try Exercise 29.*

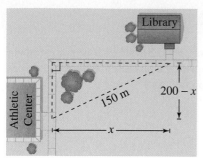

Figure 8.11

EXAMPLE 5   **An Application Involving the Pythagorean Theorem**

An L-shaped sidewalk from the athletic center to the library on a college campus is 200 meters long, as shown in Figure 8.11. By cutting diagonally across the grass, students shorten the walking distance to 150 meters. What are the lengths of the two legs of the sidewalk?

Solution

Common
Formula: $a^2 + b^2 = c^2$ Pythagorean Theorem

Labels: Length of one leg $= x$ (meters)
            Length of other leg $= 200 - x$ (meters)
            Length of diagonal $= 150$ (meters)

Equation: $x^2 + (200 - x)^2 = 150^2$

$$x^2 + 40,000 - 400x + x^2 = 22,500$$

$$2x^2 - 400x + 40,000 = 22,500$$

$$2x^2 - 400x + 17,500 = 0$$

$$x^2 - 200x + 8750 = 0$$

Using the Quadratic Formula, you can find the solutions as follows.

$$x = \frac{-(-200) \pm \sqrt{(-200)^2 - 4(1)(8750)}}{2(1)}$$ Substitute 1 for a, -200 for b, and 8750 for c.

$$= \frac{200 \pm \sqrt{5000}}{2}$$

$$= \frac{200 \pm 50\sqrt{2}}{2}$$

$$= \frac{\cancel{2}(100 \pm 25\sqrt{2})}{\cancel{2}}$$

$$= 100 \pm 25\sqrt{2}$$

Both solutions are positive, so it does not matter which you choose. If you let

$$x = 100 + 25\sqrt{2} \approx 135.4 \text{ meters}$$

the length of the other leg is

$$200 - x \approx 200 - 135.4 \approx 64.6 \text{ meters}.$$

 CHECKPOINT *Now try Exercise 31.*

In Example 5, notice that you obtain the same dimensions if you choose the other value of x. That is, if you let

$$x = 100 - 25\sqrt{2} \approx 64.6 \text{ meters}$$

the length of the other leg is

$$200 - x \approx 200 - 64.6 \approx 135.4 \text{ meters}.$$

EXAMPLE 6   **Work-Rate Problem**

An office contains two copy machines. Machine B is known to take 12 minutes longer than machine A to copy the company's monthly report. Using both machines together, it takes 8 minutes to reproduce the report. How long would it take each machine alone to reproduce the report?

Solution

Verbal Model:

$$\boxed{\begin{array}{c}\text{Work done by}\\\text{machine A}\end{array}} + \boxed{\begin{array}{c}\text{Work done by}\\\text{machine B}\end{array}} = \boxed{\begin{array}{c}\text{1 complete}\\\text{job}\end{array}}$$

$$\boxed{\begin{array}{c}\text{Rate}\\\text{for A}\end{array}} \cdot \boxed{\begin{array}{c}\text{Time}\\\text{for both}\end{array}} + \boxed{\begin{array}{c}\text{Rate}\\\text{for B}\end{array}} \cdot \boxed{\begin{array}{c}\text{Time}\\\text{for both}\end{array}} = \boxed{1}$$

Labels:

Time for machine A $= t$	(minutes)
Rate for machine A $= \dfrac{1}{t}$	(job per minute)
Time for machine B $= t + 12$	(minutes)
Rate for machine B $= \dfrac{1}{t + 12}$	(job per minute)
Time for both machines $= 8$	(minutes)
Rate for both machines $= \dfrac{1}{8}$	(job per minute)

Equation:

$$\frac{1}{t}(8) + \frac{1}{t + 12}(8) = 1 \qquad \text{Original equation}$$

$$8\left(\frac{1}{t} + \frac{1}{t + 12}\right) = 1 \qquad \text{Distributive Property}$$

$$8\left[\frac{t + 12 + t}{t(t + 12)}\right] = 1 \qquad \begin{array}{l}\text{Rewrite with common}\\\text{denominator.}\end{array}$$

$$8t(t + 12)\left[\frac{2t + 12}{t(t + 12)}\right] = t(t + 12) \qquad \begin{array}{l}\text{Multiply each side by}\\t(t + 12).\end{array}$$

$$8(2t + 12) = t^2 + 12t \qquad \text{Simplify.}$$

$$16t + 96 = t^2 + 12t \qquad \text{Distributive Property}$$

$$0 = t^2 - 4t - 96 \qquad \begin{array}{l}\text{Subtract } 16t + 96 \text{ from}\\\text{each side.}\end{array}$$

$$0 = (t - 12)(t + 8) \qquad \text{Factor right side of equation.}$$

$$t - 12 = 0 \implies t = 12 \qquad \text{Set 1st factor equal to 0.}$$

$$t + 8 = 0 \implies t = -8 \qquad \text{Set 2nd factor equal to 0.}$$

By choosing the positive value for t, you can conclude that the times for the two machines are

Time for machine A $= t = 12$ minutes

Time for machine B $= t + 12 = 12 + 12 = 24$ minutes.

Check these solutions in the original statement of the problem.

 **CHECKPOINT** *Now try Exercise 35.*

EXAMPLE 7   **The Height of a Model Rocket**

A model rocket is projected straight upward from ground level according to the height equation

$$h = -16t^2 + 192t, \, t \geq 0$$

where h is the height in feet and t is the time in seconds.

a. After how many seconds is the height 432 feet?

b. After how many seconds does the rocket hit the ground?

c. What is the maximum height of the rocket?

Solution

a.

$h = -16t^2 + 192t$	Write original equation.
$432 = -16t^2 + 192t$	Substitute 432 for h.
$16t^2 - 192t + 432 = 0$	Write in general form.
$t^2 - 12t + 27 = 0$	Divide each side by 16.
$(t - 3)(t - 9) = 0$	Factor left side of equation.
$t - 3 = 0 \implies t = 3$	Set 1st factor equal to 0.
$t - 9 = 0 \implies t = 9$	Set 2nd factor equal to 0.

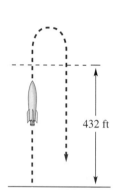

432 ft

Figure 8.12

The rocket attains a height of 432 feet at two different times—once (going up) after 3 seconds, and again (coming down) after 9 seconds. (See Figure 8.12.)

b. To find the time it takes for the rocket to hit the ground, let the height be 0.

$0 = -16t^2 + 192t$	Substitute 0 for h in original equation.
$0 = t^2 - 12t$	Divide each side by -16.
$0 = t(t - 12)$	Factor right side of equation.
$t = 0 \quad \text{or} \quad t = 12$	Solutions

The rocket hits the ground after 12 seconds. (Note that the time of $t = 0$ seconds corresponds to the time of lift-off.)

c. The maximum value of h in the equation $h = -16t^2 + 192t$ occurs when $t = -\dfrac{b}{2a}$. So, the t-coordinate is

$$t = \frac{-b}{2a} = \frac{-192}{2(-16)} = 6$$

and the h-coordinate is

$$h = -16(6)^2 + 192(6) = 576.$$

So, the maximum height of the rocket is 576 feet.

 CHECKPOINT *Now try Exercise 41.*

Concept Check

In Questions 1–4, a problem situation is given. Describe two quantities that can be set equal to each other so as to write an equation that can be used to solve the problem.

1. You know the length of the hypotenuse, and the sum of the lengths of the legs, of a right triangle. You want to find the lengths of the legs.

2. You know the area of a rectangle and you know how many units longer the length is than the width. You want to find the length and width.

3. You know the amount invested in an unknown number of product units. You know the number of units remaining when the investment is regained, and the profit per unit sold. You want to find the number of units sold and the price per unit.

4. You know the time in minutes for two machines to complete a task together and you know how many more minutes it takes one machine than the other to complete the task alone. You want to find the time to complete the task alone for each machine.

8.5 EXERCISES

Go to pages 570–571 to record your assignments.

Solving Problems

✓ 1. *Selling Price* A store owner bought a case of eggs for $21.60. By the time all but 6 dozen of the eggs had been sold at a profit of $0.30 per dozen, the original investment of $21.60 had been regained. How many dozen eggs did the owner sell, and what was the selling price per dozen? *See Example 1.*

2. *Selling Price* A computer store manager buys several computers of the same model for $12,600. The store can regain this $12,600 investment by selling all but four of the computers at a profit of $360 per computer. To do this, how many computers must be sold, and at what price?

3. *Selling Price* A flea market vendor buys a box of DVD movies for $50. After selling several of the DVDs at a profit of $3 each, the vendor still has 15 of the DVDs left by the time she regains her $50 investment. How many DVDs has the vendor sold, and at what price?

4. *Selling Price* A sorority buys a case of sweatshirts for $750 to sell at a mixer. The sorority needs to sell all but 20 of the sweatshirts at a profit of $10 per sweatshirt to regain the $750 investment. How many sweatshirts must be sold, and at what price, to do this?

▲ *Geometry* In Exercises 5–14, complete the table of widths, lengths, perimeters, and areas of rectangles.

Width	Length	Perimeter	Area
5. $1.4l$	l	54 in.	
6. w	$3.5w$	60 m	
7. w	$2.5w$		250 ft^2
8. w	$1.5w$		216 cm^2
9. $\frac{1}{3}l$	l		192 in.^{2}
10. $\frac{3}{4}l$	l		2700 in.^{2}
11. w	$w+3$	54 km	
12. $l-6$	l	108 ft	
13. $l-20$	l		12,000 m^2
14. w	$w+5$		500 ft^2

✓ 15. ▲ *Geometry* A picture frame is 4 inches taller than it is wide and has an area of 192 square inches. What are the dimensions of the picture frame? *See Example 2.*

16. ▲ *Geometry* The height of a triangle is 8 inches less than its base. The area of the triangle is 192 square inches. Find the dimensions of the triangle.

17. *Storage Area* A retail lumberyard plans to store lumber in a rectangular region adjoining the sales office (see figure). The region will be fenced on three sides, and the fourth side will be bounded by the wall of the office building. There is 350 feet of fencing available, and the area of the region is 12,500 square feet. Find the dimensions of the region.

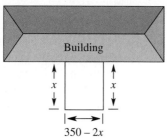

Building

x x

$350 - 2x$

18. ▲ *Geometry* Your home is on a square lot. To add more space to your yard, you purchase an additional 20 feet along one side of the property (see figure). The area of the lot is now 25,500 square feet. What are the dimensions of the new lot?

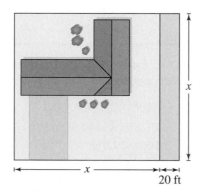

x

x

20 ft

19. *Fenced Area* A family built a fence around three sides of their property (see figure). In total, they used 550 feet of fencing. By their calculations, the lot is 1 acre (43,560 square feet). Is this correct? Explain your reasoning.

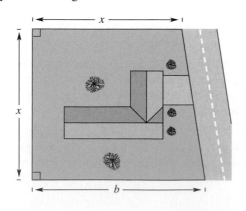

x

x

b

20. *Fenced Area* You have 100 feet of fencing. Do you have enough to enclose a rectangular region whose area is 630 square feet? Is there enough to enclose a circular area of 630 square feet? Explain.

21. *Open Conduit* An open-topped rectangular conduit for carrying water in a manufacturing process is made by folding up the edges of a sheet of aluminum 48 inches wide (see figure). A cross section of the conduit must have an area of 288 square inches. Find the width and height of the conduit.

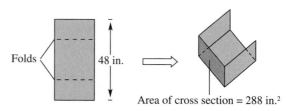

Folds   48 in.

Area of cross section = 288 in.²

22. *Photography* A photographer has a photograph that is 6 inches by 8 inches. The photographer wants to crop the photo down to half of its original area by trimming equal lengths from each side. How many inches should be trimmed from each side?

Compound Interest In Exercises 23–28, find the interest rate r. Use the formula $A = P(1 + r)^2$, where A is the amount after 2 years in an account earning r percent (in decimal form) compounded annually, and P is the original investment. ***See Example 3.***

✔ **23.** $P = \$10{,}000$
   $A = \$11{,}990.25$

24. $P = \$3000$
   $A = \$3499.20$

25. $P = \$500$
   $A = \$572.45$

26. $P = \$250$
   $A = \$280.90$

27. $P = \$6500$
   $A = \$7372.46$

28. $P = \$8000$
   $A = \$8421.41$

✔ **29.** *Reduced Rates* A service organization pays $210 for a block of tickets to a baseball game. The block contains three more tickets than the organization needs for its members. By inviting three more people to attend (and share in the cost), the organization lowers the price per person by $3.50. How many people are going to the game? ***See Example 4.***

30. *Reduced Fares* A science club charters a bus to attend a science fair at a cost of $480. To lower the bus fare per person, the club invites nonmembers to go along. When two nonmembers join the trip, the fare per person is decreased by $1. How many people are going on the excursion?

31. *Delivery Route*   You deliver pizzas to an insurance office and an apartment complex (see figure). Your total mileage in driving to the insurance office and then to the apartment complex is 12 miles. By using a direct route, you are able to drive just 9 miles to return to the pizza shop. Estimate the distance from the pizza shop to the insurance office. *See Example 5.*

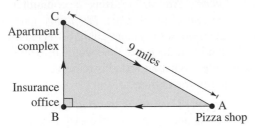

32. ▲ *Geometry*   An L-shaped sidewalk from the library (point A) to the gym (point B) on a high school campus is 100 yards long, as shown in the figure. By cutting diagonally across the grass, students shorten the walking distance to 80 yards. What are the lengths of the two legs of the sidewalk?

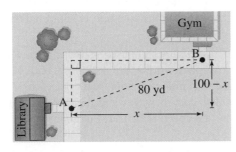

33. *Solving Graphically and Numerically*   A meteorologist is positioned 100 feet from the point where a weather balloon is launched (see figure).

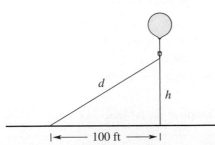

(a) Write an equation relating the distance d between the balloon and the meteorologist to the height h of the balloon.

(b) 🖩 Use a graphing calculator to graph the equation.

(c) 🖩 Use the graph to approximate the value of h when $d = 200$ feet.

(d) Complete the table.

h	0	100	200	300
d				

34. ▲ *Geometry*   An adjustable rectangular form has minimum dimensions of 3 meters by 4 meters. The length and width can be expanded by equal amounts x (see figure).

(a) Write an equation relating the length d of the diagonal to x.

(b) 🖩 Use a graphing calculator to graph the equation.

(c) 🖩 Use the graph to approximate the value of x when $d = 10$ meters.

(d) Find x algebraically when $d = 10$.

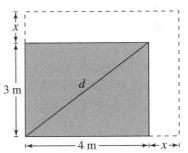

35. *Work Rate*   An office contains two printers. Machine B is known to take 3 minutes longer than machine A to produce the company's monthly financial report. Using both machines together, it takes 6 minutes to produce the report. How long would it take each machine to produce the report? *See Example 6.*

36. *Work Rate*   A builder works with two plumbing companies. Company A is known to take 3 days longer than Company B to install the plumbing in a particular style of house. Using both companies, it takes 4 days. How long would it take to install the plumbing using each company individually?

Free-Falling Object   In Exercises 37–40, find the time necessary for an object to fall to ground level from an initial height of h_0 feet if its height h at any time t (in seconds) is given by $h = h_0 - 16t^2$.

37. $h_0 = 169$ **38.** $h_0 = 729$

39. $h_0 = 1454$ (height of the Sears Tower)

40. $h_0 = 984$ (height of the Eiffel Tower)

✓ **41.** *Height*   The height h in feet of a baseball t seconds after being hit at a point 3 feet above the ground is given by $h = 3 + 75t - 16t^2$. Find the time when the ball hits the ground. *See Example 7.*

42. *Height*   You are hitting baseballs. When you toss the ball into the air, your hand is 5 feet above the ground (see figure). You hit the ball when it falls back to a height of 3.5 feet. You toss the ball with an initial velocity of 18 feet per second. The height h of the ball t seconds after leaving your hand is given by $h = 5 + 18t - 16t^2$. About how much time passes before you hit the ball?

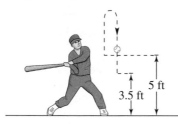

5 ft
3.5 ft

43. *Height*   A model rocket is projected straight upward from ground level according to the height equation $h = -16t^2 + 160t$, where h is the height of the rocket in feet and t is the time in seconds.

(a) After how many seconds is the height 336 feet?

(b) After how many seconds does the rocket hit the ground?

(c) What is the maximum height of the rocket?

44. *Height*   A tennis ball is tossed vertically upward from a height of 5 feet according to the height equation $h = -16t^2 + 21t + 5$, where h is the height of the tennis ball in feet and t is the time in seconds.

(a) After how many seconds is the height 11 feet?

(b) After how many seconds does the tennis ball hit the ground?

(c) What is the maximum height of the ball?

Number Problems   In Exercises 45–50, find two positive integers that satisfy the requirement.

45. The product of two consecutive integers is 182.

46. The product of two consecutive integers is 1806.

47. The product of two consecutive even integers is 168.

48. The product of two consecutive even integers is 2808.

49. The product of two consecutive odd integers is 323.

50. The product of two consecutive odd integers is 1443.

51. *Air Speed*   An airline runs a commuter flight between two cities that are 720 miles apart. If the average speed of the planes could be increased by 40 miles per hour, the travel time would be decreased by 12 minutes. What air speed is required to obtain this decrease in travel time?

52. *Average Speed*   A truck traveled the first 100 miles of a trip at one speed and the last 135 miles at an average speed of 5 miles per hour less. The entire trip took 5 hours. What was the average speed for the first part of the trip?

53. *Speed*   A company uses a pickup truck for deliveries. The cost per hour for fuel is $C = v^2/300$, where v is the speed in miles per hour. The driver is paid $15 per hour. The cost of wages and fuel for an 80-mile trip at constant speed is $36. Find the speed.

54. *Speed*   A hobby shop uses a small car for deliveries. The cost per hour for fuel is $C = v^2/600$, where v is the speed in miles per hour. The driver is paid $10 per hour. The cost of wages and fuel for a 110-mile trip at constant speed is $29.32. Find the speed.

55. *Distance*   Find any points on the line $y = 9$ that are 10 units from the point $(2, 3)$.

56. *Distance*   Find any points on the line $y = 14$ that are 13 units from the point $(1, 2)$.

57. ▲ *Geometry*   The area of an ellipse is given by $A = \pi ab$ (see figure). For a certain ellipse, it is required that $a + b = 20$.

(a) Show that $A = \pi a(20 - a)$.

(b) Complete the table.

a	4	7	10	13	16
A					

(c) Find two values of a such that $A = 300$.

(d) 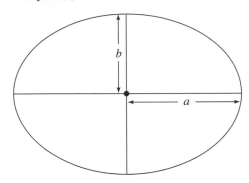 Use a graphing calculator to graph the area equation. Then use the graph to verify the results in part (c).

Figure for 57

58.  *Data Analysis* For the years 2000 through 2006, the reported numbers of boating accidents A in the United States can be approximated by

$$A = 118.52t^2 - 1140.4t + 7615, \quad 0 \le t \le 6$$

where t is the year, with $t = 0$ corresponding to 2000.   (Source: U.S. Coast Guard)

(a) Approximate the numbers of boating accidents in the years 2000 and 2006.

(b) During which year from 2000 to 2006 were there approximately 5800 boating accidents?

(c)  Use the Internet to find the Coast Guard's data for boating accidents in the years since 2006. Then use a graphing calculator to graph the model for A given in the problem. Discuss the behavior of the graph for $t > 6$. Use your data and the graph to discuss the appropriateness of the model for making predictions after 2006.

© Craig Lovell/CORBIS

Explaining Concepts

59. ✎ To solve some of the problems in this section, you wrote rational equations. Explain why these types of problems are included as applications of quadratic equations.

60. ✎ In each of Exercises 5–14, finding the area or perimeter of a rectangle is involved. The solution requires writing an equation that can be solved for the length or width of the rectangle. Explain how you can tell when this equation will be a *quadratic equation* or a *linear equation*.

61. ✎ In a *reduced rates* problem such as Example 4, does the cost per person decrease by the same amount for each additional person? Explain.

62. ✎ In a *height of an object* problem such as Example 7, suppose you try solving the height equation using a height greater than the maximum height reached by the object. What type of result will you get for t? Explain.

Cumulative Review

In Exercises 63 and 64, solve the inequality and sketch the solution on the real number line.

63. $5 - 3x > 17$

64. $-3 < 2x + 3 < 5$

In Exercises 65 and 66, solve the equation by completing the square.

65. $x^2 - 8x = 0$

66. $x^2 - 2x - 2 = 0$

8.6 Quadratic and Rational Inequalities

Will Hart/PhotoEdit, Inc.

What You Should Learn

1 ▶ Determine test intervals for polynomials.

2 ▶ Use test intervals to solve quadratic inequalities.

3 ▶ Use test intervals to solve rational inequalities.

4 ▶ Use inequalities to solve application problems.

Why You Should Learn It

Rational inequalities can be used to model and solve real-life problems. For instance, in Exercise 120 on page 569, a rational inequality is used to model the temperature of a metal in a laboratory experiment.

1 ▶ Determine test intervals for polynomials.

Finding Test Intervals

When working with polynomial inequalities, it is important to realize that the value of a polynomial can change signs only at its **zeros.** That is, a polynomial can change signs only at the x-values for which the value of the polynomial is zero. For instance, the first-degree polynomial $x + 2$ has a zero at $x = -2$, and it changes signs at that zero. You can picture this result on the real number line, as shown in Figure 8.13.

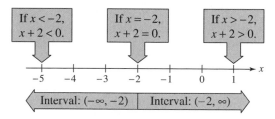

Figure 8.13

Note in Figure 8.13 that the zero of the polynomial partitions the real number line into two **test intervals.** The value of the polynomial is negative for every x-value in the first test interval $(-\infty, -2)$, and positive for every x-value in the second test interval $(-2, \infty)$. You can use the same basic approach to determine the test intervals for any polynomial.

Finding Test Intervals for a Polynomial

1. Find all real zeros of the polynomial, and arrange the zeros in increasing order. The zeros of a polynomial are called its **critical numbers.**

2. Use the critical numbers of the polynomial to determine its test intervals.

3. Choose a representative x-value in each test interval and evaluate the polynomial at that value. If the value of the polynomial is negative, the polynomial will have negative values for *every* x-value in the interval. If the value of the polynomial is positive, the polynomial will have positive values for *every* x-value in the interval.

2 ▶ Use test intervals to solve quadratic inequalities.

Quadratic Inequalities

The concepts of critical numbers and test intervals can be used to solve nonlinear inequalities, as demonstrated in Examples 1, 2, and 4.

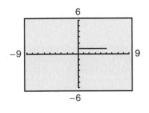
EXAMPLE 1   Solving a Quadratic Inequality

Solve the inequality $x^2 - 5x < 0$.

Solution

First find the *critical numbers* of $x^2 - 5x < 0$ by finding the solutions of the equation $x^2 - 5x = 0$.

$$x^2 - 5x = 0 \qquad \text{Write corresponding equation.}$$
$$x(x - 5) = 0 \qquad \text{Factor.}$$
$$x = 0, \; x = 5 \qquad \text{Critical numbers}$$

This implies that the test intervals are $(-\infty, 0)$, $(0, 5)$, and $(5, \infty)$. To test an interval, choose a convenient value in the interval and determine if the value satisfies the inequality.

Test interval	Representative x-value	Is inequality satisfied?
$(-\infty, 0)$	$x = -1$	$(-1)^2 - 5(-1) \overset{?}{<} 0$ $6 \not< 0$
$(0, 5)$	$x = 1$	$1^2 - 5(1) \overset{?}{<} 0$ $-4 < 0$
$(5, \infty)$	$x = 6$	$6^2 - 5(6) \overset{?}{<} 0$ $6 \not< 0$

Because the inequality $x^2 - 5x < 0$ is satisfied only by the value $x = 1$ (the value in the middle test interval), you can conclude that the solution set of the inequality is the interval $(0, 5)$, as shown in Figure 8.14.

Study Tip

In Example 1, note that you would have used the same basic procedure if the inequality symbol had been $\leq$, $>$, or $\geq$. For instance, in Figure 8.14, you can see that the solution set of the inequality $x^2 - 5x \geq 0$ consists of the union of the half-open intervals $(-\infty, 0]$ and $[5, \infty)$, which is written as $(-\infty, 0] \cup [5, \infty)$.

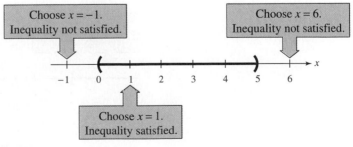

Figure 8.14

✓ **CHECKPOINT** *Now try Exercise 21.*

Just as in solving quadratic *equations*, the first step in solving a quadratic *inequality* is to write the inequality in **general form,** with the polynomial on the left and zero on the right, as demonstrated in Example 2.

EXAMPLE 2   Solving a Quadratic Inequality

Solve the inequality $2x^2 + 5x \geq 12$.

Solution

Begin by writing the inequality in the general form $2x^2 + 5x - 12 \geq 0$. Next, find the critical numbers by finding the solutions of the equation $2x^2 + 5x - 12 = 0$.

$$2x^2 + 5x - 12 = 0 \qquad \text{Write corresponding equation.}$$

$$(x + 4)(2x - 3) = 0 \qquad \text{Factor.}$$

$$x = -4, x = \frac{3}{2} \qquad \text{Critical numbers}$$

This implies that the test intervals are $(-\infty, -4)$, $\left(-4, \frac{3}{2}\right)$, and $\left(\frac{3}{2}, \infty\right)$. To test an interval, choose a convenient value in the interval and determine if the value satisfies the inequality.

Test interval	Representative x-value	Is inequality satisfied?
$(-\infty, -4)$	$x = -5$	$2(-5)^2 + 5(-5) \overset{?}{\geq} 12$ $25 \geq 12$
$\left(-4, \frac{3}{2}\right)$	$x = 0$	$2(0)^2 + 5(0) \overset{?}{\geq} 12$ $0 \not\geq 12$
$\left(\frac{3}{2}, \infty\right)$	$x = 2$	$2(2)^2 + 5(2) \overset{?}{\geq} 12$ $18 \geq 12$

From this table you can see that the inequality $2x^2 + 5x \geq 12$ is satisfied by x-values in the intervals $(-\infty, -4)$ and $\left(\frac{3}{2}, \infty\right)$. The critical numbers -4 and $\frac{3}{2}$ are also solutions to the inequality. So, the solution set of the inequality is $(-\infty, -4] \cup \left[\frac{3}{2}, \infty\right)$, as shown in Figure 8.15.

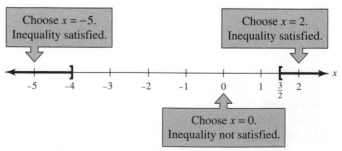

Figure 8.15

✓ **CHECKPOINT** *Now try Exercise 31.*

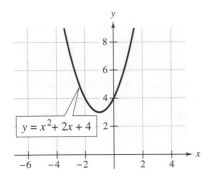

Figure 8.16

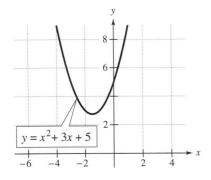

Figure 8.17

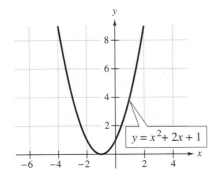

Figure 8.18

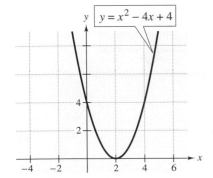

Figure 8.19

The solutions of the quadratic inequalities in Examples 1 and 2 consist, respectively, of a single interval and the union of two intervals. When solving the exercises for this section, you should watch for some unusual solution sets, as illustrated in Example 3.

EXAMPLE 3   Unusual Solution Sets

Solve each inequality.

a. The solution set of the quadratic inequality

$$x^2 + 2x + 4 > 0$$

consists of the entire set of real numbers, $(-\infty, \infty)$. This is true because the value of the quadratic $x^2 + 2x + 4$ is positive for every real value of x. You can see in Figure 8.16 that the entire parabola lies above the x-axis.

b. The solution set of the quadratic inequality

$$x^2 + 2x + 1 \le 0$$

consists of the single number $\{-1\}$. This is true because $x^2 + 2x + 1 = (x + 1)^2$ has just one critical number, $x = -1$, and it is the only value that satisfies the inequality. You can see in Figure 8.17 that the parabola meets the x-axis at $x = -1$.

c. The solution set of the quadratic inequality

$$x^2 + 3x + 5 < 0$$

is empty. This is true because the value of the quadratic $x^2 + 3x + 5$ is not less than zero for any value of x. No point on the parabola lies below the x-axis, as shown in Figure 8.18.

d. The solution set of the quadratic inequality

$$x^2 - 4x + 4 > 0$$

consists of all real numbers *except* the number 2. In interval notation, this solution set can be written as $(-\infty, 2) \cup (2, \infty)$. You can see in Figure 8.19 that the parabola lies above the x-axis *except* at $x = 2$, where it meets the x-axis.

✔ **CHECKPOINT** *Now try Exercise 35.*

Remember that checking the solution set of an inequality is not as straightforward as checking the solutions of an equation, because inequalities tend to have infinitely many solutions. Even so, you should check several x-values in your solution set to confirm that they satisfy the inequality. Also try checking x-values that are not in the solution set to verify that they do not satisfy the inequality.

For instance, the solution set of $x^2 - 5x < 0$ is the interval $(0, 5)$. Try checking some numbers in this interval to verify that they satisfy the inequality. Then check some numbers outside the interval to verify that they do not satisfy the inequality.

3 ▶ Use test intervals to solve rational inequalities.

Rational Inequalities

The concepts of critical numbers and test intervals can be extended to inequalities involving rational expressions. To do this, use the fact that the value of a rational expression can change sign only at its *zeros* (the x-values for which its numerator is zero) and its *undefined values* (the x-values for which its denominator is zero). These two types of numbers make up the **critical numbers** of a rational inequality. For instance, the critical numbers of the inequality

$$\frac{x-2}{(x-1)(x+3)} < 0$$

are $x = 2$ (the numerator is zero), and $x = 1$ and $x = -3$ (the denominator is zero). From these three critical numbers, you can see that the inequality has *four* test intervals: $(-\infty, -3)$, $(-3, 1)$, $(1, 2)$, and $(2, \infty)$.

Study Tip

When solving a rational inequality, you should begin by writing the inequality in general form, with the rational expression (as a single fraction) on the left and zero on the right. For instance, the first step in solving

$$\frac{2x}{x+3} < 4$$

is to write it as

$$\frac{2x}{x+3} - 4 < 0$$

$$\frac{2x - 4(x+3)}{x+3} < 0$$

$$\frac{-2x - 12}{x+3} < 0.$$

Try solving this inequality. You should find that the solution set is $(-\infty, -6) \cup (-3, \infty)$.

EXAMPLE 4 Solving a Rational Inequality

To solve the inequality $\dfrac{x}{x-2} > 0$, first find the critical numbers. The numerator is zero when $x = 0$, and the denominator is zero when $x = 2$. So, the two critical numbers are 0 and 2, which implies that the test intervals are $(-\infty, 0)$, $(0, 2)$, and $(2, \infty)$. To test an interval, choose a convenient value in the interval and determine if the value satisfies the inequality, as shown in the table.

Test interval	Representative x-value	Is inequality satisfied?	
$(-\infty, 0)$	$x = -1$	$\dfrac{-1}{-1-2} \overset{?}{>} 0$	$\dfrac{1}{3} > 0$
$(0, 2)$	$x = 1$	$\dfrac{1}{1-2} \overset{?}{>} 0$	$-1 \not> 0$
$(2, \infty)$	$x = 3$	$\dfrac{3}{3-2} \overset{?}{>} 0$	$3 > 0$

You can see that the inequality is satisfied for the intervals $(-\infty, 0)$ and $(2, \infty)$. So, the solution set of the inequality is $(-\infty, 0) \cup (2, \infty)$. (See Figure 8.20.)

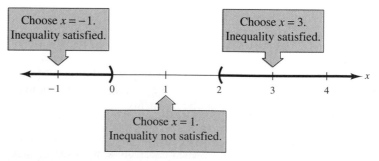

Figure 8.20

✓ **CHECKPOINT** *Now try Exercise 77.*

4 ▶ Use inequalities to solve application problems.

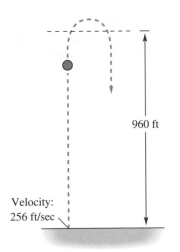

Figure 8.21

Application

EXAMPLE 5   The Height of a Projectile

A projectile is fired straight upward from ground level with an initial velocity of 256 feet per second, as shown in Figure 8.21, so that its height h at any time t is given by

$$h = -16t^2 + 256t$$

where h is measured in feet and t is measured in seconds. During what interval of time will the height of the projectile exceed 960 feet?

Solution

To solve this problem, begin by writing the inequality in general form.

$$-16t^2 + 256t > 960 \qquad \text{Write original inequality.}$$

$$-16t^2 + 256t - 960 > 0 \qquad \text{Write in general form.}$$

Next, find the critical numbers for $-16t^2 + 256t - 960 > 0$ by finding the solution to the equation $-16t^2 + 256t - 960 = 0$.

$$-16t^2 + 256t - 960 = 0 \qquad \text{Write corresponding equation.}$$

$$t^2 - 16t + 60 = 0 \qquad \text{Divide each side by } -16.$$

$$(t - 6)(t - 10) = 0 \qquad \text{Factor.}$$

$$t = 6, \, t = 10 \qquad \text{Critical numbers}$$

This implies that the test intervals are

$$(-\infty, 6), (6, 10), \text{ and } (10, \infty). \qquad \text{Test intervals}$$

To test an interval, choose a convenient value in the interval and determine if the value satisfies the inequality.

Test interval	Representative t-value	Is inequality satisfied?
$(-\infty, 6)$	$t = 0$	$-16(0)^2 + 256(0) \overset{?}{>} 960$ $0 \not> 960$
$(6, 10)$	$t = 7$	$-16(7)^2 + 256(7) \overset{?}{>} 960$ $1008 > 960$
$(10, \infty)$	$t = 11$	$-16(11)^2 + 256(11) \overset{?}{>} 960$ $880 \not> 960$

So, the height of the projectile will exceed 960 feet for values of t such that $6 < t < 10$.

 **CHECKPOINT** *Now try Exercise 111.*

Smart Study Strategy

Go to page 510 for ways to *Make Mental Cheat Sheets.*

_____ **Concept Check** _____

1. The test intervals of a polynomial are $(-\infty, -1)$, $(-1, 3), (3, \infty)$. What are the critical numbers of the polynomial?

2. In your own words, describe a procedure for solving quadratic inequalities.

3. Give a verbal description of the solution set $(-\infty, -5] \cup [10, \infty)$ of the inequality $x^2 - 5x - 50 \geq 0$.

4. How is the procedure for finding the critical numbers of a quadratic inequality different from the procedure for finding the critical numbers of a rational inequality?

8.6 EXERCISES

Go to pages 570–571 to record your assignments.

_____ **Developing Skills** _____

In Exercises 1–10, find the critical numbers.

1. $x(2x - 5)$
2. $5x(x - 3)$
3. $4x^2 - 81$
4. $9y^2 - 16$
5. $x(x + 3) - 5(x + 3)$
6. $y(y - 4) - 3(y - 4)$
7. $x^2 - 4x + 3$
8. $3x^2 - 2x - 8$
9. $6x^2 + 13x - 15$
10. $4x^2 - 4x - 3$

In Exercises 11–20, determine the intervals for which the polynomial is entirely negative and entirely positive.

11. $x - 4$
12. $3 - x$
13. $3 - \frac{1}{2}x$
14. $\frac{2}{3}x - 8$
15. $4x(x - 5)$
16. $7x(3 - x)$
17. $4 - x^2$
18. $x^2 - 36$
19. $x^2 - 4x - 5$
20. $2x^2 - 4x - 3$

In Exercises 21–60, solve the inequality and graph the solution on the real number line. (Some of the inequalities have no solutions.) *See Examples 1–3.*

✓ 21. $3x(x - 2) < 0$
22. $5x(x - 8) > 0$
23. $3x(2 - x) \geq 0$

24. $5x(8 - x) > 0$
25. $x^2 > 4$
26. $z^2 \leq 9$
27. $x^2 - 3x - 10 \geq 0$
28. $x^2 + 8x + 7 < 0$
29. $x^2 + 4x > 0$
30. $x^2 - 5x \geq 0$
✓ 31. $x^2 + 5x \leq 36$
32. $t^2 - 4t > 12$
33. $u^2 + 2u - 2 > 1$
34. $t^2 - 15t + 50 < 0$
✓ 35. $x^2 + 4x + 5 < 0$
36. $x^2 + 6x + 10 > 0$
37. $x^2 + 2x + 1 \geq 0$
38. $y^2 - 5y + 6 > 0$
39. $x^2 - 4x + 2 > 0$
40. $-x^2 + 8x - 11 \leq 0$
41. $x^2 - 6x + 9 \geq 0$
42. $x^2 + 14x + 49 < 0$
43. $u^2 - 10u + 25 < 0$
44. $y^2 + 16y + 64 \leq 0$
45. $3x^2 + 2x - 8 \leq 0$
46. $2t^2 - 3t - 20 \geq 0$
47. $-6u^2 + 19u - 10 > 0$
48. $4x^2 - 4x - 63 < 0$
49. $-2u^2 + 7u + 4 < 0$
50. $-3x^2 - 4x + 4 \leq 0$
51. $4x^2 + 28x + 49 \leq 0$

52. $9x^2 - 24x + 16 \geq 0$

53. $(x - 2)^2 < 0$

54. $(y + 3)^2 \geq 0$

55. $6 - (x - 2)^2 < 0$

56. $(y + 3)^2 - 6 \geq 0$

57. $16 \leq (u + 5)^2$

58. $25 \geq (x - 3)^2$

59. $x(x - 2)(x + 2) > 0$

60. $x(x - 1)(x + 4) \leq 0$

In Exercises 61–68, use a graphing calculator to solve the inequality. Verify your result algebraically.

61. $x^2 - 6x < 0$

62. $2x^2 + 5x > 0$

63. $0.5x^2 + 1.25x - 3 > 0$

64. $\frac{1}{3}x^2 - 3x < 0$

65. $x^2 + 6x + 5 \geq 8$

66. $x^2 - 6x + 9 < 16$

67. $9 - 0.2(x - 2)^2 < 4$

68. $8x - x^2 > 12$

Graphical Analysis In Exercises 69–72, use a graphing calculator to graph the function. Use the graph to approximate the values of x that satisfy the specified inequalities.

Function	Inequalities
69. $f(x) = x^2 - 2x + 3$	(a) $f(x) \geq 0$
	(b) $f(x) \leq 6$
70. $f(x) = -3x^2 + 6x + 2$	(a) $f(x) \leq 0$
	(b) $f(x) \geq 5$
71. $f(x) = -2x^2 + 6x - 9$	(a) $f(x) > -11$
	(b) $f(x) < 10$
72. $f(x) = 4x^2 - 10x - 7$	(a) $f(x) > -1$
	(b) $f(x) < 8$

In Exercises 73–76, find the critical numbers.

73. $\frac{5}{x - 3}$

74. $\frac{-6}{x + 2}$

75. $\frac{2x}{x + 5}$

76. $\frac{x - 2}{x - 10}$

In Exercises 77–98, solve the inequality and graph the solution on the real number line. *See Example 4.*

77. $\frac{5}{x - 3} > 0$

78. $\frac{3}{4 - x} > 0$

79. $\frac{-5}{x - 3} > 0$

80. $\frac{-3}{4 - x} > 0$

81. $\frac{3}{y - 1} \leq -1$

82. $\frac{2}{x - 3} \geq -1$

83. $\frac{x + 4}{x - 2} > 0$

84. $\frac{x - 5}{x + 2} < 0$

85. $\frac{y - 4}{y - 1} \leq 0$

86. $\frac{y + 7}{y + 3} \geq 0$

87. $\frac{4x - 2}{2x - 4} > 0$

88. $\frac{3x + 4}{2x - 1} < 0$

89. $\frac{x + 2}{4x + 6} \leq 0$

90. $\frac{u - 6}{3u - 5} \leq 0$

91. $\frac{3(u - 3)}{u + 1} < 0$

92. $\frac{2(4 - t)}{4 + t} > 0$

93. $\frac{2}{x - 5} \geq 3$

94. $\frac{1}{x + 2} > -3$

95. $\frac{4x}{x + 2} < -1$

96. $\frac{6x}{x - 4} < 5$

97. $\frac{x - 3}{x - 6} \leq 4$

98. $\frac{x + 4}{x - 5} \geq 10$

In Exercises 99–106, use a graphing calculator to solve the rational inequality. Verify your result algebraically.

99. $\frac{1}{x} - x > 0$

100. $\frac{1}{x} - 3 < 0$

101. $\dfrac{x + 6}{x + 1} - 2 < 0$

102. $\dfrac{x + 12}{x + 2} - 3 \geq 0$

103. $\dfrac{6x - 3}{x + 5} < 2$

104. $\dfrac{3x - 4}{x - 4} < -5$

105. $x + \dfrac{1}{x} > 3$

106. $4 - \dfrac{1}{x^2} > 1$

Graphical Analysis In Exercises 107–110, use a graphing calculator to graph the function. Use the graph to approximate the values of x that satisfy the specified inequalities.

Function	Inequalities	
107. $f(x) = \dfrac{3x}{x - 2}$	(a) $f(x) \leq 0$	(b) $f(x) \geq 6$
108. $f(x) = \dfrac{2(x - 2)}{x + 1}$	(a) $f(x) \leq 0$	(b) $f(x) \geq 8$
109. $f(x) = \dfrac{2x^2}{x^2 + 4}$	(a) $f(x) \geq 1$	(b) $f(x) \leq 2$
110. $f(x) = \dfrac{5x}{x^2 + 4}$	(a) $f(x) \geq 1$	(b) $f(x) \geq 0$

Solving Problems

✓ 111. *Height* A projectile is fired straight upward from ground level with an initial velocity of 128 feet per second, so that its height h at any time t is given by $h = -16t^2 + 128t$, where h is measured in feet and t is measured in seconds. During what interval of time will the height of the projectile exceed 240 feet?

112. *Height* A projectile is fired straight upward from ground level with an initial velocity of 88 feet per second, so that its height h at any time t is given by $h = -16t^2 + 88t$, where h is measured in feet and t is measured in seconds. During what interval of time will the height of the projectile exceed 50 feet?

113. *Compound Interest* You are investing $1000 in a certificate of deposit for 2 years, and you want the interest for that time period to exceed $150. The interest is compounded annually. What interest rate should you have? [*Hint:* Solve the inequality $1000(1 + r)^2 > 1150$.]

114. *Compound Interest* You are investing $500 in a certificate of deposit for 2 years, and you want the interest for that time to exceed $50. The interest is compounded annually. What interest rate should you have? [*Hint:* Solve the inequality $500(1 + r)^2 > 550$.]

115. ▲ *Geometry* You have 64 feet of fencing to enclose a rectangular region. Determine the interval for the length such that the area will exceed 240 square feet.

116. ▲ *Geometry* A rectangular playing field with a perimeter of 100 meters is to have an area of at least 500 square meters. Within what bounds must the length of the field lie?

117. *Cost, Revenue, and Profit* The revenue and cost equations for a computer desk are given by

$$R = x(50 - 0.0002x) \text{ and } C = 12x + 150,000$$

where R and C are measured in dollars and x represents the number of desks sold. How many desks must be sold to obtain a profit of at least $1,650,000?

118. *Cost, Revenue, and Profit* The revenue and cost equations for a digital camera are given by

$$R = x(125 - 0.0005x) \text{ and } C = 3.5x + 185,000$$

where R and C are measured in dollars and x represents the number of cameras sold. How many cameras must be sold to obtain a profit of at least $6,000,000?

119. *Antibiotics* The concentration C (in milligrams per liter) of an antibiotic 30 minutes after it is administered is given by

$$C(t) = \frac{21.9 - 0.043t}{1 + 0.005t}, \quad 30 \le t \le 500$$

where t is the time (in minutes).

(a) ▦ Use a graphing calculator to graph the concentration function.

(b) How long does it take for the concentration of the antibiotic to fall below 5 milligrams per liter?

120. *Data Analysis* The temperature T (in degrees Fahrenheit) of a metal in a laboratory experiment was recorded every 2 minutes for a period of 16 minutes. The table shows the experimental data, where t is the time in minutes.

t	0	2	4	6	8
T	250	290	338	410	498

t	10	12	14	16
T	560	530	370	160

A model for this data is

$$T = \frac{248.5 - 13.72t}{1.0 - 0.13t + 0.005t^2}.$$

(a) ▦ Use a graphing calculator to plot the data and graph the model in the same viewing window. Does the model fit the data well?

(b) Use the graph to approximate the times when the temperature was at least $400°\text{F}$.

Explaining Concepts

121. ✎ Explain why the critical numbers of a polynomial are not included in its test intervals.

122. ✎ Explain the difference in the solution sets of $x^2 - 4 < 0$ and $x^2 - 4 \le 0$.

123. ✎ The graph of a quadratic function g lies completely above the x-axis. What is the solution set of the inequality $g(x) < 0$? Explain your reasoning.

124. ✎ Explain how you can use the graph of $f(x) = x^2 - x - 6$ to check the solution of $x^2 - x - 6 > 0$.

Cumulative Review

In Exercises 125–130, perform the operation and simplify.

125. $\dfrac{4xy^3}{x^2y} \cdot \dfrac{y}{8x}$

126. $\dfrac{2x^2 - 2}{x^2 - 6x - 7} \cdot (x^2 - 10x + 21)$

127. $\dfrac{x^2 - x - 6}{4x^3} \cdot \dfrac{x + 1}{x^2 + 5x + 6}$

128. $\dfrac{32x^3y}{y^9} \div \dfrac{8x^4}{y^6}$

129. $\dfrac{x^2 + 8x + 16}{x^2 - 6x} \div (3x - 24)$

130. $\dfrac{x^2 + 6x - 16}{3x^2} \div \dfrac{x + 8}{6x}$

In Exercises 131–134, evaluate the expression for the specified value. Round your result to the nearest hundredth, if necessary.

131. $x^2; \ x = -\dfrac{1}{3}$

132. $1000 - 20x^3; \ x = 4.02$

133. $\dfrac{100}{x^4}; \ x = 1.06$

134. $\dfrac{50}{1 - \sqrt{x}}; \ x = 0.1024$

What Did You Learn?

Use these two pages to help prepare for a test on this chapter. Check off the key terms and key concepts you know. You can also use this section to record your assignments.

Plan for Test Success

Date of test: ☐ / /   **Study dates and times:** ☐ / / at ☐ : A.M./P.M.

☐ / / at ☐ : A.M./P.M.

Things to review:

☐ Key Terms, *p. 570*
☐ Key Concepts, *pp. 570–571*
☐ Your class notes
☐ Your assignments

☐ Study Tips, *pp. 512, 516, 522, 529, 530, 531, 532, 540, 541, 561, 562, 564*
☐ Technology Tips, *pp. 513, 515, 522, 533, 561*
☐ Mid-Chapter Quiz, *p. 538*

☐ Review Exercises, *pp. 572–574*
☐ Chapter Test, *p. 575*
☐ Video Explanations Online
☐ Tutorial Online

Key Terms

☐ double or repeated solution, *p.512*
☐ Square Root Property, *p.513*
☐ extracting square roots, *p.513*
☐ quadratic form, *p.515*
☐ completing the square, *p.521*
☐ Quadratic Formula, *p.529*

☐ discriminant, *p.532*
☐ parabola, *p.539*
☐ standard form of a quadratic function, *p.539*
☐ vertex of a parabola, *p.539*
☐ axis of a parabola, *p.539*

☐ zeros of a polynomial, *p.560*
☐ test intervals, *p.560*
☐ critical numbers of a polynomial, *p.560*
☐ general form of an inequality, *p.562*
☐ critical numbers of a rational inequality, *p.564*

Key Concepts

8.1 Solving Quadratic Equations: Factoring and Special Forms

Assignment: _____ Due date: _____

☐ **Solve quadratic equations by factoring.**
1. Write the equation in general form.
2. Factor the left side.
3. Set each factor equal to zero and solve for x.

☐ **Solve nonquadratic equations that are of quadratic form.**

An equation is said to be of quadratic form if it has the form $au^2 + bu + c = 0$, where u is an algebraic expression.

☐ **Use the Square Root Property to solve quadratic equations.**
1. The equation $u^2 = d$, where $d > 0$, has exactly two solutions: $u = \pm\sqrt{d}$. This solution process is called *extracting square roots*.
2. The equation $u^2 = d$, where $d < 0$, has exactly two solutions: $u = \pm\sqrt{|d|}\,i$.

8.2 Completing the Square

Assignment: _____ Due date: _____

☐ **Write an expression in completed square form.**

To complete the square for the expression $x^2 + bx$, add $(b/2)^2$, which is the square of half the coefficient of x. Consequently,

$$x^2 + bx + \left(\frac{b}{2}\right)^2 = \left(x + \frac{b}{2}\right)^2.$$

☐ **Solve quadratic equations by completing the square.**
1. Prior to completing the square, the coefficient of the second-degree term must be 1.
2. Preserve the equality by adding the same constant to each side of the equation.
3. Use the Square Root Property to solve the quadratic equation.

8.3 The Quadratic Formula

Assignment: _____ Due date: _____

☐ **Use the Quadratic Formula.**

The solutions of $ax^2 + bx + c = 0$, $a \neq 0$, are given by the *Quadratic Formula*

$$x = \frac{-b \pm \sqrt{b^2 - 4ac}}{2a}.$$

The expression inside the radical, $b^2 - 4ac$, is called the *discriminant*.

☐ **Write quadratic equations from their solutions.**

You can use the Zero-Factor Property in reverse to find a quadratic equation given its solutions.

☐ **Use the discriminant.**

Let a, b, and c be rational numbers such that $a \neq 0$. The discriminant of the quadratic equation $ax^2 + bx + c = 0$ is given by $b^2 - 4ac$, and can be used to classify the solutions of the equation as follows.

Discriminant	Solution Type
1. Perfect square	Two distinct rational solutions
2. Positive nonperfect square	Two distinct irrational solutions
3. Zero	One repeated rational solution
4. Negative number	Two distinct complex solutions

8.4 Graphs of Quadratic Functions

Assignment: _____ Due date: _____

☐ **Recognize graphs of quadratic functions.**

The graph of $f(x) = ax^2 + bx + c$, $a \neq 0$, is a *parabola*. The completed square form $f(x) = a(x - h)^2 + k$ is the *standard form* of the function. The *vertex* of the parabola occurs at the point (h, k), and the vertical line passing through the vertex is the *axis* of the parabola.

☐ **Sketch a parabola.**

1. Determine the vertex and axis of the parabola by completing the square or by using the formula $x = -b/(2a)$.
2. Plot the vertex, axis, x- and y-intercepts, and a few additional points on the parabola. (Using the symmetry about the axis can reduce the number of points you need to plot.)
3. Use the fact that the parabola opens upward if $a > 0$ and opens downward if $a < 0$ to complete the sketch.

8.5 Applications of Quadratic Equations

Assignment: _____ Due date: _____

☐ **Use quadratic equations to solve a wide variety of real-life problems.**

The following are samples of applications of quadratic equations.

1. Investment   2. Interest
3. Height of a projectile   4. Geometric dimensions
5. Work rate   6. Structural design

8.6 Quadratic and Rational Inequalities

Assignment: _____ Due date: _____

☐ **Find test intervals for inequalities.**

1. For a polynomial expression, find all the real zeros. For a rational expression, find all the real zeros and those x-values for which the function is undefined.
2. Arrange the numbers found in Step 1 in increasing order. These numbers are called *critical numbers*.

3. Use the critical numbers to determine the test intervals.
4. Choose a representative x-value in each test interval and evaluate the expression at that value. If the value of the expression is negative, the expression will have negative values for every x-value in the interval. If the value of the expression is positive, the expression will have positive values for every x-value in the interval.

Review Exercises

8.1 Solving Quadratic Equations: Factoring and Special Forms

1 ▶ Solve quadratic equations by factoring.

In Exercises 1–10, solve the equation by factoring.

1. $x^2 + 12x = 0$ **2.** $u^2 - 18u = 0$

3. $3y^2 - 27 = 0$ **4.** $2z^2 - 72 = 0$

5. $4y^2 + 20y + 25 = 0$

6. $x^2 + \frac{8}{3}x + \frac{16}{9} = 0$

7. $2x^2 - 2x - 180 = 0$

8. $9x^2 + 18x - 135 = 0$

9. $6x^2 - 12x = 4x^2 - 3x + 18$

10. $10x - 8 = 3x^2 - 9x + 12$

2 ▶ Solve quadratic equations by the Square Root Property.

In Exercises 11–16, solve the equation by using the Square Root Property.

11. $z^2 = 144$ **12.** $2x^2 = 98$

13. $y^2 - 12 = 0$ **14.** $y^2 - 45 = 0$

15. $(x - 16)^2 = 400$ **16.** $(x + 3)^2 = 900$

3 ▶ Solve quadratic equations with complex solutions by the Square Root Property.

In Exercises 17–22, solve the equation by using the Square Root Property.

17. $z^2 = -121$ **18.** $u^2 = -225$

19. $y^2 + 50 = 0$ **20.** $x^2 + 48 = 0$

21. $(y + 4)^2 + 18 = 0$ **22.** $(x - 2)^2 + 24 = 0$

4 ▶ Use substitution to solve equations of quadratic form.

In Exercises 23–30, solve the equation of quadratic form. (Find all real *and* complex solutions.)

23. $x^4 - 4x^2 - 5 = 0$

24. $x^4 - 10x^2 + 9 = 0$

25. $x - 4\sqrt{x} + 3 = 0$

26. $x - 4\sqrt{x} + 13 = 0$

27. $(x^2 - 2x)^2 - 4(x^2 - 2x) - 5 = 0$

28. $\left(\sqrt{x} - 2\right)^2 + 2\left(\sqrt{x} - 2\right) - 3 = 0$

29. $x^{2/3} + 3x^{1/3} - 28 = 0$

30. $x^{2/5} + 4x^{1/5} + 3 = 0$

8.2 Completing the Square

1 ▶ Rewrite quadratic expressions in completed square form.

In Exercises 31–36, add a term to the expression so that it becomes a perfect square trinomial.

31. $z^2 + 18z + $ **32.** $y^2 - 80y + $

33. $x^2 - 15x + $ **34.** $x^2 + 21x + $

35. $y^2 + \frac{2}{5}y + $ **36.** $x^2 - \frac{3}{4}x + $

2 ▶ Solve quadratic equations by completing the square.

In Exercises 37–42, solve the equation by completing the square. Give the solutions in exact form and in decimal form rounded to two decimal places. (The solutions may be complex numbers.)

37. $x^2 - 6x - 3 = 0$

38. $x^2 + 12x + 6 = 0$

39. $v^2 + 5v + 4 = 0$

40. $u^2 - 5u + 6 = 0$

41. $y^2 - \frac{2}{3}y + 2 = 0$

42. $t^2 + \frac{1}{2}t - 1 = 0$

8.3 The Quadratic Formula

2 ▶ Use the Quadratic Formula to solve quadratic equations.

In Exercises 43–48, solve the equation by using the Quadratic Formula. (Find all real *and* complex solutions.)

43. $v^2 + v - 42 = 0$ **44.** $x^2 - x - 72 = 0$

45. $2y^2 + y - 21 = 0$

46. $2x^2 - 3x - 20 = 0$ **47.** $5x^2 - 16x + 2 = 0$

48. $3x^2 + 12x + 4 = 0$

3 ▶ Determine the types of solutions of quadratic equations using the discriminant.

In Exercises 49–56, use the discriminant to determine the type of solutions of the quadratic equation.

49. $x^2 + 4x + 4 = 0$

50. $y^2 - 26y + 169 = 0$

51. $s^2 - s - 20 = 0$

52. $r^2 - 5r - 45 = 0$

53. $4t^2 + 16t + 10 = 0$

54. $8x^2 + 85x - 33 = 0$

55. $v^2 - 6v + 21 = 0$

56. $9y^2 + 1 = 0$

4 ▶ Write quadratic equations from solutions of the equations.

In Exercises 57–62, write a quadratic equation having the given solutions.

57. $3, -7$

58. $-2, 8$

59. $5 + \sqrt{7}, 5 - \sqrt{7}$

60. $2 + \sqrt{2}, 2 - \sqrt{2}$

61. $6 + 2i, 6 - 2i$

62. $3 + 4i, 3 - 4i$

8.4 Graphs of Quadratic Functions

1 ▶ Determine the vertices of parabolas by completing the square.

In Exercises 63–66, write the equation of the parabola in standard form, and find the vertex of its graph.

63. $y = x^2 - 8x + 3$

64. $y = 8 - 8x - x^2$

65. $y = 2x^2 - x + 3$

66. $y = 3x^2 + 2x - 6$

2 ▶ Sketch parabolas.

In Exercises 67–70, sketch the parabola. Identify the vertex and any x-intercepts. Use a graphing calculator to verify your results.

67. $y = x^2 + 8x$ **68.** $y = -x^2 + 3x$

69. $f(x) = -x^2 - 2x + 4$ **70.** $f(x) = x^2 + 3x - 10$

3 ▶ Write the equation of a parabola given the vertex and a point on the graph.

In Exercises 71–74, write an equation of the parabola $y = a(x - h)^2 + k$ that satisfies the conditions.

71. Vertex: $(2, -5)$; Point on the graph: $(0, 3)$

72. Vertex: $(-4, 0)$; Point on the graph: $(0, -6)$

73. Vertex: $(5, 0)$; Point on the graph: $(1, 1)$

74. Vertex: $(-2, 5)$; Point on the graph: $(-4, 11)$

4 ▶ Use parabolas to solve application problems.

75. *Path of a Ball*   The height y (in feet) of a ball thrown by a child is given by $y = -\frac{1}{10}x^2 + 3x + 6$, where x is the horizontal distance (in feet) from where the ball is thrown.

(a) ▦ Use a graphing calculator to graph the path of the ball.

(b) How high is the ball when it leaves the child's hand?

(c) How high is the ball when it reaches its maximum height?

(d) How far from the child does the ball strike the ground?

76. ▦ *Graphical Estimation*   The numbers N (in thousands) of bankruptcies filed by businesses in the United States in the years 2001 through 2005 are approximated by $N = -0.736t^2 + 3.12t + 35.0$, $1 \le t \le 5$, where t is the time in years, with $t = 1$ corresponding to 2001.   (Source: Administrative Office of the U.S. Courts)

(a) Use a graphing calculator to graph the model.

(b) Use the graph from part (a) to approximate the maximum number of bankruptcies filed by businesses from 2001 through 2005. During what year did this maximum occur?

8.5 Applications of Quadratic Equations

1 ▶ Use quadratic equations to solve application problems.

77. *Selling Price*   A car dealer bought a fleet of used cars for a total of $80,000. By the time all but four of the cars had been sold, at an average profit of $1000 each, the original investment of $80,000 had been regained. How many cars were sold, and what was the average price per car?

78. *Selling Price*   A manager of a computer store bought several computers of the same model for $27,000. When all but five of the computers had been sold at a profit of $900 per computer, the original investment of $27,000 had been regained. How many computers were sold, and what was the selling price of each computer?

79. ▲ *Geometry*   The length of a rectangle is 12 inches greater than its width. The area of the rectangle is 85 square inches. Find the dimensions of the rectangle.

80. *Compound Interest*   You want to invest $35,000 for 2 years at an annual interest rate of r (in decimal form). Interest on the account is compounded annually. Find the interest rate if a deposit of $35,000 increases to $40,221.44 over a two-year period.

81. *Reduced Rates*   A Little League baseball team obtains a block of tickets to a ball game for $96. After three more people decide to go to the game, the price per ticket is decreased by $1.60. How many people are going to the game?

82. ▲ *Geometry*   A corner lot has an L-shaped sidewalk along its sides. The total length of the sidewalk is 69 feet. By cutting diagonally across the lot, the walking distance is shortened to 51 feet. What are the lengths of the two legs of the sidewalk?

83. *Work-Rate Problem*   Working together, two people can complete a task in 10 hours. Working alone, one person takes 2 hours longer than the other. How long would it take each person to do the task alone?

84. *Height*   An object is projected vertically upward at an initial velocity of 64 feet per second from a height of 192 feet, so that the height h at any time t is given by $h = -16t^2 + 64t + 192$, where t is the time in seconds.

(a) After how many seconds is the height 256 feet?

(b) After how many seconds does the object hit the ground?

8.6 Quadratic and Rational Inequalities

1 ▶ Determine test intervals for polynomials.

In Exercises 85–88, find the critical numbers.

85. $2x(x + 7)$

86. $x(x - 2) + 4(x - 2)$

87. $x^2 - 6x - 27$

88. $2x^2 + 11x + 5$

2 ▶ Use test intervals to solve quadratic inequalities.

In Exercises 89–94, solve the inequality and graph the solution on the real number line.

89. $5x(7 - x) > 0$

90. $-2x(x - 10) \le 0$

91. $16 - (x - 2)^2 \le 0$

92. $(x - 5)^2 - 36 > 0$

93. $2x^2 + 3x - 20 < 0$

94. $-3x^2 + 10x + 8 \ge 0$

3 ▶ Use test intervals to solve rational inequalities.

In Exercises 95–98, solve the inequality and graph the solution on the real number line.

95. $\dfrac{x + 3}{2x - 7} \ge 0$

96. $\dfrac{3x + 2}{x - 3} > 0$

97. $\dfrac{x + 4}{x - 1} < 0$

98. $\dfrac{2x - 9}{x - 1} \le 0$

4 ▶ Use inequalities to solve application problems.

99. *Height*   A projectile is fired straight upward from ground level with an initial velocity of 312 feet per second, so that its height h at any time t is given by $h = -16t^2 + 312t$, where h is measured in feet and t is measured in seconds. During what interval of time will the height of the projectile exceed 1200 feet?

100. *Average Cost*   The cost C of producing x notebooks is $C = 100,000 + 0.9x$, $x > 0$. Write the average cost $\overline{C} = C/x$ as a function of x. Then determine how many notebooks must be produced if the average cost per unit is to be less than $2.

Chapter Test

Take this test as you would take a test in class. After you are done, check your work against the answers in the back of the book.

In Exercises 1–6, solve the equation by the specified method.

1. Factoring:

$x(x - 3) - 10(x - 3) = 0$

2. Factoring:

$6x^2 - 34x - 12 = 0$

3. Square Root Property:

$(x - 2)^2 = 0.09$

4. Square Root Property:

$(x + 4)^2 + 100 = 0$

5. Completing the square:

$2x^2 - 6x + 3 = 0$

6. Quadratic Formula:

$2y(y - 2) = 7$

In Exercises 7 and 8, solve the equation of quadratic form.

7. $\dfrac{1}{x^2} - \dfrac{6}{x} + 4 = 0$

8. $x^{2/3} - 9x^{1/3} + 8 = 0$

9. Find the discriminant and explain what it means in terms of the type of solutions of the quadratic equation $5x^2 - 12x + 10 = 0$.

10. Find a quadratic equation having the solutions -7 and -3.

In Exercises 11 and 12, sketch the parabola. Identify the vertex and any x-intercepts. Use a graphing calculator to verify your results.

11. $y = -x^2 + 2x - 4$

12. $y = x^2 - 2x - 15$

In Exercises 13–15, solve the inequality and sketch its solution.

13. $16 \le (x - 2)^2$

14. $2x(x - 3) < 0$

15. $\dfrac{x + 1}{x - 5} \le 0$

16. The width of a rectangle is 22 feet less than its length. The area of the rectangle is 240 square feet. Find the dimensions of the rectangle.

17. An English club chartered a bus trip to a Shakespearean festival. The cost of the bus was \$1250. To lower the bus fare per person, the club invited nonmembers to go along. When 10 nonmembers joined the trip, the fare per person decreased by \$6.25. How many club members are going on the trip?

18. An object is dropped from a height of 75 feet. Its height h (in feet) at any time t is given by $h = -16t^2 + 75$, where t is measured in seconds. Find the time required for the object to fall to a height of 35 feet.

19. Two buildings are connected by an L-shaped protected walkway. The total length of the walkway is 155 feet. By cutting diagonally across the grass, the walking distance is shortened to 125 feet. What are the lengths of the two legs of the walkway?

Study Skills in Action

Making the Most of Class Time

Have you ever slumped at your desk while in class and thought "I'll just get the notes down and study later—I'm too tired"? Learning math in college is a team effort, between instructor and student. The more you understand in class, the more you will be able to learn while studying outside of class.

Approach math class with the intensity of a navy pilot during a mission briefing. The pilot has strategic plans to learn during the briefing. He or she listens intensely, takes notes, and memorizes important information. The goal is for the pilot to leave the briefing with a clear picture of the mission. It is the same with a student in a math class.

Kimberly Nolting

VP, Academic Success Press
expert in developmental education

These students are sitting in the front row, where they are more likely to pay attention.

Smart Study Strategy

Take Control of Your Class Time

1 ▶ **Sit where you can easily see and hear the instructor, and the instructor can see you.** The instructor may be able to tell when you are confused just by the look on your face, and may adjust the lesson accordingly. In addition, sitting in this strategic place will keep your mind from wandering.

2 ▶ **Pay attention to what the instructor says about the math, not just what is written on the board.** Write problems on the left side of your notes and what the instructor says about the problems on the right side.

3 ▶ **If the instructor is moving through the material too fast, ask a question.** Questions help to slow the pace for a few minutes and also to clarify what is confusing to you.

4 ▶ **Try to memorize new information while learning it.** Repeat in your head what you are writing in your notes. That way you are reviewing the information twice.

5 ▶ **Ask for clarification.** If you don't understand something at all and do not even know how to phrase a question, just ask for clarification. You might say something like, "Could you please explain the steps in this problem one more time?"

6 ▶ **Think as intensely as if you were going to take a quiz on the material at the end of class.** This kind of mindset will help you to process new information.

7 ▶ **If the instructor asks for someone to go up to the board, volunteer.** The student at the board often receives additional attention and instruction to complete the problem.

8 ▶ **At the end of class, identify concepts or problems on which you still need clarification.** Make sure you see the instructor or a tutor as soon as possible.

Chapter 9
Exponential and Logarithmic Functions

IT WORKED FOR ME!

"I failed my first college math class because, for some reason, I thought just showing up and listening would be enough. I was wrong. Now, I get to class a little early to review my notes, sit where I can see the instructor, and ask questions. I try to learn and remember as much as possible in class because I am so busy juggling work and college."

Aaron
Business

9.1 Exponential Functions

© W. Wayne Lockwood, M.D./CORBIS

What You Should Learn

1 ▶ Evaluate exponential functions.

2 ▶ Graph exponential functions.

3 ▶ Evaluate the natural base e and graph natural exponential functions.

4 ▶ Use exponential functions to solve application problems.

Why You Should Learn It

Exponential functions can be used to model and solve real-life problems. For instance, in Exercise 89 on page 589, you will use an exponential function to model the descent of a parachutist.

1 ▶ Evaluate exponential functions.

Exponential Functions

In this section, you will study a new type of function called an **exponential function.** Whereas polynomial and rational functions have terms with variable bases and constant exponents, exponential functions have terms with *constant bases* and *variable exponents.* Here are some examples.

Polynomial or Rational Function

Constant Exponents

$$f(x) = x^2, \quad f(x) = x^{-3}$$

Variable Bases

Exponential Function

Variable Exponents

$$f(x) = 2^x, \quad f(x) = 3^{-x}$$

Constant Bases

Definition of Exponential Function

The **exponential function** f **with base** a is denoted by

$$f(x) = a^x$$

where $a > 0$, $a \neq 1$, and x is any real number.

The base $a = 1$ is excluded because $f(x) = 1^x = 1$ is a constant function, *not* an exponential function.

In Chapters 5 and 7, you learned to evaluate a^x for integer and rational values of x. For example, you know that

$$a^3 = a \cdot a \cdot a, \quad a^{-4} = \frac{1}{a^4}, \quad \text{and} \quad a^{5/3} = \left(\sqrt[3]{a}\right)^5.$$

However, to evaluate a^x for any real number x, you need to interpret forms with *irrational* exponents, such as $a^{\sqrt{2}}$ or a^{π}. For the purposes of this text, it is sufficient to think of a number such as $a^{\sqrt{2}}$, where $\sqrt{2} \approx 1.414214$, as the number that has the successively closer approximations

$$a^{1.4}, a^{1.41}, a^{1.414}, a^{1.4142}, a^{1.41421}, a^{1.414214}, \ldots .$$

The rules of exponents that were discussed in Section 5.1 can be extended to cover exponential functions, as described on the following page.

Study Tip

Rule 4 of the rules of exponential functions indicates that 2^{-x} can be written as

$$2^{-x} = \frac{1}{2^x}.$$

Similarly, $\frac{1}{3^{-x}}$ can be written as

$$\frac{1}{3^{-x}} = 3^x.$$

In other words, you can move a *factor* from the numerator to the denominator (or from the denominator to the numerator) by changing the sign of its exponent.

Rules of Exponential Functions

Let a be a positive real number, and let x and y be real numbers, variables, or algebraic expressions.

1. $a^x \cdot a^y = a^{x+y}$ Product rule

2. $\dfrac{a^x}{a^y} = a^{x-y}$ Quotient rule

3. $(a^x)^y = a^{xy}$ Power-to-power rule

4. $a^{-x} = \dfrac{1}{a^x} = \left(\dfrac{1}{a}\right)^x$ Negative exponent rule

To evaluate exponential functions with a calculator, you can use the exponential key $\boxed{y^x}$ or $\boxed{\wedge}$. For example, to evaluate $3^{-1.3}$, you can use the following keystrokes.

Keystrokes	Display	
3 $\boxed{y^x}$ 1.3 $\boxed{+/-}$ $\boxed{=}$	0.239741	*Scientific*
3 $\boxed{\wedge}$ $\boxed{(}$ $\boxed{(-)}$ 1.3 $\boxed{)}$ $\boxed{ENTER}$	0.239741	*Graphing*

EXAMPLE 1   **Evaluating Exponential Functions**

Evaluate each function. Use a calculator only if it is necessary or more efficient.

Function	Values
a. $f(x) = 2^x$	$x = 3, x = -4, x = \pi$
b. $g(x) = 12^x$	$x = 3, x = -0.1, x = \frac{5}{7}$
c. $h(x) = (1.04)^{2x}$	$x = 0, x = -2, x = \sqrt{2}$

Solution

Evaluation	Comment
a. $f(3) = 2^3 = 8$	Calculator is not necessary.
$f(-4) = 2^{-4} = \dfrac{1}{2^4} = \dfrac{1}{16}$	Calculator is not necessary.
$f(\pi) = 2^\pi \approx 8.825$	Calculator is necessary.
b. $g(3) = 12^3 = 1728$	Calculator is more efficient.
$g(-0.1) = 12^{-0.1} \approx 0.780$	Calculator is necessary.
$g\left(\dfrac{5}{7}\right) = 12^{5/7} \approx 5.900$	Calculator is necessary.
c. $h(0) = (1.04)^{2 \cdot 0} = (1.04)^0 = 1$	Calculator is not necessary.
$h(-2) = (1.04)^{2(-2)} \approx 0.855$	Calculator is more efficient.
$h\left(\sqrt{2}\right) = (1.04)^{2\sqrt{2}} \approx 1.117$	Calculator is necessary.

✓ **CHECKPOINT** *Now try Exercise 17.*

2 ▸ Graph exponential functions.

Graphs of Exponential Functions

The basic nature of the graph of an exponential function can be determined by the point-plotting method or by using a graphing calculator.

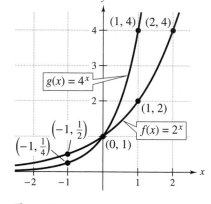

Figure 9.1

EXAMPLE 2   **The Graphs of Exponential Functions**

In the same coordinate plane, sketch the graph of each function. Determine the domain and range.

a. $f(x) = 2^x$

b. $g(x) = 4^x$

Solution

The table lists some values of each function, and Figure 9.1 shows the graph of each function. From the graphs, you can see that the domain of each function is the set of all real numbers and that the range of each function is the set of all positive real numbers.

x	-2	-1	0	1	2	3
2^x	$\frac{1}{4}$	$\frac{1}{2}$	1	2	4	8
4^x	$\frac{1}{16}$	$\frac{1}{4}$	1	4	16	64

✓ **CHECKPOINT** *Now try Exercise 31.*

You know from your study of the graphs of functions in Section 3.7 that the graph of $h(x) = f(-x) = 2^{-x}$ is a reflection of the graph of $f(x) = 2^x$ in the y-axis. This is reinforced in the next example.

EXAMPLE 3   **The Graphs of Exponential Functions**

In the same coordinate plane, sketch the graph of each function.

a. $f(x) = 2^{-x}$

b. $g(x) = 4^{-x}$

Solution

The table lists some values of each function, and Figure 9.2 shows the graph of each function.

x	-3	-2	-1	0	1	2
2^{-x}	8	4	2	1	$\frac{1}{2}$	$\frac{1}{4}$
4^{-x}	64	16	4	1	$\frac{1}{4}$	$\frac{1}{16}$

✓ **CHECKPOINT** *Now try Exercise 33.*

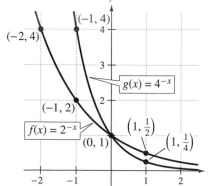

Figure 9.2

Examples 2 and 3 suggest that for $a > 1$, the values of the function $y = a^x$ increase as x increases and the values of the function $y = a^{-x} = (1/a)^x$ decrease as x increases. The graphs shown in Figure 9.3 are typical of the graphs of exponential functions. Note that each graph has a y-intercept at $(0, 1)$ and a **horizontal asymptote** of $y = 0$ (the x-axis).

Graph of $y = a^x$

- Domain: $(-\infty, \infty)$
- Range: $(0, \infty)$
- Intercept: $(0, 1)$
- Increasing
  (moves up to the right)
- Asymptote: x-axis

Graph of $y = a^{-x} = \left(\dfrac{1}{a}\right)^x$

- Domain: $(-\infty, \infty)$
- Range: $(0, \infty)$
- Intercept: $(0, 1)$
- Decreasing
  (moves down to the right)
- Asymptote: x-axis

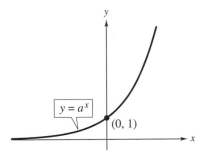
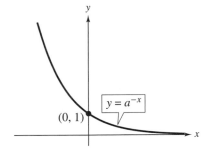

Figure 9.3   Characteristics of the exponential functions $y = a^x$ and $y = a^{-x}$ ($a > 1$)

In the next two examples, notice how the graph of $y = a^x$ can be used to sketch the graphs of functions of the form $f(x) = b \pm a^{x+c}$. Also note that the transformation in Example 4(a) keeps the x-axis as a horizontal asymptote, but the transformation in Example 4(b) yields a new horizontal asymptote of $y = -2$. Also, be sure to note how the y-intercept is affected by each transformation.

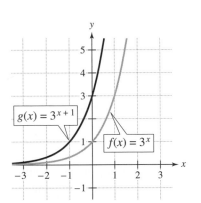

Figure 9.4

EXAMPLE 4   Transformations of Graphs of Exponential Functions

Use transformations to analyze and sketch the graph of each function.

a. $g(x) = 3^{x+1}$ **b.** $h(x) = 3^x - 2$

Solution

Consider the function $f(x) = 3^x$.

a. The function g is related to f by $g(x) = f(x + 1)$. To sketch the graph of g, shift the graph of f one unit to the left, as shown in Figure 9.4. Note that the y-intercept of g is $(0, 3)$.

b. The function h is related to f by $h(x) = f(x) - 2$. To sketch the graph of g, shift the graph of f two units downward, as shown in Figure 9.5. Note that the y-intercept of h is $(0, -1)$ and the horizontal asymptote is $y = -2$.

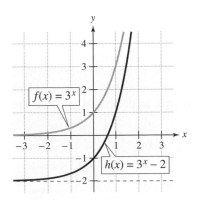

Figure 9.5

✓ **CHECKPOINT** *Now try Exercise 63.*

EXAMPLE 5   **Reflections of Graphs of Exponential Functions**

Use transformations to analyze and sketch the graph of each function.

a. $g(x) = -3^x$ **b.** $h(x) = 3^{-x}$

Solution

Consider the function $f(x) = 3^x$.

a. The function g is related to f by $g(x) = -f(x)$. To sketch the graph of g, reflect the graph of f about the x-axis, as shown in Figure 9.6. Note that the y-intercept of g is $(0, -1)$.

b. The function h is related to f by $h(x) = f(-x)$. To sketch the graph of h, reflect the graph of f about the y-axis, as shown in Figure 9.7.

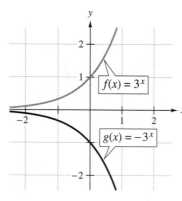

Figure 9.6

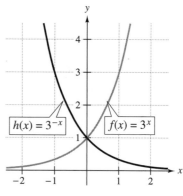

Figure 9.7

✓ **CHECKPOINT** *Now try Exercise 67.*

3 ▶ Evaluate the natural base e and graph natural exponential functions.

The Natural Exponential Function

So far, integers or rational numbers have been used as bases of exponential functions. In many applications of exponential functions, the convenient choice for a base is the following irrational number, denoted by the letter "e."

$$e \approx 2.71828 \ldots \qquad \text{Natural base}$$

This number is called the **natural base.** The function

$$f(x) = e^x \qquad \text{Natural exponential function}$$

is called the **natural exponential function.** To evaluate the natural exponential function, you need a calculator, preferably one having a natural exponential key $\boxed{e^x}$. Here are some examples of how to use such a calculator to evaluate the natural exponential function.

Value	Keystrokes	Display	
e^2	2 $\boxed{e^x}$	7.3890561	Scientific
e^2	$\boxed{e^x}$ 2 $\boxed{)}$ $\boxed{ENTER}$	7.3890561	Graphing
e^{-3}	3 $\boxed{+/-}$ $\boxed{e^x}$	0.0497871	Scientific
e^{-3}	$\boxed{e^x}$ $\boxed{(-)}$ 3 $\boxed{)}$ $\boxed{ENTER}$	0.0497871	Graphing

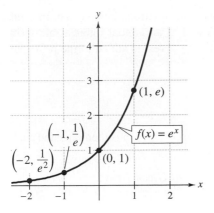

Figure 9.8

When evaluating the natural exponential function, remember that e is the constant number 2.71828 . . . ., and x is a variable. After evaluating this function at several values, as shown in the table, you can sketch its graph, as shown in Figure 9.8.

x	-2	-1.5	-1.0	-0.5	0.0	0.5	1.0	1.5
$f(x) = e^x$	0.135	0.223	0.368	0.607	1.000	1.649	2.718	4.482

From the graph, notice the following characteristics of the natural exponential function.

- Domain: $(-\infty, \infty)$
- Range: $(0, \infty)$
- Intercept: $(0, 1)$
- Increasing (moves up to the right)
- Asymptote: x-axis

Notice that these characteristics are consistent with those listed for the exponential function $y = a^x$ on page 581.

4 ▶ Use exponential functions to solve application problems.

Applications

A common scientific application of exponential functions is **radioactive decay.**

EXAMPLE 6   **Radioactive Decay**

A particular radioactive element has a half-life of 25 years. For an initial mass of 10 grams, the mass y (in grams) that remains after t years is given by

$$y = 10\left(\frac{1}{2}\right)^{t/25}, \quad t \geq 0.$$

How much of the initial mass remains after 120 years?

Solution

When $t = 120$, the mass is given by

$$y = 10\left(\frac{1}{2}\right)^{120/25} \qquad \text{Substitute 120 for } t.$$

$$= 10\left(\frac{1}{2}\right)^{4.8} \qquad \text{Simplify.}$$

$$\approx 0.359. \qquad \text{Use a calculator.}$$

So, after 120 years, the mass has decayed from an initial amount of 10 grams to only 0.359 gram. Note in Figure 9.9 that the graph of the function shows the 25-year half-life. That is, after 25 years the mass is 5 grams (half of the original), after another 25 years the mass is 2.5 grams, and so on.

✓ **CHECKPOINT** *Now try Exercise 69.*

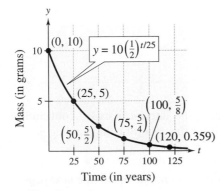

Figure 9.9

One of the most familiar uses of exponential functions involves **compound interest.** For instance, a principal P is invested at an annual interest rate r (in decimal form), compounded once a year. If the interest is added to the principal at the end of the year, the balance is

$$A = P + Pr$$

$$= P(1 + r).$$

This pattern of multiplying the previous principal by $(1 + r)$ is then repeated each successive year, as shown below.

Time in Years	*Balance at Given Time*
0	$A = P$
1	$A = P(1 + r)$
2	$A = P(1 + r)(1 + r) = P(1 + r)^2$
3	$A = P(1 + r)^2(1 + r) = P(1 + r)^3$
$\vdots$	$\vdots$
t	$A = P(1 + r)^t$

To account for more frequent compounding of interest (such as quarterly or monthly compounding), let n be the number of compoundings per year and let t be the number of years. Then the rate per compounding is r/n and the account balance after t years is

$$A = P\left(1 + \frac{r}{n}\right)^{nt}.$$

EXAMPLE 7   Finding the Balance for Compound Interest

A sum of \$10,000 is invested at an annual interest rate of 7.5%, compounded monthly. Find the balance in the account after 10 years.

Solution

Using the formula for compound interest, with $P = 10{,}000$, $r = 0.075$, $n = 12$ (for monthly compounding), and $t = 10$, you obtain the following balance.

$$A = 10{,}000\left(1 + \frac{0.075}{12}\right)^{12(10)}$$

$$\approx \$21{,}120.65$$

 CHECKPOINT *Now try Exercise 71.*

A second method that banks use to compute interest is called **continuous compounding.** The formula for the balance for this type of compounding is

$$A = Pe^{rt}.$$

The formulas for both types of compounding are summarized on the next page.

> ### Formulas for Compound Interest
>
> After t years, the balance A in an account with principal P and annual interest rate r (in decimal form) is given by one of the following formulas.
>
> **1.** For n compoundings per year: $A = P\left(1 + \dfrac{r}{n}\right)^{nt}$
>
> **2.** For continuous compounding: $A = Pe^{rt}$

Technology: Discovery

Use a graphing calculator to evaluate

$$A = 15{,}000\left(1 + \frac{0.08}{n}\right)^{n(6)}$$

for $n = 1000$, $10{,}000$, and $100{,}000$. Compare these values with those found in parts (a) and (b) of Example 8.

As n gets larger and larger, do you think that the value of A will ever exceed the value found in Example 8(c)? Explain.

EXAMPLE 8   **Comparing Three Types of Compounding**

A total of $15,000 is invested at an annual interest rate of 8%. Find the balance after 6 years for each type of compounding.

a. Quarterly

b. Monthly

c. Continuous

Solution

a. Letting $P = 15{,}000$, $r = 0.08$, $n = 4$, and $t = 6$, the balance after 6 years at quarterly compounding is

$$A = 15{,}000\left(1 + \frac{0.08}{4}\right)^{4(6)}$$

$$\approx \$24{,}126.56.$$

b. Letting $P = 15{,}000$, $r = 0.08$, $n = 12$, and $t = 6$, the balance after 6 years at monthly compounding is

$$A = 15{,}000\left(1 + \frac{0.08}{12}\right)^{12(6)}$$

$$\approx \$24{,}202.53.$$

c. Letting $P = 15{,}000$, $r = 0.08$, and $t = 6$, the balance after 6 years at continuous compounding is

$$A = 15{,}000e^{0.08(6)}$$

$$\approx \$24{,}241.12.$$

Note that the balance is greater with continuous compounding than with quarterly or monthly compounding.

 CHECKPOINT *Now try Exercise 73.*

Example 8 illustrates the following general rule. For a given principal, interest rate, and time, the more often the interest is compounded per year, the greater the balance will be. Moreover, the balance obtained by continuous compounding is larger than the balance obtained by compounding n times per year.

Concept Check

1. For the exponential function f with base a, describe the operation you must perform to evaluate f at $x = 2$.

2. Consider the graphs of the functions $f(x) = 3^x$, $g(x) = 4^x$, and $h(x) = 4^{-x}$.

   (a) What point do the graphs of f, g, and h have in common?

   (b) Describe the asymptote for each graph.

   (c) State whether each graph increases or decreases as x increases.

   (d) Compare the graphs of f and g.

3. What special function is approximately equivalent to the function $f(x) = 2.72^x$?

4. The natural base e is used in the formula for which type of interest compounding?

9.1 EXERCISES

Go to pages 646–647 to record your assignments.

Developing Skills

In Exercises 1–8, simplify the expression.

1. $3^x \cdot 3^{x+2}$

2. $e^{3x} \cdot e^{-x}$

3. $\dfrac{e^{x+2}}{e^x}$

4. $\dfrac{3^{2x+3}}{3^{x+1}}$

5. $3(e^x)^{-2}$

6. $4(e^{2x})^{-1}$

7. $\sqrt[3]{-8e^{3x}}$

8. $\sqrt{4e^{6x}}$

In Exercises 9–16, evaluate the expression. (Round your answer to three decimal places.)

9. $5^{\sqrt{2}}$

10. $4^{-\pi}$

11. $e^{1/3}$

12. $e^{-1/3}$

13. $3(2e^{1/2})^3$

14. $(9e^2)^{3/2}$

15. $\dfrac{4e^3}{12e^2}$

16. $\dfrac{6e^5}{10e^7}$

In Exercises 17–30, evaluate the function as indicated. Use a calculator only if it is necessary or more efficient. (Round your answers to three decimal places.) *See Example 1.*

✓ 17. $f(x) = 3^x$

   (a) $x = -2$

   (b) $x = 0$

   (c) $x = 1$

18. $F(x) = 3^{-x}$

   (a) $x = -2$

   (b) $x = 0$

   (c) $x = 1$

19. $g(x) = 2.2^{-x}$

   (a) $x = 1$

   (b) $x = 3$

   (c) $x = \sqrt{6}$

20. $G(x) = 4.2^x$

   (a) $x = -1$

   (b) $x = -2$

   (c) $x = \sqrt{2}$

21. $f(t) = 500\left(\frac{1}{2}\right)^t$

   (a) $t = 0$

   (b) $t = 1$

   (c) $t = \pi$

22. $g(s) = 1200\left(\frac{2}{3}\right)^s$

   (a) $s = 0$

   (b) $s = 2$

   (c) $s = \sqrt{2}$

23. $f(x) = 1000(1.05)^{2x}$

   (a) $x = 0$

   (b) $x = 5$

   (c) $x = 10$

24. $g(t) = 10{,}000(1.03)^{4t}$

   (a) $t = 1$

   (b) $t = 3$

   (c) $t = 5.5$

25. $h(x) = \dfrac{5000}{(1.06)^{8x}}$

   (a) $x = 5$

   (b) $x = 10$

   (c) $x = 20$

26. $P(t) = \dfrac{10{,}000}{(1.01)^{12t}}$

   (a) $t = 2$

   (b) $t = 10$

   (c) $t = 20$

27. $g(x) = 10e^{-0.5x}$

   (a) $x = -4$

   (b) $x = 4$

   (c) $x = 8$

28. $A(t) = 200e^{0.1t}$

   (a) $t = 10$

   (b) $t = 20$

   (c) $t = 40$

29. $g(x) = \dfrac{1000}{2 + e^{-0.12x}}$

   (a) $x = 0$

   (b) $x = 10$

   (c) $x = 50$

30. $f(z) = \dfrac{100}{1 + e^{-0.05z}}$

   (a) $z = 0$

   (b) $z = 10$

   (c) $z = 20$

In Exercises 31–46, sketch the graph of the function. Identify the horizontal asymptote. *See Examples 2 and 3.*

✓ 31. $f(x) = 3^x$

32. $h(x) = \frac{1}{2}(3^x)$

✓ **33.** $f(x) = 3^{-x} = \left(\frac{1}{3}\right)^x$

34. $h(x) = \frac{1}{2}(3^{-x})$

35. $g(x) = 3^x - 2$

36. $g(x) = 3^x + 1$

37. $g(x) = 5^{x-1}$

38. $g(x) = 5^{x+3}$

39. $f(t) = 2^{-t^2}$

40. $f(t) = 2^{t^2}$

41. $f(x) = -2^{0.5x}$

42. $h(t) = -2^{-0.5t}$

43. $f(x) = -\left(\frac{1}{3}\right)^x$

44. $f(x) = \left(\frac{3}{4}\right)^x + 1$

45. $g(t) = 200\left(\frac{1}{2}\right)^t$

46. $h(x) = 27\left(\frac{2}{3}\right)^x$

In Exercises 47–58, use a graphing calculator to graph the function.

47. $y = 7^{x/2}$

48. $y = 7^{-x/2}$

49. $y = 7^{-x/2} + 5$

50. $y = 7^{(x-3)/2}$

51. $y = 500(1.06)^t$

52. $y = 100(1.06)^{-t}$

53. $y = 3e^{0.2x}$

54. $y = 50e^{-0.05x}$

55. $P(t) = 100e^{-0.1t}$

56. $N(t) = 10,000e^{0.05t}$

57. $y = 6e^{-x^2/3}$

58. $g(x) = 7e^{(x+1)/2}$

In Exercises 59–62, match the function with its graph. [The graphs are labeled (a), (b), (c), and (d).]

(a)

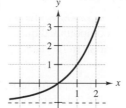

(b)

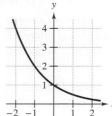

(c)

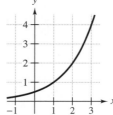

(d)

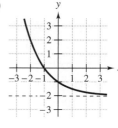

59. $f(x) = 2^{-x}$

60. $f(x) = 2^x - 1$

61. $f(x) = 2^{x-1}$

62. $f(x) = \left(\frac{1}{2}\right)^x - 2$

In Exercises 63–68, identify the transformation of the graph of $f(x) = 4^x$, and sketch the graph of h. **See Examples 4 and 5.**

✓ **63.** $h(x) = 4^x - 1$

64. $h(x) = 4^x + 2$

65. $h(x) = 4^{x+2}$

66. $h(x) = 4^{x-4}$

✓ **67.** $h(x) = -4^x$

68. $h(x) = 4^{-x}$

Solving Problems

✓ **69.** *Radioactive Decay*   After t years, 16 grams of a radioactive element with a half-life of 30 years decays to a mass y (in grams) given by

$$y = 16\left(\frac{1}{2}\right)^{t/30}, \quad t \geq 0.$$

How much of the initial mass remains after 80 years?

70.

Radioactive Substance In July of 1999, an individual bought several leaded containers from a metals recycler and found two of them labeled "radioactive." An investigation showed that the containers, originally obtained from Ohio State University, apparently had been used to store iodine-131 starting in January of 1999. Because iodine-131 has a half-life of only 8 days, no elevated radiation levels were detected. (Source: United States Nuclear Regulatory Commission)

Suppose 6 grams of iodine-131 is stored in January. The mass y (in grams) that remains after t days is given by $y = 6\left(\frac{1}{2}\right)^{t/8}$, $t \geq 0$. How much of the substance is left in July, after 180 days have passed?

✓ **71.** *Compound Interest*   A sum of $5000 is invested at an annual interest rate of 6%, compounded monthly. Find the balance in the account after 5 years.

72. *Compound Interest*   A sum of $2000 is invested at an annual interest rate of 8%, compounded quarterly. Find the balance in the account after 10 years.

Compound Interest   In Exercises 73–76, complete the table to determine the balance A for P dollars invested at rate r for t years, compounded n times per year.

n	1	4	12	365	Continuous compounding
A					

	Principal	Rate	Time
✓ **73.**	$P = \$100$	$r = 7\%$	$t = 15$ years
74.	$P = \$600$	$r = 4\%$	$t = 5$ years
75.	$P = \$2000$	$r = 9.5\%$	$t = 10$ years
76.	$P = \$1500$	$r = 6.5\%$	$t = 20$ years

Compound Interest   In Exercises 77–80, complete the table to determine the principal P that will yield a balance of A dollars when invested at rate r for t years, compounded n times per year.

n	1	4	12	365	Continuous compounding
P					

	Balance	Rate	Time
77.	$A = \$5000$	$r = 7\%$	$t = 10$ years
78.	$A = \$100{,}000$	$r = 9\%$	$t = 20$ years
79.	$A = \$1{,}000{,}000$	$r = 10.5\%$	$t = 40$ years
80.	$A = \$10{,}000$	$r = 12.4\%$	$t = 3$ years

81. *Demand*   The daily demand x and the price p for a collectible are related by $p = 25 - 0.4e^{0.02x}$. Find the prices for demands of (a) $x = 100$ units and (b) $x = 125$ units.

82. *Population Growth*   The populations P (in millions) of the United States from 1980 to 2006 can be approximated by the exponential function $P(t) = 226(1.0110)^t$, where t is the time in years, with $t = 0$ corresponding to 1980. Use the model to estimate the populations in the years (a) 2010 and (b) 2020.   (Source: U.S. Census Bureau)

83. *Property Value*   The value of a piece of property doubles every 15 years. You buy the property for $64,000. Its value t years after the date of purchase should be $V(t) = 64{,}000(2)^{t/15}$. Use the model to approximate the values of the property (a) 5 years and (b) 20 years after it is purchased.

84. *Inflation Rate*   Suppose the annual rate of inflation is expected to be 4% for the next 5 years. Then the cost C of goods or services t years from now can be approximated by $C(t) = P(1.04)^t$, $0 \le t \le 5$, where P is the present cost. Estimate the yearly cost of tuition four years from now for a college where the tuition is currently $32,000 per year.

85. *Depreciation*   After t years, the value of a car that originally cost $16,000 depreciates so that each year it is worth $\frac{3}{4}$ of its value from the previous year. Find a model for $V(t)$, the value of the car after t years. Sketch a graph of the model, and determine the value of the car 2 years and 4 years after its purchase.

86. *Depreciation*   Suppose the value of the car in Exercise 85 changes by straight-line depreciation of $3000 per year. Find a model $v(t)$ for the value of the car. Sketch a graph of $v(t)$ and the graph of $V(t)$ from Exercise 85, on the same coordinate axes used in Exercise 85. Which model do you prefer if you sell the car after 2 years? after 4 years?

87. *Match Play*   A fraternity sponsors a match play golf tournament to benefit charity. The tournament is bracketed for 1024 golfers, and half of the golfers are eliminated in each round of one-on-one matches. So, 512 golfers remain after the first round, 256 after the second round, and so on, until there is a single champion. Write an exponential function that models this problem. How many golfers remain after the eighth round?

88. *Savings Plan*   You decide to start saving pennies according to the following pattern. You save 1 penny the first day, 2 pennies the second day, 4 the third day, 8 the fourth day, and so on. Each day you save twice the number of pennies you saved on the previous day. Write an exponential function that models this problem. How many pennies do you save on the thirtieth day?

89. ▦ *Parachute Drop*   A parachutist jumps from a plane and releases her parachute 2000 feet above the ground (see figure). From there, her height h (in feet) is given by $h = 1950 + 50e^{-0.4433t} - 22t$, where t is the number of seconds after the parachute is released.

2000 ft

Not drawn to scale

(a) Use a graphing calculator to graph the function.

(b) Use a graphing calculator to find the parachutist's height every 10 seconds from $t = 0$ until the time she reaches the ground. Record your results in a table.

(c) During which 10-second period does the parachutist's height change the most? Use the context of the problem to explain why.

90. ▦ *Parachute Drop*   A parachutist jumps from a plane and releases his parachute 3000 feet above the ground. From there, his height h (in feet) is given by $h = 2940 + 60e^{-0.4021t} - 24t$, where t is the number of seconds after the parachute is released.

(a) Use a graphing calculator to graph the function.

(b) Use a graphing calculator to find the parachutist's height every 10 seconds from $t = 0$ until the time he reaches the ground. Record your results in a table.

(c) During which 10-second period does the parachutist's height change the most? Use the context of the problem to explain why.

91. ▦ *Data Analysis*   A meteorologist measures the atmospheric pressure P (in kilograms per square meter) at various altitudes h (in kilometers). The data are shown in the table.

h	0	5	10	15	20
P	10,332	5583	2376	1240	517

(a) Use a graphing calculator to plot the data points.

(b) A model for the data is given by $P = 10,958e^{-0.15h}$. Use a graphing calculator to graph the model with the data points from part (a), in the same viewing window. How well does the model fit the data?

(c) Use a graphing calculator to create a table comparing the model with the data points.

(d) Estimate the atmospheric pressure at an altitude of 8 kilometers.

(e) Use the graph to estimate the altitude at which the atmospheric pressure is 2000 kilograms per square meter.

92. *Data Analysis*   The median prices of existing one-family homes sold in the United States in the years 1999 through 2006 are shown in the table. (Source: National Association of Realtors)

Year	1999	2000	2001	2002
Price	$141,200	$147,300	$156,600	$167,600

Year	2003	2004	2005	2006
Price	$180,200	$195,200	$219,000	$221,900

A model for this data is given by $y = 73,482e^{0.0700t}$, where t is time in years, with $t = 9$ representing 1999.

(a) Use the model to complete the table below and compare the results with the actual data.

Year	1999	2000	2001	2002
Price				

Year	2003	2004	2005	2006
Price				

(b) ▦ Use a graphing calculator to graph the model with the data points from part (a), in the same viewing window.

(c) Beyond $t = 16$, do the y-values of the model increase at a *higher rate*, a *constant rate*, or a *lower rate* for increasing values of t? Do you think that the model is reliable for making predictions past 2006? Explain.

93. *Calculator Experiment*

(a) Use a calculator to complete the table.

x	1	10	100	1000	10,000
$\left(1 + \dfrac{1}{x}\right)^x$					

(b) Use the table to sketch the graph of the function

$$f(x) = \left(1 + \frac{1}{x}\right)^x.$$

Does this graph appear to be approaching a horizontal asymptote?

(c) From parts (a) and (b), what conclusions can you make about the value of

$$\left(1 + \frac{1}{x}\right)^x$$

as x gets larger and larger?

94. Identify the graphs of $y_1 = e^{0.2x}$, $y_2 = e^{0.5x}$, and $y_3 = e^x$ in the figure. Describe the effect on the graph of $y = e^{kx}$ when $k > 0$ is changed.

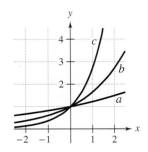

Explaining Concepts

95. ✎ Explain why 1^x is not an exponential function.

96. ✎ Compare the graphs of $f(x) = 3^x$ and $g(x) = \left(\frac{1}{3}\right)^x$.

97. ✎ Does e equal $\dfrac{271{,}801}{99{,}990}$? Explain.

98. ✎ Use the characteristics of the exponential function with base 2 to explain why $2^{\sqrt{2}}$ is greater than 2 but less than 4.

99. ✎ Consider functions of the form $f(x) = k^x$. Describe the real values of k for which the values of f will *increase*, *decrease*, and *remain constant* as x increases.

100. Look back at your answers to the compound interest problems in Exercises 73–76. In terms of the interest earned, would you say there is much difference between daily compounding and continuous compounding? Explain your reasoning.

Cumulative Review

In Exercises 101 and 102, find the domain of the function.

101. $g(s) = \sqrt{s - 4}$

102. $h(t) = \dfrac{\sqrt{t^2 - 1}}{t - 2}$

In Exercises 103 and 104, sketch a graph of the equation. Use the Vertical Line Test to determine whether y is a function of x.

103. $y^2 = x - 1$

104. $x = y^4 + 1$

9.2 Composite and Inverse Functions

Superstock, Inc.

What You Should Learn

1 ▶ Form compositions of two functions and find the domains of composite functions.

2 ▶ Use the Horizontal Line Test to determine whether functions have inverse functions.

3 ▶ Find inverse functions algebraically.

4 ▶ Graphically verify that two functions are inverse functions of each other.

Why You Should Learn It

Inverse functions can be used to model and solve real-life problems. For instance, in Exercise 109 on page 603, you will use an inverse function to determine the number of units produced for a certain hourly wage.

1 ▶ Form compositions of two functions and find the domains of composite functions.

Composite Functions

Two functions can be combined to form another function called the **composition** of the two functions. For instance, if $f(x) = 2x^2$ and $g(x) = x - 1$, the composition of f with g is denoted by $f \circ g$ and is given by

$$f(g(x)) = f(x - 1) = 2(x - 1)^2.$$

Definition of Composition of Two Functions

The **composition** of the functions f and g is given by

$$(f \circ g)(x) = f(g(x)).$$

The domain of the **composite function** $(f \circ g)$ is the set of all x in the domain of g such that $g(x)$ is in the domain of f. (See Figure 9.10.)

$f \circ g$

x　$g(x)$　$f(g(x))$

Domain of g　　Domain of f

Figure 9.10

Study Tip

A composite function can be viewed as a function within a function, where the composition

$$(f \circ g)(x) = f(g(x))$$

has f as the "outer" function and g as the "inner" function. This is reversed in the composition

$$(g \circ f)(x) = g(f(x)).$$

EXAMPLE 1　Forming the Composition of Two Functions

Given $f(x) = 2x + 4$ and $g(x) = 3x - 1$, find the composition of f with g. Then evaluate the composite function when $x = 1$ and when $x = -3$.

Solution

$$
\begin{aligned}
(f \circ g)(x) &= f(g(x)) & &\text{Definition of } f \circ g \\
&= f(3x - 1) & &g(x) = 3x - 1 \text{ is the inner function.} \\
&= 2(3x - 1) + 4 & &\text{Input } 3x - 1 \text{ into the outer function } f. \\
&= 6x - 2 + 4 & &\text{Distributive Property} \\
&= 6x + 2 & &\text{Simplify.}
\end{aligned}
$$

When $x = 1$, the value of this composite function is

$$(f \circ g)(1) = 6(1) + 2 = 8.$$

When $x = -3$, the value of this composite function is

$$(f \circ g)(-3) = 6(-3) + 2 = -16.$$

✓ **CHECKPOINT** *Now try Exercise 1.*

The composition of f with g is generally *not* the same as the composition of g with f. This is illustrated in Example 2.

EXAMPLE 2 **Comparing the Compositions of Functions**

Given $f(x) = 2x - 3$ and $g(x) = x^2 + 1$, find each composition.

a. $(f \circ g)(x)$ **b.** $(g \circ f)(x)$

Solution

a. $(f \circ g)(x) = f(g(x))$ Definition of $f \circ g$

$\qquad\qquad = f(x^2 + 1)$ $g(x) = x^2 + 1$ is the inner function.

$\qquad\qquad = 2(x^2 + 1) - 3$ Input $x^2 + 1$ into the outer function f.

$\qquad\qquad = 2x^2 + 2 - 3$ Distributive Property

$\qquad\qquad = 2x^2 - 1$ Simplify.

b. $(g \circ f)(x) = g(f(x))$ Definition of $g \circ f$

$\qquad\qquad = g(2x - 3)$ $f(x) = 2x - 3$ is the inner function.

$\qquad\qquad = (2x - 3)^2 + 1$ Input $2x - 3$ into the outer function g.

$\qquad\qquad = 4x^2 - 12x + 9 + 1$ Expand.

$\qquad\qquad = 4x^2 - 12x + 10$ Simplify.

Note that $(f \circ g)(x) \neq (g \circ f)(x)$.

✓ **CHECKPOINT** *Now try Exercise 3.*

To determine the domain of a composite function, first write the composite function in simplest form. Then use the fact that its domain *either is equal to or is a restriction of the domain of the "inner" function.* This is demonstrated in Example 3.

EXAMPLE 3 **Finding the Domain of a Composite Function**

Find the domain of the composition of f with g when $f(x) = x^2$ and $g(x) = \sqrt{x}$.

Solution

$(f \circ g)(x) = f(g(x))$ Definition of $f \circ g$

$\qquad\qquad = f(\sqrt{x})$ $g(x) = \sqrt{x}$ is the inner function.

$\qquad\qquad = (\sqrt{x})^2$ Input $\sqrt{x}$ into the outer function f.

$\qquad\qquad = x, \; x \geq 0$ Domain of $f \circ g$ is all $x \geq 0$.

The domain of the inner function $g(x) = \sqrt{x}$ is the set of all nonnegative real numbers. The simplified form of $f \circ g$ has no restriction on this set of numbers. So, the restriction $x \geq 0$ must be added to the composition of this function. The domain of $f \circ g$ is the set of all nonnegative real numbers.

✓ **CHECKPOINT** *Now try Exercise 21.*

2 ▶ Use the Horizontal Line Test to determine whether functions have inverse functions.

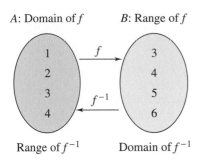

A: Domain of *f*   *B*: Range of *f*

Range of f^{-1} Domain of f^{-1}

Figure 9.11   The function *f* is one-to-one and has inverse function f^{-1}.

One-to-One and Inverse Functions

In Section 3.6, you learned that a function can be represented by a set of ordered pairs. For instance, the function $f(x) = x + 2$ from the set $A = \{1, 2, 3, 4\}$ to the set $B = \{3, 4, 5, 6\}$ can be written as follows.

$$f(x) = x + 2: \quad \{(1, 3), (2, 4), (3, 5), (4, 6)\}$$

By interchanging the first and second coordinates of each of these ordered pairs, you can form another function that is called the **inverse function** of *f*, denoted by f^{-1}. It is a function from the set *B* to the set *A*, and can be written as follows.

$$f^{-1}(x) = x - 2: \quad \{(3, 1), (4, 2), (5, 3), (6, 4)\}$$

Interchanging the ordered pairs for a function *f* will only produce another function when *f* is one-to-one. A function *f* is **one-to-one** if each value of the dependent variable corresponds to exactly one value of the independent variable. Figure 9.11 shows that the domain of *f* is the range of f^{-1} and the range of *f* is the domain of f^{-1}.

Horizontal Line Test for Inverse Functions

A function *f* has an inverse function f^{-1} if and only if *f* is one-to-one. Graphically, a function *f* has an inverse function f^{-1} if and only if no *horizontal* line intersects the graph of *f* at more than one point.

EXAMPLE 4   **Applying the Horizontal Line Test**

Use the Horizontal Line Test to determine if the function is one-to-one and so has an inverse function.

a. The graph of the function $f(x) = x^3 - 1$ is shown in Figure 9.12. Because no horizontal line intersects the graph of *f* at more than one point, you can conclude that *f is* a one-to-one function and *does* have an inverse function.

b. The graph of the function $f(x) = x^2 - 1$ is shown in Figure 9.13. Because it is possible to find a horizontal line that intersects the graph of *f* at more than one point, you can conclude that *f is not* a one-to-one function and *does not* have an inverse function.

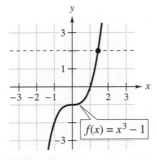

Figure 9.12

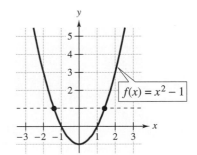

Figure 9.13

✓ **CHECKPOINT** *Now try Exercise 35.*

The formal definition of an inverse function is given as follows.

> ### Definition of Inverse Function
> Let f and g be two functions such that
> $$f(g(x)) = x \quad \text{for every } x \text{ in the domain of } g$$
> and
> $$g(f(x)) = x \quad \text{for every } x \text{ in the domain of } f.$$
> The function g is called the **inverse function** of the function f, and is denoted by f^{-1} (read "f-inverse"). So, $f(f^{-1}(x)) = x$ and $f^{-1}(f(x)) = x$. The domain of f must be equal to the range of f^{-1}, and vice versa.

Do not be confused by the use of -1 to denote the inverse function f^{-1}. Whenever f^{-1} is written, it *always* refers to the inverse function of f and *not* to the reciprocal of $f(x)$.

If the function g is the inverse function of the function f, it must also be true that the function f is the inverse function of the function g. For this reason, you can refer to the functions f and g as being *inverse functions of each other*.

EXAMPLE 5   Verifying Inverse Functions

Show that $f(x) = 2x - 4$ and $g(x) = \dfrac{x + 4}{2}$ are inverse functions of each other.

Solution

Begin by noting that the domain and range of both functions are the entire set of real numbers. To show that f and g are inverse functions of each other, you need to show that $f(g(x)) = x$ and $g(f(x)) = x$, as follows.

$$f(g(x)) = f\left(\frac{x + 4}{2}\right) \qquad g(x) = (x + 4)/2 \text{ is the inner function.}$$

$$= 2\left(\frac{x + 4}{2}\right) - 4 \qquad \text{Input } (x + 4)/2 \text{ into the outer function } f.$$

$$= x + 4 - 4 = x \qquad \text{Simplify.}$$

$$g(f(x)) = g(2x - 4) \qquad f(x) = 2x - 4 \text{ is the inner function.}$$

$$= \frac{(2x - 4) + 4}{2} \qquad \text{Input } 2x - 4 \text{ into the outer function } g.$$

$$= \frac{2x}{2} = x \qquad \text{Simplify.}$$

Note that the two functions f and g "undo" each other in the following verbal sense. The function f first *multiplies* the input x by 2 and then *subtracts* 4, whereas the function g first *adds* 4 and then *divides* the result by 2.

✓ **CHECKPOINT** *Now try Exercise 43.*

EXAMPLE 6   **Verifying Inverse Functions**

Show that the functions

$$f(x) = x^3 + 1 \text{ and } g(x) = \sqrt[3]{x - 1}$$

are inverse functions of each other.

Solution

Begin by noting that the domain and range of both functions are the entire set of real numbers. To show that f and g are inverse functions of each other, you need to show that $f(g(x)) = x$ and $g(f(x)) = x$, as follows.

$$f(g(x)) = f\left(\sqrt[3]{x - 1}\right) \qquad\qquad g(x) = \sqrt[3]{x - 1} \text{ is the inner function.}$$

$$= \left(\sqrt[3]{x - 1}\right)^3 + 1 \qquad\quad \text{Input } \sqrt[3]{x - 1} \text{ into the outer function } f.$$

$$= (x - 1) + 1 = x \qquad\quad \text{Simplify.}$$

$$g(f(x)) = g(x^3 + 1) \qquad\qquad f(x) = x^3 + 1 \text{ is the inner function.}$$

$$= \sqrt[3]{(x^3 + 1) - 1} \qquad\quad \text{Input } x^3 + 1 \text{ into the outer function } g.$$

$$= \sqrt[3]{x^3} = x \qquad\qquad \text{Simplify.}$$

Note that the two functions f and g "undo" each other in the following verbal sense. The function f first *cubes* the input x and then *adds* 1, whereas the function g first *subtracts* 1 and then *takes the cube root* of the result.

 CHECKPOINT *Now try Exercise 45.*

---

3 ▶ Find inverse functions algebraically.

Finding an Inverse Function Algebraically

You can find the inverse function of a simple function by inspection. For instance, the inverse function of $f(x) = 10x$ is $f^{-1}(x) = x/10$. For more complicated functions, however, it is best to use the following steps for finding an inverse function. The key step in these guidelines is switching the roles of x and y. This step corresponds to the fact that inverse functions have ordered pairs with the coordinates reversed.

Finding an Inverse Function Algebraically

1. In the equation for $f(x)$, replace $f(x)$ with y.

2. Interchange x and y.

3. If the new equation does not represent y as a function of x, the function f does not have an inverse function. If the new equation does represent y as a function of x, solve the new equation for y.

4. Replace y with $f^{-1}(x)$.

5. Verify that f and f^{-1} are inverse functions of each other by showing that $f(f^{-1}(x)) = x = f^{-1}(f(x))$.

EXAMPLE 7 **Finding an Inverse Function**

Determine whether each function has an inverse function. If it does, find its inverse function.

a. $f(x) = 2x + 3$ **b.** $f(x) = x^3 + 3$

Solution

a.
$f(x) = 2x + 3$ Write original function.

$y = 2x + 3$ Replace $f(x)$ with y.

$x = 2y + 3$ Interchange x and y.

$y = \dfrac{x - 3}{2}$ Solve for y.

$f^{-1}(x) = \dfrac{x - 3}{2}$ Replace y with $f^{-1}(x)$.

You can verify that $f(f^{-1}(x)) = x = f^{-1}(f(x))$, as follows.

$$f(f^{-1}(x)) = f\left(\dfrac{x-3}{2}\right) = 2\left(\dfrac{x-3}{2}\right) + 3 = (x-3) + 3 = x$$

$$f^{-1}(f(x)) = f^{-1}(2x + 3) = \dfrac{(2x+3) - 3}{2} = \dfrac{2x}{2} = x$$

b.
$f(x) = x^3 + 3$ Write original function.

$y = x^3 + 3$ Replace $f(x)$ with y.

$x = y^3 + 3$ Interchange x and y.

$\sqrt[3]{x - 3} = y$ Solve for y.

$f^{-1}(x) = \sqrt[3]{x - 3}$ Replace y with $f^{-1}(x)$.

You can verify that $f(f^{-1}(x)) = x = f^{-1}(f(x))$, as follows.

$$f(f^{-1}(x)) = f\left(\sqrt[3]{x-3}\right) = \left(\sqrt[3]{x-3}\right)^3 + 3 = (x-3) + 3 = x$$

$$f^{-1}(f(x)) = f^{-1}(x^3 + 3) = \sqrt[3]{(x^3+3) - 3} = \sqrt[3]{x^3} = x$$

 CHECKPOINT *Now try Exercise 55.*

Technology: Discovery

Use a graphing calculator to graph $f(x) = x^3 + 1$, $f^{-1}(x) = \sqrt[3]{x - 1}$, and $y = x$ in the same viewing window.

a. Relative to the line $y = x$, how do the graphs of f and f^{-1} compare?

b. For the graph of f, complete the table.

x	-1	0	1
f			

For the graph of f^{-1}, complete the table.

x	0	1	2
f^{-1}			

What can you conclude about the coordinates of the points on the graph of f compared with those on the graph of f^{-1}?

EXAMPLE 8 **A Function That Has No Inverse Function**

$f(x) = x^2$ Original equation

$y = x^2$ Replace $f(x)$ with y.

$x = y^2$ Interchange x and y.

Recall from Section 3.6 that the equation $x = y^2$ does not represent y as a function of x because you can find two different y-values that correspond to the same x-value. Because the equation does not represent y as a function of x, you can conclude that the original function f does not have an inverse function.

 **CHECKPOINT** *Now try Exercise 67.*

4 ▶ Graphically verify that two functions are inverse functions of each other.

Graphs of Inverse Functions

The graphs of f and f^{-1} are related to each other in the following way. If the point (a, b) lies on the graph of f, the point (b, a) must lie on the graph of f^{-1}, and vice versa. This means that the graph of f^{-1} is a reflection of the graph of f in the line $y = x$, as shown in Figure 9.14. This "reflective property" of the graphs of f and f^{-1} is illustrated in Examples 9 and 10.

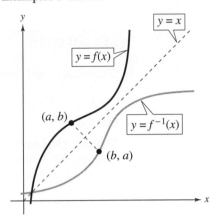

Figure 9.14   The graph of f^{-1} is a reflection of the graph of f in the line $y = x$.

EXAMPLE 9   **The Graphs of f and f^{-1}**

Sketch the graphs of the inverse functions $f(x) = 2x - 3$ and $f^{-1}(x) = \frac{1}{2}(x + 3)$ on the same rectangular coordinate system, and show that the graphs are reflections of each other in the line $y = x$.

Solution

The graphs of f and f^{-1} are shown in Figure 9.15. Visually, it appears that the graphs are reflections of each other. You can verify this reflective property by testing a few points on each graph. Note in the following list that if the point (a, b) is on the graph of f, the point (b, a) is on the graph of f^{-1}.

$f(x) = 2x - 3$	$f^{-1}(x) = \frac{1}{2}(x + 3)$
$(-1, -5)$	$(-5, -1)$
$(0, -3)$	$(-3, 0)$
$(1, -1)$	$(-1, 1)$
$(2, 1)$	$(1, 2)$
$(3, 3)$	$(3, 3)$

 CHECKPOINT *Now try Exercise 79.*

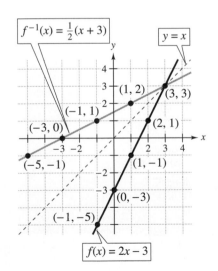

Figure 9.15

You can sketch the graph of an inverse function without knowing the equation of the inverse function. Simply find the coordinates of points that lie on the original function. By interchanging the x- and y-coordinates, you have points that lie on the graph of the inverse function. Plot these points and sketch the graph of the inverse function.

In Example 8, you saw that the function

$$f(x) = x^2$$

has no inverse function. A more complete way of saying this is "*assuming that the domain of f is the entire real line,* the function $f(x) = x^2$ has no inverse function." If, however, you *restrict* the domain of f to the nonnegative real numbers, then f does have an inverse function, as demonstrated in Example 10.

EXAMPLE 10   Verifying Inverse Functions Graphically

Graphically verify that f and g are inverse functions of each other.

$$f(x) = x^2, \quad x \geq 0 \quad \text{and} \quad g(x) = \sqrt{x}$$

Solution

You can graphically verify that f and g are inverse functions of each other by graphing the functions on the same rectangular coordinate system, as shown in Figure 9.16. Visually, it appears that the graphs are reflections of each other in the line $y = x$. You can verify this reflective property by testing a few points on each graph. Note in the following list that if the point (a, b) is on the graph of f, the point (b, a) is on the graph of g.

$f(x) = x^2, \quad x \geq 0$	$g(x) = f^{-1}(x) = \sqrt{x}$
$(0, 0)$	$(0, 0)$
$(1, 1)$	$(1, 1)$
$(2, 4)$	$(4, 2)$
$(3, 9)$	$(9, 3)$

Technology: Tip

A graphing calculator program for several models of graphing calculators that graphs the function f and its reflection in the line $y = x$ can be found at our website *www.cengage.com/math/larson/ algebra.*

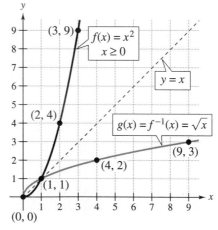

Figure 9.16

So, f and g are inverse functions of each other.

✓ **CHECKPOINT** *Now try Exercise 93.*

Concept Check

1. In general, is the composition of f with g equal to the composition of g with f? Give an example.

2. Explain the Horizontal Line Test. What is the relationship between this test and a function being one-to-one?

3. Describe how to find the inverse of a function given by an equation in x and y. Give an example.

4. Describe the relationship between the graphs of a function and its inverse function.

9.2 EXERCISES

Go to pages 646–647 to record your assignments.

Developing Skills

In Exercises 1–10, find the compositions. *See Examples 1 and 2.*

1. $f(x) = 2x + 3$, $g(x) = x - 6$
 (a) $(f \circ g)(x)$ (b) $(g \circ f)(x)$
 (c) $(f \circ g)(4)$ (d) $(g \circ f)(7)$

2. $f(x) = x - 5$, $g(x) = 3x + 2$
 (a) $(f \circ g)(x)$ (b) $(g \circ f)(x)$
 (c) $(f \circ g)(3)$ (d) $(g \circ f)(3)$

3. $f(x) = x^2 + 3$, $g(x) = x + 2$
 (a) $(f \circ g)(x)$ (b) $(g \circ f)(x)$
 (c) $(f \circ g)(2)$ (d) $(g \circ f)(-3)$

4. $f(x) = 2x + 1$, $g(x) = x^2 - 5$
 (a) $(f \circ g)(x)$ (b) $(g \circ f)(x)$
 (c) $(f \circ g)(-1)$ (d) $(g \circ f)(3)$

5. $f(x) = |x - 3|$, $g(x) = 3x$
 (a) $(f \circ g)(x)$ (b) $(g \circ f)(x)$
 (c) $(f \circ g)(1)$ (d) $(g \circ f)(2)$

6. $f(x) = |3x|$, $g(x) = x - 3$
 (a) $(f \circ g)(x)$ (b) $(g \circ f)(x)$
 (c) $(f \circ g)(-2)$ (d) $(g \circ f)(-4)$

7. $f(x) = \sqrt{x - 4}$, $g(x) = x + 5$
 (a) $(f \circ g)(x)$ (b) $(g \circ f)(x)$
 (c) $(f \circ g)(3)$ (d) $(g \circ f)(8)$

8. $f(x) = \sqrt{x + 6}$, $g(x) = 2x - 3$
 (a) $(f \circ g)(x)$ (b) $(g \circ f)(x)$
 (c) $(f \circ g)(3)$ (d) $(g \circ f)(-2)$

9. $f(x) = \dfrac{1}{x - 3}$, $g(x) = \dfrac{2}{x^2}$
 (a) $(f \circ g)(x)$ (b) $(g \circ f)(x)$
 (c) $(f \circ g)(-1)$ (d) $(g \circ f)(2)$

10. $f(x) = \dfrac{4}{x^2 - 4}$, $g(x) = \dfrac{1}{x}$
 (a) $(f \circ g)(x)$ (b) $(g \circ f)(x)$
 (c) $(f \circ g)(-2)$ (d) $(g \circ f)(1)$

In Exercises 11–14, use the functions f and g to find the indicated values.

$f = \{(-2, 3), (-1, 1), (0, 0), (1, -1), (2, -3)\}$,
$g = \{(-3, 1), (-1, -2), (0, 2), (2, 2), (3, 1)\}$

11. (a) $f(1)$ **12.** (a) $g(0)$
 (b) $g(-1)$ (b) $f(-1)$
 (c) $(g \circ f)(1)$ (c) $(f \circ g)(0)$

13. (a) $(f \circ g)(-3)$ **14.** (a) $(f \circ g)(2)$
 (b) $(g \circ f)(-2)$ (b) $(g \circ f)(2)$

In Exercises 15–18, use the functions f and g to find the indicated values.

$f = \{(0, 1), (1, 2), (2, 5), (3, 10), (4, 17)\}$,
$g = \{(5, 4), (10, 1), (2, 3), (17, 0), (1, 2)\}$

15. (a) $f(2)$ **16.** (a) $g(2)$
 (b) $g(10)$ (b) $f(0)$
 (c) $(g \circ f)(1)$ (c) $(f \circ g)(10)$

17. (a) $(g \circ f)(4)$ **18.** (a) $(f \circ g)(1)$
 (b) $(f \circ g)(2)$ (b) $(g \circ f)(0)$

In Exercises 19–26, find the compositions (a) $f \circ g$ and (b) $g \circ f$. Then find the domain of each composition. *See Example 3.*

19. $f(x) = 3x + 4$
$g(x) = x - 7$

20. $f(x) = x + 5$
$g(x) = 4x - 1$

21. $f(x) = \sqrt{x + 2}$
$g(x) = x - 4$

22. $f(x) = \sqrt{x - 5}$
$g(x) = x + 3$

23. $f(x) = x^2 + 3$
$g(x) = \sqrt{x - 1}$

24. $f(x) = \sqrt{2x - 2}$
$g(x) = x^2 - 8$

25. $f(x) = \dfrac{x}{x + 5}$
$g(x) = \sqrt{x - 1}$

26. $f(x) = \dfrac{x}{x - 4}$
$g(x) = \sqrt{x}$

In Exercises 27–34, use a graphing calculator to graph the function and determine whether the function is one-to-one.

27. $f(x) = x^3 - 1$

28. $f(x) = (2 - x)^3$

29. $f(t) = \sqrt[3]{5 - t}$

30. $h(t) = 4 - \sqrt[3]{t}$

31. $g(x) = (x - 3)^4$

32. $f(x) = x^5 + 2$

33. $h(t) = \dfrac{5}{t}$

34. $g(t) = \dfrac{5}{t^2}$

In Exercises 35–40, use the Horizontal Line Test to determine if the function is one-to-one and so has an inverse function. *See Example 4.*

35. $f(x) = x^2 - 2$

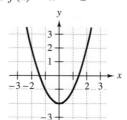

36. $f(x) = \frac{1}{5}x$

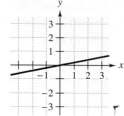

37. $f(x) = x^2, \quad x \geq 0$

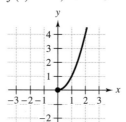

38. $f(x) = \sqrt{-x}$

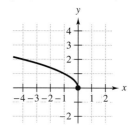

39. $g(x) = \sqrt{25 - x^2}$

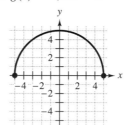

40. $g(x) = |x - 4|$

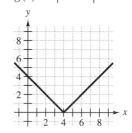

In Exercises 41–48, verify algebraically that the functions f and g are inverse functions of each other. *See Examples 5 and 6.*

41. $f(x) = -6x, \; g(x) = -\frac{1}{6}x$

42. $f(x) = \frac{2}{3}x, \; g(x) = \frac{3}{2}x$

43. $f(x) = 1 - 2x, \; g(x) = \frac{1}{2}(1 - x)$

44. $f(x) = 2x - 1, \; g(x) = \frac{1}{2}(x + 1)$

45. $f(x) = \sqrt[3]{x + 1}, \ g(x) = x^3 - 1$

46. $f(x) = x^7, \ g(x) = \sqrt[7]{x}$

47. $f(x) = \dfrac{1}{x}, \ g(x) = \dfrac{1}{x}$

48. $f(x) = \dfrac{1}{x + 1}, \ g(x) = \dfrac{1 - x}{x}$

In Exercises 49–60, find the inverse function of *f.* Verify that $f(f^{-1}(x))$ and $f^{-1}(f(x))$ are equal to the identity function. *See Example 7.*

49. $f(x) = 5x$

50. $f(x) = -8x$

51. $f(x) = -\frac{2}{5}x$

52. $f(x) = \frac{1}{3}x$

53. $f(x) = x + 10$

54. $f(x) = x - 5$

55. $f(x) = 5 - x$

56. $f(x) = 8 - x$

57. $f(x) = x^9$

58. $f(x) = x^5$

59. $f(x) = \sqrt[3]{x}$

60. $f(x) = x^{1/7}$

In Exercises 61–74, find the inverse function (if it exists). *See Examples 7 and 8.*

61. $g(x) = x + 25$

62. $f(x) = 7 - x$

63. $g(x) = 3 - 4x$

64. $g(t) = 6t + 1$

65. $g(t) = \frac{1}{4}t + 2$

66. $h(s) = 5 - \frac{3}{2}s$

67. $g(x) = x^2 + 4$

68. $h(x) = (4 - x)^2$

69. $h(x) = \sqrt{x}$

70. $h(x) = \sqrt{x + 5}$

71. $f(t) = t^3 - 1$

72. $h(t) = t^5 + 8$

73. $f(x) = \sqrt{x + 3}$

74. $f(x) = \sqrt{x^2 - 4}, \quad x \geq 2$

In Exercises 75–78, match the graph with the graph of its inverse function. [The graphs of the inverse functions are labeled (a), (b), (c), and (d).]

(a)
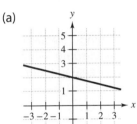

(b)
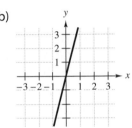

(c)
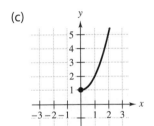

(d)
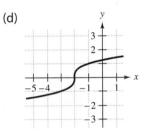

75.
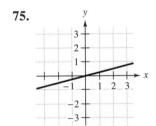

76.
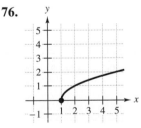

77.
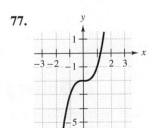

78.
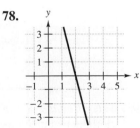

In Exercises 79–84, sketch the graphs of f and f^{-1} on the same rectangular coordinate system. Show that the graphs are reflections of each other in the line $y = x$. *See Example 9.*

79. $f(x) = x + 4$, $f^{-1}(x) = x - 4$

80. $f(x) = x - 7$, $f^{-1}(x) = x + 7$

81. $f(x) = 3x - 1$, $f^{-1}(x) = \frac{1}{3}(x + 1)$

82. $f(x) = 5 - 4x$, $f^{-1}(x) = -\frac{1}{4}(x - 5)$

83. $f(x) = x^2 - 1$, $x \geq 0$

$\quad f^{-1}(x) = \sqrt{x + 1}$

84. $f(x) = (x + 2)^2$, $x \geq -2$

$\quad f^{-1}(x) = \sqrt{x} - 2$

In Exercises 85–92, use a graphing calculator to graph the functions in the same viewing window. Graphically verify that f and g are inverse functions of each other.

85. $f(x) = \frac{1}{3}x$

$\quad g(x) = 3x$

86. $f(x) = \frac{1}{5}x - 1$

$\quad g(x) = 5x + 5$

87. $f(x) = \sqrt{x - 4}$

$\quad g(x) = x^2 + 4$, $x \geq 0$

88. $f(x) = \sqrt{4 - x}$

$\quad g(x) = 4 - x^2$, $x \geq 0$

89. $f(x) = \frac{1}{8}x^3$

$\quad g(x) = 2\sqrt[3]{x}$

90. $f(x) = \sqrt[3]{x + 2}$

$\quad g(x) = x^3 - 2$

91. $f(x) = |3 - x|$, $x \geq 3$

$\quad g(x) = 3 + x$, $x \geq 0$

92. $f(x) = |x - 2|$, $x \geq 2$

$\quad g(x) = x + 2$, $x \geq 0$

In Exercises 93–96, delete part of the graph of the function so that the remaining part is one-to-one. Find the inverse function of the remaining part and find the domain of the inverse function. (*Note:* There is more than one correct answer.) *See Example 10.*

93. $f(x) = (x - 2)^2$

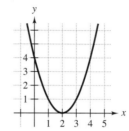

94. $f(x) = 9 - x^2$

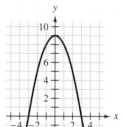

95. $f(x) = |x| + 1$

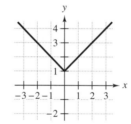

96. $f(x) = |x - 2|$

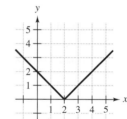

In Exercises 97 and 98, consider the function $f(x) = 3 - 2x$.

97. Find $f^{-1}(x)$.

98. Find $(f^{-1})^{-1}(x)$.

In Exercises 99–102, f is a one-to-one function such that $f(a) = b$, $f(b) = c$, and $f(c) = a$.

99. Find $f^{-1}(a)$.

100. Find $f^{-1}(b)$.

101. Find $f^{-1}(f^{-1}(c))$.

102. Find $f^{-1}(f^{-1}(a))$.

Solving Problems

103. *Sales Bonus*   You are a sales representative for a clothing manufacturer. You are paid an annual salary plus a bonus of 3% of your sales over $300,000.

Consider the two functions $s(x) = x - 300,000$ and $p(s) = 0.03s$. If x is greater than $300,000$, find and interpret $p(s(x))$.

104. *Daily Production Cost* The daily cost of producing x units in a manufacturing process is $C(x) = 8.5x + 300$. The number of units produced in t hours during a day is given by $x(t) = 12t$, $0 \le t \le 8$. Find, simplify, and interpret $(C \circ x)(t)$.

105.  *Ripples* You are standing on a bridge over a calm pond and drop a pebble, causing ripples of concentric circles in the water. The radius (in feet) of the outermost ripple is given by $r(t) = 0.6t$, where t is time in seconds after the pebble hits the water. The area of the circle is given by the function $A(r) = \pi r^2$. Find an equation for the composition $A(r(t))$. What are the input and output of this composite function? What is the area of the circle after 3 seconds?

106. *Oil Spill* An oil tanker hits a reef and begins to spill crude oil into the water. The oil forms a circular region around the ship with the radius (in feet) given by $r(t) = 15t$, where t is the time in hours after the hull is breached. The area of the circle is given by the function $A(r) = \pi r^2$. Find an equation for the composition $A(r(t))$. What are the input and output of this composite function? What is the area of the circle after 4 hours?

© Jean Louis Atlan/Sygma/CORBIS

107. *Rebate and Discount* The suggested retail price of a new car is p dollars. The dealership advertised a factory rebate of $2000 and a 5% discount.

(a) Write a function R in terms of p, giving the cost of the car after receiving the factory rebate.

(b) Write a function S in terms of p, giving the cost of the car after receiving the dealership discount.

(c) Form the composite functions $(R \circ S)(p)$ and $(S \circ R)(p)$ and interpret each.

(d) Find $(R \circ S)(26{,}000)$ and $(S \circ R)(26{,}000)$. Which yields the smaller cost for the car? Explain.

108. *Rebate and Discount* The suggested retail price of a plasma television is p dollars. The electronics store is offering a manufacturer's rebate of $200 and a 10% discount.

(a) Write a function R in terms of p, giving the cost of the television after receiving the manufacturer's rebate.

(b) Write a function S in terms of p, giving the cost of the television after receiving the 10% discount.

(c) Form the composite functions $(R \circ S)(p)$ and $(S \circ R)(p)$ and interpret each.

(d) Find $(R \circ S)(1600)$ and $(S \circ R)(1600)$. Which yields the smaller cost for the plasma television? Explain.

109. *Hourly Wage* Your wage is $9.00 per hour plus $0.65 for each unit produced per hour. So, your hourly wage y in terms of the number of units produced x is $y = 9 + 0.65x$.

(a) Find the inverse function.

(b) What does each variable represent in the inverse function?

(c) Use the context of the problem to determine the domain of the inverse function.

(d) Determine the number of units produced when your hourly wage averages $14.20.

110. *Federal Income Tax* In 2007, the function $T = 0.15(x - 7825) + 782.5$ represented the federal income tax owed by a single person whose adjusted gross income x was between $7825 and $31,850. (Source: Internal Revenue Service)

(a) Find the inverse function.

(b) What does each variable represent in the inverse function?

(c) Use the context of the problem to determine the domain of the inverse function.

(d) Determine a single person's adjusted gross income if they owed $3808.75 in federal income taxes in 2007.

111. *Exploration* Consider the functions $f(x) = 4x$ and $g(x) = x + 6$.

(a) Find $(f \circ g)(x)$.

(b) Find $(f \circ g)^{-1}(x)$.

(c) Find $f^{-1}(x)$ and $g^{-1}(x)$.

(d) Find $(g^{-1} \circ f^{-1})(x)$ and compare the result with that of part (b).

(e) Make a conjecture about $(f \circ g)^{-1}(x)$ and $(g^{-1} \circ f^{-1})(x)$.

112. *Exploration* Repeat Exercise 111 for $f(x) = x^2 + 1, x \geq 0$ and $g(x) = 3x$.

Explaining Concepts

True or False? In Exercises 113–116, decide whether the statement is true or false. If true, explain your reasoning. If false, give an example.

113. If the inverse function of f exists, the y-intercept of f is an x-intercept of f^{-1}. Explain.

114. There exists no function f such that $f = f^{-1}$.

115. If the inverse function of f exists, the domains of f and f^{-1} are the same.

116. If the inverse function of f exists and its graph passes through the point $(2, 2)$, the graph of f^{-1} also passes through the point $(2, 2)$.

117. ✎ Describe how to find the inverse of a function given by a set of ordered pairs. Give an example.

118. Give an example of a function that does not have an inverse function.

119. ✎ Why must a function be one-to-one in order for its inverse to be a function?

Cumulative Review

In Exercises 120–123, identify the transformation of the graph of $f(x) = x^2$.

120. $g(x) = (x - 4)^2$

121. $h(x) = -x^2$

122. $v(x) = x^2 + 1$

123. $k(x) = (x + 3)^2 - 5$

In Exercises 124–127, factor the expression completely.

124. $2x^3 - 6x$

125. $16 - (y + 2)^2$

126. $t^2 + 10t + 25$

127. $5 - u + 5u^2 - u^3$

In Exercises 128–131, graph the equation.

128. $y = 3 - \frac{1}{2}x$

129. $3x - 4y = 6$

130. $y = x^2 - 6x + 5$

131. $y = -(x - 2)^2 + 1$

9.3 Logarithmic Functions

A.T. Willett/Alamy

What You Should Learn

1 ▶ Evaluate logarithmic functions.

2 ▶ Graph logarithmic functions.

3 ▶ Graph and evaluate natural logarithmic functions.

4 ▶ Use the change-of-base formula to evaluate logarithms.

Why You Should Learn It

Logarithmic functions can be used to model and solve real-life problems. For instance, in Exercise 128 on page 615, you will use a logarithmic function to determine the speed of the wind near the center of a tornado.

1 ▶ Evaluate logarithmic functions.

Logarithmic Functions

In Section 9.2, you learned the concept of an inverse function. Moreover, you saw that if a function has the property that no horizontal line intersects the graph of the function more than once, the function must have an inverse function. By looking back at the graphs of the exponential functions introduced in Section 9.1, you will see that every function of the form

$$f(x) = a^x$$

passes the Horizontal Line Test and so must have an inverse function. To describe the inverse function of $f(x) = a^x$, follow the steps used in Section 9.2.

$y = a^x$ Replace $f(x)$ by y.

$x = a^y$ Interchange x and y.

At this point, there is no way to solve for y. A verbal description of y in the equation $x = a^y$ is "y equals the power to which a must be raised to obtain x." This inverse of $f(x) = a^x$ is denoted by the **logarithmic function with base a**

$$f^{-1}(x) = \log_a x.$$

Definition of Logarithmic Function

Let a and x be positive real numbers such that $a \neq 1$. The **logarithm of x with base a** is denoted by $\log_a x$ and is defined as follows.

$$y = \log_a x \quad \text{if and only if} \quad x = a^y$$

The function $f(x) = \log_a x$ is the **logarithmic function with base a.**

From the definition it is clear that

Logarithmic Equation   *Exponential Equation*

$\quad y = \log_a x \qquad$ is equivalent to $\qquad x = a^y.$

So, to find the value of $\log_a x$, *think*

"$\log_a x$ = the power to which a must be raised to obtain x."

Smart Study Strategy

Go to page 576 for ways to *Take Control of Your Class Time.*

For instance,

$$y = \log_2 8$$ Think: "The power to which 2 must be raised to obtain 8."

$$y = 3.$$

That is,

$$3 = \log_2 8.$$ This is equivalent to $2^3 = 8$.

By now it should be clear that *a logarithm is an exponent.*

EXAMPLE 1 Evaluating Logarithms

Evaluate each logarithm.

a. $\log_2 16$ **b.** $\log_3 9$ **c.** $\log_4 2$

Solution

In each case you should answer the question, "To what power must the base be raised to obtain the given number?"

a. The power to which 2 must be raised to obtain 16 is 4. That is,

$$2^4 = 16 \implies \log_2 16 = 4.$$

b. The power to which 3 must be raised to obtain 9 is 2. That is,

$$3^2 = 9 \implies \log_3 9 = 2.$$

c. The power to which 4 must be raised to obtain 2 is $\frac{1}{2}$. That is,

$$4^{1/2} = 2 \implies \log_4 2 = \frac{1}{2}.$$

✓ **CHECKPOINT** *Now try Exercise 25.*

EXAMPLE 2 Evaluating Logarithms

Evaluate each logarithm.

a. $\log_5 1$ **b.** $\log_{10} \frac{1}{10}$ **c.** $\log_3(-1)$ **d.** $\log_4 0$

Solution

a. The power to which 5 must be raised to obtain 1 is 0. That is,

$$5^0 = 1 \implies \log_5 1 = 0.$$

b. The power to which 10 must be raised to obtain $\frac{1}{10}$ is -1. That is,

$$10^{-1} = \frac{1}{10} \implies \log_{10} \frac{1}{10} = -1.$$

c. There is no power to which 3 can be raised to obtain -1. The reason for this is that for any value of x, 3^x is a positive number. So, $\log_3(-1)$ is undefined.

d. There is no power to which 4 can be raised to obtain 0. So, $\log_4 0$ is undefined.

✓ **CHECKPOINT** *Now try Exercise 35.*

The following properties of logarithms follow directly from the definition of the logarithmic function with base a.

Properties of Logarithms

Let a and x be positive real numbers such that $a \neq 1$. Then the following properties are true.

1. $\log_a 1 = 0$ because $a^0 = 1$.

2. $\log_a a = 1$ because $a^1 = a$.

3. $\log_a a^x = x$ because $a^x = a^x$.

The logarithmic function with base 10 is called the **common logarithmic function.** On most calculators, this function can be evaluated with the common logarithmic key (LOG), as illustrated in the next example.

EXAMPLE 3   Evaluating Common Logarithms

Evaluate each logarithm. Use a calculator only if necessary.

a. $\log_{10} 100$ **b.** $\log_{10} 0.01$

c. $\log_{10} 5$ **d.** $\log_{10} 2.5$

Solution

a. The power to which 10 must be raised to obtain 100 is 2. That is,

$$10^2 = 100 \quad \Longrightarrow \quad \log_{10} 100 = 2.$$

b. The power to which 10 must be raised to obtain 0.01 or $\frac{1}{100}$ is -2. That is,

$$10^{-2} = \frac{1}{100} \quad \Longrightarrow \quad \log_{10} 0.01 = -2.$$

c. There is no simple power to which 10 can be raised to obtain 5, so you should use a calculator to evaluate $\log_{10} 5$.

Keystrokes	Display	
5 (LOG)	0.69897	Scientific
(LOG) 5) (ENTER)	0.69897	Graphing

So, rounded to three decimal places, $\log_{10} 5 \approx 0.699$.

d. There is no simple power to which 10 can be raised to obtain 2.5, so you should use a calculator to evaluate $\log_{10} 2.5$.

Keystrokes	Display	
2.5 (LOG)	0.39794	Scientific
(LOG) 2.5) (ENTER)	0.39794	Graphing

So, rounded to three decimal places, $\log_{10} 2.5 \approx 0.398$.

✓ **CHECKPOINT** *Now try Exercise 47.*

Study Tip

Be sure you see that the value of a logarithm can be zero or negative, as in Example 3(b), but you *cannot* take the logarithm of zero or a negative number. This means that the logarithms $\log_{10}(-10)$ and $\log_5 0$ are undefined.

2 ▶ Graph logarithmic functions.

Graphs of Logarithmic Functions

To sketch the graph of

$$y = \log_a x$$

you can use the fact that the graphs of inverse functions are reflections of each other in the line $y = x$.

EXAMPLE 4   **Graphs of Exponential and Logarithmic Functions**

On the same rectangular coordinate system, sketch the graph of each function.

a. $f(x) = 2^x$ **b.** $g(x) = \log_2 x$

Solution

a. Begin by making a table of values for $f(x) = 2^x$.

x	-2	-1	0	1	2	3
$f(x) = 2^x$	$\frac{1}{4}$	$\frac{1}{2}$	1	2	4	8

By plotting these points and connecting them with a smooth curve, you obtain the graph shown in Figure 9.17.

b. Because $g(x) = \log_2 x$ is the inverse function of $f(x) = 2^x$, the graph of g is obtained by reflecting the graph of f in the line $y = x$, as shown in Figure 9.17.

✓ **CHECKPOINT** *Now try Exercise 57.*

Figure 9.17   Inverse Functions

Study Tip

In Example 4, the inverse property of logarithmic functions is used to sketch the graph of $g(x) = \log_2 x$. You could also use a standard point-plotting approach or a graphing calculator.

Notice from the graph of $g(x) = \log_2 x$, shown in Figure 9.17, that the domain of the function is the set of positive numbers and the range is the set of all real numbers. The basic characteristics of the graph of a logarithmic function are summarized in Figure 9.18. In this figure, note that the graph has one x-intercept at $(1, 0)$. Also note that $x = 0$ (y-axis) is a vertical asymptote of the graph.

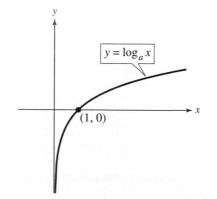

Graph of $y = \log_a x, \quad a > 1$

- Domain: $(0, \infty)$
- Range: $(-\infty, \infty)$
- Intercept: $(1, 0)$
- Increasing (moves up to the right)
- Asymptote: y-axis

Figure 9.18   Characteristics of logarithmic function $y = \log_a x \quad (a > 1)$

In the following example, the graph of $\log_a x$ is used to sketch the graphs of functions of the form $y = b \pm \log_a(x + c)$. Notice how each transformation affects the vertical asymptote.

EXAMPLE 5   Sketching the Graphs of Logarithmic Functions

The graph of each function is similar to the graph of $f(x) = \log_{10} x$, as shown in Figure 9.19. From the graph you can determine the domain of the function.

a. Because $g(x) = \log_{10}(x - 1) = f(x - 1)$, the graph of g can be obtained by shifting the graph of f one unit to the right. The vertical asymptote of the graph of g is $x = 1$. The domain of g is $(1, \infty)$.

b. Because $h(x) = 2 + \log_{10} x = 2 + f(x)$, the graph of h can be obtained by shifting the graph of f two units upward. The vertical asymptote of the graph of h is $x = 0$. The domain of h is $(0, \infty)$.

c. Because $k(x) = -\log_{10} x = -f(x)$, the graph of k can be obtained by reflecting the graph of f in the x-axis. The vertical asymptote of the graph of k is $x = 0$. The domain of k is $(0, \infty)$.

d. Because $j(x) = \log_{10}(-x) = f(-x)$, the graph of j can be obtained by reflecting the graph of f in the y-axis. The vertical asymptote of the graph of j is $x = 0$. The domain of j is $(-\infty, 0)$.

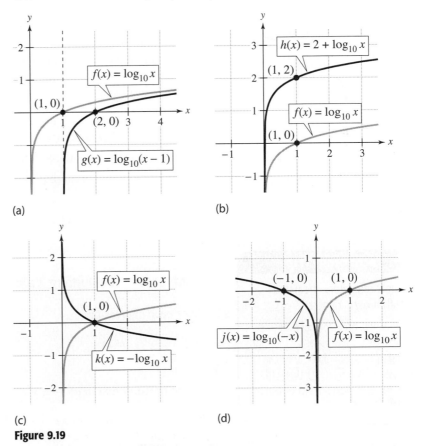

(a)

(b)

(c)

(d)

Figure 9.19

✔ **CHECKPOINT** *Now try Exercise 61.*

3 ▶ Graph and evaluate natural logarithmic functions.

The Natural Logarithmic Function

As with exponential functions, the most widely used base for logarithmic functions is the number e. The logarithmic function with base e is the **natural logarithmic function** and is denoted by the special symbol $\ln x$, which is read as "el en of x."

> ### The Natural Logarithmic Function
>
> The function defined by
>
> $$f(x) = \log_e x = \ln x$$
>
> where $x > 0$, is called the **natural logarithmic function.**

Graph of $g(x) = \ln x$
- Domain: $(0, \infty)$
- Range: $(-\infty, \infty)$
- Intercept: $(1, 0)$
- Increasing (moves up to the right)
- Asymptote: y-axis

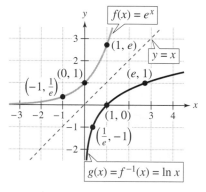

Figure 9.20 Characteristics of the natural logarithmic function $g(x) = \ln x$

The definition above implies that the natural logarithmic function and the natural exponential function are inverse functions of each other. So, every logarithmic equation can be written in an equivalent exponential form, and every exponential equation can be written in logarithmic form.

Because the functions $f(x) = e^x$ and $g(x) = \ln x$ are inverse functions of each other, their graphs are reflections of each other in the line $y = x$. This reflective property is illustrated in Figure 9.20. The figure also contains a summary of several characteristics of the graph of the natural logarithmic function.

Notice that the domain of the natural logarithmic function, as with every other logarithmic function, is the set of *positive real numbers*—be sure you see that $\ln x$ is not defined for zero or for negative numbers.

The three properties of logarithms listed earlier in this section are also valid for natural logarithms.

> ### Properties of Natural Logarithms
>
> Let x be a positive real number. Then the following properties are true.
>
> **1.** $\ln 1 = 0$ because $e^0 = 1$.
>
> **2.** $\ln e = 1$ because $e^1 = e$.
>
> **3.** $\ln e^x = x$ because $e^x = e^x$.

EXAMPLE 6   **Evaluating Natural Logarithmic Functions**

Evaluate each expression.

a. $\ln e^2$ **b.** $\ln \dfrac{1}{e}$

Solution

Using the property that $\ln e^x = x$, you obtain the following.

a. $\ln e^2 = 2$ **b.** $\ln \dfrac{1}{e} = \ln e^{-1} = -1$

✓ **CHECKPOINT**   *Now try Exercise 89.*

4 ▶ Use the change-of-base formula to evaluate logarithms.

Change of Base

Although 10 and e are the most frequently used bases, you occasionally need to evaluate logarithms with other bases. In such cases, the following **change-of-base formula** is useful.

Technology: Tip

You can use a graphing calculator to graph logarithmic functions that do not have a base of 10 by using the change-of-base formula. Use the change-of-base formula to rewrite $g(x) = \log_2 x$ in Example 4 on page 608 (with $b = 10$) and graph the function. You should obtain a graph similar to the one below.

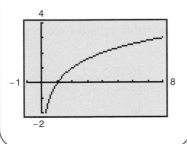

> **Change-of-Base Formula**
>
> Let a, b, and x be positive real numbers such that $a \neq 1$ and $b \neq 1$. Then $\log_a x$ is given as follows.
>
> $$\log_a x = \frac{\log_b x}{\log_b a} \qquad \text{or} \qquad \log_a x = \frac{\ln x}{\ln a}$$

The usefulness of this change-of-base formula is that you can use a calculator that has only the common logarithm key $\boxed{\text{LOG}}$ and the natural logarithm key $\boxed{\text{LN}}$ to evaluate logarithms to any base.

EXAMPLE 7   Changing Bases to Evaluate Logarithms

a. Use *common* logarithms to evaluate $\log_3 5$.

b. Use *natural* logarithms to evaluate $\log_6 2$.

Solution

Using the change-of-base formula, you can convert to common and natural logarithms by writing

$$\log_3 5 = \frac{\log_{10} 5}{\log_{10} 3} \qquad \text{and} \qquad \log_6 2 = \frac{\ln 2}{\ln 6}.$$

Now, use the following keystrokes.

a.

Keystrokes	Display	
5 $\boxed{\text{LOG}}$ $\boxed{\div}$ 3 $\boxed{\text{LOG}}$ $\boxed{=}$	1.4649735	Scientific
$\boxed{\text{LOG}}$ 5 $\boxed{)}$ $\boxed{\div}$ $\boxed{\text{LOG}}$ 3 $\boxed{)}$ $\boxed{\text{ENTER}}$	1.4649735	Graphing

So, $\log_3 5 \approx 1.465$.

b.

Keystrokes	Display	
2 $\boxed{\text{LN}}$ $\boxed{\div}$ 6 $\boxed{\text{LN}}$ $\boxed{=}$	0.3868528	Scientific
$\boxed{\text{LN}}$ 2 $\boxed{)}$ $\boxed{\div}$ $\boxed{\text{LN}}$ 6 $\boxed{)}$ $\boxed{\text{ENTER}}$	0.3868528	Graphing

So, $\log_6 2 \approx 0.387$.

✓ **CHECKPOINT** *Now try Exercise 111.*

Study Tip

In Example 7(a), $\log_3 5$ could have been evaluated using natural logarithms in the change-of-base formula.

$$\log_3 5 = \frac{\ln 5}{\ln 3} \approx 1.465$$

Notice that you get the same answer whether you use natural logarithms or common logarithms in the change-of-base formula.

At this point, you have been introduced to all the basic types of functions that are covered in this course: polynomial functions, radical functions, rational functions, exponential functions, and logarithmic functions. The only other common types of functions are *trigonometric functions*, which you will study if you go on to take a course in trigonometry or precalculus.

_____ **Concept Check** _____

In Concept Check Exercises 1 and 2, determine whether the statement is true or false. Justify your answer.

1. The statement $8 = 2^3$ is equivalent to $2 = \log_8 3$.

2. The graph of $f(x) = \ln x$ is the reflection of the graph of $f(x) = e^x$ in the x-axis.

3. Explain how to use the graph of $y = \log_{10} x$ to graph $y = 3 + \log_{10}(x - 1)$.

4. Describe and correct the error in evaluating $\log_4 10$.

$$\log_4 10 = \frac{\ln 4}{\ln 10} \approx 0.6021$$

9.3 EXERCISES

Go to pages 646–647 to record your assignments.

_____ **Developing Skills** _____

In Exercises 1–12, write the logarithmic equation in exponential form.

1. $\log_7 49 = 2$

2. $\log_{11} 121 = 2$

3. $\log_2 \frac{1}{32} = -5$

4. $\log_3 \frac{1}{27} = -3$

5. $\log_3 \frac{1}{243} = -5$

6. $\log_{10} 10{,}000 = 4$

7. $\log_{36} 6 = \frac{1}{2}$

8. $\log_{64} 4 = \frac{1}{3}$

9. $\log_8 4 = \frac{2}{3}$

10. $\log_{16} 8 = \frac{3}{4}$

11. $\log_2 5.278 \approx 2.4$

12. $\log_3 1.179 \approx 0.15$

In Exercises 13–24, write the exponential equation in logarithmic form.

13. $6^2 = 36$

14. $3^5 = 243$

15. $5^{-3} = \frac{1}{125}$

16. $6^{-4} = \frac{1}{1296}$

17. $8^{2/3} = 4$

18. $81^{3/4} = 27$

19. $25^{-1/2} = \frac{1}{5}$

20. $6^{-3} = \frac{1}{216}$

21. $4^0 = 1$

22. $6^1 = 6$

23. $5^{1.4} \approx 9.518$

24. $10^{0.36} \approx 2.291$

In Exercises 25–46, evaluate the logarithm without using a calculator. (If not possible, state the reason.) *See Examples 1 and 2.*

25. $\log_2 8$

26. $\log_3 27$

27. $\log_{10} 1000$

28. $\log_{10} 0.00001$

29. $\log_2 \frac{1}{16}$

30. $\log_3 \frac{1}{9}$

31. $\log_4 \frac{1}{64}$

32. $\log_6 \frac{1}{216}$

33. $\log_{10} \frac{1}{10{,}000}$

34. $\log_{10} \frac{1}{100}$

35. $\log_2(-3)$

36. $\log_4(-4)$

37. $\log_4 1$

38. $\log_3 1$

39. $\log_5(-6)$

40. $\log_2 0$

41. $\log_9 3$

42. $\log_{125} 5$

43. $\log_{16} 8$

44. $\log_{81} 9$

45. $\log_7 7^4$

46. $\log_5 5^3$

In Exercises 47–52, use a calculator to evaluate the common logarithm. (Round your answer to four decimal places.) *See Example 3.*

47. $\log_{10} 42$

48. $\log_{10} 7561$

49. $\log_{10} 0.023$

50. $\log_{10} 0.149$

51. $\log_{10}(\sqrt{5} + 3)$

52. $\log_{10} \frac{\sqrt{3}}{2}$

In Exercises 53–56, match the function with its graph. [The graphs are labeled (a), (b), (c), and (d).]

(a)
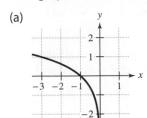

(b)
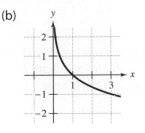

(c)
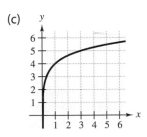

(d)
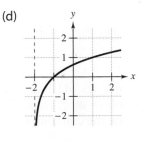

53. $f(x) = 4 + \log_3 x$

54. $f(x) = -\log_3 x$

55. $f(x) = \log_3(-x)$

56. $f(x) = \log_3(x + 2)$

In Exercises 57–60, sketch the graph of f. Then use the graph of f to sketch the graph of g. *See Example 4.*

✓ **57.** $f(x) = 3^x$
$g(x) = \log_3 x$

58. $f(x) = 4^x$
$g(x) = \log_4 x$

59. $f(x) = 6^x$
$g(x) = \log_6 x$

60. $f(x) = \left(\frac{1}{2}\right)^x$
$g(x) = \log_{1/2} x$

In Exercises 61–66, identify the transformation of the graph of $f(x) = \log_2 x$. Then sketch the graph of h. *See Example 5.*

✓ **61.** $h(x) = 3 + \log_2 x$

62. $h(x) = -5 + \log_2 x$

63. $h(x) = \log_2(x - 2)$

64. $h(x) = \log_2(x + 5)$

65. $h(x) = \log_2(-x)$

66. $h(x) = -\log_2 x$

In Exercises 67–76, sketch the graph of the function. Identify the vertical asymptote.

67. $f(x) = \log_5 x$

68. $g(x) = \log_8 x$

69. $g(t) = -\log_9 t$

70. $h(s) = -2 \log_3 s$

71. $f(x) = 2 + \log_4 x$

72. $f(x) = -2 + \log_3 x$

73. $g(x) = \log_2(x - 3)$

74. $h(x) = \log_3(x + 1)$

75. $f(x) = \log_{10}(10x)$

76. $g(x) = \log_4(4x)$

In Exercises 77–82, find the domain and vertical asymptote of the function. Then sketch its graph.

77. $f(x) = \log_4 x$

78. $g(x) = \log_6 x$

79. $h(x) = \log_5(x - 4)$

80. $f(x) = -\log_6(x + 2)$

81. $y = -\log_3 x + 2$

82. $y = \log_4(x - 2) + 3$

In Exercises 83–88, use a graphing calculator to graph the function. Determine the domain and the vertical asymptote.

83. $y = 5 \log_{10} x$

84. $y = 5 \log_{10}(x - 3)$

85. $y = -3 + 5 \log_{10} x$

86. $y = 5 \log_{10}(3x)$

87. $y = \log_{10}\left(\frac{x}{5}\right)$

88. $y = \log_{10}(-x)$

In Exercises 89–94, use a calculator to evaluate the natural logarithm. (Round your answer to four decimal places.) *See Example 6.*

✓ **89.** $\ln 38$

90. $\ln 18.6$

91. $\ln 0.15$

92. $\ln 0.002$

93. $\ln\left(\frac{3 - \sqrt{2}}{5}\right)$

94. $\ln\left(1 + \frac{0.10}{12}\right)$

In Exercises 95–98, match the function with its graph. [The graphs are labeled (a), (b), (c), and (d).]

(a)

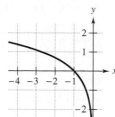

(b)

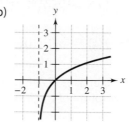

(c)

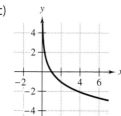

(d)

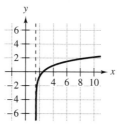

95. $f(x) = \ln(x + 1)$ **96.** $f(x) = \ln(-x)$

97. $f(x) = \ln\left(x - \frac{3}{2}\right)$ **98.** $f(x) = -\frac{3}{2}\ln x$

In Exercises 99–106, sketch the graph of the function. Identify the vertical asymptote.

99. $f(x) = -\ln x$ **100.** $f(x) = -5\ln x$

101. $f(x) = 3\ln x$ **102.** $h(t) = 4\ln t$

103. $f(x) = 3 + \ln x$ **104.** $h(x) = 2 + \ln x$

105. $g(t) = 2\ln(t - 4)$ **106.** $g(x) = -3\ln(x + 3)$

 In Exercises 107–110, use a graphing calculator to graph the function. Determine the domain and the vertical asymptote.

107. $g(x) = -\ln(x + 1)$ **108.** $h(x) = \ln(x + 4)$

109. $f(t) = 7 + 3\ln t$ **110.** $g(t) = \ln(5 - t)$

In Exercises 111–124, use a calculator to evaluate the logarithm by means of the change-of-base formula. Use (a) the common logarithm key and (b) the natural logarithm key. (Round your answer to four decimal places.) *See Example 7.*

111. $\log_9 36$ **112.** $\log_7 411$

113. $\log_5 14$ **114.** $\log_6 9$

115. $\log_2 0.72$ **116.** $\log_{12} 0.6$

117. $\log_{15} 1250$ **118.** $\log_{20} 125$

119. $\log_{1/4} 16$ **120.** $\log_{1/3} 18$

121. $\log_4 \sqrt{42}$ **122.** $\log_5 \sqrt{21}$

123. $\log_2(1 + e)$ **124.** $\log_4(2 + e^3)$

Solving Problems

125. *American Elk* The antler spread a (in inches) and shoulder height h (in inches) of an adult male American elk are related by the model

$$h = 116\log_{10}(a + 40) - 176.$$

Approximate to one decimal place the shoulder height of a male American elk with an antler spread of 55 inches.

126. *Sound Intensity* The relationship between the number of decibels B and the intensity of a sound I in watts per centimeter squared is given by

$$B = 10\log_{10}\left(\frac{I}{10^{-16}}\right).$$

Determine the number of decibels of an alarm clock with an intensity of 10^{-8} watt per centimeter squared.

127. *Compound Interest* The time t in years for an investment to double in value when compounded continuously at interest rate r is given by

$$t = \frac{\ln 2}{r}.$$

Complete the table, which shows the "doubling times" for several interest rates.

r	0.07	0.08	0.09	0.10	0.11	0.12
t						

128. *Meteorology*   Most tornadoes last less than 1 hour and travel about 20 miles. The speed of the wind S (in miles per hour) near the center of the tornado and the distance d (in miles) the tornado travels are related by the model $S = 93 \log_{10} d + 65$. On March 18, 1925, a large tornado struck portions of Missouri, Illinois, and Indiana, covering a distance of 220 miles. Approximate to one decimal place the speed of the wind near the center of this tornado.

129. *Tractrix*   A person walking along a dock (the y-axis) drags a boat by a 10-foot rope (see figure). The boat travels along a path known as a *tractrix*. The equation of the path is

$$y = 10 \ln\left(\frac{10 + \sqrt{100 - x^2}}{x}\right) - \sqrt{100 - x^2}.$$

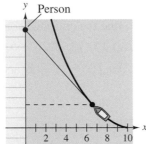

(a) 🖩 Use a graphing calculator to graph the function. What is the domain of the function?

(b) 🖩 Identify any asymptotes.

(c) Determine the position of the person when the x-coordinate of the position of the boat is $x = 2$.

130. 🖩 *Home Mortgage*   The model

$$t = 10.042 \ln\left(\frac{x}{x - 1250}\right), \quad x > 1250$$

approximates the length t (in years) of a home mortgage of $150,000 at 10% interest in terms of the monthly payment x.

(a) Use a graphing calculator to graph the model. Describe the change in the length of the mortgage as the monthly payment increases.

(b) Use the graph in part (a) to approximate the length of the mortgage when the monthly payment is $1316.35.

(c) Use the result of part (b) to find the total amount paid over the term of the mortgage. What amount of the total is interest costs?

Think About It   In Exercises 131–136, answer the question for the function $f(x) = \log_{10} x$. (Do not use a calculator.)

131. What is the domain of f?

132. Find the inverse function of f.

133. Describe the values of $f(x)$ for $1000 \le x \le 10{,}000$.

134. Describe the values of x, given that $f(x)$ is negative.

135. By what amount will x increase, given that $f(x)$ is increased by 1 unit?

136. Find the ratio of a to b when $f(a) = 3 + f(b)$.

Explaining Concepts

137. ✎ Explain the difference between common logarithms and natural logarithms.

138. ✎ Explain the relationship between the domain of the graph of $f(x) = \log_5 x$ and the range of the graph of $g(x) = 5^x$.

139. Discuss how shifting or reflecting the graph of a logarithmic function affects the domain and the range.

140. ✎ Explain why $\log_a x$ is defined only when $0 < a < 1$ and $a > 1$.

Cumulative Review

In Exercises 141–144, use the rules of exponents to simplify the expression.

141. $(-m^6 n)(m^4 n^3)$

142. $(m^2 n^4)^3 (mn^2)$

143. $\dfrac{36x^4 y}{8xy^3}$

144. $-\left(\dfrac{3x}{5y}\right)^5$

In Exercises 145–148, perform the indicated operation(s) and simplify. (Assume all variables are positive.)

145. $25\sqrt{3x} - 3\sqrt{12x}$

146. $\left(\sqrt{x} + 3\right)\left(\sqrt{x} - 3\right)$

147. $\sqrt{u}\left(\sqrt{20} - \sqrt{5}\right)$

148. $\left(2\sqrt{t} + 3\right)^2$

Mid-Chapter Quiz

Take this quiz as you would take a quiz in class. After you are done, check your work against the answers in the back of the book.

1. Given $f(x) = \left(\frac{4}{3}\right)^x$, find (a) $f(2)$, (b) $f(0)$, (c) $f(-1)$, and (d) $f(1.5)$.

2. Identify the horizontal asymptote of the graph of $g(x) = -3^{-0.5x}$.

In Exercises 3–6, sketch the graph of the function. Identify the horizontal asymptote. Use a graphing calculator for Exercises 5 and 6.

3. $y = \frac{1}{2}(4^x)$

4. $y = 5(2^{-x})$

5. ▦ $f(t) = 12e^{-0.4t}$

6. ▦ $g(x) = 100(1.08)^x$

7. Given $f(x) = 2x - 3$ and $g(x) = x^3$, find the indicated composition.

   (a) $(f \circ g)(x)$ (b) $(g \circ f)(x)$ (c) $(f \circ g)(-2)$ (d) $(g \circ f)(4)$

8. Verify algebraically and graphically that $f(x) = 5 - 2x$ and $g(x) = \frac{1}{2}(5 - x)$ are inverse functions of each other.

In Exercises 9 and 10, find the inverse function.

9. $h(x) = 10x + 3$ 10. $g(t) = \frac{1}{2}t^3 + 2$

11. Write the logarithmic equation $\log_9 \frac{1}{81} = -2$ in exponential form.

12. Write the exponential equation $2^6 = 64$ in logarithmic form.

13. Evaluate $\log_5 125$ without a calculator.

▦ In Exercises 14 and 15, use a graphing calculator to graph the function. Identify the vertical asymptote.

14. $f(t) = -2\ln(t + 3)$ 15. $h(x) = 5 + \frac{1}{2}\ln x$

16. Use the graph of f shown at the left to determine h and k if $f(x) = \log_5(x - h) + k$.

17. Use a calculator and the change-of-base formula to evaluate $\log_3 782$.

18. You deposit $1200 in an account at an annual interest rate of $6\frac{1}{4}\%$. Complete the table showing the balances A in the account after 15 years for several types of compounding.

n	1	4	12	365	Continuous compounding
A					

19. After t years, the remaining mass y (in grams) of 14 grams of a radioactive element whose half-life is 40 years is given by $y = 14\left(\frac{1}{2}\right)^{t/40}$, $t \geq 0$. How much of the initial mass remains after 125 years?

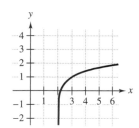

Figure for 16

9.4 Properties of Logarithms

Charles Gupton/CORBIS

What You Should Learn

1 ▶ Use the properties of logarithms to evaluate logarithms.

2 ▶ Use the properties of logarithms to rewrite, expand, or condense logarithmic expressions.

3 ▶ Use the properties of logarithms to solve application problems.

Why You Should Learn It

Logarithmic equations are often used to model scientific observations. For instance, in Example 5 on page 620, a logarithmic equation is used to model human memory.

1 ▶ Use the properties of logarithms to evaluate logarithms.

Properties of Logarithms

You know from the preceding section that the logarithmic function with base a is the *inverse function* of the exponential function with base a. So, it makes sense that each property of exponents should have a corresponding property of logarithms. For instance, the exponential property

$$a^0 = 1 \qquad \text{Exponential property}$$

has the corresponding logarithmic property

$$\log_a 1 = 0. \qquad \text{Corresponding logarithmic property}$$

In this section you will study the logarithmic properties that correspond to the following three exponential properties:

	Base a	*Natural Base*
1.	$a^m a^n = a^{m+n}$	$e^m e^n = e^{m+n}$
2.	$\dfrac{a^m}{a^n} = a^{m-n}$	$\dfrac{e^m}{e^n} = e^{m-n}$
3.	$(a^m)^n = a^{mn}$	$(e^m)^n = e^{mn}$

Properties of Logarithms

Let a be a positive real number such that $a \neq 1$, and let n be a real number. If u and v are real numbers, variables, or algebraic expressions such that $u > 0$ and $v > 0$, the following properties are true.

		Logarithm with Base a	*Natural Logarithm*
1.	Product Property:	$\log_a(uv) = \log_a u + \log_a v$	$\ln(uv) = \ln u + \ln v$
2.	Quotient Property:	$\log_a \dfrac{u}{v} = \log_a u - \log_a v$	$\ln \dfrac{u}{v} = \ln u - \ln v$
3.	Power Property:	$\log_a u^n = n \log_a u$	$\ln u^n = n \ln u$

There is no general property of logarithms that can be used to simplify $\log_a(u + v)$. Specifically,

$$\log_a(u + v) \text{ does not equal } \log_a u + \log_a v.$$

EXAMPLE 1   **Using Properties of Logarithms**

Use $\ln 2 \approx 0.693$, $\ln 3 \approx 1.099$, and $\ln 5 \approx 1.609$ to approximate each expression.

a. $\ln \dfrac{2}{3}$ **b.** $\ln 10$ **c.** $\ln 30$

Solution

a. $\ln \dfrac{2}{3} = \ln 2 - \ln 3$ Quotient Property

$\qquad \approx 0.693 - 1.099 = -0.406$ Substitute for $\ln 2$ and $\ln 3$.

b. $\ln 10 = \ln(2 \cdot 5)$ Factor.

$\qquad = \ln 2 + \ln 5$ Product Property

$\qquad \approx 0.693 + 1.609$ Substitute for $\ln 2$ and $\ln 5$.

$\qquad = 2.302$ Simplify.

c. $\ln 30 = \ln(2 \cdot 3 \cdot 5)$ Factor.

$\qquad = \ln 2 + \ln 3 + \ln 5$ Product Property

$\qquad \approx 0.693 + 1.099 + 1.609$ Substitute for $\ln 2$, $\ln 3$, and $\ln 5$.

$\qquad = 3.401$ Simplify.

 **CHECKPOINT** *Now try Exercise 25.*

When using the properties of logarithms, it helps to state the properties *verbally*. For instance, the verbal form of the Product Property

$$\ln(uv) = \ln u + \ln v$$

is: *The log of a product is the sum of the logs of the factors.* Similarly, the verbal form of the Quotient Property

$$\ln \frac{u}{v} = \ln u - \ln v$$

is: *The log of a quotient is the difference of the logs of the numerator and denominator.*

Study Tip

Remember that you can verify results such as those given in Examples 1 and 2 with a calculator.

EXAMPLE 2   **Using Properties of Logarithms**

Use the properties of logarithms to verify that $-\ln 2 = \ln \frac{1}{2}$.

Solution

Using the Power Property, you can write the following.

$\qquad -\ln 2 = (-1)\ln 2$ Rewrite coefficient as -1.

$\qquad\qquad = \ln 2^{-1}$ Power Property

$\qquad\qquad = \ln \dfrac{1}{2}$ Rewrite 2^{-1} as $\frac{1}{2}$.

 **CHECKPOINT** *Now try Exercise 41.*

2 ▸ Use the properties of logarithms to rewrite, expand, or condense logarithmic expressions.

Rewriting Logarithmic Expressions

In Examples 1 and 2, the properties of logarithms were used to rewrite logarithmic expressions involving the log of a *constant*. A more common use of these properties is to rewrite the log of a *variable expression*.

EXAMPLE 3   **Expanding Logarithmic Expressions**

Use the properties of logarithms to expand each expression.

a. $\log_{10} 7x^3 = \log_{10} 7 + \log_{10} x^3$ Product Property

$\qquad\qquad\quad = \log_{10} 7 + 3 \log_{10} x$ Power Property

b. $\log_6 \dfrac{8x^3}{y} = \log_6 8x^3 - \log_6 y$ Quotient Property

$\qquad\qquad\quad = \log_6 8 + \log_6 x^3 - \log_6 y$ Product Property

$\qquad\qquad\quad = \log_6 8 + 3 \log_6 x - \log_6 y$ Power Property

c. $\ln \dfrac{\sqrt{3x-5}}{7} = \ln\left[\dfrac{(3x-5)^{1/2}}{7}\right]$ Rewrite using rational exponent.

$\qquad\qquad\quad = \ln(3x-5)^{1/2} - \ln 7$ Quotient Property

$\qquad\qquad\quad = \dfrac{1}{2}\ln(3x-5) - \ln 7$ Power Property

✓ **CHECKPOINT** *Now try Exercise 47.*

When you rewrite a logarithmic expression as in Example 3, you are **expanding** the expression. The reverse procedure is demonstrated in Example 4, and is called **condensing** a logarithmic expression.

EXAMPLE 4   **Condensing Logarithmic Expressions**

Use the properties of logarithms to condense each expression.

a. $\ln x - \ln 3 = \ln \dfrac{x}{3}$ Quotient Property

b. $\dfrac{1}{2}\log_3 x + \log_3 5 = \log_3 x^{1/2} + \log_3 5$ Power Property

$\qquad\qquad\qquad\quad = \log_3 5\sqrt{x}$ Product Property

c. $3(\ln 4 + \ln x) = 3(\ln 4x)$ Product Property

$\qquad\qquad\quad = \ln (4x)^3$ Power Property

$\qquad\qquad\quad = \ln 64x^3$ Simplify.

✓ **CHECKPOINT** *Now try Exercise 75.*

When you expand or condense a logarithmic expression, it is possible to change the domain of the expression. For instance, the domain of the function

$$f(x) = 2 \ln x \qquad \text{Domain is the set of positive real numbers.}$$

is the set of positive real numbers, whereas the domain of

$$g(x) = \ln x^2 \qquad \text{Domain is the set of nonzero real numbers.}$$

is the set of nonzero real numbers. So, when you expand or condense a logarithmic expression, you should check to see whether the rewriting has changed the domain of the expression. In such cases, you should restrict the domain appropriately. For instance, you can write

$$f(x) = 2 \ln x$$
$$= \ln x^2, \; x > 0.$$

3 ▶ Use the properties of logarithms to solve application problems.

Application

EXAMPLE 5   Human Memory Model

In an experiment, students attended several lectures on a subject. Every month for a year after that, the students were tested to see how much of the material they remembered. The average scores for the group are given by the **human memory model**

$$f(t) = 80 - \ln(t + 1)^9, \quad 0 \le t \le 12$$

where t is the time in months. Find the average scores for the group after 8 months.

Solution

To make the calculations easier, rewrite the model using the Power Property, as follows.

$$f(t) = 80 - 9 \ln(t + 1), \quad 0 \le t \le 12$$

After 8 months, the average score was

$$f(8) = 80 - 9 \ln(8 + 1) \qquad \text{Substitute 8 for } t.$$
$$\approx 80 - 19.8 \qquad\qquad \text{Simplify.}$$
$$= 60.2. \qquad\qquad\quad \text{Average score after 8 months}$$

The graph of the function is shown in Figure 9.21.

✓ **CHECKPOINT** *Now try Exercise 113.*

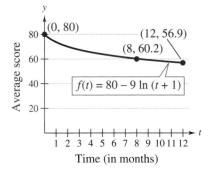

Human Memory Model
Figure 9.21

Concept Check

1. In your own words, give a verbal description of the Power Property $\ln u^n = n \ln u$.

2. Explain the steps you would take to evaluate the expression $\log_6 9 + \log_6 4$.

3. Explain the steps you would take to expand the expression

$$\log_{10} \frac{x^3}{y^5}.$$

4. For $x > 0$, is $2 \ln x$ *greater than, less than,* or *equal to* $\ln x^3 + \ln x^4 - \ln x^5$? Justify your answer.

9.4 EXERCISES

Go to pages 646–647 to record your assignments.

Developing Skills

In Exercises 1–24, use properties of logarithms to evaluate the expression without a calculator. (If not possible, state the reason.)

1. $\log_{12} 12^3$
2. $\log_3 81$
3. $\log_4 \left(\frac{1}{16}\right)^2$
4. $\log_7 \left(\frac{1}{49}\right)^3$
5. $\log_5 \sqrt[3]{5}$
6. $\ln \sqrt{e}$
7. $\ln 14^0$
8. $\ln \left(\frac{7.14}{7.14}\right)$
9. $\ln e^{-9}$
10. $\ln e^7$
11. $\log_4 8 + \log_4 2$
12. $\log_6 2 + \log_6 3$
13. $\log_8 4 + \log_8 16$
14. $\log_{10} 5 + \log_{10} 20$
15. $\log_3 54 - \log_3 2$
16. $\log_5 50 - \log_5 2$
17. $\log_6 72 - \log_6 2$
18. $\log_3 324 - \log_3 4$
19. $\log_2 5 - \log_2 40$
20. $\log_4 \left(\frac{3}{16}\right) + \log_4 \left(\frac{1}{3}\right)$
21. $\ln e^8 + \ln e^4$

22. $\ln e^5 - \ln e^2$
23. $\ln \frac{e^3}{e^2}$
24. $\ln(e^2 \cdot e^4)$

In Exercises 25–32, use $\log_4 2 = 0.5000$, $\log_4 3 \approx 0.7925$, and the properties of logarithms to approximate the expression. Do not use a calculator. *See Example 1.*

✓ 25. $\log_4 8$
26. $\log_4 24$
27. $\log_4 \frac{3}{2}$
28. $\log_4 \frac{9}{2}$
29. $\log_4 \sqrt[3]{9}$
30. $\log_4 \sqrt{3 \cdot 2^5}$
31. $\log_4 3^0$
32. $\log_4 4^3$

In Exercises 33–40, use $\ln 3 \approx 1.0986$, $\ln 5 \approx 1.6094$, and the properties of logarithms to approximate the expression. Use a calculator to verify your result.

33. $\ln 9$
34. $\ln 75$
35. $\ln \frac{5}{3}$
36. $\ln \frac{27}{125}$
37. $\ln \sqrt{45}$
38. $\ln \sqrt[3]{25}$
39. $\ln(3^5 \cdot 5^2)$
40. $\ln \sqrt{3^2 \cdot 5^3}$

In Exercises 41–46, use the properties of logarithms to verify the statement. *See Example 2.*

✓ **41.** $-3 \log_4 2 = \log_4 \frac{1}{8}$

42. $-2 \ln \frac{1}{3} = \ln 9$

43. $-3 \log_{10} 3 + \log_{10} \frac{3}{2} = \log_{10} \frac{1}{18}$

44. $-2 \ln 2 + \ln 24 = \ln 6$

45. $-\ln \frac{1}{7} = \ln 56 - \ln 8$

46. $-\log_5 10 = \log_5 10 - \log_5 100$

In Exercises 47–74, use the properties of logarithms to expand the expression. (Assume all variables are positive.) *See Example 3.*

✓ **47.** $\log_3 11x$

48. $\log_2 3x$

49. $\ln 3y$

50. $\ln 5x$

51. $\log_7 x^2$

52. $\log_3 x^3$

53. $\log_4 x^{-3}$

54. $\log_2 s^{-4}$

55. $\log_4 \sqrt{3x}$

56. $\log_3 \sqrt[3]{5y}$

57. $\log_2 \frac{z}{17}$

58. $\log_{10} \frac{7}{y}$

59. $\log_9 \frac{\sqrt{x}}{12}$

60. $\ln \frac{\sqrt{x}}{x + 9}$

61. $\ln x^2(y + 2)$

62. $\ln y(y + 1)^2$

63. $\log_4[x^6(x + 7)^2]$

64. $\log_8[(x - y)^3 z^6]$

65. $\log_3 \sqrt[3]{x + 1}$

66. $\log_5 \sqrt{xy}$

67. $\ln \sqrt{x(x + 2)}$

68. $\ln \sqrt[3]{x(x + 5)}$

69. $\ln \left(\frac{x + 1}{x + 4}\right)^2$

70. $\log_2 \left(\frac{x^2}{x - 3}\right)^3$

71. $\ln \sqrt[3]{\frac{x^2}{x + 1}}$

72. $\ln \sqrt{\frac{3x}{x - 5}}$

73. $\ln \frac{xy^2}{z^3}$

74. $\log_5 \frac{x^2 y^5}{z^7}$

In Exercises 75–102, use the properties of logarithms to condense the expression. *See Example 4.*

✓ **75.** $\log_{12} x - \log_{12} 3$

76. $\log_6 12 - \log_6 y$

77. $\log_3 5 + \log_3 x$

78. $\log_5 2x + \log_5 3y$

79. $\log_{10} 4 - \log_{10} x$

80. $\ln 10x - \ln z$

81. $4 \ln b$

82. $12 \log_4 z$

83. $-2 \log_5 2x$

84. $-5 \ln(x + 3)$

85. $7 \log_2 x + 3 \log_2 z$

86. $2 \log_{10} x + \frac{1}{2} \log_{10} y$

87. $\log_3 2 + \frac{1}{2} \log_3 y$

88. $\ln 6 - 3 \ln z$

89. $3 \ln x + \ln y - 2 \ln z$

90. $4 \ln 2 + 2 \ln x - \frac{1}{2} \ln y$

91. $4(\ln x + \ln y)$

92. $\frac{1}{3}(\ln 10 + \ln 4x)$

93. $2[\ln x - \ln(x + 1)]$

94. $5\left[\ln x - \frac{1}{2} \ln(x + 4)\right]$

95. $\log_4(x + 8) - 3 \log_4 x$

96. $5 \log_3 x + \log_3(x - 6)$

97. $\frac{1}{3} \log_5(x + 3) - \log_5(x - 6)$

98. $\frac{1}{4} \log_6(x + 1) - 5 \log_6(x - 4)$

99. $5 \log_6(c + d) - \frac{1}{2} \log_6(m - n)$

100. $2 \log_5(x + y) + 3 \log_5 w$

101. $\frac{1}{5}(3 \log_2 x - 4 \log_2 y)$

102. $\frac{1}{3}[\ln(x - 6) - 4 \ln y - 2 \ln z]$

In Exercises 103–108, simplify the expression.

103. $\ln 3e^2$

104. $\log_3(3^2 \cdot 4)$

105. $\log_5 \sqrt{50}$

106. $\log_2 \sqrt{22}$

107. $\log_8 \dfrac{8}{x^3}$

108. $\ln \dfrac{6}{e^5}$

In Exercises 109–112, use a graphing calculator to graph the two equations in the same viewing window. Use the graphs to verify that the expressions are equivalent. Assume $x > 0$.

109. $y_1 = \ln\left(\dfrac{10}{x^2 + 1}\right)^2$

$y_2 = 2[\ln 10 - \ln(x^2 + 1)]$

110. $y_1 = \ln \sqrt{x(x + 1)}$

$y_2 = \frac{1}{2}[\ln x + \ln(x + 1)]$

111. $y_1 = \ln[x^2(x + 2)]$

$y_2 = 2 \ln x + \ln(x + 2)$

112. $y_1 = \ln \dfrac{\sqrt{x}}{x - 3}$

$y_2 = \frac{1}{2} \ln x - \ln(x - 3)$

Solving Problems

113. *Sound Intensity* The relationship between the number of decibels B and the intensity of a sound I in watts per centimeter squared is given by

$$B = 10 \log_{10}\left(\dfrac{I}{10^{-16}}\right).$$

Use properties of logarithms to write the formula in simpler form, and determine the number of decibels of a thunderclap with an intensity of 10^{-3} watt per centimeter squared.

© Scott Stulberg/CORBIS

114. *Human Memory Model* Students participating in an experiment attended several lectures on a subject. Every month for a year after that, the students were tested to see how much of the material they remembered. The average scores for the group are given by the human memory model

$$f(t) = 80 - \log_{10}(t + 1)^{12}, \quad 0 \le t \le 12$$

where t is the time in months.

(a) Find the average scores for the group after 2 months and 8 months.

(b) ⌨ Use a graphing calculator to graph the function.

Molecular Transport In Exercises 115 and 116, use the following information. The energy E (in kilocalories per gram molecule) required to transport a substance from the outside to the inside of a living cell is given by

$$E = 1.4(\log_{10} C_2 - \log_{10} C_1)$$

where C_1 and C_2 are the concentrations of the substance outside and inside the cell, respectively.

115. Condense the expression.

116. The concentration of a substance inside a cell is twice the concentration outside the cell. How much energy is required to transport the substance from outside to inside the cell?

Explaining Concepts

True or False?   In Exercises 117–122, use properties of logarithms to determine whether the equation is true or false. Justify your answer.

117. $\log_2 8x = 3 + \log_2 x$

118. $\log_3(u + v) = \log_3 u + \log_3 v$

119. $\log_3(u + v) = \log_3 u \cdot \log_3 v$

120. $\dfrac{\log_6 10}{\log_6 3} = \log_6 10 - \log_6 3$

121. If $f(x) = \log_a x$, then $f(ax) = 1 + f(x)$.

122. If $f(x) = \log_a x$, then $f(a^n) = n$.

True or False?   In Exercises 123–127, determine whether the statement is true or false given that $f(x) = \ln x$. Justify your answer.

123. $f(0) = 0$

124. $f(2x) = \ln 2 + \ln x$

125. $f(x - 3) = \ln x - \ln 3, \quad x > 3$

126. $\sqrt{f(x)} = \frac{1}{2} \ln x$

127. If $f(x) > 0$, then $x > 1$.

128. *Think About It*   Without a calculator, approximate the natural logarithms of as many integers as possible between 1 and 20 using $\ln 2 \approx 0.6931$, $\ln 3 \approx 1.0986$, $\ln 5 \approx 1.6094$, and $\ln 7 \approx 1.9459$. Explain the method you used. Then verify your results with a calculator and explain any differences in the results.

129. *Think About It*   Explain how you can show that

$$\frac{\ln x}{\ln y} \neq \ln \frac{x}{y}.$$

Cumulative Review

In Exercises 130–135, solve the equation.

130. $\frac{2}{3}x + \frac{2}{3} = 4x - 6$

131. $x^2 - 10x + 17 = 0$

132. $\frac{5}{2x} - \frac{4}{x} = 3$

133. $\frac{1}{x} + \frac{2}{x - 5} = 0$

134. $|x - 4| = 3$

135. $\sqrt{x + 2} = 7$

In Exercises 136–139, sketch the parabola. Identify the vertex and any *x*-intercepts.

136. $g(x) = -(x + 2)^2$

137. $f(x) = x^2 - 16$

138. $g(x) = -2x^2 + 4x - 7$

139. $h(x) = x^2 + 6x + 14$

In Exercises 140–143, find the compositions (a) $f \circ g$ and (b) $g \circ f$. Then find the domain of each composition.

140. $f(x) = 4x + 9$
$g(x) = x - 5$

141. $f(x) = \sqrt{x}$
$g(x) = x - 3$

142. $f(x) = \frac{1}{x}$
$g(x) = x + 2$

143. $f(x) = \frac{5}{x^2 - 4}$
$g(x) = x + 1$

9.5 Solving Exponential and Logarithmic Equations

AP Photo/Tampa Tribune, Stephen McKnight

Why You Should Learn It

Exponential and logarithmic equations occur in many scientific applications. For instance, in Exercise 131 on page 633, you will use a logarithmic equation to determine a person's time of death.

1 ▶ Solve basic exponential and logarithmic equations.

What You Should Learn

1 ▶ Solve basic exponential and logarithmic equations.

2 ▶ Use inverse properties to solve exponential equations.

3 ▶ Use inverse properties to solve logarithmic equations.

4 ▶ Use exponential or logarithmic equations to solve application problems.

Exponential and Logarithmic Equations

In this section, you will study procedures for *solving equations* that involve exponential or logarithmic expressions. As a simple example, consider the exponential equation $2^x = 16$. By rewriting this equation in the form $2^x = 2^4$, you can see that the solution is $x = 4$. To solve this equation, you can use one of the following properties, which result from the fact that exponential and logarithmic functions are one-to-one functions.

> **One-to-One Properties of Exponential and Logarithmic Equations**
>
> Let a be a positive real number such that $a \neq 1$, and let x and y be real numbers. Then the following properties are true.
>
> **1.** $a^x = a^y$ if and only if $x = y$.
>
> **2.** $\log_a x = \log_a y$ if and only if $x = y$ $(x > 0, y > 0)$.

EXAMPLE 1   **Solving Exponential and Logarithmic Equations**

Solve each equation.

a. $4^{x+2} = 64$ Original equation

      $4^{x+2} = 4^3$ Rewrite with like bases.

    $x + 2 = 3$ One-to-one property

        $x = 1$ Subtract 2 from each side.

The solution is $x = 1$. Check this in the original equation.

b. $\ln(2x - 3) = \ln 11$ Original equation

      $2x - 3 = 11$ One-to-one property

          $2x = 14$ Add 3 to each side.

            $x = 7$ Divide each side by 2.

The solution is $x = 7$. Check this in the original equation.

✓ *CHECKPOINT Now try Exercise 21.*

2 ▶ Use inverse properties to solve exponential equations.

Solving Exponential Equations

In Example 1(a), you were able to use a one-to-one property to solve the original equation because each side of the equation was written in exponential form with the same base. However, if only one side of the equation is written in exponential form or if both sides cannot be written with the same base, it is more difficult to solve the equation. For example, to solve the equation $2^x = 7$, you must find the power to which 2 can be raised to obtain 7. To do this, *rewrite the exponential equation in logarithmic form* by taking the logarithm of each side, and use one of the inverse properties of exponents and logarithms listed below.

> ## Solving Exponential Equations
>
> To solve an exponential equation, first isolate the exponential expression. Then **take the logarithm of each side of the equation** (or write the equation in logarithmic form) and solve for the variable.

> ### Inverse Properties of Exponents and Logarithms
>
> | *Base a* | *Natural Base e* |
> |---|---|
> | 1. $\log_a(a^x) = x$ | $\ln(e^x) = x$ |
> | 2. $a^{(\log_a x)} = x$ | $e^{(\ln x)} = x$ |

EXAMPLE 2 Solving Exponential Equations

Solve each exponential equation.

a. $2^x = 7$ **b.** $4^{x-3} = 9$ **c.** $2e^x = 10$

Solution

a. To isolate x, take the $\log_2$ of each side of the equation or write the equation in logarithmic form, as follows.

$$2^x = 7 \qquad \text{Write original equation.}$$

$$x = \log_2 7 \qquad \text{Inverse property}$$

The solution is $x = \log_2 7 \approx 2.807$. Check this in the original equation.

b.
$$4^{x-3} = 9 \qquad \text{Write original equation.}$$

$$x - 3 = \log_4 9 \qquad \text{Inverse property}$$

$$x = \log_4 9 + 3 \qquad \text{Add 3 to each side.}$$

The solution is $x = \log_4 9 + 3 \approx 4.585$. Check this in the original equation.

c.
$$2e^x = 10 \qquad \text{Write original equation.}$$

$$e^x = 5 \qquad \text{Divide each side by 2.}$$

$$x = \ln 5 \qquad \text{Inverse property}$$

The solution is $x = \ln 5 \approx 1.609$. Check this in the original equation.

 **CHECKPOINT** *Now try Exercise 39.*

Study Tip

Remember that to evaluate a logarithm such as $\log_2 7$ you need to use the change-of-base formula.

$$\log_2 7 = \frac{\ln 7}{\ln 2} \approx 2.807$$

Similarly,

$$\log_4 9 + 3 = \frac{\ln 9}{\ln 4} + 3$$

$$\approx 1.585 + 3$$

$$= 4.585$$

Technology: Tip

Remember that you can use a graphing calculator to solve equations graphically or to check solutions that are obtained algebraically. For instance, to check the solutions in Examples 2(a) and 2(c), graph each side of each equation, as shown below.

Graph $y_1 = 2^x$ and $y_2 = 7$. Then use the *intersect* feature of the graphing calculator to approximate the intersection of the two graphs to be $x \approx 2.807$.

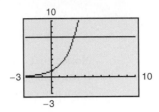

Graph $y_1 = 2e^x$ and $y_2 = 10$. Then use the *intersect* feature of the graphing calculator to approximate the intersection of the two graphs to be $x \approx 1.609$.

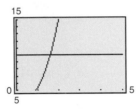

EXAMPLE 3   **Solving an Exponential Equation**

Solve $5 + e^{x+1} = 20$.

Solution

$5 + e^{x+1} = 20$	Write original equation.
$e^{x+1} = 15$	Subtract 5 from each side.
$\ln e^{x+1} = \ln 15$	Take the logarithm of each side.
$x + 1 = \ln 15$	Inverse property
$x = -1 + \ln 15$	Subtract 1 from each side.

The solution is $x = -1 + \ln 15 \approx 1.708$. You can check this as follows.

Check

$5 + e^{x+1} = 20$	Write original equation.
$5 + e^{-1 + \ln 15 + 1} \overset{?}{=} 20$	Substitute $-1 + \ln 15$ for x.
$5 + e^{\ln 15} \overset{?}{=} 20$	Simplify.
$5 + 15 = 20$	Solution checks. ✓

✓ **CHECKPOINT** *Now try Exercise 67.*

3 ▶ Use inverse properties to solve logarithmic equations.

Solving Logarithmic Equations

You know how to solve an exponential equation by *taking the logarithm of each side*. To solve a logarithmic equation, you need to **exponentiate** each side. For instance, to solve a logarithmic equation such as $\ln x = 2$, you can exponentiate each side of the equation as follows.

$$\ln x = 2 \qquad\qquad \text{Write original equation.}$$

$$e^{\ln x} = e^2 \qquad\qquad \text{Exponentiate each side.}$$

$$x = e^2 \qquad\qquad \text{Inverse property}$$

Notice that you obtain the same result by writing the equation in exponential form. This procedure is demonstrated in the next three examples. The following guideline can be used for solving logarithmic equations.

> ### Solving Logarithmic Equations
>
> To solve a logarithmic equation, first isolate the logarithmic expression. Then **exponentiate each side of the equation** (or write the equation in exponential form) and solve for the variable.

EXAMPLE 4   **Solving Logarithmic Equations**

a. $2 \log_4 x = 5 \qquad\qquad$ Original equation

$$\log_4 x = \frac{5}{2} \qquad\qquad \text{Divide each side by 2.}$$

$$4^{\log_4 x} = 4^{5/2} \qquad\qquad \text{Exponentiate each side.}$$

$$x = 4^{5/2} \qquad\qquad \text{Inverse property}$$

$$x = 32 \qquad\qquad \text{Simplify.}$$

The solution is $x = 32$. Check this in the original equation, as follows.

Check

$$2 \log_4 x = 5 \qquad\qquad \text{Original equation}$$

$$2 \log_4(32) \overset{?}{=} 5 \qquad\qquad \text{Substitute 32 for } x.$$

$$2(2.5) \overset{?}{=} 5 \qquad\qquad \text{Use a calculator.}$$

$$5 = 5 \qquad\qquad \text{Solution checks. } \checkmark$$

b. $\dfrac{1}{4} \log_2 x = \dfrac{1}{2} \qquad\qquad$ Original equation

$$\log_2 x = 2 \qquad\qquad \text{Multiply each side by 4.}$$

$$2^{\log_2 x} = 2^2 \qquad\qquad \text{Exponentiate each side.}$$

$$x = 4 \qquad\qquad \text{Inverse property}$$

The solution is $x = 4$. Check this in the original equation.

✓ *CHECKPOINT Now try Exercise 83.*

EXAMPLE 5 **Solving a Logarithmic Equation**

$3 \log_{10} x = 6$	Original equation
$\log_{10} x = 2$	Divide each side by 3.
$x = 10^2$	Exponential form
$x = 100$	Simplify.

The solution is $x = 100$. Check this in the original equation.

 CHECKPOINT *Now try Exercise 87.*

EXAMPLE 6 **Solving a Logarithmic Equation**

$20 \ln 0.2x = 30$	Original equation
$\ln 0.2x = 1.5$	Divide each side by 20.
$e^{\ln 0.2x} = e^{1.5}$	Exponentiate each side.
$0.2x = e^{1.5}$	Inverse property
$x = 5e^{1.5}$	Divide each side by 0.2.

The solution is $x = 5e^{1.5} \approx 22.408$. Check this in the original equation.

 CHECKPOINT *Now try Exercise 93.*

The next two examples use logarithmic properties as part of the solutions.

EXAMPLE 7 **Solving a Logarithmic Equation**

$\log_3 2x - \log_3(x - 3) = 1$	Original equation
$\log_3 \dfrac{2x}{x - 3} = 1$	Condense the left side.
$\dfrac{2x}{x - 3} = 3^1$	Exponential form
$2x = 3x - 9$	Multiply each side by $x - 3$.
$-x = -9$	Subtract $3x$ from each side.
$x = 9$	Divide each side by -1.

The solution is $x = 9$. Check this in the original equation.

CHECKPOINT *Now try Exercise 105.*

EXAMPLE 8   Checking For Extraneous Solutions

$\log_6 x + \log_6(x - 5) = 2$	Original equation
$\log_6 [x(x - 5)] = 2$	Condense the left side.
$x(x - 5) = 6^2$	Exponential form
$x^2 - 5x - 36 = 0$	Write in general form.
$(x - 9)(x + 4) = 0$	Factor.
$x - 9 = 0 \implies x = 9$	Set 1st factor equal to 0.
$x + 4 = 0 \implies x = -4$	Set 2nd factor equal to 0.

Check the possible solutions $x = 9$ and $x = -4$ in the original equation.

First Solution	*Second Solution*
$\log_6 (9) + \log_6 (9 - 5) \overset{?}{=} 2$	$\log_6 (-4) + \log_6 (-4 - 5) \overset{?}{=} 2$
$\log_6 (9 \cdot 4) \overset{?}{=} 2$	$\log_6 (-4) + \log_6 (-9) \neq 2$ ✗
$\log_6 36 = 2$ ✓	

Of the two possible solutions, only $x = 9$ checks. So, $x = -4$ is extraneous.

✓ **CHECKPOINT**   *Now try Exercise 109.*

---

4 ▶ Use exponential or logarithmic equations to solve application problems.

Application

EXAMPLE 9   Compound Interest

A deposit of $5000 is placed in a savings account for 2 years. The interest on the account is compounded continuously. At the end of 2 years, the balance in the account is $5416.44. What is the annual interest rate for this account?

Solution

Formula: $A = Pe^{rt}$

Labels:
Principal $= P = 5000$	(dollars)
Amount $= A = 5416.44$	(dollars)
Time $= t = 2$	(years)
Annual interest rate $= r$	(percent in decimal form)

Equation:
$5416.44 = 5000e^{2r}$	Substitute for A, P, and t.
$1.083288 = e^{2r}$	Divide each side by 5000 and simplify.
$\ln 1.083288 = \ln(e^{2r})$	Take logarithm of each side.
$0.08 \approx 2r \implies 0.04 \approx r$	Inverse property

The annual interest rate is approximately 4%. Check this solution.

✓ **CHECKPOINT**   *Now try Exercise 123.*

Concept Check

1. Can the one-to-one property of logarithms be used to solve the equation $\log_2 x = \log_3 9$? Explain.

2. Explain how to solve $5^{x+2} = 5^4$.

3. Which equation requires logarithms for its solution: $2^{x-1} = 32$ or $2^{x-1} = 30$? Explain.

4. If a solution of a logarithmic equation is negative, does this imply that the solution is extraneous? Explain.

9.5 EXERCISES

Go to pages 646–647 to record your assignments.

Developing Skills

In Exercises 1–6, determine whether each value of x is a solution of the equation.

1. $3^{2x-5} = 27$
   (a) $x = 1$
   (b) $x = 4$

2. $2^{x+5} = 16$
   (a) $x = -1$
   (b) $x = 0$

3. $e^{x+5} = 45$
   (a) $x = -5 + \ln 45$
   (b) $x \approx -2.1933$

4. $4^{x-2} = 250$
   (a) $x = 2 + \log_4 250$
   (b) $x \approx 4.9829$

5. $\log_9(6x) = \frac{3}{2}$
   (a) $x = 27$
   (b) $x = \frac{9}{2}$

6. $\ln(x + 3) = 2.5$
   (a) $x = -3 + e^{2.5}$
   (b) $x \approx 9.1825$

In Exercises 7–34, solve the equation. (Do not use a calculator.) *See Example 1.*

7. $7^x = 7^3$

8. $4^x = 4^6$

9. $e^{1-x} = e^4$

10. $e^{x+3} = e^8$

11. $5^{x+6} = 25^5$

12. $2^{x-4} = 8^2$

13. $6^{2x} = 36$

14. $5^{3x} = 25$

15. $3^{2-x} = 81$

16. $4^{2x-1} = 64$

17. $5^x = \frac{1}{125}$

18. $3^x = \frac{1}{243}$

19. $2^{x+2} = \frac{1}{16}$

20. $3^{x+2} = \frac{1}{27}$

✓ 21. $4^{x+3} = 32^x$

22. $9^{x-2} = 243^{x+1}$

23. $\ln 5x = \ln 22$

24. $\ln 4x = \ln 30$

25. $\log_6 3x = \log_6 18$

26. $\log_5 2x = \log_5 36$

27. $\ln(3 - x) = \ln 10$

28. $\ln(2x - 3) = \ln 17$

29. $\log_2(x + 3) = \log_2 7$

30. $\log_4(x - 8) = \log_4(-4)$

31. $\log_5(2x - 3) = \log_5(4x - 5)$

32. $\log_3(4 - 3x) = \log_3(2x + 9)$

33. $\log_3(2 - x) = 2$

34. $\log_2(3x - 1) = 5$

In Exercises 35–38, simplify the expression.

35. $\ln e^{2x-1}$

36. $\log_3 3^{x^2}$

37. $10^{\log_{10} 2x}$

38. $e^{\ln(x+1)}$

In Exercises 39–82, solve the exponential equation. (Round your answer to two decimal places.) *See Examples 2 and 3.*

✓ 39. $3^x = 91$

40. $4^x = 40$

41. $5^x = 8.2$

42. $2^x = 3.6$

43. $6^{2x} = 205$

44. $4^{3x} = 168$

45. $7^{3y} = 126$

46. $5^{5y} = 305$

47. $3^{2-x} = 8$

48. $5^{3-x} = 15$

49. $10^{x+6} = 250$

50. $12^{x-1} = 324$

51. $4e^{-x} = 24$

52. $6e^{-x} = 3$

53. $\frac{1}{4}e^x = 5$

54. $\frac{2}{3}e^x = 1$

55. $\frac{1}{2}e^{-2x} = 9$

56. $4e^{-3x} = 6$

57. $250(1.04)^x = 1000$

58. $32(1.5)^x = 640$

59. $300e^{x/2} = 9000$

60. $7500e^{x/3} = 1500$

61. $1000^{0.12x} = 25{,}000$

62. $1800^{0.2x} = 225$

63. $\frac{1}{5}(4^{x+2}) = 300$

64. $3(2^{t+4}) = 350$

65. $6 + 2^{x-1} = 1$

66. $5^{x+6} - 4 = 12$

✓ 67. $7 + e^{2-x} = 28$

68. $24 + e^{4-x} = 22$

69. $8 - 12e^{-x} = 7$

70. $6 - 3e^{-x} = -15$

71. $4 + e^{2x} = 10$

72. $10 + e^{4x} = 18$

73. $17 - e^{x/4} = 14$

74. $50 - e^{x/2} = 35$

75. $23 - 5e^{x+1} = 3$

76. $3e^{1-x} - 5 = 72$

77. $4(1 + e^{x/3}) = 84$

78. $50(3 - e^{2x}) = 125$

79. $\dfrac{8000}{(1.03)^t} = 6000$

80. $\dfrac{5000}{(1.05)^x} = 250$

81. $\dfrac{300}{2 - e^{-0.15t}} = 200$

82. $\dfrac{500}{1 + e^{-0.1t}} = 400$

In Exercises 83–118, solve the logarithmic equation. (Round your answer to two decimal places.) *See Examples 4–8.*

✓ 83. $\log_{10} x = -1$

84. $\log_{10} x = 3$

85. $\log_3 x = 4.7$

86. $\log_3 x = -1.8$

✓ 87. $4 \log_3 x = 28$

88. $6 \log_2 x = 18$

89. $16 \ln x = 30$

90. $12 \ln x = 20$

91. $\log_{10} 4x = 2$

92. $\log_3 6x = 4$

✓ 93. $\ln 2x = \frac{1}{5}$

94. $\ln(0.5t) = \frac{1}{4}$

95. $\ln x^2 = 6$

96. $\ln \sqrt{x} = 1.3$

97. $2 \log_4(x + 5) = 3$

98. $5 \log_{10}(x + 2) = 15$

99. $\frac{3}{4} \ln(x + 4) = -2$

100. $\frac{2}{3} \ln(x + 1) = -1$

101. $7 - 2 \log_2 x = 4$

102. $5 - 4 \log_2 x = 2$

103. $-1 + 3 \log_{10} \dfrac{x}{2} = 8$

104. $-5 + 2 \ln 3x = 5$

✓ 105. $\log_4 x + \log_4 5 = 2$

106. $\log_5 x - \log_5 4 = 2$

107. $\log_6(x + 8) + \log_6 3 = 2$

108. $\log_7(x - 1) - \log_7 4 = 1$

✓ 109. $\log_5(x + 3) - \log_5 x = 1$

110. $\log_3(x - 2) + \log_3 5 = 3$

111. $\log_{10} x + \log_{10}(x - 3) = 1$

112. $\log_{10} x + \log_{10}(x + 1) = 0$

113. $\log_2(x - 1) + \log_2(x + 3) = 3$

114. $\log_6(x - 5) + \log_6 x = 2$

115. $\log_{10} 4x - \log_{10}(x - 2) = 1$

116. $\log_2 3x - \log_2(x + 4) = 3$

117. $\log_2 x + \log_2(x + 2) - \log_2 3 = 4$

118. $\log_3 2x + \log_3(x - 1) - \log_3 4 = 1$

In Exercises 119–122, use a graphing calculator to approximate the x-intercept of the graph.

119. $y = 10^{x/2} - 5$

120. $y = 2e^x - 21$

121. $y = 6 \ln(0.4x) - 13$

122. $y = 5 \log_{10}(x + 1) - 3$

Solving Problems

✓ 123. *Compound Interest*   A deposit of $10,000 is placed in a savings account for 2 years. The interest for the account is compounded continuously. At the end of 2 years, the balance in the account is $11,051.71. What is the annual interest rate for this account?

124. *Compound Interest*   A deposit of $2500 is placed in a savings account for 2 years. The interest for the account is compounded continuously. At the end of 2 years, the balance in the account is $2847.07. What is the annual interest rate for this account?

125. *Doubling Time* Solve the exponential equation $5000 = 2500e^{0.09t}$ for t to determine the number of years for an investment of \$2500 to double in value when compounded continuously at the rate of 9%.

126. *Doubling Rate* Solve the exponential equation $10{,}000 = 5000e^{10r}$ for r to determine the interest rate required for an investment of \$5000 to double in value when compounded continuously for 10 years.

127. *Sound Intensity* The relationship between the number of decibels B and the intensity of a sound I in watts per centimeter squared is given by

$$B = 10 \log_{10}\left(\frac{I}{10^{-16}}\right).$$

Determine the intensity of a sound I if it registers 80 decibels on a decibel meter.

128. *Sound Intensity* The relationship between the number of decibels B and the intensity of a sound I in watts per centimeter squared is given by

$$B = 10 \log_{10}\left(\frac{I}{10^{-16}}\right).$$

Determine the intensity of a sound I if it registers 110 decibels on a decibel meter.

129. *Friction* In order to restrain an untrained horse, a trainer partially wraps a rope around a cylindrical post in a corral (see figure). The horse is pulling on the rope with a force of 200 pounds. The force F (in pounds) needed to hold back the horse is $F = 200e^{-0.5\pi\theta/180}$, where θ is the angle of wrap (in degrees). The trainer needs to know the smallest value of θ for which a force of 80 pounds will hold the horse.

(a) Find the smallest value of θ algebraically by letting $F = 80$ and solving the resulting equation.

(b) 🖩 Find the smallest value of θ graphically by using a graphing calculator to graph the equations $y_1 = 200e^{-0.5\pi\theta/180}$ and $y_2 = 80$ in order to find the point of intersection.

130. *Online Retail* The projected online retail sales S (in billions of dollars) in the United States for the years 2006 through 2011 are modeled by the equation $S = 59.8e^{0.1388t}$, for $6 \le t \le 11$, where t is the time in years, with $t = 6$ corresponding to 2006. You want to know the year when S is about \$210 billion.   (Source: Forrester Research, Inc.)

(a) Find the year algebraically by letting $S = 210$ and solving the resulting equation.

(b) 🖩 Find the year graphically by using a graphing calculator to graph the equations $y_1 = 59.8e^{0.1388t}$ and $y_2 = 210$ in order to find the point of intersection.

Newton's Law of Cooling In Exercises 131 and 132, use Newton's Law of Cooling

$$kt = \ln\frac{T - S}{T_0 - S}$$

where T is the temperature of a body (in °F), t is the number of hours elapsed, S is the temperature of the environment, and T_0 is the initial temperature of the body.

131. *Time of Death* A corpse was discovered in a motel room at 10 P.M., and its temperature was 85°F. Three hours later, the temperature of the corpse was 78°F. The temperature of the motel room is a constant 65°F.

(a) What is the constant k?

(b) Find the time of death using the fact that the temperature of the corpse at the time of death was 98.6°F.

(c) What is the temperature of the corpse two hours after death?

132. *Time of Death* A corpse was discovered in the bedroom of a home at 7 A.M., and its temperature was 92°F. Two hours later, the temperature of the corpse was 88°F. The temperature of the bedroom is a constant 68°F.

(a) What is the constant k?

(b) Find the time of death using the fact that the temperature of the corpse at the time of death was 98.6°F.

(c) What is the temperature of the corpse three hours after death?

Oceanography In Exercises 133 and 134, use the following information. Oceanographers use the density d (in grams per cubic centimeter) of seawater to obtain information about the circulation of water masses and the rates at which waters of different densities mix. For water with a salinity of 30%, the water temperature T (in °C) is related to the density by

$$T = 7.9 \ln(1.0245 - d) + 61.84.$$

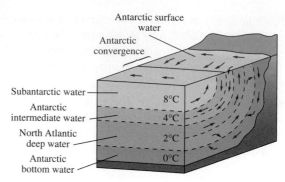

Figure for 133 and 134

This cross section shows complex currents at various depths in the South Atlantic Ocean off Antarctica.

133. Find the densities of the Subantarctic water and the Antarctic bottom water shown in the figure.

134. Find the densities of the Antarctic intermediate water and the North Atlantic deep water shown in the figure.

Explaining Concepts

135. ✎ State the three basic properties of logarithms.

136. ✎ Explain how to solve $10^{2x-1} = 5316$.

137. ✎ In your own words, state the guidelines for solving exponential and logarithmic equations.

138. ✎ Why is it possible for a logarithmic equation to have an extraneous solution?

Cumulative Review

In Exercises 139–142, solve the equation by using the Square Root Property.

139. $x^2 = -25$

140. $x^2 - 49 = 0$

141. $9n^2 - 16 = 0$

142. $(2a + 3)^2 = 18$

In Exercises 143 and 144, solve the equation of quadratic form.

143. $t^4 - 13t^2 + 36 = 0$

144. $u + 2\sqrt{u} - 15 = 0$

In Exercises 145–148, complete the table of widths, lengths, perimeters, and areas of rectangles.

	Width	Length	Perimeter	Area
145.	$2.5x$	x	42 in.	
146.	w	$1.6w$	78 ft	
147.	w	$w + 4$		192 km²
148.	$x - 3$	x		270 cm²

9.6 Applications

Blend Images/Punchstock

What You Should Learn

1 ▶ Use exponential equations to solve compound interest problems.

2 ▶ Use exponential equations to solve growth and decay problems.

3 ▶ Use logarithmic equations to solve intensity problems.

Why You Should Learn It

Exponential growth and decay models can be used in many real-life situations. For instance, in Exercise 62 on page 644, you will use an exponential growth model to represent the number of music albums downloaded in the United States.

1 ▶ Use exponential equations to solve compound interest problems.

Compound Interest

In Section 9.1, you were introduced to two formulas for compound interest. Recall that in these formulas, A is the balance, P is the principal, r is the annual interest rate (in decimal form), and t is the time in years.

n Compoundings per Year

$$A = P\left(1 + \frac{r}{n}\right)^{nt}$$

Continuous Compounding

$$A = Pe^{rt}$$

EXAMPLE 1   **Finding the Annual Interest Rate**

An investment of $50,000 is made in an account that compounds interest quarterly. After 4 years, the balance in the account is $71,381.07. What is the annual interest rate for this account?

Solution

Formula: $A = P\left(1 + \dfrac{r}{n}\right)^{nt}$

Labels:

Principal $= P = 50{,}000$	(dollars)
Amount $= A = 71{,}381.07$	(dollars)
Time $= t = 4$	(years)
Number of compoundings per year $= n = 4$	
Annual interest rate $= r$	(percent in decimal form)

Equation:

$71{,}381.07 = 50{,}000\left(1 + \dfrac{r}{4}\right)^{(4)(4)}$ Substitute for A, P, n, and t.

$1.42762 \approx \left(1 + \dfrac{r}{4}\right)^{16}$ Divide each side by 50,000.

$(1.42762)^{1/16} \approx 1 + \dfrac{r}{4}$ Raise each side to $\frac{1}{16}$ power.

$1.0225 \approx 1 + \dfrac{r}{4}$ Simplify.

$0.09 \approx r$ Subtract 1 from each side and then multiply each side by 4.

The annual interest rate is approximately 9%. Check this in the original problem.

✓ **CHECKPOINT** *Now try Exercise 1.*

Study Tip

To remove an exponent from one side of an equation, you can often raise each side of the equation to the *reciprocal* power. For instance, in Example 1, the exponent 16 is eliminated from the right side by raising each side to the reciprocal power $\frac{1}{16}$.

EXAMPLE 2   **Doubling Time for Continuous Compounding**

An investment is made in a trust fund at an annual interest rate of 8.75%, compounded continuously. How long will it take for the investment to double?

Study Tip

In "doubling time" problems, you do not need to know the value of the principal P to find the doubling time. As shown in Example 2, the factor P divides out of the equation and so does not affect the doubling time.

Solution

$A = Pe^{rt}$	Formula for continuous compounding
$2P = Pe^{0.0875t}$	Substitute known values.
$2 = e^{0.0875t}$	Divide each side by P.
$\ln 2 = 0.0875t$	Inverse property
$\dfrac{\ln 2}{0.0875} = t$	Divide each side by 0.0875.
$7.92 \approx t$	Use a calculator.

It will take approximately 7.92 years for the investment to double.

Check

$2P \stackrel{?}{=} Pe^{0.0875(7.92)}$	Substitute $2P$ for A, 0.0875 for r, and 7.92 for t.
$2P \stackrel{?}{=} Pe^{0.693}$	Simplify.
$2P \approx 1.9997P$	Solution checks. ✓

 CHECKPOINT *Now try Exercise 9.*

EXAMPLE 3   **Finding the Type of Compounding**

You deposit $1000 in an account. At the end of 1 year, your balance is $1077.63. The bank tells you that the annual interest rate for the account is 7.5%. How was the interest compounded?

Solution

If the interest had been compounded continuously at 7.5%, the balance would have been $A = 1000e^{(0.075)(1)} = \$1077.88$. Because the actual balance is slightly less than this, you should use the formula for interest that is compounded n times per year.

$$A = 1000\left(1 + \frac{0.075}{n}\right)^{n} = 1000\left(1 + \frac{0.075}{12}\right)^{12} \approx 1077.63$$

At this point, it is not clear what you should do to solve the equation for n. However, by completing a table like the one shown below, you can see that $n = 12$. So, the interest was compounded monthly.

n	1	4	12	365
$1000\left(1 + \dfrac{0.075}{n}\right)^{n}$	1075	1077.14	1077.63	1077.88

 **CHECKPOINT** *Now try Exercise 13.*

In Example 3, notice that an investment of $1000 compounded monthly produced a balance of $1077.63 at the end of 1 year. Because $77.63 of this amount is interest, the **effective yield** for the investment is

$$\text{Effective yield} = \frac{\text{Year's interest}}{\text{Amount invested}} = \frac{77.63}{1000} = 0.07763 = 7.763\%.$$

In other words, the effective yield for an investment collecting compound interest is the *simple interest rate* that would yield the same balance at the end of 1 year.

EXAMPLE 4   **Finding the Effective Yield**

An investment is made in an account that pays 6.75% interest, compounded continuously. What is the effective yield for this investment?

Solution

Notice that you do not have to know the principal or the time that the money will be left in the account. Instead, you can choose an arbitrary principal, such as $1000. Then, because effective yield is based on the balance at the end of 1 year, you can use the following formula.

$$A = Pe^{rt}$$

$$= 1000e^{0.0675(1)}$$

$$\approx 1069.83$$

Now, because the account would earn $69.83 in interest after 1 year for a principal of $1000, you can conclude that the effective yield is

$$\text{Effective yield} = \frac{69.83}{1000} = 0.06983 = 6.983\%.$$

✓ **CHECKPOINT** *Now try Exercise 19.*

 2 ▶ Use exponential equations to solve growth and decay problems.

Growth and Decay

The balance in an account earning *continuously* compounded interest is one example of a quantity that increases over time according to the **exponential growth model** $y = Ce^{kt}$.

Exponential Growth and Decay

The mathematical model for exponential growth or decay is given by

$$y = Ce^{kt}.$$

For this model, t is the time, C is the original amount of the quantity, and y is the amount after time t. The number k is a constant that is determined by the rate of growth (or decay). If $k > 0$, the model represents **exponential growth,** and if $k < 0$, it represents **exponential decay.**

One common application of exponential growth is in modeling the growth of a population. Example 5 illustrates the use of the growth model

$$y = Ce^{kt}, \quad k > 0.$$

EXAMPLE 5   Population Growth

The population of Texas was 17 million in 1990 and 21 million in 2000. What would you estimate the population of Texas to be in 2010? (Source: U.S. Census Bureau)

Solution

If you assumed a *linear growth model*, you would simply estimate the population in the year 2010 to be 25 million because the population would increase by 4 million every 10 years. However, social scientists and demographers have discovered that *exponential growth models* are better than linear growth models for representing population growth. So, you can use the exponential growth model

$$y = Ce^{kt}.$$

In this model, let $t = 0$ represent 1990. The given information about the population can be described by the following table.

t (year)	0	10	20
Ce^{kt} (million)	$Ce^{k(0)} = 17$	$Ce^{k(10)} = 21$	$Ce^{k(20)} = ?$

To find the population when $t = 20$, you must first find the values of C and k. From the table, you can use the fact that $Ce^{k(0)} = Ce^0 = 17$ to conclude that $C = 17$. Then, using this value of C, you can solve for k as follows.

$Ce^{k(10)} = 21$	From table
$17e^{10k} = 21$	Substitute value of C.
$e^{10k} = \dfrac{21}{17}$	Divide each side by 17.
$10k = \ln \dfrac{21}{17}$	Inverse property
$k = \dfrac{1}{10} \ln \dfrac{21}{17}$	Divide each side by 10.
$k \approx 0.0211$	Simplify.

Finally, you can use this value of k in the model from the table for 2010 (for $t = 20$) to estimate the population in the year 2010 to be

$$17e^{0.0211(20)} \approx 17(1.53) = 26.01 \text{ million.}$$

Figure 9.22 graphically compares the exponential growth model with a linear growth model.

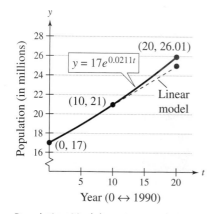

Population Models

Figure 9.22

✓ **CHECKPOINT** *Now try Exercise 47.*

EXAMPLE 6 **Radioactive Decay**

Radioactive iodine-125 is a by-product of some types of nuclear reactors. Its **half-life** is 60 days. That is, after 60 days, a given amount of radioactive iodine-125 will have decayed to half the original amount. A nuclear accident occurs and releases 20 grams of radioactive iodine-125. How long will it take for the radioactive iodine to decay to a level of 1 gram?

Solution

Use the model for exponential decay, $y = Ce^{kt}$, and the information given in the problem to set up the following table.

t (days)	0	60	?
Ce^{kt} (grams)	$Ce^{k(0)} = 20$	$Ce^{k(60)} = 10$	$Ce^{k(t)} = 1$

Because $Ce^{k(0)} = Ce^0 = 20$, you can conclude that $C = 20$. Then, using this value of C, you can solve for k as follows.

$Ce^{k(60)} = 10$	From table
$20e^{60k} = 10$	Substitute value of C.
$e^{60k} = \dfrac{1}{2}$	Divide each side by 20.
$60k = \ln \dfrac{1}{2}$	Inverse property
$k = \dfrac{1}{60} \ln \dfrac{1}{2}$	Divide each side by 60.
$k \approx -0.01155$	Simplify.

Finally, you can use this value of k in the model from the table to find the time when the amount is 1 gram, as follows.

$Ce^{kt} = 1$	From table
$20e^{-0.01155t} = 1$	Substitute values of C and k.
$e^{-0.01155t} = \dfrac{1}{20}$	Divide each side by 20.
$-0.01155t = \ln \dfrac{1}{20}$	Inverse property
$t = \dfrac{1}{-0.01155} \ln \dfrac{1}{20}$	Divide each side by -0.01155.
$t \approx 259.4$ days	Simplify.

So, 20 grams of radioactive iodine-125 will have decayed to 1 gram after about 259.4 days. This solution is shown graphically in Figure 9.23.

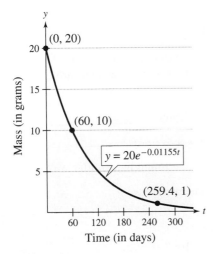

Radioactive Decay
Figure 9.23

✓ **CHECKPOINT** *Now try Exercise 65.*

EXAMPLE 7 **Website Growth**

Your college created an algebra tutoring website in 2000. The number of hits per year at the website has grown exponentially. The website had 4080 hits in 2000 and 22,440 hits in 2008. Predict the number of hits in 2014.

Solution

In the exponential growth model $y = Ce^{kt}$, let $t = 0$ represent 2000. Next, use the information given in the problem to set up the table shown at the left. Because $Ce^{k(0)} = Ce^0 = 4080$, you can conclude that $C = 4080$. Then, using this value of C, you can solve for k as follows.

t (year)	Ce^{kt}
0	$Ce^{k(0)} = 4080$
8	$Ce^{k(8)} = 22{,}440$
14	$Ce^{k(14)} = ?$

$$Ce^{k(8)} = 22{,}440 \qquad \text{From table}$$

$$4080e^{8k} = 22{,}440 \qquad \text{Substitute value of } C.$$

$$e^{8k} = 5.5 \qquad \text{Divide each side by 4080.}$$

$$8k = \ln 5.5 \qquad \text{Inverse property}$$

$$k = \tfrac{1}{8}\ln 5.5 \approx 0.2131 \qquad \text{Divide each side by 8 and simplify.}$$

Finally, you can use this value of k in the model from the table to predict the number of hits in 2014 to be $4080e^{0.2131(14)} \approx 80{,}600$.

 CHECKPOINT *Now try Exercise 61.*

3 ▶ Use logarithmic equations to solve intensity problems.

Intensity Models

On the **Richter scale**, the magnitude R of an earthquake can be measured by the **intensity model** $R = \log_{10} I$, where I is the intensity of the shock wave.

EXAMPLE 8 **Earthquake Intensity**

In 2007, an earthquake near the coast of Peru measured 8.0 on the Richter scale. Weeks later, an earthquake in San Francisco that was felt by thousands of people measured 5.4 on the Richter scale. Compare the intensities of the two earthquakes.

Solution

The intensity of Peru's earthquake is given as follows.

$$8.0 = \log_{10} I \quad \Longrightarrow \quad 10^8 = I \qquad \text{Inverse property}$$

The intensity of San Francisco's earthquake can be found in a similar way.

$$5.4 = \log_{10} I \quad \Longrightarrow \quad 10^{5.4} = I \qquad \text{Inverse property}$$

The ratio of these two intensities is

$$\frac{I \text{ for Peru}}{I \text{ for San Francisco}} = \frac{10^{8.0}}{10^{5.4}} = 10^{8.0-5.4} = 10^{2.6} \approx 398.$$

So, Peru's earthquake had an intensity that was about 398 times greater than the intensity of San Francisco's earthquake.

 **CHECKPOINT** *Now try Exercise 75.*

The intensity of an 8.0 earthquake is evidenced by the earthquake of August 15, 2007, centered off the coast of central Peru. This catastrophe left at least 35,500 buildings destroyed, 1090 people injured, and 514 people dead.

_____ Concept Check _____

1. *True or False?* If an account earns simple interest for one year, then the effective yield for the account is the simple interest rate. Justify your answer.

2. *True or False?* The exponential model $y = Ce^{0.1t}$ represents exponential decay. Justify your answer.

3. In a radioactive decay problem, you are given the initial amount and the half-life of a radioactive substance. To write the decay model $y = Ce^{kt}$ for the problem, explain how you can find the values of C and k.

4. Explain how you can use the Richter scale measurements R_1 and R_2 of two earthquakes to compare the intensities of the two earthquakes.

9.6 EXERCISES

Go to pages 646–647 to record your assignments.

_____ Solving Problems _____

Compound Interest   In Exercises 1–6, find the annual interest rate. *See Example 1.*

	Principal	Balance	Time	Compounding
✓ 1.	$500	$1004.83	10 years	Monthly
2.	$3000	$21,628.70	20 years	Quarterly
3.	$1000	$36,581.00	40 years	Daily
4.	$200	$314.85	5 years	Yearly
5.	$750	$8267.38	30 years	Continuous
6.	$2000	$4234.00	10 years	Continuous

Doubling Time   In Exercises 7–12, find the time for the investment to double. Use a graphing calculator to verify the result graphically. *See Example 2.*

	Principal	Rate	Compounding
7.	$2500	7.5%	Monthly
8.	$900	$5\frac{3}{4}\%$	Quarterly
✓ 9.	$18,000	8%	Continuous
10.	$250	6.5%	Yearly
11.	$1500	$7\frac{1}{4}\%$	Monthly
12.	$600	9.75%	Continuous

Compound Interest   In Exercises 13–18, determine the type of compounding. Solve the problem by trying the more common types of compounding. *See Example 3.*

	Principal	Balance	Time	Rate
✓ 13.	$5000	$8954.24	10 years	6%
14.	$5000	$9096.98	10 years	6%
15.	$750	$1587.75	10 years	7.5%
16.	$10,000	$73,890.56	20 years	10%
17.	$100	$141.48	5 years	7%
18.	$4000	$4788.76	2 years	9%

Effective Yield   In Exercises 19–26, find the effective yield. *See Example 4.*

	Rate	Compounding
✓ 19.	8%	Continuous
20.	9.5%	Daily
21.	7%	Monthly
22.	8%	Yearly
23.	6%	Quarterly
24.	9%	Quarterly
25.	8%	Monthly
26.	$5\frac{1}{4}\%$	Daily

27. *Doubling Time*   Is it necessary to know the principal P to find the doubling time in Exercises 7–12? Explain.

28. *Effective Yield*

(a) Is it necessary to know the principal P to find the effective yield in Exercises 19–26? Explain.

(b) When the interest is compounded more frequently, what inference can you make about the difference between the effective yield and the stated annual percentage rate?

Compound Interest In Exercises 29–36, find the principal that must be deposited under the specified conditions to obtain the given balance.

	Balance	Rate	Time	Compounding
29.	$10,000	9%	20 years	Continuous
30.	$5000	8%	5 years	Continuous
31.	$750	6%	3 years	Daily
32.	$3000	7%	10 years	Monthly
33.	$25,000	7%	30 years	Monthly
34.	$8000	6%	2 years	Monthly
35.	$1000	5%	1 year	Daily
36.	$100,000	9%	40 years	Daily

Monthly Deposits In Exercises 37–40, you make monthly deposits of P dollars in a savings account at an annual interest rate r, compounded continuously. Find the balance A after t years given that

$$A = \frac{P(e^{rt} - 1)}{e^{r/12} - 1}.$$

	Principal	Rate	Time
37.	$P = 30$	$r = 8\%$	$t = 10$ years
38.	$P = 100$	$r = 9\%$	$t = 30$ years
39.	$P = 50$	$r = 10\%$	$t = 40$ years
40.	$P = 20$	$r = 7\%$	$t = 20$ years

Monthly Deposits In Exercises 41 and 42, you make monthly deposits of $30 in a savings account at an annual interest rate of 8%, compounded continuously (see figure).

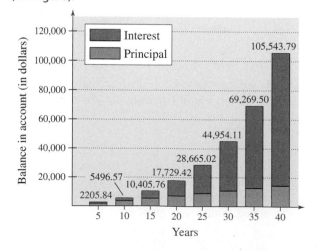

41. Find the total amount that has been deposited in the account after 20 years, and the total interest earned.

42. Find the total amount that has been deposited in the account after 40 years, and the total interest earned.

Exponential Growth and Decay In Exercises 43–46, find the constant k such that the graph of $y = Ce^{kt}$ passes through the points.

43.

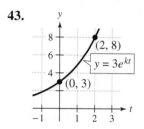

44.

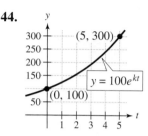

45.

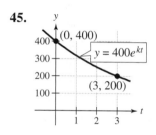

46.

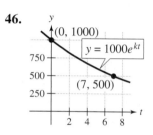

Population of a Country In Exercises 47–54, the population (in thousands) of a Caribbean locale in 2000 and the predicted population (in thousands) for 2020 are given. Find the constants C and k to obtain the exponential growth model $y = Ce^{kt}$ for the population. (Let $t = 0$ correspond to the year 2000.) Use the model to predict the population in the year 2025. *See Example 5.* (Source: United Nations)

	Country	2000	2020
✓ 47.	Aruba	90	106
48.	Bahamas	303	381
49.	Barbados	286	303
50.	Belize	245	363
51.	Jamaica	2589	2872

Country	2000	2020
52. Haiti	8573	11,584
53. Puerto Rico	3834	4252
54. Saint Lucia	153	188

55. *Rate of Growth*   Compare the values of k in Exercises 47 and 51. Which is larger? Explain.

56. *Exponential Growth Models*   What variable in the continuous compound interest formula is equivalent to k in the model for population growth? Use your answer to give an interpretation of k.

57. *World Population*   The figure shows the population P (in billions) of the world as projected by the U.S. Census Bureau. The bureau's projection can be modeled by

$$P = \frac{11.7}{1 + 1.21e^{-0.0269t}}$$

where $t = 0$ represents 1990. Use the model to predict the population in 2025.

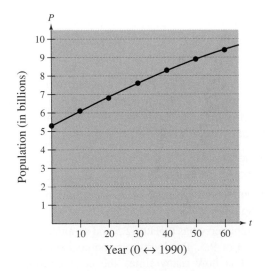

Year (0 ↔ 1990)

58. *World Population*   Use the model P given in Exercise 57 to predict the world population in 2045.

59. *Computer Virus*   In 2005, a computer worm called "Samy" interrupted the operations of a social networking website by inserting the payload message "but most of all, Samy is my hero" in the personal profile pages of the website's users. It is said that the "Samy" worm's message spread from 73 users to 1 million users within 20 hours.

(a) Find the constants C and k to obtain an exponential growth model $y = Ce^{kt}$ for the "Samy" worm.

(b) Use your model from part (a) to estimate how long it took the "Samy" worm to drop its payload message in 5300 personal profile pages.

60. *Stamp Collecting*   The most expensive stamp in the world is the "Treskilling Yellow," a stamp issued in Sweden in 1855. The Treskilling Yellow sold in 1990 for $1.3 million and again in 2008 for $2.3 million.

AP Photo/Donald Stampfli

(a) Find the constants C and k to obtain the exponential growth model $y = Ce^{kt}$ for the value of the Treskilling Yellow.

(b) Use your model from part (a) to estimate the year in which the value of the Treskilling Yellow will reach $3 million.

61. *Cellular Phones*   In 2000, there were 109,478,000 cellular telephone users in the United States. By 2006, the number had grown to 233,041,000. Use an exponential growth model to predict the number of cell phone users in 2013. (Source: CTIA–The Wireless Association®)

62. *Album Downloads* In 2004, about 4.6 million albums were purchased through downloading in the United States. In 2006, the number had increased to about 27.6 million. Use an exponential growth model to predict the number of albums that will be purchased through downloading in 2013. (Source: Recording Industry Association of America)

Radioactive Decay In Exercises 63–68, complete the table for the radioactive isotopes. *See Example 6.*

Isotope	Half-Life (Years)	Initial Quantity	Amount After 1000 Years
63. ^{226}Ra	1620	6 g	g
64. ^{226}Ra	1620	g	0.25 g
65. ^{14}C	5730	g	4.0 g
66. ^{14}C	5730	10 g	g
67. ^{239}Pu	24,100	4.2 g	g
68. ^{239}Pu	24,100	g	1.5 g

69. *Radioactive Decay* Radioactive radium (^{226}Ra) has a half-life of 1620 years. If you start with 5 grams of the isotope, how much will remain after 1000 years?

70. *Carbon 14 Dating* Carbon 14 dating assumes that all living organisms contain ^{14}C (radioactive carbon) in the same relative proportion. When an organism dies, however, its ^{14}C begins to decay according to its half-life of 5730 years. A piece of charcoal from an ancient tree contains only 15% as much ^{14}C as a piece of modern charcoal. How long ago did the ancient tree die? (Round your answer to the nearest 100 years.)

71. *Radioactive Decay* The isotope ^{230}Pu has a half-life of 24,360 years. If you start with 10 grams of this isotope, how much will remain after 10,000 years?

72. *Radioactive Decay* Carbon 14 (^{14}C) has a half-life of 5730 years. If you start with 5 grams of this isotope, how much will remain after 1000 years?

73. *Depreciation* A sport utility vehicle that cost $34,000 new has a depreciated value of $26,000 after 1 year. Find the value of the sport utility vehicle when it is 3 years old by using the exponential model $y = Ce^{kt}$.

74. *Depreciation* After x years, the value y of a recreational vehicle that cost $8000 new is given by

$$y = 8000(0.8)^x.$$

(a) Use a graphing calculator to graph the model.

(b) Graphically approximate the value of the recreational vehicle after 1 year.

(c) Graphically approximate the time when the recreational vehicle's value will be $4000.

Earthquake Intensity In Exercises 75–78, compare the intensities of the two earthquakes. *See Example 8.*

Location	Date	Magnitude
75. Chile	5/22/1960	9.5
Chile	1/22/2008	5.2
76. Southern Alaska	3/28/1964	9.2
Southern Alaska	12/31/2007	3.5
77. Fiji	1/15/2008	6.5
Philippines	1/15/2008	4.7
78. India	1/14/2008	5.8
Taiwan	1/14/2008	3.5

Acidity In Exercises 79–82, use the acidity model $pH = -\log_{10}[H^+]$, where acidity (pH) is a measure of the hydrogen ion concentration $[H^+]$ (measured in moles of hydrogen per liter) of a solution.

79. Find the pH of a solution that has a hydrogen ion concentration of 9.2×10^{-8}.

80. Compute the hydrogen ion concentration if the pH of a solution is 4.7.

81. A blueberry has a pH of 2.5 and an antacid tablet has a pH of 9.5. The hydrogen ion concentration of the fruit is how many times the concentration of the tablet?

82. If the pH of a solution is decreased by 1 unit, the hydrogen ion concentration is increased by what factor?

83. *Population Growth*   The population p of a species of wild rabbit t years after it is introduced into a new habitat is given by

$$p(t) = \frac{5000}{1 + 4e^{-t/6}}.$$

(a) ⊞ Use a graphing calculator to graph the population function.

(b) Determine the size of the population of rabbits that was introduced into the habitat.

(c) Determine the size of the population of rabbits after 9 years.

(d) After how many years will the size of the population of rabbits be 2000?

84. ⊞ *Sales Growth*   Annual sales y of a personal digital assistant x years after it is introduced are approximated by

$$y = \frac{2000}{1 + 4e^{-x/2}}.$$

(a) Use a graphing calculator to graph the model.

(b) Use the graph in part (a) to approximate annual sales of this personal digital assistant model when $x = 4$.

(c) Use the graph in part (a) to approximate the time when annual sales of this personal digital assistant model are $y = 1100$ units.

(d) Use the graph in part (a) to estimate the maximum level that annual sales of this model will approach.

85. *Advertising Effect*   The sales S (in thousands of units) of a brand of jeans after the company spent x hundred dollars in advertising are given by

$$S = 10(1 - e^{kx}).$$

(a) Write S as a function of x if 2500 pairs of jeans are sold when \$500 is spent on advertising.

(b) How many pairs of jeans will be sold if advertising expenditures are raised to \$700?

86. *Advertising Effect*   The sales S of a video game after the company spent x thousand dollars in advertising are given by

$$S = 4500(1 - e^{kx}).$$

(a) Write S as a function of x if 2030 copies of the video game are sold when \$10,000 is spent on advertising.

(b) How many copies of the video game will be sold if advertising expenditures are raised to \$25,000?

Explaining Concepts

87. ✎   Explain how to determine whether an exponential model of the form $y = Ce^{kt}$ models growth or decay.

88. ✎   The formulas for periodic and continuous compounding have the four variables A, P, r, and t in common. Explain what each variable measures.

89. For what types of compounding is the effective yield of an investment greater than the annual interest rate? Explain.

90. If the reading on the Richter scale is increased by 1, the intensity of the earthquake is increased by what factor? Explain.

Cumulative Review

In Exercises 91–94, solve the equation by using the Quadratic Formula.

91. $x^2 - 7x - 5 = 0$ **92.** $x^2 + 5x - 3 = 0$

93. $3x^2 + 9x + 4 = 0$ **94.** $3x^2 + 4x = -2x + 5$

In Exercises 95–98, solve the inequality and graph the solution on the real number line.

95. $\dfrac{4}{x - 4} > 0$ **96.** $\dfrac{x - 1}{x + 2} < 0$

97. $\dfrac{2x}{x - 3} > 1$ **98.** $\dfrac{x - 5}{x + 2} \le -1$

What Did You Learn?

Use these two pages to help prepare for a test on this chapter. Check off the key terms and key concepts you know. You can also use this section to record your assignments.

Plan for Test Success

Date of test: [/ /]   **Study dates and times:** [/ /] at [:] A.M./P.M.

[/ /] at [:] A.M./P.M.

Things to review:

☐ Key Terms, *p. 646*
☐ Key Concepts, *pp. 646–647*
☐ Your class notes
☐ Your assignments

☐ Study Tips, *pp. 579, 581, 591, 595, 606, 607, 608, 611, 618, 626, 629, 630, 635, 636*
☐ Technology Tips, *pp. 582, 598, 610, 611, 627*

☐ Mid-Chapter Quiz, *p. 616*
☐ Review Exercises, *pp. 648–652*
☐ Chapter Test, *p. 653*
☐ Video Explanations Online
☐ Tutorial Online

Key Terms

☐ exponential function, *p. 578*
☐ asymptote, *p. 581*
☐ horizontal asymptote, *p. 581*
☐ natural base, *p. 582*
☐ natural exponential function, *p. 582*
☐ composition, *p. 591*

☐ inverse function, *p. 593*
☐ one-to-one function, *p. 593*
☐ logarithmic function with base *a*, *p. 605*
☐ common logarithmic function, *p. 607*
☐ natural logarithmic function, *p. 610*

☐ change-of-base formula, *p. 611*
☐ exponentiate, *p. 628*
☐ effective yield, *p. 637*
☐ exponential growth, *p. 637*
☐ exponential decay, *p. 637*

Key Concepts

9.1 Exponential Functions

Assignment: _____ Due date: _____

☐ **Evaluate exponential functions of the form** $f(x) = a^x$.
☐ **Graph exponential functions using the characteristics of the graph of** $y = a^x$:

Domain: $(-\infty, \infty)$ Range: $(0, \infty)$
Intercept: $(0, 1)$ Increases from left to right.
The x-axis is a horizontal asymptote.

☐ **Evaluate and graph natural exponential functions of the form** $f(x) = e^x$.

The natural exponential function is simply an exponential function with a special base, the natural base $e \approx 2.71828$.

9.2 Composite and Inverse Functions

Assignment: _____ Due date: _____

☐ **Form the composition of two functions.**

$(f \circ g)(x) = f(g(x))$

The domain of $(f \circ g)$ is the set of all x in the domain of g such that $g(x)$ is in the domain of f.

☐ **Use the Horizontal Line Test to determine whether a function has an inverse.**

☐ **Find an inverse function algebraically.**

1. In the equation for $f(x)$, replace $f(x)$ with y.
2. Interchange x and y.
3. Solve for y. (If y is not a function of x, the original equation does not have an inverse.)
4. Replace y with $f^{-1}(x)$.
5. Verify that $f(f^{-1}(x)) = x = f^{-1}(f(x))$.

9.3 Logarithmic Functions

Assignment: _____   _____ Due date: _____

☐ **Evaluate logarithmic functions.**

A logarithm is an exponent. In other words, $y = \log_a x$ if and only if $x = a^y$.

Properties of logarithms   Let a and x be positive real numbers such that $a \neq 1$. Then:
1. $\log_a 1 = 0$ because $a^0 = 1$.
2. $\log_a a = 1$ because $a^1 = a$.
3. $\log_a a^x = x$ because $a^x = a^x$.

☐ **Graph logarithmic functions using the characteristics of the graph of $y = \log_a x$:**

Domain: $(0, \infty)$ Range: $(-\infty, \infty)$

Intercept: $(1, 0)$ Increases from left to right.

The y-axis is a vertical asymptote.

☐ **Use a Change-of-Base Formula.**

$$\log_a x = \frac{\log_b x}{\log_b a} \quad \text{or} \quad \log_a x = \frac{\ln x}{\ln a}$$

9.4 Properties of Logarithms

Assignment: _____   _____ Due date: _____

☐ **Use properties of logarithms.**

Let a be a positive real number such that $a \neq 1$, and let n be a real number. If u and v are real numbers, variables, or algebraic expressions such that $u > 0$ and $v > 0$, then the following properties are true.

Logarithm with base a
1. $\log_a(uv) = \log_a u + \log_a v$
2. $\log_a \frac{u}{v} = \log_a u - \log_a v$
3. $\log_a u^n = n \log_a u$

Natural logarithm
$\ln(uv) = \ln u + \ln v$
$\ln \frac{u}{v} = \ln u - \ln v$
$\ln u^n = n \ln u$

9.5 Solving Exponential and Logarithmic Equations

Assignment: _____   _____ Due date: _____

☐ **Solve exponential and logarithmic equations using the one-to-one properties.**

Let a be a positive real number such that $a \neq 1$, and let x and y be real numbers. Then the following properties are true.
1. $a^x = a^y$ if and only if $x = y$.
2. $\log_a x = \log_a y$ if and only if $x = y$ $(x > 0, y > 0)$.

☐ **Solve exponential and logarithmic equations using the inverse properties.**

Base a   *Natural base e*
1. $\log_a(a^x) = x$ $\ln(e^x) = x$
2. $a^{(\log_a x)} = x$ $e^{(\ln x)} = x$

9.6 Applications

Assignment: _____   _____ Due date: _____

☐ **Solve compound interest problems.**

The following compound interest formulas are for the balance A, principal P, annual interest rate r (in decimal form), and time t (in years).

n Compoundings per Year: $A = P\left(1 + \frac{r}{n}\right)^{nt}$

Continuous Compounding: $A = Pe^{rt}$

☐ **Solve growth and decay problems.**

The mathematical model for growth or decay is

$y = Ce^{kt}$

where y is the amount of an initial quantity C that remains after time t. The number k is a constant. The model represents growth if $k > 0$ or decay if $k < 0$.

☐ **Solve earthquake intensity problems.**

On the Richter scale, the magnitude R of an earthquake can be measured by the intensity model $R = \log_{10} I$, where I is the intensity of the shock wave.

Review Exercises

9.1 Exponential Functions

1 ▶ Evaluate exponential functions.

In Exercises 1–4, evaluate the exponential function as indicated. (Round your answer to three decimal places.)

1. $f(x) = 4^x$

   (a) $x = -3$

   (b) $x = 1$

   (c) $x = 2$

2. $g(x) = 4^{-x}$

   (a) $x = -2$

   (b) $x = 0$

   (c) $x = 2$

3. $g(t) = 5^{-t/3}$

   (a) $t = -3$

   (b) $t = \pi$

   (c) $t = 6$

4. $h(s) = 1 - 3^{0.2s}$

   (a) $s = 0$

   (b) $s = 2$

   (c) $s = \sqrt{10}$

2 ▶ Graph exponential functions.

In Exercises 5–14, sketch the graph of the function. Identify the horizontal asymptote.

5. $f(x) = 3^x$

6. $f(x) = 3^{-x}$

7. $f(x) = 3^x - 3$

8. $f(x) = 3^x + 5$

9. $f(x) = 3^{x+1}$

10. $f(x) = 3^{x-1}$

11. $f(x) = 3^{x/2}$

12. $f(x) = 3^{-x/2}$

13. $f(x) = 3^{x/2} - 2$

14. $f(x) = 3^{x/2} + 3$

⊞ In Exercises 15–18, use a graphing calculator to graph the function.

15. $f(x) = 2^{-x^2}$

16. $g(x) = 2^{|x|}$

17. $y = 10(1.09)^t$

18. $y = 250(1.08)^t$

3 ▶ Evaluate the natural base e and graph natural exponential functions.

In Exercises 19 and 20, evaluate the exponential function as indicated. (Round your answers to three decimal places.)

19. $f(x) = 3e^{-2x}$

   (a) $x = 3$

   (b) $x = 0$

   (c) $x = -19$

20. $g(x) = e^{x/5} + 11$

   (a) $x = 12$

   (b) $x = -8$

   (c) $x = 18.4$

⊞ In Exercises 21–24, use a graphing calculator to graph the function.

21. $y = 4e^{-x/3}$ **22.** $y = 6 - e^{x/2}$

23. $f(x) = e^{x+2}$ **24.** $h(t) = \dfrac{8}{1 + e^{-t/5}}$

4 ▶ Use exponential functions to solve application problems.

Compound Interest In Exercises 25 and 26, complete the table to determine the balance A for P dollars invested at interest rate r for t years, compounded n times per year.

n	1	4	12	365	Continuous compounding
A					

Principal	Rate	Time
25. $P = \$5000$	$r = 10\%$	$t = 40$ years
26. $P = \$10,000$	$r = 9.5\%$	$t = 30$ years

27. *Radioactive Decay*   After t years, the remaining mass y (in grams) of 21 grams of a radioactive element whose half-life is 25 years is given by $y = 21\left(\frac{1}{2}\right)^{t/25}$, $t \geq 0$. How much of the initial mass remains after 58 years?

28. *Depreciation*   After t years, a truck that originally cost \$38,000 depreciates in value so that each year it is worth $\frac{2}{3}$ of its value for the previous year. Find a model for $V(t)$, the value of the truck after t years. Sketch a graph of the model and determine the value of the truck 6 years after it was purchased.

9.2 Composite and Inverse Functions

1 ▶ Form compositions of two functions and find the domains of composite functions.

In Exercises 29–32, find the compositions.

29. $f(x) = x + 2$, $g(x) = x^2$
  (a) $(f \circ g)(2)$ (b) $(g \circ f)(-1)$

30. $f(x) = \sqrt[3]{x}$, $g(x) = x + 2$
  (a) $(f \circ g)(6)$ (b) $(g \circ f)(64)$

31. $f(x) = \sqrt{x + 1}$, $g(x) = x^2 - 1$
  (a) $(f \circ g)(5)$ (b) $(g \circ f)(-1)$

32. $f(x) = \dfrac{1}{x - 4}$, $g(x) = \dfrac{x + 1}{2x}$
  (a) $(f \circ g)(1)$ (b) $(g \circ f)\left(\dfrac{1}{5}\right)$

In Exercises 33 and 34, find the compositions (a) $f \circ g$ and (b) $g \circ f$. Then find the domain of each composition.

33. $f(x) = \sqrt{x + 6}$, $g(x) = 2x$

34. $f(x) = \dfrac{2}{x - 4}$, $g(x) = x^2$

2 ▶ Use the Horizontal Line Test to determine whether functions have inverse functions.

In Exercises 35–38, use the Horizontal Line Test to determine if the function is one-to-one and so has an inverse function.

35. $f(x) = x^2 - 25$

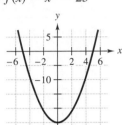

36. $f(x) = \frac{1}{4}x^3$

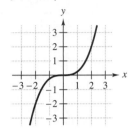

37. $h(x) = 4\sqrt[3]{x}$

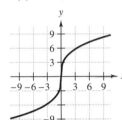

38. $g(x) = \sqrt{16 - x^2}$

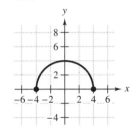

3 ▶ Find inverse functions algebraically.

In Exercises 39–44, find the inverse function.

39. $f(x) = 3x + 4$

40. $f(x) = 2x - 3$

41. $h(x) = \sqrt{5x}$

42. $g(x) = x^2 + 2$, $x \geq 0$

43. $f(t) = t^3 + 4$

44. $h(t) = \sqrt[3]{t - 1}$

4 ▶ Graphically verify that two functions are inverse functions of each other.

⊞   **In Exercises 45 and 46, use a graphing calculator to graph the functions in the same viewing window. Graphically verify that f and g are inverse functions of each other.**

45. $f(x) = 3x + 4$
  $g(x) = \frac{1}{3}(x - 4)$

46. $f(x) = \frac{1}{3}\sqrt[3]{x}$
  $g(x) = 27x^3$

In Exercises 47–50, use the graph of *f* to sketch the graph of *f*⁻¹.

47.
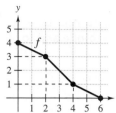

48.
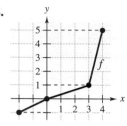

49.
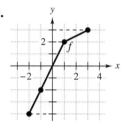

50.
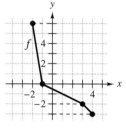

9.3 Logarithmic Functions

1 ▶ Evaluate logarithmic functions.

In Exercises 51–58, evaluate the logarithm.

51. $\log_{10} 1000$

52. $\log_{27} 3$

53. $\log_3 \frac{1}{9}$

54. $\log_4 \frac{1}{16}$

55. $\log_2 64$

56. $\log_{10} 0.01$

57. $\log_3 1$

58. $\log_2 \sqrt{4}$

2 ▶ Graph logarithmic functions.

In Exercises 59–64, sketch the graph of the function. Identify the vertical asymptote.

59. $f(x) = \log_3 x$

60. $f(x) = -\log_3 x$

61. $f(x) = -1 + \log_3 x$

62. $f(x) = 1 + \log_3 x$

63. $y = \log_2(x - 4)$

64. $y = \log_4(x + 1)$

3 ▶ Graph and evaluate natural logarithmic functions.

In Exercises 65 and 66, use your calculator to evaluate the natural logarithm. (Round your answer to four decimal places.)

65. $\ln 50$

66. $\ln\left(\dfrac{5 - \sqrt{3}}{2}\right)$

In Exercises 67–70, sketch the graph of the function. Identify the vertical asymptote.

67. $y = \ln(x - 3)$

68. $y = -\ln(x + 2)$

69. $y = 5 - \ln x$

70. $y = 3 + \ln x$

4 ▶ Use the change-of-base formula to evaluate logarithms.

In Exercises 71–74, use a calculator to evaluate the logarithm by means of the change-of-base formula. (Round your answer to four decimal places.)

71. $\log_4 9$

72. $\log_{1/2} 5$

73. $\log_8 160$

74. $\log_3 0.28$

9.4 Properties of Logarithms

1 ▶ Use the properties of logarithms to evaluate logarithms.

In Exercises 75–80, use $\log_5 2 \approx 0.4307$ and $\log_5 3 \approx 0.6826$ to approximate the expression. Do not use a calculator.

75. $\log_5 18$

76. $\log_5 \sqrt{6}$

77. $\log_5 \frac{1}{2}$

78. $\log_5 \frac{2}{3}$

79. $\log_5 (12)^{2/3}$

80. $\log_5 (5^2 \cdot 6)$

2 ▶ Use the properties of logarithms to rewrite, expand, or condense logarithmic expressions.

In Exercises 81–88, use the properties of logarithms to expand the expression. (Assume all variables are positive.)

81. $\log_4 6x^4$

82. $\log_{12} 2x^{-5}$

83. $\log_5 \sqrt{x + 2}$

84. $\ln \sqrt[3]{\dfrac{x}{5}}$

85. $\ln \dfrac{x + 2}{x + 3}$

86. $\ln x(x + 4)^2$

87. $\ln\left[\sqrt{2x}(x + 3)^5\right]$

88. $\log_3 \dfrac{a^2 \sqrt{b}}{cd^5}$

In Exercises 89–98, use the properties of logarithms to condense the expression.

89. $-\frac{2}{3} \ln 3y$

90. $5 \log_2 y$

91. $\log_8 16x + \log_8 2x^2$

92. $\log_4 6x - \log_4 10$

93. $-2(\ln 2x - \ln 3)$

94. $5(1 + \ln x + \ln 2)$

95. $4\left[\log_2 k - \log_2(k - t)\right]$

96. $\frac{1}{3}(\log_8 a + 2 \log_8 b)$

97. $3 \ln x + 4 \ln y + \ln z$

98. $\ln(x + 4) - 3 \ln x - \ln y$

True or False? **In Exercises 99–104, use properties of logarithms to determine whether the equation is true or false. If it is false, state why or give an example to show that it is false.**

99. $\log_2 4x = 2 \log_2 x$

100. $\dfrac{\ln 5x}{\ln 10x} = \ln \dfrac{1}{2}$

101. $\log_{10} 10^{2x} = 2x$

102. $e^{\ln t} = t,\ t > 0$

103. $\log_4 \dfrac{16}{x} = 2 - \log_4 x$

104. $6 \ln x + 6 \ln y = \ln(xy)^6,\ x > 0,\ y > 0$

3 ▶ Use the properties of logarithms to solve application problems.

105. *Light Intensity*   The intensity of light y as it passes through a medium is given by

$$y = \ln\left(\dfrac{I_0}{I}\right)^{0.83}.$$

Use properties of logarithms to write the formula in simpler form, and determine the intensity of light passing through this medium when $I_0 = 4.2$ and $I = 3.3$.

106. *Human Memory Model*   A psychologist finds that the percent p of retention in a group of subjects can be modeled by

$$p = \dfrac{\log_{10}(10^{68})}{\log_{10}(t + 1)^{20}}$$

where t is the time in months after the subjects' initial testing. Use properties of logarithms to write the formula in simpler form, and determine the percent of retention after 5 months.

9.5 Solving Exponential and Logarithmic Equations

1 ▶ Solve basic exponential and logarithmic equations.

In Exercises 107–112, solve the equation.

107. $2^x = 64$

108. $6^x = 216$

109. $4^{x-3} = \frac{1}{16}$

110. $3^{x-2} = 81$

111. $\log_7(x + 6) = \log_7 12$

112. $\ln(8 - x) = \ln 3$

2 ▶ Use inverse properties to solve exponential equations.

In Exercises 113–118, solve the exponential equation. (Round your answer to two decimal places.)

113. $3^x = 500$

114. $8^x = 1000$

115. $2e^{0.5x} = 45$

116. $125e^{-0.4x} = 40$

117. $12(1 - 4^x) = 18$

118. $25(1 - e^t) = 12$

3 ▶ Use inverse properties to solve logarithmic equations.

In Exercises 119–128, solve the logarithmic equation. (Round your answer to two decimal places.)

119. $\ln x = 7.25$

120. $\ln x = -0.5$

121. $\log_{10} 4x = 2.1$

122. $\log_2 2x = -0.65$

123. $\log_3(2x + 1) = 2$

124. $\log_5(x - 10) = 2$

125. $\frac{1}{3} \log_2 x + 5 = 7$

126. $4 \log_5(x + 1) = 4.8$

127. $\log_3 x + \log_3 7 = 4$

128. $2 \log_4 x - \log_4(x - 1) = 1$

4 ▶ Use exponential or logarithmic equations to solve application problems.

129. *Compound Interest*   A deposit of $5000 is placed in a savings account for 2 years. The interest for the account is compounded continuously. At the end of 2 years, the balance in the account is $5751.37. What is the annual interest rate for this account?

130. *Sound Intensity*   The relationship between the number of decibels B and the intensity of a sound I in watts per centimeter squared is given by

$$B = 10 \log_{10}\left(\frac{I}{10^{-16}}\right).$$

Determine the intensity of a firework display I if it registers 130 decibels on a decibel meter.

9.6 Applications

1 ▶ Use exponential equations to solve compound interest problems.

Annual Interest Rate   **In Exercises 131–136, find the annual interest rate.**

	Principal	Balance	Time	Compounding
131.	$250	$410.90	10 years	Quarterly
132.	$1000	$1348.85	5 years	Monthly
133.	$5000	$15,399.30	15 years	Daily
134.	$10,000	$35,236.45	20 years	Yearly
135.	$1800	$46,422.61	50 years	Continuous
136.	$7500	$15,877.50	15 years	Continuous

Effective Yield   **In Exercises 137–142, find the effective yield.**

	Rate	Compounding
137.	5.5%	Daily
138.	6%	Monthly
139.	7.5%	Quarterly
140.	8%	Yearly
141.	7.5%	Continuously
142.	3.75%	Continuously

2 ▶ Use exponential equations to solve growth and decay problems.

Radioactive Decay   **In Exercises 143–148, complete the table for the radioactive isotopes.**

	Isotope	Half-Life (Years)	Initial Quantity	Amount After 1000 Years
143.	^{226}Ra	1620	3.5 g	g
144.	^{226}Ra	1620	g	0.5 g
145.	^{14}C	5730	g	2.6 g
146.	^{14}C	5730	10 g	g
147.	^{239}Pu	24,100	5 g	g
148.	^{239}Pu	24,100	g	2.5 g

3 ▶ Use logarithmic equations to solve intensity problems.

In Exercises 149 and 150, compare the intensities of the two earthquakes.

	Location	Date	Magnitude
149.	San Francisco, California	4/18/1906	8.3
	Napa, California	9/3/2000	4.9
150.	Mexico	1/23/2008	5.8
	Puerto Rico	1/23/2008	3.3

Chapter Test

Take this test as you would take a test in class. After you are done, check your work against the answers in the back of the book.

1. Evaluate $f(t) = 54\left(\frac{2}{3}\right)^t$ when $t = -1, 0, \frac{1}{2}$, and 2.

2. Sketch a graph of the function $f(x) = 2^{x/3}$ and identify the horizontal asymptote.

3. Find the compositions (a) $f \circ g$ and (b) $g \circ f$. Then find the domain of each composition.

   $f(x) = 2x^2 + x \qquad g(x) = 5 - 3x$

4. Find the inverse function of $f(x) = 9x - 4$.

5. Verify algebraically that the functions f and g are inverse functions of each other.

   $f(x) = -\frac{1}{2}x + 3, \qquad g(x) = -2x + 6$

6. Evaluate $\log_4 \frac{1}{256}$ without a calculator.

7. Describe the relationship between the graphs of $f(x) = \log_5 x$ and $g(x) = 5^x$.

8. Use the properties of logarithms to expand $\log_8\left(4\sqrt{x}/y^4\right)$.

9. Use the properties of logarithms to condense $\ln x - 4 \ln y$.

In Exercises 10–17, solve the equation. Round your answer to two decimal places, if necessary.

10. $\log_2 x = 5$

11. $9^{2x} = 182$

12. $400e^{0.08t} = 1200$

13. $3\ln(2x - 3) = 10$

14. $12(7 - 2^x) = -300$

15. $\log_2 x + \log_2 4 = 5$

16. $\ln x - \ln 2 = 4$

17. $30(e^x + 9) = 300$

18. Determine the balance after 20 years if $2000 is invested at 7% compounded (a) quarterly and (b) continuously.

19. Determine the principal that will yield $100,000 when invested at 9% compounded quarterly for 25 years.

20. A principal of $500 yields a balance of $1006.88 in 10 years when the interest is compounded continuously. What is the annual interest rate?

21. A car that cost $20,000 new has a depreciated value of $15,000 after 1 year. Find the value of the car when it is 5 years old by using the exponential model $y = Ce^{kt}$.

In Exercises 22–24, the population p of a species of fox t years after it is introduced into a new habitat is given by

$$p(t) = \frac{2400}{1 + 3e^{-t/4}}.$$

22. Determine the size of the population that was introduced into the habitat.

23. Determine the population after 4 years.

24. After how many years will the population be 1200?

Avoiding Test-Taking Errors

For some students, the day they get their math tests back is just as nerve-racking as the day they take the test. Do you look at your grade, sigh hopelessly, and stuff the test in your book bag? This kind of response is not going to help you to do better on the next test. When professional football players lose a game, the coach does not let them just forget about it. They review all their mistakes and discuss how to correct them. That is what you need to do with every math test.

There are six types of test errors (Nolting, 2008), as listed below. Look at your test and see what types of errors you make. Then decide what you can do to avoid making them again. Many students need to do this with a tutor or instructor the first time through.

Kimberly Nolting

VP, Academic Success Press
expert in developmental education

Smart Study Strategy

Analyze Your Errors

Type of error	Corrective action
1 ▸ Misreading Directions: You do not correctly read or understand directions.	Read the instructions in the textbook exercises at least twice and make sure you understand what they mean. Make this a habit in time for the next test.
2 ▸ Careless Errors: You understand how to do a problem but make careless errors, such as not carrying a sign, miscopying numbers, and so on.	Pace yourself during a test to avoid hurrying. Also, make sure you write down every step of a solution neatly. Use a finger to move from one step to the next, looking for errors.
3 ▸ Concept Errors: You do not understand how to apply the properties and rules needed to solve a problem.	Find a tutor who will work with you on the next chapter. Visit the instructor to make sure you understand the math.
4 ▸ Application Errors: You can do numerical problems that are similar to your homework problems but struggle with problems that vary, such as application problems.	Do not just mimic the steps of solving an application problem. Explain out loud why you are doing each step. Ask the instructor or tutor for different types of problems.
5 ▸ Test-Taking Errors: You hurry too much, do not use all of the allowed time, spend too much time on one problem, and so on.	Refer to the *Ten Steps for Test-Taking* on page 368.
6 ▸ Study Errors: You do not study the right material or do not learn it well enough to remember it on a test without resources such as notes.	Take a practice test. Work with a study group. Confer with your instructor. Don't try to learn a whole chapter's worth of material in one night—cramming does not work in math!

Chapter 10
Conics

IT WORKED FOR ME!

"My instructor told me that if I just put a little more effort into studying and getting ready for the final, I could possibly get an A. So, I pulled out my old tests. That is when I noticed that most of my mistakes involved word problems and not reading directions carefully. I got help from a tutor on word problems and made sure I correctly read the instructions on the final. It worked."

Maribeth
Education

10.1 Circles and Parabolas

© 1996 CORBIS; Original courtesy of NASA/CORBIS

What You Should Learn

1 ▶ Recognize the four basic conics: circles, parabolas, ellipses, and hyperbolas.
2 ▶ Graph and write equations of circles centered at the origin.
3 ▶ Graph and write equations of circles centered at (h, k).
4 ▶ Graph and write equations of parabolas.

Why You Should Learn It

Circles can be used to model and solve scientific problems. For instance, in Exercise 93 on page 666, you will write an equation that represents the circular orbit of a satellite.

1 ▶ Recognize the four basic conics: circles, parabolas, ellipses, and hyperbolas.

The Conics

In Section 8.4, you saw that the graph of a second-degree equation of the form $y = ax^2 + bx + c$ is a parabola. A parabola is one of four types of **conics** or **conic sections.** The other three types are circles, ellipses, and hyperbolas. All four types have equations of second degree. Each figure can be obtained by intersecting a plane with a double-napped cone, as shown in Figure 10.1.

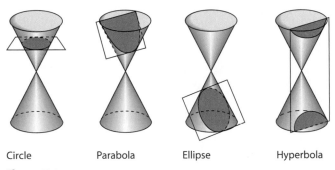

Circle   Parabola   Ellipse   Hyperbola

Figure 10.1

2 ▶ Graph and write equations of circles centered at the origin.

Conics occur in many practical applications. Reflective surfaces in satellite dishes, flashlights, and telescopes often have a parabolic shape. The orbits of planets are elliptical, and the orbits of comets are usually elliptical or hyperbolic. Ellipses and parabolas are also used in building archways and bridges.

Circles Centered at the Origin

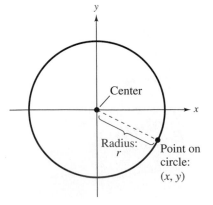

Figure 10.2

Definition of a Circle

A **circle** in the rectangular coordinate system consists of all points (x, y) that are a given positive distance r from a fixed point, called the **center** of the circle. The distance r is called the **radius** of the circle.

If the center of the circle is the origin, as shown in Figure 10.2, the relationship between the coordinates of any point (x, y) on the circle and the radius r is

$$r = \sqrt{(x - 0)^2 + (y - 0)^2} = \sqrt{x^2 + y^2}.$$ Distance Formula (See Section 3.1.)

By squaring each side of this equation, you obtain the equation below, which is called the **standard form of the equation of a circle centered at the origin.**

> ### Standard Equation of a Circle (Center at Origin)
> The **standard form of the equation of a circle centered at the origin** is
> $$x^2 + y^2 = r^2. \qquad \text{Circle with center at } (0, 0)$$
> The positive number r is called the **radius** of the circle.

EXAMPLE 1   Writing an Equation of a Circle

Write an equation of the circle that is centered at the origin and has a radius of 2, as shown in Figure 10.3.

Solution

Using the standard form of the equation of a circle (with center at the origin) and $r = 2$, you obtain

$x^2 + y^2 = r^2$	Standard form with center at $(0, 0)$
$x^2 + y^2 = 2^2$	Substitute 2 for r.
$x^2 + y^2 = 4$.	Equation of circle

✔ **CHECKPOINT** *Now try Exercise 7.*

Figure 10.3

To sketch the graph of the equation of a given circle, write the equation in standard form, which will allow you to identify the radius of the circle.

EXAMPLE 2   Sketching a Circle

Identify the radius of the circle given by the equation $4x^2 + 4y^2 - 25 = 0$. Then sketch the circle.

Solution

Begin by writing the equation in standard form.

$4x^2 + 4y^2 - 25 = 0$	Write original equation.
$4x^2 + 4y^2 = 25$	Add 25 to each side.
$x^2 + y^2 = \dfrac{25}{4}$	Divide each side by 4.
$x^2 + y^2 = \left(\dfrac{5}{2}\right)^2$	Standard form

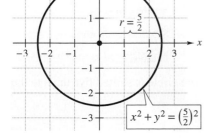

Figure 10.4

From the standard form of the equation of this circle centered at the origin, you can see that the radius is $\frac{5}{2}$. The graph of the circle is shown in Figure 10.4.

✔ **CHECKPOINT** *Now try Exercise 23.*

3 ▶ Graph and write equations of circles centered at (h, k).

Circles Centered at (h, k)

Consider a circle whose radius is r and whose center is the point (h, k), as shown in Figure 10.5. Let (x, y) be any point on the circle. To find an equation for this circle, you can use a variation of the Distance Formula and write

$$\text{Radius} = r = \sqrt{(x - h)^2 + (y - k)^2}. \qquad \text{Distance Formula (See Section 3.1.)}$$

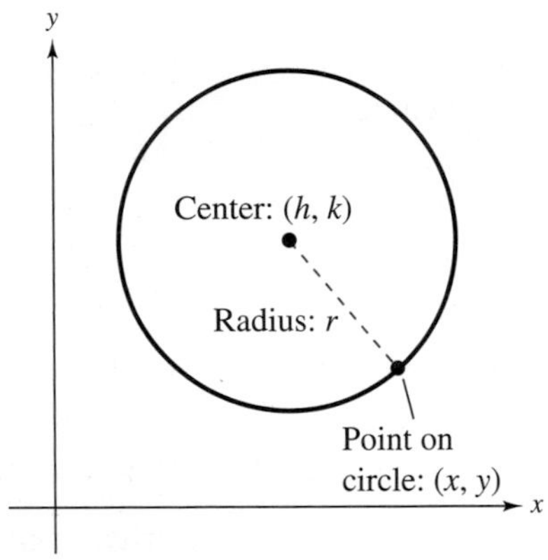

By squaring each side of this equation, you obtain the equation shown below, which is called the **standard form of the equation of a circle centered at (h, k).**

> ### Standard Equation of a Circle [Center at (h, k)]
> The **standard form of the equation of a circle centered at (h, k)** is
> $$(x - h)^2 + (y - k)^2 = r^2.$$

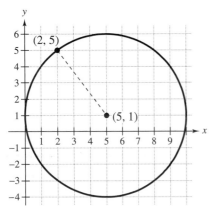

Center: (h, k)

Radius: r

Point on circle: (x, y)

Figure 10.5

When $h = 0$ and $k = 0$, the circle is centered at the origin. Otherwise, you can use the rules on horizontal and vertical shifts from Section 3.7 to shift the center of the circle h units horizontally and k units vertically from the origin.

EXAMPLE 3   Writing an Equation of a Circle

The point $(2, 5)$ lies on a circle whose center is $(5, 1)$, as shown in Figure 10.6. Write the standard form of the equation of this circle.

Solution

The radius r of the circle is the distance between $(2, 5)$ and $(5, 1)$.

$$r = \sqrt{(2 - 5)^2 + (5 - 1)^2} \qquad \text{Distance Formula}$$

$$= \sqrt{(-3)^2 + 4^2} \qquad \text{Simplify.}$$

$$= \sqrt{9 + 16} \qquad \text{Simplify.}$$

$$= \sqrt{25} \qquad \text{Simplify.}$$

$$= 5 \qquad \text{Radius}$$

Figure 10.6

Using $(h, k) = (5, 1)$ and $r = 5$, the equation of the circle is

$$(x - h)^2 + (y - k)^2 = r^2 \qquad \text{Standard form}$$

$$(x - 5)^2 + (y - 1)^2 = 5^2 \qquad \text{Substitute for } h, k, \text{ and } r.$$

$$(x - 5)^2 + (y - 1)^2 = 25. \qquad \text{Equation of circle}$$

From the graph, you can see that the center of the circle is shifted five units to the right and one unit upward from the origin.

✓ **CHECKPOINT** *Now try Exercise 15.*

Writing the equation of a circle in standard form helps you to determine both the radius and center of the circle. To write the standard form of the equation of a circle, you may need to complete the square, as demonstrated in Example 4.

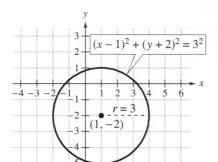

Figure 10.7

EXAMPLE 4   Writing an Equation in Standard Form

Write the equation $x^2 + y^2 - 2x + 4y - 4 = 0$ in standard form. Then sketch the circle represented by the equation.

Solution

$$x^2 + y^2 - 2x + 4y - 4 = 0 \qquad \text{Write original equation.}$$

$$\left(x^2 - 2x + \right) + \left(y^2 + 4y + \right) = 4 \qquad \text{Group terms.}$$

$$\left[x^2 - 2x + (-1)^2\right] + \left(y^2 + 4y + 2^2\right) = 4 + 1 + 4 \qquad \text{Complete the squares.}$$

$$\underbrace{}_{\text{(half)}^2} \qquad \underbrace{}_{\text{(half)}^2}$$

$$(x - 1)^2 + (y + 2)^2 = 3^2 \qquad \text{Standard form}$$

From this standard form, you can see that the circle is centered at $(1, -2)$ with a radius of 3. The graph of the equation of the circle is shown in Figure 10.7.

✓ **CHECKPOINT** *Now try Exercise 31.*

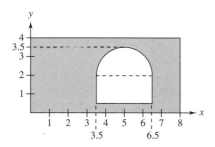

Figure 10.8

EXAMPLE 5   An Application: Mechanical Drawing

In a mechanical drawing class, you have to program a computer to model the metal piece shown in Figure 10.8. Part of the assignment is to find an equation that represents the semicircular portion of the hole in the metal piece. What is the equation?

Solution

From the drawing, you can see that the center of the circle is $(h, k) = (5, 2)$ and that the radius of the circle is $r = 1.5$. This implies that the equation of the entire circle is

$$(x - h)^2 + (y - k)^2 = r^2 \qquad \text{Standard form}$$

$$(x - 5)^2 + (y - 2)^2 = 1.5^2 \qquad \text{Substitute for } h, k, \text{ and } r.$$

$$(x - 5)^2 + (y - 2)^2 = 2.25. \qquad \text{Equation of circle}$$

To find the equation of the upper portion of the circle, solve this standard equation for y.

$$(x - 5)^2 + (y - 2)^2 = 2.25$$

$$(y - 2)^2 = 2.25 - (x - 5)^2$$

$$y - 2 = \pm\sqrt{2.25 - (x - 5)^2}$$

$$y = 2 \pm \sqrt{2.25 - (x - 5)^2}$$

Finally, take the positive square root to obtain the equation of the upper portion of the circle.

$$y = 2 + \sqrt{2.25 - (x - 5)^2}$$

✓ **CHECKPOINT** *Now try Exercise 95.*

Study Tip

In Example 5, if you had wanted the equation of the lower portion of the circle, you would have taken the negative square root

$$y = 2 - \sqrt{2.25 - (x - 5)^2}.$$

4 ▸ Graph and write equations of parabolas.

Equations of Parabolas

The second basic type of conic is a **parabola.** In Section 8.4, you studied some of the properties of parabolas. There you saw that the graph of a quadratic function of the form $y = ax^2 + bx + c$ is a parabola that opens upward if a is positive and downward if a is negative. You also learned that each parabola has a vertex and that the vertex of the graph of $y = ax^2 + bx + c$ occurs when $x = -b/(2a)$.

In this section, you will study a general definition of a parabola in the sense that it is independent of the orientation of the parabola.

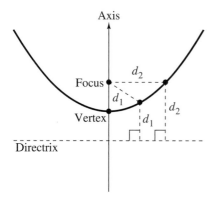

Figure 10.9

Definition of a Parabola

A **parabola** is the set of all points (x, y) that are equidistant from a fixed line (**directrix**) and a fixed point (**focus**) not on the line.

The midpoint between the focus and the directrix is called the **vertex,** and the line passing through the focus and the vertex is called the **axis** of the parabola. Note in Figure 10.9 that a parabola is symmetric with respect to its axis. Using the definition of a parabola, you can derive the **standard form of the equation of a parabola** whose directrix is parallel to the x-axis or to the y-axis.

Study Tip

If the focus of a parabola is above or to the right of the vertex, p is positive. If the focus is below or to the left of the vertex, p is negative.

Standard Equation of a Parabola

The **standard form of the equation of a parabola** with vertex at the origin $(0, 0)$ is

$$x^2 = 4py, \quad p \neq 0 \qquad \text{Vertical axis}$$

$$y^2 = 4px, \quad p \neq 0. \qquad \text{Horizontal axis}$$

The focus lies on the axis p units (*directed distance*) from the vertex. If the vertex is at (h, k), then the standard form of the equation is

$$(x - h)^2 = 4p(y - k), \quad p \neq 0 \qquad \text{Vertical axis; directrix: } y = k - p$$

$$(y - k)^2 = 4p(x - h), \quad p \neq 0. \qquad \text{Horizontal axis; directrix: } x = h - p$$

(See Figure 10.10.)

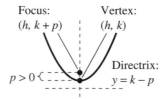

Parabola with vertical axis

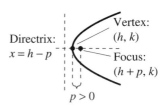

Parabola with horizontal axis

Figure 10.10

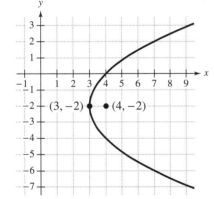

Figure 10.11

EXAMPLE 6   Writing the Standard Equation of a Parabola

Write the standard form of the equation of the parabola with vertex $(0, 0)$ and focus $(0, -2)$, as shown in Figure 10.11.

Solution

Because the vertex is at the origin and the axis of the parabola is vertical, use the equation

$$x^2 = 4py$$

where p is the directed distance from the vertex to the focus. Because the focus is two units *below* the vertex, $p = -2$. So, the equation of the parabola is

$x^2 = 4py$	Standard form
$x^2 = 4(-2)y$	Substitute for p.
$x^2 = -8y.$	Equation of parabola

✓ **CHECKPOINT** *Now try Exercise 55.*

EXAMPLE 7   Writing the Standard Equation of a Parabola

Write the standard form of the equation of the parabola with vertex $(3, -2)$ and focus $(4, -2)$, as shown in Figure 10.12.

Solution

Because the vertex is at $(h, k) = (3, -2)$ and the axis of the parabola is horizontal, use the equation

$$(y - k)^2 = 4p(x - h)$$

where $h = 3$, $k = -2$, and $p = 1$. So, the equation of the parabola is

$(y - k)^2 = 4p(x - h)$	Standard form
$[y - (-2)]^2 = 4(1)(x - 3)$	Substitute for h, k, and p.
$(y + 2)^2 = 4(x - 3).$	Equation of parabola

✓ **CHECKPOINT** *Now try Exercise 69.*

Figure 10.12

Technology: Tip

You cannot represent a circle or a parabola with a horizontal axis as a single function of x. You can, however, represent it by two functions of x. For instance, try using a graphing calculator to graph the equations below in the same viewing window. Use a viewing window in which $-1 \le x \le 10$ and $-8 \le y \le 4$. Do you obtain a parabola? Does the graphing calculator connect the two portions of the parabola?

$y_1 = -2 + 2\sqrt{x - 3}$	Upper portion of parabola
$y_2 = -2 - 2\sqrt{x - 3}$	Lower portion of parabola

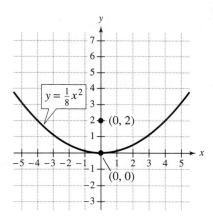

Figure 10.13

EXAMPLE 8   Analyzing a Parabola

Sketch the graph of the parabola $y = \frac{1}{8}x^2$ and identify its vertex and focus.

Solution

Because the equation can be written in the standard form $x^2 = 4py$, it is a parabola whose vertex is at the origin. You can identify the focus of the parabola by writing its equation in standard form.

$y = \frac{1}{8}x^2$	Write original equation.
$\frac{1}{8}x^2 = y$	Interchange sides of the equation.
$x^2 = 8y$	Multiply each side by 8.
$x^2 = 4(2)y$	Rewrite 8 in the form $4p$.

From this standard form, you can see that $p = 2$. Because the parabola opens upward, as shown in Figure 10.13, you can conclude that the focus lies $p = 2$ units above the vertex. So, the focus is $(0, 2)$.

✓ **CHECKPOINT** *Now try Exercise 75.*

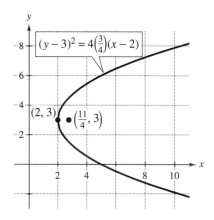

Figure 10.14

EXAMPLE 9   Analyzing a Parabola

Sketch the parabola $x = \frac{1}{3}y^2 - 2y + 5$ and identify its vertex and focus.

Solution

This equation can be written in the standard form $(y - k)^2 = 4p(x - h)$. To do this, you can complete the square, as follows.

$x = \frac{1}{3}y^2 - 2y + 5$	Write original equation.
$\frac{1}{3}y^2 - 2y + 5 = x$	Interchange sides of the equation.
$y^2 - 6y + 15 = 3x$	Multiply each side by 3.
$y^2 - 6y = 3x - 15$	Subtract 15 from each side.
$y^2 - 6y + 9 = 3x - 15 + 9$	Complete the square.
$(y - 3)^2 = 3x - 6$	Simplify.
$(y - 3)^2 = 3(x - 2)$	Factor.
$(y - 3)^2 = 4\left(\frac{3}{4}\right)(x - 2)$	Rewrite 3 in the form $4p$.

From this standard form, you can see that the vertex is $(h, k) = (2, 3)$ and $p = \frac{3}{4}$. Because the parabola opens to the right, as shown in Figure 10.14, the focus lies $p = \frac{3}{4}$ unit to the right of the vertex. So, the focus is $\left(\frac{11}{4}, 3\right)$.

✓ **CHECKPOINT** *Now try Exercise 85.*

Parabolas occur in a wide variety of applications. For instance, a parabolic reflector can be formed by revolving a parabola around its axis. The light rays emanating from the focus of a parabolic reflector used in a flashlight are all parallel to one another, as shown in Figure 10.15.

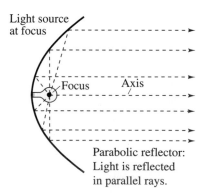

Figure 10.15

_____ **Concept Check** _____

1. Explain how you would identify the center and radius of the circle given by the equation $x^2 + y^2 - 36 = 0$.

2. Is the center of the circle given by the equation $(x + 2)^2 + (y + 4)^2 = 20$ shifted two units to the right and four units upward? Explain your reasoning.

3. Which standard form of the equation of a parabola would you use to write an equation for a parabola with vertex $(2, -3)$ and focus $(2, 1)$? Explain your reasoning.

4. Given the equation of a parabola, explain how to determine if the parabola opens upward, downward, to the right, or to the left.

10.1 EXERCISES

Go to pages 698–699 to record your assignments.

_____ **Developing Skills** _____

In Exercises 1–6, match the equation with its graph. [The graphs are labeled (a), (b), (c), (d), (e), and (f).]

5. $y = -\sqrt{4 - x^2}$

6. $y = \sqrt{4 - x^2}$

(a)

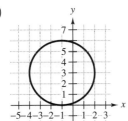

(b)

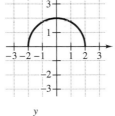

(c)

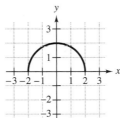

(d)

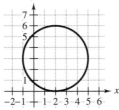

(e)

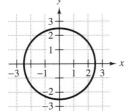

(f)

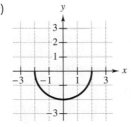

1. $x^2 + y^2 = 25$

2. $4x^2 + 4y^2 = 25$

3. $(x - 2)^2 + (y - 3)^2 = 9$

4. $(x + 1)^2 + (y - 3)^2 = 9$

In Exercises 7–14, write the standard form of the equation of the circle with center at $(0, 0)$ that satisfies the criterion. **See Example 1.**

 7. Radius: 5

8. Radius: 9

9. Radius: $\frac{2}{3}$

10. Radius: $\frac{5}{2}$

11. Passes through the point $(0, 6)$

12. Passes through the point $(-2, 0)$

13. Passes through the point $(5, 2)$

14. Passes through the point $(-1, -4)$

In Exercises 15–22, write the standard form of the equation of the circle with center at (h, k) that satisfies the criteria. **See Example 3.**

 15. Center: $(4, 3)$

   Radius: 10

16. Center: $(-4, 8)$

   Radius: 7

17. Center: $(6, -5)$

   Radius: 3

18. Center: $(-5, -2)$

Radius: $\frac{5}{2}$

19. Center: $(-2, 1)$

Passes through the point $(0, 1)$

20. Center: $(8, 2)$

Passes through the point $(8, 0)$

21. Center: $(3, 2)$

Passes through the point $(4, 6)$

22. Center: $(-3, -5)$

Passes through the point $(0, 0)$

In Exercises 23–42, identify the center and radius of the circle and sketch the circle. ***See Examples 2 and 4.***

✓ **23.** $x^2 + y^2 = 16$ **24.** $x^2 + y^2 = 1$

25. $x^2 + y^2 = 36$ **26.** $x^2 + y^2 = 15$

27. $4x^2 + 4y^2 = 1$ **28.** $9x^2 + 9y^2 = 64$

29. $25x^2 + 25y^2 - 144 = 0$

30. $\dfrac{x^2}{4} + \dfrac{y^2}{4} - 1 = 0$

✓ **31.** $(x + 1)^2 + (y - 5)^2 = 64$

32. $(x - 9)^2 + (y + 2)^2 = 144$

33. $(x - 2)^2 + (y - 3)^2 = 4$

34. $(x + 4)^2 + (y - 3)^2 = 25$

35. $\left(x + \frac{9}{4}\right)^2 + (y - 4)^2 = 16$

36. $(x - 5)^2 + \left(y + \frac{3}{4}\right)^2 = 1$

37. $x^2 + y^2 - 4x - 2y + 1 = 0$

38. $x^2 + y^2 + 6x - 4y - 3 = 0$

39. $x^2 + y^2 + 2x + 6y + 6 = 0$

40. $x^2 + y^2 - 2x + 6y - 15 = 0$

41. $x^2 + y^2 + 10x - 4y - 7 = 0$

42. $x^2 + y^2 - 14x + 8y + 56 = 0$

▦ In Exercises 43–46, use a graphing calculator to graph the circle. (*Note:* Solve for *y*. Use a square setting so that the circles display correctly.)

43. $x^2 + y^2 = 30$ **44.** $4x^2 + 4y^2 = 45$

45. $(x - 2)^2 + y^2 = 10$ **46.** $x^2 + (y - 5)^2 = 21$

In Exercises 47–52, match the equation with its graph. [The graphs are labeled (a), (b), (c), (d), (e), and (f).]

(a)
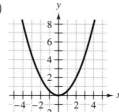

(b)
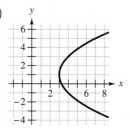

(c)
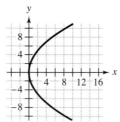

(d)
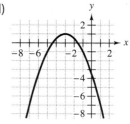

(e)
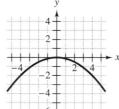

(f)
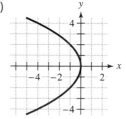

47. $y^2 = -4x$ **48.** $x^2 = 2y$

49. $x^2 = -8y$ **50.** $y^2 = 12x$

51. $(y - 1)^2 = 4(x - 3)$ **52.** $(x + 3)^2 = -2(y - 1)$

In Exercises 53–64, write the standard form of the equation of the parabola with its vertex at the origin. ***See Example 6.***

53.
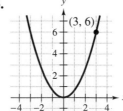

54.
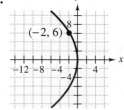

55. Focus: $\left(0, -\frac{3}{2}\right)$ **56.** Focus: $\left(\frac{5}{4}, 0\right)$

57. Focus: $(-2, 0)$ **58.** Focus: $(0, -2)$

59. Focus: $(0, 1)$ **60.** Focus: $(-3, 0)$

61. Focus: $(6, 0)$ **62.** Focus: $(0, 2)$

63. Passes through $(4, 6)$; Horizontal axis

64. Passes through $(-6, -12)$; Vertical axis

In Exercises 65–74, write the standard form of the equation of the parabola with its vertex at (h, k). *See Example 7.*

65.
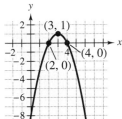

66.
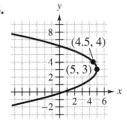

67.
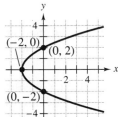

68.
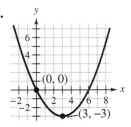

69. Vertex: $(3, 2)$;   Focus: $(1, 2)$

70. Vertex: $(-1, 2)$;   Focus: $(-1, 0)$

71. Vertex: $(0, -4)$;   Focus: $(0, -1)$

72. Vertex: $(-2, 1)$;   Focus: $(-5, 1)$

73. Vertex: $(0, 2)$;
    Passes through $(1, 3)$;   Horizontal axis

74. Vertex: $(0, 2)$;
    Passes through $(6, 0)$;   Vertical axis

In Exercises 75–88, identify the vertex and focus of the parabola and sketch the parabola. *See Examples 8 and 9.*

75. $y = \frac{1}{2}x^2$

76. $y = 2x^2$

77. $y^2 = -10x$

78. $y^2 = 3x$

79. $x^2 + 8y = 0$

80. $x + y^2 = 0$

81. $(x - 1)^2 + 8(y + 2) = 0$

82. $(x + 3) + (y - 2)^2 = 0$

83. $\left(y + \frac{1}{2}\right)^2 = 2(x - 5)$

84. $\left(x + \frac{1}{2}\right)^2 = 4(y - 3)$

85. $y = \frac{1}{3}(x^2 - 2x + 10)$

86. $4x - y^2 - 2y - 33 = 0$

87. $y^2 + 6y + 8x + 25 = 0$

88. $y^2 - 4y - 4x = 0$

In Exercises 89–92, use a graphing calculator to graph the parabola. Identify the vertex and focus.

89. $y = -\frac{1}{6}(x^2 + 4x - 2)$

90. $x^2 - 2x + 8y + 9 = 0$

91. $y^2 + x + y = 0$

92. $3y^2 - 10x + 25 = 0$

_____ **Solving Problems** _____

93. *Satellite Orbit* Write an equation that represents the circular orbit of a satellite 500 miles above the surface of Earth. Place the origin of the rectangular coordinate system at the center of Earth, and assume the radius of Earth is 4000 miles.

94. *Observation Wheel* Write an equation that represents the circular wheel of the Singapore Flyer in Singapore, which has a diameter of 150 meters. Place the origin of the rectangular coordinate system at the center of the wheel.

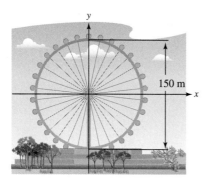

✓ **95.** *Mirror* Write an equation that represents the circular mirror, with a diameter of 3 feet, shown in the figure. The wall hangers of the mirror are shown as two points on the circle. Use the equation to determine the height of the left wall hanger.

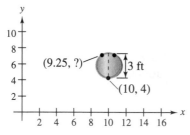

96. *Dog Leash* A leash allows a dog a semicircular boundary that has a diameter of 80 feet. Write an equation that represents the semicircle. The dog is located on the semicircle, 10 feet from the fence at the right. How far is the dog from the house?

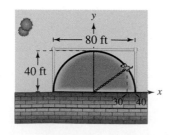

97. *Suspension Bridge* Each cable of a suspension bridge is suspended (in the shape of a parabola) between two towers that are 120 meters apart, and the top of each tower is 20 meters above the roadway. The cables touch the roadway at the midpoint between the two towers (see figure).

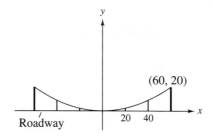

(a) Write an equation that represents the parabolic shape of each cable.

(b) Complete the table by finding the height of the suspension cables y over the roadway at a distance of x meters from the center of the bridge.

x	0	20	40	60
y				

98. *Beam Deflection* A simply supported beam is 16 meters long and has a load at the center (see figure). The deflection of the beam at its center is 3 centimeters. Assume that the shape of the deflected beam is parabolic.

(a) Write an equation of the parabola. (Assume that the origin is at the center of the deflected beam.)

(b) How far from the center of the beam is the deflection equal to 1 centimeter?

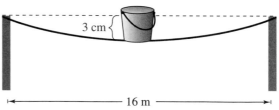

Not drawn to scale

99. 🖩 *Revenue* The revenue R generated by the sale of x video game systems is given by $R = 575x - \frac{5}{4}x^2$.

(a) Use a graphing calculator to graph the function.

(b) Use the graph to approximate the number of sales that will maximize revenue.

100. 🖩 *Path of a Softball* The path of a softball is given by $y = -0.08x^2 + x + 4$. The coordinates x and y are measured in feet, with $x = 0$ corresponding to the position from which the ball was thrown.

(a) Use a graphing calculator to graph the path of the softball.

(b) Move the cursor along the path to approximate the highest point and the range of the path.

101. *Graphical Estimation* A rectangle centered at the origin with sides parallel to the coordinate axes is placed in a circle of radius 25 inches centered at the origin (see figure). The length of the rectangle is $2x$ inches.

(a) Show that the width and area of the rectangle are given by $2\sqrt{625 - x^2}$ and $4x\sqrt{625 - x^2}$, respectively.

(b) 🖩 Use a graphing calculator to graph the area function. Approximate the value of x for which the area is maximum.

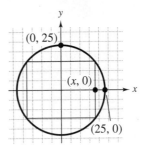

Explaining Concepts

102. ✎ The point $(-4, 3)$ lies on a circle with center $(-1, 1)$. Does the point $(3, 2)$ lie on the same circle? Explain your reasoning.

103. ✎ A student claims that

$$x^2 + y^2 - 6y = -5$$

does not represent a circle. Is the student correct? Explain your reasoning.

104. ✎ Is y a function of x in the equation $y^2 = 6x$? Explain.

105. ✎ Is it possible for a parabola to intersect its directrix? Explain.

106. ✎ If the vertex and focus of a parabola are on a horizontal line, is the directrix of the parabola vertical? Explain.

Cumulative Review

In Exercises 107–112, solve the equation by completing the square.

107. $x^2 + 4x = 6$

108. $x^2 + 6x = -4$

109. $x^2 - 2x - 3 = 0$

110. $4x^2 - 12x - 10 = 0$

111. $2x^2 + 5x - 8 = 0$ **112.** $9x^2 - 12x = 14$

In Exercises 113–116, use the properties of logarithms to expand the expression. (Assume all variables are positive.)

113. $\log_8 x^{10}$

114. $\log_{10} \sqrt{xy^3}$

115. $\ln 5x^2y$

116. $\ln \dfrac{x}{y^4}$

In Exercises 117–120, use the properties of logarithms to condense the expression.

117. $\log_{10} x + \log_{10} 6$

118. $2\log_3 x - \log_3 y$

119. $3\ln x + \ln y - \ln 9$

120. $4(\ln x + \ln y) - \ln(x^4 + y^4)$

10.2 Ellipses

AP Photo/*Billings Gazette*, Bob Zellar

What You Should Learn

1 ► Graph and write equations of ellipses centered at the origin.

2 ► Graph and write equations of ellipses centered at (h, k).

Why You Should Learn It

Equations of ellipses can be used to model and solve real-life problems. For instance, in Exercise 59 on page 676, you will use an equation of an ellipse to model a dirt race track for sprint cars.

1 ► Graph and write equations of ellipses centered at the origin.

Ellipses Centered at the Origin

The third type of conic is called an *ellipse* and is defined as follows.

Definition of an Ellipse

An **ellipse** in the rectangular coordinate system consists of all points (x, y) such that the sum of the distances between (x, y) and two distinct fixed points is a constant, as shown in Figure 10.16. Each of the two fixed points is called a **focus** of the ellipse. (The plural of focus is *foci*.)

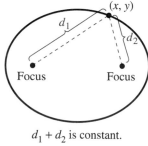

$d_1 + d_2$ is constant.

Figure 10.16

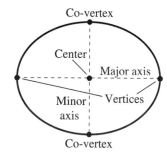

Figure 10.17

The line through the foci intersects the ellipse at two points, called the **vertices,** as shown in Figure 10.17. The line segment joining the vertices is called the **major axis,** and its midpoint is called the **center** of the ellipse. The line segment perpendicular to the major axis at the center is called the **minor axis** of the ellipse, and the points at which the minor axis intersects the ellipse are called **co-vertices.**

To trace an ellipse, place two thumbtacks at the foci, as shown in Figure 10.18. If the ends of a fixed length of string are fastened to the thumbtacks and the string is drawn taut with a pencil, the path traced by the pencil will be an ellipse.

Figure 10.18

The standard form of the equation of an ellipse takes one of two forms, depending on whether the major axis is horizontal or vertical.

Standard Equation of an Ellipse (Center at Origin)

The **standard form of the equation of an ellipse centered at the origin** with major and minor axes of lengths $2a$ and $2b$ is

$$\frac{x^2}{a^2} + \frac{y^2}{b^2} = 1 \qquad \text{or} \qquad \frac{x^2}{b^2} + \frac{y^2}{a^2} = 1, \qquad 0 < b < a.$$

The vertices lie on the major axis, a units from the center, and the co-vertices lie on the minor axis, b units from the center, as shown in Figure 10.19.

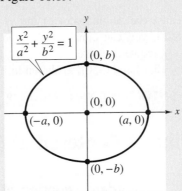

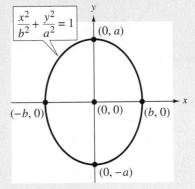

Major axis is horizontal.
Minor axis is vertical.

Major axis is vertical.
Minor axis is horizontal.

Figure 10.19

> ### Study Tip
>
> Notice that the standard form of the equation of an ellipse centered at the origin with a horizontal major axis has x-intercepts of $(\pm a, 0)$ and y-intercepts of $(0, \pm b)$.

EXAMPLE 1   **Writing the Standard Equation of an Ellipse**

Write an equation of the ellipse that is centered at the origin, with vertices $(-3, 0)$ and $(3, 0)$ and co-vertices $(0, -2)$ and $(0, 2)$.

Solution

Begin by plotting the vertices and co-vertices, as shown in Figure 10.20. The center of the ellipse is $(0, 0)$. So, the equation of the ellipse has the form

$$\frac{x^2}{a^2} + \frac{y^2}{b^2} = 1. \qquad \text{Major axis is horizontal.}$$

For this ellipse, the major axis is horizontal. So, a is the distance between the center and either vertex, which implies that $a = 3$. Similarly, b is the distance between the center and either co-vertex, which implies that $b = 2$. So, the standard form of the equation of the ellipse is

$$\frac{x^2}{3^2} + \frac{y^2}{2^2} = 1. \qquad \text{Standard form}$$

✓ **CHECKPOINT** *Now try Exercise 7.*

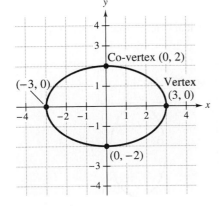

Figure 10.20

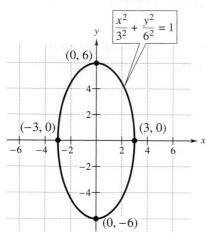

(0, 6)

(−3, 0)   (3, 0)

(0, −6)

Figure 10.21

EXAMPLE 2   **Sketching an Ellipse**

Sketch the ellipse given by $4x^2 + y^2 = 36$. Identify the vertices and co-vertices.

Solution

To sketch an ellipse, it helps first to write its equation in standard form.

$$4x^2 + y^2 = 36 \qquad \text{Write original equation.}$$

$$\frac{x^2}{9} + \frac{y^2}{36} = 1 \qquad \text{Divide each side by 36 and simplify.}$$

$$\frac{x^2}{3^2} + \frac{y^2}{6^2} = 1 \qquad \text{Standard form}$$

Because the denominator of the y^2-term is larger than the denominator of the x^2-term, you can conclude that the major axis is vertical. Moreover, because $a = 6$, the vertices are $(0, -6)$ and $(0, 6)$. Finally, because $b = 3$, the co-vertices are $(-3, 0)$ and $(3, 0)$, as shown in Figure 10.21.

✓ **CHECKPOINT** *Now try Exercise 19.*

2 ▶ Graph and write equations of ellipses centered at (h, k).

Ellipses Centered at (h, k)

Standard Equation of an Ellipse [Center at (h, k)]

The **standard form of the equation of an ellipse centered at** (h, k) with major and minor axes of lengths $2a$ and $2b$, where $0 < b < a$, is

$$\frac{(x - h)^2}{a^2} + \frac{(y - k)^2}{b^2} = 1 \qquad \text{Major axis is horizontal.}$$

or

$$\frac{(x - h)^2}{b^2} + \frac{(y - k)^2}{a^2} = 1. \qquad \text{Major axis is vertical.}$$

The foci lie on the major axis, c units from the center, with $c^2 = a^2 - b^2$.

Figure 10.22 shows the horizontal and vertical orientations for an ellipse.

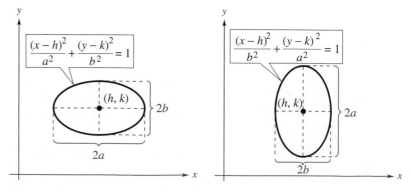

Figure 10.22

When $h = 0$ and $k = 0$, the ellipse is centered at the origin. Otherwise, you can use the rules for horizontal and vertical shifts from Section 3.7 to shift the center of the ellipse h units horizontally and k units vertically from the origin.

EXAMPLE 3   Writing the Standard Equation of an Ellipse

Write the standard form of the equation of the ellipse with vertices $(-2, 2)$ and $(4, 2)$ and co-vertices $(1, 3)$ and $(1, 1)$, as shown in Figure 10.23.

Solution

Because the vertices are $(-2, 2)$ and $(4, 2)$, the center of the ellipse is $(h, k) = (1, 2)$. The distance from the center to either vertex is $a = 3$, and the distance to either co-vertex is $b = 1$. Because the major axis is horizontal, the standard form of the equation is

$$\frac{(x - h)^2}{a^2} + \frac{(y - k)^2}{b^2} = 1. \qquad \text{Major axis is horizontal.}$$

Substitute the values of h, k, a, and b to obtain

$$\frac{(x - 1)^2}{3^2} + \frac{(y - 2)^2}{1^2} = 1. \qquad \text{Standard form}$$

From the graph, you can see that the center of the ellipse is shifted one unit to the right and two units upward from the origin.

✔ **CHECKPOINT** *Now try Exercise 37.*

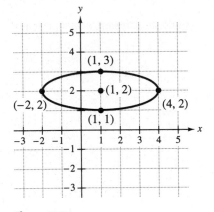

Figure 10.23

Technology: Tip

You can use a graphing calculator to graph an ellipse by graphing the upper and lower portions in the same viewing window. For instance, to graph the ellipse $x^2 + 4y^2 = 4$, first solve for y to obtain

$$y_1 = \frac{1}{2}\sqrt{4 - x^2}$$

and

$$y_2 = -\frac{1}{2}\sqrt{4 - x^2}.$$

Use a viewing window in which $-3 \le x \le 3$ and $-2 \le y \le 2$. You should obtain the graph shown below.

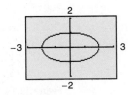

Use this procedure to graph the ellipse in Example 3 on your graphing calculator.

To write an equation of an ellipse in standard form, you must group the x-terms and the y-terms and then complete each square, as shown in Example 4.

EXAMPLE 4   Sketching an Ellipse

Sketch the ellipse given by $4x^2 + y^2 - 8x + 6y + 9 = 0$.

Solution

Begin by writing the equation in standard form. In the fourth step, note that 9 and 4 are added to *each* side of the equation.

$$4x^2 + y^2 - 8x + 6y + 9 = 0 \qquad \text{Write original equation.}$$

$$(4x^2 - 8x + \boxed{}) + (y^2 + 6y + \boxed{}) = -9 \qquad \text{Group terms.}$$

$$4(x^2 - 2x + \boxed{}) + (y^2 + 6y + \boxed{}) = -9 \qquad \text{Factor 4 out of } x\text{-terms.}$$

$$4(x^2 - 2x + 1) + (y^2 + 6y + 9) = -9 + 4(1) + 9 \qquad \text{Complete the squares.}$$

$$4(x - 1)^2 + (y + 3)^2 = 4 \qquad \text{Simplify.}$$

$$\frac{(x - 1)^2}{1} + \frac{(y + 3)^2}{4} = 1 \qquad \text{Divide each side by 4.}$$

$$\frac{(x - 1)^2}{1^2} + \frac{(y + 3)^2}{2^2} = 1 \qquad \text{Standard form}$$

You can see that the center of the ellipse is at $(h, k) = (1, -3)$. Because the denominator of the y^2-term is larger than the denominator of the x^2-term, you can conclude that the major axis is vertical. Because the denominator of the x^2-term is $b^2 = 1^2$, you can locate the endpoints of the minor axis one unit to the right of the center and one unit to the left of the center. Because the denominator of the y^2-term is $a^2 = 2^2$, you can locate the endpoints of the major axis two units up from the center and two units down from the center, as shown in Figure 10.24. To complete the graph, sketch an oval shape that is determined by the vertices and co-vertices, as shown in Figure 10.25.

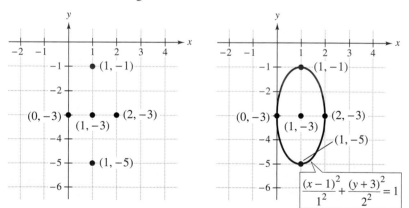

Figure 10.24   **Figure 10.25**

From Figure 10.25, you can see that the center of the ellipse is shifted one unit to the right and three units down from the origin.

✓ **CHECKPOINT** *Now try Exercise 45.*

EXAMPLE 5   An Application: Semielliptical Archway

You are responsible for designing a semielliptical archway, as shown in Figure 10.26. The height of the archway is 10 feet, and its width is 30 feet. Write an equation of the ellipse and use the equation to sketch an accurate diagram of the archway.

Solution

To make the equation simple, place the origin at the center of the ellipse. This means that the standard form of the equation is

$$\frac{x^2}{a^2} + \frac{y^2}{b^2} = 1.$$ Major axis is horizontal.

Because the major axis is horizontal, it follows that $a = 15$ and $b = 10$, which implies that the equation is

$$\frac{x^2}{15^2} + \frac{y^2}{10^2} = 1.$$ Standard form

To make an accurate sketch of the ellipse, solve this equation for y as follows.

$$\frac{x^2}{225} + \frac{y^2}{100} = 1$$ Simplify denominators.

$$\frac{y^2}{100} = 1 - \frac{x^2}{225}$$ Subtract $\frac{x^2}{225}$ from each side.

$$y^2 = 100\left(1 - \frac{x^2}{225}\right)$$ Multiply each side by 100.

$$y = 10\sqrt{1 - \frac{x^2}{225}}$$ Take the positive square root of each side.

Next, calculate several y-values for the archway, as shown in the table. Then use the values in the table to sketch the archway, as shown in Figure 10.27.

x	±15	±12.5	±10	±7.5	±5	±2.5	0
y	0	5.53	7.45	8.66	9.43	9.86	10

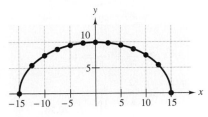

Figure 10.27

✓ **CHECKPOINT** *Now try Exercise 57.*

Figure 10.26

—————————————— **Concept Check** ——————————————

1. Define an ellipse and write the standard form of the equation of an ellipse centered at the origin.

2. What points do you need to know in order to write the equation of an ellipse?

3. From the standard equation, how can you determine the lengths of the major and minor axes of an ellipse?

4. From the standard equation, how can you determine the orientation of the major and minor axes of an ellipse?

10.2 EXERCISES

Go to pages 698–699 to record your assignments.

—————————————— **Developing Skills** ——————————————

In Exercises 1–6, match the equation with its graph. [The graphs are labeled (a), (b), (c), (d), (e), and (f).]

In Exercises 7–18, write the standard form of the equation of the ellipse centered at the origin. *See Example 1.*

(a)
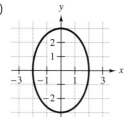

(b)
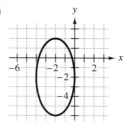

(c)
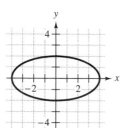

(d)
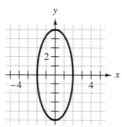

(e)
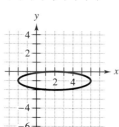

(f)
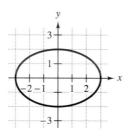

1. $\dfrac{x^2}{4} + \dfrac{y^2}{9} = 1$

2. $\dfrac{x^2}{9} + \dfrac{y^2}{4} = 1$

3. $\dfrac{x^2}{4} + \dfrac{y^2}{25} = 1$

4. $\dfrac{y^2}{4} + \dfrac{x^2}{16} = 1$

5. $\dfrac{(x-2)^2}{16} + \dfrac{(y+1)^2}{1} = 1$

6. $\dfrac{(x+2)^2}{4} + \dfrac{(y+2)^2}{16} = 1$

	Vertices	*Co-vertices*
✓ 7.	$(-4, 0), (4, 0)$	$(0, -3), (0, 3)$
8.	$(-4, 0), (4, 0)$	$(0, -1), (0, 1)$
9.	$(-2, 0), (2, 0)$	$(0, -1), (0, 1)$
10.	$(-7, 0), (7, 0)$	$(0, -4), (0, 4)$
11.	$(0, -6), (0, 6)$	$(-3, 0), (3, 0)$
12.	$(0, -5), (0, 5)$	$(-1, 0), (1, 0)$
13.	$(0, -2), (0, 2)$	$(-1, 0), (1, 0)$
14.	$(0, -8), (0, 8)$	$(-4, 0), (4, 0)$

15. Major axis (vertical) 10 units, minor axis 6 units

16. Major axis (horizontal) 24 units, minor axis 10 units

17. Major axis (horizontal) 20 units, minor axis 12 units

18. Major axis (vertical) 40 units, minor axis 30 units

In Exercises 19–32, sketch the ellipse. Identify the vertices and co-vertices. *See Example 2.*

19. $\dfrac{x^2}{16} + \dfrac{y^2}{4} = 1$ **20.** $\dfrac{x^2}{25} + \dfrac{y^2}{9} = 1$

21. $\dfrac{x^2}{4} + \dfrac{y^2}{16} = 1$ **22.** $\dfrac{x^2}{9} + \dfrac{y^2}{25} = 1$

23. $\dfrac{x^2}{25/9} + \dfrac{y^2}{16/9} = 1$ **24.** $\dfrac{x^2}{1} + \dfrac{y^2}{1/4} = 1$

25. $\dfrac{9x^2}{4} + \dfrac{25y^2}{16} = 1$ **26.** $\dfrac{36x^2}{49} + \dfrac{16y^2}{9} = 1$

27. $16x^2 + 25y^2 - 9 = 0$

28. $64x^2 + 36y^2 - 49 = 0$

29. $4x^2 + y^2 - 4 = 0$

30. $4x^2 + 9y^2 - 36 = 0$

31. $10x^2 + 16y^2 - 160 = 0$

32. $15x^2 + 3y^2 - 75 = 0$

In Exercises 33–36, use a graphing calculator to graph the ellipse. Identify the vertices. (*Note:* Solve for y.)

33. $x^2 + 2y^2 = 4$

34. $9x^2 + y^2 = 64$

35. $3x^2 + y^2 - 12 = 0$

36. $5x^2 + 2y^2 - 10 = 0$

In Exercises 37–40, write the standard form of the equation of the ellipse. *See Example 3.*

37.
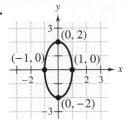

38.
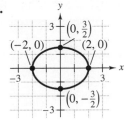

39.
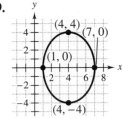

40.
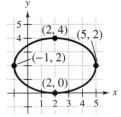

In Exercises 41–54, find the center and vertices of the ellipse and sketch the ellipse. *See Example 4.*

41. $\dfrac{(x + 5)^2}{16} + y^2 = 1$

42. $x^2 + \dfrac{(y - 3)^2}{9} = 1$

43. $\dfrac{(x - 1)^2}{9} + \dfrac{(y - 5)^2}{25} = 1$

44. $\dfrac{(x + 2)^2}{1/4} + \dfrac{(y + 4)^2}{1} = 1$

45. $4(x - 2)^2 + 9(y + 2)^2 = 36$

46. $2(x + 5)^2 + 8(y - 2)^2 = 72$

47. $12(x + 4)^2 + 3(y - 1)^2 = 48$

48. $16(x - 2)^2 + 4(y + 3)^2 = 16$

49. $9x^2 + 4y^2 + 36x - 24y + 36 = 0$

50. $9x^2 + 4y^2 - 36x + 8y + 31 = 0$

51. $25x^2 + 9y^2 - 200x + 54y + 256 = 0$

52. $25x^2 + 16y^2 - 150x - 128y + 81 = 0$

53. $x^2 + 4y^2 - 4x - 8y - 92 = 0$

54. $x^2 + 4y^2 + 6x + 16y - 11 = 0$

Solving Problems

55. *Wading Pool*   You are building a wading pool that is in the shape of an ellipse. Your plans give the following equation for the elliptical shape of the pool, measured in feet.

$$\frac{x^2}{324} + \frac{y^2}{196} = 1$$

Find the longest distance and shortest distance across the pool.

56. *Oval Office*   In the White House, the Oval Office is in the shape of an ellipse. The perimeter of the floor can be modeled in meters by the equation

$$\frac{x^2}{19.36} + \frac{y^2}{30.25} = 1.$$

Find the longest distance and shortest distance across the office.

✓ 57. *Architecture*   A semielliptical arch for a tunnel under a river has a width of 100 feet and a height of 40 feet (see figure). Determine the height of the arch 5 feet from the edge of the tunnel.

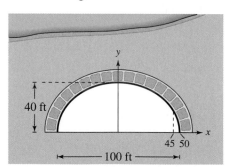

58. *Architecture*   A semielliptical arch for a fireplace has a width of 54 inches and a height of 30 inches. Determine the height of the arch 10 inches from the edge of the fireplace.

59. *Motorsports*   Most sprint car dirt tracks are elliptical in shape. Write an equation of an elliptical race track with a major axis that is 1230 feet long and a minor axis that is 580 feet long.

60. *Bicycle Chainwheel*   The pedals of a bicycle drive a chainwheel, which drives a smaller sprocket wheel on the rear axle (see figure). Many chainwheels are circular. Some, however, are slightly elliptical, which tends to make pedaling easier. Write an equation of an elliptical chainwheel with a major axis that is 8 inches long and a minor axis that is $7\frac{1}{2}$ inches long.

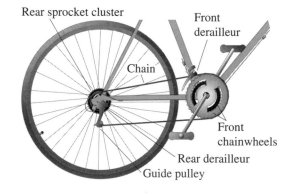

Rear sprocket cluster

Front derailleur

Chain

Front chainwheels

Rear derailleur

Guide pulley

Airplane   In Exercises 61 and 62, an airplane with enough fuel to fly 800 miles safely will take off from airport A and land at airport B. Answer the following questions given the situation in each exercise. Round answers to two decimal places, if necessary.

(a) *Explain* why the region in which the airplane can fly is bounded by an ellipse (see figure).

(b) Let $(0, 0)$ represent the center of the ellipse. Find the coordinates of each airport.

(c) Suppose the plane flies from airport A straight past airport B to a vertex of the ellipse, and then straight back to airport B. How far does the plane fly? Use your answer to find the coordinates of the vertices.

(d) Write an equation of the ellipse. (*Hint:* $c^2 = a^2 - b^2$)

(e) The area of an ellipse is given by $A = \pi ab$. Find the area of the ellipse.

61. Airport A is 500 miles from airport B.

62. Airport A is 650 miles from airport B.

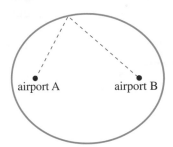

Figure for 61 and 62

───────────── **Explaining Concepts** ─────────────

63. ✎   Describe the relationship between circles and ellipses. How are they similar? How do they differ?

64. ✎   The area of an ellipse is given by $A = \pi ab$. Explain how this area is related to the area of a circle.

65. ✎   Explain the significance of the foci in an ellipse.

66. ✎   Explain how to write an equation of an ellipse if you know the coordinates of the vertices and co-vertices.

67. ✎   From the standard form of the equation, explain how you can determine if the graph of an ellipse intersects the *x*- or *y*-axis.

───────────── **Cumulative Review** ─────────────

In Exercises 68–75, evaluate the function as indicated and sketch the graph of the function.

68. $f(x) = 4^x$
  (a) $x = 3$
  (b) $x = -1$

69. $f(x) = 3^{-x}$
  (a) $x = -2$
  (b) $x = 2$

70. $g(x) = 5^{x-1}$
  (a) $x = 4$
  (b) $x = 0$

71. $g(x) = 6e^{0.5x}$
  (a) $x = -1$
  (b) $x = 2$

72. $h(x) = \log_{10} 2x$
  (a) $x = 5$
  (b) $x = 500$

73. $h(x) = \log_{16} 4x$
  (a) $x = 4$
  (b) $x = 64$

74. $f(x) = \ln(-x)$
  (a) $x = -6$
  (b) $x = 3$

75. $f(x) = \log_4(x - 3)$
  (a) $x = 3$
  (b) $x = 35$

Mid-Chapter Quiz

Take this quiz as you would take a quiz in class. After you are done, check your work against the answers in the back of the book.

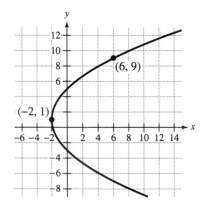

Figure for 1

1. Write the standard form of the equation of the circle shown in the figure.

2. Write the standard form of the equation of the parabola shown in the figure.

3. Write the standard form of the equation of the ellipse shown in the figure.

4. Write the standard form of the equation of the circle with center $(3, -5)$ and passing through the point $(0, -1)$.

5. Write the standard form of the equation of the parabola with vertex $(2, 3)$ and focus $(2, 1)$.

6. Write the standard form of the equation of the ellipse with vertices $(0, -10)$ and $(0, 10)$ and co-vertices $(-6, 0)$ and $(6, 0)$.

Figure for 2

In Exercises 7 and 8, write the equation of the circle in standard form. Then find the center and the radius of the circle.

7. $x^2 + y^2 + 6y - 7 = 0$

8. $x^2 + y^2 + 2x - 4y + 4 = 0$

In Exercises 9 and 10, write the equation of the parabola in standard form. Then find the vertex and the focus of the parabola.

9. $x = y^2 - 6y - 7$

10. $x^2 - 8x + y + 12 = 0$

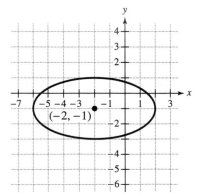

Figure for 3

In Exercises 11 and 12, write the equation of the ellipse in standard form. Then find the center and the vertices of the ellipse.

11. $4x^2 + y^2 - 16x - 20 = 0$ 12. $4x^2 + 9y^2 - 48x + 36y + 144 = 0$

In Exercises 13–18, sketch the graph of the equation.

13. $(x + 5)^2 + (y - 1)^2 = 9$ 14. $9x^2 + y^2 = 81$

15. $x = -y^2 - 4y$ 16. $x^2 + (y + 4)^2 = 1$

17. $y = x^2 - 2x + 1$ 18. $4(x + 3)^2 + (y - 2)^2 = 16$

10.3 Hyperbolas

Jonathan Nourok/PhotoEdit, Inc.

Why You Should Learn It

Equations of hyperbolas are often used in navigation. For instance, in Exercise 49 on page 686, a hyperbola is used to model long-distance radio navigation for a ship.

1 ▶ Graph and write equations of hyperbolas centered at the origin.

What You Should Learn

1 ▶ Graph and write equations of hyperbolas centered at the origin.

2 ▶ Graph and write equations of hyperbolas centered at (h, k).

Hyperbolas Centered at the Origin

The fourth basic type of conic is called a **hyperbola** and is defined as follows.

Definition of a Hyperbola

A **hyperbola** in the rectangular coordinate system consists of all points (x, y) such that the *difference* of the distances between (x, y) and two fixed points is a positive constant, as shown in Figure 10.28. The two fixed points are called the **foci** of the hyperbola. The line on which the foci lie is called the **transverse axis** of the hyperbola.

$d_2 - d_1$ is a positive constant.

Figure 10.28

Standard Equation of a Hyperbola (Center at Origin)

The **standard form of the equation of a hyperbola centered at the origin** is

$$\frac{x^2}{a^2} - \frac{y^2}{b^2} = 1 \qquad \text{or} \qquad \frac{y^2}{a^2} - \frac{x^2}{b^2} = 1$$

Transverse axis is horizontal.

Transverse axis is vertical.

where a and b are positive real numbers. The **vertices** of the hyperbola lie on the transverse axis, a units from the center, as shown in Figure 10.29.

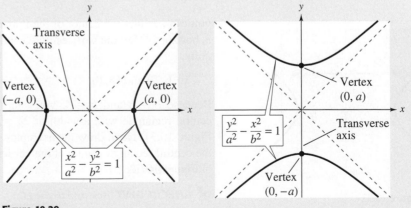

Figure 10.29

A hyperbola has two disconnected parts, each of which is called a **branch** of the hyperbola. The two branches approach a pair of intersecting lines called the **asymptotes** of the hyperbola. The two asymptotes intersect at the center of the hyperbola. To sketch a hyperbola, form a **central rectangle** that is centered at the origin and has side lengths of $2a$ and $2b$. Note in Figure 10.30 that the asymptotes pass through the corners of the central rectangle and that the vertices of the hyperbola lie at the centers of opposite sides of the central rectangle.

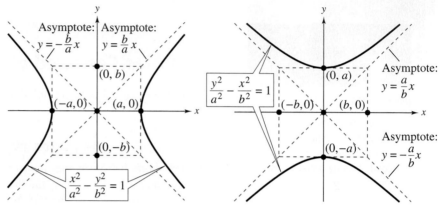

Transverse axis is horizontal.   Transverse axis is vertical.

Figure 10.30

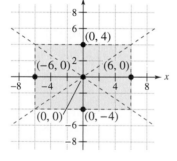

Figure 10.31

EXAMPLE 1   Sketching a Hyperbola

Identify the vertices of the hyperbola given by the equation, and sketch the hyperbola.

$$\frac{x^2}{36} - \frac{y^2}{16} = 1$$

Solution

From the standard form of the equation

$$\frac{x^2}{6^2} - \frac{y^2}{4^2} = 1$$

you can see that the center of the hyperbola is the origin and the transverse axis is horizontal. So, the vertices lie six units to the left and right of the center at the points

$$(-6, 0) \text{ and } (6, 0).$$

Because $a = 6$ and $b = 4$, you can sketch the hyperbola by first drawing a central rectangle with a width of $2a = 12$ and a height of $2b = 8$, as shown in Figure 10.31. Next, draw the asymptotes of the hyperbola through the corners of the central rectangle, and plot the vertices. Finally, draw the hyperbola, as shown in Figure 10.32.

✓ **CHECKPOINT** *Now try Exercise 11.*

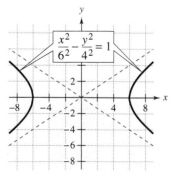

Figure 10.32

Writing the equation of a hyperbola is a little more difficult than writing equations of the other three types of conics. However, if you know the vertices and the asymptotes, you can find the values of a and b, which enable you to write the equation. Notice in Example 2 that the key to this procedure is knowing that the central rectangle has a width of $2b$ and a height of $2a$.

EXAMPLE 2   Writing the Equation of a Hyperbola

Write the standard form of the equation of the hyperbola with a vertical transverse axis and vertices $(0, 3)$ and $(0, -3)$. The equations of the asymptotes of the hyperbola are $y = \frac{3}{5}x$ and $y = -\frac{3}{5}x$.

Solution

To begin, sketch the lines that represent the asymptotes, as shown in Figure 10.33. Note that these two lines intersect at the origin, which implies that the center of the hyperbola is $(0, 0)$. Next, plot the two vertices at the points $(0, 3)$ and $(0, -3)$. You can use the vertices and asymptotes to sketch the central rectangle of the hyperbola, as shown in Figure 10.33. Note that the corners of the central rectangle occur at the points

$$(-5, 3), (5, 3), (-5, -3), \text{ and } (5, -3).$$

Because the width of the central rectangle is $2b = 10$, it follows that $b = 5$. Similarly, because the height of the central rectangle is $2a = 6$, it follows that $a = 3$. Now that you know the values of a and b, you can use the standard form of the equation of a hyperbola to write an equation.

$$\frac{y^2}{a^2} - \frac{x^2}{b^2} = 1 \qquad \text{Transverse axis is vertical.}$$

$$\frac{y^2}{3^2} - \frac{x^2}{5^2} = 1 \qquad \text{Substitute 3 for } a \text{ and 5 for } b.$$

$$\frac{y^2}{9} - \frac{x^2}{25} = 1 \qquad \text{Simplify.}$$

The graph is shown in Figure 10.34.

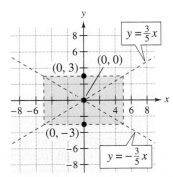

Figure 10.33

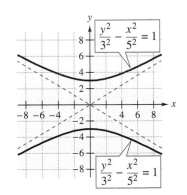

Figure 10.34

✓ **CHECKPOINT** *Now try Exercise 19.*

2 ▶ Graph and write equations of hyperbolas centered at (h, k).

Hyperbolas Centered at (h, k)

Standard Equation of a Hyperbola [Center at (h, k)]

The **standard form of the equation of a hyperbola centered at (h, k)** is

$$\frac{(x - h)^2}{a^2} - \frac{(y - k)^2}{b^2} = 1 \qquad \text{Transverse axis is horizontal.}$$

or

$$\frac{(y - k)^2}{a^2} - \frac{(x - h)^2}{b^2} = 1 \qquad \text{Transverse axis is vertical.}$$

where a and b are positive real numbers. The vertices lie on the transverse axis, a units from the center, as shown in Figure 10.35.

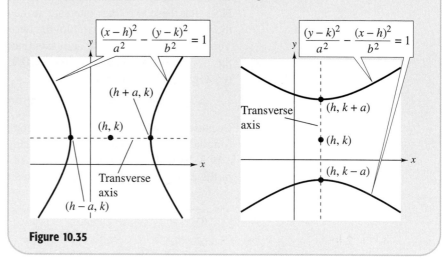

Figure 10.35

When $h = 0$ and $k = 0$, the hyperbola is centered at the origin. Otherwise, you can use the rules on horizontal and vertical shifts from Section 3.7 to shift the center of the hyperbola h units horizontally and k units vertically from the origin.

EXAMPLE 3   Sketching a Hyperbola

Sketch the hyperbola given by $\dfrac{(y - 1)^2}{9} - \dfrac{(x + 2)^2}{4} = 1$.

Solution

From the form of the equation, you can see that the transverse axis is vertical. The center of the hyperbola is $(h, k) = (-2, 1)$. Because $a = 3$ and $b = 2$, you can begin by sketching a central rectangle that is six units high and four units wide, centered at $(-2, 1)$. Then, sketch the asymptotes by drawing lines through the corners of the central rectangle. Sketch the hyperbola, as shown in Figure 10.36. From the graph, you can see that the center of the hyperbola is shifted two units to the left and one unit upward from the origin.

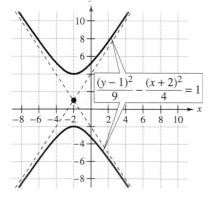

Figure 10.36

✓ **CHECKPOINT** *Now try Exercise 33.*

EXAMPLE 4   Sketching a Hyperbola

Sketch the hyperbola given by $x^2 - 4y^2 + 8x + 16y - 4 = 0$.

Solution

Complete the square to write the equation in standard form.

$$x^2 - 4y^2 + 8x + 16y - 4 = 0 \qquad \text{Write original equation.}$$

$$(x^2 + 8x +) - (4y^2 - 16y +) = 4 \qquad \text{Group terms.}$$

$$(x^2 + 8x +) - 4(y^2 - 4y +) = 4 \qquad \text{Factor 4 out of } y\text{-terms.}$$

$$(x^2 + 8x + 16) - 4(y^2 - 4y + 4) = 4 + 16 - 4(4) \qquad \text{Complete the squares.}$$

$$(x + 4)^2 - 4(y - 2)^2 = 4 \qquad \text{Simplify.}$$

$$\frac{(x + 4)^2}{4} - \frac{(y - 2)^2}{1} = 1 \qquad \text{Divide each side by 4.}$$

$$\frac{(x + 4)^2}{2^2} - \frac{(y - 2)^2}{1^2} = 1 \qquad \text{Standard form}$$

From this standard form, you can see that the transverse axis is horizontal and the center of the hyperbola is $(h, k) = (-4, 2)$. Because $a = 2$ and $b = 1$, you can begin by sketching a central rectangle that is four units wide and two units high, centered at $(-4, 2)$. Then, sketch the asymptotes by drawing lines through the corners of the central rectangle. Sketch the hyperbola, as shown in Figure 10.37. From the graph, you can see that the center of the hyperbola is shifted four units to the left and two units upward from the origin.

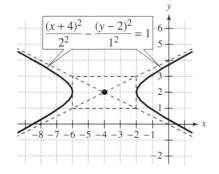

Figure 10.37

✓ **CHECKPOINT** *Now try Exercise 35.*

Technology: Tip

You can use a graphing calculator to graph a hyperbola. For instance, to graph the hyperbola $4y^2 - 9x^2 = 36$, first solve for y to obtain

$$y_1 = 3\sqrt{\frac{x^2}{4} + 1}$$

and

$$y_2 = -3\sqrt{\frac{x^2}{4} + 1}.$$

Use a viewing window in which $-6 \le x \le 6$ and $-8 \le y \le 8$. You should obtain the graph shown below.

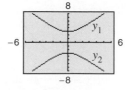

Concept Check

1. You are given the equation of a hyperbola in the standard form
$$\frac{x^2}{a^2} - \frac{y^2}{b^2} = 1.$$
Explain how you can sketch the central rectangle for the hyperbola. Explain how you can use the central rectangle to sketch the asymptotes of the hyperbola.

2. You are given the vertices and the equations of the asymptotes of a hyperbola. Explain how you can determine the values of a and b in the standard form of the equation of the hyperbola.

3. What are the dimensions of the central rectangle and the coordinates of the center of the hyperbola whose equation in standard form is
$$\frac{(y-k)^2}{a^2} - \frac{(x-h)^2}{b^2} = 1?$$

4. Given the equation of a hyperbola in the general polynomial form
$$ax^2 - by^2 + cx + dy + e = 0$$
what process can you use to find the center of the hyperbola?

10.3 EXERCISES

Go to pages 698–699 to record your assignments.

Developing Skills

In Exercises 1–6, match the equation with its graph. [The graphs are labeled (a), (b), (c), (d), (e), and (f).]

(a)
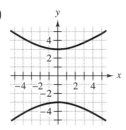

(b)
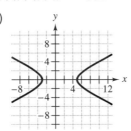

(c)
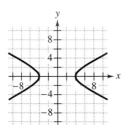

(d)
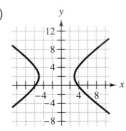

(e)
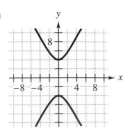

(f)
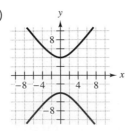

1. $\dfrac{x^2}{16} - \dfrac{y^2}{4} = 1$

2. $\dfrac{y^2}{16} - \dfrac{x^2}{4} = 1$

3. $\dfrac{y^2}{9} - \dfrac{x^2}{16} = 1$

4. $\dfrac{y^2}{16} - \dfrac{x^2}{9} = 1$

5. $\dfrac{(x-1)^2}{16} - \dfrac{y^2}{4} = 1$

6. $\dfrac{(x+1)^2}{16} - \dfrac{(y-2)^2}{9} = 1$

In Exercises 7–18, sketch the hyperbola. Identify the vertices and asymptotes. *See Example 1.*

7. $x^2 - y^2 = 9$

8. $x^2 - y^2 = 1$

9. $y^2 - x^2 = 9$

10. $y^2 - x^2 = 1$

11. $\dfrac{x^2}{9} - \dfrac{y^2}{25} = 1$

12. $\dfrac{x^2}{4} - \dfrac{y^2}{9} = 1$

13. $\dfrac{y^2}{9} - \dfrac{x^2}{25} = 1$ **14.** $\dfrac{y^2}{4} - \dfrac{x^2}{9} = 1$

15. $\dfrac{x^2}{1} - \dfrac{y^2}{9/4} = 1$ **16.** $\dfrac{y^2}{1/4} - \dfrac{x^2}{25/4} = 1$

17. $4y^2 - x^2 + 16 = 0$ **18.** $4y^2 - 9x^2 - 36 = 0$

In Exercises 19–26, write the standard form of the equation of the hyperbola centered at the origin. *See Example 2.*

	Vertices	*Asymptotes*	
19.	$(-4, 0), (4, 0)$	$y = 2x$	$y = -2x$
20.	$(-2, 0), (2, 0)$	$y = \frac{1}{3}x$	$y = -\frac{1}{3}x$
21.	$(0, -4), (0, 4)$	$y = \frac{1}{2}x$	$y = -\frac{1}{2}x$
22.	$(0, -2), (0, 2)$	$y = 3x$	$y = -3x$
23.	$(-9, 0), (9, 0)$	$y = \frac{2}{3}x$	$y = -\frac{2}{3}x$
24.	$(-1, 0), (1, 0)$	$y = \frac{1}{2}x$	$y = -\frac{1}{2}x$
25.	$(0, -1), (0, 1)$	$y = 2x$	$y = -2x$
26.	$(0, -5), (0, 5)$	$y = x$	$y = -x$

In Exercises 27–30, use a graphing calculator to graph the equation. (*Note:* Solve for *y.*)

27. $\dfrac{x^2}{16} - \dfrac{y^2}{4} = 1$ **28.** $\dfrac{y^2}{16} - \dfrac{x^2}{4} = 1$

29. $5x^2 - 2y^2 + 10 = 0$

30. $x^2 - 2y^2 - 4 = 0$

In Exercises 31–38, find the center and vertices of the hyperbola and sketch the hyperbola. *See Examples 3 and 4.*

31. $(y + 4)^2 - (x - 3)^2 = 25$

32. $(y + 6)^2 - (x - 2)^2 = 1$

33. $\dfrac{(x - 1)^2}{4} - \dfrac{(y + 2)^2}{1} = 1$

34. $\dfrac{(x - 2)^2}{4} - \dfrac{(y - 3)^2}{9} = 1$

35. $9x^2 - y^2 - 36x - 6y + 18 = 0$

36. $x^2 - 9y^2 + 36y - 72 = 0$

37. $4x^2 - y^2 + 24x + 4y + 28 = 0$

38. $25x^2 - 4y^2 + 100x + 8y + 196 = 0$

In Exercises 39–42, write the standard form of the equation of the hyperbola.

39.

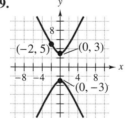

40.

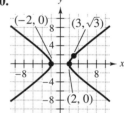

41.

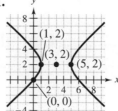

42.

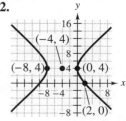

In Exercises 43–48, determine whether the graph represented by the equation is a *circle*, a *parabola*, an *ellipse*, or a *hyperbola*.

43. $\dfrac{(x-3)^2}{4^2} + \dfrac{(y-4)^2}{6^2} = 1$

44. $\dfrac{(x+2)^2}{25} + \dfrac{(y-2)^2}{25} = 1$

45. $x^2 - y^2 = 1$

46. $2x + y^2 = 0$

47. $y^2 - x^2 - 2y + 8x - 19 = 0$

48. $9x^2 + y^2 - 18x - 8y + 16 = 0$

Solving Problems

49. *Navigation*   Long-distance radio navigation for aircraft and ships uses synchronized pulses transmitted by widely separated transmitting stations. The locations of two transmitting stations that are 300 miles apart are represented by the points $(-150, 0)$ and $(150, 0)$ (see figure). A ship's location is given by $(x, 75)$. The difference in the arrival times of pulses transmitted simultaneously to the ship from the two stations is constant at any point on the hyperbola given by

$$\frac{x^2}{93^2} - \frac{y^2}{13{,}851} = 1$$

which passes through the ship's location and has the two stations as foci. Use the equation to find the x-coordinate of the ship's location.

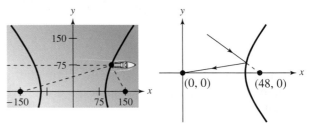

Figure for 49   Figure for 50

50. *Optics*   Hyperbolic mirrors are used in some telescopes. The figure shows a cross section of a hyperbolic mirror as the right branch of a hyperbola. A property of the mirror is that a light ray directed at the focus $(48, 0)$ is reflected to the other focus $(0, 0)$. Use the equation of the hyperbola

$$89x^2 - 55y^2 - 4272x + 31{,}684 = 0$$

to find the coordinates of the mirror's vertex.

Explaining Concepts

51. *Think About It*   Describe the part of the hyperbola

$$\frac{(x-3)^2}{4} - \frac{(y-1)^2}{9} = 1$$

given by each equation.

(a) $x = 3 - \frac{2}{3}\sqrt{9 + (y-1)^2}$

(b) $y = 1 + \frac{3}{2}\sqrt{(x-3)^2 - 4}$

52. Consider the definition of a hyperbola. How many hyperbolas have a given pair of points as foci? Explain your reasoning.

53. How many hyperbolas pass through a given point and have a given pair of points as foci? Explain your reasoning.

54. Cut cone-shaped pieces of styrofoam to demonstrate how to obtain each type of conic section: circle, parabola, ellipse, and hyperbola. Discuss how you could write directions for someone else to form each conic section. Compile a list of real-life situations and/or everyday objects in which conic sections may be seen.

Cumulative Review

In Exercises 55 and 56, determine whether the system is consistent or inconsistent.

55. $\begin{cases} -x + 3y = 8 \\ 4x - 12y = -32 \end{cases}$

56. $\begin{cases} x - 3y = 5 \\ 2x - 6y = -5 \end{cases}$

In Exercises 57 and 58, solve the system of linear equations by the method of elimination.

57. $\begin{cases} x + y = 3 \\ x - y = 2 \end{cases}$

58. $\begin{cases} 4x + 3y = 3 \\ x - 2y = 9 \end{cases}$

10.4 Solving Nonlinear Systems of Equations

StockShot/Alamy

What You Should Learn

1 ▶ Solve nonlinear systems of equations graphically.

2 ▶ Solve nonlinear systems of equations by substitution.

3 ▶ Solve nonlinear systems of equations by elimination.

4 ▶ Use nonlinear systems of equations to model and solve real-life problems.

Why You Should Learn It

Nonlinear systems of equations can be used in real-life applications. For instance, in Example 7 on page 693, nonlinear equations are used to assist rescuers in their search for victims buried by an avalanche.

1 ▶ Solve nonlinear systems of equations graphically.

Solving Nonlinear Systems of Equations by Graphing

In Chapter 4, you studied several methods for solving systems of linear equations. For instance, the following linear system has one solution, $(2, -1)$, which means that $(2, -1)$ is a point of intersection of the two lines represented by the system.

$$\begin{cases} 2x - 3y = 7 \\ x + 4y = -2 \end{cases}$$

In Chapter 4, you also learned that a linear system can have no solution, exactly one solution, or infinitely many solutions. A **nonlinear system of equations** is a system that contains at least one nonlinear equation. Nonlinear systems of equations can have no solution, one solution, or two or more solutions. For instance, the hyperbola and line in Figure 10.38(a) have no point of intersection, the circle and line in Figure 10.38(b) have one point of intersection, and the parabola and line in Figure 10.38(c) have two points of intersection.

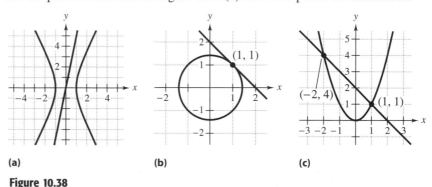

(a)   (b)   (c)

Figure 10.38

You can solve a nonlinear system of equations graphically, as follows.

Solving a Nonlinear System Graphically

1. Sketch the graph of each equation in the system.

2. Locate the point(s) of intersection of the graphs (if any) and graphically approximate the coordinates of the point(s).

3. Check the coordinates by substituting them into each equation in the original system. If the coordinates do not check, you may have to use an algebraic approach, as discussed later in this section.

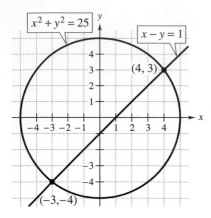

Figure 10.39

Technology: Tip

Try using a graphing calculator to solve the system in Example 1. When you do this, remember that the circle needs to be entered as two separate equations.

$y_1 = \sqrt{25 - x^2}$ — Top half of circle

$y_2 = -\sqrt{25 - x^2}$ — Bottom half of circle

$y_3 = x - 1$ — Line

EXAMPLE 1 Solving a Nonlinear System Graphically

Find all solutions of the nonlinear system of equations.

$$\begin{cases} x^2 + y^2 = 25 & \text{Equation 1} \\ x - y = 1 & \text{Equation 2} \end{cases}$$

Solution

Begin by sketching the graph of each equation. The first equation graphs as a circle centered at the origin and having a radius of 5. The second equation graphs as a line with a slope of 1 and a y-intercept of $(0, -1)$. The system appears to have two solutions: $(-3, -4)$ and $(4, 3)$, as shown in Figure 10.39.

Check

To check $(-3, -4)$, substitute -3 for x and -4 for y in each equation.

$(-3)^2 + (-4)^2 \overset{?}{=} 25$ — Substitute -3 for x and -4 for y in Equation 1.

$9 + 16 = 25$ — Solution checks in Equation 1. ✓

$(-3) - (-4) \overset{?}{=} 1$ — Substitute -3 for x and -4 for y in Equation 2.

$-3 + 4 = 1$ — Solution checks in Equation 2. ✓

To check $(4, 3)$, substitute 4 for x and 3 for y in each equation.

$4^2 + 3^2 \overset{?}{=} 25$ — Substitute 4 for x and 3 for y in Equation 1.

$16 + 9 = 25$ — Solution checks in Equation 1. ✓

$4 - 3 \overset{?}{=} 1$ — Substitute 4 for x and 3 for y in Equation 2.

$1 = 1$ — Solution checks in Equation 2. ✓

✔ **CHECKPOINT** *Now try Exercise 1.*

EXAMPLE 2 Solving a Nonlinear System Graphically

Find all solutions of the nonlinear system of equations.

$$\begin{cases} x = (y - 3)^2 & \text{Equation 1} \\ x + y = 5 & \text{Equation 2} \end{cases}$$

Solution

Begin by sketching the graph of each equation. Solve the first equation for y.

$x = (y - 3)^2$ — Write original equation.

$\pm\sqrt{x} = y - 3$ — Take the square root of each side.

$3 \pm \sqrt{x} = y$ — Add 3 to each side.

The graph of $y = 3 \pm \sqrt{x}$ is a parabola with its vertex at $(0, 3)$. The second equation graphs as a line with a slope of -1 and a y-intercept of $(0, 5)$. The system appears to have two solutions: $(4, 1)$ and $(1, 4)$, as shown in Figure 10.40. Check these solutions in the original system.

✔ **CHECKPOINT** *Now try Exercise 5.*

Figure 10.40

2 ▶ Solve nonlinear systems of equations by substitution.

Solving Nonlinear Systems of Equations by Substitution

The graphical approach to solving any type of system (linear or nonlinear) in two variables is very useful. For systems with solutions having "messy" coordinates, however, a graphical approach is usually not accurate enough to produce exact solutions. In such cases, you should use an algebraic approach. (With an algebraic approach, you should still sketch the graph of each equation in the system.)

As with systems of *linear* equations, there are two basic algebraic approaches: substitution and elimination. Substitution usually works well for systems in which one of the equations is linear, as shown in Example 3.

EXAMPLE 3   **Using Substitution to Solve a Nonlinear System**

Solve the nonlinear system of equations.

$$\begin{cases} 4x^2 + y^2 = 4 & \text{Equation 1} \\ -2x + y = 2 & \text{Equation 2} \end{cases}$$

Solution

Begin by solving for y in Equation 2 to obtain $y = 2x + 2$. Next, substitute this expression for y into Equation 1.

$4x^2 + y^2 = 4$	Write Equation 1.
$4x^2 + (2x + 2)^2 = 4$	Substitute $2x + 2$ for y.
$4x^2 + 4x^2 + 8x + 4 = 4$	Expand.
$8x^2 + 8x = 0$	Simplify.
$8x(x + 1) = 0$	Factor.
$8x = 0 \implies x = 0$	Set 1st factor equal to 0.
$x + 1 = 0 \implies x = -1$	Set 2nd factor equal to 0.

Finally, back-substitute these values of x into the revised Equation 2 to solve for y.

For $x = 0$: $y = 2(0) + 2 = 2$

For $x = -1$: $y = 2(-1) + 2 = 0$

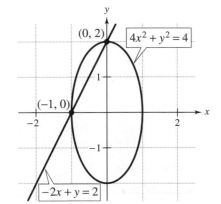

Figure 10.41

So, the system of equations has two solutions: $(0, 2)$ and $(-1, 0)$. Figure 10.41 shows the graph of the system. You can check the solutions as follows.

Check First Solution

$$4(0)^2 + 2^2 \stackrel{?}{=} 4$$
$$0 + 4 = 4 \checkmark$$
$$-2(0) + 2 \stackrel{?}{=} 2$$
$$2 = 2 \checkmark$$

Check Second Solution

$$4(-1)^2 + 0^2 \stackrel{?}{=} 4$$
$$4 + 0 = 4 \checkmark$$
$$-2(-1) + 0 \stackrel{?}{=} 2$$
$$2 = 2 \checkmark$$

✓ **CHECKPOINT** *Now try Exercise 31.*

The steps for using the method of substitution to solve a system of two equations involving two variables are summarized as follows.

Method of Substitution

To solve a system of two equations in two variables, use the steps below.
1. Solve one of the equations for one variable in terms of the other variable.
2. Substitute the expression found in Step 1 into the other equation to obtain an equation in one variable.
3. Solve the equation obtained in Step 2.
4. Back-substitute the solution from Step 3 into the expression obtained in Step 1 to find the value of the other variable.
5. Check the solution to see that it satisfies *both* of the original equations.

Example 4 shows how the method of substitution and graphing can be used to determine that a nonlinear system of equations has no solution.

EXAMPLE 4   Solving a Nonlinear System: No-Solution Case

Solve the nonlinear system of equations.

$$\begin{cases} x^2 - y = 0 & \text{Equation 1} \\ x - y = 1 & \text{Equation 2} \end{cases}$$

Solution

Begin by solving for y in Equation 2 to obtain $y = x - 1$. Next, substitute this expression for y into Equation 1.

$$x^2 - y = 0 \qquad \text{Write Equation 1.}$$

$$x^2 - (x - 1) = 0 \qquad \text{Substitute } x - 1 \text{ for } y.$$

$$x^2 - x + 1 = 0 \qquad \text{Distributive Property}$$

Use the Quadratic Formula, because this equation cannot be factored.

$$x = \frac{-(-1) \pm \sqrt{(-1)^2 - 4(1)(1)}}{2(1)} \qquad \text{Use Quadratic Formula.}$$

$$= \frac{1 \pm \sqrt{1 - 4}}{2} = \frac{1 \pm \sqrt{-3}}{2} \qquad \text{Simplify.}$$

Now, because the Quadratic Formula yields a negative number inside the radical, you can conclude that the equation $x^2 - x + 1 = 0$ has no (real) solution. So, the system has no (real) solution. Figure 10.42 shows the graph of this system. From the graph, you can see that the parabola and the line have no point of intersection.

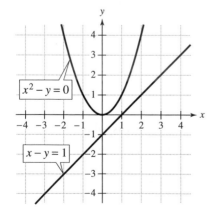

Figure 10.42

✓ **CHECKPOINT** *Now try Exercise 51.*

3 ▶ Solve nonlinear systems of equations by elimination.

Solving Nonlinear Systems of Equations by Elimination

In Section 4.2, you learned how to use the method of elimination to solve a system of linear equations. This method can also be used with special types of nonlinear systems, as demonstrated in Example 5.

EXAMPLE 5   **Using Elimination to Solve a Nonlinear System**

Solve the nonlinear system of equations.

$$\begin{cases} 4x^2 + y^2 = 64 & \text{Equation 1} \\ x^2 + y^2 = 52 & \text{Equation 2} \end{cases}$$

Solution

Because both equations have y^2 as a term (and no other terms containing y), you can eliminate y by subtracting Equation 2 from Equation 1.

$$\begin{array}{rl} 4x^2 + y^2 = & 64 \\ -x^2 - y^2 = & -52 \\ \hline 3x^2 \quad\;\; = & 12 \end{array}$$ Subtract Equation 2 from Equation 1.

After eliminating y, solve the remaining equation for x.

$$3x^2 = 12$$ Write resulting equation.

$$x^2 = 4$$ Divide each side by 3.

$$x = \pm 2$$ Take square root of each side.

By substituting $x = 2$ into Equation 2, you obtain

$$x^2 + y^2 = 52$$ Write Equation 2.

$$(2)^2 + y^2 = 52$$ Substitute 2 for x.

$$y^2 = 48$$ Subtract 4 from each side.

$$y = \pm 4\sqrt{3}.$$ Take square root of each side and simplify.

By substituting $x = -2$, you obtain the same values of y, as follows.

$$x^2 + y^2 = 52$$ Write Equation 2.

$$(-2)^2 + y^2 = 52$$ Substitute -2 for x.

$$y^2 = 48$$ Subtract 4 from each side.

$$y = \pm 4\sqrt{3}$$ Take square root of each side and simplify.

This implies that the system has four solutions:

$$\left(2, 4\sqrt{3}\right), \quad \left(2, -4\sqrt{3}\right), \quad \left(-2, 4\sqrt{3}\right), \quad \left(-2, -4\sqrt{3}\right).$$

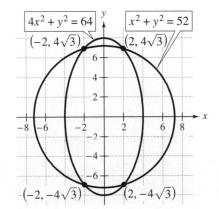

Figure 10.43

Check these solutions in the original system. Figure 10.43 shows the graph of the system. Notice that the graph of Equation 1 is an ellipse and the graph of Equation 2 is a circle.

✔ **CHECKPOINT** *Now try Exercise 59.*

Smart Study Strategy

Go to page 654 for ways to *Analyze Your Errors.*

EXAMPLE 6   Using Elimination to Solve a Nonlinear System

Solve the nonlinear system of equations.

$$\begin{cases} x^2 - 2y = 4 & \text{Equation 1} \\ x^2 - y^2 = 1 & \text{Equation 2} \end{cases}$$

Solution

Because both equations have x^2 as a term (and no other terms containing x), you can eliminate x by subtracting Equation 2 from Equation 1.

$$\begin{array}{rcl} x^2 - 2y &=& 4 \\ -x^2 + y^2 &=& -1 \\ \hline y^2 - 2y &=& 3 \end{array}$$ Subtract Equation 2 from Equation 1.

After eliminating x, solve the remaining equation for y.

$y^2 - 2y = 3$	Write resulting equation.
$y^2 - 2y - 3 = 0$	Write in general form.
$(y + 1)(y - 3) = 0$	Factor.
$y + 1 = 0 \implies y = -1$	Set 1st factor equal to 0.
$y - 3 = 0 \implies y = 3$	Set 2nd factor equal to 0.

When $y = -1$, you obtain

$x^2 - y^2 = 1$	Write Equation 2.
$x^2 - (-1)^2 = 1$	Substitute -1 for y.
$x^2 - 1 = 1$	Simplify.
$x^2 = 2$	Add 1 to each side.
$x = \pm\sqrt{2}.$	Take square root of each side.

When $y = 3$, you obtain

$x^2 - y^2 = 1$	Write Equation 2.
$x^2 - (3)^2 = 1$	Substitute 3 for y.
$x^2 - 9 = 1$	Simplify.
$x^2 = 10$	Add 9 to each side.
$x = \pm\sqrt{10}.$	Take square root of each side.

This implies that the system has four solutions:

$$\left(\sqrt{2}, -1\right), \quad \left(-\sqrt{2}, -1\right), \quad \left(\sqrt{10}, 3\right), \quad \left(-\sqrt{10}, 3\right).$$

Check these solutions in the original system. Figure 10.44 shows the graph of the system. Notice that the graph of Equation 1 is a parabola and the graph of Equation 2 is a hyperbola.

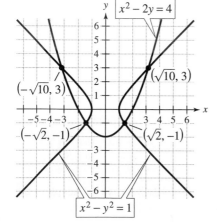

Figure 10.44

✓ **CHECKPOINT** *Now try Exercise 63.*

In Example 6, the method of elimination yields the four exact solutions $\left(\sqrt{2}, -1\right)$, $\left(-\sqrt{2}, -1\right)$, $\left(\sqrt{10}, 3\right)$, and $\left(-\sqrt{10}, 3\right)$. You can use a calculator to approximate these solutions as $(1.41, -1)$, $(-1.41, -1)$, $(3.16, 3)$, and $(-3.16, 3)$. If you use the decimal approximations to check your solutions in the original system, be aware that they may not check.

4 ▶ Use nonlinear systems of equations to model and solve real-life problems.

Application

There are many examples of the use of nonlinear systems of equations in business and science. For instance, in Example 7 a nonlinear system of equations is used to help rescue avalanche victims.

EXAMPLE 7   **Avalanche Rescue System**

RECCO® is an avalanche rescue system utilized by rescue organizations worldwide. RECCO technology enables quick directional pinpointing of a victim's exact location using harmonic radar. The two-part system consists of a detector used by rescuers, and reflectors that are integrated into apparel, helmets, protection gear, or boots. The range of the detector through snow is 30 meters. Two rescuers are 30 meters apart on the surface. What is the maximum depth of a reflector that is in range of both rescuers?

Solution

Let the first rescuer be located at the origin and let the second rescuer be located 30 meters (units) to the right. The range of each detector is circular and can be modeled by the following equations.

$$\begin{cases} x^2 + y^2 = 30^2 & \text{Range of first rescuer} \\ (x - 30)^2 + y^2 = 30^2 & \text{Range of second rescuer} \end{cases}$$

Using methods demonstrated earlier in this section, you'll find that these two equations intersect when $x = 15$ and $y \approx \pm25.98$. You are concerned only about the lower portions of the circles. So, the maximum depth in range of both rescuers is point R, as shown in Figure 10.45, which is about 26 meters beneath the surface.

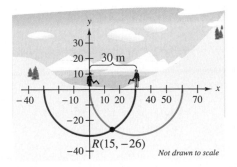

Figure 10.45

✓ **CHECKPOINT** *Now try Exercise 87.*

blickwinkel/Alamy

RECCO technology is often used in conjunction with other rescue methods such as avalanche dogs, transceivers, and probe lines.

_____ **Concept Check** _____

1. How is a system of nonlinear equations different from a system of linear equations?

2. Identify the different methods you can use to solve a system of nonlinear equations.

3. If one of the equations in a system is linear, which algebraic method usually works best for solving the system?

4. If both of the equations in a system are conics, which algebraic method usually works best for solving the system?

10.4 EXERCISES

Go to pages 698–699 to record your assignments.

_____ **Developing Skills** _____

In Exercises 1–16, graph the equations to determine whether the system has any solutions. Find any solutions that exist. *See Examples 1 and 2.*

✓ 1. $\begin{cases} y = 1 \\ x^2 + y = 0 \end{cases}$

2. $\begin{cases} x^2 + y^2 = 1 \\ x + y = 4 \end{cases}$

3. $\begin{cases} x = 0 \\ x^2 + y^2 = 9 \end{cases}$

4. $\begin{cases} y = 4 \\ x^2 - y = 0 \end{cases}$

✓ 5. $\begin{cases} x + y = 2 \\ x^2 - y = 0 \end{cases}$

6. $\begin{cases} 2x + y = 10 \\ x^2 + y^2 = 25 \end{cases}$

7. $\begin{cases} x^2 + y = 9 \\ x - y = -3 \end{cases}$

8. $\begin{cases} x - y^2 = 0 \\ x - y = 2 \end{cases}$

9. $\begin{cases} y = \sqrt{x - 2} \\ x - 2y = 1 \end{cases}$

10. $\begin{cases} x - 2y = 4 \\ x^2 - y = 0 \end{cases}$

11. $\begin{cases} x^2 + y^2 = 100 \\ x + y = 2 \end{cases}$

12. $\begin{cases} x^2 + y^2 = 169 \\ x + y = 7 \end{cases}$

13. $\begin{cases} x^2 + y^2 = 25 \\ 2x - y = -5 \end{cases}$

14. $\begin{cases} x^2 - y^2 = 16 \\ 3x - y = 12 \end{cases}$

15. $\begin{cases} 9x^2 - 4y^2 = 36 \\ 5x - 2y = 0 \end{cases}$

16. $\begin{cases} 9x^2 + 4y^2 = 36 \\ y + 1 = \sqrt{x - 1} \end{cases}$

🔲 In Exercises 17–30, use a graphing calculator to graph the equations and find any solutions of the system.

17. $\begin{cases} y = 2x^2 \\ y = -2x + 12 \end{cases}$

18. $\begin{cases} y = 5x^2 \\ y = -15x - 10 \end{cases}$

19. $\begin{cases} y = x \\ y = x^3 \end{cases}$

20. $\begin{cases} y = x^4 \\ y = 5x + 6 \end{cases}$

21. $\begin{cases} y = x^2 \\ y = -x^2 + 4x \end{cases}$

22. $\begin{cases} y = 8 - x^2 \\ y = 6 - x \end{cases}$

23. $\begin{cases} x^2 - y = 2 \\ 3x + y = 2 \end{cases}$

24. $\begin{cases} x^2 + 2y = 6 \\ x - y = -4 \end{cases}$

25. $\begin{cases} x - 3y = 1 \\ \sqrt{x} - 1 = y \end{cases}$

26. $\begin{cases} \sqrt{x} + 1 = y \\ 2x + y = 4 \end{cases}$

27. $\begin{cases} y = x^3 \\ y = x^3 - 3x^2 + 3x \end{cases}$

28. $\begin{cases} y = -2(x^2 - 1) \\ y = 2(x^4 - 2x^2 + 1) \end{cases}$

29. $\begin{cases} y = \ln x - 2 \\ y = x - 2 \end{cases}$

30. $\begin{cases} y = -x^2 + 1 \\ y = \ln(x + 1) + 1 \end{cases}$

In Exercises 31–58, solve the system by the method of substitution. *See Examples 3 and 4.*

31. $\begin{cases} y = 2x^2 \\ y = 6x - 4 \end{cases}$

32. $\begin{cases} y = 5x^2 \\ y = -5x + 10 \end{cases}$

33. $\begin{cases} x^2 + y = 5 \\ 2x + y = 5 \end{cases}$

34. $\begin{cases} x - y^2 = 0 \\ x - y = 2 \end{cases}$

35. $\begin{cases} x^2 + y^2 = 4 \\ x + y = 2 \end{cases}$

36. $\begin{cases} x^2 + y^2 = 36 \\ x = 6 \end{cases}$

37. $\begin{cases} x^2 + y^2 = 25 \\ y = 5 \end{cases}$

38. $\begin{cases} x^2 + y^2 = 1 \\ x + y = 7 \end{cases}$

39. $\begin{cases} x^2 + y^2 = 64 \\ -3x + y = 8 \end{cases}$

40. $\begin{cases} x^2 + y^2 = 81 \\ x + 3y = 27 \end{cases}$

41. $\begin{cases} 4x + y^2 = 2 \\ 2x - y = -11 \end{cases}$

42. $\begin{cases} x^2 + y^2 = 10 \\ 2x - y = 5 \end{cases}$

43. $\begin{cases} x^2 + y^2 = 9 \\ x + 2y = 3 \end{cases}$

44. $\begin{cases} x^2 + y^2 = 4 \\ x - 2y = 4 \end{cases}$

45. $\begin{cases} 2x^2 - y^2 = -8 \\ x - y = 6 \end{cases}$

46. $\begin{cases} y^2 = -x + 4 \\ x^2 + y^2 = 6 \end{cases}$

47. $\begin{cases} y = x^2 - 5 \\ 3x + 2y = 10 \end{cases}$

48. $\begin{cases} x + y = 4 \\ x^2 - y^2 = 4 \end{cases}$

49. $\begin{cases} y = \sqrt{4 - x} \\ x + 3y = 6 \end{cases}$

50. $\begin{cases} y = \sqrt{25 - x^2} \\ x + y = 7 \end{cases}$

51. $\begin{cases} x^2 - 4y^2 = 16 \\ x^2 + y^2 = 1 \end{cases}$

52. $\begin{cases} 2x^2 - y^2 = 12 \\ 3x^2 - y^2 = -4 \end{cases}$

53. $\begin{cases} y = x^2 - 3 \\ x^2 + y^2 = 9 \end{cases}$

54. $\begin{cases} x^2 + y^2 = 25 \\ x - 3y = -5 \end{cases}$

55. $\begin{cases} 16x^2 + 9y^2 = 144 \\ 4x + 3y = 12 \end{cases}$

56. $\begin{cases} 4x^2 + 16y^2 = 64 \\ x + 2y = 4 \end{cases}$

57. $\begin{cases} x^2 - y^2 = 9 \\ x^2 + y^2 = 1 \end{cases}$

58. $\begin{cases} x^2 - y^2 = 4 \\ x - y = 2 \end{cases}$

In Exercises 59–80, solve the system by the method of elimination. *See Examples 5 and 6.*

59. $\begin{cases} x^2 + 2y = 1 \\ x^2 + y^2 = 4 \end{cases}$

60. $\begin{cases} x + y^2 = 5 \\ 2x^2 + y^2 = 6 \end{cases}$

61. $\begin{cases} -x + y^2 = 10 \\ x^2 - y^2 = -8 \end{cases}$

62. $\begin{cases} x^2 + y = 9 \\ x^2 - y^2 = 7 \end{cases}$

63. $\begin{cases} x^2 + y^2 = 7 \\ x^2 - y^2 = 1 \end{cases}$

64. $\begin{cases} x^2 + y^2 = 25 \\ y^2 - x^2 = 7 \end{cases}$

65. $\begin{cases} x^2 - y^2 = 4 \\ x^2 + y^2 = 4 \end{cases}$

66. $\begin{cases} x^2 + y^2 = 25 \\ x^2 - y^2 = -36 \end{cases}$

67. $\begin{cases} x^2 + y^2 = 13 \\ 2x^2 + 3y^2 = 30 \end{cases}$

68. $\begin{cases} 3x^2 - y^2 = 4 \\ x^2 + 4y^2 = 10 \end{cases}$

69. $\begin{cases} 4x^2 + 9y^2 = 36 \\ 2x^2 - 9y^2 = 18 \end{cases}$

70. $\begin{cases} 5x^2 - 2y^2 = -13 \\ 3x^2 + 4y^2 = 39 \end{cases}$ **71.** $\begin{cases} 2x^2 + 3y^2 = 21 \\ x^2 + 2y^2 = 12 \end{cases}$

72. $\begin{cases} 2x^2 + y^2 = 11 \\ x^2 + 3y^2 = 28 \end{cases}$ **73.** $\begin{cases} -x^2 - 2y^2 = 6 \\ 5x^2 + 15y^2 = 20 \end{cases}$

74. $\begin{cases} x^2 - 2y^2 = 7 \\ x^2 + y^2 = 34 \end{cases}$ **75.** $\begin{cases} x^2 + y^2 = 9 \\ 16x^2 - 4y^2 = 64 \end{cases}$

76. $\begin{cases} 3x^2 + 4y^2 = 35 \\ 2x^2 + 5y^2 = 42 \end{cases}$

77. $\begin{cases} \dfrac{x^2}{4} + y^2 = 1 \\ x^2 + \dfrac{y^2}{4} = 1 \end{cases}$

78. $\begin{cases} x^2 - y^2 = 1 \\ \dfrac{x^2}{2} + y^2 = 1 \end{cases}$

79. $\begin{cases} y^2 - x^2 = 10 \\ x^2 + y^2 = 16 \end{cases}$

80. $\begin{cases} x^2 + y^2 = 25 \\ x^2 + 2y^2 = 36 \end{cases}$

Solving Problems

81. ▲ *Ice Rink* A rectangular ice rink has an area of 3000 square feet. The diagonal across the rink is 85 feet. Find the dimensions of the rink.

82. ▲ *Cell Phone* A cell phone has a rectangular external display that contains 19,200 pixels with a diagonal of 200 pixels. Find the resolution (the dimensions in pixels) of the external display.

83. ▲ *Dog Park* A rectangular dog park has a diagonal sidewalk that measures 290 feet. The perimeter of each triangle formed by the diagonal is 700 feet. Find the dimensions of the dog park.

84. ▲ *Sailboat* A sail for a sailboat is shaped like a right triangle that has a perimeter of 36 meters and a hypotenuse of 15 meters. Find the dimensions of the sail.

85. *Hyperbolic Mirror* In a hyperbolic mirror, light rays directed to one focus are reflected to the other focus. The mirror in the figure has the equation

$$\frac{x^2}{9} - \frac{y^2}{16} = 1.$$

At which point on the mirror will light from the point $(0, 10)$ reflect to the focus?

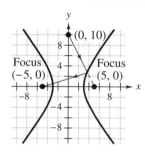

Figure for 85

86. *Miniature Golf* You are playing miniature golf and your golf ball is at $(-15, 25)$ (see figure). A wall at the end of the enclosed area is part of a hyperbola whose equation is

$$\frac{x^2}{19} - \frac{y^2}{81} = 1.$$

Using the reflective property of hyperbolas given in Exercise 85, at which point on the wall must your ball hit for it to go into the hole? (The ball bounces off the wall only once.)

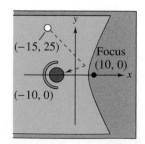

87. *Busing Boundary*   To be eligible to ride the school bus to East High School, a student must live at least 1 mile from the school (see figure). Describe the portion of Clarke Street for which the residents are *not* eligible to ride the school bus. Use a coordinate system in which the school is at $(0, 0)$ and each unit represents 1 mile.

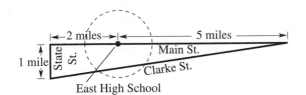

88. *Search Team*   A search team of three members splits to search an area in the woods. Each member carries a family service radio with a circular range of 3 miles. The team members agree to communicate from their bases every hour. The second member sets up base 3 miles north of the first member. Where should the third member set up base to be as far east as possible but within direct communication range of each of the other two searchers? Use a coordinate system in which the first member is at $(0, 0)$ and each unit represents 1 mile.

Explaining Concepts

89. ✎   Explain how to solve a nonlinear system of equations using the method of substitution.

90. ✎   Explain how to solve a nonlinear system of equations using the method of elimination.

91. ✎   What is the maximum number of points of intersection of a line and a hyperbola? Explain.

92. A circle and a parabola can have 0, 1, 2, 3, or 4 points of intersection. Sketch the circle given by $x^2 + y^2 = 4$. Discuss how this circle could intersect a parabola with an equation of the form $y = x^2 + C$. Then find the values of C for each of the five cases described below.

  (a) No points of intersection

  (b) One point of intersection

  (c) Two points of intersection

  (d) Three points of intersection

  (e) Four points of intersection

  Use a graphing calculator to confirm your results.

Cumulative Review

In Exercises 93–104, solve the equation and check your solution(s).

93. $\sqrt{6 - 2x} = 4$

94. $\sqrt{x + 3} = -9$

95. $\sqrt{x} = x - 6$

96. $\sqrt{x + 14} = \sqrt{x} + 3$

97. $3^x = 243$

98. $4^x = 256$

99. $5^{x-1} = 310$

100. $e^{0.5x} = 8$

101. $\log_{10} x = 0.01$

102. $\log_4 8x = 3$

103. $2 \ln(x + 1) = -2$

104. $\ln(x + 3) - \ln x = \ln 1$

What Did You Learn?

Use these two pages to help prepare for a test on this chapter. Check off the key terms and key concepts you know. You can also use this section to record your assignments.

Plan for Test Success

Date of test: ☐ / /

Study dates and times: ☐ / / at ☐ : A.M./P.M.

☐ / / at ☐ : A.M./P.M.

Things to review:

☐ Key Terms, *p. 698*
☐ Key Concepts, *pp. 698–699*
☐ Your class notes
☐ Your assignments

☐ Study Tips, *pp. 659, 660, 669, 681*
☐ Technology Tips, *pp. 661, 671, 683, 688*
☐ Mid-Chapter Quiz, *p. 678*

☐ Review Exercises, *pp. 700–702*
☐ Chapter Test, *p. 703*
☐ Video Explanations Online
☐ Tutorial Online

Key Terms

☐ conic (conic section), *p. 656*
☐ circle, *p. 656*
☐ center (of a circle), *p. 656*
☐ radius, *pp. 656, 657*
☐ standard form of equation of circle centered at $(0, 0)$, *p. 657*
☐ standard form of equation of circle centered at (h, k), *p. 658*
☐ parabola, *p. 660*
☐ directrix, *p. 660*
☐ focus (of a parabola), *p. 660*
☐ vertex (of a parabola), *p. 660*
☐ axis (of a parabola), *p. 660*

☐ ellipse, *p. 668*
☐ focus (of an ellipse), *p. 668*
☐ vertices (of an ellipse), *p. 668*
☐ major axis, *p. 668*
☐ center (of an ellipse), *p. 668*
☐ minor axis, *p. 668*
☐ co-vertices, *p. 668*
☐ standard form of equation of ellipse centered at $(0, 0)$, *p. 669*
☐ standard form of equation of ellipse centered at (h, k), *p. 670*
☐ hyperbola, *p. 679*
☐ foci (of a hyperbola), *p. 679*

☐ transverse axis (of a hyperbola), *p. 679*
☐ standard form of equation of hyperbola centered at $(0, 0)$, *p. 679*
☐ vertices (of a hyperbola), *p. 679*
☐ branch (of a hyperbola), *p. 680*
☐ asymptotes (of a hyperbola), *p. 680*
☐ central rectangle, *p. 680*
☐ standard form of equation of hyperbola centered at (h, k), *p. 682*
☐ nonlinear system of equations, *p. 687*

Key Concepts

10.1 Circles and Parabolas

Assignment: _____ Due date: _____

☐ **Graph and write equations of circles centered at the origin.**

Standard equation of a circle with radius r and center (0, 0):

$x^2 + y^2 = r^2$

☐ **Graph and write equations of circles centered at (h, k).**

Standard equation of a circle with radius r and center (h, k):

$(x - h)^2 + (y - k)^2 = r^2$

☐ **Graph and write equations of parabolas.**

Standard equation of a parabola with vertex at (0, 0):

$x^2 = 4py$ Vertical axis

$y^2 = 4px$ Horizontal axis

Standard equation of a parabola with vertex at (h, k):

$(x - h)^2 = 4p(y - k)$ Vertical axis

$(y - k)^2 = 4p(x - h)$ Horizontal axis

The focus of a parabola lies on the axis, a directed distance of p units from the vertex.

10.2 Ellipses

Assignment: _____ Due date: _____

☐ **Graph and write equations of ellipses centered at the origin.**

Standard equation of an ellipse centered at the origin:

$$\frac{x^2}{a^2} + \frac{y^2}{b^2} = 1 \qquad \text{Major axis is horizontal.}$$

$$\frac{x^2}{b^2} + \frac{y^2}{a^2} = 1 \qquad \text{Major axis is vertical.}$$

☐ **Graph and write equations of ellipses centered at (h, k).**

Standard equation of an ellipse centered at (h, k):

$$\frac{(x - h)^2}{a^2} + \frac{(y - k)^2}{b^2} = 1 \qquad \text{Major axis is horizontal.}$$

$$\frac{(x - h)^2}{b^2} + \frac{(y - k)^2}{a^2} = 1 \qquad \text{Major axis is vertical.}$$

In all of the standard equations for ellipses, $0 < b < a$. The vertices of the ellipse lie on the major axis, a units from the center (the major axis has length $2a$). The co-vertices lie on the minor axis, b units from the center.

10.3 Hyperbolas

Assignment: _____ Due date: _____

☐ **Graph and write equations of hyperbolas centered at the origin.**

Standard equation of a hyperbola centered at the origin:

$$\frac{x^2}{a^2} - \frac{y^2}{b^2} = 1 \qquad \text{Transverse axis is horizontal.}$$

$$\frac{y^2}{a^2} - \frac{x^2}{b^2} = 1 \qquad \text{Transverse axis is vertical.}$$

☐ **Graph and write equations of hyperbolas centered at (h, k).**

Standard equation of a hyperbola centered at (h, k):

$$\frac{(x - h)^2}{a^2} - \frac{(y - k)^2}{b^2} = 1 \qquad \text{Transverse axis is horizontal.}$$

$$\frac{(y - k)^2}{a^2} - \frac{(x - h)^2}{b^2} = 1 \qquad \text{Transverse axis is vertical.}$$

A hyperbola's vertices lie on the transverse axis, a units from the center. A hyperbola's central rectangle has side lengths of $2a$ and $2b$. A hyperbola's asymptotes pass through opposite corners of its central rectangle.

10.4 Solving Nonlinear Systems of Equations

Assignment: _____ Due date: _____

☐ **Solve a nonlinear system graphically.**

1. Sketch the graph of each equation in the system.
2. Graphically approximate the coordinates of any points of intersection of the graphs.
3. Check the coordinates by substituting them into each equation in the original system. If the coordinates do not check, it may be necessary to use an algebraic approach such as substitution or elimination.

☐ **Use substitution to solve a nonlinear system of two equations in two variables.**

1. Solve one equation for one variable in terms of the other.
2. Substitute the expression found in Step 1 into the other equation to obtain an equation in one variable.
3. Solve the equation obtained in Step 2.
4. Back-substitute the solution from Step 3 into the expression obtained in Step 1 to find the value of the other variable.
5. Check that the solution satisfies *both* original equations.

☐ **Use elimination to solve a nonlinear system of equations.**

Review Exercises

10.1 Circles and Parabolas

1 ▶ Recognize the four basic conics: circles, parabolas, ellipses, and hyperbolas.

In Exercises 1–6, identify the conic.

1.
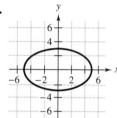

2.
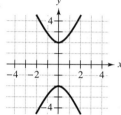

3.
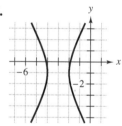

4.
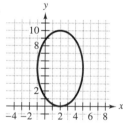

5.
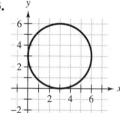

6.
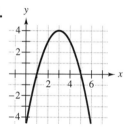

2 ▶ Graph and write equations of circles centered at the origin.

In Exercises 7 and 8, write the standard form of the equation of the circle with center at $(0, 0)$ that satisfies the criterion.

7. Radius: 6

8. Passes through the point $(-1, 3)$

In Exercises 9 and 10, identify the center and radius of the circle and sketch the circle.

9. $x^2 + y^2 = 64$

10. $9x^2 + 9y^2 - 49 = 0$

3 ▶ Graph and write equations of circles centered at (h, k).

In Exercises 11 and 12, write the standard form of the equation of the circle with center at (h, k) that satisfies the criteria.

11. Center: $(2, 6)$;   Radius: 3

12. Center: $(-2, 3)$;   Passes through the point $(1, 1)$

In Exercises 13 and 14, identify the center and radius of the circle and sketch the circle.

13. $x^2 + y^2 + 6x + 8y + 21 = 0$

14. $x^2 + y^2 - 8x + 16y + 75 = 0$

4 ▶ Graph and write equations of parabolas.

In Exercises 15–18, write the standard form of the equation of the parabola. Then sketch the parabola.

15. Vertex: $(0, 0)$;   Focus: $(-2, 0)$

16. Vertex: $(0, 5)$;   Focus: $(2, 5)$

17. Vertex: $(-1, 3)$;

Passes through the point $(-2, 5)$; Vertical axis

18. Vertex: $(8, 0)$;

Passes through the point $(2, -2)$; Horizontal axis

In Exercises 19 and 20, identify the vertex and focus of the parabola and sketch the parabola.

19. $y = \frac{1}{2}x^2 - 8x + 7$

20. $x = y^2 + 10y - 4$

10.2 Ellipses

1 ▶ Graph and write equations of ellipses centered at the origin.

In Exercises 21–24, write the standard form of the equation of the ellipse centered at the origin.

21. Vertices: $(0, -5), (0, 5)$;
Co-vertices: $(-2, 0), (2, 0)$

22. Vertices: $(-8, 0), (8, 0)$;
Co-vertices: $(0, -3), (0, 3)$

23. Major axis (vertical) 6 units, minor axis 4 units

24. Major axis (horizontal) 12 units, minor axis 2 units

In Exercises 25–28, sketch the ellipse. Identify the vertices and co-vertices.

25. $\dfrac{x^2}{64} + \dfrac{y^2}{16} = 1$

26. $\dfrac{x^2}{9} + y^2 = 1$

27. $36x^2 + 9y^2 - 36 = 0$

28. $100x^2 + 4y^2 - 4 = 0$

2 ▶ Graph and write equations of ellipses centered at (h, k).

In Exercises 29–32, write the standard form of the equation of the ellipse.

29. Vertices: $(-2, 4), (8, 4)$; Co-vertices: $(3, 0), (3, 8)$

30. Vertices: $(0, 5), (12, 5)$; Co-vertices: $(6, 2), (6, 8)$

31. Vertices: $(0, 0), (0, 8)$; Co-vertices: $(-3, 4), (3, 4)$

32. Vertices: $(5, -3), (5, 13)$;
Co-vertices: $(3, 5), (7, 5)$

In Exercises 33–36, find the center and vertices of the ellipse and sketch the ellipse.

33. $9(x + 1)^2 + 4(y - 2)^2 = 144$

34. $x^2 + 25y^2 - 4x - 21 = 0$

35. $16x^2 + y^2 + 6y - 7 = 0$

36. $x^2 + 4y^2 + 10x - 24y + 57 = 0$

10.3 Hyperbolas

1 ▶ Graph and write equations of hyperbolas centered at the origin.

In Exercises 37–40, sketch the hyperbola. Identify the vertices and asymptotes.

37. $x^2 - y^2 = 25$

38. $y^2 - x^2 = 16$

39. $\dfrac{y^2}{25} - \dfrac{x^2}{4} = 1$

40. $\dfrac{x^2}{16} - \dfrac{y^2}{25} = 1$

In Exercises 41–44, write the standard form of the equation of the hyperbola centered at the origin.

Vertices	Asymptotes	
41. $(-2, 0), (2, 0)$	$y = \frac{3}{2}x$	$y = -\frac{3}{2}x$
42. $(0, -6), (0, 6)$	$y = 3x$	$y = -3x$
43. $(0, -8), (0, 8)$	$y = \frac{4}{5}x$	$y = -\frac{4}{5}x$
44. $(-3, 0), (3, 0)$	$y = \frac{4}{3}x$	$y = -\frac{4}{3}x$

2 ▶ Graph and write equations of hyperbolas centered at (h, k).

In Exercises 45–48, find the center and vertices of the hyperbola and sketch the hyperbola.

45. $\dfrac{(x-3)^2}{9} - \dfrac{(y+1)^2}{4} = 1$

46. $\dfrac{(x+4)^2}{25} - \dfrac{(y-7)^2}{64} = 1$

47. $8y^2 - 2x^2 + 48y + 16x + 8 = 0$

48. $25x^2 - 4y^2 - 200x - 40y = 0$

In Exercises 49 and 50, write the standard form of the equation of the hyperbola.

49.

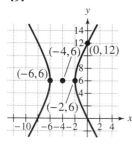

50.

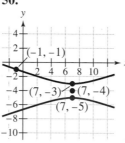

10.4 Solving Nonlinear Systems of Equations

1 ▶ Solve nonlinear systems of equations graphically.

⌨ **In Exercises 51–54, use a graphing calculator to graph the equations and find any solutions of the system.**

51. $\begin{cases} y = x^2 \\ y = 3x \end{cases}$

52. $\begin{cases} y = 2 + x^2 \\ y = 8 - x \end{cases}$

53. $\begin{cases} x^2 + y^2 = 16 \\ -x + y = 4 \end{cases}$

54. $\begin{cases} 2x^2 - y^2 = -8 \\ y = x + 6 \end{cases}$

2 ▶ Solve nonlinear systems of equations by substitution.

In Exercises 55–58, solve the system by the method of substitution.

55. $\begin{cases} y = 5x^2 \\ y = -15x - 10 \end{cases}$

56. $\begin{cases} y^2 = 16x \\ 4x - y = -24 \end{cases}$

57. $\begin{cases} x^2 + y^2 = 1 \\ x + y = -1 \end{cases}$

58. $\begin{cases} y^2 - x^2 = 9 \\ x + y = 1 \end{cases}$

3 ▶ Solve nonlinear systems of equations by elimination.

In Exercises 59 and 60, solve the system by the method of elimination.

59. $\begin{cases} 6x^2 - y^2 = 15 \\ x^2 + y^2 = 13 \end{cases}$

60. $\begin{cases} x^2 + y^2 = 16 \\ -x^2 + \dfrac{y^2}{16} = 1 \end{cases}$

4 ▶ Use nonlinear systems of equations to model and solve real-life problems.

61. ▲ *Geometry* A computer manufacturer needs a circuit board with a perimeter of 28 centimeters and a diagonal of length 10 centimeters. What should the dimensions of the circuit board be?

62. ▲ *Geometry* A home interior decorator wants to find a ceramic tile with a perimeter of 6 inches and a diagonal of length $\sqrt{5}$ inches. What should the dimensions of the tile be?

63. ▲ *Geometry* A piece of wire 100 inches long is to be cut into two pieces. Each of the two pieces is then to be bent into a square. The area of one square is to be 144 square inches greater than the area of the other square. How should the wire be cut?

64. ▲ *Geometry* You have 250 feet of fencing to enclose two corrals of equal size (see figure). The combined area of the corrals is 2400 square feet. Find the dimensions of each corral.

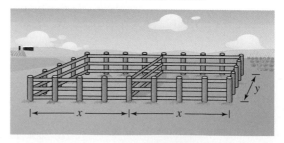

Chapter Test

Take this test as you would take a test in class. After you are done, check your work against the answers in the back of the book.

1. Write the standard form of the equation of the circle shown in the figure.

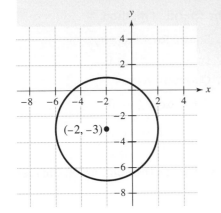

Figure for 1

In Exercises 2 and 3, write the equation of the circle in standard form. Then sketch the circle.

2. $x^2 + y^2 - 2x - 6y + 1 = 0$

3. $x^2 + y^2 + 4x - 6y + 4 = 0$

4. Identify the vertex and the focus of the parabola $x = -3y^2 + 12y - 8$. Then sketch the parabola.

5. Write the standard form of the equation of the parabola with vertex $(7, -2)$ and focus $(7, 0)$.

6. Write the standard form of the equation of the ellipse shown in the figure.

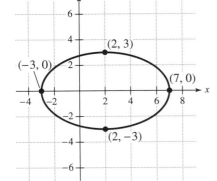

Figure for 6

In Exercises 7 and 8, find the center and vertices of the ellipse. Then sketch the ellipse.

7. $16x^2 + 4y^2 = 64$

8. $25x^2 + 4y^2 - 50x - 24y - 39 = 0$

In Exercises 9 and 10, write the standard form of the equation of the hyperbola.

9. Vertices: $(-3, 0), (3, 0)$; Asymptotes: $y = \pm\frac{2}{3}x$

10. Vertices: $(0, -5), (0, 5)$; Asymptotes: $y = \pm\frac{5}{2}x$

In Exercises 11 and 12, find the center and vertices of the hyperbola. Then sketch the hyperbola.

11. $4x^2 - 2y^2 - 24x + 20 = 0$

12. $16y^2 - 25x^2 + 64y + 200x - 736 = 0$

In Exercises 13–15, solve the nonlinear system of equations.

13. $\begin{cases} x^2/16 + y^2/9 = 1 \\ 3x + 4y = 12 \end{cases}$ **14.** $\begin{cases} x^2 + y^2 = 16 \\ x^2/16 - y^2/9 = 1 \end{cases}$ **15.** $\begin{cases} x^2 + y^2 = 10 \\ x^2 = y^2 + 2 \end{cases}$

16. Write the equation of the circular orbit of a satellite 1000 miles above the surface of Earth. Place the origin of the rectangular coordinate system at the center of Earth and assume the radius of Earth to be 4000 miles.

17. A rectangle has a perimeter of 56 inches and a diagonal of length 20 inches. Find the dimensions of the rectangle.

Cumulative Test: Chapters 8–10

Take this test as you would take a test in class. After you are done, check your work against the answers in the back of the book.

In Exercises 1–4, solve the equation by the specified method.

1. Factoring:

$4x^2 - 9x - 9 = 0$

2. Square Root Property:

$(x - 5)^2 - 64 = 0$

3. Completing the square:

$x^2 - 10x - 25 = 0$

4. Quadratic Formula:

$3x^2 + 6x + 2 = 0$

5. Solve the equation of quadratic form: $x^4 - 8x^2 + 15 = 0$.

In Exercises 6 and 7, solve the inequality and graph the solution on the real number line.

6. $3x^2 + 8x \leq 3$

7. $\dfrac{3x + 4}{2x - 1} < 0$

8. Find a quadratic equation having the solutions -2 and 6.

9. Find the compositions (a) $f \circ g$ and (b) $g \circ f$ for $f(x) = 2x^2 - 3$ and $g(x) = 5x - 1$. Then find the domain of each composition.

10. Find the inverse function of $f(x) = \dfrac{5 - 3x}{4}$.

11. Evaluate $f(x) = 7 + 2^{-x}$ when $x = 1, 0.5$, and 3.

12. Sketch the graph of $f(x) = 4^{x-1}$ and identify the horizontal asymptote.

13. Describe the relationship between the graphs of $f(x) = e^x$ and $g(x) = \ln x$.

14. Sketch the graph of $\log_3(x - 1)$ and identify the vertical asymptote.

15. Evaluate $\log_4 \frac{1}{16}$ without using a calculator.

16. Use the properties of logarithms to condense $3(\log_2 x + \log_2 y) - \log_2 z$.

17. Use the properties of logarithms to expand $\log_{10} \dfrac{\sqrt{x + 1}}{x^4}$.

In Exercises 18–21, solve the equation.

18. $\log_x\left(\frac{1}{9}\right) = -2$

19. $4 \ln x = 10$

20. $500(1.08)^t = 2000$ **21.** $3(1 + e^{2x}) = 20$

22. If the inflation rate averages 2.8% over the next 5 years, the approximate cost C of goods and services t years from now is given by

$$C(t) = P(1.028)^t, \ 0 \le t \le 5$$

where P is the present cost. The price of an oil change is presently $29.95. Estimate the price 5 years from now.

23. Determine the effective yield of an 8% interest rate compounded continuously.

24. Determine the length of time for an investment of $1500 to quadruple in value if the investment earns 7% compounded continuously.

25. Write the equation of the circle in standard form and sketch the circle:

$$x^2 + y^2 - 6x + 14y - 6 = 0.$$

26. Identify the vertex and focus of the parabola and sketch the parabola:

$$y = 2x^2 - 20x + 5.$$

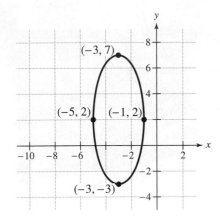

Figure for 27

27. Write the standard form of the equation of the ellipse shown in the figure.

28. Find the center and vertices of the ellipse and sketch the ellipse:

$$4x^2 + y^2 = 4.$$

29. Write the standard form of the equation of the hyperbola with vertices $(0, -3)$ and $(0, 3)$ and asymptotes $y = \pm 3x$.

30. Find the center and vertices of the hyperbola and sketch the hyperbola:

$$x^2 - 9y^2 + 18y = 153.$$

In Exercises 31 and 32, solve the nonlinear system of equations.

31. $\begin{cases} y = x^2 - x - 1 \\ 3x - y = 4 \end{cases}$ **32.** $\begin{cases} x^2 + 5y^2 = 21 \\ -x + y^2 = 5 \end{cases}$

33. A rectangle has an area of 32 square feet and a perimeter of 24 feet. Find the dimensions of the rectangle.

34. The path of a ball is given by $y = -0.1x^2 + 3x + 6$. The coordinates x and y are measured in feet, with $x = 0$ corresponding to the position from which the ball was thrown.

(a) Use a graphing calculator to graph the path of the ball.

(b) Move the cursor along the path to approximate the highest point and the range of the path.

Study Skills in Action

Preparing for the Final Exam

At the end of the semester, most students are inundated with projects, papers, and tests. Instructors may speed up the pace in lectures to get through all the material. If something unexpected is going to happen to a student, it often happens during this time.

Getting through the last couple of weeks of a math course can be challenging. This is why it is important to plan your review time for the final exam at least three weeks before the exam.

Kimberly Nolting

VP, Academic Success Press
expert in developmental education

These students are planning how they will study for the final exam.

Smart Study Strategy

Form a Final Exam Study Group

1 ▶ **Form a study group of three or four students several weeks before the final exam.** The intent of this group is to review what you have already learned while continuing to learn new material.

2 ▶ **Find out what material you must know for the final, even if the instructor has not yet covered it.** As a group, meet with the instructor outside of class. A group is likely to receive more attention and can ask more questions.

3 ▶ **Ask for or create a practice final and have the instructor look at it.** Make sure the problems are on an appropriate level of difficulty. Look for sample problems in old tests and in cumulative tests in the textbook. Review what the textbook and your notes say as you look for problems. This will refresh your memory.

4 ▶ **Have each group member take the practice final exam.** Then have each member identify what he or she needs to study. Make sure you can complete the problems with the speed and accuracy that are necessary to complete the real final exam.

5 ▶ **Decide when the group is going to meet during the next couple of weeks and what you will cover during each session.** The tutoring or learning center on campus is an ideal setting in which to meet. Many libraries have small study rooms that study groups can reserve. Set up several study times for each week. If you live at home, make sure your family knows that this is a busy time.

6 ▶ **During the study group sessions, make sure you stay on track.** Prepare for each study session by knowing what material in the textbook you are going to cover and having the class notes for that material. When you have questions, assign a group member to go to the instructor for answers. Then this member can relay the correct information to the other group members. Save socializing for after the final exam.

Chapter 11

Sequences, Series, and the Binomial Theorem

IT WORKED FOR ME!

"When I was in my math courses, I had to get an early start on getting ready for my finals. I had to get a good grade in the last math course because I was competing with other students to get into the dental hygiene program. I studied with a friend from class a couple of weeks before the final. I stuck to studying for the final more since we did it together. We both did just fine on the final."

Mindy
Dental hygiene

11.1 Sequences and Series

Macduff Everton/CORBIS

Why You Should Learn It

Sequences and series are useful in modeling sets of values in order to identify patterns. For instance, in Exercise 117 on page 718, you will use a sequence to model the depreciation of a sport utility vehicle.

1 ▶ Use sequence notation to write the terms of sequences.

What You Should Learn

1 ▶ Use sequence notation to write the terms of sequences.
2 ▶ Write the terms of sequences involving factorials.
3 ▶ Find the apparent *n*th term of a sequence.
4 ▶ Sum the terms of sequences to obtain series, and use sigma notation to represent partial sums.

Sequences

You are given the following choice of contract offers for the next 5 years of employment.

Contract A $30,000 the first year and a $3300 raise each year

Contract B $30,000 the first year and a 10% raise each year

Which contract offers the largest salary over the five-year period? The salaries for each contract are shown in the table at the left. Notice that after 4 years contract B represents a better contract offer than contract A. The salaries for each contract option represent a sequence.

A mathematical **sequence** is simply an ordered list of numbers. Each number in the list is a **term** of the sequence. A sequence can have a finite number of terms or an infinite number of terms. For instance, the sequence of positive odd integers that are less than 15 is a *finite* sequence

$$1, 3, 5, 7, 9, 11, 13 \qquad \text{Finite sequence}$$

whereas the sequence of positive odd integers is an *infinite* sequence.

$$1, 3, 5, 7, 9, 11, 13, \ldots \qquad \text{Infinite sequence}$$

Note that the three dots indicate that the sequence continues and has an infinite number of terms.

Because each term of a sequence is matched with its location, a sequence can be defined as a function whose domain is a subset of positive integers.

Year	Contract A	Contract B
1	$30,000	$30,000
2	$33,300	$33,000
3	$36,600	$36,300
4	$39,900	$39,930
5	$43,200	$43,923
Total	$183,000	$183,153

Sequences

An **infinite sequence** $a_1, a_2, a_3, \ldots, a_n, \ldots$ is a function whose domain is the set of positive integers.

A **finite sequence** $a_1, a_2, a_3, \ldots, a_n$ is a function whose domain is the finite set $\{1, 2, 3, \ldots, n\}$.

In some cases it is convenient to begin subscripting a sequence with 0 instead of 1. Then the domain of the infinite sequence is the set of nonnegative integers and the domain of the finite sequence is the set $\{0, 1, 2, \ldots, n\}$. The terms of the sequence are denoted by $a_0, a_1, a_2, a_3, a_4, \ldots, a_n, \ldots$.

$a_{()} = 2() + 1$

$a_{(1)} = 2(1) + 1 = 3$

$a_{(2)} = 2(2) + 1 = 5$

$\vdots$

$a_{(51)} = 2(51) + 1 = 103$

The subscripts of a sequence are used in place of function notation. For instance, if parentheses replaced the n in $a_n = 2n + 1$, the notation would be similar to function notation, as shown at the left.

EXAMPLE 1   Writing the Terms of a Sequence

Write the first six terms of the sequence whose nth term is

$a_n = n^2 - 1.$ Begin sequence with $n = 1$.

Solution

$a_1 = (1)^2 - 1 = 0$ $a_2 = (2)^2 - 1 = 3$ $a_3 = (3)^2 - 1 = 8$

$a_4 = (4)^2 - 1 = 15$ $a_5 = (5)^2 - 1 = 24$ $a_6 = (6)^2 - 1 = 35$

The sequence can be written as $0, 3, 8, 15, 24, 35, \ldots, n^2 - 1, \ldots$.

✓ **CHECKPOINT** *Now try Exercise 1.*

Technology: Tip

Most graphing calculators have a "sequence graphing mode" that allows you to plot the terms of a sequence as points on a rectangular coordinate system. For instance, the graph of the first six terms of the sequence given by

$a_n = n^2 - 1$

is shown below.

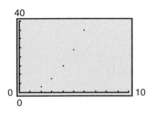

EXAMPLE 2   Writing the Terms of a Sequence

Write the first six terms of the sequence whose nth term is

$a_n = 3(2^n).$ Begin sequence with $n = 0$.

Solution

$a_0 = 3(2^0) = 3 \cdot 1 = 3$ $a_1 = 3(2^1) = 3 \cdot 2 = 6$

$a_2 = 3(2^2) = 3 \cdot 4 = 12$ $a_3 = 3(2^3) = 3 \cdot 8 = 24$

$a_4 = 3(2^4) = 3 \cdot 16 = 48$ $a_5 = 3(2^5) = 3 \cdot 32 = 96$

The sequence can be written as $3, 6, 12, 24, 48, 96, \ldots, 3(2^n), \ldots$.

✓ **CHECKPOINT** *Now try Exercise 3.*

EXAMPLE 3   A Sequence Whose Terms Alternate in Sign

Write the first six terms of the sequence whose nth term is

$a_n = \dfrac{(-1)^n}{2n - 1}.$ Begin sequence with $n = 1$.

Solution

$a_1 = \dfrac{(-1)^1}{2(1) - 1} = -\dfrac{1}{1}$ $a_2 = \dfrac{(-1)^2}{2(2) - 1} = \dfrac{1}{3}$ $a_3 = \dfrac{(-1)^3}{2(3) - 1} = -\dfrac{1}{5}$

$a_4 = \dfrac{(-1)^4}{2(4) - 1} = \dfrac{1}{7}$ $a_5 = \dfrac{(-1)^5}{2(5) - 1} = -\dfrac{1}{9}$ $a_6 = \dfrac{(-1)^6}{2(6) - 1} = \dfrac{1}{11}$

Smart Study Strategy

Go to page 706 for ways to *Form a Final Exam Study Group.*

The sequence can be written as $-1, \dfrac{1}{3}, -\dfrac{1}{5}, \dfrac{1}{7}, -\dfrac{1}{9}, \dfrac{1}{11}, \ldots, \dfrac{(-1)^n}{2n - 1}, \ldots$.

✓ **CHECKPOINT** *Now try Exercise 5.*

2 ▶ Write the terms of sequences involving factorials.

Factorial Notation

Some very important sequences in mathematics involve terms that are defined with special types of products called **factorials.**

Definition of Factorial

If n is a positive integer, n **factorial** is defined as

$$n! = 1 \cdot 2 \cdot 3 \cdot 4 \cdot \cdots \cdot (n-1) \cdot n.$$

As a special case, zero factorial is defined as $0! = 1$.

The first several factorial values are as follows.

$$0! = 1 \qquad\qquad\qquad 1! = 1$$

$$2! = 1 \cdot 2 = 2 \qquad\qquad 3! = 1 \cdot 2 \cdot 3 = 6$$

$$4! = 1 \cdot 2 \cdot 3 \cdot 4 = 24 \qquad 5! = 1 \cdot 2 \cdot 3 \cdot 4 \cdot 5 = 120$$

Many calculators have a factorial key, denoted by $\boxed{n!}$. If your calculator has such a key, try using it to evaluate $n!$ for several values of n. You will see that as n increases, the value of $n!$ becomes very large. For instance,

$$10! = 3,628,800.$$

EXAMPLE 4   A Sequence Involving Factorials

Write the first six terms of the sequence with the given nth term.

a. $a_n = \dfrac{1}{n!}$ Begin sequence with $n = 0$.

b. $a_n = \dfrac{2^n}{n!}$ Begin sequence with $n = 0$.

Solution

a. $a_0 = \dfrac{1}{0!} = \dfrac{1}{1} = 1 \qquad\qquad a_1 = \dfrac{1}{1!} = \dfrac{1}{1} = 1$

$a_2 = \dfrac{1}{2!} = \dfrac{1}{1 \cdot 2} = \dfrac{1}{2} \qquad a_3 = \dfrac{1}{3!} = \dfrac{1}{1 \cdot 2 \cdot 3} = \dfrac{1}{6}$

$a_4 = \dfrac{1}{4!} = \dfrac{1}{1 \cdot 2 \cdot 3 \cdot 4} = \dfrac{1}{24} \qquad a_5 = \dfrac{1}{5!} = \dfrac{1}{1 \cdot 2 \cdot 3 \cdot 4 \cdot 5} = \dfrac{1}{120}$

b. $a_0 = \dfrac{2^0}{0!} = \dfrac{1}{1} = 1 \qquad\qquad a_1 = \dfrac{2^1}{1!} = \dfrac{2}{1} = 2$

$a_2 = \dfrac{2^2}{2!} = \dfrac{2 \cdot 2}{1 \cdot 2} = \dfrac{4}{2} = 2 \qquad a_3 = \dfrac{2^3}{3!} = \dfrac{2 \cdot 2 \cdot 2}{1 \cdot 2 \cdot 3} = \dfrac{8}{6} = \dfrac{4}{3}$

$a_4 = \dfrac{2^4}{4!} = \dfrac{2 \cdot 2 \cdot 2 \cdot 2}{1 \cdot 2 \cdot 3 \cdot 4} = \dfrac{2}{3} \qquad a_5 = \dfrac{2^5}{5!} = \dfrac{2 \cdot 2 \cdot 2 \cdot 2 \cdot 2}{1 \cdot 2 \cdot 3 \cdot 4 \cdot 5} = \dfrac{4}{15}$

✓ **CHECKPOINT** *Now try Exercise 19.*

3 ▶ Find the apparent nth term of a sequence.

Finding the nth Term of a Sequence

Sometimes you have the first several terms of a sequence and need to find a formula (the nth term) to generate those terms. Pattern recognition is crucial in finding a form for the nth term.

Study Tip

Simply listing the first few terms is not sufficient to define a unique sequence—the nth term must be given. Consider the sequence

$$\frac{1}{2}, \frac{1}{4}, \frac{1}{8}, \frac{1}{15}, \cdots$$

The first three terms are identical to the first three terms of the sequence in Example 5(a). However, the nth term of this sequence is defined as

$$a_n = \frac{6}{(n+1)(n^2-n+6)}.$$

EXAMPLE 5   **Finding the nth Term of a Sequence**

Write an expression for the nth term of each sequence.

a. $\dfrac{1}{2}, \dfrac{1}{4}, \dfrac{1}{8}, \dfrac{1}{16}, \dfrac{1}{32}, \cdots$ **b.** $1, -4, 9, -16, 25, \ldots$

Solution

a.

n:	1	2	3	4	5	$\cdots$	n
Terms:	$\frac{1}{2}$	$\frac{1}{4}$	$\frac{1}{8}$	$\frac{1}{16}$	$\frac{1}{32}$	$\cdots$	a_n

Pattern: The numerators are 1 and the denominators are increasing powers of 2.

So, an expression for the nth term is $\dfrac{1}{2^n}$.

b.

n:	1	2	3	4	5	$\cdots$	n
Terms:	1	-4	9	-16	25	$\cdots$	a_n

Pattern: The terms have alternating signs, with those in the even positions being negative. The absolute value of each term is the square of n.

So, an expression for the nth term is $(-1)^{n+1}n^2$.

✓ **CHECKPOINT** *Now try Exercise 55.*

4 ▶ Sum the terms of sequences to obtain series, and use sigma notation to represent partial sums.

Series

In the table of salaries at the beginning of this section, the terms of the finite sequence were *added*. If you add all the terms of an *infinite* sequence, you obtain a **series.**

Definition of Series

For an infinite sequence $a_1, a_2, a_3, \ldots, a_n, \ldots$

1. the sum of the first n terms

$$S_n = a_1 + a_2 + a_3 + \cdots + a_n$$

is called a **partial sum,** and

2. the sum of all the terms

$$a_1 + a_2 + a_3 + \cdots + a_n + \cdots$$

is called an **infinite series,** or simply a **series.**

EXAMPLE 6   Finding Partial Sums

Find the indicated partial sums for each sequence.

a. Find S_1, S_2, and S_5 for $a_n = 3n - 1$.

b. Find S_2, S_3, and S_4 for $a_n = \dfrac{(-1)^n}{n + 1}$.

Solution

a. The first five terms of the sequence $a_n = 3n - 1$ are

$$a_1 = 2, a_2 = 5, a_3 = 8, a_4 = 11, \text{ and } a_5 = 14.$$

So, the partial sums are

$$S_1 = 2, S_2 = 2 + 5 = 7, \text{ and } S_5 = 2 + 5 + 8 + 11 + 14 = 40.$$

b. The first four terms of the sequence $a_n = \dfrac{(-1)^n}{n + 1}$ are

$$a_1 = -\frac{1}{2}, a_2 = \frac{1}{3}, a_3 = -\frac{1}{4}, \text{ and } a_4 = \frac{1}{5}.$$

So, the partial sums are

$$S_2 = -\frac{1}{2} + \frac{1}{3} = -\frac{1}{6}, S_3 = -\frac{1}{2} + \frac{1}{3} - \frac{1}{4} = -\frac{5}{12}, \text{ and}$$

$$S_4 = -\frac{1}{2} + \frac{1}{3} - \frac{1}{4} + \frac{1}{5} = -\frac{13}{60}.$$

✓ **CHECKPOINT** *Now try Exercise 67.*

A convenient shorthand notation for denoting a partial sum is called **sigma notation.** This name comes from the use of the uppercase Greek letter sigma, written as Σ.

Definition of Sigma Notation

The sum of the first n terms of the sequence whose nth term is a_n is

$$\sum_{i=1}^{n} a_i = a_1 + a_2 + a_3 + a_4 + \cdots + a_n$$

where i is the **index of summation,** n is the **upper limit of summation,** and 1 is the **lower limit of summation.**

Sigma (summation) notation is an instruction to add the terms of a sequence. From the definition above, the upper limit of summation tells you where to end the sum. Sigma notation helps you generate the appropriate terms of the sequence prior to finding the actual sum.

EXAMPLE 7   **Finding a Sum in Sigma Notation**

$$\sum_{i=1}^{6} 2i = 2(1) + 2(2) + 2(3) + 2(4) + 2(5) + 2(6)$$

$$= 2 + 4 + 6 + 8 + 10 + 12$$

$$= 42$$

 CHECKPOINT *Now try Exercise 71.*

Study Tip

In Example 7, the index of summation is i and the summation begins with $i = 1$. Any letter can be used as the index of summation, and the summation can begin with any integer. For instance, in Example 8, the index of summation is k and the summation begins with $k = 0$.

EXAMPLE 8   **Finding a Sum in Sigma Notation**

$$\sum_{k=0}^{8} \frac{1}{k!} = \frac{1}{0!} + \frac{1}{1!} + \frac{1}{2!} + \frac{1}{3!} + \frac{1}{4!} + \frac{1}{5!} + \frac{1}{6!} + \frac{1}{7!} + \frac{1}{8!}$$

$$= 1 + 1 + \frac{1}{2} + \frac{1}{6} + \frac{1}{24} + \frac{1}{120} + \frac{1}{720} + \frac{1}{5040} + \frac{1}{40,320}$$

$$\approx 2.71828$$

Note that this sum is approximately $e = 2.71828.\ \ldots$

 CHECKPOINT *Now try Exercise 77.*

EXAMPLE 9   **Writing a Sum in Sigma Notation**

Write each sum in sigma notation.

a. $\dfrac{2}{2} + \dfrac{2}{3} + \dfrac{2}{4} + \dfrac{2}{5} + \dfrac{2}{6}$ **b.** $1 - \dfrac{1}{3} + \dfrac{1}{9} - \dfrac{1}{27} + \dfrac{1}{81}$

Solution

a. To write this sum in sigma notation, you must find a pattern for the terms. You can see that the terms have numerators of 2 and denominators that range over the integers from 2 to 6. So, one possible sigma notation is

$$\sum_{i=1}^{5} \frac{2}{i+1} = \frac{2}{2} + \frac{2}{3} + \frac{2}{4} + \frac{2}{5} + \frac{2}{6}.$$

b. To write this sum in sigma notation, you must find a pattern for the terms. You can see that their numerators alternate in sign and their denominators are integer powers of 3, starting with 3^0 and ending with 3^4. So, one possible sigma notation is

$$\sum_{i=0}^{4} \frac{(-1)^i}{3^i} = \frac{1}{3^0} + \frac{-1}{3^1} + \frac{1}{3^2} + \frac{-1}{3^3} + \frac{1}{3^4}.$$

CHECKPOINT *Now try Exercise 95.*

_____ **Concept Check** _____

1. Explain how to find the sixth term of the sequence

$$a_n = \frac{n-3}{4}.$$

2. Explain the difference between $a_n = 5n$ and $a_n = 5n!$.

3. Determine whether the statement is true or false. Justify your answer. The only expression for the nth term of the sequence 1, 3, 9, . . . is $2n^2 + 1$.

4. How many terms are in the sum $\sum\limits_{i=1}^{5} i^2$? In the sum $\sum\limits_{i=0}^{5} 10i$?

11.1 EXERCISES

Go to pages 746–747 to record your assignments.

_____ **Developing Skills** _____

In Exercises 1–22, write the first five terms of the sequence. (Assume that n begins with 1.) *See Examples 1–4.*

✓ 1. $a_n = 2n$

2. $a_n = 3n$

✓ 3. $a_n = \left(\frac{1}{4}\right)^n$

4. $a_n = \left(\frac{1}{3}\right)^n$

✓ 5. $a_n = (-1)^n 2n$

6. $a_n = (-1)^{n+1} 3n$

7. $a_n = \left(-\frac{1}{2}\right)^{n+1}$

8. $a_n = \left(\frac{2}{3}\right)^{n-1}$

9. $a_n = 5n - 2$

10. $a_n = 2n + 3$

11. $a_n = \frac{4}{n+3}$

12. $a_n = \frac{9}{5+n}$

13. $a_n = \frac{3n}{5n-1}$

14. $a_n = \frac{2n}{6n-3}$

15. $a_n = \frac{(-1)^n}{n^2}$

16. $a_n = \frac{1}{\sqrt{n}}$

17. $a_n = 2 + \frac{1}{4^n}$

18. $a_n = 10 - \frac{1}{5^n}$

✓ 19. $a_n = \frac{(n+1)!}{n!}$

20. $a_n = \frac{n!}{(n-1)!}$

21. $a_n = \frac{2 + (-2)^n}{n!}$

22. $a_n = \frac{1 + (-1)^n}{n^2}$

In Exercises 23–26, find the indicated term of the sequence.

23. $a_n = (-1)^n(5n - 3)$

$a_{15} =$

24. $a_n = (-1)^{n+1}(3n + 10)$

$a_{20} =$

25. $a_n = \frac{n^2 - 2}{(n-1)!}$

$a_8 =$

26. $a_n = \frac{n^2}{n!}$

$a_{12} =$

In Exercises 27–38, simplify the expression.

27. $\frac{5!}{4!}$

28. $\frac{6!}{8!}$

29. $\frac{10!}{12!}$

30. $\frac{16!}{13!}$

31. $\dfrac{25!}{20!5!}$

32. $\dfrac{20!}{15!5!}$

33. $\dfrac{n!}{(n + 1)!}$

34. $\dfrac{(n + 2)!}{n!}$

35. $\dfrac{(n + 1)!}{(n - 1)!}$

36. $\dfrac{(3n)!}{(3n + 2)!}$

37. $\dfrac{(2n)!}{(2n - 1)!}$

38. $\dfrac{(5n + 2)!}{(5n)!}$

In Exercises 39–42, match the sequence with the graph of its first 10 terms. [The graphs are labeled (a), (b), (c), and (d).]

(a)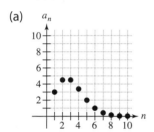

(b)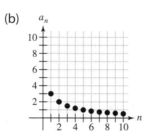

(c)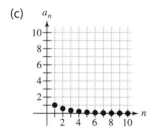

(d)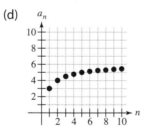

39. $a_n = \dfrac{6}{n + 1}$

40. $a_n = \dfrac{6n}{n + 1}$

41. $a_n = (0.6)^{n - 1}$

42. $a_n = \dfrac{3^n}{n!}$

⊞ In Exercises 43–48, use a graphing calculator to graph the first 10 terms of the sequence.

43. $a_n = \dfrac{4n^2}{n^2 - 2}$

44. $a_n = \dfrac{2n^2}{n^2 + 1}$

45. $a_n = 3 - \dfrac{4}{n}$

46. $a_n = 5 + \dfrac{3}{n}$

47. $a_n = 100(-0.4)^{n + 1}$

48. $a_n = 10\left(\dfrac{3}{4}\right)^{n - 1}$

In Exercises 49–66, write an expression for the *n*th term of the sequence. (Assume that *n* begins with 1.) *See Example 5.*

49. $1, 3, 5, 7, 9, \ldots$

50. $2, 4, 6, 8, 10, \ldots$

51. $2, 6, 10, 14, 18, \ldots$

52. $6, 11, 16, 21, 26, \ldots$

53. $0, 3, 8, 15, 24, \ldots$

54. $1, 8, 27, 64, 125, \ldots$

✓ **55.** $2, -4, 6, -8, 10, \ldots$

56. $1, -1, 1, -1, 1, \ldots$

57. $\dfrac{2}{3}, \dfrac{3}{4}, \dfrac{4}{5}, \dfrac{5}{6}, \dfrac{6}{7}, \ldots$

58. $\dfrac{2}{1}, \dfrac{3}{3}, \dfrac{4}{5}, \dfrac{5}{7}, \dfrac{6}{9}, \ldots$

59. $\dfrac{-1}{5}, \dfrac{1}{25}, \dfrac{-1}{125}, \dfrac{1}{625}, \dfrac{-1}{3125}, \ldots$

60. $1, \dfrac{1}{4}, \dfrac{1}{9}, \dfrac{1}{16}, \dfrac{1}{25}, \ldots$

61. $1, \dfrac{1}{2}, \dfrac{1}{4}, \dfrac{1}{8}, \ldots$

62. $\dfrac{1}{3}, \dfrac{2}{9}, \dfrac{4}{27}, \dfrac{8}{81}, \ldots$

63. $1 + \dfrac{1}{1}, 1 + \dfrac{1}{2}, 1 + \dfrac{1}{3}, 1 + \dfrac{1}{4}, 1 + \dfrac{1}{5}, \ldots$

64. $1 + \dfrac{1}{2}, 1 + \dfrac{3}{4}, 1 + \dfrac{7}{8}, 1 + \dfrac{15}{16}, 1 + \dfrac{31}{32}, \ldots$

65. $-\dfrac{1}{2}, \dfrac{1}{6}, -\dfrac{1}{24}, \dfrac{1}{120}, -\dfrac{1}{720}, \ldots$

66. $1, 2, \dfrac{2^2}{2}, \dfrac{2^3}{6}, \dfrac{2^4}{24}, \dfrac{2^5}{120}, \ldots$

In Exercises 67–70, find the indicated partial sums for the sequence. *See Example 6.*

✓ **67.** Find S_1, S_2, and S_6 for $a_n = 2n + 5$.

68. Find S_3, S_4, and S_{10} for $a_n = n^3 - 1$.

69. Find S_2, S_3, and S_9 for $a_n = \dfrac{1}{n}$.

70. Find S_1, S_3, and S_5 for $a_n = \dfrac{(-1)^{n + 1}}{n + 1}$.

In Exercises 71–86, find the partial sum. *See Examples 7 and 8.*

71. $\displaystyle\sum_{k=1}^{5} 6k$ **72.** $\displaystyle\sum_{k=1}^{4} 5k$

73. $\displaystyle\sum_{i=0}^{6} (2i + 5)$ **74.** $\displaystyle\sum_{i=0}^{4} (2i + 3)$

75. $\displaystyle\sum_{j=3}^{7} (6j - 10)$ **76.** $\displaystyle\sum_{i=2}^{7} (4i - 1)$

77. $\displaystyle\sum_{j=1}^{5} \frac{(-1)^{j+1}}{j^2}$ **78.** $\displaystyle\sum_{j=0}^{3} \frac{1}{j^2 + 1}$

79. $\displaystyle\sum_{m=1}^{8} \frac{m}{m + 1}$ **80.** $\displaystyle\sum_{k=1}^{6} \frac{k - 3}{k + 2}$

81. $\displaystyle\sum_{k=1}^{6} (-8)$ **82.** $\displaystyle\sum_{n=3}^{12} 10$

83. $\displaystyle\sum_{i=1}^{8} \left(\frac{1}{i} - \frac{1}{i + 1}\right)$ **84.** $\displaystyle\sum_{k=1}^{5} \left(\frac{2}{k} - \frac{2}{k + 2}\right)$

85. $\displaystyle\sum_{n=0}^{5} \left(-\frac{1}{3}\right)^n$ **86.** $\displaystyle\sum_{k=1}^{4} \left(\frac{5}{3}\right)^{k - 1}$

In Exercises 87–94, use a graphing calculator to find the partial sum.

87. $\displaystyle\sum_{n=1}^{8} 10n^2$ **88.** $\displaystyle\sum_{n=0}^{5} 2n^2$

89. $\displaystyle\sum_{j=2}^{6} (j! - j)$ **90.** $\displaystyle\sum_{i=0}^{4} (i! + 4)$

91. $\displaystyle\sum_{j=0}^{4} \frac{6}{j!}$ **92.** $\displaystyle\sum_{k=1}^{6} \left(\frac{1}{2k} - \frac{1}{2k - 1}\right)$

93. $\displaystyle\sum_{k=1}^{6} \ln k$ **94.** $\displaystyle\sum_{k=3}^{5} \frac{\log_{10} k}{k}$

In Exercises 95–112, write the sum using sigma notation. (Begin with $k = 0$ or $k = 1$.) *See Example 9.*

95. $1 + 2 + 3 + 4 + 5$

96. $8 + 9 + 10 + 11 + 12 + 13 + 14$

97. $5 + 10 + 15 + 20 + 25 + 30$

98. $24 + 30 + 36 + 42$

99. $\dfrac{3}{1 + 1} + \dfrac{3}{1 + 2} + \dfrac{3}{1 + 3} + \cdots + \dfrac{3}{1 + 50}$

100.

$\dfrac{1}{2(1) + 1} + \dfrac{1}{2(2) + 1} + \dfrac{1}{2(3) + 1} + \cdots + \dfrac{1}{2(30) + 1}$

101. $\dfrac{1}{2(1)} + \dfrac{1}{2(2)} + \dfrac{1}{2(3)} + \dfrac{1}{2(4)} + \cdots + \dfrac{1}{2(10)}$

102. $\dfrac{1}{1^2} + \dfrac{1}{2^2} + \dfrac{1}{3^2} + \dfrac{1}{4^2} + \cdots + \dfrac{1}{20^2}$

103. $\dfrac{1}{2^0} + \dfrac{1}{2^1} + \dfrac{1}{2^2} + \dfrac{1}{2^3} + \cdots + \dfrac{1}{2^{12}}$

104. $\left(-\dfrac{2}{3}\right)^0 + \left(-\dfrac{2}{3}\right)^1 + \left(-\dfrac{2}{3}\right)^2 + \cdots + \left(-\dfrac{2}{3}\right)^{20}$

105.

$\dfrac{1}{1(1 + 1)} + \dfrac{1}{2(2 + 1)} + \dfrac{1}{3(3 + 1)} + \cdots + \dfrac{1}{20(20 + 1)}$

106. $\dfrac{1}{2^3} - \dfrac{1}{4^3} + \dfrac{1}{6^3} - \dfrac{1}{8^3} + \cdots + \dfrac{1}{14^3}$

107. $\frac{1}{2} + \frac{2}{3} + \frac{3}{4} + \frac{4}{5} + \frac{5}{6} + \cdots + \frac{11}{12}$

108. $\frac{2}{4} + \frac{4}{7} + \frac{6}{10} + \frac{8}{13} + \frac{10}{16} + \cdots + \frac{20}{31}$

109. $\frac{2}{4} + \frac{4}{5} + \frac{6}{6} + \frac{8}{7} + \cdots + \frac{40}{23}$

110. $\left(2 + \frac{1}{1}\right) + \left(2 + \frac{1}{2}\right) + \left(2 + \frac{1}{3}\right) + \cdots + \left(2 + \frac{1}{25}\right)$

111. $1 + 1 + 2 + 6 + 24 + 120 + 720$

112. $-\frac{1}{10} + \frac{1}{20} - \frac{1}{60} + \frac{1}{240} - \frac{1}{1200} + \frac{1}{7200} - \frac{1}{50,400}$

Solving Problems

113. *Compound Interest*   A deposit of $500 is made in an account that earns 7% interest compounded yearly. The balance in the account after N years is given by

$$A_N = 500(1 + 0.07)^N, \quad N = 1, 2, 3, \ldots.$$

(a) Compute the first eight terms of the sequence.

(b) Find the balance in this account after 40 years by computing A_{40}.

(c) ▦ Use a graphing calculator to graph the first 40 terms of the sequence.

(d) The terms are increasing. Is the rate of growth of the terms increasing? Explain.

114. *Sports*   The number of degrees a_n in each angle of a regular n-sided polygon is

$$a_n = \frac{180(n - 2)}{n}, \quad n \geq 3.$$

The surface of a soccer ball is made of regular hexagons and pentagons. When a soccer ball is taken apart and flattened, as shown in the figure, the sides don't meet each other. Use the terms a_5 and a_6 to explain why there are gaps between adjacent hexagons.

115. *Stars*   Stars are formed by placing n equally spaced points on a circle and connecting each point with a second point on the circle (see figure). The measure in degrees d_n of the angle at each tip of the star is given by

$$d_n = \frac{180(n - 4)}{n}, \quad n \geq 5.$$

Write the first six terms of this sequence.

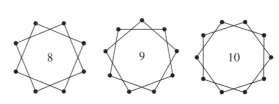

116. *Stars*   The stars in Exercise 115 were formed by placing n equally spaced points on a circle and connecting each point with the second point from it on the circle. The stars in the figure for this exercise were formed in a similar way except that each point was connected with the third point from it. For these stars, the measure in degrees d_n of the angle at each point is given by

$$d_n = \frac{180(n - 6)}{n}, \quad n \geq 7.$$

Write the first five terms of this sequence.

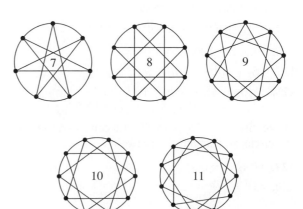

117. *Depreciation*   At the end of each year, the value of a sport utility vehicle with an initial cost of \$32,000 is three-fourths what it was at the beginning of the year. After n years, its value is given by

$$a_n = 32{,}000\left(\frac{3}{4}\right)^n, \quad n = 1, 2, 3, \ldots.$$

(a) Find the value of the sport utility vehicle 3 years after it was purchased by computing a_3.

(b) Find the value of the sport utility vehicle 6 years after it was purchased by computing a_6. Is this value half of what it was after 3 years? Explain.

118. *Annual Revenue*   The revenue a_n (in millions of dollars) of Netflix, Inc., for the years 2001 through 2007 is modeled by

$$a_n = 18.054n^2 + 51.47n - 11.1, \quad n = 1, 2, \ldots, 7$$

where n is the year, with $n = 1$ corresponding to 2001.   (Source: Netflix, Inc.)

(a) Find the terms of this finite sequence.

(b) 🖩 Use a graphing calculator to construct a bar graph that represents the sequence.

(c) Find $\sum_{i=1}^{7} a_i$. What does this number represent?

─────── **Explaining Concepts** ───────

119. ✎   In your own words, explain why a sequence is a function.

120. ✎   The nth term of a sequence is $a_n = (-1)^n n$. Which terms of the sequence are negative? Explain.

121. ✎   Explain the difference between $a_n = 4n!$ and $a_n = (4n)!$.

In Exercises 122–124, decide whether the statement is true or false. Justify your answer.

122. $\sum_{i=1}^{4} (i^2 + 2i) = \sum_{i=1}^{4} i^2 + \sum_{i=1}^{4} 2i$

123. $\sum_{k=1}^{4} 3k = 3\sum_{k=1}^{4} k$

124. $\sum_{j=1}^{4} 2^j = \sum_{j=3}^{6} 2^{j-2}$

─────── **Cumulative Review** ───────

In Exercises 125–128, evaluate the expression for the specified value of the variable.

125. $-2n + 15; n = 3$

126. $-20n + 100; n = 4$

127. $25 - 3(n + 4); n = 8$

128. $-\frac{3}{2}(n - 1) + 6; n = 10$

In Exercises 129–132, identify the center and radius of the circle, and sketch the circle.

129. $x^2 + y^2 = 36$

130. $4x^2 + 4y^2 = 9$

131. $x^2 + y^2 + 4x - 12 = 0$

132. $x^2 + y^2 - 10x - 2y - 199 = 0$

In Exercises 133–136, identify the vertex and focus of the parabola, and sketch the parabola.

133. $x^2 = 6y$

134. $y^2 = 9x$

135. $x^2 + 8y + 32 = 0$

136. $y^2 - 10x + 6y + 29 = 0$

11.2 Arithmetic Sequences

Lynn Goldsmith/CORBIS

What You Should Learn

1 ▶ Recognize, write, and find the nth terms of arithmetic sequences.

2 ▶ Find the nth partial sum of an arithmetic sequence.

3 ▶ Use arithmetic sequences to solve application problems.

Why You Should Learn It

An arithmetic sequence can reduce the amount of time it takes to find the sum of a sequence of numbers with a common difference. For instance, in Exercise 125 on page 725, you will use an arithmetic sequence to determine how much to charge for tickets to a concert at an outdoor arena.

1 ▶ Recognize, write, and find the nth terms of arithmetic sequences.

Arithmetic Sequences

A sequence whose consecutive terms have a common difference is called an **arithmetic sequence.**

Definition of Arithmetic Sequence

A sequence is called **arithmetic** if the differences between consecutive terms are the same. So, the sequence

$$a_1, a_2, a_3, a_4, \ldots, a_n, \ldots$$

is arithmetic if there is a number d such that

$$a_2 - a_1 = d, \quad a_3 - a_2 = d, \quad a_4 - a_3 = d$$

and so on. The number d is the **common difference** of the sequence.

| EXAMPLE 1 | Examples of Arithmetic Sequences |

a. The sequence whose nth term is $3n + 2$ is arithmetic. For this sequence, the common difference between consecutive terms is 3.

$$5, 8, 11, 14, \ldots, 3n + 2, \ldots \qquad \text{Begin with } n = 1.$$

$$8 - 5 = 3$$

b. The sequence whose nth term is $7 - 5n$ is arithmetic. For this sequence, the common difference between consecutive terms is -5.

$$2, -3, -8, -13, \ldots, 7 - 5n, \ldots \qquad \text{Begin with } n = 1.$$

$$-3 - 2 = -5$$

c. The sequence whose nth term is $\frac{1}{4}(n + 3)$ is arithmetic. For this sequence, the common difference between consecutive terms is $\frac{1}{4}$.

$$1, \frac{5}{4}, \frac{3}{2}, \frac{7}{4}, \ldots, \frac{1}{4}(n + 3), \ldots \qquad \text{Begin with } n = 1.$$

$$\frac{5}{4} - 1 = \frac{1}{4}$$

✔ **CHECKPOINT** *Now try Exercise 1.*

Study Tip

The nth term of an arithmetic sequence can be derived from the following pattern.

$a_1 = a_1$ 1st term

$a_2 = a_1 + d$ 2nd term

$a_3 = a_1 + 2d$ 3rd term

$a_4 = a_1 + 3d$ 4th term

$a_5 = a_1 + 4d$ 5th term

1 less

$\vdots$ $\vdots$

$a_n = a_1 + (n - 1)d$ nth term

1 less

The nth Term of an Arithmetic Sequence

The nth term of an arithmetic sequence has the form

$$a_n = a_1 + (n - 1)d$$

where d is the common difference between the terms of the sequence, and a_1 is the first term.

EXAMPLE 2   Finding the nth Term of an Arithmetic Sequence

Find a formula for the nth term of the arithmetic sequence whose common difference is 2 and whose first term is 5.

Solution

You know that the formula for the nth term is of the form $a_n = a_1 + (n - 1)d$. Moreover, because the common difference is $d = 2$ and the first term is $a_1 = 5$, the formula must have the form

$$a_n = 5 + 2(n - 1).$$

So, the formula for the nth term is $a_n = 2n + 3$, and the sequence has the following form.

$$5, 7, 9, 11, 13, \ldots, 2n + 3, \ldots$$

 **CHECKPOINT** *Now try Exercise 53.*

If you know the nth term and the common difference of an arithmetic sequence, you can find the $(n + 1)$th term by using the **recursion formula**

$$a_{n+1} = a_n + d.$$

EXAMPLE 3   Using a Recursion Formula

The 12th term of an arithmetic sequence is 52 and the common difference is 3.

a. What is the 13th term of the sequence?   **b.** What is the first term?

Solution

a. You know that $a_{12} = 52$ and $d = 3$. So, using the recursion formula $a_{13} = a_{12} + d$, you can determine that the 13th term of the sequence is

$$a_{13} = 52 + 3 = 55.$$

b. Using $n = 12$, $d = 3$, and $a_{12} = 52$ in the formula $a_n = a_1 + (n - 1)d$ yields

$$52 = a_1 + (12 - 1)(3)$$

$$19 = a_1.$$

 **CHECKPOINT** *Now try Exercise 71.*

2 ▸ Find the *n*th partial sum of an arithmetic sequence.

The Partial Sum of an Arithmetic Sequence

The sum of the first *n* terms of an arithmetic sequence is called the **nth partial sum** of the sequence. For instance, the fifth partial sum of the arithmetic sequence whose *n*th term is $3n + 4$ is

$$\sum_{i=1}^{5} (3i + 4) = 7 + 10 + 13 + 16 + 19 = 65.$$

To find a formula for the *n*th partial sum S_n of an arithmetic sequence, write out S_n forwards and backwards and then add the two forms, as follows.

$$S_n = a_1 + (a_1 + d) + (a_1 + 2d) + \cdots + [a_1 + (n - 1)d] \quad \text{Forwards}$$

$$S_n = a_n + (a_n - d) + (a_n - 2d) + \cdots + [a_n - (n - 1)d] \quad \text{Backwards}$$

$$2S_n = (a_1 + a_n) + (a_1 + a_n) + (a_1 + a_n) + \cdots + [a_1 + a_n] \quad \begin{array}{l}\text{Sum of two}\\\text{equations}\end{array}$$

$$\qquad = n(a_1 + a_n) \qquad\qquad\qquad\qquad\qquad\qquad \begin{array}{l}n \text{ groups of}\\(a_1 + a_n)\end{array}$$

Dividing each side by 2 yields the following formula.

Study Tip

You can use the formula for the *n*th partial sum of an arithmetic sequence to find the sum of consecutive numbers. For instance, the sum of the integers from 1 to 100 is

$$\sum_{i=1}^{100} i = \frac{100}{2}(1 + 100)$$

$$= 50(101)$$

$$= 5050.$$

The *n*th Partial Sum of an Arithmetic Sequence

The *n*th partial sum of the arithmetic sequence whose *n*th term is a_n is

$$\sum_{i=1}^{n} a_i = a_1 + a_2 + a_3 + a_4 + \cdots + a_n$$

$$= \frac{n}{2}(a_1 + a_n).$$

Or, equivalently, you can find the sum of the first *n* terms of an arithmetic sequence by multiplying the average of the first and *n*th terms by *n*.

EXAMPLE 4   Finding the *n*th Partial Sum

Find the sum of the first 20 terms of the arithmetic sequence whose *n*th term is $4n + 1$.

Solution

The first term of this sequence is $a_1 = 4(1) + 1 = 5$ and the 20th term is $a_{20} = 4(20) + 1 = 81$. So, the sum of the first 20 terms is given by

$$\sum_{i=1}^{n} a_i = \frac{n}{2}(a_1 + a_n) \qquad \textit{nth partial sum formula}$$

$$\sum_{i=1}^{20} (4i + 1) = \frac{20}{2}(a_1 + a_{20}) \qquad \textit{Substitute 20 for n.}$$

$$= 10(5 + 81) \qquad \textit{Substitute 5 for } a_1 \textit{ and 81 for } a_{20}.$$

$$= 10(86) \qquad \textit{Simplify.}$$

$$= 860. \qquad \textit{nth partial sum}$$

✓ **CHECKPOINT** *Now try Exercise 79.*

EXAMPLE 5   Finding the *n*th Partial Sum

Find the sum of the even integers from 2 to 100.

Solution

Because the integers

$$2, 4, 6, 8, \ldots, 100$$

form an arithmetic sequence, you can find the sum as follows.

$$\sum_{i=1}^{n} a_i = \frac{n}{2}(a_1 + a_n) \qquad \textit{n}\text{th partial sum formula}$$

$$\sum_{i=1}^{50} 2i = \frac{50}{2}(a_1 + a_{50}) \qquad \text{Substitute 50 for } n.$$

$$= 25(2 + 100) \qquad \text{Substitute 2 for } a_1 \text{ and 100 for } a_{50}.$$

$$= 25(102) \qquad \text{Simplify.}$$

$$= 2550 \qquad \textit{n}\text{th partial sum}$$

✔ **CHECKPOINT** *Now try Exercise 89.*

3 ▶ Use arithmetic sequences to solve application problems.

Application

EXAMPLE 6   Total Sales

Your business sells $100,000 worth of handmade furniture during its first year. You have a goal of increasing annual sales by $25,000 each year for 9 years. If you meet this goal, how much will you sell during your first 10 years of business?

Solution

The annual sales during the first 10 years form the following arithmetic sequence.

$100,000, $125,000, $150,000, $175,000, $200,000,
$225,000, $250,000, $275,000, $300,000, $325,000

Using the formula for the *n*th partial sum of an arithmetic sequence, you can find the total sales during the first 10 years as follows.

$$\text{Total sales} = \frac{n}{2}(a_1 + a_n) \qquad \textit{n}\text{th partial sum formula}$$

$$= \frac{10}{2}(100,000 + 325,000) \qquad \text{Substitute for } n, a_1, \text{ and } a_n.$$

$$= 5(425,000) \qquad \text{Simplify.}$$

$$= \$2,125,000 \qquad \text{Simplify.}$$

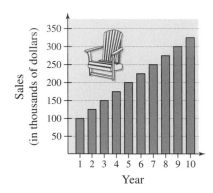

Figure 11.1

From the bar graph shown in Figure 11.1, notice that the annual sales for your company follow a *linear growth* pattern. In other words, saying that a quantity increases arithmetically is the same as saying that it increases linearly.

✔ **CHECKPOINT** *Now try Exercise 123.*

_____ **Concept Check** _____

1. In an arithmetic sequence, the common difference between consecutive terms is 3. How can you use the value of one term to find the value of the next term in the sequence?

2. Explain how you can use the first two terms of an arithmetic sequence to write a formula for the nth term of the sequence.

3. Explain how you can use the average of the first term and the nth term of an arithmetic sequence to find the nth partial sum of the sequence.

4. In an arithmetic sequence, you know the common difference d between consecutive terms. What else do you need to know to find a_6?

11.2 EXERCISES

Go to pages 746–747 to record your assignments.

_____ **Developing Skills** _____

In Exercises 1–10, find the common difference of the arithmetic sequence. *See Example 1.*

✓ 1. $2, 5, 8, 11, \ldots$ 2. $-8, 0, 8, 16, \ldots$

3. $100, 94, 88, 82, \ldots$

4. $3200, 2800, 2400, 2000, \ldots$

5. $10, -2, -14, -26, -38, \ldots$

6. $4, \frac{9}{2}, 5, \frac{11}{2}, 6, \ldots$ 7. $1, \frac{5}{3}, \frac{7}{3}, 3, \ldots$

8. $\frac{1}{2}, \frac{5}{4}, 2, \frac{11}{4}, \ldots$

9. $\frac{7}{2}, \frac{9}{4}, 1, -\frac{1}{4}, -\frac{3}{2}, \ldots$

10. $\frac{5}{2}, \frac{11}{6}, \frac{7}{6}, \frac{1}{2}, -\frac{1}{6}, \ldots$

In Exercises 11–16, find the common difference of the arithmetic sequence with the given nth term. *See Example 1.*

11. $a_n = 4n + 5$ 12. $a_n = 7n + 6$

13. $a_n = 8 - 3n$ 14. $a_n = 12 - 4n$

15. $a_n = \frac{1}{2}(n + 1)$ 16. $a_n = \frac{1}{3}(n + 4)$

In Exercises 17–32, determine whether the sequence is arithmetic. If so, find the common difference.

17. $2, 4, 6, 8, \ldots$ 18. $1, 2, 4, 8, 16, \ldots$

19. $10, 8, 6, 4, 2, \ldots$ 20. $2, 6, 10, 14, \ldots$

21. $32, 16, 0, -16, \ldots$ 22. $32, 16, 8, 4, \ldots$

23. $3.2, 4, 4.8, 5.6, \ldots$

24. $8, 4, 2, 1, 0.5, 0.25, \ldots$

25. $2, \frac{7}{2}, 5, \frac{13}{2}, \ldots$ 26. $3, \frac{5}{2}, 2, \frac{3}{2}, 1, \ldots$

27. $\frac{1}{3}, \frac{2}{3}, \frac{4}{3}, \frac{8}{3}, \frac{16}{3}, \ldots$ 28. $\frac{9}{4}, 2, \frac{7}{4}, \frac{3}{2}, \frac{5}{4}, \ldots$

29. $1, \sqrt{2}, \sqrt{3}, 2, \sqrt{5}, \ldots$

30. $1, 4, 9, 16, 25, \ldots$

31. $\ln 4, \ln 8, \ln 12, \ln 16, \ldots$

32. $e, e^2, e^3, e^4, \ldots$

In Exercises 33–42, write the first five terms of the arithmetic sequence.

33. $a_1 = 7, d = 5$ 34. $a_1 = 8, d = 3$

35. $a_1 = 11, d = 4$ 36. $a_1 = 18, d = 10$

37. $a_1 = 20, d = -4$ 38. $a_1 = 16, d = -3$

39. $a_1 = 6, a_2 = 11$ 40. $a_1 = 9, a_2 = 11$

41. $a_1 = 22, a_2 = 18$ 42. $a_1 = 30, a_2 = 20$

In Exercises 43–52, write the first five terms of the arithmetic sequence. (Assume that n begins with 1.)

43. $a_n = 3n + 4$ 44. $a_n = 5n - 4$

45. $a_n = -2n + 8$ 46. $a_n = -10n + 100$

47. $a_n = \frac{5}{2}n - 1$ 48. $a_n = \frac{2}{3}n + 2$

49. $a_n = \frac{3}{5}n + 1$

50. $a_n = \frac{3}{4}n - 2$

51. $a_n = -\frac{1}{4}(n - 1) + 4$

52. $a_n = 4(n + 2) + 24$

In Exercises 53–70, find a formula for the *n*th term of the arithmetic sequence. *See Example 2.*

✓ 53. $a_1 = 4$, $d = 3$

54. $a_1 = 7$, $d = 2$

55. $a_1 = \frac{1}{2}$, $d = \frac{3}{2}$

56. $a_1 = \frac{5}{3}$, $d = \frac{1}{3}$

57. $a_1 = 100$, $d = -5$

58. $a_1 = -6$, $d = -1$

59. $a_3 = 6$, $d = \frac{3}{2}$

60. $a_6 = 5$, $d = \frac{3}{2}$

61. $a_1 = 5$, $a_5 = 15$

62. $a_2 = 93$, $a_6 = 65$

63. $a_3 = 16$, $a_4 = 20$

64. $a_5 = 30$, $a_4 = 25$

65. $a_1 = 50$, $a_3 = 30$

66. $a_{10} = 32$, $a_{12} = 48$

67. $a_2 = 10$, $a_6 = 8$

68. $a_7 = 8$, $a_{13} = 6$

69. $a_1 = 0.35$, $a_2 = 0.30$

70. $a_1 = 0.08$, $a_2 = 0.082$

In Exercises 71–78, write the first five terms of the arithmetic sequence defined recursively. *See Example 3.*

✓ 71. $a_1 = 14$
    $a_{k+1} = a_k + 6$

72. $a_1 = 3$
    $a_{k+1} = a_k - 2$

73. $a_1 = 23$
    $a_{k+1} = a_k - 5$

74. $a_1 = 12$
    $a_{k+1} = a_k + 6$

75. $a_1 = -16$
    $a_{k+1} = a_k + 5$

76. $a_1 = -22$
    $a_{k+1} = a_k - 4$

77. $a_1 = 3.4$
    $a_{k+1} = a_k - 1.1$

78. $a_1 = 10.9$
    $a_{k+1} = a_k + 0.7$

In Exercises 79–88, find the partial sum. *See Example 4.*

✓ 79. $\displaystyle\sum_{k=1}^{20} k$

80. $\displaystyle\sum_{k=1}^{30} 4k$

81. $\displaystyle\sum_{k=1}^{50} (k + 3)$

82. $\displaystyle\sum_{n=1}^{30} (n + 2)$

83. $\displaystyle\sum_{k=1}^{10} (5k - 2)$

84. $\displaystyle\sum_{k=1}^{100} (4k - 1)$

85. $\displaystyle\sum_{n=1}^{500} \frac{n}{2}$

86. $\displaystyle\sum_{n=1}^{300} \frac{n}{3}$

87. $\displaystyle\sum_{n=1}^{30} \left(\frac{1}{3}n - 4\right)$

88. $\displaystyle\sum_{n=1}^{75} (0.3n + 5)$

In Exercises 89–100, find the *n*th partial sum of the arithmetic sequence. *See Example 5.*

✓ 89. 5, 12, 19, 26, 33, . . . , $n = 12$

90. 2, 12, 22, 32, 42, . . . , $n = 20$

91. 2, 8, 14, 20, . . . , $n = 25$

92. 500, 480, 460, 440, . . . , $n = 20$

93. 200, 175, 150, 125, 100, . . . , $n = 8$

94. 800, 785, 770, 755, 740, . . . , $n = 25$

95. $-50, -38, -26, -14, -2, . . . , n = 50$

96. $-16, -8, 0, 8, 16, . . . , n = 30$

97. 1, 4.5, 8, 11.5, 15, . . . , $n = 12$

98. 2.2, 2.8, 3.4, 4.0, 4.6, . . . , $n = 12$

99. $a_1 = 0.5$, $a_4 = 1.7$, . . . , $n = 10$

100. $a_1 = 15$, $a_{100} = 307$, . . . , $n = 100$

In Exercises 101–106, match the arithmetic sequence with its graph. [The graphs are labeled (a), (b), (c), (d), (e), and (f).]

(a)

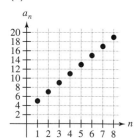

(b)

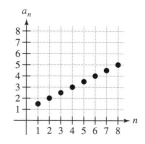

(c)

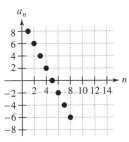

(d)

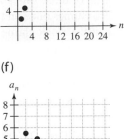

(e)

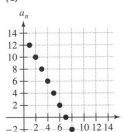

(f)

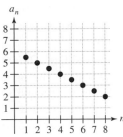

101. $a_n = \frac{1}{2}n + 1$

102. $a_n = -\frac{1}{2}n + 6$

103. $a_n = -2n + 10$

104. $a_n = 2n + 3$

105. $a_1 = 12$

$a_{n+1} = a_n - 2$

106. $a_1 = 2$

$a_{n+1} = a_n + 3$

In Exercises 107–112, use a graphing calculator to graph the first 10 terms of the sequence.

107. $a_n = -2n + 21$ **108.** $a_n = -25n + 500$

109. $a_n = \frac{3}{5}n + \frac{3}{2}$ **110.** $a_n = \frac{3}{2}n + 1$

111. $a_n = 2.5n - 8$ **112.** $a_n = 6.2n + 3$

In Exercises 113–118, use a graphing calculator to find the partial sum.

113. $\displaystyle\sum_{j=1}^{25} (750 - 30j)$ **114.** $\displaystyle\sum_{n=1}^{40} (1000 - 25n)$

115. $\displaystyle\sum_{i=1}^{60} \left(300 - \frac{8}{3}i\right)$ **116.** $\displaystyle\sum_{n=1}^{20} \left(500 - \frac{1}{10}n\right)$

117. $\displaystyle\sum_{n=1}^{50} (2.15n + 5.4)$ **118.** $\displaystyle\sum_{n=1}^{60} (200 - 3.4n)$

Solving Problems

119. *Number Problem*   Find the sum of the first 75 positive integers.

120. *Number Problem*   Find the sum of the integers from 35 to 100.

121. *Number Problem*   Find the sum of the first 50 positive odd integers.

122. *Number Problem*   Find the sum of the first 100 positive even integers.

✓ **123.** *Salary*   In your new job as an actuary, your starting salary will be $54,000 with an increase of $3000 at the end of each of the first 5 years. How much will you be paid through the end of your first 6 years of employment with the company?

124. *Wages*   You earn 5 dollars on the first day of the month, 10 dollars on the second day, 15 dollars on the third day, and so on. Determine the total amount that you will earn during a 30-day month.

125. *Ticket Prices*   There are 20 rows of seats on the main floor of an outdoor arena: 20 seats in the first row, 21 seats in the second row, 22 seats in the third row, and so on (see figure). How much should you charge per ticket in order to obtain $15,000 for the sale of all the seats on the main floor?

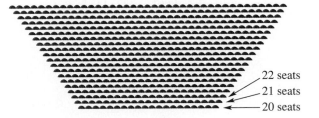

22 seats
21 seats
20 seats

126. *Pile of Logs*   Logs are stacked in a pile as shown in the figure. The top row has 15 logs and the bottom row has 21 logs. How many logs are in the pile?

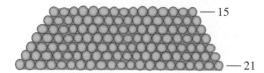

127. *Baling Hay*   In the first two trips baling hay around a large field (see figure), a farmer obtains 93 bales and 89 bales, respectively. The farmer estimates that the same pattern will continue. Estimate the total number of bales obtained if there are another six trips around the field.

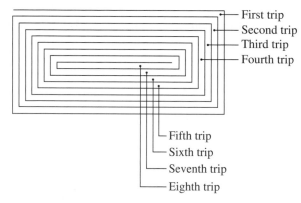

128. *Baling Hay*   In the first two trips baling hay around a field (see figure), a farmer obtains 64 bales and 60 bales, respectively. The farmer estimates that the same pattern will continue. Estimate the total number of bales obtained if there are another four trips around the field.

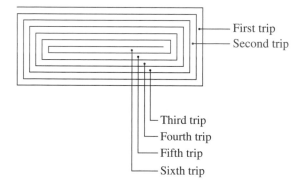

129. *Clock Chimes*   A clock chimes once at 1:00, twice at 2:00, three times at 3:00, and so on. The clock also chimes once at 15-minute intervals that are not on the hour. How many times does the clock chime in a 12-hour period?

130. *Clock Chimes*   A clock chimes once at 1:00, twice at 2:00, three times at 3:00, and so on. The clock also chimes once on the half-hour. How many times does the clock chime in a 12-hour period?

131. *Free-Falling Object*   A free-falling object falls 16 feet during the first second, 48 feet during the second second, 80 feet during the third second, and so on. What total distance does the object fall in 8 seconds?

132. *Free-Falling Object*   A free-falling object falls 4.9 meters during the first second, 14.7 meters during the second second, 24.5 meters during the third second, and so on. What total distance does the object fall in 5 seconds?

133. *Craft Beer Sales*   Craft beers are produced by small, independent brewers using innovative techniques. The amount of craft beer (in barrels) sold in the United States each year from 2003 through 2007 can be modeled by the arithmetic sequence whose nth term is given by

$$a_n = 4.41 + 0.59n$$

where $n = 1$ corresponds to 2003. What is the total amount of craft beer sold in these five years? (Source: Brewer's Association)

AP Photo/*The Journal Record*, Mark Hancock

Explaining Concepts

134. *Growth Rate*   A growth chart shows how a typical child with certain characteristics is expected to grow. A growth chart for Chinese girls shows that a 6-year-old Chinese girl is in the 50th percentile for her weight. From the data in the chart, the girl's expected yearly weight gains (in pounds) over the next five years can be modeled by the arithmetic sequence whose nth term is given by

$$a_n = 2.3 + 0.86n$$

where $n = 1$ corresponds to the weight gained from age 6 to age 7. How much total weight is the girl expected to gain in the next five years?

135. ✎   Explain what a recursion formula does.

136. ✎   Explain how to use the nth term a_n and the common difference d of an arithmetic sequence to write a recursion formula for the term a_{n+2} of the sequence.

137. ✎   Is it possible to use the nth term a_n and the common difference d of an arithmetic sequence to write a recursion formula for the term a_{2n}? Explain.

138. ✎   Explain why you cannot use the formula $a_n = a_1 + (n - 1)d$ to find the nth term of a sequence whose first term is a_0. Discuss the changes that can be made in the formula to create a new formula that can be used.

139. *Pattern Recognition*

(a) Compute the sums of positive odd integers.

$$1 + 3 = \quad\rule{1cm}{0.4pt}$$
$$1 + 3 + 5 = \quad\rule{1cm}{0.4pt}$$
$$1 + 3 + 5 + 7 = \quad\rule{1cm}{0.4pt}$$
$$1 + 3 + 5 + 7 + 9 = \quad\rule{1cm}{0.4pt}$$
$$1 + 3 + 5 + 7 + 9 + 11 = \quad\rule{1cm}{0.4pt}$$

(b) Do the partial sums of the positive odd integers form an arithmetic sequence? Explain.

(c) Use the sums in part (a) to make a conjecture about the sums of positive odd integers. Check your conjecture for the sum

$$1 + 3 + 5 + 7 + 9 + 11 + 13 = \quad\rule{1cm}{0.4pt}.$$

(d) Verify your conjecture in part (c) analytically.

140. ✎   Each term of an arithmetic sequence is multiplied by a constant C. Is the resulting sequence arithmetic? If so, how does the common difference compare with the common difference of the original sequence?

Cumulative Review

In Exercises 141–144, find the center and vertices of the ellipse.

141. $\dfrac{(x - 4)^2}{25} + \dfrac{(y + 5)^2}{9} = 1$

142. $\dfrac{(x + 2)}{4} + (y - 8)^2 = 1$

143. $9x^2 + 4y^2 - 18x + 24y + 9 = 0$

144. $x^2 + 4y^2 - 8x + 12 = 0$

In Exercises 145–148, write the sum using sigma notation. (Begin with $k = 1$.)

145. $3 + 4 + 5 + 6 + 7 + 8 + 9$

146. $3 + 6 + 9 + 12 + 15$

147. $12 + 15 + 18 + 21 + 24$

148. $2 + 2^2 + 2^3 + 2^4 + 2^5$

Mid-Chapter Quiz

Take this quiz as you would take a quiz in class. After you are done, check your work against the answers in the back of the book.

In Exercises 1–4, write the first five terms of the sequence. (Assume that n begins with 1.)

1. $a_n = 4n$

2. $a_n = 2n + 5$

3. $a_n = 32\left(\dfrac{1}{4}\right)^{n-1}$

4. $a_n = \dfrac{(-3)^n n}{n + 4}$

In Exercises 5–10, find the sum.

5. $\displaystyle\sum_{k=1}^{4} 10k$

6. $\displaystyle\sum_{i=1}^{10} 4$

7. $\displaystyle\sum_{j=1}^{5} \dfrac{60}{j + 1}$

8. $\displaystyle\sum_{n=1}^{4} \dfrac{12}{n}$

9. $\displaystyle\sum_{n=1}^{5} (3n - 1)$

10. $\displaystyle\sum_{k=1}^{4} (k^2 - 1)$

In Exercises 11–14, write the sum using sigma notation. (Begin with $k = 1$.)

11. $\dfrac{2}{3(1)} + \dfrac{2}{3(2)} + \dfrac{2}{3(3)} + \cdots + \dfrac{2}{3(20)}$

12. $\dfrac{1}{1^3} - \dfrac{1}{2^3} + \dfrac{1}{3^3} - \cdots + \dfrac{1}{25^3}$

13. $0 + \dfrac{1}{2} + \dfrac{2}{3} + \dfrac{3}{4} + \cdots + \dfrac{19}{20}$

14. $\dfrac{1}{2} + \dfrac{4}{2} + \dfrac{9}{2} + \cdots + \dfrac{100}{2}$

In Exercises 15 and 16, find the common difference of the arithmetic sequence.

15. $1, \frac{3}{2}, 2, \frac{5}{2}, 3, \ldots$

16. $100, 94, 88, 82, 76, \ldots$

In Exercises 17 and 18, find a formula for the nth term of the arithmetic sequence.

17. $a_1 = 20, \quad a_4 = 11$

18. $a_1 = 32, \quad d = -4$

19. Find the sum of the first 200 positive even numbers.

20. You save $.50 on one day, $1.00 the next day, $1.50 the next day, and so on. How much will you have accumulated at the end of one year (365 days)?

11.3 Geometric Sequences and Series

Paul A. Souders/CORBIS

What You Should Learn

1 ▶ Recognize, write, and find the nth terms of geometric sequences.

2 ▶ Find the nth partial sum of a geometric sequence.

3 ▶ Find the sum of an infinite geometric series.

4 ▶ Use geometric sequences to solve application problems.

Why You Should Learn It

A geometric sequence can reduce the amount of time it takes to find the sum of a sequence of numbers with a common ratio. For instance, in Exercise 121 on page 737, you will use a geometric sequence to find the total distance traveled by a bungee jumper.

1 ▶ Recognize, write, and find the nth terms of geometric sequences.

Geometric Sequences

In Section 11.2, you studied sequences whose consecutive terms have a common *difference*. In this section, you will study sequences whose consecutive terms have a common *ratio*.

Definition of Geometric Sequence

A sequence is called **geometric** if the ratios of consecutive terms are the same. So, the sequence $a_1, a_2, a_3, a_4, \ldots, a_n, \ldots$ is geometric if there is a number r, with $r \neq 0$, such that

$$\frac{a_2}{a_1} = r, \quad \frac{a_3}{a_2} = r, \quad \frac{a_4}{a_3} = r$$

and so on. The number r is the **common ratio** of the sequence.

| EXAMPLE 1 | Examples of Geometric Sequences |

a. The sequence whose nth term is 2^n is geometric. For this sequence, the common ratio between consecutive terms is 2.

$$2, 4, 8, 16, \ldots, 2^n, \ldots \qquad \text{Begin with } n = 1.$$

$$\frac{4}{2} = 2$$

b. The sequence whose nth term is $4(3^n)$ is geometric. For this sequence, the common ratio between consecutive terms is 3.

$$12, 36, 108, 324, \ldots, 4(3^n), \ldots \qquad \text{Begin with } n = 1.$$

$$\frac{36}{12} = 3$$

c. The sequence whose nth term is $\left(-\frac{1}{3}\right)^n$ is geometric. For this sequence, the common ratio between consecutive terms is $-\frac{1}{3}$.

$$-\frac{1}{3}, \frac{1}{9}, -\frac{1}{27}, \frac{1}{81}, \ldots, \left(-\frac{1}{3}\right)^n, \ldots \qquad \text{Begin with } n = 1.$$

$$\frac{1/9}{-1/3} = -\frac{1}{3}$$

✔ **CHECKPOINT** *Now try Exercise 1.*

The nth Term of a Geometric Sequence

The nth term of a geometric sequence has the form

$$a_n = a_1 r^{n-1}$$

where r is the common ratio of consecutive terms of the sequence. So, every geometric sequence can be written in the following form.

$$a_1, a_1 r, a_1 r^2, a_1 r^3, a_1 r^4, \ldots, a_1 r^{n-1}, \ldots$$

EXAMPLE 2   Finding the nth Term of a Geometric Sequence

a. Find a formula for the nth term of the geometric sequence whose common ratio is 3 and whose first term is 1.

b. What is the eighth term of the sequence found in part (a)?

Solution

a. The formula for the nth term is of the form $a_n = a_1 r^{n-1}$. Moreover, because the common ratio is $r = 3$ and the first term is $a_1 = 1$, the formula must have the form

$$a_n = a_1 r^{n-1} \qquad \text{Formula for geometric sequence}$$

$$= (1)(3)^{n-1} \qquad \text{Substitute 1 for } a_1 \text{ and 3 for } r.$$

$$= 3^{n-1}. \qquad \text{Simplify.}$$

The sequence has the form $1, 3, 9, 27, 81, \ldots, 3^{n-1}, \ldots$.

b. The eighth term of the sequence is $a_8 = 3^{8-1} = 3^7 = 2187$.

✓ **CHECKPOINT** *Now try Exercise 39.*

EXAMPLE 3   Finding the nth Term of a Geometric Sequence

Find a formula for the nth term of the geometric sequence whose first two terms are 4 and 2.

Solution

Because the common ratio is

$$r = \frac{a_2}{a_1} = \frac{2}{4} = \frac{1}{2}$$

the formula for the nth term must be

$$a_n = a_1 r^{n-1} \qquad \text{Formula for geometric sequence}$$

$$= 4\left(\frac{1}{2}\right)^{n-1}. \qquad \text{Substitute 4 for } a_1 \text{ and } \tfrac{1}{2} \text{ for } r.$$

The sequence has the form $4, 2, 1, \dfrac{1}{2}, \dfrac{1}{4}, \ldots, 4\left(\dfrac{1}{2}\right)^{n-1}, \ldots$.

✓ **CHECKPOINT** *Now try Exercise 47.*

2 ▶ Find the *n*th partial sum of a geometric sequence.

The Partial Sum of a Geometric Sequence

The *n*th Partial Sum of a Geometric Sequence

The *n*th partial sum of the geometric sequence whose *n*th term is $a_n = a_1 r^{n-1}$ is given by

$$\sum_{i=1}^{n} a_1 r^{i-1} = a_1 + a_1 r + a_1 r^2 + a_1 r^3 + \cdots + a_1 r^{n-1} = a_1 \left(\frac{r^n - 1}{r - 1} \right).$$

EXAMPLE 4   **Finding the *n*th Partial Sum**

Find the sum $1 + 2 + 4 + 8 + 16 + 32 + 64 + 128$.

Solution

This is a geometric sequence whose common ratio is $r = 2$. Because the first term of the sequence is $a_1 = 1$, it follows that the sum is

$$\sum_{i=1}^{8} 2^{i-1} = (1)\left(\frac{2^8 - 1}{2 - 1} \right) = \frac{256 - 1}{2 - 1} = 255.$$ Substitute 1 for a_1 and 2 for r.

✓ **CHECKPOINT** *Now try Exercise 71.*

EXAMPLE 5   **Finding the *n*th Partial Sum**

Find the sum of the first five terms of the geometric sequence whose *n*th term is $a_n = \left(\frac{2}{3} \right)^n$.

Solution

$$\sum_{i=1}^{5} \left(\frac{2}{3} \right)^i = \frac{2}{3} \left[\frac{(2/3)^5 - 1}{(2/3) - 1} \right]$$ Substitute $\frac{2}{3}$ for a_1 and $\frac{2}{3}$ for r.

$$= \frac{2}{3} \left[\frac{(32/243) - 1}{-1/3} \right]$$ Simplify.

$$= \frac{422}{243} \approx 1.737$$ Use a calculator to simplify.

✓ **CHECKPOINT** *Now try Exercise 77.*

---

3 ▶ Find the sum of an infinite geometric series.

Geometric Series

Suppose that in Example 5, you were to find the sum of all the terms of the infinite geometric sequence

$$\frac{2}{3}, \frac{4}{9}, \frac{8}{27}, \frac{16}{81}, \cdots, \left(\frac{2}{3} \right)^n, \cdots.$$

The sum of all the terms of an infinite geometric sequence is called an **infinite geometric series,** or simply a **geometric series.**

In your mind, would this sum be infinitely large or would it be a finite number? Consider the formula for the nth partial sum of a geometric sequence.

$$S_n = a_1\left(\frac{r^n - 1}{r - 1}\right) = a_1\left(\frac{1 - r^n}{1 - r}\right)$$

Suppose that $|r| < 1$. As you let n become larger, it follows that r^n approaches 0, so that the term r^n drops out of the formula above. The sum becomes

$$S = a_1\left(\frac{1}{1 - r}\right) = \frac{a_1}{1 - r}.$$

Because n is not involved, you can use this formula to evaluate the sum. In the case of Example 5, $r = \left(\frac{2}{3}\right) < 1$, and so the sum of the infinite geometric sequence is

$$S = \sum_{i=1}^{\infty}\left(\frac{2}{3}\right)^i = \frac{a_1}{1 - r} = \frac{2/3}{1 - (2/3)} = \frac{2/3}{1/3} = 2.$$

Sum of an Infinite Geometric Series

If $a_1, a_1r, a_1r^2, \ldots, a_1r^n, \ldots$ is an infinite geometric sequence and $|r| < 1$, the sum of the terms of the corresponding infinite geometric series is

$$S = \sum_{i=0}^{\infty} a_1r^i = \frac{a_1}{1 - r}.$$

EXAMPLE 6   **Finding the Sum of an Infinite Geometric Series**

Find each sum.

a. $\displaystyle\sum_{i=1}^{\infty} 5\left(\frac{3}{4}\right)^{i-1}$ **b.** $\displaystyle\sum_{n=0}^{\infty} 4\left(\frac{3}{10}\right)^n$ **c.** $\displaystyle\sum_{i=0}^{\infty}\left(-\frac{3}{5}\right)^i$

Solution

a. The series is geometric, with $a_1 = 5\left(\frac{3}{4}\right)^{1-1} = 5$ and $r = \frac{3}{4}$. So,

$$\sum_{i=1}^{\infty} 5\left(\frac{3}{4}\right)^{i-1} = \frac{5}{1 - (3/4)}$$

$$= \frac{5}{1/4} = 20.$$

b. The series is geometric, with $a_1 = 4\left(\frac{3}{10}\right)^0 = 4$ and $r = \frac{3}{10}$. So,

$$\sum_{n=0}^{\infty} 4\left(\frac{3}{10}\right)^n = \frac{4}{1 - (3/10)} = \frac{4}{7/10} = \frac{40}{7}.$$

c. The series is geometric, with $a_1 = \left(-\frac{3}{5}\right)^0 = 1$ and $r = -\frac{3}{5}$. So,

$$\sum_{i=0}^{\infty}\left(-\frac{3}{5}\right)^i = \frac{1}{1 - (-3/5)} = \frac{1}{1 + (3/5)} = \frac{5}{8}.$$

✓ **CHECKPOINT** *Now try Exercise 93.*

4 ► Use geometric sequences to solve application problems.

Applications

 EXAMPLE 7   A Lifetime Salary

You have accepted a job as a meteorologist that pays a salary of $45,000 the first year. During the next 39 years, suppose you receive a 6% raise each year. What will your total salary be over the 40-year period?

Solution

Using a geometric sequence, your salary during the first year will be $a_1 = 45,000$. Then, with a 6% raise each year, your salary for the next 2 years will be as follows.

$$a_2 = 45,000 + 45,000(0.06) = 45,000(1.06)^1$$

$$a_3 = 45,000(1.06) + 45,000(1.06)(0.06) = 45,000(1.06)^2$$

From this pattern, you can see that the common ratio of the geometric sequence is $r = 1.06$. Using the formula for the nth partial sum of a geometric sequence, you will find that the total salary over the 40-year period is given by

$$\text{Total salary} = a_1\left(\frac{r^n - 1}{r - 1}\right)$$

$$= 45,000\left[\frac{(1.06)^{40} - 1}{1.06 - 1}\right]$$

$$= 45,000\left[\frac{(1.06)^{40} - 1}{0.06}\right] \approx \$6,964,288.$$

✓ **CHECKPOINT** *Now try Exercise 107.*

AP Photo/The Journal Record, Mark Hancock

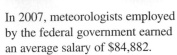

In 2007, meteorologists employed by the federal government earned an average salary of $84,882.

 EXAMPLE 8   Increasing Annuity

You deposit $100 in an account each month for 2 years. The account pays an annual interest rate of 9%, compounded monthly. What is your balance at the end of 2 years? (This type of savings plan is called an **increasing annuity.**)

Solution

The first deposit would earn interest for the full 24 months, the second deposit would earn interest for 23 months, the third deposit would earn interest for 22 months, and so on. Using the formula for compound interest, you can see that the total of the 24 deposits would be

$$\text{Total} = a_1 + a_2 + \cdots + a_{24}$$

$$= 100\left(1 + \frac{0.09}{12}\right)^1 + 100\left(1 + \frac{0.09}{12}\right)^2 + \cdots + 100\left(1 + \frac{0.09}{12}\right)^{24}$$

$$= 100(1.0075)^1 + 100(1.0075)^2 + \cdots + 100(1.0075)^{24}$$

$$= 100(1.0075)\left(\frac{1.0075^{24} - 1}{1.0075 - 1}\right) = \$2638.49.$$

✓ **CHECKPOINT** *Now try Exercise 109.*

Concept Check

1. Explain the difference between an arithmetic sequence and a geometric sequence.

2. How can you determine whether a sequence is geometric?

3. What is the general formula for the nth term of a geometric sequence?

4. Can you find the sum of an infinite geometric series if the common ratio is $\frac{3}{2}$? Explain.

11.3 EXERCISES

Go to pages 746–747 to record your assignments.

Developing Skills

In Exercises 1–12, find the common ratio of the geometric sequence. *See Example 1.*

 1. $3, 6, 12, 24, \ldots$

2. $2, 6, 18, 54, \ldots$

3. $5, -5, 5, -5, \ldots$

4. $-5, -0.5, -0.05, -0.005, \ldots$

5. $\frac{1}{2}, -\frac{1}{4}, \frac{1}{8}, -\frac{1}{16}, \ldots$

6. $\frac{2}{3}, -\frac{4}{3}, \frac{8}{3}, -\frac{16}{3}, \ldots$

7. $75, 15, 3, \frac{3}{5}, \ldots$

8. $12, -4, \frac{4}{3}, -\frac{4}{9}, \ldots$

9. $1, \pi, \pi^2, \pi^3, \ldots$

10. $e, e^2, e^3, e^4, \ldots$

11. $50(1.04), 50(1.04)^2, 50(1.04)^3, 50(1.04)^4, \ldots$

12. $25(1.07), 25(1.07)^2, 25(1.07)^3, 25(1.07)^4, \ldots$

In Exercises 13–24, determine whether the sequence is geometric. If so, find the common ratio.

13. $64, 32, 16, 8, \ldots$ **14.** $64, 32, 0, -32, \ldots$

15. $10, 15, 20, 25, \ldots$ **16.** $10, 20, 40, 80, \ldots$

17. $5, 10, 20, 40, \ldots$ **18.** $270, 90, 30, 10, \ldots$

19. $1, 8, 27, 64, 125, \ldots$

20. $2, 4, 8, 14, 22, \ldots$

21. $1, -\frac{2}{3}, \frac{4}{9}, -\frac{8}{27}, \ldots$ **22.** $\frac{1}{3}, -\frac{2}{3}, \frac{4}{3}, -\frac{8}{3}, \ldots$

23. $1, 1.1, 1.21, 1.331, \ldots$

24. $1, 0.2, 0.04, 0.008, \ldots$

In Exercises 25–38, write the first five terms of the geometric sequence. If necessary, round your answers to two decimal places.

25. $a_1 = 4, \quad r = 2$ **26.** $a_1 = 2, \quad r = 4$

27. $a_1 = 6, \quad r = \frac{1}{2}$ **28.** $a_1 = 90, \quad r = \frac{1}{3}$

29. $a_1 = 5, \quad r = -2$

30. $a_1 = -12, \quad r = -1$

31. $a_1 = -4, \quad r = -\frac{1}{2}$

32. $a_1 = 3, \quad r = -\frac{3}{2}$

33. $a_1 = 10, \quad r = 1.02$

34. $a_1 = 200, \quad r = 1.07$

35. $a_1 = 10, \quad r = \frac{3}{5}$

36. $a_1 = 36, \quad r = \frac{2}{3}$

37. $a_1 = \frac{3}{2}, \quad r = \frac{2}{3}$

38. $a_1 = \frac{4}{5}, \quad r = \frac{1}{2}$

In Exercises 39–52, find a formula for the nth term of the geometric sequence. (Assume that n begins with 1.) *See Examples 2 and 3.*

39. $a_1 = 1, \quad r = 2$ **40.** $a_1 = 5, \quad r = 4$

41. $a_1 = 2, \quad r = 2$ **42.** $a_1 = 25, \quad r = 5$

43. $a_1 = 10, \quad r = -\frac{1}{5}$ **44.** $a_1 = 1, \quad r = -\frac{4}{3}$

45. $a_1 = 4, \quad r = -\frac{1}{2}$ **46.** $a_1 = 9, \quad r = \frac{2}{3}$

47. $a_1 = 8, \quad a_2 = 2$ **48.** $a_1 = 18, \quad a_2 = 8$

49. $a_1 = 14, \quad a_2 = \frac{21}{2}$ **50.** $a_1 = 36, \quad a_2 = \frac{27}{2}$

51. $4, -6, 9, -\frac{27}{2}, \ldots$ **52.** $1, \frac{3}{2}, \frac{9}{4}, \frac{27}{8}, \ldots$

In Exercises 53–66, find the specified term of the geometric sequence. Round to the nearest hundredth, if necessary.

53. $a_1 = 6, \quad r = \frac{1}{2}, \quad a_{10} = $
54. $a_1 = 8, \quad r = \frac{3}{4}, \quad a_8 = $
55. $a_1 = 3, \quad r = \sqrt{2}, \quad a_{10} = $
56. $a_1 = 5, \quad r = \sqrt{3}, \quad a_9 = $
57. $a_1 = 200, \quad r = 1.2, \quad a_{12} = $
58. $a_1 = 500, \quad r = 1.06, \quad a_{40} = $
59. $a_1 = 120, \quad r = -\frac{1}{3}, \quad a_{10} = $
60. $a_1 = 240, \quad r = -\frac{1}{4}, \quad a_{13} = $
61. $a_1 = 4, \quad a_2 = 3, \quad a_5 = $
62. $a_1 = 1, \quad a_2 = 9, \quad a_5 = $
63. $a_3 = 3, \quad a_4 = 6, \quad a_5 = $
64. $a_2 = 5, \quad a_3 = 7, \quad a_4 = $
65. $a_2 = 12, \quad a_3 = 16, \quad a_5 = $
66. $a_4 = 100, \quad a_5 = -25, \quad a_7 = $

In Exercises 67–70, match the geometric sequence with its graph. [The graphs are labeled (a), (b), (c), and (d).]

(a)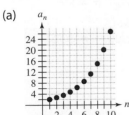
(b)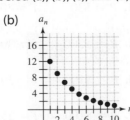
(c) (d)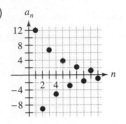

67. $a_n = 12\left(\frac{3}{4}\right)^{n-1}$ **68.** $a_n = 12\left(-\frac{3}{4}\right)^{n-1}$
69. $a_n = 2\left(-\frac{4}{3}\right)^{n-1}$ **70.** $a_n = 2\left(\frac{4}{3}\right)^{n-1}$

In Exercises 71–80, find the partial sum. Round to the nearest hundredth, if necessary. *See Examples 4 and 5.*

71. $\sum_{i=1}^{10} 2^{i-1}$

72. $\sum_{i=1}^{6} 3^{i-1}$

73. $\sum_{i=1}^{12} 3\left(\frac{3}{2}\right)^{i-1}$

74. $\sum_{i=1}^{20} 12\left(\frac{2}{3}\right)^{i-1}$

75. $\sum_{i=1}^{15} 3\left(-\frac{1}{3}\right)^{i-1}$

76. $\sum_{i=1}^{8} 8\left(-\frac{1}{4}\right)^{i-1}$

77. $\sum_{i=1}^{6} \left(\frac{3}{4}\right)^{i}$

78. $\sum_{i=1}^{4} \left(\frac{5}{6}\right)^{i}$

79. $\sum_{i=1}^{8} 6(0.1)^{i-1}$

80. $\sum_{i=1}^{12} 1000(1.06)^{i-1}$

In Exercises 81–92, find the nth partial sum of the geometric sequence. Round to the nearest hundredth, if necessary.

81. $1, -3, 9, -27, 81, \ldots, n = 10$
82. $3, -6, 12, -24, 48, \ldots, n = 12$
83. $8, 4, 2, 1, \frac{1}{2}, \ldots, n = 15$
84. $9, 6, 4, \frac{8}{3}, \frac{16}{9}, \ldots, n = 10$
85. $4, 12, 36, 108, \ldots, n = 8$
86. $\frac{1}{36}, -\frac{1}{12}, \frac{1}{4}, -\frac{3}{4}, \ldots, n = 20$
87. $60, -15, \frac{15}{4}, -\frac{15}{16}, \ldots, n = 12$

88. $40, -10, \frac{5}{2}, -\frac{5}{8}, \frac{5}{32}, \ldots, n = 10$

89. $30, 30(1.06), 30(1.06)^2, 30(1.06)^3, \ldots, n = 20$

90. $100, 100(1.08), 100(1.08)^2, 100(1.08)^3, \ldots,$
     $n = 40$

91. $1, \sqrt{3}, 3, 3\sqrt{3}, 9, \ldots, n = 18$

92. $1, \sqrt{2}, 2, 2\sqrt{2}, 4, \ldots, n = 12$

In Exercises 93–100, find the sum. *See Example 6.*

✓ 93. $\sum_{n=1}^{\infty} \left(\frac{1}{2}\right)^{n-1}$

94. $\sum_{n=1}^{\infty} \left(-\frac{1}{2}\right)^{n-1}$

95. $\sum_{n=0}^{\infty} 2\left(-\frac{2}{3}\right)^n$

96. $\sum_{n=0}^{\infty} 2\left(\frac{2}{3}\right)^n$

97. $\sum_{n=0}^{\infty} \left(\frac{1}{10}\right)^n$

98. $\sum_{n=0}^{\infty} 4\left(\frac{1}{4}\right)^n$

99. $8 + 6 + \frac{9}{2} + \frac{27}{8} + \cdots$

100. $3 - 1 + \frac{1}{3} - \frac{1}{9} + \cdots$

In Exercises 101–104, use a graphing calculator to graph the first 10 terms of the sequence.

101. $a_n = 20(-0.6)^{n-1}$

102. $a_n = 4(1.4)^{n-1}$

103. $a_n = 15(0.6)^{n-1}$

104. $a_n = 8(-0.6)^{n-1}$

Solving Problems

105. *Depreciation*   A company buys a machine for $250,000. During the next 5 years, the machine depreciates at the rate of 25% per year. (That is, at the end of each year, the depreciated value is 75% of what it was at the beginning of the year.)

(a) Find a formula for the nth term of the geometric sequence that gives the value of the machine n full years after it was purchased.

(b) Find the depreciated value of the machine at the end of 5 full years.

(c) During which year did the machine depreciate the most?

106. *Population Increase*   A city of 350,000 people is growing at the rate of 1% per year. (That is, at the end of each year, the population is 1.01 times the population at the beginning of the year.)

(a) Find a formula for the nth term of the geometric sequence that gives the population after n years.

(b) Estimate the population after 10 years.

(c) During which year did the population grow the least?

✓ 107. *Salary*   You accept a job as an archaeologist that pays a salary of $30,000 the first year. During the next 39 years, you receive a 5% raise each year. What would your total salary be over the 40-year period?

108. *Salary*   You accept a job as a marine biologist that pays a salary of $45,000 the first year. During the next 39 years, you receive a 5.5% raise each year. What would your total salary be over the 40-year period?

Increasing Annuity   In Exercises 109–114, find the balance A in an increasing annuity in which a principal of P dollars is invested each month for t years, compounded monthly at rate r.

✓ 109. $P = \$50$ $t = 10$ years $r = 9\%$

110. $P = \$50$ $t = 5$ years $r = 7\%$

111. $P = \$30$ $t = 40$ years $r = 8\%$

112. $P = \$200$ $t = 30$ years $r = 10\%$

113. $P = \$100$ $t = 30$ years $r = 6\%$

114. $P = \$100$ $t = 25$ years $r = 8\%$

115. *Wages* You start work at a company that pays $0.01 for the first day, $0.02 for the second day, $0.04 for the third day, and so on. The daily wage keeps doubling. What would your total income be for working (a) 29 days and (b) 30 days?

116. *Wages* You start work at a company that pays $0.01 for the first day, $0.03 for the second day, $0.09 for the third day, and so on. The daily wage keeps tripling. What would your total income be for working (a) 25 days and (b) 26 days?

117. *Power Supply* The electrical power for an implanted medical device decreases by 0.1% each day.

(a) Find a formula for the *n*th term of the geometric sequence that gives the percent of the initial power *n* days after the device is implanted.

(b) What percent of the initial power is still available 1 year after the device is implanted?

(c) 🖩 The power supply needs to be changed when half the power is depleted. Use a graphing calculator to graph the first 750 terms of the sequence. Estimate when the power source should be changed.

118. *Cooling* The temperature of water in an ice cube tray is 70°F when it is placed in a freezer. Its temperature *n* hours after being placed in the freezer is 20% less than 1 hour earlier.

(a) Find a formula for the *n*th term of the geometric sequence that gives the temperature of the water *n* hours after being placed in the freezer.

(b) Find the temperature of the water 6 hours after it is placed in the freezer.

(c) 🖩 Use a graphing calculator to estimate the time when the water freezes. Explain how you found your answer.

119. *Geometry* An equilateral triangle has an area of 1 square unit. The triangle is divided into four smaller triangles and the center triangle is shaded (see figure). Each of the three unshaded triangles is then divided into four smaller triangles and each center triangle is shaded. This process is repeated one more time. What is the total area of the shaded region?

120. *Geometry* A square has an area of 1 square unit. The square is divided into nine smaller squares and the center square is shaded (see figure). Each of the eight unshaded squares is then divided into nine smaller squares and each center square is shaded. This process is repeated one more time. What is the total area of the shaded region?

121. *Bungee Jumping* A bungee jumper jumps from a bridge and stretches a cord 100 feet. Each successive bounce stretches the cord 75% of its length for the preceding bounce (see figure). Find the total distance traveled by the bungee jumper during 10 bounces.

$$100 + 2(100)(0.75) + \cdots + 2(100)(0.75)^{10}$$

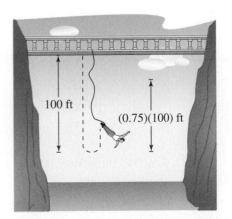

100 ft   (0.75)(100) ft

122. *Distance* A ball is dropped from a height of 16 feet. Each time it drops h feet, it rebounds $0.81h$ feet.

(a) Find the total distance traveled by the ball.

(b) The ball takes the following times for each fall.

$$s_1 = -16t^2 + 16, \qquad s_1 = 0 \text{ if } t = 1$$
$$s_2 = -16t^2 + 16(0.81), \qquad s_2 = 0 \text{ if } t = 0.9$$
$$s_3 = -16t^2 + 16(0.81)^2, \qquad s_3 = 0 \text{ if } t = (0.9)^2$$
$$s_4 = -16t^2 + 16(0.81)^3, \qquad s_4 = 0 \text{ if } t = (0.9)^3$$
$$\vdots \qquad\qquad\qquad \vdots$$
$$s_n = -16t^2 + 16(0.81)^{n-1}, \; s_n = 0 \text{ if } t = (0.9)^{n-1}$$

Beginning with s_2, the ball takes the same amount of time to bounce up as it does to fall, and so the total time elapsed before it comes to rest is

$$t = 1 + 2\sum_{n=1}^{\infty} (0.9)^n.$$

Find this total.

Explaining Concepts

123. The second and third terms of a geometric sequence are 6 and 3, respectively. What is the first term?

124. Give an example of a geometric sequence whose terms alternate in sign.

125. ✎ Explain why the terms of a geometric sequence decrease when $a_1 > 0$ and $0 < r < 1$.

126. ✎ In your own words, describe an increasing annuity.

127. ✎ Explain what is meant by the nth partial sum of a sequence.

128. ✎ A unit square is divided into two equal rectangles. One of the resulting rectangles is then divided into two equal rectangles, as shown in the figure. This process is repeated indefinitely.

(a) Explain why the areas of the rectangles (from largest to smallest) form a geometric sequence.

(b) Find a formula for the nth term of the geometric sequence.

(c) Use the formula for the sum of an infinite geometric series to show that the combined area of the rectangles is 1.

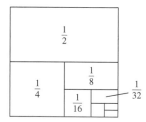

Cumulative Review

In Exercises 129 and 130, solve the system.

129. $\begin{cases} y = 2x^2 \\ y = 2x + 4 \end{cases}$ **130.** $\begin{cases} x^2 + y^2 = 1 \\ x^2 + y = 1 \end{cases}$

In Exercises 131–134, find the annual interest rate.

	Principal	Balance	Time	Compounding
131.	$1000	$2219.64	10 years	Monthly
132.	$2000	$3220.65	8 years	Quarterly
133.	$2500	$10,619.63	20 years	Yearly
134.	$3500	$25,861.70	40 years	Continuous

In Exercises 135 and 136, sketch the hyperbola. Identify the vertices and asymptotes.

135. $\dfrac{x^2}{16} - \dfrac{y^2}{9} = 1$ **136.** $\dfrac{y^2}{1} - \dfrac{x^2}{4} = 1$

11.4 The Binomial Theorem

What You Should Learn

1 ▶ Use the Binomial Theorem to calculate binomial coefficients.

2 ▶ Use Pascal's Triangle to calculate binomial coefficients.

3 ▶ Expand binomial expressions.

Binomial Coefficients

Why You Should Learn It

You can use the Binomial Theorem to expand quantities used in probability. See Exercises 55–58 on page 745.

1 ▶ Use the Binomial Theorem to calculate binomial coefficients.

Recall that a **binomial** is a polynomial that has two terms. In this section, you will study a formula that provides a quick method of raising a binomial to a power. To begin, let's look at the expansion of $(x + y)^n$ for several values of n.

$$(x + y)^0 = 1$$

$$(x + y)^1 = x + y$$

$$(x + y)^2 = x^2 + 2xy + y^2$$

$$(x + y)^3 = x^3 + 3x^2y + 3xy^2 + y^3$$

$$(x + y)^4 = x^4 + 4x^3y + 6x^2y^2 + 4xy^3 + y^4$$

$$(x + y)^5 = x^5 + 5x^4y + 10x^3y^2 + 10x^2y^3 + 5xy^4 + y^5$$

There are several observations you can make about these expansions.

1. In each expansion, there are $n + 1$ terms.

2. In each expansion, x and y have symmetrical roles. The powers of x decrease by 1 in successive terms, whereas the powers of y increase by 1.

3. The sum of the powers of each term is n. For instance, in the expansion of $(x + y)^5$, the sum of the powers of each term is 5.

$$4 + 1 = 5 \quad 3 + 2 = 5$$

$$(x + y)^5 = x^5 + 5x^4y^1 + 10x^3y^2 + 10x^2y^3 + 5xy^4 + y^5$$

4. The coefficients increase and then decrease in a symmetrical pattern.

The coefficients of a binomial expansion are called **binomial coefficients.** To find them, you can use the **Binomial Theorem.**

Study Tip

Other notations that are commonly used for $_nC_r$ are

$\binom{n}{r}$ and $C(n, r)$.

The Binomial Theorem

In the expansion of $(x + y)^n$

$$(x + y)^n = x^n + nx^{n-1}y + \cdots + {_nC_r}x^{n-r}y^r + \cdots + nxy^{n-1} + y^n$$

the coefficient of $x^{n-r}y^r$ is given by

$$_nC_r = \frac{n!}{(n - r)!r!}.$$

EXAMPLE 1   **Finding Binomial Coefficients**

Find each binomial coefficient.

a. $_8C_2$ **b.** $_{10}C_3$ **c.** $_7C_0$ **d.** $_8C_8$

Solution

a. $_8C_2 = \dfrac{8!}{6! \cdot 2!} = \dfrac{(8 \cdot 7) \cdot \cancel{6!}}{\cancel{6!} \cdot 2!} = \dfrac{8 \cdot 7}{2 \cdot 1} = 28$

b. $_{10}C_3 = \dfrac{10!}{7! \cdot 3!} = \dfrac{(10 \cdot 9 \cdot 8) \cdot \cancel{7!}}{\cancel{7!} \cdot 3!} = \dfrac{10 \cdot 9 \cdot 8}{3 \cdot 2 \cdot 1} = 120$

c. $_7C_0 = \dfrac{7!}{7! \cdot 0!} = 1$ **d.** $_8C_8 = \dfrac{8!}{0! \cdot 8!} = 1$

✔ **CHECKPOINT** *Now try Exercise 1.*

When $r \neq 0$ and $r \neq n$, as in parts (a) and (b) of Example 1, there is a simple pattern for evaluating binomial coefficients that results from dividing a common factorial expression out of the numerator and denominator.

$$_8C_2 = \overbrace{\underbrace{\dfrac{8 \cdot 7}{2 \cdot 1}}_{\text{2 factors}}}^{\text{2 factors}} \quad \text{and} \quad _{10}C_3 = \overbrace{\underbrace{\dfrac{10 \cdot 9 \cdot 8}{3 \cdot 2 \cdot 1}}_{\text{3 factors}}}^{\text{3 factors}}$$

EXAMPLE 2   **Finding Binomial Coefficients**

Find each binomial coefficient.

a. $_7C_3$ **b.** $_7C_4$ **c.** $_{12}C_1$ **d.** $_{12}C_{11}$

Solution

a. $_7C_3 = \dfrac{7 \cdot 6 \cdot 5}{3 \cdot 2 \cdot 1} = 35$

b. $_7C_4 = \dfrac{7 \cdot 6 \cdot 5 \cdot 4}{4 \cdot 3 \cdot 2 \cdot 1} = 35$ $_7C_4 = _7C_3$

c. $_{12}C_1 = \dfrac{12!}{11! \cdot 1!} = \dfrac{(12) \cdot \cancel{11!}}{\cancel{11!} \cdot 1!} = \dfrac{12}{1} = 12$

d. $_{12}C_{11} = \dfrac{12!}{1! \cdot 11!} = \dfrac{(12) \cdot \cancel{11!}}{1! \cdot \cancel{11!}} = \dfrac{12}{1} = 12$ $_{12}C_{11} = _{12}C_1$

✔ **CHECKPOINT** *Now try Exercise 7.*

In Example 2, it is not a coincidence that the answers to parts (a) and (b) are the same and that the answers to parts (c) and (d) are the same. In general,

$$_nC_r = _nC_{n-r}.$$

This shows the symmetric property of binomial coefficients.

Technology: Tip

The formula for the binomial coefficient is the same as the formula for combinations in the study of probability. Most graphing calculators have the capability to evaluate a binomial coefficient. Consult the user's guide for your graphing calculator.

2 ▶ Use Pascal's Triangle to calculate binomial coefficients.

Pascal's Triangle

There is a convenient way to remember a pattern for binomial coefficients. By arranging the coefficients in a triangular pattern, you obtain the following array, which is called **Pascal's Triangle.** This triangle is named after the famous French mathematician Blaise Pascal (1623–1662).

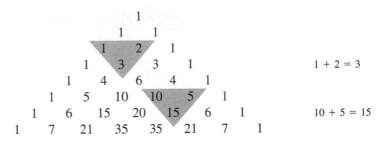

$$1 + 2 = 3$$

$$10 + 5 = 15$$

Study Tip

The top row in Pascal's Triangle is called the *zeroth row* because it corresponds to the binomial expansion

$$(x + y)^0 = 1.$$

Similarly, the next row is called the *first row* because it corresponds to the binomial expansion

$$(x + y)^1 = 1(x) + 1(y).$$

In general, the *nth row* in Pascal's Triangle gives the coefficients of $(x + y)^n$.

The first and last numbers in each row of Pascal's Triangle are 1. Every other number in each row is formed by adding the two numbers immediately above the number. Pascal noticed that numbers in this triangle are precisely the same numbers that are the coefficients of binomial expansions, as follows.

$$(x + y)^0 = 1 \qquad \text{0th row}$$
$$(x + y)^1 = 1x + 1y \qquad \text{1st row}$$
$$(x + y)^2 = 1x^2 + 2xy + 1y^2 \qquad \text{2nd row}$$
$$(x + y)^3 = 1x^3 + 3x^2y + 3xy^2 + 1y^3 \qquad \text{3rd row}$$
$$(x + y)^4 = 1x^4 + 4x^3y + 6x^2y^2 + 4xy^3 + 1y^4 \qquad \vdots$$
$$(x + y)^5 = 1x^5 + 5x^4y + 10x^3y^2 + 10x^2y^3 + 5xy^4 + 1y^5$$
$$(x + y)^6 = 1x^6 + 6x^5y + 15x^4y^2 + 20x^3y^3 + 15x^2y^4 + 6xy^5 + 1y^6$$
$$(x + y)^7 = 1x^7 + 7x^6y + 21x^5y^2 + 35x^4y^3 + 35x^3y^4 + 21x^2y^5 + 7xy^6 + 1y^7$$

Use the seventh row to find the binomial coefficients of the eighth row.

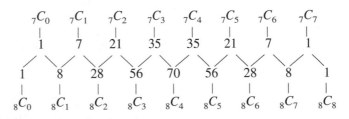

EXAMPLE 3   Using Pascal's Triangle

Use the fifth row of Pascal's Triangle to evaluate $_5C_2$.

Solution

$$
\begin{array}{cccccc}
1 & 5 & 10 & 10 & 5 & 1 \\
| & | & | & | & | & | \\
_5C_0 & _5C_1 & _5C_2 & _5C_3 & _5C_4 & _5C_5
\end{array}
$$

So, $_5C_2 = 10$.

✓ **CHECKPOINT** *Now try Exercise 17.*

3 ▶ Expand binomial expressions.

Binomial Expansions

As mentioned at the beginning of this section, when you write out the coefficients of a binomial raised to a power, you are **expanding a binomial.** The formulas for binomial coefficients give you an easy way to expand binomials.

EXAMPLE 4   Expanding a Binomial

Write the expansion of the expression $(x + 1)^5$.

Solution

The binomial coefficients from the fifth row of Pascal's Triangle are

$$1, 5, 10, 10, 5, 1.$$

So, the expansion is as follows.

$$(x + 1)^5 = (1)x^5 + (5)x^4(1) + (10)x^3(1^2) + (10)x^2(1^3) + (5)x(1^4) + (1)(1^5)$$
$$= x^5 + 5x^4 + 10x^3 + 10x^2 + 5x + 1$$

✔ **CHECKPOINT** *Now try Exercise 23.*

To expand binomials representing *differences*, rather than sums, you alternate signs. Here are two examples.

$$(x - 1)^3 = x^3 - 3x^2 + 3x - 1$$
$$(x - 1)^4 = x^4 - 4x^3 + 6x^2 - 4x + 1$$

EXAMPLE 5   Expanding a Binomial

Write the expansion of each expression.

a. $(x - 3)^4$ **b.** $(2x - 1)^3$

Solution

a. The binomial coefficients from the fourth row of Pascal's Triangle are

$$1, 4, 6, 4, 1.$$

So, the expansion is as follows.

$$(x - 3)^4 = (1)x^4 - (4)x^3(3) + (6)x^2(3^2) - (4)x(3^3) + (1)(3^4)$$
$$= x^4 - 12x^3 + 54x^2 - 108x + 81$$

b. The binomial coefficients from the third row of Pascal's Triangle are

$$1, 3, 3, 1.$$

So, the expansion is as follows.

$$(2x - 1)^3 = (1)(2x)^3 - (3)(2x)^2(1) + (3)(2x)(1^2) - (1)(1^3)$$
$$= 8x^3 - 12x^2 + 6x - 1$$

 CHECKPOINT *Now try Exercise 25.*

EXAMPLE 6   **Expanding a Binomial**

Write the expansion of the expression $(x - 2y)^4$.

Solution

Use the fourth row of Pascal's Triangle, as follows.

$$(x - 2y)^4 = (1)x^4 - (4)x^3(2y) + (6)x^2(2y)^2 - (4)x(2y)^3 + (1)(2y)^4$$

$$= x^4 - 8x^3y + 24x^2y^2 - 32xy^3 + 16y^4$$

✓ **CHECKPOINT** *Now try Exercise 27.*

EXAMPLE 7   **Expanding a Binomial**

Write the expansion of the expression $(x^2 + 4)^3$.

Solution

Use the third row of Pascal's Triangle, as follows.

$$(x^2 + 4)^3 = (1)(x^2)^3 + (3)(x^2)^2(4) + (3)x^2(4^2) + (1)(4^3)$$

$$= x^6 + 12x^4 + 48x^2 + 64$$

✓ **CHECKPOINT** *Now try Exercise 31.*

Sometimes you will need to find a specific term in a binomial expansion. Instead of writing out the entire expansion, you can use the fact that from the Binomial Theorem, the $(r + 1)$th term is $_nC_r\, x^{n-r}y^r$.

EXAMPLE 8   **Finding a Term in a Binomial Expansion**

a. Find the sixth term in the expansion of $(a + 2b)^8$.

b. Find the coefficient of the term a^6b^5 in the expansion of $(3a - 2b)^{11}$.

Solution

a. In this case, $6 = r + 1$ means that $r = 5$. Because $n = 8$, $x = a$, and $y = 2b$, the sixth term in the binomial expansion is

$$_8C_5 a^{8-5}(2b)^5 = 56 \cdot a^3 \cdot (2b)^5$$

$$= 56(2^5)a^3b^5$$

$$= 1792\, a^3b^5.$$

b. In this case, $n = 11$, $r = 5$, $x = 3a$, and $y = -2b$. Substitute these values to obtain

$$_nC_r\, x^{n-r}y^r = {}_{11}C_5(3a)^6(-2b)^5$$

$$= 462(729a^6)(-32b^5)$$

$$= -10{,}777{,}536a^6b^5.$$

So, the coefficient is $-10{,}777{,}536$.

✓ **CHECKPOINT** *Now try Exercise 43.*

Concept Check

1. How many terms are in the expansion of $(x + y)^{10}$?

2. In the expansion of $(x + y)^{10}$, is $_6C_4$ the coefficient of the x^4y^6 term? Explain.

3. Which row of Pascal's Triangle would you use to evaluate $_{10}C_3$?

4. When finding the seventh term of a binomial expansion by evaluating $_nC_r x^{n-r} y^r$, what value should you substitute for r? Explain.

11.4 EXERCISES

Go to pages 746–747 to record your assignments.

Developing Skills

In Exercises 1–10, evaluate the binomial coefficient $_nC_r$. *See Examples 1 and 2.*

✓ **1.** $_6C_4$ **2.** $_9C_3$

3. $_{10}C_5$ **4.** $_{12}C_9$

5. $_{12}C_{12}$ **6.** $_8C_1$

✓ **7.** $_{20}C_6$ **8.** $_{15}C_{10}$

9. $_{20}C_{14}$ **10.** $_{15}C_5$

In Exercises 11–16, use a graphing calculator to evaluate $_nC_r$.

11. $_{30}C_6$ **12.** $_{40}C_8$

13. $_{52}C_5$ **14.** $_{100}C_4$

15. $_{800}C_{797}$ **16.** $_{1000}C_2$

In Exercises 17–22, use Pascal's Triangle to evaluate $_nC_r$. *See Example 3.*

✓ **17.** $_6C_2$ **18.** $_9C_3$

19. $_7C_3$ **20.** $_9C_5$

21. $_8C_4$ **22.** $_{10}C_6$

In Exercises 23–32, use Pascal's Triangle to expand the expression. *See Examples 4–7.*

✓ **23.** $(t + 5)^3$

24. $(y + 2)^4$

✓ **25.** $(m - n)^5$

26. $(r - s)^7$

✓ **27.** $(3a - 1)^5$

28. $(1 - 4b)^3$

29. $(2y + z)^6$

30. $(3c + d)^6$

✓ **31.** $(x^2 + 2)^4$

32. $(5 + y^2)^5$

In Exercises 33–42, use the Binomial Theorem to expand the expression.

33. $(x + 3)^6$

34. $(m - 4)^4$

35. $(u - 2v)^3$

36. $(2x + y)^5$

37. $(3a + 2b)^4$

38. $(4u - 3v)^3$

39. $\left(x + \dfrac{2}{y}\right)^4$

40. $\left(s + \dfrac{1}{t}\right)^5$

41. $(2x^2 - y)^5$

42. $(x - 4y^3)^4$

In Exercises 43–46, find the specified term in the expansion of the binomial. *See Example 8.*

✓ **43.** $(x + y)^{10}$, 4th term   **44.** $(x - y)^6$, 7th term

45. $(a + 6b)^9$, 5th term   **46.** $(3a - b)^{12}$, 10th term

In Exercises 47–50, find the coefficient of the given term in the expansion of the binomial. *See Example 8.*

Expression	Term
47. $(x + 1)^{10}$	x^7
48. $(x + 3)^{12}$	x^9
49. $(x^2 - 3)^4$	x^4
50. $(3 - y^3)^5$	y^9

In Exercises 51–54, use the Binomial Theorem to approximate the quantity rounded to three decimal places. For example:

$$(1.02)^{10} = (1 + 0.02)^{10} \approx 1 + 10(0.02) + 45(0.02)^2.$$

51. $(1.02)^8$ **52.** $(2.005)^{10}$

53. $(2.99)^{12}$ **54.** $(1.98)^9$

_____ **Solving Problems** _____

Probability   In Exercises 55–58, use the Binomial Theorem to expand the expression. In the study of probability, it is sometimes necessary to use the expansion $(p + q)^n$, where $p + q = 1$.

55. $\left(\frac{1}{2} + \frac{1}{2}\right)^5$

56. $\left(\frac{2}{3} + \frac{1}{3}\right)^4$

57. $\left(\frac{1}{4} + \frac{3}{4}\right)^4$

58. $\left(\frac{2}{5} + \frac{3}{5}\right)^3$

59. *Pascal's Triangle*   Rows 0 through 6 of Pascal's Triangle are shown. Find the sum of the numbers in each row. Describe the pattern.

$$
\begin{array}{ccccccccccccc}
 & & & & & & 1 & & & & & & \\
 & & & & & 1 & & 1 & & & & & \\
 & & & & 1 & & 2 & & 1 & & & & \\
 & & & 1 & & 3 & & 3 & & 1 & & & \\
 & & 1 & & 4 & & 6 & & 4 & & 1 & & \\
 & 1 & & 5 & & 10 & & 10 & & 5 & & 1 & \\
1 & & 6 & & 15 & & 20 & & 15 & & 6 & & 1 \\
\end{array}
$$

60. *Pascal's Triangle*   Use each encircled group of numbers to form a 2×2 matrix. Find the determinant of each matrix. Describe the pattern.

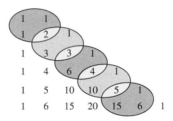

_____ **Explaining Concepts** _____

61. How do the expansions of $(x + y)^n$ and $(x - y)^n$ differ?

62. Which of the following is equal to $_{11}C_5$? Explain.

(a) $\dfrac{11 \cdot 10 \cdot 9 \cdot 8 \cdot 7}{5 \cdot 4 \cdot 3 \cdot 2 \cdot 1}$ (b) $\dfrac{11 \cdot 10 \cdot 9 \cdot 8 \cdot 7}{6 \cdot 5 \cdot 4 \cdot 3 \cdot 2 \cdot 1}$

63. ✎   In your own words, explain how to form the rows in Pascal's Triangle.

64. In the expansion of $(x + 2)^9$, are the coefficients of the x^3-term and the x^6-term identical? Explain.

_____ **Cumulative Review** _____

In Exercises 65 and 66, find the partial sum of the arithmetic sequence.

65. $\displaystyle\sum_{i=1}^{15}(2 + 3i)$ **66.** $\displaystyle\sum_{k=1}^{25}(9k - 5)$

In Exercises 67 and 68, find the partial sum of the geometric sequence. Round to the nearest hundredth.

67. $\displaystyle\sum_{k=1}^{8} 5^{k-1}$ **68.** $\displaystyle\sum_{i=1}^{20} 10\left(\frac{3}{4}\right)^{i-1}$

What Did You Learn?

Use these two pages to help prepare for a test on this chapter. Check off the key terms and key concepts you know. You can also use this section to record your assignments.

Plan for Test Success

Date of test: [/ /]   **Study dates and times:** [/ /] at [:] A.M./P.M.

[/ /] at [:] A.M./P.M.

Things to review:

- ☐ Key Terms, *p. 746*
- ☐ Key Concepts, *pp. 746–747*
- ☐ Your class notes
- ☐ Your assignments

- ☐ Study Tips, *pp. 711, 713, 720, 721, 730, 739, 741*
- ☐ Technology Tips, *pp. 709, 712, 740*
- ☐ Mid-Chapter Quiz, *p. 728*

- ☐ Review Exercises, *pp. 748–750*
- ☐ Chapter Test, *p. 751*
- ☐ Video Explanations Online
- ☐ Tutorial Online

Key Terms

- ☐ sequence, *p. 708*
- ☐ term (of a sequence), *p. 708*
- ☐ infinite sequence, *p. 708*
- ☐ finite sequence, *p. 708*
- ☐ factorials, *p. 710*
- ☐ series, *p. 711*
- ☐ partial sum, *p. 711*
- ☐ infinite series, *p. 711*

- ☐ sigma notation, *p. 712*
- ☐ index of summation, *p. 712*
- ☐ upper limit of summation, *p. 712*
- ☐ lower limit of summation, *p. 712*
- ☐ arithmetic sequence, *p. 719*
- ☐ common difference, *p. 719*
- ☐ recursion formula, *p. 720*
- ☐ nth partial sum, *pp. 721, 731*

- ☐ geometric sequence, *p. 729*
- ☐ common ratio, *p. 729*
- ☐ infinite geometric series, *p. 731*
- ☐ increasing annuity, *p. 733*
- ☐ binomial coefficients, *p. 739*
- ☐ Pascal's Triangle, *p. 741*
- ☐ expanding a binomial, *p. 742*

Key Concepts

11.1 Sequences and Series

Assignment: _____   Due date: _____

☐ **Write the terms of a sequence.**

An infinite sequence $a_1, a_2, a_3, \ldots, a_n, \ldots$ is a function whose domain is the set of positive integers.

A finite sequence $a_1, a_2, a_3, \ldots, a_n$ is a function whose domain is the finite set $\{1, 2, 3, \ldots, n\}$.

☐ **Write the terms of a sequence involving factorials.**

If n is a positive integer, n factorial is defined as

$$n! = 1 \cdot 2 \cdot 3 \cdot 4 \cdot \ldots \cdot (n-1) \cdot n.$$

As a special case, zero factorial is defined as $0! = 1$.

☐ **Sum the terms of a sequence to obtain a series.**

For an infinite sequence $a_1, a_2, a_3, \ldots, a_n, \ldots$

1. The sum of the first n terms is called a partial sum.
2. The sum of all the terms is called an infinite series, or simply a series.

☐ **Use sigma notation to represent a partial sum.**

The sum of the first n terms of the sequence whose nth term is a_n is

$$\sum_{i=1}^{n} a_i = a_1 + a_2 + a_3 + a_4 + \cdots + a_n$$

where i is the index of summation, n is the upper limit of summation, and 1 is the lower limit of summation.

11.2 Arithmetic Sequences

Assignment: _____ Due date:_____

☐ **Recognize, write, and find the *n*th term of an arithmetic sequence.**

The *n*th term of an arithmetic sequence has the form

$a_n = a_1 + (n - 1)d$

where d is the common difference between the terms of the sequence, and a_1 is the first term.

☐ **Find the *n*th partial sum of an arithmetic sequence.**

The *n*th partial sum of the arithmetic sequence whose *n*th term is a_n is

$$\sum_{i=1}^{n} a_i = a_1 + a_2 + a_3 + a_4 + \cdots + a_n$$

$$= \frac{n}{2}(a_1 + a_n).$$

11.3 Geometric Sequences and Series

Assignment: _____ Due date:_____

☐ **Recognize, write, and find the *n*th term of a geometric sequence.**

The *n*th term of a geometric sequence has the form

$a_n = a_1 r^{n-1}$

where r is the common ratio of consecutive terms of the sequence. So, every geometric sequence can be written in the following form.

$a_1, a_1 r, a_1 r^2, a_1 r^3, a_1 r^4, \ldots, a_1 r^{n-1}, \ldots$

☐ **Find the *n*th partial sum of a geometric sequence.**

The *n*th partial sum of the geometric sequence whose *n*th term is $a_n = a_1 r^{n-1}$ is given by

$$\sum_{i=1}^{n} a_1 r^{i-1} = a_1 + a_1 r + a_1 r^2 + a_1 r^3 + \cdots + a_1 r^{n-1}$$

$$= a_1\left(\frac{r^n - 1}{r - 1}\right).$$

☐ **Find the sum of an infinite geometric series.**

If $a_1, a_1 r, a_1 r^2, \ldots, a_1 r^n, \ldots$ is an infinite geometric sequence and $|r| < 1$, the sum of the terms of the corresponding infinite geometric series is

$$S = \sum_{i=0}^{\infty} a_1 r^i = \frac{a_1}{1 - r}.$$

11.4 The Binomial Theorem

Assignment: _____ Due date:_____

☐ **Use the Binomial Theorem to calculate binomial coefficients.**

In the expansion of $(x + y)^n$

$(x + y)^n = x^n + nx^{n-1}y + \cdots + {}_nC_r x^{n-r}y^r +$

$\cdots + nxy^{n-1} + y^n$

the coefficient of $x^{n-r}y^r$ is given by

$${}_nC_r = \frac{n!}{(n - r)!r!}.$$

☐ **Use Pascal's Triangle to calculate binomial coefficients.**

The first and last numbers in each row of Pascal's Triangle are 1. Every other number in each row is the sum of the two numbers immediately above it.

```
            1
          1   1
        1   2   1
      1   3   3   1
    1   4   6   4   1
  1   5  10  10   5   1
1   6  15  20  15   6   1
```

Review Exercises

11.1 Sequences and Series

1 ► Use sequence notation to write the terms of sequences.

In Exercises 1–4, write the first five terms of the sequence. (Assume that n begins with 1.)

1. $a_n = 3n + 5$

2. $a_n = \frac{1}{2}n - 4$

3. $a_n = \dfrac{n}{3n - 1}$

4. $a_n = 3^n + n$

2 ► Write the terms of sequences involving factorials.

In Exercises 5–8, write the first five terms of the sequence. (Assume that n begins with 1.)

5. $a_n = (n + 1)!$

6. $a_n = (-1)^n n!$

7. $a_n = \dfrac{n!}{2n}$

8. $a_n = \dfrac{(n + 1)!}{(2n)!}$

3 ► Find the apparent nth term of a sequence.

In Exercises 9–16, write an expression for the nth term of the sequence. (Assume that n begins with 1.)

9. $4, 7, 10, 13, 16, \ldots$

10. $3, -6, 9, -12, 15, \ldots$

11. $\frac{1}{2}, \frac{1}{5}, \frac{1}{10}, \frac{1}{17}, \frac{1}{26}, \ldots$

12. $\frac{0}{2}, \frac{1}{3}, \frac{2}{4}, \frac{3}{5}, \frac{4}{6}, \ldots$

13. $3, 1, -1, -3, -5, \ldots$

14. $3, 7, 11, 15, 19, \ldots$

15. $\frac{3}{2}, \frac{12}{5}, \frac{27}{10}, \frac{48}{17}, \frac{75}{26}, \ldots$

16. $-1, \frac{1}{2}, -\frac{1}{4}, \frac{1}{8}, -\frac{1}{16}, \ldots$

4 ► Sum the terms of sequences to obtain series, and use sigma notation to represent partial sums.

In Exercises 17–20, find the partial sum.

17. $\displaystyle\sum_{k=1}^{4} 7$

18. $\displaystyle\sum_{k=1}^{4} \frac{(-1)^k}{k}$

19. $\displaystyle\sum_{i=1}^{5} \frac{i - 2}{i + 1}$

20. $\displaystyle\sum_{n=1}^{4} \left(\frac{1}{n} - \frac{1}{n + 2} \right)$

In Exercises 21–24, write the sum using sigma notation. (Begin with $k = 0$ or $k = 1$.)

21. $[5(1) - 3] + [5(2) - 3] + [5(3) - 3] + [5(4) - 3]$

22. $(1)[(1) - 5] + (2)[(2) - 5] + (3)[(3) - 5] + (4)[(4) - 5] + (5)[(5) - 5]$

23. $\dfrac{1}{3(1)} + \dfrac{1}{3(2)} + \dfrac{1}{3(3)} + \dfrac{1}{3(4)} + \dfrac{1}{3(5)} + \dfrac{1}{3(6)}$

24. $\left(-\frac{1}{3} \right)^0 + \left(-\frac{1}{3} \right)^1 + \left(-\frac{1}{3} \right)^2 + \left(-\frac{1}{3} \right)^3 + \left(-\frac{1}{3} \right)^4$

11.2 Arithmetic Sequences

1 ► Recognize, write, and find the nth terms of arithmetic sequences.

In Exercises 25 and 26, find the common difference of the arithmetic sequence.

25. $50, 44.5, 39, 33.5, 28, \ldots$

26. $9, 12, 15, 18, 21, \ldots$

In Exercises 27–32, write the first five terms of the arithmetic sequence. (Assume that n begins with 1.)

27. $a_n = 132 - 5n$

28. $a_n = 2n + 3$

29. $a_n = \frac{1}{3}n + \frac{5}{3}$

30. $a_n = -\frac{3}{5}n + 1$

31. $a_1 = 80$
$a_{k+1} = a_k - \frac{5}{2}$

32. $a_1 = 30$
$a_{k+1} = a_k - 12$

In Exercises 33–36, find a formula for the nth term of the arithmetic sequence.

33. $a_1 = 10, \quad d = 4$

34. $a_1 = 32, \quad d = -2$

35. $a_1 = 1000, \quad a_2 = 950$

36. $a_2 = 150, \quad a_5 = 201$

2▶ Find the nth partial sum of an arithmetic sequence.

In Exercises 37–40, find the partial sum.

37. $\displaystyle\sum_{k=1}^{12} (7k - 5)$ **38.** $\displaystyle\sum_{k=1}^{10} (100 - 10k)$

39. $\displaystyle\sum_{j=1}^{120} \left(\frac{1}{4}j + 1\right)$ **40.** $\displaystyle\sum_{j=1}^{50} \frac{3j}{2}$

In Exercises 41 and 42, use a graphing calculator to find the partial sum.

41. $\displaystyle\sum_{i=1}^{60} (1.25i + 4)$ **42.** $\displaystyle\sum_{i=1}^{150} \frac{i + 4}{2}$

3▶ Use arithmetic sequences to solve application problems.

43. *Number Problem*   Find the sum of the first 50 positive integers that are multiples of 4.

44. *Number Problem*   Find the sum of the integers from 225 to 300.

45. *Auditorium Seating*   Each row in a small auditorium has three more seats than the preceding row. The front row seats 22 people and there are 12 rows of seats. Find the seating capacity of the auditorium.

46. *Wages*   You earn $25 on the first day of the month and $100 on the last day of the month. Each day you are paid $2.50 more than the previous day. How much do you earn in a 31-day month?

11.3 Geometric Sequences and Series

1▶ Recognize, write, and find the nth terms of geometric sequences.

In Exercises 47 and 48, find the common ratio of the geometric sequence.

47. $8, 20, 50, 125, \frac{625}{2}, \ldots$

48. $27, -18, 12, -8, \frac{16}{3}, \ldots$

In Exercises 49–54, write the first five terms of the geometric sequence.

49. $a_1 = 10, \quad r = 3$ **50.** $a_1 = 2, \quad r = -5$

51. $a_1 = 100, \quad r = -\frac{1}{2}$

52. $a_1 = 20, \quad r = \frac{1}{5}$

53. $a_1 = 4, \quad r = \frac{3}{2}$ **54.** $a_1 = 32, \quad r = -\frac{3}{4}$

In Exercises 55–60, find a formula for the nth term of the geometric sequence. (Assume that n begins with 1.)

55. $a_1 = 1, \quad r = -\frac{2}{3}$

56. $a_1 = 100, \quad r = 1.07$

57. $a_1 = 24, \quad a_2 = 72$

58. $a_1 = 16, \quad a_2 = -4$

59. $a_1 = 12, \quad a_4 = -\frac{3}{2}$

60. $a_2 = 1, \quad a_3 = \frac{1}{3}$

2▶ Find the nth partial sum of a geometric sequence.

In Exercises 61–66, find the partial sum. Round to the nearest thousandth, if necessary.

61. $\displaystyle\sum_{n=1}^{12} 2^n$ **62.** $\displaystyle\sum_{n=1}^{12} (-2)^n$

63. $\displaystyle\sum_{k=1}^{8} 5\left(-\frac{3}{4}\right)^k$ **64.** $\displaystyle\sum_{k=1}^{12} (-0.6)^{k-1}$

65. $\displaystyle\sum_{n=1}^{120} 500(1.01)^n$ **66.** $\displaystyle\sum_{n=1}^{40} 1000(1.1)^n$

In Exercises 67 and 68, use a graphing calculator to find the partial sum. Round to the nearest thousandth, if necessary.

67. $\displaystyle\sum_{k=1}^{75} 200(1.4)^{k-1}$ **68.** $\displaystyle\sum_{j=1}^{60} 25(0.9)^{j-1}$

3▶ Find the sum of an infinite geometric series.

In Exercises 69–72, find the sum.

69. $\displaystyle\sum_{i=1}^{\infty} \left(\frac{7}{8}\right)^{i-1}$ **70.** $\displaystyle\sum_{i=1}^{\infty} \left(\frac{3}{5}\right)^{i-1}$

71. $\sum_{k=0}^{\infty} 4\left(\frac{2}{3}\right)^k$ **72.** $\sum_{k=0}^{\infty} 1.3\left(\frac{1}{10}\right)^k$

4 ▶ Use geometric sequences to solve application problems.

73. *Depreciation*   A company pays $120,000 for a machine. During the next 5 years, the machine depreciates at the rate of 30% per year. (That is, at the end of each year, the depreciated value is 70% of what it was at the beginning of the year.)

(a) Find a formula for the nth term of the geometric sequence that gives the value of the machine n full years after it was purchased.

(b) Find the depreciated value of the machine at the end of 5 full years.

74. *Population Increase*   A city of 85,000 people is growing at the rate of 1.2% per year. (That is, at the end of each year, the population is 1.012 times what it was at the beginning of the year.)

(a) Find a formula for the nth term of the geometric sequence that gives the population after n years.

(b) Estimate the population after 50 years.

75. *Internet*   On its first day, a website has 1000 visits. During the next 89 days, the number of visits increases by 12.5% each day. What is the total number of visits during the 90-day period?

76. *Increasing Annuity*   You deposit $200 in an account each month for 10 years. The account pays an annual interest rate of 8%, compounded monthly. What is your balance at the end of 10 years?

11.4 The Binomial Theorem

1 ▶ Use the Binomial Theorem to calculate binomial coefficients.

In Exercises 77–80, evaluate $_nC_r$.

77. $_8C_3$ **78.** $_{12}C_2$

79. $_{15}C_4$ **80.** $_{100}C_1$

⌨ **In Exercises 81–84, use a graphing calculator to evaluate $_nC_r$.**

81. $_{40}C_4$ **82.** $_{32}C_8$

83. $_{25}C_6$ **84.** $_{48}C_5$

2 ▶ Use Pascal's Triangle to calculate binomial coefficients.

In Exercises 85–88, use Pascal's Triangle to evaluate $_nC_r$.

85. $_5C_3$ **86.** $_9C_9$

87. $_8C_4$ **88.** $_7C_2$

3 ▶ Expand binomial expressions.

In Exercises 89–92, use Pascal's Triangle to expand the expression.

89. $(x - 5)^4$

90. $(x + y)^7$

91. $(5x + 2)^3$

92. $(x - 3y)^4$

In Exercises 93–98, use the Binomial Theorem to expand the expression.

93. $(x + 1)^{10}$

94. $(y - 2)^6$

95. $(3x - 2y)^4$

96. $(4u + v)^5$

97. $(u^2 + v^3)^5$

98. $(x^4 - y^5)^4$

In Exercises 99 and 100, find the specified term in the expansion of the binomial.

99. $(x + 2)^{10}$, 7th term

100. $(2x - 3y)^5$, 4th term

In Exercises 101 and 102, find the coefficient of the given term in the expansion of the binomial.

Expression	*Term*
101. $(x - 3)^{10}$	x^5
102. $(3x + 4y)^6$	x^2y^4

Chapter Test

Take this test as you would take a test in class. After you are done, check your work against the answers in the back of the book.

1. Write the first five terms of the sequence $a_n = \left(-\frac{3}{5}\right)^{n-1}$. (Assume that n begins with 1.)

2. Write the first five terms of the sequence $a_n = 3n^2 - n$. (Assume that n begins with 1.)

In Exercises 3–5, find the partial sum.

3. $\displaystyle\sum_{n=1}^{12} 5$

4. $\displaystyle\sum_{k=0}^{8} (2k - 3)$

5. $\displaystyle\sum_{n=1}^{5} (3 - 4n)$

6. Use sigma notation to write $\dfrac{2}{3(1) + 1} + \dfrac{2}{3(2) + 1} + \cdots + \dfrac{2}{3(12) + 1}$.

7. Use sigma notation to write

$$\left(\frac{1}{2}\right)^0 + \left(\frac{1}{2}\right)^2 + \left(\frac{1}{2}\right)^4 + \left(\frac{1}{2}\right)^6 + \left(\frac{1}{2}\right)^8 + \left(\frac{1}{2}\right)^{10}.$$

8. Write the first five terms of the arithmetic sequence whose first term is $a_1 = 12$ and whose common difference is $d = 4$.

9. Find a formula for the nth term of the arithmetic sequence whose first term is $a_1 = 5000$ and whose common difference is $d = -100$.

10. Find the sum of the first 50 positive integers that are multiples of 3.

11. Find the common ratio of the geometric sequence: $-4, 3, -\frac{9}{4}, \frac{27}{16}, \ldots$.

12. Find a formula for the nth term of the geometric sequence whose first term is $a_1 = 4$ and whose common ratio is $r = \frac{1}{2}$.

In Exercises 13 and 14, find the partial sum.

13. $\displaystyle\sum_{n=1}^{8} 2(2^n)$

14. $\displaystyle\sum_{n=1}^{10} 3\left(\frac{1}{2}\right)^n$

In Exercises 15 and 16, find the sum of the infinite geometric series.

15. $\displaystyle\sum_{i=1}^{\infty} \left(\frac{1}{2}\right)^i$

16. $\displaystyle\sum_{i=1}^{\infty} 10(0.4)^{i-1}$

17. Evaluate: $_{20}C_3$

18. Use Pascal's Triangle to expand $(x - 2)^5$.

19. Find the coefficient of the term x^3y^5 in the expansion of $(x + y)^8$.

20. A free-falling object will fall 4.9 meters during the first second, 14.7 more meters during the second second, 24.5 more meters during the third second, and so on. What is the total distance the object will fall in 10 seconds if this pattern continues?

21. You deposit $80 each month in an increasing annuity that pays 4.8% compounded monthly. What is the balance after 45 years?

Appendix A Introduction to Graphing Calculators

Introduction

You previously studied the point-plotting method for sketching the graph of an equation. One of the disadvantages of the point-plotting method is that to get a good idea about the shape of a graph, you need to plot *many* points. By plotting only a few points, you can badly misrepresent the graph.

For instance, consider the equation $y = x^3$. To graph this equation, suppose you calculated only the following three points.

x	-1	0	1
$y = x^3$	-1	0	1
Solution point	$(-1, -1)$	$(0, 0)$	$(1, 1)$

By plotting these three points, as shown in Figure A.1, you might assume that the graph of the equation is a line. This, however, is not correct. By plotting several more points, as shown in Figure A.2, you can see that the actual graph is not straight at all.

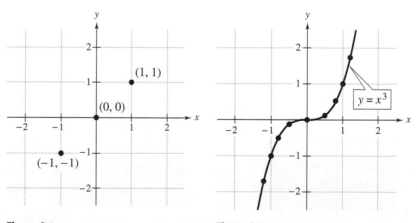

Figure A.1   Figure A.2

So, the point-plotting method leaves you with a dilemma. On the one hand, the method can be very inaccurate if only a few points are plotted. But, on the other hand, it is very time-consuming to plot a dozen (or more) points. Technology can help you solve this dilemma. Plotting several points (or even hundreds of points) on a rectangular coordinate system is something that a computer or graphing calculator can do easily.

Using a Graphing Calculator

There are many different graphing utilities: some are graphing programs for computers and some are hand-held graphing calculators. In this appendix, the steps used to graph an equation with a *TI-84* or *TI-84 Plus* graphing calculator are described. Keystroke sequences are often given for illustration; however, these sequences may not agree precisely with the steps required by *your* calculator.*

Graphing an Equation with a *TI-84* or *TI-84 Plus* Graphing Calculator

Before performing the following steps, set your calculator so that all of the standard defaults are active. For instance, all of the options at the left of the (MODE) screen should be highlighted.

1. Set the viewing window for the graph. (See Example 3.) To set the standard viewing window, press (ZOOM) 6.

2. Rewrite the equation so that y is isolated on the left side of the equation.

3. Press the (Y=) key. Then enter the right side of the equation on the first line of the display. (The first line is labeled $Y_1 = .$)

4. Press the (GRAPH) key.

EXAMPLE 1   Graphing a Linear Equation

Use a graphing calculator to graph $2y + x = 4$.

Solution

To begin, solve the equation for y in terms of x.

$$2y + x = 4 \qquad \text{Write original equation.}$$

$$2y = -x + 4 \qquad \text{Subtract } x \text{ from each side.}$$

$$y = -\frac{1}{2}x + 2 \qquad \text{Divide each side by 2.}$$

Press the (Y=) key, and enter the following keystrokes.

(−) (X,T,θ,n) (÷) 2 (+) 2

The top row of the display should now be as follows.

$$Y_1 = \text{-X/2} + 2$$

Press the (GRAPH) key, and the screen should look like the one in Figure A.3.

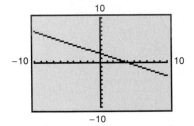

Figure A.3

---

*The graphing calculator keystrokes given in this appendix correspond to the *TI-84* and *TI-84 Plus* graphing calculators by Texas Instruments. For other graphing calculators, the keystrokes may differ. Consult your user's guide.

In Figure A.3, notice that the calculator screen does not label the tick marks on the x-axis or the y-axis. To see what the tick marks represent, you can press (WINDOW). If you set your calculator to the standard graphing defaults before working Example 1, the screen should show the following values.

Xmin $= -10$	The minimum x-value is -10.
Xmax $= 10$	The maximum x-value is 10.
Xscl $= 1$	The x-scale is 1 unit per tick mark.
Ymin $= -10$	The minimum y-value is -10.
Ymax $= 10$	The maximum y-value is 10.
Yscl $= 1$	The y-scale is 1 unit per tick mark.
Xres $= 1$	Sets the pixel resolution

These settings are summarized visually in Figure A.4.

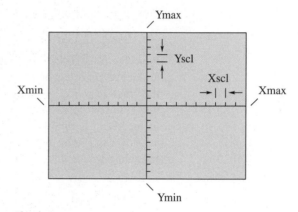

Figure A.4

EXAMPLE 2   **Graphing an Equation Involving Absolute Value**

Use a graphing calculator to graph

$$y = |x - 3|.$$

Solution

This equation is already written so that y is isolated on the left side of the equation. Press the (Y=) key, and enter the following keystrokes.

(MATH) (▶) 1 (X,T,θ,n) (−) 3 ())

The top row of the display should now be as follows.

$$Y_1 = \text{abs}(X - 3)$$

Press the (GRAPH) key, and the screen should look like the one shown in Figure A.5.

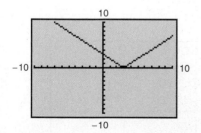

Figure A.5

Using Special Features of a Graphing Calculator

To use your graphing calculator to its best advantage, you must learn to set the viewing window, as illustrated in the next example.

EXAMPLE 3   Setting the Viewing Window

Use a graphing calculator to graph

$$y = x^2 + 12.$$

Solution

Press (Y=) and enter $x^2 + 12$ on the first line.

(X,T,θ,n) (x²) (+) 12

Press the (GRAPH) key. If your calculator is set to the standard viewing window, nothing will appear on the screen. The reason for this is that the lowest point on the graph of $y = x^2 + 12$ occurs at the point (0, 12). Using the standard viewing window, you obtain a screen whose largest y-value is 10. In other words, none of the graph is visible on a screen whose y-values vary between -10 and 10, as shown in Figure A.6. To change these settings, press (WINDOW) and enter the following values.

Xmin = -10	The minimum x-value is -10.
Xmax = 10	The maximum x-value is 10.
Xscl = 1	The x-scale is 1 unit per tick mark.
Ymin = -10	The minimum y-value is -10.
Ymax = 30	The maximum y-value is 30.
Yscl = 5	The y-scale is 5 units per tick mark.
Xres = 1	Sets the pixel resolution

Press (GRAPH) and you will obtain the graph shown in Figure A.7. On this graph, note that each tick mark on the y-axis represents five units because you changed the y-scale to 5. Also note that the highest point on the y-axis is now 30 because you changed the maximum value of y to 30.

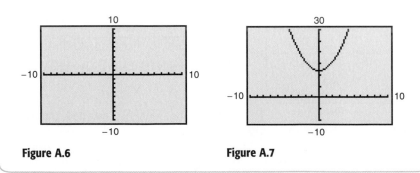

Figure A.6   Figure A.7

If you changed the y-maximum and y-scale on your calculator as indicated in Example 3, you should return to the standard setting before working Example 4. To do this, press (ZOOM) 6.

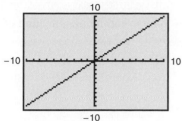

Figure A.8

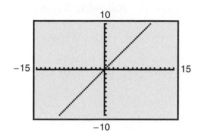

Figure A.9

EXAMPLE 4   **Using a Square Setting**

Use a graphing calculator to graph $y = x$. The graph of this equation is a line that makes a 45° angle with the x-axis and with the y-axis. From the graph on your calculator, does the angle appear to be 45°?

Solution

Press $\boxed{Y=}$ and enter x on the first line.

$$Y_1 = X$$

Press the $\boxed{GRAPH}$ key and you will obtain the graph shown in Figure A.8. Notice that the angle the line makes with the x-axis doesn't appear to be 45°. The reason for this is that the screen is wider than it is tall. This makes the tick marks on the x-axis farther apart than the tick marks on the y-axis. To obtain the same distance between tick marks on both axes, you can change the graphing settings from "standard" to "square." To do this, press the following keys.

$\boxed{ZOOM}$ 5   Square setting

The screen should look like the one shown in Figure A.9. Note in this figure that the square setting has changed the viewing window so that the x-values vary from -15 to 15.

There are many possible square settings on a graphing calculator. To create a square setting, you need the following ratio to be $\frac{2}{3}$.

$$\frac{\text{Ymax} - \text{Ymin}}{\text{Xmax} - \text{Xmin}}$$

For instance, the setting in Example 4 is square because

$$\frac{\text{Ymax} - \text{Ymin}}{\text{Xmax} - \text{Xmin}} = \frac{10 - (-10)}{15 - (-15)} = \frac{20}{30} = \frac{2}{3}.$$

EXAMPLE 5   **Graphing More than One Equation**

Use a graphing calculator to graph each equation in the same viewing window.

$$y = -x + 4, \quad y = -x, \quad \text{and} \quad y = -x - 4$$

Solution

To begin, press $\boxed{Y=}$ and enter all three equations on the first three lines. The display should now be as follows.

$Y_1 = -X + 4$ $\boxed{(-)}$ $\boxed{X,T,\theta,n}$ $\boxed{+}$ 4

$Y_2 = -X$ $\boxed{(-)}$ $\boxed{X,T,\theta,n}$

$Y_3 = -X - 4$ $\boxed{(-)}$ $\boxed{X,T,\theta,n}$ $\boxed{-}$ 4

Press the $\boxed{GRAPH}$ key and you will obtain the graph shown in Figure A.10. Note that the graph of each equation is a line and that the lines are parallel to each other.

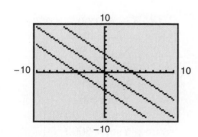

Figure A.10

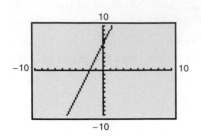

Figure A.11

EXAMPLE 6   Using the Trace Feature

Approximate the x- and y-intercepts of $y = 3x + 6$ by using the *trace* feature of a graphing calculator.

Solution

Press $\boxed{Y=}$ and enter $3x + 6$ on the first line.

$$3 \; \boxed{X,T,\theta,n} \; \boxed{+} \; 6$$

Press the $\boxed{GRAPH}$ key and you will obtain the graph shown in Figure A.11. Then press the $\boxed{TRACE}$ key and use the $\boxed{\blacktriangleleft}$ $\boxed{\blacktriangleright}$ keys to move along the graph. To get a better approximation of a solution point, you can use the following keystrokes repeatedly.

$$\boxed{ZOOM} \; 2 \; \boxed{ENTER}$$

As you can see in Figures A.12 and A.13, the x-intercept is $(-2, 0)$ and the y-intercept is $(0, 6)$.

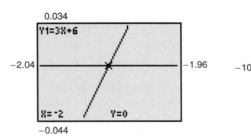

Figure A.12   **Figure A.13**

Appendix A Exercises

📊 In Exercises 1–12, use a graphing calculator to graph the equation. (Use a standard setting.) *See Examples 1 and 2.*

1. $y = -3x$
2. $y = x - 4$
3. $y = \frac{3}{4}x - 6$
4. $y = -3x + 2$
5. $y = \frac{1}{2}x^2$
6. $y = -\frac{2}{3}x^2$
7. $y = x^2 - 4x + 2$
8. $y = -0.5x^2 - 2x + 2$
9. $y = |x - 5|$
10. $y = |x + 4|$
11. $y = |x^2 - 4|$
12. $y = |x - 2| - 5$

📊 In Exercises 13–16, use a graphing calculator to graph the equation using the given window settings. *See Example 3.*

13. $y = 27x + 100$

| Xmin = 0 |
| Xmax = 5 |
| Xscl = .5 |
| Ymin = 75 |
| Ymax = 250 |
| Yscl = 25 |
| Xres = 1 |

14. $y = 50{,}000 - 6000x$

| Xmin = 0 |
| Xmax = 7 |
| Xscl = .5 |
| Ymin = 0 |
| Ymax = 50000 |
| Yscl = 5000 |
| Xres = 1 |

15. $y = 0.001x^2 + 0.5x$ **16.** $y = 100 - 0.5|x|$

Xmin = -500
Xmax = 200
Xscl = 50
Ymin = -100
Ymax = 100
Yscl = 20
Xres = 1

Xmin = -300
Xmax = 300
Xscl = 60
Ymin = -100
Ymax = 100
Yscl = 20
Xres = 1

In Exercises 17–20, find a viewing window that shows the important characteristics of the graph.

17. $y = 15 + |x - 12|$ **18.** $y = 15 + (x - 12)^2$
19. $y = -15 + |x + 12|$ **20.** $y = -15 + (x + 12)^2$

Geometry   In Exercises 21–24, use a graphing calculator to graph the equations in the same viewing window. Using a "square setting," determine the geometrical shape bounded by the graphs. *See Example 4.*

21. $y = -4, \quad y = -|x|$
22. $y = |x|, \quad y = 5$
23. $y = |x| - 8, \quad y = -|x| + 8$
24. $y = -\frac{1}{2}x + 7, \quad y = \frac{8}{3}(x + 5), \quad y = \frac{2}{7}(3x - 4)$

In Exercises 25–28, use a graphing calculator to graph both equations in the same viewing window. Are the graphs identical? If so, what basic rule of algebra is being illustrated? *See Example 5.*

25. $y_1 = 2x + (x + 1)$
   $y_2 = (2x + x) + 1$

26. $y_1 = \frac{1}{2}(3 - 2x)$
   $y_2 = \frac{3}{2} - x$

27. $y_1 = 2\left(\frac{1}{2}\right)$
   $y_2 = 1$

28. $y_1 = x(0.5x)$
   $y_2 = (0.5x)x$

In Exercises 29–36, use the *trace* feature of a graphing calculator to approximate the *x*- and *y*-intercepts of the graph. *See Example 6.*

29. $y = 9 - x^2$
30. $y = 3x^2 - 2x - 5$
31. $y = 6 - |x + 2|$
32. $y = (x - 2)^2 - 3$
33. $y = 2x - 5$
34. $y = 4 - |x|$
35. $y = x^2 + 1.5x - 1$
36. $y = x^3 - 4x$

Modeling Data   In Exercises 37 and 38, use the following models, which give the number of pieces of first-class mail and the number of pieces of standard mail handled by the U.S. Postal Service.

First Class

$y = 0.5x^2 - 5.06x + 110.3, \quad 2 \le x \le 6$

Standard

$y = -0.221x^2 + 5.88x + 75.8, \quad 2 \le x \le 6$

In these models, *y* is the number of pieces handled (in billions) and *x* is the year, with $x = 2$ corresponding to 2002.   (Source: U.S. Postal Service)

37. Use the following window settings to graph both models in the same viewing window on a graphing calculator.

Xmin = 2
Xmax = 6
Xscl = 1
Ymin = 50
Ymax = 150
Yscl = 25
Xres = 1

38. (a) Were the numbers of pieces of first-class mail and standard mail increasing or decreasing over time?

   (b) In what year were the numbers of pieces of first-class mail and standard mail the same?

   (c) After the year in part (b), was more first-class mail or more standard mail handled?

Answers to Odd-Numbered Exercises, Quizzes, and Tests

CHAPTER 1

Section 1.1   (page 9)

1. (a) $\{1, 2, 6\}$ (b) $\{-6, 0, 1, 2, 6\}$

(c) $\left\{-6, -\frac{4}{3}, 0, \frac{5}{8}, 1, 2, 6\right\}$ (d) $\left\{-\sqrt{6}, \sqrt{2}, \pi\right\}$

3. (a) $\left\{\sqrt{4}\right\}$ (b) $\left\{\sqrt{4}, 0\right\}$

(c) $\left\{-4.2, \sqrt{4}, -\frac{1}{9}, 0, \frac{3}{11}, 5.\overline{5}, 5.543\right\}$ (d) $\left\{\sqrt{11}\right\}$

5. $0.\overline{2}$ **7.** $2.\overline{12}$

9. $-5, -4, -3, -2, -1, 0, 1, 2, 3$ **11.** $5, 7, 9$

13. (a)   (b)

(c)   (d)

15. $-1 < \frac{1}{2}$ **17.** $-\frac{9}{2} < -2$ **19.** $<$ **21.** $<$

23. $<$ **25.** $>$ **27.** $>$ **29.** 6 **31.** 19

33. 50 **35.** 8 **37.** 35 **39.** 3 **41.** 10

43. 225 **45.** -85 **47.** -16 **49.** $-\frac{3}{4}$

51. -3.5 **53.** π **55.** $>$ **57.** $>$

59. $=$ **61.** $>$ **63.** $-34, 34$ **65.** $160, 160$

67. $\frac{3}{11}, \frac{3}{11}$ **69.** $-\frac{5}{4}, \frac{5}{4}$ **71.** $-4.7, 4.7$

73.   **75.**

7   5

77.   **79.**

$\frac{3}{5}$ $\frac{5}{3}$

81.

4.25

83. $x < 0$ **85.** $u \geq 16$ **87.** $16 \leq s \leq 28$

89. $p < 225$ **91.** $-4, 4$ **93.** $-2, 8$

95. True. If a number can be written as the ratio of two integers, it is rational. If not, the number is irrational.

97. 0.15 is a terminating rational number and $0.\overline{15}$ is a repeating rational number.

Section 1.2   (page 19)

1. 45 **3.** 4 **5.** -2.7 **7.** 7 **9.** -25.9

11. -20 **13.** -15.2 **15.** 24 **17.** -3.8

19. 22 **21.** $\frac{5}{4}$ **23.** $\frac{1}{2}$ **25.** $\frac{1}{10}$ **27.** $\frac{1}{24}$

29. $\frac{63}{8}$ **31.** $\frac{35}{8}$ **33.** 60 **35.** 45.95 **37.** $-\frac{121}{8}$

39. $4 \cdot 9$ **41.** $6\left(\frac{1}{4}\right)$ **43.** $4\left(-\frac{1}{5}\right)$ **45.** -30

47. 48 **49.** -40 **51.** 36 **53.** $\frac{1}{2}$ **55.** $-\frac{12}{5}$

57. $\frac{1}{12}$ **59.** $-\frac{1}{3}$ **61.** $-\frac{1}{2}$ **63.** $\frac{1}{6}$ **65.** $\frac{3}{2}$

67. $-\frac{7}{9}$ **69.** 6 **71.** -3 **73.** -9 **75.** $-\frac{5}{2}$

77. $\frac{2}{5}$ **79.** $\frac{34}{45}$ **81.** $\frac{11}{12}$ **83.** $(-7)^3$ **85.** $\left(\frac{1}{4}\right)^4$

87. -7^3 **89.** 32 **91.** 16 **93.** -64 **95.** $\frac{64}{125}$

97. $\frac{1}{32}$ **99.** 0.027 **101.** -0.32 **103.** 0

105. 4 **107.** 22 **109.** 6 **111.** 12

113. 27 **115.** 135 **117.** -6 **119.** -6

121. 1 **123.** 30 **125.** 161 **127.** $14,425$

129. 171.36 **131.** $\frac{17}{180}$ **133.** $\$2533.56$

135. (a)

Day	Daily gain or loss
Tuesday	$5
Wednesday	$8
Thursday	$-5
Friday	$16

(b) The stock gained $24 in value during the week. Find the difference between the first bar (Monday) and the last bar (Friday).

137. (a) $\$10,800$ (b) $\$15,832.22$ (c) $\$5032.22$

139. 15 square meters   **141.** 20 square inches

143. 6.125 cubic feet

145. True. A nonzero rational number is of the form $\frac{a}{b}$, where a and b are integers and $a \neq 0$, $b \neq 0$. The reciprocal will be $\frac{b}{a}$, which is also rational.

147. True. When a negative number is raised to an even power, the result is positive.

149. False. $6 \div 3 = 2 \neq \frac{1}{2} = 3 \div 6$

151. If the numbers have like signs, the product or quotient is positive. If the numbers have unlike signs, the product or quotient is negative.

153. (a) $40 - (10 + 3) = 27$

(b) $5^2 + \frac{1}{2} \cdot 4 = 27$

(c) $(8 \cdot 3 + 30) \div 2 = 27$

(d) $75 \div (2 + 1) + 2 = 27$

155. Only common factors (not terms) of the numerator and denominator can be divided out.

Section 1.3   *(page 28)*

1. Additive Inverse Property

3. Multiplicative Inverse Property

5. Commutative Property of Addition

7. Associative Property of Addition

9. Distributive Property

11. Associative Property of Multiplication

13. Multiplicative Inverse Property

15. Additive Inverse Property   **17.** Distributive Property

19. Additive Inverse Property

21. $-3(15)$ **23.** $5 \cdot 6 + 5 \cdot z$

25. $-x + 25$ **27.** $x + 8$

29. (a) -10 (b) $\frac{1}{10}$ **31.** (a) 19 (b) $-\frac{1}{19}$

33. (a) $-\frac{1}{2}$ (b) 2 **35.** (a) $\frac{5}{8}$ (b) $-\frac{8}{5}$

37. (a) $-6z$ (b) $\dfrac{1}{6z}$

39. (a) $-(x - 2)$ or $-x + 2$ (b) $\dfrac{1}{x - 2}$

41. $(32 + 4) + y$ **43.** $(9 \cdot 6)m$

45. $20 \cdot 2 + 20 \cdot 5$ **47.** $x(-2) + 6(-2)$ or $-2x - 12$

49. $-6(2y) + (-6)(-5)$ or $-12y + 30$

51. $(7 + 2)x = 9x$ **53.** $\dfrac{x}{8}(7 - 5) = \dfrac{x}{4}$

55. $3x + 15$ **57.** $-2x - 16$

59. Answers will vary.   **61.** Answers will vary.

63. $x + 5 = 3$

Original equation

$(x + 5) + (-5) = 3 + (-5)$

Addition Property of Equality

$x + [5 + (-5)] = -2$

Associative Property of Addition

$x + 0 = -2$

Additive Inverse Property

$x = -2$

Additive Identity Property

65. $2x - 5 = 6$

Original equation

$(2x - 5) + 5 = 6 + 5$

Addition Property of Equality

$2x + (-5 + 5) = 11$

Associative Property of Addition

$2x + 0 = 11$

Additive Inverse Property

$2x = 11$

Additive Identity Property

$\frac{1}{2}(2x) = \frac{1}{2}(11)$

Multiplication Property of Equality

$\left(\frac{1}{2} \cdot 2\right)x = \frac{11}{2}$

Associative Property of Multiplication

$1 \cdot x = \frac{11}{2}$

Multiplicative Inverse Property

$x = \frac{11}{2}$

Multiplicative Identity Property

67. 28   **69.** 434   **71.** 62.82

73. $a(b + c) = ab + ac$

75. $(4) + (x + 5) + (3x + 2)$; $4x + 11$

77. (a) $2(x + 6) + 2(2x)$; $6x + 12$

(b) $(x + 6)(2x)$; $2x^2 + 12x$

79. The additive inverse of a real number is its opposite. The sum of a number and its additive inverse is the additive identity zero. For example, $-3.2 + 3.2 = 0$.

81. Given two real numbers a and b, the sum a plus b is the same as the sum b plus a.

83. Sample answer: $4 \odot 7 = 15 \neq 18 = 7 \odot 4$

$3 \odot (4 \odot 7) = 21 \neq 27 = (3 \odot 4) \odot 7$

Mid-Chapter Quiz   *(page 31)*

1.

$-4.5 > -6$

2.

$\frac{3}{4} < \frac{3}{2}$

3. 22   **4.** 6.5   **5.** 7.6   **6.** -9.8 **7.** 14

8. 5   **9.** $\frac{5}{2}$ **10.** $\frac{1}{2}$ **11.** 60   **12.** $-\frac{3}{8}$

13. $\frac{7}{10}$ **14.** $-\frac{27}{8}$ **15.** 4   **16.** 2

17. (a) Distributive Property

(b) Additive Inverse Property

18. (a) Associative Property of Addition

(b) Multiplicative Identity Property

19. $1068.20   **20.** $8640

21. $\frac{7}{24}$; The sum of the parts of the circle graph is equal to 1.

Section 1.4   *(page 37)*

1. $10x, 5$; $10, 5$ **3.** $-6x^2, 12$; $-6, 12$

5. $-3y^2, 2y, -8$; $-3, 2, -8$ **7.** $-4a^3, 1.2a$; $-4, 1.2$

9. $4x^2, -3y^2, -5x, 21$; $4, -3, -5, 21$

11. $-5x^2y, 2y^2, xy$; $-5, 2, 1$ **13.** $\frac{1}{4}x^2, -\frac{3}{8}x, 5$; $\frac{1}{4}, -\frac{3}{8}, 5$

15. Commutative Property of Addition

17. Associative Property of Multiplication

19. Distributive Property

21. $5x + 30$ or $5x + 5 \cdot 6$

23. $(x + 6)5$ **25.** $7x$ **27.** $2x^2$ **29.** $-4x$

31. $8y$ **33.** $8x + 18y$ **35.** $6x^2 - 2x$

37. $-2z^4 - 4z^2 + 5z + 8$ **39.** $-x^2 + 3xy + y$

41. $8x^2 + 4x - 12$ **43.** $-18y^2 + 3y + 6$

45. $-3x^2 + 2x - 4$ **47.** $5x^2 + 2x$

49. $-12x^2 + 51x$ **51.** $10t^2 - 35t$ **53.** $12x - 35$

55. $-4x - 9$ **57.** $a + 3$ **59.** $-7y - 7$

61. $y^3 + 6$ **63.** $x^3 - 12$ **65.** $44a - 22$

67. $-6x + 96$ **69.** $12x^2 + 2x$ **71.** $-2b^2 + 4b - 36$

73. (a) 3   (b) -10 **75.** (a) 6   (b) 9

77. (a) 7   (b) 11

79. (a) Not possible; undefined   (b) $\frac{28}{9}$

81. (a) 13   (b) -36 **83.** (a) 7   (b) 7

85. (a) Not possible; undefined   (b) $\frac{1}{2}$

87. (a) 3   (b) 0   **89.** (a) 210   (b) 140

91. 252 ft^3 **93.** 3888 in.^3 **95.** $1.21

97. $7.23   **99.** $\frac{1}{2}b^2 - \frac{3}{2}b$; 90

101. $23,500 million; $23,775 million

103. $22 billion; $22.3 billion   **105.** 1440 square feet

107. No. When $y = 3$, the expression is undefined.

109. To remove a set of parentheses preceded by a minus sign, distribute -1 to each term inside the parentheses. For example: $13 - (-10 + 5) = 13 + 10 - 5 = 18$.

111. A factor can consist of a sum of terms; the term x is part of the sum $x + y$, which is a factor of $(x + y) \cdot z$.

113. No. There are an infinite number of values of x and y that would satisfy $8y - 5x = 14$. For example, $x = 10$ and $y = 8$ would be a solution, and so would $x = 2$ and $y = 3$.

Section 1.5   *(page 47)*

1. $23 + n$ **3.** $12 + 2n$ **5.** $n - 6$ **7.** $4n - 10$

9. $\frac{1}{2}n$ **11.** $\frac{x}{6}$ **13.** $8 \cdot \frac{N}{5}$ **15.** $4c + 10$

17. $0.30L$ **19.** $\frac{n + 5}{10}$ **21.** $|n - 8|$ **23.** $3x^2 - 4$

25. A number decreased by 2

27. A number increased by 50

29. Two decreased by three times a number

31. The ratio of a number and 2

33. Four-fifths of a number

35. Eight times the difference of a number and 5

37. The sum of a number and 10, divided by 3

39. The square of a number, decreased by 3

41. $0.25n$ **43.** $0.10m$ **45.** $5m + 10n$

47. $55t$ **49.** $\frac{320}{r}$ **51.** $0.45y$

53. $0.0125I$ **55.** $L - 0.20L = 0.80L$

57. $8.25 + 0.60q$ **59.** $n + 5n = 6n$

61. $(2n + 1) + (2n + 3) + (2n + 5) = 6n + 9$

63. $\dfrac{2n(2n + 2)}{4} = n^2 + n$ **65.** s^2

67. $\frac{1}{2}b(0.75b) = 0.375b^2$

69. Perimeter: $2(2w) + 2w = 6w$; Area: $2w(w) = 2w^2$

71. Perimeter: $6 + 2x + 3 + x + 3 + x = 4x + 12$

Area: $3x + 6x = 9x$ or $6(2x) - 3(x) = 9x$

73. $b(b - 50) = b^2 - 50b$; square meters

75.

n	0	1	2	3	4	5
$5n - 3$	-3	2	7	12	17	22
Differences		5	5	5	5	5

The differences are constant.

77. a **79.** a and c

81. Using a specific case may make it easier to see the form of the expression for the general case.

Review Exercises   *(page 52)*

1. (a) $\left\{52, \sqrt{9}\right\}$ (b) $\left\{-4, 0, 52, \sqrt{9}\right\}$

(c) $\left\{\frac{3}{5}, -4, 0, 52, -\frac{1}{8}, \sqrt{9}\right\}$ (d) $\left\{\sqrt{2}\right\}$

3. $\{1, 2, 3, 4, 5, 6\}$

5. (a)

(b)

(c)

(d)

7. $<$ **9.** $<$ **11.** 14   **13.** 7.3   **15.** 5

17. -7.2 **19.** 11   **21.** 230   **23.** -41.8

25. $\frac{11}{21}$ **27.** $\frac{1}{6}$ **29.** $\frac{17}{8}$ **31.** -28

33. -4200 **35.** $-\frac{1}{20}$ **37.** 14   **39.** 2

41. $(-3) + (-3) + (-3) + (-3) + (-3) + (-3) + (-3)$

43. $8(8)$ **45.** 6^7 **47.** 1296   **49.** -16 **51.** $\frac{1}{8}$

53. 20   **55.** 98   **57.** 1,165,469.01   **59.** $800

61. Additive Inverse Property   **63.** Distributive Property

65. Associative Property of Addition

67. Commutative Property of Multiplication

69. Distributive Property   **71.** $u - 3v$ **73.** $-8a + 3a^2$

75. $4y^3, -y^2, \frac{17}{2}y; 4, -1, \frac{17}{2}$

77. $-1.2x^3, \frac{1}{x}, 52; -1.2, 1, 52$

79. $9x$ **81.** $5v$ **83.** $5x - 10$ **85.** $5x - y$

87. $-15a + 18b$ **89.** (a) 0 (b) -3

91. (a) 19 (b) -8 **93.** $12 - 2n$ **95.** $y^2 + 49$

97. The sum of two times a number and 7

99. The difference of a number and 5, divided by 4

101. $0.18l$ **103.** $l(l - 5) = l^2 - 5l$

Chapter Test   *(page 55)*

1. (a) $<$ (b) $>$ **2.** 11.3   **3.** -20

4. $-\frac{1}{2}$ **5.** -150 **6.** 60   **7.** $\frac{1}{6}$

8. $\frac{4}{27}$ **9.** $-\frac{27}{125}$ **10.** 15

11. (a) Associative Property of Multiplication

 (b) Multiplicative Inverse Property

12. $-12x + 6$ **13.** $-2x^2 + 5x - 1$ **14.** $-x^2 + 26$

15. a^2 **16.** $11t + 7$

17. Evaluating an expression is solving the expression when values are provided for its variables.
 (a) 23 (b) 7

18. 6 inches   **19.** 640 cubic feet   **20.** $5n - 8$

21. $2n + (2n + 2) = 4n + 2$

22. Perimeter: $2l + 2(0.6l) = 3.2l$, 144;

 Area: $l(0.6l) = 0.6l^2$, 1215

CHAPTER 2

Section 2.1   *(page 65)*

1. (a) Not a solution   (b) Solution

3. (a) Solution   (b) Not a solution

5. (a) Not a solution   (b) Solution

7. No solution   **9.** Conditional

11. Original equation

 Subtract 15 from each side.

 Combine like terms.

 Divide each side by 3.

 Simplify.

13. Equivalent   **15.** Not equivalent

17. Equivalent   **19.** Not equivalent

21. 3   **23.** 4   **25.** $-\frac{2}{3}$ **27.** $-\frac{10}{3}$

29. No solution, because $-3 \neq 0$.

31. No solution, because $-4 \neq 0$.   **33.** $\frac{1}{3}$

35. 15   **37.** 11   **39.** $-\frac{9}{2}$

41. Infinitely many, because $-12 = -12$.

43. -3 **45.** Infinitely many, because $42 = 42$.

47. $\frac{19}{10}$ **49.** $-\frac{10}{3}$ **51.** $-\frac{20}{9}$ **53.** 23

55. 12   **57.** $\frac{1}{5}$ **59.** 125, 126   **61.** 2.5 hours

63. 6 hours   **65.** 1.5 seconds

67. (a)

t	1	1.5	2
Width	250	200	166.7
Length	250	300	333.4
Area	62,500	60,000	55,577.8

t	3	4	5
Width	125	100	83.3
Length	375	400	416.5
Area	46,875	40,000	34,694.5

 (b) Because the length is t times the width and the perimeter is fixed, as t increases, the length increases and the width and area decrease. The maximum area occurs when the length and width are equal.

69. 2001

Graphically: Find the year that has a bar corresponding to the value.

Numerically: Plug in different values of t until you arrive at the correct value of y.

Algebraically: Set y equal to the value and solve for t.

71. False. This does not follow the Multiplication Property of Equality.

73. No. To write an identity in standard form, $ax + b = 0$, the values of a and b must both be zero. Because the equation is no longer in standard form, the identity cannot be written as a linear equation in standard form.

75. Conditional equation; Sample answer: 20% tip on a restaurant bill

77. Identity; Sample answer: value of 1 roll of quarters plus x additional quarters

79. $\frac{6}{5}$ **81.** -2 **83.** 22; 29

85. 15; 8   **87.** $n - 8$ **89.** $2(n + 3)$

Section 2.2   *(page 75)*

1. (A number) $+ 24 = 68$

Equation: $x + 24 = 68$; Solution: 44

3. 26(biweekly pay) $+$ (bonus) $= 37{,}120$

Equation: $26x + 2800 = 37{,}120$; Solution: $1320

Percent	Parts out of 100	Decimal	Fraction
5. 30%	30	0.30	$\frac{3}{10}$
7. 7.5%	7.5	0.075	$\frac{3}{40}$
9. $66\frac{2}{3}\%$	$66\frac{2}{3}$	$0.\overline{6}$	$\frac{2}{3}$
11. 100%	100	1.00	1

13. 87.5   **15.** 346.8   **17.** 128   **19.** 600

21. 80   **23.** 350   **25.** 35   **27.** 2750

29. 62%   **31.** 0.5%   **33.** 175%   **35.** $\frac{2}{3}$

37. $\frac{3}{4}$ **39.** $\frac{1}{25}$ **41.** $\frac{10}{3}$ **43.** 4   **45.** $\frac{216}{7}$

47. $\frac{15}{2}$ **49.** 6   **51.** $\frac{20}{3}$ **53.** 8072 students

55. 2 students   **57.** 15.625%   **59.** $4.89

61. About 21%   **63.** About 9.8%   **65.** 7%

67. 200 parts   **69.** 177.78%, 56.25%

71. Houston: 39.83%; San Antonio: 24.08%; Dallas: 22.90%; Austin: 13.19%

73. $\approx$ \$37 per 100 pounds; 33%   **75.** 33%

77. $\frac{1}{50}$ **79.** $\frac{85}{4}$ **81.** $\frac{4}{9}$ **83.** $0.06

85. $0.11   **87.** $14\frac{1}{2}$-ounce bag

89. six-ounce tube   **91.** $5.\overline{09}$ **93.** 3

95. $\frac{516}{11} \approx 46.9$ feet   **97.** 17.1 gallons   **99.** \$2400

101. 2667 units   **103.** 6 units   **105.** 46,400 votes

107. To convert a percent to a decimal, divide by 100. For example: $24\% = \frac{24}{100} = 0.24$.

To convert a decimal to a percent, multiply by 100. For example: $0.18 = 0.18(100)\% = 18\%$.

109. Mathematical modeling is the use of mathematics to solve problems that occur in real-life situations. For examples, refer to the real-life problems in the exercise set.

111. $-\frac{1}{4}$ **113.** -27

115. Commutative Property of Addition

117. Distributive Property

119. 14   **121.** 0   **123.** -200

Section 2.3   *(page 88)*

	Cost	Selling Price	Markup	Markup Rate
1.	\$45.97	\$64.33	\$18.36	40%
3.	\$152.00	\$250.80	\$98.80	65%
5.	\$22,250.00	\$26,922.50	\$4672.50	21%
7.	\$225.00	\$416.70	\$191.70	85.2%

	List Price	Sale Price	Discount	Discount Rate
9.	\$49.95	\$25.74	\$24.21	48.5%
11.	\$300.00	\$111.00	\$189.00	63%
13.	\$45.00	\$27.00	\$18.00	40%
15.	\$1155.50	\$831.96	\$323.54	28%

17. \$26.19   **19.** 80%   **21.** \$30   **23.** 20%

25. 9 minutes, \$2.06   **27.** \$54.15   **29.** 2.5 hours

31. \$37

33. 50 gallons of solution 1; 50 gallons of solution 2

35. 8 quarts of solution 1; 16 quarts of solution 2

37. 75 pounds of type 1; 25 pounds of type 2

39. 100 children's tickets   **41.** $\frac{5}{6}$ gallon

	Distance, d	Rate, r	Time, t
43.	2275 mi	650 mi/hr	$3\frac{1}{2}$ hr
45.	1000 km	110 km/hr	$\frac{100}{11} \approx 9.1$ hr
47.	385 mi	55 mi/hr	7 hr

49. 25 miles   **51.** $1\frac{1}{4}$ hours   **53.** 6 miles per hour

55. 1440 miles   **57.** 17.14 minutes

59. 3 hours at 58 miles per hour; $2\frac{3}{4}$ hours at 52 miles per hour

61. (a) $\frac{1}{5}, \frac{1}{8}$ (b) $\frac{40}{13} = 3\frac{1}{13}$ hours   **63.** 10 minutes

65. $R = \dfrac{E}{I}$　**67.** $L = \dfrac{S}{1-r}$　**69.** $a = \dfrac{2h - 96t}{t^2}$

71. $a = \dfrac{2h - 72t - 100}{t^2}$　**73.** $h = \dfrac{S}{2\pi r} - r$

75. $147\pi \approx 461.8$ cubic centimeters　**77.** 0.926 foot

79. 43 centimeters　**81.** 30 degrees Celsius

83. \$1950　**85.** \$3571.43

87. (a) 2002, yes　(b) \$0.26; Explanations will vary.

89. The sale price is the list price times the quantity 1 minus the discount rate.

91. No, it quadruples. The area of a square with side s is s^2. If the length of the side is $2s$, the area is $(2s)^2 = 4s^2$.

93. (a) -21　(b) $\dfrac{1}{21}$　**95.** (a) $5x$　(b) $-\dfrac{1}{5x}$

97. $2x^2 - 8x + 3$　**99.** $x^3 - 6x^2$

101. 130　**103.** 13%

Mid-Chapter Quiz　*(page 93)*

1. 2　**2.** 2　**3.** 2　**4.** Identity　**5.** $\frac{28}{5}$

6. $\frac{10}{3}$　**7.** $\frac{33}{2}$　**8.** $-\frac{29}{144}$　**9.** 0.55

10. 1.41　**11.** $\frac{9}{20}$, 45%　**12.** 200

13. \$0.40 per ounce　**14.** 5000 defective units

15. The computer from the store　**16.** 7 hours

17. 40 gallons of 25% solution; 10 gallons of 50% solution

18. 1.5 hours at 62 miles per hour; 4.5 hours at 46 miles per hour

19. $\frac{15}{8} = 1.875$ hours　**20.** 169 square inches

Section 2.4　*(page 103)*

1. (a) Yes　(b) No　(c) Yes　(d) No

3. (a) No　(b) No　(c) Yes　(d) No

5. a　**6.** e　**7.** d　**8.** b　**9.** f　**10.** c

11.
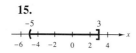

13.

15.

17.

19.

21.

23.
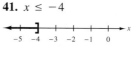

25. $-15 + x < -24$

27. Not equivalent　**29.** Not equivalent

31. Equivalent　**33.** Not equivalent

35. $x \geq 4$
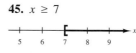

37. $x \leq 2$

39. $x < 4$
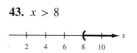

41. $x \leq -4$

43. $x > 8$

45. $x \geq 7$
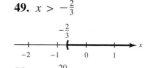

47. $x > 7.55$
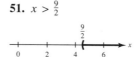

49. $x > -\frac{2}{3}$

51. $x > \frac{9}{2}$
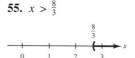

53. $x > \frac{20}{11}$
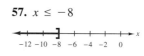

55. $x > \frac{8}{3}$

57. $x \leq -8$

59. $x > -15$
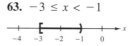

61. $\frac{5}{2} < x < 7$

63. $-3 \leq x < -1$
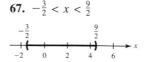

65. $-5 < x < 5$
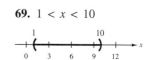

67. $-\frac{3}{2} < x < \frac{9}{2}$

69. $1 < x < 10$

71. $-1 < x \leq 4$

73. No solution

75. $-\infty < x < \infty$

77. $x < -\frac{8}{3}$ or $x \geq \frac{5}{2}$

79. $y \le -10$

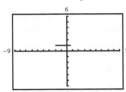

81. $-5 < x \le 0$

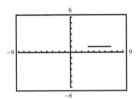

83. $x < -3$ or $x \ge 2$, $\{x|x < -3\} \cup \{x|x \ge 2\}$

85. $-5 \le x < 4$, $\{x|x \ge -5\} \cap \{x|x < 4\}$

87. $x \le -2.5$ or $x \ge -0.5$, $\{x|x \le -2.5\} \cup \{x|x \ge -0.5\}$

89. $\{x|x \ge -7\} \cap \{x|x < 0\}$

91. $\left\{x|x > -\frac{9}{2}\right\} \cap \left\{x|x \le -\frac{3}{2}\right\}$

93. $\left\{x|x < 0\right\} \cup \left\{x|x \ge \frac{2}{3}\right\}$

95. $x \ge 0$ **97.** $z \ge 8$ **99.** $10 \le n \le 16$

101. x is at least $\frac{5}{2}$.   **103.** y is at least 3 and less than 5.

105. \$2600

107. The average temperature in Miami is greater than the average temperature in New York.

109. 26,000 miles   **111.** $x \ge 31$

113. $23 \le x \le 38$ **115.** $3 \le n \le \frac{15}{2}$

117. $12.50 < 8 + 0.75n$; $n > 6$ **119.** 1999, 2000, 2001

121. The multiplication and division properties differ. The inequality symbol is reversed if both sides of an inequality are multiplied or divided by a negative real number.

123. The solution set of a linear inequality is bounded if the solution is written as a double inequality. Otherwise, the solution set is unbounded.

125. $a < x < b$; A double inequality is always bounded.

127. $x > a$ or $x < b$; $x > a$ includes all real numbers between a and b and greater than or equal to b, all the way to ∞. $x < b$ adds all real numbers less than or equal to a, all the way to $-\infty$.

129. $<$ **131.** $=$ **133.** No; Yes

135. Yes; No   **137.** $\frac{17}{2}$ **139.** $-\frac{1}{4}$

Section 2.5   *(page 113)*

1. Not a solution   **3.** Solution

5. $x - 10 = 17$; $x - 10 = -17$

7. $4x + 1 = \frac{1}{2}$; $4x + 1 = -\frac{1}{2}$ **9.** $|3x| = 1$

11. $|2x| = 2$ **13.** $4, -4$ **15.** No solution

17. 0   **19.** $3, -3$ **21.** $4, -6$ **23.** $11, -14$

25. $\frac{4}{3}$ **27.** $\frac{17}{5}, -\frac{11}{5}$ **29.** 2   **31.** $\frac{15}{4}, -\frac{1}{4}$

33. $2, 3$ **35.** $7, -3$ **37.** $\frac{1}{3}$ **39.** $\frac{1}{2}$

41. $|x - 4| = 9$ **43.** Solution

45. Not a solution   **47.** $-3 < y + 5 < 3$

49. $7 - 2h \ge 9$ or $7 - 2h \le -9$ **51.** $-4 < y < 4$

53. $x \le -6$ or $x \ge 6$ **55.** $-7 < x < 7$

57. $-1 \le y \le 1$ **59.** $x < -16$ or $x > 4$

61. $-3 \le x \le 4$ **63.** No solution

65. $-104 < y < 136$ **67.** $-5 < x < 35$

69. $-\infty < x < \infty$

71. $-2 < x < \frac{2}{3}$ **73.** $x < -6$ or $x > 3$

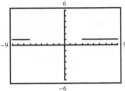

75. $3 \le x \le 7$

77. d   **78.** c   **79.** b   **80.** a   **81.** $|x| \le 2$

83. $|x - 19| > 2$ **85.** $|x| < 3$ **87.** $|2x - 3| > 5$

89.

Fastest: 41.826 seconds; Slowest: 42.65 seconds

91. (a) $|s - x| \le \frac{3}{16}$ (b) $4\frac{15}{16} \le x \le 5\frac{5}{16}$

93. All real numbers that are within one unit of 4

95. $|2x - 6| \le 6$ **97.** $4(n + 3)$

99. Selling price: \$112, Markup: \$32

101. $x > 20$ **103.** $x \ge 4$

Review Exercises   *(page 118)*

1. (a) Not a solution   (b) Solution

3. (a) Not a solution   (b) Solution

5. -7 **7.** 24   **9.** 5   **11.** -12 **13.** -24

15. 3   **17.** 2   **19.** 4   **21.** 4   **23.** 14

25. No solution   **27.** 2   **29.** -4.2 **31.** 57 and 58

33. (Total pay) = (Weekly wage)(Number of weeks) + (Training session pay)

Equation: $2635 = 320x + 75$; Solution: 8 weeks

| | Parts out | | |
Percent	of 100	Decimal	Fraction
35. 68%	68	0.68	$\frac{68}{100}$
37. 60%	60	0.6	$\frac{3}{5}$

39. 65   **41.** 3000   **43.** 125%   **45.** 6%

47. 375 parts   **49.** 8-ounce can   **51.** $\frac{1}{40}$ **53.** $\frac{7}{2}$

55. $\frac{10}{3}$ **57.** 3   **59.** $2000   **61.** 50 feet

	Cost	Selling Price	Markup	Markup Rate
63.	$99.95	$149.93	$49.98	50%

	List Price	Sale Price	Discount	Discount Rate
65.	$71.95	$53.96	$17.99	25%

67. Sales tax: $167.70; Total bill: $2962.70;
Amount financed: $2162.70

69. $3\frac{1}{3}$ liters of 30% solution; $6\frac{2}{3}$ liters of 60% solution

71. 3500 miles   **73.** 43.6 miles per hour

75. $\frac{18}{7} \approx 2.57$ hours   **77.** $340   **79.** $52,631.58

81. $30,000   **83.** 20 feet $\times$ 12 feet

85. 50.9 degrees Fahrenheit

87.
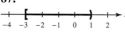

89.

91. $x \le 4$ **93.** $x > 4$

95. $x > 3$ **97.** $x \le 6$

99. $y > -\frac{70}{3}$ **101.** $x \le -3$
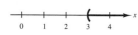

103. $-7 \le x < -2$ **105.** $-16 < x < -1$

107. $-3 < x < 2$
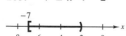

109. At least $800   **111.** $6, -6$ **113.** $4, -\frac{4}{3}$

115. $0, -\frac{8}{5}$ **117.** $\frac{1}{2}, 3$ **119.** $x < 1$ or $x > 7$

121. $-4 < x < 4$ **123.** $-4 < x < 11$

125. $b < -9$ or $b > 5$

127. $x \le 1$ or $x \ge 5$

129. $|x - 3| < 2$

131.
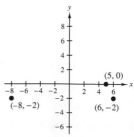

Maximum: 116.6 degrees Fahrenheit

Minimum: 40 degrees Fahrenheit

Chapter Test   *(page 123)*

1. 4   **2.** 3   **3.** 4   **4.** 24   **5.** 4000

6. 0.4%   **7.** $11,125

8. 15-ounce package; It has a lower unit price.

9. $2175   **10.** $2\frac{1}{2}$ hours

11. 15 pounds at $2.60 per pound; 25 pounds at $3.80 per pound

12. 40 minutes   **13.** $2000

14. (a) $5, -1$ (b) $\frac{2}{3}, -\frac{4}{3}$ (c) No solution

15. (a) $x \ge -6$ (b) $x > 2$
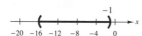

(c) $-7 < x \le 1$ (d) $-1 \le x < \frac{5}{4}$
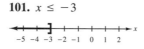

16. $t \ge 8$

17. (a) $1 \le x \le 5$ (b) $x < -\frac{9}{5}$ or $x > 3$

(c) $-8.8 < x < -7.2$

18. 25,000 miles

CHAPTER 3

Section 3.1   *(page 135)*

1.
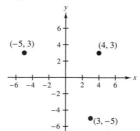

3.

5.

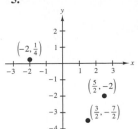

7.

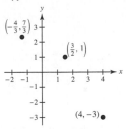

9. A: $(-2, 4)$, B: $(0, -2)$, C: $(4, -2)$

11. A: $(4, -2)$, B: $\left(-3, -\frac{5}{2}\right)$, C: $(3, 0)$

13.

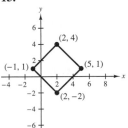

15.

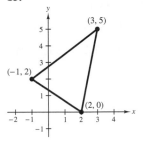

17.

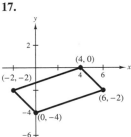

19.

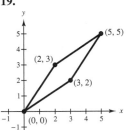

21. $(1, 4)$ **23.** $(10, -4)$ **25.** $(10, 0)$

27. $(-8, -8)$ **29.** Quadrant III   **31.** Quadrant II

33. Quadrant III   **35.** Quadrant IV

37. Quadrant I or II   **39.** Quadrant II or III

41. Quadrant I or III

43.

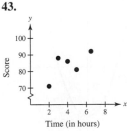

45.

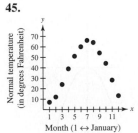

47. $(-3, -4) \Rightarrow (-1, 1)$
  $(1, -3) \Rightarrow (3, 2)$
  $(-2, -1) \Rightarrow (0, 4)$

49.

x	-2	0	2	4	6
$y = 5x + 3$	-7	3	13	23	33

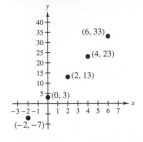

51.

x	-4	0	3	5	10
$y = \lvert 2x - 7 \rvert + 2$	17	9	3	5	15

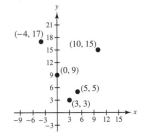

53.

x	-2	0	2	4	6
$y = x^2 + 2x + 5$	5	5	13	29	53

55. (a) Not a solution   (b) Solution
  (c) Solution   (d) Not a solution

57. (a) Solution   (b) Solution
  (c) Not a solution   (d) Not a solution

59. (a) Not a solution   (b) Solution
  (c) Not a solution   (d) Solution

61. 7; Vertical line   **63.** 7; Horizontal line

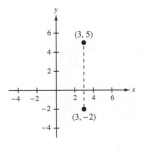

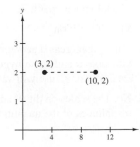

65. $\frac{3}{4}$; Vertical line   **67.** $\frac{13}{2}$; Horizontal line

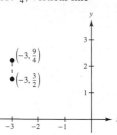

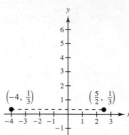

69. 5   **71.** $\sqrt{5}$ **73.** $\sqrt{58}$ **75.** $\sqrt{61}$ **77.** $\sqrt{17}$

79. $\left(\sqrt{13}\right)^2 + \left(\sqrt{13}\right)^2 = \left(\sqrt{26}\right)^2$

81. $\left(\sqrt{8}\right)^2 + \left(\sqrt{72}\right)^2 = \left(\sqrt{80}\right)^2$

83. Not collinear   **85.** Collinear

87. $3 + \sqrt{26} + \sqrt{29} \approx 13.48$

89. $(1, 4)$ **91.** $\left(\frac{7}{2}, \frac{9}{2}\right)$

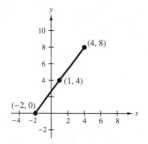

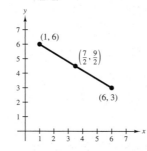

93.

x	1	2	3	4	5
$y = 150x + 425$	575	725	875	1025	1175

The cost of installation is \$425, plus \$150 for every window installed.

95. $\sqrt{1525} \approx 39.05$ yards

97. The faster the car travels, up to 60 kilometers per hour, the less gas it uses. As the speed increases past 60 kilometers per hour, the car uses progressively more gas; 10 liters (per 100 km traveled) is used when the car is traveling at 120 kilometers per hour.

99. \$13,797 million

101. The Pythagorean Theorem states that, for a right triangle with sides a and b and hypotenuse c, $a^2 + b^2 = c^2$. Examples of its use will vary.

103. No. The scales on the x- and y-axes are determined by the magnitudes of the quantities being measured by x and y.

105.

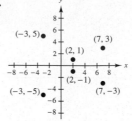

Reflection in the x-axis

107. 30   **109.** 35   **111.** 6

	List Price	Sale Price	Discount	Discount Rate
113.	\$55.00	\$35.75	\$19.25	35%
115.	\$258.50	\$134.42	\$124.08	48%

Section 3.2   *(page 145)*

1. e   **2.** b   **3.** f   **4.** a   **5.** d   **6.** c

7.   **9.**

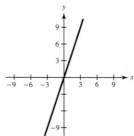

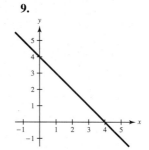

11.   **13.**

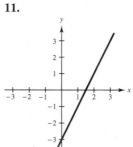

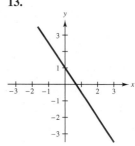

15.   **17.**

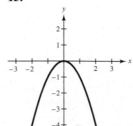

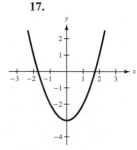

19.

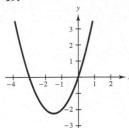

21.

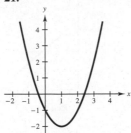

55. $(-2, 0), (-1, 0), (0, 4)$

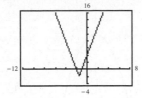

23.

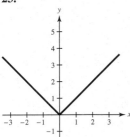

25.

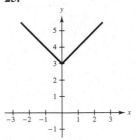

57.

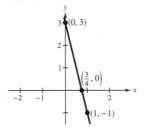

59.

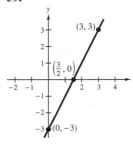

27.

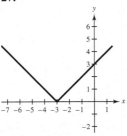

29.

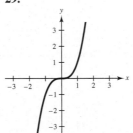

61.

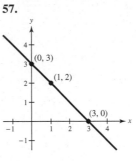

63.

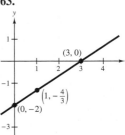

65.

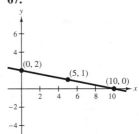

67.

31. $\left(\frac{1}{2}, 0\right), (0, -3)$ **33.** $(30, 0), (0, 12)$

35. $(10, 0), (0, 5)$ **37.** $\left(-\frac{3}{4}, 0\right), (0, 3)$

39. $(1, 0), (-1, 0), (0, -1)$ **41.** $(-5, 0), (0, -5)$

43. $(-2, 0), (4, 0), (0, -2)$ **45.** $(3, 0), (0, 2)$

47. $(0, 3)$ **49.** $(0, -2)$

51. $(2, 0), (0, 8)$ **53.** $(1, 0), (6, 0), (0, 6)$

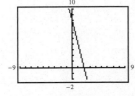

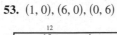

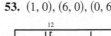

69.

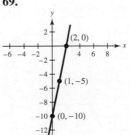

71.

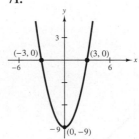

73.

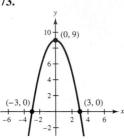

75.

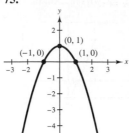

93.

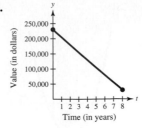

95. (a) $y = 40{,}000 - 5000t, \ 0 \le t \le 7$

(b)

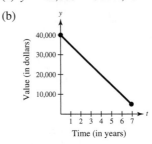

(c) $(0, 40{,}000)$; This represents the value of the delivery van when purchased.

77.

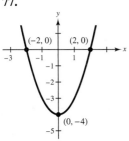

79.

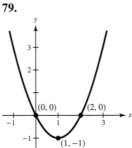

97. (a)

x	0	3	6	9	12
F	0	4	8	12	16

(b)

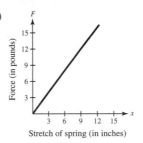

(c) F doubles because F is directly proportional to x.

81.

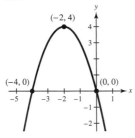

83.

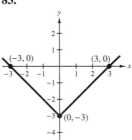

99. The scales on the y-axes are different. From graph (a) it appears that sales have not increased. From graph (b) it appears that sales have increased dramatically.

85.

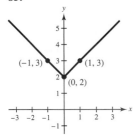

87.

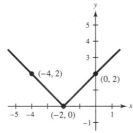

101.

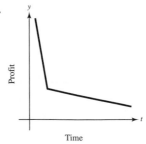

89.

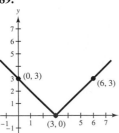

91.

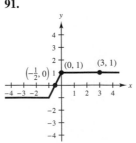

103. A horizontal line has no x-intercept. The y-intercept of a horizontal line is $(0, b)$, where b is any real number. So, a horizontal line has one y-intercept.

105. $\frac{1}{7}$ **107.** $\frac{5}{4}$ **109.** 8 **111.** 2

113.

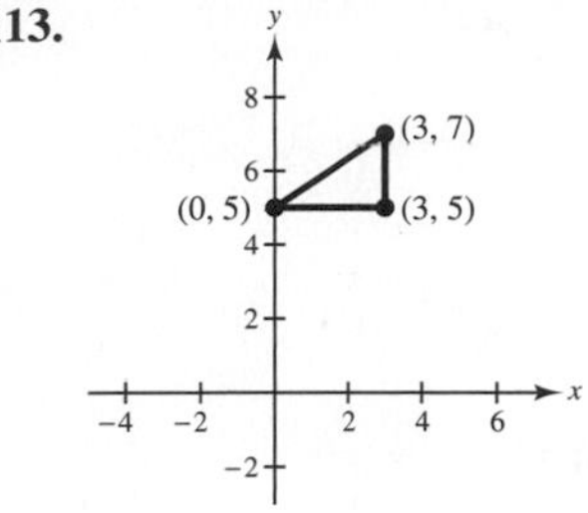

Section 3.3 *(page 158)*

1. -2 **3.** $\frac{2}{5}$ **5.** Undefined

7. (a) L_3 (b) L_2 (c) L_1

9. $m = 2$

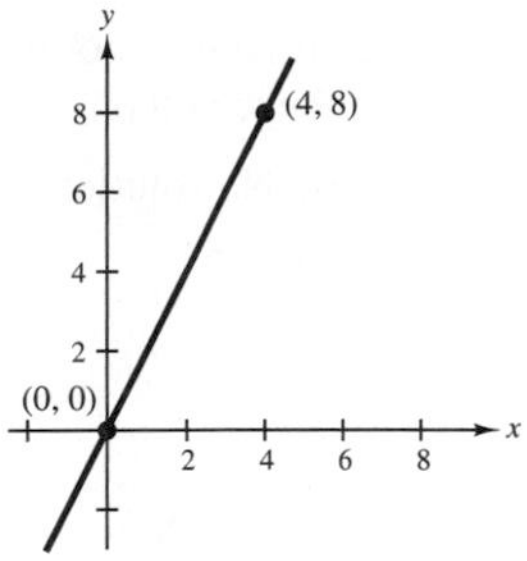

11. $m = 0$

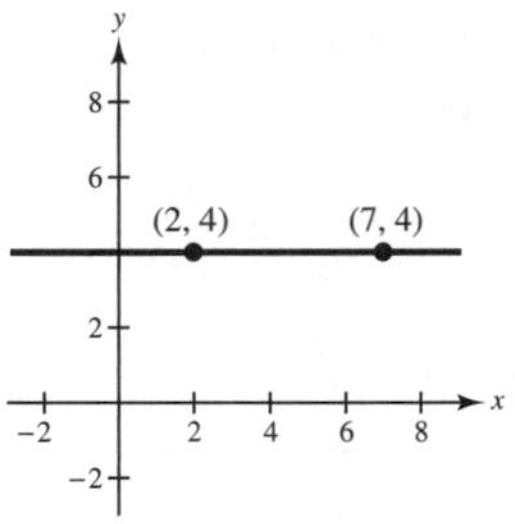

13. $m = 1$; rises

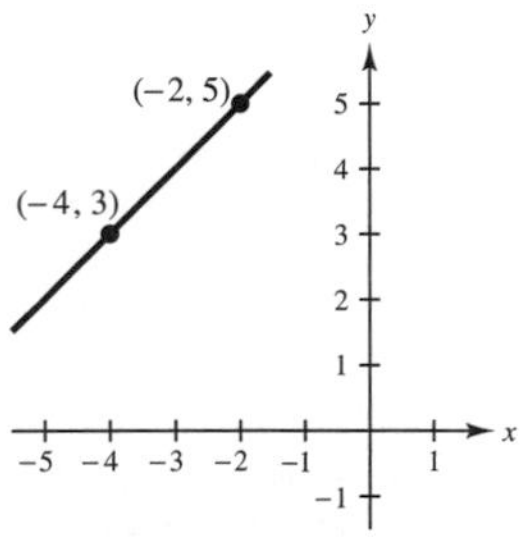

15. m is undefined; vertical

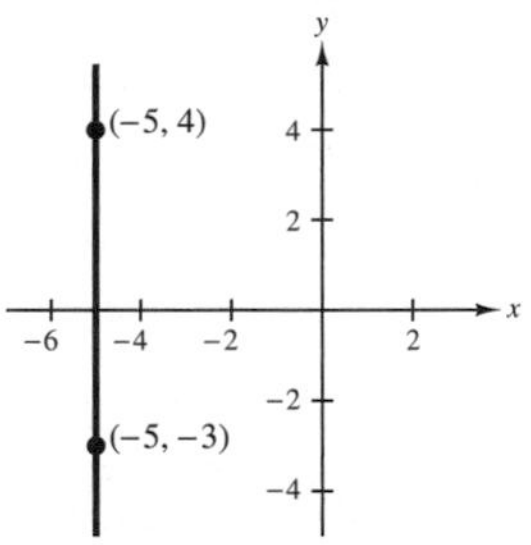

17. $m = 0$; horizontal

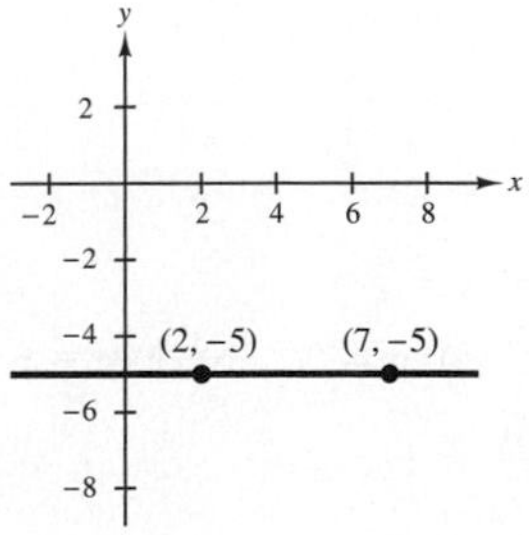

19. $m = -\frac{18}{17}$; falls

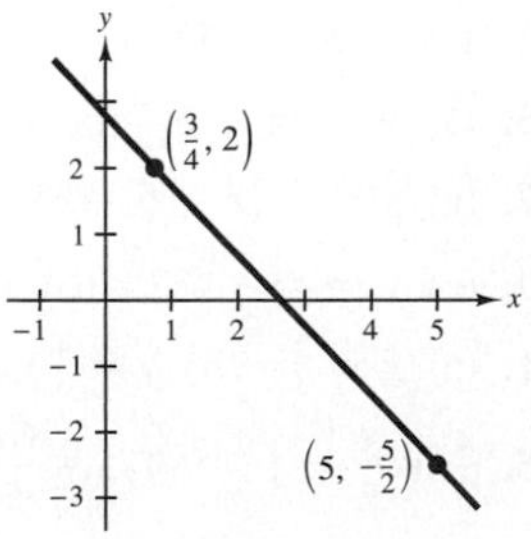

21. $m = \frac{1}{18}$; rises

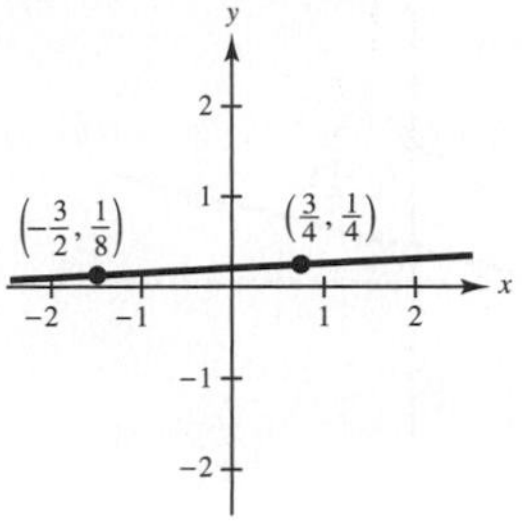

23. $m = -\frac{5}{6}$; falls

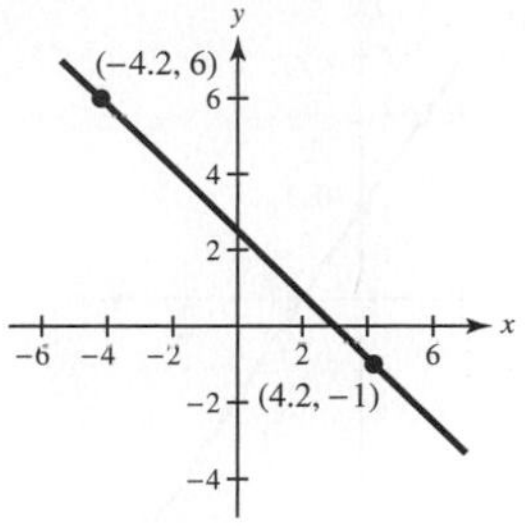

25. $m = 2$

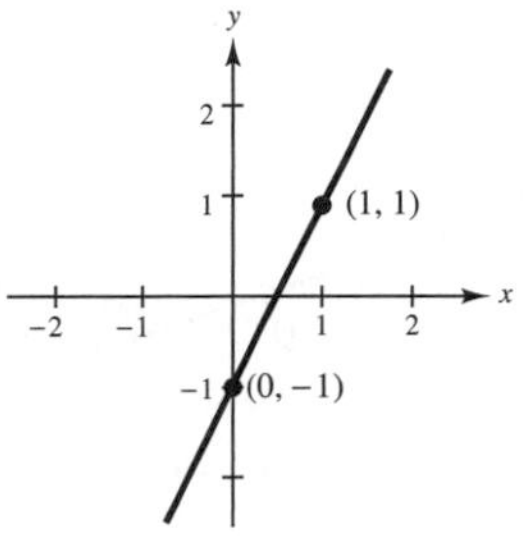

27. $m = -\frac{1}{2}$

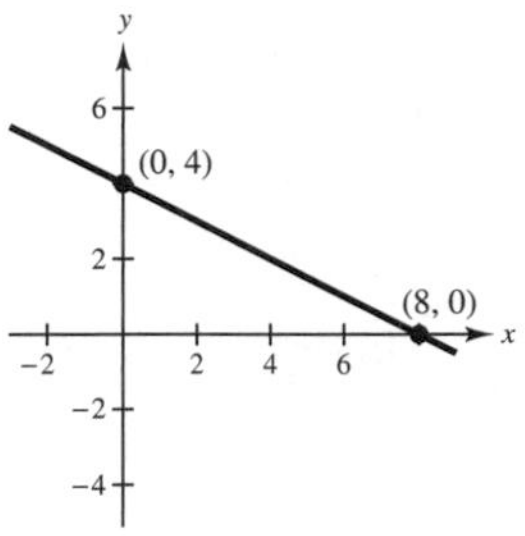

29. $m = \frac{1}{3}$

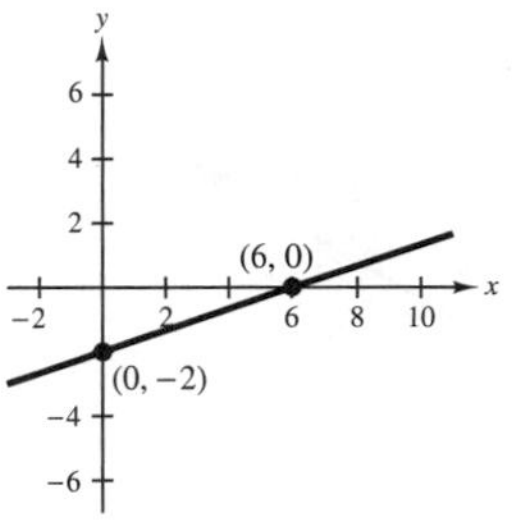

31. $x = 1$ **33.** $y = -15$ **35.** $(6, 2), (10, 2)$

37. $(4, -1), (5, 2)$ **39.** $(-3, -2), (-1, -4)$

41. $(-2, 4), (1, 8)$ **43.** $y = 2x - 3$ **45.** $y = \frac{1}{4}x - 1$

47. $y = -\frac{2}{5}x + \frac{3}{5}$ **49.** $y = \frac{1}{2}x + 2$

51. $m = 3$; $(0, -2)$ **53.** $m = \frac{2}{3}$; $(0, -4)$

55. $m = -\frac{5}{3}$; $\left(0, \frac{2}{3}\right)$

57. $y = -x$

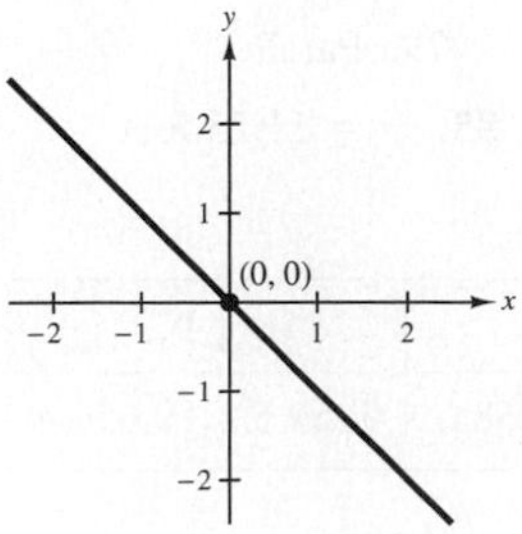

59. $y = 3x - 2$

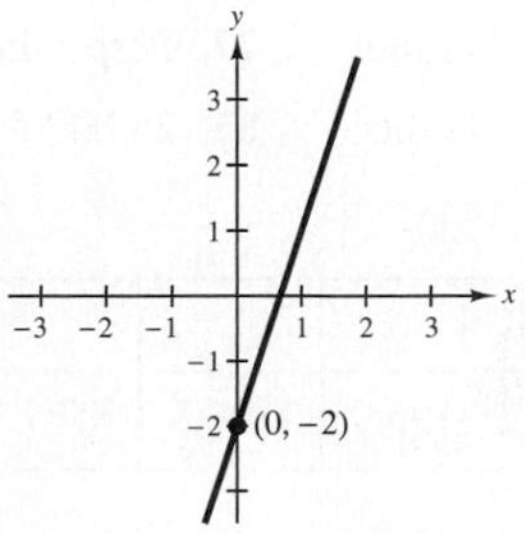

61. $y = -\frac{3}{2}x + 1$

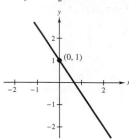

63. $y = \frac{1}{4}x + \frac{1}{2}$

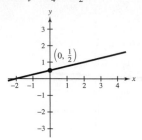

(b)

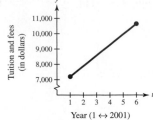

(c) $690.90   (d) $16,868.50

89. Yes. When different pairs of points are selected, the change in y and the change in x are the lengths of the sides of similar triangles. Corresponding sides of similar triangles are proportional.

91. The x-coordinate of the x-intercept is the same as the solution of the equation when $y = 0$.

93. No. Two lines are perpendicular if their slopes, not their y-intercepts, are negative reciprocals of each other.

95. $m \geq -3$ **97.** $16 \leq n < 20$ **99.** No solution

101. $-10, 10$ **103.** $7, -4$ **105.** $-12, -2$

65. $y = 0.25x - 5$

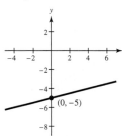

67.

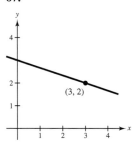

Section 3.4   (page 168)

1. b   **2.** d   **3.** a   **4.** c   **5.** $y = 3x - 9$

7. $y = -\frac{1}{2}x - \frac{1}{2}$ **9.** $y = \frac{4}{5}x - \frac{8}{5}$ **11.** $2; (3, 1)$

13. $-5; (8, -1)$ **15.** $\frac{1}{2}; (-6, -3)$ **17.** $-8; (0, 0)$

19. $y = -\frac{1}{2}x$ **21.** $y = 3x - 4$ **23.** $y = -\frac{3}{4}x + 6$

25. $y = -2x + 4$ **27.** $y = \frac{5}{4}x - 2$

29. $y = -4x - \frac{9}{2}$ **31.** $y = \frac{4}{3}x + \frac{3}{2}$ **33.** $y = -1$

35. $y = -\frac{1}{5}x + 3$ **37.** $y = -\frac{20}{3}x - \frac{2}{3}$

39. $4x + 3y = 0$ **41.** $x + y - 4 = 0$

43. $x - 2y + 7 = 0$ **45.** $2x + 5y = 0$

47. $2x - 6y + 15 = 0$ **49.** $5x + 34y - 67 = 0$

51. $52x + 15y - 395 = 0$ **53.** $4x + 5y - 11 = 0$

55. $y = 3x + 4$ **57.** $y = 3$ **59.** $x = -1$

61. $y = -5$ **63.** $x = -7$

69.

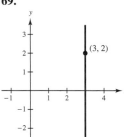

71.

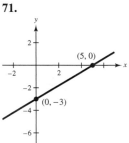

65. (a) $y = 3x - 5$ (b) $y = -\frac{1}{3}x + \frac{5}{3}$

67. (a) $y = -\frac{5}{4}x - \frac{9}{4}$ (b) $y = \frac{4}{5}x + 8$

69. (a) $y = 4x - 23$ (b) $y = -\frac{1}{4}x - \frac{7}{4}$

71. (a) $x = \frac{2}{3}$ (b) $y = \frac{4}{3}$ **73.** (a) $y = 2$ (b) $x = -1$

73.

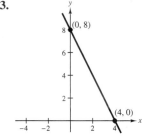

75. $\frac{x}{3} + \frac{y}{2} = 1$ **77.** $-\frac{6x}{5} - \frac{3y}{7} = 1$

75. Parallel   **77.** Perpendicular   **79.** Parallel

81. Neither   **83.** 25,000 feet   **85.** $\frac{45}{4} = 11.25$ feet

87. (a)

t	1	2	3	4	5	6
y	7195.9	7886.8	8577.7	9268.6	9959.5	10,650.4

79. $C = 20x + 5000$; $13,000

81. $S = 100,000t$; $600,000

83. $S = 0.03M + 1500$; 3%

85. (a) $S = 0.70L$

(b)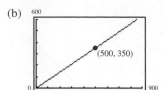

(c) $300

87. (a) $V = 7400 - 1475t$ (b) $4450

89. (a) $N = 60t + 1200$ (b) 2700 students

(c) 1800 students

91. (a) and (b)

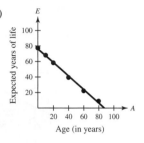

(c) Answers will vary.

Sample answer: $E = -0.87A + 76.3$

(d) 50 years

93. $x - 8y = 0$

Distance from the deep end	0	8	16	24	32	40
Depth of water	9	8	7	6	5	4

95. Point-slope form: $y - y_1 = m(x - x_1)$

Slope-intercept form: $y = mx + b$

General form: $ax + by + c = 0$

97. Any point on a vertical line is independent of y.

99.

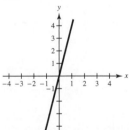

101.

103. 2   **105.** 2   **107.** 3   **109.** -8

Mid-Chapter Quiz   *(page 173)*

1. Quadrant I or II

2. (a) Not a solution   (b) Solution

(c) Solution   (d) Solution

3. Distance: 5

Midpoint: $\left(1, \frac{7}{2}\right)$

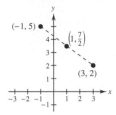

4. Distance: $10\sqrt{2}$

Midpoint: $(1, -2)$

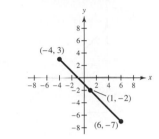

5.

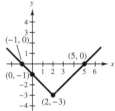

6.

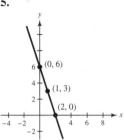

7.

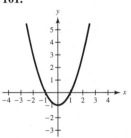

8. $m = 0$; Horizontal   **9.** $m = \frac{5}{3}$; Rises

10. $m = -\frac{4}{3}$; Falls

11. $y = -\frac{1}{2}x + 1$

$m = -\frac{1}{2}$

$(0, 1)$

12. $y = \frac{3}{2}x - 3$

$m = \frac{3}{2}$

$(0, -3)$

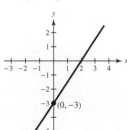

13. Perpendicular   **14.** Parallel   **15.** $x - 2y - 8 = 0$

16. $V = 124{,}000 - 12{,}000t, \; 0 \le t \le 10$

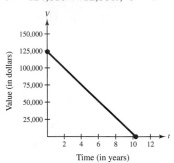

Section 3.5   *(page 179)*

1. (a) Solution   (b) Not a solution
    (c) Solution   (d) Solution
3. (a) Not a solution   (b) Not a solution
    (c) Solution   (d) Solution
5. (a) Solution   (b) Not a solution
    (c) Not a solution   (d) Solution
7. (a) Not a solution   (b) Solution
    (c) Not a solution   (d) Solution
9. b   **10.** a   **11.** d   **12.** e   **13.** f   **14.** c

15.

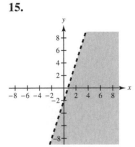

17.

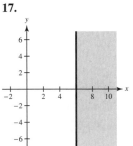

19.

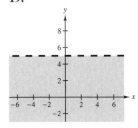

21.

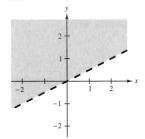

23.

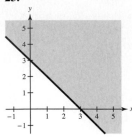

25.

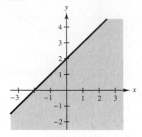

27.

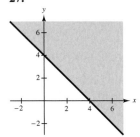

29.

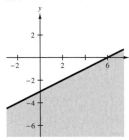

31.

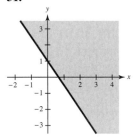

33.

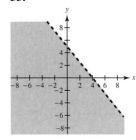

35.

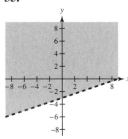

37.

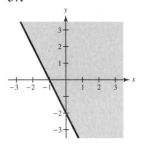

39.

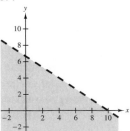

41.

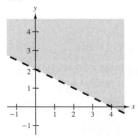

43.

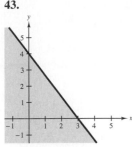

45.

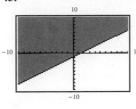

(b)

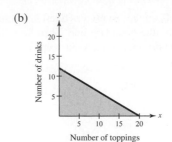

Number of drinks

Number of toppings

(c) $(6, 6)$; yes

65. (a) $12x + 16y \geq 250$; $x \geq 0$, $y \geq 0$

(b)

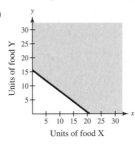

Units of food Y

Units of food X

Sample answer: (x, y): $(1, 15)$, $(10, 9)$, $(15, 5)$

47.

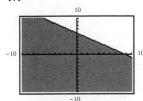

49.

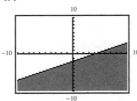

51.

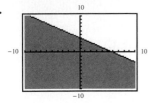

53. $3x + 4y > 17$ **55.** $y < 2$ **57.** $y \geq 2x$

59. (a) $2x + 2y \leq 500$ or $y \leq -x + 250$; $x \geq 0$, $y \geq 0$

(b)

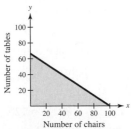

61. (a) $10x + 15y \leq 1000$ or $y \leq -\frac{2}{3}x + \frac{200}{3}$; $x \geq 0$, $y \geq 0$

(b)

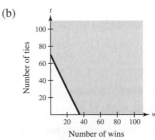

Number of tables

20 40 60 80 100
Number of chairs

63. (a) $0.60x + 1.00y \leq 12$; $x \geq 0$, $y \geq 0$

67. (a) $11x + 9y \geq 240$; $x \geq 0$, $y \geq 0$

(b)

Time mowing lawns (in hours)

Time at grocery store (in hours)

Sample answer: (x, y): $(2, 25)$, $(4, 22)$, $(10, 15)$

69. (a) $2w + t \geq 70$

(b)

Number of ties

20 40 60 80 100
Number of wins

Sample answer: (w, t): $(10, 50)$, $(20, 30)$, $(30, 10)$

71. The inequality is true when x_1 and y_1 are substituted for x and y, respectively.

73. To represent the other points in the plane, change the inequality symbol so that it is $\leq$ rather than $>$.

75. $(0, 0)$; If the point $(0, 0)$ cannot be used as a test point, the point $(0, 0)$ must lie on the boundary line. Therefore, the line passes through the origin, and the y-intercept is $(0, 0)$.

77. $x = 6, x = -6$ **79.** $2x + 3 = 9, 2x + 3 = -9$

81.

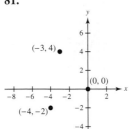

83.

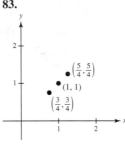

85. (a) $y = -2x + 6$ (b) $y = \frac{1}{2}x - \frac{3}{2}$

Section 3.6   *(page 191)*

1. Domain: $\{-2, 0, 1\}$ **3.** Domain: $\{0, 2, 4, 5, 6\}$
Range: $\{-1, 0, 1, 4\}$ Range: $\{-3, 0, 5, 8\}$

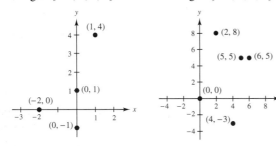

5. $(1, 1), (2, 8), (3, 27), (4, 64), (5, 125), (6, 216), (7, 343)$

7. (2004, Boston Red Sox), (2005, Chicago White Sox), (2006, St. Louis Cardinals), (2007, Boston Red Sox)

9. $(3, 9), (1, 3), (2, 6), (8, 24), (7, 21)$

11. Not a function   **13.** Function   **15.** Not a function

17. Not a function   **19.** Function   **21.** Not a function

23. (a) Function from A to B
(b) Not a function from A to B
(c) Function from A to B
(d) Not a function from A to B

25. Function   **27.** Not a function

29. There are two values of y associated with one value of x.

31. There are two values of y associated with one value of x.

33. There is one value of y associated with each value of x.

35. There is one value of y associated with each value of x.

37. There is one value of y associated with each value of x.

39. Not a function   **41.** Function

43. (a) $2; 11$ (b) $-2; -1$
(c) $k; 3k + 5$ (d) $k + 1; 3k + 8$

45. (a) $0; 3$ (b) $-3; -6$
(c) $m; 3 - m^2$ (d) $2t; 3 - 4t^2$

47. (a) $3, 3; \frac{3}{5}$ (b) $-4, -4; 2$
(c) $s, s; \dfrac{s}{s + 2}$ (d) $s - 2, s - 2; \dfrac{s - 2}{s}$

49. (a) 29 (b) 11
(c) $12a - 2$ (d) $12a + 5$

51. (a) 2 (b) 2
(c) $4y^2 - 8y + 2$ (d) 16

53. (a) 2 (b) 3
(c) $\sqrt{z}$ (d) $\sqrt{5z + 5}$

55. (a) 4 (b) 4
(c) -7 (d) $8 - |x - 6|$

57. (a) 0 (b) $-\dfrac{3}{2}$
(c) $-\dfrac{5}{2}$ (d) $\dfrac{3x + 12}{x - 1}$

59. (a) 2 (b) -2 (c) 10 (d) -8

61. (a) 0 (b) $\frac{7}{4}$ (c) 3 (d) 0

63. (a) 2 (b) $\dfrac{2x - 12}{x}$

65. All real numbers x

67. All real numbers t such that $t \neq 0, -2$

69. All real numbers x such that $x \geq -4$

71. All real numbers x such that $x \geq \frac{1}{2}$

73. All real numbers t

75. Domain: $\{0, 2, 4, 6\}$; Range: $\{0, 1, 8, 27\}$

77. Domain: $\{-3, -1, 4, 10\}$; Range: $\left\{-\frac{17}{2}, -\frac{5}{2}, 2\right\}$

79. Domain: All real numbers r such that $r > 0$
Range: All real numbers C such that $C > 0$

81. Domain: All real numbers r such that $r > 0$
Range: All real numbers A such that $A > 0$

83. $P = 4x$ **85.** $V = x^3$

87. $C(x) = 1.95x + 8000, \ x > 0$ **89.** $d = 120t$

91. $d = 65t$; 260 miles   **93.** $A = (32 - 2x)^2, \ x > 0$

95. $V = x(24 - 2x)^2, \ x > 0$ **97.** (a) \$700   (b) \$750

99. (a) \$360, \$480, \$570, \$660
(b) No. $h < 0$ is not in the domain of W.

101. Yes. For each year, there is only one value associated with the enrollment in public schools; 13 million

103. Correct

105. Yes. If a subset of A includes only input values that are mapped to only one output in the range, then the subset of A is a function.

107. Sample answer: Statement: The number of ceramic tiles required to floor a kitchen is a function of the area of the floor. Domain: The area of the floor; Range: The number of ceramic tiles. This is a function.

109. Multiplicative Identity Property

111. Multiplicative Inverse Property

113.

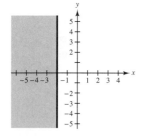

115.

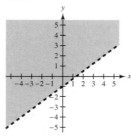

117.

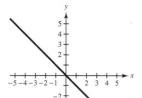

119.

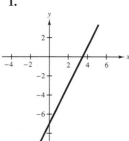

Section 3.7   *(page 203)*

1.

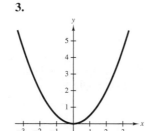

Domain: $-\infty < x < \infty$

Range: $-\infty < y < \infty$

3.

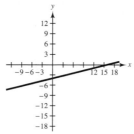

Domain: $-\infty < x < \infty$

Range: $0 \le y < \infty$

5.

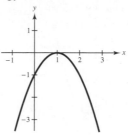

Domain: $-\infty < x < \infty$

Range: $-\infty < y \le 0$

7.

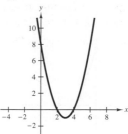

Domain: $-\infty < x < \infty$

Range: $-1 \le y < \infty$

9.

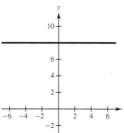

Domain: $0 \le x < \infty$

Range: $2 \le y < \infty$

11.

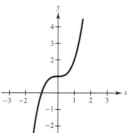

Domain: $2 \le t < \infty$

Range: $0 \le y < \infty$

13.

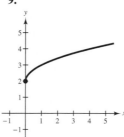

Domain: $-\infty < x < \infty$

Range: $y = 8$

15.

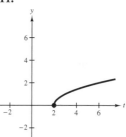

Domain: $-\infty < s < \infty$

Range: $-\infty < y < \infty$

17.

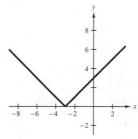

Domain: $-\infty < x < \infty$

Range: $0 \le y < \infty$

19.

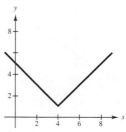

Domain: $-\infty < s < \infty$

Range: $1 \le y < \infty$

21.

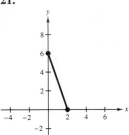

Domain: $0 \le x \le 2$

Range: $0 \le y \le 6$

23.

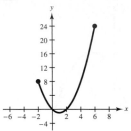

Domain: $-2 \le x \le 6$

Range: $-1 \le y \le 24$

25.

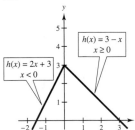

$h(x) = 3 - x$
$x \ge 0$

$h(x) = 2x + 3$
$x < 0$

Domain: $-\infty < x < \infty$

Range: $-\infty < y \le 3$

27.

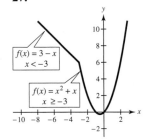

$f(x) = 3 - x$
$x < -3$

$f(x) = x^2 + x$
$x \ge -3$

Domain: $-\infty < x < \infty$

Range: $-\frac{1}{4} \le y < \infty$

29.

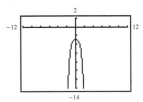

Domain: $-\infty < x < \infty$

Range: $-\infty < y \le -3$

31.

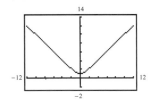

Domain: $-\infty < x < \infty$

Range: $1 \le y < \infty$

33. Function   **35.** Function   **37.** Not a function

39. y is a function of x.

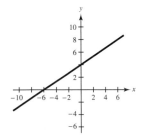

41. y is not a function of x.

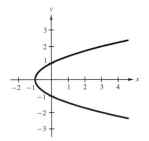

43. b   **44.** d   **45.** a   **46.** c

47. (a) Vertical shift
2 units upward

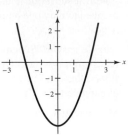

(b) Vertical shift
4 units downward

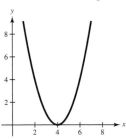

(c) Horizontal shift
2 units to the left

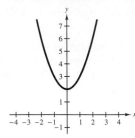

(d) Horizontal shift
4 units to the right

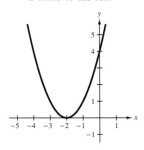

(e) Horizontal shift
3 units to the right
and a vertical shift
1 unit upward

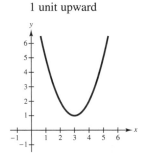

(f) Reflection in the
x-axis and a vertical
shift 4 units upward

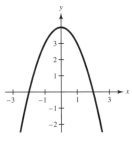

49. Horizontal shift 3 units
to the right

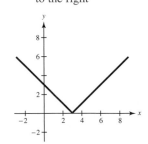

51. Vertical shift 4 units
downward

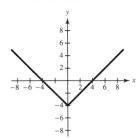

53. Reflection in the x-axis

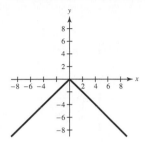

55. $h(x) = (x + 3)^2$ **57.** $h(x) = -x^2$

59. $h(x) = x^2 + 3$ **61.** $h(x) = -\sqrt{x}$

63. $h(x) = \sqrt{x + 2}$ **65.** $h(x) = \sqrt{-x}$

67. (a) $f(x) = (x + 2)^2 + 1$

    (b) $f(x) = (x - 1)^2 - 4$

    (c) $f(x) = (x - 1)^2 + 2$

    (d) $f(x) = (x - 3)^2 + 1$

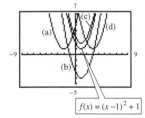

$f(x) = (x - 1)^2 + 1$

69. Horizontal shift of $y = x^3$ (cubing function)

$y = (x - 2)^3$

71. Reflection in the x-axis and horizontal and vertical shifts of $y = x^2$ (squaring function)

$y = -(x + 1)^2 + 1$

73. Vertical or horizontal shift of $y = x$ (identity function)

$y = x + 3$

75. (a)

(b)

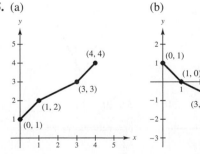

(c)

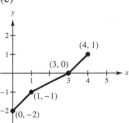

(d)

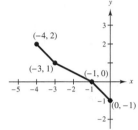

(e)

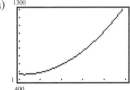

(f)

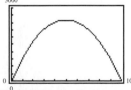

77. (a)

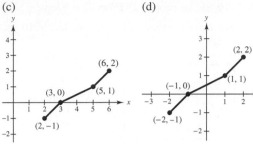

(b) 2003

79. (a) Perimeter: $P = 2l + 2w$

               $200 = 2l + 2w$

         $200 - 2l = 2w$

          $100 - l = w$

    Area: $A = lw$

           $A = l(100 - l)$

(b)

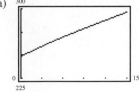

(c) $l = 50$. The figure is a square.

81. (a)

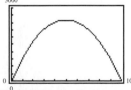

(b) Yes. For each year, there is only one population.

(c) 265.9 million people; 281.7 million people

(d) 2000. $P_1(t)$ represents a horizontal shift of 10 years to the right.

(e)
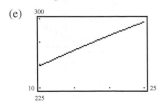

83. If the graph of an equation has the property that no vertical line intersects the graph at two (or more) points, the equation represents y as a function of x.

85. The range changes from $0 \le y \le 4$ to $0 \le y \le 8$.

87. The sum of four times a number and 1

89. The ratio of two times a number and 3

91. $y = -2x + 4$ **93.** $y = \frac{4}{3}x - 1$

95. Not a solution   **97.** Solution

Review Exercises   *(page 210)*

1.
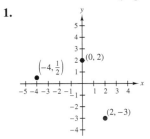

3. Quadrant I   **5.** Quadrant I or IV

7. (a) Solution   (b) Not a solution

(c) Not a solution   (d) Solution

9. 5   **11.** $3\sqrt{5}$

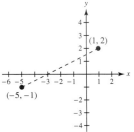

13. Collinear

15. $(4, 3)$
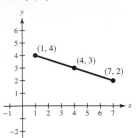

17. $\left(1, \frac{3}{2}\right)$
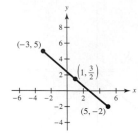

19.
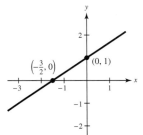

21.
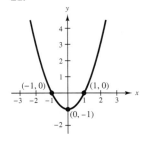

23.
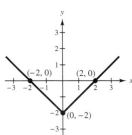

25. $\left(-\frac{1}{2}, 0\right), (0, 2)$
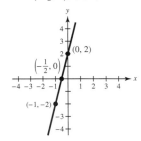

27. $(-5, 0), (5, 0), (0, 5)$
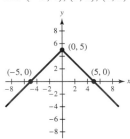

29. $(-3, 0), (2, 0), (0, -4)$
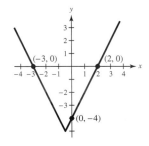

31. $(-4, 0), (0, -8), (2, 0)$
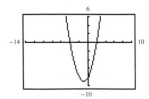

33. $(0, -11)$
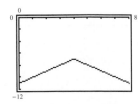

35. $(3, 0), (0, 1.73)$

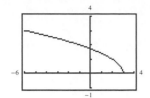

37. (a) $y = 35{,}000 - 4000t, \ 0 \le t \le 5$

(b)

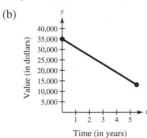

(c) $(0, 35{,}000)$; Initial purchase price

39. $\frac{3}{7}$ **41.** 0 **43.** $-\frac{3}{4}$ **45.** $(1, -1), (0, 2)$

47. $(-6, 0), (2, 2)$ **49.** $\sqrt{2320} \approx 48.17$ feet

51. $y = \frac{5}{2}x - 2$ **53.** $y = -3x - \frac{5}{2}$

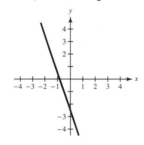

55. Neither   **57.** Perpendicular

59. $\$0.23$ **61.** $4x + y = 0$ **63.** $4y - x - 26 = 0$

65. $y = -\frac{1}{2}x - 3$ **67.** $y = \frac{3}{2}x$ **69.** $y = -9$

71. $x = -5$ **73.** (a) $y = -3x + 1$ (b) $y = \frac{1}{3}x - 1$

75. (a) $S = 1050t + 29{,}900$ (b) $\$40{,}400$ (c) $\$35{,}150$

77. (a) Not a solution   (b) Solution

(c) Solution   (d) Not a solution

79.   **81.**

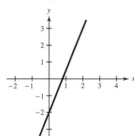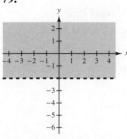

83.   **85.**

87.   **89.**

91. (a) $x + y \le 400; x \ge 0, y \ge 0$

(b)

93. Domain: $\{-3, -1, 0, 1\}$; Range: $\{0, 1, 4, 5\}$

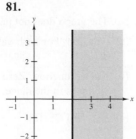

95. Not a function   **97.** Function

99. (a) 3 (b) 0

(c) $\sqrt{2}$ (d) $\sqrt{5 - 5z}$

101. (a) -8 (b) 2

(c) 0 (d) -7

103. (a) -2 (b) $\dfrac{12 - 2x}{x}$

105. All real numbers x

107. All real numbers x such that $x \ge -\frac{10}{3}$

109. $A = x(75 - x); \ 0 < x < \frac{75}{2}$

111.

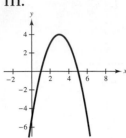

113.

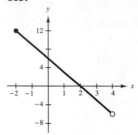

2.

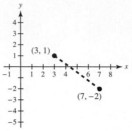

3. $(-1, 0), (0, -3)$

Domain: $-\infty < x < \infty$

Range: $-\infty < y \le 4$

Domain: $-2 \le x < 4$

Range: $-6 < y \le 12$

Distance: 5; Midpoint: $\left(5, -\frac{1}{2}\right)$

4.

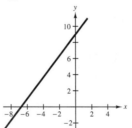

115.

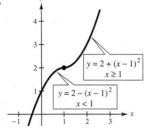

$y = 2 + (x-1)^2$
$x \ge 1$

$y = 2 - (x-1)^2$
$x < 1$

Domain: $-\infty < x < \infty$

Range: $-\infty < y < \infty$

117. c   **118.** a   **119.** b   **120.** d

121. Not a function

123. Reflection in the x-axis

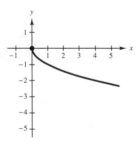

125. Horizontal shift 2 units to the right and 1 unit downward

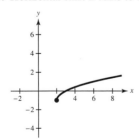

Chapter Test *(page 215)*

1. Quadrant IV

5. (a) $-\frac{2}{3}$ (b) Undefined

6.

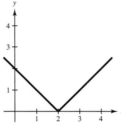

7.

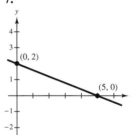

$(0, 2)$

$(5, 0)$

8. $3x - 2y - 18 = 0$ **9.** $x + 2 = 0$

10. (a) $y = \frac{3}{5}x + \frac{21}{5}$ (b) $y = -\frac{5}{3}x - \frac{1}{3}$

11.

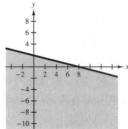

12. No. The graph does not pass the Vertical Line Test.

13. (a) Function; For each value of x there is only one value of y.

(b) Not a function; Zero is matched with two different elements in the range.

14. (a) -2 (b) 7   (c) $\dfrac{x + 2}{x - 1}$

15. (a) All real values of t such that $t \le 9$

(b) All real values of x such that $x \ne 4$

16.

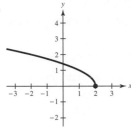

17. Reflection in the x-axis, horizontal shift 2 units to the right, vertical shift 1 unit upward

18. $V = 26{,}000 - 4000t$; 2.5 years since the car was purchased

19. (a) $y = |x - 2|$ (b) $y = |x| - 2$ (c) $y = 2 - |x|$

CHAPTER 4

Section 4.1   *(page 226)*

1. (a) Solution   (b) Not a solution

3. (a) Not a solution   (b) Solution

5. (a) Not a solution   (b) Solution

7. (a) Solution   (b) Not a solution

9. No solution   **11.** One solution

13. Infinitely many solutions   **15.** Inconsistent

17. Inconsistent   **19.** Consistent   **21.** Consistent

23.   **25.**

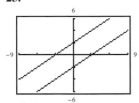

Inconsistent   Consistent; One solution

27. No solution   **29.** $\left(1, \frac{1}{3}\right)$

31. Infinitely many solutions   **33.** $(1, 2)$ **35.** $(1, 2)$

37. $(2, 0)$ **39.** $(3, 1)$ **41.** Infinitely many solutions

43. $(10, 0)$ **45.** No solution   **47.** $(3, -1)$

49.   **51.**

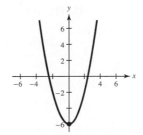

53. $(2, 1)$ **55.** $(4, 3)$ **57.** $(1, 2)$ **59.** $(4, -2)$

61. No solution   **63.** $(7, 2)$ **65.** $(10, 4)$

67. $(-2, -1)$ **69.** $\left(\frac{3}{2}, \frac{3}{2}\right)$ **71.** $\left(\frac{20}{3}, \frac{40}{3}\right)$ **73.** $(-4, 3)$

75. No solution   **77.** No solution

79.   **81.**

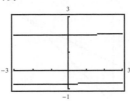

$\left(2992, \frac{798}{25}\right)$ $(50, 4)$

83. $a = 2, b = 3$ **85.** $a = 1, b = -6$

87. $\begin{cases} 2x - 3y = -7 \\ x + y = 9 \end{cases}$ **89.** $\begin{cases} 7x + y = -9 \\ -x + 3y = -5 \end{cases}$

91. $\begin{cases} 3x + 10y = 7 \\ 6x + 20y = 14 \end{cases}$ **93.** $\begin{cases} y + x = 4 \\ y + x = 3 \end{cases}$

95. 30 tons at \$125 per ton, 70 tons at \$75 per ton

97. 10,000 candy bars   **99.** 6250 bottles

101. \$4000 at 8.5%, \$8000 at 10%   **103.** 31, 49

105. 12, 15   **107.** 20, 32   **109.** 28, 43

111. 10 feet $\times$ 15 feet   **113.** 14 yards $\times$ 20 yards

115. 2001

117. A system of equations has no solution when a false statement, such as $2 = 0$, is produced by the substitution process.

119. One advantage of the substitution method is that it generates exact answers, whereas the graphical method usually yields approximate solutions.

121. The system will have exactly one solution because the system is consistent.

123. The system has no solution. Because the lines have different slopes, each pair of lines intersects at one point. Because exactly two of the lines share the same y-intercept, it is not possible for all three lines to intersect at the same point.

125. 5   **127.** 7

129.   **131.**

133. There are two values of y associated with one value of x.

Section 4.2 *(page 236)*

1.

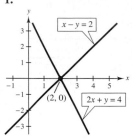

$(2, 0)$

3.

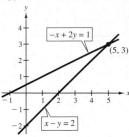

$(5, 3)$

5.

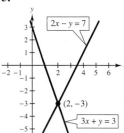

$(2, -3)$

7.

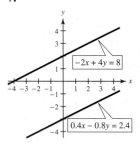

No solution

9.

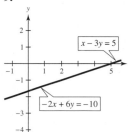

Infinitely many solutions

11.

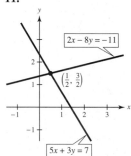

$\left(\frac{1}{2}, \frac{3}{2}\right)$

13. $\left(\frac{47}{6}, \frac{11}{3}\right)$ **15.** $(2, -2)$ **17.** $\left(3, \frac{1}{2}\right)$ **19.** $(3, -4)$

21. $(-1, -1)$ **23.** $(-4, 2)$ **25.** $(7, -2)$

27. No solution   **29.** $\left(\frac{3}{2}, 1\right)$ **31.** $(6, 3)$

33. $(-2, -1)$ **35.** Infinitely many solutions   **37.** $(4, 3)$

39. $\left(\frac{68}{5}, \frac{32}{5}\right)$ **41.** $(8, -2)$ **43.** $(2, 7)$ **45.** $(15, 10)$

47. $(4, 3)$ **49.** Consistent   **51.** Consistent

53. Inconsistent   **55.** 4

57. $\begin{cases} -2x + 3y = -24 \\ 2x + y = 0 \end{cases}$ **59.** $\begin{cases} 4x - y = -6 \\ 2x + y = 3 \end{cases}$

61. 95 weeks   **63.** 12 hours

65. $8000 at 8%, $12,000 at 9.5%   **67.** 4 hours

69. 550 miles per hour is the speed of the plane; 50 miles per hour is the wind speed.

71. 400 adult tickets, 100 children's tickets

73. $3.78 per gallon of regular unleaded, $3.89 per gallon of premium unleaded

75. 12 liters of 40% solution, 8 liters of 65% solution

77. 12 fluid ounces of 50% solution, 20 fluid ounces of 90% solution

79. 24 liters of 40% solution, 6 liters of 70% solution

81. 6.1 pounds of peanuts, 3.9 pounds of cashews

83. (a) $y = \frac{3}{2}x - \frac{1}{6}$

(b)

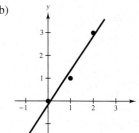

85. (a) and (b)

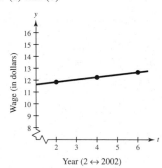

Year (2 ↔ 2002)

(b) $y = 0.208x + 11.42$

(c) The average annual increase in hourly wage is $0.208.

87. When a nonzero multiple of one equation is added to another equation to eliminate a variable and the result is $0 = 0$, the system has infinitely many solutions.

89. When a nonzero multiple of one equation is added to another equation to eliminate a variable and the result is a false statement, such as $0 = 5$, the system has no solution.

$\begin{cases} x - y = 3 \\ x - y = 8 \end{cases} \Rightarrow \begin{aligned} -x + y &= -3 \\ \underline{x - y} &= \underline{8} \\ 0 &= 5 \end{aligned}$

91. $x - 2y = 0$ **93.** $y - 2 = 0$

95. Function   **97.** Not a function

99. (a) Solution   (b) Not a solution

Section 4.3   *(page 249)*

1. (a) Not a solution   (b) Solution

(c) Solution   (d) Not a solution

3. $(22, -1, -5)$ **5.** $(14, 3, -1)$

7. $\begin{cases} x - 2y = 8 & \text{Eliminated the } x\text{-term in Equation 2} \\ \quad\quad y = 14 \end{cases}$

9. Yes. The first equation was multiplied by -2 and added to the second equation. Then the first equation was multiplied by -3 and added to the third equation.

11. $(1, 2, 3)$ **13.** $(1, 2, 3)$ **15.** $(2, -3, -2)$

17. No solution   **19.** $(-4, 8, 5)$ **21.** $(5, -2, 0)$

23. $\left(\frac{3}{10}, \frac{2}{5}, 0\right)$ **25.** $(-4, 2, 3)$ **27.** $\left(-\frac{1}{2}a + \frac{1}{2}, \frac{3}{5}a + \frac{2}{5}, a\right)$

29. No solution   **31.** $(1, -1, 2)$

33. $\left(\frac{6}{13}a + \frac{10}{13}, \frac{5}{13}a + \frac{4}{13}, a\right)$ **35.** $\begin{cases} x + 2y - z = -4 \\ \quad\quad y + 2z = 1 \\ 3x + y + 3z = 15 \end{cases}$

37. $s = -16t^2 + 144$ **39.** $s = -16t^2 + 48t$

41. $88°, 32°, 60°$

43. \$17,404 at 6%, \$31,673 at 10%, \$30,923 at 15%

45. \$4000 at 6%, \$8000 at 8%, \$3000 at 9%

47. 20 gallons of spray X, 18 gallons of spray Y, 16 gallons of spray Z

49. 84 hot dogs at \$1.50, 38 hot dogs at \$2.50, 21 hot dogs at \$3.25

51. (a) Not possible

(b) 10% solution: 0 gallons; 15% solution: 6 gallons; 25% solution: 6 gallons

(c) 10% solution: 4 gallons; 15% solution: 0 gallons; 25% solution: 8 gallons

53. 50 students on strings, 20 students in winds, 8 students in percussion

55. The solution is apparent because the row-echelon form is

$\begin{cases} x \quad\quad\quad = 1 \\ \quad y \quad\quad = -3. \\ \quad\quad z = 4 \end{cases}$

57. Three planes have no point in common when two of the planes are parallel and the third plane intersects the other two planes.

59. The graphs are three planes with three possible situations. If all three planes intersect in one point, there is one solution. If all three planes intersect in one line, there are an infinite number of solutions. If each pair of planes intersects in a line, but the three lines of intersection are all parallel, there is no solution.

61. $3x, 2; 3, 2$ **63.** $14t^5, -t, 25; 14, -1, 25$

65. $(4, 3)$ **67.** $(-2, 6)$

Mid-Chapter Quiz   *(page 253)*

1. (a) Not a solution   (b) Solution

2. 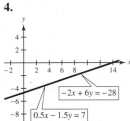   **3.**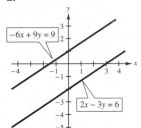

No solution   One solution

4. 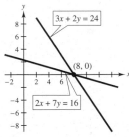   **5.**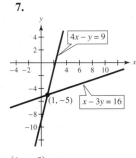

Infinitely many solutions $(4, 4)$

6. 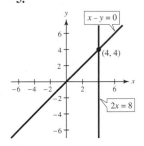   **7.**

$(8, 0)$ $(1, -5)$

8. $(5, 2)$ **9.** $\left(\frac{90}{13}, \frac{34}{13}\right)$ **10.** No solution   **11.** $(8, 1)$

12. $(3, -1)$ **13.** $\left(\frac{1}{2}, -\frac{1}{2}, 1\right)$ **14.** $(5, -1, 3)$

15. $\begin{cases} x + y = -2 \\ 2x - y = 32 \end{cases}$ **16.** $\begin{cases} x + y + z = 7 \\ 2x - y \quad\quad = 9 \\ -2x + y + 3z = 21 \end{cases}$

17. $13\frac{1}{3}$ gallons of 20% solution, $6\frac{2}{3}$ gallons of 50% solution

18. $68°, 41°, 71°$

Section 4.4   *(page 262)*

1. 4×2 **3.** 2×2 **5.** 4×1 **7.** 1×1

9. 1×4 **11.** (a) $\begin{bmatrix} 4 & -5 \\ -1 & 8 \end{bmatrix}$ (b) $\begin{bmatrix} 4 & -5 & \vdots & -2 \\ -1 & 8 & \vdots & 10 \end{bmatrix}$

13. (a) $\begin{bmatrix} 1 & 1 & 0 \\ 5 & -2 & -2 \\ 2 & 4 & 1 \end{bmatrix}$ (b) $\begin{bmatrix} 1 & 1 & 0 & \vdots & 0 \\ 5 & -2 & -2 & \vdots & 12 \\ 2 & 4 & 1 & \vdots & 5 \end{bmatrix}$

15. (a) $\begin{bmatrix} 5 & 1 & -3 \\ 0 & 2 & 4 \end{bmatrix}$ (b) $\begin{bmatrix} 5 & 1 & -3 & \vdots & 7 \\ 0 & 2 & 4 & \vdots & 12 \end{bmatrix}$

17. $\begin{cases} 4x + 3y = 8 \\ x - 2y = 3 \end{cases}$ **19.** $\begin{cases} x \quad\quad + 2z = -10 \\ \quad 3y - z = 5 \\ 4x + 2y \quad\quad = 3 \end{cases}$

21. $\begin{cases} 5x + 8y + 2z \quad\quad = -1 \\ -2x + 15y + 5z + w = 9 \\ x + 6y - 7z \quad\quad = -3 \end{cases}$

23. $\begin{cases} 13x + y + 4z - 2w = -4 \\ 5x + 4y \quad\quad - w = 0 \\ x + 2y + 6z + 8w = 5 \\ -10x + 12y + 3z + w = -2 \end{cases}$

25. Interchange the first and second rows.

27. Multiply the first row by $-\frac{1}{3}$.

29. Add 5 times the first row to the third row.

31. $\begin{bmatrix} 1 & 1 & -4 & 2 \\ 0 & 4 & 5 & 5 \\ 0 & 0 & 8 & 3 \end{bmatrix}$ **33.** $\begin{bmatrix} 1 & -2 & 3 \\ 3 & 4 & 5 \end{bmatrix}$

35. $\begin{bmatrix} 1 & 4 & 3 \\ 0 & 0 & 0 \end{bmatrix}$ **37.** $\begin{bmatrix} 1 & 2 & 3 \\ 0 & 1 & 2 \end{bmatrix}$

39. $\begin{bmatrix} 1 & 0 & -\frac{7}{5} \\ 0 & 1 & \frac{11}{10} \end{bmatrix}$ **41.** $\begin{bmatrix} 1 & 1 & 0 & 5 \\ 0 & 1 & 2 & 0 \\ 0 & 0 & 1 & -1 \end{bmatrix}$

43. $\begin{bmatrix} 1 & -1 & -1 & 1 \\ 0 & 1 & 6 & 3 \\ 0 & 0 & 1 & 0.8 \end{bmatrix}$ **45.** $\begin{bmatrix} 1 & 1 & -1 & 3 \\ 0 & 1 & -4 & 1 \\ 0 & 0 & 0 & 0 \end{bmatrix}$

47. $\begin{cases} x - 2y = 4 \\ y = -3 \end{cases}$ **49.** $\begin{cases} x + 5y = 3 \\ y = -2 \end{cases}$

$(-2, -3)$ $(13, -2)$

51. $\begin{cases} x - y + 2z = 4 \\ y - z = 2 \\ z = -2 \end{cases}$ **53.** $(5, 1)$ **55.** $(1, 1)$

$(8, 0, -2)$

57. $(-2, 1)$ **59.** No solution   **61.** $(2, -3, 2)$

63. $(2a + 1, 3a + 2, a)$ **65.** $(1, 2, -1)$ **67.** $(1, -1, 2)$

69. $(34, -4, -4)$ **71.** No solution

73. $(-12a - 1, 4a + 1, a)$ **75.** $\left(\frac{1}{2}, 2, 4\right)$ **77.** $\left(2, 5, \frac{5}{2}\right)$

79. \$800,000 at 8%, \$500,000 at 9%, \$200,000 at 12%

81. Theater A: 300 tickets; Theater B: 600 tickets; Theater C: 600 tickets

83. 5, 8, 20

85. 15 computer chips, 10 resistors, 10 transistors

87. Certificates of deposit: $250,000 - \frac{1}{2}s$

Municipal bonds: $125,000 + \frac{1}{2}s$

Blue-chip stocks: $125,000 - s$

Growth stocks: s

If $s = \$100,000$, then:

Certificates of deposit: \$200,000

Municipal bonds: \$175,000

Blue-chip stocks: \$25,000

Growth stocks: \$100,000

89. 3×5. There are 15 entries in the matrix, so the order is 3×5, 5×3, or 15×1. Because there are more columns than rows, the second number in the order must be larger than the first.

91. There will be a row in the matrix with all zero entries except in the last column.

93. The first entry in the first column is 1, and the other two are zero. In the second column, the first entry is a nonzero real number, the second number is 1, and the third number is zero. In the third column, the first two entries are nonzero real numbers and the third entry is 1.

95. -42 **97.** 26   **99.** $(4, -2, 1)$

Section 4.5   (page 275)

1. 5   **3.** 27   **5.** 0   **7.** 0   **9.** -24

11. -0.33 **13.** -24 **15.** -2 **17.** -30

19. 3   **21.** 0   **23.** 0   **25.** 102   **27.** -0.22

29. $x - 5y + 2$ **31.** 248   **33.** 105.625   **35.** 4.32

37. $(1, 2)$ **39.** $(2, -2)$ **41.** $\left(\frac{3}{4}, -\frac{1}{2}\right)$

43. Not possible, $D = 0$ **45.** $\left(\frac{2}{3}, \frac{1}{2}\right)$ **47.** $(-1, 3, 2)$

49. $\left(\frac{1}{2}, -\frac{2}{3}, \frac{1}{4}\right)$ **51.** Not possible, $D = 0$ **53.** $\left(\frac{22}{27}, \frac{22}{9}\right)$

55. $\left(\frac{51}{16}, -\frac{7}{16}, -\frac{13}{16}\right)$ **57.** -2 **59.** 16   **61.** 21

63. $\frac{31}{2}$ **65.** $\frac{33}{8}$ **67.** 16   **69.** $\frac{53}{2}$

71. 250 square miles   **73.** Collinear

75. Not collinear   **77.** Not collinear

79. $x - 2y = 0$ **81.** $7x - 6y - 28 = 0$

83. $9x + 10y + 3 = 0$ **85.** $16x - 15y + 22 = 0$

87. $I_1 = 1, I_2 = 2, I_3 = 1$ **89.** $I_1 = 2, I_2 = 3, I_3 = 5$

91. (a) $\left(\dfrac{13}{2k + 6}, \dfrac{3k - 4}{-2k^2 - 6k}\right)$ (b) $0, -3$

93. A determinant is a real number associated with a square matrix.

95. The determinant is zero. Because two rows are identical, each term is zero when expanding by minors along the other row. Therefore, the sum is zero.

97.

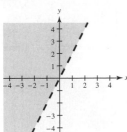

99.

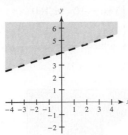

101. The graph of $h(x)$ has a vertical shift c units upward.

103.

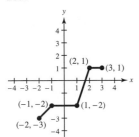

105.

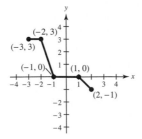

Section 4.6   *(page 284)*

1. d   **2.** b   **3.** f   **4.** c   **5.** a   **6.** e

7. (a) Not a solution   (b) Solution

9. (a) Not a solution   (b) Not a solution

11.

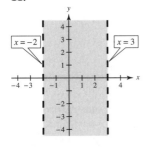

13.

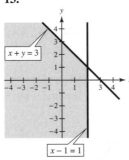

15.

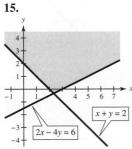

17.

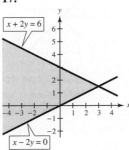

19.

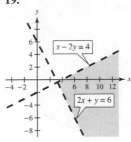

21.

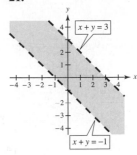

23.

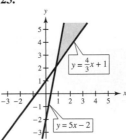

25.

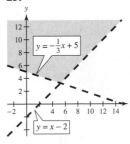

27.

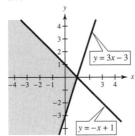

29.

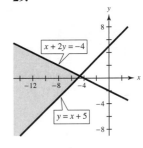

31.

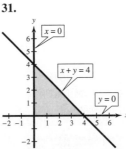

33.

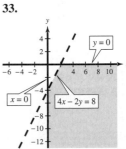

35.

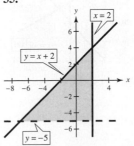

37.

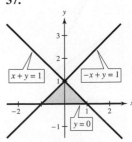

39. No solution

41.

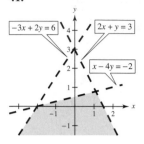

43.

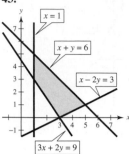

45.

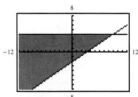

47.

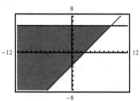

49.

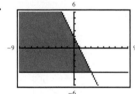

51. $\begin{cases} x \geq \quad 1 \\ y \geq \quad x - 3 \\ y \leq -2x + 6 \end{cases}$

53. $\begin{cases} x \geq -2 \\ x \leq \quad 2 \\ y \geq -\frac{1}{2}x - 4 \\ y \leq -\frac{1}{2}x + 2 \end{cases}$

55. $\begin{cases} y \leq \frac{9}{10}x + \frac{42}{5} \\ y \geq 3x \\ y \geq \frac{2}{3}x + 7 \end{cases}$

57. $\begin{cases} x + \frac{3}{2}y \leq 12 \\ \frac{4}{3}x + \frac{3}{4}y \leq 16 \\ x \qquad \geq 0 \\ y \qquad \geq 0 \end{cases}$

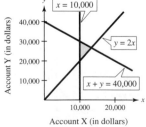

59. $\begin{cases} x + y \leq 40{,}000 \\ x \qquad \geq 10{,}000 \\ y \geq 2x \end{cases}$

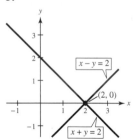

61. $\begin{cases} 20x + 10y \geq 280 \\ 15x + 10y \geq 160 \\ 10x + 20y \geq 180 \\ x \qquad\qquad \geq 0 \\ y \geq 0 \end{cases}$

63. $\begin{cases} x + \quad y \geq \ 15{,}000 \\ 30x + 45y \geq 525{,}000 \\ x \qquad\qquad \geq \quad 8{,}000 \\ y \geq \quad 4{,}000 \end{cases}$

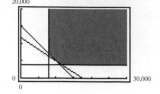

65. $\begin{cases} x \leq \quad 90 \\ y \leq \quad 0 \\ y \geq -10 \\ y \geq -\frac{1}{7}x \end{cases}$

67. The graph of a linear equation splits the xy-plane into two parts, each of which is a half-plane. The graph of $y < 5$ is a half-plane.

69. To determine the vertices of the region, find all intersections between the lines corresponding to the inequalities. The vertices are the intersection points that satisfy each inequality.

71. $\left(-\frac{1}{2}, 0\right), (0, 2)$ **73.** $(3, 0), (0, -1)$

75. $(-2, 0), (0, 2)$

77. (a) -10 (b) -5 **79.** (a) 0 (b) $6m - 4m^2$

Review Exercises   *(page 290)*

1. (a) Not a solution   (b) Solution

3. (a) Not a solution   (b) Not a solution

5.

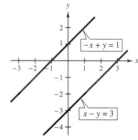

$(2, 0)$

7.

No solution

9.

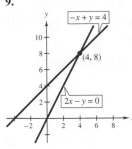

(4, 8)

11.

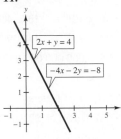

Infinitely many solutions

13. No solution

15.

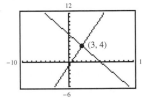

(3, 4)

17. (4, 3)   **19.** No solution   **21.** $(-10, -5)$

23. $(-1, 5)$ **25.** Infinitely many solutions

27. 5556 cameras   **29.** (0, 0)   **31.** $\left(\frac{5}{2}, 3\right)$

33. $(-2, 5)$ **35.** $(-0.5, 0.8)$

37. 16 gallons of 75% solution, 24 gallons of 50% solution

39. 45 minutes   **41.** (3, 2, 5)   **43.** (0, 3, 6)

45. $(2, -3, 3)$ **47.** $(0, 1, -2)$

49. $8000 at 7%, $5000 at 9%, $7000 at 11%

51. 1×4 **53.** 2×3

55. (a) $\begin{bmatrix} 7 & -5 \\ 1 & -1 \end{bmatrix}$ (b) $\begin{bmatrix} 7 & -5 & \vdots & 11 \\ 1 & -1 & \vdots & -5 \end{bmatrix}$

57. $\begin{cases} 4x - y = 2 \\ 6x + 3y + 2z = 1 \\ y + 4z = 0 \end{cases}$ **59.** $(10, -12)$ **61.** (0.6, 0.5)

63. $(3, -6, 7)$ **65.** $\left(\frac{1}{2}, -\frac{1}{3}, 1\right)$ **67.** $(1, 0, -4)$

69. 10   **71.** -51 **73.** 1   **75.** $(-3, 7)$

77. $(2, -3, 3)$ **79.** 16   **81.** 7   **83.** Not collinear

85. $x - 2y + 4 = 0$

87.

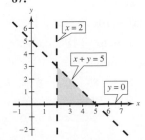

89.

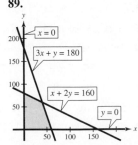

91. $\begin{cases} x + y \le 500 \\ x \ge 150 \\ y \ge 220 \end{cases}$

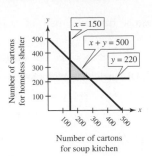

Number of cartons
for soup kitchen

Chapter Test   *(page 294)*

1. (a) Solution   (b) Not a solution

2. (3, 2)   **3.** (1, 3)   **4.** (2, 3)   **5.** $(-2, 2)$

6. $(2, 2a - 1, a)$ **7.** $(-1, 3, 3)$ **8.** $(2, 1, -2)$

9. $\left(4, \frac{1}{7}\right)$ **10.** $(5, 1, -1)$ **11.** $\left(-\frac{11}{5}, \frac{56}{25}, \frac{32}{25}\right)$

12. -16 **13.** 12

14.

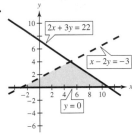

15. 16 feet $\times$ 18 feet

16. $13,000 at 4.5%, $9000 at 5%, $3000 at 8%

17. $\begin{cases} 30x + 40y \ge 300,000 \\ x \le 9,000 \\ y \le 4,000 \end{cases}$

where x is the number of reserved tickets and y is the number of floor tickets.

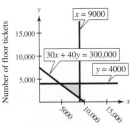

Number of reserved tickets

Cumulative Test: Chapters 1–4   *(page 295)*

1. (a) $>$ (b) $>$ (c) $=$ **2.** $3n - 8$

3. $3t^2 - 3t - 8$ **4.** $4x^3 - 2x^2 - 24$ **5.** $\frac{3}{2}$

6. $-\frac{3}{2}$ **7.** $x \le -1$ or $x \ge 5$ **8.** $-\frac{3}{2} \le x < 4$

9. $1380   **10.** 6.5

11. Because a fractional part of a unit cannot be sold, $x \geq 103$.

12. No   **13.** $2 \leq x < \infty$ **14.** (a) 3 (b) $9c^2 + 6c$

15. $m = \frac{3}{4}$; Distance: 10; Midpoint: $(0, 3)$

16. (a) $y = 2x + 5$ (b) $y = \frac{2}{3}x + \frac{7}{3}$

17.   **18.**

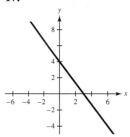
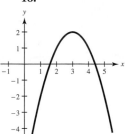

19. $(3, 3)$ **20.** $(4, -2)$ **21.** $(4, -1, 3)$

CHAPTER 5

Section 5.1   *(page 304)*

1. (a) $-3x^8$ (b) $9x^7$ **3.** (a) $-125z^6$ (b) $25z^8$

5. (a) $2u^3v^3$ (b) $-4u^9v$ **7.** (a) $-15u^8$ (b) $64u^5$

9. (a) $-m^{19}n^7$ (b) $-m^7n^3$

11. (a) $3m^4n^3$ (b) $3m^2n^3$ **13.** (a) $\dfrac{9x^2}{16y^2}$ (b) $\dfrac{125u^3}{27v^3}$

15. (a) $\dfrac{8x^4y}{9}$ (b) $-\dfrac{2x^2y^4}{3}$ **17.** (a) $\dfrac{25u^8v^2}{4}$ (b) $\dfrac{u^8v^2}{4}$

19. (a) $x^{2n-1}y^{2n-1}$ (b) $x^{2n-2}y^{n-12}$ **21.** $\frac{1}{25}$

23. $-\frac{1}{1000}$ **25.** 1   **27.** 64   **29.** -32 **31.** $\frac{3}{2}$

33. 1   **35.** 1   **37.** 729   **39.** 100,000   **41.** $\frac{1}{16}$

43. $\frac{1}{64}$ **45.** $\frac{3}{16}$ **47.** $\frac{64}{121}$ **49.** $\frac{16}{15}$ **51.** y^2 **53.** z^2

55. $\dfrac{7}{x^4}$ **57.** $\dfrac{1}{64x^3}$ **59.** x^6 **61.** $\dfrac{4a}{3}$ **63.** t^2

65. $\dfrac{1}{4x^4}$ **67.** $-\dfrac{12}{xy^3}$ **69.** $\dfrac{y^4}{9x^4}$ **71.** $\dfrac{10}{x}$ **73.** $\dfrac{x^5}{2y^4}$

75. $\dfrac{81v^8}{u^6}$ **77.** $\dfrac{b^5}{a^5}$ **79.** $\dfrac{1}{2x^8y^3}$ **81.** $6u$ **83.** x^8y^{12}

85. $\dfrac{2b^{11}}{25a^{12}}$ **87.** $\dfrac{v^2}{uv^2 + 1}$ **89.** b **91.** 4   **93.** 144

95. 1   **97.** $-\frac{4}{5}$ **99.** $\frac{1}{144}$ **101.** 3.6×10^6

103. 4.762×10^7 **105.** 3.1×10^{-4} **107.** 3.81×10^{-8}

109. 5.73×10^7 **111.** 9.4608×10^{12} **113.** 8.99×10^{-2}

115. 720,000,000   **117.** 0.0000001359

119. 34,659,000,000   **121.** 15,000,000

123. 0.00000000048   **125.** 6.8×10^5 **127.** 2.5×10^9

129. 6×10^6 **131.** 9×10^{15} **133.** 1.6×10^{12}

135. 3.46×10^{10} **137.** 4.70×10^{11} **139.** 4.43×10^{25}

141. 2.74×10^{20} **143.** 9.3×10^7 miles

145. 84,830,000,000,000,000,000,000 free electrons

147. 1.59×10^{-5} year ≈ 8 minutes

149. 8.99×10^{17} meters

151. $\$7.87 \times 10^{12} = \$7,870,000,000,000$

153. Scientific notation makes it easier to multiply or divide very large or very small numbers because the properties of exponents make it more efficient.

155. False. $\dfrac{1}{3^{-3}} = 27$, which is greater than 1.

157. The product rule can be applied only to exponential expressions with the same base.

159. The power-to-power rule applied to the expression raises the base to the *product* of the exponents.

161. $6x$ **163.** $a^2 + 3ab + 3b^2$ **165.** No solution

Section 5.2   *(page 313)*

1. $4y + 16$; 1; 4   **3.** $x^2 + 2x - 6$; 2; 1

5. $-42x^3 - 10x^2 + 3x + 5$; 3; -42

7. $t^5 - 14t^4 - 20t + 4$; 5; 1   **9.** -4; 0; -4

11. $-16t^2 + v_0t$; 2; -16 **13.** Binomial

15. Trinomial   **17.** Monomial   **19.** $3x^2$

21. $8x^3 + 5$

23. The first term is not of the form ax^k (k must be nonnegative).

25. The second term is not of the form ax^k (k must be an integer).

27. $3x + 7$ **29.** $7x^2 + 3$ **31.** $4y^2 - y + 3$

33. $-2y^4 - 5y + 4$ **35.** $2x^3 + 2x^2$

37. $6x^2 - 7x + 8$ **39.** $\frac{1}{6}x^3 + 3x + \frac{2}{5}$

41. $2.69t^2 + 7.35t - 4.2$ **43.** $2x^2 - 3x$

45. $4x^3 + 2x^2 + 9x - 6$ **47.** $2p^2 - 2p - 5$

49. $0.6b^2 - 0.6b + 7.1$ **51.** $-2y^3$ **53.** $x^2 - 3x + 2$

55. $7t^3 - t - 10$ **57.** $\frac{7}{4}y^2 - 9y - 12$

59. $9.37t^5 + 10.4t^4 - 5.4t^2 + 7.35t - 2.6$

61. $-2x^3 + x^2 + 2x$ **63.** $x^2 - 2x + 5$

65. $-2x^2 - 9x + 11$ **67.** $-11x^7 - 10x^5 + 8x^4 + 16$

69. $2x^3 - 2x + 3$ **71.** $-2x^3 - x^2 + 6x - 11$

73. 0   **75.** $7x^3 + 2x$ **77.** $3x^3 + 5x^2 + 2$

79. $3t^2 + 29$ **81.** $3v^2 + 78v + 27$ **83.** $29s + 8$

85. $5x^{2r} - 8x^r + 3$ **87.** $2x^{2m} + 6x^m - 11$

89. $3x^{4n} - x^{3n} + 3x^{2n} - 1$

91. 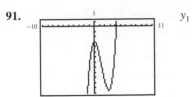 $y_1 = y_2$

93. $-x^3 - 4x^2 - x + 16$ **95.** $x^3 + 4x^2 + x - 16$

97. (a) 64 feet (b) 60 feet (c) 48 feet (d) 0 feet

At time $t = 0$, the object is dropped from a height of 64 feet and continues to fall, reaching the ground at time $t = 2$.

99. Dropped; 100 feet   **101.** Thrown upward; 12 feet

103. 1008 feet; 784 feet; 48 feet   **105.** $15,000

107. $14x + 10$ **109.** $5x + 72$ **111.** $36x$

113. (a) $T = 0.0475x^2 + 1.099x$

(b) 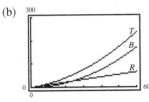   (c) 57.2 feet; 173.7 feet

115. The degree of the term ax^k is k. The degree of a polynomial is the degree of its highest-degree term.

117. (a) Not always false. A polynomial is only a trinomial when it has three terms.

(b) Always true

119. No. $x^3 + 2x + 3$ **121.** 2×3 **123.** Yes

125. $\begin{bmatrix} 1 & 0 & -1 \\ 0 & 1 & -5 \\ 0 & 0 & 1 \end{bmatrix}$ **127.** $(1, 2)$ **129.** -28

Section 5.3   *(page 323)*

1. $16a^3$ **3.** $-2y^2 + 10y$ **5.** $8x^3 - 12x^2 + 20x$

7. $-4m^4 + 8m^3 - 14m^2$ **9.** $-x^7 + 2x^6 - 5x^4 + 6x^3$

11. $75x^3 + 30x^2$ **13.** $3u^6v - 5u^4v^2 + 6u^3v^4$

15. $x^2 + 6x + 8$ **17.** $x^2 - 16$ **19.** $x^2 - 6x + 9$

21. $2x^2 + 7x - 15$ **23.** $10x^2 - 34x + 12$

25. $2x^3 + 4x^2 - x - 2$ **27.** $48y^2 + 32y - 3$

29. $6x^2 + 7xy + 2y^2$ **31.** $4t^2 - 6t + 4$

33. $x^3 - 5x^2 + 10x - 6$ **35.** $3a^3 + 11a^2 + 9a + 2$

37. $8u^3 + 22u^2 - u - 20$

39. $x^4 - 2x^3 - 3x^2 + 8x - 4$

41. $5x^4 + 20x^3 - 3x^2 + 8x - 2$ **43.** $t^4 - t^2 + 4t - 4$

45. $14x^3 - 21x^2 + 4x + 9$

47. $2u^3 + u^2 - 7u - 6$ **49.** $-2x^3 + 3x^2 - 1$

51. $t^4 - t^2 + 4t - 4$ **53.** $x^2 - 4$ **55.** $x^2 - 64$

57. $4 - 49y^2$ **59.** $9 - 4x^4$ **61.** $4a^2 - 25b^2$

63. $36x^2 - 81y^2$ **65.** $4x^2 - \frac{1}{16}$ **67.** $0.04t^2 - 0.25$

69. $x^2 + 10x + 25$ **71.** $x^2 - 20x + 100$

73. $4x^2 + 20x + 25$ **75.** $36x^2 - 12x + 1$

77. $4x^2 - 28xy + 49y^2$

79. $x^2 + 2xy + y^2 + 4x + 4y + 4$

81. $u^2 - v^2 + 6v - 9$ **83.** $k^3 + 15k^2 + 75k + 125$

85. $u^3 + 3u^2v + 3uv^2 + v^3$ **87.** $15x^{3r} + 12x^{4r-1}$

89. $12x^{3m} - 10x^{2m} - 18x^m + 15$ **91.** $x^{m^2-n^2}$

93. 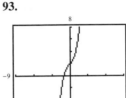   **95.**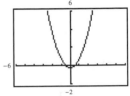

$y_1 = y_2$ $y_1 = y_2$

97. (a) $w^2 + 2w$ (b) $a^2 - 10a + 27$

99. (a) $V(n) = n^3 + 6n^2 + 8n$ (b) 105 cubic inches

(c) $A(n) = n^2 + 2n$ (d) $A(n + 5) = n^2 + 12n + 35$

101. $8x^2 + 26x$ **103.** $1.2x^2$ **105.** (a) $14w$ (b) $10w^2$

107. $R = -0.02x^2 + 175x$; $3450

109. $5000r^2 + 10,000r + 5000$

111. $(x + a)(x + b) = x^2 + bx + ax + ab$; FOIL Method

113. (a) $x^2 - 1$ (b) $x^3 - 1$ (c) $x^4 - 1$

Yes, $x^5 - 1$

115. When two polynomials are multiplied together, the Distributive Property is used because each term of the first polynomial is multiplied by each term of the second polynomial.

117. $m + n$

119. 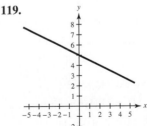   **121.**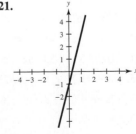

Function   Function

123.

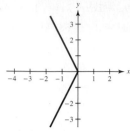

Not a function.

125. $\frac{1}{32}$　　**127.** 64　　**129.** 1

Mid-Chapter Quiz　(page 327)

1. $4; -2$

2. The first term is not of the form ax^k. (k must be nonnegative.)

3. $-10y^9$　　**4.** $-54x^5$　　**5.** $10n^5$　　**6.** $-18m^{10}$

7. $-\dfrac{48}{x}$　　**8.** $\dfrac{25x^2}{16y^4}$　　**9.** $\dfrac{9}{a^4 b^{10}}$　　**10.** $\dfrac{8y^{33}}{125x^6}$

11. $3t^3 + 3t^2 + 7$　　**12.** $7y^2 - 5y$　　**13.** $9x^3 - 4x^2 + 1$

14. $2u^2 - u + 1$　　**15.** $28y - 21y^2$

16. $k^2 + 13k + 40$　　**17.** $24x^2 - 26xy + 5y^2$

18. $2z^2 + 3z - 35$　　**19.** $36r^2 - 25$

20. $4x^2 - 12x + 9$　　**21.** $x^3 + 1$

22. $x^4 + 2x^3 - 23x^2 + 40x - 20$

23. $\frac{1}{2}(x+2)^2 - \frac{1}{2}x^2 = 2(x+1) = 2x + 2$

24. 112 feet; 52 feet　　**25.** $26,250

Section 5.4　(page 334)

1. $2 \cdot 3$　　**3.** $2 \cdot 2 \cdot 2$　　**5.** $2 \cdot 3 \cdot 5$　　**7.** $3 \cdot 3 \cdot 3$

9. 8　　**11.** 2　　**13.** x^3　　**15.** $3x$　　**17.** $8ab^2$

19. $3(x-2)^2$　　**21.** $4(x+1)$　　**23.** $2(3y-10)$

25. $12(2t^2 - 3)$　　**27.** $x(x+9)$　　**29.** $8t(t+1)$

31. No common factor other than 1　　**33.** $3y(x^2 y - 5)$

35. $4(7x^2 + 4x - 2)$　　**37.** $15(3x^2 - x + 2)$

39. $x^2(14x^2 y^3 + 21xy^2 + 9)$　　**41.** $-7(2x - 1)$

43. $-(x-6)$　　**45.** $-(y^2 - 7)$

47. $-(x^2 - x - 4)$　　**49.** $-2(3y^2 - y + 1)$

51. $10y - 3$　　**53.** $6x + 5$

55. $(y-4)(2y+5)$　　**57.** $(3x+2)(5x-3)$

59. $(7a+6)(2-3a^2)$　　**61.** $(4t-1)^2(8t^3 + 3)$

63. $(4x+9)(-2x-9)$　　**65.** $(x+25)(x+1)$

67. $(y-6)(y+2)$　　**69.** $(x+2)(x^2 + 1)$

71. $(a-4)(3a^2 - 2)$　　**73.** $(z+3)(z^3 - 2)$

75. $(x-2y)(5x^2 + 7y^2)$　　**77.** $(x+3)(x-3)$

79. $(1+a)(1-a)$　　**81.** $(4y+3)(4y-3)$

83. $(9+2x)(9-2x)$　　**85.** $(2z+y)(2z-y)$

87. $(6x+5y)(6x-5y)$　　**89.** $\left(u + \frac{1}{4}\right)\left(u - \frac{1}{4}\right)$

91. $\left(\frac{2}{3}x + \frac{4}{5}y\right)\left(\frac{2}{3}x - \frac{4}{5}y\right)$　　**93.** $(x+3)(x-5)$

95. $(14+z)(4-z)$　　**97.** $(x+9)(3x+1)$

99. $(x-2)(x^2 + 2x + 4)$　　**101.** $(y+4)(y^2 - 4y + 16)$

103. $(2t-3)(4t^2 + 6t + 9)$　　**105.** $(3u+1)(9u^2 - 3u + 1)$

107. $(4a+b)(16a^2 - 4ab + b^2)$

109. $(x+3y)(x^2 - 3xy + 9y^2)$　　**111.** $2(2+5x)(2-5x)$

113. $8(x+2)(x^2 - 2x + 4)$

115. $(y-3)(y+3)(y^2 + 9)$　　**117.** $3x^2(x+10)(x-10)$

119. $6(x^2 - 2y^2)(x^4 + 2x^2 y^2 + 4y^4)$

121. $(2x^n + 5)(2x^n - 5)$　　**123.** $2x^r(x^{2r} + 2x^r + 4)$

125. $y^{m+n}(7y^m - y^n + 4)$

127.

$y_1 = y_2$

129.

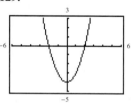

$y_1 = y_2$

131. $x^2(3x+4) - (3x+4) = (3x+4)(x+1)(x-1)$

$3x(x^2 - 1) + 4(x^2 - 1) = (x+1)(x-1)(3x+4)$

133. $p = 800 - 0.25x$　　**135.** $P(1 + rt)$

137. $w = 45 - l$　　**139.** $S = 2x(x + 2h)$

141. $\pi(R - r)(R + r)$

143. The polynomial is written as a product of polynomials.

145. To check the result, multiply the factors to obtain the original polynomial.

147. $x^2 + 2x = x(x + 2)$

149. The three binomial factors of the expression are the sum of the squares of the monomials, the sum of the monomials, and the difference of the monomials.

$a^4 - b^4 = (a^2 + b^2)(a^2 - b^2)$

$\qquad = (a^2 + b^2)(a + b)(a - b)$

151. -5　　**153.** 36　　**155.** $x^2 - 49$

157. $4x^2 - 12x + 9$

Section 5.5　(page 346)

1. $(x+2)^2$　　**3.** $(a-5)^2$　　**5.** $(5y-1)^2$

7. $(3b+2)^2$　　**9.** $(u+4v)^2$　　**11.** $(6x-5y)^2$

13. $5(x+3)^2$　　**15.** $3m(m-3)^2$　　**17.** $5v^2(2v-3)^2$

19. $\frac{1}{36}(3x - 4)^2$ **21.** ± 18 **23.** ± 12 **25.** 16

27. 9   **29.** $a + 2$ **31.** $y - 5$ **33.** $x + 6$

35. $z - 2$ **37.** $(x + 1)(x + 5)$ **39.** $(x - 3)(x - 2)$

41. $(y + 10)(y - 3)$ **43.** $(t - 8)(t + 2)$

45. $(x - 8)(x - 12)$ **47.** $(x - 7y)(x + 5y)$

49. $(x + 12y)(x + 18y)$ **51.** $\pm 6, \pm 9$

53. $\pm 4, \pm 20$ **55.** $\pm 12, \pm 36$ **57.** $-16, 8$

59. $-18, 2$ **61.** $5x + 3$ **63.** $5a - 3$ **65.** $2y - 9$

67. $(3x + 5)(2x - 5)$ **69.** $(5y + 4)(2y - 3)$

71. $(4x - 1)(3x - 1)$ **73.** $(5z - 3)(z + 1)$

75. $(2t + 1)(t - 4)$ **77.** $(3b - 1)(2b + 7)$

79. $-(2x - 3)(x + 2)$ **81.** $-(3d - 2)(5d - 3)$

83. $-(3x - 2)(4x + 1)$ **85.** Prime

87. $5y(3y - 2)(4y + 5)$ **89.** $(a + 2b)(10a + 3b)$

91. $(4x - 3y)(6x + y)$ **93.** $(3x + 4)(x + 2)$

95. $(5x + 3)(x - 3)$ **97.** $(3x - 1)(5x - 2)$

99. $3x(x + 1)(x - 1)$ **101.** $2t(5t - 9)(t + 2)$

103. $2(3x - 1)(9x^2 + 3x + 1)$

105. $9ab^2(3ab + 2)(ab - 1)$ **107.** $(x + 2)(x + 4)(x - 4)$

109. $-(r + 5)(r - 9)$ **111.** $(x - 5 + y)(x - 5 - y)$

113. $(x^4 + 1)(x^2 + 1)(x + 1)(x - 1)$

115. $(x^n - 8)(x^n + 3)$ **117.** $(x^n + 5)(x^n - 2)$

119. $(2y^n + 3)(3y^n + 2)$

121.   **123.**

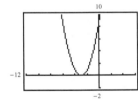   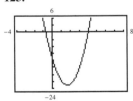

$y_1 = y_2$ $y_1 = y_2$

125. c   **126.** a   **127.** b   **128.** d

129. $4(6 + x)(6 - x)$

131. (a) $2n(2n + 2)(2n + 4)$ (b) 20, 22, 24

133. Begin by finding the factors of 6 whose sum is -5. They are -2 and -3. The factorization is $(x - 2)(x - 3)$.

135. $x^2 + x + 1$

137. No. $x(x + 2) - 2(x + 2) = (x + 2)(x - 2)$

139. Problems will vary. It is possible to create factorable polynomials by working backward: first list several factors, and then multiply them to form a single polynomial.

141. 425   **143.** 36.25%

145. 7.5 liters of 20% solution, 2.5 liters of 60% solution

147. 20 gallons of 60% solution, 100 gallons of 90% solution

Section 5.6   *(page 356)*

1. 0, 4   **3.** $-10, 3$ **5.** $-4, 2$ **7.** $-\frac{5}{2}, -\frac{1}{3}$

9. $-\frac{25}{2}, 0, \frac{3}{2}$ **11.** $-4, -\frac{1}{2}, 3$ **13.** 0, 5   **15.** $-\frac{5}{3}, 0$

17. 0, 16   **19.** 0, 3   **21.** ± 5 **23.** ± 4

25. $-2, 5$ **27.** 4, 6   **29.** $-5, \frac{5}{4}$ **31.** $-\frac{1}{2}, 7$

33. $-1, \frac{2}{3}$ **35.** 4, 9   **37.** 4   **39.** -8 **41.** $\frac{3}{2}$

43. $-2, 10$ **45.** ± 3 **47.** $-4, 9$ **49.** $-12, 6$

51. $-1, \frac{5}{2}$ **53.** $-7, 0$ **55.** $-6, 5$ **57.** $-2, 6$

59. $-5, 1$ **61.** $-2, 8$ **63.** $-13, 5$ **65.** 0, 7, 12

67. $-\frac{1}{3}, 0, \frac{1}{2}$ **69.** ± 2 **71.** $\pm 3, -2$ **73.** ± 3

75. $\pm 1, 0, 3$ **77.** $\pm 2, -\frac{3}{2}, 0$

79. $(-3, 0), (3, 0)$; The x-intercepts are solutions of the polynomial equation.

81. $(0, 0), (3, 0)$; The x-intercepts are solutions of the polynomial equation.

83.   **85.**

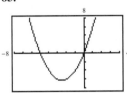   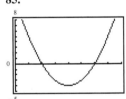

$(-5, 0), (0, 0)$ $(2, 0), (6, 0)$

87.   **89.**

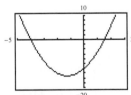   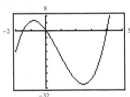

$(-4, 0), (\frac{3}{2}, 0)$ $(-\frac{3}{2}, 0), (0, 0), (4, 0)$

91. $-\dfrac{b}{a}, 0$ **93.** $x^2 - 4x - 12 = 0$ **95.** 15

97. 20 feet × 27 feet

99. Base: 6 inches; Height: 9 inches

101. 5 seconds   **103.** 2 seconds   **105.** 9.75 seconds

107. 40 units, 50 units

109. (a) $-6, -\frac{1}{2}$ (b) $-6, -\frac{1}{2}$ (c) Answers will vary.

111. (a) Length $= 5 - 2x$; Width $= 4 - 2x$; Height $= x$

Volume $= $ (Length)(Width)(Height)

$$V = (5 - 2x)(4 - 2x)(x)$$

(b) $0, 2, \frac{5}{2}; \ 0 < x < 2$

(c)

x	0.25	0.50	0.75	1.00	1.25	1.50	1.75
V	3.94	6	6.56	6	4.69	3	1.31

(d) 1.50

(e) 0.74

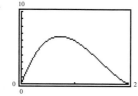

113. Maximum number: n. The third-degree equation $(x + 1)^3 = 0$ has only one real solution: $x = -1$.

115. When a quadratic equation has a repeated solution, the graph of the equation has one x-intercept, which is the vertex of the graph.

117. A solution to a polynomial equation is the value of x when y is zero. If a polynomial is not factorable, the equation can still have real number solutions for x when y is zero.

119. \$0.0625 per ounce   **121.** \$0.071 per ounce

123. All real numbers x such that $x \neq -1$

125. All real numbers x such that $x \leq 3$

Review Exercises   (page 362)

1. x^9 **3.** u^6 **5.** $-8z^3$ **7.** $4u^7v^3$ **9.** $2z^3$

11. $\dfrac{5g^2}{16}$ **13.** $144x^4$ **15.** $\dfrac{1}{72}$ **17.** $\dfrac{64}{27}$ **19.** $12y$

21. $\dfrac{2}{x^3}$ **23.** 1   **25.** $\dfrac{a^7}{2b^6}$ **27.** $\dfrac{4x^6}{y^5}$ **29.** $\dfrac{405u^5}{v}$

31. 3.19×10^{-5} **33.** 1.735×10^7 **35.** 1,950,000

37. 0.0000205   **39.** 3.6×10^7 **41.** 500

43. $-x^4 + 6x^3 + 5x^2 - 4x; 4; -1$

45. $3x^5 + x^4 - 6x - 4; 5; 3$ **47.** $-6x^5 + 2x - 4$

49. $x^2 + 13x + 8$ **51.** $-3x^3 - x^2 + 16$

53. $-2y^2 + 3y - 5$ **55.** $-9x^3 + 9x - 4$

57. $-2y - 15$ **59.** $x^5 - 4x^3 + 7x^2 - 9x + 3$

61. $-x^2 + 4x + 6$ **63.** $-t^2 + 4t$ **65.** $18x + 10$

67. \$19,800   **69.** $-8x^4 - 32x^3$ **71.** $6x^3 - 15x^2 + 9x$

73. $x^2 + 5x - 14$ **75.** $15x^2 - 11x - 12$

77. $24x^4 + 22x^2 + 3$ **79.** $4x^3 - 5x + 6$

81. $u^2 - 8u + 7$ **83.** $16x^2 - 56x + 49$

85. $36v^2 - 81$ **87.** $u^2 - v^2 - 6u + 9$

89. $14x + 3$ **91.** \$1123.60   **93.** $6(4x^2 - 3)$

95. $-b(3b - 1)$ **97.** $3x(2x + 5x^2 - 1)$

99. $-42(x + 5)$ **101.** $(v + 1)(v - 1)(v - 2)$

103. $(t + 3)(t^2 + 3)$ **105.** $(x + 6)(x - 6)$

107. $(u - 3)(u + 15)$ **109.** $(u - 1)(u^2 + u + 1)$

111. $(2x + 3)(4x^2 - 6x + 9)$ **113.** $x(x + 1)(x - 1)$

115. $3(u + 2)(u^2 - 2u + 4)$ **117.** $(x - 9)^2$

119. $(2s + 10t)^2$ **121.** $(x + 7)(x - 5)$

123. $(2x - 3)(x - 2)$ **125.** $(3x + 2)(6x + 5)$

127. $(4x + 1)(x - 1)$ **129.** $(5x - 7)(x - 1)$

131. $(7s - 4)(s + 2)$ **133.** $4a(1 + 4a)(1 - 4a)$

135. $(z^2 + 3)(z + 1)$ **137.** $\left(\tfrac{1}{2}x + y\right)^2$

139. $(x + y - 5)(x - y - 5)$ **141.** $0, 2$ **143.** $-\tfrac{1}{2}, 3$

145. $-10, -\tfrac{9}{5}, \tfrac{1}{4}$ **147.** $-\tfrac{4}{3}, 2$ **149.** $-3, \tfrac{1}{2}$

151. $-4, 9$ **153.** ± 10 **155.** $-4, 0, 3$ **157.** $0, 2, 9$

159. $\pm 1, 6$ **161.** $\pm 3, 0, 5$ **163.** $9, 11$

165. 45 inches $\times$ 20 inches   **167.** 8 inches $\times$ 8 inches

169. 15 seconds

Chapter Test   (page 367)

1. Degree: 3; Leading coefficient: 8.2

2. The variable appears in the denominator.

3. (a) $\dfrac{x^7}{8y^5}$ (b) $\dfrac{1}{4x^4y^2z^6}$ **4.** (a) $-24u^9v^5$ (b) $\dfrac{27x^6}{2y^4}$

5. (a) $6a^2 - 3a$ (b) $-2y^2 - 2y$

6. (a) $8x^2 - 4x + 10$ (b) $11t + 7$

7. (a) $-3x^2 + 12x$ (b) $2x^2 + 7xy - 15y^2$

8. (a) $3x^2 - 6x + 3$ (b) $6s^3 - 17s^2 + 26s - 21$

9. (a) $4w^2 - 28w + 49$ (b) $16 - a^2 - 2ab - b^2$

10. $6y(3y - 2)$ **11.** $\left(v - \tfrac{4}{3}\right)\left(v + \tfrac{4}{3}\right)$

12. $(x + 2)(x - 2)(x - 3)$ **13.** $(3u - 1)^2$

14. $2(x - 5)(3x + 2)$ **15.** $(x + 3)(x^2 - 3x + 9)$

16. $5, -4$ **17.** $1, -5$ **18.** $3, -\tfrac{4}{3}$

19. $-4, -1, 0$ **20.** $x^2 + 26x$

21. 6 centimeters $\times$ 9 centimeters

22. Base: 4 feet; Height: 10 feet

23. 50 computer desks

CHAPTER 6

Section 6.1   (page 377)

1. $(-\infty, \infty)$ 3. $(-\infty, 3) \cup (3, \infty)$

5. $(-\infty, 9) \cup (9, \infty)$ 7. $(-\infty, -10) \cup (-10, \infty)$

9. $(-\infty, \infty)$ 11. $(-\infty, -3) \cup (-3, 0) \cup (0, \infty)$

13. $(-\infty, 0) \cup (0, 1) \cup (1, \infty)$

15. $(-\infty, -4) \cup (-4, 4) \cup (4, \infty)$

17. $(-\infty, 0) \cup (0, 3) \cup (3, \infty)$

19. $(-\infty, 2) \cup (2, 3) \cup (3, \infty)$

21. $(-\infty, -1) \cup \left(-1, \frac{5}{3}\right) \cup \left(\frac{5}{3}, \infty\right)$

23. (a) 1   (b) -8 (c) Undefined (division by 0)   (d) 0

25. (a) 0   (b) 0   (c) Undefined (division by 0)
    (d) Undefined (division by 0)

27. (a) $\frac{25}{22}$ (b) 0   (c) Undefined (division by 0)
    (d) Undefined (division by 0)

29. $(0, \infty)$ 31. $\{1, 2, 3, 4, \ldots\}$ 33. $[0, 100]$

35. $x + 3$ 37. $3(x + 16)^2$ 39. $x(x - 2)$

41. $x + 3$ 43. $\dfrac{x}{5}$ 45. $x, \ x \neq 0$ 47. $\dfrac{6x}{5y^3}, \ x \neq 0$

49. $\dfrac{x - 3}{4x}$ 51. $x, \ x \neq 8, \ x \neq 0$ 53. $\dfrac{1}{2}, \ x \neq \dfrac{3}{2}$

55. $\dfrac{y + 7}{2}, \ y \neq 7$ 57. $\dfrac{1}{a + 3}$ 59. $\dfrac{x}{x - 7}$

61. $\dfrac{y(y + 2)}{y + 6}, \ y \neq 2$ 63. $\dfrac{y^2(y - 4)}{y - 3}, \ y \neq -4$

65. $-\dfrac{3x + 5}{x + 3}, \ x \neq 4$ 67. $\dfrac{x + 8}{x - 3}, \ x \neq -\dfrac{3}{2}$

69. $\dfrac{3x - 1}{5x - 4}, \ x \neq -\dfrac{4}{5}$ 71. $\dfrac{3y^2}{y^2 + 1}, \ x \neq 0$

73. $\dfrac{y - 8x}{15}, \ y \neq -8x$ 75. $\dfrac{5 + 3xy}{y^2}, \ x \neq 0$

77. $\dfrac{u - 2v}{u - v}, \ u \neq -2v$ 79. $\dfrac{3(m - 2n)}{m + 2n}$

81.

x	-2	-1	0	1	2		3	4
$\dfrac{x^2 - x - 2}{x - 2}$	-1	0	1	2	Undef.		4	5
$x + 1$	-1	0	1	2	3		4	5

$$\dfrac{x^2 - x - 2}{x - 2} = \dfrac{(x - 2)(x + 1)}{x - 2} = x + 1, \ x \neq 2$$

83. $\dfrac{x}{x + 3}, \ x > 0$ 85. $\dfrac{1}{4}, \ x > 0$

87. (a) $C = 2500 + 9.25x$ (b) $\overline{C} = \dfrac{2500 + 9.25x}{x}$
    (c) $\{1, 2, 3, 4, \ldots\}$ (d) \$34.25

89. (a) Van: $45(t + 3)$; Car: $60t$ (b) $d = |15(9 - t)|$
    (c) $\dfrac{4t}{3(t + 3)}$

91. $\pi, \ d > 0$ 93. $\dfrac{1189.2t + 25{,}266}{-0.35t + 67.1}, \ 1 \leq t \leq 5$

95. The rational expression is in simplified form if the numerator and denominator have no factors in common (other than ± 1).

97. $\dfrac{1}{x^2 + 1}$

99. The student incorrectly divided (the denominator may not be split up) and the domain is not restricted.
    Correct solution: $\dfrac{x^2 + 7x}{x + 7} = \dfrac{x(x + 7)}{x + 7} = x, \ x \neq -7$

101. To write the polynomial $g(x)$, multiply $f(x)$ by $(x - 2)$ and divide by $(x - 2)$.
    $$g(x) = \dfrac{f(x)(x - 2)}{(x - 2)} = f(x), \ x \neq 2$$

103. $\frac{3}{16}$ 105. 1   107. $4a^4$

109. $-3b^3 + 9b^2 - 15b$

Section 6.2   (page 387)

1. x^2 3. $(x + 2)^2$ 5. $u + 1$ 7. $-(2 + x)$

9. $\dfrac{7}{3}, \ x \neq 0$ 11. $\dfrac{s^3}{6}, \ s \neq 0$ 13. $24u^2, \ u \neq 0$

15. $24, \ x \neq -\dfrac{3}{4}$ 17. $\dfrac{2uv(u + v)}{3(3u + v)}, \ u \neq 0$

19. $-1, \ r \neq 12$ 21. $-\dfrac{x + 8}{x^2}, \ x \neq \dfrac{3}{2}$

23. $4(r + 2), \ r \neq 3, \ r \neq 2$ 25. $2t + 5, \ t \neq 3, \ t \neq -2$

27. $-\dfrac{xy(x + 2y)}{x - 2y}$ 29. $\dfrac{(x - y)^2}{x + y}, \ x \neq -3y$

31. $\dfrac{(x - 1)(2x + 1)}{(3x - 2)(x + 2)}, \ x \neq \pm 5, \ x \neq -1$

33. $-\dfrac{x^2(x + 3)(x - 3)(2x + 5)(3x - 1)}{2(2x + 1)(x + 3)(2x - 3)}, \ x \neq 0, \ x \neq \dfrac{1}{2}$

35. $\dfrac{(x + 3)^2}{x}, \ x \neq 3, \ x \neq \pm 2$ 37. $\dfrac{x(x + 1)}{3(x + 2)}, \ x \neq -1$

39. $\dfrac{4x}{3}, \ x \neq 0$ 41. $\dfrac{6}{x}$ 43. $\dfrac{3y^2}{2ux^2}, \ v \neq 0$

45. $\dfrac{3}{2(a + b)}$ 47. $\frac{4}{3}, \ x \neq -2, \ x \neq 0, \ x \neq 1$

49. $x^4y(x + 2y)$, $x \neq 0$, $y \neq 0$, $x \neq -2y$

51. $\dfrac{x - 4}{x - 5}$, $x \neq -6$, $x \neq -5$, $x \neq 3$

53. $\dfrac{x + 4}{3}$, $x \neq -2$, $x \neq 0$ **55.** $\dfrac{1}{4}$, $x \neq -1$, $x \neq 0$, $y \neq 0$

57. $\dfrac{(x + 1)(2x - 5)}{x}$, $x \neq -5$, $x \neq -1$, $x \neq -\dfrac{2}{3}$

59. $\dfrac{x^4}{(x^n + 1)^2}$, $x^n \neq -3$, $x^n \neq 3$, $x \neq 0$

61.

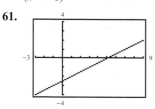

63. $\dfrac{w(2w + 3)}{6}$ **65.** $\dfrac{x}{4(2x + 1)}$ **67.** $\dfrac{\pi x}{4(2x + 1)}$

69. $Y = \dfrac{-0.696t + 8.94}{(-0.092t + 1)(0.352t + 15.97)}$, $1 \leq t \leq 6$

71. In simplifying a product of rational expressions, you divide the common factors out of the numerator and denominator.

73. The domain needs to be restricted, $x \neq a$, $x \neq b$.

75. The first expression needs to be multiplied by the reciprocal of the second expression, and the domain needs to be restricted.

$$\frac{x^2 - 4}{5x} \div \frac{x + 2}{x - 2} = \frac{x^2 - 4}{5x} \cdot \frac{x - 2}{x + 2}$$

$$= \frac{(x - 2)^2(x + 2)}{5x(x + 2)}$$

$$= \frac{(x - 2)^2}{5x}, \quad x \neq \pm 2$$

77. $\dfrac{9}{8}$ **79.** $\dfrac{13}{15}$ **81.** $-3, 0$ **83.** $\pm\dfrac{5}{2}$

Section 6.3   *(page 396)*

1. $\dfrac{3x}{2}$ **3.** $-\dfrac{3}{a}$ **5.** $-\dfrac{2}{9}$ **7.** $\dfrac{2z^2 - 2}{3}$ **9.** $\dfrac{x + 6}{3x}$

11. 1, $y \neq 6$ **13.** $\dfrac{1}{x - 3}$, $x \neq 0$ **15.** $\dfrac{1}{w - 2}$, $w \neq -2$

17. $\dfrac{1}{c + 4}$, $c \neq 1$ **19.** $-\dfrac{4}{3}$ **21.** $x - 6$, $x \neq 3$

23. $20x^3$ **25.** $36y^3$ **27.** $15x^2(x + 5)$

29. $126z^2(z + 1)^4$ **31.** $56t(t + 2)(t - 2)$

33. $2y(y + 1)(2y - 1)$ **35.** x^2 **37.** $u + 1$

39. $-(x + 2)$ **41.** $\dfrac{2n^2(n + 8)}{6n^2(n - 4)}$, $\dfrac{10(n - 4)}{6n^2(n - 4)}$

43. $\dfrac{2(x + 3)}{x^2(x + 3)(x - 3)}$, $\dfrac{5x(x - 3)}{x^2(x + 3)(x - 3)}$

45. $\dfrac{3v^2}{6v^2(v + 1)}$, $\dfrac{8(v + 1)}{6v^2(v + 1)}$

47. $\dfrac{(x - 8)(x - 5)}{(x + 5)(x - 5)^2}$, $\dfrac{9x(x + 5)}{(x + 5)(x - 5)^2}$ **49.** $\dfrac{-12x + 25}{20x}$

51. $\dfrac{7(a + 2)}{a^2}$ **53.** $\dfrac{5(5x + 22)}{x + 4}$ **55.** 0, $x \neq 4$

57. $\dfrac{3(x + 2)}{x - 8}$ **59.** 1, $x \neq \dfrac{2}{3}$ **61.** $\dfrac{3(8v - 3)}{5v(v - 1)}$

63. $\dfrac{x^2 - 7x - 15}{(x + 3)(x - 2)}$ **65.** $-\dfrac{2}{x + 3}$, $x \neq 3$

67. $\dfrac{5(x + 1)}{(x + 5)(x - 5)}$ **69.** $\dfrac{4}{x^2(x^2 + 1)}$

71. $\dfrac{6}{(x - 6)(x + 5)}$ **73.** $\dfrac{4x}{(x - 4)^2}$

75. $\dfrac{y - x}{xy}$, $x \neq -y$ **77.** $\dfrac{2(4x^2 + 5x - 3)}{x^2(x + 3)}$

79. $-\dfrac{u^2 - uv - 5u + 2v}{(u - v)^2}$ **81.** $\dfrac{x}{x - 1}$, $x \neq -6$

83.

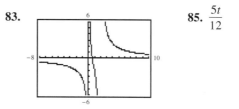

85. $\dfrac{5t}{12}$

87. $A = -4$, $B = 2$, $C = 2$

89. $T = \dfrac{633.50t^2 + 5814.42t + 13,159.79}{(0.183t + 1)(0.205t + 1)}$, $0 \leq t \leq 5$

91. When the numerators are subtracted, the result should be $(x - 1) - (4x - 11) = x - 1 - 4x + 11$.

93. Yes. The LCD of $4x + 1$ and $\dfrac{x}{x + 2}$ is $x + 2$.

95. $2v + 4$ **97.** $x^2 - 2x - 3$ **99.** $(x - 3)(x - 4)$

101. $(a - 6)(2a + 3)$

Section 6.4   *(page 404)*

1. $\dfrac{1}{4}$ **3.** $6xz^3$, $x \neq 0$, $y \neq 0$, $z \neq 0$

5. $\dfrac{2xy^2}{5}$, $x \neq 0$, $y \neq 0$ **7.** $-\dfrac{1}{y}$, $y \neq 3$

9. $-\dfrac{5x(x + 1)}{2}$, $x \neq -1$, $x \neq 0$, $x \neq 5$

11. $\dfrac{x + 5}{3(x + 4)}$, $x \neq 2$ **13.** $\dfrac{2(x + 3)}{x - 2}$, $x \neq -3$, $x \neq 7$

15. $\dfrac{(2x - 5)(3x + 1)}{3x(x + 1)}$, $x \neq \pm\dfrac{1}{3}$

17. $\dfrac{(x + 3)(4x + 1)}{(3x - 1)(x - 1)}$, $x \neq -3$, $x \neq -\dfrac{1}{4}$

19. $x + 2$, $x \neq \pm 2$, $x \neq -3$

21. $-\dfrac{(x - 5)(x + 2)^2}{(x - 2)^2(x + 3)}$, $x \neq -2$, $x \neq 7$ **23.** $\dfrac{y + 4}{y^2}$

25. $-\dfrac{3x + 4}{3x - 4}$, $x \neq 0$ **27.** $\dfrac{x^2}{2(2x + 3)}$, $x \neq 0$

29. $\dfrac{3}{4}$, $x \neq 0$, $x \neq 3$ **31.** $\dfrac{5(x + 3)}{2x(5x - 2)}$

33. $y - x$, $x \neq 0$, $y \neq 0$, $x \neq -y$ **35.** $\dfrac{x + y}{x - y}$, $x \neq 2y$

37. $-\dfrac{(y - 1)(y - 3)}{y(4y - 1)}$, $y \neq 3$ **39.** $\dfrac{20}{7}$, $x \neq -1$

41. $\dfrac{1}{x}$, $x \neq -1$ **43.** $\dfrac{x(x + 6)}{3x^3 + 10x - 30}$, $x \neq 0$, $x \neq 3$

45. $\dfrac{y(2y^2 - 1)}{10y^2 - 1}$, $y \neq 0$ **47.** $\dfrac{x^2(7x^3 + 2)}{x^4 + 5}$, $x \neq 0$

49. $\dfrac{y + x}{y - x}$, $x \neq 0$, $y \neq 0$ **51.** $\dfrac{y - x}{x^2 y^2 (y + x)}$

53. $-\dfrac{1}{2(h + 2)}$, $h \neq 0$ **55.** $\dfrac{11x}{60}$ **57.** $\dfrac{11x}{24}$

59. $\dfrac{b^2 + 5b + 8}{8b}$ **61.** $\dfrac{x}{8}, \dfrac{5x}{36}, \dfrac{11x}{72}$

63. $\dfrac{R_1 R_2}{R_1 + R_2}$, $R_1 \neq 0$, $R_2 \neq 0$

65. (a)

(b) $B = \dfrac{250(-487.42t^2 + 4510.08t + 60{,}227.5)}{3(-257.34t^2 + 1992.05t + 111{,}039.2)}$,

$0 \leq t \leq 5$

67. No. A complex fraction can be written as the division of two rational expressions, so the simplified form will be a rational expression.

69. In the second step, the parentheses cannot be moved because division is not associative.

$\dfrac{(a/b)}{b} = \dfrac{a}{b} \cdot \dfrac{1}{b} = \dfrac{a}{b^2}$

71. $72y^5$ **73.** $(3x - 1)(x + 2)$ **75.** $\dfrac{x}{8}$, $x \neq 0$

77. $(x + 1)(x + 2)^2$, $x \neq -2$, $x \neq -1$

Mid-Chapter Quiz   *(page 408)*

1. $(-\infty, -1) \cup (-1, 0) \cup (0, \infty)$

2. (a) 0 (b) $\dfrac{9}{2}$ (c) Undefined   (d) $\dfrac{8}{9}$

3. $\dfrac{3}{2}y$, $y \neq 0$ **4.** $\dfrac{2u^3}{5}$, $u \neq 0$, $v \neq 0$

5. $-\dfrac{2x + 1}{x}$, $x \neq \dfrac{1}{2}$ **6.** $\dfrac{z + 3}{2z - 1}$, $z \neq -3$

7. $\dfrac{5a + 3b^2}{ab}$ **8.** $\dfrac{n^2}{m + n}$, $2m \neq n$ **9.** $\dfrac{t}{2}$, $t \neq 0$

10. $\dfrac{5x}{x - 2}$, $x \neq -2$ **11.** $\dfrac{8x}{3(x + 3)(x - 1)^2}$

12. $\dfrac{2x^3 y}{z}$, $x \neq 0$, $y \neq 0$ **13.** $\dfrac{(a + 1)^2}{9(a + b)^2}$, $a \neq b$, $a \neq -1$

14. $\dfrac{4(u - v)^2}{5uv}$, $u \neq \pm v$ **15.** $\dfrac{7x - 11}{x - 2}$

16. $-\dfrac{4x^2 - 25x + 36}{(x - 3)(x + 3)}$ **17.** 0, $x \neq 2$, $x \neq -1$

18. $-\dfrac{3t}{2}$, $t \neq 0$, $t \neq 3$ **19.** $\dfrac{2(x + 1)}{3x}$, $x \neq -2$, $x \neq -1$

20. $\dfrac{(3y - x)(x - y)}{xy}$, $x \neq y$

21. (a) $C = 25{,}000 + 144x$ (b) $\overline{C} = \dfrac{25{,}000 + 144x}{x}$

(c) $\$194$

22. $\dfrac{13x}{18}$

Section 6.5   *(page 415)*

1. $7x^2 - 2x$, $x \neq 0$ **3.** $-4x + 2$, $x \neq 0$

5. $m^3 + 2m - \dfrac{7}{m}$ **7.** $-10z^2 - 6$, $z \neq 0$

9. $v^2 + \dfrac{5}{2}v - 2$, $v \neq 0$ **11.** $x^3 - \dfrac{3}{2}x^2 + 3x - 2$, $x \neq 0$

13. $\dfrac{5}{2}x - 4 + \dfrac{7}{2}y$, $x \neq 0$, $y \neq 0$ **15.** $112 + \dfrac{5}{9}$

17. $215 + \dfrac{2}{3}$ **19.** $x - 5$, $x \neq 3$

21. $x + 10$, $x \neq -5$ **23.** $x - 3 + \dfrac{2}{x - 2}$

25. $x + 7$, $x \neq 3$ **27.** $5x - 8 + \dfrac{19}{x + 2}$

29. $4x + 3 - \dfrac{11}{3x + 2}$ **31.** $3t - 4$, $t \neq \dfrac{3}{2}$

33. $y + 3$, $y \neq -\dfrac{1}{2}$ **35.** $x^2 + 4$, $x \neq 2$

37. $3x^2 - 3x + 1 + \dfrac{2}{3x + 2}$ **39.** $2 + \dfrac{5}{x + 2}$

41. $x - 4 + \dfrac{32}{x + 4}$ **43.** $\dfrac{6}{5}z + \dfrac{41}{25} + \dfrac{41}{25(5z - 1)}$

45. $4x - 1$, $x \neq -\dfrac{1}{4}$ **47.** $x^2 - 5x + 25$, $x \neq -5$

49. $x + 2 + \dfrac{1}{x^2 + 2x + 3}$

51. $4x^2 + 12x + 25 + \dfrac{52x - 55}{x^2 - 3x + 2}$

53. $x^3 + x^2 + x + 1,\ x \neq 1$ **55.** $x^3 - x + \dfrac{x}{x^2 + 1}$

57. $7uv,\ u \neq 0,\ v \neq 0$ **59.** $3x + 4,\ x \neq -2$

61. $x + 3,\ x \neq 2$ **63.** $x^2 - x + 4 - \dfrac{17}{x + 4}$

65. $x^3 - 2x^2 - 4x - 7 - \dfrac{4}{x - 2}$

67. $5x^2 + 14x + 56 + \dfrac{232}{x - 4}$

69. $10x^3 + 10x^2 + 60x + 360 + \dfrac{1360}{x - 6}$

71. $0.1x + 0.82 + \dfrac{1.164}{x - 0.2}$ **73.** $(x - 3)(x + 4)(x - 2)$

75. $(x - 1)(2x + 1)^2$ **77.** $(x + 3)^2(x - 3)(x + 4)$

79. $5\left(x - \tfrac{4}{5}\right)(3x + 2)$ **81.** -8

83.

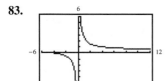

85. $x^{2n} + x^n + 4,\ x^n \neq -2$ **87.** $x^3 - 5x^2 - 5x - 10$

89. $f(k)$ equals the remainder when dividing by $(x - k)$.

k	$f(k)$	Divisor $(x - k)$	Remainder
-2	-8	$x + 2$	-8
-1	0	$x + 1$	0
0	0	x	0
$\tfrac{1}{2}$	$-\tfrac{9}{8}$	$x - \tfrac{1}{2}$	$-\tfrac{9}{8}$
1	-2	$x - 1$	-2
2	0	$x - 2$	0

91. $x^2 + 2x + 1$ **93.** $2x + 8$

95. x is not a factor of the numerator.

$$\dfrac{6x + 5y}{x} = \dfrac{6x}{x} + \dfrac{5y}{x} = 6 + \dfrac{5y}{x}$$

97. $\dfrac{x^2 + 4}{x + 1} = x - 1 + \dfrac{5}{x + 1}$

Dividend: $x^2 + 4$; Divisor: $x + 1$;
Quotient: $x - 1$; Remainder: 5

99.

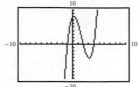

The x-intercepts are $(-1, 0), (2, 0),$ and $(4, 0)$. The polynomials in parts (a), (b), and (c) are all equivalent.

101. $x < \tfrac{3}{2}$ **103.** $1 < x < 5$ **105.** $x \leq -8$ or $x \geq 16$

107. Quadrant II or III   **109.** Located on the y-axis

Section 6.6   *(page 424)*

1. (a) Not a solution   (b) Not a solution
    (c) Not a solution   (d) Solution

3. (a) Not a solution   (b) Solution
    (c) Solution   (d) Not a solution

5. 10 **7.** 1 **9.** 0 **11.** 8 **13.** $-\tfrac{40}{9}$

15. $-3, \tfrac{8}{3}$ **17.** $-\tfrac{2}{9}$ **19.** $\tfrac{7}{4}$ **21.** $\tfrac{43}{8}$ **23.** 61

25. $\tfrac{18}{5}$ **27.** $-\tfrac{26}{5}$ **29.** 3 **31.** 3 **33.** $-\tfrac{11}{5}$

35. $\tfrac{4}{3}$ **37.** ± 6 **39.** ± 4 **41.** $-9, 8$ **43.** $3, 6$

45. No solution   **47.** -5 **49.** 8 **51.** 3 **53.** 5

55. $-\tfrac{11}{10}, 2$ **57.** 20 **59.** $\tfrac{3}{2}$ **61.** $3, -1$ **63.** $\tfrac{17}{4}$

65. $-3, 4$ **67.** (a) and (b) $(-2, 0)$

69. (a) and (b) $(-1, 0), (1, 0)$

71. (a)   (b) $(4, 0)$

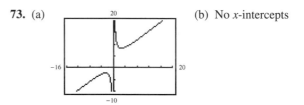

73. (a)   (b) No x-intercepts

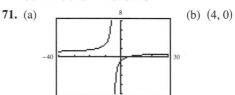

75. (a)   (b) $(-3, 0), (2, 0)$

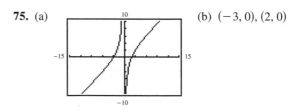

77. -12 **79.** $\dfrac{x^2 + 2x + 8}{(x+4)(x-4)}$ **81.** $6, \dfrac{1}{6}$

83. 2.4 hours $=$ 2 hours 24 minutes

85. 40 miles per hour   **87.** 121 saves

89. An extraneous solution is a "trial solution" that does not satisfy the original equation.

91. When the equation is solved, the solution is $x = 0$. However, if $x = 0$, then there is division by zero, so the equation has no solution.

93. $(x+9)(x-9)$ **95.** $\left(2x - \tfrac{1}{2}\right)\left(2x + \tfrac{1}{2}\right)$

97. $(-\infty, \infty)$

Section 6.7   *(page 435)*

1. $I = kV$ **3.** $V = kt$ **5.** $u = kv^2$ **7.** $p = k/d$

9. $A = k/t^4$ **11.** $A = klw$ **13.** $P = k/V$

15. Area varies jointly as the base and the height.

17. Volume varies jointly as the square of the radius and the height.

19. Average speed varies directly as the distance and inversely as the time.

21. $s = 5t$ **23.** $F = \tfrac{5}{16}x^2$ **25.** $n = 48/m$

27. $g = 4/\sqrt{z}$ **29.** $F = \tfrac{25}{6}xy$ **31.** $d = 120x^2/r$

33.

x	2	4	6	8	10
$y = kx^2$	4	16	36	64	100

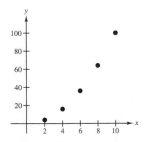

35.

x	2	4	6	8	10
$y = kx^2$	2	8	18	32	50

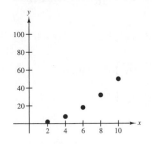

37.

x	2	4	6	8	10
$y = \dfrac{k}{x^2}$	$\dfrac{1}{2}$	$\dfrac{1}{8}$	$\dfrac{1}{18}$	$\dfrac{1}{32}$	$\dfrac{1}{50}$

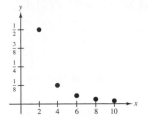

39.

x	2	4	6	8	10
$y = \dfrac{k}{x^2}$	$\dfrac{5}{2}$	$\dfrac{5}{8}$	$\dfrac{5}{18}$	$\dfrac{5}{32}$	$\dfrac{1}{10}$

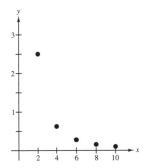

41. $y = k/x$ with $k = 4$

43. 7.5 miles per hour; 9 miles per hour   **45.** 10 people

47. 180 minutes   **49.** 85%

51. (a) $\{1, 2, 3, 4, \ldots\}$

(b)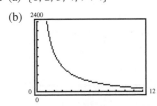

(c) $10d$ (d) $10d$

53. $5983; Price per unit

55. (a) 2 inches   (b) 15 pounds   **57.** 18 pounds

59. -128 feet per second   **61.** 192 feet

63. 1350 watts   **65.** 667 boxes   **67.** $p = \dfrac{114}{t}$; 18%

69. $6 per pizza; Answers will vary.   **71.** $22.50

73. (a) $P = \dfrac{kWD^2}{L}$

(b) Unchanged

(c) Increases by a factor of 8

(d) Increases by a factor of 4

(e) Decreases by a factor of $\dfrac{1}{4}$

(f) 3125 pounds

75. False. If x increases, then z and y do not both necessarily increase.

77. The variable y will be one-fourth as great. If $y = k/x^2$ and x is replaced with $2x$, the result is

$$y = \dfrac{k}{(2x)^2} = \dfrac{k}{4x^2}.$$

79. 6^4 **81.** $\left(\frac{1}{5}\right)^5$ **83.** $x - 7$, $x \neq -2$

85. $4x^4 - 2x^3$, $x \neq 3$

Review Exercises   (page 442)

1. $(-\infty, 8) \cup (8, \infty)$ **3.** $(-\infty, \infty)$

5. $(-\infty, 1) \cup (1, 6) \cup (6, \infty)$ **7.** $(0, \infty)$

9. $\dfrac{2x^3}{5}$, $x \neq 0$, $y \neq 0$ **11.** $\dfrac{b - 3}{6(b - 4)}$ **13.** -9, $x \neq y$

15. $\dfrac{x}{2(x + 5)}$, $x \neq 5$ **17.** $\dfrac{1}{2}$, $x > 0$ **19.** $\dfrac{x}{3}$, $x \neq 0$

21. $\dfrac{y}{8x}$, $y \neq 0$ **23.** $12z(z - 6)$, $z \neq -6$

25. $-\frac{1}{4}$, $u \neq 0$, $u \neq 3$ **27.** $20x^3$, $x \neq 0$

29. $\dfrac{125y}{x}$, $y \neq 0$ **31.** $\dfrac{1}{3x - 2}$, $x \neq -2$, $x \neq -1$

33. $\dfrac{x(x - 1)}{x - 7}$, $x \neq -1$, $x \neq 1$ **35.** $3x$ **37.** $\dfrac{4}{x}$

39. $-\dfrac{x - 13}{4x}$ **41.** $\dfrac{5y + 11}{2y + 1}$ **43.** $\dfrac{7x - 16}{x + 2}$

45. $\dfrac{2x + 3}{5x^2}$ **47.** $\dfrac{4x + 3}{(x + 5)(x - 12)}$

49. $\dfrac{5x^3 - 5x^2 - 31x + 13}{(x - 3)(x + 2)}$ **51.** $\dfrac{2x + 17}{(x - 5)(x + 4)}$

53. $\dfrac{6(x - 9)}{(x + 3)^2(x - 3)}$

55.

57. $3x^2$, $x \neq 0$

59. $-\dfrac{1}{2}$, $x \neq 0$, $x \neq 2$ **61.** $\dfrac{6(x + 5)}{x(x + 7)}$, $x \neq \pm 5$

63. $\dfrac{3t^2}{5t - 2}$, $t \neq 0$ **65.** $x - 1$, $x \neq 0$, $x \neq 2$

67. $\dfrac{-a^2 + a + 16}{(4a^2 + 16a + 1)(a - 4)}$, $a \neq -4$, $a \neq 0$

69. $2x^2 - \frac{1}{2}$, $x \neq 0$ **71.** $3xy - y + 1$, $x \neq 0$, $y \neq 0$

73. $2x^2 + \dfrac{4}{3}x - \dfrac{8}{9} + \dfrac{10}{9(3x - 1)}$ **75.** $x^2 - 2$, $x \neq \pm 1$

77. $x^2 - x - 3 - \dfrac{3x^2 - 2x - 3}{x^3 - 2x^2 + x - 1}$

79. $x + 2 + \dfrac{3}{x + 1}$ **81.** $x^2 + 5x - 7$, $x \neq -2$

83. $x^3 + 3x^2 + 6x + 18 + \dfrac{29}{x - 3}$

85. $(x - 2)(x + 1)(x + 3)$ **87.** 6   **89.** -120

91. $2, -\frac{3}{2}$ **93.** $\frac{36}{23}$ **95.** 5   **97.** $-4, 6$

99. $-\frac{16}{3}, 3$ **101.** $-4, -2$ **103.** No solution

105. $-2, 2$ **107.** $-\frac{9}{5}, 3$

109. 16 miles per hour, 10 miles per hour   **111.** 4 people

113. 8 years   **115.** 150 pounds   **117.** 2.44 hours

119. $61.49; Answers will vary.   **121.** $922.50

Chapter Test   (page 447)

1. $(-\infty, 1) \cup (1, 5) \cup (5, \infty)$ **2.** -2, $x \neq 2$

3. $\dfrac{2a + 3}{5}$, $a \neq 4$ **4.** $3x^3(x + 4)^2$ **5.** $\dfrac{5z}{3}$, $z \neq 0$

6. $\dfrac{4}{y + 4}$, $y \neq 2$ **7.** $\dfrac{14y^6}{15}$, $x \neq 0$

8. $(2x - 3)^2(x + 1)$, $x \neq -\frac{3}{2}$, $x \neq -1$, $x \neq \frac{3}{2}$

9. $\dfrac{x + 1}{x - 3}$ **10.** $\dfrac{-2x^2 + 2x + 1}{x + 1}$ **11.** $\dfrac{5x^2 - 15x - 2}{(x - 3)(x + 2)}$

12. $\dfrac{5x^3 + x^2 - 7x - 5}{x^2(x + 1)^2}$ **13.** $\dfrac{x^3}{4}$, $x \neq -2$, $x \neq 0$

14. $-(3x + 1)$, $x \neq 0$, $x \neq \frac{1}{3}$

15. $\dfrac{(3y + x^2)(x + y)}{x^2y}$, $x \neq -y$ **16.** $3x - 2 + \dfrac{4}{x}$

17. $t^2 + 3 - \dfrac{6t - 6}{t^2 - 2}$ **18.** $2x^3 + 6x^2 + 3x + 9 + \dfrac{20}{x - 3}$

19. 16   **20.** $-1, -\frac{15}{2}$ **21.** No solution

22. $v = \frac{1}{4}\sqrt{u}$ **23.** 240 cubic meters

CHAPTER 7

Section 7.1   (page 457)

1. 8   **3.** -7 **5.** -3 **7.** Not a real number

9. Perfect square   **11.** Perfect cube   **13.** Neither

15. ± 5 **17.** $\pm\frac{3}{4}$ **19.** $\pm\frac{1}{7}$ **21.** ± 0.4 **23.** 2

25. $-\frac{1}{3}$ **27.** $\frac{1}{10}$ **29.** 0.1   **31.** Irrational

33. Rational   **35.** Rational   **37.** Irrational

39. 8   **41.** 10   **43.** Not a real number   **45.** $-\frac{2}{3}$

47. Not a real number   **49.** 5   **51.** -23 **53.** 5

55. 10   **57.** 6   **59.** $-\frac{1}{4}$ **61.** 11   **63.** -24

65. 3   **67.** -2

Radical Form	*Rational Exponent Form*
69. $\sqrt{36} = 6$	$36^{1/2} = 6$
71. $\sqrt[4]{256^3} = 64$	$256^{3/4} = 64$

73. 5   **75.** -6 **77.** $\frac{1}{4}$ **79.** $\frac{1}{9}$ **81.** $\frac{4}{9}$ **83.** $\frac{3}{11}$

85. 9   **87.** -64 **89.** $t^{1/2}$ **91.** x^3 **93.** $u^{7/3}$

95. $x^{-1} = \dfrac{1}{x}$ **97.** $t^{-9/4} = \dfrac{1}{t^{9/4}}$ **99.** x^3

101. $y^{13/12}$ **103.** $x^{3/4}y^{1/4}$ **105.** $y^{5/2}z^4$ **107.** 3

109. $2^{1/3}$ **111.** $\dfrac{1}{2}$ **113.** $c^{1/2}$ **115.** $\dfrac{3y^2}{4z^{4/3}}$

117. $\dfrac{9y^{3/2}}{x^{2/3}}$ **119.** $x^{1/4}$ **121.** $y^{1/8}$ **123.** $x^{3/8}$

125. $(x+y)^{1/2}$ **127.** $\dfrac{1}{(3u-2v)^{5/6}}$ **129.** 5.9161

131. 9.9845   **133.** 0.0367   **135.** 3.8158

137. 66.7213   **139.** 1.0420   **141.** 0.7915

143. (a) 3   (b) 5   (c) Not a real number   (d) 9

145. (a) 2   (b) 3   (c) -2 (d) -4

147. (a) 2   (b) Not a real number   (c) 3   (d) 1

149. $0 \le x < \infty$ **151.** $-\infty < x \le \frac{4}{9}$

153. $-\infty < x < \infty$ **155.** $-\frac{9}{2} \le x < \infty$

157. $0 < x < \infty$

159. Domain: $(0, \infty)$ **161.** Domain: $(-\infty, \infty)$

163. $2x^{3/2} - 3x^{1/2}$ **165.** $1 + 5y$ **167.** 0.128

169. 23 feet $\times$ 23 feet   **171.** 104.64 inches

173. If a and b are real numbers, n is an integer greater than or equal to 2, and $a = b^n$, then b is the nth root of a.

175. No. $\sqrt{2}$ is an irrational number. Its decimal representation is a nonterminating, nonrepeating decimal.

177. 0, 1, 4, 5, 6, 9; Yes   **179.** 5   **181.** $-\frac{4}{5}$

183. $s = kt^2$ **185.** $a = kbc$

Section 7.2   (page 466)

1. $3\sqrt{2}$ **3.** $3\sqrt{5}$ **5.** $4\sqrt{6}$ **7.** $3\sqrt{17}$

9. $13\sqrt{7}$ **11.** 0.2   **13.** $0.06\sqrt{2}$ **15.** $2\sqrt{5}$

17. $\dfrac{\sqrt{13}}{5}$ **19.** $3x^2\sqrt{x}$ **21.** $4y^2\sqrt{3}$ **23.** $3\sqrt{13}|y^3|$

25. $2|x|y\sqrt{30y}$ **27.** $8a^2b^3\sqrt{3ab}$ **29.** $2\sqrt[3]{6}$

31. $2\sqrt[3]{14}$ **33.** $2x\sqrt[3]{5x^2}$ **35.** $3|y|\sqrt[4]{4y^2}$ **37.** $xy\sqrt[3]{x}$

39. $|xy|\sqrt[4]{4y^2}$ **41.** $2xy\sqrt[5]{y}$ **43.** $\dfrac{\sqrt[3]{35}}{4}$

45. $|y|\sqrt{13}$ **47.** $\dfrac{4a^2\sqrt{2}}{|b|}$ **49.** $\dfrac{2\sqrt[5]{x^2}}{y}$ **51.** $\dfrac{3a\sqrt[3]{2a}}{b^3}$

53. $-\dfrac{w\sqrt[3]{3w}}{2z}$ **55.** $\dfrac{\sqrt{3}}{3}$ **57.** $\dfrac{\sqrt{7}}{7}$ **59.** $\dfrac{\sqrt[4]{20}}{2}$

61. $\dfrac{3\sqrt[3]{2}}{2}$ **63.** $\dfrac{\sqrt{y}}{y}$ **65.** $\dfrac{2\sqrt{x}}{x}$ **67.** $\dfrac{\sqrt{2}}{2x}$

69. $\dfrac{2\sqrt{3b}}{b^2}$ **71.** $\dfrac{\sqrt[3]{18xy^2}}{3y}$ **73.** $3\sqrt{5}$

75. 89.44 cycles per second   **77.** $2\sqrt{194} \approx 27.86$ feet

79. $\sqrt{6} \cdot \sqrt{15} = \sqrt{90} = \sqrt{9} \cdot \sqrt{10} = 3\sqrt{10}$

81. $\left(\dfrac{5}{\sqrt{3}}\right)^2 = \dfrac{25}{3}$

No. Rationalizing the denominator produces an expression equivalent to the original expression, whereas squaring a number does not.

83. To find a perfect nth root factor, first factor the radicand completely. If the same factor appears at least n times, the perfect nth root factor is the common factor to the nth power.

85.

$(3, 2)$

87. No solution   **89.** $(4, -8, 10)$

Section 7.3   *(page 471)*

1. $2\sqrt{2}$ **3.** $7\sqrt{6}$ **5.** Cannot combine   **7.** $3\sqrt[3]{5}$

9. $13\sqrt[3]{y}$ **11.** $14\sqrt[4]{s}$ **13.** $9\sqrt{2}$

15. $4\sqrt[4]{5} - 7\sqrt[4]{13}$ **17.** $13\sqrt[3]{7} + \sqrt{3}$ **19.** $21\sqrt{3}$

21. $23\sqrt{5}$ **23.** $30\sqrt[3]{2}$ **25.** $12\sqrt{x}$ **27.** $13\sqrt{x+1}$

29. $13\sqrt{y}$ **31.** $(10 - z)\sqrt[3]{z}$ **33.** $(6a + 1)\sqrt{5a}$

35. $(x + 2)\sqrt[3]{6x}$ **37.** $4\sqrt{x-1}$ **39.** $(x + 2)\sqrt{x-1}$

41. $5a\sqrt[3]{ab^2}$ **43.** $3r^3s^2\sqrt{rs}$ **45.** 0 **47.** $\dfrac{2\sqrt{5}}{5}$

49. $\dfrac{9\sqrt{2}}{2}$ **51.** $\dfrac{5\sqrt{3y}}{3}$ **53.** $\dfrac{(3x + 2)\sqrt{3x}}{3x}$

55. $\dfrac{\sqrt{7y}(7y^3 - 3)}{7y^2}$ **57.** $>$ **59.** $<$ **61.** $12\sqrt{6x}$

63. $9x\sqrt{3} + 5\sqrt{3x}$

65. (a) $5\sqrt{10} \approx 15.8$ feet

(b) $400\sqrt{10} \approx 1264.9$ square feet

67. $T = -64 + 278.8\sqrt{t} - 214.1t + 51.8\sqrt{t^3}$; 45,756 people

69. No; $\sqrt{5} + \left(-\sqrt{5}\right) = 0$

71. Yes. $\sqrt{2x} + \sqrt{2x} = 2\sqrt{2x} = \sqrt{4} \cdot \sqrt{2x} = \sqrt{8x}$

73. (a) The student combined terms with unlike radicands; the radical expressions can be simplified no further.

(b) The student combined terms with unlike indices; the radical expressions can be simplified no further.

75. $\dfrac{3(z - 1)}{2z}$ **77.** $-3,\ x \neq 3$ **79.** $\dfrac{2v + 3}{v - 5}$

81. $\dfrac{15a^2}{2c^3},\ a \neq 0,\ b \neq 0$ **83.** $\dfrac{3(w + 1)}{w - 7},\ w \neq -1,\ w \neq 3$

Mid-Chapter Quiz   *(page 474)*

1. 15 **2.** $\frac{3}{2}$ **3.** 7 **4.** 9

5. (a) Not a real number   (b) 1 (c) 5

6. (a) 4 (b) 2 (c) 0

7. $-\infty < x < 0$ and $0 < x < \infty$ **8.** $-\dfrac{10}{3} \leq x < \infty$

9. $3|x|\sqrt{3}$ **10.** $2x^2\sqrt[4]{2}$ **11.** $\dfrac{2u\sqrt{u}}{3}$ **12.** $\dfrac{2\sqrt[3]{2}}{u^2}$

13. $5x|y|z^2\sqrt{5x}$ **14.** $4a^2b\sqrt[3]{2b^2}$ **15.** $4\sqrt{3}$

16. $3x\sqrt{7x},\ x \neq 0$ **17.** $3\sqrt{3} - 4\sqrt{7}$ **18.** $4\sqrt{2y}$

19. $7\sqrt{3}$ **20.** $4\sqrt{x + 2}$ **21.** $6x\sqrt[3]{5x^2} + 4x\sqrt[3]{5x}$

22. $4xy^2z^2\sqrt{xz}$ **23.** $23 + 8\sqrt{2} \approx 34.3$ inches

Section 7.4   *(page 479)*

1. 4 **3.** $3\sqrt{5}$ **5.** $2\sqrt[3]{9}$ **7.** 2 **9.** $3\sqrt{7} - 7$

11. $2\sqrt{10} + 8\sqrt{2}$ **13.** $3\sqrt{2}$ **15.** $12 - 4\sqrt{15}$

17. $y + 4\sqrt{y}$ **19.** $4\sqrt{a} - a$ **21.** $2 - 7\sqrt[3]{4}$

23. $\sqrt{15} - 5\sqrt{5} + 3\sqrt{3} - 15$ **25.** $8\sqrt{5} + 24$

27. $2\sqrt[3]{3} + 3\sqrt[3]{6} - 3\sqrt[3]{4} - 9$ **29.** -1 **31.** 29

33. $8 - 2\sqrt{15}$ **35.** $100 + 20\sqrt{2x} + 2x$

37. $45x - 17\sqrt{x} - 6$ **39.** $8x - 5$

41. $\sqrt[3]{4x^2} + 10\sqrt[3]{2x} + 25$ **43.** $y - 5\sqrt[3]{y} + 2\sqrt[3]{y^2} - 10$

45. $t + 5\sqrt[3]{t^2} + \sqrt[3]{t} - 3$ **47.** $x^2y^2(2y - |x|\sqrt{y})$

49. $4xy^3(x^4\sqrt[3]{x} + y\sqrt[3]{2x^2y^2})$ **51.** $x + 3$ **53.** $4 - 3x$

55. $2u + \sqrt{2u}$ **57.** $2 - \sqrt{5},\ -1$

59. $\sqrt{11} + \sqrt{3},\ 8$ **61.** $\sqrt{15} - 3,\ 6$

63. $\sqrt{x} + 3,\ x - 9$ **65.** $\sqrt{2u} + \sqrt{3},\ 2u - 3$

67. $2\sqrt{2} - \sqrt{4},\ 4$ **69.** $\sqrt{x} - \sqrt{y},\ x - y$

71. (a) $2\sqrt{3} - 4$ (b) 0 **73.** (a) 0 (b) $4 - 4\sqrt{3}$

75. $\dfrac{6\left(\sqrt{11} + 2\right)}{7}$ **77.** $\dfrac{7\left(5 - \sqrt{3}\right)}{22}$ **79.** $\dfrac{5 + 2\sqrt{10}}{5}$

81. $\dfrac{\sqrt{6} - \sqrt{2}}{2}$ **83.** $2\left(2\sqrt{3} + \sqrt{7}\right)$ **85.** $\dfrac{4\sqrt{7} + 11}{3}$

87. $\dfrac{2x - 9\sqrt{x} - 5}{4x - 1}$ **89.** $\dfrac{\left(\sqrt{15} + \sqrt{3}\right)x}{4}$

91. $\dfrac{5\sqrt{t} + t\sqrt{5}}{5 - t}$ **93.** $4\left(\sqrt{3a} - \sqrt{a}\right),\ a \neq 0$

95. $\dfrac{3(x - 4)\left(x^2 + \sqrt{x}\right)}{x(x - 1)(x^2 + x + 1)}$

97. $-\dfrac{\sqrt{u + v}\left(\sqrt{u - v} + \sqrt{u}\right)}{v}$

99.   **101.**

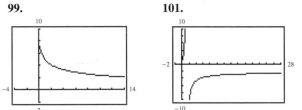

103. $\dfrac{2}{7\sqrt{2}}$ **105.** $\dfrac{10}{\sqrt{30x}}$ **107.** $\dfrac{4}{5\left(\sqrt{7} - \sqrt{3}\right)}$

109. $\dfrac{y - 25}{\sqrt{3}\left(\sqrt{y} + 5\right)}$ **111.** $192\sqrt{2}$ square inches

113. $\dfrac{500k\sqrt{k^2 + 1}}{k^2 + 1}$

115. (a) If either a or b (or both) equal zero, the expression is zero and therefore rational.

(b) If the product of a and b is a perfect square, then the expression is rational.

117. $\sqrt{a} - \sqrt{b}; \left(\sqrt{a} + \sqrt{b}\right)\left(\sqrt{a} - \sqrt{b}\right) = a - b;$
$\sqrt{b} - \sqrt{a}; \left(\sqrt{b} + \sqrt{a}\right)\left(\sqrt{b} - \sqrt{a}\right) = b - a;$

When the order of the terms is changed, the conjugate and the product both change by a factor of -1.

119. 6   **121.** No solution   **123.** $-12, 12$

125. $-5, 3$ **127.** $4|x|y^2\sqrt{2y}$ **129.** $2y\sqrt[4]{2x^2y}$

Section 7.5 *(page 489)*

1. (a) Not a solution (b) Not a solution

    (c) Not a solution (d) Solution

3. (a) Not a solution (b) Solution

    (c) Not a solution (d) Not a solution

5. 144   **7.** 49   **9.** 27   **11.** 49

13. No solution   **15.** 64   **17.** 90   **19.** -27

21. $\frac{4}{5}$ **23.** 5   **25.** No solution   **27.** 215

29. $\frac{577}{16}$ **31.** 4   **33.** No solution   **35.** 7

37. -15 **39.** $-\frac{9}{4}$ **41.** 8   **43.** $1, 3$ **45.** 1

47. $\frac{1}{4}$ **49.** $\frac{1}{2}$ **51.** 4   **53.** 7   **55.** 4

57. 216   **59.** $4, -12$ **61.** -16

63.   **65.**

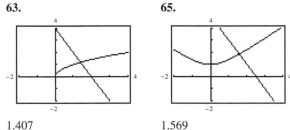

1.407   1.569

67.   **69.**

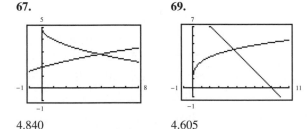

4.840   4.605

71.

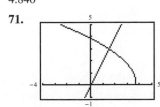

1.500

73. 25   **75.** $2, 6$ **77.** $\frac{4}{9}$ **79.** $2, 6$

81. 9.00   **83.** 12.00

85.

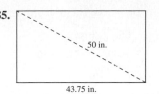

$\dfrac{25\sqrt{15}}{4} \approx 24.21$ inches

87. $2\sqrt{10} \approx 6.32$ meters   **89.** 15 feet

91. 30 inches $\times$ 16 inches

93. $h = \dfrac{\sqrt{S^2 - \pi^2 r^4}}{\pi r}$; 34 centimeters

95. $r = \dfrac{\sqrt{\pi A}}{\pi}$

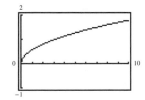

97. 64 feet   **99.** $32\sqrt{5} \approx 71.55$ feet per second

101. 39.06 feet   **103.** 1.82 feet   **105.** 500 units

107. (a)

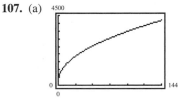

    (b) 92 months

109. $\left(\sqrt{x} + \sqrt{6}\right)^2 \neq \left(\sqrt{x}\right)^2 + \left(\sqrt{6}\right)^2$

111. Substitute $x = 20$ into the equation, and then choose any value of a such that $a \le 20$ and solve the resulting equation for b.

113. Parallel   **115.** Perpendicular   **117.** $(3, 2)$

119. $a^{4/5}$ **121.** $x^{3/2}$

Section 7.6 *(page 499)*

1. $2i$ **3.** $-12i$ **5.** $\frac{2}{5}i$ **7.** $-\frac{6}{11}i$ **9.** $2\sqrt{2}i$

11. $\sqrt{7}i$ **13.** 2   **15.** $\dfrac{3\sqrt{2}}{5}i$ **17.** $0.3i$ **19.** $10i$

21. $2i$ **23.** $3\sqrt{2}i$ **25.** $3\sqrt{3}i$ **27.** $-2\sqrt{6}$

29. $-3\sqrt{6}$ **31.** -0.44 **33.** $-3 - 2\sqrt{3}$

35. $-4\sqrt{5} + 5\sqrt{2}$ **37.** $4 + 3\sqrt{2}i$ **39.** -16

41. $-8i$ **43.** Equal   **45.** Not equal

47. $a = 3, b = -4$ **49.** $a = 2, b = -3$

51. $a = -4, b = -2\sqrt{2}$ **53.** $a = 64, b = 7$

55. $10 + 4i$ **57.** $-14 - 40i$ **59.** $-14 + 20i$

61. $9 - 7i$ **63.** $3 + 6i$ **65.** $\frac{13}{6} + \frac{3}{2}i$

67. $-6.15 - 9.3i$ **69.** $-3 + 49i$ **71.** -36

73. 24 **75.** $-35i$ **77.** $27i$ **79.** -9

81. $-65 - 10i$ **83.** $20 - 12i$ **85.** $4 + 18i$

87. $-40 - 5i$ **89.** $-14 + 42i$ **91.** 9

93. $-7 - 24i$ **95.** $-21 + 20i$ **97.** $18 + 26i$

99. $-i$ **101.** 1 **103.** -1 **105.** i **107.** -1

109. 5 **111.** 68 **113.** 31 **115.** 100 **117.** 144

119. 4 **121.** 2.5 **123.** $-10i$ **125.** $-\frac{1}{5} + \frac{2}{5}i$

127. $2 + 2i$ **129.** $1 - 2i$ **131.** $-\frac{24}{53} + \frac{84}{53}i$

133. $\frac{6}{29} + \frac{15}{29}i$ **135.** $\frac{12}{13} - \frac{5}{13}i$ **137.** $-\frac{23}{58} + \frac{43}{58}i$

139. $\frac{9}{5} - \frac{2}{5}i$ **141.** $\frac{47}{26} + \frac{27}{26}i$ **143.** $\frac{14}{29} - \frac{35}{29}i$

145–147. (a) Solution and (b) Solution

149. (a) $\left(\dfrac{-5 + 5\sqrt{3}i}{2}\right)^3 = 125$

   (b) $\left(\dfrac{-5 - 5\sqrt{3}i}{2}\right)^3 = 125$

151. (a) $1, \dfrac{-1 + \sqrt{3}i}{2}, \dfrac{-1 - \sqrt{3}i}{2}$

   (b) $2, \dfrac{-2 + 2\sqrt{3}i}{2} = -1 + \sqrt{3}i,$

   $\dfrac{-2 - 2\sqrt{3}i}{2} = -1 - \sqrt{3}i$

   (c) $4, \dfrac{-4 + 4\sqrt{3}i}{2} = -2 + 2\sqrt{3}i,$

   $\dfrac{-4 - 4\sqrt{3}i}{2} = -2 - 2\sqrt{3}i$

153. $2a$ **155.** $2bi$

157. Exercise 153: The sum of complex conjugates of the form $a + bi$ and $a - bi$ is twice the real number a, or $2a$.

Exercise 154: The product of complex conjugates of the form $a + bi$ and $a - bi$ is the sum of the squares of a and b, or $a^2 + b^2$.

Exercise 155: The difference of complex conjugates of the form $a + bi$ and $a - bi$ is twice the imaginary number bi, or $2bi$.

Exercise 156: The sum of the squares of complex conjugates of the form $a + bi$ and $a - bi$ is the difference of twice the squares of a and b, or $2a^2 - 2b^2$.

159. The numbers must be written in i-form first.

$\sqrt{-3}\sqrt{-3} = (\sqrt{3}i)(\sqrt{3}i) = 3i^2 = -3$

161. To simplify the quotient, multiply the numerator and the denominator by $-bi$. This will yield a positive real number in the denominator. The number i can also be used to simplify the quotient. The denominator will be the opposite of b, but the resulting number will be the same.

163. $-7, 5$ **165.** $-4, 0, 3$ **167.** 81 **169.** 25

Review Exercises   *(page 504)*

1. -9 **3.** -4 **5.** $-\frac{3}{4}$ **7.** $-\frac{1}{5}$

9. Not a real number

Radical Form	Rational Exponent Form
11. $\sqrt[3]{27} = 3$	$27^{1/3} = 3$
13. $\sqrt[3]{216} = 6$	$216^{1/3} = 6$

15. 81 **17.** Not a real number   **19.** $\frac{1}{16}$

21. $x^{7/12}, x \neq 0$ **23.** $z^{5/3}$ **25.** $\dfrac{1}{x^{5/4}}$

27. $ab^{2/3}$ **29.** $x^{1/8}$ **31.** $(3x + 2)^{1/3}, x \neq -\frac{2}{3}$

33. 0.0392 **35.** 10.6301

37. (a) Not a real number   (b) 7

39. (a) -1 (b) 3 **41.** $\left(-\infty, \frac{9}{2}\right]$ **43.** $6u^2|v|\sqrt{u}$

45. $0.5x^2\sqrt{y}$ **47.** $2ab\sqrt[3]{6b}$ **49.** $\dfrac{\sqrt{30}}{6}$ **51.** $\dfrac{\sqrt[3]{4x^2}}{x}$

53. $\sqrt{145}$ **55.** $8\sqrt{6}$ **57.** $14\sqrt{x} - 9\sqrt[3]{x}$

59. $7\sqrt[4]{y} + 3$ **61.** $3x\sqrt[3]{3x^2y}$ **63.** $6\sqrt{x} + 2\sqrt{3x}$ feet

65. $10\sqrt{3}$ **67.** $2\sqrt{5} + 5\sqrt{2}$ **69.** $3 - x$

71. $3 + \sqrt{7}; 2$ **73.** $\sqrt{x} - 20; x - 400$

75. $-\dfrac{(\sqrt{2} - 1)(\sqrt{3} + 4)}{13}$ **77.** $\dfrac{(\sqrt{x} + 10)^2}{x - 100}$

79. 32 **81.** No real solution   **83.** 5

85. $-5, -3$ **87.** $\frac{3}{32}$ **89.** 8 inches $\times$ 15 inches

91. 2.93 feet   **93.** 64 feet   **95.** $4\sqrt{3}i$

97. $10 - 9\sqrt{3}i$ **99.** $\frac{3}{4} - \sqrt{3}i$ **101.** $15i$

103. $\sqrt{70} - 2\sqrt{10}$ **105.** $a = 10, b = -4$

107. $a = 4, b = 7$ **109.** $8 - 3i$

111. 25 **113.** $11 - 60i$ **115.** $-\frac{7}{3}i$

117. $\frac{9}{26} - \frac{3}{13}i$ **119.** $\frac{13}{37} - \frac{33}{37}i$

Chapter Test   *(page 507)*

1. (a) 64 (b) 10 **2.** (a) $\frac{1}{25}$ (b) 6

3. $f(-8) = 7, f(0) = 3$ **4.** $\left[\frac{3}{7}, \infty\right)$

5. (a) $x^{1/3}, x \neq 0$ (b) 25 **6.** (a) $\dfrac{4\sqrt{2}}{3}$ (b) $2\sqrt[3]{3}$

7. (a) $2x\sqrt{6x}$ (b) $2xy^2\sqrt[4]{x}$ **8.** $\dfrac{2\sqrt[3]{3y^2}}{3y}$

9. $\dfrac{5\left(\sqrt{6} + \sqrt{2}\right)}{2}$ **10.** $6\sqrt{2x}$ **11.** $5\sqrt{3x} + 3\sqrt{5}$

12. $16 - 8\sqrt{2x} + 2x$ **13.** $3 + 4y$ **14.** 24

15. No solution   **16.** 9   **17.** $2 - 2i$

18. $-16 - 30i$ **19.** $-8 + 4i$ **20.** $13 + 13i$

21. $\dfrac{13}{10} - \dfrac{11}{10}i$ **22.** 144 feet

Cumulative Test: Chapters 5–7   (page 508)

1. $-\dfrac{y^2}{2x^5}$, $y \neq 0, z \neq 0$ **2.** $\dfrac{3s^7}{5t}$, $s \neq 0$

3. $\dfrac{9x^{18}}{4y^{12}}$, $x \neq 0, z \neq 0$ **4.** 2.5×10^7

5. $x^5 + 2x^2 - 11x + 4$ **6.** $-3x^3 + 15x^2 - 6x$

7. $3x^2 + 22x - 16$ **8.** $9x^3 + 3x^2 + x + 2$

9. $(2x - 5)(x - 3)$ **10.** $9(x + 4)(x - 4)$

11. $(y - 3)^2(y + 3)$ **12.** $2t(2t - 5)^2$ **13.** $-3, \dfrac{8}{3}$

14. $0, \pm 9$ **15.** $\dfrac{x(x + 2)(x + 4)}{9(x - 4)}$, $x \neq -4, x \neq 0$

16. $\dfrac{x}{(2x - 1)(x - 4)}$, $x \neq -4, x \neq 3$

17. $\dfrac{5x^2 - 15x - 2}{(x + 2)(x - 3)}$ **18.** $\dfrac{3x + 5}{x(x + 3)}$

19. $\dfrac{x^3}{4}$, $x \neq -2, x \neq 0$ **20.** $x + y$, $x \neq 0, y \neq 0, x \neq y$

21. $2x^2 - x + 4 - \dfrac{21}{x + 4}$ **22.** $2x^3 - 2x^2 - x - \dfrac{4}{2x - 1}$

23. $2, 5$ **24.** $2, 9$ **25.** $2|x|y\sqrt{6y}$ **26.** $2a^5b^2\sqrt[3]{10b^2}$

27. $\dfrac{2|b^3|\sqrt{3}}{a^2}$ **28.** $\sqrt{t}$, $t \neq 0$ **29.** $35\sqrt{5x}$

30. $2x - 6\sqrt{2x} + 9$ **31.** $\dfrac{3\left(\sqrt{10} + \sqrt{x}\right)}{10 - x}$ **32.** 41

33. 2   **34.** 11   **35.** 29   **36.** $-4 + 3\sqrt{2}\,i$

37. $-7 + 16i$ **38.** $21 + 20i$ **39.** $\dfrac{3}{20} + \dfrac{11}{20}i$

40. $4(2x + 3)$ **41.** $\dfrac{1}{2}, \dfrac{2}{7}; \dfrac{14}{11}$ hours   **42.** 6.9 years

43. $90\sqrt{5} \approx 201.25$ feet   **44.** 400 feet

CHAPTER 8

Section 8.1   (page 517)

1. $6, 9$ **3.** $-5, 6$ **5.** $-9, 5$ **7.** 8   **9.** $-\dfrac{8}{9}, 2$

11. $0, 3$ **13.** $9, 12$ **15.** $-\dfrac{9}{2}, 5$ **17.** $1, 6$

19. $-\dfrac{5}{6}, \dfrac{1}{2}$ **21.** ± 7 **23.** ± 3 **25.** $\pm \dfrac{4}{5}$ **27.** ± 14

29. $\pm \dfrac{5}{2}$ **31.** $\pm \dfrac{15}{2}$ **33.** $-12, 4$ **35.** $2.5, 3.5$

37. $2 \pm \sqrt{7}$ **39.** $-\dfrac{1}{2} \pm \dfrac{5\sqrt{2}}{2}$ **41.** $\dfrac{2}{9} \pm \dfrac{2\sqrt{3}}{3}$

43. $\pm 6i$ **45.** $\pm 2i$ **47.** $\pm\dfrac{\sqrt{17}}{3}i$ **49.** $3 \pm 5i$

51. $-\dfrac{4}{3} \pm 4i$ **53.** $-\dfrac{1}{4} \pm \sqrt{5}i$ **55.** $-3 \pm \dfrac{5}{3}i$

57. $1 \pm 3\sqrt{3}\,i$ **59.** $-1 \pm 0.2i$ **61.** $\dfrac{2}{3} \pm \dfrac{1}{3}i$

63. $-\dfrac{7}{3} \pm \dfrac{\sqrt{38}}{3}i$ **65.** $0, \dfrac{5}{2}$ **67.** $-4, \dfrac{3}{2}$ **69.** ± 30

71. $\pm 30i$ **73.** ± 3 **75.** $2 \pm 6\sqrt{3}$ **77.** $2 \pm 6\sqrt{3}\,i$

79. $-2 \pm 3\sqrt{2}\,i$

81.

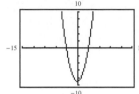

83.

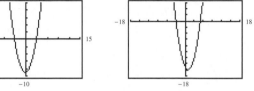

$(-3, 0), (3, 0)$;

The result is the same.

$(-3, 0), (5, 0)$;

The result is the same.

85.

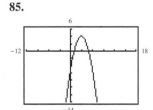

87.

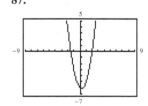

$(1, 0), (5, 0)$;

The result is the same.

$(2, 0), \left(-\dfrac{3}{2}, 0\right)$;

The result is the same.

89.

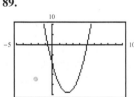

91.

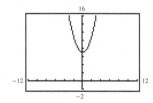

$\left(-\dfrac{2}{3}, 0\right), (5, 0)$;

The result is the same.

$\pm \sqrt{7}i$; complex solutions

93.

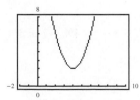

95.

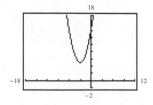

$4 \pm \sqrt{2}\,i$; complex solutions

$-3 \pm \sqrt{5}\,i$; complex solutions

97. $f(x) = \sqrt{4 - x^2}$

$g(x) = -\sqrt{4 - x^2}$

99. $f(x) = \frac{1}{2}\sqrt{4 - x^2}$

$g(x) = -\frac{1}{2}\sqrt{4 - x^2}$

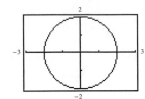

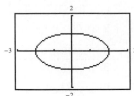

101. $\pm 1, \pm 2$ **103.** $\pm\sqrt{2}, \pm\sqrt{3}$ **105.** $\pm 1, \pm\sqrt{5}$

107. 16   **109.** 4, 25   **111.** $-8, 27$ **113.** $1, \frac{125}{8}$

115. 1, 32   **117.** $\frac{1}{32}, 243$ **119.** 729   **121.** 1, 16

123. $\frac{1}{2}, 1$ **125.** $-1, \frac{4}{5}$ **127.** $\pm 1, 2, 4$ **129.** $\frac{12}{5}$

131. 120 feet   **133.** 4 seconds

135. $2\sqrt{2} \approx 2.83$ seconds   **137.** 9 seconds

139. 6%   **141.** 2001

143. If $a = 0$, the equation would not be quadratic because it would be of degree 1, not 2.

145. Write the equation in the form $u^2 = d$, where u is an algebraic expression and d is a positive constant. Take the square root of each side of the equation to obtain the solutions $u = \pm\sqrt{d}$.

147. $x > 4$ **149.** $x \le 5$

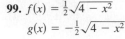

151. $(3, 2, 1)$ **153.** $3\sqrt{3}$ **155.** $16\sqrt[3]{y} - 9\sqrt[3]{x}$

157. $(4n + 1)m^2\sqrt{n}$

Section 8.2   *(page 525)*

1. 16   **3.** 100   **5.** 49   **7.** $\frac{25}{4}$ **9.** $\frac{81}{4}$ **11.** $\frac{1}{36}$

13. $\frac{16}{25}$ **15.** 0.04   **17.** 0, 20   **19.** $-6, 0$

21. 0, 5   **23.** 1, 7   **25.** $-4, -3$ **27.** $-3, 6$

29. 3   **31.** $-\frac{5}{2}, \frac{3}{2}$

33. $2 + \sqrt{7} \approx 4.65$

$2 - \sqrt{7} \approx -0.65$

35. $-2 + \sqrt{7} \approx 0.65$

$-2 - \sqrt{7} \approx -4.65$

37. $-7, 1$

39. $6 + \sqrt{26} \approx 11.10$

$6 - \sqrt{26} \approx 0.90$

41. $-7, -1$ **43.** 3, 7

45. $-\frac{5}{2} + \frac{\sqrt{13}}{2} \approx -0.70$

$-\frac{5}{2} - \frac{\sqrt{13}}{2} \approx -4.30$

47. $3 \pm i$ **49.** $-2 \pm 3i$

51. $\frac{1}{2} + \frac{\sqrt{3}}{2}i \approx 0.5 + 0.87i$

$\frac{1}{2} - \frac{\sqrt{3}}{2}i \approx 0.5 - 0.87i$

53. $-\frac{7}{2} + \frac{\sqrt{5}}{2} \approx -2.38$

$-\frac{7}{2} - \frac{\sqrt{5}}{2} \approx -4.62$

55. $\frac{1}{3} + \frac{2\sqrt{7}}{3} \approx 2.10$

$\frac{1}{3} - \frac{2\sqrt{7}}{3} \approx -1.43$

57. $-\frac{3}{8} + \frac{\sqrt{137}}{8} \approx 1.09$

$-\frac{3}{8} - \frac{\sqrt{137}}{8} \approx -1.84$

59. $-2 + \frac{\sqrt{10}}{2} \approx -0.42$

$-2 - \frac{\sqrt{10}}{2} \approx -3.58$

61. $-\frac{3}{2} + \frac{\sqrt{21}}{6} \approx -0.74$

$-\frac{3}{2} - \frac{\sqrt{21}}{6} \approx -2.26$

63. $-\frac{1}{2} + \frac{\sqrt{10}}{2} \approx 1.08$

$-\frac{1}{2} - \frac{\sqrt{10}}{2} \approx -2.08$

65. $\frac{3}{10} + \frac{\sqrt{191}}{10}i \approx 0.30 + 1.38i$

$\frac{3}{10} - \frac{\sqrt{191}}{10}i \approx 0.30 - 1.38i$

67. $\frac{1}{3} + \frac{\sqrt{127}}{3} \approx 4.09$

$\frac{1}{3} - \frac{\sqrt{127}}{3} \approx -3.42$

69. $-\frac{5}{2} + \frac{\sqrt{17}}{2} \approx -0.44$

$-\frac{5}{2} - \frac{\sqrt{17}}{2} \approx -4.56$

71. $-\frac{5}{6} + \frac{\sqrt{47}}{6}i \approx -0.83 + 1.14i$

$-\frac{5}{6} - \frac{\sqrt{47}}{6}i \approx -0.83 - 1.14i$

73. $1 \pm \sqrt{3}$ **75.** $-2, 6$ **77.** $4 + 2\sqrt{2}$

79.

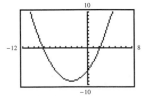

$\left(-2 \pm \sqrt{5}, 0\right)$

The result is the same.

81.

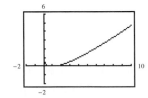

$\left(1 \pm \sqrt{6}, 0\right)$

The result is the same.

83.

$\left(-3 \pm 3\sqrt{3}, 0\right)$

The result is the same.

85.

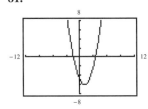

$(1, 0)$

The result is the same.

87. (a) $x^2 + 8x$ (b) $x^2 + 8x + 16$ (c) $(x + 4)^2$

89. 15 meters $\times 46\frac{2}{3}$ meters or 20 meters $\times$ 35 meters

91. 6 inches $\times$ 10 inches $\times$ 14 inches

93. 50 pairs, 110 pairs

95. Use the method of completing the square to write the quadratic equation in the form $u^2 = d$. Then use the Square Root Property to simplify.

97. (a) $d = 0$ (b) d is positive and is a perfect square.

    (c) d is positive and is not a perfect square.   (d) $d < 0$

99. 150   **101.** 7   **103.** $11 + 6\sqrt{2}$

105. $\dfrac{4\sqrt{10}}{5}$ **107.** $\sqrt{a} \cdot \sqrt{b}$

Section 8.3   *(page 534)*

1. $2x^2 + 2x - 7 = 0$ **3.** $-x^2 + 10x - 5 = 0$

5. $4, 7$ **7.** $-2, -4$ **9.** $-\dfrac{1}{4}$ **11.** $-\dfrac{3}{2}$

13. $-15, 20$ **15.** $1 \pm \sqrt{5}$ **17.** $-2 \pm \sqrt{3}$

19. $5 \pm \sqrt{2}$ **21.** $-\dfrac{3}{4} \pm \dfrac{\sqrt{15}}{4}i$ **23.** $-\dfrac{1}{3}, 1$

25. $-1 \pm \dfrac{\sqrt{10}}{2}$ **27.** $-\dfrac{3}{4} \pm \dfrac{\sqrt{21}}{4}$ **29.** $\dfrac{3}{8} \pm \dfrac{\sqrt{7}}{8}i$

31. $\dfrac{5}{4} \pm \dfrac{\sqrt{73}}{4}$ **33.** $\dfrac{1}{2} \pm \dfrac{\sqrt{13}}{6}$ **35.** $-\dfrac{1}{4} \pm \dfrac{\sqrt{13}}{4}$

37. $\dfrac{1}{5} \pm \dfrac{\sqrt{5}}{5}$ **39.** $-\dfrac{1}{5} \pm \dfrac{\sqrt{10}}{5}$

41. Two distinct complex solutions

43. Two distinct rational solutions

45. One repeated rational solution

47. Two distinct complex solutions   **49.** ± 13

51. $-3, 0$ **53.** $\dfrac{9}{5}, \dfrac{21}{5}$ **55.** $-\dfrac{3}{2}, 18$ **57.** $-4 \pm 3i$

59. $\dfrac{13}{6} \pm \dfrac{13\sqrt{11}}{6}i$ **61.** $-\dfrac{8}{5} \pm \dfrac{\sqrt{3}}{5}$

63. $-\dfrac{11}{8} \pm \dfrac{\sqrt{41}}{8}$ **65.** $x^2 - 3x - 10 = 0$

67. $x^2 - 8x + 7 = 0$ **69.** $x^2 - 2x - 1 = 0$

71. $x^2 + 25 = 0$ **73.** $x^2 - 24x + 144 = 0$

75.

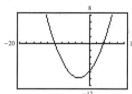

$(0.18, 0), (1.82, 0)$

The result is the same.

77.

$(1, 0), (3, 0)$

The result is the same.

79.

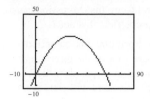

$(0.20, 0), (66.47, 0)$

The result is the same.

81.

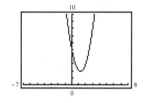

No real solutions

83.

Two real solutions

85. $\dfrac{7}{4} \pm \dfrac{\sqrt{17}}{4}$ **87.** No real values   **89.** $\dfrac{4}{3} \pm \dfrac{2\sqrt{13}}{3}$

91. $\dfrac{3}{2} + \dfrac{\sqrt{17}}{2}$ **93.** (a) $c < 9$ (b) $c = 9$ (c) $c > 9$

95. 5.1 inches $\times$ 11.4 inches

97. (a) 2.5 seconds   (b) $\dfrac{5}{4} + \dfrac{5\sqrt{3}}{4} \approx 3.42$ seconds

    (c) No. In order for the discriminant to be greater than or equal to zero, the value of c must be greater than or equal to -25. Therefore, the height cannot exceed 75 feet, or the value of c would be less than -25 when the equation is set equal to zero.

99. 2.15 or 4.65 hours

101. (a)

(b) 31.3 or 64.8 miles per hour

103.

	x_1, x_2	$x_1 + x_2$	$x_1 x_2$
(a)	$-2, 3$	1	-6
(b)	$-3, \dfrac{1}{2}$	$-\dfrac{5}{2}$	$-\dfrac{3}{2}$
(c)	$-\dfrac{3}{2}, \dfrac{3}{2}$	0	$-\dfrac{9}{4}$
(d)	$5 + 3i, 5 - 3i$	10	34

105. The Square Root Property would be convenient because the equation is of the form $u^2 = d$.

107. The Quadratic Formula would be convenient because the equation is already in general form, the expression cannot be factored, and the leading coefficient is not 1.

109. When the Quadratic Formula is applied to $ax^2 + bx + c = 0$, the square root of the discriminant is evaluated. When the discriminant is positive, the square root of the discriminant is positive and will yield two real solutions (or x-intercepts). When the discriminant is zero, the equation has one real solution (or x-intercept). When the discriminant is negative, the square root of the discriminant is negative and will yield two complex solutions (no x-intercepts).

111. Collinear   **113.** Not collinear

115.

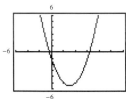

117.

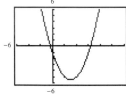

11. $y = (x + 3)^2 - 4$, $(-3, -4)$

13. $y = -(x - 3)^2 - 1$, $(3, -1)$

15. $y = -(x + 4)^2 + 21$, $(-4, 21)$

17. $y = 2\left(x + \frac{3}{2}\right)^2 - \frac{5}{2}$, $\left(-\frac{3}{2}, -\frac{5}{2}\right)$ **19.** $(4, -1)$

21. $(-1, 2)$ **23.** $\left(-\frac{1}{2}, 3\right)$ **25.** Upward, $(0, 2)$

27. Downward, $(10, 4)$ **29.** Upward, $(0, -6)$

31. Downward, $(3, 0)$ **33.** Downward, $(3, 9)$

35. $(\pm 5, 0)$, $(0, 25)$ **37.** $(0, 0)$, $(9, 0)$

39. $(-7, 0)$, $(1, 0)$, $(0, 7)$ **41.** $\left(\frac{3}{2}, 0\right)$, $(0, 9)$

43. $(0, 3)$ **45.** $\left(-\frac{3}{2} \pm \frac{\sqrt{19}}{2}, 0\right)$, $(0, 5)$

47.

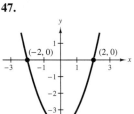

49.

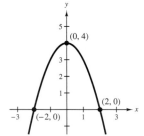

Mid-Chapter Quiz   *(page 538)*

1. ± 6 **2.** $-4, \frac{5}{2}$ **3.** $\pm 2\sqrt{3}$ **4.** $-1, 7$

5. $-\frac{7}{2} \pm \frac{\sqrt{41}}{2}$ **6.** $-\frac{3}{2} \pm \frac{\sqrt{19}}{2}$ **7.** $-2 \pm \sqrt{10}$

8. $\frac{1}{4} \pm \frac{\sqrt{105}}{12}$ **9.** $-\frac{5}{2} \pm \frac{\sqrt{3}}{2}i$ **10.** $-2, 10$

11. $-3, 10$ **12.** $-2, 5$ **13.** $\frac{3}{2}$ **14.** $-\frac{5}{3} \pm \frac{\sqrt{10}}{3}$

15. 49 **16.** $\pm 2i, \pm \sqrt{3}i$ **17.** $1, 9$

18. $\pm \sqrt{2}, \pm 2\sqrt{3}$

51.

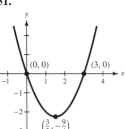

53.

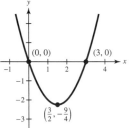

55.

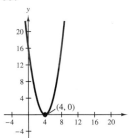

57.

19.

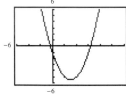

20.

$(-0.32, 0)$, $(6.32, 0)$ $(-2.24, 0)$, $(1.79, 0)$

The result is the same.   The result is the same.

21. 60 video games   **22.** 35 meters $\times$ 65 meters

Section 8.4   *(page 544)*

1. e   **2.** f   **3.** b   **4.** c   **5.** d   **6.** a

7. $y = (x - 1)^2 - 1$, $(1, -1)$

9. $y = (x - 2)^2 + 3$, $(2, 3)$

59.

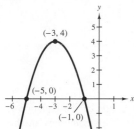

61.

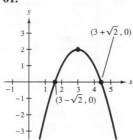

63.

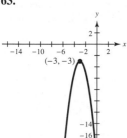

65.

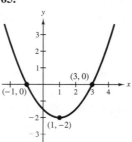

67.

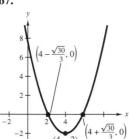

69.

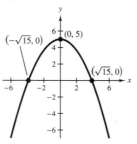

71. Vertical shift

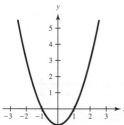

73. Horizontal shift

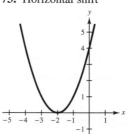

75. Horizontal shift and reflection in the x-axis

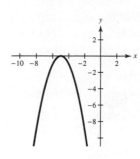

77. Horizontal and vertical shifts, reflection in the x-axis

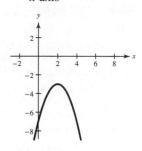

79.

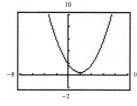

Vertex: $(2, 0.5)$

81.

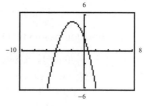

Vertex: $(-1.9, 4.9)$

83. $y = -x^2 + 4$ **85.** $y = 2(x + 1)^2 - 2$

87. $y = (x - 2)^2 + 1$ **89.** $y = (x - 2)^2 - 4$

91. $y = (x + 2)^2 - 1$ **93.** $y = \frac{2}{3}(x + 1)^2 + 1$

95. (a) 4 feet   (b) 16 feet   (c) $12 + 8\sqrt{3} \approx 25.9$ feet

97. (a) 3 feet   (b) 48 feet   (c) $15 + 4\sqrt{15} \approx 30.5$ feet

99. (a) 0 yards   (b) 30 yards   (c) 240 yards

101. 14 feet

103.

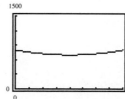

$x = 20$ when C is minimum.

105.

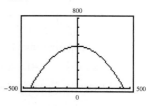

480 feet

107. $y = \frac{1}{60}x^2$

109. If the discriminant is positive, the parabola has two x-intercepts; if it is zero, the parabola has one x-intercept; and if it is negative, the parabola has no x-intercepts.

111. Find the y-coordinate of the vertex of the graph of the function.

113. $y = -\frac{1}{2}x$ **115.** $y = 2x$ **117.** $y = -\frac{11}{8}x + \frac{161}{16}$

119. $y = 8$ **121.** $8i$ **123.** $0.09i$

Section 8.5   *(page 555)*

1. 18 dozen, $1.20 per dozen   **3.** 10 DVDs at $5 per DVD

Width	Length	Perimeter	Area
5. $1.4l$	l	54 in.	177.19 in.^{2}
7. w	$2.5w$	70 ft	250 ft^2
9. $\frac{1}{3}l$	l	64 in.	192 in.^{2}
11. w	$w + 3$	54 km	180 km^2
13. $l - 20$	l	440 m	12,000 m^2

15. 12 inches × 16 inches

17. 50 feet × 250 feet or 100 feet × 125 feet

19. No.

$$\text{Area} = \frac{1}{2}(b_1 + b_2)h = \frac{1}{2}x[x + (550 - 2x)] = 43,560$$

This equation has no real solution.

21. Height: 12 inches; Width: 24 inches   **23.** 9.5%

25. 7%   **27.** $\approx 6.5\%$ **29.** 15 people

31. 3.9 miles or 8.1 miles

33. (a) $d = \sqrt{h^2 + 100^2}$

(b)
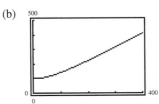

(c) $h \approx 173.2$ when $d = 200$.

(d)
h	0	100	200	300
d	100	141.4	223.6	316.2

35. 10.7 minutes, 13.7 minutes   **37.** $3\frac{1}{4}$ seconds

39. ≈ 9.5 seconds   **41.** 4.7 seconds

43. (a) 3 seconds, 7 seconds   (b) 10 seconds

(c) 400 feet

45. 13, 14   **47.** 12, 14   **49.** 17, 19

51. 400 miles per hour

53. 60 miles per hour or 75 miles per hour

55. $(-6, 9), (10, 9)$

57. (a) $b = 20 - a$; $A = \pi ab$; $A = \pi a(20 - a)$

(b)
a	4	7	10	13	16
A	201.1	285.9	314.2	285.9	201.1

(c) 7.9, 12.1

(d)
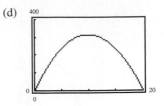

59. To solve a rational equation, each side of the equation is multiplied by the LCD. The resulting equations in this section are quadratic equations.

61. No. For each additional person, the cost-per-person decrease gets smaller because the discount is distributed to more people.

63. $x < -4$ **65.** 0, 8

Section 8.6   *(page 566)*

1. $0, \frac{5}{2}$ **3.** $\pm\frac{9}{2}$ **5.** $-3, 5$ **7.** $1, 3$ **9.** $-3, \frac{5}{6}$

11. Negative: $(-\infty, 4)$; Positive: $(4, \infty)$

13. Negative: $(6, \infty)$; Positive: $(-\infty, 6)$

15. Negative: $(0, 5)$; Positive: $(-\infty, 0) \cup (5, \infty)$

17. Negative: $(-\infty, -2) \cup (2, \infty)$; Positive: $(-2, 2)$

19. Negative: $(-1, 5)$; Positive: $(-\infty, -1) \cup (5, \infty)$

21. $(0, 2)$ **23.** $[0, 2]$

25. $(-\infty, -2) \cup (2, \infty)$ **27.** $(-\infty, -2] \cup [5, \infty)$

29. $(-\infty, -4) \cup (0, \infty)$ **31.** $[-9, 4]$

33. $(-\infty, -3) \cup (1, \infty)$ **35.** No solution

37. $(-\infty, \infty)$

39. $\left(-\infty, 2 - \sqrt{2}\right) \cup \left(2 + \sqrt{2}, \infty\right)$

41. $(-\infty, \infty)$

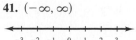

43. No solution

45. $\left[-2, \frac{4}{3}\right]$

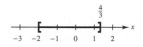

47. $\left(\frac{2}{3}, \frac{5}{2}\right)$

49. $\left(-\infty, -\frac{1}{2}\right) \cup (4, \infty)$

51. $-\frac{7}{2}$

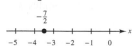

53. No solution

55. $\left(-\infty, 2 - \sqrt{6}\right) \cup \left(2 + \sqrt{6}, \infty\right)$

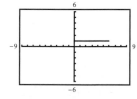

57. $(-\infty, -9] \cup [-1, \infty)$

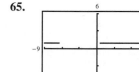

59. $(-2, 0) \cup (2, \infty)$

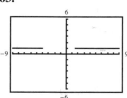

61.

$(0, 6)$

63.

$(-\infty, -4) \cup \left(\frac{3}{2}, \infty\right)$

65.

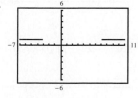

$\left(-\infty, -3 - 2\sqrt{3}\right] \cup \left[-3 + 2\sqrt{3}, \infty\right)$

67.

$(-\infty, -3) \cup (7, \infty)$

69.

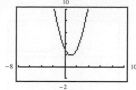

(a) $(-\infty, \infty)$ (b) $[-1, 3]$

71.

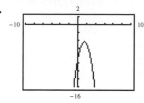

(a) $(-0.303, 3.303)$ (b) $(-\infty, \infty)$

73. 3   **75.** $0, -5$

77. $(3, \infty)$

79. $(-\infty, 3)$

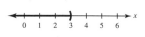

81. $[-2, 1)$

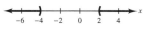

83. $(-\infty, -4) \cup (2, \infty)$

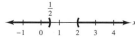

85. $(1, 4]$

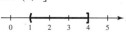

87. $\left(-\infty, \frac{1}{2}\right) \cup (2, \infty)$

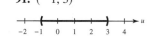

89. $\left[-2, -\frac{3}{2}\right)$

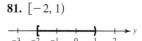

91. $(-1, 3)$

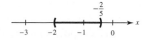

93. $\left(5, \frac{17}{3}\right)$

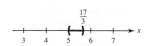

95. $\left(-2, -\frac{2}{5}\right)$

97. $(-\infty, 6) \cup [7, \infty)$

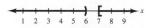

99.

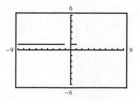

$(-\infty, -1) \cup (0, 1)$

101.

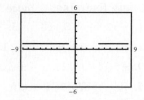

$(-\infty, -1) \cup (4, \infty)$

103.

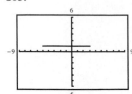

105.

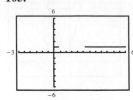

$\left(-5, \frac{13}{4}\right)$

$(0, 0.382) \cup (2.618, \infty)$

107.

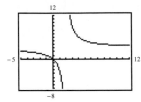

(a) $[0, 2)$ (b) $(2, 4]$

109.

(a) $(-\infty, -2] \cup [2, \infty)$

(b) $(-\infty, \infty)$

111. $(3, 5)$ **113.** $r > 7.24\%$ **115.** $(12, 20)$

117. $90{,}000 \le x \le 100{,}000$

119. (a)

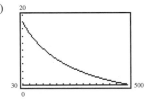

(b) $\frac{4225}{17} \approx 248.5$ minutes

121. The critical numbers of a polynomial are its zeros, so the value of the polynomial is zero at its critical numbers.

123. No solution. The value of the polynomial is positive for every real value of x, so there are no values that would make the polynomial negative.

125. $\frac{y^3}{2x^2}$, $y \ne 0$ **127.** $\frac{(x - 3)(x + 1)}{4x^3(x + 3)}$, $x \ne -2$

129. $\frac{(x + 4)^2}{3x(x - 6)(x - 8)}$ **131.** $\frac{1}{9}$ **133.** 79.21

Review Exercises *(page 572)*

1. $-12, 0$ **3.** ± 3 **5.** $-\frac{5}{2}$ **7.** $-9, 10$ **9.** $-\frac{3}{2}, 6$

11. ± 12 **13.** $\pm 2\sqrt{3}$ **15.** $-4, 36$ **17.** $\pm 11i$

19. $\pm 5\sqrt{2}\,i$ **21.** $-4 \pm 3\sqrt{2}\,i$ **23.** $\pm \sqrt{5}, \pm i$

25. $1, 9$ **27.** $1, 1 \pm \sqrt{6}$ **29.** $-343, 64$ **31.** 81

33. $\frac{225}{4}$ **35.** $\frac{1}{25}$

37. $3 + 2\sqrt{3} \approx 6.46$; $3 - 2\sqrt{3} \approx -0.46$ **39.** $-4, -1$

41. $\frac{1}{3} + \frac{\sqrt{17}}{3}i \approx 0.33 + 1.37i$; $\frac{1}{3} - \frac{\sqrt{17}}{3}i \approx 0.33 - 1.37i$

43. $-7, 6$ **45.** $-\frac{7}{2}, 3$ **47.** $\frac{8}{5} \pm \frac{3\sqrt{6}}{5}$

49. One repeated rational solution

51. Two distinct rational solutions

53. Two distinct irrational solutions

55. Two distinct complex solutions

57. $x^2 + 4x - 21 = 0$

59. $x^2 - 10x + 18 = 0$ **61.** $x^2 - 12x + 40 = 0$

63. $y = (x - 4)^2 - 13$; Vertex: $(4, -13)$

65. $y = 2\left(x - \frac{1}{4}\right)^2 + \frac{23}{8}$; Vertex: $\left(\frac{1}{4}, \frac{23}{8}\right)$

67.

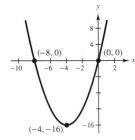

69.

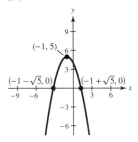

71. $y = 2(x - 2)^2 - 5$ **73.** $y = \frac{1}{16}(x - 5)^2$

75. (a)

(b) 6 feet

(c) 28.5 feet

(d) 31.9 feet

77. 16 cars; $5000   **79.** 5 inches $\times$ 17 inches

81. 15 people

83. $9 + \sqrt{101} \approx 19$ hours, $11 + \sqrt{101} \approx 21$ hours

85. $-7, 0$ **87.** $-3, 9$

89. $(0, 7)$

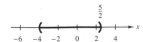

91. $(-\infty, -2] \cup [6, \infty)$

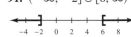

93. $\left(-4, \frac{5}{2}\right)$

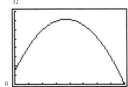

95. $(-\infty, -3] \cup \left(\frac{7}{2}, \infty\right)$

97. $(-4, 1)$ **99.** $(5.3, 14.2)$

Chapter Test   *(page 575)*

1. $3, 10$ **2.** $-\frac{1}{3}, 6$ **3.** $1.7,\ 2.3$ **4.** $-4 \pm 10i$

5. $\frac{3}{2} \pm \frac{\sqrt{3}}{2}$ **6.** $1 \pm \frac{3\sqrt{2}}{2}$ **7.** $\frac{3}{4} \pm \frac{\sqrt{5}}{4}$ **8.** $1, 512$

9. -56; A negative discriminant tells us the equation has two imaginary solutions.

10. $x^2 + 10x + 21 = 0$

11. 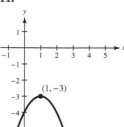   **12.**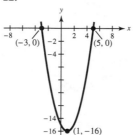

13. $(-\infty, -2] \cup [6, \infty)$ **14.** $(0, 3)$

15. $[-1, 5)$ **16.** $8 \text{ feet} \times 30 \text{ feet}$

17. 40 members   **18.** $\frac{\sqrt{10}}{2} \approx 1.58$ seconds

19. 35 feet, 120 feet

CHAPTER 9

Section 9.1   *(page 586)*

1. 3^{2x+2} **3.** e^2 **5.** $\frac{3}{e^{2x}}$ **7.** $-2e^x$ **9.** 9.739

11. 1.396 **13.** 107.561 **15.** 0.906

17. (a) $\frac{1}{9}$ (b) 1 (c) 3

19. (a) 0.455 (b) 0.094 (c) 0.145

21. (a) 500 (b) 250 (c) 56.657

23. (a) 1000 (b) 1628.895 (c) 2653.298

25. (a) 486.111 (b) 47.261 (c) 0.447

27. (a) 73.891 (b) 1.353 (c) 0.183

29. (a) 333.333 (b) 434.557 (c) 499.381

31.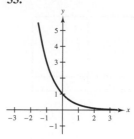

Horizontal asymptote: $y = 0$

33.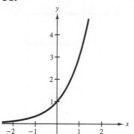

Horizontal asymptote: $y = 0$

35.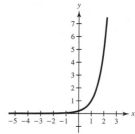

Horizontal asymptote: $y = -2$

37.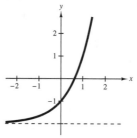

Horizontal asymptote: $y = 0$

39.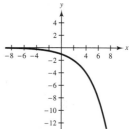

Horizontal asymptote: $y = 0$

41.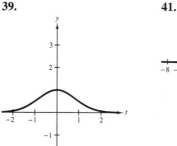

Horizontal asymptote: $y = 0$

43.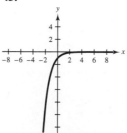

Horizontal asymptote: $y = 0$

45.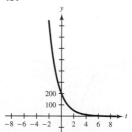

Horizontal asymptote: $y = 0$

47.

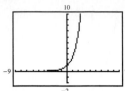

49.

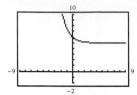

51.

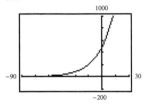

53.

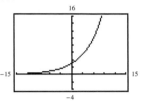

55.

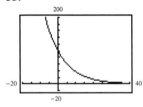

57.

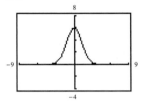

59. b   **60.** a   **61.** c   **62.** d

63. Vertical shift

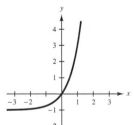

65. Horizontal shift

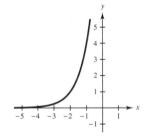

67. Reflection in the *x*-axis

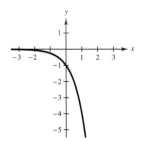

69. 2.520 grams   **71.** $6744.25

73.

n	1	4	12	365	Continuous
A	$275.90	$283.18	$284.89	$285.74	$285.77

75.

n	1	4	12
A	$4956.46	$5114.30	$5152.11

n	365	Continuous
A	$5170.78	$5171.42

77.

n	1	4	12
P	$2541.75	$2498.00	$2487.98

n	365	Continuous
P	$2483.09	$2482.93

79.

n	1	4	12
P	$18,429.30	$15,830.43	$15,272.04

n	365	Continuous
P	$15,004.64	$14,995.58

81. (a) $22.04   (b) $20.13

83. (a) $80,634.95   (b) $161,269.89

85. $V(t) = 16,000\left(\frac{3}{4}\right)^t$

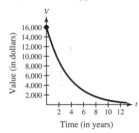

$9000 after 2 years; $5062.50 after 4 years

87. $f(x) = 1024\left(\frac{1}{2}\right)^x; f(8) = 4$ golfers

89. (a)

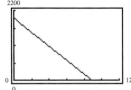

(b)

Time (in seconds)	0	10	20	30	40
Height (in feet)	2000	1731	1510	1290	1070

Time (in seconds)	50	60	70	80	90
Height (in feet)	850	630	410	190	0

(c) The height changes the most within the first 10 seconds because after the parachute is released, a few seconds pass before the descent becomes constant.

91. (a)

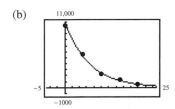

(b)

11,000

The model is a good fit for the data.

(c)

h	0	5	10	15	20
P	10,332	5583	2376	1240	517
Approx.	10,958	5176	2445	1155	546

(d) 3300 kilograms per square meter

(e) 11.3 kilometers

93. (a)

x	1	10	100	1000	10,000
$\left(1 + \dfrac{1}{x}\right)^x$	2	2.5937	2.7048	2.7169	2.7181

(b)

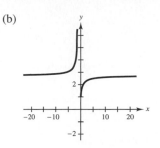

Yes, the graph appears to be approaching a horizontal asymptote.

(c) The value approaches e.

95. By definition, the base of an exponential function must be positive and not equal to 1. If the base is 1, the function simplifies to the constant function $y = 1$.

97. No; e is an irrational number.

99. When $k > 1$, the values of f will increase. When $0 < k < 1$, the values of f will decrease. When $k = 1$, the values of f will remain constant.

101. $[4, \infty)$

103.

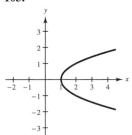

y is not a function of x.

Section 9.2   *(page 599)*

1. (a) $2x - 9$ (b) $2x - 3$ (c) -1 (d) 11

3. (a) $x^2 + 4x + 7$ (b) $x^2 + 5$ (c) 19 (d) 14

5. (a) $|3x - 3|$ (b) $3|x - 3|$ (c) 0 (d) 3

7. (a) $\sqrt{x + 1}$ (b) $\sqrt{x - 4} + 5$ (c) 2 (d) 7

9. (a) $\dfrac{x^2}{2 - 3x^2},\ x \neq 0$

(b) $2(x - 3)^2,\ x \neq 3$

(c) -1

(d) 2

11. (a) -1 (b) -2 (c) -2 **13.** (a) -1 (b) 1

15. (a) 5 (b) 1 (c) 3 **17.** (a) 0 (b) 10

19. (a) $(f \circ g)(x) = 3x - 17$

Domain: $(-\infty, \infty)$

(b) $(g \circ f)(x) = 3x - 3$

Domain: $(-\infty, \infty)$

21. (a) $(f \circ g)(x) = \sqrt{x - 2}$

Domain: $[2, \infty)$

(b) $(g \circ f)(x) = \sqrt{x + 2} - 4$

Domain: $[-2, \infty)$

23. (a) $(f \circ g)(x) = x + 2$

Domain: $[1, \infty)$

(b) $(g \circ f)(x) = \sqrt{x^2 + 2}$

Domain: $(-\infty, \infty)$

25. (a) $(f \circ g)(x) = \dfrac{\sqrt{x - 1}}{\sqrt{x - 1} + 5}$

Domain: $[1, \infty)$

(b) $(g \circ f)(x) = \sqrt{-\dfrac{5}{x + 5}}$

Domain: $(-\infty, -5)$

27.

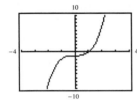

Yes

29.

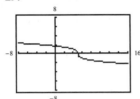

Yes

31.

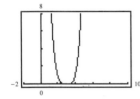

No

33.

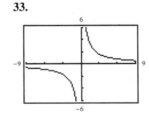

Yes

35. No   **37.** Yes   **39.** No

41. $f(g(x)) = f\left(-\frac{1}{6}x\right) = -6\left(-\frac{1}{6}x\right) = x$

$g(f(x)) = g(-6x) = -\frac{1}{6}(-6x) = x$

43. $f(g(x)) = f\left[\frac{1}{2}(1 - x)\right] = 1 - 2\left[\frac{1}{2}(1 - x)\right]$

$= 1 - (1 - x) = x$

$g(f(x)) = g(1 - 2x) = \frac{1}{2}[1 - (1 - 2x)] = \frac{1}{2}(2x) = x$

45. $f(g(x)) = f(x^3 - 1) = \sqrt[3]{(x^3 - 1) + 1} = \sqrt[3]{x^3} = x$

$g(f(x)) = g\left(\sqrt[3]{x + 1}\right) = \left(\sqrt[3]{x + 1}\right)^3 - 1$

$= x + 1 - 1 = x$

47. $f(g(x)) = f\left(\dfrac{1}{x}\right) = \dfrac{1}{(1/x)} = x$

$g(f(x)) = g\left(\dfrac{1}{x}\right) = \dfrac{1}{(1/x)} = x$

49. $f^{-1}(x) = \frac{1}{5}x$ **51.** $f^{-1}(x) = -\frac{5}{2}x$

53. $f^{-1}(x) = x - 10$ **55.** $f^{-1}(x) = 5 - x$

57. $f^{-1}(x) = \sqrt[9]{x}$ **59.** $f^{-1}(x) = x^3$

61. $g^{-1}(x) = x - 25$ **63.** $g^{-1}(x) = \dfrac{3 - x}{4}$

65. $g^{-1}(t) = 4t - 8$ **67.** Inverse does not exist.

69. $h^{-1}(x) = x^2,\ x \geq 0$ **71.** $f^{-1}(t) = \sqrt[3]{t + 1}$

73. $f^{-1}(x) = x^2 - 3,\ x \geq 0$

75. b   **76.** c   **77.** d   **78.** a

79.

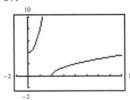

81.

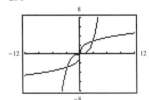

83.

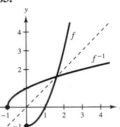

85.

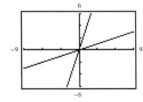

87.

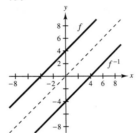

89.

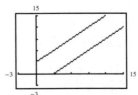

91.

93. Domain of f: $x \geq 2$; $f^{-1}(x) = \sqrt{x} + 2$;

Domain of f^{-1}: $x \geq 0$

95. Domain of f: $x \geq 0$; $f^{-1}(x) = x - 1$;

Domain of f^{-1}: $x \geq 1$

97. $f^{-1}(x) = \frac{1}{2}(3 - x)$ **99.** c   **101.** a

103. $p(s(x)) = 0.03x - 9000,\ x > 300,000$

This function represents the bonus earned for sales over $300,000.

105. $A(r(t)) = 0.36\pi t^2$

Input: time; Output: area; $A(r(3)) = 10.2$ square feet

107. (a) $R = p - 2000$ (b) $S = 0.95p$

(c) $(R \circ S)(p) = 0.95p - 2000$;

    5% discount followed by the $2000 rebate

    $(S \circ R)(p) = 0.95(p - 2000)$;

    5% discount after the price is reduced by the rebate

(d) $(R \circ S)(26,000) = 22,700$; $(S \circ R)(26,000) = 22,800$

    $R \circ S$ yields the smaller cost because the dealer discount is calculated on a larger base.

109. (a) $y = \frac{20}{13}(x - 9)$

(b) x: hourly wage; y: number of units produced

(c) $x \geq 9$ (d) 8 units

111. (a) $(f \circ g)(x) = 4x + 24$

(b) $(f \circ g)^{-1}(x) = \dfrac{x - 24}{4} = \dfrac{1}{4}x - 6$

(c) $f^{-1}(x) = \dfrac{1}{4}x;\ g^{-1}(x) = x - 6$

(d) $(g^{-1} \circ f^{-1})(x) = \dfrac{1}{4}x - 6$; The results are the same.

(e) $(f \circ g)^{-1}(x) = (g^{-1} \circ f^{-1})(x)$

113. True. If the point (a, b) lies on the graph of f, the point (b, a) must lie on the graph of f^{-1}, and vice versa.

115. False. $f(x) = \sqrt{x - 1}$; Domain: $[1, \infty)$;

    $f^{-1}(x) = x^2 + 1,\ x \geq 0$; Domain: $[0, \infty)$

117. Interchange the coordinates of each ordered pair. The inverse of the function defined by $\{(3, 6), (5, -2)\}$ is $\{(6, 3), (-2, 5)\}$.

119. A function can have only one input for every output, so the inverse will have one output for every input and is therefore a function.

121. Reflection in the x-axis

123. Horizontal and vertical shifts   **125.** $(6 + y)(2 - y)$

127. $-(u^2 + 1)(u - 5)$

129.

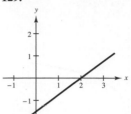

131.

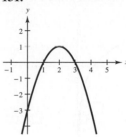

Section 9.3   *(page 612)*

1. $7^2 = 49$ **3.** $2^{-5} = \frac{1}{32}$ **5.** $3^{-5} = \frac{1}{243}$

7. $36^{1/2} = 6$ **9.** $8^{2/3} = 4$ **11.** $2^{2.4} \approx 5.278$

13. $\log_6 36 = 2$ **15.** $\log_5 \frac{1}{125} = -3$ **17.** $\log_8 4 = \frac{2}{3}$

19. $\log_{25} \frac{1}{5} = -\frac{1}{2}$ **21.** $\log_4 1 = 0$

23. $\log_5 9.518 \approx 1.4$ **25.** 3   **27.** 3   **29.** -4

31. -3 **33.** -4

35. There is no power to which 2 can be raised to obtain -3.

37. 0

39. There is no power to which 5 can be raised to obtain -6.

41. $\frac{1}{2}$ **43.** $\frac{3}{4}$ **45.** 4   **47.** 1.6232   **49.** -1.6383

51. 0.7190   **53.** c   **54.** b   **55.** a   **56.** d

57.

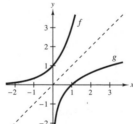

59.

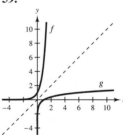

61. Vertical shift

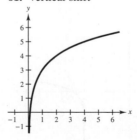

63. Horizontal shift

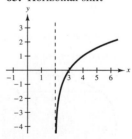

65. Reflection in the y-axis

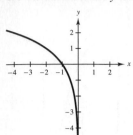

67.

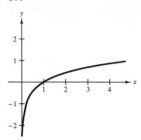

Vertical asymptote: $x = 0$

81. Domain: $(0, \infty)$

Vertical asymptote: $x = 0$

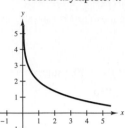

83. Domain: $(0, \infty)$

Vertical asymptote: $x = 0$

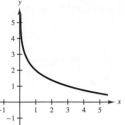

69.

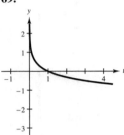

Vertical asymptote: $t = 0$

71.

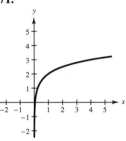

Vertical asymptote: $x = 0$

85. Domain: $(0, \infty)$

Vertical asymptote: $x = 0$

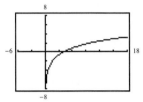

87. Domain: $(0, \infty)$

Vertical asymptote: $x = 0$

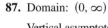

89. 3.6376   **91.** -1.8971 **93.** -1.1484 **95.** b

96. a   **97.** d   **98.** c

73.

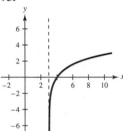

Vertical asymptote: $x = 3$

75.

Vertical asymptote: $x = 0$

99. Vertical asymptote:
$x = 0$

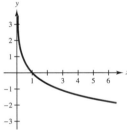

101. Vertical asymptote:
$x = 0$

77. Domain: $(0, \infty)$

Vertical asymptote:
$x = 0$

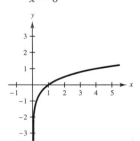

79. Domain: $(4, \infty)$

Vertical asymptote:
$x = 4$

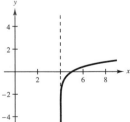

103. Vertical asymptote:
$x = 0$

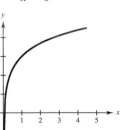

105. Vertical asymptote:
$t = 4$

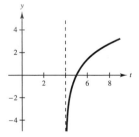

107.

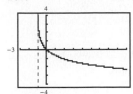

109.

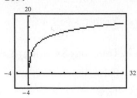

3.

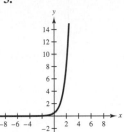

4.

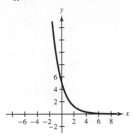

Domain: $(-1, \infty)$ Domain: $(0, \infty)$

Vertical asymptote: $x = -1$ Vertical asymptote: $t = 0$

111. 1.6309   **113.** 1.6397   **115.** -0.4739

117. 2.6332   **119.** -2 **121.** 1.3481   **123.** 1.8946

125. 53.4 inches

127.

r	0.07	0.08	0.09	0.10	0.11	0.12
t	9.9	8.7	7.7	6.9	6.3	5.8

129. (a)

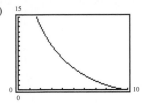

Domain: $(0, 10]$

(b) $x = 0$ (c) $(0, 22.9)$

131. $(0, \infty)$ **133.** $3 \le f(x) \le 4$ **135.** A factor of 10

137. Logarithmic functions with base 10 are common logarithms. Logarithmic functions with base e are natural logarithms.

139. A vertical shift or reflection in the x-axis of a logarithmic graph does not affect the domain or range. A horizontal shift or reflection in the y-axis of a logarithmic graph affects the domain, but the range stays the same.

141. $-m^{10}n^4$ **143.** $\dfrac{9x^3}{2y^2}$, $x \ne 0$

145. $19\sqrt{3x}$ **147.** $\sqrt{5u}$

Mid-Chapter Quiz   *(page 616)*

1. (a) $\frac{16}{9}$ (b) 1   (c) $\frac{3}{4}$ (d) 1.540

2. Horizontal asymptote: $y = 0$

Horizontal asymptote: $y = 0$ Horizontal asymptote: $y = 0$

5.

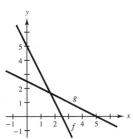

6.

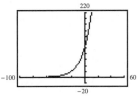

Horizontal asymptote: $y = 0$ Horizontal asymptote: $y = 0$

7. (a) $2x^3 - 3$ (b) $(2x - 3)^3 = 8x^3 - 36x^2 + 54x - 27$
    (c) -19 (d) 125

8. $f(g(x)) = 5 - 2\left[\frac{1}{2}(5 - x)\right]$
    $= 5 - 5 + x = x$
$g(f(x)) = \frac{1}{2}[5 - (5 - 2x)]$
    $= \frac{1}{2}(2x) = x$

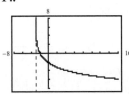

9. $h^{-1}(x) = \dfrac{x - 3}{10}$ **10.** $g^{-1}(t) = \sqrt[3]{2t - 4}$

11. $9^{-2} = \frac{1}{81}$ **12.** $\log_2 64 = 6$ **13.** 3

14.

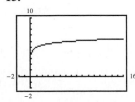

15.

Vertical asymptote: $t = -3$ Vertical asymptote: $x = 0$

16. $h = 2, k = 1$ **17.** 6.0639

18.

n	1	4	12
A	\$2979.31	\$3042.18	\$3056.86

n	365	Continuous
A	\$3064.06	\$3064.31

19. 1.60 grams

Section 9.4   *(page 621)*

1. 3   **3.** -4 **5.** $\frac{1}{3}$ **7.** 0   **9.** -9 **11.** 2

13. 2   **15.** 3   **17.** 2   **19.** -3 **21.** 12   **23.** 1

25. 1.5000   **27.** 0.2925   **29.** 0.5283   **31.** 0

33. 2.1972   **35.** 0.5108   **37.** 1.9033   **39.** 8.7118

41. $-3\log_4 2 = \log_4 2^{-3} = \log_4 \frac{1}{8}$

43. $-3\log_{10} 3 + \log_{10} \frac{3}{2} = \log_{10} 3^{-3} + \log_{10} \frac{3}{2}$
$$= \log_{10} \tfrac{1}{27} + \log_{10} \tfrac{3}{2}$$
$$= \log_{10}\left(\tfrac{1}{27} \cdot \tfrac{3}{2}\right)$$
$$= \log_{10} \tfrac{1}{18}$$

45. $-\ln \frac{1}{7} = \ln\left(\frac{1}{7}\right)^{-1} = \ln 7 = \ln \frac{56}{8} = \ln 56 - \ln 8$

47. $\log_3 11 + \log_3 x$ **49.** $\ln 3 + \ln y$ **51.** $2\log_7 x$

53. $-3\log_4 x$ **55.** $\frac{1}{2}(\log_4 3 + \log_4 x)$

57. $\log_2 z - \log_2 17$ **59.** $\frac{1}{2}\log_9 x - \log_9 12$

61. $2\ln x + \ln(y + 2)$ **63.** $6\log_4 x + 2\log_4(x + 7)$

65. $\frac{1}{3}\log_3(x + 1)$ **67.** $\frac{1}{2}[\ln x + \ln(x + 2)]$

69. $2[\ln(x + 1) - \ln(x + 4)]$ **71.** $\frac{1}{3}[2\ln x - \ln(x + 1)]$

73. $\ln x + 2\ln y - 3\ln z$ **75.** $\log_{12} \dfrac{x}{3}$ **77.** $\log_3 5x$

79. $\log_{10} \dfrac{4}{x}$ **81.** $\ln b^4$ **83.** $\log_5 \dfrac{1}{4x^2}$ **85.** $\log_2 x^7 z^3$

87. $\log_3 2\sqrt{y}$ **89.** $\ln \dfrac{x^3 y}{z^2}$ **91.** $\ln x^4 y^4$

93. $\ln\left(\dfrac{x}{x + 1}\right)^2$ **95.** $\log_4 \dfrac{x + 8}{x^3}$ **97.** $\log_5 \dfrac{\sqrt[3]{x + 3}}{x - 6}$

99. $\log_6 \dfrac{(c + d)^5}{\sqrt{m - n}}$ **101.** $\log_2 \sqrt[5]{\dfrac{x^3}{y^4}}$ **103.** $2 + \ln 3$

105. $1 + \frac{1}{2}\log_5 2$ **107.** $1 - 3\log_8 x$

109.

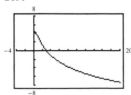

111.

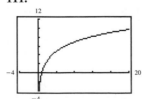

113. $B = 10(\log_{10} I + 16)$; 130 decibels

115. $E = \log_{10}\left(\dfrac{C_2}{C_1}\right)^{1.4}$

117. True; $\log_2 8x = \log_2 8 + \log_2 x = 3 + \log_2 x$

119. False; $\log_3(u + v)$ does not simplify.

121. True;
$$f(ax) = \log_a ax$$
$$= \log_a a + \log_a x$$
$$= 1 + \log_a x$$
$$= 1 + f(x)$$

123. False; 0 is not in the domain of f.

125. False; $f(x - 3) = \ln(x - 3)$

127. True; $f(1) = 0$, so when $f(x) > 0$, $x > 1$.

129. Evaluate when $x = e$ and $y = e$.

131. $5 \pm 2\sqrt{2}$ **133.** $\frac{5}{3}$ **135.** 47

137.   **139.**

141. (a) $(f \circ g)(x) = \sqrt{x - 3}$
    Domain: $[3, \infty)$
  (b) $(g \circ f)(x) = \sqrt{x} - 3$
    Domain: $[0, \infty)$

143. (a) $(f \circ g)(x) = \dfrac{5}{x^2 + 2x - 3}$
    Domain: $(-\infty, -3) \cup (-3, 1) \cup (1, \infty)$
  (b) $(g \circ f)(x) = \dfrac{5}{x^2 - 4} + 1 = \dfrac{x^2 + 1}{x^2 - 4}$
    Domain: $(-\infty, -2) \cup (-2, 2) \cup (2, \infty)$

Section 9.5   *(page 631)*

1. (a) Not a solution   (b) Solution

3. (a) Solution   (b) Not a solution

5. (a) Not a solution   (b) Solution

7. 3   **9.** -3 **11.** 4   **13.** 1   **15.** -2

17. -3 **19.** -6 **21.** 2   **23.** $\frac{22}{5}$ **25.** 6

27. -7 **29.** 4   **31.** No solution   **33.** -7

35. $2x - 1$ **37.** $2x, \; x > 0$ **39.** 4.11   **41.** 1.31

43. 1.49   **45.** 0.83   **47.** 0.11   **49.** -3.60

51. -1.79 **53.** 3.00   **55.** -1.45 **57.** 35.35

59. 6.80   **61.** 12.22   **63.** 3.28   **65.** No solution

67. -1.04 **69.** 2.48   **71.** 0.90   **73.** 4.39

75. 0.39   **77.** 8.99   **79.** 9.73   **81.** 4.62

83. 0.10   **85.** 174.77   **87.** 2187.00   **89.** 6.52

91. 25.00   **93.** 0.61   **95.** ± 20.09 **97.** 3.00

99. -3.93 **101.** 2.83   **103.** 2000.00   **105.** 3.20

107. 4.00   **109.** 0.75   **111.** 5.00   **113.** 2.46

115. 3.33   **117.** 6.00

119. $(1.40, 0)$ **121.** $(21.82, 0)$

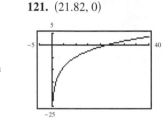

123. 5%   **125.** 7.7 years

127. 10^{-8} watt per square centimeter

129. (a) $105°$

(b)

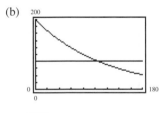

$105°$

131. (a) -0.144 (b) 6:24 P.M.   (c) $90.2°$F

133. 1.0234 and 1.0241 grams per cubic centimeter

135. $\log_a(uv) = \log_a u + \log_a v; \; \log_a \dfrac{u}{v} = \log_a u - \log_a v;$

$\log_a u^n = n \log_a u$

137. To solve an exponential equation, first isolate the exponential expression, then take the logarithm of each side of the equation and solve for the variable.

To solve a logarithmic equation, first isolate the logarithmic expression, then exponentiate each side of the equation and solve for the variable.

139. $\pm 5i$ **141.** $\pm \frac{4}{3}$ **143.** $\pm 2, \pm 3$

	Width	Length	Perimeter	Area
145.	$2.5x$	x	42 in.	90 in.^{2}
147.	w	$w + 4$	56 km	192 km^2

Section 9.6   *(page 641)*

1. 7%   **3.** 9%   **5.** 8%   **7.** 9.27 years

9. 8.66 years   **11.** 9.59 years   **13.** Yearly

15. Continuous   **17.** Quarterly   **19.** 8.33%

21. 7.23%   **23.** 6.14%   **25.** 8.30%

27. No. Each time the amount is divided by the principal, the result is always 2.

29. $1652.99   **31.** $626.46   **33.** $3080.15

35. $951.23   **37.** $5496.57   **39.** $320,250.81

41. Total deposits: $7200.00; Total interest: $10,529.42

43. $k = \frac{1}{2} \ln \frac{8}{3} \approx 0.4904$ **45.** $k = \frac{1}{3} \ln \frac{1}{2} \approx -0.2310$

47. $y = 90e^{0.0082t}$; 110,000 people

49. $y = 286e^{0.0029t}$; 308,000 people

51. $y = 2589e^{0.0052t}$; 2,948,000 people

53. $y = 3834e^{0.0052t}$; 4,366,000 people

55. k is larger in Exercise 47, because the population of Aruba is increasing faster than the population of Jamaica.

57. 7.949 billion people

59. (a) $y = 73e^{0.4763t}$ (b) 9 hours

61. $y = 109,478,000e^{0.1259t}$, where $t = 0 \leftrightarrow 2000$; 562,518,000 users

Isotope	Half-Life (Years)	Initial Quantity	Amount After 1000 Years
63. ^{226}Ra	1620	6 g	3.91 g
65. ^{14}C	5730	4.51 g	4.0 g
67. ^{239}Pu	24,100	4.2 g	4.08 g

69. 3.3 grams   **71.** 7.5 grams   **73.** $15,204

75. The Chilean earthquake in 1960 was about 19,953 times greater.

77. The earthquake in Fiji was about 63 times greater.

79. 7.04   **81.** 10^7 times

83. (a)

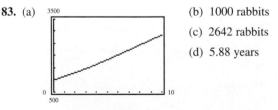

(b) 1000 rabbits

(c) 2642 rabbits

(d) 5.88 years

85. (a) $S = 10(1 - e^{-0.0575x})$ (b) 3314 pairs

87. If $k > 0$, the model represents exponential growth, and if $k < 0$, the model represents exponential decay.

89. When the investment is compounded more than once in a year (quarterly, monthly, daily, continuously), the effective yield is greater than the interest rate.

91. $\dfrac{7}{2} \pm \dfrac{\sqrt{69}}{2}$ **93.** $-\dfrac{3}{2} \pm \dfrac{\sqrt{33}}{6}$

95. $(4, \infty)$

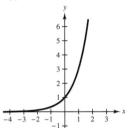

97. $(-\infty, -3) \cup (3, \infty)$

Review Exercises   *(page 648)*

1. (a) $\dfrac{1}{64}$ (b) 4 (c) 16   **3.** (a) 5 (b) 0.185 (c) $\dfrac{1}{25}$

5.

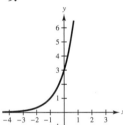

Horizontal asymptote:
$y = 0$

7.

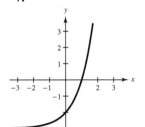

Horizontal asymptote:
$y = -3$

9.

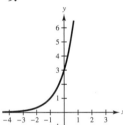

Horizontal asymptote:
$y = 0$

11.

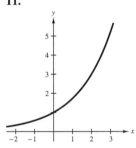

Horizontal asymptote:
$y = 0$

13.

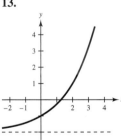

Horizontal asymptote: $y = -2$

15.

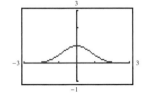

17.

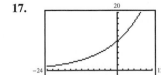

19. (a) 0.007 (b) 3 (c) 9.557×10^{16}

21.

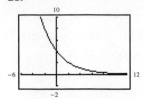

23.

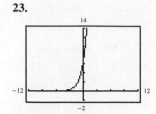

25.

n	1	4	12
A	\$226,296.28	\$259,889.34	\$268,503.32

n	365	Continuous
A	\$272,841.23	\$272,990.75

27. 4.21 grams   **29.** (a) 6 (b) 1   **31.** (a) 5 (b) -1

33. (a) $(f \circ g)(x) = \sqrt{2x + 6}$

   Domain: $[-3, \infty)$

   (b) $(g \circ f)(x) = 2\sqrt{x + 6}$

   Domain: $[-6, \infty)$

35. No   **37.** Yes   **39.** $f^{-1}(x) = \dfrac{1}{3}(x - 4)$

41. $h^{-1}(x) = \dfrac{1}{5}x^2, \ x \ge 0$ **43.** $f^{-1}(t) = \sqrt[3]{t - 4}$

45.

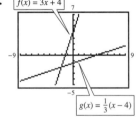

$f(x) = 3x + 4$

$g(x) = \dfrac{1}{3}(x - 4)$

47.

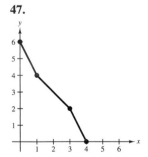

49.

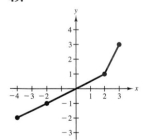

51. 3   **53.** -2 **55.** 6   **57.** 0

59.

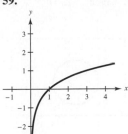

61.

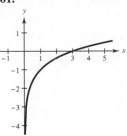

Vertical asymptote: $x = 0$ Vertical asymptote: $x = 0$

63.

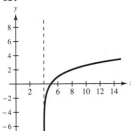

Vertical asymptote: $x = 4$

65. 3.9120

67.

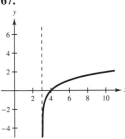

69.

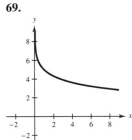

Vertical asymptote: $x = 3$ Vertical asymptote: $x = 0$

71. 1.5850   **73.** 2.4406   **75.** 1.7959   **77.** -0.4307

79. 1.0293   **81.** $\log_4 6 + 4 \log_4 x$ **83.** $\frac{1}{2} \log_5(x + 2)$

85. $\ln(x + 2) - \ln(x + 3)$

87. $\frac{1}{2}(\ln 2 + \ln x) + 5 \ln(x + 3)$ **89.** $\ln\left(\dfrac{1}{3y}\right)^{2/3}$, $y > 0$

91. $\log_8 32x^3$ **93.** $\ln \dfrac{9}{4x^2}$, $x > 0$

95. $\log_2\left(\dfrac{k}{k - t}\right)^4$, $k > t, k > 0$

97. $\ln(x^3 y^4 z)$, $x > 0$, $y > 0$, $z > 0$

99. False; $\log_2 4x = \log_2 4 + \log_2 x = 2 + \log_2 x$

101. True   **103.** True

105. $y = 0.83(\ln I_0 - \ln I)$; 0.20   **107.** 6

109. 1   **111.** 6   **113.** 5.66   **115.** 6.23

117. No solution   **119.** 1408.10   **121.** 31.47

123. 4.00   **125.** 64.00   **127.** 11.57   **129.** 7%

131. 5%   **133.** 7.5%   **135.** 6.5%   **137.** 5.65%

139. 7.71%   **141.** 7.79%

	Isotope	Half-Life (Years)	Initial Quantity	Amount After 1000 Years
143.	^{226}Ra	1620	3.5 g	2.282 g
145.	^{14}C	5730	2.934g	2.6 g
147.	^{239}Pu	24,100	5 g	4.858 g

149. The earthquake in San Francisco was about 2512 times greater.

Chapter Test   *(page 653)*

1. $f(-1) = 81$;

   $f(0) = 54$;

   $f\left(\frac{1}{2}\right) = 18\sqrt{6} \approx 44.09$;

   $f(2) = 24$

2.

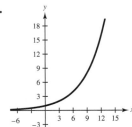

Horizontal asymptote: $y = 0$

3. (a) $(f \circ g)(x) = 18x^2 - 63x + 55$;

   Domain: $(-\infty, \infty)$

   (b) $(g \circ f)(x) = -6x^2 - 3x + 5$;

   Domain: $(-\infty, \infty)$

4. $f^{-1}(x) = \frac{1}{9}(x + 4)$

5. $(f \circ g)(x) = -\frac{1}{2}(-2x + 6) + 3$

   $= (x - 3) + 3$

   $= x$

   $(g \circ f)(x) = -2\left(-\frac{1}{2}x + 3\right) + 6$

   $= (x - 6) + 6$

   $= x$

6. -4

7. $g = f^{-1}$

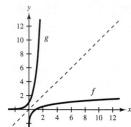

8. $\dfrac{2}{3} + \dfrac{1}{2}\log_8 x - 4\log_8 y$ **9.** $\ln\dfrac{x}{y^4}$, $y > 0$

10. 32   **11.** 1.18   **12.** 13.73   **13.** 15.52

14. 5   **15.** 8   **16.** 109.20   **17.** 0

18. (a) \$8012.78   (b) \$8110.40   **19.** \$10,806.08

20. 7%   **21.** \$4746.09   **22.** 600 foxes

23. 1141 foxes   **24.** 4.4 years

CHAPTER 10

Section 10.1   *(page 663)*

1. e   **2.** c   **3.** d   **4.** a   **5.** f   **6.** b

7. $x^2 + y^2 = 25$ **9.** $x^2 + y^2 = \frac{4}{9}$ **11.** $x^2 + y^2 = 36$

13. $x^2 + y^2 = 29$ **15.** $(x - 4)^2 + (y - 3)^2 = 100$

17. $(x - 6)^2 + (y + 5)^2 = 9$

19. $(x + 2)^2 + (y - 1)^2 = 4$

21. $(x - 3)^2 + (y - 2)^2 = 17$

23. Center: $(0, 0)$; $r = 4$ **25.** Center: $(0, 0)$; $r = 6$

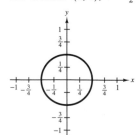   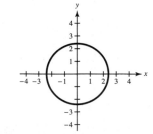

27. Center: $(0, 0)$; $r = \frac{1}{2}$ **29.** Center: $(0, 0)$; $r = \frac{12}{5}$

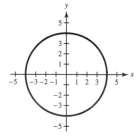   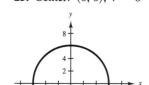

31. Center: $(-1, 5)$; $r = 8$ **33.** Center: $(2, 3)$; $r = 2$

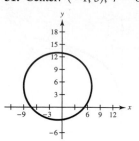   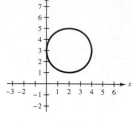

35. Center: $\left(-\frac{9}{4}, 4\right)$; $r = 4$ **37.** Center: $(2, 1)$; $r = 2$

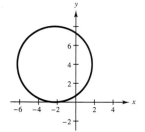   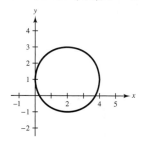

39. Center: $(-1, -3)$; $r = 2$ **41.** Center: $(-5, 2)$; $r = 6$

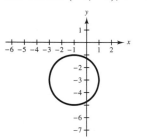   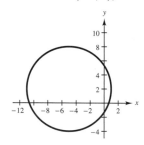

43.   **45.**

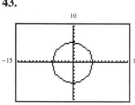   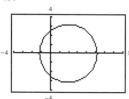

47. f   **48.** a   **49.** e   **50.** c   **51.** b   **52.** d

53. $x^2 = \frac{3}{2}y$ **55.** $x^2 = -6y$ **57.** $y^2 = -8x$

59. $x^2 = 4y$ **61.** $y^2 = 24x$ **63.** $y^2 = 9x$

65. $(x - 3)^2 = -(y - 1)$ **67.** $y^2 = 2(x + 2)$

69. $(y - 2)^2 = -8(x - 3)$ **71.** $x^2 = 12(y + 4)$

73. $(y - 2)^2 = x$

75. Vertex: $(0, 0)$
   Focus: $\left(0, \frac{1}{2}\right)$

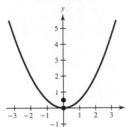

77. Vertex: $(0, 0)$
   Focus: $\left(-\frac{5}{2}, 0\right)$

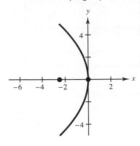

79. Vertex: $(0, 0)$
   Focus: $(0, -2)$

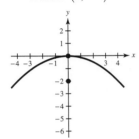

81. Vertex: $(1, -2)$
   Focus: $(1, -4)$

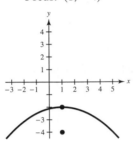

83. Vertex: $\left(5, -\frac{1}{2}\right)$
   Focus: $\left(\frac{11}{2}, -\frac{1}{2}\right)$

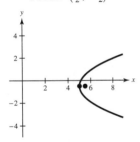

85. Vertex: $(1, 3)$
   Focus: $\left(1, \frac{15}{4}\right)$

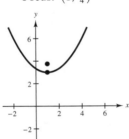

87. Vertex: $(-2, -3)$
   Focus: $(-4, -3)$

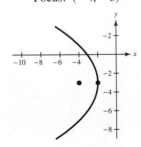

89. Vertex: $(-2, 1)$
   Focus: $\left(-2, -\frac{1}{2}\right)$

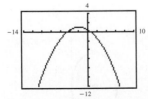

91. Vertex: $\left(\frac{1}{4}, -\frac{1}{2}\right)$
   Focus: $\left(0, -\frac{1}{2}\right)$

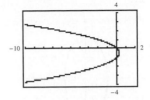

93. $x^2 + y^2 = 4500^2$

95. $(x - 10)^2 + (y - 5.5)^2 = 2.25$; 6.8 feet

97. (a) $x^2 = 180y$

(b)

x	0	20	40	60
y	0	$2\frac{2}{9}$	$8\frac{8}{9}$	20

99. (a)

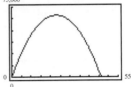

(b) 230 video game systems

101. (a) Answers will vary.

(b)

Maximum area when $x \approx 17.68$

103. No. The equation in standard form is $x^2 + (y - 3)^2 = 4$, which does represent a circle with center $(0, 3)$ and radius of 2.

105. No. If the graph intersected the directrix, there would exist points closer to the directrix than the focus.

107. $-2 \pm \sqrt{10}$ **109.** $-1, 3$ **111.** $-\frac{5}{4} \pm \frac{\sqrt{89}}{4}$

113. $10 \log_8 x$ **115.** $\ln 5 + 2 \ln x + \ln y$

117. $\log_{10} 6x$ **119.** $\ln \dfrac{x^3 y}{9}$

Section 10.2   *(page 674)*

1. a   **2.** f   **3.** d   **4.** c   **5.** e   **6.** b

7. $\dfrac{x^2}{16} + \dfrac{y^2}{9} = 1$ **9.** $\dfrac{x^2}{4} + \dfrac{y^2}{1} = 1$ **11.** $\dfrac{x^2}{9} + \dfrac{y^2}{36} = 1$

13. $\dfrac{x^2}{1} + \dfrac{y^2}{4} = 1$ **15.** $\dfrac{x^2}{9} + \dfrac{y^2}{25} = 1$

17. $\dfrac{x^2}{100} + \dfrac{y^2}{36} = 1$

19.

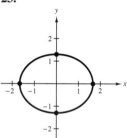

Vertices: $(\pm 4, 0)$

Co-vertices: $(0, \pm 2)$

21.

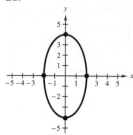

Vertices: $(0, \pm 4)$

Co-vertices: $(\pm 2, 0)$

23.

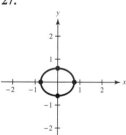

Vertices: $\left(\pm \frac{5}{3}, 0 \right)$

Co-vertices: $\left(0, \pm \frac{4}{3} \right)$

25.

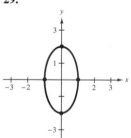

Vertices: $\left(0, \pm \frac{4}{5} \right)$

Co-vertices: $\left(\pm \frac{2}{3}, 0 \right)$

27.

Vertices: $\left(\pm \frac{3}{4}, 0 \right)$

Co-vertices: $\left(0, \pm \frac{3}{5} \right)$

29.

Vertices: $(0, \pm 2)$

Co-vertices: $(\pm 1, 0)$

31.

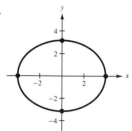

Vertices: $(\pm 4, 0)$

Co-vertices: $\left(0, \pm \sqrt{10} \right)$

33.

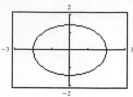

Vertices: $(\pm 2, 0)$

35.

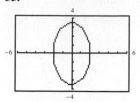

Vertices: $\left(0, \pm 2\sqrt{3} \right)$

37. $\dfrac{x^2}{1} + \dfrac{y^2}{4} = 1$ **39.** $\dfrac{(x-4)^2}{9} + \dfrac{y^2}{16} = 1$

41. Center: $(-5, 0)$

Vertices: $(-9, 0), (-1, 0)$

43. Center: $(1, 5)$

Vertices: $(1, 0), (1, 10)$

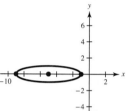

45. Center: $(2, -2)$

Vertices:

$(-1, -2), (5, -2)$

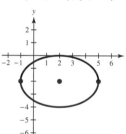

47. Center: $(-4, 1)$

Vertices:

$(-4, -3), (-4, 5)$

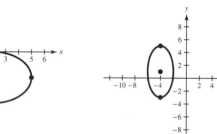

49. Center: $(-2, 3)$

Vertices:

$(-2, 6), (-2, 0)$

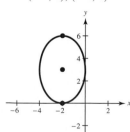

51. Center: $(4, -3)$

Vertices:

$(4, -8), (4, 2)$

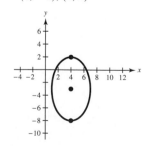

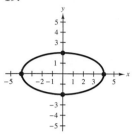
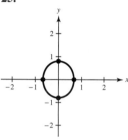

53. Center: $(2, 1)$

Vertices: $(-8, 1), (12, 1)$

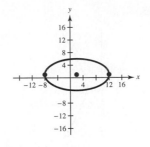

55. 36 feet; 28 feet   **57.** $4\sqrt{19} \approx 17.4$ feet

59. $\dfrac{x^2}{615^2} + \dfrac{y^2}{290^2} = 1$ or $\dfrac{x^2}{290^2} + \dfrac{y^2}{615^2} = 1$

61. (a) Every point on the ellipse represents the maximum distance (800 miles) that the plane can safely fly with enough fuel to get from airport A to airport B.

(b) Airport A: $(-250, 0)$; Airport B: $(250, 0)$

(c) 800 miles; Vertices: $(\pm 400, 0)$

(d) $\dfrac{x^2}{400^2} + \dfrac{y^2}{(50\sqrt{39})^2} = 1$

(e) $20{,}000\sqrt{39}\,\pi \approx 392{,}385$ square miles

63. A circle is an ellipse in which the major axis and the minor axis have the same length. Both circles and ellipses have foci; however, a circle has a single focus located at the center, and an ellipse has two foci that lie on the major axis.

65. The sum of the distances between each point on the ellipse and the two foci is a constant.

67. The graph of an ellipse written in the standard form

$$\dfrac{(x-h)^2}{a^2} + \dfrac{(y-k)^2}{b^2} = 1$$

intersects the y-axis if $|h| > a$ and intersects the x-axis if $|k| > b$. Similarly, the graph of

$$\dfrac{(x-h)^2}{b^2} + \dfrac{(y-k)^2}{a^2} = 1$$

intersects the y-axis if $|h| > b$ and intersects the x-axis if $|k| > a$.

69. (a) $f(-2) = 9$ **71.** (a) $g(-1) \approx 3.639$

(b) $f(2) = \frac{1}{9}$ (b) $g(2) \approx 16.310$

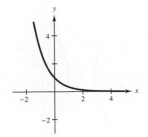

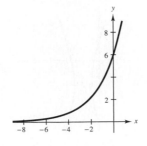

73. (a) $h(4) = 1$ **75.** (a) $f(3)$ does not exist.

(b) $h(64) = 2$ (b) $f(35) = \frac{5}{2}$

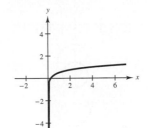

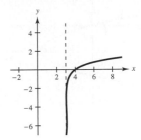

Mid-Chapter Quiz   *(page 678)*

1. $x^2 + y^2 = 25$ **2.** $(y-1)^2 = 8(x+2)$

3. $\dfrac{(x+2)^2}{16} + \dfrac{(y+1)^2}{4} = 1$

4. $(x-3)^2 + (y+5)^2 = 25$

5. $(x-2)^2 = -8(y-3)$ **6.** $\dfrac{x^2}{36} + \dfrac{y^2}{100} = 1$

7. $x^2 + (y+3)^2 = 16$; Center: $(0, -3)$; $r = 4$

8. $(x+1)^2 + (y-2)^2 = 1$; Center: $(-1, 2)$; $r = 1$

9. $(y-3)^2 = x + 16$; Vertex: $(-16, 3)$; Focus: $\left(-\frac{63}{4}, 3\right)$

10. $(x-4)^2 = -(y-4)$; Vertex: $(4, 4)$; Focus: $\left(4, \frac{15}{4}\right)$

11. $\dfrac{(x-2)^2}{9} + \dfrac{y^2}{36} = 1$

Center: $(2, 0)$

Vertices: $(2, -6), (2, 6)$

12. $\dfrac{(x-6)^2}{9} + \dfrac{(y+2)^2}{4} = 1$

Center: $(6, -2)$

Vertices: $(3, -2), (9, -2)$

13.

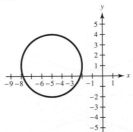

14.

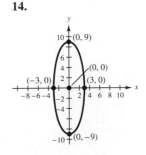

15.

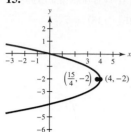

16.

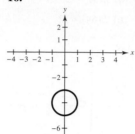

15.

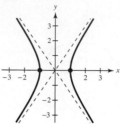

Vertices: $(\pm 1, 0)$

Asymptotes: $y = \pm\frac{3}{2}x$

17.

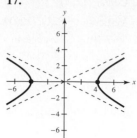

Vertices: $(\pm 4, 0)$

Asymptotes: $y = \pm\frac{1}{2}x$

19. $\dfrac{x^2}{16} - \dfrac{y^2}{64} = 1$ **21.** $\dfrac{y^2}{16} - \dfrac{x^2}{64} = 1$

23. $\dfrac{x^2}{81} - \dfrac{y^2}{36} = 1$ **25.** $\dfrac{y^2}{1} - \dfrac{x^2}{1/4} = 1$

27.

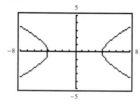

29.

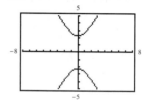

17.

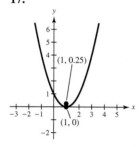

18.

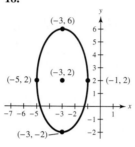

Section 10.3   (page 684)

1. c   **2.** e   **3.** a   **4.** f   **5.** b   **6.** d

7.

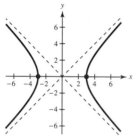

9.

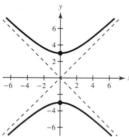

Vertices: $(\pm 3, 0)$

Asymptotes: $y = \pm x$

Vertices: $(0, \pm 3)$

Asymptotes: $y = \pm x$

31. Center: $(3, -4)$

Vertices:

$(3, 1), (3, -9)$

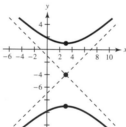

33. Center: $(1, -2)$

Vertices:

$(-1, -2), (3, -2)$

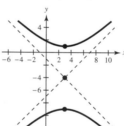

11.

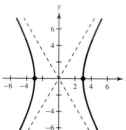

13.

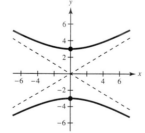

Vertices: $(\pm 3, 0)$

Asymptotes: $y = \pm\frac{5}{3}x$

Vertices: $(0, \pm 3)$

Asymptotes: $y = \pm\frac{3}{5}x$

35. Center: $(2, -3)$

Vertices:

$(1, -3), (3, -3)$

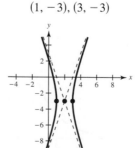

37. Center: $(-3, 2)$

Vertices:

$(-4, 2), (-2, 2)$

39. $\dfrac{y^2}{9} - \dfrac{x^2}{9/4} = 1$ **41.** $\dfrac{(x-3)^2}{4} - \dfrac{(y-2)^2}{16/5} = 1$

43. Ellipse   **45.** Hyperbola   **47.** Hyperbola

49. 110.3   **51.** (a) Left half   (b) Top half

53. One. The difference in the distances between the given point and the given foci is constant, so only one branch of one hyperbola can pass through the point.

55. Consistent   **57.** $\left(\frac{5}{2}, \frac{1}{2}\right)$

Section 10.4   *(page 694)*

1.

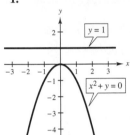

No real solution

3.

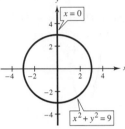

$(0, 3), (0, -3)$

5.

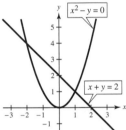

$(-2, 4), (1, 1)$

7.

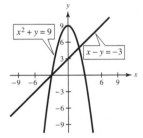

$(2, 5), (-3, 0)$

9.

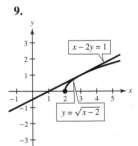

$(3, 1)$

11.

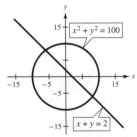

$(-6, 8), (8, -6)$

13.

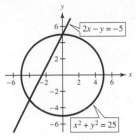

$(0, 5), (-4, -3)$

15.

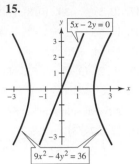

No real solution

17.

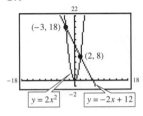

$(-3, 18), (2, 8)$

19.

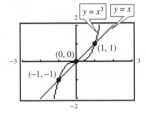

$(0, 0), (1, 1), (-1, -1)$

21.

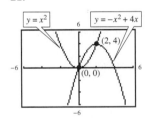

$(0, 0), (2, 4)$

23.

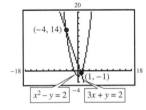

$(-4, 14), (1, -1)$

25.

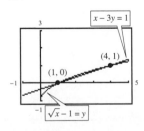

$(1, 0), (4, 1)$

27.

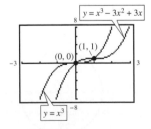

$(0, 0), (1, 1)$

29.

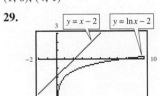

No real solution

31. $(1, 2), (2, 8)$ **33.** $(0, 5), (2, 1)$ **35.** $(0, 2), (2, 0)$

37. $(0, 5)$ **39.** $(0, 8), \left(-\frac{24}{5}, -\frac{32}{5}\right)$

41. $\left(-\frac{17}{2}, -6\right), \left(-\frac{7}{2}, 4\right)$ **43.** $\left(-\frac{9}{5}, \frac{12}{5}\right), (3, 0)$

45. $(-14, -20), (2, -4)$ **47.** $(-4, 11), \left(\frac{5}{2}, \frac{5}{4}\right)$

49. $(0, 2), (3, 1)$ **51.** No real solution

53. $\left(\pm\sqrt{5}, 2\right), (0, -3)$ **55.** $(0, 4), (3, 0)$

57. No real solution   **59.** $\left(\pm\sqrt{3}, -1\right)$

61. $\left(2, \pm 2\sqrt{3}\right), (-1, \pm 3)$ **63.** $\left(\pm 2, \pm\sqrt{3}\right)$

65. $(\pm 2, 0)$ **67.** $(\pm 3, \pm 2)$ **69.** $(\pm 3, 0)$

71. $\left(\pm\sqrt{6}, \pm\sqrt{3}\right)$ **73.** No real solution

75. $\left(\pm\sqrt{5}, \pm 2\right)$ **77.** $\left(\pm\dfrac{2\sqrt{5}}{5}, \pm\dfrac{2\sqrt{5}}{5}\right)$

79. $\left(\pm\sqrt{3}, \pm\sqrt{13}\right)$ **81.** 40 feet $\times$ 75 feet

83. 200 feet $\times$ 210 feet   **85.** $(3.633, 2.733)$

87. Between points $\left(-\frac{3}{5}, -\frac{4}{5}\right)$ and $\left(\frac{4}{5}, -\frac{3}{5}\right)$

89. Solve one of the equations for one variable in terms of the other. Substitute that expression into the other equation and solve. Back-substitute the solution into the first equation to find the value of the other variable. Check the solution to see that it satisfies both of the original equations.

91. Two. The line can intersect a branch of the hyperbola at most twice, and it can intersect only one point on each branch at the same time.

93. -5 **95.** 9   **97.** 5   **99.** 4.564

101. 1.023   **103.** -0.632

Review Exercises (page 700)

1. Ellipse   **3.** Circle   **5.** Hyperbola

7. $x^2 + y^2 = 36$

9. Center: $(0, 0)$; $r = 8$

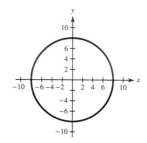

11. $(x - 2)^2 + (y - 6)^2 = 9$

13. Center: $(-3, -4)$; $r = 2$

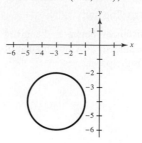

15. $y^2 = -8x$

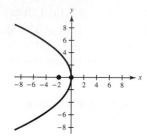

17. $(x + 1)^2 = \frac{1}{2}(y - 3)$

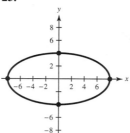

19. Vertex: $(8, -25)$

Focus: $\left(8, -\frac{49}{2}\right)$

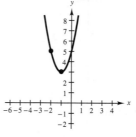

21. $\dfrac{x^2}{4} + \dfrac{y^2}{25} = 1$ **23.** $\dfrac{x^2}{4} + \dfrac{y^2}{9} = 1$

25.

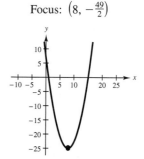

Vertices: $(\pm 8, 0)$

Co-vertices: $(0, \pm 4)$

27.

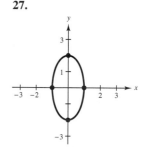

Vertices: $(0, \pm 2)$

Co-vertices: $(\pm 1, 0)$

29. $\dfrac{(x - 3)^2}{25} + \dfrac{(y - 4)^2}{16} = 1$ **31.** $\dfrac{x^2}{9} + \dfrac{(y - 4)^2}{16} = 1$

33. Center: $(-1, 2)$

Vertices:

$(-1, -4), (-1, 8)$

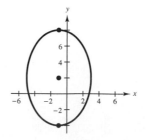

35. Center: $(0, -3)$

Vertices:

$(0, -7), (0, 1)$

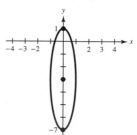

37.

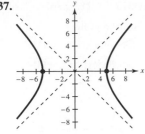

Vertices: $(\pm 5, 0)$

Asymptotes: $y = \pm x$

39.

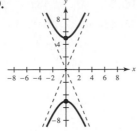

Vertices: $(0, \pm 5)$

Asymptotes: $y = \pm \frac{5}{2}x$

41. $\dfrac{x^2}{4} - \dfrac{y^2}{9} = 1$ **43.** $\dfrac{y^2}{64} - \dfrac{x^2}{100} = 1$

45. Center: $(3, -1)$

Vertices: $(0, -1), (6, -1)$

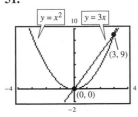

47. Center: $(4, -3)$

Vertices: $(4, -1), (4, -5)$

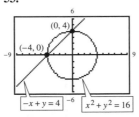

49. $\dfrac{(x+4)^2}{4} - \dfrac{(y-6)^2}{12} = 1$

51.

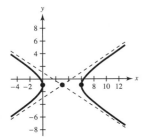

$(0, 0), (3, 9)$

53.

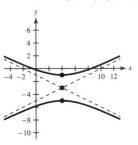

$(-4, 0), (0, 4)$

55. $(-1, 5), (-2, 20)$ **57.** $(-1, 0), (0, -1)$

59. $(\pm 2, \pm 3)$ **61.** 6 centimeters $\times$ 8 centimeters

63. Piece 1: 38.48 inches; Piece 2: 61.52 inches

Chapter Test *(page 703)*

1. $(x + 2)^2 + (y + 3)^2 = 16$

2. $(x - 1)^2 + (y - 3)^2 = 9$ **3.** $(x + 2)^2 + (y - 3)^2 = 9$

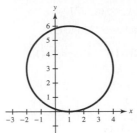   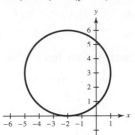

4. Vertex: $(4, 2)$; Focus: $\left(\frac{47}{12}, 2\right)$

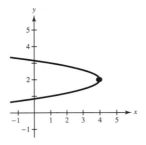

5. $(x - 7)^2 = 8(y + 2)$ **6.** $\dfrac{(x-2)^2}{25} + \dfrac{y^2}{9} = 1$

7. Center: $(0, 0)$ **8.** Center: $(1, 3)$

Vertices: $(0, \pm 4)$ Vertices: $(1, -2), (1, 8)$

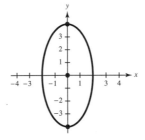

9. $\dfrac{x^2}{9} - \dfrac{y^2}{4} = 1$ **10.** $\dfrac{y^2}{25} - \dfrac{x^2}{4} = 1$

11. Center: $(3, 0)$ **12.** Center: $(4, -2)$

Vertices: $(1, 0), (5, 0)$ Vertices: $(4, -7), (4, 3)$

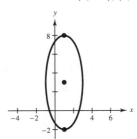   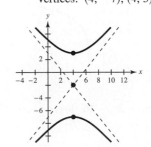

13. $(0, 3), (4, 0)$ **14.** $(\pm 4, 0)$

15. $(\sqrt{6}, 2), (\sqrt{6}, -2), (-\sqrt{6}, 2), (-\sqrt{6}, -2)$

16. $x^2 + y^2 = 5000^2$ **17.** 16 inches × 12 inches

Cumulative Test: Chapters 8–10   (page 704)

1. $-\frac{3}{4}, 3$ **2.** $-3, 13$ **3.** $5 \pm 5\sqrt{2}$

4. $-1 \pm \dfrac{\sqrt{3}}{3}$ **5.** $\pm\sqrt{3}, \pm\sqrt{5}$

6. $\left[-3, \dfrac{1}{3}\right]$

7. $\left(-\frac{4}{3}, \frac{1}{2}\right)$

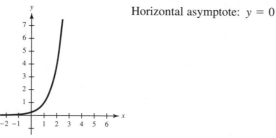

8. $x^2 - 4x - 12 = 0$

9. (a) $(f \circ g)(x) = 50x^2 - 20x - 1$; Domain: $(-\infty, \infty)$

(b) $(g \circ f)(x) = 10x^2 - 16$; Domain: $(-\infty, \infty)$

10. $f^{-1}(x) = -\frac{4}{3}x + \frac{5}{3}$

11. $f(1) = \frac{15}{2}; f(0.5) \approx 7.707; f(3) = \frac{57}{8}$

12.

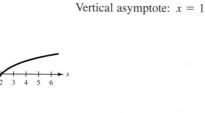

Horizontal asymptote: $y = 0$

13. The graphs are reflections of each other in the line $y = x$.

14.

Vertical asymptote: $x = 1$

15. -2 **16.** $\log_2 \dfrac{x^3 y^3}{z}$ **17.** $\dfrac{1}{2}\log_{10}(x + 1) - 4\log_{10} x$

18. 3   **19.** 12.182   **20.** 18.013   **21.** 0.867

22. \$34.38   **23.** 8.33%   **24.** 19.8 years

25. $(x - 3)^2 + (y + 7)^2 = 64$ **26.** Vertex: $(5, -45)$

Focus: $\left(5, -\dfrac{359}{8}\right)$

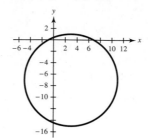

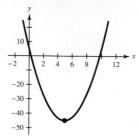

27. $\dfrac{(x + 3)^2}{4} + \dfrac{(y - 2)^2}{25} = 1$

28. Center: $(0, 0)$ **29.** $\dfrac{y^2}{9} - x^2 = 1$

Vertices: $(0, \pm 2)$

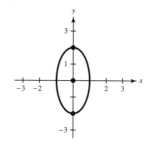

30. Center: $(0, 1)$ **31.** $(1, -1), (3, 5)$

Vertices: $(\pm 12, 1)$

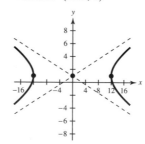

32. $(-4, \pm 1), (-1, \pm 2)$ **33.** 8 feet × 4 feet

34. (a)

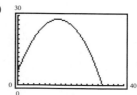

(b) Highest point: 28.5 feet; Range: $[0, 28.5]$

CHAPTER 11

Section 11.1   *(page 714)*

1. $2, 4, 6, 8, 10$ **3.** $\frac{1}{4}, \frac{1}{16}, \frac{1}{64}, \frac{1}{256}, \frac{1}{1024}$

5. $-2, 4, -6, 8, -10$ **7.** $\frac{1}{4}, -\frac{1}{8}, \frac{1}{16}, -\frac{1}{32}, \frac{1}{64}$

9. $3, 8, 13, 18, 23$ **11.** $1, \frac{4}{5}, \frac{2}{3}, \frac{4}{7}, \frac{1}{2}$ **13.** $\frac{3}{4}, \frac{2}{3}, \frac{9}{14}, \frac{12}{19}, \frac{5}{8}$

15. $-1, \frac{1}{4}, -\frac{1}{9}, \frac{1}{16}, -\frac{1}{25}$ **17.** $\frac{9}{4}, \frac{33}{16}, \frac{129}{64}, \frac{513}{256}, \frac{2049}{1024}$

19. $2, 3, 4, 5, 6$ **21.** $0, 3, -1, \frac{3}{4}, -\frac{1}{4}$ **23.** -72

25. $\frac{31}{2520}$ **27.** 5 **29.** $\frac{1}{132}$ **31.** $53,130$

33. $\dfrac{1}{n+1}$ **35.** $n(n+1)$ **37.** $2n$ **39.** b

40. d   **41.** c   **42.** a

43.   **45.**

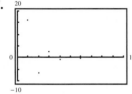

47.   **49.** $a_n = 2n - 1$

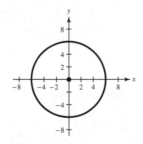

51. $a_n = 4n - 2$ **53.** $a_n = n^2 - 1$

55. $a_n = (-1)^{n+1}2n$ **57.** $a_n = \dfrac{n+1}{n+2}$

59. $a_n = \left(-\dfrac{1}{5}\right)^n$ **61.** $a_n = \dfrac{1}{2^{n-1}}$ **63.** $a_n = 1 + \dfrac{1}{n}$

65. $a_n = \dfrac{(-1)^n}{(n+1)!}$ **67.** $S_1 = 7; S_2 = 16; S_6 = 72$

69. $S_2 = \frac{3}{2}; S_3 = \frac{11}{6}; S_9 = \frac{7129}{2520}$ **71.** 90 **73.** 77

75. 100 **77.** $\frac{3019}{3600}$ **79.** $\frac{15,551}{2520}$ **81.** -48 **83.** $\frac{8}{9}$

85. $\frac{182}{243}$ **87.** 2040 **89.** 852 **91.** 16.25

93. 6.5793 **95.** $\displaystyle\sum_{k=1}^{5} k$ **97.** $\displaystyle\sum_{k=1}^{6} 5k$ **99.** $\displaystyle\sum_{k=1}^{50} \dfrac{3}{1+k}$

101. $\displaystyle\sum_{k=1}^{10} \dfrac{1}{2k}$ **103.** $\displaystyle\sum_{k=0}^{12} \dfrac{1}{2^k}$ **105.** $\displaystyle\sum_{k=1}^{20} \dfrac{1}{k(k+1)}$

107. $\displaystyle\sum_{k=1}^{11} \dfrac{k}{k+1}$ **109.** $\displaystyle\sum_{k=1}^{20} \dfrac{2k}{k+3}$ **111.** $\displaystyle\sum_{k=0}^{6} k!$

113. (a) $\$535, \$572.45, \$612.52, \$655.40, \$701.28, \$750.37,$
$\$802.89, \$859.09$

(b) $\$7487.23$

(c)

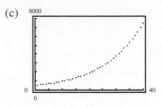

(d) Yes. Any investment earning compound interest increases at an increasing rate.

115. $36°, 60°, 77.1°, 90°, 100°, 108°$

117. (a) $\$13,500$

(b) $\$5695.31$; No. The SUV loses
$1 - \left(\frac{3}{4}\right)^3 = 1 - \frac{27}{64} \approx 58\%$ (a little more than half) of its value every three years.

119. A sequence is a function because there is only one value for each term of the sequence.

121. $a_n = 4n! = 4(1 \cdot 2 \cdot 3 \cdot 4 \cdots \cdot (n-1) \cdot n)$

$a_n = (4n)!$
$= 1 \cdot 2 \cdot 3 \cdot 4 \cdots \cdot 4(n-1) \cdot (4n)$

123. True

$$\sum_{k=1}^{4} 3k = 3 + 6 + 9 + 12$$
$$= 3(1 + 2 + 3 + 4) = 3\sum_{k=1}^{4} k$$

125. 9 **127.** -11

129. Center: $(0, 0); r = 6$ **131.** Center: $(-2, 0); r = 4$

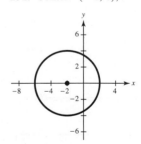

133. Vertex: $(0, 0)$ **135.** Vertex: $(0, -4)$

Focus: $\left(0, \frac{3}{2}\right)$ Focus: $(0, -6)$

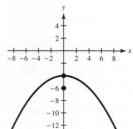

Section 11.2   *(page 723)*

1. 3   **3.** -6 **5.** -12 **7.** $\frac{2}{3}$ **9.** $-\frac{5}{4}$

11. 4   **13.** -3 **15.** $\frac{1}{2}$ **17.** Arithmetic, 2

19. Arithmetic, -2 **21.** Arithmetic, -16

23. Arithmetic, 0.8   **25.** Arithmetic, $\frac{3}{2}$

27. Not arithmetic   **29.** Not arithmetic

31. Not arithmetic   **33.** 7, 12, 17, 22, 27

35. 11, 15, 19, 23, 27   **37.** 20, 16, 12, 8, 4

39. 6, 11, 16, 21, 26   **41.** 22, 18, 14, 10, 6

43. 7, 10, 13, 16, 19   **45.** 6, 4, 2, 0, -2

47. $\frac{3}{2}, 4, \frac{13}{2}, 9, \frac{23}{2}$ **49.** $\frac{8}{5}, \frac{11}{5}, \frac{14}{5}, \frac{17}{5}, 4$ **51.** $4, \frac{15}{4}, \frac{7}{2}, \frac{13}{4}, 3$

53. $a_n = 3n + 1$ **55.** $a_n = \frac{3}{2}n - 1$

57. $a_n = -5n + 105$ **59.** $a_n = \frac{3}{2}n + \frac{3}{2}$

61. $a_n = \frac{5}{2}n + \frac{5}{2}$ **63.** $a_n = 4n + 4$

65. $a_n = -10n + 60$ **67.** $a_n = -\frac{1}{2}n + 11$

69. $a_n = -0.05n + 0.40$ **71.** 14, 20, 26, 32, 38

73. 23, 18, 13, 8, 3   **75.** $-16, -11, -6, -1, 4$

77. 3.4, 2.3, 1.2, 0.1, -1 **79.** 210   **81.** 1425

83. 255   **85.** 62,625   **87.** 35   **89.** 522

91. 1850   **93.** 900   **95.** 12,200   **97.** 243   **99.** 23

101. b   **102.** f   **103.** e   **104.** a

105. c   **106.** d

107.   **109.**

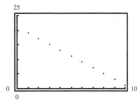 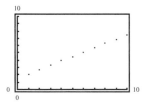

111.   **113.** 9000

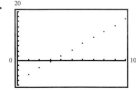

115. 13,120   **117.** 3011.25   **119.** 2850   **121.** 2500

123. $369,000   **125.** $25.43   **127.** 632 bales

129. 114 times   **131.** 1024 feet   **133.** 30.9 barrels

135. A recursion formula gives the relationship between the terms a_{n+1} and a_n.

137. Yes. Because a_{2n} is n terms away from a_n, add n times the difference d to a_n.

$$a_{2n} = a_n + nd$$

139. (a) 4; 9; 16; 25; 36

(b) No. There is no common difference between consecutive terms of the sequence.

(c) $\displaystyle\sum_{k=1}^{n} (2k - 1) = n^2;$ 49

(d) Answers will vary.

141. Center: $(4, -5)$

Vertices: $(-1, -5), (9, -5)$

143. Center: $(1, -3)$

Vertices: $(1, -6), (1, 0)$

145. $\displaystyle\sum_{k=1}^{7} (k + 2)$ **147.** $\displaystyle\sum_{k=1}^{5} (3k + 9)$

Mid-Chapter Quiz   *(page 728)*

1. 4, 8, 12, 16, 20   **2.** 7, 9, 11, 13, 15

3. $32, 8, 2, \frac{1}{2}, \frac{1}{8}$ **4.** $-\frac{3}{5}, 3, -\frac{81}{7}, \frac{81}{2}, -135$ **5.** 100

6. 40   **7.** 87   **8.** 25   **9.** 40   **10.** 26

11. $\displaystyle\sum_{k=1}^{20} \frac{2}{3k}$ **12.** $\displaystyle\sum_{k=1}^{25} \frac{(-1)^{k+1}}{k^3}$ **13.** $\displaystyle\sum_{k=1}^{20} \frac{k-1}{k}$

14. $\displaystyle\sum_{k=1}^{10} \frac{k^2}{2}$ **15.** $\frac{1}{2}$ **16.** -6 **17.** $a_n = -3n + 23$

18. $a_n = -4n + 36$ **19.** 40,200   **20.** $33,397.50

Section 11.3   *(page 734)*

1. 2   **3.** -1 **5.** $-\frac{1}{2}$ **7.** $\frac{1}{5}$ **9.** π **11.** 1.04

13. Geometric, $\frac{1}{2}$ **15.** Not geometric   **17.** Geometric, 2

19. Not geometric   **21.** Geometric, $-\frac{2}{3}$

23. Geometric, 1.1   **25.** 4, 8, 16, 32, 64

27. $6, 3, \frac{3}{2}, \frac{3}{4}, \frac{3}{8}$ **29.** $5, -10, 20, -40, 80$

31. $-4, 2, -1, \frac{1}{2}, -\frac{1}{4}$ **33.** 10, 10.2, 10.40, 10.61, 10.82

35. $10, 6, \frac{18}{5}, \frac{54}{25}, \frac{162}{125}$ **37.** $\frac{3}{2}, 1, \frac{2}{3}, \frac{4}{9}, \frac{8}{27}$ **39.** $a_n = 2^{n-1}$

41. $a_n = 2(2)^{n-1}$ **43.** $a_n = 10\left(-\frac{1}{5}\right)^{n-1}$

45. $a_n = 4\left(-\frac{1}{2}\right)^{n-1}$ **47.** $a_n = 8\left(\frac{1}{4}\right)^{n-1}$

49. $a_n = 14\left(\frac{3}{4}\right)^{n-1}$ **51.** $a_n = 4\left(-\frac{3}{2}\right)^{n-1}$

53. $\frac{3}{256}$ **55.** $48\sqrt{2}$ **57.** 1486.02

59. $-\frac{40}{6561}$ **61.** $\frac{81}{64}$ **63.** 12   **65.** $\frac{256}{9}$

67. b   **68.** d   **69.** c   **70.** a   **71.** 1023

73. 772.48   **75.** 2.25   **77.** 2.47   **79.** 6.67

81. $-14,762$ **83.** 16.00   **85.** 13,120   **87.** 48.00

89. 1103.57   **91.** 26,886.11   **93.** 2   **95.** $\frac{6}{5}$

97. $\frac{10}{9}$ **99.** 32

101.   **103.**

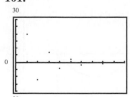   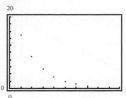

105. (a) There are many correct answers.

$$a_n = 187{,}500(0.75)^{n-1} \text{ or } a_n = 250{,}000(0.75)^n$$

(b) $59,326.17   (c) The first year

107. $3,623,993.23   **109.** $9748.28   **111.** $105,428.44

113. $100,953.76

115. (a) $5,368,709.11   (b) $10,737,418.23

117. (a) $P = (0.999)^n$ (b) 69.4%

(c)   693 days

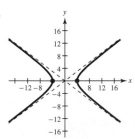

119. $\frac{37}{64} \approx 0.578$ square unit

121. 666.21 feet   **123.** $a_1 = 12$. Answers may vary.

125. When a positive number is multiplied by a number between 0 and 1, the result is a smaller positive number, so the terms of the sequence decrease.

127. The nth partial sum of a sequence is the sum of the first n terms of the sequence.

129. $(-1, 2), (2, 8)$ **131.** 8%   **133.** 7.5%

135.

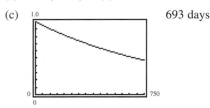

Vertices: $(\pm 4, 0)$

Asymptotes: $\pm \frac{3}{4}x$

Section 11.4   *(page 744)*

1. 15   **3.** 252   **5.** 1   **7.** 38,760   **9.** 38,760

11. 593,775   **13.** 2,598,960   **15.** 85,013,600

17. 15   **19.** 35   **21.** 70

23. $t^3 + 15t^2 + 75t + 125$

25. $m^5 - 5m^4n + 10m^3n^2 - 10m^2n^3 + 5mn^4 - n^5$

27. $243a^5 - 405a^4 + 270a^3 - 90a^2 + 15a - 1$

29. $64y^6 + 192y^5z + 240y^4z^2 + 160y^3z^3 + 60y^2z^4$
$\qquad + 12yz^5 + z^6$

31. $x^8 + 8x^6 + 24x^4 + 32x^2 + 16$

33. $x^6 + 18x^5 + 135x^4 + 540x^3 + 1215x^2 + 1458x + 729$

35. $u^3 - 6u^2v + 12uv^2 - 8v^3$

37. $81a^4 + 216a^3b + 216a^2b^2 + 96ab^3 + 16b^4$

39. $x^4 + \dfrac{8x^3}{y} + \dfrac{24x^2}{y^2} + \dfrac{32x}{y^3} + \dfrac{16}{y^4}$

41. $32x^{10} - 80x^8y + 80x^6y^2 - 40x^4y^3 + 10x^2y^4 - y^5$

43. $120x^7y^3$ **45.** $163{,}296a^5b^4$ **47.** 120

49. 54   **51.** 1.172   **53.** 510,568.785

55. $\frac{1}{32} + \frac{5}{32} + \frac{10}{32} + \frac{10}{32} + \frac{5}{32} + \frac{1}{32}$

57. $\frac{1}{256} + \frac{12}{256} + \frac{54}{256} + \frac{108}{256} + \frac{81}{256}$

59. The sum of the numbers in each row is a power of 2. Because the sum of the numbers in Row 2 is $1 + 2 + 1 = 4 = 2^2$, the sum of the numbers in Row n is 2^n.

61. The signs of the terms alternate in the expansion of $(x - y)^n$.

63. The first and last numbers in each row are 1. Every other number in the row is formed by adding the two numbers immediately above the number.

65. 390   **67.** 97,656

Review Exercises   *(page 748)*

1. 8, 11, 14, 17, 20   **3.** $\frac{1}{2}, \frac{2}{5}, \frac{3}{8}, \frac{4}{11}, \frac{5}{14}$

5. 2, 6, 24, 120, 720   **7.** $\frac{1}{2}, \frac{1}{2}, 1, 3, 12$ **9.** $a_n = 3n + 1$

11. $a_n = \dfrac{1}{n^2 + 1}$ **13.** $a_n = -2n + 5$

15. $a_n = \dfrac{3n^2}{n^2 + 1}$ **17.** 28   **19.** $\dfrac{13}{20}$ **21.** $\displaystyle\sum_{k=1}^{4} (5k - 3)$

23. $\displaystyle\sum_{k=1}^{6} \dfrac{1}{3k}$ **25.** -5.5 **27.** 127, 122, 117, 112, 107

29. $2, \frac{7}{3}, \frac{8}{3}, 3, \frac{10}{3}$ **31.** $80, \frac{155}{2}, 75, \frac{145}{2}, 70$

33. $a_n = 4n + 6$ **35.** $a_n = -50n + 1050$

37. 486   **39.** 1935   **41.** 2527.5   **43.** 5100

45. 462 seats   **47.** $\frac{5}{2}$ **49.** 10, 30, 90, 270, 810

51. $100, -50, 25, -\frac{25}{2}, \frac{25}{4}$ **53.** $4, 6, 9, \frac{27}{2}, \frac{81}{4}$

55. $a_n = \left(-\frac{2}{3}\right)^{n-1}$ **57.** $a_n = 24(3)^{n-1}$

59. $a_n = 12\left(-\frac{1}{2}\right)^{n-1}$ **61.** 8190   **63.** -1.928

65. 116,169.538   **67.** 4.556×10^{13} **69.** 8   **71.** 12

73. (a) There are many correct answers. $a_n = 120{,}000(0.70)^n$

(b) $20,168.40

75. 321,222,672 visits   **77.** 56   **79.** 1365

81. 91,390   **83.** 177,100   **85.** 10   **87.** 70

89. $x^4 - 20x^3 + 150x^2 - 500x + 625$

91. $125x^3 + 150x^2 + 60x + 8$

93. $x^{10} + 10x^9 + 45x^8 + 120x^7 + 210x^6 + 252x^5$
$\quad + 210x^4 + 120x^3 + 45x^2 + 10x + 1$

95. $81x^4 - 216x^3y + 216x^2y^2 - 96xy^3 + 16y^4$

97. $u^{10} + 5u^8v^3 + 10u^6v^6 + 10u^4v^9 + 5u^2v^{12} + v^{15}$

99. $13,440x^4$ **101.** $-61,236$

Chapter Test *(page 751)*

1. $1, -\frac{3}{5}, \frac{9}{25}, -\frac{27}{125}, \frac{81}{625}$ **2.** 2, 10, 24, 44, 70   **3.** 60

4. 45   **5.** -45 **6.** $\sum_{k=1}^{12} \frac{2}{3k+1}$ **7.** $\sum_{k=1}^{6} \left(\frac{1}{2}\right)^{2k-2}$

8. 12, 16, 20, 24, 28   **9.** $a_n = -100n + 5100$

10. 3825   **11.** $-\frac{3}{4}$ **12.** $a_n = 4\left(\frac{1}{2}\right)^{n-1}$ **13.** 1020

14. $\frac{3069}{1024}$ **15.** 1   **16.** $\frac{50}{3}$ **17.** 1140

18. $x^5 - 10x^4 + 40x^3 - 80x^2 + 80x - 32$

19. 56   **20.** 490 meters   **21.** $153,287.87

APPENDIX

Appendix A *(page A6)*

1. 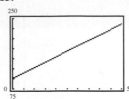   **3.**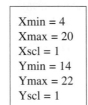

5. 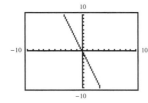   **7.**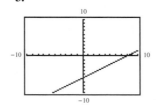

9. 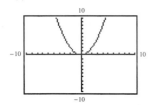   **11.**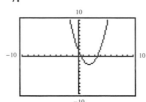

13. 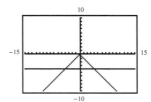   **15.**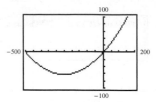

17. Sample answer:

Xmin = 4
Xmax = 20
Xscl = 1
Ymin = 14
Ymax = 22
Yscl = 1

19. Sample answer:

Xmin = -20
Xmax = -4
Xscl = 1
Ymin = -16
Ymax = -8
Yscl = 1

21. Triangle 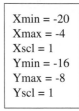   **23.** Square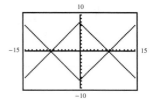

25. Yes, Associative Property of Addition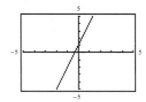

27. Yes, Multiplicative Inverse Property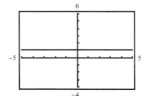

29. $(-3, 0), (3, 0), (0, 9)$ **31.** $(-8, 0), (4, 0), (0, 4)$

33. $\left(\frac{5}{2}, 0\right), (0, -5)$ **35.** $(-2, 0), \left(\frac{1}{2}, 0\right), (0, -1)$

37.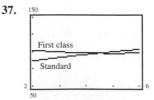

Index of Applications

Property
  Additive, 23
  Multiplicative, 23
i-form, 493
Imaginary number(s), 495
  pure, 495
Imaginary part of a complex number, 495
Imaginary unit *i*, 493
Implied domain, 189
  restrictions, 372
Inconsistent system of equations, 219
Increasing annuity, 733
Independent variable, 186
Index
  of a radical, 450
  of summation, 712
Inequality (inequalities)
  absolute value, solving, 110
  algebraic, 94
  compound, 99
  conjunctive, 99
  disjunctive, 99
  double, 99
  equivalent, 96
  graph of, 94
  linear (one variable), 97
  linear (two variables), 174
    graph of, 175
    solution of, 174
    systems of, 279
  properties of, 96
  quadratic, general form of, 562
  rational, critical numbers of, 564
  satisfy an, 94
  solution set of, 94
  solutions of, 94
  solve an, 94
  symbols, 5
Infinite
  geometric series, 731
    sum of, 732
  interval, 94
  sequence, 708
  series, 711
Infinitely many solutions, 64
Infinity
  negative, 95
  positive, 95
Initial
  height, 311
  velocity, 311
Input-output ordered pairs, 184
Integers, 2
  consecutive, 45
  even, 45
  labels for, 45
  negative, 2
  odd, 45
  positive, 2
Intensity model, 640

Intercepts, 143
  x-intercept, 143
  y-intercept, 143
Intercept form of the equation of a line, 169
Interest formulas
  compound, 585
  simple, 85
Intersection, 100, 371
Intervals, test, 560
Intervals on the real number line
  bounded, 94
  closed, 94
  endpoints, 94
  infinite, 94
  length of, 94
  open, 94
  unbounded, 94, 95
Inverse, additive, 7
Inverse function, 593, 594
  finding algebraically, 595
  Horizontal Line Test for, 593
Inverse properties
  of exponents and logarithms, 626
  of *n*th powers and *n*th roots, 452
Inverse property
  Additive, 23
  Multiplicative, 23
Inverse variation, 432
Inversely proportional, 432
Irrational numbers, 3
Isolate *x*, 60

J
Joint variation, 434
Jointly proportional, 434

K
Key words and phrases, translating, 41
Kirchhoff's Laws, 277, 278

L
Labels for integers, 45
  consecutive, 45
  even, 45
  odd, 45
Law of Trichotomy, 8
Leading 1, 257
Leading coefficient of a polynomial, 308
Least common denominator (LCD), 13, 393
Least common multiple (LCM), 13, 392
Left-to-Right Rule, 17
Length of an interval, 94
Less than, 5
  or equal to, 5
Like
  denominators, 13
  radicals, 468
  signs, 11, 14

  terms, 33
Line(s)
  horizontal, 151, 166
  parallel, 156, 166
  perpendicular, 156, 166
  slope of, 151, 166
  vertical, 151, 166
Linear equation in one variable, 60
  standard form of, 60
Linear equation in two variables, 140
  general form of, 163, 166
  intercept form of, 169
  point-slope form of, 163, 166
  slope-intercept form of, 154, 166
  summary of, 166
  two-point form of, 164, 274
Linear extrapolation, 167
Linear inequality in one variable, 97
Linear inequality in two variables, 174
  graph of, 175
    half-plane, 175
  solution of, 174
  system of, 279
    graphing, 280
    solution of, 279
    solution set of, 279
Linear interpolation, 167
Linear system, number of solutions of, 245
Logarithm(s)
  inverse property of, 626
  natural, properties of, 610, 617
    power, 617
    product, 617
    quotient, 617
  properties of, 607, 617
    power, 617
    product, 617
    quotient, 617
  of *x* with base *a*, 605
Logarithmic equations
  one-to-one property of, 625
  solving, 628
Logarithmic expression
  condensing, 619
  expanding, 619
Logarithmic function, 605
  common, 607
  natural, 610
  with base *a*, 605
Long division of polynomials, 411
Lower limit of summation, 712

M
Major axis of an ellipse, 668
Markup, 80
  rate, 80
Mathematical
  model, constructing a, 43
  modeling, 69
  system, 24